Elements of Chemical Reaction Engineering

Fifth Edition

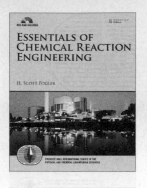

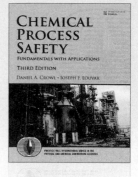

Elements of Chemical Reaction Engineering

Fifth Edition

H. SCOTT FOGLER

Ame and Catherine Vennema Professor of Chemical Engineering
and the Arthur F. Thurnau Professor
The University of Michigan, Ann Arbor

PRENTICE
HALL

Boston • Columbus • Indianapolis • New York • San Francisco • Amsterdam • Cape Town
Dubai • London • Madrid • Milan • Munich • Paris • Montreal • Toronto • Delhi • Mexico City
São Paulo • Sidney • Hong Kong • Seoul • Singapore • Taipei • Tokyo

Many of the designations used by manufacturers and sellers to distinguish their products are claimed as trademarks. Where those designations appear in this book, and the publisher was aware of a trademark claim, the designations have been printed with initial capital letters or in all capitals.

The author and publisher have taken care in the preparation of this book, but make no expressed or implied warranty of any kind and assume no responsibility for errors or omissions. No liability is assumed for incidental or consequential damages in connection with or arising out of the use of the information or programs contained herein.

For information about buying this title in bulk quantities, or for special sales opportunities (which may include electronic versions; custom cover designs; and content particular to your business, training goals, marketing focus, or branding interests), please contact our corporate sales department at corpsales@pearsoned.com or (800) 382-3419.

For government sales inquiries, please contact governmentsales@pearsoned.com.

For questions about sales outside the United States, please contact international@pearsoned.com.

Visit us on the Web: informit.com/ph

Library of Congress Cataloging-in-Publication Data
Fogler, H. Scott, author.
  Elements of chemical reaction engineering / H. Scott Fogler.—Fifth edition.
     pages   cm
  Includes index.
  ISBN 978-0-13-388751-8 (hardcover : alk. paper)
1. Chemical reactors. I. Title.
  TP157.F65 2016
  660'.2832—dc23
                                                    2015032892

ISBN-13: 978-0-13-388751-8
ISBN-10: 0-13-388751-0

Text printed in the United States on recycled paper at RR Donnelley in Kendallville, Indiana.
First printing, January 2016

Dedicated to

Janet Meadors Fogler

*For her companionship, encouragement,
sense of humor, love, and support throughout the years*

Contents

248- Linear plot of each Order of rxn (handwritten annotation)

Preface

The man who has ceased to learn ought not to be allowed
to wander around loose in these dangerous days.

M. M. Coady

A. Who Is the Intended Audience?

This book and interactive Web site is intended for use as both an undergradu-
ate-level and a graduate-level text in chemical reaction engineering. The level
will depend on the choice of chapters, the *Professional Reference Shelf (PRS)*
material (from the companion Web site) to be covered, and the type and degree
of difficulty of problems assigned. It was written with today's students in
mind. It provides instantaneous access to information; does not waste time on
extraneous details; cuts right to the point; uses more bullets to make information easier to access; and includes new, novel problems on chemical reaction
engineering (e.g., solar energy). It gives more emphasis to chemical reactor
safety (Chapters 12 and 13) and alternative energy sources—solar (Chapters 3,
8, and 10) and biofuel production (Chapter 9). The graduate material on topics
such as effectiveness factors, non-ideal reactors, and residence time distribution is in Chapters 14–18 and now includes the software solutions for maximum mixedness and segregation models. A draft version of this book was class
tested at the University of Michigan and other leading universities; then the
text was further revised, taking into account the suggestions of more than
100 students. Much of the material was revised and reworked based on that
feedback.

B. What Are the Goals of This Book?

B.1   To Have Fun Learning Chemical Reaction Engineering (CRE)

Chemical reaction engineering (CRE) is one of two core courses that is unique
to chemical engineering and that separates the chemical engineer from other

engineers. CRE is a great subject that is fun to learn and is the heart of chemical engineering. I have tried to provide a little Michigan humor as we go. Take a look at the humorous YouTube videos (e.g., "Black Widow" or "Chemical Engineering Gone Wrong") that illustrate certain principles in the text. These videos were made by chemical engineering students at the universities of Alabama and Michigan. In addition, I have found that students very much enjoy the Interactive Computer Games (ICGs) that, along with the videos, are linked from the CRE homepage (*http://www.umich.edu/~elements/5e*).

B.2   To Develop a Fundamental Understanding of Reaction Engineering

The second goal of this book is to help the reader clearly understand the fundamentals of CRE. This goal is achieved by presenting a structure that allows the reader to solve reaction engineering problems through reasoning rather than through memorization and recall of numerous equations and the restrictions and conditions under which each equation applies. The algorithms presented in the text for reactor design provide this framework, and the homework problems give the reader practice using the algorithms described in Figures P-1 and P-2, shown in Section C. The conventional homework problems at the end of each chapter are designed to reinforce the principles in the chapter. These problems are about equally divided between those that can be solved with a calculator and those that require a personal computer with a numerical software package, such as Polymath, AspenTech, or COMSOL.

    To give a reference point as to the level of understanding of CRE required in the profession, a number of reaction engineering problems from the California Board of Registration for Civil and Professional Engineers—Chemical Engineering Examinations (PECEE) are included in the text.[1] Typically, these problems should each require approximately 30 minutes to solve.

    Finally, the companion Web site should greatly facilitate learning the fundamentals of CRE because it includes Summary Notes of the material in each chapter, PowerPoint slides of class lecture notes, additional examples, expanded derivations, and self-tests. A complete description of these learning resources is in Appendix I.

B.3.   To Enhance Thinking Skills

A third goal of this text is to enhance *critical thinking skills* and *creative thinking skills*. How does the book help enhance your critical and creative thinking skills? We discuss ways to achieve this enhancement in Section I of the Preface.

---

[1] The permission for use of these problems—which, incidentally, may be obtained from the Documents Section, California Board of Registration for Civil and Professional Engineers—Chemical Engineering, 1004 6th Street, Sacramento, CA 95814, is gratefully acknowledged. (Note: These problems have been copyrighted by the California Board of Registration and may not be reproduced without its permission.)

C. What Is the Structure of CRE?

C.1   What Are the Concepts that Form the Foundation of CRE?

The strategy behind the presentation of material is to build continually on a few basic ideas in CRE to solve a wide variety of problems. These ideas, referred to as the Pillars of Chemical Reaction Engineering (Figure P-1), are the foundation on which different applications rest. They represent not only components of chemical reaction analysis, but also the physical phenomena of diffusion and contacting that affect chemical reactor design.

Figure P-2 shows the first building blocks of CRE and the primary algorithm that allows us to solve CRE problems through logic rather than memorization. We start with the *Mole Balance Building Block* (Chapter 1) and then place the other blocks one at a time on top of the others until we reach the *Evaluate Block* (Chapter 5), by which time we can solve a multitude of isothermal CRE problems. As we study each block we need to make sure we understand everything in that block and don't leave anything out so we don't wind up with a cylindrical block. A tower containing cylindrical blocks would be unstable and would fall apart as we study later chapters. See the end of Chapter 1 lecture notes on the CRE Web site to see the tower of CRE fall if you have blocks with rounded edges.

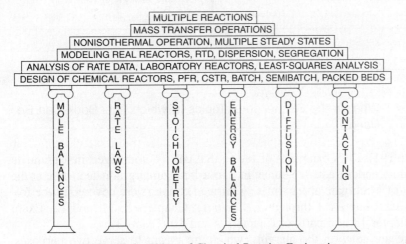

Figure P-1   Pillars of Chemical Reaction Engineering.

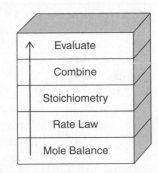

Figure P-2   Building blocks.

From these pillars and building blocks, we construct our CRE algorithm:

Mole Balance + Rate Laws + Stoichiometry + Energy Balance + Combine → Solution

With a few restrictions, the contents of this book can be studied in virtually any order after students have mastered the first six chapters. A flow diagram showing the possible paths is shown in Figure P-3.

The reader will observe that although metric units are used primarily in this text (e.g., $kmol/m^3$, J/mol), English units are also employed (e.g., lb_m/ft^3, Btu). This choice is intentional! We believe that whereas most papers published today use the metric system, a significant amount of reaction engineering data exists in the older literature in English units. Because engineers will be faced with extracting information and reaction rate data from older literature as well as from the current literature, they should be equally at ease with both English and metric units.

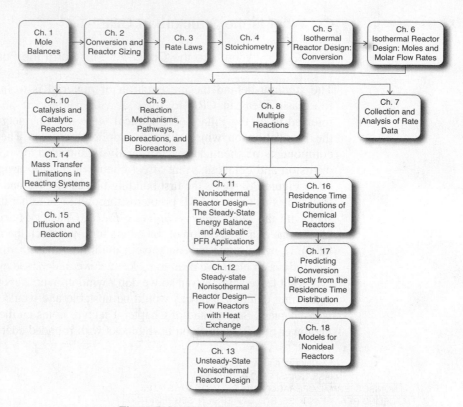

Figure P-3   Sequences for study using this text.

C.2   **What Is the Sequence of Topics in which This Book Can Be Used?**

Table P-1 shows examples of topics that can be converged in a graduate course and an undergraduate course. In a four-hour undergraduate course at the University of Michigan, approximately thirteen chapters are covered in the following order: Chapters 1 through 7 (Exam 1); Chapters 8, 11, and 12 (Exam 2); and Chapter 13 and parts of Chapters 9 and 10 (Exam 3).

There are notes in the margins, which are meant to serve two purposes. First, they act as guides or commentary as one reads through the material. Second, they identify key equations and relationships that are used to solve CRE problems.

D.   What Are the Components of the CRE Web Site?

The companion, interactive Web site material is an updated version of the CRE Web site and is a novel and unique part of this book. The main purposes of the Web site are to serve as an enrichment resource and as a "professional reference shelf." The home page for the CRE Web site (*http://www.umich.edu/~elements/5e/index.html*) is shown in Figure P-4. For discussion of how to use the Web site and text interactively, see Appendix I.

TABLE P-1   UNDERGRADUATE/GRADUATE COVERAGE OF CRE

Undergraduate Material/Course	Graduate Material/Course
Mole Balances (Ch. 1)	Short Review (Ch. 1–8, 11–12)
Smog in Los Angeles Basin (PRS Ch. 1)	Collision Theory (PRS Ch. 3)
Reactor Staging (Ch. 2)	Transition State Theory (PRS Ch. 3)
Hippopotamus Stomach (PRS Ch. 2)	Molecular Dynamics (PRS Ch. 3)
Rate Laws (Ch. 3)	Aerosol Reactors (PRS Ch. 4)
Stoichiometry (Ch. 4)	Multiple Reactions (Ch. 8):
Reactors (Ch. 5):	Fed Membrane Reactors
Batch, PFR, CSTR, PBR	Bioreactions and Reactors (Ch. 9, PRS 9.3–9.5)
Reactors (Ch. 6):	Polymerization (PRS Ch. 9)
Semibatch, Membrane	Co- and Countercurrent Heat Exchange
Data Analysis: Regression (Ch. 7)	(Ch. 12)
Multiple Reactions (Ch. 8)	Radial and Axial Gradients in a PFR
Bioreaction Engineering (Ch. 9)	COMSOL (Ch. 12)
Adiabatic Reactor (Ch. 11)	Reactor Stability and Safety (Ch. 12, 13, PRS
Steady-State Heat Effects (Ch. 12):	12.3)
PFR and CSTR with and without a	Runaway Reactions (PRS, Ch. 12)
Heat Exchanger	Catalyst Deactivation (Ch. 10)
Multiple Steady States	Residence Time Distribution (Ch. 16, 17)
Unsteady-State Heat Effects (Ch. 13)	Models of Real Reactors (Ch. 18)
Reactor Safety	Applications (PRS): Multiphase Reactors,
Catalysis (Ch. 10)	CVD Reactors, Bioreactors

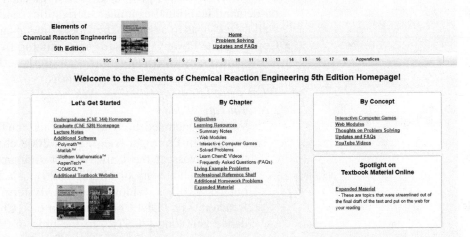

Figure P-4   Screen shot of the book's companion Web site *(http://www.umich.edu/~elements/5e/index.html)*.

The objectives of the Web site are fourfold:

(1) To facilitate the learning of CRE by using the companion Web site to actively address the Felder/Solomon Inventory of Learning Styles 7 discussed in Web Appendix I

(2) To provide additional technical material

(3) To provide tutorial information and self-assessment exercises

(4) To make the learning of CRE fun through the use of interactive games

The following sections (D.1 through D.4) are listed at the end of most chapters and can be accessed from each chapter in the companion Web site.[2]

D.1   Expanded Material

The expanded material consists of material that was removed from the printed text and moved to the Web site to reduce the size and weight of the physical textbook. Reducing the weight makes it easier for the students to carry the book with them at all times, such as while on the campus shuttle or while eating at the cafeteria or the student union. The expanded material includes derivations, examples, and novel applications of CRE principles.

D.2   Learning Resources

The Learning Resources give an overview of the material in each chapter and provide extra explanations, examples, and applications to reinforce the basic concepts of CRE; the Learning Resources are described in Appendix I. The CRE Web site includes the following additional resources:

1. ***Summary Notes and PowerPoint Slides***

    The Summary Notes give an overview of each chapter and provide on-demand additional examples, derivations, and audio comments, as well as self-tests to assess each reader's understanding of the material. Copies of the PowerPoint slides are available from this course taught at the University of Michigan as well as those from Professor Mary Kraft's class at the University of Illinois.

2. ***What Entertainment Is on the Web Site?***

    2.A. *YouTube Videos*

    We have included links to humorous YouTube Videos made by students in Professor Alan Lane's 2008 chemical reaction engineering class at the University of Alabama, as well as videos from the University of Michigan 2011 class. Specifically, check out "The Black Widow Murder Mystery" (Chapter 3), "CRF Reactor," and "Diet Coke and Mentos" (Chapter 4); learn a new dance and song ("CSTR" to the tune of "YMCA"); hear a rap song ("Find Your Rhythm," an "Ice Ice Baby" remix (Chapter 5)); and watch University of Michigan videos, including the ever-popular chemical engineering classic, "Reaction Engineering Gone Wrong."

    2.B. *Interactive Computer Games (ICGs)*

    Students have found the Interactive Computer Games to be both fun and extremely useful to review the important chapter concepts and then apply them to real problems in a unique and

---

[2] *http://www.ncsu.edu/felder-public/ILSdir/styles.htm*

entertaining fashion. The following ICGs are available on the Web site:

- Quiz Show I (Ch. 1)
- Reactor Staging (Ch. 2)
- Quiz Show II (Ch. 4)
- Murder Mystery (Ch. 5)
- Tic Tac (Ch. 5)
- Ecology (Ch. 7)
- The Great Race (Ch. 8)
- Enzyme Man (Ch. 9)
- Catalysis (Ch. 10)
- Heat Effects I (Ch. 12)
- Heat Effects II (Ch. 12)

As you play these interactive games, you will be asked a number of questions related to the corresponding material in the textbook. The ICG keeps track of all the correct answers and at the end of the game displays a coded performance number that reflects how well you mastered the material in the text. Instructors have a manual to decode the performance number.

2.C. *Web Modules*
The Web Modules are a number of examples that apply key CRE concepts to both standard and nonstandard reaction engineering problems (e.g., glow sticks, the use of wetlands to degrade toxic chemicals, and pharmacokinetics of death from a cobra bite). The Web Modules can be loaded directly from the CRE Web site (*http://www.umich.edu/~elements/5e/web_mod/index.html*).

3. *Solved Problems*
A number of solved problems are presented along with problem-solving heuristics. Problem-solving strategies and additional worked example problems are available in the Problem Solving section of the CRE Web site.

D.3 Professional Reference Shelf

This section of the CRE Web site contains

1. Material from the fifth edition of *Elements of Chemical Reaction Engineering* that is not included in the printed textbook.
2. Material that is important to the practicing engineer, such as details of the industrial reactor design for the oxidation of SO_2 and design of spherical reactors and other material that is typically not included in the majority of chemical reaction engineering courses but is included here.

Reference Shelf

E. Why Do We Assign Homework Problems?

The working of homework problems facilitates a true understanding of CRE. After reading a chapter the student may feel they have an understanding of the material. However, when attempting a new or slightly different application of CRE in a homework problem, students sometimes need to go back and re-read

different parts of the chapter to get the level of understanding needed to eventually solve the homework problem.

The end-of-chapter problems numbered "2" (e.g., P3-2$_A$, P11-2$_B$) ask questions about the example problems in that chapter. These example problems are a key resource. These number-2-level problems should be worked before tackling the more challenging homework problems in a given chapter. The subscript letter (A, B, C, or D) after each problem number denotes the difficulty of the problem (i.e., A = easy; D = difficult).

F. What Is a Living Example Problem (LEP)?

Living Example Problem

The example problems that use an Ordinary Differential Equation (ODE) solver (e.g., Polymath) are referred to as "Living Example Problems" or "LEPs" because students can load the Polymath program directly onto their own computers in order to study the problem. Students are encouraged to change parameter values and to "play with" the key variables and assumptions. Using the LEPs to explore the problem and asking "what if…?" questions provide students with the opportunity to practice critical and creative thinking skills.

G. What Software Is Available to Solve the LEPs?

Polymath. Polymath was developed by Professors Michael Cutlip and Mordechai Shacham. The Polymath software includes an ordinary differential equation (ODE) solver, a nonlinear equation solver, and nonlinear regression. As with previous editions of this book, Polymath is used to explore the example problems and to solve the homework problems. Polymath tutorials with screen shots are given on the CRE Web site Summary Notes in Chapter 1 and can also be accessed from the Home Page by going to Living Example Problems (LEPs) and then clicking on Polymath. Most chemical engineering departments in the United States have site licenses for Polymath. If your department does not have a site license and would like one, have your instructor e-mail the CACHE Corporation at cache@uts.cc.utexas.edu to learn how to obtain one. The LEPs need to be copied from the CRE Web site and pasted into the Polymath software. The Polymath software used in the examples in the text is available in most department computer labs in the United States. If you want to have Polymath on your personal laptop computer, you will need to purchase the program. An educational version of the software has been made available by Professors Cutlip and Shacham to students for $20 for a 4-month license, $30 for a 12-month license, or for $39 for a perpetual-use license. Polymath 6.1 is compatible with Windows XP, Windows Vista, Windows 7, and Windows 8. See the Polymath Web site (*http://www.polymath-software.com*) to obtain the laptop version. A special Polymath Web site (*http://www.polymath-software.com/fogler*) has been set up for this book by Polymath authors Professors Cutlip and Shacham.

AspenTech. AspenTech is a process flow sheet simulator used in most senior chemical engineering design courses. It is now routinely introduced in earlier chemical engineering courses, such as thermodynamics, separations, and now in CRE. See the AspenTech Web site (*http://www.aspentech.com*) for more information. Like Polymath, AspenTech site licenses are available in most chemical engineering departments in the United States. Four AspenTech simulation examples specific to CRE are provided on the CRE Web site with step-by-step tutorial screen shots.

As with Polymath programs, the input parameters in AspenTech can be varied to learn how they change the temperature and concentration profiles. Further details are given in Appendix D.

COMSOL Multiphysics. The COMSOL Multiphysics software is a partial differential equation solver that is used with Chapters 12 and 18 to view both axial and radial temperature and concentration profiles. For users of this text, COMSOL has provided a special Web site that includes a step-by-step tutorial, along with examples. See *http://www.comsol.com/ecre*. Further details are given in Appendix D.

Further details of these three software packages can be found in Appendix D.

H. Are There Other Web Site Resources?

FAQs. The Frequently Asked Questions (FAQs) page on the CRE Web site contains a compilation of questions collected over the years from undergraduate students taking reaction engineering.

Visual Encyclopedia of Equipment (*http://encyclopedia.che.engin.umich.edu*). This section was developed by Dr. Susan Montgomery at the University of Michigan. Here, a wealth of photographs and descriptions of real and ideal reactors are given. Students with visual, active, sensing, and intuitive learning styles of the Felder/Solomon Index will particularly benefit from this section.

Reactor Lab (*http://www.ReactorLab.net*). Developed by Professor Richard Herz at the University of California at San Diego, this interactive tool will allow students not only to test their comprehension of the CRE material, but also to explore different situations and combinations of reaction orders and types of reactions.

CRE Web Site. The CRE Web site (*http://www.umich.edu/~elements/5e/index.html*) will be used to update the text and identify typographical and other errors in the first and later printings of this text—available under **Updates and FAQs** on the CRE Web site home page. Additional material may also be added to include more solved problems, as well as additional Web Modules, which will also be found under **Updates and FAQs**.

I. How Can Critical Thinking and Creative Thinking Skills Be Enhanced?

I.1.   Enhance Critical Thinking Skills

A third goal of this book is to enhance critical thinking skills. How does one enhance their critical thinking skills? Answer by learning how to ask the critical thinking questions in Table P-2 and carry out the actions in Table P-3. A number of homework problems have been included that are designed for this purpose. Socratic questioning is at the heart of critical thinking, and a number of homework problems draw from R. W. Paul's six types of Socratic questions,[3] shown in Table P-2 and given in the expanded material on the Web site.

TABLE P-2   SIX TYPES OF SOCRATIC QUESTIONS USED IN CRITICAL THINKING

(1) *Questions for clarification:* Why do you say that? How does this relate to our discussion?

"Are you going to include diffusion in your mole balance equations?"

(2) *Questions that probe assumptions:* What could we assume instead? How can you verify or disprove that assumption?

"Why are you neglecting radial diffusion and including only axial diffusion?"

(3) *Questions that probe reasons and evidence:* What would be an example?

"Do you think that diffusion is responsible for the lower conversion?"

(4) *Questions about viewpoints and perspectives:* What would be an alternative?

"With all the bends in the pipe, from an industrial/practical perspective, do you think diffusion and dispersion will be large enough to affect the conversion?"

(5) *Questions that probe implications and consequences:* What generalizations can you make? What are the consequences of that assumption?

"How would the results be affected if you neglected diffusion?"

(6) *Questions about the question:* What was the point of this question? Why do you think I asked this question?

"Why do you think diffusion is important?"

It is important to know these six types and be able to apply them when investigating a problem such as "Is there a chance the reactor will run away and explode?" or "Why did the reactor explode?"

Critical thinking skills are like any skill, they must be practiced. Scheffer and Rubenfeld[4,5] describe how to practice critical thinking skills using the activities, statements, and questions shown in Table P-3. The reader should try to practice using some or all of these actions every day, as well as asking the critical thinking questions in Table P-1 and on the Web site.

---

[3] R. W. Paul, *Critical Thinking* (Santa Rosa, CA: Foundation for Critical Thinking, 1992).

[4] Courtesy of B. K. Scheffer and M. G. Rubenfeld, "A Consensus Statement on Critical Thinking in Nursing," *Journal of Nursing Education*, 39, 352–359 (2000).

[5] Courtesy of B. K. Scheffer and M. G. Rubenfeld, "Critical Thinking: What Is It and How Do We Teach It?" *Current Issues in Nursing* (2001).

TABLE P-3   CRITICAL THINKING ACTIONS[6]

Analyzing: separating or breaking a whole into parts to discover their nature, function, and relationships
"I studied it piece by piece."
"I sorted things out."

Applying Standards: judging according to established personal, professional, or social rules or criteria
"I judged it according to...."

Discriminating: recognizing differences and similarities among things or situations and distinguishing carefully as to category or rank
"I rank ordered the various...."
"I grouped things together."

Information Seeking: searching for evidence, facts, or knowledge by identifying relevant sources and gathering objective, subjective, historical, and current data from those sources
"I knew I needed to look up/study...."
"I kept searching for data."

Logical Reasoning: drawing inferences or conclusions that are supported in or justified by evidence
"I deduced from the information that...."
"My rationale for the conclusion was...."

Predicting: envisioning a plan and its consequences
"I envisioned the outcome would be...."
"I was prepared for...."

Transforming Knowledge: changing or converting the condition, nature, form, or function of concepts among contexts
"I improved on the basics by...."
"I wondered if that would fit the situation of ...."

I have found that the best way to develop and practice critical thinking skills is to use Tables P-2 and P-3 to help students write a question on any assigned homework problem and then to explain why the question involves critical thinking.

More information on critical thinking can be found on the CRE Web site in the section on Problem Solving (*http://www.umich.edu/~elements/5e/probsolv/index.htm*).

I.2   Enhance Creative Thinking Skills

The fourth goal of this book is to help enhance creative thinking skills. This goal is achieved by using a number of problems that are open-ended to various degrees. With these, students can practice their creative skills by exploring the example problems, as outlined at the beginning of the home problems of each chapter, and by making up and solving an original problem. Problem P5-1 in the text gives some guidelines for developing original problems. A number of techniques that can aid students in practicing and enhancing their creativity

---

[6] R. W. Paul, *Critical Thinking* (Santa Rosa, CA: Foundation for Critical Thinking, 1992); B. K. Scheffer and M. G. Rubenfeld, "A Consensus Statement on Critical Thinking in Nursing," *Journal of Nursing Education*, 39, 352–359 (2000).

can be found in Fogler, LeBlanc, and Rizzo[7] (and its companion Web site), *Strategies for Creative Problem Solving, Third Edition.* The Web site for that book can be accessed from the CRE Web site home page. We use these techniques, such as Osborn's checklist and de Bono's lateral thinking (which involves considering other people's views and responding to random stimulation) to answer add-on questions such as those in Table P-4.

TABLE P-4   PRACTICING CREATIVE THINKING

(1) Brainstorm ideas to ask another question or suggest another calculation that can be made for this homework problem.

(2) Brainstorm ways you could work this homework problem incorrectly.

(3) Brainstorm ways to make this problem easier or more difficult or more exciting.

(4) Brainstorm a list of things you learned from working this homework problem and what you think the point of the problem is.

(5) Brainstorm the reasons why your calculations overpredicted the conversion that was measured when the reactor was put on stream. Assume you made no numerical errors in your calculations.

(6) "What if..." questions: The "What if..." questions are particularly effective when used with the *Living Example Problems,* where one varies the parameters to explore the problem and to carry out a sensitivity analysis. For example, *what if someone suggested that you should double the catalyst particle diameter, what would you say?*

One of the major goals at the undergraduate level is to bring students to the point where they can solve complex reaction problems, such as multiple reactions with heat effects, and then ask "What if . . . ?" questions and look for optimum operating conditions and unsafe operating conditions. The solution to one problem exemplifies this goal: the Manufacture of Styrene (Chapter 12, Problem P12-26$_C$). This problem is particularly interesting because two reactions are endothermic and one is exothermic.

(1) Ethylbenzene → Styrene + Hydrogen: Endothermic
(2) Ethylbenzene → Benzene + Ethylene: Endothermic
(3) Ethylbenzene + Hydrogen → Toluene + Methane: Exothermic

The student could get further practice in critical and creative thinking skills by adding any of the following exercises (x), (y), and (z) to any of the end-of-chapter homework problems.

(x) How could you make this problem easier? More difficult?
(y) Critique your answer by writing a critical thinking question.
(z) Describe two ways you could work this problem incorrectly.

To summarize, it is this author's experience that both critical and creative thinking skills can be enhanced by using Tables P-2, P-3, and P-4 to extend any of the homework problems at the end of each chapter.

---

[7] H. S. Fogler, S. E. LeBlanc, with B. Rizzo, *Strategies for Creative Problem Solving,* 3rd Ed. (Upper Saddle River, N.J.: Prentice Hall, 2014).

J. What's New in This Edition?

J.1   Pedagogy

This book maintains all the strengths of the fourth edition of *Elements of Chemical Reaction Engineering* by using algorithms that allow students to learn chemical reaction engineering through logic rather than memorization. It has the added strength of breaking down the material into smaller bites, as there are now 18 chapters to cover the same concepts as the 14 chapters in the fourth edition.

At the same time, this edition provides new resources that allow students to go beyond solving equations in order to get an intuitive feel and understanding of how reactors behave under different situations. This understanding is achieved through more than 80 interactive simulations (LEPs) provided on the Web site. The Web site has been greatly expanded to address the Felder/Solomon Inventory of Different Learning Styles[8] through interactive Summary Notes and new and updated Interactive Computer Games (ICGs). For example, as discussed in Appendix I the Global Learner can get an overview of the chapter material from the Summary Notes; the Sequential Learner can use all the ⬛ Self Test hot buttons; and the active learner can interact with the ICGs and use the ⬛ Derive hot buttons in the Summary Notes.

A new pedagogical concept is introduced in this text through expanded emphasis on the example problems. Here, the students simply load the Living Example Problems (LEPs) onto their computers and then explore the problems to obtain a deeper understanding of the implications and generalizations before working the homework problems for that chapter. This exploration helps students get an innate feel for reactor behavior and operation, as well as develop and practice their creative thinking skills.

To develop critical thinking skills, instructors can assign one of the new homework problems on troubleshooting, as well as ask the students to expand homework problems by asking a related question that involves critical thinking using Tables P-2 and P-3.

Creative thinking skills can be enhanced by exploring the example problems and asking "What if . . . ?" questions, by using one or more of the brainstorming exercises in Table P-4 to extend any of the homework problems, and by solving the open-ended problems. For example, in the case study on safety, students can use the LEP on the CRE Web site to carry out a postmortem analysis on the nitroaniline explosion in Example 13-2 to learn what would have happened if the cooling had failed for five minutes instead of ten minutes. To this end, a new feature in the text is an Analysis paragraph at the end of each example problem. Significant effort has been devoted to developing example and homework problems that foster critical and creative thinking.

---

[8] *http://www.ncsu.edu/felder-public/ILSdir/styles.htm*

J.2   Content

The following areas have an increased emphasis in this new edition over previous CRE editions by including thorough example problems and homework
problems:

1. Safety: Three industrial explosions are discussed and modeled.
   a. Ammonium Nitrate CSTR Explosion (Chapters 12 and 13)
   b. Nitroaniline Batch Reactor Runaway (Chapter 13)
   c. T2 Laboratories Batch Reactor Runaway (Chapter 13)
   d. Resources from SAChE and CCPS (Chapter 12)
2. Solar Energy: Three examples of solar energy conversion are discussed.
   a. Solar Chemical Reactions (Chapter 3)
   b. Solar Thermal Reactors (Chapter 8)
   c. Solar Catalytic Water Splitting (Chapter 10)
3. Alternative Fuels:
   a. Production of Algae for Biomass (Chapter 9)
4. AspenTech: An AspenTech tutorial for chemical reaction engineering
   and four example problems are provided on the CRE Web site. The
   example problems are
   a. Production of Ethylene from Ethane
   b. The Pyrolysis of Benzene
   c. Adiabatic Liquid Phase Isomerization of Normal Butane
   d. Adiabatic Production of Acetic Anhydride

*However, all intensive laws tend often to have exceptions. Very important con-
cepts take orderly, responsible statements. Virtually all laws intrinsically are
natural thoughts. General observations become laws under experimentation.*

K.   How Do I Say Thank You?

There are so many colleagues and students who contributed to this book that it
would require another chapter to thank them all in an appropriate manner.
I again acknowledge all my friends, students, and colleagues for their contributions to the fifth edition of *Elements of Chemical Reaction Engineering*.
I would like to give special recognition as follows.

First of all, I am indebted to Ame and Catherine Vennema, whose gift of
an endowed chair greatly facilitated the completion of this project. My colleague Dr. Nihat Gürmen coauthored the original Web site during the writing
of the fourth edition of this book. He has been a wonderful colleague to work
with. I also would like to thank University of Michigan undergraduate students
Arthur Shih, Maria Quigley, and Brendan Kirchner, who worked on earlier
versions of the Web site. Their hard work and suggestions are greatly appreciated. Ben Griessmann was instrumental in making everything come together
for the Web site for the fifth edition, including converting the fourth edition's
physical CD-ROM to online-only content for this new edition.

The many stimulating discussions on activation energy with Professor Michael Stamatakis in the Chemical Engineering Department at University College London are greatly appreciated. Michael B. Cutlip, coauthor of Polymath, not only gave suggestions and a critical reading of many sections, but also, most importantly, provided continuous support and encouragement throughout the course of this project. Professor Chau-Chyun Chen provided two AspenTech examples. Ed Fontes at COMSOL Mutiphysic not only provided encouragement, but also provided a COMSOL Web site containing a tutorial with CRE examples. Bernard Goodwin and Laura Lewin, editors at Prentice Hall, were extremely encouraging, helpful, and supportive throughout. Julie Nahil, full-service production manager at Prentice Hall, was fantastic throughout. She provided encouragement, attention to detail, and a great sense of humor, which were greatly appreciated. Indian Institute of Technology (IIT) students Darshan Shah, Anamika Singh, and Sravya Jangareddy, along with Fan Zhang, a University of Michigan student, and Keyvan Edrisi from Swedish Royal Institute of Technology, not only participated in the preparation of the solutions manual, but along with Maithri Venkat worked on the Web site to place many of the LEPs in Wolfram. Richa Motwani from IIT Guwahati, and Gunish Handa and Prafful Bhansali from IIT Bombay, did an extraordinary job in proofreading the galley proofs of the manuscript and making helpful suggestions for changes as well as putting the solution manual in final form. Thank you to students Krittin Binabdullah, Apirak Hanpan, and Thanaphoom Khrutphisit from Chulalongkorn University in Bangkok, along with Ph.D. candidate Cláudio Vilas Bôas Fávero for help in meeting the final deadline for this manuscript.

I very much appreciated the patience of all my Ph.D. students during the period in which this book was written, Michael Senra, Zhenyu Huang, Michael Hoepfner, Nasim Haji Akbari Balou, Claudio Vilas Boas Favero, and Mark Sheng Zheng. Mark helped proofread a number of chapters of the page proofs; Professor Michael Senra class-tested the draft version of the fifth edition, and he and his students gave many valuable suggestions to this edition. There are others I would like to thank for a variety of different reasons: David Bogle, Lee Brown, Brice Carnahan, John Chen, Stu Churchill, Rane Curl, Jim Duderstadt, Tom Edgar, John Falconer, Rich Felder, Asterios Gavriilidis, Joe Goddard, Jay Jorgenson, Costas Kravaris, Steve LeBlanc, Joe Martin, Susan Montgomery, Max Peters, Phil Savage, Johannes Schwank, Mordechai Shacham, Klaus Timmerhaus, Ron West, Jim Wilkes, June Wispelwey, Max, Joe (aka "Jofo"), Sophia, Nicolas, and to the Starbucks staff at Plymouth Road Mall, where most of my final editing of this book was accomplished.

Laura Bracken is very much a part of this book. I appreciate her excellent deciphering of equations and scribbles, her organization, her discovery of mistakes and inconsistencies, and her attention to detail in working with the galleys and proofs. Through all this was her ever-present wonderful disposition. Thanks, Radar!!

Finally, to my wife Janet, love and thanks. Not only did she type the first edition of this book—on a Royal Select typewriter!—she also was a sounding board for so many things in this edition. She was always willing to help with the wording and sentence structure. For example, I often asked her, "Is this the correct phrase or word to use here?" or "Should I mention Jofostan here?" Jan also helped me learn that creativity also involves knowing what to leave out. Without her enormous help and support the project would never have been possible.

HSF
Ann Arbor, Michigan
November 2015

For updates and new and exciting applications, go to the Web site:

http://www.umich.edu/~elements/5e/index.html

For typographical errors, click on Updates & FAQ on the Home page to find

http://www.umich.edu/~elements/5e/updates/index.html

About the Author

**H. Scott Fogler** is the Ame and Catherine Vennema professor of chemical engineering and the Arthur F. Thurnau professor at the University of Michigan in Ann Arbor, and was the 2009 National President of the American Institute of Chemical Engineers, a 50,000-member organization. He received his B.S. from the University of Illinois and his M.S. and Ph.D. from the University of Colorado. He is also the author of the *Essentials of Chemical Reaction Engineering* and co-author, with Steven LeBlanc and Benjamin Rizzo, of *Strategies for Creative Problem Solving, Third Edition.*

Professor Fogler's research interests include flow and reaction in porous media, wax and asphaltene deposition, asphaltene flocculation kinetics, gellation kinetics, colloidal phenomena, and catalyzed dissolution. He has been research advisor to more than forty-five Ph.D. students and has more than two hundred thirty-five refereed publications in these areas. Fogler has chaired ASEE's Chemical Engineering Division, served as director of the American Institute of Chemical Engineers, and earned the Warren K. Lewis Award from AIChE for contributions to chemical engineering education. He also received the Chemical Manufacturers Association's National Catalyst Award and the 2010 Malcom E. Pruitt Award from the Council for Chemical Research (CCR). He is the recipient of 11 named lectureships and is associate editor of *Energy & Fuels.*

Mole Balances 1

The first step to knowledge
is to know that we are ignorant.
—Socrates (470–399 B.C.)

The Wide Wild World of Chemical Reaction Engineering

Chemical kinetics is the study of chemical reaction rates and reaction mechanisms. The study of chemical reaction engineering (CRE) combines the study of chemical kinetics with the reactors in which the reactions occur. Chemical kinetics and reactor design are at the heart of producing almost all industrial chemicals, such as the manufacture of phthalic anhydride shown in Figure 1-1. It is primarily a knowledge of chemical kinetics and reactor design that distinguishes the chemical engineer from other engineers. The selection of a reaction system that operates in the safest and most efficient manner can be the key to the economic success or failure of a chemical plant. For example, if a reaction system produces a large amount of undesirable product, subsequent purification and separation of the desired product could make the entire process economically unfeasible.

How is a chemical
engineer different
from other
engineers?

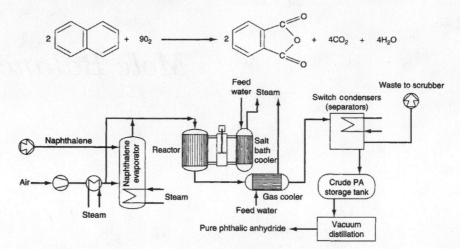

Figure 1-1 Manufacture of phthalic anhydride.

The chemical reaction engineering (**CRE**) principles learned here can also be applied in many areas, such as waste treatment, microelectronics, nano-particles, and living systems, in addition to the more traditional areas of the manufacture of chemicals and pharmaceuticals. Some of the examples that illustrate the wide application of CRE principles in this book are shown in Figure 1-2. These examples include modeling smog in the Los Angeles (L.A.) basin (Chapter 1), the digestive system of a hippopotamus (Chapter 2 on the CRE Web site, *www.umich.edu/~elements/5e/index.html*), and molecular CRE (Chapter 3). Also shown are the manufacture of ethylene glycol (antifreeze), where three of the most common types of industrial reactors are used (Chapters 5 and 6), and the use of wetlands to degrade toxic chemicals (Chapter 7 on the CRE Web site). Other examples shown are the solid-liquid kinetics of acid-rock interactions to improve oil recovery (Chapter 7); pharmacokinetics of cobra bites (Chapter 8 Web Module); free-radical scavengers used in the design of motor oils (Chapter 9); enzyme kinetics (Chapter 9) and drug delivery pharmacokinetics (Chapter 9 on the CRE Web site); heat effects, runaway reactions, and plant safety (Chapters 11 through 13); and increasing the octane number of gasoline and the manufacture of computer chips (Chapter 10).

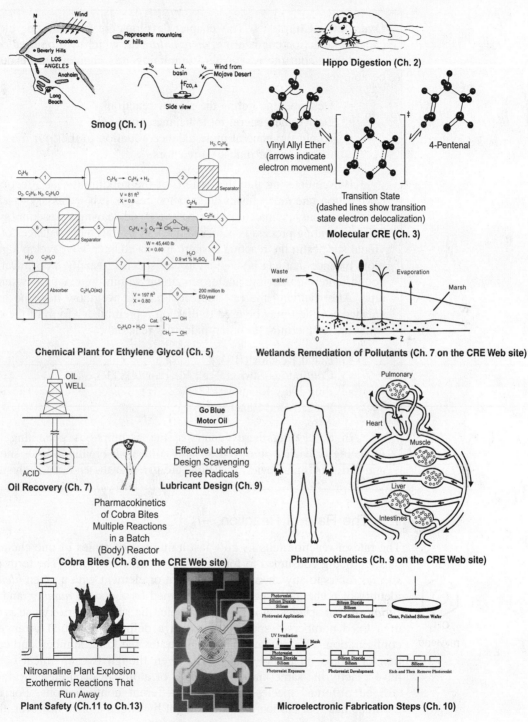

Smog (Ch. 1)

Hippo Digestion (Ch. 2)

Vinyl Allyl Ether (arrows indicate electron movement)

Transition State (dashed lines show transition state electron delocalization)

4-Pentenal

Molecular CRE (Ch. 3)

Chemical Plant for Ethylene Glycol (Ch. 5)

Wetlands Remediation of Pollutants (Ch. 7 on the CRE Web site)

Oil Recovery (Ch. 7)

Pharmacokinetics of Cobra Bites Multiple Reactions in a Batch (Body) Reactor

Cobra Bites (Ch. 8 on the CRE Web site)

Effective Lubricant Design Scavenging Free Radicals

Lubricant Design (Ch. 9)

Pharmacokinetics (Ch. 9 on the CRE Web site)

Nitroanaline Plant Explosion Exothermic Reactions That Run Away

Plant Safety (Ch. 11 to Ch.13)

Microelectronic Fabrication Steps (Ch. 10)

Figure 1-2 The wide world of CRE applications.

Overview—Chapter 1. This chapter develops the first building block of chemical reaction engineering, *mole balances*, which will be used continually throughout the text. After completing this chapter, the reader will be able to:

- Describe and define the rate of reaction
- Derive the general mole balance equation
- Apply the general mole balance equation to the four most common types of industrial reactors

Before entering into discussions of the conditions that affect chemical reaction rate mechanisms and reactor design, it is necessary to account for the various chemical species entering and leaving a reaction system. This accounting process is achieved through overall mole balances on individual species in the reacting system. In this chapter, we develop a general mole balance that can be applied to any species (usually a chemical compound) entering, leaving, and/or remaining within the reaction system volume. After defining the rate of reaction, $-r_A$, we show how the general balance equation may be used to develop a preliminary form of the design equations of the most common industrial reactors:

- Batch Reactor (BR)
- Continuous-Stirred Tank Reactor (CSTR)
- Plug-Flow Reactor (PFR)
- Packed-Bed Reactor (PBR)

In developing these equations, the assumptions pertaining to the modeling of each type of reactor are delineated. Finally, a brief summary and series of short review questions are given at the end of the chapter.

1.1 The Rate of Reaction, $-r_A$

CH₃

CH₃

p-xylene

The rate of reaction tells us how fast a number of moles of one chemical species are being consumed to form another chemical species. The term *chemical species* refers to any chemical component or element with a given *identity*. The identity of a chemical species is determined by the *kind, number,* and *configuration* of that species' atoms. For example, the species para-xylene is made up of a fixed number of specific atoms in a definite molecular arrangement or configuration. The structure shown illustrates the kind, number, and configuration of atoms on a molecular level. Even though two chemical compounds have exactly the same kind and number of atoms of each element, they could still be different species because of different configurations. For example, 2-butene has four carbon atoms and eight hydrogen atoms; however, the atoms in this compound can form two different arrangements.

$$
\begin{array}{ccc}
\text{H} \qquad\quad \text{H} & & \text{H} \qquad\quad \text{CH}_3 \\
\text{C}=\text{C} & \text{and} & \text{C}=\text{C} \\
\text{CH}_3 \qquad \text{CH}_3 & & \text{CH}_3 \qquad\quad \text{H}
\end{array}
$$

cis-2-butene   *trans*-2-butene

As a consequence of the different configurations, these two isomers display different chemical and physical properties. Therefore, we consider them as two different species, even though each has the same number of atoms of each element.

We say that a *chemical reaction* has taken place when a detectable number of molecules of one or more species have lost their identity and assumed a new form by a change in the kind or number of atoms in the compound and/or by a change in structure or configuration of these atoms. In this classical approach to chemical change, it is assumed that the total mass is neither created nor destroyed when a chemical reaction occurs. The mass referred to is the total collective mass of all the different species in the system. However, when considering the individual species involved in a particular reaction, we do speak of the rate of disappearance of mass of a particular species. *The rate of disappearance of a species, say species A, is the number of A molecules that lose their chemical identity per unit time per unit volume through the breaking and subsequent re-forming of chemical bonds during the course of the reaction.* In order for a particular species to "appear" in the system, some prescribed fraction of another species must lose its chemical identity.

There are three basic ways a species may lose its chemical identity: decomposition, combination, and isomerization. In *decomposition*, the molecule loses its identity by being broken down into smaller molecules, atoms, or atom fragments. For example, if benzene and propylene are formed from a cumene molecule,

the cumene molecule has lost its identity (i.e., disappeared) by breaking its bonds to form these molecules. A second way that a molecule may lose its chemical identity is through *combination* with another molecule or atom. In the above reaction, the propylene molecule would lose its chemical identity if the reaction were carried out in the reverse direction, so that it combined with benzene to form cumene. The third way a species may lose its chemical identity is through *isomerization,* such as the reaction

Here, although the molecule neither adds other molecules to itself nor breaks into smaller molecules, it still loses its identity through a change in configuration.

To summarize this point, we say that a given number of molecules (i.e., moles) of a particular chemical species have reacted or disappeared when the molecules have lost their chemical identity.

The rate at which a given chemical reaction proceeds can be expressed in several ways. To illustrate, consider the reaction of chlorobenzene and chloral to produce the banned insecticide DDT (dichlorodiphenyl-trichloroethane) in the presence of fuming sulfuric acid.

$$CCl_3CHO + 2C_6H_5Cl \longrightarrow (C_6H_4Cl)_2CHCCl_3 + H_2O$$

Letting the symbol A represent chloral, B be chlorobenzene, C be DDT, and D be H_2O, we obtain

$$A + 2B \longrightarrow C + D$$

The numerical value of the rate of disappearance of reactant A, $-r_A$, is a positive number.

What is $-r_A$?

> The rate of reaction, $-r_A$, is the number of moles of A (e.g., chloral) reacting (disappearing) per unit time per unit volume (mol/dm^3·s).

Example 1–1

Chloral is being consumed at a rate of 10 moles per second per m^3 when reacting with chlorobenzene to form DDT and water in the reaction described above. In symbol form, the reaction is written as

$$A + 2B \longrightarrow C + D$$

Write the rates of disappearance and formation (i.e., generation) for each species in this reaction.

Solution

(a) *Chloral[A]:* The rate of reaction of chloral [A] ($-r_A$) is given as 10 mol/m^3·s

Rate of disappearance of A = $-r_A$ = 10 mol/m^3·s

Rate of formation of A = r_A = –10 mol/m^3·s

(b) *Chlorobenzene[B]:* For every mole of chloral that disappears, two moles of chlorobenzene [B] also disappear.

Rate of disappearance of B = $-r_B$ = 20 mol/m^3·s

Rate of formation of B = r_B = –20 mol/m^3·s

(c) *DDT[C]:* For every mole of chloral that disappears, one mole of DDT[C] appears.

Rate of formation of C = r_C = 10 mol/m^3·s

Rate of disappearance of C = $-r_C$ = –10 mol/m^3·s

(d) *Water[D]:* Same relationship to chloral as the relationship to DDT

Rate of formation of D = r_D = 10 mol/m^3·s

Rate of disappearance of D = $-r_D$ = –10 mol/m^3·s

$$A + 2B \rightarrow C + D$$
The convention

$-r_A = 10$ mol A/m³·s
$r_A = -10$ mol A/m³·s
$-r_B = 20$ mol B/m³·s
$r_B = -20$ mol B/m³·s
$r_C = 10$ mol C/m³·s

Analysis: The purpose of this example is to better understand the convention for the rate of reaction. The symbol r_j is the rate of formation (generation) of species j. If species j is a reactant, the numerical value of r_j will be a negative number. If species j is a product, then r_j will be a positive number. The rate of reaction, $-r_A$, is the rate of disappearance of reactant A and must be a positive number. A mnemonic relationship to help remember how to obtain relative rates of reaction of A to B, etc., is given by Equation (3-1) on page 71.

In Chapter 3, we will delineate the prescribed relationship between the rate of formation of one species, r_j (e.g., DDT [C]), and the rate of disappearance of another species, $-r_i$ (e.g., chlorobenzene [B]), in a chemical reaction.

Heterogeneous reactions involve more than one phase. In heterogeneous reaction systems, the rate of reaction is usually expressed in measures other than volume, such as reaction surface area or catalyst weight. For a gas-solid catalytic reaction, the gas molecules must interact with the solid catalyst surface for the reaction to take place, as described in Chapter 10.

What is $-r'_A$?

The dimensions of this heterogeneous reaction rate, $-r'_A$ (prime), *are the number of moles of A reacting per unit time per unit mass of catalyst* (mol/s·g catalyst).

Definition of r_j

Most of the introductory discussions on chemical reaction engineering in this book focus on homogeneous systems, in which case we simply say that r_j *is the rate of formation of species j per unit volume.* It is the number of moles of species j generated per unit volume per unit time.

We can say four things about the reaction rate r_j. The reaction rate law for r_j is

The rate law does not depend on the type of reactor used!!

- **The rate of formation of species *j* (mole/time/volume)**
- **An algebraic equation**
- **Independent of the type of reactor (e.g., batch or continuous flow) in which the reaction is carried out**
- **Solely a function of the properties of the reacting materials and reaction conditions (e.g., species concentration, temperature, pressure, or type of catalyst, if any) at a point in the system**

What is $-r_A$ a function of?

However, because the properties and reaction conditions of the reacting materials may vary with position in a chemical reactor, r_j can in turn be a function of position and can vary from point to point in the system.

The chemical reaction rate law is essentially an algebraic equation involving concentration, not a differential equation.[1] For example, the algebraic form of the rate law for $-r_A$ for the reaction

$$A \longrightarrow products$$

may be a linear function of concentration,

$$-r_A = kC_A \qquad (1-1)$$

or, as shown in Chapter 3, it may be some other algebraic function of concentration, such as

---

[1] For further elaboration on this point, see *Chem. Eng. Sci., 25,* 337 (1970); B. L. Crynes and H. S. Fogler, eds., *AIChE Modular Instruction Series E: Kinetics,* 1, 1 (New York: AIChE, 1981); and R. L. Kabel, "Rates," *Chem. Eng. Commun., 9,* 15 (1981).

$$-r_A = kC_A^2 \qquad (1\text{-}2)$$

or

$$-r_A = \frac{k_1 C_A}{1 + k_2 C_A}$$

For a given reaction, the particular concentration dependence that the rate law follows (i.e., $-r_A = kC_A$ or $-r_A = kC_A^2$ or ...) must be determined from *experimental observation*. Equation (1-2) states that the rate of disappearance of A is equal to a rate constant k (which is a function of temperature) times the square of the concentration of A. As noted earlier, by convention, r_A is the rate of formation of A; consequently, $-r_A$ is the rate of disappearance of A. Throughout this book, the phrase *rate of generation* means exactly the same as the phrase *rate of formation*, and these phrases are used interchangeably.

The convention

1.2 The General Mole Balance Equation

To perform a mole balance on any system, the system boundaries must first be specified. The volume enclosed by these boundaries is referred to as the *system volume*. We shall perform a mole balance on species j in a system volume, where species j represents the particular chemical species of interest, such as water or NaOH (Figure 1-3).

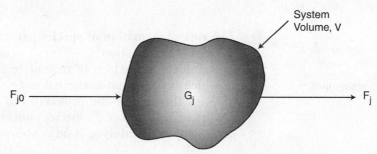

Figure 1-3   Mole balance on species j in a system volume, V.

A mole balance on species j at any instant in time, t, yields the following equation:

$$
\begin{bmatrix}
\text{Rate of flow} \\
\text{of } j \text{ into} \\
\text{the system} \\
\text{(moles/time)}
\end{bmatrix}
-
\begin{bmatrix}
\text{Rate of flow} \\
\text{of } j \text{ out of} \\
\text{the system} \\
\text{(moles/time)}
\end{bmatrix}
+
\begin{bmatrix}
\text{Rate of generation} \\
\text{of } j \text{ by chemical} \\
\text{reaction within} \\
\text{the system} \\
\text{(moles/time)}
\end{bmatrix}
=
\begin{bmatrix}
\text{Rate of} \\
\text{accumulation} \\
\text{of } j \text{ within} \\
\text{the system} \\
\text{(moles/time)}
\end{bmatrix}
$$

Mole balance   **In**   –   **Out**   +   **Generation**   =   **Accumulation**

$$F_{j0} \quad - \quad F_j \quad + \quad G_j \quad = \quad \frac{dN_j}{dt} \qquad (1\text{-}3)$$

In this equation, N_j represents the number of moles of species j in the system at time t. If all the system variables (e.g., temperature, catalytic activity, and concentration of the chemical species) are spatially uniform throughout the system volume, the rate of generation of species j, G_j, is just the product of the reaction volume, V, and the rate of formation of species j, r_j.

$$G_j = r_j \cdot V$$

$$\frac{moles}{time} = \frac{moles}{time \cdot volume} \cdot volume$$

Now suppose that the rate of formation of species j for the reaction varies with position in the system volume. That is, it has a value r_{j1} at location 1, which is surrounded by a small volume, ΔV_1, within which the rate is uniform; similarly, the reaction rate has a value r_{j2} at location 2 and an associated volume, ΔV_2, and so on (Figure 1-4).

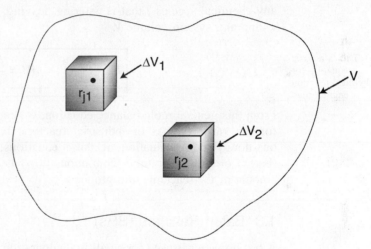

Figure 1-4   Dividing up the system volume, V.

The rate of generation, ΔG_{j1}, in terms of r_{j1} and subvolume ΔV_1, is

$$\Delta G_{j1} = r_{j1} \Delta V_1$$

Similar expressions can be written for ΔG_{j2} and the other system subvolumes, ΔV_i. The total rate of generation within the system volume is the sum of all the rates of generation in each of the subvolumes. If the total system volume is divided into M subvolumes, the total rate of generation is

$$G_j = \sum_{i=1}^{M} \Delta G_{ji} = \sum_{i=1}^{M} r_{ji} \, \Delta V_i$$

By taking the appropriate limits (i.e., let $M \rightarrow \infty$ and $\Delta V \rightarrow 0$) and making use of the definition of an integral, we can rewrite the foregoing equation in the form

$$G_j = \int^V r_j \, dV$$

From this equation, we see that r_j will be an indirect function of position, since the properties of the reacting materials and reaction conditions (e.g., concentration, temperature) can have different values at different locations in the reactor volume.

We now replace G_j in Equation (1-3)

$$F_{j0} - F_j + G_j = \frac{dN_j}{dt} \tag{1-3}$$

by its integral form to yield a form of the general mole balance equation for any chemical species j that is entering, leaving, reacting, and/or accumulating within any system volume V.

This is a basic equation for chemical reaction engineering.

$$F_{j0} - F_j + \int^V r_j \, dV = \frac{dN_j}{dt} \tag{1-4}$$

From this general mole balance equation, we can develop the design equations for the various types of industrial reactors: batch, semibatch, and continuous-flow. Upon evaluation of these equations, we can determine the time (batch) or reactor volume (continuous-flow) necessary to convert a specified amount of the reactants into products.

1.3 Batch Reactors (BRs)

When is a batch reactor used?

A batch reactor is used for small-scale operation, for testing new processes that have not been fully developed, for the manufacture of expensive products, and for processes that are difficult to convert to continuous operations. The reactor can be charged (i.e., filled) through the holes at the top (see Figure 1-5(a)). The batch reactor has the advantage of high conversions that can be obtained by leaving the reactant in the reactor for long periods of time, but it also has the disadvantages of high labor costs per batch, the variability of products from batch to batch, and the difficulty of large-scale production (see Industrial Reactor Photos in *Professional Reference Shelf [PRS]* on the CRE Web site, *www.umich.edu/~elements/ 5e/index.html*).

Reference Shelf

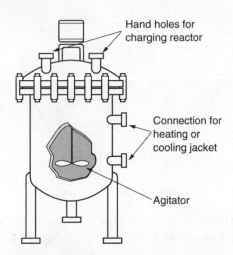

Figure 1-5(a)   Simple batch homogeneous batch reactor (BR). [Excerpted by special permission from *Chem. Eng., 63*(10), 211 (Oct. 1956). Copyright 1956 by McGraw-Hill, Inc., New York, NY 10020.]

Figure 1-5(b)   Batch reactor mixing patterns. Further descriptions and photos of the batch reactors can be found in both the *Visual Encyclopedia of Equipment* and in the *Professional Reference Shelf* on the CRE Web site.

A batch reactor has neither inflow nor outflow of reactants or products while the reaction is being carried out: $F_{j0} = F_j = 0$. The resulting general mole balance on species j is

$$\frac{dN_j}{dt} = \int^{V} r_j \ dV$$

If the reaction mixture is perfectly mixed (Figure 1-5(b)) so that there is no variation in the rate of reaction throughout the reactor volume, we can take r_j out of the integral, integrate, and write the mole balance in the form

Perfect mixing

$$\boxed{\frac{dN_j}{dt} = r_j V} \tag{1-5}$$

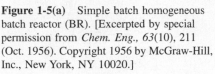

Batch Reactor

Let's consider the isomerization of species A in a batch reactor

$$A \longrightarrow B$$

As the reaction proceeds, the number of moles of A decreases and the number of moles of B increases, as shown in Figure 1-6.

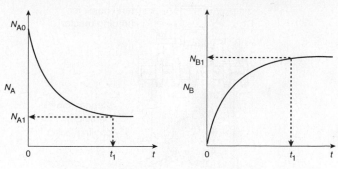

Figure 1-6 Mole-time trajectories.

We might ask what time, t_1, is necessary to reduce the initial number of moles from N_{A0} to a final desired number N_{A1}. Applying Equation (1-5) to the isomerization

$$\frac{dN_A}{dt} = r_A V$$

rearranging,

$$dt = \frac{dN_A}{r_A V}$$

and integrating with limits that at $t = 0$, then $N_A = N_{A0}$, and at $t = t_1$, then $N_A = N_{A1}$, we obtain

$$t_1 = \int_{N_{A1}}^{N_{A0}} \frac{dN_A}{-r_A V} \tag{1-6}$$

This equation is the integral form of the mole balance on a batch reactor. It gives the time, t_1, necessary to reduce the number of moles from N_{A0} to N_{A1} and also to form N_{B1} moles of B.

1.4 Continuous-Flow Reactors

Reference Shelf

Continuous-flow reactors are almost always operated at steady state. We will consider three types: the *continuous-stirred tank reactor* (CSTR), the *plug-flow reactor* (PFR), and the *packed-bed reactor* (PBR). Detailed physical descriptions of these reactors can be found in both the *Professional Reference Shelf* (*PRS*) for Chapter 1 and in the *Visual Encyclopedia of Equipment*, *encyclopedia.che.engin.umich.edu*, and on the CRE Web site.

1.4.1 Continuous-Stirred Tank Reactor (CSTR)

What is a CSTR used for?

A type of reactor commonly used in industrial processing is the stirred tank operated continuously (Figure 1-7). It is referred to as the *continuous-stirred tank reactor* (CSTR) or *vat*, or *backmix reactor,* and is primarily used for

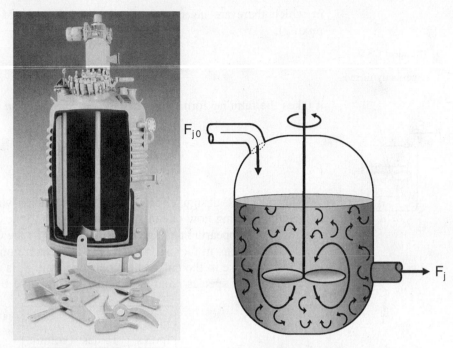

Figure 1-7(a)   CSTR/batch reactor. (Photo courtesy of Pfaudler, Inc.)

Figure 1-7(b)   CSTR mixing patterns. Also see the *Visual Encyclopedia of Equipment* on the CRE Web site.

liquid-phase reactions. It is normally operated **at steady state** and is assumed to be **perfectly mixed**; consequently, there is no time dependence or position dependence of the temperature, concentration, or reaction rate inside the CSTR. That is, every variable is the same at every point inside the reactor. Because the temperature and concentration are identical everywhere within the reaction vessel, they are the same at the *exit point* as they are elsewhere in the tank. Thus, the temperature and concentration in the exit stream are modeled as being the same as those inside the reactor. In systems where mixing is highly nonideal, the well-mixed model is inadequate, and we must resort to other modeling techniques, such as residence time distributions, to obtain meaningful results. This topic of nonideal mixing is discussed in Chapters 16, 17, and 18 on nonideal reactors.

When the general mole balance equation

$$F_{j0} - F_j + \int^V r_j \, dV = \frac{dN_j}{dt} \tag{1-4}$$

is applied to a CSTR operated at steady state (i.e., conditions do not change with time),

$$\frac{dN_j}{dt} = 0$$

in which there are no spatial variations in the rate of reaction (i.e., perfect mixing),

The ideal CSTR is
assumed to be
perfectly mixed.

$$\int^{V} r_j \, dV = V r_j$$

it takes the familiar form known as the *design equation* for a CSTR

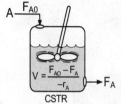

CSTR

$$V = \frac{F_{j0} - F_j}{-r_j} \tag{1-7}$$

The CSTR design equation gives the reactor volume V necessary to reduce the entering flow rate of species j from F_{j0} to the exit flow rate F_j, when species j is disappearing at a rate of $-r_j$. We note that the CSTR is modeled such that the conditions in the exit stream (e.g., concentration and temperature) **are identical** to those in the tank. The molar flow rate F_j is just the product of the concentration of species j and the volumetric flow rate v

$$F_j = C_j \cdot v$$
$$\frac{\text{moles}}{\text{time}} = \frac{\text{moles}}{\text{volume}} \cdot \frac{\text{volume}}{\text{time}} \tag{1-8}$$

Similarly, for the entrance molar flow rate we have $F_{j0} = C_{j0} \cdot v_0$. Consequently, we can substitute for F_{j0} and F_j into Equation (1-7) to write a balance on species A as

$$V = \frac{v_0 C_{A0} - v C_A}{-r_A} \tag{1-9}$$

The ideal CSTR mole balance equation is an algebraic equation, not a differential equation.

1.4.2 Tubular Reactor

When is a tubular
reactor most
often used?

In addition to the CSTR and batch reactors, another type of reactor commonly used in industry is the *tubular reactor*. It consists of a cylindrical pipe and is normally operated at steady state, as is the CSTR. Tubular reactors are used most often for gas-phase reactions. A schematic and a photograph of industrial tubular reactors are shown in Figure 1-8.

In the tubular reactor, the reactants are continually consumed as they flow down the length of the reactor. In modeling the tubular reactor, we assume that the concentration varies continuously in the axial direction through the reactor. Consequently, the reaction rate, which is a function of concentration for all but zero-order reactions, will also vary axially. For the purposes of the material presented here, we consider systems in which the flow field may be modeled by that of a plug-flow profile (e.g., uniform velocity as

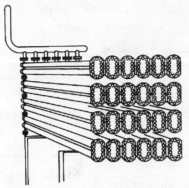

Figure 1-8(a)   Tubular reactor schematic. Longitudinal tubular reactor. [Excerpted by special permission from *Chem. Eng.*, 63(10), 211 (Oct. 1956). Copyright 1956 by McGraw-Hill, Inc., New York, NY 10020.]

Figure 1-8(b)   Tubular reactor photo. Tubular reactor for production of Dimersol G. (Photo courtesy of Editions Techniq Institut français du pétrole.)

in turbulent flow), as shown in Figure 1-9. That is, there is no radial variation in reaction rate, and the reactor is referred to as a *plug-flow reactor* (PFR). (The laminar-flow reactor is discussed in Chapters 16 through 18 on nonideal reactors.)

Also see PRS and *Visual Encyclopedia of Equipment.*

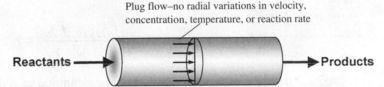

Plug flow–no radial variations in velocity, concentration, temperature, or reaction rate

Reactants → → Products

Figure 1-9   Plug-flow tubular reactor.

The general mole balance equation is given by Equation (1-4)

$$F_{j0} - F_j + \int^V r_j \, dV = \frac{dN_j}{dt} \tag{1-4}$$

The equation we will use to design PFRs at steady state can be developed in two ways: (1) directly from Equation (1-4) by differentiating with respect to volume V, and then rearranging the result or (2) from a mole balance on species j in a differential segment of the reactor volume ΔV. Let's choose the second way to arrive at the differential form of the PFR mole balance. The differential volume, ΔV, shown in Figure 1-10, will be chosen sufficiently small such that there are no spatial variations in reaction rate within this volume. Thus the generation term, ΔG_j, is

$$\Delta G_j = \int^{\Delta V} r_j \, dV = r_j \, \Delta V$$

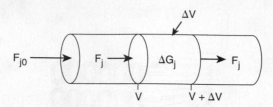

Figure 1-10   Mole balance on species j in volume ΔV.

$$
\begin{bmatrix} \text{Molar flow} \\ \text{rate of species } j \\ \textit{In } \text{at } V \\ \textit{moles/time} \end{bmatrix} - \begin{bmatrix} \text{Molar flow} \\ \text{rate of species } j \\ \textit{Out } \text{at } (V+\Delta V) \\ \textit{moles/time} \end{bmatrix} + \begin{bmatrix} \text{Molar rate of} \\ \textit{Generation} \\ \text{of species } j \\ \text{within } \Delta V \\ \textit{moles/time} \end{bmatrix} = \begin{bmatrix} \text{Molar rate of} \\ \textit{Accumulation} \\ \text{of species } j \\ \text{within } \Delta V \\ \textit{moles/time} \end{bmatrix}
$$

$$
\textbf{In} \qquad - \qquad \textbf{Out} \qquad + \textbf{Generation} \; = \textbf{Accumulation}
$$

$$
F_j\big|_V \qquad - \qquad F_j\big|_{V+\Delta V} \qquad + \qquad r_j\Delta V \qquad = \qquad 0 \qquad (1\text{-}10)
$$

Dividing by ΔV and rearranging

$$
\left[\frac{F_j\big|_{V+\Delta V} - F_j\big|_V}{\Delta V} \right] = r_j
$$

the term in brackets resembles the definition of a derivative

$$
\lim_{\Delta x \to 0} \left[\frac{f(x+\Delta x) - f(x)}{\Delta x} \right] = \frac{df}{dx}
$$

Taking the limit as ΔV approaches zero, we obtain the differential form of steady state mole balance on a PFR

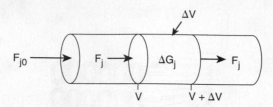

Tubular reactor

$$
\boxed{ \frac{dF_j}{dV} = r_j } \qquad (1\text{-}11)
$$

We could have made the cylindrical reactor on which we carried out our mole balance an irregularly shaped reactor, such as the one shown in Figure 1-11 for reactant species A.

However, we see that by applying Equation (1-10), the result would yield the same equation (i.e., Equation (1-11)). For species A, the mole balance is

$$
\boxed{ \frac{dF_A}{dV} = r_A } \qquad (1\text{-}12)
$$

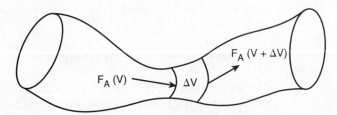

Picasso's
reactor

Figure 1-11   Pablo Picasso's reactor.

Consequently, we see that Equation (1-11) applies equally well to our model of tubular reactors of variable and constant cross-sectional area, although it is doubtful that one would find a reactor of the shape shown in 1-11 unless it were designed by Pablo Picasso.

The conclusion drawn from the application of the design equation to Picasso's reactor is an important one: the degree of completion of a reaction achieved in an ideal plug-flow reactor (PFR) does not depend on its shape, only on its total volume.

Again consider the isomerization A $\rightarrow$ B, this time in a PFR. As the reactants proceed down the reactor, A is consumed by chemical reaction and B is produced. Consequently, the molar flow rate F_A decreases, while F_B increases as the reactor volume V increases, as shown in Figure 1-12.

$$V = \int_{F_A}^{F_{A0}} \frac{dF_A}{-r_A}$$

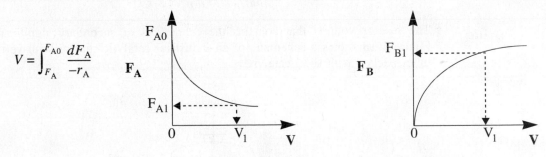

Figure 1-12   Profiles of molar flow rates in a PFR.

We now ask what is the reactor volume V_1 necessary to reduce the entering molar flow rate of A from F_{A0} to F_{A1}. Rearranging Equation (1-12) in the form

$$dV = \frac{dF_A}{r_A}$$

and integrating with limits at $V = 0$, then $F_A = F_{A0}$, and at $V = V_1$, then $F_A = F_{A1}$

$$V_1 = \int_{F_{A0}}^{F_{A1}} \frac{dF_A}{r_A} = \int_{F_{A1}}^{F_{A0}} \frac{dF_A}{-r_A} \qquad (1-13)$$

V_1 is the volume necessary to reduce the entering molar flow rate F_{A0} to some specified value F_{A1} and also the volume necessary to produce a molar flow rate of B of F_{B1}.

1.4.3 Packed-Bed Reactor (PBR)

The principal difference between reactor design calculations involving homogeneous reactions and those involving fluid-solid heterogeneous reactions is that for the latter, the reaction takes place on the surface of the catalyst (see Chapter 10). The greater the mass of a given catalyst, the greater the reactive surface area. Consequently, the reaction rate is based on mass of solid catalyst, W, rather than on reactor volume, V. For a fluid–solid heterogeneous system, the rate of reaction of a species A is defined as

$$-r_A' = \text{mol A reacted/(time} \times \text{mass of catalyst)}$$

The mass of solid catalyst is used because the amount of catalyst is what is important to the rate of product formation. We note that by multiplying the heterogeneous reaction rate, $-r_A'$, by the bulk catalyst density, $\rho_b \left(\dfrac{mass}{volume}\right)$, we can obtain the homogeneous reaction rate, $-r_A$

$$-r_A = \rho_b \left(-r_A'\right)$$

$$\left(\frac{mol}{dm^3 \cdot s}\right) = \left(\frac{g}{dm^3}\right)\left(\frac{mol}{g \cdot s}\right)$$

The reactor volume that contains the catalyst is of secondary significance. Figure 1-13 shows a schematic of an industrial catalytic reactor with vertical tubes packed with solid catalyst.

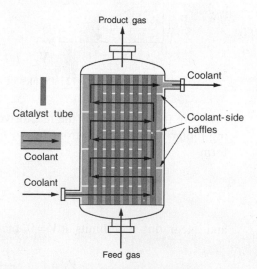

Figure 1-13   Longitudinal catalytic packed-bed reactor. [From Cropley, *American Institute of Chemical Engineers*, 86(2), 34 (1990). Reproduced with permission of the American Institute of Chemical Engineers, Copyright © 1990 AIChE. All rights reserved.]

In the three idealized types of reactors just discussed (the perfectly mixed batch reactor, the plug-flow tubular reactor [PFR]), and the perfectly mixed continuous-stirred tank reactor [CSTR]), the design equations (i.e., mole balances) were developed based on reactor volume. The derivation of the design equation for a packed-bed catalytic reactor (PBR) will be carried out in a manner analogous to the development of the tubular design equation. To accomplish this derivation, we simply replace the volume coordinate in Equation (1-10) with the catalyst mass (i.e., weight) coordinate W (Figure 1-14).

PBR
Mole Balance

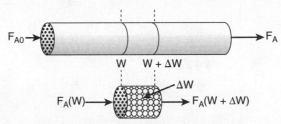

Figure 1-14   Packed-bed reactor schematic.

As with the PFR, the PBR is assumed to have no radial gradients in concentration, temperature, or reaction rate. The generalized mole balance on species A over catalyst weight ΔW results in the equation

$$
\textbf{\textit{In}} \quad – \quad \textbf{\textit{Out}} \quad + \quad \textbf{\textit{Generation}} \quad = \quad \textbf{\textit{Accumulation}}
$$

$$
F_{A|W} \quad – \quad F_{A|(W+\Delta W)} \quad + \quad r'_A \,\Delta W \quad = \quad 0 \qquad (1\text{-}14)
$$

The dimensions of the generation term in Equation (1-14) are

$$
(r'_A)\,\Delta W \equiv \frac{moles\ A}{(time)(mass\ of\ catalyst)} \cdot (mass\ of\ catalyst) \equiv \frac{moles\ A}{time}
$$

which are, as expected, the same dimensions of the molar flow rate F_A. After dividing by ΔW and taking the limit as $\Delta W \to 0$, we arrive at the differential form of the mole balance for a packed-bed reactor

Use the differential form of design equation for catalyst decay and pressure drop.

$$
\boxed{\frac{dF_A}{dW} = r'_A} \qquad (1\text{-}15)
$$

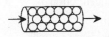

When pressure drop through the reactor (see Section 5.5) and catalyst decay (see Section 10.7 on the CRE Web site Chapter 10) are neglected, the integral form of the packed-catalyst-bed design equation can be used to calculate the catalyst weight

You can use the integral form *only* when there is no ΔP and no catalyst decay.

$$
W = \int_{F_{A0}}^{F_A} \frac{dF_A}{r'_A} = \int_{F_A}^{F_{A0}} \frac{dF_A}{-r'_A} \qquad (1\text{-}16)
$$

W is the catalyst weight necessary to reduce the entering molar flow rate of species A, F_{A0}, down to a flow rate F_A.

For some insight into things to come, consider the following example of how one can use the tubular reactor design in Equation (1-11).

Example 1–2 How Large Is It?

Consider the liquid phase *cis – trans* isomerization of 2–butene

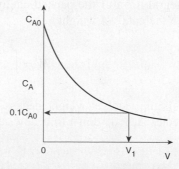

cis-2-butene   *trans*-2-butene

which we will write symbolically as

$$A \longrightarrow B$$

The reaction is first order in A ($-r_A = kC_A$) and is carried out in a tubular reactor in which the volumetric flow rate, v, is constant, i.e., $v = v_0$.

1. Sketch the concentration profile.
2. Derive an equation relating the reactor volume to the entering and exiting concentrations of A, the rate constant k, and the volumetric flow rate v_0.
3. Determine the reactor volume, V_1, necessary to reduce the exiting concentration to 10% of the entering concentration, i.e., $C_A = 0.1C_{A0}$, when the volumetric flow rate is 10 dm³/min (i.e., liters/min) and the specific reaction rate, k, is 0.23 min⁻¹.

Solution

1. Sketch C_A as a function of *V*.

   Species A is consumed as we move down the reactor, and as a result, both the molar flow rate of A and the concentration of A will decrease as we move. Because the volumetric flow rate is constant, $v = v_0$, one can use Equation (1-8) to obtain the concentration of A, $C_A = F_A/v_0$, and then by comparison with the Figure 1-12 plot, obtain the concentration of A as a function of reactor volume, as shown in Figure E1-2.1.

Figure E1-2.1   Concentration profile.

2. Derive an equation relating V, v_0, k, C_{A0}, and C_A.

For a tubular reactor, the mole balance on species A ($j = A$) was shown to be given by Equation (1-11). Then for species A ($j = A$)

Mole Balance: $$\frac{dF_A}{dV} = r_A$$ (1-12)

For a first-order reaction, the rate law (discussed in Chapter 3) is

Rate Law: $$-r_A = kC_A$$ (E1-2.1)

Because the volumetric flow rate, v, is constant ($v = v_0$), as it is for most all liquid-phase reactions,

$$\frac{dF_A}{dV} = \frac{d(C_A v)}{dV} = \frac{d(C_A v_0)}{dV} = v_0 \frac{dC_A}{dV} = r_A$$ (E1-2.2)

Multiplying both sides of Equation (E1-2.2) by minus one and then substituting Equation (E1-2.1) yields

Combine: $$-\frac{v_0 dC_A}{dV} = -r_A = kC_A$$ (E1-2.3)

Separating the variables and rearranging gives

$$-\frac{v_0}{k}\left(\frac{dC_A}{C_A}\right) = dV$$

Using the conditions at the entrance of the reactor that when $V = 0$, then $C_A = C_{A0}$

$$-\frac{v_0}{k}\int_{C_{A0}}^{C_A} \frac{dC_A}{C_A} = \int_0^V dV$$ (E1-2.4)

Carrying out the integration of Equation (E1-2.4) gives

Solve: $$\boxed{V = \frac{v_0}{k}\, ln\, \frac{C_{A0}}{C_A}}$$ (E1-2.5)

We can also rearrange Equation (E1-2.5) to solve for the concentration of A as a function of reactor volume to obtain

$$C_A = C_{A0}\exp\left(-kV/v_0\right)$$

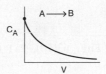

Reactor sizing (margin note)

Concentration Profile (margin note)

3. Calculate V. We want to find the volume, V_1, at which $C_A = \frac{1}{10}C_{A0}$ for $k = 0.23$ min^{-1} and $v_0 = 10$ dm^3/min.

Evaluate:

Substituting C_{A0}, C_A, v_0, and k in Equation (E1-2.5), we have

$$V = \frac{10 \text{ dm}^3/\text{min}}{0.23 \text{ min}^{-1}} \ln \frac{C_{A0}}{0.1 C_{A0}} = \frac{10 \text{ dm}^3}{0.23}\ln 10 = 100 \text{ dm}^3 \text{ (i.e., 100 L; 0.1 m}^3\text{)}$$

Let's calculate the volume to reduce the entering concentration to $C_A = 0.01$ C_{A0}. Again using Equation (E1-2.5)

$$V = \frac{10 \text{ dm}^3/\text{min}}{0.23 \text{ min}^{-1}} \ln \frac{C_{A0}}{0.01 C_{A0}} = \frac{10 \text{ dm}^3}{0.23}\ln 100 = 200 \text{ dm}^3$$

Note: We see that a larger reactor (200 dm^3) is needed to reduce the exit concentration to a smaller fraction of the entering concentration (e.g., $C_A = 0.01\ C_{A0}$).

We see that a reactor volume of 0.1 m^3 is necessary to convert 90% of species A entering into product B for the parameters given.

<u>Analysis:</u> For this irreversible liquid-phase first order reaction (i.e., $-r_A = kC_A$) being carried out in a PFR, the concentration of the reactant decreases exponentially down the length (i.e., volume V) of the reactor. The more species A consumed and converted to product B, the larger must be the reactor volume V. The purpose of the example was to give a vision of the types of calculations we will be carrying out as we study chemical reaction engineering (CRE).

1.5 Industrial Reactors[2]

Reference Shelf

Be sure to view the actual photographs of industrial reactors on the CRE Web site. There are also links to view reactors on different Web sites. The CRE Web site also includes a portion of the *Visual Encyclopedia of Equipment*, *encyclopedia.che.engin.umich.edu*, "Chemical Reactors" developed by Dr. Susan Montgomery and her students at the University of Michigan. Also see *Professional Reference Shelf* on the CRE Web site for "Reactors for Liquid-Phase and Gas-Phase Reactions," along with photos of industrial reactors, and Expanded Material on the CRE Web site.

In this chapter, and on the CRE Web site, we've introduced each of the major types of industrial reactors: batch, stirred tank, tubular, and fixed bed (packed bed). Many variations and modifications of these commercial reactors (e.g., semibatch, fluidized bed) are in current use; for further elaboration, refer to the detailed discussion of industrial reactors given by Walas.[3]

---

[2] *Chem. Eng.,* 63(10), 211 (1956). See also *AIChE Modular Instruction Series E, 5* (1984).

[3] S. M. Walas, *Reaction Kinetics for Chemical Engineers* (New York: McGraw-Hill, 1959), Chapter 11.

Solved Problems

The CRE Web site describes industrial reactors, along with typical feed and operating conditions. In addition, two solved example problems for Chapter 1 can be found on the CRE Web site.

Closure. The goal of this text is to weave the fundamentals of chemical reaction engineering into a structure or algorithm that is easy to use and apply to a variety of problems. We have just finished the first building block of this algorithm: mole balances.

Mole Balance

This algorithm and its corresponding building blocks will be developed and discussed in the following chapters:

- Mole Balance, Chapters 1 and 2
- Rate Law, Chapter 3
- Stoichiometry, Chapter 4
- Combine, Chapter 5
- Evaluate, Chapter 5
- Energy Balance, Chapters 11 through 13

With this algorithm, one can approach and solve chemical reaction engineering problems through logic rather than memorization.

SUMMARY

Each chapter summary gives the key points of the chapter that need to be remembered and carried into succeeding chapters.

1. A mole balance on species j, which enters, leaves, reacts, and accumulates in a system volume V, is

$$F_{j0} - F_j + \int^V r_j \, dV = \frac{dN_j}{dt} \tag{S1-1}$$

If, and only if, the contents of the reactor are well mixed will the mole balance (Equation (S1-1)) on species A give

$$F_{A0} - F_A + r_A V = \frac{dN_A}{dt} \tag{S1-2}$$

2. The kinetic rate law for r_j is
   - The rate of formation of species j per unit volume (e.g., mol/s·dm^3)
   - Solely a function of the properties of reacting materials and reaction conditions (e.g., concentration [activities], temperature, pressure, catalyst, or solvent [if any]) and does not depend on reactor type
   - An intensive quantity (i.e., it does not depend on the total amount)
   - An algebraic equation, not a differential equation (e.g., $-r_A = kC_A$, $-r_A = kC_A^2$)

For homogeneous catalytic systems, typical units of $-r_j$ may be gram moles per second per liter; for heterogeneous systems, typical units of r_j' may be gram moles per second per gram of catalyst. By convention, $-r_A$ is the rate of disappearance of species A and r_A is the rate of formation of species A.

3. Mole balances on species A in four common reactors are shown in Table S1-1.

TABLE S1-1   SUMMARY OF REACTOR MOLE BALANCES

Reactor	Comment	Mole Balance Differential Form	Algebraic Form	Integral Form
BR	No spatial variations	$\dfrac{dN_A}{dt} = r_A V$	—	$t_1 = \displaystyle\int_{N_{A1}}^{N_{A0}} \dfrac{dN_A}{-r_A V}$
CSTR	No spatial variations, steady state	—	$V = \dfrac{F_{A0} - F_A}{-r_A}$	—
PFR	Steady state	$\dfrac{dF_A}{dV} = r_A$		$V_1 = \displaystyle\int_{F_{A1}}^{F_{A0}} \dfrac{dF_A}{-r_A}$
PBR	Steady state	$\dfrac{dF_A}{dW} = r_A'$		$W_1 = \displaystyle\int_{F_{A1}}^{F_{A0}} \dfrac{dF_A}{-r_A'}$

CRE WEB SITE MATERIALS

- **Expanded Materials**
  1. *Industrial Reactors*
- **Learning Resources**
  1. *Summary Notes*
  2. *Web Material*
     A. Problem-Solving Algorithm
     B. Getting Unstuck on a Problem
        This Web site gives tips on how to overcome mental barriers in problem solving.
     C. Smog in L.A. Web module includes a *Living Example Problem*.

B. Getting Unstuck   **C. Smog in L.A.**

Fotografiert von ©2002 Hank Good.

3. *Interactive Computer Games*
   A. Quiz Show I

4. *Solved Problems*
   CDP1-A$_B$ Batch Reactor Calculations: A Hint of Things to Come

- **FAQ [Frequently Asked Questions]**—In Updates/FAQ icon section
- **Professional Reference Shelf**
  R1.1 *Photos of Real Reactors*

R1.2 *Reactor Section of the Visual Encyclopedia of Equipment (encyclopedia.che.engin.umich.edu)*
This section of the CRE Web site shows industrial equipment and discusses its operation. The reactor portion of this encyclopedia is included on the CRE Web site.

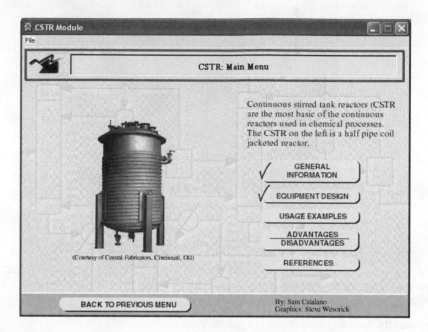

R1.3 *Industrial Reactors*
    A. Liquid Phase
        • Reactor sizes and costs
        • Battery of stirred tanks
        • Semibatch
    B. Gas Phase
        • Costs
        • Fluidized bed schematic
R1.4 *Top Ten List of Chemical Products and Chemical Companies*

QUESTIONS AND PROBLEMS

> I wish I had an answer for that, because I'm getting tired of answering that question.
>
> —Yogi Berra, New York Yankees
> *Sports Illustrated,* June 11, 1984

The subscript to each of the problem numbers indicates the level of difficulty, i.e., A, least difficult; B, moderate difficulty; C, fairly difficult; D, (double black diamond), most difficult. A = ●   B = ■   C = ◆   D = ◆◆ For example, P1-5$_B$ means "1" is the Chapter number, "5" is the problem number, "$_B$" is the problem difficulty, in this case B means moderate difficulty.

    Before solving the problems, state or sketch qualitatively the expected results or trends.

Questions

Q1-1$_A$ Read through the Preface. Write a paragraph describing both the content goals and the intellectual goals of the course and text. Also describe what's on the Web site and how the Web site can be used with the text and course.

Q1-2$_A$ View the photos and schematics on the Web site under Essentials of Chemical Reaction Engineering—Chapter 1. Look at the QuickTime videos. Write a paragraph describing two or more of the reactors. What similarities and differences do you observe between the reactors on the Web (e.g., *www.loe-bequipment.com*), on the Web site, and in the text? How do the used reactor prices compare with those in Table 1-1?

Q1-3$_A$ Surf the Web and the CRE Web site (*www.umich.edu/~elements/5e/index.html*). Go on a scavenger hunt using the summary notes for Chapter 1 on the Web site. Take a quick look at the Web Modules and list the ones that you feel are the most novel applications of CRE.

Q1-4$_A$ What does a negative number for the rate of formation of species (e.g., species A), $r_A = -3 \, mol/dm^3 \cdot s$, signify? What does a positive number signify? Explain.

Q1-5$_A$ What assumptions were made in the derivation of the design equation for:
(a) The batch reactor (BR)?
(b) The CSTR?
(c) The plug-flow reactor (PFR)?
(d) The packed-bed reactor (PBR)?
(e) State in words the meanings of $-r_A$ and $-r'_A$.

Q1-6$_A$ Use the mole balance to derive an equation analogous to Equation (1-7) for a fluidized CSTR containing catalyst particles in terms of the catalyst weight, W, and other appropriate terms.

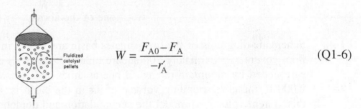

$$W = \frac{F_{A0} - F_A}{-r'_A} \tag{Q1-6}$$

Figure Q1-6 Fluidized Bed CSTR.

Problems

P1-1$_A$ **(a) Revisit Example 1-1.** Rework this example using Equation (3-1) on page 71.
(b) Revisit Example 1-2. Calculate the volume of a CSTR for the conditions used to calculate the plug-flow reactor volume in Example 1-2. Which volume is larger, the PFR or the CSTR? Explain why. Suggest two ways to work this problem incorrectly.
(c) Revisit Example 1-2. Calculate the time to reduce the number of moles of A to 1% of its initial value in a constant-volume batch reactor for the same reaction and data in Example 1-2. Suggest two ways to work this problem incorrectly.

P1-2$_A$ **(a)** Find the Interactive Computer Games (ICG) on the CRE Web site. Play this game and then record your performance number, which indicates your mastery of the material.

ICG Quiz Show		
Mole Balance	Reactions	Rate Laws
100	100	100
200	200	200
300	300	300

ICG Kinetics Challenge 1 Performance # _____

P1-3$_A$ The reaction

$$A + B \longrightarrow 2C$$

takes place in an unsteady CSTR. The feed is only A and B in equimolar proportions. Which of the following sets of equations gives the correct set of mole balances on A, B, and C? Species A and B are disappearing and species C is being formed. Circle the correct answer where <u>all</u> the mole balances are correct.

(a)
$$F_{B0} - F_A - \int^V r_A dV = \frac{dN_A}{dt}$$

$$F_{B0} - F_B - \int^V r_A dV = \frac{dN_B}{dt}$$

$$-F_C + 2\int^V r_A dV = \frac{dN_C}{dt}$$

(b)
$$F_{A0} - F_A + \int^V r_A dV = \frac{dN_A}{dt}$$

$$F_{A0} - F_B + \int^V r_A dV = \frac{dN_B}{dt}$$

$$-F_C - 2\int^V r_A dV = \frac{dN_C}{dt}$$

(c)
$$F_{A0} - F_A + \int_0^V r_A dV = \frac{dN_A}{dt}$$

$$F_{A0} - F_B + \int^V r_A dV = \frac{dN_B}{dt}$$

$$F_C + \int^V r_C dV = \frac{dN_C}{dt}$$

(d)
$$F_{B0} - F_A - \int^V r_A dV = \frac{dN_A}{dt}$$

$$F_{B0} - F_{A0} - \int^V r_A dV = \frac{dN_B}{dt}$$

$$-F_C + \int^V r_C dV = \frac{dN_C}{dt}$$

(e) None of the above.

P1-4$_B$ Schematic diagrams of the Los Angeles basin are shown in Figure P1-4$_B$. The basin floor covers approximately 700 square miles (2×10^{10} ft^2) and is almost completely surrounded by mountain ranges. If one assumes an inversion height in the basin of 2,000 ft, the corresponding volume of air in the basin is 4×10^{13} ft^3. We shall use this system volume to model the accumulation and depletion of air pollutants. As a very rough first approximation, we shall treat the Los Angeles basin as a well-mixed container (analogous to a CSTR) in which there are no spatial variations in pollutant concentrations.

Hall of Fame

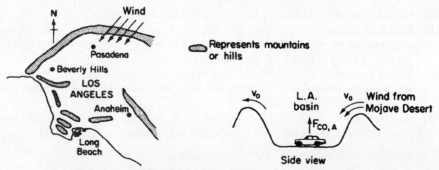

Figure P1-4$_B$ Schematic diagrams of the Los Angeles basin.

We shall perform an unsteady-state mole balance (Equation (1–4)) on CO as it is depleted from the basin area by a Santa Ana wind. Santa Ana winds are high-velocity winds that originate in the Mojave Desert just to the northeast of Los Angeles. Load the **Smog in Los Angeles Basin Web Module**. Use the data in the module to work parts 1–12 (a) through (h) given in the module. Load the **Living Example Polymath code** and explore the problem. For part (i), vary the parameters v_0, a, and b, and write a paragraph describing what you find.

There is heavier traffic in the L.A. basin in the mornings and in the evenings as workers go to and from work in downtown L.A. Consequently, the flow of CO into the L.A. basin might be better represented by the sine function over a 24-hour period.

P1-5$_B$ The reaction

$$A \longrightarrow B$$

is to be carried out isothermally in a continuous-flow reactor. The entering volumetric flow rate v_0 is 10 dm^3/h. (*Note*: $F_A = C_A v$. For a constant volumetric flow rate $v = v_0$, then $F_A = C_A v_0$. Also, $C_{A0} = F_{A0}/v_0 = ([5 \text{ mol/h}]/[10 \text{ dm}^3/\text{h}])$ 0.5 mol/dm^3.)

Calculate both the CSTR and PFR reactor volumes necessary to consume 99% of A (i.e., $C_A = 0.01 C_{A0}$) when the entering molar flow rate is 5 mol/h, assuming the reaction rate $-r_A$ is

(a) $-r_A = k$ with $k = 0.05 \dfrac{\text{mol}}{\text{h} \cdot \text{dm}^3}$ [*Ans.*: $V_{CSTR} = 99$ dm^3]

(b) $-r_A = kC_A$ with $k = 0.0001$ s^{-1}

(c) $-r_A = kC_A^2$ with $k = 300 \dfrac{\text{dm}^3}{\text{mol} \cdot \text{h}}$ [*Ans.*: $V_{CSTR} = 660$ dm^3]

(d) Repeat **(a)**, **(b)**, and/or **(c)** to calculate the time necessary to consume 99.9% of species A in a 1000 dm^3 constant-volume batch reactor with $C_{A0} = 0.5$ mol/dm^3.

P1-6$_B$ This problem focuses on using Polymath, an ordinary differential equation (ODE) solver, and also a nonlinear equation (NLE) solver. These equation solvers will be used extensively in later chapters. Information on how to obtain and load the Polymath Software is given in **Appendix D** and on the CRE Web site.

(a) There are initially 500 rabbits (x) and 200 foxes (y) on Farmer Oat's property near Riça, Jofostan. Use Polymath or MATLAB to plot the concentration of foxes and rabbits as a function of time for a period of up to 500 days. The predator–prey relationships are given by the following set of coupled ordinary differential equations:

$$\frac{dx}{dt} = k_1 x - k_2 x \cdot y$$

$$\frac{dy}{dt} = k_3 x \cdot y - k_4 y$$

Constant for growth of rabbits $k_1 = 0.02$ day^{-1}
Constant for death of rabbits $k_2 = 0.00004/(\text{day} \times \text{no. of foxes})$
Constant for growth of foxes after eating rabbits $k_3 = 0.0004/(\text{day} \times \text{no. of rabbits})$
Constant for death of foxes $k_4 = 0.04$ day^{-1}
What do your results look like for the case of $k_3 = 0.00004/(\text{day} \times \text{no. of rabbits})$ and $t_{final} = 800$ days? Also, plot the number of foxes versus the number of rabbits. Explain why the curves look the way they do.

(b) Use Polymath or MATLAB to solve the following set of nonlinear algebraic equations

$$x^3 y - 4y^2 + 3x = 1$$

$$6y^2 - 9xy = 5$$

with inital guesses of $x = 2$, $y = 2$. Try to become familiar with the edit keys in Polymath and MATLAB. See the CRE Web site for instructions
    Screen shots on how to run Polymath are shown at the end of Summary Notes for Chapter 1 or on the CRE Web site, *www.umich.edu/~elements/5e/software/polymath-tutorial.html.*

P1-7$_A$ **Enrico Fermi (1901–1954) Problems (EFP).** Enrico Fermi was an Italian physicist who received the Nobel Prize for his work on nuclear processes. Fermi was famous for his "Back of the Envelope Order of Magnitude Calculation" to obtain an estimate of the answer through *logic* and then to make reasonable assumptions. He used a process to set bounds on the answer by saying it is probably larger than one number and smaller than another, and arrived at an answer that was within a factor of 10. See *http://mathforum.org/workshops/sum96/interdisc/sheila2.html.*

Enrico Fermi Problem

(a) **EFP #1.** How many piano tuners are there in the city of Chicago? Show the steps in your reasoning.

1. Population of Chicago _____

2. Number of people per household _____

3. Etc. _____

   An answer is given on the CRE Web site under Summary Notes for Chapter 1.

(b) **EFP #2.** How many square meters of pizza were eaten by an undergraduate student body population of 20,000 during the Fall term 2016?

(c) **EFP #3.** How many bathtubs of water will the average person drink in a lifetime?

P1-8$_A$ **What is wrong with this solution?** The irreversible liquid phase second order reaction $(-r_A = kC_A^2)$

$$2A \xrightarrow{k_1} B \quad k = 0.03\,\text{dm}^3/\text{mol}\cdot\text{s}$$

is carried out in a CSTR. The entering concentration of A, C_{A0}, is 2 molar. and the exit concentration of A, C_A is 0.1 molar. The volumetric flow rate, v_0, is constant at 3 dm³/s. What is the corresponding reactor volume?

Solution

1. Mole Balance

$$V = \frac{F_{A0} - F_A}{-r_A}$$

2. Rate Law (2nd order)

$$-r_A = kC_A^2$$

3. Combine

$$V = \frac{F_{A0} - F_A}{kC_A^2}$$

4. $F_{A0} = v_o C_{A0} = \dfrac{3\ \text{dm}^3}{s} \cdot \dfrac{2\ \text{molA}}{\text{dm}^3} = \dfrac{6\ \text{molA}}{s}$

5. $F_A = v_o C_A = \dfrac{3\ \text{dm}^3}{s} \cdot \dfrac{0.1\ \text{molA}}{\text{dm}^3} = \dfrac{0.3\ \text{molA}}{s}$

6. $V = \dfrac{(6 - 0.3)\dfrac{\text{mol}}{s}}{\left(0.03\dfrac{\text{dm}^3}{\text{mol}\cdot s}\right)\left(2\dfrac{\text{mol}}{\text{dm}^3}\right)^2} = 47.5\ \text{dm}^3$

For more puzzles on what's wrong with this solution, see additional material for each chapter on the CRE Web site home page, under "Expanded Material."

Solved Problems

NOTE TO INSTRUCTORS: Additional problems (cf. those from the preceding editions) can be found in the solutions manual and on the CRE Web site These problems could be photocopied and used to help reinforce the fundamental principles discussed in this chapter.

SUPPLEMENTARY READING

1. For further elaboration of the development of the general balance equation, see not only the Web site *www.umich.edu/~elements/5e/index.html* but also

   FELDER, R. M., and R. W. ROUSSEAU, *Elementary Principles of Chemical Processes*, 3rd ed. New York: Wiley, 2000, Chapter 4.

   SANDERS, R. J., *The Anatomy of Skiing*. Denver, CO: Golden Bell Press, 1976.

2. A detailed explanation of a number of topics in this chapter can be found in the tutorials.

   CRYNES, B. L., and H. S. FOGLER, eds., *AIChE Modular Instruction Series E: Kinetics*, Vols. 1 and 2. New York: AIChE, 1981.

3. A discussion of some of the most important industrial processes is presented by

   AUSTIN, G. T., *Shreve's Chemical Process Industries*, 5th ed. New York: McGraw-Hill, 1984.

Conversion and Reactor Sizing 2

Be more concerned with your character than with your
reputation, because character is what you really are
while reputation is merely what others think you are.
— John Wooden, head coach, UCLA Bruins

Overview. In the first chapter, the general mole balance equation was
derived and then applied to the four most common types of industrial
reactors. A balance equation was developed for each reactor type and
these equations are summarized in Table S1-1 in Chapter 1. In Chapter 2,
we will show how to size and arrange these reactors conceptually, so that
the reader may see the structure of CRE design and will not get lost in
the mathematical details.

In this chapter, we

- Define conversion
- Rewrite all balance equations for the four types of industrial
  reactors in Chapter 1 in terms of conversion, X
- Show how to size (i.e., determine the reactor volume) these
  reactors once the relationship between the reaction rate and conversion is known—i.e., given $-r_A = f(X)$
- Show how to compare CSTR and PFR sizes
- Show how to decide the best arrangements for reactors in series,
  a most important principle

In addition to being able to determine CSTR and PFR sizes given the
rate of reaction as a function of conversion, you will be able to calculate
the overall conversion and reactor volumes for reactors arranged in
series.

2.1 Definition of Conversion

In defining conversion, we choose one of the reactants as the basis of calculation and then relate the other species involved in the reaction to this basis. In virtually all instances we must choose the limiting reactant as the basis of calculation. We develop the stoichiometric relationships and design equations by considering the general reaction

$$a\text{A} + b\text{B} \longrightarrow c\text{C} + d\text{D} \tag{2-1}$$

The uppercase letters represent chemical species, and the lowercase letters represent stoichiometric coefficients. We shall choose species A as our limiting reactant and, thus, our *basis of calculation*. The limiting reactant is the reactant that will be completely consumed first after the reactants have been mixed. Next, we divide the reaction expression through by the stoichiometric coefficient of species A, in order to arrange the reaction expression in the form

$$\text{A} + \frac{b}{a}\,\text{B} \longrightarrow \frac{c}{a}\,\text{C} + \frac{d}{a}\,\text{D} \tag{2-2}$$

to put every quantity on a "per mole of A" basis, our limiting reactant.

Now we ask such questions as "How can we quantify how far a reaction [e.g., Equation (2-2)] proceeds to the right?" or "How many moles of C are formed for every mole of A consumed?" A convenient way to answer these questions is to define a parameter called *conversion*. The conversion X_A is the number of moles of A that have reacted per mole of A fed to the system:

Definition of X
$$X_A = \frac{\text{Moles of A reacted}}{\text{Moles of A fed}}$$

Because we are defining conversion with respect to our basis of calculation [A in Equation (2-2)], we eliminate the subscript A for the sake of brevity and let $X \equiv X_A$. For irreversible reactions, the maximum conversion is 1.0, i.e., complete conversion. For reversible reactions, the maximum conversion is the equilibrium conversion X_e (i.e., $X_{max} = X_e$). We will take a closer look at equilibrium conversion in Chapter 4.

2.2 Batch Reactor Design Equations

In most batch reactors, the longer a reactant stays in the reactor, the more the reactant is converted to product until either equilibrium is reached or the reactant is exhausted. Consequently, in batch systems the conversion X is a function of the time the reactants spend in the reactor. If N_{A0} is the number of moles of A initially present in the reactor (i.e., $t = 0$), then the total number of moles of A that have reacted (i.e., have been consumed) after a time t is $[N_{A0}X]$.

$$[\text{Moles of A reacted (consumed)}] = [\text{Moles of A fed}] \cdot \left[\frac{\text{Moles of A reacted}}{\text{Moles of A fed}} \right]$$

$$\begin{bmatrix} \text{Moles of A} \\ \text{reacted} \\ \text{(consumed)} \end{bmatrix} \quad = \quad [N_{A0}] \quad \cdot \quad [X] \qquad (2\text{-}3)$$

Now, the number of moles of A that remain in the reactor after a time t, N_A, can be expressed in terms of N_{A0} and X:

$$\begin{bmatrix} \text{Moles of A} \\ \text{in reactor} \\ \text{at time } t \end{bmatrix} = \begin{bmatrix} \text{Moles of A} \\ \text{initially fed} \\ \text{to reactor at} \\ t = 0 \end{bmatrix} - \begin{bmatrix} \text{Moles of A that} \\ \text{have been con-} \\ \text{sumed by chemical} \\ \text{reaction} \end{bmatrix}$$

$$[N_A] \quad = \quad [N_{A0}] \quad - \quad [N_{A0}X]$$

The number of moles of A in the reactor after a conversion X has been achieved is

Moles of A in the
reactor at a time t

$$\boxed{N_A = N_{A0} - N_{A0}X = N_{A0}(1 - X)} \qquad (2\text{-}4)$$

When no spatial variations in reaction rate exist, the mole balance on species A for a batch system is given by the following equation [cf. Equation (1-5)]:

$$\frac{dN_A}{dt} = r_A V \qquad (2\text{-}5)$$

This equation is valid whether or not the reactor volume is constant. In the general reaction, Equation (2-2), reactant A is disappearing; therefore, we multiply both sides of Equation (2-5) by –1 to obtain the mole balance for the batch reactor in the form

$$-\frac{dN_A}{dt} = (-r_A)V$$

The rate of disappearance of A, $-r_A$, in this reaction might be given by a rate law similar to Equation (1-2), such as $-r_A = kC_A C_B$.

For batch reactors, we are interested in determining how long to leave the reactants in the reactor to achieve a certain conversion X. To determine this length of time, we write the mole balance, Equation (2-5), in terms of conversion by differentiating Equation (2-4) with respect to time, remembering that N_{A0} is the number of moles of A initially present in the reactor and is therefore a constant with respect to time.

$$\frac{dN_A}{dt} = 0 - N_{A0}\frac{dX}{dt}$$

Combining the above with Equation (2-5) yields

$$-N_{A0}\frac{dX}{dt} = r_A V$$

For a batch reactor, the design equation in differential form is

Batch reactor (BR)
design equation

$$\boxed{N_{A0}\frac{dX}{dt} = -r_A V} \tag{2-6}$$

We call Equation (2-6) the differential form of the **design equation** for a batch reactor because we have written the mole balance in terms of conversion. The differential forms of the batch reactor mole balances, Equations (2-5) and (2-6), are often used in the interpretation of reaction rate data (Chapter 7) and for reactors with heat effects (Chapters 11–13), respectively. Batch reactors are frequently used in industry for both gas-phase and liquid-phase reactions. The laboratory bomb calorimeter reactor is widely used for obtaining reaction rate data. Liquid-phase reactions are frequently carried out in batch reactors when small-scale production is desired or operating difficulties rule out the use of continuous-flow systems.

To determine the time to achieve a specified conversion X, we first separate the variables in Equation (2-6) as follows:

$$dt = N_{A0}\frac{dX}{-r_A V}$$

Batch time t
to achieve a
conversion X

This equation is now integrated with the limits that the reaction begins at time equals zero where there is no conversion initially (when $t = 0$, $X = 0$) and ends at time t when a conversion X is achieved (i.e., when $t = t$, then $X = X$). Carrying out the integration, we obtain the time t necessary to achieve a conversion X in a batch reactor

Batch Design
Equation

$$\boxed{t = N_{A0}\int_0^X \frac{dX}{-r_A V}} \tag{2-7}$$

The longer the reactants are left in the reactor, the greater the conversion will be. Equation (2-6) is the differential form of the design equation, and Equation (2-7) is the integral form of the design equation for a batch reactor.

2.3 Design Equations for Flow Reactors

For a batch reactor, we saw that conversion increases with time spent in the reactor. For continuous-flow systems, this time usually increases with increasing reactor volume, e.g., the bigger/longer the reactor, the more time it will take the reactants to flow completely through the reactor and thus, the more time to react. Consequently, the conversion X is a function of reactor volume V. If F_{A0} is the molar flow rate of species A fed to a system operated at steady state, the molar rate at which species A is reacting *within* the entire system will be $F_{A0}X$.

$$[F_{A0}] \cdot [X] = \frac{\text{Moles of A fed}}{\text{time}} \cdot \frac{\text{Moles of A reacted}}{\text{Moles of A fed}}$$

$$[F_{A0} \cdot X] = \frac{\text{Moles of A reacted}}{\text{time}}$$

The molar feed rate of A *to* the system *minus* the rate of reaction of A within the system *equals* the molar flow rate of A leaving the system F_A. The preceding sentence can be expressed mathematically as

$$\begin{bmatrix} \text{Molar flow rate} \\ \text{at which A is} \\ \text{fed to the system} \end{bmatrix} - \begin{bmatrix} \text{Molar rate at} \\ \text{which A is} \\ \text{consumed within} \\ \text{the system} \end{bmatrix} = \begin{bmatrix} \text{Molar flow rate} \\ \text{at which A leaves} \\ \text{the system} \end{bmatrix}$$

$$[F_{A0}] \qquad - \qquad [F_{A0}X] \qquad = \qquad [F_A]$$

Rearranging gives

$$\boxed{F_A = F_{A0}(1 - X)} \tag{2-8}$$

The entering molar flow rate of species A, F_{A0} (mol/s), is just the product of the entering concentration, C_{A0} (mol/dm^3), and the entering volumetric flow rate, v_0 (dm^3/s).

$$\boxed{F_{A0} = C_{A0} v_0} \tag{2-9}$$

Liquid phase   *For liquid systems*, the volumetric flow rate, v, is constant and equal to v_0, and C_{A0} is commonly given in terms of molarity, for example, $C_{A0} = 2$ mol/dm^3.

For gas systems, C_{A0} can be calculated from the entering mole fraction, y_{A0}, the temperature, T_0, and pressure, P_0, using the ideal gas law or some other gas law. For an ideal gas (see Appendix B):

Gas phase
$$C_{A0} = \frac{P_{A0}}{RT_0} = \frac{y_{A0}P_0}{RT_0} \qquad (2\text{-}10)$$

Now that we have a relationship [Equation (2-8)] between the molar flow rate and conversion, it is possible to express the design equations (i.e., mole balances) in terms of conversion for the *flow* reactors examined in Chapter 1.

2.3.1 CSTR (Also Known as a Backmix Reactor or a Vat)

Recall that the CSTR is modeled as being well mixed such that there are no spatial variations in the reactor. For the general reaction

$$A + \frac{b}{a}\,B \longrightarrow \frac{c}{a}\,C + \frac{d}{a}\,D \qquad (2\text{-}2)$$

the CSTR mole balance Equation (1-7) can be arranged to

$$V = \frac{F_{A0} - F_A}{-r_A} \qquad (2\text{-}11)$$

We now substitute for F_A in terms of F_{A0} and X

$$F_A = F_{A0} - F_{A0}X \qquad (2\text{-}12)$$

and then substitute Equation (2-12) into (2-11)

$$V = \frac{F_{A0} - (F_{A0} - F_{A0}X)}{-r_A}$$

Simplifying, we see that the CSTR volume necessary to achieve a specified conversion X is

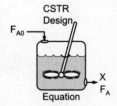

CSTR
Design
Equation

Perfect mixing

F_{A0}

X
F_A

$$\boxed{V = \frac{F_{A0}X}{(-r_A)_{exit}}} \qquad (2\text{-}13)$$

Evaluate $-r_A$ at
the CSTR exit
conditions!!

Because the reactor is *perfectly mixed*, the exit composition from the reactor is identical to the composition inside the reactor, and, therefore, the rate of reaction, $-r_A$, is evaluated at the exit conditions.

2.3.2 Tubular Flow Reactor (PFR)

We model the tubular reactor as having the fluid flowing in plug flow—i.e., no radial gradients in concentration, temperature, or reaction rate.[1] As the reactants enter and flow axially down the reactor, they are consumed and the

---

[1] This constraint can be removed when we extend our analysis to nonideal (industrial) reactors in Chapters 16 through 18.

conversion increases along the length of the reactor. To develop the PFR design equation, we first multiply both sides of the tubular reactor design equation (1-12) by –1. We then express the mole balance equation for species A in the reaction as

$$\frac{-dF_A}{dV} = -r_A \tag{2-14}$$

For a flow system, F_A has previously been given in terms of the entering molar flow rate F_{A0} and the conversion X

$$F_A = F_{A0} - F_{A0}X \tag{2-12}$$

Differentiating

$$dF_A = -F_{A0}dX$$

and substituting into (2-14) gives the differential form of the design equation for a plug-flow reactor (PFR)

$$\boxed{F_{A0}\frac{dX}{dV} = -r_A} \tag{2-15}$$

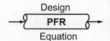

Design
PFR
Equation

We now separate the variables and integrate with the limits $V = 0$ when $X = 0$ to obtain the plug-flow reactor volume necessary to achieve a specified conversion X

$$\boxed{V = F_{A0}\int_0^X \frac{dX}{-r_A}} \tag{2-16}$$

To carry out the integrations in the batch and plug-flow reactor design equations (2-7) and (2-16), as well as to evaluate the CSTR design equation (2-13), we need to know how the reaction rate $-r_A$ varies with the concentration (hence conversion) of the reacting species. This relationship between reaction rate and concentration is developed in Chapter 3.

2.3.3 Packed-Bed Reactor (PBR)

Packed-bed reactors are tubular reactors filled with catalyst particles. In PBRs it is the weight of catalyst W that is important, rather than the reactor volume. The derivation of the differential and integral forms of the design equations for packed-bed reactors are analogous to those for a PFR [cf. Equations (2-15) and (2-16)]. That is, substituting Equation (2-12) for F_A in Equation (1-15) gives

PBR design
equation

$$\boxed{F_{A0}\frac{dX}{dW} = -r'_A} \tag{2-17}$$

The differential form of the design equation [i.e., Equation (2-17)] **must** be used when analyzing reactors that have a pressure drop along the length of the reactor. We discuss pressure drop in packed-bed reactors in Chapter 5.

In the *absence* of pressure drop, i.e., $\Delta P = 0$, we can integrate (2-17) with limits $X = 0$ at $W = 0$, and when $W = W$ then $X = X$ to obtain

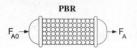

PBR

$$W = F_{A0} \int_0^X \frac{dX}{-r_A'} \qquad (2\text{-}18)$$

Equation (2-18) can be used to determine the catalyst weight W (i.e., mass) necessary to achieve a conversion X when the total pressure remains constant.

2.4 Sizing Continuous-Flow Reactors

In this section, we are going to show how we can size CSTRs and PFRs (i.e., determine their reactor volumes) from knowledge of the rate of reaction, $-r_A$, as a function of conversion, X [i.e., $-r_A = f(X)$]. The rate of disappearance of A, $-r_A$, is almost always a function of the concentrations of the various species present (see Chapter 3). When only one reaction is occurring, each of the concentrations can be expressed as a function of the conversion X (see Chapter 4); consequently, $-r_A$ can be expressed as a function of X.

A particularly simple functional dependence, yet one that occurs often, is the first-order dependence

$$-r_A = kC_A = kC_{A0}(1 - X)$$

Here, k is the specific reaction rate and is a function only of temperature, and C_{A0} is the entering concentration of A. We note in Equations (2-13) and (2-16) that the reactor volume is a function of the reciprocal of $-r_A$. For this first-order dependence, a plot of the reciprocal rate of reaction ($1/-r_A$) as a function of conversion yields a curve similar to the one shown in Figure 2-1, where

$$\frac{1}{-r_A} = \frac{1}{kC_{A0}}\left(\frac{1}{1 - X}\right)$$

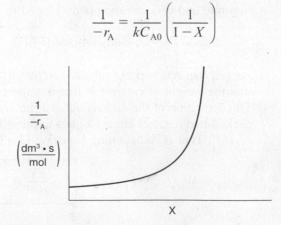

Figure 2-1 Reciprocal rate as a function of conversion.

We can use Figure 2-1 to size CSTRs and PFRs for different entering flow rates. By *sizing* we mean either determine the reactor volume for a specified conversion or determine the conversion for a specified reactor volume. Before sizing flow reactors, let's consider some insights. If a reaction is carried out isothermally, the rate is usually greatest at the start of the reaction when the concentration of reactant is greatest (i.e., when there is negligible conversion $[X \cong 0]$). Hence, the reciprocal rate $(1/-r_A)$ will be small. Near the end of the reaction, when the reactant has been mostly used up and thus the concentration of A is small (i.e., the conversion is large), the reaction rate will be small. Consequently, the reciprocal rate $(1/-r_A)$ is large.

For all irreversible reactions of greater than zero order (see Chapter 3 for zero-order reactions), as we approach complete conversion where all the limiting reactant is used up, i.e., $X = 1$, the reciprocal rate approaches infinity as does the reactor volume, i.e.

$A \rightarrow B + C$ $\qquad$ As $X \rightarrow 1, -r_A \rightarrow 0$, thus, $\dfrac{1}{-r_A} \rightarrow \infty$ and therefore $V \rightarrow \infty$

"To infinity and beyond" —Buzz Lightyear

Consequently, we see that an infinite reactor volume is necessary to reach complete conversion, $X = 1.0$.

For reversible reactions (e.g., A $\rightleftarrows$ B), the maximum conversion is the equilibrium conversion X_e. At equilibrium, the reaction rate is zero ($r_A \equiv 0$). Therefore,

$A \rightleftarrows B + C$ $\qquad$ as $X \rightarrow X_e, -r_A \rightarrow 0$, thus, $\dfrac{1}{-r_A} \rightarrow \infty$ and therefore $V \rightarrow \infty$

and we see that an infinite reactor volume would also be necessary to obtain the exact equilibrium conversion, $X = X_e$. We will discuss X_e further in Chapter 4.

Examples of Reactor Design and Staging Given $-r_A = f(X)$

To illustrate the design of continuous-flow reactors (i.e., CSTRs and PFRs), we consider the isothermal gas-phase isomerization

$$A \longrightarrow B$$

We are going to the laboratory to determine the rate of chemical reaction as a function of the conversion of reactant A. The laboratory measurements given in Table 2-1 show the chemical reaction rate as a function of conversion. The temperature was 500 K (440°F), the total pressure was 830 kPa (8.2 atm), and the initial charge to the reactor was pure A. The entering molar flow of A rate is $F_{A0} = 0.4$ mol/s.

If we know $-r_A$ as a function of X, we can size any isothermal reaction system.

TABLE 2-1   RAW DATA[†]

X	0	0.1	0.2	0.4	0.6	0.7	0.8
$-r_A$ (mol/m$^3 \cdot$ s)	0.45	0.37	0.30	0.195	0.113	0.079	0.05

[†] Proprietary coded data courtesy of Jofostan Central Research Laboratory, Çölow, Jofostan, and published in *Jofostan Journal of Chemical Engineering Research*, Volume 21, page 73 (1993).

Recalling the CSTR and PFR design equations, (2-13) and (2-16), we see that the reactor volume varies directly with the molar flow rate F_{A0} and with the reciprocal of $-r_A$, $\left(\dfrac{1}{-r_A}\right)$, e.g., $V = \left(\dfrac{F_{A0}}{-r_A}\right)X$. Consequently, to size reactors, we first convert the raw data in Table 2-1, which gives $-r_A$ as a function of X first to $\left(\dfrac{1}{-r_A}\right)$ as a function of X. Next, we multiply by the entering molar flow rate, F_{A0}, to obtain $\left(\dfrac{F_{A0}}{-r_A}\right)$ as a function of X as shown in Table 2-2 of the processed data for $F_{A0} = 0.4$ mol/s.

TABLE 2-2   PROCESSED DATA

We shall use the data in this table for the next five Example Problems.

X	0.0	0.1	0.2	0.4	0.6	0.7	0.8
$-r_A\left(\dfrac{\text{mol}}{\text{m}^3 \cdot \text{s}}\right)$	0.45	0.37	0.30	0.195	0.113	0.079	0.05
$(1/-r_A)\left(\dfrac{\text{m}^3 \cdot \text{s}}{\text{mol}}\right)$	2.22	2.70	3.33	5.13	8.85	12.7	20
$(F_{A0}/-r_A)(\text{m}^3)$	0.89	1.08	1.33	2.05	3.54	5.06	8.0

To size reactors for different entering molar flow rates, F_{A0}, we would use rows 1 and 3 in Table 2-2 to construct the following figure:

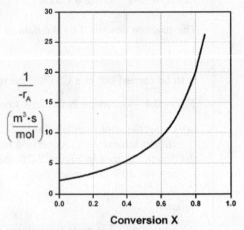

Figure 2-2A   Processed data 1.

However, for a given F_{A0}, rather than use Figure 2-2A to size reactors, it is often more advantageous to plot $\left(\dfrac{F_{A0}}{-r_A}\right)$ as a function of X, which is called a Levenspiel plot. We are now going to carry out a number of examples where we have specified the flow rate F_{A0} at 0.4 mol A/s.

Plotting $\left(\dfrac{F_{A0}}{-r_A}\right)$ as a function of X using the data in Table 2-2 we obtain the plot shown in Figure 2-2B.

Levenspiel plot

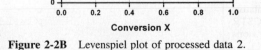

Figure 2-2B   Levenspiel plot of processed data 2.

We are now going to use the Levenspiel plot of the processed data (Figure 2-2B) to size a CSTR and a PFR.

Example 2–1   Sizing a CSTR

The reaction described by the data in Table 2-2

$$A \rightarrow B$$

is to be carried out in a CSTR. Species A enters the reactor at a molar flow rate of $F_{A0} = 0.4\,\dfrac{mol}{s}$, which is the flow rate used to construct Figure 2-2B.

(a) Using the data in either Table 2-2 or Figure 2-2B, calculate the volume necessary to achieve 80% conversion in a CSTR.

(b) Shade the area in Figure 2-2B that would give the CSTR volume necessary to achieve 80% conversion.

Solutions

CSTR
Design

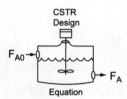

Equation

(a) Equation (2-13) gives the volume of a CSTR as a function of F_{A0}, X, and $-r_A$

$$V = \frac{F_{A0}X}{(-r_A)_{exit}} \tag{2-13}$$

In a CSTR, the composition, temperature, and conversion of the effluent stream are identical to that of the fluid within the reactor, because perfect mixing is assumed. Therefore, we need to find the value of $-r_A$ (or reciprocal thereof) at $X = 0.8$. From either Table 2-2 or Figure 2-2A, we see that when $X = 0.8$, then

$$\left(\frac{1}{-r_A}\right)_{X=0.8} = 20\,\frac{m^3 \cdot s}{mol}$$

Substitution into Equation (2-13) for an entering molar flow rate, F_{A0}, of 0.4 mol A/s and $X = 0.8$ gives

$$V = 0.4\frac{mol}{s}\left(\frac{20\ m^3 \cdot s}{mol}\right)(0.8) = 6.4\ m^3 \tag{E2-1.1}$$

$$V = 6.4\ m^3 = 6400\ dm^3 = 6400\ \text{liters}$$

(b) Shade the area in Figure 2-2B that yields the CSTR volume. Rearranging Equation (2-13) gives

$$V = \left[\frac{F_{A0}}{-r_A}\right]X \tag{2-13}$$

In Figure E2-1.1, the volume is equal to the area of a rectangle with a height $(F_{A0}/-r_A = 8\ m^3)$ and a base $(X = 0.8)$. This rectangle is shaded in the figure.

Representative
Industrial CSTR
Dimensions

$$V = \left[\frac{F_{A0}}{-r_A}\right]_{X=0.8}(0.8) \tag{E2-1.2}$$

$$V = \text{Levenspiel rectangle area} = \text{height} \times \text{width}$$

$$V = [8\ m^3][0.8] = 6.4\ m^3 = 6400\ dm^3 = 6400\ L$$

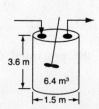

The CSTR volume necessary to achieve 80% conversion is 6.4 m³ when operated at 500 K, 830 kPa (8.2 atm), and with an entering molar flow rate of A of 0.4 mol/s. This volume corresponds to a reactor about 1.5 m in diameter and 3.6 m high. It's a large CSTR, but this is a gas-phase reaction, and CSTRs are normally not used for gas-phase reactions. CSTRs are used primarily for liquid-phase reactions.

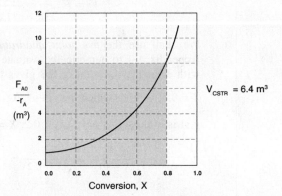

Figure E2-1.1   Levenspiel CSTR plot.

Plots of $(F_{A0}/-r_A)$ vs. X are sometimes referred to as Levenspiel plots (after Octave Levenspiel).

$V_{CSTR} = 6.4$ m^3

Analysis: Given the conversion, the rate of reaction as a function of conversion along with the molar flow of the species A, we saw how to calculate the volume of a CSTR. From the data and information given, we calculated the volume to be 6.4 m^3 for 80% conversion. We showed how to carry out this calculation using the design equation (2-13) and also using a Levenspiel plot.

Example 2–2 Sizing a PFR

The reaction described by the data in Tables 2-1 and 2-2 is to be carried out in a PFR. The entering molar flow rate of A is again 0.4 mol/s.

(a) First, use one of the integration formulas given in Appendix A.4 to determine the PFR reactor volume necessary to achieve 80% conversion.

(b) Next, shade the area in Figure 2-2B that would give the PFR volume necessary to achieve 80% conversion.

(c) Finally, make a qualitative sketch of the conversion, X, and the rate of reaction, $-r_A$, down the length (volume) of the reactor.

Solution

We start by repeating rows 1 and 4 of Table 2-2 to produce the results shown in Table 2-3.

TABLE 2-3   PROCESSED DATA 2

X	0.0	0.1	0.2	0.4	0.6	0.7	0.8
$(F_{A0}/-r_A)$(m^3)	0.89	1.08	1.33	2.05	3.54	5.06	8.0

(a) Numerically evaluate PFR volume. For the PFR, the differential form of the mole balance is

$$F_{A0}\frac{dX}{dV} = -r_A \tag{2-15}$$

Rearranging and integrating gives

$$V = F_{A0} \int_0^{0.8} \frac{dX}{-r_A} = \int_0^{0.8} \frac{F_{A0}}{-r_A} dX \qquad (2\text{-}16)$$

We shall use the *five-point quadrature* formula [Equation (A-23)] given in Appendix A.4 to numerically evaluate Equation (2-16). The five-point formula with a final conversion of 0.8 gives four equal segments between $X = 0$ and $X = 0.8$, with a segment length of $\Delta X = \dfrac{0.8}{4} = 0.2$. The function inside the integral is evaluated at $X = 0$, $X = 0.2$, $X = 0.4$, $X = 0.6$, and $X = 0.8$.

$$V = \frac{\Delta X}{3}\left[\frac{F_{A0}}{-r_A(X=0)} + \frac{4F_{A0}}{-r_A(X=0.2)} + \frac{2F_{A0}}{-r_A(X=0.4)} + \frac{4F_{A0}}{-r_A(X=0.6)} + \frac{F_{A0}}{-r_A(X=0.8)}\right]$$

$$(E2\text{-}2.1)$$

Using values of $[F_{A0}/(-r_A)]$ corresponding to the different conversions in Table 2-3 yields

$$V = \left(\frac{0.2}{3}\right)[0.89 + 4(1.33) + 2(2.05) + 4(3.54) + 8.0]\,\mathrm{m}^3 = \left(\frac{0.2}{3}\right)(32.47\ \mathrm{m}^3)$$

$$\boxed{V = 2.165\ \mathrm{m}^3 = 2165\ \mathrm{dm}^3}$$

The PFR reactor volume necessary to achieve 80% conversion is 2165 dm³. This volume could result from a bank of 100 PFRs that are each 0.1 m in diameter with a length of 2.8 m (e.g., see margin figure or Figures 1-8(a) and (b)).

(b) The PFR volume, i.e., the integral in Equation (2-16), can also be evaluated from the area under the curve of a plot of $(F_{A0}/-r_A)$ versus X.

$$V = \int_0^{0.8} \frac{F_{A0}}{-r_A} dX = \begin{array}{l}\text{Area under the curve between } X = 0 \text{ and } X = 0.8\\ \text{(see shaded area in Figure E2-2.1)}\end{array}$$

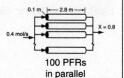

100 PFRs
in parallel

PFR

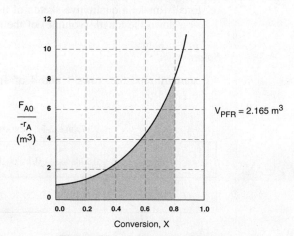

Figure E2-2.1 Levenspiel PFR plot.

The area under the curve will give the tubular reactor volume necessary to achieve the specified conversion of A. For 80% conversion, the shaded area is roughly equal to 2165 dm^3 (2.165 m^3).

(c) Sketch the profiles of $-r_A$ and X down the length of the reactor.

We know that as we proceed down the reactor, the conversion increases as more and more reactant is converted to product. Consequently, as the reactant is consumed, the concentration of reactant decreases, as does the rate of disappearance of A for isothermal reactions.

(i) For $\underline{X = 0.2}$, we calculate the corresponding reactor volume using Simpson's rule [given in Appendix A.4 as Equation (A-21)] with increment $\Delta X = 0.1$ and the data in rows 1 and 4 in Table 2-2.

$$V = F_{A0}\int_0^{0.2} \frac{dX}{-r_A} = \frac{\Delta X}{3}\left[\frac{F_{A0}}{-r_A(X=0)} + \frac{4F_{A0}}{-r_A(X=0.1)} + \frac{F_{A0}}{-r_A(X=0.2)}\right] \quad \text{(E2-2.2)}$$

$$= \left[\frac{0.1}{3}\left[0.89 + 4(1.08) + 1.33\right]\right]m^3 = \frac{0.1}{3}(6.54\ m^3) = 0.218\ m^3 = 218\ dm^3$$

$$= 218\ dm^3$$

This volume (218 dm^3) is the volume at which $X = 0.2$. From Table 2-3, we see the corresponding rate of reaction at $X = 0.2$ is $-r_A = 0.3\ \dfrac{mol}{dm^3 \cdot s}$.

Therefore at $X = 0.2$, then $-r_A = 0.3\ \dfrac{mol}{dm^3 \cdot s}$ and $V = 218\ dm^3$.

(ii) For $\underline{X = 0.4}$, we can again use Table 2-3 and Simpson's rule with $\Delta X = 0.2$ to find the reactor volume necessary for a conversion of 40%.

$$V = \frac{\Delta X}{3}\left[\frac{F_{A0}}{-r_A(X=0)} + \frac{4F_{A0}}{-r_A(X=0.2)} + \frac{F_{A0}}{-r_A(X=0.4)}\right]$$

$$= \left[\frac{0.2}{3}\left[0.89 + 4(1.33) + 2.05\right]\right]m^3 = 0.551\ m^3$$

$$= 551\ dm^3$$

From Table 2-3 we see that at $X = 0.4$, $-r_A = 0.195\dfrac{mol}{dm^3 \cdot s}$ and $V = 551$ dm^3.

We can continue in this manner to arrive at Table E2-2.1.

TABLE E2-2.1   CONVERSION AND REACTION RATE PROFILES

X	0	0.2	0.4	0.6	0.8
$-r_A \left(\dfrac{mol}{m^3 \cdot s} \right)$	0.45	0.30	0.195	0.113	0.05
V (dm^3)	0	218	551	1093	2165

The data in Table E2-2.1 are plotted in Figures E2-2.2(a) and (b).

For isothermal reactions, the conversion increases and the rate decreases as we move down the PFR.

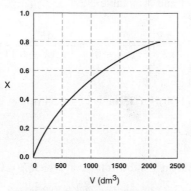

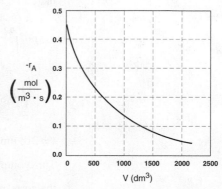

Figure E2-2.2(a)   Conversion profile.   **Figure E2-2.2(b)**   Reaction rate profile.

<u>***Analysis:***</u> One observes that the reaction rate, $-r_A$, decreases as we move down the reactor while the conversion increases. These plots are typical for reactors operated isothermally.

Example 2–3   Comparing CSTR and PFR Sizes

Compare the volumes of a CSTR and a PFR required for the same conversion using the data in Figure 2-2B. Which reactor would require the smaller volume to achieve a conversion of 80%: a CSTR or a PFR? The entering molar flow rate and the feed conditions are the same in both cases.

Solution

We will again use the data in Table 2-3.

TABLE 2-3   PROCESSED DATA 2

X	0.0	0.1	0.2	0.4	0.6	0.7	0.8
$(F_{A0}/-r_A)$(m^3)	0.89	1.08	1.33	2.05	3.54	5.06	8.0

The CSTR volume was 6.4 m³ and the PFR volume was 2.165 m³. When we combine Figures E2-1.1 and E2-2.1 on the same graph, Figure 2-3.1(a), we see that the crosshatched area above the curve is the difference in the CSTR and PFR reactor volumes.

For isothermal reactions *greater* than zero order (see Chapter 3), the CSTR volume will always be greater than the PFR volume for the same conversion and reaction conditions (temperature, flow rate, etc.).

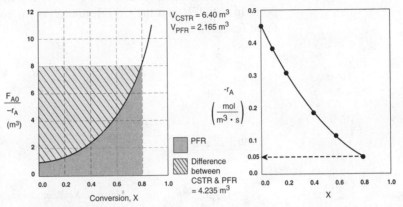

Figure E2-3.1(a)   Comparison of CSTR and PFR reactor sizes for $X = 0.8$.

Figure E2-3.1(b) $-r_A$ as a function of X obtained from Table 2-2.1.

___

Analysis: We see that the reason the isothermal CSTR volume is usually greater than the PFR volume is that the CSTR is always operating at the lowest reaction rate (e.g., $-r_A = 0.05 \text{ mol/m}^3 \cdot \text{s}$ in Figure E2-3.1(b)). The PFR, on the other hand, starts at a high rate at the entrance and gradually decreases to the exit rate, thereby requiring less volume because the volume is inversely proportional to the rate. However, there are exceptions such as autocatalytic reactions, product-inhibited reactions, and nonisothermal exothermic reactions; these trends will not always be the case, as we will see in Chapters 9 and 11.

2.5 Reactors in Series

Many times, reactors are connected in series so that the exit stream of one reactor is the feed stream for another reactor. When this arrangement is used, it is often possible to speed calculations by defining conversion in terms of location at a point downstream rather than with respect to any single reactor. That is, the conversion X is the *total number of moles* of A that have reacted up to that point per mole of A fed to the *first* reactor.

Only valid for **NO** side streams!!

For reactors in series

$$X_i = \frac{\text{Total moles of A reacted up to point } i}{\text{Moles of A fed to the first reactor}}$$

However, this definition can *only* be used when the feed stream only enters the first reactor in the series and there are *no* side streams either fed or withdrawn. The molar flow rate of A at point *i* is equal to the moles of A fed to the first reactor, minus all the moles of A reacted up to point *i*.

$$F_{Ai} = F_{A0} - F_{A0}X_i$$

For the reactors shown in Figure 2-3, X_1 at point $i = 1$ is the conversion achieved in the PFR, X_2 at point $i = 2$ is the total conversion achieved at this point in the PFR and the CSTR, and X_3 is the total conversion achieved by all three reactors.

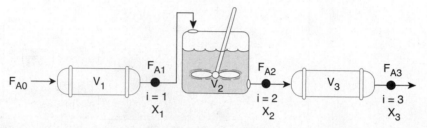

Figure 2-3   Reactors in series.

To demonstrate these ideas, let us consider three different schemes of reactors in series: two CSTRs, two PFRs, and then a combination of PFRs and CSTRs in series. To size these reactors, we shall use laboratory data that give the reaction rate at different conversions.

2.5.1 CSTRs in Series

The first scheme to be considered is the two CSTRs in series shown in Figure 2-4.

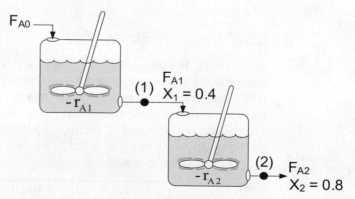

Figure 2-4   Two CSTRs in series.

For the first reactor, the rate of disappearance of A is $-r_{A1}$ at conversion X_1.
A mole balance on reactor 1 gives

$$\textbf{In} - \textbf{Out} + \textbf{Generation} = 0$$

Reactor 1: $\quad F_{A0} - F_{A1} + \quad r_{A1}V_1 \quad = 0$ \hfill (2-19)

The molar flow rate of A at point 1 is

$$F_{A1} = F_{A0} - F_{A0}X_1 \hfill (2\text{-}20)$$

Combining Equations (2-19) and (2-20), or rearranging

Reactor 1 $\qquad\qquad \boxed{V_1 = \dfrac{F_{A0}X_1}{-r_{A1}}}$ \hfill (2-21)

In the second reactor, the rate of disappearance of A, $-r_{A2}$, is evaluated at
the conversion of the exit stream of reactor 2, X_2. A steady-state mole balance
on the second reactor is

$$\textbf{In} - \textbf{Out} + \textbf{Generation} = 0$$

Reactor 2: $\quad F_{A1} - F_{A2} + \quad r_{A2}V_2 \quad = 0$ \hfill (2-22)

The molar flow rate of A at point 2 is

$$F_{A2} = F_{A0} - F_{A0}X_2 \hfill (2\text{-}23)$$

Combining and rearranging

$$V_2 = \frac{F_{A1} - F_{A2}}{-r_{A2}} = \frac{(F_{A0} - F_{A0}X_1) - (F_{A0} - F_{A0}X_2)}{-r_{A2}}$$

Reactor 2 $\qquad\qquad \boxed{V_2 = \dfrac{F_{A0}}{-r_{A2}}(X_2 - X_1)}$ \hfill (2-24)

For the second CSTR, recall that $-r_{A2}$ is evaluated at X_2 and then use (X_2-X_1)
to calculate V_2.

In the examples that follow, we shall again use the molar flow rate of A
used in Example 2-1 (i.e., $F_{A0} = 0.4$ mol A/s) and the reaction conditions given in
Table 2-3.

Example 2–4 Comparing Volumes for CSTRs in Series

For the two CSTRs in series, 40% conversion is achieved in the first reactor. What
is the volume of each of the two reactors necessary to achieve 80% overall conversion of the entering species A? (See Table 2-3.)

TABLE 2-3   PROCESSED DATA 2

X	0.0	0.1	0.2	0.4	0.6	0.7	0.8
$(F_{A0}/-r_A)(m^3)$	0.89	1.09	1.33	2.05	3.54	5.06	8.0

Solution

For Reactor 1, we observe from either Table 2-3 or Figure 2-2B that when $X = 0.4$, then

$$\left(\frac{F_{A0}}{-r_{A1}}\right)_{X=0.4} = 2.05 \text{ m}^3$$

Then, using Equation (2-13)

$$V_1 = \left(\frac{F_{A0}}{-r_{A1}}\right)_{X_1} X_1 = \left(\frac{F_{A0}}{-r_{A1}}\right)_{0.4} X_1 = (2.05)(0.4) = 0.82 \text{ m}^3 = 820 \text{ dm}^3$$

For Reactor 2, when $X_2 = 0.8$, then $\left(\frac{F_{A0}}{-r_A}\right)_{X=0.8} = 8.0 \text{ m}^3$

using Equation (2-24)

$$V_2 = \left(\frac{F_{A0}}{-r_{A2}}\right)(X_2 - X_1) \qquad (2\text{-}24)$$

$$V_2 = (8.0 \text{ m}^3)(0.8 - 0.4) = 3.2 \text{ m}^3 = 3200 \text{ dm}^3$$

$$V_2 = 3200 \text{ dm}^3 \text{ (liters)}$$

To achieve the same overall conversion, the total volume for two CSTRs in series is less than that required for one CSTR.

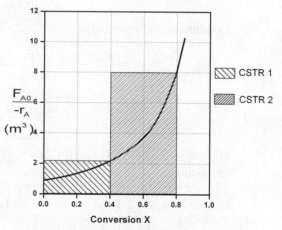

Figure E2-4.1   Two CSTRs in series.

The shaded areas in Figure E2-4.1 can also be used to determine volumes of CSTR 1 and CSTR 2.

Note again that for CSTRs in series, the rate $-r_{A1}$ is evaluated at a conversion of 0.4 and rate $-r_{A2}$ is evaluated at a conversion of 0.8. The total volume for these two reactors in series is

$$V = V_1 + V_2 = 0.82 \text{ m}^3 + 3.2 \text{ m}^3 = 4.02 \text{ m}^3 = 4020 \text{ dm}^3$$

We need only
$-r_A = f(X)$ and
F_{A0} to size
reactors.

By comparison, the volume necessary to achieve 80% conversion in **one** CSTR is

$$V = \left(\frac{F_{A0}}{-r_{A1}} \right) X = (8.0)(0.8) = 6.4 \text{ m}^3 = 6400 \text{ dm}^3$$

Notice in Example 2-5 that the sum of the two CSTR reactor volumes (4.02 m³) in series is less than the volume of one CSTR (6.4 m³) to achieve the same overall conversion.

Analysis: When we have reactors in series, we can speed our analysis and calculations by defining an overall conversion at a point in the series, rather than the conversion of each individual reactor. In this example, we saw that 40% was achieved at point 1, the exit to the first reactor, and that a total of 80% conversion was achieved by the time we exit the second reactor.

Approximating a PFR by a Large Number of CSTRs in Series

Consider approximating a PFR with a number of small, equal-volume CSTRs of V_i in series (Figure 2-5). We want to compare the _total volume_ of all the CSTRs with the volume of one plug-flow reactor for the same conversion, say 80%.

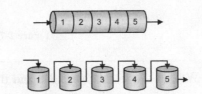

Figure 2-5   Modeling a PFR with CSTRs in series.

From Figure 2-6, we note a very important observation! The total volume to achieve 80% conversion for five CSTRs of equal volume in series is

The fact that we can model a PFR with a large number of CSTRs is an important result.

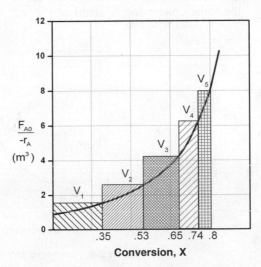

Figure 2-6   Levenspiel plot showing comparison of CSTRs in series with one PFR.

"roughly" the same as the volume of a PFR. As we make the volume of each CSTR smaller and increase the number of CSTRs, the total volume of the CSTRs in series and the volume of the PFR will become identical. *That is, we can model a PFR with a large number of CSTRs in series.* This concept of using many CSTRs in series to model a PFR will be used later in a number of situations, such as modeling catalyst decay in packed-bed reactors or transient heat effects in PFRs.

2.5.2 PFRs in Series

We saw that two CSTRs in series gave a smaller total volume than a single CSTR to achieve the same conversion. This case does not hold true for the two plug-flow reactors connected in series shown in Figure 2-7.

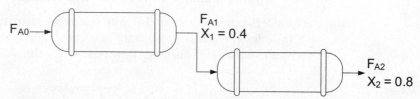

Figure 2-7 Two PFRs in series.

PFRs in series   We can see from Figure 2-8 and from the following equation

$$\int_0^{X_2} F_{A0}\frac{dX}{-r_A} \equiv \int_0^{X_1} F_{A0}\frac{dX}{-r_A} + \int_{X_1}^{X_2} F_{A0}\frac{dX}{-r_A}$$

that it is immaterial whether you place two plug-flow reactors in series or have one continuous plug-flow reactor; the total reactor volume required to achieve the same conversion is identical!

The overall conversion of two PFRs in series is the same as one PFR with the same total volume.

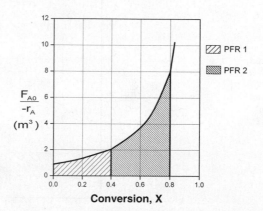

Figure 2-8   Levenspiel plot for two PFRs in series.

2.5.3 Combinations of CSTRs and PFRs in Series

The final sequences we shall consider are combinations of CSTRs and PFRs in series. An industrial example of reactors in series is shown in the photo in Figure 2-9. This sequence is used to dimerize propylene (A) into olefins (B), e.g.,

$$2CH_3-CH=CH_2 \longrightarrow CH_3\overset{\overset{\textstyle CH_3}{\textstyle |}}{C}=CH-CH_2-CH_3$$

$$2\,A \longrightarrow B$$

Not sure if the size of these CSTRs is in the Guiness Book of World Records

Figure 2-9   Dimersol G (an organometallic catalyst) unit (two CSTRs and one tubular reactor in series) to dimerize propylene into olefins. Institut Français du Pétrole process. Photo courtesy of Editions Technip (Institut Français du Pétrole).

A schematic of the industrial reactor system in Figure 2-9 is shown in Figure 2-10.

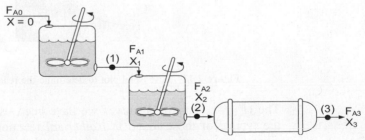

Figure 2-10   Schematic of a real system.

For the sake of illustration, let's assume that the reaction carried out in the reactors in Figure 2-10 follows the same $\left(\dfrac{F_{A0}}{-r_A}\right)$ vs. X curve given by Table 2-3.

The volumes of the first two CSTRs in series (see Example 2-5) are:

In this series arrangement, $-r_{A2}$ is evaluated at X_2 for the second CSTR.

Reactor 1:
$$V_1 = \frac{F_{A0}X_1}{-r_{A1}} \tag{2-13}$$

Reactor 2:
$$V_2 = \frac{F_{A0}(X_2 - X_1)}{-r_{A2}} \tag{2-24}$$

Starting with the differential form of the PFR design equation

$$F_{A0}\frac{dX}{dV} = -r_A \tag{2-15}$$

rearranging and integrating between limits, when $V = 0$, then $X = X_2$, and when $V = V_3$, then $X = X_3$ we obtain

Reactor 3:
$$V_3 = \int_{X_2}^{X_3} \frac{F_{A0}}{-r_A} dX \tag{2-25}$$

The corresponding reactor volumes for each of the three reactors can be found from the shaded areas in Figure 2-11.

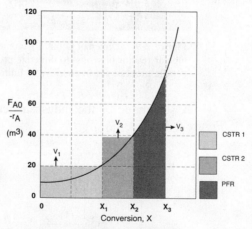

Figure 2-11   Levenspiel plot to determine the reactor volumes V_1, V_2, and V_3.

The $(F_{A0}/-r_A)$ versus X curves we have been using in the previous examples are typical of those found in *isothermal* reaction systems. We will now consider a real reaction system that is carried out *adiabatically*. Isothermal reaction systems are discussed in Chapter 5 and adiabatic systems in Chapter 11.

Example 2–5   An Adiabatic Liquid-Phase Isomerization

The isomerization of butane

$$n\text{-}C_4H_{10} \rightleftharpoons i\text{-}C_4H_{10}$$

was carried out adiabatically in the liquid phase. The data for this reversible reaction are given in Table E2-5.1. (Example 11.3 shows how the data in Table E2-5.1 were generated.)

TABLE E2-5.1   RAW DATA

X	0.0	0.2	0.4	0.6	0.65
$-r_A(\text{kmol/m}^3 \cdot \text{h})$	39	53	59	38	25

Don't worry how we got this data or why the $(1/-r_A)$ looks the way it does; we will see how to construct this table in Chapter 11, Example 11-3. It is *real data* for a *real reaction* carried out adiabatically, and the reactor scheme shown below in Figure E2-5.1 is used.

Real Data
for a
Real Reaction

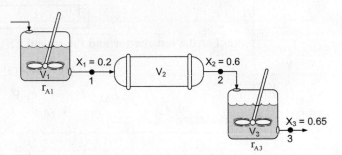

Figure E2-5.1   Reactors in series.

Calculate the volume of each of the reactors for an entering molar flow rate of n-butane of 50 kmol/hr.

Solution

Taking the reciprocal of $-r_A$ and multiplying by F_{A0}, we obtain Table E2-5.2.

$$\text{E.g., at } X = 0: \quad \frac{F_{A0}}{-r_A} = \frac{50 \text{ kmol/h}}{39 \text{ kmol/h} \cdot \text{m}^3} = 1.28 \text{ m}^3$$

TABLE E2-5.2   PROCESSED DATA

X	0.0	0.2	0.4	0.6	0.65
$-r_A \ (\text{kmol/m}^3 \cdot \text{h})$	39	53	59	38	25
$[F_{A0}/-r_A] \ (\text{m}^3)$	1.28	0.94	0.85	1.32	2.0

(a) For the first CSTR,
when $X = 0.2$, then $\dfrac{F_{A0}}{-r_A} = 0.94 \text{ m}^3$

$$V_1 = \frac{F_{A0}}{-r_A} X_1 = (0.94 \text{ m}^3)(0.2) = 0.188 \text{ m}^3 \qquad \text{(E2-5.1)}$$

$$\boxed{V_1 = 0.188 \text{ m}^3 = 188 \text{ dm}^3} \qquad \text{(E2-5.2)}$$

(b) For the PFR,

$$V_2 = \int_{0.2}^{0.6} \left(\frac{F_{A0}}{-r_A}\right) dX$$

Using Simpson's three-point formula with $\Delta X = (0.6 - 0.2)/2 = 0.2$, and $X_1 = 0.2$, $X_2 = 0.4$, and $X_3 = 0.6$

$$V_2 = \int_{0.2}^{0.6} \frac{F_{A0}}{-r_A} dX = \frac{\Delta X}{3}\left[\left.\frac{F_{A0}}{-r_A}\right)_{X=0.2} + 4\left.\frac{F_{A0}}{-r_A}\right)_{X=0.4} + \left.\frac{F_{A0}}{-r_A}\right)_{X=0.6}\right]$$

$$= \frac{0.2}{3}[0.94 + 4(0.85) + 1.32]\text{m}^3 \qquad \text{(E2-5.3)}$$

$$\boxed{V_2 = 0.38 \text{ m}^3 = 380 \text{ dm}^3} \qquad \text{(E2-5.4)}$$

(c) For the last reactor and the second CSTR, mole balance on A for the CSTR:

$$\textbf{In} - \textbf{Out} + \textbf{Generation} = \textbf{0}$$

$$F_{A2} - F_{A3} + r_{A3}V_3 = 0 \qquad \text{(E2-5.5)}$$

Rearranging

$$V_3 = \frac{F_{A2} - F_{A3}}{-r_{A3}} \qquad \text{(E2-5.6)}$$

$$F_{A2} = F_{A0} - F_{A0}X_2$$

$$F_{A3} = F_{A0} - F_{A0}X_3$$

$$V_3 = \frac{(F_{A0} - F_{A0}X_2) - (F_{A0} - F_{A0}X_3)}{-r_{A3}}$$

Simplifying

$$\boxed{V_3 = \left(\frac{F_{A0}}{-r_{A3}}\right)(X_3 - X_2)} \qquad \text{(E2-5.7)}$$

We find from Table E2-5.2 that at $X_3 = 0.65$, then $\dfrac{F_{A0}}{-r_{A3}} = 2.0 \text{ m}^3$

$$V_3 = 2 \text{ m}^3 \, (0.65 - 0.6) = 0.1 \text{ m}^3$$

$$\boxed{V_3 = 0.1 \text{ m}^3 = 100 \text{ dm}^3} \qquad \text{(E2-5.8)}$$

A Levenspiel plot of $(F_{A0}/-r_A)$ vs. X is shown in Figure E2-5.2.

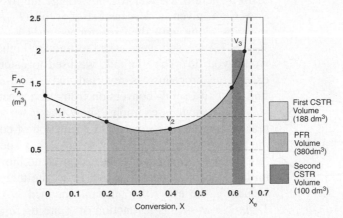

Figure E2-5.2 Levenspiel plot for adiabatic reactors in series.

For this adiabatic reaction the three reactors in series resulted in an overall conversion of 65%. The maximum conversion we can achieve is the equilibrium conversion, which is 68%, and is shown by the dashed line in Figure E2-5.2. Recall that at equilibrium, the rate of reaction is zero and an infinite reactor volume is required to reach equilibrium $\left(V \sim \dfrac{1}{-r_A} \sim \dfrac{1}{0} = \infty \right)$.

Analysis: For exothermic reactions that are not carried out isothermally, the rate usually increases at the start of the reaction because reaction temperature increases. However, as the reaction proceeds the rate eventually decreases as the conversion increases as the reactants are consumed. These two competing effects give the bowed shape of the curve in Figure E2-5.2, which will be discussed in detail in Chapter 12. Under these circumstances, we saw that a CSTR will require a smaller volume than a PFR at low conversions.

2.5.4 Comparing the CSTR and PFR Reactor Volumes and Reactor Sequencing

If we look at Figure E2-5.2, the area under the curve (PFR volume) between $X = 0$ and $X = 0.2$, we see that the PFR area is greater than the rectangular area corresponding to the CSTR volume, i.e., $V_{PFR} > V_{CSTR}$. However, if we compare the areas under the curve between $X = 0.6$ and $X = 0.65$, we see that the area under the curve (PFR volume) is smaller than the rectangular area corresponding to the CSTR volume, i.e., $V_{CSTR} > V_{PFR}$. This result often occurs when the reaction is carried out adiabatically, which is discussed when we look at heat effects in Chapter 11.

In the *sequencing of reactors,* one is often asked, "Which reactor should go first to give the highest overall conversion? Should it be a PFR followed by a CSTR, or two CSTRs, then a PFR, or …?" The answer is **"It depends."** It depends not only on the shape of the Levenspiel plot $(F_{A0}/-r_A)$ versus X, but

Which arrangement is best?

also on the relative reactor sizes. As an exercise, examine Figure E2-5.2 to learn if there is a better way to arrange the two CSTRs and one PFR. Suppose you were given a Levenspiel plot of $(F_{A0}/-r_A)$ vs. X for three reactors in series, along with their reactor volumes $V_{CSTR1} = 3$ m^3, $V_{CSTR2} = 2$ m^3, and $V_{PFR} = 1.2$ m^3, and were asked to find the highest possible conversion X. What would you do? The methods we used to calculate reactor volumes all apply, except the procedure is reversed and a *trial-and-error solution* is needed to find the exit overall conversion from each reactor (see Problem P2-5$_B$).

Given V and $\left[\dfrac{1}{-r_A} \text{ vs. } X\right]$ find X

The previous examples show that *if* we know the molar flow rate to the reactor and the reaction rate as a function of conversion, *then* we can calculate the reactor volume necessary to achieve a specified conversion. The reaction rate does not depend on conversion alone, however. It is also affected by the initial concentrations of the reactants, the temperature, and the pressure. Consequently, the experimental data obtained in the laboratory and presented in Table 2-1 as $-r_A$ as a function of X are useful only in the design of full-scale reactors that are to be operated at the *identical conditions* as the laboratory experiments (temperature, pressure, and initial reactant concentrations). However, such circumstances are **seldom** encountered and we must revert to the methods we describe in Chapters 3 and 4 to obtain $-r_A$ as a function of X.

Only need $-r_A = f(X)$ to size flow reactors

It is important to understand that if the rate of reaction is available or can be obtained solely as a function of conversion, $-r_A = f(X)$, or if it can be generated by some intermediate calculations, one can design a variety of reactors and combinations of reactors.

Chapter 3 shows how to find $-r_A = f(X)$.

Ordinarily, laboratory data are used to formulate a rate law, and then the reaction rate–conversion functional dependence is determined using the rate law. The preceding sections show that with the reaction rate–conversion relationship, different reactor schemes can readily be sized. In Chapters 3 and 4, we show how we obtain this relationship between reaction rate and conversion from rate law and reaction stoichiometry.

2.6 Some Further Definitions

Before proceeding to Chapter 3, some terms and equations commonly used in reaction engineering need to be defined. We also consider the special case of the plug-flow design equation when the volumetric flow rate is constant.

2.6.1 Space Time

The space time tau, τ, is obtained by dividing the reactor volume by the volumetric flow rate entering the reactor

τ is an important quantity!

$$\tau \equiv \frac{V}{v_0} \tag{2-26}$$

The space time is the time necessary to process one reactor volume of fluid based on entrance conditions. For example, consider the tubular reactor shown in Figure 2-12, which is 20 m long and 0.2 m^3 in volume. The dashed line in Figure 2-12 represents 0.2 m^3 of fluid directly upstream of the reactor. The time it takes for this fluid to enter the reactor completely is called the *space time tau*. It is also called the *holding time* or *mean residence time*.

Space time or
mean residence time
$\tau = V/v_0$

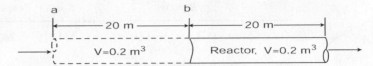

Figure 2-12   Tubular reactor showing identical volume upstream.

For example, if the reactor volume is 0.2 m^3 and the inlet volumetric flow rate is 0.01 m^3/s, it would take the upstream equivalent reactor volume (V = 0.2 m^3), shown by the dashed lines, a time τ equal to

$$\tau = \frac{0.2 \text{ m}^3}{0.01 \text{ m}^3/\text{s}} = 20 \text{ s}$$

to enter the reactor (V = 0.2 m^3). In other words, it would take 20 s for the fluid molecules at point a to move to point b, which corresponds to a space time of 20 s. We can substitute for $F_{A0} = v_0 C_{A0}$ in Equations (2-13) and (2-16) and then divide both sides by v_0 to write our mole balance in the following forms:

$$\text{For a PFR} \quad \tau_p = \left(\frac{V_p}{v_0}\right) = C_{A0} \int_0^X \frac{dX}{-r_A}$$

and

$$\text{For a CSTR} \quad \tau = \left(\frac{V}{v_0}\right) = \frac{C_{A0}}{-r_A} X$$

For plug flow, the space time is equal to the mean residence time in the reactor, t_m (see Chapter 16). This time is the average time the molecules spend in the reactor. A range of typical processing times in terms of the space time (residence time) for industrial reactors is shown in Table 2-4.

TABLE 2-4   TYPICAL SPACE TIME FOR INDUSTRIAL REACTORS[2]

Practical
guidelines

Reactor Type	Mean Residence Time Range	Production Capacity
Batch	15 min to 20 h	Few kg/day to 100,000 tons/year
CSTR	10 min to 4 h	10 to 3,000,000 tons/year
Tubular	0.5 s to 1 h	50 to 5,000,000 tons/year

[2] Trambouze, Landeghem, and Wauquier, *Chemical Reactors* (Paris: Editions Technip, 1988; Houston: Gulf Publishing Company, 1988), p. 154.

Table 2-5 shows an order of magnitude of the space times for six industrial reactions and associated reactors.

TABLE 2-5   SAMPLE INDUSTRIAL SPACE TIMES[3]

	Reaction	Reactor	Temperature	Pressure atm	Space Time
(1)	$C_2H_6 \rightarrow C_2H_4 + H_2$	PFR[†]	860°C	2	1 s
(2)	$CH_3CH_2OH + HCH_3COOH \rightarrow$ $CH_3CH_2COOCH_3 + H_2O$	CSTR	100°C	1	2 h
(3)	Catalytic cracking	PBR	490°C	20	$1\ s < \tau < 400\ s$
(4)	$C_6H_5CH_2CH_3 \rightarrow C_6H_5CH = CH_2 + H_2$	PBR	600°C	1	0.2 s
(5)	$CO + H_2O \rightarrow CO_2 + H_2$	PBR	300°C	26	4.5 s
(6)	$C_6H_6 + HNO_3 \rightarrow$ $C_6H_5NO_2 + H_2O$	CSTR	50°C	1	20 min

Typical industrial reaction space times

[†]The reactor is tubular but the flow may or may not be ideal plug flow.

Reference Shelf

Table 2-6 gives typical sizes for batch and CSTR reactors (along with the comparable size of a familiar object) and the costs associated with those sizes. All reactors are glass lined and the prices include heating/cooling jacket, motor, mixer, and baffles. The reactors can be operated at temperatures between 20 and 450°F, and at pressures up to 100 psi.

TABLE 2-6   REPRESENTATIVE PFAUDLER CSTR/BATCH REACTOR SIZES AND PRICES[†]

Volume	Price
20 dm³ (Waste Basket)	$30,000
200 dm³ (Garbage Can)	$40,000
2,000 dm³ (Jacuzzi)	$75,000
30,000 dm³ (Gasoline Tanker)	$300,000

[†]Doesn't include instrumentation costs.

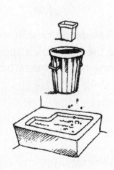

2.6.2 Space Velocity

The space velocity (SV), which is defined as

$$SV \equiv \frac{v_0}{V} \qquad SV = \frac{1}{\tau} \tag{2-27}$$

[3] Walas, S. M. Chemical Reactor Data, *Chemical Engineering*, 79 (October 14, 1985).

might be regarded at first sight as the reciprocal of the space time. However, there can be a difference in the two quantities' definitions. For the space time, the entering volumetric flow rate is measured at the entrance conditions, but for the space velocity, other conditions are often used. The two space velocities commonly used in industry are the *liquid-hourly* and *gas-hourly space velocities*, LHSV and GHSV, respectively. The entering volumetric flow rate, v_0, in the LHSV is frequently measured as that of a liquid feed rate at 60°F or 75°F, even though the feed to the reactor may be a vapor at some higher temperature. Strange but true. The gas volumetric flow rate, v_0, in the GHSV is normally reported at standard temperature and pressure (STP).

$$LHSV = \frac{v_0|_{liquid}}{V} \qquad\qquad (2\text{-}28)$$

$$GHSV = \frac{v_0|_{STP}}{V} \qquad\qquad (2\text{-}29)$$

Example 2–6   Reactor Space Times and Space Velocities

Calculate the space time, τ, and space velocities for the reactor in Examples 2-1 and 2-3 for an entering volumetric flow rate of 2 dm³/s.

Solution
The entering volumetric flow is 2 dm³/s (0.002 m³/s).

From Example 2-1, the CSTR volume was 6.4 m³ and the corresponding space time, τ, and space velocity, SV are

$$\tau = \frac{V}{v_0} = \frac{6.4\ \text{m}^3}{0.002\ \text{m}^3/\text{s}} = 3200\ \text{s} = 0.89\ \text{h}$$

It takes 0.89 hours to put 6.4 m³ into the reactor.

$$SV = \frac{1}{\tau} = \frac{1}{0.89\ \text{h}} = 1.125\ \text{h}^{-1}$$

From Example 2-3, the PFR volume was 2.165 m³, and the corresponding space time and space velocity are

$$\tau = \frac{V}{v_0} = \frac{2.165\ \text{m}^3}{0.002\ \text{m}^3/\text{s}} = 1083\ \text{s} = 0.30\ \text{h}$$

$$SV = \frac{1}{\tau} = \frac{1}{0.30\ \text{h}} = 3.3\ \text{h}^{-1}$$

<u>*Analysis:*</u> This example gives an *important industrial concept*. These space times are the times for each of the reactors to take the volume of fluid equivalent to one reactor volume and put it into the reactor.

Summary

In these last examples we have seen that in the design of reactors that are to be operated at conditions (e.g., temperature and initial concentration) identical to those at which the reaction rate data were obtained, we can size (determine the reactor volume) both CSTRs and PFRs alone or in various combinations. In principle, it may be possible to scale up a laboratory-bench or pilot-plant reaction system solely from knowledge of $-r_A$ as a function of X or C_A. However, for most reactor systems in industry, a scale-up process cannot be achieved in this manner because knowledge of $-r_A$ solely as a function of X is seldom, if ever, available under identical conditions. By combining the information in Chapters 3 and 4, we shall see how we can obtain $-r_A = f(X)$ from information obtained either in the laboratory or from the literature. This relationship will be developed in a two-step process. In Step 1, we will find the rate law that gives the rate as a function of concentration (Chapter 3) and in Step 2, we will find the concentrations as a function of conversion (Chapter 4). Combining Steps 1 and 2 in Chapters 3 and 4, we obtain $-r_A = f(X)$. We can then use the methods developed in this chapter, along with integral and numerical methods, to size reactors.

Coming attractions in Chapters 3 and 4

The CRE Algorithm
- Mole Balance, Ch 1
- Rate Law, Ch 3
- Stoichiometry, Ch 4
- Combine, Ch 5
- Evaluate, Ch 5
- Energy Balance, Ch 11

Closure

In this chapter, we have shown that if you are given the rate of reaction as a function of conversion, i.e., $-r_A = f(X)$, you will be able to size CSTRs and PFRs, and arrange the order of a given set of reactors to determine the maximum overall conversion. After completing this chapter, the reader should be able to

a. Define the parameter *conversion* and rewrite the mole balances in terms of conversion

b. Show that by expressing $-r_A$ as a function of conversion X, a number of reactors and reaction systems can be sized or a conversion calculated from a given reactor size

c. Arrange reactors in series to achieve the maximum conversion for a given Levenspiel plot

SUMMARY

1. The conversion X is the moles of A reacted per mole of A fed.

For batch systems:
$$X = \frac{N_{A0} - N_A}{N_{A0}} \tag{S2-1}$$

For flow sytems:
$$X = \frac{F_{A0} - F_A}{F_{A0}} \tag{S2-2}$$

For reactors in series with no side streams, the conversion at point i is

$$X_i = \frac{\text{Total moles of A reacted up to point } i}{\text{Moles A fed to the first reactor}} \tag{S2-3}$$

2. In terms of the conversion, the differential and integral forms of the reactor design equations become:

TABLE S2-1   MOLE BALANCE FOR BR, CSTR, PFR, AND PBR TERMS OF CONVERSION

	Differential Form	Algebraic Form	Integral Form
Batch	$N_{A0}\dfrac{dX}{dt} = -r_A V$		$t = N_{A0}\displaystyle\int_0^X \dfrac{dX}{-r_A V}$
CSTR		$V = \dfrac{F_{A0}(X_{out} - X_{in})}{(-r_A)_{out}}$	
Fluidized CSTR		$W = \dfrac{F_{A0}(X_{out} - X_{in})}{(-r'_A)_{out}}$	
PFR	$F_{A0}\dfrac{dX}{dV} = -r_A$		$V = F_{A0}\displaystyle\int_{X_{in}}^{X_{out}} \dfrac{dX}{-r_A}$
PBR	$F_{A0}\dfrac{dX}{dW} = -r'_A$		$W = F_{A0}\displaystyle\int_{X_{in}}^{X_{out}} \dfrac{dX}{-r'_A}$

3. If the rate of disappearance of A is given as a function of conversion, the following graphical techniques can be used to size a CSTR and a plug-flow reactor.

A.   Graphical Integration Using Levenspiel Plots

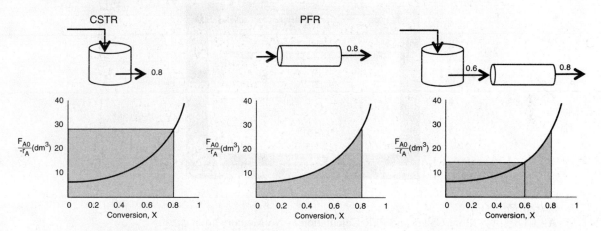

The PFR integral could also be evaluated by

B.   Numerical Integration

See Appendix A.4 for quadrature formulas such as the five-point quadrature formula with $\Delta X = 0.8/4$ of five equally spaced points, $X_1 = 0$, $X_2 = 0.2$, $X_3 = 0.4$, $X_4 = 0.6$, and $X_5 = 0.8$.

4. Space time, τ, and space velocity, SV, are given by

$$\tau = \frac{V}{v_0} \tag{S2-4}$$

$$SV = \frac{v_0}{V} \text{ (at STP)} \tag{S2-5}$$

CRE WEB SITE

- **Expanded Material**
  1. *Web P2-1$_A$ Reactor Sizing for Reversible Reactions*
  2. *Web P2-2$_A$ Puzzle Problem "What's Wrong with this Solution?"*
- **Learning Resources**
  1. *Summary Notes for Chapter 2*
  2. *Web Module*
     Hippopotamus Digestive System

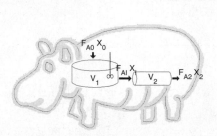
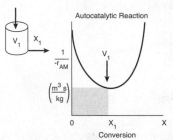

Levenspiel Plot for Autocatalytic Digestion in a CSTR

  3. *Interactive Computer Games*
     Reactor Staging

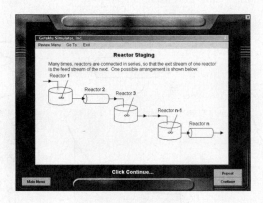

  4. *Solved Problems*
     A. CDP2-A$_B$ More CSTR and PFR Calculations—No Memorization
- **FAQ (Frequently Asked Questions)**
- **Professional Reference Shelf**

QUESTIONS AND PROBLEMS

The subscript to each of the problem numbers indicates the level of difficulty: A, least difficult; D, most difficult.

$$A = \bullet \quad B = \blacksquare \quad C = \blacklozenge \quad D = \blacklozenge\blacklozenge$$

Questions

Q2-1   (a) Without referring back, make a list of the most important items you learned in this chapter.
   (b) What do you believe was the overall purpose of the chapter?
   Before solving the problems, state or sketch qualitatively the expected results or trends.

Q2-2   Go to the Web site *www.engr.ncsu.edu/learningstyles/ilsweb.html.*
   (a) Take the Inventory of Learning Style test, and record your learning style according to the Solomon/Felder inventory.
   Global/Sequential_____
   Active/Reflective_____
   Visual/Verbal_____
   Sensing/Intuitive_____
   (b) After checking the CRE Web site, *www.umich.edu/~elements/asyLearn/learningstyles.htm*, suggest two ways to facilitate your learning style in each of the four categories.

Problems

P2-1$_A$ (a) Revisit **Examples 2-1** through **2-3.** How would your answers change if the flow rate, F_{A0}, were cut in half? If it were doubled? What conversion can be achieved in a 4.5 m^3 PFR and in a 4.5 m^3 CSTR?
   (b) Revisit **Example 2-2.** Being a company about to go bankrupt, you can only afford a 2.5 m^3 CSTR. What conversion can you achieve?
   (c) Revisit **Example 2-3.** What conversion could you achieve if you could convince your boss, Dr. Pennypincher, to spend more money to buy a 1.0 m^3 PFR to attach to a 2.40 CSTR?
   (d) Revisit **Example 2-4.** How would your answers change if the two CSTRs (one 0.82 m^3 and the other 3.2 m^3) were placed in parallel with the flow, F_{A0}, divided equally between the reactors.
   (e) Revisit **Example 2-5.** (1) What would be the reactor volumes if the two intermediate conversions were changed to 20% and 50%, respectively? (2) What would be the conversions, X_1, X_2, and X_3, if all the reactors had the same volume of 100 dm^3 and were placed in the same order? (3) What is the worst possible way to arrange the two CSTRs and one PFR?

   (f) Revisit **Example 2-6.** If the term $C_{A0}\displaystyle\int_0^X \frac{dX}{-r_A}$ is 2 seconds for 80% conversion, how much fluid (m^3/min) can you process in a 3 m^3 reactor?

P2-2$_A$ **ICG Staging.** Download the Interactive Computer Game (ICG) from the CRE Web site. Play this game and then record your performance number, which indicates your mastery of the material. Your professor has the key to decode your performance number. Note: To play this game you *must* have Windows 2000 or a later version.
   ICG Reactor Staging Performance # _____

P2-3$_B$ You have two CSTRs and two PFRs, each with a volume of 1.6 m^3. Use Figure 2-2B on page 41 to calculate the conversion for each of the reactors in the following arrangements.
   (a) Two CSTRs in series.
   (b) Two PFRs in series.
   (c) Two CSTRs in parallel with the feed, F_{A0}, divided equally between the two reactors.
   (d) Two PFRs in parallel with the feed divided equally between the two reactors.

(e) **Caution:** Part (e) is a C level problem. A CSTR and a PFR in parallel with the flow equally divided. Calculate the overall conversion, X_{ov}

$$X_{ov} = \frac{F_{A0} - F_{ACSTR} - F_{APFR}}{F_{A0}}, \text{ with } F_{ACSTR} = \frac{F_{A0}}{2} - \frac{F_{A0}}{2} X_{CSTR}, \text{ and } F_{APFR} = \frac{F_{A0}}{2}(1 - X_{PFR})$$

(f) A PFR followed by a CSTR.

(g) A CSTR followed by a PFR.

(h) A PFR followed by two CSTRs. Is this arrangement a good arrangement or is there a better one?

P2-4$_B$ The exothermic reaction of stillbene (A) to form the economically important trospophene (B) and methane (C), i.e.,

$$A \longrightarrow B + C$$

was carried out adiabatically and the following data recorded:

X	0	0.2	0.4	0.45	0.5	0.6	0.8	0.9
$-r_A$ (mol/dm^3·min)	1.0	1.67	5.0	5.0	5.0	5.0	1.25	0.91

The entering molar flow rate of A was 300 mol/min.

(a) What are the PFR and CSTR volumes necessary to achieve 40% conversion? (V_{PFR} = 72 dm^3, V_{CSTR} = 24 dm^3)

(b) Over what range of conversions would the CSTR and PFR reactor volumes be identical?

(c) What is the maximum conversion that can be achieved in a 105-dm^3 CSTR?

(d) What conversion can be achieved if a 72-dm^3 PFR is followed in series by a 24-dm^3 CSTR?

(e) What conversion can be achieved if a 24-dm^3 CSTR is followed in a series by a 72-dm^3 PFR?

(f) Plot the conversion and rate of reaction as a function of PFR reactor volume up to a volume of 100 dm^3.

P2-5$_B$ The financially important reaction to produce the valuable product B (not the real name) was carried out in Jesse Pinkman's garage. This breaking bad, fly-by-night company is on a shoestring budget and has very little money to purchase equipment. Fortunately, cousin Bernie has a reactor surplus company and get reactors for them. The reaction

$$A \longrightarrow B + C$$

takes place in the liquid phase. Below is the Levenspiel plot for this reaction.

You have up to $10,000 to use to purchase reactors from those given below.

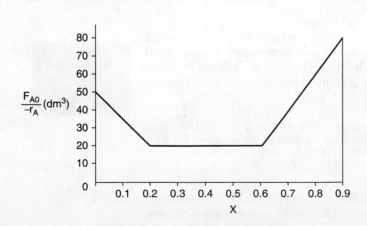

Reactor Type	Number	Volume (dm^3)	Cost
CSTR	2	2	$1,000
CSTR	1	4	$2,000
PFR	2	4	$2,000
CSTR	2	6	$4,000
CSTR	3	12	$8,000
PFR	2	12	$6,000

What reactors do you choose, how do you arrange them, and what is the highest conversion you can get for $10,000? Approximately what is the corresponding highest conversion with your arrangement of reactors?

Scheme and sketch your reactor volumes.

P2-6$_B$ Read the Web Module "Chemical Reaction Engineering of Hippopotamus Stomach" on the CRE Web site.

(a) Write five sentences summarizing what you learned from the Web Module.
(b) Work problems (1) and (2) in the Hippo Web Module.
(c) The hippo has picked up a river fungus, and now the effective volume of the CSTR stomach compartment is only 0.2 m^3. The hippo needs 30% conversion to survive. Will the hippo survive?
(d) The hippo had to have surgery to remove a blockage. Unfortunately, the surgeon, Dr. No, accidentally reversed the CSTR and the PFR during the operation. **Oops!!** What will be the conversion with the new digestive arrangement? Can the hippo survive?

P2-7$_B$ The adiabatic exothermic irreversible gas-phase reaction

$$2A + B \longrightarrow 2C$$

is to be carried out in a flow reactor for an equimolar feed of A and B. A Levenspiel plot for this reaction is shown in Figure P2-7$_B$.

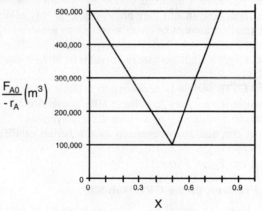

Figure P2-7$_B$ Levenspiel plot.

(a) What PFR volume is necessary to achieve 50% conversion?
(b) What CSTR volume is necessary to achieve 50% conversion?
(c) What is the volume of a second CSTR added in series to the first CSTR (**Part b**) necessary to achieve an overall conversion of 80%?
(d) What PFR volume must be added to the first CSTR (**Part b**) to raise the conversion to 80%?
(e) What conversion can be achieved in a 6 × 10^4 m^3 CSTR? In a 6 × 10^4 m^3 PFR?
(f) Think critically (cf. Preface, Section I, page xxviii) to critique the answers (numbers) to this problem.

P2-8$_A$ Estimate the reactor volumes of the two CSTRs and the PFR shown in the photo in Figure 2-9. [*Hint:* Use the dimensions of the door as a scale.]

P2-9$_D$ Don't calculate anything. Just go home and relax.

P2-10$_C$ The curve shown in Figure 2-1 is typical of a reaction carried out isothermally, and the curve shown in Figure P2-10$_B$ (see Example 11-3) is typical of a gas-solid catalytic exothermic reaction carried out adiabatically.

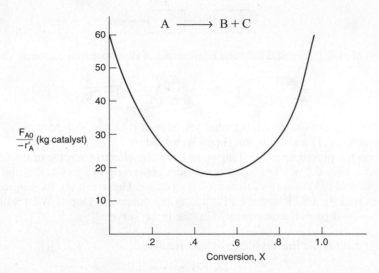

Figure P2-10$_B$ Levenspiel plot for an adiabatic exothermic heterogeneous reaction.

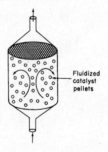

(a) Assuming that you have a fluidized CSTR and a PBR containing equal weights of catalyst, how should they be arranged for this adiabatic reaction? Use the smallest amount of catalyst weight to achieve 80% conversion of A.

(b) What is the catalyst weight necessary to achieve 80% conversion in a fluidized CSTR?

(c) What fluidized CSTR weight is necessary to achieve 40% conversion?

(d) What PBR weight is necessary to achieve 80% conversion?

(e) What PBR weight is necessary to achieve 40% conversion?

(f) Plot the rate of reaction and conversion as a function of PBR catalyst weight, W.

Additional information: $F_{A0} = 2$ mol/s.

- **Additional Homework Problems on the CRE Web Site**

CDP2-A$_A$ *An ethical dilemma* as to how to determine the reactor size in a competitor's chemical plant. [*ECRE,* 2nd Ed. P2-18$_B$]

SUPPLEMENTARY READING

Further discussion of the proper staging of reactors in series for various rate laws, in which a plot of $-1/r_A$ versus X is given, may or may not be presented in

BURGESS, THORNTON W., *The Adventures of Poor Mrs. Quack*, New York: Dover Publications, Inc., 1917.

KARRASS, CHESTER L., *Effective Negotiating: Workbook and Discussion Guide,* Beverly Hills, CA: Karrass Ltd., 2004.

LEVENSPIEL, O., *Chemical Reaction Engineering*, 3rd ed. New York: Wiley, 1999, Chapter 6, pp. 139–156.

Rate Laws   3

Success is measured not so much by the position
one has reached in life, as by the obstacles one has
overcome while trying to succeed.

—Booker T. Washington

Overview.   In Chapter 2, we showed that if we had the rate of reaction
as a function of conversion, $-r_A = f(X)$, we could calculate reactor volumes necessary to achieve a specified conversion for flow systems and
the time to achieve a given conversion in a batch system. Unfortunately,
one is seldom, if ever, given $-r_A = f(X)$ directly from raw data. Not to
fear, in the next two chapters we will show how to obtain the rate of
reaction as a function of conversion. This relationship between reaction
rate and conversion will be obtained in two steps.

- In Step 1, described in Chapter 3, we define the rate law, which
relates the rate of reaction to temperature and the concentrations
of the reacting species.
- In Step 2, which will be described in Chapter 4, we define
concentrations for flow and batch systems and develop a stoichiometric table so that one can write concentrations as a function
of conversion.
- Combining Steps 1 and 2, we see that one can then write the rate
as a function of conversion and use the techniques in Chapter 2
to design reaction systems.

After completing this chapter, you will be able to

- relate the rates of reaction of species in a reaction to each other,
- write the rate law in terms of concentrations,
- use the Arrhenius Equation to find the rate constant as a function of temperature, and
- describe the energy barrier height (activation energy) and the
fraction of molecular collisions that have the energy to pass
over it and react.

3.1 Basic Definitions

A *homogeneous reaction* is one that involves only one phase. A *heterogeneous reaction* involves more than one phase, and the reaction usually occurs at the interface between the phases. An *irreversible reaction* is one that proceeds in only one direction and continues in that direction until one of the reactants is exhausted. A *reversible reaction*, on the other hand, can proceed in either direction, depending on the concentrations of reactants and products relative to the corresponding equilibrium concentrations. An irreversible reaction behaves as if no equilibrium condition exists. Strictly speaking, no chemical reaction is completely irreversible. However, for many reactions, the equilibrium point lies so far to the product side that these reactions are treated as irreversible reactions.

Types of reactions

The *molecularity* of a reaction is the number of atoms, ions, or molecules involved (colliding) in a reaction step. The terms *unimolecular*, *bimolecular*, and *termolecular* refer to reactions involving, respectively, one, two, and three atoms (or molecules) interacting or colliding in any one reaction step. The most common example of a *unimolecular* reaction is radioactive decay, such as the spontaneous emission of an alpha particle from uranium-238 to give thorium and helium

$$_{92}U^{238} \rightarrow {}_{90}Th^{234} + {}_2He^4$$

The rate of disappearance of uranium (U) is given by the rate law

$$-r_U = kC_U$$

The only true bimolecular reactions are those that involve the collision with free radicals (i.e., unpaired electrons, e.g., Br•), such as

$$Br• + C_2H_6 \rightarrow HBr + C_2H_5•$$

with the rate of disappearance of bromine given by the rate law

$$-r_{Br•} = kC_{Br•}C_{C_2H_6}$$

The probability of a *termolecular* reaction, where three molecules collide all at once, is almost nonexistent, and in most instances the reaction pathway follows a series of *bimolecular* reactions, as in the case of the reaction

$$2NO + O_2 \rightarrow 2NO_2$$

The reaction pathway for this "Hall of Fame" reaction is quite interesting and is discussed in Chapter 9, along with similar reactions that form active intermediate complexes in their reaction pathways.

In discussing *Rates of Chemical Reaction*, we have three rates to consider

- Relative Rates
- Rate Laws
- Net Rates

Relative Rates tell us how fast one species is disappearing or appearing relative to the other species in the given reaction.

Rate Laws are the algebraic equations that apply to a given reaction.

Net Rates of formation of a given species (e.g., A) is the sum of the rate of reactions of A all the reactions in which A is either a reactant or product in the system.

A thorough discussion of net rates is given in Chapter 8 where we discuss multiple reactions.

3.1.1 Relative Rates of Reaction

The relative rates of reaction of the various species involved in a reaction can be obtained from the ratio of the stoichiometric coefficients. For Reaction (2-2),

$$A + \frac{b}{a}B \rightarrow \frac{c}{a}C + \frac{d}{a}D \tag{2-2}$$

we see that for every mole of A that is consumed, c/a moles of C appear. In other words,

$$\text{Rate of formation of C} = \frac{c}{a} \text{ (Rate of disappearance of A)}$$

$$r_C = \frac{c}{a}(-r_A) = -\frac{c}{a} r_A$$

Similarly, the relationship between the rates of formation of C and D is

$$r_C = \frac{c}{d} r_D$$

The relationship can be expressed directly from the stoichiometry of the reaction

$$aA + bB \rightarrow cC + dD \tag{2-1}$$

for which

$$\boxed{\frac{-r_A}{a} = \frac{-r_B}{b} = \frac{r_C}{c} = \frac{r_D}{d}} \tag{3-1}$$

Reaction stoichiometry   or

$$\boxed{\frac{r_A}{-a} = \frac{r_B}{-b} = \frac{r_C}{c} = \frac{r_D}{d}}$$

For example, in the reaction

$$2NO + O_2 \rightleftharpoons 2NO_2$$

we have

$$\frac{r_{NO}}{-2} = \frac{r_{O_2}}{-1} = \frac{r_{NO_2}}{2}$$

If NO_2 is being formed at a rate of 4 mol/m^3/s, i.e.,

$$r_{NO_2} = 4 \text{ mol/m}^3/\text{s}$$

then the rate of formation of NO is

$$r_{NO} = \frac{-2}{2} \, r_{NO_2} = -4 \text{ mol/m}^3/\text{s}$$

the rate of disappearance of NO is

$$-r_{NO} = 4 \text{ mol/m}^3/\text{s}$$

and the rate of disappearance of oxygen, O_2, is

$$-r_{O_2} = \frac{-1}{-2} \, r_{NO_2} = 2 \text{ mol/m}^3/\text{s}$$

3.2 The Reaction Order and the Rate Law

In the chemical reactions considered in the following paragraphs, we take as the basis of calculation a species A, which is one of the reactants that is disappearing as a result of the reaction. The limiting reactant is usually chosen as our basis for calculation. The rate of disappearance of A, $-r_A$, depends on temperature and concentration. For many irreversible reactions, it can be written as the product of a *reaction rate constant, k_A,* and a function of the concentrations (activities) of the various species involved in the reaction:

$$\boxed{-r_A = [k_A(T)][\text{fn}(C_A, C_B, \ldots)]} \tag{3-2}$$

Concept 1. Law of Mass Action. The rate of reaction increases with increasing concentration of reactants owing the corresponding increase in the number of molecular collisions. The rate law for bimolecular collisions is derived in the collision theory section of the *Professional Reference Shelf R3.1.* A schematic of the reaction of **A** and **B** molecules colliding and reacting is shown in Figure R3.1 on page 97.

The rate law gives the relationship between reaction rate and concentration.

The algebraic equation that relates $-r_A$ to the species concentrations is called the kinetic expression or **rate law**. The specific rate of reaction (also called the rate constant), k_A, like the reaction rate, $-r_A$, always refers to a particular species in the reaction and normally should be subscripted with respect to that species. However, for reactions in which the stoichiometric coefficient is 1 for all species involved in the reaction, for example,

$$1 \text{ NaOH} + 1 \text{ HCl} \rightarrow 1 \text{ NaCl} + 1 \text{ H}_2\text{O}$$

we shall delete the subscript on the specific reaction rate, (e.g., A in k_A), to let

$$k = k_{NaOH} = k_{HCl} = k_{NaCl} = k_{H_2O}$$

3.2.1 Power Law Models and Elementary Rate Laws

The dependence of the reaction rate, $-r_A$, on the concentrations of the species present, $\text{fn}(C_j)$, is almost without exception determined by experimental observation. Although the functional dependence on concentration may be postulated from theory, experiments are necessary to confirm the proposed form. One of the most common general forms of this dependence is the *power law model*. Here the rate law is the product of concentrations of the individual reacting species, each of which is raised to a power, for example,

$$\boxed{-r_A = k_A C_A^{\alpha} C_B^{\beta}} \qquad (3\text{-}3)$$

The exponents of the concentrations in Equation (3-3) lead to the concept of *reaction order*. The **order of a reaction** refers to the powers to which the concentrations are raised in the kinetic rate law.[1] In Equation (3-3), the reaction is *α order with respect to reactant* A, *and β order with respect to reactant* B. The overall order of the reaction, *n*, is

Overall reaction order

$$n = \alpha + \beta$$

The units of $-r_A$ are always in terms of concentration per unit time, while the units of the specific reaction rate, k_A, will vary with the order of the reaction. Consider a reaction involving only one reactant, such as

$$A \rightarrow \text{Products}$$

with an overall reaction order *n*. The units of rate, $-r_A$, and the specific reaction rate, *k* are

$$\{-r_A\} = [\text{concentration/time}]$$

$$\text{and } \{k\} = \frac{[\text{concentration}]^{1-n}}{\text{time}}$$

Generally, the overall reaction order can be deduced from the units of the specific reaction rate constant *k*. For example, the rate laws corresponding to a zero-, first-, second-, and third-order reaction, together with typical units for the corresponding rate constants, are

Zero-order ($n = 0$): $\qquad\qquad -r_A = k_A$:

$$\{k\} = \text{mol/dm}^3/\text{s} \qquad (3\text{-}4)$$

First-order ($n = 1$): $\qquad\qquad -r_A = k_A C_A$:

$$\{k\} = \text{s}^{-1} \qquad (3\text{-}5)$$

---

[1] Strictly speaking, the reaction rates should be written in terms of the activities, a_i, ($a_i = \gamma_i C_i$, where γ_i is the activity coefficient). Kline and Fogler, *JCIS, 82*, 93 (1981); ibid., p. 103; and *Ind. Eng. Chem Fundamentals* 20, 155 (1981).

$$-r_A = k_A' a_A^{\alpha} a_B^{\beta}$$

However, for many reacting systems, the activity coefficients, γ_i, do not change appreciably during the course of the reaction, and they are absorbed into the specific reaction rate constant, k_A

$$-r_A = k_A' a_A^{\alpha} a_B^{\beta} = k_A'(\gamma_A C_A)^{\alpha}(\gamma_B C_B)^{\beta} = \overbrace{\left(k_A' \gamma_A^{\alpha} \gamma_B^{\beta}\right)}^{k_A} C_A^{\alpha} C_B^{\beta} = k_A C_A^{\alpha} C_B^{\beta}$$

Second-order ($n = 2$): $\qquad -r_A = k_A C_A^2$:

$$\{k\} = dm^3/mol/s \qquad (3\text{-}6)$$

Third-order ($n = 3$): $\qquad -r_A = k_A C_A^3$:

$$\{k\} = (dm^3/mol)^2/s \qquad (3\text{-}7)$$

An *elementary reaction* is one that involves a single reaction step, such as the bimolecular reaction between an oxygen free radical and methanol molecule

$$O\bullet + CH_3OH \rightarrow CH_3O\bullet + OH\bullet$$

The stoichiometric coefficients in this reaction are *identical* to the powers in the rate law. Consequently, the rate law for the disappearance of molecular oxygen is

$$-r_{O\bullet} = k C_{O\bullet} C_{CH_3OH}$$

Reference Shelf

Collision theory

The reaction is first order in an oxygen free radical and first order in methanol; therefore, we say that both the reaction and the rate law are *elementary*. This form of the rate law can be derived from *Collision Theory,* as shown in the *Professional Reference Shelf R3.1* on the CRE Web site (*www.umich. edu/~ elements/5e/index.html*). There are many reactions where the stoichiometric coefficients in the reaction are identical to the reaction orders, but the reactions are not elementary, owing to such things as pathways involving active intermediates and series reactions. For these reactions that are not elementary but whose stoichiometric coefficients are identical to the reaction orders in the rate law, we say the reaction *follows an elementary rate law.* For example, the oxidation reaction of nitric oxide discussed earlier

$$2NO + O_2 \rightarrow 2NO_2$$

is not really an elementary reaction, but follows an elementary rate law; therefore,

Note: The rate constant, k, is defined with respect to NO.

$$-r_{NO} = k_{NO} C_{NO}^2 C_{O_2}$$

Another nonelementary reaction that follows an elementary rate law is the gas-phase reaction between hydrogen and iodine

$$H_2 + I_2 \rightarrow 2HI$$

with

$$-r_{H_2} = k_{H_2} C_{H_2} C_{I_2}$$

In summary, for many reactions involving multiple steps and pathways, the powers in the rate laws surprisingly agree with the stoichiometric coefficients. Consequently, to facilitate describing this class of reactions, we say that a reaction *follows an elementary rate law* when the reaction orders are identical to the stoichiometric coefficients of the reacting species for the ***reaction as written***. It is important to remember that the rate laws **are determined by experimental observation!** Chapter 7 describes how these and other rate laws can be developed from experimental data. They are a function of the reaction chemistry and not the type of reactor in which the reactions occur. Table 3-1 gives examples of rate laws for a number of reactions. By saying a reaction follows an elementary rate law as written gives us a quick way to look at the reaction stoichiometry and then write the mathematical form of the rate law. The values of specific reaction rates for these and a number of other reactions can be found in the *Database* found on the CRE Web site.

Where do you find rate laws?

The rate constants and the reaction orders for a large number of gas- and liquid-phase reactions can be found in the National Bureau of Standards' circulars and supplements.[2] Also consult the journals listed at the end of Chapter 1.

We note in Table 3-1 that Reaction Number (3) in the First-Order Rate Laws and Reaction Number (1) in the Second-Order Rate Laws do not follow elementary reaction rate laws. We know this because the reaction orders are not the same as the stoichiometric coefficients for the reactions as they are written.[†]

TABLE 3-1   EXAMPLES OF REACTION RATE LAWS

A. First-Order Rate Laws

(1) $C_2H_6 \longrightarrow C_2H_4 + H_2$ $\boxed{-r_A = kC_{C_2H_6}}$

(2)   [φN=NCl benzene ring] $\xrightarrow{}$ [φCl benzene ring] $+ N_2$ $\boxed{-r_A = kC_{\phi N = NCl}}$

(3) $\underset{CH_2 - CH_2}{\overset{O}{\triangle}} + H_2O \xrightarrow{H_2SO_4} \underset{CH_2OH}{\overset{CH_2OH}{|}}$ $\boxed{-r_{CH_2OCH_2} = kC_{CH_2OCH_2}}$

(4) $CH_3COCH_3 \longrightarrow CH_2CO + CH_4$ $\boxed{-r_A = kC_{CH_3COCH_3}}$

(5) $nC_4H_{10} \rightleftharpoons iC_4H_{10}$ $\boxed{-r_n = k[C_{nC_4} - C_{iC_4}/K_C]}$

Very important references. You should also look in the other literature before *going to the lab.*

[2] Kinetic data for a larger number of reactions can be obtained on CD-ROMs provided by *National Institute of Standards and Technology (NIST).* Standard Reference Data 221/A320 Gaithersburg, MD 20899; phone: (301) 975-2208. Additional sources are *Tables of Chemical Kinetics: Homogeneous Reactions*, National Bureau of Standards Circular 510 (Sept. 28, 1951); Suppl. 1 (Nov. 14, 1956); Suppl. 2 (Aug. 5, 1960); Suppl. 3 (Sept. 15, 1961) (Washington, DC: U.S. Government Printing Office). *Chemical Kinetics and Photochemical Data for Use in Stratospheric Modeling*, Evaluate No. 10, JPL Publication 92-20 (Pasadena, CA: Jet Propulsion Laboratories, Aug. 15, 1992).

[†] Just as an aside, Prof. Dr. Sven Köttlov, a prominent resident of Riça, Jofostan, was once jailed and questioned for 12 hours for breaking a reaction rate law. He stated it was second order when experiments clearly showed it was first order.

TABLE 3-1   EXAMPLES OF REACTION RATE LAWS (CONTINUED)

B. Second-Order Rate Laws

(1)

$$-r_{\text{ONCB}} = k_{\text{ONCB}} C_{\text{ONCB}} C_{\text{NH}_3}$$

(2) $CNBr + CH_3NH_2 \longrightarrow CH_3Br + NCNH_2$

$$-r_{\text{CNBr}} = k C_{\text{CNBr}} C_{\text{CH}_3\text{NH}_2}$$

(3)

$$CH_3COOC_2H_5 + C_4H_9OH \rightleftarrows CH_3COOC_4H_9 + C_2H_5OH$$
$$\quad A \quad + \quad B \quad \rightleftarrows \quad C \quad + \quad D$$

$$-r_A = k[C_A C_B - C_C C_D / K_C]$$

C. Nonelementary Rate Laws

(1) Homogeneous

$$CH_3CHO \xrightarrow{\text{cat}} CH_4 + CO$$

$$-r_{\text{CH}_3\text{CHO}} = k C_{\text{CH}_3\text{CHO}}^{3/2}$$

(2) Heterogeneous

$CH(CH_3)_2$

$$\xrightarrow{\text{cat}} \qquad + C_3H_6$$

$$-r'_C = \frac{k[P_C - P_B P_P / K_P]}{1 + K_B P_B + K_C P_C}$$

Cumene (C) $\longrightarrow$ Benzene (B) + Propylene (P)

D. Enzymatic Reactions (Urea (U) + Urease (E))

$$NH_2CONH_2 + \text{Urease} + H_2O \xrightarrow{\quad} 2NH_3 + CO_2 + \text{Urease}$$

$$-r_U = \frac{k C_U}{K_M + C_U}$$

E. Biomass Reactions

Substrate (S) + Cells (C) $\rightarrow$ More Cells + Product

$$-r_S = \frac{k C_S C_C}{K_S + C_S}$$

Note: The rate constants, k, and activation energies for a number of the reactions in these examples are given in the Databases in Appendix E.

3.2.2 Nonelementary Rate Laws

A large number of both homogeneous and heterogeneous reactions do not follow simple rate laws. Examples of reactions that don't follow simple elementary rate laws are discussed below.

Homogeneous Reactions. The overall order of a reaction does not have to be an integer, nor does the order have to be an integer with respect to any individual component. As an example, consider the gas-phase synthesis of phosgene,

$$CO + Cl_2 \rightarrow COCl_2$$

in which the kinetic *rate law* is

$$-r_{CO} = kC_{CO}C_{Cl_2}^{3/2}$$

This reaction is first order with respect to carbon monoxide, three-halves order with respect to chlorine, and five-halves order overall.

Sometimes reactions have complex rate expressions that cannot be separated into solely temperature-dependent and concentration-dependent portions. In the decomposition of nitrous oxide,

$$2N_2O \rightarrow 2N_2 + O_2$$

the kinetic *rate law* is

$$-r_{N_2O} = \frac{k_{N_2O}C_{N_2O}}{1 + k'C_{O_2}}$$

Both k_{N_2O} and k' are strongly temperature dependent. When a rate expression such as the one given above occurs, we cannot state an overall reaction order. Here, we can only speak of reaction orders under certain limiting conditions. For example, at very low concentrations of oxygen, the second term in the denominator would be negligible with respect to 1 ($1 \gg k'C_{O_2}$), and the reaction would be "apparent" first order with respect to nitrous oxide and first order overall. However, if the concentration of oxygen were large enough so that the number 1 in the denominator were insignificant in comparison with the second term, $k'C_{O_2}$ ($k'C_{O_2} \gg 1$), the *apparent* reaction order would be –1 with respect to oxygen and first order with respect to nitrous oxide, giving an overall *apparent* zero order. Rate expressions of this type are very common for liquid and gaseous reactions promoted by solid catalysts (see Chapter 10). They also occur in homogeneous reaction systems with reactive intermediates (see Chapter 9).

It is interesting to note that although the reaction orders often correspond to the stoichiometric coefficients, as evidenced for the reaction between hydrogen and iodine, just discussed to form HI, the rate expression for the reaction between hydrogen and another halogen, bromine, is quite complex. This non-elementary reaction

$$H_2 + Br_2 \rightarrow 2HBr$$

proceeds by a free-radical mechanism, and its reaction rate law is

$$-r_{Br_2} = \frac{k_{Br_2}C_{H_2}C_{Br_2}^{1/2}}{k' + C_{HBr}/C_{Br_2}} \tag{3-8}$$

Apparent First-Order Reactions. Because the law of mass action in collision theory shows that two molecules must collide giving a second-order dependence on the rate, you are probably wondering how rate laws such as (3-8) as well as the rate law for first-order reactions come about. An example of first-order reaction not involving radioactive decay is the decomposition of ethanol to form ethylene and hydrogen.

$$C_2H_6 \longrightarrow C_2H_4 + H_2$$

$$-r_{C_2H_6} = kC_{C_2H_6}$$

In terms of symbols

$$A \longrightarrow B + C$$

$$-r_A = kC_A$$

Rate laws of this form usually involve a number of elementary reactions and at least one active intermediate. An *active intermediate is a high-energy molecule that reacts virtually as fast as it is formed. As a result,* it is present in very small concentrations. Active intermediates (e.g., $A*$) can be formed by collision or interaction with other molecules (M) such as inerts or reactants

$$A + M \xrightarrow{\ k_1\ } A^* + M$$

Here, the activation occurs when translational kinetic energy is transferred into energy stored in internal degrees of freedom, particularly vibrational degrees of freedom.[3] An unstable molecule (i.e., active intermediate) is not formed solely as a consequence of the molecule moving at a high velocity (high-translational kinetic energy). The translational kinetic energy must be absorbed into the chemical bonds where high-amplitude oscillations will lead to bond ruptures, molecular rearrangement, and decomposition. In the absence of photochemical effects or similar phenomena, the transfer of translational energy to vibrational energy to produce an active intermediate can occur only as a consequence of molecular collision or interaction. Collision theory is discussed in the *Professional Reference Shelf* on the CRE Web site for Chapter 3.

As will be shown in Chapter 9, the mole A becomes activated to $A*$ by collision with another molecule M. The activated molecule can become deactivated by collision with another molecule or the activated molecule can decompose to B and C.

$$A + M \underset{k_2}{\overset{k_1}{\rightleftharpoons}} A^* + M$$
$$\downarrow k_3$$
$$B + C$$

Using this mechanism, we show in Section 9.1.1 that at high concentrations of M the rate law for this mechanism becomes

$$-r_A = k_A C_A$$

In Chapter 9, we will discuss reaction mechanisms and pathways that lead to nonelementary rate laws, such as the rate of formation of HBr shown in Equation (3-8).

---

[3] W. J. Moore, *Physical Chemistry* (Reading, MA: Longman Publishing Group, 1998).

Heterogeneous Reactions. Historically, it has been the practice in many gas-solid catalyzed reactions to write the rate law in terms of partial pressures rather than concentrations. In heterogeneous catalysis it is the weight of catalyst that is important, rather than the reactor volume. Consequently, we use $-r'_A$ in order to write the rate law in terms of mol per kg of catalyst per time in order to design PBRs. An example of a heterogeneous reaction and corresponding rate law is the hydrodemethylation of toluene (T) to form benzene (B) and methane (M) carried out over a solid catalyst

$$C_6H_5CH_3 + H_2 \underset{cat}{\rightarrow} C_6H_6 + CH_4$$

The rate of disappearance of toluene per mass of catalyst, $-r'_T$, i.e., (mol/mass/time) follows Langmuir-Hinshelwood kinetics (discussed in Chapter 10), and the rate law was found experimentally to be

$$-r'_T = \frac{k'P_{H_2}P_T}{1 + K_B P_B + K_T P_T}$$

where the prime in $-r'_T$ notes typical units are in per kilogram of catalyst (mol/kg-cat/s), P_T, P_{H_2}, and P_B are partial pressures of toluene, hydrogen, and benzene in (kPa or atm), and K_B and K_T are the adsorption constants for benzene and toluene respectively, with units of kPa^{-1} (or atm^{-1}). The specific reaction rate k has units of

$$\{k'\} = \frac{\text{mol toluene}}{\text{kg-cat} \cdot \text{s} \cdot \text{kPa}^2}$$

You will find that almost all heterogeneous catalytic reactions will have a term such as $(1 + K_A P_A + \dots)$ or $(1 + K_A P_A + \dots)^2$ in the denominator of the rate law (cf. Chapter 10).

To express the rate of reaction in terms of concentration rather than partial pressure, we simply substitute for P_i using the ideal gas law

$$\boxed{P_i = C_i RT} \qquad (3\text{-}9)$$

The rate of reaction per unit weight (i.e., mass) catalyst, $-r'_A$ (e.g., $-r'_T$), and the rate of reaction per unit volume, $-r_A$, are related through the bulk density ρ_b (mass of solid/volume) of the *catalyst particles* in the fluid media:

Relating rate per unit volume and rate by per unit mass of catalyst

$$\boxed{-r_A = (\rho_b)(-r'_A)}$$

$$\frac{\text{moles}}{\text{time} \cdot \text{volume}} = \left(\frac{\text{mass}}{\text{volume}}\right)\left(\frac{\text{moles}}{\text{time} \cdot \text{mass}}\right)$$

In fluidized catalytic beds, the bulk density, ρ_b, is normally a function of the volumetric flow rate through the bed.

Consequently, using the above equations for P_i and $-r'_T$ we can write the rate law for the hydromethylation of toluene in terms of concentration and in (mole/dm^3) and the rate, $-r_T$ in terms of reactor volume, i.e.,

$$-r_T = \frac{\overbrace{\left[\rho_b k'(RT)^2\right]}^{k} C_{H_2} C_T}{1 + K_B RTC_B + K_T RTC_T} \quad \left(\frac{mol}{dm^3 \bullet s}\right)$$

$$\{k\} = \left(\frac{dm^3}{mol \bullet s}\right)$$

or as we will see in Chapter 4 leave it in terms of partial pressures.

In summary on reaction orders, they **cannot** be deduced from reaction stoichiometry. Even though a number of reactions follow elementary rate laws, at least as many reactions do not. One **must** determine the reaction order from the literature or from experiments.

3.2.3 Reversible Reactions

All rate laws for reversible reactions *must* reduce to the thermodynamic relationship relating the reacting species concentrations at equilibrium. At equilibrium, the rate of reaction is identically zero for all species (i.e., $-r_A \equiv 0$). That is, for the general reaction

$$aA + bB \; \underset{}{\overset{}{\rightleftharpoons}} \; cC + dD \tag{2-1}$$

the concentrations at equilibrium are related by the thermodynamic relationship for the equilibrium constant K_C (see Appendix C).

Thermodynamic equilibrium relationship

$$K_C = \frac{C_{Ce}^c C_{De}^d}{C_{Ae}^a C_{Be}^b} \tag{3-10}$$

The units of the thermodynamic equilibrium constant, K_C, are $(mol/dm^3)^{d+c-b-a}$.

To illustrate how to write rate laws for reversible reactions, we will use the combination of two benzene molecules to form one molecule of hydrogen and one of diphenyl. In this discussion, we shall consider this gas-phase reaction to be elementary and reversible

$$2C_6H_6 \; \underset{k_{-B}}{\overset{k_B}{\rightleftharpoons}} \; C_{12}H_{10} + H_2$$

or, symbolically,

$$2B \; \underset{k_{-B}}{\overset{k_B}{\rightleftharpoons}} \; D + H_2$$

The forward and reverse specific reaction rate constants, k_B and k_{-B}, respectively, will *be defined with respect to benzene*.

Benzene (B) is being depleted by the forward reaction

$$2C_6H_6 \; \overset{k_B}{\longrightarrow} \; C_{12}H_{10} + H_2$$

in which the rate of disappearance of benzene is

$$-r_{B, forward} = k_B C_B^2$$

If we multiply both sides of this equation by -1, we obtain the expression for the rate of formation of benzene for the forward reaction

$$r_{B, \text{forward}} = -k_B C_B^2 \tag{3-11}$$

For the reverse reaction between diphenyl (D) and hydrogen (H_2),

$$C_{12}H_{10} + H_2 \xrightarrow{k_{-B}} 2C_6H_6$$

the rate of formation of benzene is given as

$$r_{B, \text{reverse}} = k_{-B} C_D C_{H_2} \tag{3-12}$$

> The specific reaction rate constant, k_i, must be defined with respect to a particular species.

Again, both the rate constants k_B and k_{-B} are *defined with respect to benzene!!!*

The net rate of formation of benzene is the sum of the rates of formation from the forward reaction [i.e., Equation (3-11)] and the reverse reaction [i.e., Equation (3-12)]

Net rate

$$r_B \equiv r_{B, \text{net}} = r_{B, \text{forward}} + r_{B, \text{reverse}}$$

$$r_B = -k_B C_B^2 + k_{-B} C_D C_{H_2} \tag{3-13}$$

Multiplying both sides of Equation (3-13) by -1, and then factoring out k_B, we obtain the rate law for the rate of disappearance of benzene, $-r_B$

> **Elementary reversible**
> $A \rightleftarrows B$
> $$-r_A = k\left(C_A - \frac{C_B}{K_C}\right)$$

$$-r_B = k_B C_B^2 - k_{-B} C_D C_{H_2} = k_B \left(C_B^2 - \frac{k_{-B}}{k_B} C_D C_{H_2} \right)$$

Replacing the ratio of the reverse to forward rate law constants by the reciprocal of the concentration equilibrium constant, K_C, we obtain

$$\boxed{-r_B = k_B \left(C_B^2 - \frac{C_D C_{H_2}}{K_C} \right)} \tag{3-14}$$

where

$$\frac{k_B}{k_{-B}} = K_C = \text{Concentration equilibrium constant}$$

The equilibrium constant decreases with increasing temperature for exothermic reactions and increases with increasing temperature for endothermic reactions.

Let's write the rate of formation of diphenyl, r_D, in terms of the concentrations of hydrogen, H_2, diphenyl, D, and benzene, B. The rate of formation of diphenyl, r_D, **must** have the same functional dependence on the reacting species concentrations as does the rate of disappearance of benzene, $-r_B$. The rate of formation of diphenyl is

$$r_D = k_D \left(C_B^2 - \frac{C_D C_{H_2}}{K_C} \right) \tag{3-15}$$

Using the relationship given by Equation (3-1) for the general reaction

Relative rates

$$\boxed{\frac{r_A}{-a} = \frac{r_B}{-b} = \frac{r_C}{c} = \frac{r_D}{d}} \tag{3-1}$$

we can obtain the relationship between the various specific reaction rates, k_B, k_D

$$\frac{r_D}{1} = \frac{r_B}{-2} = \frac{-k_B\,[C_B^2 - C_D C_{H_2}/K_C]}{-2} = \frac{k_B}{2}\left[C_B^2 - \frac{C_D C_{H_2}}{K_C}\right] \tag{3-16}$$

Comparing Equations (3-15) and (3-16), we see the relationship between the specific reaction rate with respect to diphenyl, k_D, and the specific reaction rate with respect to benzene, k_B, is

$$k_D = \frac{k_B}{2}$$

Consequently, we see the need to define the rate constant, k, with respect to a particular species.

Finally, we need to check to see if the rate law given by Equation (3-14) is thermodynamically consistent at equilibrium. Applying Equation (3-10) (and Appendix C) to the diphenyl reaction and substituting the appropriate species concentration and exponents, thermodynamics tells us that

$$K_C = \frac{C_{De} C_{H_2 e}}{C_{Be}^2} \tag{3-17}$$

At equilibrium, the rate law must reduce to an equation consistent wth thermodynamic equilibrium.

Now let's look at the rate law. At equilibrium, $-r_B \equiv 0$, and the rate law given by Equation (3-14) becomes

$$-r_B \equiv 0 = k_B\left[C_{Be}^2 - \frac{C_{De} C_{H_2 e}}{K_C}\right]$$

Rearranging, we obtain, as expected, the equilibrium expression

$$K_C = \frac{C_{De} C_{H_2 e}}{C_{Be}^2}$$

that is identical to Equation (3-17) obtained from thermodynamics.

From Appendix C, Equation (C-9), we know that when there is no change in the total number of moles and the heat capacity term, $\Delta C_P = 0$, the temperature dependence of the concentration equilibrium constant is

$$\boxed{K_C(T) = K_C(T_1)\exp\left[\frac{\Delta H_{Rx}^\circ}{R}\left(\frac{1}{T_1} - \frac{1}{T}\right)\right]} \tag{C-9}$$

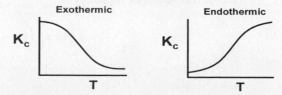

Therefore, if we know the equilibrium constant at one temperature, T_1 [i.e., K_C (T_1)], and the heat of reaction, ΔH°_{Rx}, we can calculate the equilibrium constant at any other temperature T. For endothermic reactions, the equilibrium constant, K_C, increases with increasing temperature; for exothermic reactions, K_C decreases with increasing temperature. A further discussion of the equilibrium constant and its thermodynamic relationship is given in Appendix C. For large values of the equilibrium constant, K_C, the reaction behaves as if it were irreversible.

3.3 Rates and the Reaction Rate Constant

There are three molecular concepts relating to the rate of reaction we will discuss:

Concept 1. Law of Mass Action. The rate of reaction increases with increasing concentration of reactants owing to the increased number of molecular collisions at the higher reactant concentrations. In Section 3.2, we have just discussed **Concept 1**: rate laws, and the dependence of the rate reactant concentration. We now discuss **Concept 2**, potential energy surfaces and energy barriers, and **Concept 3**, the energy needed for crossing the barriers.

3.3.1 The Rate Constant k

The reaction rate constant k is not truly a constant; it is merely independent of the concentrations of the species involved in the reaction. The quantity k is referred to as either the **specific reaction rate or the rate constant**. It is almost always strongly dependent on temperature. It also depends on whether or not a catalyst is present, and in gas-phase reactions, it may be a function of total pressure. In liquid systems it can also be a function of other parameters, such as ionic strength and choice of solvent. These other variables normally exhibit much less effect on the specific reaction rate than does temperature, with the exception of supercritical solvents, such as supercritical water. Consequently, for the purposes of the material presented here, it will be assumed that k_A depends *only on* temperature. This assumption is valid in most laboratory and industrial reactions, and seems to work quite well.

It was the great Nobel Prize–winning Swedish chemist Svante Arrhenius (1859–1927) who first suggested that the temperature dependence of the specific reaction rate, k_A, could be correlated by an equation of the type

Arrhenius equation

$$\boxed{k_A(T) = Ae^{-E/RT}}\tag{3-18}$$

where A = pre-exponential factor or frequency factor
         E = activation energy, J/mol or cal/mol
         R = gas constant = 8.314 J/mol • K = 1.987 cal/mol • K
         T = absolute temperature, K

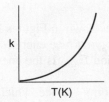

Equation (3-18), known as the *Arrhenius equation*, has been verified empirically to give the correct temperature behavior of most reaction rate constants within experimental accuracy over fairly large temperature ranges. The Arrhenius equation is derived in the *Professional Reference Shelf R3.A: Collision Theory* on the CRE Web site.

Reference Shelf

Additionally, one can view the activation energy in terms of collision theory (*Professional Reference Shelf R3.1*). By increasing the temperature, we increase the kinetic energy of the reactant molecules. This kinetic energy can in turn be transferred through molecular collisions to internal energy to increase the stretching and bending of the bonds, causing them to reach an activated state, vulnerable to bond breaking and reaction.

Why is there an activation energy? If the reactants are free radicals that essentially react immediately on collision, there usually isn't an activation energy. However, for most atoms and molecules undergoing reaction, there is an activation energy. A couple of the reasons are that in order to react

1. The molecules need energy to distort or stretch their bonds so that they break and now can form new bonds.
2. The molecules need energy to overcome the steric and electron repulsive forces as they come close together.

The activation energy can be thought of as a <u>barrier</u> to energy transfer (from kinetic energy to potential energy) between reacting molecules that must be overcome. The activation energy is the minimum increase in potential energy of the reactants that must be provided to transform the reactants into products. This increase can be provided by the kinetic energy of the colliding molecules.

In addition to the concentrations of the reacting species, there are two other factors that affect the rate of reaction,

- the height of the barrier, i.e., activation energy, and
- the fraction of molecular collisions that have sufficient energy to cross over the barrier (i.e., react when the molecules collide).

If we have a small barrier height, the molecules colliding will need only low kinetic energies to cross over the barrier. For reactions of molecules with small barrier heights occurring at room temperatures, a greater fraction of molecules will have this energy at low temperatures. However, for larger barrier heights, we require higher temperatures where a higher fraction of colliding molecules will have the necessary energy to cross over the barrier and react. We will discuss each of these concepts separately.

Concept 2. Potential Energy Surfaces and Energy Barriers. One way to view the barrier to a reaction is through the use of potential energy surfaces and the reaction coordinates. These coordinates denote the minimum potential energy of the system as a function of the progress along the reaction path as we go from reactants to an intermediate to products. For the exothermic reaction

$$A + BC \rightleftharpoons A - B - C \longrightarrow AB + C$$

the potential energy surface and the reaction coordinate are shown in Figures 3-1 and 3-2. Here E_A, E_C, E_{AB}, and E_{BC} are the potential energy surface energies of the reactants A, BC and product molecules (AB and C), and E_{ABC} is the energy of the complex A–B–C at the top of the barrier.

Figure 3-1(a) shows the 3–D plot of the potential energy surface, which is analogous to a mountain pass where we start out in a valley and then climb up to pass over the top of the pass, i.e., the <u>col</u> or <u>saddle point</u>, and proceed down into the next valley. Figure 3-1(b) shows a contour plot of the pass and valleys and the reaction coordinate as we pass over the col from valley to valley.

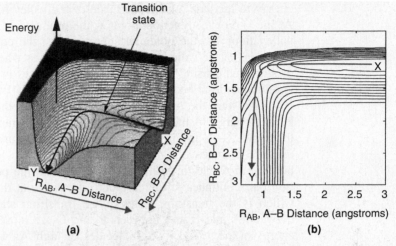

(a)   **(b)**

Figure 3-1   A potential energy surface for the $H + CH_3OH \longrightarrow H_2 + CH_2OH$ from the calculations of Blowers and Masel. The lines in the figure are contours of constant energy. The lines are spaced 5 kcal/mol. Richard I. Masel, *Chemical Kinetics and Catalysis*, p. 370, Fig.7.6 (Wiley, 2001).

Side view at point X in **Figure 3-1(b)** showing the valley

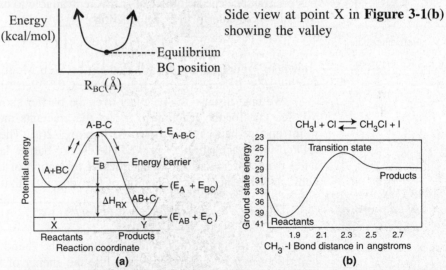

(a)   **(b)**

Figure 3-2   Progress along reaction path. **(a)** Symbolic reaction; **(b)** Calculated from computational software on the CRE Web site, Chapter 3 Web Module.

Energy changes as we move within the potential energy surfaces.

At point X in Figure 3-1(b), species A and BC are far apart and are not interacting; R_{BC} is just the equilibrium bond length. Consequently, the potential energy is just the BC bond energy. Here, A and BC are in their minimum potential energy in the valley and the steep rise up from the valley to the col from X would correspond to increases in the potential energy as A comes close to BC. If the BC bond is stretched from its equilibrium position at X, the potential energy increases up one side of the valley hill (R_{BC} increases). The potential energy is now greater because of the attractive forces trying to bring the B–C distance back to its equilibrium position. If the BC bond is compressed at X from its equilibrium position, the repulsive forces cause the potential energy to increase on the other side of the valley hill, i.e., R_{BC} decreases at X. At the end of the reaction, point Y, the products AB and C are far apart on the valley floor and the potential energy is just the equilibrium AB bond energy. The sides of the valley at Y represent the cases where AB is either compressed or stretched causing the corresponding increases in potential energy at point Y and can be described in an analogous manner to BC at point X.

We want the minimum energy path across the barrier for converting the kinetic energy of the molecules into potential energy. This path is shown by the curve X→Y in Figure 3-1(a) and also by Figure 3-2(a). As we move along the A–B distance axis in Figure 3-2(a), A comes closer to BC and B begins to bond with A and to push BC apart such that the potential energy of the reaction pair continues to increase until we arrive at the top of the energy barrier, which is the transition state. In the transition state the molecular distances between A and B and between B and C are close. As a result, the potential energy of the three atoms (molecules) is high. As we proceed further along the arc length of the reaction coordinate depicted in Figure 3-1(a), the AB bond strengthens and the BC bond weakens and C moves away from AB and the energy of the reacting pair decreases until we arrive at the valley floor where AB is far apart from C. The reaction coordinate quantifies how far the reaction has progressed. The commercial software available to carry out calculations for the transition state for the real reaction

Reference Shelf

$$CH_3I + Cl \rightarrow CH_3Cl + I$$

shown in Figure 3-2(b) is discussed in the Web Module *Molecular Modeling in Chemical Reaction Engineering* on the CRE Web site.

We next discuss the pathway over the barrier shown along the line Y–X. We see that for the reaction to occur, the reactants must overcome the minimum energy barrier, E_B, shown in Figure 3-2(a). The energy barrier, E_B, is related to the activation energy, E. The energy barrier height, E_B, can be calculated from differences in the energies of formation of the transition-state molecule and the energy of formation of the reactants; that is,

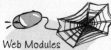

Web Modules

$$\boxed{E_B = E_{fA-B-C}^\circ - (E_{fA}^\circ + E_{fB-C}^\circ)} \tag{3-19}$$

The energy of formation of the reactants can be found in the literature or calculated from quantum mechanics, while the energy of formation of the transition state can also be calculated from quantum mechanics using a number of software packages, such as Gaussian (*http://www.gaussian.com/*) and Dacapo (*https://wiki.fysik.dtu.dk/dacapo*). The activation energy, E, is often approximated by the barrier height, E_B, which is a good approximation in the absence of quantum mechanical tunneling.

Now that we have the general idea for a reaction coordinate, let's consider another real reaction system

$$H\bullet + C_2H_6 \rightarrow H_2 + C_2H_5\bullet$$

The energy-reaction coordinate diagram for the reaction between a hydrogen atom and an ethane molecule is shown in Figure 3-3 where the bond distortions, breaking, and forming are identified. This figure shows schematically how the hydrogen molecule, H, moves into CH_3–CH_3 molecule, distorting the C–H bond and pushing the C–H bond off the methyl hydrogen to arrive at the transition state. As one continues along the reaction coordinate, the methyl hydrogen moves out of the C–H bond into the H–H bond to form the products CH_3CH_2 and H_2.

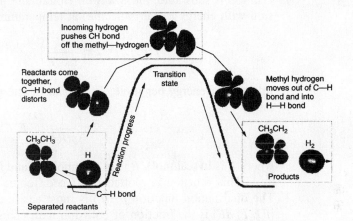

Figure 3-3   A diagram of the orbital distortions during the reaction

$$H \bullet + CH_3CH_3 \rightarrow H_2 + CH_2CH_3 \bullet$$

The diagram shows only the interaction with the energy state of ethane (the C–H bond). Other molecular orbitals of the ethane also distort. Courtesy of Richard I. Masel, *Chemical Kinetics and Catalysis*, p. 594 (Wiley, 2001).

Concept 3. Fraction of Molecular Collisions That Have Sufficient Energy to React. Now that we have established a barrier height, we need to know what fraction of molecular collisions have sufficient energy to cross over the barrier and react. To discuss this issue, we consider reactions in the gas phase where the reacting molecules will not have only one velocity, U, but a distribution of velocities, $f(U,T)$. Some will have high velocities and some will have low velocities as they move around and collide. These velocities are not defined with respect to a fixed coordinate system; these velocities are defined with respect to the other reactant molecules. The Maxwell-Boltzmann distribution of relative velocities is given by the probability function, $f(U, T)$

$$f(U,T) = 4\pi \left(\frac{m}{2\pi k_B T} \right)^{3/2} \exp\left[\frac{-mU^2}{2k_B T} \right] U^2$$

k_B = Boltzmann's constant = 3.29×10^{-24} cal/molecule/K
m = Reduced mass, g
U = Relative velocity, *m/s*
T = Absolute Temperature, K
e = Energy kcal/molecule
E = Kinetic energy kcal/mol

We usually interpret $f(U, T)$ with dU, i.e., $f(U, T)\ dU$ = fraction of reactant molecules with velocities between U and $(U + dU)$.

Rather than using velocities to discuss the fraction of molecules with sufficient energy to cross the barrier, we convert these velocities to energies using the equation for kinetic energy in making this conversion

$$e = \frac{1}{2}mU^2$$

Using this substitute, the Maxwell Boltzmann probability distribution of collision with energy e (cal/molecule) at temperature T is

$$f(e,T) = 2\pi\left(\frac{1}{\pi k_B T}\right)^{3/2} e^{1/2} \; exp\left[\frac{-e}{k_B T}\right] \tag{3-19}$$

In terms of energy per mole, E, *instead* of energy per molecule, e, we have

$$f(E,T) = 2\pi\left(\frac{1}{\pi RT}\right)^{3/2} E^{1/2} \; exp\left[\frac{-E}{RT}\right] \tag{3-20}$$

where E is in (cal/mol), R is in (cal/mol/K), and $f(E, T)$ is in mol/cal.

This function is plotted for two different temperatures in Figure 3-4(a). The distribution function $f(E,T)$ is most easily interpreted by recognizing that $[f(E,T) \, dE]$ is the fraction of collision with energies between E and $E + dE$.

$$f(E,T)dE = 2\pi\left(\frac{1}{\pi k_B T}\right)^{3/2} E^{1/2} \exp\left[-\frac{E}{k_B T}\right]dE \tag{3-21}$$

For example, the fraction of collisions with energies between 0.25 kcal/mol and 0.35 kcal/mol would be

$$\left(\text{Fraction with energies to between } (0.25 \text{ and } 0.35) \; \frac{\text{kcal}}{\text{mol}}\right) = \int_{0.25}^{0.35} f(E,T)\,dE$$

This fraction is shown by the shaded area in Figure 3-4(a) and is approximated by the average value of $f(E,T)$ at $E = 0.3$ kcal/mole is 0.81 mol/kcal.

$$f(E,T)dE = f(0.3, \; 300K)\Delta E = \frac{0.81 \text{ mol}}{\text{kcal}}\left(0.35\frac{\text{kcal}}{\text{mol}} - 0.25\frac{\text{kcal}}{\text{mol}}\right) = 0.081$$

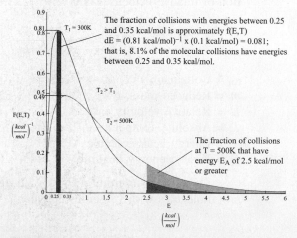

Figure 3-4(a)   Energy distribution of reacting molecules.

Thus 8.1% of the molecular collisions have energies between 0.25 and 0.35 kcal/mol.

We can also determine the fraction of collision that have energies greater than a certain value, E_A

$$\left(\begin{array}{c}\text{Fraction of} \\ \text{Molecules} \\ \text{with}\end{array}\quad E > E_A\right) = F(E > E_A, T) = \int_{E_A}^{\infty} f(E, T)\, dE \qquad (3\text{-}22)$$

This fraction is shown by the shaded area for $E_A = 2.5$ kcal/mole for $T = 300$ K (heavier shade) and for $T = 500$ K (lighter shade). One can easily see that for $T = 500$ K a greater fraction of the collisions that cross the barrier with $E_A = 2.5$ kcal/mol and react, which is consistent with our observation that the rates of reaction increase with increasing temperature.

For $E_A > 3RT$, we can obtain an analytical approximation for the fraction of molecules of collision with energies greater than E_A by combining Equations (3-21) and (3-22) and integrating to get

$$\left[\begin{array}{c}\text{Fraction of collision} \\ \text{with energies greater} \\ \text{than } E_A\end{array}\right] = F(E > E_A,\ T) \cong \frac{2}{\sqrt{\pi}}\left(\frac{E_A}{RT}\right)^{1/2}\exp\left(-\frac{E_A}{RT}\right) \quad (3\text{-}23)$$

Figure 3.4(b) shows the fraction of collisions that have energies greater than E_A as as function of the choose E_A at two different temperatures. One observes for an activation energy E_A of 20 kcal/mol and a temperature of 300 K the fraction of collisions with energies greater than 20 kcal/mol is 1.76×10^{-14} while at 600 K, the fraction increases to 2.39×10^{-7}, which is a 7 orders of magnitude difference.

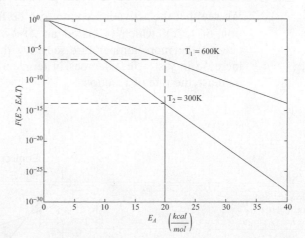

Figure 3-4(b)   Fraction of collision with energies greater than E_A.

To summarize this discussion on the three concepts

Concept 1   The rate increases with increasing reactant concentration,

Concept 2   The rate is related to the potential barrier height and to the conversion of translational energy into potential energy, and

Concept 3   The rate increases with the increasing fraction of collisions that have sufficient energy to cross over the barrier and form products.

To carry this discussion to the next level is beyond the scope of this text as it involves averaging over a collection of pairs of molecules to find an average rate with which they pass over the transition state to become products.

Recapping the last section, the energy of the individual molecules falls within a distribution of energies some molecules have more energy than others. One such distribution is the Boltzmann distribution, shown in Figure 3-4, where $f(E, T)$ is the energy distribution function for the kinetic energies of the reacting molecules. It is interpreted most easily by recognizing the product $[f(E, T) \, dE]$ as the fraction of molecular collisions that have an energy between E and $(E + dE)$. For example, in Figure 3-4(a), the fraction of collisions that have energies between 0.25 kcal/mol and 0.35 kcal/mol is 0.081, as shown by the shaded area on the left. The activation energy has been equated with a minimum energy that must be possessed by reacting molecules before the reaction will occur. The fraction of the molecular collisions that have an energy E_A or greater is shown by the shaded areas at the right in Figure 3-4(a). Figure 3-4(b) shows the fraction of collisions with energies greater than as a function of E_A at two different temperatures.

3.3.2 The Arrhenius Plot

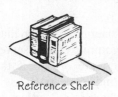

Reference Shelf

Postulation of the Arrhenius equation, Equation (3-18), remains the greatest single advancement in chemical kinetics, and retains its usefulness today, more than a century later. The activation energy, E, is determined experimentally by measuring the reaction rate at several different temperatures. After taking the natural logarithm of Equation (3-18), we obtain

$$\ln k_A = \ln A - \frac{E}{R}\left(\frac{1}{T}\right) \tag{3-24}$$

We see that the activation energy can be found from a plot of $\ln k_A$ as a function of $(1/T)$, which is called an *Arrhenius plot*. The larger the activation energy, the more temperature-sensitive the reaction. That is, for large E, an increase in just a few degrees in temperature can greatly increase k and thus increase the rate of reaction.

Calculation of the
activation energy

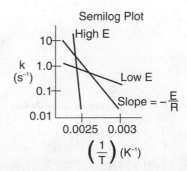

Figure 3-5 Calculation of the activation energy from an Arrhenius plot.

Example 3–1 Determination of the Activation Energy

Calculate the activation energy for the decomposition of benzene diazonium chloride to give chlorobenzene and nitrogen

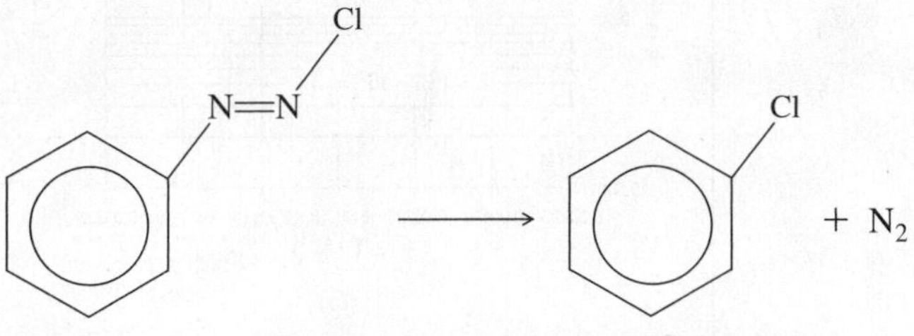

using the information in Table E3-1.1 for this first-order reaction.

TABLE E3-1.1 DATA

k (s^{-1})	0.00043	0.00103	0.00180	0.00355	0.00717
T (K)	313.0	319.0	323.0	328.0	333.0

Solution

We start by recalling Equation (3-20)

$$\ln k_A = \ln A - \frac{E}{R}\left(\frac{1}{T}\right) \tag{3-24}$$

We can use the data in Table E3-1.1 to determine the activation energy, E, and frequency factor, A, in two different ways. One way is to make a semilog plot of k vs. $(1/T)$ and determine E from the slope $(-E/R)$ of an Arrhenius plot. Another way is to use Excel or Polymath to regress the data. The data in Table E3-1.1 was entered in Excel and is shown in Figure E3-1.1, which was then used to obtain Figure E3-1.2.

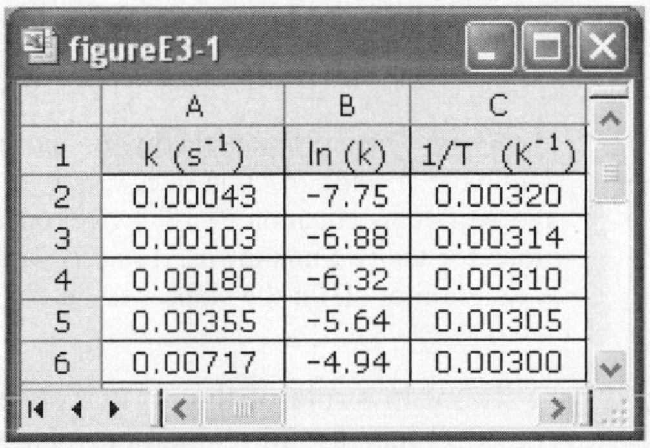

Summary Notes

	A	B	C
1	k (s^{-1})	ln (k)	1/T (K^{-1})
2	0.00043	-7.75	0.00320
3	0.00103	-6.88	0.00314
4	0.00180	-6.32	0.00310
5	0.00355	-5.64	0.00305
6	0.00717	-4.94	0.00300

Figure E3-1.1 Excel spreadsheet.

Tutorials

A step-by-step tutorial to construct both an Excel and a Polymath spreadsheet is given in the Chapter 3 Summary Notes on the CRE Web site.

The equation for the best fit of the data

$$\ln k = \frac{-14{,}017}{T} + 37.12 \tag{E3-1.1}$$

$$k = k_1 \exp\left[\frac{E}{R}\left(\frac{1}{T_1} - \frac{1}{T}\right)\right]$$

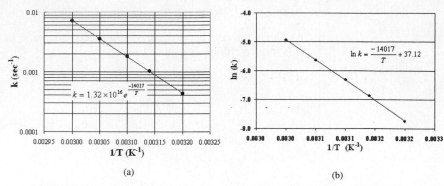

(a)   (b)

Figure E3-1.2   (a) Excel semilog plot; (b) Excel normal plot after taking ln(k).

is also shown in Figure E3-1.2(b). From the slope of the line given in Figure 3-1.2(b) and Equation (3-20), we obtain

$$-\frac{E}{R} = -14,017 \text{ K}$$

$$E = (14,017 \text{ K})R = (14,017 \text{ K})\left(8.314 \frac{\text{J}}{\text{mol} \cdot \text{K}}\right)$$

$$\boxed{E = 116.5 \frac{\text{kJ}}{\text{mol}}}$$

From Figure E3-1.2(b) and Equation (E3-1.1), we see

$$\ln A = 37.12$$

Taking the antilog, we find the frequency factor to be

$$A = 1.32 \times 10^{16} \text{ s}^{-1}$$

$$k = 1.32 \times 10^{16} \, exp\left[-\frac{14,017 \text{ K}}{\text{T}}\right] \tag{E3-1.2}$$

Analysis: The activation energy, E, and frequency factor, A, can be calculated if we know the specific reaction rate, k, at two temperatures, T_1 and T_2. We can either use the Arrhenius equation (3-18) twice, once at T_1 and once at T_2, to solve two equations for the two unknowns, A and E, *or* we can take the slope of a plot of (ln k) as a function of (1/T); the slope will be equal to (−E/R).

The rate does not always double for a temperature increase of 10°C.

There is a rule of thumb that states that the rate of reaction doubles for every 10°C increase in temperature. However, this rule is true only for specific combinations of activation energies and temperatures. For example, if the activation energy is 53.6 kJ/mol, the rate will double only if the temperature is raised from 300 K to 310 K. If the activation energy is 147 kJ/mol, the rule will be valid only if the temperature is raised from 500 K to 510 K (see Problem P3-7$_\text{B}$ for the derivation of this relationship).

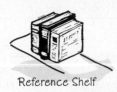

Reference Shelf

The larger the activation energy, the more temperature-sensitive is the rate of reaction. While there are no typical values of the frequency factor and activation energy for a first-order gas-phase reaction, if one were forced to make a guess, values of A and E might be 10^{13} s^{-1} and 100 kJ/mol. However, for families of reactions (e.g., halogenation), a number of correlations can be used to estimate the activation energy. One such correlation is the *Polanyi-Semenov equation*, which relates activation energy to the heat of reaction (see *Professional Reference Shelf R3.1*). Another correlation relates the activation energy to differences in bond strengths between products and reactants.[4] While the activation energy cannot be currently predicted *a priori*, significant research efforts are under way to calculate activation energies from first principles.[5]

One final comment on the Arrhenius equation, Equation (3-18). It can be put in a most useful form by finding the specific reaction rate at a temperature T_0; that is,

$$k(T_0) = Ae^{-E/RT_0}$$

and at a temperature T

$$k(T) = Ae^{-E/RT}$$

and taking the ratio to obtain

A most useful form of $k(T)$

$$\boxed{k(T) = k(T_0)e^{\frac{E}{R}\left(\frac{1}{T_0} - \frac{1}{T}\right)}} \qquad (3\text{-}25)$$

This equation says that if we know the specific reaction rate $k(T_0)$ at a temperature, T_0, and we know the activation energy, E, we can find the specific reaction rate $k(T)$ at any other temperature, T, for that reaction.

3.4 Present Status of Our Approach to Reactor Sizing and Design

Where are we?

In Chapter 2, we combined the different reactor mole balances with the definition of conversion to arrive at the design equation for each of four types of reactors, as shown in Table 3-2. Next we showed that *if* the rate of disappearance is known as a function of the conversion X

$$-r_A = g(X)$$

then it is possible to size CSTRs, PFRs, and PBRs operated at the same conditions under which $-r_A = g(X)$ was obtained.

---

[4] M. Boudart, *Kinetics of Chemical Processes* (Upper Saddle River, NJ: Prentice Hall, 1968), p. 168. J. W. Moore and R. G. Pearson, *Kinetics and Mechanisms,* 3rd ed. (New York: Wiley, 1981), p. 199. S. W. Benson, *Thermochemical Kinetics,* 2nd ed. (New York: Wiley, 1976).

[5] R. Masel, *Chemical Kinetics and Catalysis,* New York: Wiley, 2001, p. 594.

TABLE 3-2   DESIGN EQUATIONS

	Differential Form		Algebraic Form	Integral Form	
Batch	$N_{A0}\dfrac{dX}{dt} = -r_A V$	(2-6)		$t = N_{A0}\displaystyle\int_0^X \dfrac{dX}{-r_A V}$	(2-9)
Backmix (CSTR)			$V = \dfrac{F_{A0}X}{-r_A}$ (2-13)		
Fluidized CSTR			$W = \dfrac{F_{A0}X}{-r_A'}$		
Tubular (PFR)	$F_{A0}\dfrac{dX}{dV} = -r_A$	(2-15)		$V = F_{A0}\displaystyle\int_0^X \dfrac{dX}{-r_A}$	(2-16)
Packed bed (PBR)	$F_{A0}\dfrac{dX}{dW} = -r_A'$	(2-17)		$W = F_{A0}\displaystyle\int_0^X \dfrac{dX}{-r'_A}$	(2-18)

The design equations

In general, information in the form $-r_A = g(X)$ is not available. However, we have seen in Section 3.2 that the rate of disappearance of A, $-r_A$, is normally expressed in terms of the concentration of the reacting species. This functionality

$$-r_A = [k_A(T)][\text{fn}(C_A, C_B, \ldots)] \tag{3-2}$$

is called a *rate law*. In Chapter 4, we show how the concentration of the reacting species may be written in terms of the conversion X

$$C_j = h_j(X) \tag{3-26}$$

> $-r_A = f(C_j)$
> $+$
> $C_j = h_j(X)$
> $\downarrow$
> $-r_A = g(X)$
> and then we can design isothermal reactors

With these additional relationships, one observes that if the rate law is given and the concentrations can be expressed as a function of conversion, *then in fact we have $-r_A$ as a function of X and this is all that is needed to evaluate the isothermal design equations*. One can use either the numerical techniques described in Chapter 2 or, as we shall see in Chapter 5, a table of integrals, and/or software programs (e.g., Polymath).

> **Closure.** Having completed this chapter, you should be able to write the rate law in terms of concentration and the Arrhenius temperature dependence. We have now completed the first two basic building blocks in our algorithm to study isothermal chemical reactions and reactors.
>
> **Rate Law**
> **Mole Balance**
>
> In Chapter 4, we focus on the third building block, **stoichiometry**, as we use the stoichiometric table to write the concentrations in terms of conversion to finally arrive at a relationship between the rate of reaction and conversion.

> The CRE Algorithm
> • Mole Balance, Ch 1
> • Rate Law, Ch 3
> • Stoichiometry, Ch 4
> • Combine, Ch 5
> • Evaluate, Ch 5
> • Energy Balance, Ch 11

SUMMARY

1. Relative rates of reaction for the generic reaction:

$$A + \frac{b}{a}B \rightarrow \frac{c}{a}C + \frac{d}{a}D \tag{S3-1}$$

The relative rates of reaction can be written either as

$$\boxed{\frac{-r_A}{a} = \frac{-r_B}{b} = \frac{r_C}{c} = \frac{r_D}{d}} \text{ or } \boxed{\frac{r_A}{-a} = \frac{r_B}{-b} = \frac{r_C}{c} = \frac{r_D}{d}} \tag{S3-2}$$

2. *Reaction order* is determined from experimental observation:

$$A + B \longrightarrow C \tag{S3-3}$$

$$-r_A = kC_A^\alpha C_B^\beta$$

The reaction in Equation (S3-3) is α order with respect to species A and β order with respect to species B, whereas the overall order, n, is ($\alpha + \beta$). If $\alpha = 1$ and $\beta = 2$, we would say that the reaction is first order with respect to A, second order with respect to B, and overall third order. We say a reaction follows an elementary rate law if the reaction orders agree with the stoichiometric coefficients for the reaction as written.

Examples of reactions that follow an elementary rate law:

Irreversible reactions

First order

$$C_2H_6 \longrightarrow C_2H_4 + H_2 \qquad\qquad \boxed{-r_{C_2H_6} = kC_{C_2H_6}}$$

Second order

$$CNBr + CH_3NH_2 \longrightarrow CH_3Br + NCNH_2 \qquad \boxed{-r_{CNBr} = kC_{CNBr}C_{CH_3NH_2}}$$

Reversible reactions

Second order

$$2C_6H_6 \rightleftharpoons C_{12}H_{10} + H_2 \qquad \boxed{-r_{C_2H_6} = k_{C_2H_6}\left(C_{C_6H_6}^2 - \frac{C_{C_{12}H_{10}}C_{H_2}}{K_C}\right)}$$

Examples of reactions that follow nonelementary rate laws:

Homogeneous

$$CH_3CHO \longrightarrow CH_4 + CH_2 \qquad\qquad \boxed{-r_{CH_3CHO} = kC_{CH_3CHO}^{3/2}}$$

Heterogeneous reactions

$$C_2H_4 + H_2 \xrightarrow[cat]{} C_2H_6 \qquad\qquad \boxed{-r_{C_2H_4} = k\frac{P_{C_2H_4}P_{H_2}}{1 + K_{C_2H_4}P_{C_2H_4}}}$$

3. The temperature dependence of a specific reaction rate is given by the *Arrhenius equation*

$$k = Ae^{-E/RT} \tag{S3-4}$$

where A is the frequency factor and E the activation energy.

If we know the specific reaction rate, k, at a temperature, T_0, and the activation energy, we can find k at any temperature, T

$$k(T) = k(T_0)\exp\left[\frac{E}{R}\left(\frac{1}{T_0} - \frac{1}{T}\right)\right] \tag{S3-5}$$

Concept 1   The rate increases with increasing reactant concentrations,
Concept 2   The rate is related to the potential barrier height and to the conversion of translational energy into potential energy, and
Concept 3   The rate increases as the fraction of collisions that have sufficient energy to cross over the barrier and form products increases.

Similarly from Appendix C, Equation (C-9), if we know the partial pressure equilibrium constant K_P at a temperature, T_1, and the heat of reaction, we can find the equilibrium constant at any other temperature

$$K_P(T) = K_P(T_1)\exp\left[\frac{\Delta H^\circ_{Rx}}{R}\left(\frac{1}{T_1} - \frac{1}{T}\right)\right] \tag{C-9}$$

CRE WEB SITE MATERIALS

- **Expanded Material**
  1. *Collision Theory*
  2. *Transition State*
  3. *Molecular Simulation*
  4. *How to Estimate the Activation Energy from the Polanyi Equation*
  5. *Web P3-1_A Puzzle Problem "What's Wrong with this Solution?"*
  6. *Additional Homework Problems*
- **Learning Resources**
  1. *Summary Notes for Chapter 3*
  2. *Web Modules*
  A. Cooking a Potato
  Chemical reaction engineering is applied to cooking a potato

$$\text{Starch (crystalline)} \xrightarrow{k} \text{Starch amorphous}$$

with

$$k = Ae^{-E/RT}$$

      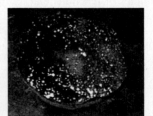

8 minutes at 400° F   12 minutes at 400° F   16 minutes at 400° F

B. Molecular Reaction Engineering
   Molecular simulators (e.g., Gaussian) are used to make predictions of the activation energy.

- **FAQs (Frequently Asked Questions)**
- **Professional Reference Shelf**

    R3.1. *Collision Theory*

    In this section, the fundamentals of collision theory

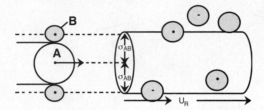

Figure R3.1   Schematic of collision cross section.

are applied to the reaction

$$A + B \rightarrow C + D$$

to arrive at the following rate law

$$-r_A = \underbrace{\pi \sigma_{AB}^2 \left(\frac{8 \pi k_B T}{\mu \pi} \right)^{1/2} N_{Avo} e^{-E_A/RT}}_{A} C_A C_B = A e^{-E_A/RT} C_A C_B$$

The activation energy, E_A, can be estimated from the Polanyi equation

$$E_A = E_A^\circ + \gamma_P \Delta H_{Rx}$$

    R3.2. *Transition-State Theory*

    In this section, the rate law and rate-law parameters are derived for the reaction

$$A + BC \rightleftarrows ABC^{\#} \rightarrow AB + C$$

using transition-state theory. Figure PW3B-1 shows the energy of the molecules along the reaction coordinate, which measures the progress of the reaction.

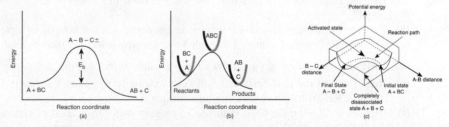

Figure PW3B-1   Reaction coordinate for (a) S_{N2} reaction, and (b) generalized reaction. (c) 3-D energy surface for generalized reaction.

    R3.3. *Molecular Dynamics Simulations*

    The reaction trajectories are calculated to determine the reaction cross section of the reacting molecules. The reaction probability is found by counting up the number of reactive trajectories after Karplus.[6]

---

[6] M. Karplus, R.N. Porter, and R.D. Sharma, *J. Chem. Phys.*, **43** (9), 3259 (1965).

Nonreactive Trajectory

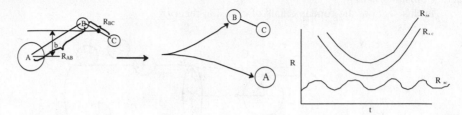

Reactive Trajectory

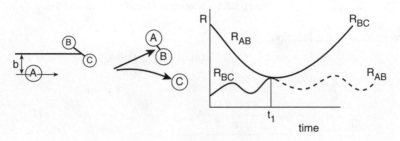

QUESTIONS AND PROBLEMS

The subscript to each of the problem numbers indicates the level of difficulty: A, least difficult; D, most difficult.

$$A = \bullet \quad B = \blacksquare \quad C = \blacklozenge \quad D = \blacklozenge\blacklozenge$$

Homework Problems

Questions

Q3-1$_C$ **(a)** List the important concepts that you learned from this chapter. What concepts are you not clear about?

(b) Explain the strategy to evaluate reactor design equations and how this chapter expands on Chapter 2.

(c) Choose a FAQ from Chapters 1 through 3 and say why it was the most helpful.

(d) Listen to the audios 🎧 on the CRE Web site. Select a topic and explain it.

(e) Read through the Self-Tests and Self-Assessments in Chapters 1 through 3 in the *Summary Notes* on the CRE Web site. Select one and critique it.

(f) Which example on the CRE Web site's *Summary Notes* for Chapters 1 through 3 was most helpful?

Problems

P3-1$_A$ **(a)** **Example 3-1.** Make a plot of k versus T and $ln\ k$ versus $(1/T)$ for $E = 240$ kJ/mol and for $E = 60$ kJ/mol. (1) Write a couple of sentences describing what you find. (2) Next, write a paragraph describing the activation, how it affects chemical reaction rates, and what its origins are.

(b) **Collision Theory**—Professional Reference Shelf. Make an outline of the steps that were used to derive

$$-r_A = Ae^{-E/RT}C_A C_B$$

(c) The rate law for the reaction $(2A + B \rightarrow C)$ is $-r_A = k_A C_A^2 C_B$ with $k_A = 25(dm^3/mol)^2/s$. What are k_B and k_C?

P3-2$_B$ Molecular collision energies—refer to Figure 3-3.

(a) What fraction of molecular collisions have energies less than or equal to 2.5 kcal at 300 K? At 500 K?

(b) What fraction of molecular collisions have energies between 3 and 4 kcal/mol at T = 300 K? At T = 500K?

(c) What fraction of molecular collisions have energies greater than the activation energy E_A = 25 kcal at T = 300 K? At T = 500K?

P3-3$_B$ (a) Use Figure 3-1(b) to sketch the trajectory over the saddle point when the BC and AB molecules vibrate with the minimum separation distance being 0.20 Angstroms and the maximum separation being 0.4 Angstroms.

(b) At point Y, R_{AB} = 2.8 Angstroms, sketch the potential energy as a function of the distance R_{BC} noting the minimum at valley floor.

(c) At Point X, R_{BC} = 2.8 Angstroms, sketch the potential energy as a function of R_{AB} noting the minimum on the valley floor.

P3-4$_B$ Use Equation (3-20) to make a plot of f(E,T) as a function of E for T = 300, 500, 800, and 1200 K.

(a) What is the fraction of molecules that have sufficient energy to pass over a energy barrier of 25 kcal at 300, 500, 800, and 1200 K?

(b) For a temperature of 300 K, what is the ratio of the faction of energies between 15 and 25 kcal to the same energy range (15-25 kcal) at 1200 K?

(c) Make a plot of $f(E > E_A, T)$ as a function of (E_A/RT) for $(E_A/RT) > 3$. What is the fraction of collisions that have energies greater than 25 kcal/mole at 700 K?

(d) What fraction of molecules have energies greater than 15 kcal/mol at 500 K?

(e) Construct a plot of $f(E > E_A, T)$ versus T for E_A = 3, 10, 25, and 40 kcal/mole. Describe what you find.

P3-5$_A$ The following figures show the energy distribution function at 300 K for the reaction $A + B \rightarrow C$

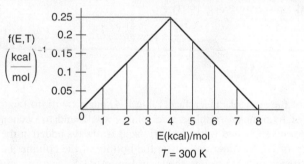

Figure P3-5(a)

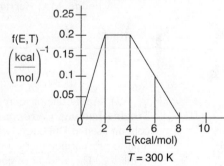

Figure P3-5(b)

For each figure, determine the following:

(a) What fraction of the collisions have energies between 3 and 5 kcal/mol?

(b) What fraction of collisions have energies greater than 5 kcal/mol?

(c) What is the fraction with energies greater than 0 kcal/mol?

(d) What is the fraction with energies 8 kcal/mol or greater?

(e) If the activation energy for Figure P3-5(b) is 8 kcal/mol, what fraction of molecules have an energy greater than E_A?

(f) Guess what (sketch) the shape of the curve $f(E,T)$ versus E shown in Figure P3-5(b) would look like if the temperature were increased to 400 K. (Remember: $\int_0^\infty f(E,T)dE = 1.$)

P3-6$_A$ The frequency of fireflies flashing and the frequency of crickets chirping as a function of temperature follow. Source: Keith J. Laidler, "Unconventional applications of the Arrhenius law." J. Chem. Educ., 5, 343 (1972). Copyright (c) 1972, American Chemical Society. Reprinted by permission.

For fireflies:				For crickets:			
T (°C)	21.0	25.00	30.0	T (°C)	14.2	20.3	27.0
Flashes/min	9.0	12.16	16.2	*Chirps/min*	80	126	200

The running speed of ants and the flight speed of honeybees as a function of temperature are given below. Source: B. Heinrich, *The Hot-Blooded Insects* (Cambridge, MA: Harvard University Press, 1993).

For ants:					For honeybees:				
T (°C)	10	20	30	38	T (°C)	25	30	35	40
V (cm/s)	0.5	2	3.4	6.5	V (cm/s)	0.7	1.8	3	?

(a) What do the firefly and cricket have in common? What are their differences?

(b) What is the velocity of the honeybee at 40°C? At –5°C?

(c) Nicolas wants to know if the bees, ants, crickets, and fireflies have anything in common. If so, what is it? You may also do a pair-wise comparison.

(d) Would more data help clarify the relationships among frequency, speed, and temperature? If so, in what temperature should the data be obtained? Pick an insect, and explain how you would carry out the experiment to obtain more data. For an alternative to this problem, see CDP3-A$_B$.

(e) Data on the tenebrionid beetle whose body mass is 3.3 g show that it can push a 35-g ball of dung at 6.5 cm/s at 27° C, 13 cm/s at 37° C, and 18 cm/s at 40° C.

   (1) How fast can it push dung at 41.5° C? Source: B. Heinrich. *The Hot-Blooded Insects* (Cambridge, MA: Harvard University Press, 1993).

   (2) Apply one or more of the six ideas in Preface Table P-4, page xxviii, to this problem.

P3-7$_B$ **Troubleshooting.** Corrosion of high-nickel stainless steel plates was found to occur in a distillation column used at DuPont to separate HCN and water. Sulfuric acid is always added at the top of the column to prevent polymerization of HCN. Water collects at the bottom of the column and HCN collects at the top. The amount of corrosion on each tray is shown in Figure P3-5$_B$ as a function of plate location in the column.

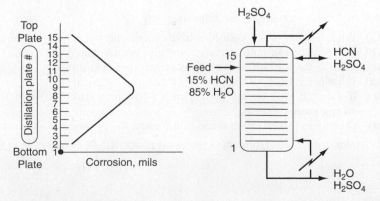

Figure P3-5$_B$ Corrosion in a distillation column.

The bottom-most temperature of the column is approximately 125°C and the topmost is 100°C. The corrosion rate is a function of temperature and the concentration of an HCN–H_2SO_4 complex. Suggest an explanation for the observed corrosion-plate profile in the column. What effect would the column operating conditions have on the corrosion-plate profile?

P3-8$_B$ **Inspector Sgt. Ambercromby of Scotland Yard.** It is believed, although never proven, that Bonnie murdered her first husband, Lefty, by poisoning the tepid brandy they drank together on their first anniversary. Lefty was unaware she had coated her glass with an antidote before she filled both glasses with the poisoned brandy. Bonnie married her second husband, Clyde, and some years later when she had tired of him, she called him one day to tell him of her new promotion at work and to suggest that they celebrate with a glass of brandy that evening. She had the fatal end in mind for Clyde. However, Clyde suggested that instead of brandy, they celebrate with ice-cold Russian vodka and they down it Cossack style, in one gulp. She agreed and decided to follow her previously successful plan and to put the poison in the vodka and the antidote in her glass. The next day, both were found dead. Sgt. Ambercromby arrives. What are the first three questions he asks? What are two possible explanations? Based on what you learned from this chapter, what do you feel Sgt. Ambercromby suggested as the most logical explanation?

Source: Professor Flavio Marin Flores, ITESM, Monterrey, Mexico.

[*Hint:* View the YouTube video (*www.youtube.com*) made by the chemical reaction engineering students at the University of Alabama, entitled *The Black Widow.* You can access it from the CRE Web site (*www.umich.edu/~elements/5e*) using the YouTube Videos link under *By Concepts* on the home page.]

P3-9$_B$ **Activation Energy**

    (a) The rule of thumb that the rate of reaction doubles for a 10°C increase in temperature occurs only at a specific temperature for a given activation energy. Develop a relationship between the temperature and activation energy for which the rule of thumb holds. Neglect any variations in concentrations.

    (b) Write a paragraph explaining activation energy, *E*, and how it affects the chemical reaction rate. Refer to Section 3.3 and especially the *Professional Reference Shelf sections R3.1, R3.2,* and *R3.3* if necessary.

P3-10$_C$ The initial reaction rate for the elementary reaction

$$2A + B \rightarrow 4C$$

was measured as a function of temperature when the concentration of A was 2 M and that of B was 1.5 M.

$-r_A$(mol/dm$^3 \cdot s$):	0.002	0.046	0.72	8.33
T(K):	300	320	340	360

    (a) What is the activation energy?

    (b) What is the frequency factor?

    (c) What is the rate constant as a function of temperature using $T = 300$K as the base case?

P3-11$_B$ Determine the rate law for the reaction described in each of the cases below involving species A, B, and C. The rate laws should be elementary as written for reactions that are either of the form A $\rightarrow$ B or A + B $\rightarrow$ C.

    (a) The units of the specific reaction rate are $k = \left[\dfrac{dm^3}{mol \cdot h}\right]$: Rate Law _____

    (b) The units of the specific reaction rate are $k = \left[\dfrac{mol}{kg\text{-}cat \cdot h(atm)^2}\right]$: Rate Law _____

(c) The units of the specific reaction rate are $k = \left[\dfrac{1}{h}\right]$: Rate Law _____

(d) The units of a nonelementary reaction rate are $k = \left[\dfrac{mol}{dm^3 \cdot h}\right]$: Rate Law _____

P3-12$_A$ **(a)** Write the rate law for the following reactions assuming each reaction follows an elementary rate law. Give the units of k_A for each, keeping in mind some are homogeneous and some reactants are heterogeneous.

$$\text{(a)} \qquad\qquad C_2H_6 \longrightarrow C_2H_4 + H_2$$

$$\text{(b)} \qquad\qquad C_2H_4 + \tfrac{1}{2}O_2 \rightarrow \overset{\displaystyle O}{\overset{\displaystyle \diagup\ \ \diagdown}{CH_2 - CH_2}}$$

$$\text{(c)} \qquad (CH_3)_3COOC(CH_3)_3 \rightleftharpoons C_2H_6 + 2CH_3COCH_3$$

$$\text{(d)} \qquad\qquad nC_4H_{10} \rightleftharpoons iC_4H_{10}$$

$$\text{(e)} \qquad CH_3COOC_2H_5 + C_4H_9OH \rightleftharpoons CH_3COOC_4H_9 + C_2H_5OH$$

$$\text{(f)} \qquad\qquad 2CH_3NH_2 \underset{\text{cat}}{\rightleftharpoons} (CH_3)_2 NH + NH_3$$

$$\text{(g)} \qquad\qquad (CH_3CO)_2O + H_2O \rightleftharpoons 2CH_3COOH$$

P3-13$_A$ **(a)** Write the rate law for the reaction

$$2A + B \rightarrow C$$

if the reaction
(1) is second order in B and overall third order, $-r_A =$ _____
(2) is zero order in A and first order in B, $-r_A =$ _____
(3) is zero order in both A and B, $-r_A =$ _____
(4) is first order in A and overall zero order. $-r_A =$ _____

(b) Find and write the rate laws for the following reactions
(1) $H_2 + Br_2 \rightarrow 2HBr$
(2) $H_2 + I_2 \rightarrow 2HI$

P3-14$_B$ The rate laws for each of the reactions listed below were obtained at low temperatures. The reactions are highly exothermic and therefore reversible at high temperatures. Suggest a rate law for each of the reactions [(**a**), (**b**), and (**c**)] at high temperatures, which may or may not be elementary.

Web Hint

(a) The reaction

$$A \rightarrow B$$

is irreversible at low temperatures and the rate law is

$$-r_A = kC_A$$

(b) The reaction

$$A + 2B \rightarrow 2D$$

is irreversible at low temperatures and the rate law is

$$-r_A = kC_A^{1/2} C_B$$

(c) The gas-solid catalyzed reaction

$$A + B \underset{cat}{\rightarrow} C + D$$

is irreversible at low temperatures and the rate law is

$$-r_A = \frac{kP_A P_B}{1 + K_A P_A + K_B P_B}$$

In each case, make sure that the rate laws at high temperatures are thermodynamically consistant at equilibrium (cf. Appendix C).

P3-15$_B$ **Solar Chemical Energy Storage.** The major ways to utilize, capture, or store the sun's energy are solar thermal (see P8-16$_B$), solar voltaic, biomass conversion, solar water splitting (P10-13$_B$), and solar chemical. *Solar chemical* refers to processes that harness and store solar energy by adsorbing light in a reversible chemical reaction; see *http://en.wikipedia.org/wiki/Solar_chemical*. For example, the photodimerization of anthracene absorbs and stores solar energy that can be released when the reverse reaction takes place.

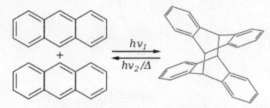

Figure P3-15.1 Anthracene dimerization.

Another reaction of interest is the Norbornadiene–Quadricyclane (NQ) couple, where solar energy is adsorbed and stored in one direction and released in the other.

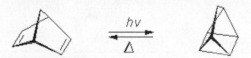

Figure P3-15.2 The Norbornadiene–
Quadricyclane (NQ) couple is of potential interest
for sunlight energy storage.

(a) Suggest a rate law for the reversible photodimerization of anthracene.
(b) Suggest a rate law for the reversible storage of energy by the NQ couple.

SUPPLEMENTARY READING

1. Two references relating to the discussion of activation energy have already been cited in this chapter. Activation energy is usually discussed in terms of either collision theory or transition-state theory. A concise and readable account of these two theories may or may not be found in

BURGESS, THORNTON W., *The Adventures of Reddy Fox*. New York: Dover Publications, Inc., 1913.
LAIDLER, K. J. *Chemical Kinetics*. New York: Harper & Row, 1987, Chap. 3.
MASEL, R., *Chemical Kinetics and Catalysis*. New York: Wiley, 2001, p. 594.

2. Prof. Dr. Sven Köttlov of Jofostan University has pointed out that in addition to the books listed above, the rate laws and activation energies can be found in NBS circulars, such as those shown below Table 3-1 on page 75. Also check chemistry literature, e.g., *Journal of Physical Chemistry*.

Stoichiometry   4

If you are thinking about someone while reading this book, you are most definitely in LOVE.

—Tim Newberger
Undergraduate ChE Student W2013

Overview. In Chapter 2 we carried out the sizing and sequencing of flow reactors given the reaction rate, $-r_A$, as a function of conversion X. To find $-r_A = f(X)$ is a two-step process. In Chapter 3, we described (*Step 1*) how the rate of reaction, $-r_A$, is related to concentration and temperature. In this chapter, we show how concentration can be related to conversion (*Step 2*), and once we do that we will have $-r_A = f(X)$ and can design a multitude of reaction systems. We will use stoichiometric tables, along with the definitions of concentration, to find the concentration as a function of conversion.

Batch   Flow

$$C_A = \frac{N_A}{V} = \frac{N_{A0}(1-X)}{V} \qquad C_A = \frac{F_A}{v} = \frac{F_{A0}(1-X)}{v}$$

$$\downarrow$$

$$\text{Liquid, } v = v_0 \quad \text{Gas, } v = v_0(1+\varepsilon X)\frac{P_0}{P}\frac{T}{T_0}$$

$$V = V_0$$

$$\downarrow \qquad\qquad\qquad \downarrow$$

$$C_A = C_{A0}(1-X) \qquad C_A = C_{A0}(1-X) \quad C_A = C_{A0}\frac{(1-X)}{(1+\varepsilon X)}\frac{P}{P_0}\frac{T_0}{T}$$

- For batch systems the reactor is rigid, so $V = V_0$, and one then uses the stoichiometric table to express concentration as a function of conversion: $C_A = N_A/V_0 = C_{A0}(1-X)$

- For liquid-phase flow systems, the volumetric flow rate is constant, $v = v_0$, and $C_A = (F_{A0}/v_0)(1 - X) = C_{A0}(1 - X)$.
- For gas-phase flow systems, the process becomes more complicated, as the volumetric flow rate for gases can vary with conversion, and we need to develop the relationship relating v and X, i.e., $v = v_0(1 + \varepsilon X)(P_0/P)(T/T_0)$ and thus

$$C_A = \frac{C_{A0}(1 - X)}{(1 + \varepsilon X)} \frac{P}{P_0} \frac{T_0}{T} = C_{A0} \frac{(1 - X)}{(1 + \varepsilon X)} p \frac{T_0}{T}$$

- For most gas-solid catalyst reactions, the rate laws are written in terms of partial pressures, which can also be written in terms of conversion.

$$P_A = C_A RT = P_{A0} \frac{(1 - X)}{(1 + \varepsilon X)} \frac{P}{P_0} = P_{A0} \frac{(1 - X)}{(1 + \varepsilon X)} p$$

After completing this chapter, you will be able to write the rate of reaction as a function of conversion and to calculate the equilibrium conversion for both batch and flow reactors.

Now that we have shown how the rate law can be expressed as a function of concentrations, we need only express concentration as a function of conversion in order to carry out calculations similar to those presented in Chapter 2 to size reactors. If the rate law depends on more than one species, we must relate the concentrations of the different species to each other. This relationship is most easily established with the aid of a stoichiometric table. This table presents the stoichiometric relationships between reacting molecules for a single reaction. That is, it tells us how many molecules of one species will be formed during a chemical reaction when a given number of molecules of another species disappears. These relationships will be developed for the general reaction

$$a\mathrm{A} + b\mathrm{B} \;\rightleftharpoons\; c\mathrm{C} + d\mathrm{D} \tag{2-1}$$

Recall that we have already used stoichiometry to relate the *relative rates of reaction* for Equation (2-1):

This stoichiometric relationship relating reaction rates will be used in Chapters 6 and 8.

$$\boxed{\dfrac{-r_A}{a} = \dfrac{-r_B}{b} = \dfrac{r_C}{c} = \dfrac{r_D}{d}} \tag{3-1}$$

In formulating our stoichiometric table, we shall take species A as our basis of calculation (i.e., the limiting reactant) and then divide through by the stoichiometric coefficient of A

$$\mathrm{A} + \frac{b}{a}\,\mathrm{B} \;\longrightarrow\; \frac{c}{a}\,\mathrm{C} + \frac{d}{a}\,\mathrm{D} \tag{2-2}$$

in order to put everything on a basis of "per mole of A."

Next, we develop the stoichiometric relationships for reacting species that give the change in the number of moles of each species (i.e., A, B, C, and D).

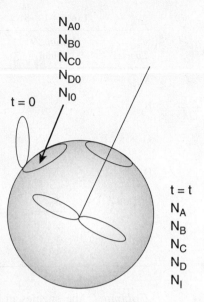

Figure 4-1   Batch reactor. (Schematic with special permission by Renwahr.)

4.1 Batch Systems

Batch reactors are primarily used for the production of specialty chemicals and to obtain reaction rate data in order to determine reaction rate laws and rate-law parameters such as k, the specific reaction rate.

Figure 4-1 shows a starving artist's rendition of a batch system in which we will carry out the reaction given by Equation (2-2). At time $t = 0$, we will open the reactor and place a number of moles of species A, B, C, and D, and inerts I (N_{A0}, N_{B0}, N_{C0}, N_{D0}, and N_{I0}, respectively) into the reactor.

Species A is our basis of calculation, and N_{A0} is the number of moles of A initially present in the reactor. After a time t, $N_{A0}X$ moles of A are consumed in the system as a result of the chemical reaction, leaving ($N_{A0} - N_{A0}X$) moles of A in the system. That is, the number of moles of A remaining in the reactor after a conversion X has been achieved is

$$N_A = N_{A0} - N_{A0}X = N_{A0}(1 - X)$$

We now will use conversion in a similar fashion to express the number of moles of B, C, and D in terms of conversion.

To determine the number of moles of each species remaining after $N_{A0}X$ moles of A have reacted, we form the stoichiometric table (Table 4-1). This stoichiometric table presents the following information:

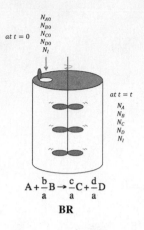

$at\ t = 0$
N_{A0}
N_{B0}
N_{C0}
N_{D0}
N_I

$at\ t = t$
N_A
N_B
N_C
N_D
N_I

$$A + \frac{b}{a}B \rightarrow \frac{c}{a}C + \frac{d}{a}D$$

BR

TABLE 4-1   STOICHIOMETRIC TABLE FOR A BATCH SYSTEM

Species	Initially (mol)	Change (mol)	Remaining (mol)
A	N_{A0}	$-(N_{A0}X)$	$N_A = N_{A0} - N_{A0}X$
B	N_{B0}	$-\dfrac{b}{a}(N_{A0}X)$	$N_B = N_{B0} - \dfrac{b}{a}N_{A0}X$
C	N_{C0}	$\dfrac{c}{a}(N_{A0}X)$	$N_C = N_{C0} + \dfrac{c}{a}N_{A0}X$
D	N_{D0}	$\dfrac{d}{a}(N_{A0}X)$	$N_D = N_{D0} + \dfrac{d}{a}N_{A0}X$
I (inerts)	N_{I0}	–	$N_I = N_{I0}$
Totals	N_{T0}		$N_T = N_{T0} + \underbrace{\left(\dfrac{d}{a} + \dfrac{c}{a} - \dfrac{b}{a} - 1\right)}_{\delta} N_{A0}X$

Components of the
stoichiometric table

Column 1: the species in the reaction
Column 2: the number of moles of each species initially present
Column 3: the change in the number of moles brought about by reaction
Column 4: the number of moles remaining in the system at time t

To calculate the number of moles of species B remaining at time t, we recall that at time t the number of moles of A that have reacted is $N_{A0}X$. For every mole of A that reacts, b/a moles of B must react; therefore, the total number of moles of B that have reacted is

$$\text{Moles B reacted} = \frac{\text{Moles B reacted}}{\text{Moles A reacted}} \cdot \text{Moles A reacted}$$

$$= \frac{b}{a}(N_{A0}X)$$

Because B is disappearing from the system, the sign of the "change" is negative. N_{B0} is the number of moles of B initially in the system. Therefore, the number of moles of B remaining in the system, N_B, at a time t, is given in the last column of Table 4-1 as

$$N_B = N_{B0} - \frac{b}{a}N_{A0}X$$

The complete stoichiometric table delineated in Table 4-1 is for all species in the general reaction

$$\boxed{A + \frac{b}{a}B \longrightarrow \frac{c}{a}C + \frac{d}{a}D} \qquad (2\text{-}2)$$

Let's take a look at the totals in the last column of Table 4-1. The stoichiometric coefficients in parentheses $(d/a + c/a - b/a - 1)$ represent the change in the total number of moles per mole of A reacted. Because this term occurs so often in our calculations, it is given the symbol δ:

$$\boxed{\delta = \frac{d}{a} + \frac{c}{a} - \frac{b}{a} - 1}$$ (4-1)

Definition of δ   The parameter δ

$$\boxed{\delta = \frac{\text{Change in the total number of moles}}{\text{Mole of A reacted}}}$$

The total number of moles can now be calculated from the equation

$$N_T = N_{T0} + \delta(N_{A0}X)$$

We recall from Chapter 1 and Chapter 3 that the kinetic rate law (e.g., $-r_A = kC_A^2$) is a function solely of the intensive properties of the reacting system (e.g., temperature, pressure, concentration, and catalysts, if any). The reaction rate, $-r_A$, usually depends on the concentration of the reacting species raised to some power. Consequently, to determine the reaction rate as a function of conversion, X, we need to know the concentrations of the reacting species as a function of conversion, X.

We want
$C_j = h_j(X)$

4.1.1 Batch Concentrations for the Generic Reaction, Equation (2-2)

The concentration of A is the number of moles of A per unit volume

Batch
concentration

$$C_A = \frac{N_A}{V}$$

After writing similar equations for B, C, and D, we use the stoichiometric table to express the concentration of each component in terms of the conversion X:

$$C_A = \frac{N_A}{V} = \frac{N_{A0}(1-X)}{V}$$ (4-2)

$$C_B = \frac{N_B}{V} = \frac{N_{B0} - (b/a)N_{A0}X}{V}$$ (4-3)

$$C_C = \frac{N_C}{V} = \frac{N_{C0} + (c/a)N_{A0}X}{V}$$ (4-4)

$$C_D = \frac{N_D}{V} = \frac{N_{D0} + (d/a)N_{A0}X}{V}$$ (4-5)

Because almost all batch reactors are solid vessels, the reactor volume is constant, so we can take $V = V_0$, then

$$C_A = \frac{N_A}{V_0} = \frac{N_{A0}(1-X)}{V_0}$$

$$C_A = C_{A0}(1-X)$$ (4-6)

We will soon see that Equation (4-6) also applies to liquid systems.

We further simplify these equations by defining the parameter Θ_i, which allows us to factor out N_{A0} in each of the expressions for concentration

$$\boxed{\Theta_i = \frac{N_{i0}}{N_{A0}} = \frac{C_{i0}}{C_{A0}} = \frac{y_{i0}}{y_{A0}}},$$

$$\boxed{\Theta_i = \frac{\text{Moles of species "i" initially}}{\text{Moles of species A initially}}}$$

$$C_B = \frac{N_{A0}\left[N_{B0}/N_{A0} - (b/a)X\right]}{V_0} = \frac{N_{A0}\left[\Theta_B - (b/a)X\right]}{V_0}$$

$$C_B = C_{A0}\left(\Theta_B - \frac{b}{a}X\right) \tag{4-7}$$

$$\text{with } \Theta_B = \frac{N_{B0}}{N_{A0}}$$

Feed

Equalmolar:
$\quad \Theta_B = 1$

Stoichiometric:

$\quad \Theta_B = \dfrac{b}{a}$

for an equalmolar feed $\Theta_B = 1$ and for a stoichiometric feed $\Theta_B = b/a$.
Continuing for species C and D

$$C_C = \frac{N_{A0}\left[\Theta_C + (c/a)X\right]}{V_0}$$

$$C_C = C_{A0}\left(\Theta_C + \frac{c}{a}X\right) \tag{4-8}$$

Constant-
volume batch
concentrations

$$\text{with } \Theta_C = \frac{N_{C0}}{N_{A0}}$$

$$C_D = \frac{N_{A0}\left[\Theta_D + (d/a)X\right]}{V_0}$$

$$C_D = C_{A0}\left(\Theta_D + \frac{d}{a}X\right) \tag{4-9}$$

$$\text{with } \Theta_D = \frac{N_{D0}}{N_{A0}}$$

For constant-volume batch reactors, e.g., steel containers $V = V_0$, we now have concentration as a function of conversion. If we know the rate law, we can now obtain $-r_A = f(X)$ to couple with the differential mole balance in terms of conversion in order to solve for the reaction time, t.

For liquid-phase reactions taking place in solution, the solvent usually dominates the situation. For example, most liquid-phase organic reactions do not change density during the course of the reaction and represent still another case for which the constant-volume simplifications apply. As a result, changes

For liquids
$V = V_0$
and
$v = v_0$

in the density of the solute **do not** affect the overall density of the solution significantly and therefore it is essentially a constant-volume reaction process, $V = V_0$ and $v = v_0$. Consequently, Equations (4-6) through (4-9) can be used for liquid-phase reactions as well. An important exception to this general rule exists for polymerization processes.

To summarize for constant-volume batch systems and for liquid-phase reactions, we can use a rate law for reaction (2-2) such as $-r_A = k_A C_A C_B$ to obtain $-r_A = f(X)$; that is,

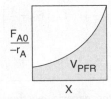

$$-r_A = kC_A C_B = kC_{A0}^2 (1-X)\left(\Theta_B - \frac{b}{a}X\right) = f(X)$$

Substituting for the given parameters k, C_{A0}, and Θ_B, we can now use the techniques in Chapter 2 to size the CSTRs and PFRs for liquid-phase reactions.

Example 4–1   Expressing $C_j = h_j(X)$ for a Liquid-Phase Batch Reaction

Soap consists of the sodium and potassium salts of various fatty acids, such as oleic, stearic, palmitic, lauric, and myristic acids. The saponification for the formation of soap from aqueous caustic soda and glyceryl stearate is

$$3NaOH(aq) + (C_{17}H_{35}COO)_3C_3H_5 \longrightarrow 3C_{17}H_{35}COONa + C_3H_5(OH)_3$$

Letting X represent the conversion of sodium hydroxide (the moles of sodium hydroxide reacted per mole of sodium hydroxide initially present), set up a stoichiometric table expressing the concentration of each species in terms of its initial concentration and the conversion, X.

Solution

Because we have taken sodium hydroxide as our basis of calculation, we divide through by the stoichiometric coefficient of sodium hydroxide to put the reaction expression in the form

Choosing a basis of
calculation

$$NaOH + \tfrac{1}{3}(C_{17}H_{35}COO)_3C_3H_5 \longrightarrow C_{17}H_{35}COONa + \tfrac{1}{3}C_3H_5(OH)_3$$

$$A \quad + \quad \tfrac{1}{3}B \quad \longrightarrow \quad C \quad + \quad \tfrac{1}{3}D$$

We may then perform the calculations shown in Table E4-1.1. Because this is a liquid-phase reaction, the density ρ is considered to be constant; therefore, $V = V_0$.

$$C_A = \frac{N_A}{V} = \frac{N_A}{V_0} = \frac{N_{A0}(1-X)}{V_0} = C_{A0}(1-X)$$

$$\Theta_B = \frac{C_{B0}}{C_{A0}} \quad \Theta_C = \frac{C_{C0}}{C_{A0}} \quad \Theta_D = \frac{C_{D0}}{C_{A0}}$$

TABLE E4-1.1   STOICHIOMETRIC TABLE FOR LIQUID-PHASE SOAP REACTION

Species	Symbol	Initially	Change	Remaining	Concentration
NaOH	A	N_{A0}	$-N_{A0}X$	$N_{A0}(1-X)$	$C_{A0}(1-X)$
$(C_{17}H_{35}COO)_3C_3H_5$	B	N_{B0}	$-\frac{1}{3}N_{A0}X$	$N_{A0}\left(\Theta_B - \frac{X}{3}\right)$	$C_{A0}\left(\Theta_B - \frac{X}{3}\right)$
$C_{17}H_{35}COONa$	C	N_{C0}	$N_{A0}X$	$N_{A0}(\Theta_C + X)$	$C_{A0}(\Theta_C + X)$
$C_3H_5(OH)_3$	D	N_{D0}	$\frac{1}{3}N_{A0}X$	$N_{A0}\left(\Theta_D + \frac{X}{3}\right)$	$C_{A0}\left(\Theta_D + \frac{X}{3}\right)$
Water (inert)	I	N_{10}	–	N_{10}	C_{10}
Totals		N_{T0}	0	$N_T = N_{T0}$	

Stoichiometric table (batch)

Analysis: The purpose of this example was to show how the generic reaction in Table 4-1 is applied to a real reaction.

Example 4–2 What Is the Limiting Reactant?

Having set up the stoichiometric table in Example 4-1, one can now readily use it to calculate the concentrations at a given conversion. If the initial mixture consists of sodium hydroxide at a concentration of 10 mol/dm³ (i.e., 10 mol/L or 10 kmol/m³) and glyceryl stearate at a concentration of 2 mol/dm³, what are the concentrations of glycerol stearate, B, and of glycerine, D, when the conversion of sodium hydroxide is (a) 20% and (b) 90%?

Solution

Only the reactants NaOH and $(C_{17}H_{35}COO)_3C_3H_5$ are initially present; therefore, $\Theta_C = \Theta_D = 0$.

(a) For 20% conversion of NaOH

$$C_D = C_{A0}\left(\frac{X}{3}\right) = (10)\left(\frac{0.2}{3}\right) = 0.67 \text{ mol/L} = 0.67 \text{ mol/dm}^3$$

$$C_B = C_{A0}\left(\Theta_B - \frac{X}{3}\right) = 10\left(\frac{2}{10} - \frac{0.2}{3}\right) = 10(0.133) = 1.33 \text{ mol/dm}^3$$

(b) For 90% conversion of NaOH

$$C_D = C_{A0}\left(\frac{X}{3}\right) = 10\left(\frac{0.9}{3}\right) = 3 \text{ mol/dm}^3$$

Let us find C_B

$$C_B = 10\left(\frac{2}{10} - \frac{0.9}{3}\right) = 10(0.2 - 0.3) = -1 \text{ mol/dm}^3$$

Oops!! Negative concentration—impossible! What went wrong?

The basis of
calculation **must**
be the limiting
reactant.

Analysis: We chose the wrong basis of calculation! We **must** choose the limiting reactant as our basis of calculation. Ninety-percent conversion of NaOH is not possible because glyceryl stearate is the limiting reactant and is used up before 90% of the NaOH can be reacted. Glyceryl stearate should have been our basis of calculation and therefore we should not have divided the reaction as written by the stoichiometric coefficient of 3.

4.2 Flow Systems

The form of the stoichiometric table for a continuous-flow system (see Figure 4-2) is virtually identical to that for a batch system (Table 4-1), except that we replace N_{j0} by F_{j0} and N_j by F_j (Table 4-2). Again taking A as the basis, we divide Equation (2-1) through by the stoichiometric coefficient of A to obtain

$$A + \frac{b}{a} B \longrightarrow \frac{c}{a} C + \frac{d}{a} D \tag{2-2}$$

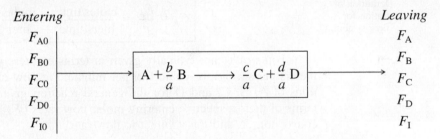

Figure 4-2   Flow reactor.

TABLE 4-2   STOICHIOMETRIC TABLE FOR A FLOW SYSTEM

Stoichiometric
table for a
flow system

Species	Feed Rate to Reactor (mol/time)	Change within Reactor (mol/time)	Effluent Rate from Reactor (mol/time)
A	F_{A0}	$-F_{A0}X$	$F_A = F_{A0}(1-X)$
B	$F_{B0} = \Theta_B F_{A0}$	$-\dfrac{b}{a}F_{A0}X$	$F_B = F_{A0}\left(\Theta_B - \dfrac{b}{a}X\right)$
C	$F_{C0} = \Theta_C F_{A0}$	$\dfrac{c}{a}F_{A0}X$	$F_C = F_{A0}\left(\Theta_C + \dfrac{c}{a}X\right)$
D	$F_{D0} = \Theta_D F_{A0}$	$\dfrac{d}{a}F_{A0}X$	$F_D = F_{A0}\left(\Theta_D + \dfrac{d}{a}X\right)$
I	$F_{I0} = \Theta_I F_{A0}$	–	$F_I = F_{A0}\Theta_I$
Totals	F_{T0}		$F_T = F_{T0} + \left(\dfrac{d}{a} + \dfrac{c}{a} - \dfrac{b}{a} - 1\right)F_{A0}X$ $F_T = F_{T0} + \delta F_{A0}X$

where

$$\Theta_B = \frac{F_{B0}}{F_{A0}} = \frac{C_{B0} v_0}{C_{A0} v_0} = \frac{C_{B0}}{C_{A0}} = \frac{y_{B0}}{y_{A0}}$$

and Θ_C, Θ_D, and Θ_I are defined similarly

and where

$$\delta = \frac{d}{a} + \frac{c}{a} - \frac{b}{a} - 1 \qquad (4-1)$$

4.2.1 Equations for Concentrations in Flow Systems

For a flow system, the concentration C_A at a given point can be determined from the molar flow rate F_A and the volumetric flow rate v at that point:

Definition of
concentration for a
flow system

$$C_A = \frac{F_A}{v} = \frac{\text{moles/time}}{\text{liters/time}} = \frac{\text{moles}}{\text{liter}} \qquad (4-10)$$

Units of v are typically given in terms of liters per second, cubic decimeters per second, or cubic feet per minute. We now can write the concentrations of A, B, C, and D for the general reaction given by Equation (2-2) in terms of their respective entering molar flow rates (F_{A0}, F_{B0}, F_{C0}, F_{D0}), the conversion, X, and the volumetric flow rate, v.

$$C_A = \frac{F_A}{v} = \frac{F_{A0}}{v}(1 - X) \qquad C_B = \frac{F_B}{v} = \frac{F_{B0} - (b/a)F_{A0}X}{v}$$

$$C_C = \frac{F_C}{v} = \frac{F_{C0} + (c/a)F_{A0}X}{v} \qquad C_D = \frac{F_D}{v} = \frac{F_{D0} + (d/a)F_{A0}X}{v} \qquad (4-11)$$

4.2.2 Liquid-Phase Concentrations

For liquids, the fluid volume change with reaction is negligible when no phase changes are taking place. Consequently, we can take

$$v = v_0$$

For liquids
$C_A = C_{A0}(1 - X)$
$C_B = C_{A0}\left(\Theta_B - \dfrac{b}{a} X\right)$
Therefore, for a given rate law we have
$-r_A = g(X)$

Then

$$C_A = \frac{F_{A0}}{v_0}(1 - X) = C_{A0}(1 - X) \qquad (4-12)$$

$$C_B = C_{A0}\left(\Theta_B - \frac{b}{a} X\right) \qquad (4-13)$$

and so forth for C_C and C_D.

Consequently, using any one of the rate laws in Chapter 3, we can now find $-r_A = f(X)$ for liquid-phase reactions. **However,** for gas-phase reactions the volumetric flow rate most often changes during the course of the reaction because of a change in the total number of moles or a change in temperature or pressure. Hence, one cannot always use Equation (4-13) to express concentration as a function of conversion for gas-phase reactions.

4.2.3 Gas-Phase Concentrations

In our previous discussions, we considered primarily systems in which the reaction volume or volumetric flow rate did not vary as the reaction progressed. Most batch and liquid-phase and some gas-phase systems fall into this category. There are other systems, though, in which either V or v **does** vary, and these will now be considered.

A situation where one encounters a varying flow rate occurs quite frequently in gas-phase reactions that do not have an equal number of product and reactant moles. For example, in the synthesis of ammonia

$$N_2 + 3H_2 \;\rightleftharpoons\; 2NH_3$$

4 mol of reactants gives 2 mol of product. In flow systems where this type of reaction occurs, the molar flow rate will be changing as the reaction progresses. Because equal numbers of moles occupy equal volumes in the gas phase at the same temperature and pressure, the volumetric flow rate will also change.

In the stoichiometric tables presented on the preceding pages, it was not necessary to make assumptions concerning a volume change in the first four columns of the table (i.e., the species, initial number of moles or molar feed rate, change within the reactor, and the remaining number of moles or the molar effluent rate). All of these columns of the stoichiometric table are independent of the volume or density, and they are *identical* for constant-volume (constant-density) and varying-volume (varying-density) situations. Only when concentration is expressed as a function of conversion does variable density enter the picture.

Flow Reactors with Variable Volumetric Flow Rate. To derive the concentrations of each species in terms of conversion for a gas-phase flow system, we shall use the relationships for the total concentration. The total concentration, C_T, at any point in the reactor is the total molar flow rate, F_T, divided by the volumetric flow rate, v [cf. Equation (4-10)]. In the gas phase, the total concentration is also found from the gas law, $C_T = P/ZRT$. Equating these two relationships gives

$$C_T = \frac{F_T}{v} = \frac{P}{ZRT} \tag{4-14}$$

At the entrance to the reactor

$$C_{T0} = \frac{F_{T0}}{v_0} = \frac{P_0}{Z_0 R T_0} \tag{4-15}$$

Taking the ratio of Equation (4-14) to Equation (4-15) and assuming negligible changes in the compressibility factor, i.e., $Z \cong Z_0$, during the course of the reaction we have upon rearrangement

Gas-phase reactions

$$v = v_0 \left(\frac{F_T}{F_{T0}} \right) \frac{P_0}{P} \left(\frac{T}{T_0} \right) \qquad (4\text{-}16)$$

We can now express the concentration of species j for a flow system in terms of its flow rate, F_j, the temperature, T, and total pressure, P.

$$C_j = \frac{F_j}{v} = \frac{F_j}{v_0 \left(\frac{F_T}{F_{T0}} \frac{P_0}{P} \frac{T}{T_0} \right)} = \left(\frac{F_{T0}}{v_0} \right) \left(\frac{F_j}{F_T} \right) \left(\frac{P}{P_0} \right) \left(\frac{T_0}{T} \right)$$

Use this concentration equation for membrane reactors (Chapter 6) and for multiple reactions (Chapter 8).

$$C_j = C_{T0} \left(\frac{F_j}{F_T} \right) \left(\frac{P}{P_0} \right) \left(\frac{T_0}{T} \right) \qquad (4\text{-}17)$$

The total molar flow rate is just the sum of the molar flow rates of each of the species in the system and is

$$F_T = F_A + F_B + F_C + F_D + F_I + \cdots = \sum_{j=1}^{n} F_j \qquad (4\text{-}18)$$

We can also write equation (4-17) in terms of the mole fraction of species j, y_j, and the pressure ratio, p, with respect to the initial or entering conditions, i.e., sub "0"

$$y_j = \frac{F_j}{F_T}$$

$$p = \frac{P}{P_0}$$

$$C_j = C_{T0} y_j p \frac{T_0}{T} \qquad (4\text{-}19)$$

The molar flow rates, F_j, are found by solving the mole balance equations. The concentration given by Equation (4-17) will be used for measures other than conversion when we discuss membrane reactors (Chapter 6) and multiple gas-phase reactions (Chapter 8).

Now, let's express the concentration in terms of conversion for gas flow systems. From Table 4-2, the total molar flow rate can be written in terms of conversion and is

$$F_T = F_{T0} + F_{A0} \, \delta X$$

We divide this equation through by F_{T0}

$$\frac{F_T}{F_{T0}} = 1 + \frac{F_{A0}}{F_{T0}}\delta X = 1 + \overset{\varepsilon}{\overbrace{y_{A0}\delta}} X$$

Then

$$\frac{F_T}{F_{T0}} = 1 + \varepsilon X \qquad (4\text{-}20)$$

where y_{A0} is the mole fraction of A at the inlet (i.e., (F_{A0}/F_{T0})), and where δ is given by Equation (4-1) and ε is given by

Relationship
between
δ and ε

$$\varepsilon = \left(\frac{d}{a} + \frac{c}{a} - \frac{b}{a} - 1\right)\frac{F_{A0}}{F_{T0}} = y_{A0}\delta$$

$$\boxed{\varepsilon = y_{A0}\delta} \qquad (4\text{-}21)$$

Equation (4-21) holds for both batch and flow systems. To interpret ε, let's rearrange Equation (4-20) at complete conversion (i.e., $X = 1$ and $F_T = F_{Tf}$)

$$\varepsilon = \frac{F_{Tf} - F_{T0}}{F_{T0}}$$

Interpretation of ε

$$\boxed{\varepsilon = \frac{\text{Change in total number of moles for complete conversion}}{\text{Total moles fed}}} \qquad (4\text{-}22)$$

Substituting for (F_T/F_{T0}) in Equation (4-16) for the volumetric flow rate, v, we have

Gas-phase
volumetric flow
rate

$$\boxed{v = v_0(1 + \varepsilon X)\frac{P_0}{P}\left(\frac{T}{T_0}\right)} \qquad (4\text{-}23)$$

The concentration of species j in a flow system is

$$C_j = \frac{F_j}{v} \qquad (4\text{-}24)$$

The molar flow rate of species j is

$$F_j = F_{j0} + v_j(F_{A0}X) = F_{A0}(\Theta_j + v_j X)$$

where v_i is the stoichiometric coefficient, which is negative for reactants and positive for products. For example, for the reaction

$$A + \frac{b}{a}B \longrightarrow \frac{c}{a}C + \frac{d}{a}D \qquad (2\text{-}2)$$

$$\boxed{v_A = -1, \quad v_B = -b/a, \quad v_C = c/a, \quad v_D = d/a, \text{ and } \Theta_j = F_{j0}/F_{A0}.}$$

Substituting for v using Equation (4-23) and for F_j, we have

$$C_j = \frac{F_{A0}(\Theta_j + v_j X)}{v_0\left((1 + \varepsilon X)\dfrac{P_0}{P}\dfrac{T}{T_0}\right)}$$

Rearranging

Gas-phase
concentration as a
function of
conversion

$$\boxed{C_j = \frac{C_{A0}(\Theta_j + v_j X)}{1 + \varepsilon X}\left(\frac{P}{P_0}\right)\frac{T_0}{T}} \qquad (4\text{-}25)$$

Recall that $y_{A0} = F_{A0}/F_{T0}$, $C_{A0} = y_{A0}C_{T0}$, and ε is given by Equation (4-21) (i.e., $\varepsilon = y_{A0}\delta$).

The stoichiometric table for the gas-phase reaction (2-2) is given in Table 4-3.

TABLE 4-3   CONCENTRATIONS IN A VARIABLE-VOLUME GAS FLOW SYSTEM

$$C_A = \frac{F_A}{v} = \frac{F_{A0}(1-X)}{v} = \frac{F_{A0}(1-X)}{v_0(1+\varepsilon X)}\left(\frac{T_0}{T}\right)\frac{P}{P_0} = C_{A0}\left(\frac{1-X}{1+\varepsilon X}\right)\frac{T_0}{T}\left(\frac{P}{P_0}\right)$$

$$C_B = \frac{F_B}{v} = \frac{F_{A0}[\Theta_B - (b/a)X]}{v} = \frac{F_{A0}[\Theta_B - (b/a)X]}{v_0(1+\varepsilon X)}\left(\frac{T_0}{T}\right)\frac{P}{P_0} = C_{A0}\left(\frac{\Theta_B - (b/a)X}{1+\varepsilon X}\right)\frac{T_0}{T}\left(\frac{P}{P_0}\right)$$

$$C_C = \frac{F_C}{v} = \frac{F_{A0}[\Theta_C + (c/a)X]}{v} = \frac{F_{A0}[\Theta_C + (c/a)X]}{v_0(1+\varepsilon X)}\left(\frac{T_0}{T}\right)\frac{P}{P_0} = C_{A0}\left(\frac{\Theta_C + (c/a)X}{1+\varepsilon X}\right)\frac{T_0}{T}\left(\frac{P}{P_0}\right)$$

$$C_D = \frac{F_D}{v} = \frac{F_{A0}[\Theta_D + (d/a)X]}{v} = \frac{F_{A0}[\Theta_D + (d/a)X]}{v_0(1+\varepsilon X)}\left(\frac{T_0}{T}\right)\frac{P}{P_0} = C_{A0}\left(\frac{\Theta_D + (d/a)X}{1+\varepsilon X}\right)\frac{T_0}{T}\left(\frac{P}{P_0}\right)$$

$$C_I = \frac{F_I}{v} = \frac{F_{A0}\Theta_I}{v} = \frac{F_{A0}\Theta_I}{v_0(1+\varepsilon X)}\left(\frac{T_0}{T}\right)\frac{P}{P_0} = \frac{C_{A0}\Theta_I}{1+\varepsilon X}\left(\frac{T_0}{T}\right)\frac{P}{P_0}$$

At last!
We now have
$C_j = h_j(X)$
and
$-r_A = g(X)$
for variable-volume
gas-phase reactions.

One of the major objectives of this chapter is to learn how to express any given rate law $-r_A$ as a function of conversion. The schematic diagram in Figure 4-3 helps to summarize our discussion on this point. The concentration of the reactant species B in the generic reaction is expressed as a function of conversion in both flow and batch systems for various conditions of temperature, pressure, and volume.

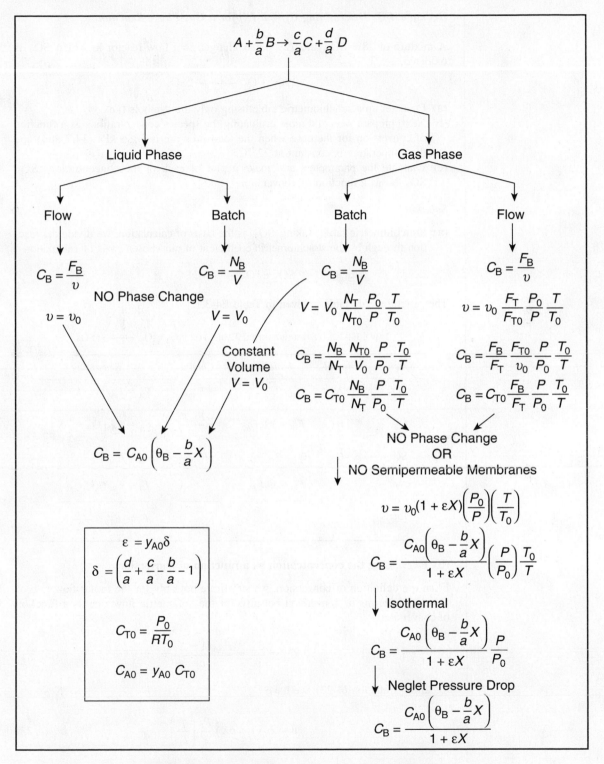

Figure 4-3 Expressing concentration as a function of conversion.

Example 4–3   Determining $C_j = h_j(X)$ for a Gas-Phase Reaction

A mixture of 28% SO_2 and 72% air is charged to a flow reactor in which SO_2 is oxidized.

$$2SO_2 + O_2 \longrightarrow 2SO_3$$

(a) First, set up a stoichiometric table using only the symbols (i.e., Θ_i, F_i).
(b) Next, prepare a second table evaluating the species concentrations as a function of conversion for the case when the total pressure is 1485 kPa (14.7 atm) and the temperature is constant at 227°C.
(c) Evaluate the parameters and make a plot of each of the concentrations SO_2, SO_3, N_2 as a function of conversion

Solution

(a) **Stoichiometric table.** Taking SO_2 as the basis of calculation, we divide the reaction through by the stoichiometric coefficient of our chosen basis of calculation

$$SO_2 + \tfrac{1}{2}O_2 \longrightarrow SO_3$$

The stoichiometric table is given in Table E4-3.1.

TABLE E4-3.1   STOICHIOMETRIC TABLE FOR $SO_2 + \tfrac{1}{2}O_2 \longrightarrow SO_3$

Species	Symbol	Initially	Change	Remaining
SO_2	A	F_{A0}	$-F_{A0}X$	$F_A = F_{A0}(1-X)$
O_2	B	$F_{B0} = \Theta_B F_{A0}$	$-\dfrac{F_{A0}X}{2}$	$F_B = F_{A0}\left(\Theta_B - \dfrac{1}{2}X\right)$
SO_3	C	0	$+F_{A0}X$	$F_C = F_{A0}X$
N_2	I	$F_{I0} = \Theta_I F_{A0}$	—	$F_I = F_{I0} = \Theta_I F_{A0}$
Totals		F_{T0}		$F_T = F_{T0} - \dfrac{F_{A0}X}{2}$

(b) **Expressing the concentration as a function of conversion.**

From the definition of conversion, we substitute not only for the molar flow rate of SO_2 (A) in terms of conversion but also for the volumetric flow rate as a function of conversion

$$C_A = \frac{F_A}{v} = \frac{F_{A0}(1-X)}{v} \tag{E4-3.1}$$

Recalling Equation (4-23), we have

$$v = v_0(1 + \varepsilon X)\frac{P_0}{P}\left(\frac{T}{T_0}\right) \tag{4-23}$$

Neglecting pressure drop in the reaction, $P = P_0$, yields

$$v = v_0(1 + \varepsilon X)\frac{T}{T_0} \tag{E4-3.2}$$

If the reaction is also carried out isothermally, $T = T_0$, we obtain

$$v = v_0(1 + \varepsilon X)$$

Isothermal operation, $T = T_0$

$$C_A = \frac{F_{A0}(1-X)}{v_0(1+\varepsilon X)} = C_{A0}\left(\frac{1-X}{1+\varepsilon X}\right)$$

Similarly for B also with $T = T_0$ and $P = P_0$, (i.e., $p = 1$)

$$C_B = C_{A0}\frac{\left(\Theta_B - \frac{b}{a}X\right)}{1+\varepsilon X}p\frac{T_0}{T} = \frac{C_{A0}\left(\Theta_B - \frac{1}{2}X\right)}{1+\varepsilon X}$$

(c) Parameter evaluation and concentrations of a function of conversion plot.
The inlet concentration of A is equal to the inlet mole fraction of A multiplied by the total inlet molar concentration. The total concentration can be calculated from an equation of state such as the ideal gas law. Recall that $y_{A0} = 0.28$, $T_0 = 500$ K, and $P_0 = 1485$ kPa.

$$C_{A0} = y_{A0}C_{T0} = y_{A0}\left(\frac{P_0}{RT_0}\right)$$

$$= 0.28\left[\frac{1485\ \text{kPa}}{8.314\ \text{kPa} \cdot \text{dm}^3/(\text{mol} \cdot \text{K}) \times 500\ \text{K}}\right]$$

$$= 0.1\ \text{mol/dm}^3$$

The total concentration at constant temperature and pressure is

$$C_T = \frac{F_T}{v} = \frac{F_{T0}+\overbrace{y_{A0}\delta}^{\varepsilon}XF_{T0}}{v_0(1+\varepsilon X)} = \frac{F_{T0}(1+\varepsilon X)}{v_0(1+\varepsilon X)} = \frac{F_{T0}}{v_0} = C_{T0} \qquad \text{(E4-3.3)}$$

$$C_{T0} = \frac{P_0}{RT_0} = \frac{1485\ \text{kPa}}{[8.314\ \text{kPa} \cdot \text{dm}^3/(\text{mol} \cdot \text{K})](500\ \text{K})} = 0.357\frac{\text{mol}}{\text{dm}^3} \qquad \text{(E4-3.4)}$$

We now evaluate ε.

$$\varepsilon = y_{A0}\ \delta = (0.28)(1 - 1 - \tfrac{1}{2}) = -0.14 \qquad \text{(E4-3.5)}$$

Initially, 72% of the total number of moles is air containing 21% O_2 and 79% N_2, along with 28% SO_2.

$$F_{A0} = (0.28)(F_{T0})$$

$$F_{B0} = (0.72)(0.21)(F_{T0})$$

$$\Theta_B = \frac{F_{B0}}{F_{A0}} = \frac{(0.72)(0.21)}{0.28} = 0.54$$

$$\Theta_I = \frac{F_{I0}}{F_{A0}} = \frac{(0.72)(0.79)}{0.28} = 2.03$$

Substituting for C_{A0} and ε in the species concentrations:

SO$_2$ $$C_A = C_{A0}\left(\frac{1-X}{1+\varepsilon X}\right) = 0.1\left(\frac{1-X}{1-0.14X}\right)\ \text{mol/dm}^3 \qquad \text{(E4-3.6)}$$

O₂ $$C_B = C_{A0} \left(\frac{\Theta_B - \frac{1}{2}X}{1 + \varepsilon X} \right) = \frac{0.1\,(0.54 - 0.5X)}{1 - 0.14X} \text{ mol/dm}^3 \qquad \text{(E4-3.7)}$$

SO₃ $$C_C = \frac{C_{A0}X}{1 + \varepsilon X} = \frac{0.1X}{1 - 0.14X} \text{ mol/dm}^3 \qquad \text{(E4-3.8)}$$

N₂ $$C_I = \frac{C_{A0}\Theta_I}{1 + \varepsilon X} = \frac{(0.1)(2.03)}{1 - 0.14X} \text{ mol/dm}^3 \qquad \text{(E4-3.9)}$$

The concentrations of different species at various conversions are calculated in Table E4-3.2 and plotted in Figure E4-3.1. *Note* that the concentration of N₂ is changing even though it is an inert species in this reaction!!

TABLE E4-3.2   CONCENTRATION AS A FUNCTION OF CONVERSION

Species		C_i (mol/dm³)				
		$X = 0.0$	$X = 0.25$	$X = 0.5$	$X = 0.75$	$X = 1.0$
SO₂	$C_A =$	0.100	0.078	0.054	0.028	0.000
O₂	$C_B =$	0.054	0.043	0.031	0.018	0.005
SO₃	$C_C =$	0.000	0.026	0.054	0.084	0.116
N₂	$C_I =$	0.203	0.210	0.218	0.227	0.236
Total	$C_T =$	0.357	0.357	0.357	0.357	0.357

Note: Because the volumetic flow rate varies with conversion, $v = v_0(1 - 0.14X)$, the concentration of inerts (N₂) is *not* constant.

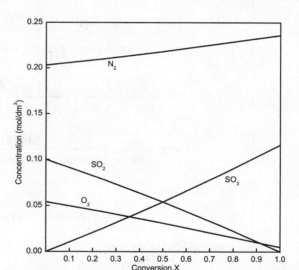

Figure E4-3.1   Concentration as a function of conversion.

Now use techniques presented in Chapter 2 to size reactors.

We are now in a position to express $-r_A$ as a function of X and use the techniques in Chapter 2. However, we will use a better method to solve CRE problems, namely a Table of Integrals (Appendix A) or the Polymath software, discussed in the next chapter.

Analysis: In this example, we formed a stoichiometric table in terms of molar flow rates. We then showed how to express the concentrations of each species in a gas-phase reaction in which there is a change in the total number of moles. Next, we plotted each species concentration as a function of conversion and noted that the concentration of the inert, N_2, was not constant but increased with increasing conversion because of the decrease in the total molar flow rate, F_T, with conversion.

As previously mentioned many, if not most, rate laws for catalyst reactions are given in terms of partial pressures. Luckily, partial pressures are easily related to conversion with the aid of the ideal gas law and Equation (4-17).

$$P_i = C_i RT = C_{T0} \frac{F_i}{F_{T0}} \overbrace{\left(\frac{P}{P_0} \right)}^{p} \frac{T_0}{T} RT = \overbrace{C_{T0} RT_0}^{P_{T0}} \overbrace{\frac{F_i}{F_{T0}}}^{y_i} p$$

The following equation is used when the mole balance is written in terms of molar flow rates

$$\boxed{P_i = P_{T0} \left(\frac{F_i}{F_{T0}} \right) p} \qquad (4\text{-}26)$$

However, when the mole balance is written in terms of conversion, we use Equation (4-25)

$$P_i = C_i RT = C_{A0} \frac{\left(\Theta_j + \upsilon_j X \right)}{\left(1 + \varepsilon X \right)} \frac{P}{P_0} \frac{T_0}{T} RT$$

$$= \overbrace{C_{A0} RT_0}^{P_{A0}} \frac{\left(\Theta_j + \upsilon_j X \right)}{\left(1 + \varepsilon X \right)} p$$

$$\boxed{P_j = P_{A0} \frac{\left(\Theta_j + \upsilon_j X \right)}{1 + \varepsilon X} p} \qquad (4\text{-}27)$$

For example, the rate law for the hydrodemethylation of toluene (T) to form methane (M) and benzene (B) given by Equation (10-80) on page 440 can now be written in terms of conversion.

T + H$_2$ → M + B

$$-r_T' = \frac{k P_{A0}^2 (1 - X)\left(\Theta_{H_2} - X \right)}{1 + K_B P_{A0} X + K_B P_{A0}(1 - X)} (p)^2$$

If you haven't decided which computer to buy or borrow, or don't have your integral tables, you could resort to the graphical techniques in Chapter 2 and use a Levenspiel plot, $(F_{A0}/-r_A')$ versus X, to achieve a specified conversion of toluene.

Example 4–4   Expressing the Rate Law for SO$_2$ Oxidation in Terms of Partial Pressures and Conversions

The SO$_2$ oxidation discussed in Example 4-3 is to be carried out over a solid platinum catalyst. As with almost all gas-solid catalytic reactions, the rate law is expressed in terms of partial pressures instead of concentrations. The rate law for this SO$_2$ oxidation was found experimentally to be[1]

$$-r_{SO_2}' = \frac{k \left[P_{SO_2} \sqrt{P_{O_2}} - \dfrac{P_{SO_3}}{K_P} \right]}{\left(1 + \sqrt{P_{O_2} K_{O_2}} + P_{SO_2} K_{SO_2} \right)^2}, \text{ mol } SO_2 \text{ oxidized/(h)(g-cat)} \qquad (E4\text{-}4.1)$$

[1]Uychara, O.A. and K. M. Watson, *Ind. Engrg. Chem. 35* p.541.

where P_i (atm) is the partial pressure of species i.

The reaction is to be carried out isothermally at 400°C. At this temperature the rate constant k per gram of catalyst (g-cat), the adsorption constants for O_2 (K_{O_2}) and SO_2 (K_{SO_2}), and the pressure equilibrium constant, K_P, were experimentally found to be:

$$k = 9.7 \text{ mol } SO_2/atm^{3/2}/h/g\text{-cat}$$

$$K_{O_2} = 38.5 \text{ atm}^{-1}, \ K_{SO_2} = 42.5 \text{ atm}^{-1}, \text{ and } K_P = 930 atm^{-1/2}$$

The total pressure and the feed composition (e.g., 28% SO_2) are the same as in Example 4-3. Consequently, the entering partial pressure of SO_2 is 4.1 atm. There is no pressure drop.

Write the rate law as a function of conversion.

Solution

No Pressure Drop and Isothermal Operation

For SO$_2$

First we need to recall the relationship between partial pressure and concentration, followed by the relationship between concentration and conversion. Because we know how to express concentration as a function of conversion, we know how to express partial pressure as a function of conversion.

$$P_i = P_{A0}\frac{(\Theta_i + v_i X)}{(1+\varepsilon X)}\frac{P}{P_0} = P_{A0}\frac{(\Theta + v_i X)}{(1+\varepsilon X)}p$$

$$P_{SO_2} = C_{SO_2}RT = \frac{F_{SO_2}}{v}RT = \frac{F_{SO_2,0}(1-X)RT}{v_0(1+\varepsilon X)\frac{T}{T_0}\frac{P_0}{P}} = \frac{F_{SO_2,0}}{v_0}\frac{RT_0(1-X)\frac{P}{P_0}}{(1+\varepsilon X)}$$

$$P_{SO_2} = \frac{P_{SO_2,0}(1-X)\frac{P}{P_0}}{(1+\varepsilon X)} = \frac{P_{SO_2,0}(1-X)p}{(1+\varepsilon X)} \qquad (E4\text{-}4.2)$$

For no pressure drop $P = P_0$, i.e., $p = 1$

$$\boxed{P_{SO_2} = \frac{P_{SO_2,0}(1-X)}{(1+\varepsilon X)}}$$

$$P_{SO_2,0} = y_{SO_2,0}P_0 = 4.1 \text{ atm} \qquad (E4\text{-}4.3)$$

For SO$_3$

$$\boxed{P_{SO_3} = C_{SO_3}RT = \frac{C_{SO_2,0}RT_0 X}{(1+\varepsilon X)} = \frac{P_{SO_2,0}X}{1+\varepsilon X}} \qquad (E4\text{-}4.4)$$

For O$_2$

$$P_{O_2} = C_{O_2}RT = C_{SO_2,0}\frac{\left(\Theta_B - \frac{1}{2}X\right)RT_0}{(1+\varepsilon X)} = P_{SO_2,0}\frac{\left(\Theta_B - \frac{1}{2}X\right)}{(1+\varepsilon X)} \qquad (E4\text{-}4.5)$$

From Example 4-3

$$\Theta_B = 0.54$$

Factoring out $\frac{1}{2}$ in Equation (E4-4.5) gives

$$P_{O_2} = P_{SO_2,0} \frac{\left(\Theta_B - \frac{1}{2}X\right)}{(1+\varepsilon X)} = \frac{P_{SO_2,0}(1.08 - X)}{2(1+\varepsilon X)} \tag{E4-4.6}$$

From Equation (E4-3.5)

$$\varepsilon = -0.14 \tag{E4-3.5}$$

Substitute for the partial pressure in the rate-law equation (E4-4.1)

$$-r'_{SO_2} = k \left[\frac{P_{SO_2,0}^{3/2}\left(\dfrac{1-X}{1-0.14X}\right)\sqrt{\dfrac{(1.08-X)}{2(1-0.14X)}} - \dfrac{P_{SO_2,0}X}{(1-0.14X)}\left(\dfrac{1}{930\text{ atm}^{-1/2}}\right)}{\left(1+\sqrt{\dfrac{38.5\ P_{SO_2,0}(1.08-X)}{2(1-0.14X)}} + \dfrac{42.5\ P_{SO_2,0}(1-X)}{(1-0.14X)}\right)^2} \right] \tag{E4-4.7}$$

with k = 9.7 mol SO$_2$/atm$^{3/2}$/h/g-cat $P_{SO_2,0} = 4.1$ atm, $P_{SO_2,0}^{3/2} = 8.3$ atm$^{3/2}$

$$-r'_{SO_2} = 9.7 \frac{\text{mol}}{\text{h g-cat atm}^{3/2}} \left[\frac{\dfrac{8.3\text{ atm}^{3/2}(1-X)}{(1-0.14X)}\sqrt{\dfrac{1.08-X}{2(1-0.14X)}} - \dfrac{0.0044\text{ atm}^{3/2}\,X}{(1-0.14X)}}{\left(1+\sqrt{\dfrac{79(1.08-X)}{(1-0.14X)}} + \dfrac{174(1-X)}{1-0.14X}\right)^2} \right] \tag{E4-4.8}$$

We could now use a Levenspiel plot to find the catalyst weight W in a packed-bed reactor (PBR) to achieve a specified conversion.

$$F_{A0}\frac{dX}{dW} = -r'_A \tag{2-17}$$

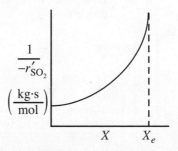

Figure E4-4.1 Reciprocal rate of SO_2 oxidation as a function of conversion.

However, we will see in the next chapter there is a much, much better way to solve for the catalysis weight, W, by using numerical software packages. For example, we would couple Equation (E4-4.8) with Equation (2-17) and use an ordinary differential equation (ODE) solver, such as Polymath, to find the conversion X as a function of catalyst weight W. So, be sure to buy or borrow a laptop computer before attempting to solve the problems in Chapter 5 and beyond.

**Analysis:** In most heterogeneous catalytic reactions, rate laws are expressed in terms of partial pressures instead of concentration. However, we see that through the use of the ideal gas law we could _easily_ express the partial pressure as a function of concentration then conversion in order to express the rate law as a function of conversion. In addition, for most all heterogeneous reactions you will usually find a term like $(1 + K_A P_A + K_B P_B + \ldots)$ in the denominator of the rate law, as will be explained in Chapter 10.

4.3 Reversible Reactions and Equilibrium Conversion

Thus far in this chapter, we have focused mostly on irreversible reactions. The procedure one uses for the isothermal reactor design of reversible reactions is virtually the same as that for irreversible reactions, with one notable exception: the maximum conversion that can be achieved at the reaction temperature is the equilibrium conversion, X_e. In the following example, it will be shown how our algorithm for reactor design is easily extended to reversible reactions.

Need to first calculate X_e

Example 4–5 Calculating the Equilibrium Conversion

The reversible gas-phase decomposition of nitrogen tetroxide, N_2O_4, to nitrogen dioxide, NO_2,

$$N_2O_4 \rightleftharpoons 2NO_2$$

is to be carried out at constant temperature. The feed consists of pure N_2O_4 at 340 K and 202.6 kPa (2 atm). The concentration equilibrium constant, K_C, at 340 K is 0.1 mol/dm^3 and the rate constant $k_{N_2O_4}$ is 0.5min^{-1}.

(a) Set up a stoichiometric table and then calculate the equilibrium conversion of N_2O_4 in a constant-volume batch reactor.

(b) Calculate the equilibrium conversion of N_2O_4 in a flow reactor.

(c) Assuming the reaction is elementary, express the rate of reaction solely as a function of conversion for a flow system and for a batch system.

(d) Determine the CSTR volume necessary to achieve 80% of the equilibrium conversion.

Solution

$$N_2O_4 \rightleftharpoons 2NO_2$$
$$A \rightleftharpoons 2B$$

At equilibrium, the concentrations of the reacting species are related by the relationship dictated by thermodynamics (see Equation (3-10) and Appendix C).

$$K_C = \frac{C_{Be}^2}{C_{Ae}} \tag{E4-5.1}$$

(a) Batch system—constant volume, $V = V_0$.

TABLE E4-5.1   STOICHIOMETRIC TABLE

Species	Symbol	Initial	Change	Remaining
N_2O_4	A	N_{A0}	$-N_{A0}X$	$N_A = N_{A0}(1-X)$
NO_2	B	0	$+2N_{A0}X$	$N_B = 2N_{A0}X$
		$N_{T0} = N_{A0}$		$N_T = N_{T0} + N_{A0}X$

Living Example Problem

For batch systems $C_i = N_i/V$

$$C_A = \frac{N_A}{V} = \frac{N_A}{V_0} = \frac{N_{A0}(1-X)}{V_0} = C_{A0}(1-X) \qquad \text{(E4-5.2)}$$

$$C_B = \frac{N_B}{V} = \frac{N_B}{V_0} = \frac{2N_{A0}X}{V_0} = 2C_{A0}X \qquad \text{(E4-5.3)}$$

$$C_{A0} = \frac{y_{A0}P_0}{RT_0} = \frac{(1)(2 \text{ atm})}{(0.082 \text{ atm} \cdot \text{dm}^3/\text{mol} \cdot \text{K})(340 \text{ K})}$$

$$= 0.07174 \text{ mol}/\text{dm}^3$$

At equilibrium, $X = X_e$, we substitute Equations (E4-5.2) and (E4-5.3) into Equation (E4-5.1)

$$K_C = \frac{C_{Be}^2}{C_{Ae}} = \frac{4C_{A0}^2 X_e^2}{C_{A0}(1-X_e)} = \frac{4C_{A0}X_e^2}{1-X_e}$$

(math-math-math-math) to get

$$X_e = \sqrt{\frac{K_C(1-X_e)}{4C_{A0}}} \qquad \text{(E4-5.4)}$$

We will use the software package Polymath to solve for the equilibrium conversion and let Xeb represent the equilibrium conversion in a constant-volume batch reactor. Equation (E4-5.4) written in Polymath format becomes

$$f(Xeb) = Xeb - [kc*(1-Xeb)/(4*cao)]\ ^{\wedge}0.5$$

The Polymath program and solution are given in Table E4-5.2.

TABLE E4-5.2   POLYMATH PROGRAM AND SOLUTION FOR BOTH BATCH AND FLOW SYSTEMS

Nonlinear equations
1 f(Xef) = Xef-(Kc*(1-Xef)*(1+epsilon*Xef)/(4*Cao))^0.5 = 0
2 f(Xeb) = Xeb-(Kc*(1-Xeb)/(4*Cao))^0.5 = 0

Explicit equations
1 Cao = 0.07174
2 epsilon = 1.0
3 Kc = 0.1

Calculated values of NLE variables

	Variable	Value	f(x)	Initial Guess
1	Xeb	0.4412597	7.266E-09	0.4
2	Xef	0.5083548	2.622E-10	0.5

	Variable	Value
1	Cao	0.07174
2	epsilon	1.
3	Kc	0.1

When looking at Equation (E4-5.4), you probably asked yourself, *"Why not use the quadratic formula to solve for the equilibrium conversion in both batch and flow systems?"* That is

There is a Polymath
tutorial in the
Summary Notes for
Chapter 1 on the
CRE Web site.

$$\text{Batch:} \quad X_e = \frac{1}{8}[(-1 + \sqrt{1 + 16C_{A0}/K_C})/(C_{A0}/K_C)]$$

$$\text{Flow:} \quad X_e = \frac{[(\varepsilon - 1) + \sqrt{(\varepsilon - 1)^2 + 4(\varepsilon + 4C_{A0}/K_C)}]}{2(\varepsilon + 4C_{A0}/K_C)}$$

The answer is that future problems will be nonlinear and require Polymath solutions; therefore, this simple exercise increases the reader's ease in using Polymath.

The equilibrium conversion in a constant-volume batch reactor is

$$\boxed{X_{eb} = 0.44}$$

Note: A tutorial on Polymath can be found in the summary notes for Chapter 1 on the CRE Web site (*www.umich.edu/~elements/5e/index.html*).

Polymath Tutorial
Chapter 1

Summary Notes

(b) Flow system. The stoichiometric table is the same as that for a batch system except that the number of moles of each species, N_i, is replaced by the molar flow rate of that species, F_i. For constant temperature and pressure, the volumetric flow rate is $v = v_0(1 + \varepsilon X)$, and the resulting concentrations of species A and B are

$$C_A = \frac{F_A}{v} = \frac{F_{A0}(1 - X)}{v} = \frac{F_{A0}(1 - X)}{v_0(1 + \varepsilon X)} = \frac{C_{A0}(1 - X)}{1 + \varepsilon X} \qquad \text{(E4-5.5)}$$

$$C_B = \frac{F_B}{v} = \frac{2F_{A0}X}{v_0(1 + \varepsilon X)} = \frac{2C_{A0}X}{1 + \varepsilon X} \qquad \text{(E4-5.6)}$$

At equilibrium, $X = X_e$, we can substitute Equations (E4-5.5) and (E4-5.6) into Equation (E4-5.1) to obtain the expression

$$K_C = \frac{C_{Be}^2}{C_{Ae}} = \frac{[2C_{A0}X_e/(1 + \varepsilon X_e)]^2}{C_{A0}(1 - X_e)/(1 + \varepsilon X_e)}$$

Simplifying gives

$$K_C = \frac{4C_{A0}X_e^2}{(1 - X_e)(1 + \varepsilon X_e)} \qquad \text{(E4-5.7)}$$

Rearranging to use Polymath yields

$$X_e = \sqrt{\frac{K_C(1 - X_e)(1 + \varepsilon X_e)}{4C_{A0}}} \qquad \text{(E4-5.8)}$$

For a flow system with pure N_2O_4 feed, $\varepsilon = y_{A0} \delta = 1(2 - 1) = 1$.

We shall let Xef represent the equilibrium conversion in a flow system. Equation (E4-5.8) written in the Polymath format becomes

$$f(Xef) = Xef - [kc*(1 - Xef)*(1 + eps*Xef)/4/cao]^0.5$$

This solution is also shown in Table E4-5.2 $\boxed{X_{ef} = 0.51}$.

Note that the equilibrium conversion in a flow reactor (i.e., $X_{ef} = 0.51$), with no pressure drop, is greater than the equilibrium conversion in a constant-volume batch reactor ($X_{eb} = 0.44$). Recalling Le Châtelier's principle, can you suggest an explanation for this difference in X_e?

(c) Rate laws. Assuming that the reaction follows an elementary rate law, then

$$-r_A = k_A \left[C_A - \frac{C_B^2}{K_C} \right] \tag{E4-5.9}$$

1. *For a constant volume ($V = V_0$) batch system*

   Here, $C_A = N_A / V_0$ and $C_B = N_B / V_0$. Substituting Equations (E4-5.2) and (E4-5.3) into the rate law, we obtain the rate of disappearance of A as a function of conversion

$$\boxed{-r_A = k_A \left[C_A - \frac{C_B^2}{K_C} \right] = k_A \left[C_{A0}(1-X) - \frac{4C_{A0}^2 X^2}{K_C} \right]} \tag{E4-5.10}$$

2. *For a flow system*

   Here, $C_A = F_A/v$ and $C_B = F_B/v$ with $v = v_0 (1 + \varepsilon X)$. Consequently, we can substitute Equations (E4-5.5) and (E4-5.6) into Equation (E4-5.9) to obtain

$$\boxed{-r_A = k_A \left[\frac{C_{A0}(1-X)}{1 + \varepsilon X} - \frac{4C_{A0}^2 X^2}{K_C(1 + \varepsilon X)^2} \right]} \tag{E4-5.11}$$

As expected, the dependence of reaction rate on conversion for a constant-volume batch system [i.e., Equation (E4-5.10)] is different than that for a flow system [Equation (E4-5.11)] for gas-phase reactions.

If we substitute the values for C_{A0}, K_C, ε, and $k_A = 0.5$ min^{-1} in Equation (E4-5.11), we obtain $-r_A$ solely as a function of X for the flow system.

$$-r_A = \frac{0.5}{\text{min}} \left[0.072 \frac{\text{mol}}{\text{dm}^3} \frac{(1-X)}{(1+X)} - \frac{4(0.072 \ \text{mol/dm}^3)^2 X^2}{0.1 \ \text{mol/dm}^3 (1+X)^2} \right]$$

$$-r_A = 0.036 \left[\frac{(1-X)}{(1+X)} - \frac{2.88 \ X^2}{(1+X)^2} \right] \left(\frac{\text{mol}}{\text{dm}^3 \cdot \text{min}} \right) \tag{E4-5.12}$$

We can now prepare our Levenspiel plot.

We see as we approach equilibrium, $-r_A$ goes to zero and $(1/-r_A)$ goes to infinity as X approaches X_e.

(d) CSTR volume. Just for fun (and this *really* is fun), let's calculate the CSTR reactor volume necessary to achieve 80% of the equilibrium conversion of 51% (i.e., $X = 0.8X_e = (0.8)(0.51) = 0.4$) for a molar feed rate of A of 3 mol/min.

(margin notes)

$-r_A = f(X)$ for a batch reactor with $V = V_0$

$-r_A = f(X)$ for a flow reactor

Solution

Evaluating Equation E4-5.12 at $X = 0.4$

$$-r_A = 0.036 \left[\frac{(1-0.4)}{(1+0.4)} - \frac{2.88\,(0.4)^2}{(1+(0.4))^2} \right]$$

$$= 0.0070 \text{ mol/dm}^3/\text{min}$$

$$V = \frac{F_{A0}X}{-r_A|_X} = \frac{F_{A0}(0.4)}{-r_A|_{0.4}} = \frac{(3 \text{ mol/min})(0.4)}{0.0070 \dfrac{\text{mol}}{\text{dm}^3 \cdot \text{min}}}$$

$$V = 171 \text{ dm}^3 = 0.171 \text{ m}^3$$

The CSTR volume necessary to achieve 40% conversion is 0.171 m³.

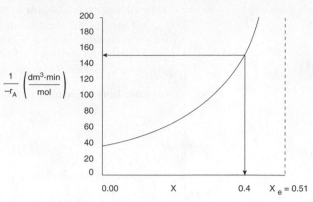

$\dfrac{1}{-r_A} \left(\dfrac{\text{dm}^3 \cdot \text{min}}{\text{mol}} \right)$

Figure E4-5.1 Levenspiel plot for a flow system.

Analysis: The purpose of this example was to calculate the equilibrium conversion first for a constant volume batch reactor in part (**a**), and then for a constant pressure flow reactor in part (**b**). One notes that there is a change in the total number of moles in this reaction and, as a result, these two equilibrium conversions are not the same!! We next showed in part (**c**) how to express $-r_A = f(X)$ for a reversible gas-phase reaction. Finally, in Part (**d**) having $-r_A = f(X)$, we specified a molar flow rate of A (i.e., 3.0 mol A/min) and calculated the CSTR volume necessary to achieve 40% conversion. We did this calculation to give insight to the types of analyses we, as chemical reaction engineers, will carry out as we move into similar but more complex calculations in Chapters 5 and 6.

Closure. Having completed this chapter, you should be able to write the rate law solely in terms of conversion and the reaction-rate parameters, (e.g., k, K_C) for both liquid-phase and gas-phase reactions. Once expressing $-r_A = f(X)$ is accomplished, you can proceed to use the techniques in Chapter 2 to calculate reactor sizes and conversion for single CSTRs, PFRs, and PBRs, as well as those connected in series. However, in the next chapter we will show you how to carry out these calculations much more easily by instead using a Table of Integrals or Polymath without having to resort to Levenspiel plots. After studying this chapter you should also be able to calculate the equilibrium conversion for both constant-volume batch reactors and for constant-pressure flow reactors. We have now completed the following building blocks of our CRE tower. In Chapter 5 we will focus on the fourth and fifth blocks: Combine and Evaluate.

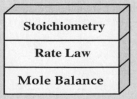

In Chapter 5, we will focus on the **combine** and **evaluation** building blocks, which will then complete our algorithm for isothermal chemical reactor design.

The CRE Algorithm

- Mole Balance, Ch 1
- Rate Law, Ch 3
- Stoichiometry, Ch 4
- Combine, Ch 5
- Evaluate, Ch 5
- Energy Balance, Ch 11

SUMMARY

1. The *stoichiometric table* for the reaction given by Equation (S4-1) being carried out in a flow system is

$$A + \frac{b}{a}B \rightarrow \frac{c}{a}C + \frac{d}{a}D \qquad \text{(S4-1)}$$

2. In the case of ideal gases, Equation (S4-3) relates volumetric flow rate to conversion.

Batch constant volume: $V = V_0$ (S4-2)

Flow systems: Gas: $v = v_0 \left(\dfrac{P_0}{P}\right)(1 + \varepsilon X)\dfrac{T}{T_0}$ (S4-3)

Liquid: $v = v_0$ (S4-4)

For the general reaction given by (S4-1), we have

$$\boxed{\delta = \frac{d}{a} + \frac{c}{a} - \frac{b}{a} - 1} \qquad \text{(S4-5)}$$

Species	Entering	Change	Leaving
A	F_{A0}	$-F_{A0}X$	$F_{A0}(1 - X)$
B	F_{B0}	$-\left(\dfrac{b}{a}\right)F_{A0}X$	$F_{A0}\left(\Theta_B - \dfrac{b}{a}X\right)$
C	F_{C0}	$\left(\dfrac{c}{a}\right)F_{A0}X$	$F_{A0}\left(\Theta_C + \dfrac{c}{a}X\right)$
D	F_{D0}	$\left(\dfrac{d}{a}\right)F_{A0}X$	$F_{A0}\left(\Theta_D + \dfrac{d}{a}X\right)$
I	F_{I0}	- - -	F_{I0}
Totals	F_{T0}	$\delta F_{A0}X$	$F_T = F_{T0} + \delta F_{A0}X$

Definitions of δ and ε

$$\delta = \frac{\text{Change in total number of moles}}{\text{Mole of A reacted}}$$

and

$$\boxed{\varepsilon = y_{A0}\delta}$$
(S4-6)

$$\varepsilon = \frac{\text{Change in total number of moles for complete conversion}}{\text{Total number of moles fed to the reactor}}$$

3. For incompressible liquids or for batch gas phase reactions taking place in a constant volume, $V = V_0$, the concentrations of species A and C in the reaction given by Equation (S4-1) can be written as

$$C_A = \frac{F_A}{v} = \frac{F_{A0}}{v_0}(1-X) = C_{A0}(1-X)$$
(S4-7)

$$C_C = C_{A0}\left(\Theta_C + \frac{c}{a}X\right)$$
(S4-8)

Equations (S4-7) and (S4-8) also hold for gas-phase reactions carried out in constant-volume batch reactors.

4. For gas-phase reactions, we use the definition of concentration ($C_A = F_A/v$) along with the stoichiometric table and Equation (S4-3) to write the concentration of A and C in terms of conversion.

$$C_A = \frac{F_A}{v} = \frac{F_{A0}(1-X)}{v} = C_{A0}\left[\frac{1-X}{1+\varepsilon X}\right]\frac{P}{P_0}\left(\frac{T_0}{T}\right)$$
(S4-9)

$$C_C = \frac{F_C}{v} = C_{A0}\left[\frac{\Theta_C + (c/a)X}{1+\varepsilon X}\right]\frac{P}{P_0}\left(\frac{T_0}{T}\right)$$
(S4-10)

with $\Theta_C = \dfrac{F_{C0}}{F_{A0}} = \dfrac{C_{C0}}{C_{A0}} = \dfrac{y_{C0}}{y_{A0}}$

5. In terms of gas-phase molar flow rates, the concentration of species i is

$$C_i = C_{T0}\frac{F_i}{F_T}\frac{P}{P_0}\frac{T_0}{T}$$
(S4-11)

Equation (S4-11) must be used for membrane reactors (Chapter 6) and for multiple reactions (Chapter 8).

6. Many catalytic rate laws are given in terms of partial pressure, e.g.,

$$-r_A' = \frac{k_A P_A}{1 + K_A P_A}$$
(S4-12)

The partial pressure is related to conversion through the stoichiometric table. For any species "i" in the reaction

$$P_i = P_{A0}\frac{(\Theta_i + v_i X)}{(1 + \varepsilon X)}p$$
(S4-13)

For species A

$$P_A = P_{A0}\frac{(1-X)}{(1+\varepsilon X)}p$$

Substituting in the rate law

$$-r'_A = \frac{k_A P_{A0} \dfrac{(1-X)}{(1+\varepsilon X)} p}{1 + K_A P_{A0} \dfrac{(1-X)}{(1+\varepsilon X)} p} = \frac{k P_{A0}(1-X) p}{(1+\varepsilon X) + K_A P_{A0}(1-X) p} \tag{S4-14}$$

CRE WEB SITE MATERIALS

- **Expanded Material**
    1. *Web P4-1$_B$ Puzzle Problem "What Four Things Are Wrong with This Solution?"*
- **Learning Resources**
    1. *Summary Notes for Chapter 4*
    2. *Interactive Computer Games*
       Quiz Show II

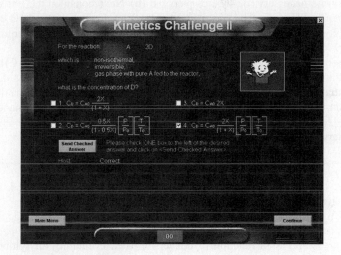

    3. *Solved Problems*
       CDP4-B$_B$ Microelectronics Industry and the Stoichiometric Table
- **Living Example Problems**
    1. *Example 4-5 Calculating the Equilibrium Conversion*
- **FAQ (Frequently Asked Questions)—In Updates/FAW icon section**
- **Professional Reference Shelf**

QUESTIONS AND PROBLEMS

The subscript to each of the problem numbers indicates the level of difficulty: A, least difficult; D, most difficult.

$$A = \bullet \quad B = \blacksquare \quad C = \blacklozenge \quad D = \blacklozenge\blacklozenge$$

Questions

Q4-1$_A$ **(a)** List the important concepts that you learned from this chapter. Make a list of concepts that you are not clear about and ask your instructor or colleague about them.
  (b) Explain the strategy to evaluate reactor design equations and how this chapter expands on Chapters 2 and 3.

Problems

P4-1$_A$ (a) **Example 4-1.** Would the example be correct if water were considered an inert? Explain.

(b) **Example 4-2.** How would the answer change if the initial concentration of glyceryl stearate were 3 mol/dm^3? Rework Example 4-2 correctly using the information given in the problem statement.

(c) **Example 4-3.** Under what conditions will the concentration of the inert nitrogen be constant? Plot Equation (E4-5.2) in terms of $(1/-r_A)$ as a function of X up to value of $X = 0.99$. What did you find?

(e) **Example 4-4.** Why is the equilibrium conversion lower for the batch system than the flow system? Will this always be the case for constant-volume batch systems? For the case in which the total concentration C_{T0} is to remain constant as the inerts are varied, plot the equilibrium conversion as a function of the mole fraction of inerts for both a PFR and a constant-volume batch reactor. The pressure and temperature are constant at 2 atm and 340 K. Only N_2O_4 and inert I are to be fed. Go to the Living Example Problems and load Wolfram. (1) What values of K_C, C_{A0}, and ε cause X_{ef} to be the farthest away from X_{eb}? (2) At what value C_{A0} will X_{eb} and X_{ef} be closest together?

(d) **Example 4-5. (a)** Using the molar flow rate of A of 3 mol/minute and Figure E4-5.1, calculate the PBR volume necessary for 40% conversion. **(b)** Next consider the entering flow rate of SO_2 is 1,000 mol/h. Plot $\left(F_{A0}/-r_A'\right)$ as a function of X to determine the PBR catalyst weight to achieve (1) 30% conversion, (2) 40% conversion, and (3) 99% of the equilibrium conversion, i.e., $X = 0.99\ X_e$.

P4-2$_A$ Load the Interactive Computer Games (ICG) Kinetic Challenge from the CRE Web site. Play the game and then record your performance number for the module that indicates your mastering of the material. Your professor has the key to decode your performance number. ICG Kinetics Challenge Performance # _____.

Kinetics Challenge II		
Rate	Law	Stoich
100	100	100
200	200	200
300	300	300

P4-3$_A$ The elementary reversible reaction

$$2A \rightleftharpoons B$$

is carried out in a flow reactor where pure A is fed at a concentration of 4.0 mol/dm^3. If the equilibrium conversion is found to be 60%,

(a) What is the equilibrium constant, K_C if the reaction is a gas phase reaction?

(b) What is the K_C if the reaction is a liquid-phase reaction?

(c) Write $-r_A$ solely as a function of conversion (i.e., evaluating all symbols) when the reaction is an elementary, reversible, gas-phase, isothermal reaction with no pressure drop with $k_A = 2$ dm^6/mol•s and $K_C = 0.5$ all in proper units.

(d) Repeat (c) for a constant-volume batch reactor.

P4-4$_B$ **Stoichiometry.** The elementary gas reaction

$$2A + B \rightarrow C$$

is carried out isothermally in a PFR with no pressure drop. The feed is equal molar in A and B, and the entering concentration of A is 0.1 mol/dm^3. Set up a stoichiometric table and then determine the following.

(a) What is the entering concentration (mol/dm^3) of B?

(b) What are the concentrations of A and C (mol/dm^3) at 25% conversion of A?

(c) What is the concentration of B (mol/dm^3) at 25% conversion of A?

(d) What is the concentration of B (mol/dm^3) at 100% conversion of A?

(e) If at a particular conversion the rate of formation of C is 2 mol/min/dm^3, what is the rate of formation of A at the same conversion?

(f) Write $-r_A$ solely as a function of conversion (i.e., evaluating all symbols) when the reaction is an elementary, irreversible, gas-phase, isothermal reaction with no pressure drop with an equal molar feed and with $C_{A0} = 2.0$ mol/dm^3 at, $k_A = 2$ dm^6/mol•s.

(g) What is the rate of reaction at $X = 0.5$?

P4-5$_A$ Set up a stoichiometric table for each of the following reactions and express the concentration of each species in the reaction as a function of conversion, evaluating all constants (e.g., ε, Θ). Next, assume the reaction follows an elementary rate law, and write the reaction rate solely as a function of conversion, i.e., $-r_A = f(X)$.

(a) For the liquid-phase reaction

$$\underset{\substack{O \\ \diagdown \\ CH_2{-}CH_2}}{} + H_2O \xrightarrow{H_2SO_4} \underset{\substack{CH_2{-}OH \\ | \\ CH_2{-}OH}}{}$$

the entering concentrations of ethylene oxide and water, after mixing the inlet streams, are 16.13 mol/dm³ and 55.5 mol/dm³, respectively. The specific reaction rate is $k = 0.1$ dm³/mol·s at 300 K with $E = 12{,}500$ cal/mol.

(1) After finding $-r_A = f(X)$, calculate the CSTR space-time, τ, for 90% conversion at 300 K and also at 350 K.

(2) If the volumetric flow rate is 200 liters per second, what are the corresponding reactor volume?

(b) For the isothermal, isobaric gas-phase pyrolysis

$$C_2H_6 \longrightarrow C_2H_4 + H_2$$

pure ethane enters a flow reactor at 6 atm and 1100 K. Set up a stoichiometric table and then write $-r_A = f(X)$. How would your equation for the concentration and reaction rate, i.e., $-r_A = f(X)$, change if the reaction were to be carried out in a constant-volume batch reactor?

(c) For the isothermal, isobaric, catalytic gas-phase oxidation

$$C_2H_4 + \tfrac{1}{2}O_2 \longrightarrow \underset{\substack{O \\ \diagdown \\ CH_2{-}CH_2}}{}$$

the feed enters a PBR at 6 atm and 260°C, and is a stoichiometric mixture of only oxygen and ethylene. Set up a stoichiometric table and then write $-r_A'$ as a function of partial pressures. Express the partial pressures and $-r_A'$ as a function of conversion for (1) a fluidized batch reactor and (2) a PBR. Finally, write $-r_A'$ solely as a function of the rate constant and conversion.

(d) Set up a stoichiometric table for the isothermal, isobaric, catalytic gas-phase reaction carried out in a fluidized CSTR.

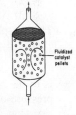

Fluidized CSTR

The feed is stoichiometric and enters at 6 atm and 170°C. What catalyst weight is required to reach 80% conversion in a fluidized CSTR at 170°C and at 270°C? The rate constant is defined with respect to benzene and $v_0 = 50$ dm³/min.

$$k_B = \frac{53 \ \text{mol}}{\text{kgcat} \cdot \text{min} \cdot \text{atm}^3} \quad \text{at 300 K with } E = 80 \ \text{kJ/mol}$$

First write the rate law in terms of partial pressures and then express the rate law as a function of conversion.

P4-6$_A$ *Ortho*nitroanaline (an important intermediate in dyes—called *fast orange*) is formed from the reaction of *ortho*nitrochlorobenzene (ONCB) and aqueous ammonia (see explosion in Figure E13-2.1 in Example 13-2).

The liquid-phase reaction is first order in both ONCB and ammonia with $k = 0.0017$ m^3/kmol · min at 188°C with $E = 11,273$ cal/mol. The initial entering concentrations of ONCB and ammonia are 1.8 kmol/m^3 and 6.6 kmol/m^3, respectively (more on this reaction in Chapter 13).

(a) Set up a stoichiometric table for this reaction for a flow system.

(b) Write the rate law for the rate of disappearance of ONCB in terms of concentration.

(c) Explain how parts (a) and (b) would be different for a batch system.

(d) Write $-r_A$ solely as a function of conversion. $-r_A =$ _____

(e) What is the initial rate of reaction ($X = 0$)   at 188°C? $-r_A =$ _____
                                                        at 25°C? $-r_A =$ _____
                                                        at 288°C? $-r_A =$ _____

(f) What is the rate of reaction when $X = 0.90$ at 188°C? $-r_A =$ _____
                                                    at 25°C? $-r_A =$ _____
                                                    at 288°C? $-r_A =$ _____

(g) What would be the corresponding CSTR reactor volume at 25°C to achieve 90% conversion and at 288°C for a feed rate of 2 dm^3/min

at 25°C? $V =$ _____
at 288°C? $V =$ _____

P4-7$_B$ Consider the following elementary gas-phase reversible reaction to be carried out isothermally with no pressure drop and for an equal molar feed of A and B with $C_{A0} = 2.0$ mol/dm^3.

$$2A + B \rightleftharpoons C$$

(a) What is the concentration of B initially? $C_{B0} =$ ____ (mol/dm^3)

(b) What is the limiting reactant? _____

(c) What is the exit concentration of B when the conversion of A is 25%? $C_B =$ _____ (mol/dm^3)

(d) Write $-r_A$ solely as a function of conversion (i.e., evaluating all symbols) when the reaction is an elementary, reversible, gas-phase, isothermal reaction with no pressure drop with an equal molar feed and with $C_{A0} = 2.0$ mol/dm^3, $k_A = 2$ dm^6/mol^2·s, and $K_C = 0.5$ all in proper units $-r_A =$ ____.

(e) What is the equilibrium conversion?

(f) What is the rate when the conversion is
    (1) 0%?
    (2) 50%?
    (3) 0.99 X_e?

P4-8$_B$ The gas-phase reaction

$$\tfrac{1}{2}N_2 + \tfrac{3}{2}H_2 \longrightarrow NH_3$$

is to be carried out isothermally first in a flow reactor. The molar feed is 50% H$_2$ and 50% N$_2$, at a pressure of 16.4 atm and at a temperature of 227°C.

(a) Construct a complete stoichiometric table.

(b) Express the concentrations in mol/dm^3 of each for the reacting species as a function of conversion. Evaluate C_{A0}, δ, and ε, and then calculate the concentrations of ammonia and hydrogen when the conversion of H$_2$ is 60%. (*Ans:* $C_{H_2} = 0.1$ mol/dm^3)

(c) Suppose by chance the reaction is elementary with $k_{N_2} = 40$ dm^3/mol/s. Write the rate of reaction *solely* as a function of conversion for (1) a flow reactor and for (2) a constant-volume batch reactor.

P4-9$_B$ Calculate the equilibrium conversion and concentrations for each of the following reactions:

(a) The reversible reaction

$$2A \rightleftharpoons B$$

is carried out in a flow reactor where pure A is fed at a concentration of 4.0 mol/dm^3. If the equilibrium conversion is found to be 60%.

(1) What is the equilibrium constant, K_C, if the reaction is a gas phase reaction?

(2) What is the K_C if the reaction is a liquid-phase reaction?

(b) The gas-phase reaction

$$A \xrightleftharpoons{} 3C$$

is carried out in a flow reactor with no pressure drop. Pure A enters at a temperature of 400 K and a pressure of 10 atm. At this temperature, $K_C = 0.25(\text{mol/dm}^3)^2$.

(3) The gas-phase reaction is carried out in a constant-volume batch reactor.

(4) The gas-phase reaction is carried out in a constant-pressure batch reactor.

P4-10$_C$ Consider a *cylindrical batch reactor* that has one end fitted with a frictionless piston attached to a spring (Figure P4-10$_C$). The reaction

$$A + B \longrightarrow 8C$$

with the rate law

$$-r_A = k_1 C_A^2 C_B$$

Hall of Fame

is taking place in this type of reactor.

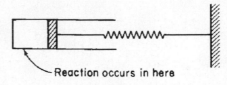

Reaction occurs in here

Figure P4-10$_C$

(a) Write the rate law solely as a function of conversion, numerically evaluating all possible symbols. (*Ans.:* $-r_A = 5.03 \times 10^{-9}\,[(1 - X)^3/(1 + 3X)^{3/2}]$ lb mol/ft^3·s.)

(b) What is the conversion and rate of reaction when $V = 0.2$ ft^3? (*Ans.:* $X = 0.259$, $-r_A = 8.63 \times 10^{-10}$ lb mol/ft^3·s.)

Additional information:

Equal moles of A and B are present at $t = 0$

Initial volume: 0.15 ft^3

Value of k_1: 1.0 (ft^3/lb mol)^2·s^{-1}

The spring constant is such that the relationship between the volume of the reactor and pressure within the reactor is

$$V = (0.1)(P) \qquad (V \text{ in ft}^3, P \text{ in atm})$$

Temperature of system (considered constant): 140°F

Gas constant: 0.73 ft^3·atm/lb mol·°R

SUPPLEMENTARY READING

Further elaboration of the development of the general balance equation may be found on the CRE Web site *www.umich.edu/~elements/5e/index.html* and also may or may not be found in

KEILLOR, GARRISON and TIM RUSSELL, *Dusty and Lefty: The Lives of the Cowboys* (Audio CD), St. Paul, MN: Highbridge Audio, 2006.

FELDER, R. M., and R. W. ROUSSEAU, *Elementary Principles of Chemical Processes*, 4th ed. New York: Wiley, 2015, Chapter 4.

HIMMELBLAU, D. M., and J. D. RIGGS, *Basic Principles and Calculations in Chemical Engineering*, 7th ed. Upper Saddle River, NJ: Prentice Hall, 2004, Chapters 2 and 6.

Isothermal Reactor Design: Conversion — 5

Why, a four-year-old child could understand this.
Someone get me a four-year-old child.

—Groucho Marx

Tying everything
together

Overview. Chapters 1 and 2 discussed mole balances on reactors and the manipulation of these balances to predict reactor sizes. Chapter 3 discussed reactions and reaction-rate laws, and Chapter 4 discussed reaction stoichiometry. In Chapters 5 and 6, we combine reactions and reactors as we bring all the material in the preceding four chapters together to arrive at a logical structure for the design of various types of reactors. By using this structure, one should be able to solve reactor engineering problems by reasoning, rather than by memorizing numerous equations together with the various restrictions and conditions under which each equation applies (e.g., whether or not there is a change in the total number of moles).

In this chapter, we use the mole balances written in terms of conversion, shown in Chapter 2, Table S2-1, to study isothermal reactor designs. Conversion is the preferred parameter to measure progress for single reactions occurring in batch reactors (BRs), CSTRs, PFRs, and PBRs. Both batch reactor times and flow reactor volumes to achieve a given conversion will be calculated.

We have chosen four different reactions and four different reactors to illustrate the salient principles of isothermal reactor design using conversion as a variable, namely

- The use of a laboratory batch reactor to determine the specific reaction rate constant, k, for the liquid-phase reaction to form ethylene glycol.
- The design of an industrial CSTR to produce ethylene glycol using k from the batch experiment.
- The design of a PFR for the gas-phase pyrolysis of ethane to form ethylene.

> • The design of a packed-bed reactor with pressure drop to form
>   ethylene oxide from the partial oxidation of ethylene.
>
> When we put all these reactions and reactors together, we will see
> we have designed a chemical plant to produce 200 million pounds per
> year of ethylene glycol.

5.1 Design Structure for Isothermal Reactors

Logic vs. memorization

One of the primary goals of this chapter is to solve chemical reaction engineering (CRE) problems by using logic rather than memorizing which equation applies where. It is the author's experience that following this structure, shown in Figure 5-1, will lead to a greater understanding of isothermal reactor design. We begin by applying our general mole balance equation (level ①) to a specific reactor to arrive at the design equation for that reactor (level ②). If the feed conditions are specified (e.g., N_{A0} or F_{A0}), all that is required to evaluate the design equation is the rate of reaction as a function of conversion at the same conditions as those at which the reactor is to be operated (e.g., temperature and pressure). When $-r_A = f(X)$ is known or given, one can go directly

Use the algorithm rather than memorizing equations.

from level ③ to the last level, level ⑨, to determine either the batch time or reactor volume necessary to achieve the specified conversion.

When the rate of reaction is not given explicitly as a function of conversion, we must proceed to level ④, where the rate law must be determined by either finding it in books or journals or by determining it experimentally in the laboratory. Techniques for obtaining and analyzing rate data to determine the reaction order and rate constant are presented in Chapter 7. After the rate law has been established, one has only to use stoichiometry (level ⑤) together with the conditions of the system (e.g., constant volume, temperature) to express concentration as a function of conversion.

For liquid-phase reactions and for gas-phase reactions with no pressure drop ($P = P_0$), one can combine the information in levels ④ and ⑤ to express the rate of reaction as a function of conversion and arrive at level ⑥. It is now possible to determine either the time or reactor volume necessary to achieve the desired conversion by substituting the relationship linking conversion and rate of reaction into the appropriate design equation (level ⑨).

For gas-phase reactions in packed beds where there is a pressure drop, we need to proceed to level ⑦ to evaluate the pressure ratio p (that is, ($p = P / P_0$)) in the concentration term using the Ergun equation (Section 5.5). In level ⑧, we combine the equations for pressure drop in level ⑦ with the information in levels ④ and ⑤ to proceed to level ⑨, where the equations are then evaluated in the appropriate manner (i.e., analytically using a table of integrals, or numerically using an ODE solver). Although this structure emphasizes the determination of a reaction time or reactor volume for a specified conversion, it can also readily be used for other types of reactor calculations, such as determining the conversion for a specified volume. Different manipulations can be performed in level ⑨ to answer the different types of questions mentioned here.

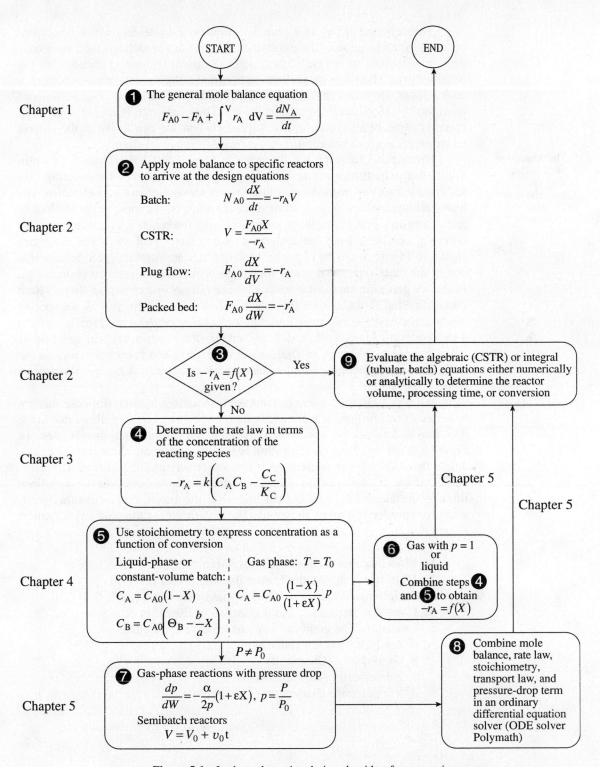

Figure 5-1   Isothermal-reaction design algorithm for conversion.

The structure shown in Figure 5-1 allows one to develop a few basic concepts and then to arrange the parameters (equations) associated with each concept in a variety of ways. Without such a structure, one is faced with the possibility of choosing or perhaps memorizing the correct equation from a *multitude of equations* that can arise for a variety of different combinations of reactions, reactors, and sets of conditions. The challenge is to put everything together in an orderly and logical fashion so that we can arrive at the correct equation for a given situation.

Fortunately, by using the algorithm to formulate CRE problems shown in Figure 5-2, which happens to be analogous to the algorithm for ordering dinner from a fixed-price menu in a fine French restaurant, we can eliminate virtually all memorization. In both of these algorithms, we must make **choices** in each category. For example, in ordering from a French menu, we begin by choosing one dish from the *appetizers* listed. Step 1 of the CRE algorithm shown in Figure 5-2 is to begin by choosing the appropriate mole balance for one of the three types of reactors shown. After making our reactor choice (e.g., PFR), we go to the small oval ready to make our second choice in Step 2, Rate Laws. In Step 2 we choose the rate law (*entrée*), and in Step 3 we specify whether the reaction is gas *or* liquid phase (*cheese* or *dessert*). Finally, in Step 4 we combine Steps 1, 2, and 3 and either obtain an analytical solution or solve the equations using an ODE solver. The complete French menu is on the CRE Web site (*www.umich.edu/~elements/5e/index.html*), Chapter 5 *Summary Notes*.

We now will apply this algorithm to a specific situation. Suppose that we have, as shown in Figure 5-2, mole balances for three reactors, three rate laws, and the equations for concentrations for both liquid and gas phases. In Figure 5-2, we see how the algorithm is used to formulate the equation to calculate the *PFR reactor volume for a first-order gas-phase reaction*. The pathway to arrive at this equation is shown by the ovals connected to the dark lines through the algorithm. The dashed lines and the boxes represent other pathways for solutions to other situations. The algorithm for the pathway shown in Figure 5-2 is

1. **Mole balances,** choose species A reacting in a PFR
2. **Rate laws,** choose the irreversible first-order reaction
3. **Stoichiometry,** choose the gas-phase concentration
4. **Combine** steps 1, 2, and 3 to arrive at Equation A
5. **Evaluate.** The combine step can be evaluated either
   a. Analytically (Appendix Al)
   b. Graphically (Chapter 2)
   c. Numerically (Appendix A4)
   d. Via software (Polymath)

The Algorithm
1. Mole balance
2. Rate law
3. Stoichiometry
4. Combine
5. Evaluate

French Menu Analogy

Choices

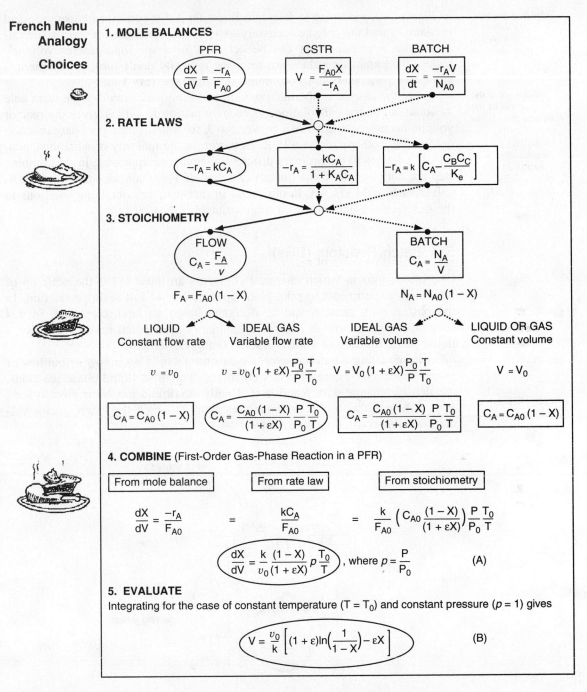

1. **MOLE BALANCES**

PFR
$$\frac{dX}{dV} = \frac{-r_A}{F_{A0}}$$

CSTR
$$V = \frac{F_{A0}X}{-r_A}$$

BATCH
$$\frac{dX}{dt} = \frac{-r_A V}{N_{A0}}$$

2. **RATE LAWS**

$$-r_A = kC_A$$

$$-r_A = \frac{kC_A}{1 + K_A C_A}$$

$$-r_A = k\left[C_A - \frac{C_B C_C}{K_e}\right]$$

3. **STOICHIOMETRY**

FLOW
$$C_A = \frac{F_A}{v}$$

BATCH
$$C_A = \frac{N_A}{V}$$

$$F_A = F_{A0}(1 - X)$$

$$N_A = N_{A0}(1 - X)$$

LIQUID
Constant flow rate

IDEAL GAS
Variable flow rate

IDEAL GAS
Variable volume

LIQUID OR GAS
Constant volume

$$v = v_0$$

$$v = v_0(1 + \varepsilon X)\frac{P_0}{P}\frac{T}{T_0}$$

$$V = V_0(1 + \varepsilon X)\frac{P_0}{P}\frac{T}{T_0}$$

$$V = V_0$$

$$C_A = C_{A0}(1 - X)$$

$$C_A = \frac{C_{A0}(1 - X)}{(1 + \varepsilon X)}\frac{P}{P_0}\frac{T_0}{T}$$

$$C_A = \frac{C_{A0}(1 - X)}{(1 + \varepsilon X)}\frac{P}{P_0}\frac{T_0}{T}$$

$$C_A = C_{A0}(1 - X)$$

4. **COMBINE** (First-Order Gas-Phase Reaction in a PFR)

From mole balance

From rate law

From stoichiometry

$$\frac{dX}{dV} = \frac{-r_A}{F_{A0}} = \frac{kC_A}{F_{A0}} = \frac{k}{F_{A0}}\left(C_{A0}\frac{(1-X)}{(1+\varepsilon X)}\right)\frac{P}{P_0}\frac{T_0}{T}$$

$$\frac{dX}{dV} = \frac{k}{v_0}\frac{(1-X)}{(1+\varepsilon X)}p\frac{T_0}{T} \text{, where } p = \frac{P}{P_0} \qquad \text{(A)}$$

5. **EVALUATE**

Integrating for the case of constant temperature ($T = T_0$) and constant pressure ($p = 1$) gives

$$V = \frac{v_0}{k}\left[(1 + \varepsilon)\ln\left(\frac{1}{1 - X}\right) - \varepsilon X\right] \qquad \text{(B)}$$

Figure 5-2 Algorithm for isothermal reactors.

Substitute parame-
ter values in
steps 1–4 *only* if
they are zero.

In Figure 5-2, we chose to integrate Equation A for constant temperature and pressure to find the volume necessary to achieve a specified conversion (or calculate the conversion that can be achieved in a specified reactor volume). Unless the parameter values are zero, we typically don't substitute numerical values for parameters in the combine step until the very end.

For the case of isothermal operation with no pressure drop, we were able to obtain an analytical solution, given by equation B, which gives the reactor volume necessary to achieve a conversion X for a first-order gas-phase reaction carried out isothermally in a PFR. However, in the majority of situations, analytical solutions to the ordinary differential equations appearing in the combine step are not possible. Consequently, we include Polymath, or some other ODE solver such as MATLAB, in our menu in that it makes obtaining solutions to the differential equations much more palatable.

We can solve the
equations in the
combine step either
1. Analytically
   (Appendix A1)
2. Graphically
   (Chapter 2)
3. Numerically
   (Appendix A4)
4. Using software
   (Polymath).

5.2 Batch Reactors (BRs)

One of the jobs in which chemical engineers are involved is the scale-up of laboratory experiments to pilot-plant operation or to full-scale production. In the past, a pilot plant would be designed based on laboratory data. In this section, we show how to analyze a laboratory-scale batch reactor in which a liquid-phase reaction of known order is being carried out.

In modeling a batch reactor, we assume there is no inflow or outflow of material and that the reactor is well mixed. For most liquid-phase reactions, the density change with reaction is usually small and can be neglected (i.e., $V = V_0$). In addition, for *gas-phase* reactions in which the batch reactor volume remains constant, we also have $V = V_0$.

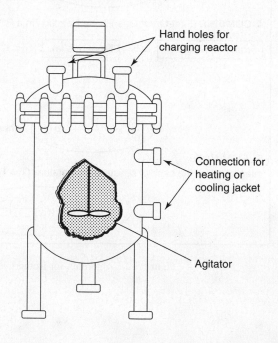

Hand holes for
charging reactor

Connection for
heating or
cooling jacket

Agitator

5.2.1 Batch Reaction Times

The time necessary to achieve a specific conversion depends upon how fast the reaction takes place, which in turn is dependent on the rate constant and the reactant concentration. To get a feel of how long it takes to carry a batch reaction, we shall calculate the batch reaction times for different values of the reaction rate constant, k, for a first- and for a second-order reaction. First, let's solve for the time to achieve a conversion X for the second-order reaction

$$2A \rightarrow B + C$$

The Algorithm

1. The **mole balance** on a constant-volume, $V = V_0$, batch reactor is

Mole balance
$$N_{A0}\frac{dX}{dt} = -r_A V_0 \qquad (2\text{-}6)$$

Dividing by N_{A0} and recognizing $C_{A0} = N_{A0}/V_0$ we obtain

$$\frac{dX}{dt} = -\frac{r_A}{C_{A0}} \qquad (5\text{-}1)$$

2. The **rate law** is

Rate Law
$$-r_A = k_2 C_A^2 \qquad (5\text{-}2)$$

3. From **stoichiometry** for a constant-volume batch reactor, we obtain

Stoichiometry
$$C_A = C_{A0}(1 - X) \qquad (4\text{-}12)$$

4. **Combining** the **mole balance, rate law,** and **stoichiometry** we obtain

$$-r_A = kC_{A0}^2(1 - X)^2$$

Next, we substitute for $-r_A$ in Equation (5-1)

Combine
$$\frac{dX}{dt} = k_2 C_{A0}(1 - X)^2 \qquad (5\text{-}3)$$

5. To **evaluate,** we separate the variables and integrate

$$\frac{dX}{(1 - X)^2} = k_2 C_{A0}\,dt$$

Initially, if $t = 0$, then $X = 0$. If the reaction is carried out isothermally, k will be constant; we can integrate this equation (see Appendix A.1 for a table of integrals used in CRE applications) to obtain

Evaluate
$$\int_0^t dt = \frac{1}{k_2 C_{A0}} \int_0^X \frac{dX}{(1 - X)^2}$$

Second-order,
isothermal,
constant-volume
batch reaction

$$t_R = \frac{1}{k_2 C_{A0}}\left(\frac{X}{1-X}\right) \qquad (5\text{-}4)$$

This time is the reaction time t (i.e., t_R) needed to achieve a conversion X for a second-order reaction in a batch reactor. In a similar fashion, we can apply the CRE algorithm to a first-order reaction to obtain the reaction time, t_R, needed to achieve a conversion X

$$t_R = \frac{1}{k_1} ln \frac{1}{1-X} \qquad (5\text{-}5)$$

It is important to have a grasp of the order of magnitudes of batch reaction times, t_R, to achieve a given conversion, say 90%, for different values of the product of specific reaction rate, k, and initial concentration, C_{A0}. Table 5-1 shows the algorithm to find the batch reaction times, t_R, for both first- and second-order reactions carried out isothermally. We can obtain these estimates of t_R by considering the first- and second-order irreversible reactions of the form

$$2A \rightarrow B + C$$

TABLE 5-1   ALGORITHM TO ESTIMATE REACTION TIMES

Mole Balance	$\dfrac{dX}{dt_R} = \dfrac{-r_A}{N_{A0}} V$	
Rate Law	*First-Order*	*Second-Order*
	$-r_A = k_1 C_A$	$-r_A = k_2 C_A^2$
Stoichiometry $(V = V_0)$	$C_A = \dfrac{N_A}{V_0} = C_{A0}(1-X)$	
Combine	$\dfrac{dX}{dt_R} = k_1(1-X)$	$\dfrac{dX}{dt_R} = k_2 C_{A0}(1-X)^2$
Evaluate (Integrate)	$t_R = \dfrac{1}{k_1} \ln \dfrac{1}{1-X}$	$t_R = \dfrac{X}{k_2 C_{A0}(1-X)}$

For *first-order reactions*, the reaction time to reach 90% conversion (i.e., $X = 0.9$) in a constant-volume batch reactor scales as

$$t_R = \frac{1}{k_1} \ln \frac{1}{1-X} = \frac{1}{k_1} \ln \frac{1}{1-0.9} = \frac{2.3}{k_1}$$

If $k_1 = 10^{-4} \text{ s}^{-1}$,

$$t_R = \frac{2.3}{10^{-4} \text{ s}^{-1}} = 23,000 \text{ s} = 6.4 \text{ h}$$

The time necessary to achieve 90% conversion in a batch reactor for an irreversible first-order reaction in which the specific reaction rate, k_1, is (10^{-4} s^{-1}) is 6.4 h.

For *second-order reactions*, we have

$$t_R = \frac{1}{k_2 C_{A0}} \frac{X}{1-X} = \frac{0.9}{k_2 C_{A0}(1-0.9)} = \frac{9}{k_2 C_{A0}}$$

If $k_2 C_{A0} = 10^{-3} \text{ s}^{-1}$,

$$t_R = \frac{9}{10^{-3} \text{ s}^{-1}} = 9000 \text{ s} = 2.5 \text{ h}$$

We note that if 99% conversion had been required for this value of kC_{A0}, the reaction time, t_R, would jump to 27.5 h.

Table 5-2 gives the *order of magnitude* of time to achieve 90% conversion for first- and second-order irreversible batch reactions. Flow reactors would be used for reactions with *characteristic reaction times, t_R,* of minutes or less.

Estimating reaction times

TABLE 5-2   BATCH REACTION TIMES

First-Order k_1 (s^{-1})	Second-Order $k_2 C_{A0}$ (s^{-1})	Reaction Time t_R
10^{-4}	10^{-3}	Hours
10^{-2}	10^{-1}	Minutes
1	10	Seconds
1000	10,000	Milliseconds

The times in Table 5-2 are the reaction time to achieve 90% conversion (i.e., to reduce the concentration from C_{A0} to $0.1\ C_{A0}$). The total cycle time in any batch operation is considerably longer than the reaction time, t_R, as one must account for the time necessary to fill (t_f) and heat (t_e) the reactor together with the time necessary to clean the reactor between batches, t_c. In some cases, the reaction time calculated from Equations (5-4) and (5-5) may be only a small fraction of the total cycle time, t_t.

$$t_t = t_f + t_e + t_c + t_R$$

Typical cycle times for a batch polymerization process are shown in Table 5-3. Batch polymerization reaction times may vary between 5 and 60 hours. Clearly, decreasing the reaction time with a 60-hour reaction is a promising endeavor. As the reaction time is reduced (e.g., 2.5 h for a second-order reaction with $k_2 C_{A0} = 10^{-3} \text{ s}^{-1}$), it becomes important to use large lines and pumps to achieve rapid transfers and to utilize efficient sequencing to minimize the cycle time.

TABLE 5-3   TYPICAL CYCLE TIME FOR A BATCH POLYMERIZATION PROCESS

Activity	Time (h)
1. Charge feed to the reactor and agitate, t_f	0.5–2.0
2. Heat to reaction temperature, t_e	0.5–2.0
3. Carry out reaction, t_R	(varies)
4. Empty and clean reactor, t_c	1.5–3.0
Total time excluding reaction	2.5–7.0

Batch operation times

Usually, one has to optimize the reaction time with the processing times listed in Table 5-3 to produce the maximum number of batches (i.e., pounds or kilograms of product) in a day.

In the next four examples, we will describe the various reactors needed to produce 200 million pounds per year of ethylene glycol from a feedstock of ethane. We begin by finding the rate constant, k, for the hydrolysis of ethylene oxide to form ethylene glycol.

Example 5–1   Determining k *from Batch Data*

It is desired to design a CSTR to produce 200 million pounds of ethylene glycol per year by hydrolyzing ethylene oxide. However, before the design can be carried out, it is necessary to perform and analyze a batch-reactor experiment to determine the specific reaction-rate constant, k. Because the reaction will be carried out isothermally, the specific reaction rate will need to be determined only at the reaction temperature of the CSTR. At temperatures above 80°C, there is a significant by-product formation, while at temperatures below 40°C, the reaction does not proceed at a significant rate; consequently, a temperature of 55°C has been chosen. Because water is present in excess, its concentration (55.5 mol/dm³) may be considered constant during the course of the reaction. The reaction is first-order in ethylene oxide.

$$\underset{A}{\underset{\big|}{\overset{O}{CH_2-CH_2}}} + \underset{B}{H_2O} \xrightarrow[\text{catalyst}]{H_2SO_4} \underset{C}{\overset{CH_2-OH}{\underset{|}{CH_2-OH}}}$$

In the laboratory experiment, 500 mL of a 2 M solution (2 kmol/m³) of ethylene oxide (A) in water was mixed with 500 mL of water (B) containing 0.9 wt % sulfuric acid, which is a catalyst. The temperature was maintained at 55°C. The concentration of ethylene glycol (C) was recorded as a function of time (Table E5-1.1).

(a) Derive an equation for the concentration of ethylene glycol as a function of time.

(b) Rearrange the equation derived in (a) to obtain a linear plot of a function concentration versus time.

(c) Using the data in Table E5-1.1, determine the specific reaction rate at 55°C.

TABLE E5-1.1   CONCENTRATION-TIME DATA

Time (min)	Concentration of Ethylene Glycol (C) (kmol/m³)*
0.0	0.000
0.5	0.145
1.0	0.270
1.5	0.376
2.0	0.467
3.0	0.610
4.0	0.715
6.0	0.848
10.0	0.957

*1 kmol/m³ = 1 mol/dm³ = 1 mol/L.

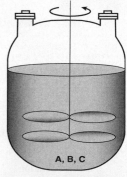

Batch Reactor

> Check 10 types of homework problems on the CRE Web site for more solved examples using this algorithm.

Solution

Part (a)

1. The **mole balance** on ethylene oxide (A) given in Equation (1-5) for a constant volume, V_0, well-mixed batch reactor can be written as

$$\frac{1}{V_0}\frac{dN_A}{dt} = r_A \qquad (E5\text{-}1.1)$$

Following the Algorithm

Mole Balance

Rate Law

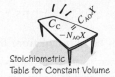

Stoichiometric Table for Constant Volume

Taking V_0 inside the differential and recalling that the concentration is

$$C_A = \frac{N_A}{V_0}$$

then the differential mole balance becomes

$$\frac{d(N_A/V_0)}{dt} = \frac{dC_A}{dt} = r_A \qquad \text{(E5-1.2)}$$

2. The **rate law** for the ethylene oxide hydrolysis is

$$-r_A = kC_A \qquad \text{(E5-1.3)}$$

Because water is present in such excess, the concentration of water at any time t is virtually the same as the initial concentration, and the rate law is independent of the concentration of H_2O ($C_B \cong C_{B0}$).

3. **Stoichiometry.** Liquid phase, no volume change, $V = V_0$ (Table E5-1.2):

TABLE E5-1.2   STOICHIOMETRIC TABLE

Species	Symbol	Initial	Change	Remaining	Concentration
CH_2CH_2O	A	N_{A0}	$-N_{A0}X$	$N_A = N_{A0}(1-X)$	$C_A = C_{A0}(1-X)$
H_2O	B	$\Theta_B N_{A0}$	$-N_{A0}X$	$N_B = N_{A0}(\Theta_B - X)$	$C_B = C_{A0}(\Theta_B - X)$
					$C_B \approx C_{A0}\Theta_B = C_{B0}$
$(CH_2OH)_2$	C	0	$N_{A0}X$	$N_C = N_{A0}X$	$C_C = C_{A0}X = C_{A0} - C_A$
		$\overline{N_{T0}}$		$\overline{N_T = N_{T0} - N_{A0}X}$	

Recall that Θ_B is the ratio of the initial number of moles of B to A (i.e.,

$$\Theta_B = \frac{N_{B0}}{N_{A0}}).$$

For species B, i.e., water,

$$C_B = C_{A0}(\Theta_B - X)$$

We quickly see that water is in excess, as the molarity of water is 55 moles per liter. The initial concentration of A after mixing the two volumes together is 1 molar. Therefore,

$$\Theta_B = \frac{55\,\text{mol}/\text{dm}^3}{1\,\text{mol}/\text{dm}^3} = 55$$

The maximum value of X is 1, and $\Theta_B \gg 1$, therefore C_B is virtually constant

$$C_B \cong C_{A0}\Theta = C_{B0}$$

For species C, i.e., ethylene glycol, the concentration is

$$C_C = \frac{N_C}{V_0} = \frac{N_{A0}X}{V_0} = \frac{N_{A0} - N_A}{V_0} = C_{A0} - C_A \qquad \text{(E5-1.4)}$$

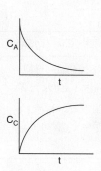

Combining
mole balance,
rate law, and
stoichiometry

4. **Combining** the rate law and the mole balance, we have

$$-\frac{dC_A}{dt} = kC_A \tag{E5-1.5}$$

5. **Evaluate.** For isothermal operation, k is constant, so we can integrate this equation (E5-1.5)

$$-\int_{C_{A0}}^{C_A} \frac{dC_A}{C_A} = \int_0^t k\, dt = k\int_0^t dt$$

using the initial condition that when $t = 0$, then $C_A = C_{A0} = 1$ mol/dm^3 = 1 kmol/m^3.

Integrating yields

$$\ln \frac{C_{A0}}{C_A} = kt \tag{E5-1.6}$$

The concentration of *ethylene oxide* (A) at any time t is

$$C_A = C_{A0}e^{-kt} \tag{E5-1.7}$$

The concentration of *ethylene glycol* (C) at any time t can be obtained from the reaction stoichiometry

$$C_C = C_{A0} - C_A = C_{A0}(1 - e^{-kt}) \tag{E5-1.8}$$

Part (b)

We are now going to rearrange Equation (E5-1.8) for the concentration of ethylene glycol in such a manner that we can easily use the data in Table E5-1.1 to determine the rate constant k

$$\ln \frac{C_{A0} - C_C}{C_{A0}} = -kt \tag{E5-1.9}$$

Part (c)

We see that a plot of $\ln[(C_{A0} - C_C)/C_{A0}]$ as a function of t will be a straight line with a slope $-k$. Using Table E5-1.1, we can construct Table E5-1.3 and use Excel to plot $\ln(C_{A0} - C_C)/C_{A0}$ as a function of t.

TABLE E5-1.3   PROCESSED DATA

t (min)	C_C (kmol/m^3)	$\dfrac{C_{A0} - C_C}{C_{A0}}$	$\ln\left(\dfrac{C_{A0} - C_C}{C_{A0}}\right)$
0.0	0.000	1.000	0.0000
0.5	0.145	0.855	−0.1570
1.0	0.270	0.730	−0.3150
1.5	0.376	0.624	−0.4720
2.0	0.467	0.533	−0.6290
3.0	0.610	0.390	−0.9420
4.0	0.715	0.285	−1.2550
6.0	0.848	0.152	−1.8840
10.0	0.957	0.043	−3.1470

Evaluating the
specific reaction
rate from
batch-reactor
concentration–
time data

From the slope of a plot of $\ln[(C_{A0} - C_C)/C_{A0}]$ versus t, we can find k, as shown in the Excel plot in Figure E5-1.1.

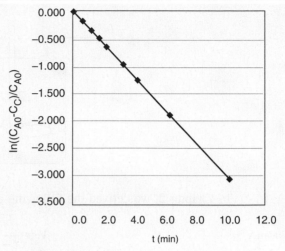

Figure E5-1.1   Excel plot of data.

$$\text{Slope} = -k = -0.311 \text{ min}^{-1}$$

$$k = 0.311 \text{ min}^{-1}$$

The rate law becomes

$$-r_A = 0.311 \text{ min}^{-1} C_A$$

The rate law can now be used in the design of an industrial CSTR. For those who prefer to find k using semilog graph paper, this type of analysis can be found at *www.physics.uoguelph.ca/tutorials/GLP*. Tutorials are also given in the *Summary Notes* for Chapter 3 (last example) and Chapter 7 (Excel).

Summary Notes

Analysis: In this example, we used our CRE algorithm

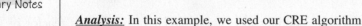

(mole balance → rate law → stoichiometry → combine)

to calculate the concentration of species C, C_C, as a function of time, t. We then used experimental batch data of C_C versus t to verify the reaction as a first-order reaction and to determine the specific reaction-rate constant, k.

5.3 Continuous-Stirred Tank Reactors (CSTRs)

Continuous-stirred tank reactors (CSTRs), such as the one shown here schematically, are typically used for liquid-phase reactions.

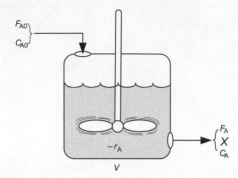

In Chapter 2, we derived the following design equation for a CSTR

Mole balance

$$V = \frac{F_{A0}X}{(-r_A)_{\text{exit}}} \qquad (2\text{-}13)$$

which gives the volume V necessary to achieve a conversion X. As we saw in Chapter 2, the space time, τ, is a characteristic time of a reactor. To obtain the space time, τ, as a function of conversion, we first substitute for $F_{A0} = v_0 C_{A0}$ in Equation (2-13)

$$V = \frac{v_0 C_{A0}X}{(-r_A)_{\text{exit}}} \qquad (5\text{-}6)$$

and then divide by v_0 to obtain the space time, τ, to achieve a conversion X in a CSTR

$$\tau = \frac{V}{v_0} = \frac{C_{A0}X}{(-r_A)_{\text{exit}}} \qquad (5\text{-}7)$$

This equation applies to a single CSTR or to the first reactor of CSTRs connected in series.

5.3.1 A Single CSTR

5.3.1.1   *First-Order Reaction*
Let's consider a first-order irreversible reaction for which the rate law is

Rate law

$$-r_A = kC_A$$

For liquid-phase reactions, there is no volume change during the course of the reaction, so we can use Equation (4-12) to relate concentration and conversion

Stoichiometry

$$C_A = C_{A0}(1 - X) \qquad (4\text{-}12)$$

Combine   We can combine the mole balance equation (5-7), the rate law, and the concentration equation (4-12) to obtain

$$\tau = \frac{1}{k}\left(\frac{X}{1-X}\right)$$

CSTR relationship between space time and conversion for a first-order liquid-phase reaction

Rearranging

$$\boxed{X = \frac{\tau k}{1 + \tau k}} \tag{5-8}$$

A plot of conversion as a function of τk using Equation (5-8) is shown in Figure 5-3.

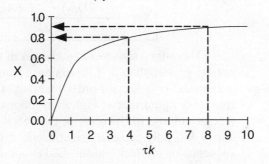

Figure 5-3   First-order reaction in a CSTR.

We can increase τk by either increasing the temperature to increase k or increasing the space time τ by increasing the volume V or decreasing the volumetric flow rate v_0. For example, when we increase the reactor volume, V, by a factor of 2 (or decrease the volumetric flow rate v_0 by a factor of 2) as we go from $\tau k = 4$ to $\tau k = 8$, the conversion only increases from 0.8 to 0.89.

We could also combine Equations (4-12) and (5-8) to find the exit reactor concentration of A, C_A,

$$C_A = \frac{C_{A0}}{1 + \tau k} \tag{5-9}$$

5.3.1.2 A Second-Order Reaction in a CSTR

For a second-order liquid-phase reaction being carried out in a CSTR, the **combination** of the **rate law** and the **design equation** yields

$$V = \frac{F_{A0}X}{-r_A} = \frac{F_{A0}X}{kC_A^2} \tag{5-10}$$

Using our stoichiometric table for constant density $v = v_0$, $C_A = C_{A0}(1 - X)$, and $F_{A0}X = v_0 C_{A0} X$, then

$$V = \frac{v_0 C_{A0} X}{kC_{A0}^2(1 - X)^2}$$

Dividing by v_0

$$\tau = \frac{V}{v_0} = \frac{X}{kC_{A0}(1-X)^2} \tag{5-11}$$

We solve Equation (5-11) for the conversion X

Conversion for
a second-order
liquid-phase
reaction
in a CSTR

$$X = \frac{(1 + 2\tau k C_{A0}) - \sqrt{(1 + 2\tau k C_{A0})^2 - (2\tau k C_{A0})^2}}{2\tau k C_{A0}}$$

$$= \frac{(1 + 2\tau k C_{A0}) - \sqrt{1 + 4\tau k C_{A0}}}{2\tau k C_{A0}}$$

$$\boxed{X = \frac{(1 + 2\text{Da}_2) - \sqrt{1 + 4\text{Da}_2}}{2\text{Da}_2}} \tag{5-12}$$

The minus sign must be chosen in the quadratic equation because X cannot be greater than 1. Conversion is plotted as a function of the Damköhler parameter for a second-order reaction, $\text{Da}_2 = \tau k C_{A0}$, in Figure 5-4. Observe from this figure that at high conversions (say 67%), a 10-fold increase in the reactor volume (or increase in the specific reaction rate by raising the temperature) will only increase the conversion up to 88%. This observation is a consequence of the fact that the CSTR operates under the condition of the lowest reactant concentration (i.e., the exit concentration), and consequently the smallest value of the rate of reaction.

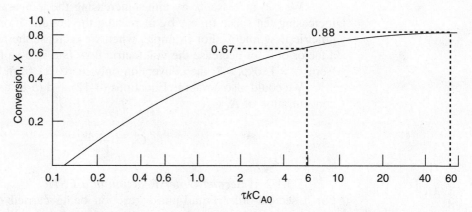

Figure 5-4   Conversion as a function of the Damköhler number $(\tau k C_{A0})$ for a second-order reaction in a CSTR.

5.3.1.3 *The Damköhler Number*

$$\text{Da} = \frac{-r_{A0}V}{F_{A0}}$$

For a first-order reaction, the product τk is often referred to as the reaction **Damköhler number,** Da_1, which is a dimensionless number that can give us a quick estimate of the degree of conversion that can be achieved in continuous-flow reactors. The Damköhler number is the ratio of the rate of reaction of A to the rate of convective transport of A evaluated at the entrance to the reactor.

$$\boxed{\text{Da} = \frac{-r_{A0}V}{F_{A0}} = \frac{\text{Rate of reaction at entrance}}{\text{Entering flow rate of A}} = \frac{\text{``A reaction rate''}}{\text{``A convection rate''}}}$$

The Damköhler number for a first-order irreversible reaction is

$$\text{Da}_1 = \frac{-r_{A0}V}{F_{A0}} = \frac{k_1 C_{A0} V}{v_0 C_{A0}} = \tau k_1$$

For a second-order irreversible reaction, the Damköhler number is

$$\text{Da}_2 = \frac{-r_{A0}V}{F_{A0}} = \frac{k_2 C_{A0}^2 V}{v_0 C_{A0}} = \tau k_2 C_{A0}$$

It is important to know what values of the Damköhler number, Da, give high and low conversion in continuous-flow reactors. For irreversible reactions, a value of Da = 0.1 or less will usually give less than 10% conversion, and a value of Da = 10.0 or greater will usually give greater than 90% conversion; that is, the rule of thumb is

$0.1 \leq \text{Da} \leq 10$

$$\boxed{\begin{array}{l} \text{if Da} < 0.1, \text{ then } X < 0.1 \\ \text{if Da} > 10, \text{ then } X > 0.9 \end{array}}$$

Equation (5-8) for a first-order liquid-phase reaction in a CSTR can also be written in terms of the Damköhler number

$$X = \frac{\text{Da}_1}{1 + \text{Da}_1}$$

5.3.2 CSTRs in Series

A first-order reaction with no change in the volumetric flow rate ($v = v_0$) is to be carried out in two CSTRs placed in series (Figure 5-5).

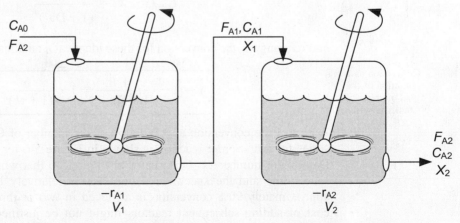

Figure 5-5 Two CSTRs in series.

The effluent concentration of reactant A from the first CSTR can be found using Equation (5-9)

$$C_{A1} = \frac{C_{A0}}{1 + \tau_1 k_1}$$

with $\tau_1 = V_1/v_0$. From a mole balance on reactor 2

$$V_2 = \frac{F_{A1} - F_{A2}}{-r_{A2}} = \frac{v_0(C_{A1} - C_{A2})}{k_2 C_{A2}}$$

Solving for C_{A2}, the concentration exiting the second reactor, we obtain

First-order reaction

$$C_{A2} = \frac{C_{A1}}{1 + \tau_2 k_2} = \frac{C_{A0}}{(1 + \tau_2 k_2)(1 + \tau_1 k_1)} \qquad (5\text{-}13)$$

If both reactors are of equal size ($\tau_1 = \tau_2 = \tau$) and operate at the same temperature ($k_1 = k_2 = k$), then

$$C_{A2} = \frac{C_{A0}}{(1 + \tau k)^2}$$

If, instead of two CSTRs in series, we had n equal-sized CSTRs connected in series ($\tau_1 = \tau_2 = \cdots = \tau_n = \tau_i = (V_i/v_0)$) operating at the same temperature ($k_1 = k_2 = \cdots = k_n = k$), the concentration leaving the last reactor would be

$$C_{An} = \frac{C_{A0}}{(1 + \tau k)^n} = \frac{C_{A0}}{(1 + \mathrm{Da}_1)^n} \qquad (5\text{-}14)$$

Substituting for C_{An} in terms of conversion

CSTRs in series

$$C_{A0}(1 - X) = \frac{C_{A0}}{(1 + \mathrm{Da}_1)^n}$$

and rearranging, the conversion for these identical n tank reactors in series will be

Conversion as a function of the number of tanks in series

$$\boxed{X = 1 - \frac{1}{(1 + \mathrm{Da}_1)^n} \equiv 1 - \frac{1}{(1 + \tau k)^n}} \qquad (5\text{-}15)$$

A plot of the conversion as a function of the number of CSTRs (i.e., tanks) in series for a first-order reaction is shown in Figure 5-6 for various values of the Damköhler number τk. Observe from Figure 5-6 that when the product of the space time and the specific reaction rate is relatively large, say, $\mathrm{Da}_1 \geq 1$, approximately 90% conversion is achieved in two or three reactors; thus, the cost of adding subsequent reactors might not be justified. When the product τk is small, $\mathrm{Da}_1 \sim 0.1$, the conversion continues to increase significantly with each reactor added.

Economics

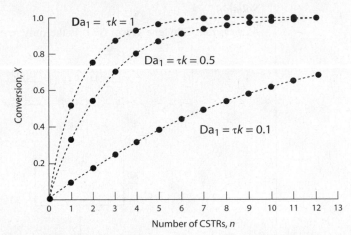

Figure 5-6 Conversion as a function of the number of CSTRs (i.e., tanks) in series for different Damköhler numbers for a first-order reaction.

The rate of disappearance of A in the nth reactor is

$$-r_{An} = kC_{An} = k\,\frac{C_{A0}}{(1 + \tau k)^n} \qquad (5\text{-}16)$$

Uses and
economics

Scale-up
of
batch reactor
data

Example 5–2 Producing 200 Million Pounds per Year in a CSTR

Close to 16 billion pounds of ethylene glycol (EG) were produced in 2013. It previously ranked as the twenty-sixth most produced chemical in the nation on a total pound basis. About one-half of the ethylene glycol is used for *antifreeze,* while the other half is used in the manufacture of polyesters. In the polyester category, 88% was used for fibers and 12% for the manufacture of bottles and films. The 2013 selling price for ethylene glycol was $0.60 per pound.

It is desired to produce 200 million pounds per year of EG. The reactor is to be operated isothermally. A 16.1 mol/dm^3 solution of ethylene oxide (EO) in water is mixed (see Figure E5-2.1) with an equal volumetric solution of water containing 0.9 wt % of the catalyst H_2SO_4 and fed to a CSTR. The specific reaction-rate constant is 0.311 min^{-1}, as determined in Example 5-1. Practical guidelines for reactor scale-up are given by Mukesh.[1]

(a) If 80% conversion is to be achieved, determine the necessary CSTR volume.

(b) If two 800-gal reactors were arranged in parallel with the feed equally divided, what would be the corresponding conversion?

(c) If two 800-gal reactors were arranged in series, what would be the corresponding conversion?

---

[1] D. Mukesh, *Chemical Engineering,* 46 (January 2002), *www.CHE.com.*

Solution

Assumption: Ethylene glycol (EG) is the only reaction product formed.

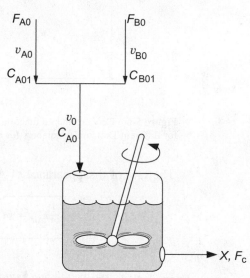

$$A \quad + \quad B \quad \xrightarrow{\text{catalyst}} \quad C$$

Figure E5-2.1   Single CSTR.

The specified ethylene glycol (EG) production rate in mol/s is

$$F_C = 2 \times 10^8 \, \frac{\text{lb}_m}{\text{yr}} \times \frac{1 \, \text{yr}}{365 \, \text{days}} \times \frac{1 \, \text{day}}{24 \, \text{h}} \times \frac{1 \, \text{h}}{3600 \, \text{s}} \times \frac{454 \, \text{g}}{\text{lb}_m} \times \frac{1 \, \text{mol}}{62 \, \text{g}} = 46.4 \, \frac{\text{mol}}{\text{sec}}$$

From the reaction stoichiometry

$$F_C = F_{A0} X$$

we find the required molar flow rate of ethylene oxide for 80% conversion to be

$$F_{A0} = \frac{F_C}{X} = \frac{46.4 \ \text{mol/s}}{0.8} = 58.0 \, \text{mol/s}$$

(a) We now calculate the single CSTR volume to achieve 80% conversion using the **CRE algorithm**.

1. **CSTR Mole Balance:**

$$V = \frac{F_{A0} X}{-r_A} \tag{E5-2.1}$$

2. **Rate Law:**

$$-r_A = k C_A \tag{E5-2.2}$$

Following the Algorithm

3. **Stoichiometry:** Liquid phase $(v = v_0)$:

$$C_A = \frac{F_A}{v_0} = \frac{F_{A0}(1-X)}{v_0} = C_{A0}(1-X) \qquad \text{(E5-2.3)}$$

4. **Combining:**

$$V = \frac{F_{A0}X}{kC_{A0}(1-X)} = \frac{v_0 X}{k(1-X)} \qquad \text{(E5-2.4)}$$

5. **Evaluate:**

The entering volumetric flow rate of stream A, with $C_{A01} = 16.1$ mol/dm^3 before mixing, is

$$v_{A0} = \frac{F_{A0}}{C_{A01}} = \frac{58 \text{ mol/s}}{16.1 \text{ mol/dm}^3} = 3.6 \frac{\text{dm}^3}{\text{s}}$$

From the problem statement $v_{B0} = v_{A0}$

$$F_{B0} = v_{B0}C_{B01} = 3.62\frac{\text{dm}^3}{\text{s}} \times \left[\frac{1,000\text{g}}{\text{dm}^3} \times \frac{1\text{mol}}{18\text{g}}\right] = 201\frac{\text{mol}}{\text{s}}$$

The total entering volumetric flow rate of liquid is

$$v_0 = v_{A0} + v_{B0} = 3.62\frac{\text{dm}^3}{\text{s}} + 3.62\frac{\text{dm}^3}{\text{s}} = 7.2\frac{\text{dm}^3}{\text{s}}$$

Substituting in Equation (E5-2.4), recalling that $k = 0.311$ min^{-1}, yields

$$k_1 = \frac{0.311}{min} \times \frac{1\text{min}}{60\text{s}} = \frac{0.0052}{\text{s}}$$

$$V = \frac{v_0 X}{k(1-X)} = \frac{7.2\,\text{dm}^3/\text{s}}{0.0052/\text{s}}\frac{0.8}{1-0.8} = 5538 \text{ dm}^3$$

$$\boxed{V = 5.538\,\text{m}^3 = 197\,\text{ft}^3 = 1463 \text{ gal}}$$

A tank 5 ft in diameter and approximately 10 ft tall is necessary to achieve 80% conversion.

(b) **CSTRs in parallel.** What would you guess would happen if two 800-gal CSTRs arranged in parallel as shown in Figure E5-2.2 had a volumetric flow rate $v_0 = 3.62$ dm^3/s fed to each reactor? Would the conversion increase, decrease, or remain the same? To find out, let's begin by rearranging Equation (E5-2.4)

$$\frac{V}{v_0}k = \tau k = \frac{X}{1-X}$$

to obtain

$$X = \frac{\tau k}{1 + \tau k} \qquad \text{(E5-2.5)}$$

where

$$\tau = \frac{V}{v_0/2} = 800\text{gal} \times \frac{3.785\text{dm}^3}{\text{gal}}\frac{1}{3.62\,\text{dm}^3/\text{s}} = 836.5\text{s}$$

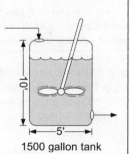

1500 gallon tank

The ***Damköhler number*** for a first-order reaction is

$$Da_1 = \tau k = 836.5\,\text{s} \times 0.0052\,\text{s}^{-1} = 4.35$$

Substituting into Equation (E5-2.5) gives us

$$X = \frac{Da_1}{1 + Da_1} = \frac{4.35}{1 + 4.35} = 0.81$$

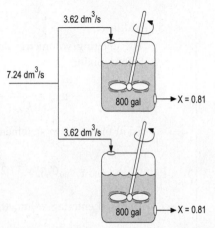

Figure E5-2.2   CSTRs in parallel.

The conversion exiting both of the CSTRs in parallel is 81%.

Problem P5-2(b) asks you to generalize the result for n equal-size reactors V_i in parallel with equal feed rates (F_{A0}/n) and show that the conversion would also be the same if everything were fed to one big reactor of volume $V = nV_i$

(c) CSTRs in series. Recalling what we found in Chapter 2 when we sequenced reactors, what do you guess will happen to the conversion in Part **(a)** if the reactors are placed in series? Will it increase or decrease? If the 800-gal reactors are arranged in series, the conversion in the *first reactor* [cf. Equation (E5-2.5)] is

$$X_1 = \frac{\tau_1 k}{1 + \tau_1 k} \tag{E5-2.6}$$

where

$$\tau = \frac{V_1}{v_0} = \left(800\ \text{gal} \times \frac{3.785\ \text{dm}^3}{\text{gal}}\right) \times \frac{1}{7.24\ \text{dm}^3/\text{s}} = 418.2\ \text{s}$$

First CSTR

The Damköhler number is

$$Da_1 = \tau_1 k = 418.2\ \text{s} \times \frac{0.0052}{\text{s}} = 2.167$$

$$X_1 = \frac{2.167}{1 + 2.167} = \frac{2.167}{3.167} = 0.684$$

To calculate the conversion exiting the second reactor, we recall that $V_1 = V_2 = V$ and $v_{01} = v_{02} = v_0$; then

$$\tau_1 = \tau_2 = \tau$$

Conversion in the series arrangement is greater than in parallel for CSTRs. From our discussion of reactor staging in Chapter 2, we could have predicted that the series arrangement would have given the higher conversion.

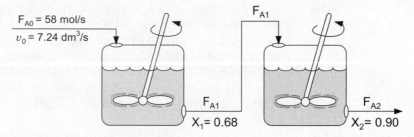

Figure E5-2.3   CSTRs in series.

A mole balance on the *second reactor* is

$$\text{In}\quad-\quad\text{Out}\quad+\quad\text{Generation}\quad=\quad0$$

$$\overbrace{F_{A1}}\quad-\quad\overbrace{F_{A2}}\quad+\quad\overbrace{r_{A2}V}\quad=\quad0$$

Basing the conversion on the total number of moles reacted up to a point per mole of A fed to the first reactor

Second CSTR

$$F_{A1} = F_{A0}(1-X_1)\quad\text{and}\quad F_{A2} = F_{A0}(1-X_2)$$

Rearranging

$$V = \frac{F_{A1}-F_{A2}}{-r_{A2}} = F_{A0}\frac{X_2-X_1}{-r_{A2}}$$

$$-r_{A2} = kC_{A2} = k\frac{F_{A2}}{v_0} = \frac{kF_{A0}(1-X_2)}{v_0} = kC_{A0}(1-X_2)$$

Combining the mole balance on the second reactor [cf. Equation (2-24)] with the rate law, we obtain

$$V = \frac{F_{A0}(X_2-X_1)}{-r_{A2}} = \frac{C_{A0}v_0(X_2-X_1)}{kC_{A0}(1-X_2)} = \frac{v_0}{k}\left(\frac{X_2-X_1}{1-X_2}\right) \qquad \text{(E5-2.7)}$$

Solving for the conversion exiting the second reactor yields

$$X_2 = \frac{X_1 + Da_1}{1 + Da_1} = \frac{X_1 + \tau k}{1 + \tau k} = \frac{0.684 + 2.167}{1 + 2.167} = 0.90$$

The same result could have been obtained from Equation (5-15)

$$X_2 = 1 - \frac{1}{(1+\tau k)^n} = 1 - \frac{1}{(1+2.167)^2} = 0.90$$

Over two hundred million pounds of EG per year can be produced using two 800-gal (3.0-m^3) reactors in series.

Analysis: The CRE algorithm was applied to a first-order irreversible liquid-phase reaction carried out isothermally in a single CSTR, two CSTRs in series, and also two CSTRs in parallel. The equations were solved algebraically for each case. When the entering molar flow rate was equally divided between the two CSTRs in parallel, the overall conversion was the same as that for a single CSTR. For two CSTRs in series, the overall conversion was greater than that of a single CSTR. This result will always be the case for isothermal reactions with power-law rate laws with reaction orders greater than zero.

Safety considerations   We can find information online about the safety of ethylene glycol and other chemicals from Table 5-4. One source is the Vermont Safety Information Resources, Inc., Web site (Vermont SIRI, *www.siri.org*). For example, we can learn from the *Control Measures* that we should use neoprene gloves when handling the material, and that we should avoid breathing the vapors. If we click on "Dow Chemical USA" and scroll the *Reactivity Data*, we would find that ethylene glycol will ignite in air at 413°C.

Safety information MSDS

TABLE 5-4   ACCESSING SAFETY INFORMATION
_____

1. Type in: *www.avantormaterials.com*
2. When the first screen appears, go to Products in the upper left of the screen, then drag down to MSDS for material safety data sheet.
3. When the next page appears, type in the chemical you want to find.

    Example: Search Text Ethylene Glycol
4. Next page shows the list of companies with product number that provide the data on ethylene glycol.

| 5001-19 Macron Fine Chemicals | Ethylene Glycol | 107-21-1 | VIEW SDS |
| 5387-03 J.T.Baker | Ethylene Glycol | 107-21-1 | VIEW SDS |

Let's click on view MSDS for Macron Fine Chemicals. The material safety data sheet provided will appear.

5. Scroll down for information you desire.

    1.   *Product Identification*
    2.   *Composition/Information on Ingredients*
    3.   *Hazards Identification*
    4.   *First Aid Measures*
    5.   *Fire Fighting Measures*
    6.   *Accidental Release Measures*
    7.   *Handling and Storage*
    8.   *Exposure Controls/Personal Protection*
    9.   *Physical and Chemical Properties*
    10-16. *Other Information*
_____

5.4 Tubular Reactors

Gas-phase reactions are carried out primarily in tubular reactors where the flow is generally turbulent. By assuming that there is no dispersion and there are no radial gradients in either temperature, velocity, concentration, or reaction rate, we can model the flow in the reactor as plug flow.[2]

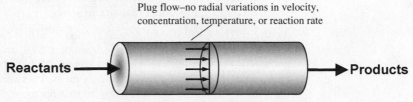

Plug flow–no radial variations in velocity, concentration, temperature, or reaction rate

Reactants → Products

Figure 1-9   (Revisited) tubular reactor.

_____

[2] Laminar flow reactors (LFTs) and dispersion effects are discussed in Chapter 17. As a general rule, the conversion calculated for a PFR will not be significantly different than that for an LFR.

The *differential form* of the PFR design equation such as

Use this differential
form of the
PFR/PBR **mole
balances** when
there is ∆P.

$$F_{A0} \frac{dX}{dV} = -r_A \tag{2-15}$$

must be used when there is a pressure drop in the reactor or heat exchange between the PFR and the surroundings. In the absence of pressure drop or heat exchange, the integral form of the *plug-flow design* equation can be used,

$$V = F_{A0} \int_0^X \frac{dX}{-r_A} \tag{2-16}$$

As an example, consider the elementary reaction

$$2A \longrightarrow \text{Products}$$

for which the rate law is

Rate Law

$$-r_A = kC_A^2$$

We shall first consider the reaction to take place as a liquid-phase reaction and then as a gas-phase reaction.

Liquid Phase $v = v_0$

The combined PFR mole balance and rate law is

$$\frac{dX}{dV} = \frac{kC_A^2}{F_{A0}}$$

If the reaction is carried out in the liquid phase, the concentration of A is

Stoichiometry
(liquid phase)

$$C_A = C_{A0}(1 - X)$$

and for isothermal operation, we can bring k outside the integral

Combine

$$V = \frac{F_{A0}}{kC_{A0}^2} \int_0^X \frac{dX}{(1-X)^2} = \frac{v_0}{kC_{A0}}\left(\frac{X}{1-X}\right)$$

This equation gives the reactor volume to achieve a conversion X. Dividing by v_0 ($\tau = V/v_0$) and solving for conversion, we find

Evaluate

$$\boxed{X = \frac{\tau k C_{A0}}{1 + \tau k C_{A0}} = \frac{\text{Da}_2}{1 + \text{Da}_2}}$$

where Da_2 is the Damköhler number for a second-order reaction, i.e. $\tau k C_{A0}$.

Gas Phase $v = v_0 (1 + \varepsilon X)(T/T_0)(P_0/P)$

For constant-temperature ($T = T_0$) and constant-pressure ($P = P_0$) *gas-phase reactions,* the concentration is expressed as a function of conversion

Stoichiometry
(gas phase)

$$C_A = \frac{F_A}{v} = \frac{F_A}{v_0(1+\varepsilon X)} = \frac{F_{A0}(1-X)}{v_0(1+\varepsilon X)} = C_{A0}\frac{(1-X)}{(1+\varepsilon X)}$$

and then combining the PFR mole balance, rate law, and stoichiometry

Combine

$$V = F_{A0}\int_0^X \frac{(1+\varepsilon X)^2}{kC_{A0}^2(1-X)^2}\,dX$$

The entering concentration C_{A0} can be taken outside the integral sign since it is not a function of conversion. Because the reaction is carried out isothermally, the specific reaction rate constant, k, can also be taken outside the integral sign.

For an isothermal reaction, k is constant.

$$V = \frac{F_{A0}}{kC_{A0}^2}\int_0^X \frac{(1+\varepsilon X)^2}{(1-X)^2}\,dX$$

From the integral equations in Appendix A.1, we find that

Evaluate
Reactor volume for a second-order gas-phase reaction

$$V = \frac{v_0}{kC_{A0}}\left[2\varepsilon(1+\varepsilon)\ln(1-X) + \varepsilon^2 X + \frac{(1+\varepsilon)^2 X}{1-X}\right] \qquad (5\text{-}17)$$

Effect of ε on Conversion

We now look at the effect of the change in the total number of moles in the gas phase on the relationship between conversion and volume. For constant temperature and pressure, Equation (4-23) becomes

$$v = v_0(1+\varepsilon X)$$

Let's now consider three types of reactions, one in which $\varepsilon = 0$ ($\delta = 0$), one in which $\varepsilon < 0$ ($\delta < 0$), and one in which $\varepsilon > 0$ ($\delta > 0$). When there is no change in the number of moles with reaction, (e.g., A $\rightarrow$ B) $\delta = 0$ and $\varepsilon = 0$, then the fluid moves through the reactor at a constant volumetric flow rate ($v = v_0$) as the conversion increases.

When there is a decrease in the number of moles ($\delta < 0$, $\varepsilon < 0$) in the gas phase, the volumetric gas flow rate decreases and the conversion increases. For example, when pure A enters for the reaction 2A $\rightarrow$ B, and taking A as the basis of calculation, then A $\rightarrow$ B/2 and we have $\varepsilon = y_{A0}\delta = 1(\tfrac{1}{2}-1) = -0.5$

$$v = v_0(1-0.5X)$$

Consequently, the gas molecules will spend more time in the reactor than they would if the flow rate were constant, $v = v_0$. As a result, this longer residence time would result in a higher conversion than if the flow were constant at v_0.

On the other hand, if there is an increase in the total number of moles ($\delta > 0$, $\varepsilon > 0$) in the gas phase, then the volumetric flow rate will increase as the conversion increases. For example, for the reaction A $\rightarrow$ 2B, then $\varepsilon = y_{A0}\delta = 1(2-1) = 1$

$$v = v_0 (1 + X)$$

and the molecules will spend less time in the reactor than they would if the volumetric flow rate were constant. As a result of this smaller residence time in the reactor, the conversion will be less than what would result if the volumetric flow rate were constant at v_0.

The importance of changes in volumetric flow rate (i.e., $\varepsilon \neq 0$) with reaction

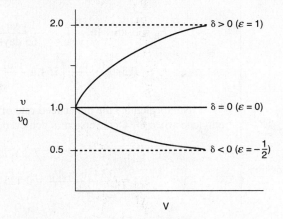

Figure 5-7   Change in gas-phase volumetric flow rate down the length of the reactor.

Figure 5-7 shows the volumetric flow rate profiles for the three cases just discussed. We note that, at the end of the reactor, virtually complete conversion has been achieved.

Example 5–3   Producing 300 Million Pounds per Year of Ethylene in a Plug-Flow Reactor: Design of a Full-Scale Tubular Reactor

The economics

The uses

Ethylene ranks first in the United States in total pounds of organic chemicals produced each year, and it is the number-one organic chemical produced each year. Over 60 billion pounds were produced in 2010, and it sold for $0.37 per pound. Sixty-five percent of the ethylene produced is used in the manufacture of fabricated plastics, 20% for ethylene oxide, 16% for ethylene dichloride and ethylene glycol, 5% for fibers, and 5% for solvents.

Determine the plug-flow reactor volume necessary to produce 300 million pounds of ethylene a year by cracking a feed stream of pure ethane. The reaction is irreversible and follows an elementary rate law. We want to achieve 80% conversion of ethane, operating the reactor isothermally at 1100 K and at a pressure of 6 atm. The specific reaction rate at 1,000 K is 0.072 s^{-1} and the activation energy is 82,000 cal/mol.

Solution

$$C_2H_6 \longrightarrow C_2H_4 + H_2$$

Let A = C_2H_6, B = C_2H_4, and C = H_2. In symbols,

$$A \longrightarrow B + C$$

Because we want the reader to be familiar with both metric units **and** English units, we will work some of the examples using English units. Trust me, a number of old timers still use concentrations in lb-mol/ft³. To help you relate English and metric units, the corresponding metric units will be given in parenthesis next to the English units. The only step in the algorithm that is different is the evaluation step.

The molar flow rate of ethylene exiting the reactor is

$$F_B = 300 \times 10^6 \frac{\text{lb}_m}{\text{year}} \times \frac{1 \text{ year}}{365 \text{ days}} \times \frac{1 \text{ day}}{24 \text{ h}} \times \frac{1 \text{ h}}{3600 \text{ s}} \times \frac{\text{lb-mol}}{28 \text{ lb}_m}$$

$$= 0.340 \frac{\text{lb-mol}}{\text{s}} \left(154.4 \frac{\text{mol}}{\text{s}} \right)$$

Next, calculate the molar feed rate of ethane, F_{A0}, to produce 0.34 lb mol/s of ethylene when 80% conversion is achieved,

$$F_B = F_{A0} X$$

$$F_{A0} = \frac{0.34 \text{ lb mol/s}}{0.8} = 0.425 \frac{\text{lb-mol}}{\text{s}} \quad (F_{A0} = 193 \text{ mol/s})$$

Following the Algorithm

1. **Plug-Flow Mole Balance:**

Mole balance

$$F_{A0} \frac{dX}{dV} = -r_A \tag{2-15}$$

Rearranging and integrating for the case of **no pressure drop** and isothermal operation yields

$$\boxed{V = F_{A0} \int_0^X \frac{dX}{-r_A}} \tag{E5-3.1}$$

2. **Rate Law:**[3]

Rate Law

$$\boxed{-r_A = kC_A} \quad \text{with} \quad k = 0.072 \text{ s}^{-1} \text{ at } 1000 \text{ K} \tag{E5-3.2}$$

The activation energy is 82 kcal/g mol.

3. **Stoichiometry:** For isothermal operation and negligible pressure drop, the concentration of ethane is calculated as follows:

Gas phase, constant T and P:

Stoichiometry

$$v = v_0 \frac{F_T}{F_{T0}} = v_0 (1 + \varepsilon X)$$

$$\boxed{C_A = \frac{F_A}{v} = \frac{F_{A0}(1-X)}{v_0(1+\varepsilon X)} = C_{A0} \left(\frac{1-X}{1+\varepsilon X} \right)} \tag{E5-3.3}$$

[3] *Ind. Eng. Chem. Process Des. Dev.*, 14, 218 (1975); *Ind. Eng. Chem.*, 59(5), 70 (1967).

$$C_C = \frac{C_{A0}X}{(1 + \varepsilon X)} \tag{E5-3.4}$$

4. **Combine** Equations (E5-3.1) through (E5-3.3) to obtain

Combining the design equation, rate law, and stoichiometry

$$V = F_{A0} \int_0^X \frac{dX}{kC_{A0}(1-X)/(1+\varepsilon X)} = F_{A0} \int_0^X \frac{(1 + \varepsilon X)\,dX}{kC_{A0}(1-X)}$$

$$= \frac{F_{A0}}{C_{A0}} \int_0^X \frac{(1 + \varepsilon X)\,dX}{k(1-X)} \tag{E5-3.5}$$

5. **Evaluate:**

Since the reaction is carried out isothermally, we can take k outside the integral sign and use Appendix A.1 to carry out our integration.

Analytical solution

$$V = \frac{F_{A0}}{kC_{A0}} \int_0^X \frac{(1 + \varepsilon X)\,dX}{1-X} = \frac{F_{A0}}{kC_{A0}}\left[(1+\varepsilon)\ln\frac{1}{1-X} - \varepsilon X\right] \tag{E5-3.6}$$

6. **Parameter evaluation:**

Evaluate

$$C_{A0} = y_{A0}C_{T0} = \frac{y_{A0}P_0}{RT_0} = (1.0)\left(\frac{6\ \text{atm}}{(0.73\ \text{ft}^3 \cdot \text{atm/lb-mol} \cdot \text{°R}) \times (1980\text{°R})}\right)$$

$$= 0.00415\ \frac{\text{lb-mol}}{\text{ft}^3}\quad (0.066\ \text{mol/dm}^3)$$

$$\varepsilon = y_{A0}\delta = (1)(1 + 1 - 1) = 1$$

Oops! The rate constant k is given at 1000 K, and we need to calculate k at reaction conditions, which is 1100 K.

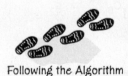

Following the Algorithm

$$k(T_2) = k(T_1)\exp\left[\frac{E}{R}\left(\frac{1}{T_1} - \frac{1}{T_2}\right)\right]$$

$$= k(T_1)\exp\left[\frac{E}{R}\left(\frac{T_2 - T_1}{T_1 T_2}\right)\right] \tag{E5-3.7}$$

$$= \frac{0.072}{\text{s}}\exp\left[\frac{82,000\ \text{cal/mol}(1100 - 1000)\ \text{K}}{1.987\ \text{cal/(mol} \cdot \text{K})(1000\ \text{K})(1100\ \text{K})}\right]$$

$$= 3.07\ \text{s}^{-1}$$

Substituting into Equation (E5-3.6) yields

$$V = \frac{0.425\ \text{lb mol/s}}{(3.07/\text{s})(0.00415\ \text{lb-mol/ft}^3)}\left[(1+1)\ln\frac{1}{1-X} - (1)X\right] \tag{E5-3.8}$$

$$= 33.36\ \text{ft}^3\left[2\ln\left(\frac{1}{1-X}\right) - X\right]$$

For $X = 0.8$,

$$V = 33.36 \text{ ft}^3 \left[2 \ln \left(\frac{1}{1 - 0.8} \right) - 0.8 \right]$$

$$= 80.7 \text{ ft}^3 = (2280 \text{ dm}^3 = 2.28 \text{ m}^3)$$

It was decided to use a bank of 2-inch schedule 80 pipes in parallel that are 40 feet in length. For schedule 80 pipe, the cross-sectional area, A_C, is 0.0205 ft². The number of pipes necessary is

The number of PFRs in parallel

100 pipes in parallel

$$n = \frac{80.7 \text{ ft}^3}{(0.0205 \text{ ft}^2)(40 \text{ ft})} = 98.4 \qquad \text{(E5-3.9)}$$

To determine the concentrations and conversion profiles down the length of the reactor, z, we divide the volume equation (E5-3.8) by the cross-sectional area, A_C,

$$z = \frac{V}{A_C} \qquad \text{(E5-3.10)}$$

Equation (E5-3.9) was used along with $A_C = 0.0205$ ft², and Equations (E5-3.8) and (E5-3.3) were used to obtain Figure E5-3.1. Using a bank of 100 pipes will give us the reactor volume necessary to make 300 million pounds per year of ethylene from ethane. The concentration and conversion profiles down any one of the pipes are shown in Figure E5-3.1.

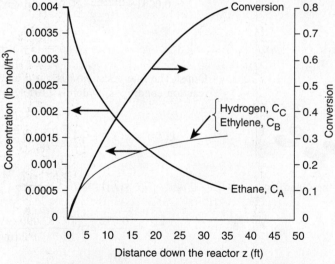

Figure E5-3.1   Conversion and concentration profiles.

Analysis: The CRE algorithm was applied to a gas-phase reaction that had a change in the total number of moles during the reaction. A bank of 100 PFRs in parallel, each with a volume of 0.81 ft³, will give the same conversion as 1 PFR with a volume of 81 ft³. The conversion and concentration profiles are shown in Figure E5-3.1. You will note that the profiles change more rapidly near the entrance to the reactor where the reactant concentrations are high and change more slowly near the exit where most of the reactants have been consumed, resulting in a smaller rate of reaction.

5.5 Pressure Drop in Reactors

Pressure drop is ignored for liquid-phase kinetics calculations.

In liquid-phase reactions, the concentration of reactants is insignificantly affected by even relatively large changes in the total pressure. Consequently, we can totally ignore the effect of pressure drop on the rate of reaction when sizing liquid-phase chemical reactors. However, in gas-phase reactions, the concentration of the reacting species is proportional to the total pressure; therefore, proper accounting for the effects of pressure drop on the reaction system can, in many instances, be a key factor in the success or failure of the reactor operation. This fact is especially true in microreactors packed with solid catalyst. Here, the channels are so small (see Problem 5-21$_B$) that pressure drop can limit the throughput and conversion for gas-phase reactions.

5.5.1 Pressure Drop and the Rate Law

For gas-phase reactions, pressure drop may be very important.

We now focus our attention on accounting for the pressure drop in the rate law. For an ideal gas, we recall Equation (4-25) to write the concentration of reacting species i as

$$C_i = C_{A0} \left(\frac{\Theta_i + v_i X}{1 + \varepsilon X} \right) \frac{P}{P_0} \frac{T_0}{T} \tag{5-18}$$

where $\Theta_i = \dfrac{F_{i0}}{F_{A0}}$, $\varepsilon = y_{A0}\delta$, and v_i is the stoichiometric coefficient (e.g., $v_A = -1$, $v_B = -b/a$). We now must determine the ratio pressure (P/P_0) as a function of the PFR reactor volume, V, or the PBR catalyst weight, W, to account for pressure drop. We then can combine the concentration, rate law, and design equation. However, whenever accounting for the effects of pressure drop, *the differential form of the mole balance (design equation) must be used*.

If, for example, the second-order reaction

$$2A \longrightarrow B + C$$

When $P \neq P_0$, one must use the differential forms of the PFR/PBR design equations.

is being carried out in a packed-bed reactor, the **differential form of the mole balance** equation in terms of catalyst weight is

$$F_{A0} \frac{dX}{dW} = -r'_A \qquad \left(\frac{\text{Gram moles}}{\text{Gram catalyst} \cdot \text{min}} \right) \tag{2-17}$$

The **rate law** is

$$-r'_A = kC_A^2 \tag{5-19}$$

From **stoichiometry** for gas-phase reactions (Table 3-5)

$$C_A = \frac{C_{A0}(1 - X)}{1 + \varepsilon X} \frac{P}{P_0} \frac{T_0}{T}$$

and the rate law can be written as

$$-r_A' = k \left[\frac{C_{A0}(1-X)}{1+\varepsilon X} \frac{P}{P_0} \frac{T_0}{T} \right]^2 \tag{5-20}$$

Note from Equation (5-20) that the larger the pressure drop (i.e., the smaller P) from frictional losses, the smaller the reaction rate!

Combining Equation (5-20) with the mole balance (2-17) and assuming isothermal operation ($T = T_0$) gives

$$F_{A0} \frac{dX}{dW} = k \left[\frac{C_{A0}(1-X)}{1+\varepsilon X} \right]^2 \left(\frac{P}{P_0} \right)^2$$

Dividing by F_{A0} (i.e., $v_0 C_{A0}$) yields

$$\frac{dX}{dW} = \frac{k C_{A0}}{v_0} \left(\frac{1-X}{1+\varepsilon X} \right)^2 \left(\frac{P}{P_0} \right)^2$$

For isothermal operation ($T = T_0$), the right-hand side is a function of only conversion and pressure

Another equation
is needed
(e.g., $P = f(W)$).

$$\frac{dX}{dW} = F_1(X, P) \tag{5-21}$$

We now need to relate the pressure drop to the catalyst weight in order to determine the conversion as a function of catalyst weight (i.e., catalyst mass).

5.5.2 Flow Through a Packed Bed

The majority of gas-phase reactions are catalyzed by passing the reactant through a packed bed of catalyst particles.

The equation used most often to calculate pressure drop in a packed porous bed is the **Ergun equation**:[4,5]

---

[4] R. B. Bird, W. E. Stewart, and E. N. Lightfoot, *Transport Phenomena,* 2nd ed. (New York: Wiley, 2002), p. 191.

[5] A slightly different set of constants for the Ergun Equation (e.g., 1.8G instead of 1.75G) can be found in *Ind. Eng. Chem. Fundamentals,* 18 (1979), p. 199.

Ergun equation

$$\frac{dP}{dz} = -\frac{G}{\rho g_c D_P}\left(\frac{1-\phi}{\phi^3}\right)\left[\overbrace{\frac{150(1-\phi)\mu}{D_P}}^{\text{Term 1}} + \overbrace{1.75G}^{\text{Term 2}}\right] \qquad (5\text{-}22)$$

Term 1 is dominant for laminar flow, and Term 2 is dominant for turbulent flow, where

P = pressure, lb_f/ft^2 or (kPa)

ϕ = porosity = $\dfrac{\text{volume of void}}{\text{total bed volume}}$ = void fraction

$1 - \phi = \dfrac{\text{volume of solid}}{\text{total bed volume}}$

g_c = 32.174 $\text{lb}_m \cdot \text{ft/s}^2 \cdot \text{lb}_f$ (conversion factor)

   = 4.17×10^8 $\text{lb}_m \cdot \text{ft/h}^2 \cdot \text{lb}_f$

(Recall that for the metric system $g_c = 1.0$)

D_P = diameter of particle in the bed, ft or (m)

μ = viscosity of gas passing through the bed, $\text{lb}_m/\text{ft} \cdot \text{h}$ or (kg/m $\cdot$ s)

z = length down the packed bed of pipe, ft or (m)

u = superficial velocity = volumetric flow rate $\div$ cross-sectional
   area of pipe, ft/h or (m/s)

ρ = gas density, lb_m/ft^3 or (kg/m^3)

$G = \rho u$ = superficial mass velocity, $(\text{lb}_m/\text{ft}^2 \cdot \text{h})$ or (kg/m$^2 \cdot$ s)

In calculating the pressure drop using the Ergun equation, the only parameter that varies with pressure on the right-hand side of Equation (5-22) is the gas density, ρ. We are now going to calculate the pressure drop through a packed-bed reactor.

Because the PBR is operated at steady state, the mass flow rate at any point down the reactor, $\dot{m}$ (kg/s), is equal to the entering mass flow rate, $\dot{m}_0$ (i.e., equation of continuity)

$$\dot{m}_0 = \dot{m}$$

$$\rho_0 v_0 = \rho v$$

Recalling Equation (4-16), we have

$$v = v_0 \frac{P_0}{P}\left(\frac{T}{T_0}\right)\frac{F_T}{F_{T0}} \qquad (4\text{-}16)$$

$$\rho = \rho_0 \frac{v_0}{v} = \rho_0 \frac{P}{P_0}\left(\frac{T_0}{T}\right)\frac{F_{T0}}{F_T} \qquad (5\text{-}23)$$

Combining Equations (5-22) and (5-23) gives

$$\frac{dP}{dz} = - \underbrace{\frac{G(1-\phi)}{\rho_0 g_c D_P \phi^3} \left[\frac{150(1-\phi)\mu}{D_P} + 1.75G \right]}_{\beta_0} \frac{P_0}{P} \left(\frac{T}{T_0} \right) \frac{F_T}{F_{T0}}$$

Simplifying yields

$$\boxed{\frac{dP}{dz} = -\beta_0 \frac{P_0}{P} \left(\frac{T}{T_0} \right) \frac{F_T}{F_{T0}}} \tag{5-24}$$

where β_0 is a constant that depends only on the properties of the packed bed (ϕ, D_P) and the fluid properties at the entrance conditions (i.e., μ, G, ρ_0, T_0, P_0). Typical units of β_0 are (atm/ft) or (Pa/m).

$$\boxed{\beta_0 = \frac{G(1-\phi)}{\rho_0 g_c D_P \phi^3} \left[\frac{150(1-\phi)\mu}{D_P} + 1.75G \right] \quad \left(\text{e.g., } \frac{\text{kPa}}{\text{m}}, \frac{\text{atm}}{\text{ft}} \right)} \tag{5-25}$$

For tubular packed-bed reactors, we are more interested in catalyst weight rather than the distance z down the reactor. The catalyst weight up to a distance of z down the reactor is

$$\underbrace{W}_{\begin{bmatrix} \text{Weight of} \\ \text{catalyst} \end{bmatrix}} = \underbrace{(1-\phi)A_c z}_{\begin{bmatrix} \text{Volume of} \\ \text{solids} \end{bmatrix}} \times \underbrace{\rho_c}_{\begin{bmatrix} \text{Density of} \\ \text{solid catalyst} \end{bmatrix}} \tag{5-26}$$

where A_c is the cross-sectional area. The *bulk density* of the catalyst, ρ_b (mass of catalyst per volume of reactor bed), is just the product of the density of the solid catalyst particles, ρ_c, and the fraction of solids, $(1-\phi)$:

Bulk density
$$\rho_b = \rho_c (1-\phi)$$

Using the relationship between z and W [Equation (5-26)], we can change our variables to express the Ergun equation in terms of catalyst weight:

Use this form for multiple reactions and membrane reactors.

$$\frac{dP}{dW} = -\frac{\beta_0}{A_c(1-\phi)\rho_c} \frac{P_0}{P} \left(\frac{T}{T_0} \right) \frac{F_T}{F_{T0}}$$

Further simplification yields

$$\frac{dP}{dW} = -\frac{\alpha}{2} \frac{T}{T_0} \frac{P_0}{P/P_0} \left(\frac{F_T}{F_{T0}} \right) \tag{5-27}$$

Let $p = (P / P_0)$, then

Used for
multiple reactions

$$\frac{dp}{dW} = -\frac{\alpha}{2p} \frac{T}{T_0} \frac{F_T}{F_{T0}}$$

(5-28)

where

$$\alpha = \frac{2\beta_0}{A_c \rho_c (1 - \phi) P_0}$$

(5-29)

and where typical units of α might be (kg^{-1}) or (lb_m^{-1}).

Differential form of
Ergun equation for
the pressure drop in
packed beds

We will use Equation (5-28) when multiple reactions are occurring or when there is pressure drop in a membrane reactor. However, for single reactions in packed-bed reactors, it is more convenient to express the Ergun equation in terms of the conversion X. Recalling Equation (4-20) for F_T,

$$\frac{F_T}{F_{T0}} = 1 + \varepsilon X$$

(4-20)

where, as before

$$\varepsilon = y_{A0}\delta = \frac{F_{A0}}{F_{T0}}\delta$$

(4-22)

Substituting for the ratio (F_T/F_{T0}), Equation (5-28) can now be written as

Use for
single reactions

$$\frac{dp}{dW} = -\frac{\alpha}{2p}(1 + \varepsilon X)\frac{T}{T_0}$$

(5-30)

We note that when ε is negative, the pressure drop ΔP will be less (i.e., higher pressure) than that for $\varepsilon = 0$. When ε is positive, the pressure drop ΔP will be greater than when $\varepsilon = 0$.

For isothermal operation, Equation (5-30) is only a function of conversion and pressure

$$\frac{dp}{dW} = F_2(X, p)$$

(5-31)

Two coupled
equations to be
solved numerically

Recalling Equation (5-21), for the combined mole balance, rate law, and stoichiometry[†]

$$\frac{dX}{dW} = F_1(X, p)$$

(5-21)

we see that we have two coupled first-order differential equations, (5-31) and (5-21), that must be solved simultaneously. A variety of software packages (e.g., Polymath) and numerical integration schemes are available for this purpose.

[†] For relaxation, see YouTube video on pressure drop, "Chemical Engineering Gone Wrong," accessible through the CRE Web site home page.

Analytical Solution. If $\varepsilon = 0$, *or* if we can neglect (εX) with respect to 1.0 (i.e., $1 \gg \varepsilon X$), we can obtain an analytical solution to Equation (5-30) for isothermal operation (i.e., $T = T_0$). For isothermal operation with $\varepsilon = 0$, Equation (5-30) becomes

Isothermal with
$\varepsilon = 0$

$$\frac{dp}{dW} = -\frac{\alpha}{2p} \tag{5-32}$$

Rearranging gives

$$\frac{2p\,dp}{dW} = -\alpha$$

Taking p inside the derivative, we have

$$\frac{dp^2}{dW} = -\alpha$$

Integrating with $p = 1$ ($P = P_0$) at $W = 0$ yields

$$p^2 = (1 - \alpha W)$$

Taking the square root of both sides gives

Pressure ratio
only for $\varepsilon = 0$
and isothermal

$$\boxed{p = \frac{P}{P_0} = (1 - \alpha W)^{1/2}} \tag{5-33}$$

Caution

Be sure ***not*** to use this equation if $\varepsilon \neq 0$ or if the reaction is not carried out isothermally. The pressure drop parameter α is

$$\boxed{\alpha = \frac{2\beta_0}{A_c(1 - \phi)\rho_c P_0}} \quad (\text{kg}^{-1} \text{ or } \text{lb}_m^{-1}) \tag{5-29}$$

Equation (5-33) can be used to substitute for the pressure in the rate law, in which case the mole balance can then be written solely as a function of conversion and catalyst weight. The resulting equation can readily be solved either analytically or numerically.

  If we wish to express the pressure in terms of reactor length z, we can use Equation (5-26) to substitute for W in Equation (5-33). Then

$$p = \frac{P}{P_0} = \left(1 - \frac{2\beta_0 z}{P_0}\right)^{1/2} \tag{5-34}$$

5.5.3 Pressure Drop in Pipes

Normally, the pressure drop for gases flowing through pipes without packing can be neglected. For flow in pipes, the pressure drop along the length of the pipe can be approximated by

$$p = (1 - \alpha_p V)^{1/2} \qquad (5\text{-}35)$$

where

$$\alpha_p = \frac{4fG^2}{A_c \rho_0 P_0 D} \qquad (5\text{-}36)$$

and where f is the Fanning friction factor, D is the pipe diameter, and the other parameters are the same as previously defined.

For the flow conditions given in Example 5-4 in a 1000-ft length of $1\frac{1}{2}$-inch schedule 40 pipe ($\alpha_p = 0.0118$ ft^{-3}), the pressure drop is less than 10%. However, for high volumetric flow rates through microreactors, the pressure drop could be significant.

Example 5–4 Calculating Pressure Drop in a Packed Bed

Plot the pressure drop in a 60-ft length of $1\frac{1}{2}$-inch schedule 40 pipe packed with catalyst pellets $\frac{1}{4}$ inch in diameter. There is 104.4 lb$_m$/h of gas passing through the bed. The temperature is constant along the length of pipe at 260°C. The void fraction is 45% and the properties of the gas are similar to those of air at this temperature. The entering pressure is 10 atm.

Solution

(a)   First let's calculate the total pressure drop.

At the end of the reactor, $z = L$, Equation (5-34) becomes

$$\frac{P}{P_0} = \left(1 - \frac{2\beta_0 L}{P_0} \right)^{1/2} \qquad (E5\text{-}4.1)$$

$$\beta_0 = \frac{G(1 - \phi)}{g_c \rho_0 D_p \phi^3} \left[\frac{150(1 - \phi)\mu}{D_p} + 1.75G \right] \qquad (5\text{-}25)$$

Evaluating
the pressure-drop
parameters

$$G = \frac{\dot{m}}{A_c} \qquad (E5\text{-}4.2)$$

We now look up the cross-sectional area for $1\frac{1}{2}$-inch schedule 40 pipe, to find $A_c = 0.01414$ ft^2

$$G = \frac{104.4 \ \text{lb}_m/\text{h}}{0.01414 \ \text{ft}^2} = 7383.3 \ \frac{\text{lb}_m}{\text{h} \cdot \text{ft}^2}$$

For air at 260°C and 10 atm

$$\mu = 0.0673 \ \text{lb}_m/\text{ft} \cdot \text{h}$$
$$\rho_0 = 0.413 \ \text{lb}_m/\text{ft}^3$$

$$v_0 = \frac{\dot{m}}{\rho_0} = \frac{104.4 \text{ lb}_m/\text{h}}{0.413 \text{ lb}_m/\text{ft}^3} = 252.8 \text{ ft}^3/\text{h} \ (7.16 \text{ m}^3/\text{h})$$

From the problem statement

$$D_p = 1/4 \text{ in.} = 0.0208 \text{ ft}, \quad \phi = 0.45 \text{ and}$$

$$g_c = 4.17 \times 10^8 \frac{\text{lb}_m \cdot \text{ft}}{\text{lb}_f \cdot \text{h}^2}$$

Evaluating the parameters in the Ergun equation

Substituting these values into Equation (5-25) gives

$$\beta_0 = \left[\frac{7383.3 \text{ lb}_m/\text{ft}^2 \cdot \text{h}(1-0.45)}{(4.17 \times 10^8 \text{ lb}_m \cdot \text{ft}/\text{lb}_f \cdot \text{h}^2)(0.413 \text{ lb}_m/\text{ft}^3)(0.0208 \text{ ft})(0.45)^3} \right] \quad \text{(E5-4.3)}$$

$$\times \left[\frac{150(1-0.45)(0.0673 \text{ lb}_m/\text{ft} \cdot \text{h})}{0.0208 \text{ ft}} + 1.75(7383.3) \frac{\text{lb}_m}{\text{ft}^2 \cdot \text{h}} \right]$$

$$\beta_0 = 0.01244 \frac{\text{lb}_f \cdot \text{h}}{\text{ft} \cdot \text{lb}_m} [\overbrace{266.9}^{\text{Term 1}} + \overbrace{12{,}920.8}^{\text{Term 2}}] \frac{\text{lb}_m}{\text{ft}^2 \cdot \text{h}} = 164.1 \frac{\text{lb}_f}{\text{ft}^3} \quad \text{(E5-4.4)}$$

We note that the turbulent flow term, Term 2, is dominant

$$\beta_0 = 164.1 \frac{\text{lb}_f}{\text{ft}^3} \times \frac{1 \text{ ft}^2}{144 \text{ in.}^2} \times \frac{1 \text{ atm}}{14.7 \text{ lb}_f/\text{in.}^2}$$

Unit conversion for β_0:
$$\frac{1 \text{ atm}}{\text{ft}} = 333\frac{\text{kPa}}{\text{m}}$$

$$\boxed{\beta_0 = 0.0775 \ \frac{\text{atm}}{\text{ft}} = 25.8 \ \frac{\text{kPa}}{\text{m}}} \quad \text{(E5-4.5)}$$

We are now in a position to calculate the total pressure drop ΔP

$$p = \frac{P}{P_0} = \left(1 - \frac{2\beta_0 L}{P_0}\right)^{1/2} = \left(1 - \frac{\overbrace{2 \times 0.0775}^{0.155} \text{ atm/ft} \times 60 \text{ ft}}{10 \text{ atm}}\right)^{1/2} \quad \text{(E5-4.6)}$$

$$p = 0.265$$

$$P = 0.265 P_0 = 2.65 \text{ atm} \ (268 \ \text{kPa})$$
$$\Delta P = P_0 - P = 10 - 2.65 = 7.35 \text{ atm} \ (744 \ \text{kPa}) \qquad \text{(E5-4.7)}$$

(b) Now let's use the data to plot the pressure and the volumetric flow rate profiles. Recalling Equation (5-23) for the case $\varepsilon = 0$ and $T = T_0$

$$\boxed{v = v\frac{P_0}{P} = \frac{v_0}{p}} \qquad \text{(E5-4.8)}$$

Equations (5-34) and (E5-4.8) were used in the construction of Table E5-4.1.

TABLE E5-4.1   *P* AND *v* PROFILES

z (ft)	0	10	20	30	40	50	60
P (atm)	10	9.2	8.3	7.3	6.2	4.7	2.65
v (ft³/h)	253	275	305	347	408	538	955

For $\rho_c = 120\ \text{lb}_m/\text{ft}^3$

$$\alpha = \frac{2\beta_0}{\rho_c(1-\phi)A_cP_0} = \frac{2(0.0775)\text{atm/ft}}{120\ \text{lb}_m/\text{ft}^3(1-0.45)(0.01414\text{ft}^2)10\,\text{atm}}$$

Typical value of α

$$\boxed{\alpha = 0.0165\ \text{lb}_m^{-1} = 0.037\ \text{kg}^{-1}} \qquad\qquad \text{(E5-4.9)}$$

Equations (E5-34.1) and (E5-4.8), along with the values in Table E5-4.1, were used to obtain Figure E5-4.1.

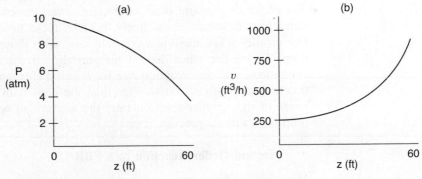

Figure E5-4.1   Pressure and volumetric flow rate profiles $\left(z = W/\left(A_c\rho_c(1-\phi)\right)\right)$.

Note how the volumetric flow rate increases dramatically as we move down the reactor.

Analysis: This example showed how to calculate the pressure drop and pressure-drop parameters (α and β_0) for a gas flowing through a packed-bed reactor. The pressure and volumetric flow rate profiles were calculated as a function of z (i.e., catalyst weight), as shown in Figure E5-4.1. One thing I bet you did not expect in this figure is how much the volumetric flow rate increases as the pressure decreases as one moves down the length of the PBR.

5.5.4 Analytical Solution for Reaction with Pressure Drop

How will pressure drop, ΔP, affect our calculations?

Let's reason out how pressure drop affects our CRE algorithm. Figure 5-8 shows qualitatively the effects of pressure drop on reactor design. Starting with Figure 5-8(**a**), we see how the Ergun equation predicts the pressure decrease

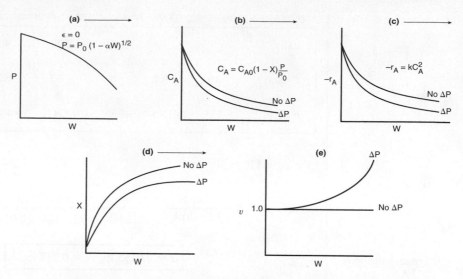

Figure 5-8   Effect of pressure drop on P (**a**), C_A (**b**), $-r_A$ (**c**), X (**d**), and v (**e**).

down the packed-bed reactor. The subsequent figures, (**b**) through (**e**), show this effect of pressure drop on concentration, reaction rate, conversion, and volumetric flow rate, respectively. Each of these figures compares the respective profiles when there is a pressure drop with those profiles for *no* pressure drop. We see that when there is pressure drop in the reactor, the reactant concentrations, and thus reaction rate for reaction (for reaction orders greater than 0 order), will always be smaller than the case with no pressure drop. As a result of this smaller reaction rate, the conversion will be less with pressure drop than without pressure drop.

Second-Order Reaction in a PBR

Now that we have expressed pressure as a function of catalyst weight [Equation (5-33) for $\varepsilon = 0$], we can return to the second-order isothermal reaction

$$A \longrightarrow B$$

to relate conversion and catalyst weight. Recall our mole balance, rate law, and stoichiometry.

1. Mole Balance:
$$F_{A0} \frac{dX}{dW} = -r_A'$$
(2-17)

2. Rate Law:
$$-r_A' = kC_A^2$$
(5-19)

3. Stoichiometry: Gas-phase isothermal reaction ($T = T_0$) with $\varepsilon = 0$. From Equation (5-23), $v = v_0/p$

Only
for
$\varepsilon = 0$

$$C_A = \frac{F_A}{v} = \frac{F_{A0}(1 - X)}{v} = C_{A0}(1 - X)\, p$$
(5-37)

$$p = \frac{P}{P_0} = (1 - \alpha W)^{1/2} \qquad (5\text{-}33)$$

Following the Algorithm

Using Equation (5-33) to substitute for y in terms of the catalyst weight, we obtain

$$C_A = C_{A0}(1 - X)(1 - \alpha W)^{1/2}$$

4. Combining: $\quad \dfrac{dX}{dW} = \dfrac{kC_{A0}^2}{F_{A0}}(1 - X)^2 \, [(1 - \alpha W)^{1/2}]^2$

5. Separating Variables: $\quad \dfrac{F_{A0}}{kC_{A0}^2}\dfrac{dX}{(1 - X)^2} = (1 - \alpha W)\, dW$

Integrating with limits $X = 0$ when $W = 0$ and substituting for $F_{A0} = C_{A0}v_0$ yields

$$\frac{v_0}{kC_{A0}}\left(\frac{X}{1 - X}\right) = W\left(1 - \frac{\alpha W}{2}\right)$$

6.A Solving for conversion gives

$$\boxed{X = \dfrac{\dfrac{kC_{A0}W}{v_0}\left(1 - \dfrac{\alpha W}{2}\right)}{1 + \dfrac{kC_{A0}W}{v_0}\left(1 - \dfrac{\alpha W}{2}\right)}} \qquad (5\text{-}38)$$

6.B Solving for the catalyst weight, we have

Catalyst weight
for
second-order
reaction in PBR
with ΔP

$$\boxed{W = \dfrac{1 - \{1 - [(2v_0\alpha)/kC_{A0}][X/(1 - X)]\}^{1/2}}{\alpha}} \qquad (5\text{-}39)$$

Example 5–5   Effect of Pressure Drop on the Conversion Profile

Reconsider the packed-bed reactor in Example 5-4 for the case where a second-order reaction

$$2A \rightarrow B + C$$

is taking place in 20 meters of a $1\frac{1}{2}$-inch schedule 40 pipe packed with catalyst. The flow and packed-bed conditions in the example remain the same except that they are converted to SI units; that is, $P_0 = 10$ atm $= 1013$ kPa, and

We need to be able
to work either
metric, S.I., or
English units.

Entering volumetric flow rate: $v_0 = 7.15$ m^3/h (252 ft^3/h)
Catalyst pellet size: $D_p = 0.006$ m (ca. $\frac{1}{4}$-inch)

Solid catalyst density: ρ_c = 1923 kg/m³ or (120 lb$_m$/ft³)
Cross-sectional area of 1½-inch schedule 40 pipe: A_C = 0.0013 m²
Pressure-drop parameter: β_0 = 25.8 kPa/m
Reactor length: L = 20 m

We will change the particle size to learn its effect on the conversion profile. However, we will assume that the specific reaction rate, k, is unaffected by particle size, an assumption that we know from Chapters 14 and 15 is valid only for small particles.

(a)   First, calculate the conversion in the absence of pressure drop.

(b)   Next, calculate the conversion accounting for pressure drop.

(c)   Finally, determine how your answer to (b) would change if the catalyst particle diameter were doubled.

The entering concentration of A is 0.1 kmol/m³ and the specific reaction rate is

$$k = \frac{12\,\text{m}^6}{\text{kmol}\cdot\text{kg-cat}\cdot\text{h}}$$

Solution

Using Equation (5-38)

$$X = \frac{\dfrac{kC_{A0}W}{v_0}\left(1 - \dfrac{\alpha W}{2}\right)}{1 + \dfrac{kC_{A0}W}{v_0}\left(1 - \dfrac{\alpha W}{2}\right)} \tag{5-38}$$

For the bulk catalyst density

$$\rho_b = \rho_c(1 - \phi) = (1923)(1 - 0.45) = 1058 \text{ kg/m}^3$$

The weight of catalyst in the 20 m of 1½-inch schedule 40 pipe is

$$W = A_c\rho_b L = (0.0013 \text{ m}^2)\left(1058\ \frac{\text{kg}}{\text{m}^3}\right)(20\ \text{m})$$

$$W = 27.5 \text{ kg}$$

$$\frac{kC_{A0}W}{v_0} = \frac{12\,\text{m}^6}{\text{kmol}\cdot\text{kg-cat}\cdot\text{h}}\cdot 0.1\frac{\text{kmol}}{\text{m}^3}\cdot\frac{27.5\ \text{kg}}{7.15\ \text{m}^3/\text{h}} = 4.6$$

(a)   First, calculate the conversion for $\Delta P = 0$ (i.e., $\alpha = 0$)

$$X = \frac{\dfrac{kC_{A0}W}{v_0}}{1 + \dfrac{kC_{A0}W}{v_0}} = \frac{4.6}{1 + 4.6} = 0.82 \tag{E5-5.1}$$

$$\boxed{X = 0.82}$$

(b)   Next, calculate the conversion with pressure drop. Recalling Equation (5-29) and substituting the bulk density $\rho_b = (1 - \phi)\,\rho_c = 1058$ kg/m³

In Equation (E5-4.9) in Example 5-4 on page 177, we calculated α to be

$$\alpha = \frac{2\beta_0}{P_0 A_c \rho_b} = 0.037\,\text{kg}^{-1} \qquad \text{(E5-5.2)}$$

then

$$\left(1 - \frac{\alpha W}{2}\right) = 1 - \frac{(0.037)(27.5)}{2} = 0.49 \qquad \text{(E5-5.3)}$$

$$X = \frac{\dfrac{kC_{A0}W}{v_0}\left(1 - \dfrac{\alpha W}{2}\right)}{1 + \dfrac{kC_{A0}W}{v_0}\left(1 - \dfrac{\alpha W}{2}\right)} = \frac{(4.6)(0.49)}{1 + (4.6)(0.49)} = \frac{2.26}{3.26} \qquad \text{(E5-5.4)}$$

$$\boxed{X = 0.693}$$

Watch out for under design!

Analysis: We see the predicted conversion dropped from 82.2% to 69.3% because of pressure drop. It would be not only embarrassing but also an economic disaster if we had neglected pressure drop and the actual conversion had turned out to be significantly smaller.

5.5.5 Robert the Worrier Wonders: _What If..._

Robert

Robert is one of the most important members of any engineering team. He is always trying to anticipate changes and the effect of those changes on the process, not only qualitatively, but quantitatively—a skill <u>all</u> engineers should try to hone. He uses basic engineering analysis to predict the new condition if a change is made or occurs unexpectedly.

Let's continue Example 5-5 in Example 5-6 where we ask a couple of _What if..._ questions. To illustrate engineering analysis, we will address Robert's worry... _What if we decrease the catalyst size by a factor of 4 and increase the entering pressure by a factor of 3?_

Let's see if we can help Robert out.

First, we need to see how the pressure-drop parameter, α, varies with the system parameters. Combining Equations (5-29) and (5-25), we obtain

$$\alpha = \frac{2}{A_C(1-\phi)\rho_C P_0}\beta_0 = \frac{2}{A_C(1-\phi)\rho_C P_0}\left[\frac{G(1-\phi)}{\rho_0 g_C D_P \phi^3}\left[\overset{\text{Laminar}}{\frac{150(1-\phi)\mu}{D_P}} + \overset{\text{Turbulent}}{1.75G}\right]\right] \qquad \text{(5-39)}$$

Using an average molecular weight of the gas and the ideal gas law, we can express the entering density as

$$\rho_0 = \left(\frac{1}{MW}\right)\frac{P_0}{RT_0} \qquad \text{(5-40)}$$

Next, substitute for ρ_0 in Equation (5-39) to obtain

$$\alpha = \frac{2RT_0(MW)}{A_C \rho_c g_c P_0^2 D_P \phi^3} G \left[\overbrace{\frac{150(1-\phi)\mu}{D_P}}^{\text{Term 1}} + \overbrace{1.75G}^{\text{Term 2}} \right] \qquad (5\text{-}41)$$

We are now going to consider a *What if..* analysis for laminar and turbulent flow separately. We will only consider changing the superficial mass flow rate, entering pressure, catalyst particle diameter, and cross-sectional area of the tube in which the catalyst is placed. We keep the catalyst pellet density, ρ_c and void fraction ϕ unchanged and we will leave it as an exercise for the reader to change the temperature and analyze the results.

A. *Laminar Flow Dominant (Term 1 >> Term 2)*

   We see the pressure drop parameter α varies as

$$\alpha \sim \frac{GT_0}{A_C D_P^2 P_0^2} \qquad (5\text{-}42)$$

We now consider the original condition, Case 1, and the condition after the change is made, Case 2, and take the ratio of Case 2 to Case 1.

Case 1 / Case 2

$$\alpha_2 = \alpha_1 \left(\frac{G_2}{G_1}\right)\left(\frac{A_{C1}}{A_{C2}}\right)\left(\frac{D_{P1}}{D_{P2}}\right)^2 \left(\frac{P_{01}}{P_{02}}\right)^2 \left(\frac{T_{02}}{T_{01}}\right) \qquad (5\text{-}43)$$

For constant mass flow rate for Case 1 and Case 2, we can substitute for G

$$G = \frac{\dot{m}}{A_C}$$

to obtain

$$\boxed{\alpha_2 = \alpha_1 \left(\frac{A_{C1}}{A_{C2}}\right)^2 \left(\frac{D_{P1}}{D_{P2}}\right)^2 \left(\frac{P_{01}}{P_{02}}\right)^2 \left(\frac{T_{02}}{T_{01}}\right)} \qquad (5\text{-}44)$$

We now proceed for turbulent flow to learn how α varies with the system parameters.

B. *Turbulent Flow Dominant (Term 2 >> Term 1)*

   We see the pressure-drop parameter varies

$$\alpha \sim \frac{G^2 T_0}{A_C D_P P_0^2}$$

Taking the ratio of Case 1 to Case 2

$$\boxed{\alpha_2 = \alpha_1 \left(\frac{G_2}{G_1}\right)^2 \left(\frac{A_{C1}}{A_{C2}}\right)^2 \left(\frac{P_{01}}{P_{02}}\right)^2 \left(\frac{D_{P1}}{D_{P2}}\right)\left(\frac{T_{02}}{T_{01}}\right)} \qquad (5\text{-}45)$$

If the mass flow rate, $\dot{m}$, is the same for the two cases, we can substitute $G = \dot{m}/A_C$ to obtain

$$\alpha_2 = \left(\frac{A_{C1}}{A_{C2}}\right)^3 \left(\frac{P_{01}}{P_{02}}\right)^2 \left(\frac{D_{P1}}{D_{P2}}\right)\left(\frac{T_{02}}{T_{01}}\right) \tag{5-46}$$

We observe that for turbulent flow the pressure drop parameter α is more sensitive to changes to the cross-sectional area A_C and less sensitive to changes in the particle diameter D_P than that for laminar flow.

His friends
sometimes call him
"Robert What If"

Example 5–6 Robert Worries what if...

How will the pressure drop parameter (e.g., α) and conversion change if you decrease the particle diameter by a factor of 4 and increase entering pressure by a factor of 3, keeping everything else the same? To answer this question we need to consider the flow regime, laminar or turbulent.

(a)   **Laminar Flow**. First, let's consider how the conversion would change if the flow were laminar.

$$\alpha_2 = \alpha_1 \left(\frac{A_{C1}}{A_{C2}}\right)^2 \left(\frac{D_{P1}}{D_{P2}}\right)^2 \left(\frac{P_{01}}{P_{02}}\right)^2 \left(\frac{T_{02}}{T_{01}}\right) \tag{5-43}$$

Here $G_1 = G_2$, $A_{C1} = A_{C2}$ and $T_{02} = T_{01}$, but

$$D_{P2} = \frac{1}{4}D_{P1} \text{ and } P_{02} = 3P_{01}$$

Substituting into Equation (5-44)

$$\alpha_2 = \alpha_1 \left(\frac{D_{P1}}{\frac{1}{4}D_{P1}}\right)^2 \left(\frac{P_{01}}{3P_{01}}\right)^2 = \frac{16}{9}\alpha_1 \tag{E5-6.1}$$

Substitute α_2 into Equation (5-38)

$$\alpha = 0.037\frac{16}{9}kg^{-1} = 0.066\ kg^{-1}$$

$$\left(1 - \frac{\alpha_2 W}{2}\right) = \left(1 - \frac{\left(0.066\ kg^{-1}\right)}{2}27.5\ kg\right) = 0.093$$

$$X = \frac{\dfrac{kC_{A0}W}{v_0}\left(1 - \dfrac{\alpha_2 W}{2}\right)}{1 + \dfrac{kC_{A0}W}{v_0}\left(1 - \dfrac{\alpha_2 W}{2}\right)} = \frac{(4.6)(0.096)}{1 + (4.6)(0.096)} = 0.31 \tag{E5-6.2}$$

(b)   Turbulent flow

$$\alpha_2 = \left(\frac{A_{C1}}{A_{C2}}\right)^3 \left(\frac{P_{01}}{P_{02}}\right)^2 \left(\frac{D_{P1}}{D_{P2}}\right)\left(\frac{T_{02}}{T_{01}}\right) \tag{5-46}$$

As above, $A_{C1} = A_{C2}$ and $T_{02} = T_{01}$, but $D_{P2} = \frac{1}{4}D_{P1}$ and $P_{02} = 3P_{01}$

$$\alpha_2 = \alpha_1 \left(\frac{D_{P1}}{\frac{1}{4}D_{P1}}\right)\left(\frac{P_{01}}{3P_{01}}\right)^2 = \frac{4}{9}\alpha_1 \tag{E5-6.3}$$

Solving for our new α and then substituting into Equation (5-38)

$$\alpha_2 = 0.037\,\frac{4}{9} = 0.0164\ kg^{-1}$$

$$\left(1 - \frac{\alpha W}{2}\right) = \left(1 - \frac{\left(0.0164\ kg^{-1}\right)}{2}\right)(27.5\ kg) = 0.77 \tag{E5-6.4}$$

$$X = \frac{\dfrac{kC_{A0}W}{\upsilon_0}\left(1 - \dfrac{\alpha W}{2}\right)}{1 + \dfrac{kC_{A0}W}{\upsilon_0}\left(1 - \dfrac{\alpha W}{2}\right)} = \frac{(4.6)(0.77)}{1 + (4.6)(0.77)} = \frac{3.56}{4.56} = 0.78 \tag{E5-6.5}$$

Analysis: The following table gives a summary of the conversion for the four cases presented in Examples 5-5 and 5-6.

1. No pressure drop	$X = 0.82$
2. Pressure drop (base case)	$X = 0.69$
3. Pressure drop laminar flow	$X = 0.31$
4. Pressure drop turbulent flow	$X = 0.78$

In both laminar and turbulent flow, increasing the pressure, P_0, decreases the pressure-drop parameter α. Decreasing the particle diameter D_p increases the pressure-drop parameter α for both laminar and turbulent flow, but more so for the laminar where $\alpha \sim 1/D_p^2$. The smaller the pressure-drop parameter, α, the smaller the pressure drop and thus the greater the conversion. For laminar flow, the negative effects of reducing the particle size by a factor of 4 are stronger than the effect of increasing pressure by a factor of 3.

Robert the Worrier, like the author of this book, thinks it is important to be able to carry out an *engineering analysis* using Case 1 and Case 2 and then taking ratios to estimate the effect of changing parameters on conversion and reactor operation.

Additionally Noted: Because there is no change in the net number of moles during this isothermal gas-phase reaction occurring in a PBR, one can obtain an analytical solution to our CRE algorithm instead of using the Polymath software. Now let's look at what we could expect by changing the particle diameter of the catalyst pellets.

By increasing the particle diameter, we decrease the pressure-drop parameter and thus increase the reaction rate and the conversion. However, Chapter 10 and

Chapter 15 explain that when interparticle diffusion effects are important in the catalyst pellet, this increase in conversion with increasing particle size will not always be the case. For larger particles, it takes a longer time for a given number of reactant and product molecules to diffuse in and out of the catalyst particle where they undergo reaction (see Figure 10-5). Consequently, the specific reaction rate decreases with increasing particle size $k \sim 1/D_P$ [see the Chapter 15 *Summary Notes* on the CRE Web site, Equation (15-33)], which in turn decreases the conversion. At small particle diameters, the rate constant, k, is large and at its maximum value, but the pressure drop is also large, resulting in a low rate of reaction. At large particle diameters, the pressure drop is small, but so is the rate constant, k, and the rate of reaction, resulting in low conversion. Thus, we see a low conversion at both large and small particle diameters with an optimum in between. This optimum is shown in Figure E5-6.1.

The variation

$$k \sim \frac{1}{D_p}$$

is discussed in detail in Chapter 15. Also, see Chapter 5 *Summary Notes*.

Summary Notes

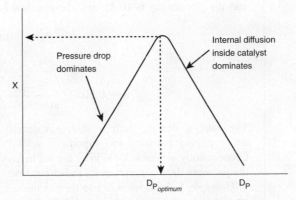

Figure E5-6.1   Finding the optimum particle diameter.

Problems with large-diameter tubes

(1) Bypassing of catalyst
(2) Smaller heat-transfer area

If pressure drop is to be minimized, *why not pack the catalyst into a larger diameter tube* to decrease the superficial velocity, G, thereby reducing ΔP? There are two reasons for *not* increasing the tube diameter: (1) There is an increased chance the gas could channel and bypass most of the catalyst, resulting in little conversion; (2) the ratio of the heat-transfer surface area to reactor volume (catalyst weight) will be decreased, thereby making heat transfer more difficult for highly exothermic and endothermic reactions.

We now proceed in Example 5-7 to combine pressure drop with reaction in a packed bed when we have volume change with reaction and therefore cannot obtain an analytical solution.[6]

Example 5–7   Calculating X in a Reactor with Pressure Drop

The economics

The uses

Approximately 8.5 billion pounds of ethylene oxide were produced in the United States. The 2010 selling price was $0.53 a pound, amounting to a commercial value of $4.0 billion. Over 60% of the ethylene oxide produced is used to make ethylene glycol. The major end uses of ethylene oxide are antifreeze (30%), polyester (30%), surfactants (10%), and solvents (5%). We want to calculate the catalyst weight necessary to achieve 60% conversion when ethylene oxide is to be made by the vapor-phase catalytic oxidation of ethylene with air.

---

[6]*Ind. Eng. Chem., 45*, 234.

$$C_2H_4 + \frac{1}{2}O_2 \longrightarrow CH_2\!-\!CH_2$$

$$A \; + \frac{1}{2}B \longrightarrow \quad C$$

Ethylene and oxygen are fed in stoichiometric proportions to a packed-bed reactor operated isothermally at 260°C. Ethylene is fed at a rate of 136.21 mol/s at a pressure of 10 atm (1013 kP$_a$). It is proposed to use 10 banks of $1\frac{1}{2}$-inch-diameter schedule 40 tubes packed with catalyst with 100 tubes per bank. Consequently, the molar flow rate to each tube is to be 0.1362 mol/s. The properties of the reacting fluid are to be considered identical to those of air at this temperature and pressure. The density of the 1/4-inch-catalyst particles is 1925 kg/m³, the bed void fraction is 0.45, and the gas density is 16 kg/m³. The rate law is

$$-r_A' = kP_A^{1/3}P_B^{2/3} \qquad \text{mol/kgcat}\cdot s$$

with

$$k = 0.00392 \; \frac{\text{mol}}{\text{atm}\cdot\text{kg-cat}\cdot\text{s}} \; \text{at } 260°C$$

The catalyst density, particle size, gas density, void fraction, pipe cross-sectional area, entering pressure, and superficial velocity are the same as in Example E5-4. Consequently we are in luck. Why are we in luck? Because we don't have to calculate the pressure-drop parameters β_0 and α because they are the same as those calculated in Example 5-4 and we will use these values, i.e., $\beta_0 = 25.8$ atm/m and $\alpha = 0.0367$ kg⁻¹ in this example.

Solution

1. **Differential Mole Balance:**

$$\boxed{F_{A0}\frac{dX}{dW} = -r_A'} \tag{E5-7.1}$$

Following the Algorithm

2. **Rate Law:**

$$-r_A' = kP_A^{1/3}P_B^{2/3} = k(C_ART)^{1/3}(C_BRT)^{2/3} \tag{E5-7.2}$$

$$= kRTC_A^{1/3}C_B^{2/3} \tag{E5-7.3}$$

The algorithm

3. **Stoichiometry:** Gas-phase, isothermal $v = v_0(1 + \varepsilon X)(P_0/P)$:

$$C_A = \frac{F_A}{v} = \frac{C_{A0}(1-X)}{1+\varepsilon X}\left(\frac{P}{P_0}\right) = \frac{C_{A0}(1-X)p}{1+\varepsilon X} \quad \text{where } p = \frac{P}{P_0} \tag{E5-7.4}$$

$$C_B = \frac{F_B}{v} = \frac{C_{A0}(\Theta_B - X/2)}{1+\varepsilon X}p \tag{E5-7.5}$$

For stoichiometric feed $\Theta_B = \dfrac{F_{B0}}{F_{A0}} = \dfrac{1}{2}$

$$C_B = \frac{C_{A0}}{2}\frac{(1-X)}{(1+\varepsilon X)}p$$

For isothermal operation, Equation (5-30) becomes

$$\frac{dy}{dW} = -\frac{\alpha}{2p}(1 + \varepsilon X)$$ (E5-7.6)

4. **Combining** the rate law and concentrations

$$-r'_A = kRT_0 \left[\frac{C_{A0}(1-X)}{1+\varepsilon X}(p)\right]^{1/3}\left[\frac{C_{A0}(1-X)}{2(1+\varepsilon X)}(p)\right]^{2/3}$$ (E5-7.7)

We can evaluate
the **combine** step
either
1. Analytically
2. Graphically
3. Numerically, or
4. Using software

Factoring $\left(\frac{1}{2}\right)^{2/3}$ and recalling $P_{A0} = C_{A0}RT_0$, we can simplify Equation (E5-7.7) to

$$\boxed{-r'_A = k'\left(\frac{1-X}{1+\varepsilon X}\right)p}$$ (E5-7.8)

where $k' = kP_{A0}\left(\frac{1}{2}\right)^{2/3} = 0.00392\dfrac{\text{mol}}{\text{atm kg-cat}\cdot\text{s}} \times 3\text{atm} \times 0.63 = 0.0074\dfrac{\text{mol}}{\text{kg-cat}\cdot\text{s}}$.

5. **Parameter evaluation per tube** (i.e., divide feed rates by 1000)

Ethylene: $F_{A0} = 0.1362$ mol/s

Oxygen: $F_{B0} = 0.068$ mol/s

Inerts (N_2): $F_I = 0.068\text{ mol/s} \times \dfrac{79\text{ mol }N_2}{21\text{ mol }O_2} = 0.256\dfrac{\text{mol}}{\text{s}}$

Summary: $F_{T0} = F_{A0} + F_{B0} + F_I = 0.460\dfrac{\text{mol}}{\text{s}}$

$$y_{A0} = \frac{F_{A0}}{F_{T0}} = \frac{0.1362}{0.460} = 0.30$$

$$\varepsilon = y_{A0}\delta = (0.3)\left(1 - \frac{1}{2} - 1\right) = -0.15$$

$$P_{A0} = y_{A0}P_0 = 3.0 \text{ atm}$$

As noted in the problem statement, $\beta_0 = 25.8$ kPa/m and $\alpha = 0.0367kg^{-1}$.

6. **Summary:** Combining Equation (E5-6.1) and (E5-6.8) and summarizing

$$\frac{dX}{dW} = \frac{-r'_A}{F_{A0}}$$ (E5-7.9)

$$\frac{dp}{dW} = -\alpha\frac{(1+\varepsilon X)}{2p}$$ (E5-7.10)

$$r'_A = -\frac{k'(1-X)}{1+\varepsilon X}p$$

$$k' = 0.0074\left(\frac{\text{mol}}{\text{kg}\cdot\text{s}}\right)$$ (E5-7.11)

$$F_{A0} = 0.1362 \left(\frac{\text{mol}}{\text{s}} \right) \qquad \text{(E5-7.12)}$$

$$\alpha = 0.0367 \left(\text{kg}^{-1} \right) \qquad \text{(E5-7.13)}$$

$$\varepsilon = -0.15 \qquad \text{(E5-7.14)}$$

We will guess the final catalyst weight to achieve 60% conversion to be 27 kg and use this value as the integration end point in our Polymath program.

$$W_f = 27 \text{ kg}$$

We have the boundary conditions $W = 0$, $X = 0$, $p = 1.0$, and $W_f = 27$kg. Here, we are guessing an upper limit of the integration to be 27 kg, with the expectation that 60% conversion will be achieved *within* this catalyst weight. If 60% conversion is not achieved, we will guess a higher weight and redo the calculation.

Living Example Problems (**LEPs**). A large number of ordinary differential equation solver software packages (i.e., ODE solvers), which are extremely user friendly, have become available. We shall use Polymath[7] to solve the examples in this book. With Polymath, one simply enters Equations (E5-7.9) and (E5-7.10) and the corresponding parameter values [Equations (5-7.11) through (5-7.14)] into the computer with the boundary conditions and they are solved and displayed as shown in Figures E5-7.1 and E5-7.2. Equations (E5-7.9) and (E5-7.10) are entered as differential equations and the parameter values are set using explicit equations. The rate law may be entered as an explicit equation in order to generate a plot of reaction rate as it changes down the length of the reactor, using Polymath's graphing function. The CRE Web site contains all of the MATLAB and Polymath solution programs used to solve the example problems, as well as an example using AspenTech. Consequently, one can download the Polymath program directly from the CRE Web site (*www.umich.edu/~elements/5e/index.html*), which has programmed Equations (E5-7.9) through (E5-7.14), and run the program for different parameter values.

It is also interesting to learn what happens to the volumetric flow rate along the length of the reactor. Recalling Equation (4-23),

$$v = v_0 (1 + \varepsilon X) \frac{P_0}{P} \frac{T}{T_0} = \frac{v_0 (1 + \varepsilon X)(T/T_0)}{p} \qquad \text{(4-23)}$$

We let f be the ratio of the volumetric flow rate, v, to the entering volumetric flow rate, v_0, at any point down the reactor. For isothermal operation, Equation (4-23) becomes

Volumetric flow rate increases with increasing pressure drop.

$$f = \frac{v}{v_0} = \frac{1 + \varepsilon X}{p} \qquad \text{(E5-7.15)}$$

The Polymath program and output are shown in Tables E5-7.1 and E5-7.2.

TABLE E5-6.1 POLYMATH PROGRAM

(Information on how to obtain and load the Polymath software can be found in Appendix D. Tutorials can be found on the CRE Web site home page under Living Examples, Polymath.)

Polymath and MATLAB program examples can be downloaded from the CRE Web site M (see the Introduction).

---

[7] Developed by Professor M. Cutlip of the University of Connecticut, and Professor M. Shacham of Ben Gurion University. Available from the CACHE Corporation, P.O. Box 7939, Austin, TX 78713. See Appendix D for more details.

Living Example Problem

For all Living Example
Problems, Polymath
and MATLAB can
be loaded from the
CRE Web site (see the
Introduction).

ODE REPORT (STIFF)

Differential equations

1 d(X)/d(W) = -raprime/Fao

2 d(p)/d(W) = -alpha*(1+eps*X)/2/p

Explicit equations

1 eps = -0.15

2 kprime = 0.0266

3 Fao = 1.08

4 alpha = 0.0166

5 raprime = -kprime*(1-X)/(1+eps*X)*p

6 f = (1+eps*X)/p

7 rate = -raprime

Table E5-7.1 Polymath program.

Calculated values of DEQ variables

	Variable	Initial value	Minimal value	Maximal value	Final value
1	alpha	0.0166	0.0166	0.0166	0.0166
2	eps	-0.15	-0.15	-0.15	-0.15
3	f	1.	1.	3.418753	3.418753
4	Fao	1.08	1.08	1.08	1.08
5	kprime	0.0266	0.0266	0.0266	0.0266
6	p	1.	0.263302	1.	0.263302
7	raprime	-0.0266	-0.0266	-0.0026021	-0.0026021
8	rate	0.0266	0.0026021	0.0266	0.0026021
9	W	0	0	60.	60.
10	X	0	0	0.6655698	0.6655698

Table E5-7.2 Numerical output.

Figure E5-7.3(a) shows X, p (i.e., $p = P/P_0$), and f down the length of the reactor. We see that both the conversion and the volumetric flow increase along the length of the reactor, while the pressure decreases. Figure 5-7.3(b) shows how the rate of reaction, $-r'_A$, decreases as we move down the reactor. For gas-phase reactions with orders greater than zero, the decrease in pressure will cause the reaction rate to be less than in the case of no pressure drop.

From either the conversion profile (shown in Figure E5-7.3) or the Polymath table of results (not shown in the text, but available on the CRE Web site), we find 60% conversion is achieved with 20 kg catalyst in each tube.

Effect of added
catalyst on
conversion

We note from Figure E5-7.3 that the catalyst weight necessary to raise the conversion the last 1%, from 65% to 66% (0.9 kg), is 8.5 times more than that required to raise the conversion 1% at the reactor's entrance. Also, during the last 5% increase in conversion, the pressure decreases from 3.8 atm to 2.3 atm.

This catalyst weight of 20 kg/tube corresponds to a pressure drop of approximately 5 atm. If we had erroneously neglected pressure drop, the catalyst weight would have been found by integrating equation (E5-7.9) with $p = 1$ to give

Neglecting
pressure drop
results in poor
design (here, 53%
vs. 60% conversion)

$$W = \frac{F_{A0}}{k'}\left[(1+\varepsilon)\,\ln\!\left(\frac{1}{1-X}\right) - \varepsilon X\right] \tag{E5-7.16}$$

$$= \frac{0.1362}{0.0074} \times \left[(1-0.15)\,\ln\frac{1}{1-0.6} - (-0.15)(0.6)\right] \tag{E5-7.17}$$

= 16 kg of catalyst per tube (neglecting pressure drop)

Analysis: If we had used this 16 kg per tube catalyst weight in our reactor, we would have had insufficient catalyst to achieve the desired conversion. For this 16 kg catalyst weight, Figure E5-7.3(a) shows that for the case of pressure drop, only 53% conversion would have been achieved and this would have been _embarrassing!_ In arriving at this job-saving conclusion, we applied the CRE algorithm to a gas-phase reaction with a change in the total number of moles carried out in a PBR. The only small change from the previous example is that we had to use the ODE solver Polymath to combine and solve all the steps to obtain the profiles of the reaction rate ($-r_A$), conversion (X), pressure ratio (P/P_0), and volumetric flow rate ratio (f) as a function of catalyst weight down the length of the PBR.

Embarrassing!

$$f = \frac{v}{v_0}$$

$$p = \frac{P}{P_0}$$

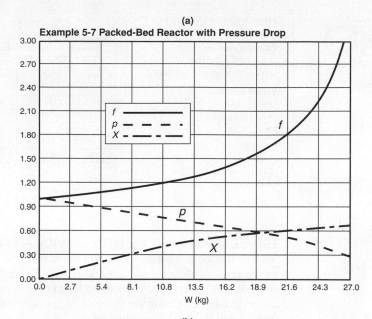

(a)

Example 5-7 Packed-Bed Reactor with Pressure Drop

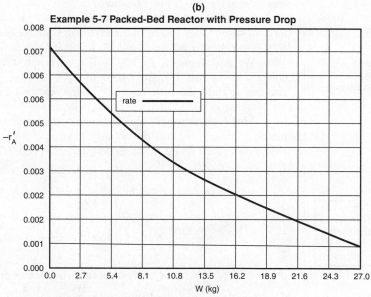

(b)

Example 5-7 Packed-Bed Reactor with Pressure Drop

Figure E5-7.3   Output in graphical form from Polymath.

5.6 Synthesizing the Design of a Chemical Plant

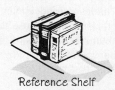

Reference Shelf

Synthesizing a
chemical plant

Careful study of the various reactions, reactors, and molar flows of the reactants and products used in the example problems in this chapter reveals that they can be arranged to form a chemical plant to produce 200 million pounds of ethylene glycol from a feedstock of 402 million pounds per year of ethane. The flowsheet for the arrangement of the reactors, together with the molar flow rates, is shown in Figure 5-9. Here, 0.425 lb mol/s of ethane is fed to 100 tubular plug-flow reactors connected in parallel; the total volume is 81 ft^3 to

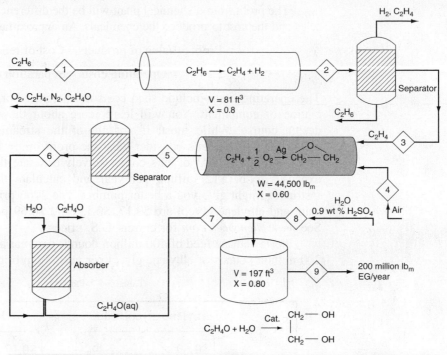

Stream	Component[a]	Flow Rate (lb mol/s)	Stream	Component[a]	Flow Rate (lb mol/s)
1	C_2H_6	0.425	6	EO	0.150
2	C_2H_4	0.340	7	EO	0.128
3	C_2H_4	0.300	8	H_2O	0.443
4	Air	0.714	9	EG	0.102
5	EO	0.180			

[a]EG, ethylene glycol; EO, ethylene oxide.

Figure 5-9   Production of ethylene glycol.

Always challenge the assumptions, constraints, and boundaries of the problem.

produce 0.34 lb mol/s of ethylene (see Example 5-3). The reaction mixture is then fed to a separation unit where 0.04 lb mol/s of ethylene is lost in the separation process in the ethane and hydrogen streams that exit the separator. This process provides a molar flow rate of ethylene of 0.3 lb mol/s, which enters the packed-bed catalytic reactor together with 0.15 lb mol/s of O_2 and 0.564 lb mol/s of N_2. There are 0.18 lb mol/s of ethylene oxide (see Example 5-6) produced in the 1000 pipes arranged in parallel and packed with silver-coated catalyst pellets. There is 60% conversion achieved in each pipe and the total catalyst weight in all the pipes is 44,500 lb_m. The effluent stream is passed to a separator where 0.03 lb mol/s of ethylene oxide is lost. The ethylene oxide stream is then contacted with water in a gas absorber to produce a 1-lb mol/ft^3 solution of ethylene oxide in water. In the absorption process, 0.022 lb mol/s of ethylene oxide is lost. The ethylene oxide solution is fed to a 197-ft^3 CSTR, together with a stream of 0.9 wt % H_2SO_4 solution, to produce ethylene glycol at a rate of 0.102 lb mol/s (see Example 5-2). This rate is equivalent to approximately 200 million pounds of ethylene glycol per year.

The profit from a chemical plant will be the difference between income from sales and the cost to produce the chemicals. An approximate formula might be

$$\$\$\$\$ \quad \text{Profit} = \text{Value of products} - \text{Cost of reactants}$$

$$- \text{Operating costs} - \text{Separation costs}$$

The operating costs include such costs as energy, labor, overhead, and depreciation of equipment. You will learn more about these costs in your senior design course. While most, if not all, of the streams from the separators could be recycled, let's consider what the profit might be if the streams were to go unrecovered. Also, let's conservatively estimate the operating and other expenses to be $12 million per year and calculate the profit. Your design instructor might give you a better number. The 2006 prices of ethane, sulfuric acid, and ethylene glycol are $0.17, $0.15, and $0.69 per pound, respectively. See *www.chemweek.com* for current U.S. prices.

For an ethane feed of 400 million pounds per year and a production rate of 200 million pounds of ethylene glycol per year, the profit is shown in Table 5-4.

TABLE 5-4   PROFITS

$$\text{Profit} = \left[\left(\overbrace{\frac{\$0.69}{lb_m} \times 2 \times 10^8 \frac{lb_m}{year}}^{\text{Ethylene glycol value}} \right) = \left(\overbrace{\frac{\$0.17}{lb_m} \times 4 \times 10^8 \frac{lb_m}{year}}^{\text{Ethane cost}} \right) \right.$$

$$\left. - \left(\overbrace{\frac{\$0.15}{lb_m} \times 2.26 \times 10^6 \frac{lb_m}{year}}^{\text{Sulfuric acid cost}} \right) - \overbrace{\$12,000,000}^{\text{Operating cost}} \right]$$

$$= \$138,000,000 - \$68,000,000 - 340,000 - \$12,000,000$$

$$\cong \$57.7 \text{ million}$$

You will learn more economics of chemical processing in your senior design class.

Using $58 million a year as a rough estimate of the profit, you can now make different approximations about the conversion, separations, recycle streams, and operating costs to learn how they affect the profit.

Closure. This chapter presents the heart of chemical reaction engineering for isothermal reactors. After completing this chapter, the reader should be able to apply the algorithm building blocks

The CRE algorithm

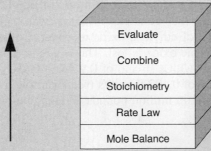

Evaluate

Combine

Stoichiometry

Rate Law

Mole Balance

to any of the reactors discussed in this chapter: batch reactor, CSTR, PFR, and PBR. The reader should be able to account for pressure drop and describe the effects of the system variables such as catalyst particle size on the PBR conversion, and explain why there is an optimum in the conversion when the catalyst particle size is varied. The reader should be able to use conversion to solve chemical reaction engineering problems. Finally, after completing this chapter, the reader should be able to work the California Professional Engineers' Exam Problems in approximately 30 minutes (cf. P5-15$_B$ through P5-20$_B$) and to diagnose and troubleshoot malfunctioning reactors (cf. P5-8$_B$).

SUMMARY

1. **Solution algorithm**
   a. Mole balances (BR, CSTR, PFR, PBR):

$$N_{A0}\frac{dX}{dt} = -r_A V, \quad V = \frac{F_{A0}X}{-r_A}, \quad F_{A0}\frac{dX}{dV} = -r_A, \quad F_{A0}\frac{dX}{dW} = -r_A' \tag{S5-1}$$

   b. Rate law: For example,

$$-r_A' = kC_A^2 \tag{S5-2}$$

   c. Stoichiometry: $\quad A + \frac{b}{a}B \rightarrow \frac{c}{a}C + \frac{d}{a}D$

   (1) *Liquid phase:* $v = v_0$ $\tag{S5-3}$
   $$C_A = C_{A0}(1-X)$$

   (2) *Gas phase:* $v = v_0(1+\varepsilon X)\left(\frac{P_0}{P}\right)\left(\frac{T}{T_0}\right)$, where $\varepsilon = y_{A0}\delta = y_{A0}\left(\frac{d}{a}+\frac{c}{a}-\frac{b}{a}-1\right)$

   $$p = \frac{P}{P_0}$$

   $$f = v/v_0 \tag{S5-4}$$

$$C_A = \frac{F_A}{v} = \frac{F_{A0}(1-X)}{v} = \frac{F_{A0}(1-X)}{v_0(1+\varepsilon X)}\left(\frac{P}{P_0}\right)\frac{T_0}{T} = C_{A0}\left(\frac{1-X}{1+\varepsilon X}\right)p\frac{T_0}{T} \tag{S5-5}$$

   For a **PBR**

$$\frac{dp}{dW} = -\frac{\alpha(1+\varepsilon X)}{2p}\left(\frac{T}{T_0}\right) \tag{S5-6}$$

$$\alpha = \frac{2\beta_0}{A_c(1-\phi)\rho_c P_0} \text{ and } \beta_0 = \frac{G(1-\phi)}{\rho_0 g_c D_p \phi^3}\left[\frac{150(1-\phi)\mu}{D_p}+1.75G\right]$$

   Variable density with $\varepsilon = 0$ or $\varepsilon X \ll 1$ and isothermal operation:

   IFF ε = 0 $\qquad\qquad p = \frac{P}{P_0} = (1-\alpha W)^{1/2}$ $\tag{S5-7}$

d. Combining the rate law and stoichiometry for isothermal operation
in a PBR

$$Liquid: \quad -r_A' = kC_{A0}^2(1-X)^2 \tag{S5-8}$$

$$Gas: \quad -r_A' = kC_{A0}^2 \frac{(1-X)^2}{(1+\varepsilon X)^2}p^2 \tag{S5-9}$$

e. Solution techniques:
(1) Numerical integration—Simpson's rule
(2) Table of integrals
(3) Software packages
    (a) Polymath
    (b) MATLAB
An ODE solver (e.g., Polymath) will combine all the equations for you.

ODE SOLVER ALGORITHM

When using an ordinary differential equation (ODE) solver such as Polymath or MATLAB, it is usually easier to leave the mole balances, rate laws, and concentrations as separate equations, rather than combining them into a single equation as we did to obtain an analytical solution. Writing the equations separately leaves it to the computer to combine them and produce a solution. The formulations for a packed-bed reactor with pressure drop are given below for an elementary reversible reaction carried out isothermally.

Reversible Gas Phase

$$A + B \rightleftarrows 3C$$

Packed-Bed Reactor

$$\frac{dX}{dW} = \frac{-r_A'}{F_{A0}} \qquad\qquad k = 1000(\text{dm}^3/\text{mol})/\text{min}/\text{kg}$$

$$\qquad\qquad\qquad\qquad \alpha = 0.01 \text{ kg}^{-1}$$

$$r_A' = -k\left[C_A C_B - \frac{C_C^3}{K_C}\right] \qquad \theta_B = 2.0$$

$$\qquad\qquad\qquad\qquad \varepsilon = 0.33$$

$$C_A = C_{A0}\frac{1-X}{1+\varepsilon X}p \qquad C_{A0} = 0.01 \text{ mol/dm}^3$$

$$\qquad\qquad\qquad\qquad F_{A0} = 15.0 \text{ mol/min}$$

$$C_B = C_{A0}\frac{\theta_B - X}{1+\varepsilon X}p \qquad\qquad K_C = 0.05 \text{ mol/dm}^3$$

$$C_C = \frac{3C_{A0}X}{(1+\varepsilon X)}p \qquad W_{\text{final}} = 90 \text{ kg}$$

$$\frac{dp}{dW} = -\frac{\alpha(1+\varepsilon X)}{2p}$$

(where $p = P/P_0$)

Polymath will combine and solve the above equations and then allow you to plot the variables (e.g., p, $-r_A$, C_A) as a function of W or each other. The Polymath solution to the above equation is given on the CRE Web site (*www.umich.edu/~elements/5e/index.html*) in the **Chapter 5** *Summary Notes*.

CRE WEB SITE MATERIALS

- **Expanded Material**

    1. *Web P5-1$_B$ Puzzle Problem "What Five Things Are Wrong with This Solution?"*
    2. *Web P5-2$_B$ Gas Phase Reaction $A \rightarrow 3B$*
    3. *Web P5-3$_B$ Catalytic Rate Law*
    4. *Web P5-4$_B$ Fluidized Bed Catalytic Reactor*

- **Learning Resources**

    1. *Summary Notes*
    2. *Web Modules*
        A. Wetlands
        B. Membrane Reactors
        C. Reactive Distillation
        D. Aerosol Reactors
    3. *Interactive Computer Games*

        A. Murder Mystery   B. AspenTech solution to Problem 5-3

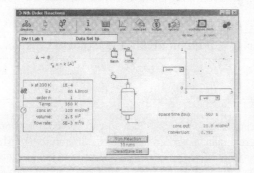
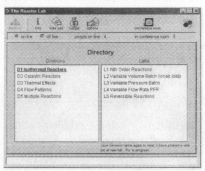

    C. Reactor Lab Modules

    Interactive reactor Lab Modules (*http://reactorlab.net/quiz*) have been developed by Professor Richard Herz in the Chemical Engineering Department at the University of California, San Diego. They are copyrighted and are used here with permission.

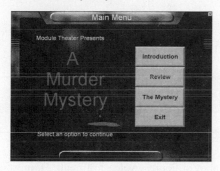

    4. *Solved Problems*
        A. A sinister-looking gentlemen is interested in producing methyl perchlorate in a batch reactor. The reactor has a strange and unsettling rate law (*ECRE*, 2nd Ed. P4-28).
        B. Solution to California Professional Engineers' Exam Problem
        C. Ten Types of Home Problems: 20 Solved Problems

     5. *Analogy of CRE Algorithms to a Menu in a Fine French Restaurant*
     6. *Algorithm for Gas-Phase Reaction*
- **Living Example Problems**
  1. *Example 5-6 Calculating Conversion in a Reactor with Pressure Drop*
     Programs in Polymath, MatLab, and Wolfram
  2. AspenTech solution to *Example Problem 5-3*
- **Professional Reference Shelf**
  R5.1. *Spherical Packed-Bed Reactors*
      When small catalyst pellets are required, the pressure drop can be significant. One type of reactor that minimizes pressure drop and is also inexpensive to build is the *spherical reactor*, shown here. In this reactor, called an *ultraformer*, dehydrogenation reactions such as

$$\text{Paraffin} \longrightarrow \text{Aromatic} + 3H_2$$

  are carried out.
  R5.2 *Recycle Reactors*
      Recycle reactors are used (1) when conversion of unwanted (toxic) products is required and they are recycled to extinction, (2) the reaction is autocatalytic, or (3) it is necessary to maintain isothermal operation. To design recycle reactors, one simply follows the procedure developed in this chapter and then adds a little additional bookkeeping.

QUESTIONS AND PROBLEMS

Homework Problems

The subscript to each of the problem numbers indicates the level of difficulty: A, least difficult; D, most difficult.

$$A = \bullet \quad B = \blacksquare \quad C = \blacklozenge \quad D = \blacklozenge\blacklozenge$$

In each of the following questions and problems, rather than just drawing a box around your answer, write a sentence or two describing how you solved the problem, the assumptions you made, the reasonableness of your answer, what you learned, and any other facts that you want to include. You may wish to refer to W. Strunk and E. B. White, *The Elements of Style*, 4th ed. (New York: Macmillan, 2000) and Joseph M. Williams, *Style: Ten Lessons in Clarity & Grace,* 6th ed. (Glenview, IL: Scott, Foresman, 1999) to enhance the quality of your sentences.

> **Before** solving the problems, state or sketch qualitatively the expected results or trends.

Questions

Q5-1ₐ Read through all the problems at the end of this chapter. Make up and solve an *original* problem based on the material in this chapter. **(a)** Use real data and reactions from the literature. **(b)** Make up a reaction and data. **(c)** Use an example from everyday life (e.g., making toast or cooking spaghetti). In preparing your original problem, first list the principles you want to get across and why the problem is important. Ask yourself how your example will be different from those in the text or lecture. Other things for you to consider when choosing a problem are relevance, interest, impact of the solution, time required to obtain a solution, and degree of difficulty. Look through some of the journals for data, or to get some ideas for industrially important reactions, or for novel applications of reaction engineering principles (the environment, food processing, etc.). At the end of the problem and solution, describe the creative process used to generate the idea for the problem. **(d)** Write a question based on the material in this chapter that requires critical thinking. Explain why your question requires critical thinking. (*Hint:* See Preface, Section B.2) **(e)** Listen to the audios on the CRE Web site (*www.umich.edu/~elements/5e/index.html*) Lecture Notes, pick one, and describe how you might explain it differently.

Q5-2$_A$ If it takes 11 minutes to cook spaghetti in Ann Arbor, Michigan, and 14 minutes in Boulder, Colorado, how long would it take in Cuzco, Peru? Discuss ways to make the spaghetti more tasty. If you prefer to make a creative spaghetti dinner for family or friends rather than answering this question, that's OK, too; you'll get full credit—but **only if** you turn in your recipe and bring your instructor a taste.

Problems

P5-1$_B$ **What if...** you were asked to explore the example problems in this chapter to learn the effects of varying the different parameters? This sensitivity analysis can be carried out by downloading the examples from the CRE Web site (*www.umich.edu/~elements/ 5e/index.html*, under Living Example Problems (**LEPs**)). For each of the example problem you investigate, write a paragraph describing your findings.

    (a) **Example 5-1.** (1) What would be the error in k if the batch reactor were only 80% filled with the same concentrations of reactants, instead of being completely filled as in the example? (2) What generalizations can you draw from this example?

    (b) **Example 5-2.** (1) What conversion would be achieved if three 800-gallon CSTRs were placed in series? In parallel with the feed equally divided? (2) What are the advantages and disadvantages of adding this third reactor? (3) Show that for n equal-size CSTRs, V_i, placed in parallel with equal feed to each, $F_{Ai0} = F_{A0}/n$, that conversion achieved in any one of the reactors will be identical to what would be achieved if the reactor were fed in one stream, $F_{A0} = nF_{Ai0}$, to one large reactor of volume $V = nV_i$.

    (c) **Example 5-3.** (1) How would your reactor volume and number of reactors change if you only needed 50% conversion to produce the 200 million pounds per year required? (2) What generalizations can you draw from this example?

    (d) **Example 5-4.** (1) How would the pressure drop and pressure-drop parameters, α and β_0, change if the particle diameter were reduced by 25%? (2) Plot α as a function of ϕ, keeping constant the other parameters in the example. (3) What generalizations can you draw from this example?

    (e) **Example 5-5.** (1) What would be the reactor volume for $X = 0.8$ if the pressure were increased by a factor of 10, assuming everything else remains the same? (2) Plot and analyze $-r_A$ as a function of V. (3) What generalizations can you draw from this example?

    (f) **Example 5-6.** Load the *Living Example Problem 5-6* from the CRE Web site. How much would the catalyst weight change if the pressure were increased by a factor of 5 and the particle size decreased by a factor of 5 (recall α is also a function of P_0)? Use plots and figures to describe what you find.

    (g) *AspenTech* **Example 5-3.** (1) Using $F_{A0} = 0.425$ lb$_m$ mol/s, run the AspenTech simulation at 1000 K and at 1200 K and compare with the specified temperature of 1100 K. (2) Explore what a small change in activation energy can make in your results by changing E from 82 kcal/mol to 74 kcal/mol and then to 90 kcal/mol and compare your results with the base case of 82 kcal/mol. (3) Double both the flow rate of A and the pressure, and describe what you find.

    (h) **High Finance.** (1) How would your profit/numbers change in Table 5-4 if you used the following 2010 prices? (2) Ethylene glycol $0.54/kg, ethylene $0.76/kg, ethylene oxide $1.17/kg, ethane $0.31/kg, sulfuric acid $0.10/kg (98 wgt %), and propylene glycol $1.70/kg. (3) What pops out at you?

    (i) **Learn a New Dance.** View the YouTube video (*www.youtube.com*) made by the chemical reaction engineering students at the University of Alabama, entitled *CSTR* to the tune of "YMCA." Type in "chemicalreactor" to narrow your search. You can also access it directly from a link in Chapter 5 *Summary Notes* on the CRE Web site (*www.umich.edu/~elements/5e/index.html*); scroll down until you find the YouTube "CSTR."

(j) Download Reactor Lab (*www.reactorlab.net*) onto your computer and call up *D1 Isothermal Reactors*. Detailed instructions with screen shots are given in Chapter 4 *Summary Notes*. **(1)** For **L1** Nth Order Reactions, vary the parameters n, E, T for a batch, CSTR, and PFR. Write a paragraph discussing the trends (e.g., first order versus second order) and describe what you find. **(2)** Next, choose the "Quiz" at the top of the screen, find the reaction order, **(3)** and turn in your performance number.

Performance number: _____

(k) Work the Chapter 5 Self-Tests in *Summary Notes* **on the Web.** Write a question for this problem that involves critical thinking and explain why it involves critical thinking.

P5-2$_B$ Load the Interactive Computer Games (ICG) from the CRE Web site. Run the games and then record your performance number, which indicates your mastery of the material. Your instructor has the key to decode your performance number.

ICG—Mystery Theater—A real "who done it?," see *Pulp and Paper*, 25 (January 1993) and also *Pulp and Paper*, 9 (July 1993). The outcome of the murder trial is summarized in the December 1995 issue of *Papermaker*, page 12. You will use fundamental chemical engineering from Sections 5.1 to 5.3 to identify the victim and the murderer.

Performance number: _____

P5-3$_A$ **Multiple Choice.** In each case, you will need to explain the reason you chose the answer you did.

(a) An irreversible, liquid-phase, second-order reaction, A → Product(s), proceeds to 50% conversion in a PFR operating isothermally, isobarically, and at steady state. What conversion would be obtained if the PFR operated at half the original pressure (with all else unchanged)?
   (1) >50%   (2) <50%   (3) 50%   (4) Insufficient information to answer definitively

(b) An irreversible, gas-phase, second-order reaction, A → Product(s), proceeds to 50% conversion in a PFR operating isothermally, isobarically, and at steady state. What conversion would be obtained if the PFR operated at half the original pressure (with all else unchanged)?
   (1) >50%   (2) <50%   (3) 50%   (4) insufficient information to answer definitively

(c) The rate constant for an irreversible, heterogeneously catalyzed, gas-phase, second-order reaction, A → Product(s), was determined to be 0.234 from experimental data in a packed-bed reactor. The person analyzing the experimental data failed to include the large pressure drop in the reactor in his analysis. If the pressure drop were properly accounted for, the rate constant would be
   (1) >0.234   (2) <0.234   (3) 0.234   (4) insufficient information to answer definitively

P5-4$_B$ **Multiple Choice.** In each of the cases below, **(a)** through **(e)**, you will need to explain why you chose the answer you did.

The elementary isomerization exothermic reaction

$$A \underset{cat}{\overset{}{\rightleftarrows}} B$$

is carried out isothermally at 400 K in a PBR in which pressure drop plays a role, with $\alpha = 0.001$ kg^{-1}. Currently, 50% conversion is achieved. The equilibrium constant at this temperature is 3.0.

(a) For a fixed mass flow rate $\dot{m}$, if the reactor diameter is increased by a factor of 4, the conversion will
   (1) X > 0.5   (2) X < 0.5   (3) X = 0.5   (4) insufficient information to tell.

(b) For a fixed mass flow rate $\dot{m}$, the equilibrium conversion is
   (1) $X_e = 0.5$ (2) $X_e = 0.667$ (3) $X_e = 0.75$ (4) insufficient information to tell

(c) For a fixed mass flow rate $\dot{m}$, if the reactor diameter is increased by a factor of 2, the equilibrium conversion X_e will
   (1) increase   (2) decrease   (3) remain the same   (4) insufficient information to tell

(d) For a fixed mass flow rate $\dot{m}$, if the particle size is increased, the equilibrium conversion will
   (1) increase   (2) decrease   (3) remain the same   (4) insufficient information to tell

(e) For a fixed mass flow rate $\dot{m}$, if the particle size is increased, the conversion will
   (1) increase   (2) decrease   (3) remain the same   (4) insufficient information to tell

P5-5$_A$ The elementary gas-phase reaction

$$2A \rightarrow B$$

is carried out in a constant-volume batch reactor where 50% conversion is achieved in 1 hour. Pure A is charged to the reactor at an initial concentration of 0.2 mol/dm³. If the same reaction is carried out in a CSTR, what volume would be necessary to achieve 50% conversion for a feed molar flow rate of 500 mol/h and an entering concentration of A of 0.2 mol/dm³?

P5-6$_C$ A reversible liquid-phase isomerization A $\rightleftarrows$ B is carried out *isothermally* in a 1000-gal CSTR. *The reaction is second order in both the forward and reverse directions.* The liquid enters at the top of the reactor and exits at the bottom. Experimental data taken in a batch reactor shows the CSTR conversion to be 40%. The reaction is reversible with K_C = 3.0 at 300 K, and ΔH_{Rx}° = –25,000 cal/mol. Assuming that the batch data taken at 300 K are accurate and that E = 15,000 cal/mol, what CSTR temperature do you recommend to obtain maximum conversion? [*Hint:* Read Appendix C and assume ΔC_P = 0 in the appendix Equation (C-9)]:

$$K_C(T) = K_C(T_0)\exp\left[\frac{\Delta H_{Rx}^{\circ}}{R}\left(\frac{1}{T_0} - \frac{1}{T}\right)\right]$$

Use Polymath to make a plot of X versus T. Does it go through a maximum? If so, explain why.

P5-7$_B$ The gas-phase reaction

$$A \rightarrow B + C$$

follows an elementary rate law and is to be carried out first in a PFR and then in a separate experiment in a CSTR. When pure A is fed to a 10 dm³ PFR at 300 K and a volumetric flow rate of 5 dm³/s, the conversion is 80%. When a mixture of 50% A and 50% inert (I) is fed to a 10 dm³ CSTR at 320 K and a volumetric flow rate of 5 dm³/s, the conversion is also 80%. What is the activation energy in cal/mol?

P5-8$_B$ The elementary gas-phase reaction

$$A \rightarrow B$$

takes place isobarically and isothermally in a PFR where 63.2% conversion is achieved. The feed is pure A. It is proposed to put a CSTR of equal volume upstream of the PFR. Based on the entering molar flow rate to A to the first reactor, what will be the intermediate from the CSTR, X_1, and exit conversion from the PFR, X_2, based on the feed to first reactor? The entering flow rates and all other variables remain the same as that for the single PFR. Source: Modified California Registration Exam Problem.

P5-9$_A$ The liquid-phase reaction

$$A + B \longrightarrow C$$

follows an elementary rate law and is carried out isothermally in a flow system. The concentrations of the A and B feed streams are 2 *M* before mixing. The volumetric flow rate of each stream is 5 dm³/min, and the entering temperature is 300 K. The streams are mixed immediately before entering. Two reactors are available. One is a gray, 200.0-dm³ CSTR that can be heated to 77°C or cooled to 0°C, and the other is a white, 800.0-dm³ PFR operated at 300 K that cannot be heated or cooled but can be painted red or black. Note that k = 0.07 dm³/mol·min at 300 K and E = 20 kcal/mol.

(a) Which reactor and what conditions do you recommend? Explain the reason for your choice (e.g., color, cost, space available, weather conditions). Back up your reasoning with the appropriate calculations.

(b) How long would it take to achieve 90% conversion in a 200-dm³ batch reactor with $C_{A0} = C_{B0} = 1$ *M* after mixing at a temperature of 77°C?

(c) What would your answer to part **(b)** be if the reactor were cooled to 0°C? (*Ans.* 2.5 days)

(d) What conversion would be obtained if the CSTR and PFR were operated at 300 K and connected in series? In parallel with 5 mol/min to each?

(e) Keeping Table 4-3 in mind, what batch reactor volume would be necessary to process the same amount of species A per day as the flow reactors, while achieving 90% conversion? Referring to Table 1-1, estimate the cost of the batch reactor.

(f) Write a couple of sentences describing what you learned from the problem and what you believe to be the point of the problem.

(g) Apply one or more of the six ideas in Preface Table P-4, page xxviii, to this problem.

P5-10$_B$ Troubleshooting

(a) A liquid-phase isomerization A $\longrightarrow$ B is carried out in a 1000-gal CSTR that has a single impeller located halfway down the reactor. The liquid enters at the top of the reactor and exits at the bottom. The reaction is second order. Experimental data taken in a batch reactor predicted the CSTR conversion should be 50%. However, the conversion measured in the actual CSTR was 57%. Suggest reasons for the discrepancy and suggest something, e.g., a model, that would give closer agreement between the predicted and measured conversions. Back your suggestions with calculations. P.S. It was raining *very* hard that day.

(b) The liquid-phase reaction

$$A \longrightarrow B$$

was carried out in a CSTR. For an entering concentration of 2 mol/dm^3, the conversion was 40%. For the same reactor volume and entering conditions as the CSTR, the expected PFR conversion is 48.6%. However, the PFR conversion was, amazingly, 52.6% exactly. Brainstorm reasons for the disparity. Quantitatively show how these conversions came about (i.e., the expected conversion and the actual conversion). Part (b) is C-level difficulty.

(c) The gas-phase reaction

$$A + B \longrightarrow C + D$$

is carried out in a packed-bed reactor. When the particle size was decreased by 15%, the conversion remained unchanged. When the particle size was decreased by 20%, the conversion decreased. When the original particle size was increased by 15%, the conversion also decreased. In all cases, the temperature, the total catalyst weight, and all other conditions remained unchanged. What's going on here?

P5-11$_B$ The irreversible elementary gas-phase reaction

$$A + B \longrightarrow C + D$$

is carried out isothermally at 305 K in a packed-bed reactor with 100 kg of catalyst.

The entering pressure was 20 atm and the exit pressure is 2 atm. The feed is equal molar in A and B and the flow is in the turbulent flow regime, with $F_{A0} = 10$ mol/min and $C_{A0} = 0.4$ mol/dm^3. Currently 80% conversion is achieved. What would be the conversion if the catalyst particle size were doubled and everything else remained the same?

P5-12$_B$ The reversible gas-phase elementary reaction

(*ethylbenzene* $\leftrightarrow$ *styrene* + *H$_2$*)

is carried out in an isothermal CSTR with no pressure drop. The feed enters at a volumetric flow rate of $v_0 = 5000\dfrac{dm^3}{hour}$. The feed consists of half ethyl benzene (i.e., A) and half inerts on a molar basis and is well mixed before it enters the reactor (I). The pressure in the reactor is 6 atm (so $P_{A0} = 3$ atm and $P_{I0} = 3$ atm, making the entering concentration of ethyl benzene, A, $C_{A0} = 0.04 \dfrac{mol}{dm^3}$). The molar flow rate of A is $F_{A0} = 200 \dfrac{mol}{hr}$. At the reaction temperature of 640°C, the rate constant, k_A, is 5.92 $\dfrac{mol}{dm^3 \cdot hr \cdot atm}$. The equilibrium constant, K_P, is 9 atm and the corresponding equilibrium conversion is $X_e = 0.84$.

Reference: Won Jae Lee and Gilbert F. Froment. *Ind. Eng. Chem. Res.* 2008, 47, pp. 9183–9194.

(a) Write out each step of the algorithm.
(b) Write the rate of reaction, $-r_A$, solely as a function of P_{A0} X, K_P and k.
(c) Calculate the reactor volume necessary to achieve 90% of the equilibrium conversion, X_e.
(d) How would the conversion from part (a) be affected if the reactor diameter increased and height decreased but total volume remained the same? Explain.

P5-13$_B$ Walter White, Jesse Pinkman, and Mike Ehrmantraut stole 1000 gallons of methylamine during an episode of the TV series *Breaking Bad*. Shortly thereafter, Jesse and Mike decided they would get out of the cooking business and sell their shares of the methylamine. Walter wanted to keep all of the methylamine for future meth cooks and not let Jesse and Mike have their shares. Suppose Jesse and Mike decided to sabotage Walter's cooking operation by ruining the methylamine using the following gas phase reaction:

$$2\,CH_3NH_2 \leftrightarrow (CH_3)_2NH + NH_3$$

or, in short:

$$2A \rightleftarrows B + C$$

This reaction converts the methylamine to dimethylamine, using a silica-alumina catalyst. The molar flow rate into a packed-bed reactor is 23.6 mol/s and the entering pressure is 18 atm. Assume there is no pressure drop or temperature change in the reactor. The reaction rate follows an elementary rate law in terms of partial pressure.

$$k_1 = 4.25 \times 10^{-6} \ \dfrac{mol}{atm^2 \bullet gcatalyst \bullet s} \ \text{ and } K_e = 2.5$$

(a) Write the mole balance.
(b) Write the rate law in terms of partial pressures.
(c) Set up a stoichiometric table for this reaction.
(d) Write the partial pressures in terms of conversion.
(e) Write the rate law solely in terms of conversion.
(f) What is the equilibrium conversion, X_e?
(g) Write out your algorithm in terms of conversion.
(h) How many kilograms of catalyst would Jesse need to load in a PBR to obtain a conversion of $0.9*X_e$? For $X = 0.75\ X_e$?
(i) How many kilograms of catalyst would be needed to obtain 90% of the equilibrium conversion in a fluidized-bed reactor? If this weight is very, very large, what might you suggest to reduce the weight?
(j) What conversion would be achieved in a 100 kg PBR with pressure drop and $\alpha = 0.0098$ kg^{-1}? At what catalyst weight does the exit pressure fall below 1.0 atm?
(k) Repeat (j) when pressure drop is accounted for with $\alpha = 6 \times 10^{-4}$ kg^{-1}.

Ph.D. student Julia Faeth created this problem using modified data from J. W. Mitchell, et al., *Ind. Eng. Chem. Res. 33*, 1994, pp. 181–184.

P5-14$_B$ The dehydration butanol of alumina is carried out over a silica-alumina catalyst at 680 K.

$$CH_3CH_2CH_2CH_2OH \xrightarrow{\text{cat}} CH_3CH = CHCH_3 + H_2O$$

The rate law is

$$-r'_{Bu} = \frac{kP_{Bu}}{\left(1 + K_{Bu}P_{Bu}\right)^2}$$

with $k = 0.054$ mol/gcat•h•atm and $K_{Bu} = 0.32$ atm^{-1}. Pure butanol enters a thin-tubed, packed-bed reactor at a molar flow rate of 50 kmol/hr and a pressure of 10 atm (1013•kPa).

(a) What PBR catalyst weight is necessary to achieve 80% conversion in the absence of pressure drop? Plot and analyze X, p, f (i.e., (v/v_0)) and reaction rate, $-r'_A$, as a function of catalyst weight.

(b) What "fluidized CSTR" catalyst weight is necessary to achieve 80% conversion?

(c) Repeat (a) when there is pressure drop, with the pressure-drop parameter $\alpha = 0.0006$ kg^{-1}. Do you observe a maximum in the rate of reaction, and if so, why? What catalyst weight is necessary to achieve 70% conversion? Compare this weight with that for no pressure drop to achieve the same conversion.

(d) What generalizations can you make about this problem?

(e) Write a question for this problem that requires critical thinking and then explain why your question requires critical thinking. (*Hint:* See Preface Section I.2)

P5-15$_B$ The gaseous reaction A → B has a unimolecular reaction rate constant of 0.0015 min^{-1} at 80°F. This reaction is to be carried out in *parallel tubes* 10 ft long and 1 in. inside diameter, under a pressure of 132 psig at 260°F. A production rate of 1000 lb/h of B is required. Assuming an activation energy of 25,000 cal/mol, how many tubes are needed if the conversion of A is to be 90%? Assume perfect gas laws. A and B each have molecular weights of 58. Source: From California Professional Engineers' Exam.

P5-16$_B$ (a) The irreversible elementary reaction 2A → B takes place in the gas phase in an *isothermal tubular (plug-flow) reactor*. Reactant A and a diluent C are fed in equimolar ratio, and conversion of A is 80%. If the molar feed rate of A is cut in half, what is the conversion of A assuming that the feed rate of C is left unchanged? Assume ideal behavior and that the reactor temperature remains unchanged. What was the point of this problem? Source: From California Professional Engineers' Exam.

(b) Write a question that requires critical thinking, and explain why it involves critical thinking. (See Preface Section I.3.)

P5-17$_B$ Compound A undergoes a reversible isomerization reaction, A → B, over a supported metal catalyst. Under pertinent conditions, A and B are liquid, miscible, and of nearly identical density; the equilibrium constant for the reaction (in concentration units) is 5.8. In a *fixed-bed isothermal flow reactor* in which backmixing is negligible (i.e., plug flow), a feed of pure A undergoes a net conversion to B of 55%. The reaction is elementary. If a second, identical flow reactor at the same temperature is placed downstream from the first, what overall conversion of A would you expect if:

(a) The reactors are directly connected in series? [*Ans.:* $X = 0.74$.]

(b) The products from the first reactor are separated by appropriate processing and only the unconverted A is fed to the second reactor?

(c) Apply one or more of the six ideas in Preface Table P-4, page xxviii, to this problem.

P5-18$_B$ A total of 2500 gal/h of metaxylene is being isomerized to a mixture of orthoxylene, metaxylene, and paraxylene in a reactor containing 1000 ft^3 of catalyst. The reaction is being carried out at 750°F and 300 psig. Under these conditions, 37% of the metaxylene fed to the reactor is isomerized. At a flow rate of 1667 gal/h, 50% of the metaxylene is isomerized at the same temperature and pressure. Energy changes are negligible.

It is now proposed that a second plant be built to process 5500 gal/h of metaxylene at the same temperature and pressure as described earlier. What size reactor (i.e., what volume of catalyst) is required if conversion in the new plant is to be 46% instead of 37%? Justify any assumptions made for the scale-up calculation. [*Ans.:* 2931 ft^3 of catalyst.] Source: From California Professional Engineers' Exam. Make a list of the things you learned from this problem.

P5-19$_B$ It is desired to carry out the gaseous reaction A → B in an existing *tubular reactor* consisting of 50 parallel tubes 40 ft long with a 0.75-in. inside diameter. Bench-scale experiments have given the reaction rate constant for this first-order reaction as 0.00152 s^{-1} at 200°F and 0.0740 s^{-1} at 300°F. At what temperature should the reactor be operated to give a conversion of A of 80% with a feed rate of 500 lb m/h of pure A and an operating pressure of 100 psig? A has a molecular weight of 73. Departures from perfect gas behavior may be neglected, and the reverse reaction is insignificant at these conditions. [*Ans.: T* = 275°F.] Source: From California Professional Engineers' Exam.

P5-20$_B$ The irreversible first-order (wrt partial pressure of A) gas-phase reaction

$$A \rightarrow B$$

is carried out isothermally in a "fluidized" catalytic CSTR containing 50 kg of catalyst.

   Currently, 50% conversion is realized for pure A entering at a pressure of 20 atm. There is virtually no pressure drop in the CSTR. It is proposed to put a PBR containing the same catalyst weight in series with the CSTR. The pressure drop parameter for the PBR, α, given by Equation (5-29) is $\alpha = 0.018$ kg^{-1}. The particle size is 0.2 mm, the bed porosity is 40%, and the viscosity is the same as that of air at 200°C.

Fluidized CSTR

(a) Should the PBR be placed upstream or downstream of the CSTR in order to achieve the highest conversion? Explain qualitatively using concepts you learned in Chapter 2.

(b) What is the conversion coming out of the first reactor?

(c) What is the conversion exiting the last reactor?

(d) What is the pressure at the exit of the packed bed?

(e) How would your answers change if the catalyst diameter were decreased by a factor of 2 and the PBR diameter were increased by 50%, assuming turbulent flow?

P5-21$_B$ A microreactor from the MIT group similar to the one shown in Figure P5-21$_B$ was used to produce phosgene in the gas phase. We continue our discussion on micro reactors in Chapter 6.

$$CO + Cl_2 \rightarrow COCl_2$$

$$A + B \rightarrow C$$

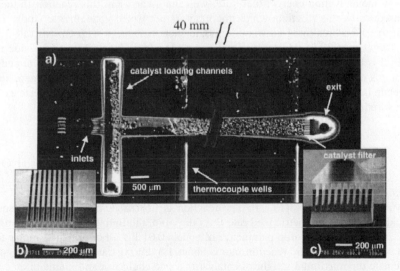

Figure P5-21$_B$ Microreactor. Source: Courtesy of S. K. Ajmera, M. W. Losey, K. F. Jensen, and M. A. Schmidt, *AIChE J. 47*, 1639 (2001). (Article titled "Microfabricated cross-flow chemical reactor for catalyst testing.")

The microreactor is 20 mm long, 500 μm in diameter, and packed with catalyst particles 35 μm in diameter. The entering partial pressure of A is 231 kPa (2.29 atm), and the entering flow to each microreactor is equimolar. The molar flow rate of CO is 2×10^{-5} mol/s and the volumetric flow is 2.83×10^{-7} m^3/s. The weight of catalyst in one microreactor: $W = 3.5 \times 10^{-6}$ kg. The reactor is kept isothermal at 120°C. Because the catalyst is also slightly different than the one in Figure P5-21$_B$, the rate law is different as well:

$$-r'_A = k_A C_A C_B$$

Additional information:

$\alpha = 3.55 \times 10^5$/kg catalyst (based on properties of air and $\phi = 0.4$)

$k = 0.004$ m^6/(mol · s · kg catalyst) at 120°C

$v_0 = 2.83 \cdot 10^{-7}$ m^3/s, $\rho = 7$ kg/m^3, $\mu = 1.94 \cdot 10^{-5}$ kg/m · s

$A_c = 1.96 \cdot 10^{-7}$ m^2, $G = 10.1$ kg/m^2 · s

(a) Plot the molar flow rates F_A, F_B, and F_C, the conversion X, and pressure ratio p along the length (i.e., catalyst weight, W) of the reactor.

(b) Calculate the number of microreactors in parallel needed to produce 10,000 kg/year phosgene.

(c) Repeat part (a) for the case when the catalyst weight remains the same but the particle diameter is cut in half. If possible, compare your answer with part (a) and describe what you find, noting anything unusual.

(d) How would your answers to part (a) change if the reaction were reversible with $K_C = 0.4$ dm^3/mol? Describe what you find.

(e) What are the advantages and disadvantages of using an array of microreactors over using one conventional packed-bed reactor that provides the same yield and conversion?

(f) Write a question that involves critical thinking, and explain why it involves critical thinking. (See Preface, Tables P-3 and P-4.)

(g) Apply one or more of the six ideas in Preface Table P-4, page xxviii, to this problem.

P5-22$_A$ A very proprietary industrial-waste reaction, which we'll code as A→B+S, is to be carried out in a 10-dm^3 CSTR followed by a 10-dm^3 PFR. The reaction is elementary, but A, which enters at a concentration of 0.001 mol/dm^3 and a molar flow rate of 20 mol/min, has trouble decomposing. The specific reaction rate at 42°C (i.e., room temperature in the Mojave desert) is 0.0001 s^{-1}. However, we don't know the activation energy; therefore, we cannot carry out this reaction in the winter in Michigan. Consequently, this reaction, while important, is not worth your time to study. Therefore, perhaps you want to take a break and go watch a movie such as *Dances with Wolves* (all-time favorite of the author), *The Dark Knight*, *Marigold Hotel II*, or *Mad Max: Fury Road*. As a side note, the Jofostan film festival is held the last week in January in Riça, Jofostan. But don't try to attend the Jofostan film festival as it's only "A-List" celebrities such as Denzel Washington, Meryl Streep, and Sven Köttlov.

P5-23$_B$ Ethyl acetate is an extensively used solvent and can be formed by the vapor-phase esterification of acetic acid and ethanol.

$$\underset{\parallel}{\overset{\displaystyle O}{}}\quad\quad\quad\quad\quad\quad\underset{\parallel}{\overset{\displaystyle O}{}}$$

CH$_3$– C – OOH + CH$_3$CH$_2$OH $\longrightarrow$ CH$_3$–C – OCH$_2$CH$_3$ + H$_2$O

The reaction was studied using a microporous resin as a catalyst in a *packed-bed microreactor* [*Ind. Eng. Chem. Res.*, 26(2), 198(1987)]. The reaction is first-order in ethanol and pseudo-zero-order in acetic acid. The total volumetric feed rate is 25 dm^3/min, the initial pressure is 10 atm, the temperature is 223°C, and the pressure-drop parameter, α, equals 0.01 kg^{-1}. For an equal molar feed rate of acetic acid and ethanol, the specific reaction rate is about 1.3 dm^3/kg-cat -min.

(a) Calculate the maximum weight of catalyst that one could use and maintain an exit pressure above 1 atm. [*Ans.*: $W = 99$ kg]

(b) Write out the CRE algorithm and then solve these equations analytically to determine the catalyst weight necessary to achieve 90% conversion.

(c) Write a Polymath program to plot and analyze X, p, and $f = v/v_0$ as a function of catalyst weight down the packed-bed reactor. You can either use your analytical equations for x, p, and f or you can plot these quantities using the Polymath program.

(d) What is the ratio of catalyst needed to achieve the last 5% (85% to 90%) conversion to the weight necessary to achieve the first 5% conversion (0% to 5%) in the reactor? [Note: You can use the results in part **(c)** to also answer this part.]

P5-24$_B$ The gas-phase reaction

$$A + B \rightarrow C + D$$

takes place isothermally at 300 K in a packed-bed reactor in which the feed is equal molar in A and B with $C_{A0} = 0.1$ mol/dm^3. The reaction is second order in A and zero order in B. Currently, 50% conversion is achieved in a reactor with 100 kg of catalysts for a volumetric flow rate 100 dm^3/min. The pressure-drop parameter, α, is $\alpha = 0.0099$ kg^{-1}. If the activation energy is 10,000 cal/mol, what is the specific reaction rate constant at 400 K?

P5-25$_B$ Go to Professor Herz's Reactor Lab on the Web at *www.reactorlab.net*. Click on Download and then click on the English version link. Provide the information asked and then click Download. Select Division 2, Lab 2 of The Reactor Lab concerning a packed-bed reactor (labeled PFR) in which a gas with the physical properties of air flows over spherical catalyst pellets. Perform experiments here to get a feeling for how pressure drop varies with input parameters such as reactor diameter, pellet diameter, gas flow rate, and temperature. In order to get significant pressure drop, you may need to change some of the input values substantially from those shown when you enter the lab. If you get a notice that you can't get the desired flow, then you need to increase the inlet pressure. In Chapter 10, you will learn how to analyze the conversion results in such a reactor.

SUPPLEMENTARY READING

KEILLOR, GARRISON, *Pretty Good Joke Book, A Prairie Home Companion*. St. Paul, MN: Highbridge Co., 2000.

BUTT, JOHN B. *Reaction Kinetics and Reactor Design,* 2nd ed. Revised and Expanded. New York: Marcel Dekker, Inc., 1999.

FOGLER, H. S., *Current Fashions: What the Well-Dressed Chemical Reaction Engineer Will Be Wearing in Spring 2016*. Available Fall 2016.

LEVENSPIEL, O., *Chemical Reaction Engineering*, 3rd ed. New York: Wiley, 1998, Chaps. 4 and 5.

Recent information on reactor design can usually be found in the following journals: *Chemical Engineering Science, Chemical Engineering Communications, Industrial and Engineering Chemistry Research, Canadian Journal of Chemical Engineering, AIChE Journal, Chemical Engineering Progress.*

Isothermal Reactor Design: 6
Moles and Molar Flow Rates

Don't let your fears . . .
Get in the way of your dreams

Anonymous

Overview. In the last chapter we used conversion to design a number of isothermal reactors for single reactions. While in many situations writing the mole balances in terms of conversion is an extremely effective strategy, there are many instances where it is more convenient, and in some cases absolutely necessary, to write the mole balance in terms of moles (N_A, N_B) or molar flow rates (F_A, F_B), as shown in Table S-1 in Chapter 1. In this chapter we show how to make small changes in our algorithm to analyze these situations. Using our algorithm, we first write a mole balance on each and every species, and second, we need to relate the rates of reaction of each species to one another using the relative rates described in Chapter 3, i.e., Equation (3-1).

We will use molar flow rates in our mole balance to analyze

- A microreactor with the reaction

$$2NOCl \rightarrow 2NO + Cl_2$$

- A membrane reactor used for the dehydrogenation of ethylbenzene

$$C_6H_5 CH_2CH_3 \rightarrow C_6H_5CH = CH_2 + H_2$$

- A semibatch reactor used for the reaction

$$CNBr + CH_3NH_2 \rightarrow CH_3Br + NCNH_2$$

We will again use mole balances in terms of these variables (N_i, F_i) for multiple reactions in Chapter 8 and for heat effects in Chapters 11 through 13.

6.1 The Molar Flow Rate Balance Algorithm

Used for:
• Multiple rxns
• Membranes
• Unsteady state

There are many instances when it is much more convenient to work in terms of the number of moles (N_A, N_B) or molar flow rates (F_A, F_B, etc.) rather than conversion. Membrane reactors and multiple reactions taking place in the gas phase are two such cases where molar flow rates are necessary rather than conversion. We now modify our algorithm by using concentrations for liquids and molar flow rates for gases as our dependent variables. The main difference between the conversion algorithm and the molar flow rate/concentration algorithm is that, in the conversion algorithm, we needed to write a mole balance on only *one species,* whereas in the molar flow rate and concentration algorithm, we must write a mole balance on *each and every species.* This algorithm is shown in Figure 6-1. First, we write the mole balances on all species present, as shown in Step ①. Next, we write the rate law, Step ②, and then we relate the mole balances to one another through the relative rates of reaction, as shown in Step ③. Steps ④ and ⑤ are used to relate the concentrations in the rate law to the molar flow rates. In Step ⑥, all the steps are combined by the ODE solver (e.g., Polymath).

6.2 Mole Balances on CSTRs, PFRs, PBRs, and Batch Reactors

6.2.1 Liquid Phase

For liquid-phase reactions, the density remains constant and consequently there is no change in either the volume V or the volumetric flow rate $v = v_0$ during the course of the reaction. Therefore, concentration is the preferred design variable. The mole balances derived in Chapter 1 (Table S-1) are now applied to each species for the generic reaction

$$a\mathrm{A} + b\mathrm{B} \rightarrow c\mathrm{C} + d\mathrm{D} \qquad (2\text{-}1)$$

The mole balances are then coupled to one another using the relative rates of reaction

Used to couple the mole balances

$$\frac{r_A}{-a} = \frac{r_B}{-b} = \frac{r_C}{c} = \frac{r_D}{d} \qquad (3\text{-}1)$$

to arrive at Table 6-1, which gives the balance equations in terms of concentration for the four types of reactors we have been discussing. We see from Table 6-1 that we have *only* to specify the parameter values for the system (C_{A0}, v_0, etc.) and for the rate-law parameters (e.g., k_A, α, β) to solve the coupled ordinary differential equations for either BRs, PFRs, or PBRs, or to solve the coupled algebraic equations for a CSTR.

$$A + 2B \rightleftharpoons C$$

Mole Balance

① Write mole balance on each species.[†]

e.g., $\dfrac{dF_A}{dV} = r_A$, $\dfrac{dF_B}{dV} = r_B$, $\dfrac{dF_C}{dV} = r_C$

Rate Law

② Write rate law in terms of concentration.

e.g., $-r_A = k_A\left(C_A C_B^2 - \dfrac{C_C}{K_C}\right)$

Relative Rates

③ Relate the rates of reaction of each species to one another.

$$\dfrac{-r_A}{1} = \dfrac{-r_B}{2} = \dfrac{r_C}{1}$$

e.g., $r_B = 2r_A$, $r_C = -r_A$

Stoichiometry

④ **(a)** Write the concentrations in terms of molar flow rates for **gas-phase** reactions.

e.g., $C_A = C_{T0}\dfrac{F_A}{F_T}\dfrac{T_0}{T}\dfrac{P}{P_0}$, $C_B = C_{T0}\dfrac{F_B}{F_T}\dfrac{T_0}{T}\dfrac{P}{P_0}$

with $F_T = F_A + F_B + F_C$

(b) For **batch** or **liquid-phase** reactions, use concentration, e.g., C_A, C_B

Pressure Drop

⑤ Write the *gas-phase* pressure drop term in terms of molar flow rates.

$$\dfrac{dp}{dW} = -\dfrac{\alpha}{2p}\left(\dfrac{T}{T_0}\right)\dfrac{F_T}{F_{T_0}}, \quad \text{with } p = \dfrac{P}{P_0}$$

Combine

⑥ For isothermal operation, $T = T_0$, we can use an ODE solver or a nonlinear equation solver (e.g., Polymath) to combine Steps ① through ⑤ to solve for, for example, the profiles of molar flow rates, concentration, and pressure.

[†] For a PBR, use $\dfrac{dF_A}{dW} = r_A'$, $\dfrac{dF_B}{dW} = r_B'$, and $\dfrac{dF_C}{dW} = r_C'$.

Figure 6-1 Isothermal-reaction design algorithm for mole balances.

LIQUIDS

TABLE 6-1   MOLE BALANCES FOR LIQUID-PHASE REACTIONS

Batch $\dfrac{dC_A}{dt} = r_A$ and $\dfrac{dC_B}{dt} = \dfrac{b}{a} r_A$

CSTR $V = \dfrac{v_0(C_{A0} - C_A)}{-r_A}$ and $V = \dfrac{v_0(C_{B0} - C_B)}{-(b/a)r_A}$

PFR $v_0 \dfrac{dC_A}{dV} = r_A$ and $v_0 \dfrac{dC_B}{dV} = \dfrac{b}{a} r_A$

PBR $v_0 \dfrac{dC_A}{dW} = r'_A$ and $v_0 \dfrac{dC_B}{dW} = \dfrac{b}{a} r'_A$

6.2.2 Gas Phase

The mole balances for gas-phase reactions are given in Table 6-2 in terms of the number of moles (batch) or molar flow rates for the generic rate law for the generic reaction, Equation (2-1). The molar flow rates for each species F_j are obtained from a mole balance on each species (i.e., A, B, C, and D), as given in Table 6-2. For example, for a plug-flow reactor

Must write a mole balance on each species

$$\frac{dF_j}{dV} = r_j \tag{1-11}$$

The generic power-law rate law for species A is

Rate Law

$$-r_A = k_A C_A^{\alpha} C_B^{\beta} \tag{3-3}$$

The rate law wrt A is coupled with the equation for relative rates,

$$\frac{-r_A}{a} = \frac{-r_B}{b} = \frac{r_C}{c} = \frac{r_D}{d} \tag{3-1}$$

Given the rate law for species A, we use Equation (3-1) to substitute for species j, rj, in Equation (1-11), the PFR mole balance.

To relate concentrations to molar flow rates, recall Equation (4-17), with $p = P/P_0$

Stoichiometry

$$\boxed{C_j = C_{T0} \frac{F_j}{F_T} \frac{T_0}{T}\, p} \tag{4-17}$$

The pressure-drop equation, Equation (5-28), for isothermal operation ($T = T_0$) is

$$\frac{dp}{dW} = \frac{-\alpha}{2p} \frac{F_T}{F_{T0}} \tag{5-28}$$

TABLE 6-2   ALGORITHM FOR GAS-PHASE REACTIONS

$$a\text{A} + b\text{B} \longrightarrow c\text{C} + d\text{D}$$

1. Mole balances:

CSTR	PFR	PBR
$V = \dfrac{F_{A0} - F_A}{-r_A}$	$\dfrac{dF_A}{dV} = r_A$	$\dfrac{dF_A}{dW} = r_A{}'$
$V = \dfrac{F_{B0} - F_B}{-r_B}$	$\dfrac{dF_B}{dV} = r_B$	$\dfrac{dF_B}{dW} = r_B{}'$
$V = \dfrac{F_{C0} - F_C}{-r_C}$	$\dfrac{dF_C}{dV} = r_C$	$\dfrac{dF_C}{dW} = r_C{}'$
$V = \dfrac{F_{D0} - F_D}{-r_D}$	$\dfrac{dF_D}{dV} = r_D$	$\dfrac{dF_D}{dW} = r_D{}'$

We shall continue the algorithm using a **PBR** as an example.

2. Rates:

Rate Law

$$-r_A{}' = k_A C_A^{\alpha} C_B^{\beta}$$

Relative Rates

$$\frac{r_A'}{-a} = \frac{r_B'}{-b} = \frac{r_C'}{c} = \frac{r_D'}{d}$$

then

$$r_B' = \frac{b}{a} r_A' \qquad r_C' = -\frac{c}{a} r_A' \qquad r_D' = -\frac{d}{a} r_A'$$

3. Stoichiometry:

Concentrations

$$C_A = C_{T0}\frac{F_A T_0}{F_T T}\, p \qquad C_B = C_{T0}\frac{F_B T_0}{F_T T}\, p$$

$$C_C = C_{T0}\frac{F_C T_0}{F_T T}\, p \qquad C_D = C_{T0}\frac{F_D T_0}{F_T T}\, p$$

$$\frac{dp}{dW} = \frac{-\alpha}{2p}\frac{F_T}{F_{T0}}\frac{T}{T_0}, \quad p = \frac{P}{P_0}$$

Total molar flow rate: $F_T = F_A + F_B + F_C + F_D + F_I$

4. Combine:

    Appropriate reactor mole balance on each species
    Rate law
    Concentration for each species
    Pressure-drop equation

5. Evaluate:

    1. Specify and enter parameter values: $k_A, C_{T0}, \alpha, \beta, T_0, a, b, c, d$
    2. Specify and enter entering molar flow rates: $F_{A0}, F_{B0}, F_{C0}, F_{D0}$, and final volume, V_{final}

6. Use an ODE solver.

Many times we will let the ODE solver replace **Step 4, Combine**.

Gas phase

The total molar flow rate is given by the sum of the flow rates of the individual species

$$F_T = \sum_{j=1}^{n} F_j$$

when species A, B, C, D, and inert I are the only ones present. Then

$$F_T = F_A + F_B + F_C + F_D + F_I$$

We now combine all the preceding information, as shown in Table 6-2.[1]

6.3 Application of the PFR Molar Flow Rate Algorithm to a Microreactor

A photo of a micro reactor is shown in Figure P5-21$_B$ on page 203.

Microreactors are emerging as a new technology in CRE. Microreactors are characterized by their high surface-area-to-volume ratios in their microstructured regions that contain tubes or channels. A typical channel width might be 100 μm with a length of 20,000 μm (2 cm). The resulting high surface-area-to-volume ratio (ca. 10,000 m^2/m^3) reduces or even eliminates heat and mass transfer resistances often found in larger reactors. Consequently, surface-catalyzed reactions can be greatly facilitated, hot spots in highly exothermic reactions can be minimized, and in many cases highly exothermic reactions can be carried out isothermally. These features provide the opportunity for microreactors to be used to study the intrinsic kinetics of reactions. Another advantage of microreactors is their use in the production of toxic or explosive intermediates where a leak or microexplosion for a single unit will do minimal damage because of the small quantities of material involved. Other advantages include shorter residence times and narrower residence time distributions.

Advantages of microreactors

Figure 6-2 shows (a) a microreactor with heat exchanger and (b) a microplant with reactor, valves, and mixers. Heat, $\dot{Q}$, is added or taken away by the fluid flowing perpendicular to the reaction channels, as shown in Figure 6-2(a). Production in microreactor systems can be increased simply by adding more units in parallel. For example, the catalyzed reaction

$$R-CH_2OH + \tfrac{1}{2}O_2 \xrightarrow{\text{Ag}} R-CHO + H_2O$$

required only 32 microreaction systems in parallel to produce 2000 tons/yr of acetate!

Microreactors are also used for the production of specialty chemicals, combinatorial chemical screening, lab-on-a-chip, and chemical sensors. In

---

[1] View the YouTube video made by the chemical reaction engineering students at the University of Alabama, entitled *Chemical Goodtime Rhyme*. Videos can be accessed directly from the home page of the CRE Web site, *www.umich.edu/~elements/5e/index.html*.

(b)

(a)

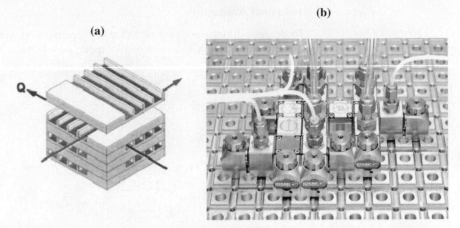

Figure 6-2   Microreactor (a) and microplant (b). (Photo courtesy of Ehrfeld, Hessel, and Löwe, *Microreactors: New Technology for Modern Chemistry*, Weinheim, Germany: Wiley-VCH, 2000.)

modeling microreactors, we will assume they are either in plug flow for which the mole balance is

$$\frac{dF_A}{dV} = r_A \tag{1-12}$$

or in laminar flow, in which case we will use the segregation model discussed in Chapter 17. For the plug-flow case, the algorithm is described in Figure 6-1.

Example 6–1   Gas-Phase Reaction in a Microreactor—Molar Flow Rates

The gas-phase reaction

$$2NOCl \longrightarrow 2NO + Cl_2$$

is carried out at 425°C and 1641 kPa (16.2 atm). Pure NOCl is to be fed, and the reaction follows an elementary rate law.[2] It is desired to produce 20 tons of NO per year in a microreactor system using a bank of ten microreactors in parallel. Each microreactor has 100 channels with each channel 0.2 mm square and 250 mm in length.

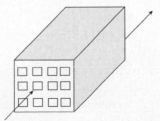

(a)   Plot and analyze the molar flow rates as a function of volume down the length of the reactor. The volume of each channel is 10^{-5} dm^3.
(b)   What is the reactor volume necessary to achieve 85% conversion?

---

[2] J. B. Butt, *Reaction Kinetics and Reactor Design*, 2nd ed. (New York: Marcel Dekker, 2001), p. 153.

Additional Information

To produce 20 tons per year of NO at 85% conversion would require a feed rate of 0.0226 mol/s of NOCl, or 2.26×10^{-5} mol/s per channel. The rate constant is

$$k = 0.29 \frac{dm^3}{mol \cdot s} \text{ at 500 K with } E = 24 \frac{kcal}{mol}$$

Solution

For one channel

Find V.

$$F_{A0} = \frac{22.6 \ \mu mol}{s}$$

$$F_B = \frac{19.2 \ \mu mol}{s}, \quad X = 0.85, \quad V = ?$$

Although this particular problem could be solved using conversion, we shall illustrate how it can also be solved using molar flow rates as the variable in the mole balance. Why do we do this? We do this to give practice using molar flow rates as the variables in order to help prepare the reader for the more complex problems where conversion cannot be used as a variable.

We first write the reaction in symbolic form and then divide by the stoichiometric coefficient of the limiting reactant, NOCl.

Chemistry:	$2NOCl \rightarrow 2NO + Cl_2$
Symbolic Form:	$2A \rightarrow 2B + C$
Limiting Reactant:	$A \rightarrow B + \frac{1}{2}C$

1. Mole balances on species A, B, and C:

$$\frac{dF_A}{dV} = r_A \tag{E6-1.1}$$

$$\frac{dF_B}{dV} = r_B \tag{E6-1.2}$$

$$\frac{dF_C}{dV} = r_C \tag{E6-1.3}$$

Following the Algorithm

2. Rates:

(a) *Rate Law*

$$-r_A = kC_A^2, \text{ with } k = 0.29 \frac{dm^3}{mol \cdot s} \text{ at 500 K} \tag{E6-1.4}$$

(b) *Relative Rates*

$$\frac{r_A}{-1} = \frac{r_B}{1} = \frac{r_C}{\frac{1}{2}}$$

$$r_B = -r_A$$

$$r_C = -\tfrac{1}{2}r_A$$

3. Stoichiometry: Gas phase with $T = T_0$ and $P = P_0$, then $v = v_0 \dfrac{F_T}{F_{T0}}$
Concentration Gas Phase

$$C_j = C_{T0}\left(\frac{F_j}{F_T}\right)\left(\frac{P}{P_0}\right)\left(\frac{T_0}{T}\right) \tag{4-17}$$

Applying Equation (4-17) to species A, B, and C, for isothermal operation $T = T_0$, and for no pressure drop, $P = P_0$ ($P = 1$), the concentrations are

$$C_A = C_{T0}\frac{F_A}{F_T}, \quad C_B = C_{T0}\frac{F_B}{F_T}, \quad C_C = C_{T0}\frac{F_C}{F_T} \tag{E6-1.5}$$

$$\text{with } F_T = F_A + F_B + F_C$$

4. Combine: The rate law in terms of molar flow rates is

$$-r_A = kC_{T0}^2\left(\frac{F_A}{F_T}\right)^2$$

combining all

$$\frac{dF_A}{dV} = -kC_{T0}^2\left(\frac{F_A}{F_T}\right)^2 \tag{E6-1.6}$$

$$\frac{dF_B}{dV} = kC_{T0}^2\left(\frac{F_A}{F_T}\right)^2 \tag{E6-1.7}$$

$$\frac{dF_C}{dV} = \frac{k}{2}C_{T0}^2\left(\frac{F_A}{F_T}\right)^2 \tag{E6-1.8}$$

5. Evaluate:

$$C_{T0} = \frac{P_0}{RT_0} = \frac{(1641 \text{ kPa})}{\left(8.314\ \dfrac{\text{kPa} \cdot \text{dm}^3}{\text{mol} \cdot \text{K}}\right)698 \text{ K}} = 0.286\frac{\text{mol}}{\text{dm}^3} = \frac{0.286 \text{ mmol}}{\text{cm}^3}$$

When using Polymath or another ODE solver, one does not have to actually combine the mole balances, rate laws, and stoichiometry, as was done in the Combine step previously in Chapter 5. The ODE solver will do that for you. Thanks, ODE solver! The Polymath program and output are shown in Table E6-1.1 and Figure E6-1.1. Note that explicit Equation #6 in the Polymath program calculates the reaction-rate constant k at the specified temperature of 425°C (i.e., 698 K).

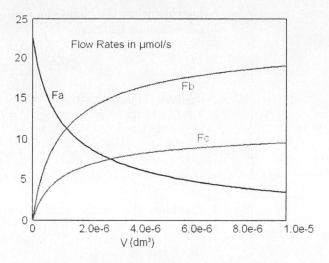

Figure E6-1.1 Profiles of microreactor molar flow rates.

TABLE E6-1.1   POLYMATH PROGRAM

Information on how to obtain and load the Polymath software can be found in Appendix E.

Differential equations

1 d(Fa)/d(V) = ra

2 d(Fb)/d(V) = rb

3 d(Fc)/d(V) = rc

Explicit equations

1   T = 698

2   Cto = 1641/8.314/T

3   E = 24000

4   Ft = Fa+Fb+Fc

5   Ca = Cto*Fa/Ft

6   k = 0.29*exp(E/1.987*(1/500-1/T))

7   Fao = 0.0000226

8   vo = Fao/Cto

9   Tau = V/vo

10 ra = -k*Ca^2

11 X = 1-Fa/Fao

12 rb = -ra

13 rc = -ra/2

Calculated values of DEQ variables

	Variable	Initial value	Final value
1	Ca	0.2827764	0.0307406
2	Cto	0.2827764	0.2827764
3	E	2.4E+04	2.4E+04
4	Fa	2.26E-05	3.495E-06
5	Fao	2.26E-05	2.26E-05
6	Fb	0	1.91E-05
7	Fc	0	9.552E-06
8	Ft	2.26E-05	3.215E-05
9	k	274.4284	274.4284
10	ra	-21.94397	-0.2593304
11	rateA	21.94397	0.2593304
12	rb	21.94397	0.2593304
13	rc	10.97199	0.1296652
14	T	698.	698.
15	Tau	0	0.1251223
16	V	0	1.0E-05
17	vo	7.992E-05	7.992E-05
18	X	0	0.8453416

**Analysis:** This gas-phase reaction in a PFR example could just as easily have been solved using conversion as a basis. However, membrane reactors and multiple reactions **cannot** be solved using conversion. You will note we just wrote out the equations in Steps 1 through 5 of our reaction algorithm (Table 6-2) and then typed them directly into our ODE solver, Polymath, to obtain the molar flow rate profiles shown in Figure E6-1.1. Notice the profiles change rapidly near the reactor entrance and then there is very little change after 6×10^{-6} dm^3 down the reactor. Other interesting variables you will want to plot when you download this program from the *Living Example Problem* (i.e., **LEP**) file are the total molar flow rate, F_T, the concentrations of the reacting species, C_A, C_B, and C_C (for C_B and C_C you will need to type in two additional equations), and the rates $-r_A$, r_B and r_C.

6.4 Membrane Reactors

Membrane reactors can be used to increase conversion when the reaction is *thermodynamically limited*, as well as to increase the selectivity when multiple reactions are occurring. Thermodynamically limited reactions are reactions where the equilibrium lies far to the left (i.e., reactant side) and as a result there is little conversion. If the reaction is exothermic, increasing the temperature will only drive the reaction further to the left, and decreasing the temperature will result in a reaction rate so slow that there is very little conversion. If the reaction is endothermic, increasing the temperature will move the reaction to the right to favor a higher conversion; however, for many reactions these higher temperatures cause the catalyst to become deactivated.

The term *membrane reactor* describes a number of different types of reactor configurations that contain a membrane. The membrane can either provide a barrier to certain components while being permeable to others, prevent certain components such as particulates from contacting the catalyst, or contain reactive sites and be a catalyst in itself. Like reactive distillation, the membrane reactor is another technique for driving reversible reactions to the right toward completion in order to achieve very high conversions. These high conversions can be achieved by having one of the reaction products diffuse out through a semipermeable membrane surrounding the reacting mixture. As a result, the reverse reaction will not be able to take place, and the reaction will continue to proceed to the right toward completion.

> By having one of the products pass through the membrane, we drive the reaction toward completion.

Two of the main types of catalytic membrane reactors are shown in Figure 6-3. The reactor in Figure 6-3(b) is called an *inert membrane reactor with catalyst pellets on the feed side* (IMRCF). Here, the membrane is inert and serves as a barrier to the reactants and some of the products. The reactor in Figure 6-3(c) is a *catalytic membrane reactor* (CMR). The catalyst is deposited directly on the membrane, and only specific reaction products are able to exit the permeate side. As an example, consider the reversible reaction where

$$C_6H_{12} \; \rightleftharpoons \; 3H_2 + C_6H_6$$

$$A \; \rightleftharpoons \; 3B + C$$

> H_2 diffuses through the membrane, while C_6H_6 does not.

the hydrogen molecule is small enough to diffuse through the small pores of the membrane, while C_6H_{12} and C_6H_6 cannot. Consequently, the reaction continues to proceed to the right even for a small value of the equilibrium constant.

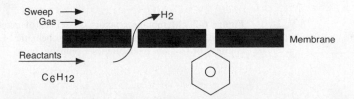

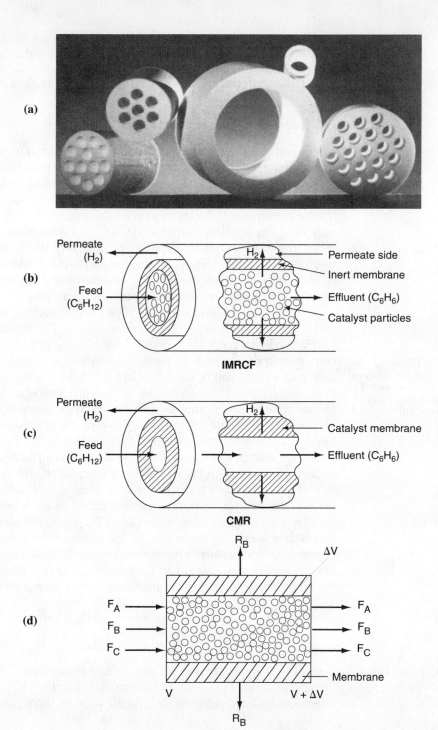

Figure 6-3 Membrane reactors. (a) Photo of ceramic reactors, (b) cross section of IMRCF, (c) cross section of CMR, (d) schematic of IMRCF for mole balance. (Photo courtesy of Golden Technologies Company, Inc.)

Hydrogen, species B, flows out through the sides of the reactor as it flows down the reactor with the other products that cannot leave until they exit the reactor.

In analyzing membrane reactors, we only need to make a small change to the algorithm shown in Figure 6-1. We shall choose the reactor volume rather than catalyst weight as our independent variable for this example. The catalyst weight, W, and reactor volume, V, are easily related through the bulk catalyst density, ρ_b (i.e., $W = \rho_b V$). The mole balances on the chemical species that stay *within* the reactor, namely A and C, are shown in Figure 6-3(d).

$$\boxed{\frac{dF_A}{dV} = r_A} \tag{1-11}$$

$$V = \frac{W}{\rho_b}$$

The mole balance on C is carried out in an identical manner to A, and the resulting equation is

$$\boxed{\frac{dF_C}{dV} = r_C} \tag{6-1}$$

However, the mole balance on B (H_2) must be modified because hydrogen leaves through both the sides of the reactor and at the end of the reactor.

First, we shall perform mole balances on the volume element ΔV shown in Figure 6-3(d). The mole balance on hydrogen (B) is over a differential volume ΔV shown in Figure 6-3(d) and it yields

Balance on B in the catalytic bed:

Now there are two "OUT" terms for species B.

$$\begin{bmatrix} \text{In} \\ \text{by flow} \end{bmatrix} - \begin{bmatrix} \text{Out} \\ \text{by flow} \end{bmatrix} - \begin{bmatrix} \text{Out} \\ \text{by diffusion} \end{bmatrix} + [\text{Generation}] = [\text{Accumulation}]$$

$$\overbrace{F_B|_V} \quad - \quad \overbrace{F_B|_{V+\Delta V}} \quad - \quad \overbrace{R_B \Delta V} \quad + \quad \overbrace{r_B \Delta V} \quad = 0 \tag{6-2}$$

where R_B is the molar rate of B leaving through the sides of the reactor per unit volume of reactor (mol/m^3·s). Dividing by ΔV and taking the limit as $\Delta V \to 0$ gives

$$\boxed{\frac{dF_B}{dV} = r_B - R_B} \tag{6-3}$$

The rate of transport of B out through the membrane R_B is the product of the molar flux of B normal to the membrane, W_B (mol/m^2/s), and the surface area per unit volume of reactor, a (m^2/m^3). The molar flux of B, W_B in (mol/m^2/s) out through the sides of the reactor is the product of the mass transfer coefficient, k_C' (m/s), and the concentration driving force across the membrane.

$$W_B = k_C'(C_B - C_{BS}) \tag{6-4}$$

Here, k'_C is the overall mass transfer coefficient in m/s and C_{BS} is the concentration of B in the sweep gas channel (mol/m^3). The overall mass transfer coefficient accounts for all resistances to transport: the tube-side resistance of the membrane, the membrane itself, and on the shell- (sweep gas) side resistance. Further elaboration of the mass transfer coefficient and its correlations can be found in the literature and in Chapter 14. In general, this coefficient can be a function of the membrane and fluid properties, the fluid velocity, and the tube diameters.

To obtain the rate of removal of B per unit volume of reactor, R_B (mol/m^3/s), we need to multiply the flux through the membrane, W_B (mol/m^2·s), by the membrane surface area per volume of reactor, a (m^2/m^3); that is

$$R_B = W_B a = k'_C a(C_B - C_{BS}) \tag{6-5}$$

The membrane surface area per unit volume of reactor is

$$a = \frac{\text{Area}}{\text{Volume}} = \frac{\pi D L}{\frac{\pi D^2}{4} L} = \frac{4}{D}$$

Letting $k_C = k'_C\, a$ and assuming the concentration in the sweep gas is essentially zero (i.e., $C_{BS} \approx 0$), we obtain

Rate of B out through the sides.

$$\boxed{R_B = k_C C_B} \tag{6-6}$$

where the units of k_C are s^{-1}.

More detailed modeling of the transport and reaction steps in membrane reactors is beyond the scope of this text but can be found in *Membrane Reactor Technology*.[3] The salient features, however, can be illustrated by the following example. When analyzing membrane reactors, we must use molar flow rates because expressing the molar flow rate of B in terms of conversion will not account for the amount of B that has left the reactor through the sides.

According to the DOE, 10 trillion Btu/yr could be saved by using membrane reactors.

Example 6–2   Membrane Reactor

According to the Department of Energy (DOE), an energy saving of 10 trillion Btu per year could result from the use of catalytic membrane reactors as replacements for conventional reactors for dehydrogenation reactions such as the dehydrogenation of ethylbenzene to styrene

---

[3] R. Govind and N. Itoh, eds., *Membrane Reactor Technology,* AIChE Symposium Series 85, no. 268 (1989). T. Sun and S. Khang, *Ind. Eng. Chem. Res.,* 27, 1136 (1988).

and of butane to butene:

$$C_4H_{10} \longrightarrow C_4H_8 + H_2$$

The dehydrogenation of propane is another reaction that has proven successful with a membrane reactor.[4]

$$C_3H_8 \longrightarrow C_3H_6 + H_2$$

All the preceding elementary dehydrogenation reactions described above can be represented symbolically as

$$A \rightleftharpoons B + C$$

and will take place on the catalyst side of an IMRCF. The equilibrium constant for this reaction is quite small at 227°C (e.g., $K_C = 0.05$ mol/dm^3). The membrane is permeable to B (e.g., H$_2$) but not to A and C. Pure gaseous A enters the reactor at 8.2 atm and 227°C ($C_{T0} = 0.2$ mol/dm^3) at a molar flow rate of 10 mol/min.

The rate of diffusion of B out of the reactor per unit volume of reactor, R_B, is proportional to the concentration of B (i.e., $R_B = k_C C_B$).

(a) Perform differential mole balances on A, B, and C to arrive at a set of coupled differential equations to solve.

(b) Plot and analyze the molar flow rates of each species as a function of reactor volume.

(c) Calculate the conversion of A at V = 500 dm^3.

Additional information: Even though this reaction is a gas–solid catalytic reaction, we will use the bulk catalyst density in order to write our balances in terms of reactor volume rather than catalyst weight (recall $-r_A = -r_A' \rho_b$). For the bulk catalyst density of $\rho_b = 1.5$ g/cm^3 and a 2-cm inside-diameter tube containing the catalyst pellets, the specific reaction rate, k, and the transport coefficient, k_C, are $k = 0.7$ min^{-1} and $k_C = 0.2$ min^{-1}, respectively.

Solution

We shall choose reactor volume rather than catalyst weight as our independent variable for this example. The catalyst weight, W, and reactor volume, V, are easily related through the bulk catalyst density, ρ_b, (i.e., $W = \rho_b V$). First, we shall perform mole balances on the volume element ΔV shown in Figure 6-3(d).

1. **Mole balances:**

Balance on A in the catalytic bed

Mole balance
on each and
every species

$$\left[\begin{array}{c} \text{In} \\ \text{by flow} \end{array} \right] - \left[\begin{array}{c} \text{Out} \\ \text{by flow} \end{array} \right] + \left[\text{Generation} \right] = \left[\text{Accumulation} \right]$$

$$\overbrace{F_A|_V} - \overbrace{F_A|_{V+\Delta V}} + \overbrace{r_A \Delta V} = 0$$

---

[4] *J. Membrane Sci.*, 77, 221 (1993).

Dividing by ΔV and taking the limit as $\Delta V \rightarrow 0$ gives

$$\boxed{\frac{dF_A}{dV} = r_A}$$

(E6-2.1)

Balance on B in the catalytic bed
The balance on B is given by Equation (6-3)

$$\boxed{\frac{dF_B}{dV} = r_B - R_B}$$

(E6-2.2)

where R_B is the molar flow of B out through the membrane per unit volume of the reactor.

The mole balance on C is carried out in an identical manner to A, and the resulting equation is

$$\boxed{\frac{dF_C}{dV} = r_C}$$

(E6-2.3)

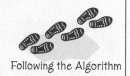

Following the Algorithm

2. **Rates:**
Rate Law

$$\boxed{-r_A = k\left(C_A - \frac{C_B C_C}{K_C}\right)}$$

(E6-2.4)

Relative Rates

$$\frac{r_A}{-1} = \frac{r_B}{1} = \frac{r_C}{1}$$

(E6-2.5)

$$\boxed{r_B = -r_A}$$

(E6-2.6)

$$\boxed{r_C = -r_A}$$

(E6-2.7)

3. **Transport out of the reactor:** We apply Equation (6-5) for the case in which the concentration of B of the sweep side is zero, $C_{BS} = 0$, to obtain

$$\boxed{R_B = k_C C_B}$$

(E6-2.8)

where k_C is a transport coefficient. In this example, we shall assume that the resistance to species B out of the membrane is a constant and, consequently, k_C is a constant.

4. **Stoichiometry:** Recalling Equation (4-17) for the case of constant temperature and pressure, we have for isothermal operation, $T = T_0$, and no pressure drop, $P = P_0$, (i.e., $p = 1$).

Concentrations

$$C_A = C_{T0} \frac{F_A}{F_T} \tag{E6-2.9}$$

$$C_B = C_{T0} \frac{F_B}{F_T} \tag{E6-2.10}$$

$$C_C = C_{T0} \frac{F_C}{F_T} \tag{E6-2.11}$$

$$F_T = F_A + F_B + F_C \tag{E6-2.12}$$

5. **Combining and summarizing:**

Summary of
equations
describing flow and
reaction in a
membrane
reactor

$$\frac{dF_A}{dV} = r_A$$

$$\frac{dF_B}{dV} = -r_A - k_C C_{T0} \left(\frac{F_B}{F_T} \right)$$

$$\frac{dF_C}{dV} = -r_A$$

$$-r_A = k C_{T0} \left[\left(\frac{F_A}{F_T} \right) - \frac{C_{T0}}{K_C} \left(\frac{F_B}{F_T} \right) \left(\frac{F_C}{F_T} \right) \right]$$

$$F_T = F_A + F_B + F_C$$

6. **Parameter evaluation:**

$$C_{T0} = \frac{P_0}{RT_0} = \frac{830.6 \text{ kPa}}{[8.314 \text{ k Pa} \cdot \text{dm}^3/(\text{mol} \cdot \text{K})] (500 \text{ K})} = 0.2 \frac{\text{mol}}{\text{dm}^3}$$

$$k = 0.7 \text{ min}^{-1}, K_C = 0.05 \text{ mol/dm}^3, k_C = 0.2 \text{ min}^{-1}$$

$$F_{A0} = 10 \text{ mol/min}$$

$$F_{B0} = F_{C0} = 0$$

7. **Numerical solution:** Equations (E6-2.1) through (E6-2.11) were solved using Polymath and MATLAB, another ODE solver. The profiles of the molar flow rates are shown here. Table E6-2.1.1 shows the Polymath programs, and Figure E6-2.1 shows the results of the numerical solution for the entering conditions.

$$V = 0: \quad F_A = F_{A0}, \quad F_B = 0, \quad F_C = 0$$

Information on how to obtain and download the Polymath software can be found in Appendix E.

TABLE E6-2.1   POLYMATH PROGRAM

Differential equations

1 d(Fa)/d(V) = ra

2 d(Fb)/d(V) = -ra-kc*Cto*(Fb/Ft)

3 d(Fc)/d(V) = -ra

Explicit equations

1 Kc = 0.05

2 Ft = Fa+Fb+Fc

3 k = 0.7

4 Cto = 0.2

5 ra = -k*Cto*((Fa/Ft)-Cto/Kc*(Fb/Ft)*(Fc/Ft))

6 kc = 0.2

Calculated values of DEQ variables

	Variable	Initial value	Final value
1	Cto	0.2	0.2
2	Fa	10.	3.995179
3	Fb	0	1.832577
4	Fc	0	6.004821
5	Ft	10.	11.83258
6	k	0.7	0.7
7	Kc	0.05	0.05
8	kc	0.2	0.2
9	ra	-0.14	-0.0032558
10	V	0	500.

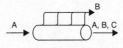

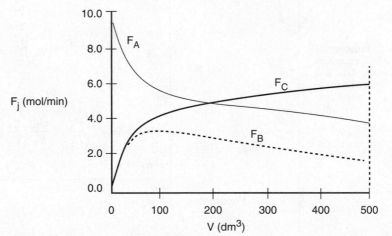

Figure E6-2.1   Polymath solution.

We note that F_B goes through a maximum as a result of the competition between the rate of B being formed from A and the rate of B being removed through the sides of the reactor.

(c) From Figure E6-2.1, we see that the exit molar flow rate of A at 500 dm³ is 4 mol/min, for which the corresponding conversion is

$$X = \frac{F_{A0} - F_A}{F_{A0}} = \frac{10 - 4}{10} = 0.60$$

__Analysis:__ The molar flow rate of A drops rapidly until about 100 dm³, where the reaction approaches equilibrium. At this point the reaction will only proceed to the right at the rate at which B is removed through the sides of the membrane, as noted by the similar slopes of F_A and F_B in this plot. You will want to use Problem 6-2$_A$(b) to show that if B is removed rapidly, F_B will close to zero and the reaction behaves as if it is irreversible, and that if B is removed slowly, F_B will be large throughout the reactor and the rate of reaction, $-r_A$, will be small.

Use of Membrane Reactors to Enhance Selectivity.   In addition to species leaving through the sides of the membrane reactor, species can also be fed to the reactor through the sides of the membrane. For example, for the reaction

$$A + B \rightarrow C + D$$

species A will be fed only to the entrance, and species B will be fed only through the membrane as shown here.

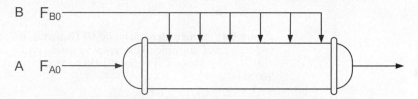

As we will see in Chapter 8, this arrangement is often used to improve selectivity when multiple reactions take place. Here, B is usually fed uniformly through the membrane along the length of the reactor. The balance on B is

$$\frac{dF_B}{dV} = r_B + R_B \tag{6-7}$$

where $R_B = F_{B0}/V_t$ with F_{B0}, representing the total molar feed rate of B through the sides and V_t the total reactor volume. The feed rate of B can be controlled by controlling the pressure drop across the reactor membrane.[5] This arrangement will keep the concentration of A high and the concentration of B low to maximize the selectivity given by Equation (E8-2.2) for the reactions given in Section 8.6, e.g., Example 8-8.

6.5 Unsteady-State Operation of Stirred Reactors

In Chapter 5 we discussed the unsteady operation of one type of reactor, the batch reactor. In this section, we discuss two other aspects of unsteady operation: startup of a CSTR and of semibatch reactors. First, the startup of a CSTR is examined to determine the time necessary to reach steady-state operation (see Figure 6-4(a)), and then semibatch reactors are discussed. In each of these cases, we are interested in predicting the concentration and conversion as a function of time. Closed-form analytical solutions to the differential equations arising from the mole balance of these reaction types can be obtained only for zero- and first-order reactions. ODE solvers must be used for other reaction orders.

---

[5] The velocity of B through the membrane, U_B, is given by Darcy's law

$$U_B = K(P_s - P_r)$$

where K is the membrane permeability, P_s is the shell-side pressure, and P_r the reactor-side pressure

$$F_{B0} = \overbrace{C_{B0}aU_B}^{R_B} V_t = R_B V_t$$

where, as before, a is the membrane surface area per unit volume, C_{B0} is the entering concentration of B, and V_t is the total reactor volume.

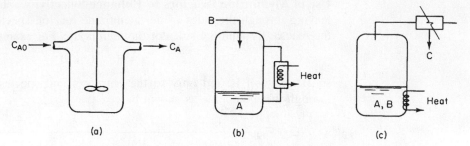

Figure 6-4   Semibatch reactors: **(a)** CSTR startup, **(b)** semibatch with cooling, and **(c)** reactive distillation. (Excerpted by special permission from *Chem. Eng.*, 63 (10) 211 (Oct. 1956). Copyright © 1956 by McGraw-Hill, Inc., New York, NY 10020.)

An unsteady-state analysis can be used to determine the startup time for a CSTR (Figure 6-4(a)) and this analysis is given in the *Expanded Material* for Chapter 6 on the CRE Web site. Here, we show the time to steady state for a first-order reaction is approximately

$$t_s = \frac{4.6\tau}{1 + \tau k}$$

For most first-order systems, the time to reach steady state is 3 to 4 space times.

There are two basic types of semibatch operations. In one type, one of the reactants in the reaction

$$A + B \rightarrow C + D$$

(e.g., B) is slowly fed to a reactor containing the other reactant (e.g., A), which has already been charged to a reactor such as that shown in Figure 6-4(b). This type of reactor is generally used when unwanted side reactions occur at high concentrations of B (see Section 8.1) or when the reaction is highly exothermic (Chapter 11). In some reactions, the reactant B is a gas and is bubbled continuously through liquid reactant A. Examples of reactions used in this type of semibatch reactor operation include *ammonolysis*, *chlorination*, and *hydrolysis*. The other type of semibatch reactor is reactive distillation and is shown schematically in Figure 6-4(c). Here, reactants A and B are charged simultaneously and one of the products vaporizes and is withdrawn continuously. Removal of one of the products in this manner (e.g., C) shifts the equilibrium toward the right, increasing the final conversion above that which would be achieved had C not been removed. In addition, removal of one of the products further concentrates the reactant, thereby producing an increased rate of reaction and decreased processing time. This type of reaction operation is called *reactive distillation*. Examples of reactions carried out in this type of reactor include *acetylation reactions* and *esterification reactions* in which water is removed.

6.6 Semibatch Reactors

6.6.1 Motivation for Using a Semibatch Reactor

One of the best reasons to use semibatch reactors is to enhance selectivity in liquid-phase reactions. For example, consider the following two simultaneous reactions. One reaction produces the desired product D

$$A + B \xrightarrow{\;k_D\;} D$$

with the rate law

$$r_D = k_D C_A^2 C_B$$

and the other reaction produces an undesired product U

$$A + B \xrightarrow{\;k_U\;} U$$

with the rate law

$$r_U = k_U C_A C_B^2$$

The instantaneous selectivity $S_{D/U}$ is the ratio of these two rates

We want $S_{D/U}$ as large as possible.

$$S_{D/U} = \frac{r_D}{r_U} = \frac{k_D C_A^2 \, C_B}{k_U C_A C_B^2} = \frac{k_D}{k_U} \frac{C_A}{C_B} \tag{6-15}$$

and guides us in how to produce the most amount of our desired product and the least amount of our undesired product (see Section 8.1). We see from the instantaneous selectivity that we can increase the formation of D and decrease the formation of U by keeping the concentration of A high and the concentration of B low. This result can be achieved through the use of the semibatch reactor, which is charged with pure A and to which B is fed slowly to A in the vat.

6.6.2 Semibatch Reactor Mole Balances

Of the two types of semibatch reactors, we focus attention primarily on the one with constant molar feed. A schematic diagram of this semibatch reactor is shown in Figure 6-5. We shall consider the elementary liquid-phase reaction

$$A + B \rightarrow C$$

in which reactant B is slowly added to a well-mixed vat containing reactant A. A **mole balance on species A yields**

Mole balance on species A

$$\begin{bmatrix} \text{Rate} \\ \text{in} \end{bmatrix} - \begin{bmatrix} \text{Rate} \\ \text{out} \end{bmatrix} + \begin{bmatrix} \text{Rate of} \\ \text{generation} \end{bmatrix} = \begin{bmatrix} \text{Rate of} \\ \text{accumulation} \end{bmatrix} \tag{6-16}$$

$$\overbrace{0}^{} \;\; - \;\; \overbrace{0}^{} \;\; + \;\; \overbrace{r_A V(t)}^{} \;\; = \;\; \overbrace{\frac{dN_A}{dt}}^{}$$

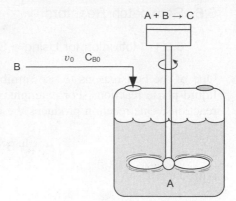

Figure 6-5 Semibatch reactor.

Three variables can be used to formulate and solve semibatch reactor problems: the concentrations, C_j, the number of moles, N_j, and the conversion, X.

We shall use concentration as our preferred variable, leaving the analysis of semibatch reactors using the number of moles, N_j, and conversion X to the CRE Web site *Summary Notes* and *Professional Reference Shelf* for Chapter 6.

Recalling that the number of moles of A, N_A, is just the product of the concentration of A, C_A, and the volume, V, [i.e., $(N_A = C_A V)$], we can rewrite Equation (6-16) as

$$r_A V = \frac{dN_A}{dt} = \frac{d(C_A V)}{dt} = \frac{V dC_A}{dt} + C_A \frac{dV}{dt} \qquad (6\text{-}17)$$

We note that since the reactor is being filled, the volume, V, varies with time. The reactor volume at any time t can be found from an **overall mass balance** of all species. The mass flow rate into the reactor, $\dot{m}_0$, is just the product of the liquid density, ρ_0, and volumetric flow rate v_0. The mass of liquid inside the reactor, m, is just the product of the liquid density ρ and the volume of liquid V in the reactor, i.e., $m = \rho V$. There is no mass flow out and no generation of mass.

Overall mass
balance

$$\begin{bmatrix} \text{Mass} \\ \text{rate} \\ \text{in} \end{bmatrix} - \begin{bmatrix} \text{Mass} \\ \text{rate} \\ \text{out} \end{bmatrix} + \begin{bmatrix} \text{Rate of} \\ \text{mass} \\ \text{generation} \end{bmatrix} = \begin{bmatrix} \text{Rate of} \\ \text{mass} \\ \text{accumulation} \end{bmatrix}$$

$$\dot{m}_0 \quad - \quad 0 \quad + \quad 0 \quad = \quad \frac{dm}{dt}$$

$$\overbrace{\rho_0 v_0} \quad - \quad \overbrace{0} \quad + \quad \overbrace{0} \quad = \quad \overbrace{\frac{d(\rho V)}{dt}} \qquad (6\text{-}18)$$

For a constant-density system, $\rho_0 = \rho$, and

$$\frac{dV}{dt} = v_0 \qquad (6\text{-}19)$$

with the initial condition $V = V_0$ at $t = 0$, integrating for the case of constant volumetric flow rate v_0 yields

Semibatch
reactor
volume as a
function of time

$$\boxed{V = V_0 + v_0 t} \qquad (6\text{-}20)$$

Substituting Equation (6-19) into the right-hand side of Equation (6-17) and rearranging gives us

$$-v_0 C_A + V r_A = \frac{V d C_A}{dt}$$

The **balance on A**, seen in Equation (6-17), can be rewritten as

Mole balance on A

$$\boxed{\frac{dC_A}{dt} = r_A - \frac{v_0}{V} C_A}$$

(6-21)

A **mole balance on B** that is fed to the reactor at a rate F_{B0} is

In	+	Out	+	Generation	=	Accumulation

$$\overbrace{F_{B0}} \quad - \quad \overbrace{0} \quad + \quad \overbrace{r_B V} \quad = \quad \overbrace{\frac{dN_B}{dt}}$$

Rearranging gives

$$\frac{dN_B}{dt} = r_B V + F_{B0}$$

(6-22)

Substituting for N_B in terms of concentration and reactor volume ($N_B = C_B V$), differentiating, and then using Equation (6-19) to substitute for (dV/dt), and $F_{B0} = C_{B0} v_0$, the mole balance on B given in Equation (6-23) becomes

$$\frac{dN_B}{dt} = \frac{d(V C_B)}{dt} = \frac{dV}{dt} C_B + \frac{V d C_B}{dt} = r_B V + F_{B0} = r_B V + v_0 C_{B0}$$

Rearranging gives

Mole balance on B

$$\boxed{\frac{dC_B}{dt} = r_B + \frac{v_0(C_{B0} - C_B)}{V}}$$

(6-23)

Similarly, for species C we have

$$\frac{dN_C}{dt} = r_C V = -r_A V$$

(6-24)

$$\frac{dN_C}{dt} = \frac{d(C_C V)}{dt} = V \frac{dC_C}{dt} + C_C \frac{dV}{dt} = V \frac{dC_C}{dt} + v_0 C_C$$

(6-25)

Combining Equations (6-24) and (6-25) and rearranging we obtain

Mole balance on C

$$\boxed{\frac{dC_C}{dt} = r_C - \frac{v_0 C_C}{V}}$$

(6-26)

Following the same procedure for species D

Mole balance on D

$$\boxed{\frac{dC_D}{dt} = r_D - \frac{v_0 C_D}{V}}$$

(6-27)

At time $t = 0$, the initial concentrations of B, C, and D in the vat are zero, $C_{Bi} = C_{Ci} = C_{Di} = 0$. The concentration of B in the feed is C_{B0}. If the reaction order is other than zero- or first-order, or if the reaction is nonisothermal, we *must* use numerical techniques to determine the concentrations and conversion as a function of time. Equations (6-21), (6-23), (6-26), and (6-27) are easily solved with an ODE solver.

Example 6–3   Isothermal Semibatch Reactor with Second-Order Reaction

The production of methyl bromide is an irreversible liquid-phase reaction that follows an elementary rate law. The reaction

$$CNBr + CH_3NH_2 \rightarrow CH_3Br + NCNH_2$$

is carried out isothermally in a semibatch reactor. An aqueous solution of methyl amine (B) at a concentration of 0.025 mol/dm^3 is to be fed at a volumetric rate of 0.05 dm^3/s to an aqueous solution of bromine cyanide (A) contained in a glass-lined reactor. The initial volume of liquid in the vat is to be 5 dm^3 with a bromine-cyanide concentration of 0.05 mol/dm^3. The specific reaction rate constant is

$$k = 2.2 \text{ dm}^3/\text{s} \cdot \text{mol}$$

Solve for the concentrations of bromine cyanide (A), methyl amine (B), methyl bromide (C), and cyanamide (D), and the rate of reaction as a function of time, and then analyze your results.

Solution

Symbolically, we write the reaction as

$$A + B \rightarrow C + D$$

1. Mole Balances:

Mole Balance on every species

$$\frac{dC_A}{dt} = r_A - \frac{v_0 C_A}{V} \tag{6-21}$$

$$\frac{dC_B}{dt} = \frac{v_0(C_{B0} - C_B)}{V} + r_B \tag{6-23}$$

$$\frac{dC_C}{dt} = r_C - \frac{v_0 C_C}{V} \tag{6-25}$$

$$\frac{dC_D}{dt} = r_D - \frac{v_0 C_D}{V} \tag{6-26}$$

2. Rates:

Rates
Law
Relative

(a) *Rate Law (Elementary)*

$$-r_A = kC_A C_B \tag{E6-3.1}$$

(b) *Relative Rates*

$$-r_A = -r_B = r_C = r_D \tag{E6-3.2}$$

3. Combine:

Before going to the stoichiometry step, normally Step 3, let's combine the mole balances from Equations (6-21), (6-23), (6-26), and (6-27), the rate law Equation (E6-3.1), and the relative rates Equation (E6-3.2), to arrive at the following forms of the mole balances on A, B, C, and D solely in terms of concentrations

$$\frac{dC_A}{dt} = -kC_AC_B - \frac{v_0C_A}{V} \tag{E6-3.3}$$

Combined mole balances and rate laws on A, B, C, and D

$$\frac{dC_B}{dt} = -kC_AC_B + \frac{v_0(C_{B0} - C_B)}{V} \tag{E6-3.4}$$

$$\frac{dC_C}{dt} = kC_AC_B - \frac{v_0C_C}{V} \tag{E6-3.5}$$

$$\frac{dC_D}{dt} = kC_AC_D - \frac{v_0C_D}{V} \tag{E6-3.6}$$

4. Stoichiometry:

The volume of liquid in the reactor at any time t is

$$V = V_0 + v_0t \tag{E6-3.7}$$

These coupled equations are easily solved with an ODE solver such as Polymath.
We could also calculate the conversion of A:

$$X = \frac{N_{A0} - N_A}{N_{A0}} \tag{E6-3.8}$$

Substituting for N_{A0} and N_A

$$\boxed{X = \frac{C_{A0}V_0 - C_AV}{C_{A0}V_0}} \tag{E6-3.9}$$

5. Evaluate:

The initial conditions are $t = 0$, $C_{A0} = 0.05$ mol/dm^3, $C_B = C_C$, $= C_D = 0$, and $V_0 = 5$ dm^3.

Equations (E6-3.2) through (E6-3.9) are easily solved with the aid of an ODE solver such as Polymath (Table E6-3.1).

TABLE E6-3.1 POLYMATH PROGRAM

ODE REPORT (RKF45)

Living Example Problem

Differential equations

1 d(Ca)/d(t) = ra- vo*Ca/V
2 d(Cb)/d(t) = ra+ (Cbo-Cb)*vo/V
3 d(Cc)/d(t) = -ra-vo*Cc/V
4 d(Cd)/d(t) = -ra-vo*Cd/V

Explicit equations

1 vo = 0.05
2 Vo = 5
3 V = Vo+vo*t
4 k = 2.2
5 Cbo = 0.025
6 ra = -k*Ca*Cb
7 Cao = 0.05
8 rate = -ra
9 X = (Cao*Vo-Ca*V)/(Cao*Vo)

Calculated values of DEQ variables

	Variable	Initial value	Final value
1	Ca	0.05	7.731E-06
2	Cao	0.05	0.05
3	Cb	0	0.0125077
4	Cbo	0.025	0.025
5	Cc	0	0.0083256
6	Cd	0	0.0083256
7	k	2.2	2.2
8	ra	0	-2.127E-07
9	rate	0	2.127E-07
10	t	0	500.
11	V	5.	30.
12	vo	0.05	0.05
13	Vo	5.	5.
14	X	0	0.9990722

The concentrations of bromine cyanide (A), methyl amine (B), and methyl bromide (C) are shown as a function of time in Figure E6-3.1, and the rate is shown in Figure E6-3.2.

Why does the
concentration of
CH_3Br (C) go
through a
maximum wrt
time?

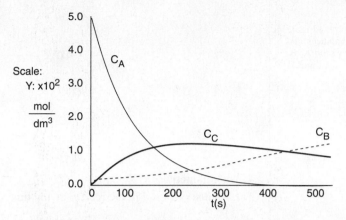

Figure E6-3.1   Polymath output: concentration–time trajectories.

We note that the concentration of methyl bromide (C), which is identical to the concentration of cynanamide (D) goes through a maximum. The maximum occurs because once all of A has been consumed, then no more C will be formed and the continual flow of B into the reactor will thus dilute the moles of C produced and hence the concentration of C.

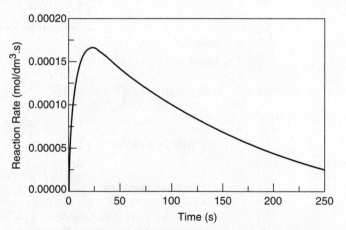

Figure E6-3.2   Reaction rate–time trajectory.

<u>**Analysis:**</u> Let's look at the trends. The concentration of A falls close to zero at about 250 seconds, as does the reaction rate. Consequently, very little C and D are formed after this time, and what has been formed begins to be diluted as B continues to be added to the reactor and stops just before it overflows. Now what do you think of the time to carry out this reaction? It's about 5 minutes, hardly enough time to turn the valves on and off. Take-away lessons: While this example showed how to analyze a semibatch reactor, you would not use a semibatch reactor to carry out this reaction at this temperature because the times are too short. Instead, you would use a tubular reactor with B fed through the sides or a number of CSTRs in series with A fed to the first reactor and small amounts of B fed to each of the following reactors. We will discuss this further in Chapter. 8.

Equilibrium Conversion.   For reversible reactions carried out in a semibatch reactor, the maximum attainable conversion (i.e., the *equilibrium conversion*) will change as the reaction proceeds because more reactant is continuously added to the reactor. This addition shifts the equilibrium continually to the right toward more product.

An outline of what is given on the CRE Web site follows:

At Equilibrium

$$K = \frac{N_{Ce}N_{De}}{N_{Ae}N_{Be}} \tag{6-28}$$

Using the number of moles in terms of conversion

$$N_{Ae} = N_{A0}(1 - X_e) \qquad N_{Ce} = N_{A0}X_e$$
$$N_{De} = N_{A0}X_e \qquad N_{Be} = F_{B0}t - N_{A0}X_e \tag{6-29}$$

Substituting

$$K_C = \frac{N_{A0}X_e^2}{(1 - X_e)(F_{B0}t - N_{A0}X_e)} \tag{6-30}$$

Solving for X_e

$$X_e = \frac{K_C\left(1 + \dfrac{F_{B0}t}{N_{A0}}\right) - \sqrt{\left[K_C\left(1 + \dfrac{F_{B0}t}{N_{A0}}\right)\right]^2 - 4(K_C - 1)K_C \dfrac{tF_{B0}}{N_{A0}}}}{2(K_C - 1)} \tag{6-31}$$

One notes the equilibrium conversion, X_e, changes with time. Further discussion on this point and calculation of the equilibrium conversion can be found in *Professional Reference Shelf R6.1*, in the example problem on the CRE Web site.

Closure. Chapters 5 and 6 present the heart of chemical reaction engineering for isothermal reactors. After completing these chapters, the reader should be able to apply the algorithm building blocks

The CRE algorithm

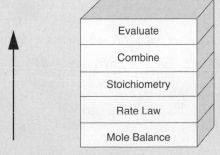

to any of the reactors discussed in this chapter: batch reactor, CSTR, PFR, PBR, membrane reactor, and semibatch reactor. The reader should be able to account for pressure drop and describe the effects of the system variables. The reader should also be able to use either conversions (Chapter 5) or concentration and molar flow rates (Chapter 6) to solve chemical reaction engineering problems.

SUMMARY

1. Solution algorithm—measures other than conversion

When using measures other than conversion for reactor design, the mole balances are written for each species in the reacting mixture:

Mole balances on each and every species

$$\frac{dF_A}{dV} = r_A, \quad \frac{dF_B}{dV} = r_B, \quad \frac{dF_C}{dV} = r_C, \quad \frac{dF_D}{dV} = r_D \qquad (S6\text{-}1)$$

The mole balances are then coupled through their relative rates of reaction. If

Rate Law

$$-r_A = k C_A^\alpha C_B^\beta \qquad (S6\text{-}2)$$

for $aA + bB \rightarrow cC + dD$, then

Relative Rates

$$r_B = \frac{b}{a} r_A, \quad r_C = -\frac{c}{a} r_A, \quad r_D = -\frac{d}{a} r_A \qquad (S6\text{-}3)$$

Concentration can also be expressed in terms of the number of moles (batch) and in terms of molar flow rates).

Gas:
$$C_A = C_{T0} \frac{F_A}{F_T} \frac{P}{P_0} \frac{T_0}{T} = C_{T0} \frac{F_A}{F_T} \frac{T_0}{T} p \qquad (S6\text{-}4)$$

$$C_B = C_{T0} \frac{F_B}{F_T} \frac{T_0}{T} p \qquad (S6\text{-}5)$$

$$p = \frac{P}{P_0}$$

Stoichiometry
$$F_T = F_A + F_B + F_C + F_D + F_I \qquad (S6\text{-}6)$$

$$\frac{dp}{dW} = \frac{-\alpha}{2p} \left(\frac{F_T}{F_{T0}} \right) \left(\frac{T}{T_0} \right) \qquad (S6\text{-}7)$$

Liquid:
$$C_A = \frac{F_A}{v_0} \qquad (S6\text{-}8)$$

2. For membrane reactors, the mole balances for the reaction

$$A \rightleftharpoons B + C$$

when reactant A and product C do not diffuse out the membrane

Mole Balance
$$\frac{dF_A}{dV} = r_A, \quad \frac{dF_B}{dV} = r_B - R_B, \quad \text{and} \quad \frac{dF_C}{dV} = r_C \qquad (S6\text{-}9)$$

with

Tansport Law
$$R_B = k_c C_B \qquad (S6\text{-}10)$$

and k_c is the overall mass transfer coefficient.

3. For semibatch reactors, reactant B is fed continuously to a vat initially containing only A

$$A + B \rightleftharpoons C + D$$

Mole Balances

$$\frac{dC_A}{dt} = r_A - \frac{v_0}{V}C_A$$

(S6-11)

$$\frac{dC_B}{dt} = r_B + \frac{v_0(C_{B0} - C_B)}{V}$$

(S6-12)

ODE SOLVER ALGORITHM

When using an ordinary differential equation (ODE) solver such as Polymath or MATLAB, it is usually easier to leave the mole balances, rate laws, and concentrations as separate equations, rather than combining them into a single equation as we did to obtain an analytical solution. Writing the equations separately leaves it to the computer to combine them and produce a solution. The formulations for a packed-bed reactor with pressure drop and a semibatch reactor are given below for two elementary reactions carried out isothermally.

Gas Phase

$A + B \rightarrow 3C$

Packed-Bed Reactor

$$\frac{dF_A}{dW} = r_A'$$

$$\frac{dF_B}{dW} = r_B'$$

$$\frac{dF_C}{dW} = r_C'$$

$$r_A' = -kC_A C_B$$

$$r_B' = r_A'$$

$$r_C' = 3(-r_A')$$

$$C_A = C_{T0}\frac{F_A}{F_T}p$$

$$C_B = C_{T0}\frac{F_B}{F_T}p$$

$$\frac{dp}{dW} = -\frac{\alpha}{2p}\frac{F_T}{F_{T0}}$$

$F_{T0} = 30, \quad C_{T0} = 0.02, \quad C_{A0} = 0.01,$

$C_{B0} = 0.01, \quad k = 5000, \quad \alpha = 0.009$

$W_{\text{final}} = 80$

Liquid Phase

$A + B \rightleftarrows 2C$

Semibatch Reactor

$$\frac{dC_A}{dt} = r_A - \frac{v_0 C_A}{V}$$

$$\frac{dC_B}{dt} = r_A + \frac{v_0(C_{B0} - C_B)}{V}$$

$$\frac{dC_C}{dt} = -2r_A - \frac{v_0 C_C}{V}$$

$$r_A = -k\left[C_A C_B - \frac{C_C^2}{K_C}\right]$$

$$V = V_0 + v_0 t$$

$k = 0.15, \quad K_C = 16.0, \quad V_0 = 10.0$

$v_0 = 0.1, \quad C_{B0} = 0.1, \quad C_{Ai} = 0.04$

$t_{\text{final}} = 200$

The Polymath solutions to the above equations are given on the CRE Web site in the **Chapter 6** *Summary Notes* under "Link" PBR ODE Solver.

CRE WEB SITE MATERIALS

- **Expanded Material**

    1. *Start-Up of a CSTR*
    2. *Puzzle Problem "What's Wrong with this Solution?"*
    3. *Additional Homework Problems*

- **Learning Resources**

    1. *Summary Notes*
    2. *Modules and Games*

    A. Wetlands Web Module

    B. Tic-Tac Interactive Game

- **Living Example Problems**
    Example 6-1 Gas-Phase Reaction in Microreactor—Molar Flow Rate
    Example 6-2 Membrane Reactor
    Example 6-3 Isothermal Semibatch Reactor
- **Professional Reference Shelf**
    R6.1 *Unsteady CSTRs and Semibatch Reactors*
        R6.1A *Startup of a CSTR*
        R6.1B *Semibatch Reactor Balances in Terms of Number of Moles*
        R6.1C *Semibatch Reactor Balance in Terms of Conversion*
        R6.1D *Equilibrium Conversion*
    R6.2 *The Practical Side*
        A number of practical guidelines for operating chemical reactors are given.
    R6.3 *Aerosol Reactors*
        Aerosol reactors are used to synthesize nano-size particles. Owing to their size, shape, and high specific surface area, nanoparticles can be used in a number of applications such as pigments in cosmetics, membranes, photocatalytic reactors, catalysts and ceramics, and catalytic reactors.
        We use the production of aluminum particles as an example of an aerosol plug-flow reactor (APFR) operation. A stream of argon gas saturated with Al vapor is cooled.

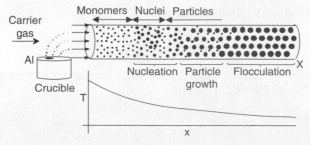

Aerosol reactor and temperature profile.

As the gas is cooled, it becomes supersaturated, leading to the nucleation of particles. This nucleation is a result of molecules colliding and agglomerating until a critical nucleus size is reached and a particle is formed. As these particles move down the reactor, the supersaturated gas molecules condense on the particles, causing them to grow in size and then to flocculate. In the development on the CRE Web site in the Web Modules category, we will model the formation and growth of aluminum nanoparticles in an AFPR.

QUESTIONS AND PROBLEMS

The subscript to each of the problem numbers indicates the level of difficulty: A, least difficult; D, most difficult.

$$A = \bullet \quad B = \blacksquare \quad C = \blacklozenge \quad D = \blacklozenge\blacklozenge$$

In each of the following questions and problems, rather than just drawing a box around your answer, write a sentence or two describing how you solved the problem, the assumptions you made, the reasonableness of your answer, what you learned, and any other facts that you want to include. You may wish to refer to W. Strunk and E. B. White, *The Elements of Style*, 4th ed. (New York: Macmillan, 2000) and Joseph M. Williams, *Style: Ten Lessons in Clarity & Grace,* 6th ed. (Glenview, IL: Scott, Foresman, 1999) to enhance the quality of your sentences. See the preface for additional generic parts (x), (y), and (z) to the home problems.

Homework Problems

Before solving the problems, state or sketch qualitatively the expected results or trends.

Questions

Q6-1$_A$ Read through all the problems at the end of this chapter. Make up and solve an *original* problem based on the material in this chapter. **(a)** Use real data and reactions for further instructions. **(b)** Make up a reaction and data. **(c)** Use an example from everyday life (e.g., cooking spaghetti; see P5-1$_A$).

Problems

P6-1$_B$ **What if...** you were asked to explore the example problems in this chapter to learn the effects of varying the different parameters? This sensitivity analysis can be carried out by downloading the examples from the CRE Web site. For each of the example problems you investigate, write a paragraph describing your findings.

    (a) **Example 6-1.** Download the *Living Example Problem 6-1* from the CRE Web site. (1) What would be the conversion if the pressure were doubled and the temperature were decreased by 20°C? (2) Compare Figure E6-1.1 profiles with those for a reversible reaction with $K_C = 0.02$ mol/dm^3 and describe the differences in the profiles. (3) How would your profiles change for the case of an irreversible reaction with pressure drop when $\alpha_p = 99 \times 10^3$ dm^{-3} for each tube?

    (b) **Example 6-2.** Download the *Living Example Problem 6-2* from the CRE Web site. You might find it easiest to answer most of the following questions using Wolfram. (1) Starting with all values of K_C, k, C_{T0}, and k_C in the middle range ($K_C = 0.25$, $k = 2.1$, $C_{T0} = 0.52$, and $k_C = 1.06$), vary each parameter individually and describe what you find. Note and explain any maximum or minimum values of your plots down the length (i.e., volume = 500 dm^3) of your reactor. *Hint*: Go to the extremes of the range. (2) Repeat (1) but set K_C at its maximum value and then vary k and k_C, and describe what you find. (3) Use Polymath to find the parameter values at which R_B is a maximum and the concentrations/flow rates are at the maximum. (4) Vary ratios of parameters such as (k/k_C) and $(k \ \tau \ C_{A0}/K_C)$ [Note: $\tau = 400$ min] and write a paragraph describing what you find. What ratio of parameters has the greatest effect on the conversion $X = (F_{A0} - F_{A0})/F_{A0}$? (5) Write a summary paragraph of all the trends and your results. (6) Make up a question/problem on membrane reactors with a solution in which Wolfram must be used to obtain the answer. *Hint*: See Preface Table P-4, page xxviii. Also comment on what types of questions would you ask when using Wolfram.

(c) **Example 6-3.** Download the *Living Example Problem 6-3* from the CRE Web site. The temperature is to be lowered by 35°C so that the reaction-rate constant is now one-tenth its original value. (1) If the concentration of B is to be maintained at 0.01 mol/dm³ or below, what is the maximum feed rate of B? (2) How would your answer change if the concentration of A were tripled? (3) Redo this problem when the reaction is reversible with $K_C = 0.1$ and compare with the irreversible case. (Only a couple of changes in the Polymath program are necessary.)

(d) *Web Module on Wetlands* on the CRE Web site. Download the Polymath program and vary a number of parameters such as rainfall, evaporation rate, atrazine concentration, and liquid flow rate, and write a paragraph describing what you find. This topic is a hot Ch.E. research area.

(e) *Web Module on Aerosol Reactors* on the CRE Web site. Download the Polymath program and (1) vary the parameters, such as cooling rate and flow rate, and describe their effect on each of the regimes: nucleation, growth, and flocculation. Write a paragraph describing what you find. (2) It is proposed to replace the carrier gas by helium.

  (i)   Compare your plots (He versus Ar) of the number of Al particles as a function of time. Explain the shape of the plots.

  (ii) How does the final value of d_p compare with that when the carrier gas was argon? Explain.

  (iii) Compare the time at which the rate of nucleation reaches a peak in the two cases (carrier gas = Ar and He). Discuss the comparison.

  Data for a He molecule: mass = 6.64×10^{-27} kg, volume = 1.33×10^{-29} m³, surface area = 2.72×10^{-19} m², bulk density = 0.164 kg/m³, at normal temperature (25°C) and pressure (1 atm).

(f) **The Work Self-Tests on the Web.** Write a question for this problem that involves critical thinking and explain why it involves critical thinking. See examples on the CRE Web site, *Summary Notes* for Chapter 6.

P6-2$_B$ Download the Interactive Computer Games (ICG) from the CRE Web site. Play the game and then record your performance number, which indicates your mastery of the material. Your instructor has the key to decode your performance number. Knowledge of all sections is necessary to pit your wit against the computer adversary in playing a game of Tic-Tac-Toe.

Performance number: _____

P6-3$_C$ The second-order liquid phase reaction

$$C_6H_5COCH_2Br + C_6H_5N \longrightarrow C_6H_5COCH_2NC_5H_5Br$$

is carried out in a batch reactor at 35°C. The specific reaction-rate constant is 0.0445 dm³/mol/min. Reactor 1 is charged with 1,000 dm³, where the concentration of each reactant after mixing is 2M.

(a) What is the conversion after 10, 50, and 100 minutes?

Now, consider the case when, after filling reactor 1, the drain at the bottom of reactor 1 is left open and it drains into reactor 2, mounted below it, at a volumetric rate of 10 dm³/min.

(b) What will be the conversion and concentration of each species in reactor 1 after 10, 50, and 80 minutes in the reactor that is being drained?

(c) What is the conversion and concentration of each species in reactor 2 that is filling up with the liquid from reactor 1 after 10 and after 50 minutes?

(d) At the end of 50 minutes, the contents of the two reactors are added together. What is the overall conversion after mixing?

(e) Apply one or more of the six ideas in Preface Table P-4, page xxviii, to this problem.

P6-4$_B$ The elementary gas-phase reaction

$$(CH_3)_3COOC(CH_3)_3 \rightarrow C_2H_6 + 2CH_3COCH_3$$
$$A \rightarrow B + 2C$$

is carried out isothermally at 400 K in a flow reactor with no pressure drop. The specific reaction rate at 50°C is 10^{-4} min^{-1} (from pericosity data) and the activation energy is 85 kJ/mol. Pure di-*tert*-butyl peroxide enters the reactor at 10 atm and 127°C and a molar flow rate of 2.5 mol/min, i.e., $F_A = 2.5$ mol/min.

(a) Use the algorithm for molar flow rates to formulate and solve the problem. Plot F_A, F_B, F_C, and then X as a function of plug-flow reactor volume and space time to achieve 90% conversion.

(b) Calculate the plug-flow volume and space time for a CSTR for 90% conversion.

P6-5$_B$ For the reaction and data in P6-4$_B$, we now consider the case when the reaction is reversible with $K_C = 0.025$ dm^6/mol^2 and the reaction is carried out at 300 K in a membrane reactor where C_2H_6 is diffusing out. The membrane transport coefficient is $k_C = 0.08$ s^{-1}.

(a) What is the equilibrium conversion and what is the exit conversion in a conventional PFR?

(b) Plot and analyze the conversion and molar flow rates in the membrane reactor as a function of reactor volume up to the point where 80% conversion of di-tert-butyl peroxide is achieved. Note any maxima in the flow rates.

(c) Apply one or more of the six ideas in Preface Table P-4, page xxviii, to this problem.

P6-6$_C$ (*Membrane reactor*) The first-order, gas-phase, reversible reaction

$$A \; \underset{\longleftarrow}{\overset{\longrightarrow}{\rightleftharpoons}} \; B + 2C$$

is taking place in a membrane reactor. Pure A enters the reactor, and B diffuses out through the membrane. Unfortunately, a small amount of the reactant A also diffuses through the membrane.

(a) Plot and analyze the flow rates of A, B, and C and the conversion X down the reactor, as well as the flow rates of A and B through the membrane.

(b) Next, compare the conversion profiles in a conventional PFR with those of a membrane reactor from part (a). What generalizations can you make?

(c) Would the conversion of A be greater or smaller if C were diffusing out instead of B?

(d) Discuss qualitatively how your curves would change if the temperature were increased significantly or decreased significantly for an exothermic reaction. Repeat the discussion for an endothermic reaction.

Additional information:

$k = 10$ min^{-1}	$F_{A0} = 100$ mol/min
$K_C = 0.01$ mol^2/dm^6	$v_0 = 100$ dm^3/min
$k_{CA} = 1$ min^{-1}	$V_{reactor} = 20$ dm^3
$k_{CB} = 40$ min^{-1}	

P6-7$_B$ **Fuel Cells Rationale.** With the focus on alternative clean-energy sources, we are moving toward an increased use of fuel cells to operate appliances ranging from computers to automobiles. For example, the hydrogen/oxygen fuel cell produces *clean energy* as the products are water and electricity, which may lead to a hydrogen-based economy instead of a petroleum-based economy.

A large component in the processing train for fuel cells is the water-gas shift membrane reactor. (M. Gummala, N. Gupla, B. Olsomer, and Z. Dardas, *Paper 103c*, 2003, AIChE National Meeting, New Orleans, LA.)

Fuel Cell

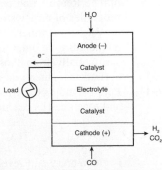

$$CO + H_2O \; \rightleftharpoons \; CO_2 + H_2$$

Here, CO and water are fed to the membrane reactor containing the catalyst. Hydrogen can diffuse out the sides of the membrane, while CO, H_2O, and CO_2 cannot. Based on the following information, plot the concentrations and molar flow rates of each of the reacting species down the length of the membrane reactor. Assume the following: The volumetric feed is 10 dm^3 /min at 10 atm, and the equimolar feed of CO and water vapor with $C_{T0} = 0.4$ mol/dm^3.

The equilibrium constant is $K_e = 1.44$, with $k = 1.37$ dm^6/mol kg-cat $\cdot$ min, and the mass transfer coefficient $k_{H_2} = 0.1$ dm^3/kg-cat $\cdot$ min (*Hint:* First calculate the entering molar flow rate of CO and then relate F_A and X.)

    (a) What is the membrane reactor volume necessary to achieve 85% conversion of CO?

    (b) Sophia wants you to compare the MR with a conventional PFR. What will you tell her?

    (c) For that same membrane reactor volume, Nicolas wants to know what would be the conversion of CO if the feed rate were doubled?

P6-8$_C$ The production of ethylene glycol from ethylene chlorohydrin and sodium bicarbonate

$$CH_2OHCH_2Cl + NaHCO_3 \rightarrow (CH_2OH)_2 + NaCl + CO_2\uparrow$$

is carried out in a semibatch reactor. A 1.5-molar solution of ethylene chlorohydrin is fed at a rate of 0.1 mole/minute to 1500 dm^3 of a 0.75-molar solution of sodium bicarbonate. The reaction is elementary and carried out isothermally at 30°C where the specific reaction rate is 5.1 dm^3/mol/h. Higher temperatures produce unwanted side reactions. The reactor can hold a maximum of 2500 dm^3 of liquid. Assume constant density.

    (a) Plot and analyze the conversion, reaction rate, concentration of reactants and products, and number of moles of glycol formed as a function of time.

    (b) Suppose you could vary the flow rate between 0.01 and 200 mol/min. What flow rate and holding time would you choose to make the greatest number of moles of ethylene glycol in 24 hours, keeping in mind the downtimes for cleaning, filling, etc., shown in Table 5-3?

    (c) Suppose the ethylene chlorohydrin is fed at a rate of 0.15 mol/min until the reactor is full and then shut in. Plot the conversion as a function of time.

    (d) Discuss what you learned from this problem and what you believe to be the point of this problem.

P6-9$_C$ The following elementary reaction is to be carried out in the liquid phase

$$NaOH + CH_3COOC_2H_5 \longrightarrow CH_3COO^-Na^+ + C_2H_5OH$$

The initial concentrations are 0.2 M in NaOH and 0.25 M in $CH_3COOC_2H_5$ with $k = 5.2 \times 10^{-5}$ dm^3/mol$\cdot$s at 20°C with $E = 42{,}810$ J/mol. Design a set of operating conditions (e.g., v_0, T, . . .) to produce 200 mol/day of ethanol in a semibatch reactor and not operate above 37°C and below a concentration of NaOH of 0.02 molar.[6] The semibatch reactor you have available is 1.5 m in diameter and 2.5 m tall. The reactor down time is $(t_c + t_e + t_f) = 3$h.

P6-10$_B$ Go to Professor Herz's Reactor Lab Web site at *www.reactorlab.net*. From the menu at the top of the page, select Download and then click on the English version link. Provide the required information and then download, install, and open the software. Select Division D2, Lab L2 and there the labeled PFR of The Reactor Lab concerning a packed-bed reactor (labeled PFR) in which a gas with the physical properties of airflows over spherical catalyst pellets. Perform experiments here to get a feeling for how pressure drop varies with input parameters such as reactor diameter, pellet diameter, gas-flow rate, and temperature. In order to get significant pressure drop, you may need to change some of the input values substantially from those shown when you enter the lab. If you get a notice that you can't get the desired flow, then you need to increase the inlet pressure.

P6-11$_B$ Pure butanol is to be fed into a *semibatch reactor* containing pure ethyl acetate to produce butyl acetate and ethanol. The reaction

$$CH_3COOC_2H_5 + C_4H_9OH \rightleftharpoons CH_3COOC_4H_9 + C_2H_5OH$$

is elementary and reversible. The reaction is carried out isothermally at 300 K. At this temperature, the equilibrium constant is 1.08 and the specific reaction rate is 9×10^{-5} dm^3/mol$\cdot$s. Initially, there is 200 dm^3 of ethyl acetate in the vat, and butanol is fed at a volumetric rate of 0.05 dm^3/s. The feed

---

[6] Manual of Chemical Engineering Laboratory, University of Nancy, Nancy, France, 1994 (*eric@ist.uni-stuttgart.de*; *www.sysbio.del/AICHE*).

and initial concentrations of butanol and ethyl acetate are 10.93 mol/dm^3 and 7.72 mol/dm^3, respectively.

(a) Plot and analyze the equilibrium conversion of ethyl acetate as a function of time.

(b) Plot and analyze the conversion of ethyl acetate, the rate of reaction, and the concentration of butanol as a function of time.

(c) Rework part (b), assuming that ethanol evaporates (reactive distillation) as soon as it forms. (This is a graduate level question.)

(d) Use Polymath or some other ODE solver to learn the sensitivity of conversion to various combinations of parameters (e.g., vary F_{B0}, N_{A0}, v_0).

(e) Apply one or more of the six ideas in Preface Table P-4, page xxviii, to this problem.

(f) Write a question that requires critical thinking and then explain why your question requires critical thinking. (*Hint*: See Preface Section I.)

P6-12$_C$ Use the reaction data in P6-11$_B$ and the molar flow rate algorithm to carry out the following problems:

(a) Calculate the CSTR reactor volume to achieve 80% of the equilibrium conversion for an equal molar feed and for a volumetric feed of 0.05 dm^3/s.

(b) *Safety*. Now consider the case where we want to shut down the reactor by feeding water at a volumetric rate of 0.05 dm^3/s. How long will it take to reduce the rate to 1% of the rate in the CSTR under the conditions of part (a)?

P6-13$_C$ An isothermal reversible reaction A $\rightleftarrows$ B is carried out in an aqueous solution. The reaction is first-order in both directions. The forward rate constant is 0.4 h^{-1} and the equilibrium constant is 4.0. The feed to the plant contains 100 kg/m^3 of A and enters at the rate of 12 m^3/h. Reactor effluents pass to a separator, where B is completely recovered. The reactor is a *stirred tank* of volume 60 m^3. A fraction, f_1, of the unreacted effluent is recycled as a solution containing 100 kg/m^3 of A and the remainder is discarded. Product B is worth \$2 per kilogram and operating costs are \$50 per cubic meter of solution entering the separator. What value of f maximizes the operational profit of the plant? What fraction A fed to the plant is converted at the optimum? Source: H. S. Shankar, IIT Mumbai.

SUPPLEMENTARY READING

ANTHONY, MAXWELL, presidential inauguration address, *"The economic future of Jofostan and the chemical reaction industry and one's ability to deal with multiple reactions."* Riça, Jofostan, January 1, 2017.

KEILLOR, GARRISON and TIM RUSSELL, *Dusty and Lefty: The Lives of the Cowboys* (Audio CD). St. Paul, MN: Highbridge Audio, 2006.

FROMENT, G. F., and K. B. BISCHOFF, *Chemical Reactor Analysis and Design*, 2nd ed. New York: Wiley, 1990.

Recent information on reactor design can usually be found in the following journals: *Chemical Engineering Science, Chemical Engineering Communications, Industrial and Engineering Chemistry Research, Canadian Journal of Chemical Engineering, Jofostan Journal of Chemical Engineering, AIChE Journal,* and *Chemical Engineering Progress.*

Collection and Analysis of Rate Data   7

> You can observe a lot just by watching.
>
> Yogi Berra, New York Yankees

Overview. In Chapters 5 and 6 we have shown that once the rate law is known, it can be substituted into the appropriate mole balance, and then through the use of the appropriate stoichiometric relationships, we can apply the CRE algorithm to analyze any isothermal reaction system. In this chapter we focus on ways of obtaining and analyzing reaction-rate data to obtain the rate law for a specific reaction.

We discuss two common types of reactors used for obtaining rate data: the batch reactor, which is used primarily for homogeneous reactions, and the differential reactor, which is used for solid-fluid heterogeneous reactions. In batch-reactor experiments, concentration, pressure, and/or volume are usually measured and recorded at different times during the course of the reaction. Data are collected from the batch reactor during transient operation, whereas measurements on the differential reactor are made during steady-state operation. In experiments with a differential reactor, the product concentration is usually monitored for different sets of feed conditions.

Three different methods of analyzing the data collected are used:
- The integral method
- The differential method
- Nonlinear regression

The integral and differential methods are used primarily in analyzing batch-reactor data. Because a number of software packages (e.g., Polymath, MATLAB) are now available to analyze data, a rather extensive discussion of nonlinear regression is included.

243

7.1 The Algorithm for Data Analysis

For batch systems, the usual procedure is to collect concentration-time data, which we then use to determine the rate law. Table 7-1 gives the seven-step procedure we will emphasize in analyzing reaction engineering data.

Data for homogeneous reactions is most often obtained in a batch reactor. After postulating a rate law in Step 1 and combining it with a mole balance in Step 2, we next use any or all of the methods in Step 5 to process the data and arrive at the reaction orders and specific reaction-rate constants.

Analysis of heterogeneous reactions is shown in Step 6. For gas–solid heterogeneous reactions, we need to have an understanding of the reaction and possible mechanisms in order to postulate the rate law in Step 6B. After studying Chapter 10 on heterogeneous reactions, one will be able to postulate different rate laws and then use Polymath nonlinear regression to choose the "best" rate-law and reaction-rate-law parameters (see Example 10-3 on page 452).

The procedure we should use to delineate the rate law and rate-law parameters is given in Table 7-1.

TABLE 7-1   STEPS IN ANALYZING RATE DATA

---

1. **Postulate a rate law.**
   A.   Power-law models for homogeneous reactions

   $$-r_A = kC_A^{\alpha} \,, \quad -r_A = kC_A^{\alpha} \, C_B^{\beta}$$

   B.   Langmuir-Hinshelwood models for heterogeneous reactions

   $$-r_A' = \frac{kP_A}{1 + K_A P_A} \,, \quad -r_A' = \frac{kP_A P_B}{(1 + K_A P_A + P_B)^2}$$

2. **Select reactor type and corresponding mole balance.**
   A.   If batch reactor (Section 7.2), use mole balance on Reactant A

   $$-r_A = -\frac{dC_A}{dt} \qquad\qquad \text{(TE7-1.1)}$$

   B.   If differential PBR (Section 7.6), use mole balance on Product P (A $\rightarrow$ P)

   $$-r_A' = \frac{F_P}{\Delta W} = C_P v_0 / \Delta W \qquad\qquad \text{(TE7-1.2)}$$

3. **Process your data in terms of the measured variable** (e.g., N_A, C_A, or P_A). If necessary, rewrite your mole balance in terms of the measured variable (e.g., P_A).
4. **Look for simplifications.** For example, if one of the reactants is in excess, assume its concentration is constant. If the gas-phase mole fraction of reactant A is small, set $\varepsilon \approx 0$.
5. **For a batch reactor, calculate $-r_A$ as a function of concentration C_A to determine the reaction order.**
   A.   *Differential analysis* (Section 7.4)
         Combine the mole balance (TE7-1.1) and power law model (TE7-1.3)

   $$-r_A = kC_A^{\alpha} \qquad\qquad \text{(TE7-1.3)}$$

TABLE 7-1   STEPS IN ANALYZING RATE DATA (CONTINUED)

$$-\frac{dC_A}{dt} = kC_A^{\alpha} \tag{TE7-1.4}$$

and then take the natural log

$$\ln\left(-\frac{dC_A}{dt}\right) = \ln(-r_A) = \ln k + \alpha \ln C_A \tag{TE7-1.5}$$

(1) Find $-\dfrac{dC_A}{dt}$ from C_A versus t data by either the

    (a) Graphical differential
    (b) Finite differential method or
    (c) Polynomial fit

(2) Either plot $\left[\ln\left(-\dfrac{dC_A}{dt}\right)\right]$ versus $\ln C_A$ to find reaction order α, which is the

    slope of the line fit to the data *or*
(3) Use nonlinear regression to find α and k simultaneously

B.   *Integral method* (Section 7.3)
    For $-r_A = kC_A^{\alpha}$, the combined mole balance and rate law is

$$-\frac{dC_A}{dt} = kC_A^{\alpha} \tag{TE7-1.4}$$

    Guess α and integrate Equation (TE7-1.4). Rearrange your equation to obtain the appropriate function of C_A, which when plotted as a function of time should be linear. If it is linear, then the guessed value of α is correct and the slope is the specific reaction rate, k. If it is not linear, guess again for α. If you guess $\alpha = 0$, 1, and 2, and none of these orders fit the data, proceed to nonlinear regression.

C.   *Nonlinear regression* (Polymath) (Section 7.5):
    Integrate Equation (TE7-1.4) to obtain

$$t = \frac{1}{k}\left[\frac{C_{A0}^{(1-\alpha)} - C_A^{(1-\alpha)}}{(1-\alpha)}\right] \text{ for } \alpha \neq 1 \tag{TE7-1.6}$$

    Use Polymath regression to find α and k. A Polymath tutorial on regression with screen shots is shown in the Chapter 7 *Summary Notes* on the CRE Web site, *www.umich.edu/~elements/5e/index.html*.

Summary Notes

6. **For differential PBR, calculate $-r'_A$ as a function of C_A or P_A (Section 7.6)**

    A.   Calculate $-r'_A = \dfrac{v_0 C_P}{\Delta W}$ as a function of reactant concentration, C_A or partial pressure P_A.

    B.   Choose a model (see Chapter 10), e.g.,

$$-r'_A = \frac{kP_A}{1 + K_A P_A}$$

    C.   Use nonlinear regression to find the best model and model parameters. See example on the CRE Web site *Summary Notes* for Chapter 10, using data from heterogeneous catalysis.

7. **Analyze your rate law model for "goodness of fit."** Calculate a correlation coefficient.

7.2 Determining the Reaction Order for Each of Two Reactants Using the Method of Excess

Batch reactors are used primarily to determine rate-law parameters for homogeneous reactions. This determination is usually achieved by measuring concentration as a function of time and then using either the integral, differential, or nonlinear regression method of data analysis to determine the reaction order, α, and specific reaction-rate constant, k. If some reaction parameter other than concentration is monitored, such as pressure, the mole balance must be rewritten in terms of the measured variable (e.g., pressure, as shown in the example in *Solved Problems* on the CRE Web site).

Process data in terms of the measured variable.

When a reaction is *irreversible*, it is possible in many cases to determine the reaction order α and the specific rate constant by either nonlinear regression or by numerically differentiating *concentration versus time data*. This latter method is most applicable when reaction conditions are such that the rate is essentially a function of the concentration of only one reactant; for example, if, for the decomposition reaction

$$A \rightarrow \text{Products}$$

Assume that the rate law is of the form
$-r_A = k_A C_A^{\alpha}$.

$$-r_A = k_A C_A^{\alpha} \tag{7-1}$$

then the differential method may be used.

However, by utilizing the method of excess, it is also possible to determine the relationship between $-r_A$ and the concentration of other reactants. That is, for the irreversible reaction

$$A + B \rightarrow \text{Products}$$

with the rate law

$$-r_A = k_A C_A^{\alpha} C_B^{\beta} \tag{7-2}$$

where α and β are both unknown, the reaction could first be run in an excess of B so that C_B remains essentially unchanged during the course of the reaction (i.e., $C_B \approx C_{B0}$) and

$$-r_A = k' C_A^{\alpha} \tag{7-3}$$

where

Method of excess

$$k' = k_A C_B^{\beta} \approx k_A C_{B0}^{\beta}$$

After determining α, the reaction is carried out in an excess of A, for which the rate law is approximated as

$$-r_A = k'' C_B^{\beta} \tag{7-4}$$

where $k'' = k_A C_A^{\alpha} \approx k_A C_{A0}^{\alpha}$

Once α and β are determined, k_A can be calculated from the measurement of $-r_A$ at known concentrations of A and B

$$k_A = \frac{-r_A}{C_A^\alpha C_B^\beta} = \frac{(\text{dm}^3/\text{mol})^{\alpha+\beta-1}}{s} \tag{7-5}$$

Both α and β can be determined by using the method of excess, coupled with a differential analysis of data for batch systems.

7.3 Integral Method

The integral method is the quickest method to use to determine the rate law if the order turns out to zero, first, or second order. In the integral method, we guess the reaction order, α, in the combined batch reactor mole balance and rate law equation

The integral method uses a trial-and-error procedure to find the reaction order.

$$\frac{dC_A}{dt} = -kC_A^\alpha \tag{7-6}$$

and integrate the differential equation to obtain the concentration as a function of time. If the order we assume is correct, the appropriate plot (determined from this integration) of the concentration–time data should be linear. The integral method is used most often when the reaction order is known and it is desired to evaluate the specific reaction rate constant at different temperatures to determine the activation energy.

In the integral method of analysis of rate data, we are looking for the appropriate function of concentration corresponding to a particular rate law that is linear with time. You should be thoroughly familiar with the methods of obtaining these linear plots for reactions of *zero, first, and second order.*

For the reaction

$$A \rightarrow \text{Products}$$

It is important to know how to generate linear plots of functions of C_A versus t for zero-, first-, and second-order reactions.

carried out in a constant-volume batch reactor, the mole balance is

$$\frac{dC_A}{dt} = r_A$$

For a zero-order reaction, $r_A = -k$, and the combined rate law and mole balance is

$$\frac{dC_A}{dt} = -k \tag{7-7}$$

Integrating with $C_A = C_{A0}$ at $t = 0$, we have

Zero order

$$\boxed{C_A = C_{A0} - kt} \tag{7-8}$$

A plot of the concentration of A as a function of time will be linear (Figure 7-1) with slope $(-k)$ for a zero-order reaction carried out in a constant-volume batch reactor.

If the reaction is first order (Figure 7-2), integration of the combined *mole balance and the rate law*

$$-\frac{dC_A}{dt} = kC_A$$

with the limit $C_A = C_{A0}$ at $t = 0$ gives

First order

$$\boxed{\ln\frac{C_{A0}}{C_A} = kt} \tag{7-9}$$

Consequently, we see that the slope of a plot of $[\ln(C_{A0}/C_A)]$ as a function of time is linear with slope k.

If the reaction is second order (Figure 7-3), then

$$-\frac{dC_A}{dt} = kC_A^2$$

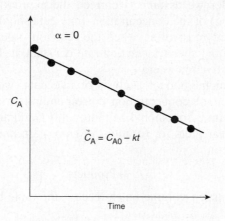

Figure 7-1   Zero-order reaction.

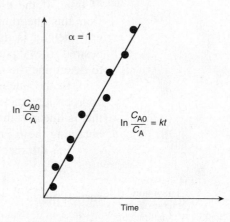

Figure 7-2   First-order reaction.

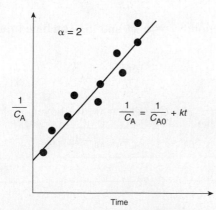

Figure 7-3   Second-order reaction.

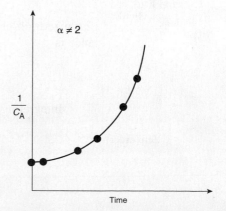

Figure 7-4   Plot of reciprocal concentration as a function of time.

Integrating, with $C_A = C_{A0}$ initially (i.e., $t = 0$), yields

Second order

$$\boxed{\frac{1}{C_A} - \frac{1}{C_{A0}} = kt}$$

(7-10)

We see that for a second-order reaction a plot of $(1/C_A)$ as a function of time should be linear with slope k.

In Figures 7-1, 7-2, and 7-3, we saw that when we plotted the appropriate function of concentration (i.e., C_A, $\ln C_A$, or $1/C_A$) versus time, the plots were linear, and we concluded that the reactions were zero, first, or second order, respectively. However, if the plots of concentration data versus time had turned out **not to be linear**, such as shown in Figure 7-4, we would say that the proposed reaction order did not fit the data. In the case of Figure 7-4, we would conclude that the reaction is not second order. After finding that the integral method for first, second, and third orders do not fit the data, one should use one of the other methods discussed in Table 7-1.

The idea is to arrange the data so that a linear relationship is obtained.

It is important to restate that, given a reaction-rate law, you should be able to quickly choose the appropriate function of concentration or conversion that yields a straight line when plotted against time or space time. The goodness-of-fit of such a line may be assessed statistically by calculating the linear correlation coefficient, r^2, which should be as close to 1 as possible. The value of r^2 is given in the output of Polymath's nonlinear regression.

Example 7–1   Integral Method of CRE Data Analysis

The liquid-phase reaction

Trityl (A) + Methanol (B) → Products (C)

was carried out in a batch reactor at 25°C in a solution of benzene and pyridine in an excess of methanol ($C_{B0} = 0.5 \dfrac{\text{mol}}{\text{dm}^3}$). (We need to point out that this batch reactor was purchased at the Sunday market in Riça, Jofostan.) Pyridine reacts with HCl, which then precipitates as pyridine hydro-chloride thereby making the reaction irreversible. The reaction is first order in methanol. The concentration of triphenyl methyl chloride (A) was measured as a function of time and is shown below

TABLE E7-1.1   RAW DATA

t (min)	0	50	100	150	200	250	300
C_A (mol/dm³)	0.05	0.038	0.0306	0.0256	0.0222	0.0195	0.0174

Use the integral method to confirm that the reaction is second order with regard to triphenyl methyl chloride

Solution

We use the power-law model, Equation (7-2), along with information from the problem statement that the reaction is first order in methanol, (B), i.e., $\beta = 1$ to obtain

$$-r_A = kC_A^\alpha C_B$$

(E7-1.1)

Excess methanol: The initial concentration of methanol (B) is 10 times that of trityl (A), so even if all A were consumed, 90% of B remains. Consequently, we will take the concentration of B as a constant and combine it with k to form

$$-r_A = kC_A^\alpha C_{B0} = k'C_A^\alpha \tag{E7-1.2}$$

where k' is the pseudo rate constant $k' = kC_{B0}$ and k is the true rate constant. Substituting $\alpha = 2$ and combining with the mole balance on a batch reactor, we obtain

$$-\frac{dC_A}{dt} = k'C_A^2 \tag{E7-1.3}$$

Integrating with $C_A = C_{A0}$ at $t = 0$

$$t = \frac{1}{k'}\left[\frac{1}{C_A} - \frac{1}{C_{A0}}\right] \tag{E7-1.4}$$

Rearranging

$$\frac{1}{C_A} = \frac{1}{C_{A0}} + k't \tag{E7-1.5}$$

We see that if the reaction is indeed second order then a plot of $(1/C_A)$ versus t should be linear. Using the data in Table E7-1.1, we calculate $(1/C_A)$ to construct Table E7-1.2.

TABLE E7-1.2   PROCESSED DATA

t (min)	0	50	100	150	200	250	300
C_A (mol/dm^3)	0.05	0.038	0.0306	0.0256	0.0222	0.0195	0.0174
$1/C_A$ (dm^3/mol)	20	26.3	32.7	39.1	45	51.3	57.5

In a graphical solution, the data in Table E7-1.2 can be used to construct a plot of $1/C_A$ as a function of t, which will yield the specific reaction rate k'. This plot is shown in Figure E7-1.1. Again, Excel or Polymath could be used to find k' from the data in Table E7-1.2. The slope of the line is the specific reaction rate k'.

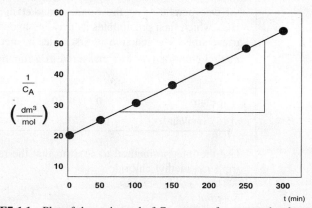

Figure E7-1.1   Plot of the reciprocal of C_A versus t for a second-order reaction.

We see from the Excel analysis and plot that the slope of the line is 0.12 dm^3/mol · min.

$$k' = 0.12\, \frac{dm^3}{mol \cdot min} \tag{E7-1.6}$$

We now use Equation (E7-1.6), along with the initial concentration of methanol, to find the true rate constant, k.

$$k = \frac{k'}{C_{B0}} = \frac{0.12}{0.5} \frac{\text{dm}^3/\text{mol}/\text{min}}{\text{mol}/\text{dm}^3} = 0.24 \left(\frac{\text{dm}^3}{\text{mol}} \right)^2 /\text{min}$$

The rate law is

$$-r_A = \left[0.24 \left(\frac{\text{dm}^3}{\text{mol}} \right)^2 /\text{min} \right] C_A^2 C_B \qquad \text{(E7-1.7)}$$

We note that the integral method tends to smooth the data.

Analysis: In this example, the reaction orders are known so that the integral method can be used to (1) verify the reaction is second order in trityl and (2) to find the specific pseudo reaction rate $k' = kC_{B0}$ for the case of excess methanol (B). Knowing k' and C_{B0}, we can then find the true rate constant k.

7.4 Differential Method of Analysis

To outline the procedure used in the differential method of analysis, we consider a reaction carried out isothermally in a constant-volume batch reactor and the concentration of A, recorded as a function of time. By combining the mole balance with the rate law given by Equation (7-1), we obtain

Constant-volume
batch reactor

$$-\frac{dC_A}{dt} = k_A C_A^\alpha$$

After taking the natural logarithm of both sides of Equation (5-6)

$$\ln\left(-\frac{dC_A}{dt} \right) = \ln k_A + \alpha \ln C_A \qquad (7\text{-}11)$$

observe that the slope of a plot of $[\ln(-dC_A/dt)]$ as a function of $(\ln C_A)$ is the reaction order, α (see Figure 7-5).

Plot
$$\ln\left(-\frac{dC_A}{dt} \right)$$
versus $\ln C_A$
to find
α and k_A

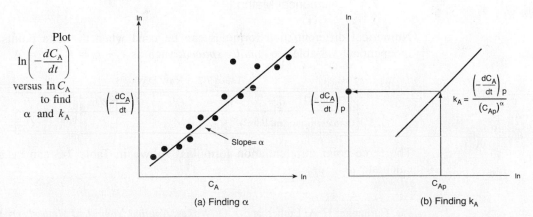

(a) Finding α (b) Finding k_A

Figure 7-5 Differential method to determine reaction order.

Figure 7-5(a) shows a plot of $[-(dC_A/dt)]$ versus $[C_A]$ on log-log paper (or use Excel to make the plot) where the slope is equal to the reaction order α. The specific reaction rate, k_A, can be found by first choosing a concentration in the plot, say C_{Ap}, and then finding the corresponding value of $[-(dC_A/dt)_p]$ on the line, as shown in Figure 7-5(b). The concentration chosen, C_{Ap}, to find the derivative at C_{Ap}, need not be a data point, it just needs to be on the line. After raising C_{Ap} to the power α, we divide it into $[-(dC_A/dt)_p]$ to determine k_A

$$k_A = \frac{-(dC_A/dt)_p}{(C_{Ap})^\alpha} \tag{7-12}$$

To obtain the derivative $(-dC_A/dt)$ used in this plot, we must differentiate the concentration–time data either numerically or graphically. Three methods to determine the derivative from data giving the concentration as a function of time are

Methods for finding $-\dfrac{dC_A}{dt}$ from concentration-time data

- Graphical differentiation
- Numerical differentiation formulas
- Differentiation of a polynomial fit to the data

We shall only discuss the graphical and numerical methods.

7.4.1 Graphical Differentiation Method

See Appendix A.2.

This method is very old (from slide rule days—"What's a slide rule, Grandfather?"), when compared with the numerous software packages. So why do we use it? Because with this method, disparities in the data are easily seen. Consequently, it is advantageous to use this technique to analyze the data before planning the next set of experiments. As explained in Appendix A.2, the graphical method involves plotting $(-\Delta C_A/\Delta t)$ as a function of t and then using equal-area differentiation to obtain $(-dC_A/dt)$. An illustrative example is also given in Appendix A.2.

In addition to the graphical technique used to differentiate the data, two other methods are commonly used: differentiation formulas and polynomial fitting.

7.4.2 Numerical Method

Numerical differentiation formulas can be used when the data points in the independent variable are *equally spaced*, such as $t_1 - t_0 = t_2 - t_1 = \Delta t$.

TABLE 7-2   RAW DATA

Time (min)	t_0	t_1	t_2	t_3	t_4	t_5
Concentration (mol/dm^3)	C_{A0}	C_{A1}	C_{A2}	C_{A3}	C_{A4}	C_{A5}

The three-point differentiation formulas[1] shown in Table 7-3 can be used to calculate dC_A/dt.

---

[1] B. Carnahan, H. A. Luther, and J. O. Wilkes, *Applied Numerical Methods* (New York: Wiley, 1969), p. 129.

TABLE 7-3   DIFFERENTIATION FORMULAS

Initial point:	$\left(\dfrac{dC_A}{dt}\right)_{t_0} = \dfrac{-3C_{A0} + 4C_{A1} - C_{A2}}{2\Delta t}$	(7-13)
Interior points:	$\left(\dfrac{dC_A}{dt}\right)_{t_i} = \dfrac{1}{2\Delta t}[(C_{A(i+1)} - C_{A(i-1)})]$ $\left[\text{e.g.,}\ \left(\dfrac{dC_A}{dt}\right)_{t_3} = \dfrac{1}{2\Delta t}[C_{A4} - C_{A2}]\right]$	(7-14)
Last point:	$\left(\dfrac{dC_A}{dt}\right)_{t_5} = \dfrac{1}{2\Delta t}[C_{A3} - 4C_{A4} + 3C_{A5}]$	(7-15)

Equations (7-13) and (7-15) are used for the first and last data points, respectively, while Equation (7-14) is used for all intermediate data points (see Step 6A.1a in Example 7-2).

7.4.3 Finding the Rate-Law Parameters

Now, using either the graphical method, differentiation formulas, or the polynomial derivative, the following table can be set up.

TABLE 7-4   PROCESSED DATA

Time	t_0	t_1	t_2	t_3
Concentration	C_{A0}	C_{A1}	C_{A2}	C_{A3}
Derivative	$\left(-\dfrac{dC_A}{dt}\right)_0$	$\left(-\dfrac{dC_A}{dt}\right)_1$	$\left(-\dfrac{dC_A}{dt}\right)_2$	$\left(-\dfrac{dC_A}{dt}\right)_3$

The reaction order can now be found from a plot of $\ln(-dC_A/dt)$ as a function of $\ln C_A$, as shown in Figure 7-5(a), since

$$\ln\left(-\frac{dC_A}{dt}\right) = \ln k_A + \alpha \ln C_A \tag{7-16}$$

Before solving the following example problems, review the steps to determine the reaction-rate law from a set of data points (Table 7-1).

Example 7–2 Determining the Rate Law

The reaction of triphenyl methyl chloride (trityl) (A) and methanol (B) discussed in Example 7-1 is now analyzed using the differential method.

$$(C_6H_5)_3CCl + CH_3OH \rightarrow (C_6H_5)_3C\overset{\overset{\displaystyle O}{\|}}{C}CH_3 + HCl$$

$$\text{A}\quad +\quad \text{B}\quad \rightarrow \quad\quad \text{C}\quad\quad +\ \text{D}$$

The concentration–time data in Table E7-2.1 was obtained in a batch reactor.

TABLE E7-2.1   RAW DATA

Time (min)	0	50	100	150	200	250	300
Concentration of A (mol/dm³) × 10³ (At $t = 0$, $C_A = 0.05$ M)	50	38	30.6	25.6	22.2	19.5	17.4

The initial concentration of methanol was 0.5 mol/dm³.

Part (1)   Determine the reaction order with respect to triphenyl methyl chloride.

Part (2)   In a separate set of experiments, the reaction order wrt methanol was found to be first order. Determine the specific reaction-rate constant.

Solution

Part (1) Find the reaction order with respect to trityl.

Step 1   Postulate a rate law.

$$-r_A = k C_A^\alpha C_B^\beta \tag{E7-2.1}$$

Step 2   Process your data in terms of the measured variable, which in this case is C_A.

Step 3   Look for simplifications. Because the concentration of methanol is 10 times the initial concentration of triphenyl methyl chloride, its concentration is essentially constant

$$C_B \cong C_{B0} \tag{E7-2.2}$$

Substituting for C_B in Equation (E7-2.1)

$$-r_A = \underbrace{k C_{B0}^\beta}_{k'} C_A^\alpha$$

$$-r_A = k' C_A^\alpha \tag{E7-2.3}$$

Step 4   Apply the CRE algorithm.
Mole Balance

$$\frac{dN_A}{dt} = r_A V \tag{E7-2.4}$$

Rate Law:

$$-r_A = k' C_A^\alpha \tag{E7-2.3}$$

Stoichiometry: Liquid $V = V_0$

$$C_A = \frac{N_A}{V_0}$$

Following the Algorithm

Combine: Mole balance, rate law, and stoichiometry

$$-\frac{dC_A}{dt} = k' C_A^\alpha \tag{E7-2.5}$$

Evaluate: Taking the natural log of both sides of Equation (E7-2.5)

$$\ln\left[-\frac{dC_A}{dt}\right] = \ln k' + \alpha \ln C_A \tag{E7-2.6}$$

The slope of a plot of $\ln \left[-\dfrac{dC_A}{dt} \right]$ versus $\ln C_A$ will yield the reaction order α with respect to triphenyl methyl chloride (A).

Step 5 Find $\left[-\dfrac{dC_A}{dt} \right]$ as a function of C_A from concentration–time data.

Step 5A.1a *Graphical Method.* We now show how to construct Table E7-2.2. The derivative $(-dC_A/dt)$ is determined by calculating and plotting $(-\Delta C_A/\Delta t)$ as a function of time, t, and then using the equal-area differentiation technique (Appendix A.2) to determine $(-dC_A/dt)$ as a function of C_A. First, we calculate the ratio $(-\Delta C_A/\Delta t)$ from the first two columns of Table E7-2.2; the result is written in the third column.

TABLE E7-2.2   PROCESSED DATA

t (min)	$C_A \times 10^3$ (mol/dm³)	$-\dfrac{\Delta C_A}{\Delta t} \times 10^4$ (mol/dm³ · min)	$-\dfrac{dC_A}{dt} \times 10^4$ (mol/dm³ · min)
0	50		3.0
		2.40†	
50	38		1.86
		1.48	
100	30.6		1.2
		1.00	
150	25.6		0.8
		0.68	
200	22.2		0.5
		0.54	
250	19.5		0.47
		0.42	
300	17.4		

$$\dagger \quad -\frac{\Delta C_A}{\Delta t} = -\frac{C_{A2} - C_{A1}}{t_2 - t_1} = -\left(\frac{38 - 50}{50 - 0} \right) \times 10^{-3} = 0.24 \times 10^{-3} = 2.4 \times 10^{-4} \, (\text{mol/dm}^3 \cdot \text{min})$$

Next, we use Table E7-2.2 to plot the third column as a function of the first column in Figure E7-1.1 [i.e., $(-\Delta C_A/\Delta t)$ versus t]. Using equal-area differentiation, the value of $(-dC_A/dt)$ is read off the figure (represented by the arrows); then it is used to complete the fourth column of Table E7-2.2.

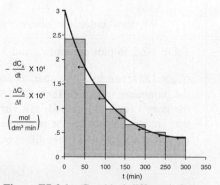

Figure E7-2.1   Graphical differentiation.

The results to find ($-dC_A/dt$) at each time, t, and concentration, C_A, are summarized in Table E7-2.3.

Step 6A.1a Finite Difference Method. We now show how to calculate (dC_A/dt) using the finite difference formulas [i.e., Equations (7-13) through (7-15)].

$$t = 0 \quad \left(\frac{dC_A}{dt}\right)_{t=0} = \frac{-3C_{A0} + 4C_{A1} - C_{A2}}{2\Delta t}$$

$$= \frac{[-3(50) + 4(38) - 30.6] \times 10^{-3}}{100}$$

$$= -2.86 \times 10^{-4} \text{ mol/dm}^3 \cdot \text{min}$$

$$-\frac{dC_A}{dt} \times 10^4 = 2.86 \text{ mol/dm}^3 \cdot \text{min}$$

$$t = 50 \quad \left(\frac{dC_A}{dt}\right)_1 = \frac{C_{A2} - C_{A0}}{2\Delta t} = \frac{(30.6 - 50) \times 10^{-3}}{100}$$

$$= -1.94 \times 10^{-4} \text{ mol/dm}^3 \cdot \text{min}$$

$$t = 100 \quad \left(\frac{dC_A}{dt}\right)_2 = \frac{C_{A3} - C_{A1}}{2\Delta t} = \frac{(25.6 - 38) \times 10^{-3}}{100}$$

$$= -1.24 \times 10^{-4} \text{ mol/dm}^3 \cdot \text{min}$$

$$t = 150 \quad \left(\frac{dC_A}{dt}\right)_3 = \frac{C_{A4} - C_{A2}}{2\Delta t} = \frac{(22.2 - 30.6) \times 10^{-3}}{100}$$

$$= -0.84 \times 10^{-4} \text{ mol/dm}^3 \cdot \text{min}$$

$$t = 200 \quad \left(\frac{dC_A}{dt}\right)_4 = \frac{C_{A5} - C_{A3}}{2\Delta t} = \frac{(19.5 - 25.6) \times 10^{-3}}{100}$$

$$= -0.61 \times 10^{-4} \text{ mol/dm}^3 \cdot \text{min}$$

$$t = 250 \quad \left(\frac{dC_A}{dt}\right)_5 = \frac{C_{A6} - C_{A4}}{2\Delta t} = \frac{(17.4 - 22.2) \times 10^{-3}}{100}$$

$$= -0.48 \times 10^{-4} \text{ mol/dm}^3 \cdot \text{min}$$

$$t = 300 \quad \left(\frac{dC_A}{dt}\right)_6 = \frac{C_{A4} - 4C_{A5} + 3C_{A6}}{2\Delta t} = \frac{[22.2 - 4(19.5) + 3(17.4)] \times 10^{-3}}{100}$$

$$= -0.36 \times 10^{-4} \text{ mol/dm}^3 \cdot \text{min}$$

We now enter the above values for ($-dC_A/dt$) in Table E7-2.3 and use Table 7-2.2 to plot columns 2 and 3 $\left(-\dfrac{dC_A}{dt} \times 10{,}000\right)$ as a function of column 4 ($C_A \times 1{,}000$) on log-log paper, as shown in Figure E7-2.2. We could also substitute the parameter values in Table E7-2.3 into Excel to find α and k'. Note that most of the points for both methods fall virtually on top of one another. This table is, in a way, redundant because it is not necessary to always find ($-dC_A dt$) by both techniques, graphical and finite difference.

TABLE E7-2.3   SUMMARY OF PROCESSED DATA

	Graphical	Finite Difference	
	$-\dfrac{dC_A}{dt} \times 10{,}000$	$-\dfrac{dC_A}{dt} \times 10{,}000$	$C_A \times 1{,}000$
t (min)	(mol/dm³ · min)	(mol/dm³ · min)	(mol/dm³)
0	3.0	2.86	50
50	1.86	1.94	38
100	1.20	1.24	30.6
150	0.80	0.84	25.6
200	0.68	0.61	22.2
250	0.54	0.48	19.5
300	0.42	0.36	17.4

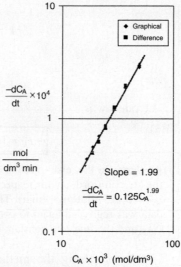

Figure E7-2.2   Excel plot to determine α and k.

From Figure E7-2.2, we found the slope to be 1.99, so that the reaction is said to be second order ($\alpha = 2.0$) with respect to triphenyl methyl chloride. To evaluate k', we can evaluate the derivative in Figure E7-2.2 at $C_{Ap} = 20 \times 10^{-3}$ mol/dm³, which is

$$\left(-\frac{dC_A}{dt}\right)_p = 0.5 \times 10^{-4} \text{ mol/dm}^3 \cdot \text{min} \tag{E7-2.7}$$

then

$$k' = \frac{\left(-\dfrac{dC_A}{dt}\right)_p}{C_{Ap}^2} \tag{E7-2.8}$$

$$= \frac{0.5 \times 10^{-4} \text{mol/dm}^3 \cdot \text{min}}{\left(20 \times 10^{-3} \text{mol/dm}^3\right)^2} = 0.125 \text{ dm}^3 / \text{mol} \cdot \text{min}$$

As will be shown in Section 7-5, we could also use nonlinear regression on Equation (E7-1.5) to find k'

$$k' = 0.122 \text{ dm}^3/\text{mol} \cdot \text{min} \tag{E7-2.9}$$

Summary Notes

The Excel graph shown in Figure E7-2.2 gives $\alpha = 1.99$ and $k' = 0.13$ dm³/mol · min. We now set $\alpha = 2$ and regress again to find $k' = 0.122$ dm³/mol · min.

ODE Regression. There are techniques and software becoming available whereby an ODE solver can be combined with a regression program to solve differential equations, such as

$$-\frac{dC_A}{dt} = k_A' C_A^{\alpha}$$

(E5-7-2.5)

to find k_A and α from concentration–time data.

<u>Part (2) The reaction was said to be first order with respect to methanol, $\beta = 1$,</u>

$$k' = C_{B0}^{\beta} k = C_{B0} k$$

(E7-2.10)

Assuming C_{B0} is constant at 0.5 mol/dm³ and solving for k yields[2]

$$k = \frac{k'}{C_{B0}} = \frac{0.122 \dfrac{\text{dm}^3}{\text{mol} \cdot \text{min}}}{0.5 \dfrac{\text{mol}}{\text{dm}^3}}$$

$$k = 0.244 \ (\text{dm}^3/\text{mol})^2 \ / \ \text{min}$$

The rate law is

$$\boxed{-r_A = [0.244 (\text{dm}^3/\text{mol})^2 / \text{min}] C_A^2 C_B}$$

(E7-2.11)

<u>*Analysis:*</u> In this example, the differential method of data analysis was used to find the reaction order with respect to trityl ($\alpha = 1.99$) and the pseudo rate constant ($k' = 0.125$ (dm³/mol)/min). The reaction order was rounded up to $\alpha = 2$ and the data was regressed again to obtain $k' = 0.122$ (dm³/mol)/min, again knowing k' and C_{B0}, and the true rate constant is $k = 0.244$ (dm³/mol)²/min.

Integral method
normally used to
find k when order
is known

By comparing the methods of analysis of the rate data presented in Examples 7-1 and 7-2, we note that the differential method tends to accentuate the uncertainties in the data, while the integral method tends to smooth the data, thereby disguising the uncertainties in it. In most analyses, it is imperative that the engineer know the limits and uncertainties in the data. This prior knowledge is necessary to provide for a safety factor when scaling up a process from laboratory experiments to design either a pilot plant or full-scale industrial plant.

7.5 Nonlinear Regression

In nonlinear regression analysis, we search for those parameter values that minimize the sum of the squares of the differences between the measured values and the calculated values for all the data points. Not only can nonlinear

---

[2] M. Hoepfner and D. K. Roper, "Describing Temperature Increases in Plasmon-Resonant Nanoparticle Systems," *Journal of Thermal Analysis and Calorimetry*, 98(1), (2009), pp. 197–202.

regression find the best estimates of parameter values, it can also be used to discriminate between different rate-law models, such as the Langmuir-Hinshelwood models discussed in Chapter 10. Many software programs are available to find these parameter values so that all one has to do is enter the data. The Polymath software will be used to illustrate this technique. In order to carry out the search efficiently, in some cases one has to enter initial estimates of the parameter values close to the actual values. These estimates can be obtained using the linear-least-squares technique discussed on the CRE Web site *Professional Reference Shelf R7.3*.

We will now apply nonlinear regression to reaction-rate data to determine the rate-law parameters. Here, we make initial estimates of the parameter values (e.g., reaction order, specific rate constant) in order to calculate the concentration for each data point, C_{ic}, obtained by solving an integrated form of the combined mole balance and rate law. We then compare the measured concentration at that point, C_{im}, with the calculated value, C_{ic}, for the parameter values chosen. We make this comparison by calculating the sum of the squares of the differences at each point $\Sigma(C_{im} - C_{ic})^2$. We then continue to choose new parameter values and search for those values of the rate law that will minimize the sum of the squared differences of the measured concentrations, C_{im}, and the calculated concentrations values, C_{ic}. That is, we want to find the rate-law parameters for which the sum of all data points $\Sigma(C_{im} - C_{ic})^2$ is a minimum. If we carried out N experiments, we would want to find the parameter values (e.g., E, activation energy, reaction orders) that minimize the quantity

$$\sigma^2 = \frac{s^2}{N-K} = \sum_{i=1}^{N} \frac{(C_{im} - C_{ic})^2}{N-K} \tag{7-17}$$

where

$$s^2 = \sum_{i=1}^{i=N} (C_{im} - C_{ic})^2$$

N = number of runs

K = number of parameters to be determined

C_{im} = measured concentration rate for run i

C_{ic} = calculated concentration rate for run i

One notes that if we minimize s^2 in Equation (7-17), we minimize σ^2.

To illustrate this technique, let's consider the reaction

$$A \longrightarrow Product$$

for which we want to learn the reaction order, α, and the specific reaction rate, k,

$$-r_A = kC_A^{\alpha}$$

The reaction rate will be measured at a number of different concentrations. We now choose values of k and α, and calculate the rate of reaction (C_{ic}) at each concentration at which an experimental point was taken. We then subtract the calculated value (C_{ic}) from the measured value (C_{im}), square the result, and sum the squares for all the runs for the values of k and α that we have chosen.

This procedure is continued by further varying α and k until we find those values of k and α that minimize the sum of the squares. Many well-known searching techniques are available to obtain the minimum value $\sigma^2_{\min}$.[3] Figure 7-6 shows a hypothetical plot of the sum of the squares as a function of the parameters α and k:

$$\sigma^2 = f(k, \alpha) \tag{7-18}$$

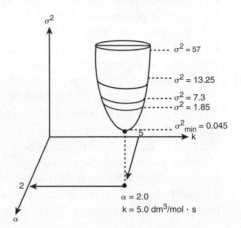

Figure 7-6   Minimum sum of squares.

Look at the top circle. We see that there are many combinations of α and k (e.g., $\alpha = 2.2$, $k = 4.8$ or $\alpha = 1.8$, $k = 5.3$) that will give a value of $\sigma^2 = 57$. The same is true for $\sigma^2 = 1.85$. We need to find the combination of α and k that gives the lowest value of σ^2.

In searching to find the parameter values that give the minimum of the sum of squares σ^2, one can use a number of optimization techniques or software packages. The searching procedure begins by guessing parameter values and then calculating ($C_{im} - C_{ic}$) and then σ^2 for these values. Next, a few sets of parameters are chosen around the initial guess, and σ^2 is calculated for these sets as well. The search technique looks for the smallest value of σ^2 in the vicinity of the initial guess and then proceeds along a trajectory in the direction of decreasing σ^2 to choose different parameter values and determine the corresponding σ^2. The trajectory is continually adjusted so as to always proceed in the direction of decreasing σ^2 until the minimum value of σ^2 is reached. For example, in Figure 7-6 the search technique keeps choosing combinations of α and k until a minimum value of $\sigma^2 = 0.045$ $(mol/dm^3)^2$ is reached.

---

[3] (a) B. Carnahan and J. O. Wilkes, *Digital Computing and Numerical Methods* (New York: Wiley, 1973), p. 405. (b) D. J. Wilde and C. S. Beightler, *Foundations of Optimization,* 2nd ed. (Upper Saddle River, NJ: Prentice Hall, 1979). (c) D. Miller and M. Frenklach, *Int. J. Chem. Kinet.*, 15 (1983), p. 677.

The combination that gives that minimum is $\alpha = 2$ and $k = 5.0$ dm^3/mol·min as shown in Figure 7-6. If the equations are highly nonlinear, the initial guesses of α and k are very important.

Reference Shelf

A number of software packages are available to carry out the procedure to determine the best estimates of the parameter values and the corresponding confidence limits. All we have to do is to enter the experimental values into the computer, specify the model, enter the initial guesses of the parameters, and then push the "compute" button, and the best estimates of the parameter values along with 95% confidence limits appear. If the confidence limits for a given parameter are larger than the parameter itself, the parameter is probably not significant and should be dropped from the model. After the appropriate model parameters are eliminated, the software is run again to determine the best fit with the new model equation.

Concentration–Time Data.   We will now use nonlinear regression to determine the rate-law parameters from concentration–time data obtained in batch experiments. We recall that the combined rate-law stoichiometry mole balance for a constant-volume batch reactor is

$$\frac{dC_A}{dt} = -kC_A^{\alpha} \tag{7-19}$$

We now integrate Equation (7-6) to give

$$C_{A0}^{1-\alpha} - C_A^{1-\alpha} = (1 - \alpha)kt$$

Rearranging to obtain the concentration as a function of time, we obtain

$$C_A = [C_{A0}^{1-\alpha} - (1 - \alpha)kt]^{1/(1-\alpha)} \tag{7-20}$$

Now we could use either Polymath or MATLAB to find the values of α and k that would minimize the sum of squares of the differences between the measured concentrations, C_{Aim}, and calculated concentrations, C_{Aic}. That is, for N data points,

$$s^2 = \sum_{i=1}^{N} (C_{Aim} - C_{Aic})^2 = \sum_{i=1}^{N} \left[C_{Aim} - [C_{A0}^{1-\alpha} - (1 - \alpha)kt_i]^{1/(1-\alpha)} \right]^2 \tag{7-21}$$

we want the values of α and k that will make s^2 a minimum.

If Polymath is used, one should use the absolute value for the term in brackets in Equation (7-16), that is,

$$s^2 = \sum_{i=1}^{n} \left[C_{Aim} - \{(\text{abs}[C_{A0}^{1-\alpha} - (1 - \alpha)kt_i]\}^{1/(1-\alpha)} \right]^2 \tag{7-22}$$

Another, and perhaps easier, way to solve for the parameter values is to use time rather than concentrations, rearranging Equation (7-20) to get

$$\boxed{t_c = \frac{C_{A0}^{1-\alpha} - C_A^{1-\alpha}}{k(1 - \alpha)}} \tag{7-23}$$

That is, we find the values of k and α that minimize

$$s^2 = \sum_{i=1}^{N} (t_{im} - t_{ic})^2 = \sum_{i=1}^{N} \left[t_{im} - \frac{C_{A0}^{1-\alpha} - C_{Ai}^{1-\alpha}}{k(1-\alpha)} \right]^2 \qquad (7\text{-}24)$$

Reference Shelf

Finally, a discussion of *weighted least squares* as applied to a first-order reaction is provided in the *Professional Reference Shelf R7.5* on the CRE Web site.

Example 7–3 Use of Regression to Find the Rate-Law Parameters

We shall use the reaction and data in Examples E7-1 and E7-2 to illustrate how to use regression to find α and k'.

$$(C_6H_5)_3CCl + CH_3OH \rightarrow (C_6H_5)_3COCH_3 + HCl$$

$$A \quad + \quad B \quad \rightarrow \quad C \quad + \quad D$$

The Polymath regression program is included on the CRE Web site. Recalling Equation (E5-1.5)

$$-\frac{dC_A}{dt} = k' C_A^{\alpha} \qquad (E7\text{-}2.5)$$

and integrating with the initial condition when $t = 0$ and $C_A = C_{A0}$ for $\alpha \neq 1.0$

$$t = \frac{1}{k'} \frac{C_{A0}^{(1-\alpha)} - C_A^{(1-\alpha)}}{(1-\alpha)} \qquad (E7\text{-}3.1)$$

Given or assuming k' and α, Equation (7-2.5) can be solved to calculate the time t to reach a concentration C_A or we could calculate the concentration C_A at time t. We can proceed two ways from this point, both of which will give the same result. We can search for the combination α and k that minimizes $[\sigma^2 = \Sigma(t_{im} - t_{ic})^2]$, or we could solve Equation (E7-4.3) for C_A and find α and k that minimize $[\sigma^2 = \Sigma(C_{Aim} - C_{Aic})^2]$. We shall choose the former. So, substituting for the initial concentration $C_{A0} = 0.05$ mol/dm³ into Equation (E7-3.1)

$$t = \frac{1}{k'} \frac{(0.05)^{(1-\alpha)} - C_A^{(1-\alpha)}}{(1-\alpha)} \qquad (E7\text{-}3.2)$$

Summary Notes

A brief tutorial on how to input data in Polymath is given in the Polymath link in the *Summary Notes* on the CRE Web site for Chapter 7. The Polymath tutorial on the CRE Web site shows screen shots of how to enter the raw data in Table E7-2.1 and how to carry out a nonlinear regression on Equation (E7-3.2). For $C_{A0} = 0.05$ mol/dm³, that is, Equation (E7-3.1) becomes

$$t_c = \frac{1}{k'} \frac{(0.05)^{(1-\alpha)} - C_A^{(1-\alpha)}}{(1-\alpha)} \qquad (E7\text{-}3.3)$$

We want to minimize s^2 to give α and k'.

$$s^2 = \sum_{i=1}^{N} (t_{im} - t_{ic})^2 = \sum_{i=1}^{N} \left[t_{im} - \frac{0.05^{(1-\alpha)} - C_{Aic}^{(1-\alpha)}}{k'(1-\alpha)} \right]^2 \qquad (7\text{-}25)$$

The result of the first and second Polymath regressions are shown in Tables E7-3.1 and E7-3.2.

TABLE E7-3.1   RESULTS OF 1ST REGRESSION

POLYMATH Results
Example 7-3.1 Use of Regression to Find Rate Law Parameters 08-05-2004

Nonlinear regression (L-M)

Model: t = (.05^(1-a)-Ca^(1-a))/(k*(1-a))

Variable	Ini guess	Value	95% confidence
a	3	2.04472	0.0317031
k	0.1	0.1467193	0.0164118

Nonlinear regression settings
Max # iterations = 64

Precision
R^2   = 0.9999717
R^2adj   = 0.999966
Rmsd   = 0.2011604
Variance = 0.3965618

TABLE E7-3.2   RESULTS OF 2ND REGRESSION

POLYMATH Results
Example 7-3.2 Use of Regression to Find Rate Law Parameters 08-05-2004

Nonlinear regression (L-M)

Model: t = (.05^(1-2)-Ca^(1-2))/(k*(1-2))

Variable	Ini guess	Value	95% confidence
k	0.1	0.1253404	7.022E-04

Nonlinear regression settings
Max # iterations = 64

Precision
R^2   = 0.9998978
R^2adj   = 0.9998978
Rmsd   = 0.3821581
Variance = 1.1926993

The results shown are

$$\alpha = 2.04$$
$$k' = 0.147 \ \text{dm}^3/\text{mol} \cdot \text{min}$$

$$\alpha = 2.0$$
$$k' = 0.125 \ \text{dm}^3/\text{mol} \cdot \text{min}$$

The first regression gives $\alpha = 2.04$, as shown in Table E7-3.1. We shall round off α to make the reaction second order, (i.e., $\alpha = 2.00$). Now having fixed α at 2.0, we must do another regression (cf. Table E7-3.2) on k' because the k' given in Table E.7-3.1 is for $\alpha = 2.04$. We now regress the equation

$$t = \frac{1}{k'}\left[\frac{1}{C_A} - \frac{1}{C_{A0}}\right]$$

The second regression gives $k' = 0.125 \ \text{dm}^3/\text{mol} \cdot \text{min}$. We now calculate k

$$k = \frac{k'}{C_{A0}} = 0.25\left(\frac{\text{dm}^3}{\text{mol}}\right)^2/\text{min}$$

Analysis: In this example, we showed how to use nonlinear regression to find k' and α. The first regression gave $\alpha = 2.04$, which we rounded to 2.00 and then regressed again for the best value of k' for $\alpha = 2.0$, which was $k' = 0.125$ (dm^3/mol)/min giving a value of the true specific reaction rate of $k = 0.25$ (mol/dm^3)^2/min. We note that the reaction order is the same as that in Examples 7-1 and 7-2; however, the value of k is about 8% larger. The r^2 and other statistics are in Polymath's output.

Model Discrimination.   One can also determine which model or equation best fits the experimental data by comparing the sums of the squares for each model and then choosing the equation with a smaller sum of squares and/or carrying out an F-test. Alternatively, we can compare the residual plots for each model. These plots show the error associated with each data point, and one looks to see if the error is randomly distributed or if there is a trend in the error. When the error is randomly distributed, this is an additional indication that the correct rate law has been chosen. An example of model discrimination using nonlinear regression is given in Chapter 10.

7.6 Reaction-Rate Data from Differential Reactors

Data acquisition using the method of initial rates and a differential reactor is similar in that the rate of reaction is determined for a specified number of predetermined initial or entering reactant concentrations. A differential reactor (PBR) is normally used to determine the rate of reaction as a function of either concentration or partial pressure. It consists of a tube containing a very small amount of catalyst, usually arranged in the form of a thin wafer or disk. A typical arrangement is shown schematically in Figure 7-7. The criterion for a reactor being differential is that the conversion of the reactants in the bed is extremely small, as is the change in temperature and reactant concentration through the bed. As a result, the reactant concentration through the reactor is essentially constant and approximately equal to the inlet concentration. That is, the reactor is considered to be gradientless,[4] and the reaction rate is considered spatially uniform within the bed.

Most commonly used catalytic reactor to obtain experimental data

The differential reactor is relatively easy to construct at a low cost. Owing to the low conversion achieved in this reactor, the heat release per unit volume will be small (or can be made small by diluting the bed with inert solids) so that the reactor operates essentially in an isothermal manner. When operating this reactor, precautions must be taken so that the reactant gas or liquid does not bypass or channel through the packed catalyst, but instead flows uniformly across the catalyst. If the catalyst under investigation decays rapidly, the differential reactor is not a good choice because the reaction-rate parameters at the start of a run will be different from those at the end of the run. In some cases, sampling and analysis of the product stream may be difficult for small conversions in multicomponent systems.

Limitations of the differential reactor

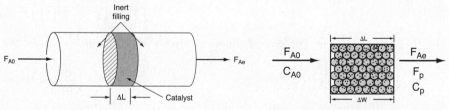

Figure 7-7 Differential reactor.   **Figure 7-8** Differential catalyst bed.

For the reaction of species A going to product (P)

$$A \rightarrow P$$

the volumetric flow rate through the catalyst bed is monitored, as are the entering and exiting concentrations (Figure 7-9). Therefore, if the weight of catalyst, ΔW, is known, the rate of reaction per unit mass of catalyst, $-r'_A$, can be calculated. Since the differential reactor is assumed to be *gradientless*, the design equation will be similar to the CSTR design equation. A steady-state mole balance on reactant A gives

---

[4] B. Anderson, ed., *Experimental Methods in Catalytic Research* (San Diego, CA: Academic Press, 1976).

$$\begin{bmatrix} \text{Flow} \\ \text{rate} \\ \text{in} \end{bmatrix} - \begin{bmatrix} \text{Flow} \\ \text{rate} \\ \text{out} \end{bmatrix} + \begin{bmatrix} \text{Rate of} \\ \text{generation} \end{bmatrix} = \begin{bmatrix} \text{Rate of} \\ \text{accumulation} \end{bmatrix}$$

$$[F_{A0}] - [F_{Ae}] + \left[\left(\frac{\text{Rate of reaction}}{\text{Mass of catalyst}} \right) (\text{Mass of catalyst}) \right] = 0$$

$$F_{A0} - F_{Ae} + (r'_A)(\Delta W) = 0$$

The subscript e refers to the exit of the reactor. Solving for $-r'_A$, we have

$$-r'_A = \frac{F_{A0} - F_{Ae}}{\Delta W} \tag{7-26}$$

The mole balance equation can also be written in terms of concentration

Differential reactor
design equation

$$\boxed{-r'_A = \frac{v_0 C_{A0} - v C_{Ae}}{\Delta W}} \tag{7-27}$$

or in terms of conversion or product flow rate F_P

$$\boxed{-r'_A = \frac{F_{A0} X}{\Delta W} = \frac{F_P}{\Delta W}} \tag{7-28}$$

The term $F_{A0} X$ gives the rate of formation of product, F_P, when the stoichiometric coefficients of A and of P are identical. Adjustments to Equation (7-28) must be made when this is not the case.

For constant volumetric flow, Equation (7-28) reduces to

$$-r'_A = \frac{v_0(C_{A0} - C_{Ae})}{\Delta W} = \frac{v_0 C_P}{\Delta W} \tag{7-29}$$

Consequently, we see that the reaction rate, $-r'_A$, can be determined by measuring the product concentration, C_P.

By using very little catalyst and large volumetric flow rates, the concentration difference, $(C_{A0} - C_{Ae})$, can be made quite small. The rate of reaction determined from Equation (7-29) can be obtained as a *function* of the reactant concentration in the catalyst bed, C_{Ab}

$$-r'_A = -r'_A(C_{Ab}) \tag{7-30}$$

by varying the inlet concentration. One approximation of the concentration of A within the bed, C_{Ab}, would be the arithmetic mean of the inlet and outlet concentrations

$$C_{Ab} = \frac{C_{A0} + C_{Ae}}{2} \tag{7-31}$$

However, since very little reaction takes place within the bed, the bed concentration is essentially equal to the inlet concentration

$$C_{Ab} \approx C_{A0}$$

so $-r_A'$ is a function of C_{A0}

$$-r_A' = -r_A'(C_{A0}) \tag{7-32}$$

As with the method of initial rates (see the CRE Web site, PRS R7.1), various numerical and graphical techniques can be used to determine the appropriate algebraic equation for the rate law. When collecting data for fluid–solid reacting systems, care must be taken that we use high flow rates through the differential reactor and small catalyst particle sizes in order to avoid mass transfer limitations. If data show the reaction to be first order with a low activation energy, say 8 kcal/mol, one should suspect the data are being collected in the mass transfer limited regime. We will expand on mass transfer limitations and how to avoid them in Chapters 10, 14, and 15.

Example 7–4 Using a Differential Reactor to Obtain Catalytic Rate Data

The formation of methane from carbon monoxide and hydrogen using a nickel catalyst was studied by Pursley.[5] The reaction

$$3H_2 + CO \rightarrow CH_4 + H_2O$$

was carried out at 500°F in a differential reactor where the effluent concentration of methane was measured. The raw data is shown in Table E7-4.1.

TABLE E7-4.1   RAW DATA

Run	P_{CO} (atm)	P_{H_2} (atm)	C_{CH_4}(mol/dm^3)
1	1	1.0	1.73×10^{-4}
2	1.8	1.0	4.40×10^{-4}
3	4.08	1.0	10.0×10^{-4}
4	1.0	0.1	1.65×10^{-4}
5	1.0	0.5	2.47×10^{-4}
6	1.0	4.0	1.75×10^{-4}

P_{H_2} is constant in Runs 1, 2, 3.
P_{CO} is constant in Runs 4, 5, 6.

The exit volumetric flow rate from a differential packed bed containing 10 g of catalyst was maintained at 300 dm^3/min for each run. The partial pressures of H$_2$ and CO were determined at the entrance to the reactor, and the methane concentration was measured at the reactor exit. Determine the rate law and rate law parameters.

(a) *Relate the rate of reaction to the exit methane concentration.* The reaction-rate law is assumed to be the product of a function of the partial pressure of CO and a function of the partial pressure of H$_2$,

$$r_{CH_4}' = f(CO) \cdot g(H_2) \tag{E7-4.1}$$

---

[5] J. A. Pursley, "An Investigation of the Reaction between Carbon Monoxide and Hydrogen on a Nickel Catalyst above One Atmosphere," Ph.D. thesis, University of Michigan.

(b) Determine the rate-law dependence on carbon monoxide, using the data generated in part **(a)**. Assume that the functional dependence of r'_{CH_4} on P_{CO} is of the form

$$r'_{CH_4} \sim P_{CO}^{\alpha} \qquad \text{(E7-4.2)}$$

(c) Determine the rate-law dependence on H_2. Generate a table of the reaction rate as a function of partial pressures of carbon monoxide and hydrogen.

Solution

(a) Calculate the Rates of Reaction. In this example the product composition, rather than the reactant concentration, is being monitored. The term $(-r'_{CO})$ can be written in terms of the flow rate of methane from the reaction

$$-r'_{CO} = r'_{CH_4} = \frac{F_{CH_4}}{\Delta W}$$

Substituting for F_{CH_4} in terms of the volumetric flow rate and the concentration of methane gives

$$-r'_{CO} = \frac{v_0 C_{CH_4}}{\Delta W} \qquad \text{(E7-4.3)}$$

Since v_0, C_{CH_4}, and ΔW are known for each run, we can calculate the rate of reaction.

For run 1

$$-r'_{CO} = \left(\frac{300 \text{ dm}^3}{\text{min}} \right) \frac{1.73 \times 10^{-4}}{10 \text{ g-cat}} \text{ mol/dm}^3 = 5.2 \times 10^{-3} \frac{\text{mol CH}_4}{\text{g-cat} \times \text{min}}$$

The rate for runs 2 through 6 can be calculated in a similar manner (Table E7-4.2).

TABLE E7-4.2   RAW AND CALCULATED DATA

Run	P_{CO} (atm)	P_{H_2} (atm)	C_{CH_4}(mol/dm³)	$r'_{CH_4} \left(\dfrac{\text{mol CH}_4}{\text{g-cat} \times \text{min}} \right)$
1	1.0	1.0	1.73×10^{-4}	5.2×10^{-3}
2	1.8	1.0	4.40×10^{-4}	13.2×10^{-3}
3	4.08	1.0	10.0×10^{-4}	30.0×10^{-3}
4	1.0	0.1	1.65×10^{-4}	4.95×10^{-3}
5	1.0	0.5	2.47×10^{-4}	7.42×10^{-3}
6	1.0	4.0	1.75×10^{-4}	5.25×10^{-3}

(b) Determining the Rate-Law Dependence in CO. For constant hydrogen concentration (runs 1, 2, and 3), the rate law

$$r'_{CH_4} = kP_{CO}^{\alpha} \cdot g(P_{H_2})$$

can be written as

$$r'_{CH_4} = k' P_{CO}^{\alpha} \qquad \text{(E7-4.4)}$$

Taking the natural log of Equation (E7-5.4) gives us

$$\ln (r'_{CH_4}) = \ln k' + \alpha \ln P_{CO}$$

We now plot $\ln(r'_{CH_4})$ versus $\ln P_{CO}$ using runs 1, 2, and 3, for which the H_2 concentration is constant, in Figure E7-4.1. We see from the Excel plot that $\alpha = 1.22$.

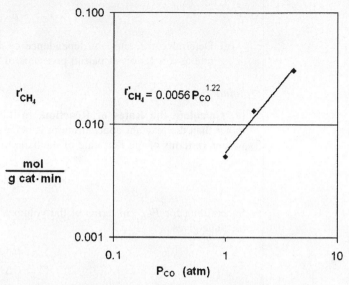

Figure E7-4.1   Reaction rate as a function of concentration.

Had we included more points, we would have found that the reaction is essentially first order with $\alpha = 1$, that is

$$-r'_{CO} = k' P_{CO} \tag{E7-4.5}$$

From the first three data points where the partial pressure of H_2 is constant, we see the rate is linear in partial pressure of CO

$$r'_{CH_4} = k' P_{CO} \cdot g(H_2)$$

Now let's look at the hydrogen dependence.

(c) Determining the Rate-Law Dependence on H_2. From Table E7-4.2 it appears that the dependence of r'_{CH_4} on P_{H_2} cannot be represented by a power law. Comparing run 4 with run 5 and then run 5 with run 6, we see that the reaction rate first increases with increasing partial pressure of hydrogen, and subsequently decreases with increasing P_{H_2}. That is, there appears to be a concentration of hydrogen at which the rate is maximum. One set of rate laws that is consistent with these observations is:

1. At low H_2 concentrations where r'_{CH_4} increases as P_{H_2} increases, the rate law may be of the form

$$r'_{CH_4} \sim P_{H_2}^{\beta_1} \tag{E7-4.6}$$

2. At high H_2 concentrations where r'_{CH_4} decreases as P_{H_2} increases, the rate law may be of the form

$$r'_{CH_4} \sim \frac{1}{P_{H_2}^{\beta_2}} \tag{E7-4.7}$$

We would like to find one rate law that is consistent with reaction-rate data at both high and low hydrogen concentrations. After we have studied heterogeneous reactions in Chapter 10, we would recognize that Equations (E7-4.6) and (E7-4.7) can be combined into the form

$$r'_{CH_4} \sim \frac{P_{H_2}^{\beta_1}}{1 + b P_{H_2}^{\beta_2}} \tag{E7-4.8}$$

We will see in Chapter 10 that this combination and similar rate laws that have reactant concentrations (or partial pressures) in the numerator and denominator are common in *heterogeneous catalysis*.

Let's see if the resulting rate law (E7-4.8) is qualitatively consistent with the rate observed.

1. *For condition 1*: At low P_{H_2}, $[b(P_{H_2})^{\beta_2} \ll 1]$ and Equation (E7-4.8) reduces to

$$r'_{CH_4} \sim P_{H_2}^{\beta_1} \tag{E7-4.9}$$

Equation (E7-4.9) is consistent with the trend in comparing runs 4 and 5.

2. *For condition 2*: At high P_{H_2}, $[b(P_{H_2})^{\beta_2} \gg 1]$ and Equation (E5-5.8) reduces to

$$r'_{CH_4} \sim \frac{(P_{H_2})^{\beta_1}}{(P_{H_2})^{\beta_2}} \sim \frac{1}{(P_{H_2})^{\beta_2 - \beta_1}} \tag{E7-4.10}$$

where $\beta_2 > \beta_1$. Equation (E7-4.10) is consistent with the trends in comparing runs 5 and 6.

Combining Equations (E7-4.8) and (E7-4.5)

Typical form of the rate law for heterogeneous catalysis

$$r'_{CH_4} = \frac{a P_{CO} P_{H_2}^{\beta_1}}{1 + b P_{H_2}^{\beta_2}} \tag{E7-4.11}$$

We now use the Polymath regression program to find the parameter values a, b, β_1, and β_2. The results are shown in Table E7-4.3.

TABLE E7-4.3   FIRST REGRESSION

Summary Notes

Polymath and Excel regression tutorial are given in the Chapter 7 Summary Notes on the CRE Web site.

POLYMATH Results

No Title 01-31-2004

Nonlinear regression (L-M)

Model: Rate = a*Pco*Ph2^beta1/(1+b*Ph2^beta2)

Variable	Ini guess	Value	95% confidence
a	1	0.0252715	0.4917749
beta1	1	0.6166542	6.9023286
b	1	2.4872569	68.002944
beta2	1	1.0262047	3.2344414

Nonlinear regression settings
Max # iterations = 64

The corresponding rate law is

$$r'_{CH_4} = \frac{0.025 P_{CO} P_{H_2}^{0.61}}{1 + 2.49 P_{H_2}} \tag{E7-4.12}$$

We could use the rate law given by Equation (E7-4.12) as is, but there are only six data points, and we should be concerned about extrapolating the rate law over a wider range of partial pressures. We could take more data, and/or we could carry out a theoretical analysis of the type discussed in Chapter 10 for heterogeneous reactions. If we assume hydrogen undergoes dissociative adsorption on the catalyst surface, we would expect a dependence on the partial pressure of hydrogen to be to the ½ power. Because 0.61 is close to 0.5, we are going to regress the data again, setting $\beta_1 = \frac{1}{2}$ and $\beta_2 = 1.0$. The results are shown in Table E7-4.4.

TABLE E7-4.4   SECOND REGRESSION

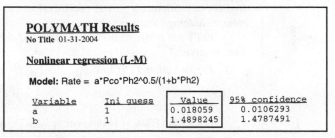

POLYMATH Results
No Title 01-31-2004

Nonlinear regression (L-M)

Model: Rate = a*Pco*Ph2^0.5/(1+b*Ph2)

Variable	Ini guess	Value	95% confidence
a	1	0.018059	0.0106293
b	1	1.4898245	1.4787491

The rate law is now

$$ r'_{CH_4} = \frac{0.018 P_{CO} P_{H_2}^{1/2}}{1 + 1.49 P_{H_2}} $$

where r'_{CH_4} is in (mol/g-cat · s) and the partial pressures are in (atm).

We could also have set $\beta_1 = \frac{1}{2}$ and $\beta_2 = 1.0$ and rearranged Equation (E7-4.11) in the form

Linearizing the rate law to determine the rate law parameters

$$ \frac{P_{CO} P_{H_2}^{1/2}}{r'_{CH_4}} = \frac{1}{a} + \frac{b}{a} P_{H_2} \tag{E7-4.13} $$

A plot of $P_{CO} P_{H_2}^{1/2}/r'_{CH_4}$ as a function of P_{H_2} should be a straight line with an intercept of $1/a$ and a slope of b/a. From the plot in Figure E7-4.2, we see that the rate law is indeed consistent with the rate-law data.

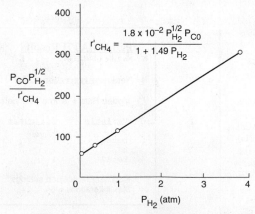

Figure E7-4.2   Linearized plot of data.

Analysis: The reaction-rate data in this example were obtained at steady state, and as a result, neither the integral method nor differential method of analysis can be used. One of the purposes of this example is to show how to reason out the form of the rate law and to then use regression to determine the rate-law parameters. Once the parameters were obtained, we showed how to linearize the rate-law [e.g., Equation (E7-4.13)] to generate a single plot of all the data, Figure (E7-4.2).

7.7 Experimental Planning

> Four to six weeks in the lab can save you an hour in
> the library.
>
> G. C. Quarderer, Dow Chemical Co.

So far, this chapter has presented various methods of analyzing rate data. It is
just as important to know in which circumstances to use each method as it is to
know the mechanics of these methods. In the Expanded Material on the CRE
Web site, we give a thumbnail sketch of a heuristic to plan experiments to generate the data necessary for reactor design. However, for a more thorough discussion, the reader is referred to the books and articles by Box and Hunter.[6]

Closure. After reading this chapter, the reader should be able to analyze
data to determine the rate law and rate-law parameters using the graphical
and numerical techniques, as well as software packages. Nonlinear regression is the easiest method to analyze rate-concentration data to determine
the parameters, but the other techniques, such as graphical differentiation,
help one get a feel for the disparities in the data. The reader should be
able to describe the care that needs to be taken in using nonlinear regression to ensure that one does not arrive at a false minimum for σ^2. Consequently, it is advisable to use more than one method to analyze the data.

SUMMARY

1. *Integral method*
    a. Guess the reaction order and integrate the mole balance equation.
    b. Calculate the resulting function of concentration for the data and plot it as a function of time. If the
       resulting plot is linear, you have probably guessed the correct reaction order.
    c. If the plot is not linear, guess another order and repeat the procedure.
2. *Differential method for constant-volume systems*

$$-\frac{dC_A}{dt} = kC_A^{\alpha} \tag{S7-1}$$

    a. Plot $-\Delta C_A/\Delta t$ as a function of t.
    b. Determine $-dC_A/dt$ from this plot.
    c. Take the ln of both sides of (S7-1) to get

$$\ln\left(-\frac{dC_A}{dt}\right) = \ln k + \alpha \ln C_A \tag{S7-2}$$

Plot $\ln(-dC_A/dt)$ versus $\ln C_A$. The slope will be the reaction order α. We could use finite-difference
formulas or software packages to evaluate $(-dC_A/dt)$ as a function of time and concentration.

---

[6] G. E. P. Box, W. G. Hunter, and J. S. Hunter, *Statistics for Experimenters: An
Introduction to Design, Data Analysis, and Model Building* (New York: Wiley, 1978).

3. *Nonlinear regression*: Search for the parameters of the rate law that will minimize the sum of the squares of the difference between the measured rate of reaction and the rate of reaction calculated from the parameter values chosen. For N experimental runs and K parameters to be determined, use Polymath.

$$\sigma^2 = \sum_{i=1}^{N} \frac{[P_i(\text{measured}) - P_i(\text{calculated})]^2}{N-K} \tag{S7-3}$$

$$s^2 = \sum_{i=1}^{N} (t_{im} - t_{ic})^2 = \sum_{i=1}^{N} \left[t_{im} - \frac{C_{A0}^{1-\alpha} - C_{Ai}^{1-\alpha}}{k(1-\alpha)} \right]^2 \tag{S7-4}$$

Caution: Be sure to avoid a false minimum in σ^2 by varying your initial guess.

4. *Modeling the differential reactor*:

The rate of reaction is calculated from the equation

$$-r'_A = \frac{F_{A0}X}{\Delta W} = \frac{F_P}{\Delta W} = \frac{v_0(C_{A0} - C_{Ae})}{\Delta W} = \frac{C_P v_0}{\Delta W} \tag{S7-5}$$

In calculating the reaction order, α

$$-r'_A = k C_A^{\alpha}$$

the concentration of A is evaluated either at the entrance conditions or at a mean value between C_{A0} and C_{Ae}. However, power-law models such as

$$-r'_A = k C_A^{\alpha} C_B^{\beta} \tag{S7-6}$$

are not the best way to describe heterogeneous reaction-rate laws. Typically, they take the form

$$-r'_A = \frac{k P_A P_B}{1 + K_A P_A + K_B P_B}$$

or a similar form, with the reactant partial pressures in the numerator *and* denominator of the rate law.

CRE WEB SITE MATERIALS

- **Expanded Material**
  1. *Evaluation of Laboratory Reactors*
  2. *Summary of Reactor-Ratings, Gas-Liquid, Powdered Catalyst Decaying Catalyst System*
  3. *Experimental Planning*
  4. *Additional Homework Problems*
- **Learning Resources**
  1. *Summary Notes*
  2. *Interactive Computer Games*

Summary Notes

A. Ecology

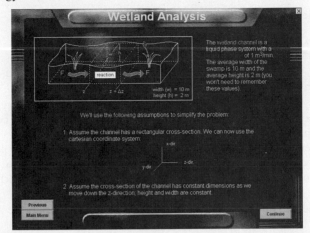

Interactive

Computer Games

B. Reactor Lab (*www.reactorlab.net*). See *Reactor Lab Chapter 7* and P7-3$_A$.
3. *Solved Problems*
    A. Example: Differential Method of Analysis of Pressure- Time Data
    B. Example: Integral Method of Analysis of Pressure- Time Data
    C. Example: Oxygenating Blood
• **Living Example Problems**
    1. *Example 7-3 Use of Regression to Find the Rate Law Parameters*
• **FAQ (Frequently Asked Questions)—In Updates/FAQ icon section**
• **Professional Reference Shelf**
    R7.1 *Method of Initial Rates*
    R7.2 *Method of Half Lives*
    R7.3 *Least-Squares Analysis of the Linearized Rate Law*
    The CRE Web site describes how the rate law

Solved Problems

Reference Shelf

$$-r_A = kC_A^\alpha \, C_B^\beta$$

is linearized

$$\ln(-r_A) = \ln k + \alpha \ln C_A + \beta \ln C_B$$

and put in the form

$$Y = a_0 + \alpha X_1 + \beta X_2$$

and used to solve for α, β, and k. The etching of a semiconductor, MnO_2, is used as an example to illustrate this technique.

R7.4 *A Discussion of Weighted Least Squares*
For the case when the error in measurement is not constant, we must use a weighted least-squares analysis.

R7.5 *Experimental Planning*
    A. Why perform the experiment?
    B. Are you choosing the correct parameters?
    C. What is the range of your experimental variables?
    D. Can you repeat the measurement? (Precision)
    E. Milk your data for all it's worth.
    F. We don't believe an experiment until it's proven by theory.
    G. Tell someone about your result.

R7.6 *Evaluation or Laboratory Reactor*

QUESTIONS AND PROBLEMS

The subscript to each of the problem numbers indicates the level of difficulty: A, least difficult; D, most difficult.

$$A = \bullet \quad B = \blacksquare \quad C = \blacklozenge \quad D = \blacklozenge\blacklozenge$$

Home Work Problems

Questions

Q7-1$_A$ **(a)** Listen to the audios on the CRE Web site and pick one and say why it could be eliminated.

(b) Create an original problem based on Chapter 7 material.

(c) Design an experiment for the undergraduate laboratory that demonstrates the principles of chemical reaction engineering and will cost less than $500 in purchased parts to build. (From 1998 AIChE National Student Chapter Competition). Rules are provided on the CRE Web site.

(d) **K-12 Experiment.** Plant a number of seeds in different pots (corn works well). The plant and soil of each pot will be subjected to different conditions. Measure the height of the plant as a function of time and fertilizer concentration. Other variables might include lighting, pH, and room temperature. (Great grade school or high school science project.)

Creative Thinking

Problems

P7-1$_B$ **(a)** Revisit **Example 7-1.** What is the error in assuming the concentration of species B is constant and what limits can you put on the calculated value of k? (I.e., $k = 0.24 \pm?$)

(b) Revisit **Example 7-3.** Explain why the regression was carried out twice to find k' and k.

(c) Revisit **Example 7-4.** Regress the data to fit the rate law

$$r_{CH_4} = k P_{CO}^{\alpha} P_{H_2}^{\beta}$$

What is the difference in the correlation and sums of squares compared with those given in Example 7-4? Why was it necessary to regress the data twice, once to obtain Table E7-4.3 and once to obtain Table E7-4.4?

P7-2$_A$ Download the Interactive Computer Game (ICG) from the CRE Web site. Play the game and then record your performance number for the module that indicates your mastery of the material. Your professor has the key to decode your performance number.

ICM Ecology Performance # _____.

P7-3$_A$ Go to Professor Herz's **Reactor Lab** on the Web at *www.reactorlab.net*. Do (a) one quiz, or (b) two quizzes from Division 1. When you first enter a lab, you see all input values and can vary them. In a lab, click on the Quiz button in the navigation bar to enter the quiz for that lab. In a quiz, you cannot see some of the input values: you need to find those with "???" hiding the values. In the quiz, perform experiments and analyze your data in order to determine the unknown values. See the bottom of the Example Quiz page at *www.reactorlab.net* for equations that relate E and k. Click on the "???" next to an input and supply your value. Your answer will be accepted if it is within ±20% of the correct value. Scoring is done with imaginary dollars to emphasize that you should design your experimental study rather than do random experiments. Each time you enter a quiz, new unknown values are assigned. To reenter an unfinished quiz at the same stage you left, click the [i] info button in the Directory for instructions. Turn in copies of your data, your analysis work, and the Budget Report.

P7-4$_A$ When arterial blood enters a tissue capillary, it exchanges oxygen and carbon dioxide with its environment, as shown in this diagram.

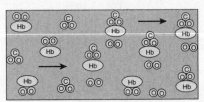

The kinetics of this deoxygenation of hemoglobin in blood was studied with the aid of a **tubular reactor** by Nakamura and Staub (*J. Physiol.*, 173, 161).

$$HbO_2 \underset{k_{-1}}{\overset{k_1}{\rightleftarrows}} Hb + O_2$$

Although this is a reversible reaction, measurements were made in the initial phases of the decomposition so that the reverse reaction could be neglected. Consider a system similar to the one used by Nakamura and Staub: the solution enters a tubular reactor (0.158 cm in diameter) that has oxygen electrodes placed at 5-cm intervals down the tube. The solution flow rate into the reactor is 19.6 cm³/s with $C_{A0} = 2.33 \times 10^{-6}$ mol/cm³.

Electrode position	1	2	3	4	5	6	7
Percent decomposition of HbO$_2$	0.00	1.93	3.82	5.68	7.48	9.25	11.00

(a) Using the method of differential analysis of rate data, determine the reaction order and the forward specific reaction-rate constant k_1 for the deoxygenation of hemoglobin.

(b) Repeat using regression.

P7-5$_B$ The liquid-phase irreversible reaction

$$A \rightarrow B + C$$

is carried out in a CSTR. To learn the rate law, the volumetric flow rate, v_0, (hence $\tau = V/v_0$) is varied and the effluent concentrations of species A are recorded as a function of the space time τ. Pure A enters the reactor at a concentration of 2 mol/dm³. Steady-state conditions exist when the measurements are recorded.

Run	1	2	3	4	5
τ (min)	15	38	100	300	1200
C_A (mol/dm³)	1.5	1.25	1.0	0.75	0.5

(a) Determine the reaction order and specific reaction-rate constant.

(b) If you were to repeat this experiment to determine the kinetics, what would you do differently? Would you run at a higher, lower, or the same temperature? If you were to take more data, where would you place the measurements (e.g., τ)?

(c) It is believed that the technician may have made a dilution factor-of-10 error in one of the concentration measurements. What do you think? How do your answers compare using regression (Polymath or other software) with those obtained by graphical methods?

Note: All measurements were taken at steady-state conditions.

P7-6$_A$ The reaction

$$A \rightarrow B + C$$

was carried out in a constant-volume batch reactor where the following concentration measurements were recorded as a function of time.

t (min)	0	5	9	15	22	30	40	60
C_A (mol/dm^3)	2	1.6	1.35	1.1	0.87	0.70	0.53	0.35

(a) Use nonlinear least squares (i.e., regression) and one other method to determine the reaction order, α, and the specific reaction rate, k.

(b) Nicolas Bellini wants to know, if you were to take more data, where would you place the points? Why?

(c) Prof. Dr. Sven Köttlov from Jofostan University always asks his students, if you were to repeat this experiment to determine the kinetics, what would you do differently? Would you run at a higher, lower, or the same temperature? Take different data points? Explain.

(d) It is believed that the technician made a dilution error in the concentration measured at 60 min. What do you think? How do your answers compare using regression (Polymath or other software) with those obtained by graphical methods?

P7-7$_A$ The following data were reported [from C. N. Hinshelwood and P. J. Ackey, *Proc. R. Soc. (Lond).*, *A115*, 215] for a gas-phase constant-volume decomposition of dimethyl ether at 504°C in a *batch reactor*. Initially, only $(CH_3)_2O$ was present.

Time (s)	390	777	1195	3155	∞
Total Pressure (mmHg)	408	488	562	799	931

(a) Why do you think the total pressure measurement at $t = 0$ is missing? Can you estimate it?

(b) Assuming that the reaction

$$(CH_3)_2O \rightarrow CH_4 + H_2 + CO$$

is irreversible and goes virtually to completion, determine the reaction order and specific reaction rate k.

(c) What experimental conditions would you suggest if you were to obtain more data?

(d) How would the data and your answers change if the reaction were run at a higher temperature? A lower temperature?

P7-8$_A$ In order to study the photochemical decay of aqueous bromine in bright sunlight, a small quantity of liquid bromine was dissolved in water contained in a glass battery jar and placed in direct sunlight. The following data were obtained at 25°C:

Time (min)	10	20	30	40	50	60
ppm Br$_2$	2.45	1.74	1.23	0.88	0.62	0.44

(a) Determine whether the reaction rate is zero, first, or second order in bromine, and calculate the reaction-rate constant in units of your choice.

(b) Assuming identical exposure conditions, calculate the required hourly rate of injection of bromine (in pounds per hour) into a sunlit body of water, 25,000 gal in volume, in order to maintain a sterilizing level of bromine of 1.0 ppm. [*Ans.*: 0.43 lb/h]

(c) Apply one or more of the six ideas in Preface Table P-4, page xxviii, to this problem.

(*Note*: ppm = parts of bromine per million parts of brominated water by weight. In dilute aqueous solutions, 1 ppm ≡ 1 milligram per liter.) (From California Professional Engineers' Exam.)

P7-9$_C$ The reactions of ozone were studied in the presence of alkenes [from R. Atkinson et al., *Int. J. Chem. Kinet., 15(8)*, 721 (1983)]. The data in Table P7-9$_C$ are for one of the alkenes studied, *cis*-2-butene. The reaction was carried out isothermally at 297 K. Determine the rate law and the values of the rate-law parameters.

TABLE P7-9$_C$ RATE AS A FUNCTION OF OZONE AND BUTENE CONCENTRATIONS

Run	Ozone Rate (mol/s·dm^3 × 10^7)	Ozone Concentration (mol/dm^3)	Butene Concentration (mol/dm^3)
1	1.5	0.01	10^{-12}
2	3.2	0.02	10^{-11}
3	3.5	0.015	10^{-10}
4	5.0	0.005	10^{-9}
5	8.8	0.001	10^{-8}
6	4.7*	0.018	10^{-9}

*Hint: Ozone also decomposes by collision with the wall.

P7-10$_A$ Tests were run on a small experimental reactor used for decomposing nitrogen oxides in an automobile exhaust stream. In one series of tests, a nitrogen stream containing various concentrations of NO$_2$ was fed to a reactor, and the kinetic data obtained are shown in Figure P7-10$_A$. Each point represents one complete run. The reactor operates essentially as an *isothermal backmix reactor (CSTR)*. What can you deduce about the apparent order of the reaction over the temperature range studied?

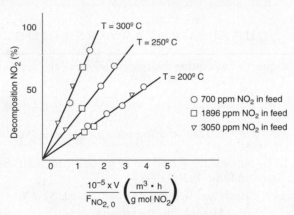

Figure P7-10$_A$ Auto exhaust data.

The plot gives the fractional decomposition of NO$_2$ fed versus the ratio of reactor volume V (in cm^3) to the NO$_2$ feed rate, $F_{NO_{2,0}}$ (g mol/h), at different feed concentrations of NO$_2$ (in parts per million by weight). Determine as many rate law parameters as you can.

P7-11$_A$ The thermal decomposition of isopropyl isocyanate was studied in a *differential packed-bed reactor*. From the data in Table P7-11$_A$, determine the reaction-rate-law parameters.

TABLE P7-11$_A$ RAW DATA[†]

Run	Rate (mol/s·dm^3)	Concentration (mol/dm^3)	Temperature (K)
1	4.9×10^{-4}	0.2	700
2	1.1×10^{-4}	0.02	750
3	2.4×10^{-3}	0.05	800
4	2.2×10^{-2}	0.08	850
5	1.18×10^{-1}	0.1	900
6	1.82×10^{-2}	0.06	950

[†] *Jofostan Journal of Chemical Engineering*, Vol. 15, page 743 (1995).

- **Additional Homework Problems are on the CRE Web site**

Green Engineering

New Problems on the Web

CDP7-New From time to time, new problems relating Chapter 7 material to everyday interests or emerging technologies will be placed on the Web. Solutions to these problems can be obtained by emailing the author. Also, one can go to the Web site, *nebula.rowan.edu:82/home.asp*, and work the home problem specific to this chapter.

Links

SUPPLEMENTARY READING

1. A wide variety of techniques for measuring the concentrations of the reacting species may or may not be found in

> BURGESS, THORNTON W., *Mr. Toad and Danny the Meadow Mouse Take a Walk.* New York: Dover Publications, Inc., 1915.

> FOGLER, H. SCOTT and STEVEN E. LEBLANC, *Strategies for Creative Problem Solving.* Englewood Cliffs, NJ: Prentice Hall, 1995.

> KARRASS, CHESTER L., *In Business As in Life, You Don't Get What You Deserve, You Get What You Negotiate.* Hill, CA: Stanford Street Press, 1996.

> ROBINSON, J. W., *Undergraduate Instrumental Analysis*, 5th ed. New York: Marcel Dekker, 1995.

> SKOOG, DOUGLAS A., F. JAMES HOLLER, and TIMOTHY A. NIEMAN, *Principles of Instrumental Analysis*, 5th ed. Philadelphia: Saunders College Publishers, Harcourt Brace College Publishers, 1998.

2. The design of laboratory catalytic reactors for obtaining rate data is presented in

> RASE, H. F., *Chemical Reactor Design for Process Plants*, Vol. 1. New York: Wiley, 1983, Chap. 5.

3. The sequential design of experiments and parameter estimation is covered in

> BOX, G. E. P., W. G. HUNTER, and J. S. HUNTER, *Statistics for Experimenters: An Introduction to Design, Data Analysis, and Model Building.* New York: Wiley, 1978.

Multiple Reactions   8

The breakfast of champions is not cereal, it's your opposition.
—Nick Seitz

Overview. Seldom is the reaction of interest the *only one* that occurs in a chemical reactor. Typically, multiple reactions will occur, some desired and some undesired. One of the key factors in the economic success of a chemical plant is the minimization of undesired side reactions that occur along with the desired reaction.

In this chapter, we discuss reactor selection and general mole balances, net rates, and relative rates for multiple reactions.

First, we describe the four basic types of multiple reactions:

- Series
- Parallel
- Independent
- Complex

Next, we define the selectivity parameter and discuss how it can be used to minimize unwanted side reactions by proper choice of operating conditions and reactor selection.

We then show how to modify our CRE algorithm to solve reaction engineering problems when multiple reactions are involved. The modification builds on the algorithm presented in Chapter 6 by numbering all reactions and expanding the *Rates Building Block* into three parts:

- Rate laws
- Relative rates
- Net rates

Finally, a number of examples are given that show how the algorithm is applied to real reactions.

8.1 Definitions

8.1.1   Types of Reactions

There are four basic types of multiple reactions: parallel, series, independent, and complex. These types of multiple reactions can occur by themselves, in pairs, or all together. When there is a combination of parallel and series reactions, they are often referred to as *complex reactions*.

Parallel reactions (also called *competing reactions*) are reactions where the reactant is consumed by two different reaction pathways to form different products:

Parallel reactions

$$
A \underset{k_2}{\overset{k_1}{<}} \begin{matrix} B \\ \\ C \end{matrix}
$$

An example of an industrially significant parallel reaction is the oxidation of ethylene to ethylene oxide while avoiding complete combustion to carbon dioxide and water:

Serious chemistry

$$
CH_2{=}CH_2 + O_2 \overset{}{<} \begin{matrix} 2CO_2 + 2H_2O \\ \\ O \\ CH_2{-}CH_2 \end{matrix}
$$

Series reactions (also called *consecutive reactions*) are reactions where the reactant forms an intermediate product, which reacts further to form another product:

Series reactions

$$
A \xrightarrow{\;k_1\;} B \xrightarrow{\;k_2\;} C
$$

An example of a series reaction is the reaction of ethylene oxide (EO) with ammonia to form mono-, di-, and triethanolamine:

$$
\overset{O}{CH_2{-}CH_2} + NH_3 \longrightarrow HOCH_2CH_2NH_2
$$

$$
\xrightarrow{\;EO\;} (HOCH_2CH_2)_2NH \xrightarrow{\;EO\;} (HOCH_2CH_2)_3N
$$

In recent years, the shift has been toward the production of diethanolamine as the *desired* product rather than triethanolamine.

Independent reactions are reactions that occur at the same time but neither the products nor the reactants react with themselves or one another.

Independent reactions

$$
A \longrightarrow B + C
$$
$$
D \longrightarrow E + F
$$

An example is the cracking of crude oil to form gasoline, where two of the many reactions occurring are

$$
C_{15}H_{32} \longrightarrow C_{12}C_{26} + C_3H_6
$$
$$
C_8H_{18} \longrightarrow C_6H_{14} + C_2H_4
$$

Complex reactions are multiple reactions that involve combinations of series and independent parallel reactions, such as

$$A+B \longrightarrow C+D$$
$$A+C \longrightarrow E$$
$$E \longrightarrow G$$

An example of a combination of parallel and series reactions is the formation of butadiene from ethanol:

$$C_2H_5OH \longrightarrow C_2H_4+H_2O$$
$$C_2H_5OH \longrightarrow CH_3CHO+H_2$$
$$C_2H_4+CH_3CHO \longrightarrow C_4H_6+H_2O$$

8.1.2 Selectivity

Desired and Undesired Reactions. Of particular interest are reactants that are consumed in the formation of a *desired product*, D, and the formation of an *undesired product*, U, in a competing or side reaction. In the parallel reaction sequence

$$A \xrightarrow{k_D} D$$
$$A \xrightarrow{k_U} U$$

or in the series sequence

$$A \xrightarrow{k_D} D \xrightarrow{k_U} U$$

we want to minimize the formation of U and maximize the formation of D because the greater the amount of undesired product formed, the greater the cost of separating the undesired product U from the desired product D (see Figure 8-1).

The economic incentive

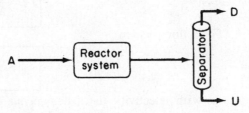

Figure 8-1 Reaction-separation system producing both desired and undesired products.

Selectivity tells us how one product is favored over another when we have multiple reactions. We can quantify the formation of D with respect to U by defining the selectivity and yield of the system. The **instantaneous selectivity** of D with respect to U is the ratio of the rate of formation of D to the rate of formation of U.

Instantaneous selectivity

$$\boxed{S_{D/U} = \frac{r_D}{r_U} = \frac{\text{rate of formation of D}}{\text{rate of formation of U}}} \qquad (8\text{-}1)$$

In the next section, we will see how evaluating $S_{D/U}$ will guide us in the design and selection of our reaction system to maximize the selectivity.

Another definition of selectivity used in the current literature, $\tilde{S}_{D/U}$, is given in terms of the flow rates leaving the reactor. $\tilde{S}_{D/U}$ is the **overall selectivity**.

Overall selectivity

$$\tilde{S}_{D/U} = \frac{F_D}{F_U} = \frac{\text{Exit molar flow rate of desired product}}{\text{Exit molar flow rate of undesired product}} \qquad (8\text{-}2a)$$

It is easily shown that for a CSTR the instantaneous and overall selectivites are identical. See P8-1$_A$(d) and *Summary Notes* on the CRE Web site, *www.umich.edu/~elements/5e/index.html*.

For a batch reactor, the overall selectivity is given in terms of the number of moles of D and U at the end of the reaction time.

$$\tilde{S}_{D/U} = \frac{N_D}{N_U} \qquad (8\text{-}2b)$$

8.1.3 Yield

Two definitions for selectivity and yield are found in the literature.

Reaction yield, like selectivity, has two definitions: one based on the ratio of reaction rates and one based on the ratio of molar flow rates. In the first case, the yield at a point can be defined as the ratio of the reaction rate of a given product to the reaction rate of the *key* reactant A, usually the basis of calculation. This yield is referred to as the *instantaneous yield Y_D*.

Instantaneous yield based on reaction rates

$$Y_D = \frac{r_D}{-r_A} \qquad (8\text{-}3)$$

The *overall yield*, $\tilde{Y}_D$, is based on molar flow rates and defined as the ratio of moles of product formed at the end of the reaction to the number of moles of the key reactant, A, that have been consumed.

For a batch system

Overall yield based on moles

$$\tilde{Y}_D = \frac{N_D}{N_{A0} - N_A} \qquad (8\text{-}4)$$

For a flow system

Overall yield based on molar flow rates

$$\tilde{Y}_D = \frac{F_D}{F_{A0} - F_A} \qquad (8\text{-}5)$$

As with selectivity, the instantaneous yield and the overall yield are identical for a CSTR (i.e., $\tilde{Y}_D = Y_D$). From an economic standpoint, the *overall* selectivities, $\tilde{S}$, and yields, $\tilde{Y}$, are important in determining profits, while the instantaneous selectivities give insights in choosing reactors and reaction schemes that will help maximize the profit. There often is a conflict between selectivity and conversion because you want to make as much as possible of your desired product (D) and at the same time minimize the undesired product (U). However, in many instances, the greater the conversion you achieve, not only do you make more D, but you also form more U.

8.2 Algorithm for Multiple Reactions

The multiple-reaction algorithm can be applied to parallel reactions, series reactions, independent reactions, and complex reactions. The availability of software packages (ODE solvers) makes it much easier to solve problems using moles N_j or molar flow rates F_j rather than conversion. For liquid systems, concentration is usually the preferred variable used in the mole balance equations.

After numbering each and every reaction involved we carry out a mole balance on each and every species. The mole balances for the various types of reactors we have been studying are shown in Table 8-1. The rates shown Table 8-1, e.g., r_A, are the net rates of formation and are discussed in detail in Table 8-2. The resulting coupled differential mole balance equations can be

TABLE 8-1   MOLE BALANCES FOR MULTIPLE REACTIONS

General Mole Balance

$$\boxed{\frac{dN_j}{dt} = F_{j0} - F_j + \int^V r_j \, dV}$$

	Molar Quantities (Gas or Liquid)	**Concentration** (Liquid)
Batch	$\dfrac{dN_A}{dt} = r_A V$	$\dfrac{dC_A}{dt} = r_A$
	$\dfrac{dN_B}{dt} = r_B V$	$\dfrac{dC_B}{dt} = r_B$
	$\vdots$	$\vdots$
PFR/PBR	$\dfrac{dF_A}{dV} = r_A$	$\dfrac{dC_A}{dV} = \dfrac{r_A}{v_0}$
	$\dfrac{dF_B}{dV} = r_B$	$\dfrac{dC_B}{dV} = \dfrac{r_B}{v_0}$
	$\vdots$	$\vdots$
CSTR	$V = \dfrac{F_{A0} - F_A}{(-r_A)_{exit}}$	$V = \dfrac{v_0[C_{A0} - C_A]}{(-r_A)_{exit}}$
	$V = \dfrac{F_{B0} - F_B}{(-r_B)_{exit}}$	$V = \dfrac{v_0[C_{B0} - C_B]}{(-r_B)_{exit}}$
	$\vdots$	$\vdots$
Membrane: C diffuses out	$\dfrac{dF_A}{dV} = r_A$	$\dfrac{dF_A}{dV} = r_A$
	$\dfrac{dF_B}{dV} = r_B$	$\dfrac{dF_B}{dV} = r_B$
	$\dfrac{dF_C}{dV} = r_C - R_C$	$\dfrac{dF_C}{dV} = r_C - R_C$
	$\vdots$	$\vdots$
Semibatch B added to **A**	$\dfrac{dN_A}{dt} = r_A V$	$\dfrac{dC_A}{dt} = r_A - \dfrac{v_0 C_A}{V}$
	$\dfrac{dN_B}{dt} = F_{B0} + r_B V$	$\dfrac{dC_B}{dt} = r_B + \dfrac{v_0[C_{B0} - C_B]}{V}$
	$\vdots$	$\vdots$

Mole balances on every species

easily solved using an ODE solver. In fact, this section has been developed to take advantage of the vast number of computational techniques now available on laptop computers (e.g., Polymath).

8.2.1 Modifications to the Chapter 6 CRE Algorithm for Multiple Reactions

Just a very few changes to our CRE algorithm for multiple reactions

There are a few small changes to the CRE algorithm presented in Table 6-2, and we will describe these changes in detail when we discuss complex reactions in Section 8.5. However, before discussing parallel and series reactions, it is necessary to point out some of the modifications to our algorithm. These changes are highlighted in Table 8-2. When analyzing multiple reactions, we first number every reaction. Next, we must perform a mole balance on each and every species, just as we did in Chapter 6 to analyze reactions in terms of the mole balances for different reactor types. The rates of formation shown in the mole

r_{ij}
$\llcorner$ Species
$\llcorner$ Reaction number

TABLE 8-2   MODIFICATION TO THE CRE ALGORITHM

Identify   **1.** Number Each and Every Reaction Separately

Mole Balance   **2.** Mole Balance on Each and Every Species

a. *Rate Law* for Every Reaction

e.g., $-r_{ij} = k_{ij} f\left(C_A, C_B, \ldots C_j\right)$

The subscript "i" refers to the reaction number and the subscript "j" refers to the species.

r_{ij}
$\llcorner$ Species
$\llcorner$ Reaction number

b. *Net Rates of Reaction* for Each Species, e.g., j

$$r_j = \sum_{i=1}^{N} r_{ij}$$

For N reactions, the net rate of formation of species A is:

$$r_A = \sum_{i=1}^{N} r_{iA} = r_{1A} + r_{2A} + \ldots$$

c. *Relative Rates* for Every Reaction

For a given reaction *i:* $a_i A + b_i B \rightarrow c_i C + d_i D$

$$\boxed{\frac{r_{iA}}{-a_i} = \frac{r_{iB}}{-b_i} = \frac{r_{iC}}{c_i} = \frac{r_{iD}}{d_i}}$$

3. Rates

The remaining steps to the algorithm in Table 6-2 remain unchanged, e.g.,

Gas Phase

$$C_j = C_{T0} \frac{F_j}{F_T} \frac{P}{P_0} \frac{T_0}{T}$$

$$F_T = \sum_{j=1}^{n} F_j$$

Stoichiometry

Liquid Phase

$$C_j = \frac{F_j}{v_0}$$

balances in Table 6-2 (e.g., r_A, r_B, r_j) are the *net rates* of formation. The main change in the CRE algorithm in Table 6-2 is that the **Rate Law** step in our algorithm has now been replaced by the step **Rates**, which includes three substeps:

- Rate Laws
- Net Rates
- Relative Rates

8.3 Parallel Reactions

8.3.1 Selectivity

In this section, we discuss various means of minimizing the undesired product, U, through the selection of reactor type and operating conditions. We also discuss the development of efficient reactor schemes.

For the competing reactions such as

$$(1) \quad A \xrightarrow{k_D} D \qquad \text{(Desired)}$$

$$(2) \quad A \xrightarrow{k_U} U \qquad \text{(Undesired)}$$

the rate laws are

Rate laws for formation of desired and undesired products

$$r_D = k_D C_A^{\alpha_1} \tag{8-6}$$

$$r_U = k_U C_A^{\alpha_2} \tag{8-7}$$

The net rate of disappearance of A for this reaction sequence is the sum of the rates of formation of U and D:

$$-r_A = r_D + r_U \tag{8-8}$$

$$-r_A = k_D C_A^{\alpha_1} + k_U C_A^{\alpha_2} \tag{8-9}$$

where α_1 and α_2 are positive reaction orders. We want the rate of formation of D, r_D, to be high with respect to the rate of formation of U, r_U. Taking the ratio of these rates [i.e., Equation (8-6) to Equation (8-7)], we obtain the **instantaneous selectivity**, $S_{D/U}$, which is to be maximized:

Instantaneous selectivity

$$S_{D/U} = \frac{r_D}{r_U} = \frac{k_D}{k_U} C_A^{\alpha_1 - \alpha_2} \tag{8-10}$$

8.3.2 Maximizing the Desired Product for One Reactant

In this section, we examine ways to maximize the instantaneous selectivity, $S_{D/U}$, for different reaction orders of the desired and undesired products.

Case 1: $\alpha_1 > \alpha_2$.   The reaction order of the desired product, α_1, is greater than the reaction order of the undesired product, α_2. Let a be a positive number that is the difference between these reaction orders ($a > 0$):

$$\alpha_1 - \alpha_2 = a$$

For $\alpha_1 > \alpha_2$, make C_A as large as possible by using a PFR or batch reactor.

Then, upon substitution into Equation (8-10), we obtain

$$S_{D/U} = \frac{r_D}{r_U} = \frac{k_D}{k_U} C_A^a \tag{8-11}$$

To make this ratio as large as possible, we want to carry out the reaction in a manner that will keep the concentration of reactant A as high as possible during the reaction. If the reaction is carried out in the gas phase, we should run it without inerts and at high pressures to keep C_A high. If the reaction is in the liquid phase, the use of diluents should be kept to a minimum.[1]

A batch or plug-flow reactor should be used in this case because, in these two reactors, the concentration of A starts at a high value and drops progressively during the course of the reaction. In a *perfectly mixed* CSTR, the concentration of reactant within the reactor is always at its lowest value (i.e., that of the outlet concentration) and therefore the CSTR should not be chosen under these circumstances.

Case 2: $\alpha_2 > \alpha_1$. The reaction order of the undesired product is greater than that of the desired product. Let $b = \alpha_2 - \alpha_1$, where b is a positive number; then

$$S_{D/U} = \frac{r_D}{r_U} = \frac{k_D C_A^{\alpha_1}}{k_U C_A^{\alpha_2}} = \frac{k_D}{k_U C_A^{\alpha_2 - \alpha_1}} = \frac{k_D}{k_U C_A^b} \tag{8-12}$$

For the ratio r_D/r_U to be high, the concentration of A should be as low as possible.

This low concentration may be accomplished by diluting the feed with inerts and running the reactor at low concentrations of species A. A CSTR should be used because the concentrations of reactants are maintained at a low level. A recycle reactor in which the product stream acts as a diluent could be used to maintain the entering concentrations of A at a low value.

For $\alpha_2 > \alpha_1$ use a CSTR and dilute the feed stream.

Because the activation energies of the two reactions in cases 1 and 2 are not given, it cannot be determined whether the reaction should be run at high or low temperatures. The sensitivity of the rate selectivity parameter to temperature can be determined from the ratio of the specific reaction rates

Effect of temperature on selectivity

$$S_{D/U} \sim \frac{k_D}{k_U} = \frac{A_D}{A_U} e^{-[(E_D - E_U)/RT]} \tag{8-13}$$

where A is the frequency factor and E the activation energy, and the subscripts D and U refer to desired and undesired product, respectively.

Case 3: $E_D > E_U$. In this case, the specific reaction rate of the desired reaction k_D (and therefore the overall rate r_D) increases more rapidly with increasing temperature, T, than does the specific rate of the undesired reaction k_U. Consequently, the reaction system should be operated at the highest possible temperature to maximize $S_{D/U}$.

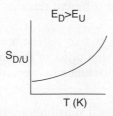

---

[1] For a number of liquid-phase reactions, the proper choice of a solvent can enhance selectivity. See, for example, *Ind. Eng. Chem.*, 62(9), 16. In gas-phase heterogeneous catalytic reactions, selectivity is an important parameter of any particular catalyst.

Case 4: $E_U > E_D$.   In this case, the reaction should be carried out at a low temperature to maximize $S_{D/U}$, *but not* so low that the desired reaction does not proceed to any significant extent.

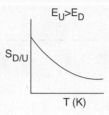

Example 8–1   Maximizing the Selectivity for the Famous Trambouze Reactions

Reactant A decomposes by three simultaneous reactions to form three products, one that is desired, B, and two that are undesired, X and Y. These gas-phase reactions, along with the appropriate rate laws, are called the *Trambouze reactions (AIChE J., 5, 384).*

1) $A \xrightarrow{k_1} X$ $-r_{1A} = r_X = k_1 = 0.0001 \dfrac{\text{mol}}{\text{dm}^3 \cdot \text{s}}$ (zero order)

2) $A \xrightarrow{k_2} B$ $-r_{2A} = r_B = k_2 C_A = (0.0015 \ s^{-1}) C_A$ (first order)

3) $A \xrightarrow{k_3} Y$ $-r_{3A} = r_Y = k_3 C_A^2 = \left(0.008 \dfrac{\text{dm}^3}{\text{mol} \cdot \text{s}} \right) C_A^2$ (second order)

The specific reaction rates are given at 300 K and the activation energies for reactions (1), (2), and (3) are $E_1 = 10,000$ kcal/mole, $E_2 = 15,000$ kcal/mole, and $E_3 = 20,000$ kcal/mole.

  (a) How, and under what conditions (e.g., reactor type(s), temperature, concentrations), should the reaction be carried out to maximize the selectivity of species B for an entering concentration of species A of 0.4 M and a volumetric flow rate of 2.0 dm³/s?

  (b) How could the conversion of B be increased and still keep selectivity relatively high?

Solution

Part (a)
The instantaneous selectivity of species B with respect to species X and Y is

$$S_{B/XY} = \frac{r_B}{r_X + r_Y} = \frac{k_2 C_A}{k_1 + k_3 C_A^2} \qquad \text{(E8-1.1)}$$

When we plot $S_{B/XY}$ vs. C_A, we see that there is a maximum, as shown in Figure E8-1.1.

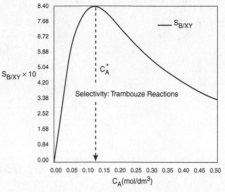

Figure E8-1.1   Selectivity as a function of the concentration of A.

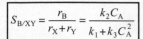

As we can see, the selectivity reaches a maximum at a concentration C_A^*. Because the concentration changes down the length of a PFR, we cannot operate at this maximum. Consequently, we will use a CSTR and design it to operate at this maximum. To find the maximum, C_A^*, we differentiate $S_{B/XY}$ with respect to C_A, set the derivative to zero, and solve for C_A^*. That is

$$\frac{dS_{B/XY}}{dC_A} = 0 = \frac{k_2[k_1 + k_3 C_A^{*2}] - k_2 C_A^*[2k_3 C_A^*]}{[k_1 + k_3 C_A^{*2}]^2} \tag{E8-1.2}$$

Solving for C_A^*

$$C_A^* = \sqrt{\frac{k_1}{k_3}} = \sqrt{\frac{0.0001(\,\text{mol/dm}^3 \cdot \text{s})}{0.008\ (\text{dm}^3/\text{mol} \cdot \text{s})}} = 0.112\ \text{mol/dm}^3 \tag{E8-1.3}$$

We see from Figure E8-1.1 that the selectivity is indeed a maximum at $C_A^* = 0.112\ \text{mol/dm}^3$.

$$\boxed{C_A^* = \sqrt{\frac{k_1}{k_3}} = 0.112\ \text{mol/dm}^3}$$

Therefore, to maximize the selectivity $S_{B/XY}$, we want to carry out our reaction in such a manner that the CSTR concentration of A is always at C_A^*. The corresponding selectivity at C_A^* is

$$S_{B/XY} = \frac{k_2 C_A^*}{k_1 + k_3 C_A^{*2}} = \frac{k_2 \sqrt{\dfrac{k_1}{k_3}}}{k_1 + k_1} = \frac{k_2}{2\sqrt{k_1 k_3}} = \frac{0.0015}{2[(0.0001)(0.008)]^{1/2}} \tag{E8-1.4}$$

$$\boxed{S_{B/XY} = 0.84}$$

We now calculate the CSTR volume when the exit concentration is C_A^*. The **net rate** of formation of A is a sum of the reaction rates from Equations (1), (2), and (3)

$$r_A = r_{1A} + r_{2A} + r_{3A} = -k_{1A} - k_{2A}C_A - k_{3A}C_A^2 \tag{E8-1.5}$$

$$-r_A = k_1 + k_2 C_A + k_3 C_A^2$$

Operate at this CSTR reactant concentration: $C_A^* = 0.112\ \text{mol/dm}^3$.

$$S_{B/XY} = \frac{r_B}{r_X + r_Y} = \frac{k_2 C_A}{k_1 + k_3 C_A^2}$$

Using Equation (E8-1.5) in the **mole balance** on a CSTR for this liquid-phase reaction ($v = v_0$) to **combine** it with the **net rate** we obtain

$$V = \frac{v_0[C_{A0}-C_A^*]}{-r_A^*} = \frac{v_0[C_{A0}-C_A^*]}{[k_1+k_2C_A^*+k_3C_A^{*2}]} \tag{E8-1.6}$$

$$\tau = \frac{V}{v_0} = \frac{C_{A0}-C_A^*}{-r_A^*} = \frac{(C_{A0}-C_A^*)}{k_1+k_2C_A^*+k_3C_A^{*2}} \tag{E8-1.7}$$

$$\tau = \frac{(0.4-0.112)}{(0.0001)+(0.0015)(0.112)+0.008(0.112)^2} = 782\,\text{s}$$

$$V = v_0\tau = (2\,\text{dm}^3/\text{s})(782\,\text{s})$$

$$\boxed{V = 1564\,\text{dm}^3 = 1.564\,\text{m}^3}$$

CSTR volume to
maximize selectivity
$\tilde{S}_{B/XY} = S_{B/XY}$

For an entering volumetric flow rate of 2 dm³/s, we must have a CSTR volume of 1564 dm³ to maximize the selectivity, $S_{B/XY}$.

Maximize the selectivity with respect to temperature.

We now substitute C_A^* in Equation (E8-1.1) and then substitute for C_A^* in terms of k_1 and k_3, cf. Equation (E8-1.3), to get $S_{B/XY}$ in terms of k_1, k_2, and k_3.

At what temperature
should we operate
the CSTR?

$$S_{B/XY} = \frac{k_2C_A^*}{k_1+k_3C_A^{*2}} = \frac{k_2\sqrt{\frac{k_1}{k_3}}}{k_1+k_1} = \frac{k_2}{2\sqrt{k_1k_3}} \tag{E8-1.4}$$

$$S_{B/XY} = \frac{A_2}{2\sqrt{A_1A_3}}\exp\left[\frac{\frac{E_1+E_3}{2}-E_2}{RT}\right] \tag{E8-1.8}$$

Case 1: If $\dfrac{E_1+E_3}{2} < E_2$ $\left\{\begin{array}{l}\text{Run at as high a temperature as possible with existing}\\ \text{equipment and watch out for other side reactions that}\\ \text{might occur at higher temperatures}\end{array}\right.$

Case 2: If $\dfrac{E_1+E_3}{2} > E_2$ $\left\{\begin{array}{l}\text{Run at low temperatures but not so low that a significant}\\ \text{conversion is not achieved}\end{array}\right.$

For the activation energies given in this example

$$\frac{E_1+E_3}{2}-E_2 = \frac{10{,}000+20{,}000}{2}-15{,}000 = 0$$

So the selectivity for this combination of activation energies is independent of temperature!

What is the conversion of A in the CSTR?

$$\boxed{X^* = \frac{C_{A0}-C_A^*}{C_{A0}} = \frac{0.4-0.112}{0.4} = 0.72}$$

Part (b)

If a greater than 72% conversion of A is required, say 90%, then the CSTR operated with a reactor concentration of 0.112 mol/dm³ should be followed by a PFR because the conversion will increase continuously as we move down the PFR (see Figure E8-1.2(**b**)). However, as can be seen in Figure E8-1.2, the concentration will decrease continuously from C_A^*, as will the selectivity $S_{B/XY}$ as we move down the PFR to an exit concentration C_{Af}. Hence the system

How can we increase the conversion and still have a high selectivity $S_{B/XY}$?

$$\left[CSTR \Big|_{C_A^*} + PFR \Big|_{C_A^*}^{C_{Af}} \right]$$

would give the highest selectivity and least total reactor volume while forming more of the desired product B, beyond what was formed at C_A^* in a CSTR.

Figure E8-1.2 illustrates how as the conversion is increased above X* by adding the PFR reactor volume; however, the selectivity decreases.

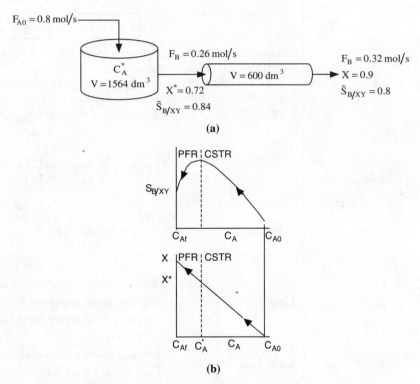

Figure E8-1.2 Effect of adding a PFR to increase conversion. (**a**) Reactor arrangement; (**b**) Selectivity and conversion trajectories.

This calculation for the PFR is carried out on the CRE Web site in Example 8-1. The results of this calculation show that at the exit of the PFR, the molar flow rates are $F_X = 0.22$ mol/s, $F_B = 0.32$ mol/s, and $F_Y = 0.18$ mol/s corresponding to a conversion of $X = 0.9$. The corresponding selectivity at a conversion of 90% is

Do you really want to add the PFR?

$$\boxed{\tilde{S}_{B/XY} = \frac{F_B}{F_X + F_Y} = 0.8}$$

Analysis: One now has to make a decision as to whether adding the PFR to increase the conversion of A from 0.72 to 0.9 and the molar flow rate of B from 0.26 to 0.32 mol/s is worth not only the added cost of the PFR, but also the decrease in selectivity from 0.84 to 0.8. In this example, we used the Trambouze reactions to show how to optimize the selectivity to species B in a CSTR. Here, we found the optimal exit conditions ($C_A = 0.112$ mol/dm^3), conversion ($X = 0.72$), and selectivity ($S_{B/XY} = 0.84$). The corresponding CSTR volume was $V = 1564$ dm^3. If we wanted to increase the conversion to 90%, we could use a PFR to follow the CSTR, and find that the selectivity decreased.

8.3.3 Reactor Selection and Operating Conditions

Next, consider two simultaneous reactions in which two reactants, A and B, are being consumed to produce a desired product, D, and an unwanted product, U, resulting from a side reaction. The rate laws for the reactions

$$A + B \xrightarrow{\ k_1\ } D$$

$$A + B \xrightarrow{\ k_2\ } U$$

are

$$r_D = k_1 C_A^{\alpha_1} C_B^{\beta_1} \tag{8-14}$$

$$r_U = k_2 C_A^{\alpha_2} C_B^{\beta_2} \tag{8-15}$$

The instantaneous selectivity

Instantaneous selectivity

$$S_{D/U} = \frac{r_D}{r_U} = \frac{k_1}{k_2} C_A^{\alpha_1 - \alpha_2} C_B^{\beta_1 - \beta_2} \tag{8-16}$$

is to be maximized. Shown in Figure 8-2 are various reactor schemes and conditions that might be used to maximize $S_{D/U}$.

The two reactors with recycle shown in (i) and (j) can be used for highly exothermic reactions. Here, the recycle stream is cooled and returned to the reactor to dilute and cool the inlet stream, thereby avoiding hot spots and runaway reactions. The PFR with recycle is used for gas-phase reactions, and the CSTR is used for liquid-phase reactions.

The last two reactors in Figure 8-2, (k) and (l), are used for thermodynamically limited reactions where the equilibrium lies far to the left (reactant side)

$$A + B \rightleftarrows C + D$$

and one of the products must be removed (e.g., C) for the reaction to continue to completion. The membrane reactor (k) is used for thermodynamically limited gas-phase reactions, while reactive distillation (l) is used for liquid-phase reactions when one of the products has a higher volatility (e.g., C) than the other species in the reactor.

In making our selection of a reactor, the criteria are *safety, selectivity, yield, temperature control,* and *cost.*

Reactor Selection

Criteria:
- Safety
- Selectivity
- Yield
- Temperature control
- Cost

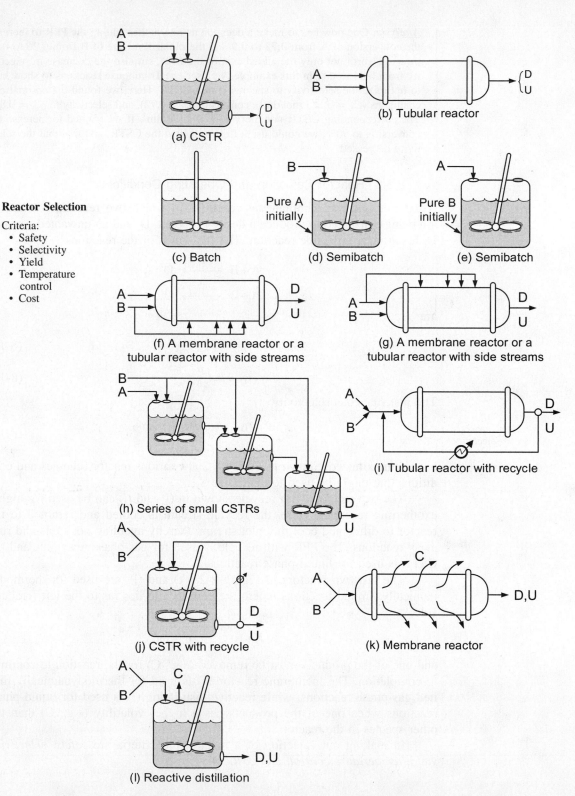

Figure 8-2 Different reactors and schemes for maximizing $S_{D/U}$ in Equation 8-16. Note unreacted A and B also exit the reactor along with D and U.

Example 8–2   Choice of Reactor and Conditions to Minimize Unwanted Products

For the parallel reactions

$$A + B \longrightarrow D : \qquad r_D = k_1 C_A^{\alpha_1} C_B^{\beta_1}$$

$$A + B \longrightarrow U : \qquad r_U = k_2 C_A^{\alpha_2} C_B^{\beta_2}$$

consider all possible combinations of reaction orders and select the reaction scheme that will maximize $S_{D/U}$.

Solution

Case 1: $\alpha_1 > \alpha_2$, $\beta_1 > \beta_2$. Let $a = \alpha_1 - \alpha_2$ and $b = \beta_1 - \beta_2$, where a and b are positive constants. Using these definitions, we can write Equation (8-16) in the form

$$\boxed{S_{D/U} = \frac{r_D}{r_U} = \frac{k_1}{k_2} \, C_A^a C_B^b} \tag{E8-2.1}$$

To maximize the ratio r_D/r_U, maintain the concentrations of both A and B as high as possible. To do this, use

- A tubular reactor, Figure 8-2(b).
- A batch reactor, Figure 8-2(c).
- High pressures (if gas phase), and reduce inerts.

Case 2: $\alpha_1 > \alpha_2$, $\beta_1 < \beta_2$. Let $a = \alpha_1 - \alpha_2$ and $b = \beta_2 - \beta_1$, where a and b are positive constants. Using these definitions, we can write Equation (8-16) in the form

$$S_{D/U} = \frac{r_D}{r_U} = \frac{k_1 C_A^a}{k_2 C_B^b} \tag{E8-2.2}$$

To make $S_{D/U}$ as large as possible, we want to make the concentration of A high and the concentration of B low. To achieve this result, use

- A semibatch reactor in which B is fed slowly into a large amount of A as in Figure 8-2(d).
- A membrane reactor or a tubular reactor with side streams of B continually fed to the reactor as in Figure 8-2(f).
- A series of small CSTRs with A fed only to the first reactor and small amounts of B fed to each reactor. In this way, B is mostly consumed before the CSTR exit stream flows into the next reactor as in Figure 8-2(h).

Case 3: $\alpha_1 < \alpha_2$, $\beta_1 < \beta_2$. Let $a = \alpha_2 - \alpha_1$ and $b = \beta_2 - \beta_1$, where a and b are positive constants. Using these definitions, we can write Equation (8-16) in the form

$$S_{D/U} = \frac{r_D}{r_U} = \frac{k_1}{k_2 C_A^a C_B^b} \tag{E8-2.3}$$

To make $S_{D/U}$ as large as possible, the reaction should be carried out at low concentrations of A and of B. Use

- A CSTR as in Figure 8-2(a).
- A tubular reactor in which there is a large recycle ratio as in Figure 8-2(i).
- A feed diluted with inerts.
- Low pressure (if gas phase).

Case 4: $\alpha_1 < \alpha_2$, $\beta_1 > \beta_2$. Let $a = \alpha_2 - \alpha_1$ and $b = \beta_1 - \beta_2$, where a and b are positive constants. Using these definitions, we can write Equation (8-16) in the form

$$S_{D/U} = \frac{r_D}{r_U} = \frac{k_1 C_B^b}{k_2 C_A^a} \tag{E8-2.4}$$

To maximize $S_{D/U}$, run the reaction at high concentrations of B and low concentrations of A. Use

- A semibatch reactor with A slowly fed to a large amount of B as in Figure 8-2(e).

- A membrane reactor or a tubular reactor with side streams of A as in Figure 8-2(g).

- A series of small CSTRs with fresh A fed to each reactor.

Analysis: In this *very* important example we showed how to use the instantaneous selectivity, $S_{D/U}$, to guide the initial selection of the type of reactor and reactor system to maximize the selectivity with respect to the desired species D. The final selection should be made after one calculates the overall selectivity $\tilde{S}_{D/U}$ for the reactors and operating conditions chosen.

8.4 Reactions in Series

In Section 8.1, we saw that the undesired product could be minimized by adjusting the reaction conditions (e.g., concentration, temperature) and by choosing the proper reactor. For series (i.e., consecutive) reactions, the most important variable is time: space time for a flow reactor and real time for a batch reactor. To illustrate the importance of the time factor, we consider the sequence

$$A \xrightarrow{\ k_1\ } B \xrightarrow{\ k_2\ } C$$

in which species B is the desired product.

If the first reaction is slow and the second reaction is fast, it will be extremely difficult to produce species B. If the first reaction (formation of B) is fast and the reaction to form C is slow, a large yield of B can be achieved. However, if the reaction is allowed to proceed for a long time in a batch reactor, or if the tubular flow reactor is too long, the desired product B will be converted to the undesired product C. In no other type of reaction is exactness in the calculation of the time needed to carry out the reaction more important than in series reactions.

Example 8–3 Series Reactions in a Batch Reactor

The elementary liquid-phase series reaction

$$A \xrightarrow{\ k_1\ } B \xrightarrow{\ k_2\ } C$$

is carried out in a batch reactor. The reaction is heated very rapidly to the reaction temperature, where it is held at this temperature until the time it is quenched.

  (a) Plot and analyze the concentrations of species A, B, and C as a function of time.

  (b) Calculate the time to quench the reaction when the concentration of B will be a maximum.

  (c) What are the overall selectivity and yields at this quench time?

Additional Information

$$C_{A0} = 2M, \ k_1 = 0.5h^{-1}, \ k_2 = 0.2h^{-1}$$

Solution

Part (a)

Number the Reactions:

The preceding series reaction can be written as two reactions

(1) Reaction 1 $A \xrightarrow{k_1} B$ $-r_{1A} = k_1 C_A$

(2) Reaction 2 $B \xrightarrow{k_2} C$ $-r_{2B} = k_2 C_B$

1. **Mole Balances:**

2A. **Mole Balance on A:**

$$\frac{dN_A}{dt} = r_A V$$

a. **Mole balance** in terms of concentration for $V = V_0$ becomes

$$\frac{dC_A}{dt} = r_A \tag{E8-3.1}$$

b. **Rate law for Reaction 1:**   Reaction is elementary

$$r_A = r_{1A} = -k_1 C_A \tag{E8-3.2}$$

Following the Algorithm

c. **Combining** the mole balance and rate law

$$\frac{dC_A}{dt} = -k_1 C_A \tag{E8-3.3}$$

Integrating with the initial condition $C_A = C_{A0}$ at t = 0

$$ln\frac{C_A}{C_{A0}} = -k_1 t \tag{E8-3.4}$$

Solving for C_A

$$\boxed{C_A = C_{A0}e^{-k_1 t}} \tag{E8-3.5}$$

2B. **Mole Balance on B:**

a. **Mole balance** for a constant-volume batch reactor becomes

$$\frac{dC_B}{dt} = r_B \tag{E8-3.6}$$

b. **Rates:**

Rate Laws

Elementary reactions

$$r_{2B} = -k_2 C_B \tag{E8-3.7}$$

Relative Rates

Rate of formation of B in Reaction 1 equals the rate of disappearance of A in Reaction 1.

$$r_{1B} = -r_{1A} = k_1 C_A \tag{E8-3.8}$$

Net Rates

The net rate of reaction of B will be the rate of formation of B in reaction (1) plus the rate of formation of B in reaction (2).

$$r_B = r_{1B} + r_{2B} \tag{E8-3.9}$$

$$r_B = k_1 C_A - k_2 C_B \tag{E8-3.10}$$

c. **Combining** the mole balance and rate law

$$\frac{dC_B}{dt} = k_1 C_A - k_2 C_B \tag{E8-3.11}$$

Rearranging and substituting for C_A

$$\frac{dC_B}{dt} + k_2 C_B = k_1 C_{A0} e^{-k_1 t} \tag{E8-3.12}$$

Using the integrating factor gives

There is a tutorial on the integrating factor in Appendix A and on the CRE Web site.

$$\frac{d\left(C_B e^{k_2 t}\right)}{dt} = k_1 C_{A0} e^{(k_2 - k_1)t} \tag{E8-3.13}$$

At time t = 0, C_B = 0. Solving Equation (E8-3.13) gives

$$\boxed{C_B = k_1 C_{A0} \left[\frac{e^{-k_1 t} - e^{-k_2 t}}{k_2 - k_1} \right]} \tag{E8-3.14}$$

2C. Mole Balance on C:

The mole balance on C is similar to Equation (E8-3.1).

$$\frac{dC_C}{dt} = r_C \tag{E8-3.15}$$

The rate of formation of C is just the rate of disappearance of B in reaction (2), i.e., $r_C = -r_{2B} = k_2 C_B$

$$\frac{dC_C}{dt} = k_2 C_B \tag{E8-3.16}$$

Substituting for C_B

$$\frac{dC_C}{dt} = \frac{k_1 k_2 C_{A0}}{k_2 - k_1} \left(e^{-k_1 t} - e^{-k_2 t} \right)$$

and integrating with C_C = 0 at t = 0 gives

$$\boxed{C_C = \frac{C_{A0}}{k_2 - k_1} \left[k_2 \left[1 - e^{-k_1 t} \right] - k_1 \left[1 - e^{-k_2 t} \right] \right]} \tag{E8-3.17}$$

Note that as $t \to \infty$, then $C_C = C_{A0}$ as expected. We also note the concentration of C, C_C, could have been obtained more easily from an overall balance.

Calculating the concentration of C the easy way

$$\boxed{C_C = C_{A0} - C_A - C_B} \tag{E8-3.18}$$

The concentrations of A, B, and C are shown as a function of time in Figure E8-3.1.

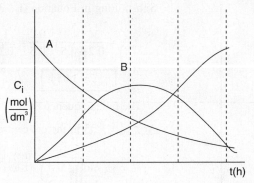

Figure E8-3.1 Concentration trajectories in a batch reactor.

Part (b)

4. **Optimum Yield**

   We note from Figure E8-3.1 that the concentration of B goes through a maximum. Consequently, to find the maximum we need to differentiate Equation E8-3.14 and set it to zero.

$$\frac{dC_B}{dt} = 0 = \frac{k_1 C_{A0}}{k_2 - k_1}\left[-k_1 e^{-k_1 t} + k_2 e^{-k_2 t}\right] \tag{E8-3.19}$$

   Solving for t_{max} gives

$$\boxed{t_{max} = \frac{1}{k_2 - k_1}\ln\frac{k_2}{k_1}} \tag{E8-3.20}$$

Series Reaction

   Substituting Equation (E8-3.20) into Equation (E8-3.5), we find the concentration of A at the maximum for C_B is

$$C_A = C_{A0}e^{-k_1\left(\frac{1}{k_2-k_1}\ln\frac{k_2}{k_1}\right)} \tag{E8-3.21}$$

$$\boxed{C_A = C_{A0}\left[\frac{k_1}{k_2}\right]^{\frac{k_1}{k_2-k_1}}} \tag{E8-3.22}$$

   Similarly, the concentration of B at the maximum is

$$\boxed{C_B = \frac{k_1 C_{A0}}{k_2 - k_1}\left[\left(\frac{k_1}{k_2}\right)^{\frac{k_1}{k_2-k_1}} - \left(\frac{k_1}{k_2}\right)^{\frac{k_2}{k_2-k_1}}\right]} \tag{E8-3.23}$$

5. **Evaluate:** Substituting for $C_{A0} = 2$ mol/dm³, $k_1 = 0.5\,\text{h}^{-1}$, and $k_2 = 0.2\,\text{h}^{-1}$ in Equations (E8-3.5), (E8-3.14), and (E8-3.18), the concentrations as a function of time are

$$C_A = 2\,\text{mol/dm}^3\left(e^{-0.5t}\right)$$

$$C_B = \frac{2\left(\text{mol/dm}^3\right)}{(0.2 - 0.5)}(0.5)\left[e^{-0.5t} - e^{-0.2t}\right]$$

$$C_B = 3.33\left(\text{mol/dm}^3\right)\left[e^{-0.2t} - e^{-0.5t}\right]$$

$$C_C = 2\,\text{mol/dm}^3 - 2\left(\text{mol/dm}^3\right)e^{-0.5t} - 3.33\,\text{mol/dm}^3\left[e^{-0.2t} - e^{-0.5t}\right]$$

Substituting in Equation (E8-3.20)

$$t_{max} = \frac{1}{0.2 - 0.5} \ln \frac{0.2}{0.5} = \frac{1}{0.3} \ln \frac{0.5}{0.2}$$

$$\boxed{t_{max} = 3.05 \text{ h}}$$

The time to quench the reaction is at 3.05 h.

<u>At $t_{max} = 3.05$ h</u>, the concentrations of A, B, and C are

$$C_A = 2 \frac{\text{mol}}{\text{dm}^3} \left[\left(\frac{0.5}{0.2} \right)^{\left(\frac{(0.5)}{0.2 - 0.5} \right)} \right] = 0.44 \frac{\text{mol}}{\text{dm}^3}$$

$$C_B = 2 \frac{\text{mol}}{\text{dm}^3} \frac{(0.5)}{(0.2 - 0.5)} \left[\left(\frac{0.5}{0.2} \right)^{\left(\frac{0.5}{0.2 - 0.5} \right)} - \left(\frac{0.5}{0.2} \right)^{\left(\frac{0.2}{0.2 - 0.5} \right)} \right]$$

$$C_B = 1.07 \frac{\text{mol}}{\text{dm}^3}$$

The concentration of C at the time we quench the reaction is

$$C_C = C_{A0} - C_A - C_B = 2 - 0.44 - 1.07 = 0.49 \text{ mol/dm}^3$$

Part (c) Calculate the overall selectivity and yield at the reaction quench time. The selectivity is

$$\tilde{S}_{B/C} = \frac{C_B}{C_C} = \frac{1.07}{0.49} = 2.2$$

The yield is

$$\tilde{Y}_B = \frac{C_B}{C_{A0} - C_A} = \frac{1.07}{2.0 - 0.44} = 0.69$$

<u>**Analysis:**</u> In this example, we applied our CRE algorithm for multiple reactions to the series reaction A → B → C. Here, we obtained an analytical solution to find the time at which the concentration of the desired product B was a maximum and, consequently, the time to quench the reaction. We also calculated the concentrations of A, B, and C at this time, along with the selectivity and yield.

We will now carry out this same series reaction in a CSTR.

Example 8–4 Series Reaction in a CSTR

The reactions discussed in Example 8-3 are now to be carried out in a CSTR.

$$A \xrightarrow{\ k_1\ } B$$
$$B \xrightarrow{\ k_2\ } C$$

(a)   Determine the exit concentrations from the CSTR.

(b)   Find the value of the space time τ that will maximize the concentration of B.

Solution

Part (a)

1. a. **Mole Balance on A:**

$$\text{IN} \quad - \quad \text{OUT} \quad + \quad \text{GENERATION} \quad = \quad 0$$

$$F_{A0} \quad - \quad F_A \quad + \quad r_A V \quad = \quad 0$$

$$v_0 C_{A0} \quad - \quad v_0 C_A \quad + \quad r_A V \quad = \quad 0$$

Dividing by v_0, rearranging and recalling that $\tau = V/v_0$, we obtain

$$C_{A0} - C_A + r_A \tau = 0 \qquad \text{(E8-4.1)}$$

b. **Rates**

The laws and net rates are the same as in Example 8-3.

$$\text{Reaction 1:} \qquad r_A = -k_1 C_A \qquad \text{(E8-4.2)}$$

c. **Combining** the mole balance of A with the rate of disappearance of A

$$C_{A0} - C_A - k_1 C_A \tau = 0 \qquad \text{(E8-4.3)}$$

Solving for C_A

$$\boxed{C_A = \frac{C_{A0}}{1 + \tau k_1}} \qquad \text{(E8-4.4)}$$

We now use the same algorithm for species B we did for species A to solve for the concentration of B.

2. a. **Mole Balance on B:**

$$\text{IN} \quad - \quad \text{OUT} \quad + \quad \text{GENERATION} \quad = \quad 0$$

$$0 \quad - \quad F_B \quad + \quad r_B V \quad = \quad 0$$

$$- \quad v_0 C_B \quad + \quad r_B V \quad = \quad 0$$

Dividing by v_0 and rearranging

$$-C_B + r_B \tau = 0 \qquad \text{(E8-4.5)}$$

b. **Rates**

The laws and net rates are the same as in Example 8-3.

Net Rates

$$r_B = k_1 C_A - k_2 C_B \qquad \text{(E8-4.6)}$$

c. **Combine**

$$-C_B + \left(k_1 C_A - k_2 C_B\right)\tau = 0$$

$$C_B = \frac{k_1 C_A \tau}{1 + k_2 \tau} \qquad \text{(E8-4.7)}$$

Substituting for C_A

$$\boxed{C_B = \frac{\tau k_1 C_{A0}}{\left(1 + k_1 \tau\right)\left(1 + k_2 \tau\right)}} \qquad \text{(E8-4.8)}$$

3. **Mole Balance on C:**

$$0 - v_0 C_C + r_C V = 0$$

$$-C_C + r_C \tau = 0 \tag{E8-4.9}$$

Rates

$$r_C = -r_{2B} = k_2 C_B$$

$$C_C = r_C \tau = k_2 C_B \tau$$

$$\boxed{C_C = \frac{\tau^2 k_1 k_2 C_{A0}}{(1 + k_1 \tau)(1 + k_2 \tau)}} \tag{E8-4.10}$$

Part (b) Optimum Concentration of B

To find the maximum concentration of B, we set the differential of Equation (E8-4.8) with respect to τ equal to zero

$$\frac{dC_B}{d\tau} = \frac{k_1 C_{A0}(1 + \tau k_1)(1 + \tau k_2) - \tau k_1 C_{A0}(k_1 + k_2 + 2\tau k_1 k_2)}{\left[(1 + k_1 \tau)(1 + k_2 \tau)\right]^2} = 0$$

Solving for τ at which the concentration of B is a maximum at

$$\boxed{\tau_{max} = \frac{1}{\sqrt{k_1 k_2}}} \tag{E8-4.11}$$

The exiting concentration of B at the optimum value of τ is

$$C_B = \frac{\tau_{max} k_1 C_{A0}}{(1 + \tau_{max} k_1)(1 + \tau_{max} k_2)} = \frac{\tau_{max} k_1 C_{A0}}{1 + \tau_{max} k_1 + \tau_{max} k_2 + \tau_{max}^2 k_1 k_2} \tag{E8-4.12}$$

Substituting Equation (E8-4.11) for τ_{max} in Equation (E8-4.12)

$$C_B = \frac{C_{A0} \dfrac{k_1}{\sqrt{k_1 k_2}}}{1 + \dfrac{k_1}{\sqrt{k_1 k_2}} + \dfrac{k_2}{\sqrt{k_1 k_2}} + 1} \tag{E8-4.13}$$

Rearranging, we find the concentration of B at the optimum space time is

$$\boxed{C_B = \frac{C_{A0} k_1}{2\sqrt{k_1 k_2} + k_1 + k_2}} \tag{E8-4.14}$$

Evaluation

$$\tau_{max} = \frac{1}{\sqrt{(0.5)(0.2)}} = 3.16 \text{h}$$

At τ_{max}, the concentrations of A, B, and C are

$$C_A = \frac{C_{A0}}{1 + \tau_{max} k_1} = \frac{2\dfrac{\text{mol}}{\text{dm}^3}}{1 + (3.16\text{h})\left(\dfrac{0.5}{h}\right)} = 0.78 \frac{\text{mol}}{\text{dm}^3}$$

$$C_B = 2 \frac{mol}{dm^3} \frac{0.5}{2\sqrt{(0.2)(0.5)} + 0.2 + 0.5} = 0.75 \frac{mol}{dm^3}$$

$$C_C = C_{A0} - C_A - C_B = \left(2 - 0.78 - 0.75 \frac{mol}{dm^3} \right) = 0.47 \frac{mol}{dm^3}$$

The conversion is

$$X = \frac{C_{A0} - C_A}{C_{A0}} = \frac{2 - 0.78}{2} = 0.61$$

The selectivity is

$$\tilde{S}_{B/C} = \frac{C_B}{C_C} = \frac{0.75}{0.47} = 1.60$$

The yield is

$$\tilde{Y}_B = \frac{C_B}{C_{A0} - C_A} = \frac{0.75}{2 - 0.78} = 0.63$$

<u>**Analysis:**</u> The CRE algorithm for multiple reactions was applied to the series reaction $A \rightarrow B \rightarrow C$ in a CSTR to find the CSTR space time necessary to maximize the concentration of B, i.e., $\tau = 3.16$ h. The conversion at this space time is 61%, the selectivity, $\tilde{S}_{B/C}$, is 1.60, and the yield, $\tilde{Y}_B$, is 0.63. The conversion and selectivity are less for the CSTR than those for the batch reactor at the time of quenching.

PFR

If the series reaction were carried out in a PFR, the results would essentially be those of a batch reactor where we replaced the time variable "t" with the space time, "τ". Data for the series reaction

$$\text{Ethanol} \xrightarrow{\ k_1\ } \text{Aldehyde} \xrightarrow{\ k_2\ } \text{Products}$$

is compared for different values of the specific reaction rates, k_1 and k_2, in Figure 8-3.

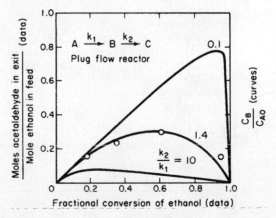

Figure 8-3   Yield of acetaldehyde as a function of ethanol conversion. Data were obtained at 518 K. Data points (in order of increasing ethanol conversion) were obtained at space velocities of 26,000, 52,000, 104,000, and 208,000 h^{-1}. The curves were calculated for a first-order series reaction in a plug-flow reactor and show yield of the intermediate species B as a function of the conversion of reactant for various ratios of rate constants k_2 and k_1. (McCabe, Robert W. and Patricia J. Mitchell. Oxidation of ethanol and acetaldehyde over alumina-supported catalysts. Industrial & Engineering Chemistry Product Research and Development 1983 22(2), 212–217. Copyright © 1963, American Chemical Society. Reprinted by permission.)

Living Example Problem

A complete analysis of this reaction carried out in a PFR is given on the CRE Web site.

Side Note: Blood Clotting

Many metabolic reactions involve a large number of sequential reactions, such as those that occur in the coagulation of blood.

$$\text{Cut} \rightarrow \text{Blood} \rightarrow \text{Clotting}$$

Blood coagulation (see Figure A) is part of an important host defense mechanism called *hemostasis*, which causes the cessation of blood loss from a damaged vessel. The clotting process is initiated when a nonenzymatic lipoprotein (called the *tissue factor*) contacts blood plasma because of cell damage. The tissue factor (TF) is normally not in contact with plasma (see Figure B) because of an intact endothelium. The rupture (e.g., cut) of the endothelium exposes the plasma to TF and a cascade* of series reactions proceeds (Figure C). These series reactions ultimately result in the conversion of fibrinogen (soluble) to fibrin (insoluble), which produces the clot. Later, as wound healing occurs, mechanisms that restrict formation of fibrin clots, necessary to maintain the fluidity of the blood, start working.

Ouch!

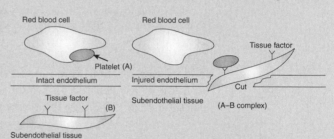

Figure A Normal clot coagulation of blood. (Dietrich Mebs, *Venomous and Poisonous Animals*, Stuttgart: Medpharm, (2002), p. 305 Reprinted by permission of the author.)

Figure B Schematic of separation of TF (A) and plasma (B) before cut occurs.

Figure C Cut allows contact of plasma to initiate coagulation. (A + B → Cascade)

*Platelets provide procoagulant, phospholipids-equivalent surfaces upon which the complex-dependent reactions of the blood coagulation cascade are localized.

Solved Problems

An abbreviated form (1) of the initiation and following cascade metabolic reactions that can capture the clotting process is

$$\text{TF} + \text{VIIa} \underset{k_{-1}}{\overset{k_1}{\rightleftharpoons}} \text{TF} - \text{VIIa(complex)} \xrightarrow[k_2]{+x} \text{Xa} \quad \Big| k_3 \qquad (1)$$

$$\text{Clot} \xleftarrow[k_5(fast)]{+\text{XIIIa}} \text{Fibrin} \xleftarrow[k_4]{+fibrinogen} \text{IIa} \xleftarrow{+\text{II}}$$

In order to maintain the fluidity of the blood, the clotting sequence (2) must be moderated. The reactions that attenuate the clotting process are

$$ATIII + Xa \xrightarrow{k_6} Xa_{inactive}$$

$$ATIII + IIa \xrightarrow{k_7} IIa_{inactive} \tag{2}$$

$$ATIII + TF - VIIa \xrightarrow{k_8} TF - VIIa_{inactive}$$

where TF = tissue factor, VIIa = factor novoseven, X = Stuart Prower factor, Xa = Stuart Prower factor activated, II = prothrombin, IIa = thrombin, ATIII = antithrombin, and XIIIa = factor XIIIa.

Symbolically, the clotting equations can be written as

$$\boxed{Cut \rightarrow A + B \rightarrow C \rightarrow D \rightarrow E \rightarrow F \rightarrow Clot}$$

One can model the clotting process in a manner identical to the series reactions by writing a mole balance and rate law for each species, such as

$$\frac{dC_{TF}}{dt} = -k_1 \cdot C_{TF} \cdot C_{VIIa} + k_{-1} \cdot C_{TF-VIIa}$$

$$\frac{dC_{VIIa}}{dt} = -k_1 \cdot C_{TF} \cdot C_{VIIa} + k_{-1} \cdot C_{TF-VIIa}$$

etc.

and then using Polymath to solve the coupled equations to predict the thrombin (shown in Figure D) and other species concentration as a function of time, as well as to determine the clotting time. Laboratory data are also shown below for a TF concentration of 5 p*M*. **One notes that when the complete set of equations is used, the Polymath output is identical to Figure E.** The complete set of equations, along with the *Polymath Living Example Problem* code, is given in the *Solved Problems* on the CRE Web site. You can load the program directly and vary some of the parameters.

The next time you have to go to the emergency room for a serious cut, discussing the clotting mechanism should impress the doctor as you are stitched up.

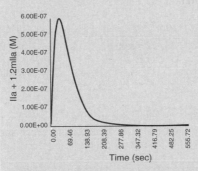

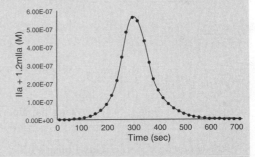

Figure D Total thrombin as a function of time with an initiating TF concentration of 25 p*M* (after running Polymath) for the abbreviated blood-clotting cascade.

Figure E Total thrombin as a function of time with an initiating TF concentration of 25 p*M*. Full blood clotting cascade. (Enzyme Catalysis and Regulation: Mathew F. Hockin, Kenneth C. Jones, Stephen J. Everse, and Kenneth G. Maan. A Model for the Stoichiometric Regulation of Blood Coagulation. J. Biol. Chem. 2002 277: 18322–18333. First published on March 13, 2002. Copyright © 2002, by the American Society for Biochemistry and Molecular Biology.)

8.5 Complex Reactions

A complex reaction system consists of a combination of interacting series and parallel reactions. Overall, this algorithm is very similar to the one given in Chapter 6 for writing the mole balances in terms of molar flow rates and concentrations (i.e., Figure 6-1). After numbering each reaction, we write a mole balance on each species, similar to those in Figure 6-1. The major difference between the two algorithms is in the rate-law step. As shown in Table 8-2, we have three steps (3, 4, and 5) to find the net rate of reaction for each species in terms of the concentration of the reacting species. As an example, we shall study the following complex reactions

$$A + 2B \longrightarrow C$$

$$2A + 3C \longrightarrow D$$

carried out in a PBR, a CSTR, and a semibatch reactor.

8.5.1 Complex Gas-Phase Reactions in a PBR

We now apply the algorithms in Tables 8-1 and 8-2 to a very important complex reaction carried out in a PBR. To protect the confidential nature of this reaction the chemicals have been given the names A, B, C, and D.

Example 8–5 Multiple Gas-Phase Reactions in a PBR

The following complex gas-phase reactions follow elementary rate laws

$$\begin{array}{llll} (1) & A + 2B \rightarrow C & -r'_{1A} = k_{1A}C_A C_B^2 \\ (2) & 2A + 3C \rightarrow D & -r'_{2C} = k_{2C}C_A^2 C_C^3 \end{array}$$

and take place isothermally in a PBR. The feed is equimolar in A and B with $F_{A0} = 10$ mol/min and the volumetric flow rate is 100 dm³/min. The catalyst weight is 1,000 kg, the pressure drop is $\alpha = 0.0019$ kg⁻¹, and the total entering concentration is $C_{T0} = 0.2$ mol/dm³.

$$k_{1A} = 100 \left(\frac{dm^9}{mol^2 \cdot kg\text{-cat} \cdot min} \right) \text{ and } k_{2C} = 1{,}500 \left(\frac{dm^{15}}{mol^4 \cdot kg\text{-cat} \cdot min} \right)$$

(a) Plot and analyze F_A, F_B, F_C, F_D, y, and $\tilde{S}_{C/D}$ as a function of catalyst weight, W.

Solution

Following the algorithm in Table 8-2 and having numbered our reactions, we now proceed to carry out a mole balance on each and every species in the reactor.

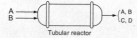

A →
B →
Tubular reactor
→ (A, B C, D)

Gas-Phase PBR

1. Mole Balances

$$(1) \quad \frac{dF_A}{dW} = r'_A \qquad \left(F_{A0} = 10 \frac{mol}{min} \right) \qquad W_f = 1{,}000 \text{ kg} \qquad (E8\text{-}5.1)$$

$$(2) \quad \frac{dF_B}{dW} = r'_B \qquad \left(F_{B0} = 10 \frac{mol}{min} \right) \qquad (E8\text{-}5.2)$$

$$(3) \quad \frac{dF_C}{dW} = r'_C \qquad (E8\text{-}5.3)$$

$$(4) \quad \frac{dF_D}{dW} = r'_D \qquad (E8\text{-}5.4)$$

2. Rates

Net Rates

$$(5) \quad r'_A = r'_{1A} + r'_{2A} \tag{E8-5.5}$$

$$(6) \quad r'_B = r'_{1B} \tag{E8-5.6}$$

$$(7) \quad r'_C = r'_{1C} + r'_{2C} \tag{E8-5.7}$$

$$(8) \quad r'_D = r'_{2D} \tag{E8-5.8}$$

Rate Laws

$$(9) \quad r'_{1A} = -k_{1A} C_A C_B^2 \tag{E8-5.9}$$

$$(10) \quad r'_{2C} = -k_{2C} C_A^2 C_C^3 \tag{E8-5.10}$$

Relative Rates

Following the Algorithm

Reaction 1: $A + 2B \rightarrow C$ $\dfrac{r'_{1A}}{-1} = \dfrac{r'_{1B}}{-2} = \dfrac{r'_{1C}}{1}$

$$(11) \quad r'_{1B} = 2\, r'_{1A} \tag{E8-5.11}$$

$$(12) \quad r'_{1C} = -r'_{1A} \tag{E8-5.12}$$

Reaction 2: $2A + 3C \rightarrow D$ $\dfrac{r'_{2A}}{-2} = \dfrac{r'_{2C}}{-3} = \dfrac{r'_{2D}}{1}$

$$(13) \quad r'_{2A} = \frac{2}{3} r'_{2C} \tag{E8-5.13}$$

$$(14) \quad r'_{2D} = -\frac{1}{3} r'_{2C} \tag{E8-5.14}$$

The net rates of reaction for species A, B, C, and D are

$$
\boxed{
\begin{aligned}
r'_A &= r'_{1A} + r'_{2A} = -k_{1A} C_A C_B^2 - \frac{2}{3} k_{2C} C_A^2 C_C^3 \\
r'_B &= r'_{1B} = -2 k_{1A} C_A C_B^2 \\
r'_C &= r'_{1C} + r'_{2C} = k_{1A} C_A C_B^2 - k_{2C} C_A^2 C_C^3 \\
r'_D &= r'_{2D} = \frac{1}{3} k_{2C} C_A^2 C_C^3
\end{aligned}
}
$$

Selectivity

$$\tilde{S}_{C/D} = \frac{F_C}{F_D}$$

At $W = 0$, $F_D = 0$ causing $S_{C/D}$ to go to infinity. Therefore, we set $S_{C/D} = 0$ between $W = 0$ and a very small number, $W = 0.0001$ kg to prevent the ODE solver, as well as other ODE solvers such as MATLAB and Excel, from crashing. In Polymath, this condition is written

$$(15) \quad \tilde{S}_{C/D} = \text{if } (W > 0.0001) \text{ then } \left(\frac{F_C}{F_D}\right) \text{ else } (0) \tag{E8-5.15}$$

3. Stoichiometry   Isothermal $T = T_0$

(16) $C_A = C_{T0}\left(\dfrac{F_A}{F_T}\right)p$ (E8-5.16)

(17) $C_B = C_{T0}\left(\dfrac{F_B}{F_T}\right)p$ (E8-5.17)

(18) $C_C = C_{T0}\left(\dfrac{F_C}{F_T}\right)p$ (E8-5.18)

(19) $C_D = C_{T0}\left(\dfrac{F_D}{F_T}\right)p$ (E8-5.19)

(20) $\dfrac{dp}{dW} = -\dfrac{\alpha}{2p}\left(\dfrac{F_T}{F_{T0}}\right)$ (E8-5.20)

(21) $F_T = F_A + F_B + F_C + F_D$ (E8-5.21)

4. Parameters

(22) $C_{T0} = 0.2 \ \text{mol/dm}^3$

(23) $\alpha = 0.0019 \ \text{kg}^{-1}$

(24) $v_0 = 100 \, \text{dm}^3/\text{min}$

(25) $k_{1A} = 100 \left(\text{dm}^3/\text{mol}\right)^2/\text{min/kg-cat}$

(26) $k_{2C} = 1{,}500\left(\text{dm}^{15}/\text{mol}^4\right)/\text{min/kg-cat}$

(27) $F_{T0} = 20 \ \text{mol/min}$

Entering the above equations into Polymath's ODE solver, we obtain the following results in Table E8-5.1 and Figures E8-5.1 and E8-5.2.

Analysis: This example is a very important one as it shows, step by step, how to handle isothermal complex reactions. For complex reactions occurring in any of the reactors shown in Figure 8-2, a numerical solution is virtually always required. The Polymath program and sample results are shown below. This problem is a Living Example Problem, _LEP_, so the reader can download the Polymath code and vary the reaction parameters (e.g., k_{1A}, α, k_{2C}, v_0) to learn how Figures E8-5.1 and E8-5.2 change.

In looking at the solution we note from Figure E8-5.2 that the selectivity reaches a maximum very close to the entrance (W ≈ 60 kg) and then drops rapidly. However, 90% of A is not consumed until 200 kg, the catalyst weight at which the desired product C reaches its maximum flow rate. If the activation energy for reaction (1) is greater than that for reaction (2), try increasing the temperature to increase the molar flow rate of C and selectivity. However, if that does not help, then one has to decide which is more important, selectivity or the molar flow rate of the desired product. In the former case, the PBR catalyst weight will be 60 kg. In the latter case, the PBR catalyst weight will be 200 kg.

TABLE E8-5.1   POLYMATH PROGRAM AND OUTPUT

Differential equations

1 d(Fa)/d(W) = ra
2 d(Fb)/d(W) = rb
3 d(Fc)/d(W) = rc
4 d(Fd)/d(W) = rd
5 d(p)/d(W) = -alpha/2/p*(Ft/Fto)

Explicit equations

1 Ft = Fa+Fb+Fc+Fd
2 k1a = 100
3 k2c = 1500
4 Cto = 0.2
5 Ca = Cto*(Fa/Ft)*p
6 Cb = Cto*(Fb/Ft)*p
7 Cc = Cto*(Fc/Ft)*p
8 r1a = -k1a*Ca*Cb^2
9 r1b = 2*r1a
10 rb = r1b
11 r2c = -k2c*Ca^2*Cc^3
12 r2a = 2/3*r2c
13 r2d = -1/3*r2c
14 r1c = -r1a
15 rd = r2d
16 ra = r1a+r2a
17 rc = r1c+r2c
18 v = 100
19 Cd = Cto*(Fd/Ft)*p
20 alpha = .0019
21 Fto = 20
22 Scd = if(W>0.0001)then(Fc/Fd)else(0)

PBR Results

POLYMATH Report
Ordinary Differential Equations

Calculated values of DEQ variables

	Variable	Initial value	Minimal value	Maximal value	Final value
1	alpha	0.0019	0.0019	0.0019	0.0019
2	Ca	0.1	0.0257858	0.1	0.0257858
3	Cb	0.1	0.0020471	0.1	0.0020471
4	Cc	0	0	0.0664046	0.0211051
5	Cd	0	0	0.0057647	0.0026336
6	Cto	0.2	0.2	0.2	0.2
7	Fa	10.	4.293413	10.	4.293413
8	Fb	10.	0.3408417	10.	0.3408417
9	Fc	0	0	4.038125	3.514068
10	Fd	0	0	0.4385037	0.4385037
11	Ft	20.	8.586827	20.	8.586827
12	Fto	20.	20.	20.	20.
13	k1a	100.	100.	100.	100.
14	k2c	1500.	1500.	1500.	1500.
15	r1a	-0.1	-0.1	-1.081E-05	-1.081E-05
16	r1b	-0.2	-0.2	-2.161E-05	-2.161E-05
17	r1c	0.1	1.081E-05	0.1	1.081E-05
18	r2a	0	-0.0022091	0	-6.251E-06
19	r2c	0	-0.0033136	0	-9.376E-06
20	r2d	0	0	0.0011045	3.125E-06
21	ra	-0.1	-0.1	-1.706E-05	-1.706E-05
22	rb	-0.2	-0.2	-2.161E-05	-2.161E-05
23	rc	0.1	-0.0015019	0.1	1.429E-06
24	rd	0	0	0.0011045	3.125E-06
25	Scd	0	0	7747.617	8.01377
26	v	100.	100.	100.	100.
27	W	0	0	1000.	1000.
28	p	1.	0.2578577	1.	0.2578577

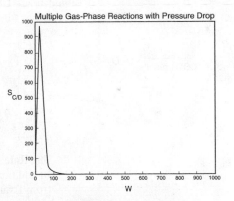

Figure E8-5.1 Molar flow rate profiles.   **Figure E8-5.2** Selectivity profile.

8.5.2 Complex Liquid-Phase Reactions in a CSTR

For a CSTR, a coupled set of algebraic equations analogous to the PFR differential equations must be solved. These equations are arrived at from a mole balance on CSTR for every species coupled with the rates step and stoichiometry. For q liquid-phase reactions occurring where N different species are present, we have the following set of algebraic equations:

$$F_{10} - F_1 = -r_1 V = V \sum_{i=1}^{q} -r_{i1} = V \cdot f_1(C_1 ..., C_N) \tag{8-17}$$

$$\vdots$$

$$F_{j0} - F_j = -r_j V = V \cdot f_j(C_1 ..., C_N) \tag{8-18}$$

$$\vdots$$

$$F_{N0} - F_N = -r_N V = V \cdot f_N(C_1 ..., C_N) \tag{8-19}$$

We can use a nonlinear algebraic equation solver (NLE) in Polymath or a similar program to solve Equations (8-17) through (8-19).

Example 8–6 Complex Reactions in a Liquid-Phase CSTR

The complex reactions discussed in Example 8-5 now take place in the *liquid phase* in a 2,500 dm^3 CSTR. The feed is equimolar in A and B with $F_{A0} = 200$ mol/min and the volumetric flow rate is 100 dm^3/min. The rate constants are

$$k_{1A} = 10 \left(\frac{dm^3}{mol} \right)^2 \Big/ min \quad \text{and} \quad k_{2C} = 15 \left(\frac{dm^3}{mol} \right)^4 \Big/ min$$

Find the concentrations of A, B, C, and D exiting the reactor, along with the exiting selectivity, $\tilde{S}_{C/D}$.

$$\tilde{S}_{C/D} = \frac{C_C}{C_D}$$

Solution

Liquid-Phase CSTR: $v = v_0$ (Polymath Formulation)

Mole Balances

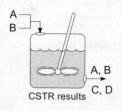

CSTR results

$$(1) \quad f(C_A) = v_0 C_{A0} - v_0 C_A + r_A V \tag{E8-6.1}$$

$$(2) \quad f(C_B) = v_0 C_{B0} - v_0 C_B + r_B V \tag{E8-6.2}$$

$$(3) \quad f(C_C) = \qquad - v_0 C_C + r_C V \tag{E8-6.3}$$

$$(4) \quad f(C_D) = \qquad - v_0 C_D + r_D V \tag{E8-6.4}$$

The **Rate Laws**, **Relative Rates**, and **Net Rates** are the same as in Example 8-5. In addition, Steps (5) through (14) (i.e., Equations (E8-5.5) through (E8-5.14) from Example 8-5) remain unchanged for this example. This fact is going to save us a lot of time in solving this example problem.

Selectivity

Note: We add a very small number (0.001 mol/min) to the term in the denominator to prevent $S_{C/D}$ going to infinity when $F_D = 0$.

$$(15) \quad \tilde{S}_{C/D} = \frac{F_C}{(F_D + 0.001)} \tag{E8-6.5}$$

Parameters

$$(16)\quad v_0 = 100 \ \text{dm}^3/\text{min} \qquad\qquad (19)\quad V = 2,500 \ \text{dm}^3$$

$$(17)\quad k_{1A} = 10 \ \left(\text{dm}^3/\text{mol}\right)^2/\text{min} \qquad (20)\quad C_{A0} = 2.0 \ \text{mol}/\text{dm}^3$$

$$(18)\quad k_{2C} = 15 \ \left(\text{dm}^3/\text{mol}\right)^4/\text{min} \qquad (21)\quad C_{B0} = 2.0 \ \text{mol}/\text{dm}^3$$

These equations are now used to solve for the exit concentrations using Polymath's nonlinear equation solver.

Liquid-Phase CSTR

TABLE E8-6.1   POLYMATH PROGRAM AND OUTPUT

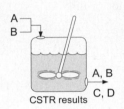

CSTR results

Nonlinear equations

1 f(Ca) = v*Cao-v*Ca+ra*V = 0

2 f(Cb) = v*Cbo-v*Cb+rb*V = 0

3 f(Cc) = -v*Cc+rc*V = 0

4 f(Cd) = -v*Cd+rd*V = 0

Explicit equations

1 k1a = 10

2 k2c = 15

3 r2c = -k2c*Ca^2*Cc^3

4 r1a = -k1a*Ca*Cb^2

5 r1b = 2*r1a

6 r2a = 2/3*r2c

7 r1c = -r1a

8 r2d = -1/3*r2c

9 rb = r1b

10 ra = r1a+r2a

11 Cao = 2

12 Cbo = 2

13 rc = r1c+r2c

14 rd = r2d

15 V = 2500

16 v = 100

17 Scd = Cc/(Cd+0.001)

POLYMATH Report
Nonlinear Equations

Calculated values of NLE variables

	Variable	Value	f(x)	Initial Guess
1	Ca	0.5326529	-3.979E-13	2.
2	Cb	0.0848008	8.527E-14	1.
3	Cc	0.1929784	-7.354E-13	1.
4	Cd	0.2548737	2.345E-13	0.1

— Results

	Variable	Value
1	Cao	2.
2	Cbo	2.
3	k1a	10.
4	k2c	15.
5	r1a	-0.038304
6	r1b	-0.076608
7	r1c	0.038304
8	r2a	-0.0203899
9	r2c	-0.0305848
10	r2d	0.0101949
11	ra	-0.0586939
12	rb	-0.076608
13	rc	0.0077191
14	rd	0.0101949
15	Scd	0.7541941	◄── Selectivity
16	v	100.
17	V	2500.

The exit concentrations are $C_A = 0.53$ M, $C_B = 0.085$ M, $C_C = 0.19$ M, and $C_D = 0.25$ M with $\tilde{S}_{C/D} = 0.75$. The corresponding conversion of A is

$$X = \frac{C_{A0} - C_A}{C_{A0}} = \frac{2 - 0.533}{2} = 0.73$$

Analysis: The CRE algorithm for a complex reaction carried out in a CSTR was solved using the nonlinear equation solver. The CSTR exit concentrations shown in the results table correspond to a selectivity $\tilde{S}_{C/D} = 0.75$, as shown in the Polymath report. While the CSTR conversion is reasonable, the selectivity is rather low. The PFR is a better choice for these reactions to maximize selectivity.

8.5.3 Complex Liquid-Phase Reactions in a Semibatch Reactor

Next up is the application of Tables 8-1 and 8-2 to a complex reaction carried out in a semibatch reactor.

Example 8–7   Complex Reactions in a Semibatch Reactor

The complex liquid-phase reactions discussed in Example 8-6 now take place in a semibatch reactor where A is fed to B with F_{A0} = 3 mol/min. The volumetric flow rate is 10 dm^3/min and the initial reactor volume is 1,000 dm^3. The rate constants are

$$A + 2B \rightarrow C$$
$$2A + 3C \rightarrow D$$

$$k_{1A} = 10 \left(\frac{dm^3}{mol}\right)^2 \bigg/ min \quad \text{and} \quad k_{2C} = 15 \left(\frac{dm^3}{mol}\right)^4 \bigg/ min$$

The maximum volume is 2,000 dm^3. The inlet concentration of A is C_{A0} = 0.3 mol/dm^3 and the initial concentration of B is C_{Bi} = 0.2 mol/dm^3.

(a) Plot and analyze N_A, N_B, N_C, N_D, and $S_{C/D}$ as a function of time.

Solution
Mole Balances

A —

B

Semibatch Reactor

$$(1) \qquad \frac{dN_A}{dt} = r_A V + F_{A0} \qquad \left(N_{Ai} = 0\right) \qquad\qquad (E8\text{-}7.1)$$

$$(2) \qquad \frac{dN_B}{dt} = r_B V \qquad \left(N_{Bi} = C_{Bi} V_0 = 200 \text{ moles}\right) \qquad (E8\text{-}7.2)$$

$$(3) \qquad \frac{dN_C}{dt} = r_C V \qquad \left(N_{Ci} = 0\right) \qquad\qquad (E8\text{-}7.3)$$

$$(4) \qquad \frac{dN_D}{dt} = r_D V \qquad \left(N_{Di} = 0\right) \qquad\qquad (E8\text{-}7.4)$$

Net Rates, Rate Laws, and Relative Rates are the same as liquid-phase CSTR.

Same as Step 2 in Example 8-5, i.e., Steps (5) through (14), Equations (E8-5.5) through (E8-5.14).

Stoichiometry

$$(15) \quad C_A = N_A/V \quad (E8\text{-}7.5) \qquad\qquad (18) \quad C_D = N_D/V \quad (E8\text{-}7.8)$$

$$(16) \quad C_B = N_B/V \quad (E8\text{-}7.6) \qquad\qquad (19) \quad V = V_0 + v_0 t \quad (E8\text{-}7.9)$$

$$(17) \quad C_C = N_C/V \quad (E8\text{-}7.7)$$

Selectivity
Tricks of the Trade. Because N_D is 0 at time t = 0, the selective goes to infinity so we again use an "if" statement.

$$(20) \qquad \tilde{S}_{C/D} = \text{if } (t > 0.0001 \text{ min}) \text{ then} \left(\frac{N_C}{N_D}\right) \text{else } (0) \qquad (E8\text{-}7.10)$$

Parameters

New Parameters

$$(21) \qquad v_0 = 10 \text{ dm}^3 / \text{min}$$

$$(22) \qquad V_0 = 1,000 \text{ dm}^3$$

$$(23) \qquad F_{A0} = 3 \text{ mol} / \text{min}$$

Putting this information into Polymath's ODE solver, we obtain the following results.

Liquid-Phase Multiple Reactions in a Semibatch Reactor

Semibatch Reactor

Differential equations

1 d(Nb)/d(t) = rb*V

2 d(Na)/d(t) = ra*V +Fao

3 d(Nd)/d(t) = rd*V

4 d(Nc)/d(t) = rc*V

Explicit equations

1 k1a = 10

2 k2c = 15

3 Vo = 1000

4 vo = 10

5 V = Vo+vo*t

6 Ca = Na/V

7 Cb = Nb/V

8 r1a = -k1a*Ca*Cb^2

9 Cc = Nc/V

10 r1b = 2*r1a

11 rb = r1b

12 r2c = -k2c*Ca^2*Cc^3

13 Fao = 3

14 r2a = 2/3*r2c

15 r2d = -1/3*r2c

16 r1c = -r1a

17 rd = r2d

18 ra = r1a+r2a

19 Cd = Nd/V

20 rc = r1c+r2c

21 Scd = if(t>0.0001)then(Nc/Nd)else(0)

POLYMATH Report
Ordinary Differential Equations

Calculated values of DEQ variables

	Variable	Initial value	Final value
1	Ca	0	0.1034461
2	Cb	0.2	0.0075985
3	Cc	0	0.0456711
4	Cd	0	0.0001766
5	Fao	3.	3.
6	k1a	10.	10.
7	k2c	15.	15.
8	Na	0	206.8923
9	Nb	200.	15.197
10	Nc	0	91.34215
11	Nd	0	0.3531159
18	ra	0	-6.992E-05
19	rb	0	-0.0001195
20	rc	0	4.444E-05
21	rd	0	5.097E-06
22	Scd	0	258.6747
23	t	0	100.
24	V	1000.	2000.
25	vo	10.	10.
26	Vo	1000.	1000.

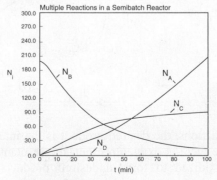

Figure E8-7.1 Number of moles as a function of time.

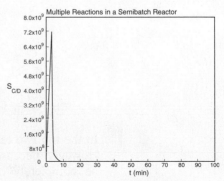

Figure E8-7.2 Selectivity as a function of time.

Analysis: The CRE algorithm for a complex reaction was applied to a semibatch reactor and solved using the ODE solver Polymath. The maximum in the selectivity occurs after only 6.5 minutes; however, very little of the desired product, C, has been formed at this time. If $E_D > E_U$, first try changing the temperature to see if that will improve the selectivity and the amount of product formed. If that does not work, an economical decision needs to be made. Are selectivity and the cost of separating C and D more important than making more C to sell?

One notes in Figure E8-7.1 that after 70 minutes, very little change in the number of moles of species A, C, and D takes place; however, the number of moles of B continue to increase because it is continually fed and there isn't sufficient A to react with B. The number of moles of D produced is so small that it is indistinguishable from the x-axis in Figure E8-7.1. Finally, we note that these times, 6.5 and 10 minutes, are **far too short to use a semibatch** reactor and, consequently, one should consider another reactor scheme such as Figure 8-2(g), where A is fed along the length of the reactor, or analogous to that shown in Figure 8-2(h), where A is fed to each of the CSTRs.

8.6 Membrane Reactors to Improve Selectivity in Multiple Reactions

In addition to using membrane reactors (MRs) to remove a reaction product in order to shift the equilibrium toward completion, we can use membrane reactors to increase selectivity in multiple reactions. This increase can be achieved by injecting one of the reactants along the length of the reactor. It is particularly effective in partial oxidation of hydrocarbons, as well as chlorination, ethoxylation, hydrogenation, nitration, and sulfonation reactions, to name a few.[2]

$$C_2H_4 + \frac{1}{2}O_2 \longrightarrow C_2H_4O \xrightarrow{+\frac{5}{2}O_2} 2CO_2 + 2H_2O$$

In the top two reactions, the desired product is the intermediate (e.g., C_2H_4O). However, because there is oxygen present, the reactants and intermediates can be completely oxidized to form undesired products, CO_2 and water. The desired product in the bottom reaction is xylene. We can enhance selectivity by keeping one of the reactants at a low concentration, which can be achieved by feeding it through the sides of a membrane reactor.

---

[2] W. J. Asher, D. C. Bomberger, and D. L. Huestis, *Evaluation of SRI's Novel Reactor Process Permix^{TM}* (New York: AIChE, 2000).

In the solved example problem in the *Expanded Material* on the CRE Web site, we have used a membrane reactor (MR) for the hydrodealkylation of mesitylene reaction. In some ways, this CRE Web site example parallels the use of MRs for partial oxidation reactions. We will now do an example for a different reaction to illustrate the advantages of an MR for certain types of reactions.

Example 8–8   Membrane Reactor (MR) to Improve Selectivity in Multiple Reactions

The reactions

$$(1)\ A + B \longrightarrow D \qquad -r_{1A} = k_{1A} C_A^2 C_B,\ \ k_{1A} = 2\ dm^6/mol^2 \cdot s$$

$$(2)\ A + B \longrightarrow U \qquad -r_{2A} = k_{2A} C_A C_B^2,\ \ k_{2A} = 3\ dm^6/mol^2 \cdot s$$

take place in the gas phase. The overall selectivities, $\tilde{S}_{D/U}$, are to be compared for a membrane reactor (MR) and a conventional PFR.[†] First, we use the instantaneous selectivity to determine which species should be fed through the membrane

$$S_{D/U} = \frac{k_1 C_A^2 C_B}{k_2 C_B^2 C_A} = \frac{k_1 C_A}{k_2 C_B}$$

We see that to maximize $S_{D/U}$, we need to keep the concentration of A high and the concentration of B low; therefore, we feed B through the membrane. The molar flow rate of A entering the reactor is 4 mol/s and that of B entering through the membrane is 4 mol/s, as shown in Figure E8-8.1. For the PFR, B enters along with A.

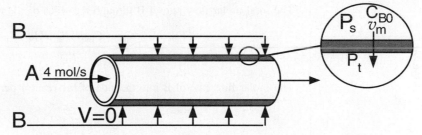

Figure E8-8.1   Membrane reactor with one reactant fed through the sides.

The reactor volume is 50 dm³ and the entering total concentration is 0.8 mol/dm³.

Plot and analyze the molar flow rates and the overall selectivity, $\tilde{S}_{D/U}$, as a function of reactor volume for both the MR and PFR.

Solution

1. Mole Balances for Both the PFR and the MR

	PFR		**MR**	
Species A:	$\dfrac{dF_A}{dV} = r_A$	(E8-8.1[a])	$\dfrac{dF_A}{dV} = r_A$	(E8-8.1[b])
Species B:	$\dfrac{dF_B}{dV} = r_B$	(E8-8.2[a])	$\dfrac{dF_B}{dV} = r_B + R_B$	(E8-8.2[b])

[†] Prof. Dr. Sven Köttlov in the chemical engineering department at Jofostan University in Riça, Jofostan, is one of the best membrane analysts in that area of the world.

Species C: $\dfrac{dF_D}{dV} = r_D$ (E8-8.3[a]) | $\dfrac{dF_D}{dV} = r_D$ (E8-8.3[b])

Species D: $\dfrac{dF_U}{dV} = r_U$ (E8-8.4[a]) | $\dfrac{dF_U}{dV} = r_U$ (E8-8.4[b])

2. Net Rates and Rate Laws (same for PFR and MR)

$$r_A = r_{1A} + r_{2A} = -k_{1A}C_A^2 C_B - k_{2A}C_A C_B^2 \tag{E8-8.5}$$

$$r_B = r_{1B} + r_{2B} = -k_{1A}C_A^2 C_B - k_{2A}C_A C_B^2 \tag{E8-8.6}$$

$$r_D = r_{1D} = k_{1A}C_A^2 C_B \tag{E8-8.7}$$

$$r_U = r_{2U} = k_{2A}C_A C_B^2 \tag{E8-8.8}$$

3. Transport Law (MR)

The volumetric flow rate through the membrane is given by Darcy's Law.[1]

$$v_m = K[P_s - P_t]A_t \tag{E8-8.9}$$

where K is the membrane permeability (m/s · kPa), P_s (kPa) and P_t (kPa) are the shell-side and tube-side pressures, and A_t is the membrane surface area (m²). The flow rate through the membrane can be controlled by adjusting the pressure drop across the membrane ($P_s - P_t$). Recall from Equation (6-5) that "a" is the membrane surface area per unit volume of reactor

$$A_t = aV_t \tag{E8-8.10}$$

The total molar flow rate of B through the sides of the reactor is

$$F_{B0} = C_{B0}v_m = \underbrace{\dfrac{C_{B0}K[P_s - P_t]a}{R_B}} \cdot V_t = R_B V_t \tag{E8-8.11}$$

The molar flow rate of B into the side of the reactor per unit volume of reactor is

$$R_B = \dfrac{F_{B0}}{V_t} \tag{E8-8.12}$$

4. Stoichiometry (same for PFR and MR)

Isothermal ($T = T_0$) and neglect pressure drop down the length of the reactor ($P = P_0$, $p = 1.0$).

For no pressure drop down the length of the reactor and isothermal operation, the concentrations for both the PFR and MR are

Here, $T = T_0$ and $\Delta P = 0$.

$$C_A = C_{T0}\dfrac{F_A}{F_T} \quad \text{(E8-8.13)} \qquad\qquad C_D = C_{T0}\dfrac{F_D}{F_T} \quad \text{(E8-8.15)}$$

$$C_B = C_{T0}\dfrac{F_B}{F_T} \quad \text{(E8-8.14)} \qquad\qquad C_U = C_{T0}\dfrac{F_U}{F_T} \quad \text{(E8-8.16)}$$

---

[1] Wilkes, J. O., *Fluid Mechanics for Chemical Engineers with Microfluidics and CFD,* 2nd Ed. (Upper Saddle River, NJ: Prentice Hall, 2006).

5. Combine

The Polymath program will combine the mole balance, net rates, and stoichiometric equations to solve for the molar flow rate and selectivity profiles for both the conventional PFR and the MR, and also the selectivity profile.

A note of caution on calculating the overall selectivity

$$\tilde{S}_{D/U} = \frac{F_D}{F_U} \qquad \text{(E8-8.17)}$$

Fool Polymath!

We have to fool Polymath because at the entrance of the reactor $F_U = 0$, Polymath will look at Equation (E8-8.17) and will not run because it will say you are dividing by zero. Therefore, we need to add a very small number to the denominator, say 0.0001 mol/s (or smaller); that is

$$\tilde{S}_{D/U} = \frac{F_D}{F_U + 0.0001} \qquad \text{(E8-8.18)}$$

Sketch the trends or results you expect before working out the details of the problem.

Table E8-8.1 shows the Polymath program and report sheet.

TABLE E8-8.1   POLYMATH PROGRAM

Differential equations		Calculated values of DEQ variables		
1 d(Fa)/d(V) = ra			Variable	Initial value	Final value

Differential equations

1 d(Fa)/d(V) = ra
2 d(Fb)/d(V) = rb+Rb
3 d(Fd)/d(V) = rd
4 d(Fu)/d(V) = ru

Explicit equations

1 Ft = Fa+Fb+Fd+Fu
2 Ct0 = 0.8
3 k1a = 2
4 k2a = 3
5 Cb = Ct0*Fb/Ft
6 Ca = Ct0*Fa/Ft
7 ra = -k1a*Ca^2*Cb-k2a*Ca*Cb^2
8 rb = ra
9 Cd = Ct0*Fd/Ft
10 Cu = Ct0*Fu/Ft
11 rd = k1a*Ca^2*Cb
12 ru = k2a*Ca*Cb^2
13 Vt = 50
14 Fbo = 4
15 Rb = Fbo/Vt
16 Sdu = Fd/(Fu+.0000000000001)

	Variable	Initial value	Final value
1	Ca	0.8	0.2020242
2	Cb	0	0.2020242
3	Cd	0	0.2855303
4	Ct0	0.8	0.8
5	Cu	0	0.1104213
6	Fa	4.	1.351387
7	Fb	0	1.351387
8	Fbo	4.	4.
9	Fd	0	1.909979
10	Ft	4.	5.351387
11	Fu	0	0.7386336
12	k1a	2.	2.
13	k2a	3.	3.
14	ra	0	-0.0412269
15	rb	0	-0.0412269
16	Rb	0.08	0.08
17	rd	0	0.0164908
18	ru	0	0.0247361
19	Sdu	0	2.585827
20	V	0	50.
21	Vt	50.	50.

Living Example Problem

We can easily modify the program, shown in Table E8-8.1, for the PFR simply by setting R_B equal to zero ($R_B = 0$) and the initial condition for B to be 4.0.

Figures E8-8.2(a) and E8-8.2(b) show the molar flow rate profiles for the conventional PFR and MR, respectively.

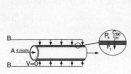

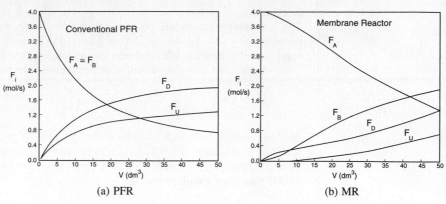

Figure E8-8.2 Molar flow rates.

Figures E8-8.3(a) and E8-8.3(b) show the selectivity for the PFR and MR. One notices the enormous enhancement in selectivity the MR has over the PFR.

Selectivities at
V = 5 dm³
MR: $S_{D/U} = 14$
PFR: $S_{D/U} = 0.65$

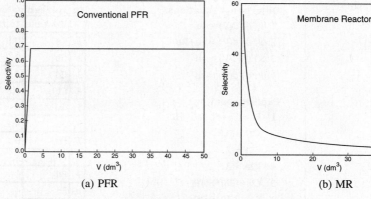

Figure E8-8.3 Selectivity.

Be sure to download this Living Example Problem (**LEP**) from the CRE Web site and play with the reactions and reactors. With minor modifications, you can explore reactions analogous to partial oxidations

$$A + B \longrightarrow D \quad r_D = k_1 C_A C_B \qquad (E8\text{-}8.19)$$

$$B + D \longrightarrow U \quad r_U = k_2 C_B C_D \qquad (E8\text{-}8.20)$$

where oxygen (B) is fed through the side of the membrane (see Problem P8-15$_C$).

Analysis: One notes that A is consumed more rapidly in the PFR than in the side-fed MR and that more of the desired product is also formed in the PFR. However, the selectivity is much higher in the MR than the PFR. One also notes, when comparing the molar flow rates, that the rates in the side-fed MR continue to change significantly after 30 dm³ down the reactor, while those in the PFR do not change significantly after 30 dm³.

8.7 Sorting It All Out

In Example 8-5 we were given the rate laws and asked to calculate the product distribution profiles. The inverse of the problem described in Example 8-5 must frequently be solved. Specifically, the rate laws often must be determined from the variation in the product distribution profiles generated by changing the feed concentrations. In some instances, this determination may not be possible without carrying out independent experiments on some of the reactions in the sequence. The best strategy to use to *sort out* all of the rate-law parameters will vary from reaction sequence to reaction sequence. Consequently, the strategy developed for one system may not be the best approach for other multiple-reaction systems. One general rule is to start an analysis by looking for species produced in only one reaction; next, study the species involved in only two reactions, then three, and so on.

When some of the intermediate products are free radicals, it may not be possible to perform independent experiments to determine the rate-law parameters. Consequently, we must deduce the rate law parameters from changes in the distribution of reaction products with feed conditions. Under these circumstances, the analysis turns into an optimization problem to estimate the best values of the parameters that will minimize the sums of the squares between the calculated variables and measured variables. This process is basically the same as that described in Section 7.5, but more complex, owing to the larger number of parameters to be determined. We begin by estimating the parameter values using some of the methods just discussed. Next, we use these estimates in nonlinear regression techniques to determine the best estimates of our parameter values from the data for all of the experiments.[3] Software packages are available for an analysis such as this one.

Nonlinear least squares

8.8 The Fun Part

I'm not talking about the fun you can have at an amusement park, but CRE fun. Now that we have an understanding on how to solve for the exit concentrations of multiple reactions in a CSTR and how to plot the species concentrations down the length of a PFR or PBR, we can address one of the most important and fun areas of chemical reaction engineering. This area, discussed in Section 8.3, is learning how to maximize the desired product and minimize the undesired product. It is this area that can make or break a chemical process financially. It is also an area that requires creativity in designing the reactor schemes and feed conditions that will maximize profits. Here, you can mix and match reactors, feed streams, and side streams, as well as vary the ratios of feed concentration in order to maximize or minimize the selectivity of a particular species. Problems of this type are what I call *digital-age problems* because we normally need to use ODE solvers, along with critical and creative thinking skills, to find the best answer.[4] A number of problems at the end of this chapter will allow you to practice these critical and creative thinking skills. These

---

[3] See, for example, Y. Bard, *Nonlinear Parameter Estimation* (San Diego, CA: Academic Press, 1974).

[4] H. Scott Fogler, *Teaching Critical Thinking, Creative Thinking, and Problem Solving in the Digital Age*, Phillips Lecture (Stillwater, OK: OSU Press, 1997).

problems offer the opportunity to explore many different solution alternatives to enhance selectivity and have fun doing it. The chemical reaction engineering students at Jofostan University in Riça, Jofostan, have often commented on how much they enjoy the LEPs and varying the parameters as they explore the problems and maximize the selective.

However, to carry CRE to the next level and to have a lot more fun solving multiple-reaction problems, we will have to be patient a little longer. The reason is that in this chapter we consider only isothermal multiple reactions, and it is nonisothermal multiple reactions where things really get interesting. Consequently, we will have to wait to carry out schemes to maximize the desired product in nonisothermal multiple reactions until we study heat effects in Chapters 11, 12, and 13. After studying these chapters, we will add a new dimension to multiple reactions, as we now have another variable, temperature, that we may or may not be able to use to affect selectivity and yield. In one particularly interesting problem (**P12-27**$_C$), we will study the production of styrene from ethylbenzene in which two side reactions, one endothermic and one exothermic, must be taken into account. Here, we may vary a whole slew of variables, such as entering temperature, diluent rate, and observed optima, in the production of styrene. However, we will have to delay gratification of the styrene study until we have mastered Chapters 11 and 12.

Multiple reactions with heat effects is unique to this book

> **Closure.** After completing this chapter the reader should be able to describe the different types of multiple reactions (series, parallel, independent, and complex) and to select a reaction system that maximizes the selectivity. The reader should be able to write down and use the algorithm for solving CRE problems with multiple reactions. The reader should also be able to point out the major differences in the CRE algorithm for the multiple reactions from that for the single reactions, and then discuss why care must be taken when writing the rate law and stoichiometric steps to account for the rate laws for each reaction, the relative rates, and the net rates of reaction.
>
> Finally, readers should feel a sense of accomplishment by knowing they have now reached a level where they can solve realistic CRE problems with complex kinetics.

SUMMARY

1. For the competing reactions

Reaction 1: $A + B \xrightarrow{k_D} D$ $r_D = A_D e^{-E_D/RT} C_A^{\alpha_1} C_B^{\beta_1}$ (S8-1)

Reaction 2: $A + B \xrightarrow{k_U} U$ $r_U = A_U e^{-E_U/RT} C_A^{\alpha_2} C_B^{\beta_2}$ (S8-2)

the instantaneous selectivity parameter is defined as

$$S_{D/U} = \frac{r_D}{r_U} = \frac{A_D}{A_U} \exp\left(-\frac{(E_D - E_U)}{RT}\right) C_A^{\alpha_1 - \alpha_2} C_B^{\beta_1 - \beta_2}$$ (S8-3)

a. If $E_D > E_U$, the selectivity parameter $S_{D/U}$ will increase with increasing temperature.

b. If $\alpha_1 > \alpha_2$ and $\beta_2 > \beta_1$, the reaction should be carried out at high concentrations of A and low concentrations of B to maintain the selectivity parameter $S_{D/U}$ at a high value. Use a semibatch reactor with pure A initially or a tubular reactor in which B is fed at different locations down the reactor. Other cases discussed in the text are $(\alpha_2 > \alpha_1,\ \beta_1 > \beta_2)$, $(\alpha_2 > \alpha_1,\ \beta_2 > \beta_1)$, and $(\alpha_1 > \alpha_2,\ \beta_1 > \beta_2)$.

The *overall selectivity*, based on molar flow rates leaving the reactor, for the reactions given by Equations (S8-1) and (S8-2) is

$$\tilde{S}_{D/U} = \frac{F_D}{F_U} \tag{S8-4}$$

2. The *overall yield* is the ratio of the number of moles of a product at the end of a reaction to the number of moles of the key reactant that have been consumed

$$\tilde{Y}_D = \frac{F_D}{F_{A0} - F_A} \tag{S8-5}$$

3. The algorithm for multiple reactions is shown in Table 8S-1. As noted earlier in this chapter, equations for the **Rates Step** are the major change in our CRE algorithm.

TABLE 8S-1   ALGORITHM FOR MULTIPLE REACTIONS

Number all reactions (1), (2), etc.

Mole balances:

Mole balance on each and every species

PFR	$\dfrac{dF_j}{dV} = r_j$	(S8-6)
CSTR	$F_{j0} - F_j = -r_j V$	(S8-7)
Batch	$\dfrac{dN_j}{dt} = r_j V$	(S8-8)
Membrane ("i" diffuses in)	$\dfrac{dF_i}{dV} = r_i + R_i$	(S8-9)
Liquid-semibatch	$\dfrac{dC_j}{dt} = r_j + \dfrac{v_0(C_{j0} - C_j)}{V}$	(S8-10)

Rates:

Laws	$r_{ij} = k_{ij} f_i(C_j, C_n)$	(S8-11)
Relative rates	$\dfrac{r_{iA}}{-a_i} = \dfrac{r_{iB}}{-b_i} = \dfrac{r_{iC}}{c_i} = \dfrac{r_{iD}}{d_i}$	(S8-12)
Net rates	$r_j = \displaystyle\sum_{i=1}^{q} r_{ij}$	(S8-13)

TABLE 8S-1   ALGORITHM FOR MULTIPLE REACTIONS (CONTINUED)

Stoichiometry:

Gas phase

$$C_j = C_{T0} \frac{F_j}{F_T} \frac{P}{P_0} \frac{T_0}{T} = C_{T0} \frac{F_j}{F_T} \frac{T_0}{T} p \qquad \text{(S8-14)}$$

$$p = \frac{P}{P_0}$$

$$F_T = \sum_{j=1}^{n} F_j \qquad \text{(S8-15)}$$

$$\frac{dp}{dW} = -\frac{\alpha}{2p} \left(\frac{F_T}{F_{T0}} \right) \frac{T}{T_0} \qquad \text{(S8-16)}$$

Liquid phase $v = v_0$

$$C_A, C_B, \ldots$$

Combine:
Polymath will combine all the equations for you. Thank you, Polymath!!

CRE WEB SITE MATERIALS

- **Expanded Material**

    1. *Hydrodealkylation of Mesitylene in a PFR*
    2. *Hydrodealkylation of Mesitylene in a CSTR*
    3. *Calculating Concentration as a Function of Position for NH$_3$ Oxidation in a PFR*
    4. *Puzzle Problem "What's Wrong with this Solution?"*

- **Learning Resources**

    1. *Summary Notes*
    2. *Web Modules*
       A. Cobra Bites   Web Modules   B. AspenTech

Summary Notes

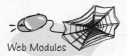

3. *Interactive Computer Games (ICG)*
   The Great Race

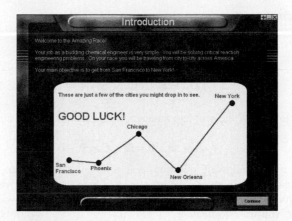

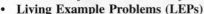

4. *Reactor Lab. See Learning Resources at the end of Chapter 5 for a description of these interactive computer exercises.*
5. *Solved Problems*

   A. Blood Coagulation; also *Living Example Problems* **LEP**
   B. Hydrodealkylation of Mesitylene in a PFR, CSTR, and Membrane Reactor; also *Living Example Problems (**LEPs**)*
   C. All You Wanted to Know About Making Malic Anhydride and More; also *Living Example Problems*
   D. Oxidation of Ammonia in a PFR; also *Living Example Problems (**LEP**)*

6. *Clarification: PFR with feed streams along the length of the reactor.*

- **Living Example Problems (LEPs)**
  The LEPs are given using POLYMATH, MATLAB, and Wolfram
   1. *Example 8-1 Trambouzi Reaction: Taking the output from the CSTR and feeding it to a PFR to increase conversion (decrease selectivity).*
   2. *Example 8-5 Complex Gas-Phase Reactions in a PFR*
   3. *Example 8-6 Complex Liquid-Phase Reactions in a CSTR*
   4. *Example 8-7 Complex Liquid-Phase Reactions in a Semibatch Reactor*
   5. *Example 8-8 Membrane Reactor to Improve Selectivity in Multiple Reactions*
   6. *Example Web/Web Modules: Calculating Concentrations as a Function of Position for NH_3 Oxidation in a PFR (See Chapter 8 Solved Problems on the CRE Web site for problem statement.)*
   7. *Example Web/Web Modules: Cobra Bite Problem*
   8. *Example Web/Web Modules: Solved Problems: Blood Coagulation*
   9. *Example Web/Web Modules: Oscillating Reactions*
   10. *AspenTech Example: Pyrolysis of Benzene*
- **FAQ (Frequently Asked Questions) — In Updates/FAQ icon section**
- **Professional Reference Shelf**
  R8.1 *Attainable Region Analysis (ARA)*

   The ARA allows one to find the optimum reaction system for certain types of rate laws. The example used is one of modified **van de Vusse kinetics**

$$A \underset{}{\overset{}{\rightleftharpoons}} B \rightarrow C$$
$$2A \rightarrow D$$

to find the optimum with respect to B using a combination of PFRs and CSTRs.

Links

R8.2   *Oxidation of Ammonia*
The coupled reactions for the oxidation of ammonia are modeled using a PFR.

QUESTIONS AND PROBLEMS

The subscript to each of the problem numbers indicates the level of difficulty: A, least difficult; D, most difficult.

$$A = \bullet \quad B = \blacksquare \quad C = \blacklozenge \quad D = \blacklozenge\blacklozenge$$

In each of the following questions and problems, rather than just drawing a box around your answer, write a sentence or two describing how you solved the problem, the assumptions you made, the reasonableness of your answer, what you learned, and any other facts that you want to include.

Homework Problems

Questions

Q8-1$_A$ **(a)**   Reactant A has been unsuccessful in courting/dating reactant B because of a completing reaction. The matchmaker advises that the only way A will succeed is to raise the temperature. Is this a <u>surefire</u> idea? Will that work?
   (b)   Make up and solve an original problem to illustrate the principles of this chapter. See Problem P5-1$_A$ for guidelines.
   (c)   Write a question based on the material in this chapter that requires critical thinking. Explain why your question requires critical thinking. (*Hint:* See Preface Section I.)
   (d)   Are the overall and instantaneous selectivities identical for a CSTR, i.e., $S_{D/U} \equiv \tilde{S}_{D/U}$? Also, are the instantaneous and overall yields for a CSTR equal, i.e., $Y_D \equiv \tilde{Y}_D$ for a CSTR?

Q8-2$_C$ Read the cobra bite **Web Module** on the **CRE Web site.**
   (a)   Determine how many cobra bites are necessary in order that no amount of antivenom will save the victim.
   (b)   Suppose the victim was bitten by a harmless snake and not bitten by a cobra and anti-venom was injected. How much antivenom would need to be injected to cause death?

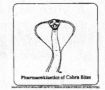
Pharmacokinetics of Cobra Bites

   (c)   What is the amount and latest possible time that anti-venom can be injected after a bite, such that the victim would not die?
   (d)   Apply one or more of the six ideas in Preface Table P-4, page xxviii, to this problem. (*Hint:* The *Living Example* Polymath program is on the CRE Web site.)

Problems

P8-1$_A$ **(a)**   **Example 8-1.** (1) What are Y_B, $\tilde{Y}_B$, C_A, C_X, and C_Y at C_A^*? (2) What would have been the selectivity $S_{B/XY}$ and conversion, X, if the reaction had been carried out in a single PFR with the same volume as the CSTR? (3) How would your answers change if the pressure were increased by a factor of 100?
   (b)   **Example 8-2.** Make a table/list for each reactor shown in Figure 8-2, identifying all the types of reactions that would be best carried out in this reactor. For example, Figure 8-2(d) Semibatch: used for (1) highly exothermic reactions and (2) increased selectivity.
   (c)   **Example 8-3.** How would t_{opt} change if $k_1 = k_2 = 0.25/h$ at 300 K?
   (d)   **Example 8-4.** (1) What are $S_{B/C}$ and Y_B? (2) What CSTR operating temperature (with $\tau = 0.5s$) would you recommend to maximize B for $C_{A0} = 5$ mol/dm^3, $k_1 = 0.4$ s^{-1} and $k_2 = 0.01$ s^{-1}, with $E_1 = 10$ kcal/mol and $E_2 = 20$ kcal/mol? (*Hint:* Plot C_B versus T. Use Wolfram if available on your computer.)

(e) **Example 8-5.** Download the *Living Example Problem (LEP)* from the CRE Web site. Explore the problem. (1) Vary the ratio of entering flow rates of A to B to learn the effect on selectivity. (2) Do the same for volumetric flow rate. (3) How would your answers change if the first reaction were reversible $A + 2B \rightleftharpoons C$ with the equilibrium constant $K_C = 0.002 \, (dm^3/mol)^2$?

(f) **Example 8-6.** Download the *Living Example Problem (LEP)* from the CRE Web site. Explore the problem and describe what you find. (*Hint:* Repeat **(e)**.)

(g) **Example 8-7.** Download the *Living Example Problem (LEP)* from the CRE Web site. Vary the flow rate to learn its effect on selectivity. Feed A to B to learn how the selectivity varies.

(h) **Example 8-8.** Download the *Living Example Problem (LEP)* from the CRE Web site. (1) How would your answers change if $F_{B0} = 2F_{A0}$? (2) If reaction (1) were $A + 2B \rightarrow D$ with the rate law remaining the same? Describe what you find.

(i) **AspenTech Benzene Pyrolysis Example.** (1) Change the activation energies to $E_1 = 28$ kcal/mol and $E_2 = 32$ kcal/mol, run the AspenTech program, and describe what you find. Compare with original data. (2) Repeat (1) by changing $E_1 = 32$ kcal/mol and $E_2 = 28$ kcal/mol, and describe what you find. (3) Double the reactor volume and compare the molar flow rate profiles. Describe what you find.

(j) **Web Example. PFR Mesitylene Reaction.** Download the *Living Example Problem (LEP)* from the CRE Web site. (1) How would your answers change if the feed were equal molar in hydrogen and mesitylene? (2) What is the effect of Θ_H on τ_{opt}? On $\tilde{S}_{X/T}$?

(k) **Web Example. CSTR Mesitylene Reaction.** Same question as P8-2(j).

(l) **Web Example. Oxidation of Ammonia.** Consider the following set of reactions:

Rate Laws Determined from Totusimetry Data (11/2/2019)

Reaction 1: $4NH_3 + 6NO \longrightarrow 5N_2 + 6H_2O \quad -r_{1NO} = k_{1NO}C_{NH_3}C_{NO}^{1.5}$

Reaction 2: $2NO \longrightarrow N_2 + O_2 \qquad\qquad r_{2N_2} = k_{2N_2}C_{NO}^2$

Reaction 3: $N_2 + 2O_2 \longrightarrow 2NO_2 \qquad\quad -r_{3O_2} = k_{3O_2}C_{N_2}C_{O_2}^2$

Use **Wolfram** to investigate the set of reactions in a PFR. Describe what you find.

(m) **Read Solved Blood-Coagulation Problem.** Download the *Living Example Problem.* (1) Plot out some of the other concentrations, such as TF-VIIa and TF-VIIaX. (2) Why do the curves look the way they do? What reaction in the cascade is most likely to be inhibited causing one to bleed to death? (3) What reactions, if eliminated, could cause one to die of a blood clot? (*Hint:* Look at ATIII and/or TFPI.)

(n) **Web Module Living Example: Oscillating Reactions.** Use the *Living Example Polymath Program* for oscillating reactions on the CRE Web site. For the (IO⁻) and (I) reactions set $k_1 = 0.0001/\text{min}^{-1}$ and for reaction (1) $C_{P0} = 0.01 \, \text{mol}/dm^3$. (1) What did you find? Look at the linearized stability analysis on the CRE Web site. (2) What factors affect the frequency and onset of the oscillations? (3) Explore and write a paragraph describing what you find. (4) Download the *Living Example Polymath Program* for the BZ reaction. Vary the parameters and write a paragraph describing what you find.

P8-2$_A$ Download the *Interactive Computer Game (ICG) The Great Race* from the CRE Web site. Play the game and then record your performance number for the module, which indicates your mastery of the material. Your professor has the key to decode your performance number.

Performance # _____.

P8-3$_B$ The following reactions

$$A \underset{}{\overset{k_1}{\rightleftharpoons}} D \qquad\qquad -r_{1A} = k_1[C_A - C_D/K_{1A}]$$

$$A \underset{}{\overset{k_2}{\rightleftharpoons}} U \qquad\qquad -r_{2A} = k_2[C_A - C_U/K_{2A}]$$

take place in a batch reactor.

Additional information:
$k_1 = 1.0$ min^{-1}, $K_{1A} = 10$
$k_2 = 100$ min^{-1}, $K_{2A} = 1.5$
$C_{A0} = 1$ mol/dm^3
(Adapted from a problem by John Falconer, University of Colorado.)

(a) Plot and analyze conversion and the concentrations of A, D, and U as a function of time. When would you stop the reaction to maximize the concentration of D? Describe what you find.

(b) When does the maximum concentration of U occur?

(c) What are the equilibrium concentrations of A, D, and U?

(d) What would be the exit concentrations from a CSTR with a space time of 1.0 min? Of 10.0 min? Of 100 min?

P8-4$_A$ Consider the following system of gas-phase reactions:

$$A \longrightarrow X \quad r_X = k_1 C_A^{1/2} \quad k_1 = 0.004 (\text{mol/dm}^3)^{1/2} \cdot \text{min}^{-1}$$

$$A \longrightarrow B \quad r_B = k_2 C_A \quad k_2 = 0.3 \text{ min}^{-1}$$

$$A \longrightarrow Y \quad r_Y = k_3 C_A^2 \quad k_3 = 0.25 \text{ dm}^3/\text{mol} \cdot \text{min}$$

B is the desired product, and X and Y are foul pollutants that are expensive to get rid of. The specific reaction rates are at 27°C. The reaction system is to be operated at 27°C and 4 atm. Pure A enters the system at a volumetric flow rate of 10 dm^3/min.

(a) Sketch the instantaneous selectivities ($S_{B/X}, S_{B/Y}$, and $S_{B/XY} = r_B/(r_X+r_Y)$) as a function of the concentration of C_A.

(b) Consider a series of reactors. What should be the volume of the first reactor?

(c) What are the effluent concentrations of A, B, X, and Y from the first reactor?

(d) What is the conversion of A in the first reactor?

(e) If 99% conversion of A is desired, what reaction scheme and reactor sizes should you use to maximize $S_{B/XY}$?

(f) Suppose that $E_1 = 20{,}000$ cal/mol, $E_2 = 10{,}000$ cal/mol, and $E_3 = 30{,}000$ cal/mol. What temperature would you recommend for a single CSTR with a space time of 10 min and an entering concentration of A of 0.1 mol/dm^3?

(g) If you could vary the pressure between 1 and 100 atm, what pressure would you choose?

P8-5$_B$ Pharmacokinetics concerns the ingestion, distribution, reaction, and elimination reaction of drugs in the body. Consider the application of pharmacokinetics to one of the major problems we have in the United States, drinking and driving. Here, we shall model how long one must wait to drive after having a tall martini. In most states, the legal intoxication limit is 0.8 g of ethanol per liter of body fluid. (In Sweden it is 0.5 g/L, and in Eastern Europe and Russia it is any value above 0.0 g/L.)

Hall of Fame

The ingestion of ethanol into the bloodstream and its subsequent elimination can be modeled as a series reaction. The rate of absorption from the gastrointestinal tract into the bloodstream and body is a first-order reaction with a specific reaction-rate constant of 10 h^{-1}. The rate at which ethanol is broken down in the bloodstream is limited by regeneration of a coenzyme. Consequently, the process may be modeled as a zero-order reaction with a specific reaction rate of 0.192 g/h·L of body fluid.

Suppose you immediately drank two tall martinis after arriving at a party. How long would you have to wait before your blood-alcohol concentration is below the legal limit in order to drive (a) in the United States, (b) in Sweden, and (c) in Russia? How would your answer change if (d) the drinks were taken $\frac{1}{2}$ hour apart? (e) and if the two drinks were consumed at a uniform rate during the first hour?

(f) Suppose that one went to a party, had two tall martinis right away, and then received a phone call saying an emergency had come up and the person needed to drive home immediately. How many minutes would the individual have to reach home before

he/she became legally intoxicated, assuming that the person had nothing further to drink? (**g**) How would your answers be different for a thin person? A heavy person? (*Hint:* Base all ethanol concentrations on the volume of body fluid. Plot the concentration of ethanol in the blood as a function of time.) What generalizations can you make? (**h**) What is the major unspoken point of this problem?

Additional information:

Ethanol in a tall martini: 40 g

Volume of body fluid: 40 L   (**SADD-MADD problem**)

(See Chapter 9 PRS R9-7 for a more in-depth look at alcohol metabolism.)

P8-6$_B$ (*Pharmacokinetics*) Tarzlon is a liquid antibiotic that is taken orally to treat infections of the spleen. It is effective only if it can maintain a concentration in the bloodstream (based on volume of body fluid) above 0.4 mg per dm^3 of body fluid. Ideally, a concentration of 1.0 mg/dm^3 in the blood should be realized. However, if the concentration in the blood exceeds 1.5 mg/dm^3, harmful side effects can occur. Once the Tarzlon reaches the stomach, it can proceed in two pathways, both of which are first order: (1) It can be absorbed into the bloodstream through the stomach walls; (2) it can pass out through the gastrointestinal tract and not be absorbed into the blood. Both these processes are first order in Tarzlon's concentration in the stomach. Once in the bloodstream, Tarzlon attacks bacterial cells and is subsequently degraded by a zero-order process. Tarzlon can also be removed from the blood and excreted in urine through a first-order process within the kidneys. In the stomach:

Hall of Fame

Absorption into blood $k_1 = 0.15$ h^{-1}

Elimination through gastrointestinal $k_2 = 0.6$ h^{-1}

In the bloodstream:

Degradation of Tarzlon $k_3 = 0.1$ mg/dm$^3 \cdot$h

Elimination through urine $k_4 = 0.2$ h^{-1}

One dose of Tarzlon is 250 mg in liquid form: Volume of body fluid = 40 dm^3.

(**a**) Plot and analyze the concentration of Tarzlon in the blood as a function of time when 1 dose (i.e., one liquid capsule) of Tarzlon is taken.

(**b**) How should the Tarzlon be administered (dosage and frequency) over a 48-h period to be most effective? (*Hint:* Recall what it says on many antibiotic prescriptions regarding the first dose.)

(**c**) Comment on the dose concentrations and potential hazards.

(**d**) How would your answers change if the drug were taken on a full or empty stomach?

P8-7$_C$ (***Reactor selection and operating conditions***) For each of the following sets of reactions, describe your reactor system and conditions to maximize the selectivity to D. Make sketches where necessary to support your choices. The rates are in (mol/dm^3·s), and concentrations are in (mol/dm^3).

(**a**) (1)   A + B → D $-r_{1A} = 10 \exp(-8{,}000 \text{ K}/T)C_A C_B$

(2)   A + B → U $-r_{2A} = 100 \exp(-1{,}000 \text{ K}/T)C_A^{1/2} C_B^{3/2}$

(**b**) (1)   A + B → D $-r_{1A} = 100 \exp(-1{,}000 \text{ K}/T)C_A C_B$

(2)   A + B → U $-r_{2A} = 10^6 \exp(-8{,}000 \text{ K}/T)C_A C_B$

(**c**) (1)   A + B → D $-r_{1A} = 10 \exp(-1{,}000 \text{ K}/T)C_A C_B$

(2)   B + D → U $-r_{2B} = 10^9 \exp(-10{,}000 \text{ K}/T)C_B C_D$

(**d**) (1)   A ⟶ D $-r_{1A} = 4280 \exp(-12{,}000 \text{ K}/T)C_A$

(2)   D ⟶ U$_1$ $-r_{2D} = 10{,}100 \exp(-15{,}000 \text{ K}/T)C_D$

(3)   A ⟶ U$_2$ $-r_{3A} = 26 \exp(-18{,}800 \text{ K}/T)C_A$

(e) (1) $A + B \rightarrow D$ $-r_{1A} = 10^9 \exp(-10{,}000 \text{ K}/T)C_A C_B$

    (2) $D \rightarrow A + B$ $-r_{2D} = 20 \exp(-2{,}000 \text{ K}/T)C_D$

    (3) $A + B \rightarrow U$ $-r_{3A} = 10^3 \exp(-3{,}000 \text{ K}/T)C_A C_B$

(f) (1) $A + B \rightarrow D$ $-r_{1A} = 800 \exp\left(\dfrac{-8{,}000 K}{T}\right) C_A^{0.5} C_B$

    (2) $A + B \rightarrow U_1$ $r_{2B} = 10 \exp\left(\dfrac{-300 K}{T}\right) C_A C_B$

    (3) $D + B \rightarrow U_2$ $r_{3D} = 10^6 \exp\left(\dfrac{-8{,}000 K}{T}\right) C_D C_B$

P8-8$_B$ Consider the reaction

$$A \xrightarrow{\ k_1\ } D \xrightarrow{\ k_2\ } U$$

Pure A is fed to a 1.0-dm³ CSTR where it reacts to form a desired product (D), which can then react further to produce an undesired product (U); both reactions are elementary and irreversible, and everything is liquid phase. The entering concentration of A is 1 mole/dm³ at a molar flow rate of 1 mol/min.

(a) Sketch the conversion of A, X, the instantaneous selectivity of D to U, $S_{D/U}$, and the instantaneous yield of D, Y_D, as a function of space time (make sure to label them on the plot). You may want to write a sentence or two of reasoning for partial credit purposes.

(b) If at $\tau = 1.0$ minutes the instantaneous selectivity, $S_{D/U}$, is (1/2) and the conversion of A is (0.5), what are the specific reactions rates k_1 and k_2?

Exam Problem

P8-9$_B$ The elementary liquid-phase series reaction

$$A \xrightarrow{\ k_1\ } B \xrightarrow{\ k_2\ } C$$

is carried out in a 500-dm³ batch reactor. The initial concentration of A is 1.6 mol/dm³. The desired product is B, and separation of the undesired product C is very difficult and costly. Because the reaction is carried out at a relatively high temperature, the reaction is easily quenched.

(a) Plot and analyze the concentrations of A, B, and C as a function of time. Assume that each reaction is irreversible, with $k_1 = 0.4$ h⁻¹ and $k_2 = 0.01$ h⁻¹.

(b) Plot and analyze the concentrations of A, B, and C as a function of time when the first reaction is reversible, with $k_{-1} = 0.3$ h⁻¹.

(c) Plot and analyze the concentrations of A, B, and C as a function of time for the case where both reactions are reversible, with $k_{-2} = 0.005$ h⁻¹.

(d) Compare **(a)**, **(b)**, and **(c)** and describe what you find.

(e) Vary k_1, k_2, k_{-1}, and k_{-2}. Explain the consequence of $k_1 > 100$ and $k_2 < 0.1$ with $k_{-1} = k_{-2} = 0$ and with $k_{-2} = 1$, $k_{-1} = 0$, and $k_{-2} = 0.25$.

(f) Apply one or more of the six ideas in Preface Table P-4, page xxviii, to this problem.

P8-10$_B$ Terephthalic acid (TPA) finds extensive use in the manufacture of synthetic fibers (e.g., Dacron) and as an intermediate for polyester films (e.g., Mylar). The formation of potassium terephthalate from potassium benzoate was studied using a tubular reactor [*Ind. Eng. Chem. Res., 26*, 1691].

It was found that the intermediates (primarily K-phthalates) formed from the dissociation of K-benzoate over a CdCl₂ catalyst reacted with K-terephthalate in an autocatalytic reaction step

$$A \xrightarrow{\ k_1\ } R \xrightarrow{\ k_2\ } S \qquad \text{Series}$$

$$R + S \xrightarrow{\ k_3\ } 2S \qquad \text{Autocatalytic}$$

where A = K-benzoate, R = lumped intermediates (K-phthalates, K-isophthalates, and K-benzenecarboxy-lates), and S = K-terephthalate. Pure A is charged to the reactor at a pressure of 110 kPa. The specific reaction rates at 410°C are $k_1 = 1.08 \times 10^{-3}$ s^{-1} with $E_1 = 42.6$ kcal/mol, $k_2 = 1.19 \times 10^{-3}$ s^{-1} with $E_2 = 48.6$ kcal/mol, and $k_3 = 1.59 \times 10^{-3}$ dm^3/mol · s with $E_3 = 32$ kcal/mol.

(a) Plot and analyze the concentrations of A, R, and S as a function of time in a batch reactor at 410°C, noting when the maximum in R occurs.

(b) Repeat (a) for temperatures of 430°C and 390°C.

(c) What would be the exit concentrations from a CSTR operated at 410°C and a space time of 1200 s?

P8-11$_A$ The following liquid-phase reactions were carried out in a CSTR at 325 K:

$$3A \longrightarrow B+C \qquad -r_{1A} = k_{1A}C_A \qquad k_{1A} = 7.0 \text{ min}^{-1}$$

$$2C+A \longrightarrow 3D \qquad r_{2D} = k_{2D}C_C^2 C_A \qquad k_{2D} = 3.0 \frac{\text{dm}^6}{\text{mol}^2 \cdot \text{min}}$$

$$4D+3C \longrightarrow 3E \qquad r_{3E} = k_{3E}C_D C_C \qquad k_{3E} = 2.0 \frac{\text{dm}^3}{\text{mol} \cdot \text{min}}$$

> Sketch the trends or results you expect **before** working out the details of the problem.

The concentrations measured *inside* the reactor were $C_A = 0.10$, $C_B = 0.93$, $C_C = 0.51$, and $C_D = 0.049$ all in mol/dm^3.

(a) What are r_{1A}, r_{2A}, and r_{3A}? ($r_{1A} = -0.7$ mol/dm^3·min)

(b) What are r_{1B}, r_{2B}, and r_{3B}?

(c) What are r_{1C}, r_{2C}, and r_{3C}? ($r_{1C} = 0.23$ mol/dm^3·min)

(d) What are r_{1D}, r_{2D}, and r_{3D}?

(e) What are r_{1E}, r_{2E}, and r_{3E}?

(f) What are the net rates of formation of A, B, C, D, and E?

(g) The entering volumetric flow rate is 100 dm^3/min and the entering concentration of A is 3 *M*. What is the CSTR reactor volume? (*Ans.:* 400 dm^3.)

(h) What are the exit molar flow rates from the 400 dm^3 CSTR?

Note: The following parts require an ODE solver and are of "B level" difficulty.

(i) **PFR.** Now assume the reactions take place in the gas phase. Use the preceding data to plot the molar flow rate's selectivity and p as a function of PFR volume up to 400 dm^3. The pressure-drop parameter is 0.001 dm^{-3}, the total concentration entering the reactor is 0.2 mol/dm^3, and $v_0 = 100$ dm^3/min. What are $\tilde{S}_{D/E}$ and $\tilde{S}_{C/D}$?

(j) **Membrane Reactor.** Repeat (i) when species C diffuses out of a membrane reactor and the transport coefficient, k_C, is 10 min^{-1}. Compare your results with part (i).

P8-12$_B$ In this problem, the complex reactions described below will first be carried out in the liquid phase (parts (a) through (d)) and then in the gas phase (parts (e) through (g)). One need not solve the liquid phase to solve the gas-phase problems.

The following reactions are carried out isothermally.

$$A+2B \longrightarrow C+D \qquad r_{1D} = k_{1D}C_A C_B$$

$$2D+3A \longrightarrow C+E \qquad r_{2E} = k_{2E}C_A C_D$$

$$B+2C \longrightarrow D+F \qquad r_{3F} = k_{3F}C_B^2 C_C$$

Additional information:

$k_{1D} = 0.25$ dm^6/mol^2· min	$v_0 = 10$ dm^3/min
$k_{2E} = 0.1$ dm^3/mol· min	$C_{A0} = 1.5$ mol/dm^3
$k_{3F} = 5.0$ dm^9/mol^2· min	$C_{B0} = 2.0$ mol/dm^3

(a) Consider the reactions to be liquid phase and plot the species concentrations and the conversion of A as a function of the distance (i.e., volume) down a 50-dm^3 PFR. Note any maxima.

(b) Consider the reactions to be liquid phase and determine the effluent concentrations and conversion from a 50-dm^3 CSTR. (*Ans.:* $C_A = 0.61$, $C_B = 0.79$, $C_F = 0.25$, and $C_D = 0.45$ mol/dm^3.)

(c) Plot and analyze the species concentrations and the conversion of A as a function of time when the reaction is carried out in a semibatch reactor initially containing 40 dm^3 of liquid. Consider two cases: (1) A is fed to B, and (2) B is fed to A. What differences do you observe for these two cases? Describe what you find.

(d) Vary the ratio of B to A $(1 < \Theta_B < 10)$ in the feed to the PFR and describe what you find. What generalizations can you make from this problem?

(e) Rework (a) for the case when the reaction is a gas-phase reaction. We will keep the constants the same so you won't have to make too many changes in your Polymath program, but we will make $v_0 = 100$ dm^3/min, $C_{T0} = 0.4$ mol/dm^3, $V = 500$ dm^3, and an equalmolar feed of A and B. Plot the molar flow rates and $S_{C/D}$ and $S_{E/F}$ down a PFR. Describe what you find.

(f) Repeat (e) when D diffuses out through the sides of a membrane reactor where the mass-transfer coefficient, k_{CD}, can be varied between 0.1 min^{-1} and 10 min^{-1}. What trends do you find?

(g) Repeat (e) when B is fed through the sides of a membrane reactor. Describe what you find.

P8-13$_B$ The gas-phase reactions take place isothermally in a membrane reactor packed with catalyst. Pure A enters the reactor at 24.6 atm and 500 K, and a flow rate of A of 10 mol/min.

$$A \rightleftharpoons B + C \qquad r'_{1C} = k_{1C}\left[C_A - \frac{C_B C_C}{K_{1C}}\right]$$

$$A \longrightarrow D \qquad r'_{2D} = k_{2D}C_A$$

$$2C + D \longrightarrow 2E \qquad r'_{3E} = k_{3E}C_C^2 C_D$$

Only species B diffuses out of the reactor through the membrane.

Additional Information:

Overall mass-transfer coefficient $k_C = 1.0$ dm^3 / kg-cat · min

$k_{1C} = 2$ dm^3 / kg-cat · min	$k_{3E} = 5.0$ dm^3 / mol^2 · kg-cat · min
$K_{1C} = 0.2$ mol / dm^3	$W_f = 100$ kg
$k_{2D} = 0.4$ dm^3 / kg-cat · min	$\alpha = 0.008$ kg^{-1}

(a) Plot and analyze the concentrations down the length of the reactor.

(b) Explain why your curves look the way they do.

(c) Describe the major differences you observe when C diffuses out instead of B, with the same mass-transfer coefficient.

(d) Vary some of the parameters (e.g., k_B, k_{1C}, K_{1C}) and write a paragraph describing what you find.

P8-14$_B$ The complex reactions involved in the oxidation of formaldehyde to formic acid over a Vanadium titanium oxide catalyst [*Ind. Eng. Chem. Res.* 28, p. 387 (1989)] are shown below. Each reaction follows an elementary rate law.

$$HCHO + \frac{1}{2}O_2 \xrightarrow{k_1} HCOOH \xrightarrow{k_3} CO + H_2O$$

$$2HCHO \xrightarrow{k_2} HCOOCH_3$$

$$HCOOCH_3 \xrightarrow{k_4} CH_3OH + HCOOH$$

Let A = HCHO, B = O$_2$, C = HCOOH, D = HCOOCH$_3$, E = CO, W = H$_2$O, and G = CH$_3$OH.

The entering flow rates are $F_{A0} = 10$ mol/s and $F_{B0} = 5$ mol/s, and $v_0 = 100$ dm^3/s. At a total entering concentration $C_{T0} = 0.147$ mol/dm^3, the suggested reactor volume is 1,000 dm^3.

Additional information:

At 300 K

$$k_1 = 0.014 \left(\frac{dm^3}{mol}\right)^{1/2}\Big/s, \quad k_2 = 0.007\frac{dm^3}{mol \cdot s}$$

$$k_3 = 0.014/s, \quad k_4 = 0.45\frac{dm^3}{mol \cdot s}$$

(a) Plot the molar flow rates of each species along the volume (length) of the reactor on the same figure and then analyze why the profiles look the way they do.

(b) Plot and analyze $\tilde{Y}_C$, $\tilde{S}_{A/E}$, $\tilde{S}_{C/D}$, and $\tilde{S}_{D/G}$ along the length of the reactor. Note and explain any maximums and the volume at which they occur.

(c) Plot and analyze the overall HCOOH yield and overall selectivity of HCOH to CO, of $HCOOCH_3$ to CH_3OH, and of HCOOH to $HCOOCH_3$ as a function of the Θ_{O_2}. Suggest some conditions to best produce formic acid. Write a paragraph describing what you find.

(d) Compare your plot in part **(a)** with a similar plot when pressure drop is taken into account with $\alpha = 0.002$ dm^{-3}. Note any unusual differences between parts **(a)** and **(d)**.

(e) Suppose that $E_1 = 10,000$ cal/mol, $E_2 = 30,000$ cal/mol, $E_3 = 20,000$ cal/mol, and $E_4 = 10,000$ cal/mol, what temperature would you recommend for a 1000-dm^3 PFR?

P8-15$_C$ The ethylene epoxydation is to be carried out using a cesium-doped silver catalyst in a packed-bed reactor.

$$(1) \quad C_2H_4 + \frac{1}{2}O_2 \rightarrow C_2H_4O \quad -r_{1E} = \frac{k_{1E}P_E P_O^{0.58}}{(1 + K_{1E}P_E)^2}$$

Along with the desired reaction, the complete combustion of ethylene also occurs

$$(2) \quad C_2H_4 + 3O_2 \rightarrow 2CO_2 + 2H_2O \quad -r_{2E} = \frac{k_{2E}P_E P_O^{0.3}}{(1 + K_{2E}P_E)^2}$$

[M. Al-Juaied, D. Lafarga, and A. Varma, *Chem. Eng. Sci. 56*, 395 (2001)].

It is proposed to replace the conventional PBR with a membrane reactor in order to improve the selectivity. As a rule of thumb, a 1% increase in the selectivity to ethylene oxide translates to an increase in profit of about $2 million/yr. The feed consists of 12% (mole) oxygen, 6% ethylene, and the remainder nitrogen at a temperature of 250°C and a pressure of 2 atm. The total molar flow rate is 0.0093 mol/s to a reactor containing 2 kg of catalyst.

Additional information:

$$k_{1E} = 0.15\frac{mol}{kg \cdot s \ atm^{1.58}} \text{ at 523 K with } E_1 = 60.7 \text{ kJ/mol}$$

$$k_{2E} = 0.0888\frac{mol}{kg \cdot s \ atm^{1.3}} \text{ at 523 K with } E_2 = 73.2 \text{ kJ/mol}$$

$$K_{1E} = 6.50 \text{ atm}^{-1}, K_{2E} = 4.33 \text{ atm}^{-1}$$

(a) What conversion and selectivity of ethylene epoxide to CO_2 are expected in a conventional PBR?

(b) What would be the conversion and selectivity if the total molar flow rate were divided and the 12% oxygen stream (no ethylene) were uniformly fed through the sides of the membrane reactor, and 6% ethylene (no oxygen) were fed at the entrance?

(c) Repeat **(b)** for a case when ethylene is fed uniformly through the sides and oxygen is fed at the entrance. Compare with parts **(a)** and **(b)**. Describe what you find.

P8-16$_B$ **Solar energy** capture has great potential to help meet the world's growing energy demand, which is 12 terawatts in 2010 and is expected to rise to 36 terawatts in 2050 (cf. **P3-15$_B$**). Professor Al Weiner and his students at the University of Colorado are engaged in developing methods of utilizing solar-thermal energy. In solar-thermal reactors, mirrors are used to focus and concentrate the sun's energy on a flow-type cavity reactor where temperatures as high as 1200°C can be realized, as shown below.

 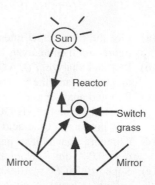

Figure P8-16.1 Solar field design. Five 265-m-tall towers with three heliostat fields/tower, on 275 acres of land in Daggett, CA. Net concentration 3,868 suns and 295 MW delivered to each solar reactor. (Melinda M. Channel, Jonathan Scheffe, Allan Lewandowski, and Alan W. Weimer, November 11, 2009. Also see: *Chemical Engineering*, 116, p. 18, March 2009).

The switch grass is fed to the 1200°C solar-thermal reactor. At these temperatures, biomass can be converted to CO and H_2, i.e., syn gas, which then can be used for liquid fuels. Switch grass, which is approximately $\frac{2}{3}$ cellulose ($C_6H_{10}O_5$) and $\frac{1}{3}$ lignin ($C_{10}H_{12}O_3$) will be fed with steam to produce CO, H_2 and a small amount of ash, which we will neglect. In order to simplify this process into a tractable home problem, we assume the switch grass is volatilized immediately upon entering the plug-flow reactor and that the reactions and postulated rate laws are

(1) Cellulose: $C_6H_{10}O_5(C) + H_2O(W) \rightarrow 6H_2 + 6CO$

(2) Lignin: $C_{10}H_{12}O_3(L) + 7H_2O(W) \rightarrow 13H_2 + 10CO$

[*AIChE J.* 55, p. 286 (2009)]. Also see *Science* p. 326, 1472 (2009).

The rate laws and constants are hypothesized to be

$$-r_{1C} = k_{1C}C_C C_W$$

$$-r_{2L} = k_{2L}C_L C_W^2$$

$$\text{with } k_{1C} = 3 \times 10^4 \left(\frac{dm^3}{mol}\right)\bigg/s \text{ and } k_{2L} = 1.4 \times 10^7 \left(\frac{dm^3}{mol}\right)\bigg/s$$

Total gas concentration in the feed and reactor with $C_{T0} = \dfrac{P_0}{RT_0} = \dfrac{1 \text{ atm}}{(0.082)(1473)} = 0.00828 \text{ mol/dm}^3$ with the entering molar flow rates of cellulose, lignin, and water are $F_{C0} = 0.00411$ mol/s and $F_{L0} = 0.00185$ mol/s, $F_{W0} = 0.02$ mol/s, respectively.

(a) Plot and analyze the molar flow rates as a function of PFR volume up to $V = 0.417$ dm³.
(b) Plot and analyze Y_C, Y_W, Y_L, and $\tilde{S}_{CO/H_2}$ down the reactor. Describe what you find.
(c) Repeat (a) for different molar flow rates of water.

P8-17$_B$ **Solar-thermal biochar gasification** has also been studied at the University of Colorado (see **P8-16$_B$**). [*Chemical Engineering and Processing: Process Intensification* 48, p. 1279 (2009) and *AIChE J. 55* p.286 (2009).*] While this process follows a shrinking core model (see the Expanded Material for Chapter 14 on the CRE Web site), for the purposes of this example, we will use the following sequence:

(1) Lignin: $C_{10}H_{12}O_3(L) + 3H_2O(W) \rightarrow 3H_2 + 3CO + Char$ (e.g., cresol)

(2) Char: $Char(Ch) + 4H_2O \rightarrow 10H_2 + 7CO$

The rate laws at 1200°C are hypothesized to be

$$-r_{1L} = k_{1L}C_L C_W^2 \text{ with } k_{1L} = 3721\left(\frac{dm^3}{mol}\right)^2 \Big/ s$$

$$-r_{2Ch} = k_{2Ch}C_{Ch}C_W^2 \text{ with } k_{2Ch} = 1,000\left(\frac{dm^3}{mol}\right)^2 \Big/ s$$

The entering molar flow rates are $F_{L0} = 0.0123$ mol/s, $F_{W0} = 0.111$ mol/s, the total entering concentration is $C_{T0} = 0.2$ mol/dm^3, and the reactor volume is 0.417 dm^3.
  (a) Plot and analyze F_{Ch}, F_L, F_W, F_{CO}, and F_{H_2} down the length of a plug-flow reactor.
  (b) Repeat **(a)** for the concentrations C_C, C_{Ch}, etc.
  (c) Plot and analyze the selectivity $\tilde{S}_{CO/H_2}$ and yields $\tilde{Y}_W$ and Y_L down the PFR.
  (d) At what point is the char molar flow rate a maximum? How does it change with changing feed conditions, such as the ratio of (F_{W0}/F_{L0}), C_{T0}, etc? Describe what you found in parts **(a)** through **(d)**.

P8-18$_A$ Go to Professor Herz's **Reactor Lab** on the CRE Web site at *www.reactorlab.net*.
  (a) Download Division 5, Lab 2 of the Reactor Lab from the CRE Web site for the selective oxidation of ethylene to ethylene oxide. Click the [i] info button to get information about the system. Perform experiments and develop rate equations for the reactions. Write a technical memo that reports your results and includes plots and statistical measurements of how well your kinetic model fits experimental data.
  (b) Download Division 5, Labs 3 and 4 of the Reactor Lab for batch reactors in which parallel and series reactions, respectively, can be carried out. Investigate how dilution with solvent affects the selectivity for different reaction orders and write a memo describing your findings.

• **Additional Homework Problems**

A number of homework problems that can be used for exams or supplementary problems or examples are found on the CRE Web site, *http://www.umich.edu/~elements/5e/index.html*.

SUPPLEMENTARY READING

1. Selectivity, reactor schemes, and staging for multiple reactions, together with evaluation of the corresponding design equations, may or may not be presented in

BURGESS, THORNTON W., *The Adventures of Chatterer the Red Squirrel*, New York: Dover Publications, Inc., 1915.

BUTT, JOHN B, *Reaction Kinetics and Reactor Design, Second Edition, Revised and Expanded*, New York: Marcel Dekker, Inc., 1999.

DENBIGH, K. G., and J. C. R. TURNER, *Chemical Reactor Theory*, 2nd ed. Cambridge: Cambridge University Press, 1971, Chap. 6.

2. Many analytical solutions for parallel, series, and combination reactions are presented in

WALAS, S. M., *Chemical Reaction Engineering Handbook of Solved Problems*. Newark, NJ: Gordon and Breach, 1995.

Reaction Mechanisms, Pathways, Bioreactions, and Bioreactors

9

The next best thing to knowing something is knowing where to find it.

—Samuel Johnson (1709–1784)

Overview. The topics of this chapter include the pseudo-steady state hypothesis (PSSH), enzyme mechanisms and reaction kinetics, and the growth of microorganisms.

We start with Section 9.1 where we discuss PSSH and the formation of active intermediate species, which disappear as fast as they are formed. Next we show how we apply the PSSH to active intermediate species to develop rate laws for chemical reactions that do not follow elementary rate laws.

To develop rate laws for nonelementary reactions we

- choose an active intermediate and a reaction mechanism,
- write the elementary rate law for each reaction in the mechanism,
- write the net rates of reaction for each species, and
- invoke the PSSH for the active intermediates in order to arrive at a rate law that is consistent with experimental observation.

In Section 9.2 we apply the PSSH to biochemical reactions, with a focus on enzymatic reaction mechanisms and kinetics. Here, we study

- Michaelis–Menten kinetics,
- Lineweaver–Burk plot and other plots to analyze data, and
- types of enzyme inhibition and the corresponding rate law.

In Section 9.3 we study the growth of microorganisms. Microorganism growth logically follows the discussion of enzyme kinetics because of the similarity of the Michaelis–Menten rate law and the Monod growth law. Here we study

- balances on the cell mass and the substrate,
- kinetics of cell growth,
- yield coefficients relating to cell growth on substrate consumption, and
- bioreactors.

We will use these principles to model both batch reactors and CSTRs, which are called chemostats. We should note that microorganism growth is becoming of increasing interest because of the use of algae as a biofuel.

9.1 Active Intermediates and Nonelementary Rate Laws

In Chapter 3 a number of simple power-law models, e.g.,

$$-r_A = kC_A^n$$

were presented, where n was an integer of 0, 1, or 2 corresponding to a zero-, first-, or second-order reaction. However, for a large number of reactions, the orders are noninteger, such as the decomposition of acetaldehyde at 500°C

$$CH_3CHO \rightarrow CH_4 + CO$$

where the rate law developed in problem P9-5$_B$(**b**) is

$$-r_{CH_3CHO} = kC_{CH_3CHO}^{3/2}$$

The rate law could also have concentration terms in both the numerator and denominator such as the formation of HBr from hydrogen and bromine

$$H_2 + Br_2 \rightarrow 2HBr$$

where the rate law developed in problem P9-5$_B$(**c**) is

$$r_{HBr} = \frac{k_1 C_{H_2} C_{Br_2}^{3/2}}{C_{HBr} + k_2 C_{Br_2}}$$

Rate laws of this form usually involve a number of elementary reactions and at least one active intermediate. An *active intermediate* is a high-energy molecule that reacts virtually as fast as it is formed. As a result, it is present in very small concentrations. Active intermediates (e.g., $A*$) can be formed by collision or interaction with other molecules.

$$A + M \rightarrow A* + M$$

Properties of an active intermediate $A*$

Here, the activation occurs when translational kinetic energy is transferred into internal energy, i.e., vibrational and rotational energy.[1] An unstable molecule

[1] W. J. Moore, *Physical Chemistry*, (Reading, MA: Longman Publishing Group, 1998).

(i.e., active intermediate) is not formed solely as a consequence of the molecule moving at a high velocity (high translational kinetic energy). The energy must be absorbed into the chemical bonds, where high-amplitude oscillations will lead to bond ruptures, molecular rearrangement, and decomposition. In the absence of photochemical effects or similar phenomena, the transfer of translational energy to vibrational energy to produce an active intermediate can occur only as a consequence of molecular collision or interaction. Collision theory is discussed in the *Professional Reference Shelf* in Chapter 3 on the CRE Web site. Other types of active intermediates that can be formed are *free radicals* (one or more unpaired electrons; e.g., $CH_3\bullet$), ionic intermediates (e.g., carbonium ion), and enzyme-substrate complexes, to mention a few.

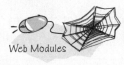
Web Modules

The idea of an active intermediate was first postulated in 1922 by F. A. Lindemann, who used it to explain changes in the reaction order with changes in reactant concentrations.[2] Because the active intermediates were so short-lived and present in such low concentrations, their existence was not really definitively confirmed until the work of Ahmed Zewail, who received the Nobel Prize in Chemistry in 1999 for femtosecond spectroscopy.[3] His work on cyclobutane showed that the reaction to form two ethylene molecules did not proceed directly, as shown in Figure 9-1(a), but formed the active intermediate shown in the small trough at the top of the energy barrier on the reaction-coordinate diagram in Figure 9-1(b). As discussed in Chapter 3, an estimation of the barrier height, E, can be obtained using computational software packages such as Spartan, Cerius[2], or Gaussian as discussed in the *Molecular Modeling Web Module* in Chapter 3 on the CRE Web site, *www.umich.edu/~elements/5e/ index.html*.

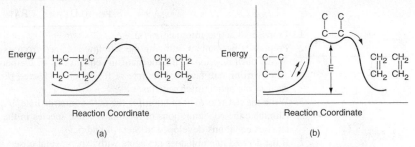

Figure 9-1 Reaction coordinate. Ivars Peterson, "Chemistry Nobel spotlights fast reactions," *Science News,* 156, 247 (1999).

9.1.1 Pseudo-Steady-State Hypothesis (PSSH)

In the theory of active intermediates, decomposition of the intermediate does not occur instantaneously after internal activation of the molecule; rather, there is a time lag, although infinitesimally small, during which the species remains activated. Zewail's work was the first definitive proof of a gas-phase active intermediate that exists for an infinitesimally short time. Because a reactive intermediate reacts virtually as fast as it is formed, the net rate of formation of an active intermediate (e.g., A*) is zero, i.e.,

$$r_{A*} \equiv 0 \qquad\qquad (9\text{-}1)$$

---

[2] F. A. Lindemann, *Trans. Faraday. Soc.,* 17, 598 (1922).

[3] J. Peterson, *Science News,* 156, 247 (1999).

This condition is also referred to as the *Pseudo-Steady-State Hypothesis* (PSSH). If the active intermediate appears in *n* reactions, then

PSSH

$$r_{A*} = \sum_{i=1}^{n} r_{iA*} = 0 \qquad (9\text{-}2)$$

To illustrate how rate laws of this type are formed, we shall first consider the gas-phase decomposition of azomethane, AZO, to give ethane and nitrogen

$$(CH_3)_2N_2 \longrightarrow C_2H_6 + N_2$$

Experimental observations show that the rate of formation of ethane is first order with respect to AZO at pressures greater than 1 atm (relatively high concentrations)[4]

$$r_{C_2H_6} \propto C_{AZO}$$

and second order at pressures below 50 mmHg (low concentrations):

Why the reaction order changes

$$r_{C_2H_6} \propto C_{AZO}^2$$

We could combine these two observations to postulate a rate law of the form

$$-r_{AZO} = \frac{k_1 C_{AZO}^2}{1 + k_2 C_{AZO}}$$

To find a mechanism that is consistent with the experimental observations, we use the steps shown in Table 9-1.

TABLE 9-1   STEPS TO DEDUCE A RATE LAW

1. Propose an active intermediate(s).
2. Propose a mechanism, utilizing the rate law obtained from experimental data, if possible.
3. Model each reaction in the mechanism sequence as an elementary reaction.
4. After writing rate laws for the rate of formation of desired product, write the rate laws for each of the active intermediates.
5. Write the net rate of formation for the active intermediate and use the PSSH.
6. Eliminate the concentrations of the intermediate species in the rate laws by solving the simultaneous equations developed in Steps 4 and 5.
7. If the derived rate law does not agree with experimental observation, assume a new mechanism and/or intermediates and go to Step 3. A strong background in organic and inorganic chemistry is helpful in predicting the activated intermediates for the reaction under consideration.

We will now follow the steps in Table 9-1 to develop the rate law for azomethane (AZO) decomposition, $-r_{AZO}$.

Step 1. **Propose an active intermediate.** We will choose as an active intermediate an azomethane molecule that has been excited through molecular collisions, to form AZO*, i.e., $[(CH_3)_2N_2]*$.

Step 2. **Propose a mechanism.**

Mechanism

Reaction 1: $(CH_3)_2N_2 + (CH_3)_2N_2 \xrightarrow{k_{1AZO*}} (CH_3)_2N_2 + [(CH_3)_2N_2]^*$

Reaction 2: $[(CH_3)_2N_2]^* + (CH_3)_2N_2 \xrightarrow{k_{2AZO*}} (CH_3)_2N_2 + (CH_3)_2N_2$

Reaction 3: $[(CH_3)_2N_2]^* \xrightarrow{k_{3AZO*}} C_2H_6 + N_2$

---

[4] H. C. Ramsperger, *J. Am. Chem. Soc.*, **49**, 912 (1927).

In *reaction 1*, two AZO molecules collide and the kinetic energy of one AZO molecule is transferred to the internal rotational and vibrational energies of the other AZO molecule, and it becomes activated and highly reactive (i.e., AZO*). In *reaction 2*, the activated molecule (AZO*) is deactivated through collision with another AZO by transferring its internal energy to increase the kinetic energy of the molecules with which AZO* collides. In *reaction 3*, this highly activated AZO* molecule, which is wildly vibrating, spontaneously decomposes into ethane and nitrogen.

Step 3. *Write rate laws.*

Because each of the reaction steps is elementary, the corresponding rate laws for the active intermediate AZO* in reactions (1), (2), and (3) are

Note: The specific reaction rates, k, are all defined wrt the active intermediate AZO*.

(1) $$r_{1\text{AZO}*} = k_{1\text{AZO}*}C_{\text{AZO}}^2 \qquad (9\text{-}3)$$

(2) $$r_{2\text{AZO}*} = -k_{2\text{AZO}*}C_{\text{AZO}*}C_{\text{AZO}} \qquad (9\text{-}4)$$

(3) $$r_{3\text{AZO}*} = -k_{3\text{AZO}*}C_{\text{AZO}*} \qquad (9\text{-}5)$$

(Let $k_1 = k_{1\text{AZO}*}$, $k_2 = k_{2\text{AZO}*}$, and $k_3 = k_{3\text{AZO}*}$)

The rate laws shown in Equations (9-3) through (9-5) are pretty much useless in the design of any reaction system because the concentration of the active intermediate AZO* is not readily measurable. Consequently, we will use the Pseudo-Steady-State-Hypothesis (PSSH) to obtain a rate law in terms of measurable concentrations.

Step 4. *Write rate of formation of product.*

We first write the rate of formation of product

$$\boxed{r_{\text{C}_2\text{H}_6} = k_3 C_{\text{AZO}*}} \qquad (9\text{-}6)$$

Step 5. ***Write net rate of formation of the active intermediate and use the PSSH.***

To find the concentration of the active intermediate AZO*, we set the net rate of formation of AZO* equal to zero,[5] $r_{\text{AZO}*} \equiv 0$.

$$r_{\text{AZO}*} = r_{1\text{AZO}*} + r_{2\text{AZO}*} + r_{3\text{AZO}*} = 0$$

$$= k_1 C_{\text{AZO}}^2 - k_2 C_{\text{AZO}*}C_{\text{AZO}} - k_3 C_{\text{AZO}*} = 0 \qquad (9\text{-}7)$$

Solving for $C_{\text{AZO}*}$

$$C_{\text{AZO}*} = \frac{k_1 C_{\text{AZO}}^2}{k_2 C_{\text{AZO}} + k_3} \qquad (9\text{-}8)$$

Step 6. ***Eliminate the concentration of the active intermediate species in the rate laws by solving the simultaneous equations developed in Steps 4 and 5.***

Substituting Equation (9-8) into Equation (9-6)

$$\boxed{r_{\text{C}_2\text{H}_6} = \frac{k_1 k_3 C_{\text{AZO}}^2}{k_2 C_{\text{AZO}} + k_3}} \qquad (9\text{-}9)$$

---

[5] For further elaboration on this section, see R. Aris, *Am. Sci.,* 58, 419 (1970).

Step 7. **Compare with experimental data.**

At low AZO concentrations,

$$k_2 C_{AZO} \ll k_3$$

for which case we obtain the following second-order rate law:

$$r_{C_2H_6} = k_1 C_{AZO}^2$$

At high concentrations

$$k_2 C_{AZO} \gg k_3$$

in which case the rate expression follows first-order kinetics

$$r_{C_2H_6} = \frac{k_1 k_3}{k_2} C_{AZO} = k C_{AZO}$$

Apparent reaction orders

In describing reaction orders for this equation, one would say that the reaction is *apparent first order* at high azomethane concentrations and *apparent second order* at low azomethane concentrations.

9.1.2 Why Is the Rate Law First Order?

The PSSH can also explain why one observes so many first-order reactions such as

$$(CH_3)_2O \rightarrow CH_4 + H_2 + CO$$

Symbolically, this reaction will be represented as A going to product P; that is,

$$A \rightarrow P$$

with

$$-r_A = k C_A$$

The reaction is first order but the reaction is not elementary. The reaction proceeds by first forming an active intermediate, A^*, from the collision of the reactant molecule and an inert molecule of M. Either this wildly oscillating active intermediate, A^*, is deactivated by collision with inert M, or it decomposes to form product. See Figure 9-2.

The mechanism consists of the three elementary reactions:

1. Activation $\qquad A + M \xrightarrow{\ k_1\ } A^* + M$

2. Deactivation $\qquad A^* + M \xrightarrow{\ k_2\ } A + M$

3. Decomposition $\qquad A^* \xrightarrow{\ k_3\ } P$

Writing the rate of formation of product

$$r_P = k_3 C_{A^*}$$

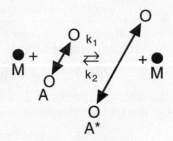

Figure 9-2   Collision and activation of a vibrating A molecule.[6]

Reaction pathways[6]   and using the PSSH to find the concentrations of A* in a manner similar to the azomethane decomposition described earlier, the rate law can be shown to be

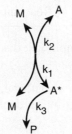

$$r_P = -r_A = \frac{k_3 k_1 C_A C_M}{k_2 C_M + k_3} \tag{9-10}$$

Because the concentration of the inert M is constant, we let

$$k = \frac{k_1 k_3 C_M}{k_2 C_M + k_3} \tag{9-11}$$

to obtain the first-order rate law

$$-r_A = k C_A$$

First-order rate law for a nonelementary reaction

Consequently, we see the reaction

$$A \rightarrow P$$

follows an elementary rate law but is not an elementary reaction.

9.1.3 Searching for a Mechanism

In many instances the rate data are correlated before a mechanism is found. It is a normal procedure to reduce the additive constant in the denominator to 1. We therefore divide the numerator and denominator of Equation (9-9) by k_3 to obtain

$$\boxed{r_{C_2H_6} = \frac{k_1 C_{AZO}^2}{1 + k' C_{AZO}}} \tag{9-12}$$

General Considerations. Developing a mechanism is a difficult and time-consuming task. The rules of thumb listed in Table 9-2 are by no means inclusive but may be of some help in the development of a simple mechanism that is consistent with the experimental rate law.

---

[6] The reaction pathway for the reaction in Figure 9-2 is shown in the margin. We start at the top of the pathway with A and M, and move down to show the pathway of A and M touching k_1 (collision) where the curved lines come together and then separate to form A* and M. The reverse of this reaction is shown starting with A* and M and moving upward along the pathway where arrows A* and M touch, k_2, and then separate to form A and M. At the bottom, we see the arrow from A* to form P with k_3.

TABLE 9-2   RULES OF THUMB FOR DEVELOPMENT OF A MECHANISM

1. Species having the concentration(s) appearing in the *denominator* of the rate law probably collide with the active intermediate; for example,

$$M + A^* \longrightarrow \text{[Collision products]} \qquad -r_{A1}^* = k_1(M)(A^*)$$

2. If a constant appears in the *denominator*, one of the reaction steps is probably the spontaneous decomposition of the active intermediate; for example,

$$A^* \longrightarrow \text{[Decomposition products]} \qquad -r_{A2}^* = k_2(A^*)$$

3. Species having the concentration(s) appearing in the *numerator* of the rate law probably produce the active intermediate in one of the reaction steps; for example,

$$\text{[Reactants]} \longrightarrow A^* + \text{[Other products]} \qquad = k_3(\text{Reactants})$$

Upon application of Table 9-2 to the azomethane example just discussed, we see the following from rate equation (9-12):

1. The active intermediate, AZO*, collides with azomethane, AZO [Reaction 2], resulting in the concentration of AZO in the denominator.
2. AZO* decomposes spontaneously [Reaction 3], resulting in a constant in the denominator of the rate expression.
3. The appearance of AZO in the numerator suggests that the active intermediate AZO* is formed from AZO. Referring to [Reaction 1], we see that this case is indeed true.

Example 9–1   The Stern–Volmer Equation

Light is given off when a high-intensity ultrasonic wave is applied to water.[6] This light results from microsize gas bubbles (0.1 mm) being formed by the ultrasonic wave and then being compressed by it. During the compression stage of the wave, the contents of the bubble (e.g., water and whatever else is dissolved in the water, e.g., CS_2, O_2, N_2) are compressed adiabatically.

This compression gives rise to high temperatures and kinetic energies of the gas molecules, which through molecular collisions generate active intermediates and cause chemical reactions to occur in the bubble.

$$M + H_2O \longrightarrow H_2O^* + M$$

The intensity of the light given off, I, is proportional to the rate of deactivation of an activated water molecule that has been formed in the microbubble.

Collapsing
cavitation
microbubble

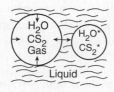

$$H_2O^* \xrightarrow{\ k\ } H_2O + h\nu$$

Light intensity (I) $\propto (-r_{H_2O^*}) = k\,C_{H_2O^*}$

---

[6] P. K. Chendke and H. S. Fogler, *J. Phys. Chem.*, 87, 1362 (1983).

An order-of-magnitude increase in the intensity of sonoluminescence is observed when either carbon disulfide or carbon tetrachloride is added to the water. The intensity of luminescence, I, for the reaction

$$CS_2^* \xrightarrow{\ k_4\ } CS_2 + h\nu$$

is

$$I \propto (-r_{CS_2^*}) = k_4 C_{CS_2^*}$$

A similar result exists for CCl_4.

However, when an aliphatic alcohol, X, is added to the solution, the intensity decreases with increasing concentration of alcohol. The data are usually reported in terms of a Stern–Volmer plot in which relative intensity is given as a function of alcohol concentration, C_X. (See Figure E9-1.1, where I_0 is the sonoluminescence intensity in the absence of alcohol and I is the sonoluminescence intensity in the presence of alcohol.)

(a) Suggest a mechanism consistent with experimental observation.

(b) Derive a rate law consistent with Figure E9-1.1.

Stern–Volmer plot

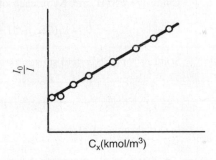

Figure E9-1.1 Ratio of luminescence intensities as a function of scavenger concentration.

Solution

(a) Mechanism
From the linear plot we know that

$$\frac{I_0}{I} = A + BC_X \equiv A + B(X) \tag{E9-1.1}$$

where $C_X \equiv (X)$. Inverting yields

$$\frac{I}{I_0} = \frac{1}{A + B(X)} \tag{E9-1.2}$$

From **Rule 1** of Table 9-2, the denominator suggests that alcohol (X) collides with the active intermediate:

$$X + \text{Intermediate} \longrightarrow \text{Deactivation products} \tag{E9-1.3}$$

Reaction pathways

The alcohol acts as what is called a *scavenger* to deactivate the active intermediate. The fact that the addition of CCl_4 or CS_2 increases the intensity of the luminescence

$$I \propto (CS_2) \tag{E9-1.4}$$

Reaction pathways

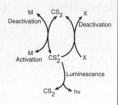

The mechanism

leads us to postulate (**Rule 3** of Table 9-2) that the active intermediate was probably formed from CS_2

$$M + CS_2 \longrightarrow CS_2^* + M \qquad (E9\text{-}1.5)$$

where M is a third body (CS_2, H_2O, N_2, etc.).

We also know that deactivation can occur by the reverse of reaction (E9-1.5). Combining this information, we have as our mechanism:

Activation: $M + CS_2 \xrightarrow{\ k_1\ } CS_2^* + M \qquad (E9\text{-}1.5)$

Deactivation: $M + CS_2^* \xrightarrow{\ k_2\ } CS_2 + M \qquad (E9\text{-}1.6)$

Deactivation: $X + CS_2^* \xrightarrow{\ k_3\ } CS_2 + X \qquad (E9\text{-}1.3)$

Luminescence: $CS_2^* \xrightarrow{\ k_4\ } CS_2 + h\nu \qquad (E9\text{-}1.7)$

$$\boxed{I = k_4(CS_2^*)} \qquad (E9\text{-}1.8)$$

(b) Rate Law

Using the PSSH on CS_2^* in each of the above elementry reactions yields

$$r_{CS_2^*} = 0 = k_1(CS_2)(M) - k_2(CS_2^*)(M) - k_3(X)(CS_2^*) - k_4(CS_2^*)$$

Solving for (CS_2^*) and substituting into Equation (E9-1.8) gives us

$$I = \frac{k_4 k_1 (CS_2)(M)}{k_2(M) + k_3(X) + k_4} \qquad (E9\text{-}1.9)$$

In the absence of alcohol

$$I_0 = \frac{k_4 k_1 (CS_2)(M)}{k_2(M) + k_4} \qquad (E9\text{-}1.10)$$

For constant concentrations of CS_2 and the third body, M, we take a ratio of Equation (E9-1.10) to (E9-1.9)

$$\boxed{\frac{I_0}{I} = 1 + \frac{k_3}{k_2(M) + k_4}(X) = 1 + k'(X)} \qquad (E9\text{-}1.11)$$

which is of the same form as that suggested by Figure E9-1.1. Equation (E9-1.11) and similar equations involving scavengers are called *Stern–Volmer equations*.

Analysis: This example showed how to use the Rules of Thumb (Table 9-2) to develop a mechanism. Each step in the mechanism is assumed to follow an elementary rate law. The PSSH was applied to the net rate of reaction for the *active intermediate* in order to find the concentration of the *active intermediate*. This concentration was then substituted into the rate law for the rate of formation of product to give the rate law. The rate law from the mechanism was found to be consistent with experimental data.

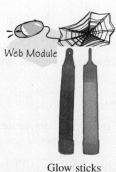

Web Module

Glow sticks
Web Module

A discussion of luminescence in the **Web Module**, **Glow Sticks** on the CRE Web site (*www.umich.edu/~elements/5e/index.html*). Here, the PSSH is applied to glow sticks. First, a mechanism for the reactions and luminescence is developed. Next, mole balance equations are written on each species and coupled with the rate law obtained using the PSSH; the resulting equations are solved and compared with experimental data.

9.1.4 Chain Reactions

A chain reaction consists of the following sequence:

Steps in a chain
reaction

1. *Initiation:* formation of an active intermediate
2. *Propagation or chain transfer:* interaction of an active intermediate with the reactant or product to produce another active intermediate
3. *Termination:* deactivation of the active intermediate to form products

Living Example Problem

An example comparing the application of the PSSH with the Polymath solution to the full set of equations is given in the *Professional Reference Shelf R9.1, Chain Reactions,* on the CRE Web site for the cracking of ethane. Also included in *Professional Reference Shelf R9.2* is a discussion of *Reaction Pathways* and the chemistry of smog formation.

9.2 Enzymatic Reaction Fundamentals

An *enzyme* is a high-molecular-weight protein or protein-like substance that acts on a substrate (reactant molecule) to transform it chemically at a greatly accelerated rate, usually 10^3 to 10^{17} times faster than the uncatalyzed rate. Without enzymes, essential biological reactions would not take place at a rate necessary to sustain life. Enzymes are usually present in small quantities and are not consumed during the course of the reaction, nor do they affect the chemical reaction equilibrium. Enzymes provide an alternate pathway for the reaction to occur, thereby requiring a lower activation energy. Figure 9-3 shows the reaction coordinate for the uncatalyzed reaction of a reactant molecule, called a *substrate* (S), to form a product (P)

$$S \rightarrow P$$

The figure also shows the catalyzed reaction pathway that proceeds through an *active intermediate* (E · S), called the *enzyme–substrate complex*, that is,

$$\boxed{S + E \rightleftharpoons E \cdot S \rightarrow E + P}$$

Because enzymatic pathways have lower activation energies, enhancements in reaction rates can be enormous, as in the degradation of urea by urease, where the degradation rate is on the order of 10^{14} higher than without the enzyme urease.

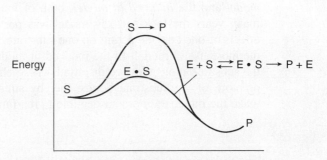

Figure 9-3   Reaction coordinate for enzyme catalysis.

An important property of enzymes is that they are specific; that is, *one* enzyme can usually catalyze only *one* type of reaction. For example, a protease hydrolyzes *only* bonds between specific amino acids in proteins, an amylase works on bonds between glucose molecules in starch, and lipase attacks fats, degrading them to fatty acids and glycerol. Consequently, unwanted products are easily controlled in enzyme-catalyzed reactions. Enzymes are produced only by living organisms, and commercial enzymes are generally produced by bacteria. Enzymes usually work (i.e., catalyze reactions) under mild conditions: pH 4 to 9 and temperatures 75°F to 160°F. Most enzymes are named in terms of the reactions they catalyze. It is a customary practice to add the suffix *-ase* to a major part of the name of the substrate on which the enzyme acts. For example, the enzyme that catalyzes the decomposition of urea is urease and the enzyme that attacks tyrosine is tyrosinase. However, there are exceptions to the naming convention, such as α-amylase. The enzyme α-amylase catalyzes the transformation of starch in the first step in the production of the controversial soft drink (e.g., Red Pop) sweetener high-fructose corn syrup (HFCS) from corn starch, which is a \$4 billion per year business.

Red Pop

$$\text{Corn starch} \xrightarrow{\text{α-amylase}} \text{Thinned starch} \xrightarrow[\text{amylase}]{\text{gluco-}} \text{Glucose} \xrightarrow[\text{isomerase}]{\text{Glucose}} \text{HFCS}$$

9.2.1 Enzyme–Substrate Complex

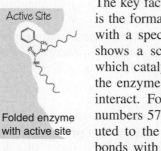

Active Site

Folded enzyme with active site

The key factor that sets enzymatic reactions apart from other catalyzed reactions is the formation of an enzyme–substrate complex, (E · S). Here, substrate binds with a specific *active site* of the enzyme to form this complex.[7] Figure 9-4 shows a schematic of the enzyme chymotrypsin (MW = 25,000 Daltons), which catalyzes the hydrolytic cleavage of polypeptide bonds. In many cases, the enzyme's active catalytic sites are found where the various folds or loops interact. For chymotrypsin, the catalytic sites are noted by the amino acid numbers 57, 102, and 195 in Figure 9-4. Much of the catalytic power is attributed to the binding energy of the substrate to the enzyme through multiple bonds with the specific functional groups on the enzyme (amino side chains, metal ions). The interactions that stabilize the enzyme–substrate complex are hydrogen bonding and hydrophobic, ionic, and London van der Waals forces. If the enzyme is exposed to extreme temperatures or pH environments (i.e., both high and low pH values), it may unfold, losing its active sites. When this occurs, the enzyme is said to be *denatured* (see Figure 9.8 and Problem P9-13$_B$).

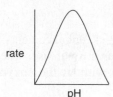

rate

pH

Two models for enzyme–substrate interactions are the *lock-and-key model* and the *induced fit model*, both of which are shown in Figure 9-5. For many years the lock-and-key model was preferred because of the sterospecific effects of one enzyme acting on one substrate. However, the induced fit model is the more useful model. In the induced fit model, both the enzyme molecule and the substrate molecules are distorted. These changes in conformation distort one or more of the substrate bonds, thereby stressing and weakening the bond to make the molecule more susceptible to rearrangement or attachment.

---

[7] M. L. Shuler and F. Kargi, *Bioprocess Engineering Basic Concepts*, 2nd ed. (Upper Saddle River, NJ: Prentice Hall, 2002).

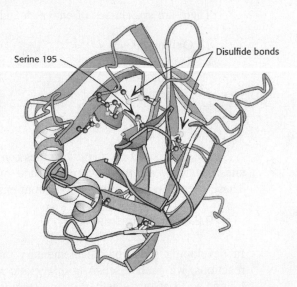

Figure 9-4   Enzyme chymotrypsin from *Biochemistry*, 7th ed., 2010 by Lubert Stryer, p. 258. Used with permission of W. H. Freeman and Company.

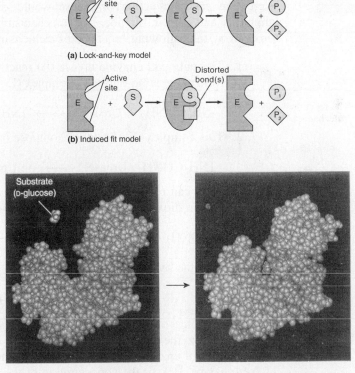

Figure 9-5   Two models for enzyme–substrate interaction.

There are six classes of enzymes and only six:

1. Oxidoreductases $\qquad$ $AH_2 + B + \mathbf{E} \rightarrow A + BH_2 + \mathbf{E}$
2. Transferases $\qquad$ $AB + C + \mathbf{E} \rightarrow AC + B + \mathbf{E}$
3. Hydrolases $\qquad$ $AB + H_2O + \mathbf{E} \rightarrow AH + BOH + \mathbf{E}$
4. Isomerases $\qquad$ $A + \mathbf{E} \rightarrow isoA + \mathbf{E}$
5. Lyases $\qquad$ $AB + \mathbf{E} \rightarrow A + B + \mathbf{E}$
6. Ligases $\qquad$ $A + B + \mathbf{E} \rightarrow AB + \mathbf{E}$

Links

More information about enzymes can be found on the following two Web sites: *http://us.expasy.org/enzyme/* and *www.chem.qmw.ac.uk/iubmb/enzyme.* These sites also give information about enzymatic reactions in general.

9.2.2 Mechanisms

In developing some of the elementary principles of the kinetics of enzyme reactions, we shall discuss an enzymatic reaction that has been suggested by Levine and LaCourse as part of a system that would reduce the size of an artificial kidney.[8] The desired result is a prototype of an artificial kidney that could be worn by the patient and would incorporate a replaceable unit for the elimination of the body's nitrogenous waste products, such as uric acid and creatinine. In the microencapsulation scheme proposed by Levine and LaCourse, the enzyme urease would be used in the removal of urea from the bloodstream. Here, the catalytic action of urease would cause urea to decompose into ammonia and carbon dioxide. The mechanism of the reaction is believed to proceed by the following sequence of elementary reactions:

1. The unbound enzyme urease (E) reacts with the substrate urea (S) to form an enzyme–substrate complex (E · S):

The reaction mechanism

$$NH_2CONH_2 + Urease \xrightarrow{\ k_1\ } [NH_2CONH_2 \cdot Urease]^* \qquad (9\text{-}13)$$

2. This complex (E · S) can decompose back to urea (S) and urease (E):

$$[NH_2CONH_2 \cdot Urease]^* \xrightarrow{\ k_2\ } Urease + NH_2CONH_2 \qquad (9\text{-}14)$$

$S + E \underset{}{\rightleftarrows} E \cdot S \xrightarrow{+H_2O} P + E$

3. Or, it can react with water (W) to give the products (P) ammonia and carbon dioxide, and recover the enzyme urease (E):

$$[NH_2CONH_2 \cdot Urease]^* + H_2O \xrightarrow{\ k_3\ } 2NH_3 + CO_2 + Urease \quad (9\text{-}15)$$

Symbolically, the overall reaction is written as

$$S + E \rightleftarrows E \cdot S \xrightarrow{+H_2O} P + E$$

We see that some of the enzyme added to the solution binds to the urea, and some of the enzyme remains unbound. Although we can easily measure the total concentration of enzyme, (E_t), it is difficult to measure either the concentration of free enzyme, (E), or the concentration of the bound enzyme (E · S).

---

[8] N. Levine and W. C. LaCourse, *J. Biomed. Mater. Res.*, 1, 275.

Letting E, S, W, E · S, and P represent the enzyme, substrate, water, the enzyme–substrate complex, and the reaction products, respectively, we can write Reactions (9-13), (9-14), and (9-15) symbolically in the forms

$$S + E \xrightarrow{\;k_1\;} E \cdot S \tag{9-16}$$

$$E \cdot S \xrightarrow{\;k_2\;} E + S \tag{9-17}$$

$$E \cdot S + W \xrightarrow{\;k_3\;} P + E \tag{9-18}$$

Here, $P = 2NH_3 + CO_2$.

The corresponding rate laws for Reactions (9-16), (9-17), and (9-18) are

$$r_{1E \cdot S} = k_1(E)(S) \tag{9-16A}$$

$$r_{2E \cdot S} = -k_2(E \cdot S) \tag{9-17A}$$

$$r_{3E \cdot S} = -k_3(E \cdot S)(W) \tag{9-18A}$$

where all the specific reaction rates are defined with respect to (E · S). The net rate of formation of product, r_P, is

$$\boxed{r_P = k_3(W)(E \cdot S)} \tag{9-19}$$

For the overall reaction

$$E + S \longrightarrow P + E$$

we know that the rate of consumption of the urea substrate equals the rate of formation of product CO_2, i.e., $-r_S = r_P$.

This rate law, Equation (9-19), is of not much use to us in making reaction engineering calculations because we cannot measure the concentration of the enzyme substrate complex (E · S). We will use the PSSH to express (E · S) in terms of measured variables.

The net rate of formation of the enzyme–substrate complex is

$$r_{E \cdot S} = r_{1E \cdot S} + r_{2E \cdot S} + r_{3E \cdot S}$$

Substituting the rate laws, we obtain

$$r_{E \cdot S} = k_1(E)(S) - k_2(E \cdot S) - k_3(W)(E \cdot S) \tag{9-20}$$

Using the PSSH, $r_{E \cdot S} = 0$, we can now solve Equation (9-20) for (E · S)

$$\boxed{(E \cdot S) = \frac{k_1(E)(S)}{k_2 + k_3(W)}} \tag{9-21}$$

and substitute for (E · S) into Equation (9-19)

$$-r_S = r_P = \frac{k_1 k_3(E)(S)(W)}{k_2 + k_3(W)} \tag{9-22}$$

We need to replace unbound enzyme concentration (E) in the rate law.

We *still* cannot use this rate law because we cannot measure the unbound enzyme concentration (E); however, we can measure the total enzyme concentration, E_t.

In the absence of enzyme denaturation, the total concentration of the enzyme in the system, (E_t), is constant and equal to the sum of the concentrations of the free or unbounded enzyme, (E), and the enzyme–substrate complex, $(E \cdot S)$:

Total enzyme concentration = Bound + Free enzyme concentration.

$$(E_t) = (E) + (E \cdot S) \tag{9-23}$$

Substituting for $(E \cdot S)$

$$(E_t) = (E) + \frac{k_1(E)(S)}{k_2 + k_3(W)}$$

solving for (E)

$$(E) = \frac{(E_t)(k_2 + k_3(W))}{k_2 + k_3(W) + k_1(S)}$$

and substituting for (E) in Equation (9-22), the rate law for substrate consumption is

$$\boxed{-r_S = \frac{k_1 k_3(W)(E_t)(S)}{k_1(S) + k_2 + k_3(W)}} \tag{9-24}$$

Note: Throughout the following text, $E_t \equiv (E_t)$ = total concentration of enzyme with typical units such as $(kmol/m^3)$ or (g/dm^3).

9.2.3 Michaelis–Menten Equation

Because the reaction of urea and urease is carried out in an aqueous solution, water is, of course, in excess, and the concentration of water (W) is therefore considered constant, ca. $55\ mol/dm^3$. Let

$$k_{cat} = k_3(W) \quad \text{and let} \quad K_M = \frac{k_{cat} + k_2}{k_1}$$

Dividing the numerator and denominator of Equation (9-24) by k_1, we obtain a form of the *Michaelis–Menten equation*:

$$\tag{9-25}$$

The final form of the rate law

$$-r_S = \frac{k_{cat}(E_t)(S)}{(S) + K_M}$$

Turnover number k_{cat}

The parameter $\textbf{\textit{k}}_{\textbf{cat}}$ is also referred to as the ***turnover number***. It is the number of substrate molecules converted to product in a given time on a single-enzyme molecule when the enzyme is saturated with substrate (i.e., all the active sites on the enzyme are occupied, $(S) \gg K_M$). For example, the turnover number for the decomposition of hydrogen-peroxide, H_2O_2, by the enzyme catalase is $40 \times 10^6\ s^{-1}$. That is, 40 million molecules of H_2O_2 are decomposed every second on a single-enzyme molecule saturated with H_2O_2. The constant K_M (mol/dm^3) is called the *Michaelis constant* and for simple systems is a measure of the attraction of the enzyme for its substrate, so it's also called the *affinity constant*. The Michaelis constant, K_M, for the decomposition of H_2O_2 discussed earlier is 1.1 M, while that for chymotrypsin is 0.1 M.[9]

Michaelis constant K_M

---

[9] D. L. Nelson and M. M. Cox, *Lehninger Principles of Biochemistry*, 3rd ed. (New York: Worth Publishers, 2000).

If, in addition, we let V_{max} represent the maximum rate of reaction for a given total enzyme concentration

$$V_{max} = k_{cat}(E_t)$$

the Michaelis–Menten equation takes the familiar form

Michaelis–Menten equation

$$-r_S = \frac{V_{max}(S)}{K_M + (S)} \tag{9-26}$$

For a given enzyme concentration, a sketch of the rate of disappearance of the substrate is shown as a function of the substrate concentration in Figure 9-6.

Michaelis–Menten plot

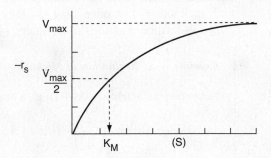

Figure 9-6   Michaelis–Menten plot identifying the parameters V_{max} and K_M.

A plot of this type is sometimes called a *Michaelis–Menten* plot. At low substrate concentration, $K_M \gg (S)$, Equation (9-26) reduces to

$$-r_S \cong \frac{V_{max}(S)}{K_M}$$

and the reaction is apparent first order in the substrate concentration. At high substrate concentrations,

$$(S) \gg K_M$$

Equation (9-26) reduces to

$$-r_S \cong V_{max}$$

and we see the reaction is apparent zero order.

Interpretation of Michaelis constant

What does K_M represent? Consider the case when the substrate concentration is such that the reaction rate is equal to one-half the maximum rate

$$-r_S = \frac{V_{max}}{2}$$

then

$$\frac{V_{max}}{2} = \frac{V_{max}(S_{1/2})}{K_M + (S_{1/2})} \tag{9-27}$$

Solving Equation (9-27) for the Michaelis constant yields

$$K_M = (S_{1/2}) \tag{9-28}$$

$K_M = (S_{1/2})$ *The Michaelis constant is equal to the substrate concentration at which the rate of reaction is equal to one-half the maximum rate*, i.e., $-r_A = V_{max}/2$. The larger the value of K_M, the higher the substrate concentration necessary for the reaction rate to reach half of its maximum value.

The parameters V_{max} and K_M characterize the enzymatic reactions that are described by Michaelis–Menten kinetics. V_{max} is dependent on total enzyme concentration, whereas K_M is not.

Two enzymes may have the same values for k_{cat} but have different reaction rates because of different values of K_M. One way to compare the catalytic efficiencies of different enzymes is to compare their ratios k_{cat}/K_M. When this ratio approaches 10^8 to 10^9 ($dm^3/mol/s$), the reaction rate approaches becoming diffusion limited. That is, it takes a long time for the enzyme and substrate to find each other, but once they do they react immediately. We will discuss diffusion-limited reactions in Chapters 14 and 15.

Example 9–2 Evaluation of Michaelis–Menten Parameters V_{max} and K_M

Determine the Michaelis–Menten parameters V_{max} and K_M for the reaction

$$\text{Urea} + \text{Urease} \underset{k_2}{\overset{k_1}{\rightleftarrows}} [\text{Urea·Urease}]^* \xrightarrow[+H_2O]{k_3} 2NH_3 + CO_2 + \text{Urease}$$

$$S + E \rightleftarrows E \cdot S \xrightarrow{+H_2O} P + E$$

The rate of reaction is given as a function of urea concentration in the table below, where $(S) \equiv C_{urea}$.

C_{urea} (kmol/m³)	0.2	0.02	0.01	0.005	0.002
$-r_{urea}$ (kmol/m³·s)	1.08	0.55	0.38	0.2	0.09

Solution

Lineweaver–Burk equation

Inverting Equation (9-26) gives us the Lineweaver–Burk equation

$$\frac{1}{-r_s} = \frac{(S) + K_M}{V_{max}(S)} = \frac{1}{V_{max}} + \frac{K_M}{V_{max}}\frac{1}{(S)} \tag{E9-2.1}$$

or

$$\boxed{\frac{1}{-r_{urea}} = \frac{1}{V_{max}} + \frac{K_M}{V_{max}}\left(\frac{1}{C_{urea}}\right)} \tag{E9-2.2}$$

A plot of the reciprocal reaction rate versus the reciprocal urea concentration should be a straight line with an intercept ($1/V_{max}$) and slope (K_M/V_{max}). This type of plot is called a *Lineweaver–Burk plot*. We shall use the data in Table E9-2.1 to make two plots: a plot of $-r_{urea}$ as a function of C_{urea} using Equation (9-26), which is called a *Michaelis–Menten plot* and is shown in Figure E9-2.1(**a**); and a plot of ($1/-r_{urea}$) as a function ($1/C_{urea}$), which is called a *Lineweaver–Burk plot* and is shown in Figure 9-2.1 (**b**).

TABLE E9-2.1   RAW AND PROCESSED DATA

C_{urea} (kmol/m³)	$-r_{urea}$ (kmol/m³·s)	$1/C_{urea}$ (m³/kmol)	$1/-r_{urea}$ (m³·s/kmol)
0.20	1.08	5.0	0.93
0.02	0.55	50.0	1.82
0.01	0.38	100.0	2.63
0.005	0.20	200.0	5.00
0.002	0.09	500.0	11.11

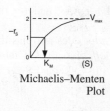

Michaelis–Menten Plot

Lineweaver–Burk Plot

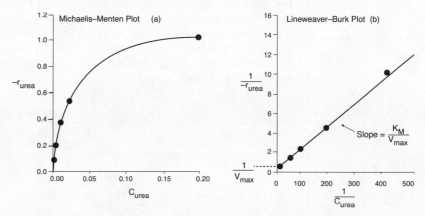

Figure E9-2.1   (a) Michaelis–Menten plot; (b) Lineweaver–Burk plot.

The intercept on Figure E9-2.1(b) is 0.75, so

$$\frac{1}{V_{max}} = 0.75 \ \text{m}^3 \cdot \text{s/kmol}$$

Therefore, the maximum rate of reaction is

$$V_{max} = 1.33 \ \text{kmol/m}^3 \cdot \text{s} = 1.33 \ \text{mol/dm}^3 \cdot \text{s}$$

From the slope, which is 0.02 s, we can calculate the Michaelis constant, K_M

$$\frac{K_M}{V_{max}} = \text{slope} = 0.02 \ \text{s}$$

For enzymatic reactions, the two key rate-law parameters are V_{max} and K_M.

Solving for K_M: $K_M = 0.0266 \ \text{kmol/m}^3$.

Substituting K_M and V_{max} into Equation (9-26) gives us

$$\boxed{-r_{urea} = \frac{1.33 C_{urea}}{0.0266 + C_{urea}}} \tag{E9-2.3}$$

where C_{urea} has units of (kmol/m³) and $-r_{urea}$ has units of (kmol/m³·s). Levine and LaCourse suggest that the total concentration of urease, (E_t), corresponding to the value of V_{max} above is approximately 5 g/dm³.

In addition to the *Lineweaver–Burk plot*, one can also use a *Hanes–Woolf plot* or an *Eadie–Hofstee plot*. Here, $S \equiv C_{urea}$, and $-r_S \equiv -r_{urea}$. Equation (9-26)

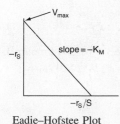

Eadie–Hofstee Plot

$$-r_S = \frac{V_{max}(S)}{K_M + (S)} \tag{9-26}$$

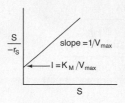

Hanes–Woolf Plot

can be rearranged in the following forms. For the *Eadie–Hofstee* form

$$-r_S = V_{max} - K_M\left(\frac{-r_S}{(S)}\right) \tag{E9-2.4}$$

For the *Hanes–Woolf* form, we can rearrange Equation (9-26) to

$$\frac{(S)}{-r_S} = \frac{K_M}{V_{max}} + \frac{1}{V_{max}}(S) \tag{E9-2.5}$$

For the *Eadie–Hofstee model*, we plot $-r_S$ as a function of $[-r_S/(S)]$ and for the *Hanes–Woolf model*, we plot $[(S)/-r_S]$ as a function of (S).

When to use the different models? The *Eadie–Hofstee plot* does not bias the points at low substrate concentrations, while the *Hanes–Woolf plot* gives a more accurate evaluation of V_{max}. In Table E9-2.2, we add two columns to Table E9-2.1 to generate these plots ($C_{urea} \equiv S$).

TABLE E9-2.2   RAW AND PROCESSED DATA

S (kmol/m³)	$-r_S$ (kmol/m³ · s)	$1/S$ (m³/kmol)	$1/-r_S$ (m³ · s/kmol)	$S/-r_s$ (s)	$-r_S/S$ (1/s)
0.20	1.08	5.0	0.93	0.185	5.4
0.02	0.55	50.0	1.82	0.0364	27.5
0.01	0.38	100.0	2.63	0.0263	38
0.005	0.20	200.0	5.00	0.0250	40
0.002	0.09	500.0	11.11	0.0222	45

The slope of the *Hanes Woolf* plot in Figure E9-2.2 (i.e., $(1/V_{max}) = 0.826$ s · m³/kmol), gives $V_{max} = 1.2$ kmol/m³ · s from the intercept, $K_M/V_{max} = 0.02$ s, and we obtain K_M to be 0.024 kmol/m³.

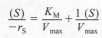

Hanes–Woolf plot

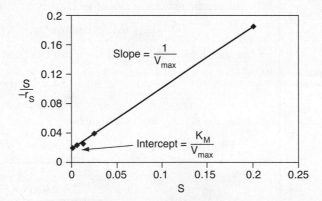

Figure E9-2.2 Hanes–Woolf plot.

From the slope at the *Eadie-Hofstee* plot in Figure E9-2.3 (–0.0244 kmol/m³), we find $K_M = 0.024$ kmol/m³. Next, using the intercept at $-r_S = 0$, i.e., $V_{max}/K_M = 50$ s⁻¹, we calculate $V_{max} = 1.22$ kmol/m³ · s.

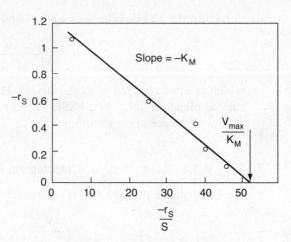

$$-r_S = V_{max} - K_M\left(\frac{-r_S}{(S)}\right)$$

Eadie–Hofstee plot

Figure E9-2.3 Eadie–Hofstee plot.

Regression

Equation (9-26) and Table E9-2.1 were used in the regression program of Polymath with the following results for V_{max} and K_M.

TABLE E9-3   REGRESSION RESULTS

Nonlinear regression (L-M)

Model: rate = Vmax*Curea/(Km+Curea)

Variable	Ini guess	Value	95% confidence
Vmax	1	1.2057502	0.0598303
Km	0.02	0.0233322	0.003295

Nonlinear regression settings
Max # iterations = 64

$$V_{max} = 1.2 \text{ mol/dm}^3 \cdot s$$
$$K_M = 0.0233 \text{ mol/dm}^3$$

Precision
R^2   = 0.9990611
R^2adj   = 0.9987481
Rmsd   = 0.0047604
Variance = 1.888E-04

These values are within experimental error of those values of V_{max} and K_m determined graphically.

Analysis: This example demonstrated how to evaluate the parameters V_{max} and K_M in the Michaelis–Menten rate law from enzymatic reaction data. Two techniques were used: a Lineweaver–Burk plot and nonlinear regression. It was also shown how the analysis could be carried out using Hanes–Woolf and Eadie–Hofstee plots.

The Product-Enzyme Complex

In many reactions the enzyme and product complex $(E \cdot P)$ is formed directly from the enzyme substrate complex $(E \cdot S)$ according to the sequence

$$E + S \rightleftharpoons E \cdot S \rightleftharpoons P \cdot E \rightleftharpoons P + E$$

Briggs–Haldane Rate Law

Applying the PSSH, after much effort and some approximations we obtain

$$-r_S = \frac{V_{max}(C_S - C_P/K_C)}{C_S + K_{max} + K_P C_P} \tag{9-29}$$

which is often referred to as the Briggs–Haldane Equation [see P9-8$_B$ **(a)**] and the application of the PSSH to enzyme kinetics, often called the *Briggs–Haldane approximation*.

9.2.4 Batch-Reactor Calculations for Enzyme Reactions

A mole balance on urea in a batch reactor gives

Mole balance

$$-\frac{dN_{urea}}{dt} = -r_{urea} V$$

Because this reaction is liquid phase, $V = V_0$, the mole balance can be put in the following form:

$$-\frac{dC_{urea}}{dt} = -r_{urea} \tag{9-30}$$

The rate law for urea decomposition is

Rate law

$$-r_{urea} = \frac{V_{max} C_{urea}}{K_M + C_{urea}} \tag{9-31}$$

Substituting Equation (9-31) into Equation (9-30) and then rearranging and integrating, we get

Combine

$$t = \int_{C_{urea}}^{C_{urea0}} \frac{dC_{urea}}{-r_{urea}} = \int_{C_{urea}}^{C_{urea0}} \frac{K_M + C_{urea}}{V_{max} C_{urea}} dC_{urea}$$

Integrate

$$\boxed{t = \frac{K_M}{V_{max}} \ln \frac{C_{urea0}}{C_{urea}} + \frac{C_{urea0} - C_{urea}}{V_{max}}} \tag{9-32}$$

We can write Equation (9-32) in terms of conversion as

$$C_{urea} = C_{urea0}(1 - X)$$

Time to achieve a conversion X in a batch enzymatic reaction

$$\boxed{t = \frac{K_M}{V_{max}} \ln \frac{1}{1 - X} + \frac{C_{urea0} X}{V_{max}}} \tag{9-32}$$

The parameters K_M and V_{max} can readily be determined from batch-reactor data by using the integral method of analysis. Dividing both sides of Equation (9-32) by $(t K_M / V_{max})$ and rearranging yields

$$\frac{1}{t} \ln \frac{1}{1 - X} = \frac{V_{max}}{K_M} - \frac{C_{urea0} X}{K_M t}$$

We see that K_M and V_{max} can be determined from the slope and intercept of a plot of $(1/t) \ln[1/(1 - X)]$ versus X/t. We could also express the Michaelis–Menten equation in terms of the substrate concentration S

$$\boxed{\frac{1}{t} \ln \frac{S_0}{S} = \frac{V_{max}}{K_M} - \frac{S_0 - S}{K_M t}} \tag{9-33}$$

where S_0 is the initial concentration of substrate. In cases similar to Equation (9-33) where there is no possibility of confusion, we shall not bother to enclose the substrate or other species in parentheses to represent concentration [i.e., $C_S \equiv (S) \equiv S$]. The corresponding plot in terms of substrate concentration is shown in Figure 9-7.

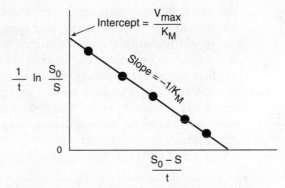

Figure 9-7   Evaluating V_{max} and K_M from batch-reactor data.

Example 9–3   Batch Enzymatic Reactors

Calculate the time needed to convert 99% of the urea to ammonia and carbon dioxide in a 0.5-dm^3 batch reactor. The initial concentration of urea is 0.1 mol/dm^3, and the urease concentration is 0.001 g/dm^3. The reaction is to be carried out isothermally at the same temperature at which the data in Table E9-2.2 were obtained.

Solution

We can use Equation (9-32)

$$t = \frac{K_M}{V_{max}} \ln \frac{1}{1 - X} + \frac{C_{urea0} X}{V_{max}} \tag{9-32}$$

From Table E9-2.3 we know $K_M = 0.0233$ mol/dm^3, $V_{max} = 1.2$ $mol/dm^3 \cdot s$. The conditions given are $X = 0.99$ and $C_{urea0} = 0.1$ mol/dm^3 (i.e., 0.1 $kmol/m^3$). However, for the conditions in the batch reactor, the enzyme concentration is only 0.001 g/dm^3, compared with 5 g/dm^3 in Example 9-2. Because $V_{max} = E_t \cdot k_3$, V_{max} for the second enzyme concentration is

$$V_{max2} = \frac{E_{t2}}{E_{t1}} V_{max1} = \frac{0.001}{5} \times 1.2 = 2.4 \times 10^{-4} \text{ mol/s} \cdot dm^3$$

$$K_M = 0.0233 \text{ mol/dm}^3 \quad \text{and} \quad X = 0.99$$

Substituting into Equation (9-32)

$$t = \frac{2.33 \times 10^{-2} \text{ mol/dm}^3}{2.4 \times 10^{-4} \text{ mol/dm}^3/\text{s}} \ln\left(\frac{1}{0.01}\right) + \frac{(0.1 \text{ mol/dm}^3)(0.99)}{2.4 \times 10^{-4} \text{ mol/dm}^3/\text{s}}$$

$$= 447 \text{ s} + 412 \text{ s}$$

$$= 859 \text{ s} \quad (14.325 \text{ min})$$

Analysis: This example shows a straightforward Chapter 5–type calculation of the batch reactor time to achieve a certain conversion X for an enzymatic reaction with a Michaelis–Menten rate law. This batch reaction time is very short; consequently, a continuous-flow reactor would be better suited for this reaction.

Effect of Temperature

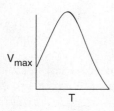

The effect of temperature on enzymatic reactions is very complex. If the enzyme structure would remain unchanged as the temperature is increased, the rate would probably follow the Arrhenius temperature dependence. However, as the temperature increases, the enzyme can unfold and/or become denatured and lose its catalytic activity. Consequently, as the temperature increases, the reaction rate, $-r_S$, increases up to a maximum and then decreases as the temperature is increased further. The descending part of this curve is called *temperature inactivation* or *thermal denaturizing*.[10] Figure 9-8 shows an example of this optimum in enzyme activity.[11]

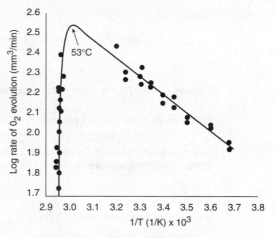

Figure 9-8   Catalytic breakdown rate of H_2O_2 depending on temperature. Courtesy of S. Aiba, A. E. Humphrey, and N. F. Mills, *Biochemical Engineering*, Academic Press (1973).

9.3 Inhibition of Enzyme Reactions

In addition to temperature and solution pH, another factor that greatly influences the rates of enzyme-catalyzed reactions is the presence of an inhibitor. Inhibitors are species that interact with enzymes and render the enzyme either

---

[10] M. L. Shuler and F. Kargi, *Bioprocess Engineering Basic Concepts*, 2nd ed. (Upper Saddle River, NJ: Prentice Hall, 2002), p. 77.

[11] S. Aiba, A. E. Humphrey, and N. F. Mills, *Biochemical Engineering* (New York: Academic Press, 1973), p. 47.

partially or totally ineffective to catalyze its specific reaction. The most dramatic consequences of enzyme inhibition are found in living organisms, where the inhibition of any particular enzyme involved in a primary metabolic pathway will render the entire pathway inoperative, resulting in either serious damage to or death of the organism. For example, the inhibition of a single enzyme, cytochrome oxidase, by cyanide will cause the aerobic oxidation process to stop; death occurs in a very few minutes. There are also beneficial inhibitors, such as the ones used in the treatment of leukemia and other neoplastic diseases. Aspirin inhibits the enzyme that catalyzes the synthesis of the module prostaglandin, which is involved in the pain-producing process. Recently the discovery of DDP-4 enzyme inhibitor Januvia has been approved for the treatment of Type 2 diabetes, a disease affecting 240 million people worldwide.

The three most common types of reversible inhibition occurring in enzymatic reactions are *competitive*, *uncompetitive*, and *noncompetitive*. The enzyme molecule is analogous to a heterogeneous catalytic surface in that it contains active sites. When *competitive* inhibition occurs, the substrate and inhibitor are usually similar molecules that compete for the same site on the enzyme. *Uncompetitive* inhibition occurs when the inhibitor deactivates the enzyme–substrate complex, sometimes by attaching itself to both the substrate and enzyme molecules of the complex. *Noncompetitive* inhibition occurs with enzymes containing at least two different types of sites. The substrate attaches only to one type of site, and the inhibitor attaches only to the other to render the enzyme inactive.

9.3.1 Competitive Inhibition

Competitive inhibition is of particular importance in pharmacokinetics (drug therapy). If a patient were administered two or more drugs that react simultaneously within the body with a common enzyme, cofactor, or active species, this interaction could lead to competitive inhibition in the formation of the respective metabolites and produce serious consequences.

In competitive inhibition, another substance, i.e., I, competes with the substrate for the enzyme molecules to form an inhibitor–enzyme complex, as shown in Figure 9-9.

In addition to the three Michaelis–Menten reaction steps, there are two additional steps as the inhibitor (I) reversely ties up the enzyme, as shown in Reaction Steps 4 and 5.

The rate law for the formation of product is the same [cf. Equations (9-18A) and (9-19)] as it was before in the absence of inhibitor

$$r_P = k_3 \, (E \cdot S) \tag{9-34}$$

Applying the PSSH, the net rate of reaction of the enzyme–substrate complex is

$$r_{E \cdot S} = 0 = k_1 \, (E)(S) - k_2 (E \cdot S) - k_3 \, (E \cdot S) \tag{9-35}$$

The net rate of formation of the inhibitor–substrate complex $(E \cdot I)$ is also zero

$$r_{E \cdot I} = 0 = k_4 \, (E)(I) - k_5 (E \cdot I) \tag{9-36}$$

Reaction Steps

(1) $E + S \xrightarrow{k_1} E \cdot S$

(2) $E \cdot S \xrightarrow{k_2} E + S$

(3) $E \cdot S \xrightarrow{k_3} P + E$

(4) $I + E \xrightarrow{k_4} E \cdot I$ (inactive)

(5) $E \cdot I \xrightarrow{k_5} E + I$

Competitive Inhibition Pathway

Competitive
inhibition pathway

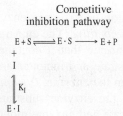

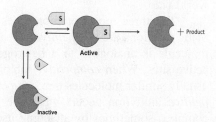

Figure 9-9   Competitive inhibition. Schematic drawing courtesy of Jofostan National Library, Lunčo, Jofostan, established 2019.

The total enzyme concentration is the sum of the bound and unbound enzyme concentrations

$$E_t = [E] + (E \cdot S) + (E \cdot I) \tag{9-37}$$

Combining Equations (9-35), (9-36), and (9-37), solving for $(E \cdot S)$ then substituting in Equation (9-34) and simplifying

Rate law for
competitive
inhibition

$$r_P = -r_S = \frac{V_{max}(S)}{(S) + K_M \left(1 + \dfrac{(I)}{K_I}\right)} \tag{9-38}$$

V_{max} and K_M are the same as before when no inhibitor is present; that is,

$$V_{max} = k_3 E_t \text{ and } K_M = \frac{k_2 + k_3}{k_1}$$

and the inhibition constant K_I (mol/dm^3) is

$$K_I = \frac{k_5}{k_4}$$

By letting $K'_M = K_M(1 + (I)/K_I)$, we can see that the effect of adding a competitive inhibitor is to increase the "apparent" Michaelis constant, K'_M. A consequence of the larger "apparent" Michaelis constant K'_M is that a larger substrate concentration is needed for the rate of substrate decomposition, $-r_S$, to reach half its maximum rate.

Rearranging Equation (9-38) in order to generate a Lineweaver–Burk plot

$$\frac{1}{-r_s} = \frac{1}{V_{max}} + \frac{1}{(S)}\left[\frac{K_M}{V_{max}}\left(1 + \frac{(I)}{K_I}\right)\right] \tag{9-39}$$

From the Lineweaver–Burk plot (Figure 9-10), we see that as the inhibitor (I) concentration is increased, the slope increases (i.e., the rate decreases), while the intercept remains fixed.

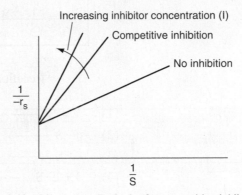

Figure 9-10   Lineweaver–Burk plot for competitive inhibition.

Side Note: Methanol Poisoning. An interesting and important example of *competitive substrate inhibition* is the enzyme alcohol dehydrogenase (ADH) in the presence of ethanol and methanol. If a person ingests methanol, ADH will convert it to formaldehyde and then formate, which causes blindness. Consequently, one treatment involves intravenously injecting ethanol (which is metabolized at a slower rate than methanol) at a controlled rate to tie up ADH to slow the metabolism of methanol to formaldehyde to formate so that the kidneys have time to filter out the methanol, which is then excreted in the urine. With this treatment, blindness may be avoided.

9.3.2 Uncompetitive Inhibition

Here, the inhibitor has no affinity for the enzyme by itself and thus does not compete with the substrate for the enzyme; instead, it ties up the enzyme–substrate complex by forming an inhibitor–enzyme–substrate complex, (I · E · S), which is inactive. In uncompetitive inhibition, the inhibitor reversibly ties up enzyme–substrate complex *after* it has been formed.

As with competitive inhibition, two additional reaction steps are added to the Michaelis–Menten kinetics for uncompetitive inhibition, as shown in Reaction Steps 4 and 5 in Figure 9-11.

Starting with the equation for the rate of formation of product, Equation (9-34), and then applying the pseudo-steady-state hypothesis to the intermediate $(I \cdot E \cdot S)$, we arrive at the rate law for uncompetitive inhibition

Rate law for uncompetitive inhibition

$$-r_s = r_p = \frac{V_{max}(S)}{K_M + (S)\left(1 + \dfrac{(I)}{K_I}\right)} \quad \text{where } K_I = \frac{k_5}{k_4} \tag{9-40}$$

The intermediate steps are shown in the Chapter 9 *Summary Notes* in Learning Resources on the CRE Web site. Rearranging Equation (9-40)

$$\frac{1}{-r_s} = \frac{1}{(S)}\frac{K_M}{V_{max}} + \frac{1}{V_{max}}\left(1 + \frac{(I)}{K_I}\right) \tag{9-41}$$

Reaction Steps

(1) $E + S \xrightarrow{\;k_1\;} E \cdot S$

(2) $E \cdot S \xrightarrow{\;k_2\;} E + S$

(3) $E \cdot S \xrightarrow{\;k_3\;} P + E$

(4) $I + E \cdot S \xrightarrow{\;k_4\;} I \cdot E \cdot S$ (inactive)

(5) $I \cdot E \cdot S \xrightarrow{\;k_5\;} I + E \cdot S$

Uncompetitive Pathway

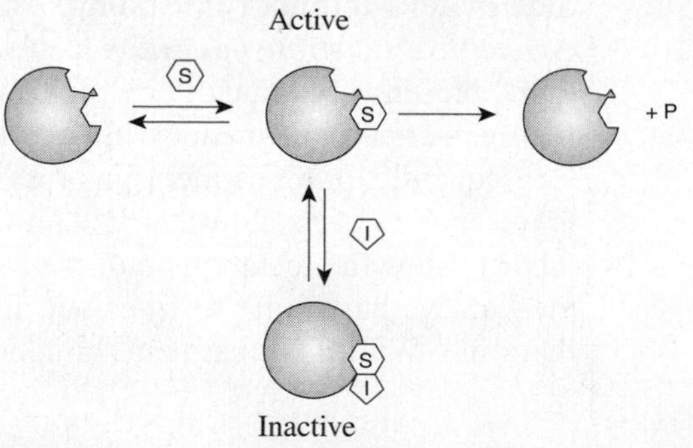

Active

Uncompetitive inhibition pathway

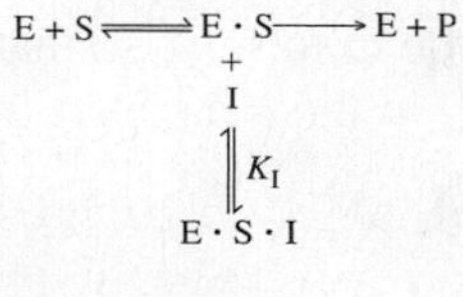

$E + S \rightleftharpoons E \cdot S \longrightarrow E + P$
       $+$
       I
       $\Big\Vert K_I$
$E \cdot S \cdot I$

Inactive

Figure 9-11   Steps in uncompetitive enzyme inhibition.

The Lineweaver–Burk plot is shown in Figure 9-12 for different inhibitor concentrations. The slope (K_M/V_{max}) remains the same as the inhibitor (I) concentration is increased, while the intercept $[(1/V_{max})(1 + (I)/K_I)]$ increases.

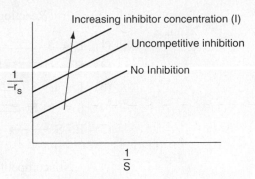

Figure 9-12   Lineweaver–Burk plot for uncompetitive inhibition.

9.3.3 Noncompetitive Inhibition (Mixed Inhibition)[12]

In noncompetitive inhibition, also sometimes called *mixed inhibition*, the substrate and inhibitor molecules react with different types of sites on the enzyme molecule. Whenever the inhibitor is attached to the enzyme, it is inactive and cannot form products. Consequently, the deactivating complex $(I \cdot E \cdot S)$ can be formed by two reversible reaction paths.

1. After a substrate molecule attaches to the enzyme molecule at the substrate site, then the inhibitor molecule attaches to the enzyme at the inhibitor site. $(E \cdot S + I \rightleftarrows I \cdot E \cdot S)$

2. After an inhibitor molecule attaches to the enzyme molecule at the inhibitor site, then the substrate molecule attaches to the enzyme at the substrate site. $(E \cdot I + S \rightleftarrows I \cdot E \cdot S)$

These paths, along with the formation of the product, P, are shown in Figure 9-13. In noncompetitive inhibition, the enzyme can be tied up in its inactive form either *before or after* forming the enzyme–substrate complex as shown in Steps 2, 3, and 4.

Again, starting with the rate law for the rate of formation of product and then applying the PSSH to the complexes $(I \cdot E)$ and $(I \cdot E \cdot S)$, we arrive at the rate law for the noncompetitive inhibition

Summary Notes

Rate law for
noncompetitive
inhibition

$$-r_{\mathrm{s}} = \frac{V_{\max}(\mathrm{S})}{((\mathrm{S}) + K_{\mathrm{M}}\left(1 + \dfrac{(I)}{K_I}\right)} \tag{9-42}$$

The derivation of the rate law is given in the *Summary Notes* on the CRE Web site. Equation (9-42) is in the form of the rate law that is given for an enzymatic reaction exhibiting noncompetitive inhibition. Heavy metal ions such as Pb^{2+}, Ag^+, and Hg^{2+}, as well as inhibitors that react with the enzyme to form chemical derivatives, are typical examples of noncompetitive inhibitors.

Rearranging

$$\frac{1}{-r_s} = \frac{1}{V_{\max}}\left(1 + \frac{(I)}{K_I}\right) + \frac{1}{(\mathrm{S})}\frac{K_{\mathrm{M}}}{V_{\max}}\left(1 + \frac{(I)}{K_I}\right) \tag{9-43}$$

---

[12] In some texts, mixed inhibition is a combination of competitive and uncompetitive inhibition.

Reaction Steps

(1) $E + S \rightleftharpoons E \cdot S$

(2) $E + I \rightleftharpoons I \cdot E$ (inactive)

(3) $I + E \cdot S \rightleftharpoons I \cdot E \cdot S$ (inactive)

(4) $S + I \cdot E \rightleftharpoons I \cdot E \cdot S$ (inactive)

(5) $E \cdot S \longrightarrow P + E$

Noncompetitive Pathway

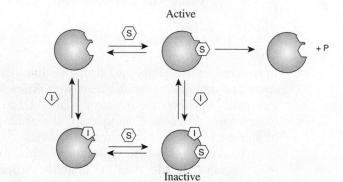

Mixed inhibition

$$\begin{array}{ccc} E+S \rightleftharpoons & E \cdot S \longrightarrow E+P \\ + & + \\ I & I \\ \Big\downarrow K_I & \Big\downarrow K_I \\ E \cdot I + S \rightleftharpoons & E \cdot S \cdot I \end{array}$$

Active

Inactive

Figure 9-13 Steps in noncompetitive enzyme inhibition.

For noncompetitive inhibition, we see in Figure 9-14 that both the slope $\left(\dfrac{K_M}{V_{max}}\left[1+\dfrac{(I)}{K_I}\right]\right)$ and intercept $\left(\dfrac{1}{V_{max}}\left[1+\dfrac{(I)}{K_I}\right]\right)$ increase with increasing inhibitor concentration. In practice, *uncompetitive inhibition* and *mixed inhibition* are generally observed only for enzymes with two or more substrates, S_1 and S_2.

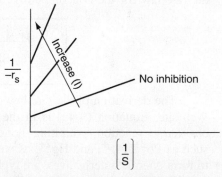

$\dfrac{1}{-r_s}$ Increase (I)   No inhibition $\left[\dfrac{1}{S}\right]$

Figure 9-14 Lineweaver–Burk plot for noncompetitive enzyme inhibition.

The three types of inhibition are compared with a reaction without inhibitors and are summarized on the Lineweaver–Burk plot shown in Figure 9-15.

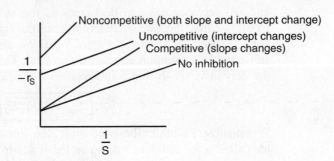

Summary plot of
types of inhibition

Figure 9-15 Summary: Lineweaver–Burk plots for three types
of enzyme inhibition.

In summary, we observe the following trends and relationships:

1. In *competitive inhibition*, the slope increases with increasing inhibitor concentration, while the intercept remains fixed.
2. In *uncompetitive inhibition*, the *y*-intercept increases with increasing inhibitor concentration, while the slope remains fixed.
3. In *noncompetitive inhibition (mixed inhibition)*, both the *y*–intercept and slope will increase with increasing inhibitor concentration.

Problem P9-12$_B$ asks you to find the type of inhibition for the enzyme catalyzed reaction of starch.

9.3.4 Substrate Inhibition

In a number of cases, the substrate itself can act as an inhibitor. In the case of uncompetitive inhibition, the inactive molecule $(S \cdot E \cdot S)$ is formed by the reaction

$$S + E \cdot S \longrightarrow S \cdot E \cdot S \quad \text{(inactive)}$$

Consequently, we see that by replacing (I) by (S) in Equation (9-40), the rate law for $-r_S$ is

$$-r_S = \frac{V_{max}(S)}{K_M + (S) + \dfrac{(S)^2}{K_I}} \tag{9-44}$$

We see that at low substrate concentrations

$$K_M >> \left((S) + \frac{(S)^2}{K_I} \right) \tag{9-45}$$

then

$$-r_S \sim \frac{V_{max}(S)}{K_M} \tag{9-46}$$

and the rate increases linearly with increasing substrate concentration.
At high substrate concentrations $((S)^2 / K_I) >> (K_M + (S))$, then

Substrate inhibition
$$-r_S = \frac{V_{max}K_I}{S} \tag{9-47}$$

and we see that the rate decreases as the substrate concentration increases. Consequently, the rate of reaction goes through a maximum in the substrate concentration, as shown in Figure 9-16. We also see that there is an optimum substrate concentration at which to operate. This maximum is found by setting the derivative of $-r_S$ wrt S in Equation (9-44) equal to 0, to obtain

$$\boxed{S_{max} = \sqrt{K_M K_I}} \tag{9-48}$$

When substrate inhibition is possible, the substrate is fed to a semibatch reactor called a *fed batch* to maximize the reaction rate and conversion.

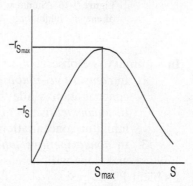

Figure 9-16   Substrate reaction rate as a function of substrate concentration for substrate inhibition.

Our discussion of enzymes is continued in the *Professional Reference Shelf* on the CRE Web site where we describe multiple enzyme and substrate systems, enzyme regeneration, and enzyme co-factors (see R9.6).

9.4 Bioreactors and Biosynthesis

A *bioreactor* is a reactor that sustains and supports life for cells and tissue cultures. Virtually all cellular reactions necessary to maintain life are mediated by enzymes as they catalyze various aspects of cell metabolism such as the transformation of chemical energy and the construction, breakdown, and digestion of cellular components. Because enzymatic reactions are involved in the growth of microorganisms (biomass), we now proceed to study microbial growth and bioreactors. Not surprisingly, the Monod equation, which describes the growth law for a number of bacteria, is similar to the Michaelis–Menten equation. Consequently, even though bioreactors are not truly homogeneous because of the presence of living cells, we include them in this chapter as a logical progression from enzymatic reactions.

The use of living cells to produce marketable chemical products is becoming increasingly important. The number of chemicals, agricultural products, and food products produced by biosynthesis has risen dramatically. In 2016, companies in this sector raised over $200 billion of new financing.[13] Both microorganisms and mammalian cells are being used to produce a variety of products, such as insulin, most antibiotics, and polymers. It is expected that in the future a number of organic chemicals currently derived from petroleum

Reference Shelf

Nutrients

Cell   Products

The growth of biotechnology

$200 billion

---

[13] *http://www.statista.com/topics/1634/biotechnology-industry/*

will be produced by living cells. The advantages of bioconversions are mild reaction conditions; high yields (e.g., 100% conversion of glucose to gluconic acid with *Aspergillus niger*); and the fact that organisms contain several enzymes that can catalyze successive steps in a reaction and, most importantly, act as stereospecific catalysts. A common example of specificity in bioconversion production of a *single* desired isomer that, when produced chemically, yields a mixture of isomers is the conversion of *cis*-proenylphosphonic acid to the antibiotic $(-)$ *cis*-1,2-epoxypropyl-phosphonic acid. Bacteria can also be modified and turned into living chemical factories. For example, using recombinant DNA, Biotechnic International engineered a bacteria to produce fertilizer by turning nitrogen into nitrates.[14]

More recently, the synthesis of biomass (i.e., cell/organisms) has become an important alternative energy source. Sapphire energy, the world's first integrated algae-oil production facility, has pilot growth ponds in New Mexico. In this process, they grow the algae, flocculate and concentrate it, extract it, and then refine it and convert it to fuel oil in a liquid-phase flow reactor (see Problems P9-20$_B$ and P9-21$_B$). In 2009, ExxonMobil invested over 600 million dollars to develop algae growth and harvest it in waste ponds. It is estimated that one acre of algae can provide 2,000 gallons of gasoline per year.

In biosynthesis, the cells, also referred to as the *biomass*, consume nutrients to grow and produce more cells and important products. Internally, a cell uses its nutrients to produce energy and more cells. This transformation of nutrients to energy and bioproducts is accomplished through a cell's use of a number of different enzymes in a series of reactions to produce metabolic products. These products can either remain in the cell (intracellular) or be secreted from the cells (extracellular). In the former case, the cells must be lysed (ruptured) and the product filtered and purified from the whole broth (reaction mixture). A schematic of a cell is shown in Figure 9-17.

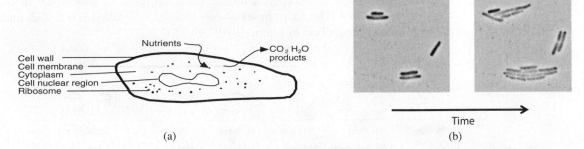

(a)   (b)

Figure 9-17   (a) Schematic of cell; (b) division of E. Coli.
Adapted from "Indole prevents *Escherichia coli* cell division by modulating membrane potential."
Catalin Chimirel, Christopher M. Field, Silvia Piñera-Fernandez, Ulrich F. Keyser, David K.
Summers. *Biochimica et Biophysica Acta–Biomembranes*, vol. 1818, issue 7, July 2012.

The cell consists of a cell wall and an outer membrane that encloses the cytoplasm containing a nuclear region and ribosomes. The cell wall protects the cell from external influences. The cell membrane provides for selective transport of materials into and out of the cell. Other substances can attach to the cell membrane to carry out important cell functions. The cytoplasm con-

---

[14] *Chem. Eng. Progr.*, August 1988, p. 18

tains the ribosomes that contain ribonucleic acid (RNA), which are important in the synthesis of proteins. The nuclear region contains deoxyribonucleic acid (DNA), which provides the genetic information for the production of proteins and other cellular substances and structures.[15]

The reactions in the cell all take place simultaneously and are classified as either class (I) nutrient degradation (fueling reactions), class (II) synthesis of small molecules (amino acids), or class (III) synthesis of large molecules (polymerization, e.g., RNA, DNA). A rough overview with only a fraction of the reactions and metabolic pathways is shown in Figure 9-18. A more detailed model is given in Figures 5.1 and 6.14 of Shuler and Kargi.[16] In the Class I reactions, adenosine triphosphate (ATP) participates in the degradation of nutrients to form products to be used in the biosynthesis reactions (Class II) of small molecules (e.g., amino acids), which are then polymerized to form RNA and DNA (Class III). ATP also transfers the energy it releases when it loses a phosphonate group to form adenosine diphosphate (ADP).

$$ATP + H_2O \rightarrow ADP + P + H_2O + Energy$$

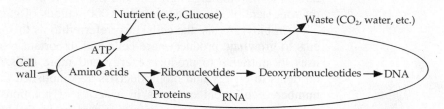

Figure 9-18   Examples of reactions occurring in the cell.

Cell Growth and Division

The cell growth and division typical of mammalian cells is shown schematically in Figure 9-19. The four phases of cell division are called G1, S, G2, and M, and are also described in Figure 9-19.

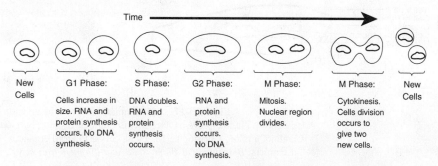

Figure 9-19   Phases of cell division.

---

[15] M. L. Shuler and F. Kargi, *Bioprocess Engineering Basic Concepts*, 2nd ed. (Upper Saddle River, NJ: Prentice Hall, 2002).

[16] M. L. Shuler and F. Kargi, *Bioprocess Engineering Basic Concepts*, 2nd ed. (Upper Saddle River, NJ: Prentice Hall, 2002), pp.135, 185.

In general, the growth of an aerobic organism follows the equation

$$[\text{Cells}] + \begin{bmatrix} \text{Carbon} \\ \text{source} \end{bmatrix} + \begin{bmatrix} \text{Nitrogen} \\ \text{source} \end{bmatrix} + \begin{bmatrix} \text{Oxygen} \\ \text{source} \end{bmatrix} + \begin{bmatrix} \text{Phosphate} \\ \text{source} \end{bmatrix} + \cdots$$

Cell multiplication $$[CO_2] + [H_2O] + [\text{Products}] + \begin{bmatrix} \text{More} \\ \text{cells} \end{bmatrix} \xleftarrow{} \begin{array}{c} \text{Culture media} \\ \text{conditions} \\ \text{(pH, temperature, etc.)} \end{array}$$

(9-49)

A more abbreviated form of Equation (9-49) generally used is that a substrate in the presence of cells produces more cells plus product, i.e.,

$$\text{Substrate} \xrightarrow{\text{Cells}} \text{More cells} + \text{Product} \qquad (9\text{-}50)$$

The products in Equation (9-50) include carbon dioxide, water, proteins, and other species specific to the particular reaction. An excellent discussion of the stoichiometry (atom and mole balances) of Equation (9-49) can be found in Shuler and Kargi,[17] Bailey and Ollis,[18] and Blanch and Clark.[19] The substrate culture medium contains all the nutrients (carbon, nitrogen, etc.) along with other chemicals necessary for growth. Because, as we will soon see, the rate of this reaction is proportional to the cell concentration, the reaction is autocatalytic. A rough schematic of a simple batch biochemical reactor and the growth of two types of microorganisms, cocci (i.e., spherical) bacteria and yeast, is shown in Figure 9-20.

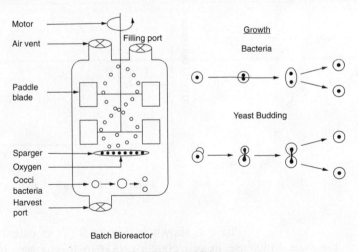

Figure 9-20   Batch bioreactor.

[17] M. L. Shuler and F. Kargi, *Bioprocess Engineering Basic Concepts*, 2nd ed. (Upper Saddle River, NJ: Prentice Hall, 2002).

[18] J. E. Bailey and D. F. Ollis, *Biochemical Engineering*, 2nd ed. (New York: McGraw-Hill, 1987).

[19] H. W. Blanch and D. S. Clark, *Biochemical Engineering* (New York: Marcel Dekker, Inc. 1996).

9.4.1 Cell Growth

Stages of cell growth in a batch reactor are shown schematically in Figures 9-21 and 9-22. Initially, a small number of cells is inoculated into (i.e., added to) the batch reactor containing the nutrients and the growth process begins, as shown in Figure 9-21. In Figure 9-22, the number of living cells is shown as a function of time.

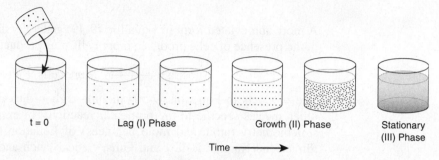

Figure 9-21 Increase in cell concentration.

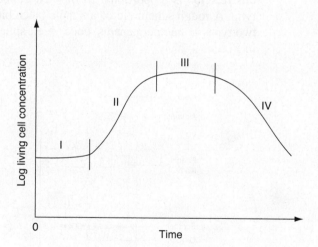

Figure 9-22 Phases of bacteria cell growth.

Lag phase

Phase I, shown in Figure 9-22, is called the *lag phase*. There is little increase in cell concentration in this phase. In the lag phase, the cells are adjusting to their new environment, carrying out such functions as synthesizing transport proteins for moving the substrate into the cell, synthesizing enzymes for utilizing the new substrate, and beginning the work for replicating the cells' genetic material. The duration of the lag phase depends upon many things, one of which is the growth medium from which the inoculum was taken relative to the reaction medium in which it is placed. If the inoculum is similar to the medium of the batch reactor, the lag phase can be almost nonexistent. If, however, the inoculum were placed in a medium with a different nutrient or other contents, or if the inoculum culture were in the stationary or death phase, the cells would have to readjust their metabolic path to allow them to consume the nutrients in their new environment.[20]

Exponential growth
phase

Phase II is called the *exponential growth phase*, owing to the fact that the cells' growth rate is proportional to the cell concentration. In this phase, the cells are dividing at the maximum rate because all of the enzyme's pathways for metabolizing the substrate are now in place (as a result of the lag phase) and the cells are able to use the nutrients most efficiently.

Phase III is the *stationary phase*, during which the cells reach a minimum biological space where the lack of one or more nutrients limits cell growth. During the stationary phase, the net cell growth rate is zero as a result of the depletion of nutrients and essential metabolites. Many important fermentation products, including many antibiotics, are produced in the stationary phase. For example, penicillin produced commercially using the fungus *Penicillium chrysogenum* is formed only after cell growth has ceased. Cell growth is also slowed by the buildup of organic acids and toxic materials generated during the growth phase.

Antibiotics
produced during
the stationary
phase

Death phase

The final phase, Phase IV, is the *death phase,* where a decrease in live cell concentration occurs. This decline is a result of the toxic by-products, harsh environments, and/or depletion of nutrient supply.

9.4.2 Rate Laws

While many laws exist for the cell growth rate of new cells, that is,

$$\text{Cells} + \text{Substrate} \longrightarrow \text{More cells} + \text{Product}$$

the most commonly used expression is the *Monod* equation for exponential growth:

$$r_g = \mu C_c \tag{9-51}$$

where r_g = cell growth rate, g/dm$^3 \cdot$s
  C_c = cell concentration, g/dm^3
  μ = specific growth rate, s^{-1}

The cell concentration is often given in terms of weight (g) of dry cells per liquid volume and is specified "grams dry weight per dm^3," i.e., (gdw/dm^3).

The specific cell growth rate can be expressed as

$$\mu = \mu_{max} \frac{C_s}{K_s + C_s} \quad \text{s}^{-1} \tag{9-52}$$

where μ_{max} = maximum specific growth reaction rate, s^{-1}
  K_s = *Monod* constant, g/dm^3
  C_s = substrate (i.e., nutrient) concentration, g/dm^3

Representative values of μ_{max} and K_s are 1.3 h^{-1} and 2.2×10^{-5} g/dm^3, respectively, which are the parameter values for the *E. coli* growth on glucose. Combining Equations (9-51) and (9-52), we arrive at the Monod equation for bacterial cell growth rate

[20] B. Wolf and H. S. Fogler, "Alteration of the Growth Rate and Lag Time of *Leuconostoc mesenteroides* NRRL-B523," *Biotechnology and Bioengineering*, 72 (6), 603 (2001). B. Wolf and H. S. Fogler, "Growth of *Leuconostoc mesenteroides* NRRL-B523, in Alkaline Medium," *Biotechnology and Bioengineering*, 89 (1), 96 (2005).

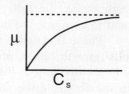

Figure 9-23 Specific cell growth rate, μ, as a function of substrate concentration C_s.

Monod equation

$$r_g = \frac{\mu_{max} C_s C_c}{K_s + C_s} \qquad (9\text{-}53)$$

For a number of different bacteria, the constant K_s is very small, with regard to typical substrate concentrations, in which case the rate law reduces to

$$r_g = \mu_{max} C_c \qquad (9\text{-}54)$$

The growth rate, r_g, often depends on more than one nutrient concentration; however, the nutrient that is limiting is usually the one used in Equation (9-53).

In many systems the product inhibits the rate of growth. A classic example of this inhibition is in winemaking, where the fermentation of glucose to produce ethanol is inhibited by the product ethanol. There are a number of different equations to account for inhibition; one such rate law takes the empirical form

$$r_g = k_{obs} \frac{\mu_{max} C_s C_c}{K_s + C_s} \qquad (9\text{-}55)$$

where

Empirical form of Monod equation for product inhibition

$$k_{obs} = \left(1 - \frac{C_p}{C_p^*}\right)^n \qquad (9\text{-}56)$$

with

C_p = product concentration (g/dm³)
C_p^* = product concentration at which all metabolism ceases, g/dm³
n = empirical constant

For the glucose-to-ethanol fermentation, typical inhibition parameters are

$$n = 0.5 \quad \text{and} \quad C_p^* = 93 \text{ g/dm}^3$$

In addition to the *Monod* equation, two other equations are also commonly used to describe the cell growth rate; they are the *Tessier* equation

$$r_g = \mu_{max}\left[1 - \exp\left(-\frac{C_s}{k}\right)\right] C_c \qquad (9\text{-}57)$$

and the *Moser* equation,

$$r_g = \frac{\mu_{max} C_c}{(1 + k C_s^{-\lambda})} \qquad (9\text{-}58)$$

where λ and k are empirical constants determined by a best fit of the data. The Moser and Tessier growth laws are often used because they have been found to better fit experimental data at the beginning or end of fermentation. Other growth equations can be found in Dean.[21]

The cell death rate is a result of harsh environments, mixing shear forces, local depletion of nutrients, and the presence of toxic substances. The rate law is

$$r_d = (k_d + k_t C_t) C_c \tag{9-59}$$

where C_t is the concentration of a substance toxic to the cell. The specific death rate constants k_d and k_t refer to the natural death and death due to a toxic substance, respectively. Representative values of k_d range from 0.1 h^{-1} to less than 0.0005 h^{-1}. The value of k_t depends on the nature of the toxin.

Doubling times   Microbial growth rates are measured in terms of *doubling times*. Doubling time is the time required for a mass of an organism to double. Typical doubling times for bacteria range from 45 minutes to 1 hour but can be as fast as 15 minutes. Doubling times for simple eukaryotes, such as yeast, range from 1.5 to 2 hours but may be as fast as 45 minutes.

Effect of Temperature.   As with enzymes (cf. Figure 9-8), there is an optimum in growth rate with temperature, owing to the competition of increased rates with increasing temperature and enzyme denaturation at high temperatures. An empirical law that describes this functionality is given in Aiba et al.[22] and is of the form

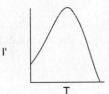

$$\mu(T) = \mu(T_m) I'$$
$$I' = \frac{aTe^{-E_1/RT}}{1 + be^{-E_2/RT}} \tag{9-60}$$

where I' is the fraction of the maximum growth rate, T_m is the temperature at which the maximum growth occurs, and $\mu(T_m)$ is the growth rate at this temperature. For the rate of oxygen uptake of *Rhizobium trifollic*, the equation takes the form

$$I' = \frac{0.0038 Te^{[21.6 - 6700/T]}}{1 + e^{[153 - 48,000/T]}} \tag{9-61}$$

The maximum growth of Rhizobium trifolic occurs at 310°K. However, experiments by Prof. Dr. Sven Köttlov of Jofostan University in Riça, Jofostan, show that this temperature should be 312°K, not 310°K.

9.4.3 Stoichiometry

The stoichiometry for cell growth is very complex and varies with the microorganism/nutrient system and environmental conditions such as pH, temperature, and redox potential. This complexity is especially true when more than one nutrient contributes to cell growth, as is usually the case. We shall focus

---

[21] A. R. C. Dean, *Growth, Function, and Regulation in Bacterial Cells* (London: Oxford University Press, 1964).

[22] S. Aiba, A. E. Humphrey, and N. F. Millis, *Biochemical Engineering* (New York: Academic Press, 1973), p. 407.

our discussion on a simplified version for cell growth, one that is limited by only one nutrient in the medium. In general, we have

$$\text{Cells} + \text{Substrate} \longrightarrow \text{More cells} + \text{Product}$$

In order to relate the substrate consumed, new cells formed, and product generated, we introduce the *yield* coefficients. The yield coefficient for cells and substrate is

$$\boxed{Y_{c/s} = \frac{\text{Mass of new cells formed}}{\text{Mass of substrate consumed}} = -\frac{\Delta C_c}{\Delta C_s}} \qquad (9\text{-}62)$$

The yield coefficient $Y_{c/s}$ is the ratio of the increase in the mass concentration of cells, ΔC_C, to the decrease in the substrate concentration $(-\Delta C_s)$, $(-\Delta C_s = C_{s0} - C_s)$, to bring about this increase in cell mass concentration. A representative value of $Y_{c/s}$ might be 0.4 (g/g).

The reciprocal of $Y_{c/s}$, i.e., $Y_{s/c}$

$$Y_{s/c} = \frac{1}{Y_{c/s}}$$

gives the ratio of $-\Delta C_s$, the substrate that must be consumed, to the increase in cell mass concentration ΔC_c.

Product formation can take place during different phases of the cell growth cycle. When product formation only occurs during the exponential growth phase, the rate of product formation is

Growth associated product formation

$$r_p = Y_{p/c} r_g = Y_{p/c} \mu C_C = Y_{p/c} \frac{\mu_{max} C_c C_s}{K_s + C_s} \qquad (9\text{-}63)$$

where

$$\boxed{Y_{p/c} = \frac{\text{Mass of product formed}}{\text{Mass of new cells formed}} = \frac{\Delta C_p}{\Delta C_c}} \qquad (9\text{-}64)$$

The product of $Y_{p/c}$ and μ, that is, $(q_P = Y_{p/c}\, \mu)$, is often called the *specific rate of product formation*, q_P, (mass product/volume/time). When the product is formed during the stationary phase where no cell growth occurs, we can relate the rate of product formation to substrate consumption by

Nongrowth associated product formation

$$r_p = Y_{p/s}\,(-r_s) \qquad (9\text{-}65)$$

The substrate in this case is usually a secondary nutrient, which we discuss in more detail later when the stationary phase is discussed.

The stoichiometric yield coefficient that relates the amount of product formed per mass of substrate consumed is

$$\boxed{Y_{p/s} = \frac{\text{Mass of product formed}}{\text{Mass of substrate consumed}} = -\frac{\Delta C_p}{\Delta C_s}} \qquad (9\text{-}66)$$

In addition to consuming substrate to produce new cells, part of the substrate must be used just to maintain a cell's daily activities. The corresponding maintenance utilization term is

Cell maintenance

$$m = \frac{\text{Mass of substrate consumed for maintenance}}{\text{Mass of cells} \cdot \text{Time}}$$

A typical value is

$$m = 0.05 \ \frac{\text{g substrate}}{\text{g dry weight}} \ \frac{1}{h} = 0.05 \ h^{-1}$$

The rate of substrate consumption for maintenance, r_{sm}, whether or not the cells are growing is

$$r_{sm} = mC_c \qquad (9\text{-}67)$$

When maintenance can be neglected, we can relate the concentration of new cells formed to the amount of substrate consumed by the equation

Neglecting cell maintenance

$$C_c = Y_{c/s}[C_{s0} - C_s] \qquad (9\text{-}68)$$

This equation can be used for both batch and continuous flow reactors.

If it were possible to sort out the substrate (S) that is consumed in the presence of cells to form new cells (C) from the substrate that is consumed to form product (P), that is

$$S \xrightarrow{\ \text{cells}\ } Y'_{c/s} C + Y'_{p/s} P$$

the yield coefficients would be written as

$$Y'_{s/c} = \frac{\text{Mass of substrate consumed to form new cells}}{\text{Mass of new cells formed}} \qquad (9\text{-}69\text{A})$$

$$Y'_{s/p} = \frac{\text{Mass of substrate consumed to form product}}{\text{Mass of product formed}} \qquad (9\text{-}69\text{B})$$

These yield coefficients will be discussed further in the substrate utilization section.

Substrate Utilization.   We now come to the task of relating the rate of nutrient (i..e., substrate) consumption, $-r_s$, to the rates of cell growth, product generation, and cell maintenance. In general, we can write

Substrate accounting

$$\begin{bmatrix} \text{Net rate of} \\ \text{substrate} \\ \text{consumption} \end{bmatrix} = \begin{bmatrix} \text{Rate of} \\ \text{substrate} \\ \text{consumed} \\ \text{by cells} \end{bmatrix} + \begin{bmatrix} \text{Rate of} \\ \text{substrate} \\ \text{consumed to} \\ \text{form product} \end{bmatrix} + \begin{bmatrix} \text{Rate of} \\ \text{substrate} \\ \text{consumed for} \\ \text{maintenance} \end{bmatrix}$$

$$-r_s \ = \ Y'_{s/c} r_g \ + \ Y'_{s/p} r_p \ + \ mC_c$$

In a number of cases, extra attention must be paid to the substrate balance. If product is produced during the growth phase, it may not be possible to separate out the amount of substrate consumed for cell growth (i.e., produce more cells) from that consumed to produce the product. Under these circumstances, all the substrate consumed for growth and for product formation is lumped into a single stoichiometric yield coefficient, $Y_{s/c}$, and the rate of substrate disappearance is

$$\boxed{-r_s = Y_{s/c} r_g + m C_c}$$ (9-70)

The corresponding rate of product formation is

$$\boxed{r_p = r_g Y_{p/c}}$$ (9-63)

The Stationary Phase.   Because there is no growth during the stationary phase, it is clear that Equation (9-70) cannot be used to account for substrate consumption, nor can the rate of product formation be related to the growth rate [e.g., Equation (9-63)]. Many antibiotics, such as penicillin, are produced in the stationary phase. In this phase, the nutrient required for growth becomes virtually exhausted, and a different nutrient, called the *secondary nutrient*, is used for cell maintenance and to produce the desired product. Usually, the rate law for product formation during the stationary phase is similar in form to the Monod equation, that is

Nongrowth-
associated product
formation
in the stationary
phase

$$\boxed{r_p = \frac{k_p C_{sn} C_c}{K_{sn} + C_{sn}}}$$ (9-71)

where k_p = specific rate constant with respect to product, $(dm^3/g \cdot s)$
   C_{sn} = concentration of the secondary nutrient, (g/dm^3)
   C_c = cell concentration, g/dm^3 (g ≡ gdw = gram dry weight)
   K_{sn} = Monod constant for secondary nutrient, (g/dm^3)
   $r_p = Y_{p/sn}(-r_{sn})$, $(g/dm^3 \cdot s)$

In the stationary
phase, the concen-
tration of live cells
is constant.

The net rate of secondary nutrient consumption, r_{sn}, during the stationary phase is

$$-r_{sn} = m C_c + Y_{sn/p} r_p$$

$$-r_{sn} = m C_c + \frac{Y_{sn/p} k_p C_{sn} C_c}{K_{sn} + C_{sn}}$$ (9-72)

Because the desired product can be produced when there is no cell growth, it is always best to relate the product concentration to the change in secondary nutrient concentration. For a batch system, the concentration of product, C_p, formed after a time t in the stationary phase can be related to the secondary nutrient concentration, C_{sn}, at that time.

Neglects cell
maintenance

$$C_p = Y_{p/sn}(C_{sn0} - C_{sn})$$ (9-73)

We have considered two limiting situations for relating substrate consumption to cell growth and product formation: product formation only during the growth phase and product formation only during the stationary phase. An example where neither of these situations applies is fermentation using *lactobacillus*, where lactic acid is produced during both the logarithmic growth and stationary phase.

The specific rate of product formation is often given in terms of the Luedeking–Piret equation, which has two parameters, α (growth) and β (non-growth)

$$q_p = \alpha\mu_g + \beta \tag{9-74}$$

Luedeking–Piret
equation for the rate
of product formation

with

$$r_p = q_p C_c$$

The assumption here in using the β-parameter is that the secondary nutrient is in excess.

Example 9–4 Estimate the Yield Coefficients

The following data was obtained from batch-reactor experiments for the yeast *Saccharomyces cerevisiae*

TABLE E9-4.1   RAW DATA

Glucose $\xrightarrow{\text{cells}}$ More cells + Ethanol

Time, t (hr)	Cells, C_c (g/dm³)	Glucose, C_s (g/dm³)	Ethanol, C_p (g/dm³)
0	1	250	0
1	1.37	245	2.14
2	1.87	238.7	5.03
3	2.55	229.8	8.96

(a) Determine the yield coefficients $Y_{s/c}$, $Y_{c/s}$, $Y_{s/p}$, $Y_{p/s}$, and $Y_{p/c}$. Assume no lag and neglect maintenance at the start of the growth phase when there are just a few cells.

(b) Describe how to find the rate-law parameters μ_{max} and K_s.

Solution

(a)   Yield coefficients

Calculate the substrate and cell *yield coefficients*, $Y_{s/c}$ and $Y_{c/s}$.

Between $t = 0$ and $t = 1$ h

$$Y_{s/c} = \frac{-\Delta C_s}{\Delta C_c} = -\frac{245-250}{1.37-1} = 13.51 \text{ g/g} \tag{E9-4.1}$$

Between $t = 2$ and $t = 3$ h

$$Y_{s/c} = -\frac{229.8-238.7}{2.55-1.87} = \frac{8.9}{0.68} = 13.1 \text{ g/g} \tag{E9-4.2}$$

Taking an average

$$\boxed{Y_{s/c} = 13.3 \text{ g/g}} \tag{E9-4.3}$$

We could also have used Polymath regression to obtain

$$Y_{c/s} = \frac{1}{Y_{s/c}} = \frac{1}{13.3 \text{ g/g}} = 0.075 \text{ g/g} \tag{E9-4.4}$$

Similarly, using the data at 1 and 2 hours, the substrate/product *yield coefficient* is

$$Y_{s/p} = -\frac{\Delta C_s}{\Delta C_P} = -\frac{238.7 - 245}{5.03 - 2.14} = \frac{6.3}{2.89} = 2.18 \text{ g/g} \tag{E9-4.5}$$

$$Y_{p/s} = \frac{1}{Y_{s/p}} = \frac{1}{2.18 \text{ g/g}} = 0.459 \text{ g/g} \tag{E9-4.6}$$

and the product/cell *yield coefficient* is

$$Y_{p/c} = \frac{\Delta C_P}{\Delta C_c} = \frac{5.03 - 2.14}{1.87 - 1.37} = 5.78 \text{ g/g} \tag{E9-4.7}$$

$$Y_{c/p} = \frac{1}{Y_{p/c}} = \frac{1}{5.78 \text{ g/g}} = 0.173 \text{ g/g} \tag{E9-4.8}$$

(b)   Rate-law parameters

We now need to determine the rate-law parameters μ_{max} and K_s in the Monod equation

$$r_g = \frac{\mu_{max} C_c C_s}{K_s + C_s} \tag{9-53}$$

For a batch system

$$r_g = \frac{dC_c}{dt} \tag{E9-4.9}$$

To find the rate-law parameters μ_{max} and K_s, we first apply the differential formulas in Chapter 7 to columns 1 and 2 of Table E9-4.1 to find r_g and add another column to Table E9-4.1.

TABLE E9-4.2   PROCESSED DATA

t	C_C	r_g	C_S
0	1		250
1	1.37	0.37	245
2	1.87	0.5	238
3	2.55	0.68	229.8

How to regress the Monod equation for μ_{max} and K_s

Because $C_s \gg K_s$ initially, it is best to regress the data using the Hanes–Woolf form of the Monod equation

$$\frac{C_c}{r_g} = \frac{K_s}{\mu_{max}}\left(\frac{1}{C_s}\right) + \frac{1}{\mu_{max}} \tag{E9-4.10}$$

We now use the newly calculated r_g along with C_c and C_s in Table E9-4.1 to prepare a table of (C_c/r_g) as a function of $(1/C_s)$. Next, we use Polymath's nonlinear regression of Equation (E9-5.10), along with more data points, to find μmax = 0.33 h^{-1} and K_s = 1.7g/dm^3.

Analysis: We first used the data in Table E9-4.1 to calculate the yield coefficients $Y_{s/c}$, $Y_{c/s}$, $Y_{s/p}$, $Y_{p/s}$, and $Y_{p/c}$. Next, we used nonlinear regression to find the Monod rate-law parameters μ_{max} and K_s.

9.4.4 Mass Balances

There are two ways that we could account for the growth of microorganisms. One is to account for the number of living cells, and the other is to account for the mass of the living cells. We shall use the latter. A mass balance on the microorganisms in a CSTR (chemostat) (e.g., margin figure that follows and Figure 9-24) of constant volume is

Cell mass balance

$$\begin{bmatrix} \text{Rate of} \\ \text{accumulation} \\ \text{of cells,} \\ \text{g/s} \end{bmatrix} = \begin{bmatrix} \text{Rate of} \\ \text{cells} \\ \text{entering,} \\ \text{g/s} \end{bmatrix} - \begin{bmatrix} \text{Rate of} \\ \text{cells} \\ \text{leaving,} \\ \text{g/s} \end{bmatrix} + \begin{bmatrix} \text{Net rate of} \\ \text{generation} \\ \text{of live cells,} \\ \text{g/s} \end{bmatrix} \quad (9\text{-}75)$$

$$V\frac{dC_c}{dt} = v_0 C_{c0} - v_0 C_c + (r_g - r_d)V$$

The corresponding substrate balance is

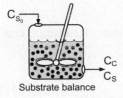

Substrate balance

$$\begin{bmatrix} \text{Rate of} \\ \text{accumulation} \\ \text{of substrate,} \\ \text{g/s} \end{bmatrix} = \begin{bmatrix} \text{Rate of} \\ \text{substrate} \\ \text{entering,} \\ \text{g/s} \end{bmatrix} - \begin{bmatrix} \text{Rate of} \\ \text{substrate} \\ \text{leaving,} \\ \text{g/s} \end{bmatrix} + \begin{bmatrix} \text{Rate of} \\ \text{substrate} \\ \text{generation,} \\ \text{g/s} \end{bmatrix} \quad (9\text{-}76)$$

$$V\frac{dC_s}{dt} = v_0 C_{s0} - v_0 C_s + r_s V$$

In most systems, the entering microorganism concentration, C_{c0}, is zero for a flow reactor.

Batch Operation

For a batch system $v = v_0 = 0$, the mass balances are as follows:

Cell Mass Balance

The mass balances

$$V\frac{dC_c}{dt} = r_g V - r_d V$$

Dividing by the reactor volume V gives

$$\frac{dC_c}{dt} = r_g - r_d \quad (9\text{-}77)$$

Substrate Mass Balance

The rate of disappearance of substrate, $-r_s$, results from substrate used for cell growth and substrate used for cell maintenance

$$V\frac{dC_s}{dt} = r_sV = Y_{s/c}(-r_g)V - mC_cV$$ (9-78)

Dividing by V yields the substrate balance for the growth phase

Growth phase

$$\frac{dC_s}{dt} = Y_{s/c}(-r_g) - mC_c$$ (9-79)

For cells in the *stationary phase*, where there is no growth in cell concentration, cell maintenance and product formation are the only reactions to consume the secondary substrate. Under these conditions the substrate balance, Equation (9-76), reduces to

Stationary phase

$$V\frac{dC_{sn}}{dt} = -mC_cV + Y_{sn/p}(-r_p)V$$ (9-80)

Typically, r_p will have the same Monod form of the rate law as r_g [e.g., Equation (9-71)]. Of course, Equation (9-79) only applies for substrate concentrations greater than zero.

Product Mass Balance

The rate of product formation, r_p, can be related to the rate of substrate consumption, $-r_s$, through the following balance when $m = 0$:

Batch stationary
growth phase

$$V\frac{dC_p}{dt} = r_pV = Y_{p/s}(-r_s)V$$ (9-81)

During the growth phase, we could also relate the rate of formation of product, r_p, to the cell growth rate, r_g, Equation (9-63), i.e., $r_p = Y_{p/c}r_g$. The coupled first-order ordinary differential equations above can be solved by a variety of numerical techniques.

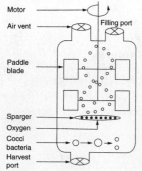

Motor
Air vent   Filling port
Paddle
blade
Sparger
Oxygen
Cocci
bacteria
Harvest
port

Example 9–5 Bacteria Growth in a Batch Reactor

Glucose-to-ethanol fermentation is to be carried out in a batch reactor using an organism such as *Saccharomyces cerevisiae*. Plot the concentrations of cells, substrate, and product and the rates for growth, death, and maintenance, i.e., r_g, r_d, and r_{sm} as functions of time. The initial cell concentration is 1.0 g/dm³, and the substrate (glucose) concentration is 250 g/dm³.

Additional data (Partial source: R. Miller and M. Melick, *Chem. Eng.*, Feb. 16, 1987, p. 113):

$$C_p^* = 93 \text{ g/dm}^3 \qquad Y_{c/s} = 0.08 \text{ g/g}$$

$$n = 0.52 \qquad Y_{p/s} = 0.45 \text{ g/g}$$

$$\mu_{max} = 0.33 \text{ h}^{-1} \qquad Y_{p/c} = 5.6 \text{ g/g}$$

$$K_s = 1.7 \text{ g/dm}^3 \qquad k_d = 0.01 \text{ h}^{-1}$$

$$m = 0.03 \text{ (g substrate)}/(\text{g cells} \cdot \text{h})$$

Solution

1. **Mass balances:**

   Cells:

$$V \frac{dC_c}{dt} = (r_g - r_d)V \qquad \text{(E9-5.1)}$$

   Substrate:

$$V \frac{dC_s}{dt} = Y_{s/c}(-r_g)V - r_{sm}V \qquad \text{(E9-5.2)}$$

   Product:

$$V \frac{dC_p}{dt} = Y_{p/c}(r_g V) \qquad \text{(E9-5.3)}$$

2. **Rate laws:**

   Growth:

$$r_g = \mu_{max} \left(1 - \frac{C_p}{C_p^*}\right)^{0.52} \frac{C_c C_s}{K_s + C_s} \qquad \text{(E9-5.4)}$$

   Death:

$$r_d = k_d C_c \qquad \text{(E9-5.5)}$$

   Maintenance:

$$r_{sm} = m C_c \qquad \text{(9-67)}$$

3. **Stoichiometry:**

$$r_p = Y_{p/c} r_g \qquad \text{(E9-5.6)}$$

4. **Combining gives**

$$\frac{dC_c}{dt} = \mu_{max} \left(1 - \frac{C_p}{C_p^*}\right)^{0.52} \frac{C_c C_s}{K_s + C_s} - k_d C_c \qquad \text{(E9-5.7)}$$

$$\frac{dC_s}{dt} = -Y_{s/c} \mu_{max} \left(1 - \frac{C_p}{C_p^*}\right)^{0.52} \frac{C_c C_s}{K_s + C_s} - m C_c \qquad \text{(E9-5.8)}$$

$$\frac{dC_p}{dt} = Y_{p/c} r_g$$

These equations were solved using an ODE equation solver (see Table E9-5.1). The results are shown in Figure E9-5.1 for the parameter values given in the problem statement.

Living Example Problem

TABLE E9-5.1   POLYMATH PROGRAM

Differential equations

1 d(Cc)/d(t) = rg-rd

2 d(Cs)/d(t) = Ysc*(-rg)-rsm

3 d(Cp)/d(t) = rg*Ypc

Explicit equations

1 rd = Cc*.01

2 Ysc = 1/.08

3 Ypc = 5.6

4 Ks = 1.7

5 m = .03

6 umax = .33

7 rsm = m*Cc

8 kobs = (umax*(1-Cp/93)^.52)

9 rg = kobs*Cc*Cs/(Ks+Cs)

Calculated values of DEQ variables

	Variable	Initial value	Final value
1	Cc	1.	16.18406
2	Cp	0	89.82293
3	Cs	250.	46.93514
4	kobs	0.33	0.0570107
5	Ks	1.7	1.7
6	m	0.03	0.03
7	rd	0.01	0.1618406
8	rg	0.3277712	0.8904142
9	rsm	0.03	0.4855217
10	t	0	12.
11	umax	0.33	0.33
12	Ypc	5.6	5.6
13	Ysc	12.5	12.5

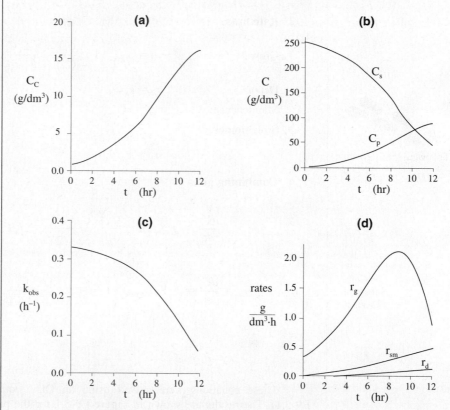

Figure E9-5.1   Concentrations and rates as a function of time.

The substrate concentration C_s can never be less than zero. However, we note that when the substrate is completely consumed, the first term on the right-hand side of Equation (E9-5.8) (and line 3 of the Polymath program) will be zero but the second term for maintenance, mC_c, will not. Consequently, **if** the integration is carried further in time, the integration program will predict a **negative** value of C_s! This inconsistency can be addressed in a number of ways, such as including an **if** statement in the Polymath program (e.g., if C_s is less than or equal to zero, then $m = 0$).

Analysis: In this example, we applied a modified CRE algorithm to biomass formation and solved the resulting equations using the ODE solver Polymath. We note in Figure E9-5.1 **(d)** the growth rate, r_g, goes through a maximum, increasing at the start of the reaction as the concentration of cells, C_c, increases then decreasing as the substrate (nutrient) and k_{obs} decrease. We see from Figures E9-5.1 **(a)** and **(b)** that the cell concentration increases dramatically with time while the product concentration does not. The reason for this difference is that part of the substrate is consumed for maintenance and part for cell growth, leaving only the remainder of the substrate to be transformed into product.

9.4.5 Chemostats

Chemostats are essentially CSTRs that contain microorganisms. A typical chemostat is shown in Figure 9-24, along with the associated monitoring equipment and pH controller. One of the most important features of the chemostat is that it allows the operator to control the cell growth rate. This control of the growth rate is achieved by adjusting the volumetric feed rate (dilution rate).

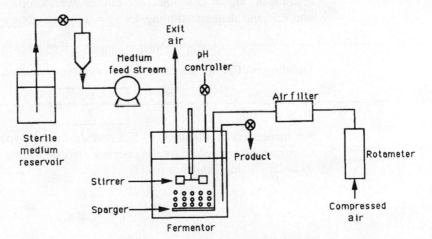

Figure 9-24 Chemostat (CSTR) system.

9.4.6 CSTR Bioreactor Operation

In this section, we return to the mass balance equations on the cells [Equation (9-75)] and substrate [Equation (9-76)], and consider the case where the volumetric flow rates in and out are the same and that no live (i.e., viable) cells enter the chemostat. We next define a parameter common to bioreactors called the dilution rate, D. The dilution rate is

$$D = \frac{v_0}{V}$$

and is simply the reciprocal of the space time τ, i.e., $D = \frac{1}{\tau}$. Dividing Equations (9-75) and (9-76) by V and using the definition of the dilution rate, we have

$$\text{Accumulation} = \text{In} \quad - \text{Out} + \text{Generation}$$

CSTR mass balances

$$\text{Cell:} \quad \frac{dC_c}{dt} = 0 \quad - DC_c + (r_g - r_d) \tag{9-82}$$

$$\text{Substrate:} \quad \frac{dC_s}{dt} = DC_{s0} - DC_s + r_s \tag{9-83}$$

Using the Monod equation, the growth rate is determined to be

Rate law

$$r_g = \mu C_c = \frac{\mu_{max} C_s C_c}{K_s + C_s} \tag{9-53}$$

For steady-state operation we have

$$DC_c = r_g - r_d \tag{9-84}$$

Steady state   and

$$D(C_{s0} - C_s) = -r_s \tag{9-85}$$

We now neglect the death rate, r_d, and combine Equations (9-51) and (9-84) for steady-state operation to obtain the mass flow rate of cells out of the chemostat, $\dot{m}_c = C_c v_0$, and the rate of generation of cells, $r_g V$. Equating $\dot{m}_c$ and $r_g V$, and then substituting for $r_g = \mu C_c$, we obtain

$$\dot{m}_c = C_c v_0 = r_g V = \mu C_c V \tag{9-86}$$

Dividing by $C_c V$ we see the cell concentration cancels to give the dilution rate D

Dilution rate

$$\boxed{D = \frac{v_0}{V} = \mu} \tag{9-87}$$

How to control cell growth

An inspection of Equation (9-87) reveals that the specific growth rate of the cells *can be controlled* by the operator by controlling the dilution rate D, i.e., $D = \frac{v_0}{V}$. Using Equation (9-52)

$$\mu = \mu_{max} \frac{C_s}{K_s + C_s} \quad\quad \text{s}^{-1} \tag{9-52}$$

to substitute for μ in terms of the substrate concentration and then solving for the steady-state substrate concentration yields

$$C_s = \frac{DK_s}{\mu_{max} - D} \tag{9-88}$$

Assuming that a single nutrient is limiting, cell growth is the only process contributing to substrate utilization, and that cell maintenance can be neglected, the stoichiometry is

$$-r_s = r_g Y_{s/c} \tag{9-89}$$

$$C_c = Y_{c/s}(C_{s0} - C_s) \tag{9-68}$$

Substituting for C_s using Equation (7-87) and rearranging, we obtain

$$C_c = Y_{c/s}\left[C_{s0} - \frac{DK_s}{\mu_{max} - D}\right] \qquad (9\text{-}90)$$

9.4.7 Wash-Out

To learn the effect of increasing the dilution rate, we combine Equations (9-82) and (9-54), and set $r_d = 0$ to get

$$\frac{dC_c}{dt} = (\mu - D)C_c \qquad (9\text{-}91)$$

We see that if $D > \mu$, then (dC_c/dt) will be negative, and the cell concentration will continue to decrease until we reach a point where all cells will be washed out:

$$C_c = 0$$

The dilution rate at which wash-out will occur is obtained from Equation (9-90) by setting $C_c = 0$.

Flow rate at which
wash-out occurs

$$D_{max} = \frac{\mu_{max}C_{s0}}{K_s + C_{s0}} \qquad (9\text{-}92)$$

We next want to determine the other extreme for the dilution rate, which is the rate of maximum cell production. The cell production rate per unit volume of reactor is the mass flow rate of cells out of the reactor (i.e., $\dot{m}_c = C_c v_0$) divided by the volume V, or

Maximum rate of
cell production
(DC_c)

$$\frac{\dot{m}_c}{V} = \frac{v_0 C_c}{V} = DC_c \qquad (9\text{-}93)$$

Using Equation (9-90) to substitute for C_c yields

$$DC_c = DY_{c/s}\left(C_{s0} - \frac{DK_s}{\mu_{max} - D}\right) \qquad (9\text{-}94)$$

Figure 9-25 shows production rate, cell concentration, and substrate concentration as functions of dilution rate.

We observe a maximum in the production rate, and this maximum can be found by differentiating the production rate, Equation (9-94), with respect to the dilution rate D:

$$\frac{d(DC_c)}{dD} = 0 \qquad (9\text{-}95)$$

Then

Maximum rate of
cell production

$$D_{maxprod} = \mu_{max}\left(1 - \sqrt{\frac{K_s}{K_s + C_{s0}}}\right) \qquad (9\text{-}96)$$

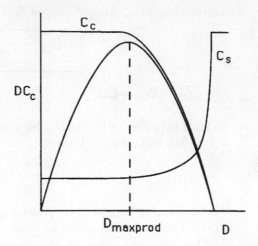

Figure 9-25   Cell and substrate concentrations and production rate as functions of dilution rate.

The organism *Streptomyces aureofaciens* was studied in a 10-dm³ chemostat using sucrose as a substrate. The cell concentration, C_c (mg/ml), the substrate concentration, C_s (mg/ml), and the production rate, DC_c (mg/ml/h), were measured at steady state for different dilution rates. The data are shown in Figure 9-26.[23] Note that the data follow the same trends as those discussed in Figure 9-25.

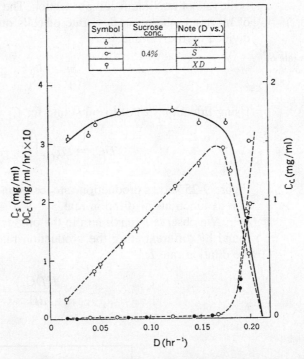

Figure 9-26   Continuous culture of *Streptomyces aureofaciens* in chemostats. (Note: $X \equiv C_c$) Courtesy of S. Aiba, A. E. Humphrey, and N. F. Millis, *Biochemical Engineering*, 2nd ed. (New York: Academic Press, 1973).

[23] B. Sikyta, J. Slezak, and M. Herold, *Appl. Microbiol.*, *9*, 233 (1961).

Closure. The theme running through most of this chapter is the pseudo-steady-state hypothesis (PSSH) as it applies to chemical reactions and to enzymatic reactions. The reader should be able to apply the PSSH to reactions in such problems as P9-4$_B$ to P9-8$_B$ in order to develop rate laws. After completing this chapter, the reader should be able to describe and analyze enzymatic reactions and the different types of inhibition as displayed on a Lineweaver–Burk plot. The reader should be able to explain the use of microorganisms to produce chemical products, along with the stages of cell growth and how the Monod equation for cell growth is coupled with mass balances on the substrate, cells, and product to obtain the concentration-time trajectories in a batch reactor. The reader should also be able to apply the growth laws and balance equations to a chemostat (CSTR) to predict the maximum product flow rate and the wash-out rate.

SUMMARY

1. In the PSSH, we set the rate of formation of the active intermediates equal to zero. If the active intermediate A* is involved in m different reactions, we set it to

$$r_{A^*, \text{net}} \equiv \sum_{i=1}^{m} r_{A^*i} = 0 \qquad (S9\text{-}1)$$

This approximation is justified when the active intermediate is highly reactive and present in low concentrations.

2. The azomethane (AZO) decomposition mechanism is

$$2\,\text{AZO} \underset{k_2}{\overset{k_1}{\rightleftharpoons}} \text{AZO} + \text{AZO*}$$

$$\text{AZO*} \xrightarrow{k_3} \text{N}_2 + \text{ethane} \qquad (S9\text{-}2)$$

$$r_{\text{N}_2} = \frac{k(\text{AZO})^2}{1 + k'(\text{AZO})} \qquad (S9\text{-}3)$$

By applying the PSSH to AZO*, we show the rate law, which exhibits first-order dependence with respect to AZO at high AZO concentrations and second-order dependence with respect to AZO at low AZO concentrations.

3. Enzyme kinetics: enzymatic reactions follow the sequence

$$\text{E} + \text{S} \underset{k_2}{\overset{k_1}{\rightleftharpoons}} \text{E} \cdot \text{S} \xrightarrow{k_3} \text{E} + \text{P}$$

Using the PSSH for (E · S) and a balance on the total enzyme, E_t, which includes both the bound (E · S) and unbound enzyme (E) concentrations

$$E_t = (\text{E}) + (\text{E} \cdot \text{S})$$

we arrive at the Michaelis–Menten equation

$$-r_s = \frac{V_{\max}(\text{S})}{K_{\text{M}} + (\text{S})} \qquad (S9\text{-}4)$$

where V_{max} is the maximum reaction rate at large substrate concentrations ($S \gg K_M$) and K_M is the Michaelis constant. K_M is the substrate concentration at which the rate is half the maximum rate ($S_{1/2} = K_M$).

4. The three different types of inhibition—competitive, uncompetitive, and noncompetitive (mixed)—are shown on the Lineweaver–Burk plot:

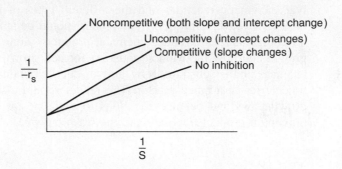

5. Bioreactors:

$$\text{Cells + Substrate} \longrightarrow \text{More cells + Product}$$

(a) Phases of bacteria growth:

I. Lag   II. Exponential   III. Stationary   IV. Death

(b) Unsteady-state **mass balance** on a chemostat

$$\frac{dC_c}{dt} = D(C_{c0} - C_c) + r_g - r_d \tag{S9-5}$$

$$\frac{dC_s}{dt} = D(C_{s0} - C_s) + r_s \tag{S9-6}$$

(c) Monod growth **rate law**

$$r_g = \mu_{max} \frac{C_c C_s}{K_s + C_s} \tag{S9-7}$$

(d) **Stoichiometry**

$$Y_{c/s} = \frac{\text{Mass of new cells formed}}{\text{Mass of substrate consumed}} \tag{S9-8}$$

$$Y_{s/c} = \frac{1}{Y_{c/s}} \tag{S9-9}$$

Substrate consumption

$$-r_s = Y_{s/c} r_g + m C_c \tag{S9-10}$$

CRE WEB SITE MATERIALS

- **Expanded Material**
    1. *Puzzle Problem "What's Wrong with this Solution?"*
    2. *Physiologically Based Pharmacokinetic Model for Alcohol Metabolism*

- **Learning Resources**
    1. *Summary Notes*
    2. *Web Modules*
        A.   Ozone Layer

B. Glow Sticks

Earth Probe TOMs Total Ozone
September 8, 2000

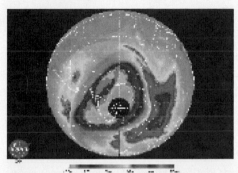

Ozone (Dotson Units)

Photo courtesy of Goddard Space Flight Center (NASA). See the Web Modules on the CRE Web site for color pictures of the ozone layer and the glow sticks.

    3. *Interactive Computer Games*
    Enzyme Man

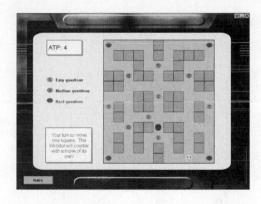

- **Living Example Problems**
    1. *Example 9-5 Bacteria Growth in a Batch Reactor*
    2. *Example Chapter 9 CRE Web site: PSSH Applied to Thermal Cracking of Ethane*
    3. *Example Chapter 9 CRE Web site: Alcohol Metabolism*
    4. *Example Web Module: Ozone*
    5. *Example Web Module: Glowsticks*
- **Professional Reference Shelf**
    R9-1. Chain Reactions Example Problem
    R9-2. Reaction Pathways
    R9-3. *Polymerization*
        A.   Step Polymerization
            Example R9-3.1 Determining the Concentration of Polymers for Step Polymerization
        B.   Chain Polymerizations
            Example R9-3.2 Parameters of MW Distribution

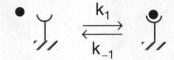

$$C_2H_5OH \underset{ADH}{\overset{}{\rightleftharpoons}} CH_3CHO \underset{AlDH}{\longrightarrow} CH_3COOH$$

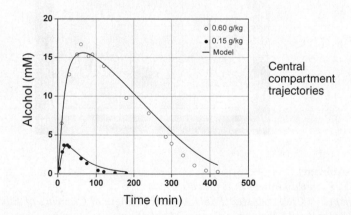

Figure R9-7.1 Blood alcohol–time trajectories from data of Wilkinson et al.[24]

[24] P. K. Wilkinson, et al., "Pharmacokinetics of Ethanol After Oral Administration in the Fasting State," *J. Phar-macoket. Biopharm.*, 5(3): 207–224 (1977).

R9-8. *Pharmacokinetics in Drug Delivery*
Pharmacokinetic models of drug delivery for medication administered either orally or intravenously are developed and analyzed.

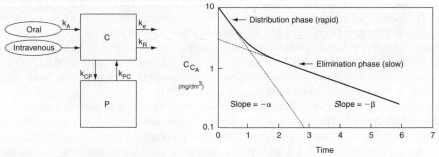

Figure A Two-compartment model.   **Figure B** Drug response curve.

PROBLEMS

In each of the following questions and problems, rather than just drawing a box around your answer, write a sentence or two describing how you solved the problem, the assumptions you made, the reasonableness of your answer, what you learned, and any other facts that you want to include.

You may wish to refer to W. Strunk and E. B. White, *The Elements of Style, 4th ed.* (New York: Macmillan, 2000) and Joseph M. Williams, *Style: Ten Lessons in Clarity & Grace, 6th ed.* (Glenview, IL: Scott, Foresman, 1999) to enhance the quality of your sentences.

P9-1$_A$ ICG Enzyme Man.
    (a) Load the ICG on your computer and carry out the exercise. Performance number = _____.
    (b) Apply one or more of the six ideas in Preface Table P-4, page xxviii, to this problem.

P9-2$_B$ (a) **Example 9-1**. How would the results change if the concentration of CS_2 and M were increased?
    (b) **Example 9-2**. (1) The following additional runs were carried out when an inhibitor was present:

$C_{urea}(kmol/m^3)$	$C_{inhibitor}(kmol/m^3)$	$-r_{urea}(kmol/m^3 \cdot s)$
0.01	0.1	0.125
0.005	0.1	0.065

What type of inhibition is taking place? (2) Sketch the curves for no inhibition as well as competitive, uncompetitive, noncompetitive (mixed), and substrate inhibition on a Woolf–Hanes plot and on an Eadie–Hofstee plot.
    (c) **Example 9-3**. (1) What would the conversion be after 10 minutes if the initial concentration of urea were decreased by a factor of 100? (2) What would be the conversion in a CSTR with the same residence time, τ, as the batch reactor time t? (3) In a PFR?
    (d) **Example 9-4**. What is the total mass of substrate consumed in grams per mass of cells plus what is consumed to form product? Is there disparity here?
    (e) **Example 9-5**. Download the *Living Example Problem*. (1) Modify the code to carry out the fermentation in a fed-batch (e.g., semibatch) reactor in which the substrate is fed at a rate of 0.5 dm^3/h and a concentration of 5 g/dm^3 to an initial liquid volume of 1.0 dm^3 containing a cell mass with an initial concentration of $C_{ci} = 0.2$ mg/dm^3 and an initial substrate concentration of $C_{ci} = 0.5$ mg/dm^3. Plot and analyze the concentration of cells, substrate, and product as a function of time, along with the mass of product up to 24 hours. (2) Repeat (1) when the growth is uncompetitively inhibited by the substrate with $K_I = 0.7$ g/dm^3. (3) Set $C_P^* = 10,000$ g/dm^3, and compare your results with the base case.

(f) **Chain Reaction Example** discussed in *Professional Reference Shelf R9.1* on the CRE Web site. Over what range of time is the PSSH not valid? Download the *Living Example Problem*. Vary the temperature ($800 < T < 1600$). What temperature gives the greatest disparity with the PSSH results? Specifically compare the PSSH solution with the full numerical solution.

(g) **Example on Alcohol Metabolism on the CRE Web site**. This problem is a **gold mine** for things to be learned about the effect of alcohol on the human body. Download the *Polymath Living Example Program* from the CRE Web site. (1) Start by varying the initial doses of alcohol. (2) Next consider individuals who are ALDH enzyme deficient, which includes about 40% to 50% of Asians and Native Americans. Set V_{max} for acetaldehydes between 10% and 50% of its normal value and compare the concentration-time trajectories with the base cases. *Hint:* Read the journal article in the **Summary Notes** [*Alcohol 35*, p.1 (2005)].

P9-3$_C$ (*Flame retardants*) Hydrogen radicals are important to sustaining combustion reactions. Consequently, if chemical compounds that can scavenge the hydrogen radicals are introduced, the flames can be extinguished. While many reactions occur during the combustion process, we shall choose CO flames as a model system to illustrate the process (S. Senkan et al., *Combustion and Flame*, 69, 113). In the absence of inhibitors

$$O_2 \longrightarrow O\cdot + O\cdot \qquad\qquad (P9\text{-}3.1)$$

$$H_2O + O\cdot \longrightarrow 2OH\cdot \qquad\qquad (P9\text{-}3.2)$$

$$CO + OH\cdot \longrightarrow CO_2 + H\cdot \qquad\qquad (P9\text{-}3.3)$$

$$H\cdot + O_2 \longrightarrow OH\cdot + O\cdot \qquad\qquad (P9\text{-}3.4)$$

The last two reactions are rapid compared to the first two. When HCl is introduced to the flame, the following additional reactions occur:

$$H\cdot + HCl \longrightarrow H_2 + Cl\cdot$$

$$H\cdot + Cl\cdot \longrightarrow HCl$$

Assume that all reactions are elementary and that the PSSH holds for the $O\cdot$, $OH\cdot$, and $Cl\cdot$ radicals.

(a) Derive a rate law for the consumption of CO when no retardant is present.

(b) Derive an equation for the concentration of $H\cdot$ as a function of time, assuming constant concentration of O_2, CO, and H_2O for both uninhibited combustion and combustion with HCl present. Sketch $H\cdot$ versus time for both cases.

P9-4$_A$ The pyrolysis of acetaldehyde is believed to take place according to the following sequence:

$$CH_3CHO \xrightarrow{\ k_1\ } CH_3\cdot + CHO\cdot$$

$$CH_3\cdot + CH_3CHO \xrightarrow{\ k_2\ } CH_3\cdot + CO + CH_4$$

$$CHO\cdot + CH_3CHO \xrightarrow{\ k_3\ } CH_3\cdot + 2CO + H_2$$

$$2CH_3\cdot \xrightarrow{\ k_4\ } C_2H_6$$

(a) Derive the rate expression for the rate of disappearance of acetaldehyde, $-r_{Ac}$.

(b) Under what conditions does it reduce to the equation at the beginning of Section 9.1 on page 334?

(c) Sketch a reaction pathway diagram for this reaction. *Hint:* See margin note on page 338.

P9-5$_B$ For each of the reactions in parts **(a)**, **(b)**, and **(c)**, suggest a mechanism and apply the PSSH to learn if the mechanism is consistent with the rate law.

(a) The gas-phase homogeneous oxidation of nitrogen monoxide (NO) to dioxide (NO_2)

$$2NO + O_2 \xrightarrow{\ k\ } 2NO_2$$

is known to have a form of third-order kinetics, which suggests that the reaction is elementary as written, at least for low partial pressures of the nitrogen oxides. However, the rate constant k actually *decreases* with increasing absolute temperature, indicating an apparently *negative* activation energy. Because the activation energy of any elementary reaction must be positive, some explanation is in order.

Hall of Fame

Provide an explanation, starting from the fact that an active intermediate species, NO_3, is a participant in some other known reactions that involve oxides of nitrogen. Draw the reaction pathway. *Hint:* See margin in Section 9.1.2.

(b) The rate law for formation of phosgene, $COCl_2$, from chlorine, Cl_2, and carbon monoxide, CO, has the rate law

$$r_{COCl_2} = k C_{CO} C_{Cl_2}^{3/2}$$

Suggest a mechanism for this reaction that is consistent with this rate law and draw the reaction pathway. *Hint:* Cl formed from the dissociation of Cl_2 is one of the two active intermediates.

(c) Suggest an active intermediate(s) and mechanism for the reaction $H_2 + Br_2 \rightarrow 2HBr$. Use the PSSH to show whether or not your mechanism is consistent with the rate law

$$r_{HBr} = \frac{k_1 C_{H_2} C_{Br}^{3/2}}{C_{HBr} + k_2 C_{Br_2}}$$

P9-6$_C$ (*Tribology*) **Why you change your motor oil?** One of the major reasons for engine-oil degradation is the oxidation of the motor oil. To retard the degradation process, most oils contain an antioxidant [see *Ind. Eng. Chem. 26*, 902 (1987)]. Without an inhibitor to oxidation present, the suggested mechanism at low temperatures is

$$I_2 \xrightarrow{k_0} 2I\cdot$$

$$I\cdot + RH \xrightarrow{k_i} R\cdot + HI$$

$$R\cdot + O_2 \xrightarrow{k_{p1}} RO_2\dot{}$$

$$RO_2\dot{} + RH \xrightarrow{k_{p2}} ROOH + R\cdot$$

$$2RO_2\dot{} \xrightarrow{k_t} inactive$$

where I_2 is an initiator and RH is the hydrocarbon in the oil.

When an antioxidant is added to retard degradation at low temperatures, the following additional termination steps occur:

$$RO_2\cdot + AH \xrightarrow{k_{A1}} ROOH + A\cdot$$

$$A\cdot + RO_2\dot{} \xrightarrow{k_{A2}} inactive$$

for example, AH = (OH–phenyl ring–CH_3),   inactive = (OH–phenyl ring–CH_2OOR)

(a) Derive a rate law for the degradation of the motor oil in the absence of an antioxidant at low temperatures.

(b) Derive a rate law for the rate of degradation of the motor oil in the presence of an antioxidant for low temperatures.

(c) How would your answer to part (a) change if the radicals $I\cdot$ were produced at a constant rate in the engine and then found their way into the oil?

(d) Sketch a reaction pathway diagram for both high and low temperatures, with and without antioxidant.

(e) See the open-ended problem G.2 in Appendix G and on the CRE Web site for more on this problem.

P9-7$_A$ **Epidemiology.** Consider the application of the PSSH to epidemiology. We shall treat each of the following steps as elementary, in that the rate will be proportional to the number of people in a particular state of health. A healthy person, H, can become ill, I, spontaneously, such as by contracting smallpox spores:

$$H \xrightarrow{\ k_1\ } I \tag{P9-7.1}$$

or the person may become ill through contact with another ill person:

$$I + H \xrightarrow{\ k_2\ } 2I \tag{P9-7.2}$$

The ill person may become healthy:

$$I \xrightarrow{\ k_3\ } H \tag{P9-7.3}$$

or the ill person may expire:

$$I \xrightarrow{\ k_4\ } D \tag{P9-7.4}$$

The reaction given in Equation (P9-7.4) is normally considered completely irreversible, although the reverse reaction has been reported to occur.

(a) Derive an equation for the death rate.

(b) At what concentration of healthy people does the death rate become critical? (*Ans.:* When [H] = $(k_3 + k_4)/k_2$.)

(c) Comment on the validity of the PSSH under the conditions of Part (b).

(d) If $k_1 = 10^{-8}$ h^{-1}, $k_2 = 10^{-16}$ (people·h)$^{-1}$, $k_3 = 5 \times 10^{-10}$ h^{-1}, $k_4 = 10^{-11}$ h^{-1}, and H$_o = 10^9$ people, use Polymath to plot H, I, and D versus time. Vary k_i and describe what you find. Check with your local *disease control center* or search online to modify the model and/or substitute appropriate values of k_i. Extend the model, taking into account what you learn from other sources (e.g., the Internet).

(e) Apply one or more of the six ideas in Preface Table P-4, page xxviii, to this problem.

P9-8$_B$ Derive the rate laws for the following enzymatic reactions and sketch and compare, where possible, with the plots shown in Figure E9-2.1.

(a) $E + S \rightleftarrows E \cdot S \rightleftarrows P + E$

(b) $E + S \rightleftarrows E \cdot S \rightleftarrows E \cdot P \rightarrow P + E$

(c) $E + S_1 \rightleftarrows E \cdot S_1$
$E \cdot S_1 + S_2 \rightleftarrows E \cdot S_1 S_2$
$E \cdot S_1 S_2 \rightarrow P + E$

(d) $E + S \rightleftarrows E \cdot S \rightarrow P$
$P + E \rightleftarrows E \cdot P$

(e) Which of the reactions (a) through (d), if any, lend themselves to analysis by a Lineweaver–Burk plot?

P9-9$_B$ Beef catalase has been used to accelerate the decomposition of hydrogen peroxide to yield water and oxygen [*Chem. Eng. Educ.*, 5, 141 (1971)]. The concentration of hydrogen peroxide is given as a function of time for a reaction mixture with a pH of 6.76 maintained at 30°C.

t (min)	0	10	20	50	100
$C_{H_2O_2}$ (mol/L)	0.02	0.01775	0.0158	0.0106	0.005

(a) Determine the Michaelis–Menten parameters V_{max} and K_M.

(b) If the total enzyme concentration is tripled, what will the substrate concentration be after 20 minutes?

(c) Apply one or more of the six ideas in Preface Table P-4, page xxviii, to this problem.

(d) List ways you can work this problem incorrectly.

P9-10$_B$ It has been observed that substrate inhibition occurs in the following enzymatic reaction:

$$E + S \longrightarrow P + E$$

(a) Show that the rate law for substrate inhibition is consistent with the plot in Figure P9-10$_B$ of $-r_s$ (mmol/L·min) versus the substrate concentration S (mmol/L).

(b) If this reaction is carried out in a CSTR that has a volume of 1000 dm^3, to which the volumetric flow rate is 3.2 dm^3/min, determine the three possible steady states, noting, if possible, which are stable. The entrance concentration of the substrate is 50 mmol/dm^3. What is the highest conversion?

(c) What would be the effluent substrate concentration if the total enzyme concentration is reduced by 33%?

(d) List ways you can work this problem incorrectly.

(e) How could you make this problem more difficult?

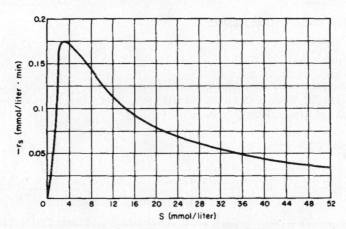

Figure P9-10$_B$ Michaelis–Menten plot for substrate inhibition.

P9-11$_B$ The following data on baker's yeast in a particular medium at 23.4°C were obtained in the presence and in the absence of an inhibitor, sulfanilamide. The reaction rate ($-r_S$) was measured in terms of the oxygen uptake rate Q_{O_2}, obtained as a function of oxygen partial pressure.

(a) Assume the rate Q_{O_2} follows Michaelis–Menten kinetics with respect to oxygen. Calculate the Q_{O_2} maximum (i.e., V_{max}), and the Michaelis–Menten constant K_M.
(*Ans.:* $V_{max} = 52.63 \ \mu L \ O_2/h \cdot mg \ cells.$)

(b) Using the Lineweaver–Burk plot, determine the type of inhibition sulfanilamide that causes the O_2 uptake to change.

P_{O_2}*	Q_{O_2} (no sulfanilamide)	Q_{O_2} (20 mg sulfanilamide/mL added to medium)
0.0	0.0	0.0
0.5	23.5	17.4
1.0	33.0	25.6
1.5	37.5	30.8
2.5	42.0	36.4
3.5	43.0	39.6
5.0	43.0	40.0

*P_{O_2} = oxygen partial pressure, mmHg; Q_{O_2} = oxygen uptake rate, μL of O_2 per hour per mg of cells.

(c) List ways you can work this problem incorrectly.

(d) Apply one or more of the six ideas in Preface Table P-4, page xxviii, to this problem.

P9-12$_B$ The enzymatic hydrolysis of starch was carried out with and without maltose and α-dextrin added. [Adapted from S. Aiba, A. E. Humphrey, and N.F. Mills, *Biochemical Engineering* (New York: Academic Press, 1973).]

$$\text{Starch} \rightarrow \alpha\text{-dextrin} \rightarrow \text{Limit dextrin} \rightarrow \text{Maltose}$$

No Inhibition

C_S (g/dm^3)	12.5	9.0	4.25	1.0
$-r_S$ (relative)	100	92	70	29

Maltose added (I = 12.7 mg/dm^3)

C_S (g/dm^3)	10	5.25	2.0	1.67
$-r_S$ (relative)	77	62	38	34

α-dextrin added (I = 3.34 mg/dm^3)

C_S (g/dm^3)	33	10	3.6	1.6
$-r_S$ (relative)	116	85	55	32

Determine the types of inhibition for maltose and for α-dextrin.

P9-13$_B$ The hydrogen ion, H$^+$, binds with the enzyme (E$^-$) to activate it in the form EH. The hydrogen ion, H$^+$, also binds with EH to deactivate it by forming EH$_2^+$.

$$\text{H}^+ + \text{E}^- \rightleftharpoons \text{EH} \qquad K_1 = \frac{(\text{EH})}{(\text{H}^+)(\text{E}^-)}$$

$$\text{H}^+ + \text{EH} \rightleftharpoons \text{EH}_2^+ \qquad K_2 = \frac{(\text{EH}_2^+)}{(\text{H}^+)(\text{EH})}$$

$$\text{EH} + \text{S} \xrightarrow{K_M} \text{EHS} \longrightarrow \text{EH} + \text{P}, \quad K_M = \frac{(\text{EHS})}{(\text{EH})(\text{S})}$$

Figure P9-13$_B$ Enzyme pH dependence.

where E$^-$ and EH$_2^+$ are inactive.

(a) Determine if the preceding sequence can explain the optimum in enzyme activity with pH shown in Figure P9-13$_B$.

(b) List ways you can work this problem incorrectly.

(c) Apply one or more of the six ideas in Preface Table P-4, page xxviii, to this problem.

P9-14$_B$ An Eadie-Hofstee plot is shown below for the different types of enzyme inhibition. Match the line with the type of inhibition.

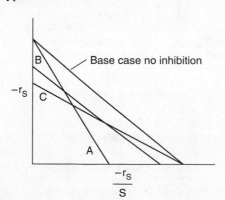

(a) Line A Inhibition Mechanism. Ans: _____

(b) Line B Inhibition Mechanism. Ans: _____

(c) Line C Inhibition Mechanism. Ans: _____

P9-15$_B$ The biomass reaction

$$\text{Substrate S + Cells} \longrightarrow \text{More cells + Product}$$

takes place in a 12-dm^3 CSTR (chemostat) where the entering concentration of substrate is 200 g/dm^3. The rate law follows the Monod equation with $\mu_{max} = 0.5s^{-1}$ and $K_S = 50$ g/dm^3. What is the volumetric flow rate, v_0 (dm^3/h), that will give the maximum production rate of cells (g/h)?

P9-16$_B$ The production of a product P from a particular gram-negative bacteria follows the Monod growth law

$$r_g = \frac{\mu_{max} C_s C_c}{K_S + C_s}$$

with $\mu_{max} = 1$ h^{-1}, $K_S = 0.25$ g/dm^3, and $Y_{c/s} = 0.5$ g/g.

(a)   The reaction is to be carried out in a batch reactor with the initial cell concentration of $C_{c0} = 0.1$ g/dm^3 and substrate concentration of $C_{s0} = 20$ g/dm^3.

$$C_c = C_{c0} + Y_{c/s}(C_{s0} - C_s)$$

Plot r_g, $-r_s$, $-r_c$, C_s, and C_c as a function of time.

(b)   The reaction is now to be carried out in a CSTR with $C_{s0} = 20$ g/dm^3 and $C_{c0} = 0$. What is the dilution rate at which wash-out occurs?

(c)   For the conditions in part **(b)**, what is the dilution rate that will give the maximum product rate (g/h) if $Y_{p/c} = 0.15$ g/g? What are the concentrations C_c, C_s, C_p, and $-r_s$ at this value of D?

(d)   How would your answers to **(b)** and **(c)** change if cell death could not be neglected with $k_d = 0.02$ h^{-1}?

(e)   How would your answers to **(b)** and **(c)** change if maintenance could not be neglected with $m = 0.2$ g/h/dm^3?

(f)   Redo part **(a)** and use a **logistic growth** law

$$r_g = \mu_{max}\left(1 - \frac{C_c}{C_\infty}\right)C_c$$

and plot C_c and r_c as a function of time. The term C_∞ is the maximum cell mass concentration and is called the *carrying capacity*, and is equal to $C_\infty = 1.0$ g/dm^3. Can you find an analytical solution for the batch reactor? Compare with part **(a)** for $C_\infty = Y_{c/s} C_{s0} + C_{c0}$.

(g)   List ways you can work this problem incorrectly.

(h)   Apply one or more of the six ideas in Preface Table P-4, page xxviii, to this problem.

P9-17$_B$ Redo Problem P9-16$_B$ **(a)**, **(c)**, and **(d)** using the Tessier equation

$$r_g = \mu_{max}[1 - e^{-C_s/k}]C_c$$

with $\mu_{max} = 1.0$ h^{-1} and $k = 8$ g/dm^3.

(a)   List ways you can work this problem incorrectly.

(b)   How could you make this problem more difficult?

P9-18$_B$ The bacteria X-II can be described by a simple Monod equation with $\mu_{max} = 0.8$ h^{-1} and $K_S = 4$ g/dm^3, $Y_{p/c} = 0.2$ g/g, and $Y_{s/c} = 2$ g/g. The process is carried out in a CSTR in which the feed rate is 1000 dm^3/h at a substrate concentration of 10 g/dm^3.

(a)   What size fermentor is needed to achieve 90% conversion of the substrate? What is the exiting cell concentration?

(b)   How would your answer to **(a)** change if all the cells were filtered out and returned to the feed stream?

(c)   Consider now two 5000-dm^3 CSTRs connected in series. What are the exiting concentrations C_s, C_c, and C_p from each of the reactors?

(d) Determine, if possible, the volumetric flow rate at which wash-out occurs and also the flow rate at which the cell production rate ($C_c v_0$) in grams per day is a maximum.

(e) Suppose you could use the two 5000-dm³ reactors as batch reactors that take two hours to empty, clean, and fill. What would your production rate be in (grams per day) if your initial cell concentration is 0.5 g/dm³? How many 500-dm³ batch reactors would you need to match the CSTR production rate?

(f) List ways you can work this problem incorrectly.

(g) Apply one or more of the six ideas in Preface Table P-4, page xxviii, to this problem.

P9-19$_A$ A CSTR is being operated at steady state. The cell growth follows the Monod growth law without inhibition. The exiting substrate and cell concentrations are measured as a function of the volumetric flow rate (represented as the dilution rate), and the results are shown below. Of course, measurements are not taken until steady state is achieved after each change in the flow rate. Neglect substrate consumption for maintenance and the death rate, and assume that $Y_{p/c}$ is zero. For run 4, the entering substrate concentration was 50 g/dm³ and the volumetric flow rate of the substrate was 2 dm³/h.

Run	C_s (g/dm³)	D (day⁻¹)	C_c (g/dm³)
1	1	1	0.9
2	3	1.5	0.7
3	4	1.6	0.6
4	10	1.8	4

(a) Determine the Monod growth parameters μ_{max} and K_S.

(b) Estimate the stoichiometric coefficients, $Y_{c/s}$ and $Y_{s/c}$.

(c) Apply one or more of the six ideas in Preface Table P-4, page xxviii, to this problem.

(d) How could you make this problem more difficult?

P9-20$_B$ **Alternative Energy Source.**[25] In the summer of 2009, ExxonMobil decided to invest 600 million dollars on developing algae as an alternative fuel. Algae would be grown and their oil extracted to provide an energy source. It is estimated that one acre of a biomass pond can provide 6,000 gallons of gasoline per year, which would require the capture of a CO_2 source more concentrated than air (e.g., fuel gas from a refinery) and also contribute to the sequestration of CO_2. The biomass biosynthesis during the day is

$$\text{Sunlight} + CO_2 + H_2O + \text{Algae} \rightarrow \text{More Algae} + O_2$$

Consider a 5,000-gallon pond with perforated pipes into which CO_2 is injected and slowly bubbled into the solution to keep the water saturated with CO_2.

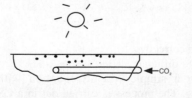 

Figure P9-20.1 Commercial microalgae production in open raceway, paddle-wheel mixed ponds. Courtesy of Cyanotech Co., Hawaii.

The doubling time during the day is 12 h at high-noon sunlight and zero during the night. As a first approximation, the growth during the 12 hours of daylight law is

$$r_g = f\mu C_C$$

---

[25] The contributions of John Benemann to this problem are appreciated.

with f = sunlight = sin (π t/12) between 6 a.m. and 6 p.m., otherwise f = 0, C_C is the algae concentration (g/dm³) and μ = 0.9 day⁻¹ (assumes constant CO_2 saturation at 1 atm is 1.69g/kg water). The pond is 30-cm deep and for effective sunlight penetration, the algae concentration cannot exceed 200 mg/dm³.

(a) Derive an equation for the ratio of the algae cell concentration C_C at time t to initial cell concentration C_{C0}, i.e., (C_C/C_{C0}). Plot and analyze (C_C/C_{C0}) versus time up to 48 hours.

(b) If the pond is initially seeded with 0.5 mg/dm³ of algae, how long will it take the algae to reach a cell density (i.e., concentration) of 200 mg/dm³, which is the concentration at which sunlight can no longer effectively penetrate the depth of the pond? Plot and analyze r_g and C_C as a function of time. As a first approximation, assume the pond is well mixed.

(c) Suppose the algae limit the sun's penetration significantly even before the concentration reaches 200 mg/dm³ with e.g., $\mu = \mu_0 (1 - C_C/200)$. Plot and analyze r_g and C_C as a function of time. How long would it take to completely stop growth at 200 mg/dm³?

(d) Now, let's consider continuous operation. Once the cell density reaches 200 mg/dm, one-half of the pond is harvested and the remaining broth is mixed with fresh nutrient. What is the steady-state algae productivity in gm/year, again assuming the pond is well mixed?

(e) Now consider a constant feed of waste water and removal of algae at a dilution rate of one reciprocal day. What is the mass flow rate of algae out of the 5,000 gallon pond (g/d)? Assume the pond is well mixed.

(f) Now consider that the reaction is to be carried out in an enclosed, transparent reactor. The reactor can be pressurized with CO_2 up to 10 atm with K_S = 2 g/dm³. Assume that after the initial pressurization, no more CO_2 can be injected. Plot and analyze the algae concentration as a function of time.

(g) An invading algae can double twice as fast as the strain you are cultivating. Assume that it initially is at 0.1 mg/l concentration. How long until it is the dominant species (over 50% of the cell density)?

P9-21$_A$ Short calculations on the algae ponds.

(a) If the pond is initially seeded with 0.5 mg/dm³ of algae, how long will it take the algae to reach a cell density (i.e., concentration) of 200 mg/dm³? Sketch a rough plot of r_g and C_C over time.

(b) Suppose the algae limit the sun's penetration significantly even before the concentration reaches 200 mg/dm³ by using $\mu = \mu_0 (1-C_C/200)$. Assume μ_0 = 0.9 day⁻¹. Qualitatively, what happens to the growth rate as the concentration of cells increases? Approximately how long would it take for the concentration to reach 200 mg/dm³? Why?

(c) An invading algae species can double twice as fast as the strain you are cultivating. Assume that it is initially at a concentration of 0.01 mg/dm³. How long until it becomes the dominant species in the pond (over 50% of the cell density)?

• **Additonal Homework Problems**

A number of homework problems that can be used for exams or supplementary problems or examples are found on the CRE Web site, *www.umich.edu/~elements/5e/index.html*.

SUPPLEMENTARY READING

Web

Review the following Web sites:
  www.enzymes.com
  www.pharmacokinetics.com

Text

1. A discussion of complex reactions involving active intermediates is given in

FROST, A. A., and R. G. PEARSON, *Kinetics and Mechanism*, 2nd ed. New York: Wiley, 1961, Chapter 10. Old but great examples.

LAIDLER, K. J., *Chemical Kinetics*, 3rd ed. New York: HarperCollins, 1987.

PILLING, M. J., *Reaction Kinetics*, New York: Oxford University Press, 1995.

2. Further discussion of enzymatic reactions:

Just about everything you want to know about basic enzyme kinetics can be found in SEGEL, I. H., *Enzyme Kinetics*. New York: Wiley-Interscience, 1975.

An excellent description of parameter estimation, biological feedback, and reaction pathways can be found in VOIT, E. O., *Computational Analysis of Biochemical Systems*. Cambridge, UK: Cambridge University Press, 2000.

CORNISH-BOWDEN, A., *Analysis of Enzyme Kinetic Data*. New York: Oxford University Press, 1995.

NELSON, D. L., and M. M. COX, *Lehninger Principles of Biochemistry*, 3rd ed. New York: Worth Publishers, 2000.

SHULER, M. L., and F. KARGI, *Bioprocess Engineering Principles*, 2nd ed. Upper Saddle River, NJ: Prentice Hall, 2002.

3. Material on bioreactors can be found in

BAILEY, T. J., and D. OLLIS, *Biochemical Engineering*, 2nd ed. New York: McGraw-Hill, 1987.

BLANCH, H. W., and D. S. CLARK, *Biochemical Engineering*. New York: Marcel Dekker, 1996.

4. Also see

BURGESS, THORNTON W., *The Adventures of Old Mr. Toad*. New York: Dover Publications, Inc., 1916.

KEILLOR, GARRISON, *Pretty Good Joke Book*: *A Prairie Home Companion*. St. Paul, MN: HighBridge Co., 2000.

MASKILL, HOWARD, *The Investigation of Organic Reactions and Their Mechanisms*. Oxford UK: Blackwell Publishing Ltd, 2006.

Catalysis **10** and Catalytic Reactors

It isn't that they can't see the solution. It is that they can't see the problem.

—G. K. Chesterton

Overview. The objectives of this chapter are to develop an understanding of catalysts, reaction mechanisms, and catalytic reactor design. Specifically, after reading this chapter one should be able to

- Define a catalyst and describe its properties.
- Describe the steps in a catalytic reaction and in chemical vapor deposition (CVD).
- Suggest a mechanism and apply the concept of a rate-limiting step to derive a rate law.
- Use nonlinear regression to determine the rate law and rate-law parameters that best fit the data.
- Use the rate law parameters to design PBRs and fluidized CSTRs.
- Describe the analogy between catalytic steps and CVD in microelectronic fabrication.
- Analyze catalyst decay and reactors that can be used to help off-set decay.

The various sections of this chapter roughly correspond to these objectives.

10.1 Catalysts

Catalysts have been used by humankind for over 2000 years.[1] The first observed uses of catalysts were in the making of wine, cheese, and bread. It was found that it was always necessary to add small amounts of the previous

---

[1] S. T. Oyama and G. A. Somorjai, *J. Chem. Educ.*, 65, 765 (1986).

batch to make the current batch. However, it wasn't until 1835 that Berzelius began to tie together observations of earlier chemists by suggesting that small amounts of a foreign substance could greatly affect the course of chemical reactions. This mysterious force attributed to the substance was called *catalytic*. In 1894, Ostwald expanded Berzelius's explanation by stating that catalysts were substances that accelerate the rate of chemical reactions without being consumed during the reaction. During the 180 years since Berzelius's work, catalysts have come to play a major economic role in the world market. In the United States alone, sales of process catalysts will reach over $20 billion by 2018, the major uses being in petroleum refining and in chemical production.

10.1.1 Definitions

A *catalyst* is a substance that affects the rate of a reaction but emerges from the process unchanged. A catalyst usually changes a reaction rate by promoting a different molecular path ("mechanism") for the reaction. For example, gaseous hydrogen and oxygen are virtually inert at room temperature, but react rapidly when exposed to platinum. The reaction coordinate (cf. Chapter 3) shown in Figure 10-1 is a measure of the progress along the reaction path as H_2 and O_2 approach each other and pass over the activation energy barrier to form H_2O. *Catalysis is the occurrence, study, and use of catalysts and catalytic processes.* Commercial chemical catalysts are immensely important. Approximately one-third of the material gross national product of the United States involves a catalytic process somewhere between raw material and finished product.[2] The development and use of catalysts is a major part of the constant search for new ways of increasing product yield and selectivity from chemical reactions. Because a catalyst makes it possible to obtain an end product by a different pathway with a lower energy barrier, it can affect both the yield and the selectivity.

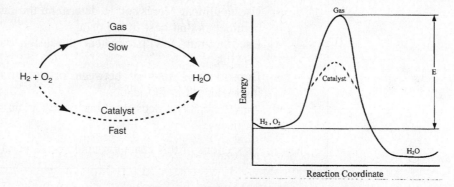

Figure 10-1   Different reaction paths.

Catalysts can accelerate the reaction rate but cannot change the equilibrium.

Normally when we talk about a catalyst, we mean one that speeds up a reaction, although strictly speaking, a catalyst can either accelerate or slow the formation of a particular product species. *A catalyst changes only the rate of a reaction; it does not affect the equilibrium.*

---

[2] V. Haensel and R. L. Burwell, Jr., *Sci. Am.*, 225(10), 46.

The 2007 Nobel Prize for Chemistry was awarded to Gerhard Ertl for his pioneering work on heterogeneous catalytic reactions. A *heterogeneous catalytic reaction* involves more than one phase; usually the catalyst is a solid and the reactants and products are in liquid or gaseous form. One example is the production of benzene, which is mostly manufactured today from the dehydrogenation of cyclohexane (obtained from the distillation of petroleum crude oil) using platinum-on-alumina as the catalyst:

The simple and complete separation of the fluid product mixture from the solid catalyst makes heterogeneous catalysis economically attractive, especially because many catalysts are quite valuable and their reuse is demanded.

A heterogeneous catalytic reaction occurs at or very near the fluid–solid interface. The principles that govern heterogeneous catalytic reactions can be applied to both catalytic and noncatalytic fluid–solid reactions. The two other types of heterogeneous reactions involve gas–liquid and gas–liquid–solid systems. Reactions between gases and liquids are usually mass-transfer limited.

10.1.2 Catalyst Properties

Because a catalytic reaction occurs at the fluid–solid interface, a large interfacial area is almost always essential in attaining a significant reaction rate. In many catalysts, this area is provided by an inner porous structure (i.e., the solid contains many fine pores, and the surface of these pores supplies the area needed for the high rate of reaction), see Figures 10-4(b) and 10-9. The area possessed by some porous catalysis materials is surprisingly large. A typical silica-alumina cracking catalyst has a pore volume of 0.6 cm^3/g and an average pore radius of 4 nm. The corresponding surface area is 300 m^2/g of these *porous catalysts*. Examples include the Raney nickel used in the hydrogenation of vegetable and animal oils, platinum-on-alumina used in the reforming of petroleum naphthas to obtain higher octane ratings, and promoted iron used in ammonia synthesis. Sometimes pores are so small that they will admit small molecules but prevent large ones from entering. Materials with this type of pore are called *molecular sieves*, and they may be derived from natural substances such as certain clays and zeolites, or they may be totally synthetic, such as some crystalline aluminosilicates (see Figure 10-2). These sieves can form the basis for quite selective catalysts; the pores can control the residence time of various molecules near the catalytically active surface to a degree that essentially allows *only* the desired molecules to react. One example of the high selectivity of zeolite catalysts is the formation of para-xylene from toluene and methane shown in Figure 10-2(b).[3] Here, benzene and toluene enter through the zeolite pore and react on the interior surface to form a mixture of ortho-, meta-, and para-xylenes. However, the size of the pore mouth is such that only

Ten grams of this catalyst possess more surface area than a U.S. football field

Catalyst types:
· Porous
· Molecular sieves
· Monolithic
· Supported
· Unsupported

Typical zeolite catalyst

High selectivity to para-xylene

[3] R. I. Masel, *Chemical Kinetics and Catalysis* (New York: Wiley Interscience, 2001), p. 741.

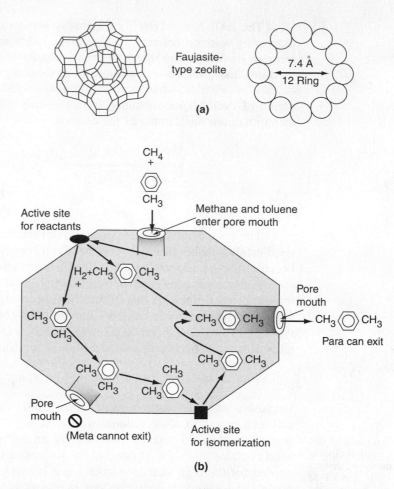

Figure 10-2   (a) Framework structures and (b) pore cross sections of two types of zeolites. (a) Faujasite-type zeolite has a three-dimensional channel system with pores at least 7.4 Å in diameter. A pore is formed by 12 oxygen atoms in a ring. (b) Schematic of reaction CH_4 and $C_6H_5CH_3$. (Note that the size of the pore mouth and the interior of the zeolite are not to scale.) [(a) from N. Y. Chen and T. F. Degnan, *Chem. Eng. Prog.*, *84*(2), 33 (1988). Reproduced by permission of the American Institute of Chemical Engineers. Copyright © 1988 AIChE. All rights reserved.]

para-xylene can exit through the pore mouth, as meta- and ortho- xylene with their methyl group on the side cannot fit through the pore mouth. There are interior sites that can isomerize ortho-xylene and meta-xylene to para-xylene. Hence, we have a very high selectivity to form para-xylene.

In some cases a catalyst consists of minute particles of an active material dispersed over a less-active substance called a *support*. The active material is frequently a pure metal or metal alloy. Such catalysts are called *supported catalysts*, as distinguished from *unsupported catalysts*. Catalysts can also have small amounts of active ingredients added called *promoters*, which increase their activity. Examples of supported catalysts are the packed-bed catalytic converter in an automobile, the platinum-on-alumina catalyst used in petroleum reforming, and the vanadium pentoxide on silica used to oxidize sulfur dioxide in manufacturing sulfuric acid. On the other hand, platinum gauze for ammonia oxidation, promoted iron for ammonia synthesis, and the silica–alumina dehydrogenation catalyst used in butadiene manufacture typify unsupported catalysts.

10.1.3 Catalytic Gas-Solid Interactions

For the moment, let us focus our attention on gas-phase reactions catalyzed by solid surfaces. For a catalytic reaction to occur, at least one and frequently all of the reactants must become attached to the surface. This attachment is known as *adsorption* and takes place by two different processes: physical adsorption and chemisorption. *Physical adsorption* is similar to condensation. The process is exothermic, and the heat of adsorption is relatively small, being on the order of 1 to 15 kcal/mol. The forces of attraction between the gas molecules and the solid surface are weak. These van der Waals forces consist of interaction between permanent dipoles, between a permanent dipole and an induced dipole, and/or between neutral atoms and molecules. The amount of gas physically adsorbed decreases rapidly with increasing temperature, and above its critical temperature only very small amounts of a substance are physically adsorbed.

The type of adsorption that affects the rate of a chemical reaction is *chemisorption*. Here, the adsorbed atoms or molecules are held to the surface by valence forces of the same type as those that occur between bonded atoms in molecules. As a result, the electronic structure of the chemisorbed molecule is perturbed significantly, causing it to be extremely reactive. Interaction with the catalyst causes bonds of the adsorbed reactant to be stretched, making them easier to break.

Figure 10-3 shows the bonding from the adsorption of ethylene on a platinum surface to form chemisorbed ethylidyne. Like physical adsorption, chemisorption is an exothermic process, but the heats of adsorption are

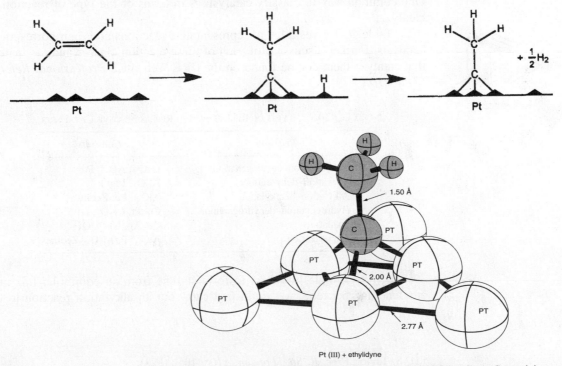

Pt (III) + ethylidyne

Figure 10-3 Ethylidyne chemisorbed on platinum. (Adapted from G. A. Somorjai, *Introduction to Surface Chemistry and Catalysis.* © 1994 John Wiley & Sons, Inc. Reprinted by permission of John Wiley & Sons, Inc. All rights reserved.)

generally of the same magnitude as the heat of a chemical reaction (i.e., 40 to 400 kJ/mol). If a catalytic reaction involves chemisorption, it must be carried out within the temperature range where chemisorption of the reactants is appreciable.

Chemisorption on active sites is what catalyzes the reaction.

In a landmark contribution to catalytic theory, Taylor suggested that a reaction is not catalyzed over the entire solid surface but only at certain *active sites* or *centers*.[4] He visualized these sites as unsaturated atoms in the solids that resulted from surface irregularities, dislocations, edges of crystals, and cracks along grain boundaries. Other investigators have taken exception to this definition, pointing out that other properties of the solid surface are also important. The active sites can also be thought of as places where highly reactive intermediates (i.e., chemisorbed species) are stabilized long enough to react. This stabilization of a reactive intermediate is key in the design of any catalyst. Consequently, for our purposes we will define an *active site* as *a point on the catalyst surface that can form strong chemical bonds with an adsorbed atom or molecule.*

One parameter used to quantify the activity of a catalyst is the *turnover frequency (TOF)*, f. It is the number of molecules reacting per active site per second at the conditions of the experiment. When a metal catalyst such as platinum is deposited on a support, the metal atoms are considered active sites. The TOFs for a number of reactions are shown in *Professional Reference Shelf R10.1*. The *dispersion, D*, of the catalyst is the fraction of the metal atoms deposited on a catalyst that are *on* the surface.

10.1.4 Classification of Catalysts

One common way to classify catalysts is in terms of the type of reaction they catalyze.

Reference Shelf

Table 10-1 gives a list of representative reactions and their corresponding catalysts. Further discussion of each of these reaction classes and the materials that catalyze them can be found on the CRE Web site's *Professional Reference Shelf R10.1*.

TABLE 10-1   TYPES OF REACTIONS AND REPRESENTATIVE CATALYSTS

Reaction	Catalysts
1. Halogenation–dehalogenation	$CuCl_2$, AgCl, Pd
2. Hydration–dehydration	Al_2O_3, MgO
3. Alkylation–dealkylation	$AlCl_3$, Pd, Zeolites
4. Hydrogenation–dehydrogenation	Co, Pt, Cr_2O_3, Ni
5. Oxidation	Cu, Ag, Ni, V_2O_5
6. Isomerization	$AlCl_3$, Pt/Al_2O_3, Zeolites

If, for example, we were to form styrene from an equimolar mixture of ethylene and benzene, we could first carry out an alkylation reaction to form

---

[4] H. S. Taylor, *Proc. R. Soc. London, A108*, 105 (1928).

ethyl benzene, which is then dehydrogenated to form styrene. We need both an alkylation catalyst and a dehydrogenation catalyst:

$$C_2H_4 + C_6H_6 \xrightarrow[\text{trace HCl}]{AlCl_3} C_6H_5C_2H_5 \xrightarrow{Ni} C_6H_5CH{=}CH_2 + H_2$$

10.2 Steps in a Catalytic Reaction

A photograph of different types and sizes of catalysts is shown in Figure 10-4a. A schematic diagram of a tubular reactor packed with catalytic pellets is shown in Figure 10-4b. The overall process by which heterogeneous catalytic reactions proceed can be broken down into the sequence of individual steps shown in Table 10-2 and pictured in Figure 10-5 for an isomerization reaction.

Figure 10-4a   Catalyst particles of different shapes (spheres, cylinders) and sizes (0.1 cm to 1 cm). (Photo courtesy of the Engelhard Corporation.)

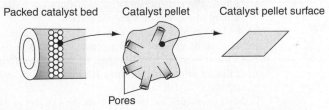

Figure 10-4b   Catalytic packed-bed reactor—schematic.

Each step in Table 10-2 is shown schematically in Figure 10-5.

The overall rate of reaction is limited by the rate of the slowest step in the sequence. When the diffusion steps (1, 2, 6, and 7 in Table 10-2) are very fast compared with the surface reaction-rate steps (3, 4, and 5), the concentrations in the immediate vicinity of the active sites are indistinguishable from those in the bulk fluid. In this situation, the transport or diffusion steps do not affect the overall rate of the reaction. In other situations, if the reaction steps are very fast compared with the diffusion steps, mass transport does affect the reaction rate. In systems where diffusion from the bulk gas or liquid to the external catalyst surface or to the mouths of catalyst pores affects the rate, i.e., steps 1 and 7, changing the flow conditions past the catalyst should change the overall reaction rate (see Chapter 14). Once inside the porous catalysts, on the other hand, diffusion within the catalyst pores, i.e., steps 2 and 6, may limit the rate of reaction and, as a result, the overall rate will be unaffected by external flow conditions even though diffusion affects the overall reaction rate (see Chapter 15).

A reaction takes place on the surface, but the species involved in the reaction must get to and from the surface.

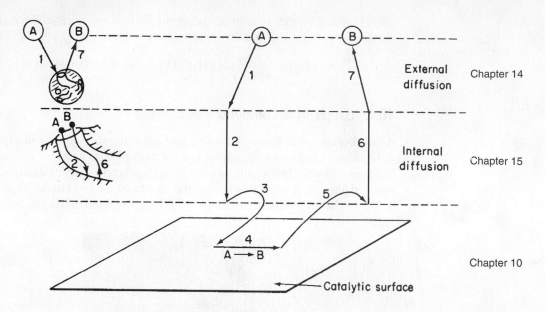

Figure 10-5   Steps in a heterogeneous catalytic reaction.

TABLE 10-2   STEPS IN A CATALYTIC REACTION

1. Mass transfer (diffusion) of the reactant(s) (e.g., species A) from the bulk fluid to the external surface of the catalyst pellet
2. Diffusion of the reactant from the pore mouth through the catalyst pores to the immediate vicinity of the internal catalytic surface
3. Adsorption of reactant A onto the catalyst surface
4. Reaction on the surface of the catalyst (e.g., A $\longrightarrow$ B)
5. Desorption of the products (e.g., B) from the surface
6. Diffusion of the products from the interior of the pellet to the pore mouth at the external surface
7. Mass transfer of the products from the external pellet surface to the bulk fluid

There are many variations of the situation described in Table 10-2. Sometimes, of course, two reactants are necessary for a reaction to occur, and both of these may undergo the steps listed above. Other reactions between two substances may have only one of them adsorbed.

In this chapter we focus on:
3. Adsorption
4. Surface reaction
5. Desorption

With this introduction, we are ready to treat individually the steps involved in catalytic reactions. In this chapter, only steps 3, 4, and 5—i.e., adsorption, surface reaction, and desorption—are considered as we assume that the diffusion steps (1, 2, 6, and 7) are very fast so that the overall reaction rate is not affected by mass transfer in any fashion. Further treatment of the effects involving diffusion limitations is provided in Chapters 14 and 15.

Where Are We Heading?[†]   As we saw in Chapter 7, one of the tasks of a chemical reaction engineer is to analyze rate data and to develop a rate law that can be used in reactor design. Rate laws in heterogeneous catalysis seldom

---

[†] "If you don't know where you are going, you'll probably wind up someplace else."
Yogi Berra

follow power-law models and hence are inherently more difficult to formulate from the data. To develop an in-depth understanding and insight as to how the rate laws are formed from heterogeneous catalytic data, we are going to proceed in somewhat of a reverse manner than what is normally done in industry when one is asked to develop a rate law. That is, we will postulate catalytic mechanisms and *then* derive rate laws for the various mechanisms. The mechanism will typically have an adsorption step, a surface reaction step, and a desorption step, one of which is usually rate-limiting. Suggesting mechanisms and rate-limiting steps is not the first thing we normally do when presented

An algorithm with data. However, by deriving equations for different mechanisms, we will observe the various forms of the rate law one can have in heterogeneous catalysis. Knowing the different forms that catalytic rate equations can take, it will be easier to view the trends in the data and deduce the appropriate rate law. This deduction is usually what is done first in industry before a mechanism is proposed. Knowing the form of the rate law, one can then numerically evaluate the rate-law parameters and postulate a reaction mechanism and rate-limiting step that are consistent with the rate data. Finally, we use the rate law to design catalytic reactors. This procedure is shown in Figure 10-6. The dashed lines represent feedback to obtain new data in specific regions (e.g., concentrations, temperature) to evaluate the rate-law parameters more precisely or to differentiate between reaction mechanisms.

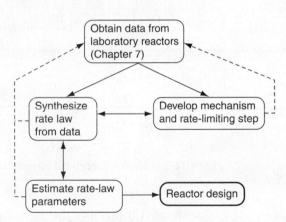

Figure 10-6   Collecting information for catalytic reactor design.

We will discuss each of the steps shown in Figure 10-5 and Table 10-2. As mentioned earlier, this chapter focuses on Steps 3, 4, and 5 (the adsorption, surface reaction and desorption steps) by assuming that Steps 1, 2, 6, and 7 are

Chance Card: very rapid. Consequently, to understand when this assumption is valid, we
Do not pass go shall give a quick overview of Steps 1, 2, 6, and 7. These steps involve diffu-
proceed directly to sion of the reactants to and within the catalyst pellet. While these diffusion
Section 10.2.3. steps are covered in detail in Chapters 14 and 15, it is worthwhile to give a brief description of these two mass-transfer steps to better understand the entire sequence of steps. If you have had the core course in *Mass Transfer* or *Transport Phenomena* you can skip Sections 10.2.1 and 10.2.2, and go directly to Section 10.2.3.

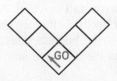

10.2.1 Step 1 Overview: Diffusion from the Bulk to the External Surface of the Catalyst

External and internal mass transfer in catalysis are covered in detail in Chapters 14 and 15.

For the moment let's assume that the transport of A from the bulk fluid to the external surface of the catalyst is the slowest step in the sequence. We lump all the resistance to transfer from the bulk fluid to the surface in the mass-transfer boundary layer surrounding the pellet. In this step the reactant A, which is at a bulk concentration C_{Ab} must travel (diffuse) through the boundary layer of thickness δ to the external surface of the pellet where the concentration is C_{As}, as shown in Figure 10-7. The rate of transfer (and hence rate of reaction, $-r'_A$) for this slowest step is

$$\text{Rate} = k_C\,(C_{Ab} - C_{As})$$

where the mass-transfer coefficient, k_C, is a function of the hydrodynamic conditions, namely the fluid velocity, U, and the particle diameter, D_p.

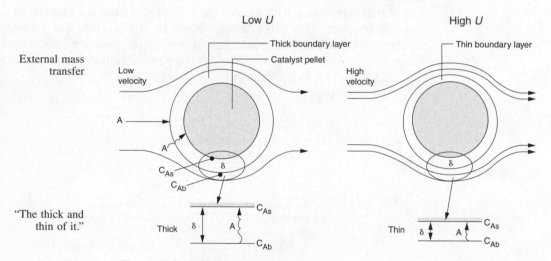

Figure 10-7 Diffusion through the external boundary layer (also see Figure 14-3).

As we see (Chapter 14), the mass-transfer coefficient is inversely proportional to the boundary layer thickness, δ, and directly proportional to the diffusion coefficient (i.e., the diffusivity D_{AB}).

$$k_C = \frac{D_{AB}}{\delta}$$

At low velocities of fluid flow over the pellet, the boundary layer across which A and B must diffuse is thick, and it takes a long time for A to travel to the surface, resulting in a small mass-transfer coefficient k_C. As a result, mass transfer across the boundary layer is slow and limits the rate of the overall reaction. As the velocity over the pellet is increased, the boundary layer becomes thinner and the mass-transfer rate is increased. At very high velocities, the boundary layer thickness, δ, is so small it no longer offers any resistance to the diffusion across the boundary layer. As a result, external mass transfer no longer limits the rate of reaction. This external resistance also decreases as the particle size is decreased. As the fluid velocity increases and/or the particle diameter decreases, the mass-transfer coefficient increases until a plateau is reached, as shown in Figure 10-8. On this plateau, $C_{Ab} \approx C_{As}$, and one of the other steps in the sequence, i.e., Steps 2 through 6, is the slowest step and limits the overall reaction rate. Further details on external mass transfer are discussed in Chapter 14.

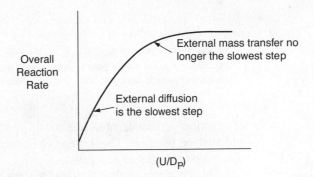

Figure 10-8   Effect of particle size and fluid velocity on the overall reaction rate.

10.2.2 Step 2 Overview: Internal Diffusion

Now consider that we are operating at a fluid velocity where external diffusion is no longer the rate-limiting step and that internal diffusion is the slowest step. In Step 2 the reactant A diffuses from the external pellet surface at a concentration C_{As} into the pellet interior, where the concentration is C_A. As A diffuses into the interior of the pellet, it reacts with catalyst deposited on the sides of the catalyst pellet's pore walls.

For large pellets, it takes a long time for the reactant A to diffuse into the interior, compared to the time that it takes for the reaction to occur on the interior pore surface. Under these circumstances, the reactant is only consumed near the exterior surface of the pellet and the catalyst near the center of the pellet is wasted catalyst. On the other hand, for very small pellets it takes very little time to diffuse into and out of the pellet interior and, as a result, internal diffusion no longer limits the rate of reaction. When internal mass transfer no longer limits the rate of reaction, the rate law can be expressed as

$$\text{Rate} = k_r \, C_{As}$$

where C_{As} is the concentration at the external pellet surface and k_r is an overall rate constant, which is a function of particle size. The overall rate constant, k_r, increases as the pellet diameter decreases. In Chapter 15, we show that Figure 15-5 can be combined with Equation (15-34) to arrive at the plot of k_r as a function of D_P, shown in Figure 10-9(b).

We see in Figure 10-9 that at small particle sizes, internal diffusion is no longer the slow step and that the surface reaction sequence of adsorption, surface reaction, and desorption (Steps 3, 4, and 5 in Figure 10-5) limit the overall rate

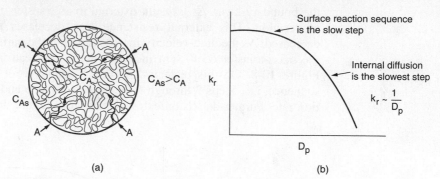

Internal mass
transfer

(a)   (b)

Figure 10-9   Effect of particle size on the overall reaction-rate constant.
(a) Branching of a single pore with deposited metal; (b) decrease in rate constant with
increasing particle diameter. (See the CRE Web site, Chapter 12.)

of reaction. Consider now one more important point about internal diffusion and surface reaction. These steps (2 through 6) are **not at all** affected by flow conditions external to the pellet.

In the material that follows, we are going to choose our pellet size and external fluid velocity such that neither external diffusion nor internal diffusion is limiting as discussed in Chapters 14 and 15. Instead, we assume that either Step 3 (adsorption), Step 4 (surface reaction), or Step 5 (desorption), or a combination of these steps, limits the overall rate of reaction.

10.2.3 Adsorption Isotherms

Because chemisorption is usually a necessary part of a catalytic process, we shall discuss it before treating catalytic reaction rates. The letter S will represent an active site; alone, it will denote a vacant site, with no atom, molecule, or complex adsorbed on it. The combination of S with another letter (e.g., $A \cdot S$) will mean that one unit of species A will be chemically adsorbed on the site S. Species A can be an atom, molecule, or some other atomic combination, depending on the circumstances. Consequently, the adsorption of A on a site S is represented by

$$A + S \rightleftharpoons A \cdot S$$

The total molar concentration of active sites per unit mass of catalyst is equal to the number of active sites per unit mass divided by Avogadro's number and will be labeled C_t (mol/g-cat). The molar concentration of vacant sites, C_v (mol/g-cat), is the number of vacant sites per unit mass of catalyst divided by Avogadro's number. In the absence of catalyst deactivation, we assume that the total concentration of active sites, C_t, remains constant. Some further definitions include

P_i = partial pressure of species i in the gas phase, (atm or kPa)
$C_{i \cdot S}$ = surface concentration of sites occupied by species i, (mol/g-cat)

A conceptual model depicting species A and B adsorbed on two different sites is shown in Figure 10-10.

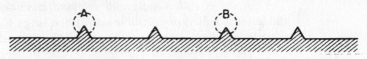

Figure 10-10   Vacant and occupied sites.

For the system shown in Figure 10-10, the total concentration of sites is

Site balance

$$C_t = C_v + C_{A \cdot S} + C_{B \cdot S} \qquad (10\text{-}1)$$

This equation is referred to as a *site balance*. A typical value for the total concentration of sites could be the order of 10^{22} sites/g-cat.

Now consider the adsorption of a nonreacting gas onto the surface of a catalyst. Adsorption data are frequently reported in the form of adsorption *isotherms*. Isotherms portray the amount of a gas adsorbed on a solid at different pressures at a given temperature.

Postulate models; then see which one(s) fit(s) the data.

First, an adsorption mechanism is proposed, and then the isotherm (see Figure 10-11, page 413) obtained from the mechanism is compared with the experimental data. If the isotherm predicted by the model agrees with the experimental data, the model may reasonably describe what is occurring physically in the real system. If the predicted curve does not agree with the experimental data, the model fails to match the physical situation in at least one important characteristic and perhaps more.

We will consider two types of adsorption: molecular adsorption and dissociative adsorption. To illustrate the difference between molecular adsorption and dissociative adsorption, we will postulate two models for the adsorption of carbon monoxide on metal surfaces. In the molecular adsorption model, CO is adsorbed as molecules, CO,

$$CO + S \rightleftharpoons CO \cdot S$$

as is the case on nickel

Two models:
1. Adsorption as CO
2. Adsorption as C and O

$$
\begin{array}{ccc}
CO & & C \diagdown \\
 & & \quad\;\; O \\
+ & & \vdots \\
-Ni-Ni-Ni- & \rightleftharpoons & -Ni-Ni-Ni-
\end{array}
$$

In the dissociative adsorption model, carbon monoxide is adsorbed as oxygen and carbon atoms instead of molecular CO

$$CO + 2S \rightleftharpoons C \cdot S + O \cdot S$$

as is the case on iron[5]

$$
\begin{array}{ccc}
CO & & C \quad O \\
+ & & \vdots \quad \vdots \\
-Fe-Fe-Fe- & \rightleftharpoons & -Fe-Fe-Fe-
\end{array}
$$

---

[5] R. I. Masel, *Principles of Adsorption and Reaction on Solid Surfaces* (New York: Wiley, 1996).

The former is called *molecular* or *nondissociated adsorption* (e.g., CO) and the latter is called *dissociative adsorption* (e.g., C and O). Whether a molecule adsorbs nondissociatively or dissociatively depends on the surface.

The adsorption of carbon monoxide molecules will be considered first. Because the carbon monoxide does not react further after being adsorbed, we need only to consider the adsorption process:

Molecular Adsorption
$$CO + S \; \rightleftharpoons \; CO \cdot S \qquad (10\text{-}2)$$

In obtaining a rate law for the rate of adsorption, the reaction in Equation (10-2) can be treated as an *elementary reaction*. The rate of attachment of the carbon monoxide molecules to the active site on the surface is proportional to the number of collisions that these molecules make with a surface active site per second. In other words, a specific fraction of the molecules that strike the surface become adsorbed. The collision rate is, in turn, directly proportional to the carbon monoxide partial pressure, P_{CO}. Because carbon monoxide molecules adsorb only on vacant sites and not on sites already occupied by other carbon monoxide molecules, the rate of attachment is also directly proportional to the concentration of vacant sites, C_v. Combining these two facts means that the rate of attachment of carbon monoxide molecules to the surface is directly proportional to the product of the partial pressure of CO and the concentration of vacant sites; that is,

$$\boxed{P_{CO} = C_{CO}RT}$$

$$\text{A} + \text{S} \longrightarrow \text{A} \cdot \text{S}$$

$$\text{Rate of attachment} = k_A P_{CO} C_v$$

The rate of detachment of molecules from the surface can be a first-order process; that is, the detachment of carbon monoxide molecules from the surface is usually directly proportional to the concentration of sites occupied by the adsorbed molecules (e.g., $C_{CO \cdot S}$):

$$\text{A} \cdot \text{S} \longrightarrow \text{A} + \text{S}$$

$$\text{Rate of detachment} = k_{-A} C_{CO \cdot S}$$

The net rate of adsorption is equal to the rate of molecular attachment to the surface minus the rate of detachment from the surface. If k_A and k_{-A} are the constants of proportionality for the attachment and detachment processes, then

$$r_{AD} = k_A P_{CO} C_v - k_{-A} C_{CO \cdot S} \qquad (10\text{-}3)$$

The ratio $K_A = k_A / k_{-A}$ is the *adsorption equilibrium constant*. Using K_A to rearrange Equation (10-3) gives

Adsorption
$$\text{A} + \text{S} \rightleftharpoons \text{A} \cdot \text{S}$$
$$r_{AD} = k_A \left(P_A C_v - \dfrac{C_{A \cdot S}}{K_A} \right)$$

$$\boxed{ r_{AD} = k_A \left(P_{CO} C_v - \dfrac{C_{CO \cdot S}}{K_A} \right) } \qquad (10\text{-}4)$$

The adsorption rate constant, k_A, for molecular adsorption is virtually independent of temperature, while the desorption constant, k_{-A}, increases exponentially with increasing temperature. Consequently, *the equilibrium adsorption constant K_A decreases exponentially with increasing temperature.*

Because carbon monoxide is the only material adsorbed on the catalyst, the site balance gives

$$C_t = C_v + C_{CO \cdot S} \qquad (10\text{-}5)$$

$$-r'_A = r_{AD} = \left(\frac{mol}{g\text{-cat}\cdot s}\right)$$

$$k_A = \left(\frac{1}{atm\cdot s}\right)$$

$$P_A = (atm)$$

$$C_v = \left(\frac{mol}{g\text{-cat}}\right)$$

$$K_A = \left(\frac{1}{atm}\right)$$

$$C_{A\cdot s} = \left(\frac{mol}{g\text{-cat}}\right)$$

At equilibrium, the net rate of adsorption equals zero, i.e., $r_{AD} \equiv 0$. Setting the left-hand side of Equation (10-4) equal to zero and solving for the concentration of CO adsorbed on the surface, we get

$$C_{CO\cdot s} = K_A C_v P_{CO} \tag{10-6}$$

Using Equation (10-5) to give C_v in terms of $C_{CO\cdot s}$ and the total number of sites C_t, we can solve for the equilibrium value of $C_{CO\cdot s}$ in terms of constants and the pressure of carbon monoxide

$$C_{CO\cdot s} = K_A C_v P_{CO} = K_A P_{CO}(C_t - C_{CO\cdot s})$$

Rearranging gives us

$$\boxed{C_{CO\cdot s} = \frac{K_A P_{CO} C_t}{1 + K_A P_{CO}}} \tag{10-7}$$

This equation thus gives the equilibrium concentration of carbon monoxide adsorbed on the surface, $C_{CO\cdot s}$, as a function of the partial pressure of carbon monoxide, and is an equation for the adsorption isotherm. This particular type of isotherm equation is called a *Langmuir isotherm*.[6] **Figure 10-11(a)** shows the Langmuir isotherm for the amount of CO adsorbed per unit mass of catalyst as a function of the partial pressure of CO. For the case of dissociative adsorption, Equation (10-11), **Figure 10-11(b)**, shows the concentration of the atoms C and O adsorbed per unit mass of catalyst.

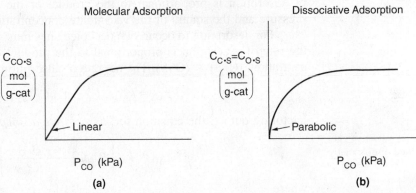

Figure 10-11 Langmuir isotherms for (a) molecular adsorption (b) dissociative adsorption of CO.

One method of checking whether a model (e.g., molecular adsorption versus dissociative adsorption) predicts the behavior of the experimental data is to linearize the model's equation and then plot the indicated variables against one another. For example, the molecular adsorption isotherm, Equation (10-7), may be arranged in the form

$$\frac{P_{CO}}{C_{CO\cdot s}} = \frac{1}{K_A C_t} + \frac{P_{CO}}{C_t} \tag{10-8}$$

---

[6] Named after Irving Langmuir (1881–1957), who first proposed it. He received the Nobel Prize in 1932 for his discoveries in surface chemistry.

and the linearity of a plot of $P_{CO}/C_{CO \cdot S}$ as a function of P_{CO} will determine if the data conform to molecular adsorption, i.e., a Langmuir single-site isotherm.

Molecular Adsorption

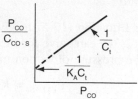

Next, we derive the isotherm for carbon monoxide disassociating into separate atoms as it adsorbs on the surface, i.e.,

Dissociative adsorption

$$CO + 2S \;\rightleftharpoons\; C \cdot S + O \cdot S$$

When the carbon monoxide molecule dissociates upon adsorption, it is referred to as the *dissociative adsorption* of carbon monoxide. As in the case of molecular adsorption, the rate of adsorption is proportional to the pressure of carbon monoxide in the system because this rate is governed by the number of gaseous collisions with the surface. For a molecule to dissociate as it adsorbs, however, two adjacent vacant active sites are required, rather than the single site needed when a substance adsorbs in its molecular form. The probability of two vacant sites occurring adjacent to one another is proportional to the square of the concentration of vacant sites. These two observations mean that the rate of adsorption is proportional to the product of the carbon monoxide partial pressure and the square of the vacant-site concentration, $P_{CO}C_v^2$.

For desorption to occur, two occupied sites must be adjacent, meaning that the rate of desorption is proportional to the product of the occupied-site concentration, $(C \cdot S) \times (O \cdot S)$. The net rate of adsorption can then be expressed as

$$r_{AD} = k_A P_{CO} C_v^2 - k_{-A} C_{O \cdot S} C_{C \cdot S} \tag{10-9}$$

Factoring out k_A, the equation for *dissociative adsorption* is

Rate of dissociative adsorption

$$r_{AD} = k_A \left(P_{CO} C_v^2 - \frac{C_{C \cdot S} C_{O \cdot S}}{K_A} \right)$$

where

$$K_A = \frac{k_A}{k_{-A}}$$

For dissociative adsorption, both k_A and k_{-A} increase exponentially with increasing temperature, while the *adsorption equilibrium constant K_A* decreases with increasing temperature.

At equilibrium, $r_{AD} \equiv 0$, and

$$k_A P_{CO} C_v^2 = k_{-A} C_{C \cdot S} C_{O \cdot S}$$

For $C_{C \cdot S} = C_{O \cdot S}$

$$(K_A P_{CO})^{1/2} C_v = C_{O \cdot S} \tag{10-10}$$

Substituting for $C_{C \cdot S}$ and $C_{O \cdot S}$ in a site balance equation (10-1),

Site balance:
$$C_t = C_v + C_{O \cdot S} + C_{C \cdot S}$$
$$= C_v + (K_{CO}P_{CO})^{1/2}C_v + (K_{CO}P_{CO})^{1/2}C_v = C_v(1 + 2(K_{CO}P_{CO})^{1/2})$$

Solving for C_v

$$C_v = C_t / (1 + 2(K_{CO}P_{CO})^{1/2})$$

This value may be substituted into Equation (10-10) to give an expression that can be solved for the equilibrium value of $C_{O \cdot S}$. The resulting equation for the isotherm shown in Figure 10-11(**b**) is

$$\boxed{C_{O \cdot S} = \frac{(K_A P_{CO})^{1/2} C_t}{1 + 2(K_A P_{CO})^{1/2}}} \tag{10-11}$$

Taking the inverse of both sides of the equation, then multiplying through by $(P_{CO})^{1/2}$, yields

$$\frac{(P_{CO})^{1/2}}{C_{O \cdot S}} = \frac{1}{C_t(K_A)^{1/2}} + \frac{2(P_{CO})^{1/2}}{C_t} \tag{10-12}$$

If dissociative adsorption is the correct model, a plot of $(P_{CO}^{1/2}/C_{O \cdot S})$ versus $P_{CO}^{1/2}$ should be linear with slope $(2/C_t)$.

Dissociative Adsorption

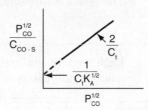

When more than one substance is present, the adsorption isotherm equations are somewhat more complex. The principles are the same, though, and the isotherm equations are easily derived. It is left as an exercise to show that the adsorption isotherm of A in the presence of another adsorbate B is given by the relationship

$$C_{A \cdot S} = \frac{K_A P_A C_t}{1 + K_A P_A + K_B P_B} \tag{10-13}$$

When the adsorption of both A and B are first-order processes, the desorptions are also first order, and both A and B are adsorbed as molecules. The derivations of other Langmuir isotherms are relatively easy.

In obtaining the Langmuir isotherm equations, several aspects of the adsorption system were presupposed in the derivations. The most important of

these, and the one that has been subject to the greatest doubt, is that a *uniform* surface is assumed. In other words, any active site has the same attraction for an impinging molecule as does any other active site. Isotherms different from the Langmuir isotherm, such as the Freundlich isotherm, may be derived based on various assumptions concerning the adsorption system, including different types of nonuniform surfaces.

10.2.4 Surface Reaction

The rate of adsorption of species A onto a solid surface

$$A + S \;\rightleftharpoons\; A \cdot S$$

is given by

$$r_{AD} = k_A \left(P_A C_v - \frac{C_{A \cdot S}}{K_A} \right) \tag{10-14}$$

Surface reaction models

After a reactant has been adsorbed onto the surface, i.e., A · S, it is capable of reacting in a number of ways to form the reaction product. Three of these ways are:

1. **Single site.** The surface reaction may be a single-site mechanism in which only the site on which the reactant is adsorbed is involved in the reaction. For example, an adsorbed molecule of A may isomerize (or perhaps decompose) directly on the site to which it is attached, such as

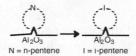

N = n-pentene   I = i-pentene

The pentane isomerization can be written in generic form as

$$A \cdot S \;\rightleftharpoons\; B \cdot S$$

Each step in the reaction mechanism is elementary, so the surface reaction rate law is

$$r_S = k_S C_{A \cdot S} - k_{-S} C_{B \cdot S} = k_S \left(C_{A \cdot S} - \frac{C_{B \cdot S}}{K_S} \right) \tag{10-15}$$

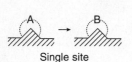

Single site

Single Site

$$k_S = \left(\frac{1}{s} \right)$$

$K_s =$ (dimensionless)

where K_S is the surface-reaction equilibrium constant $K_S = k_S/k_{-S}$

2. **Dual site.** The surface reaction may be a dual-site mechanism in which the adsorbed reactant interacts with another site (either unoccupied or occupied) to form the product.

First type of dual-site mechanism

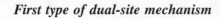

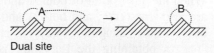

Dual site

For example, adsorbed A may react with an adjacent vacant site to yield a vacant site and a site on which the product is adsorbed, or as in the case of the dehydration of butanol, the products may adsorb on two adjacent sites.

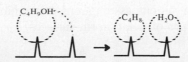

$$C_4H_9OH \cdot S + S \rightarrow C_4H_g \cdot S + H_2O \cdot S$$

$$r_S = k_S\left[(C_{C_4H_9OH \cdot S})(C_v) - \frac{(C_{C_4H_g \cdot S})(C_{H_2O \cdot S})}{K_S}\right]$$

For the generic reaction

$$A \cdot S + S \; \rightleftharpoons \; B \cdot S + S$$

the corresponding surface-reaction rate law is

Dual Site

$$r_S = \left(\frac{mol}{g\text{-cat} \cdot s}\right)$$

$$k_S = \left(\frac{g\text{-cat}}{mol \cdot s}\right)$$

$$K_S = (dimensionless)$$

$$\boxed{r_S = k_S \left(C_{A \cdot S}C_v - \frac{C_{B \cdot S}C_v}{K_S}\right)} \qquad (10\text{-}16)$$

A second dual-site mechanism is the reaction between two adsorbed species, such as the reaction of CO with O.

$$CO \cdot S + O \cdot S \; \rightleftharpoons \; CO_2 \cdot S + S$$

$$r_S = k_S\left[(C_{CO \cdot S})(C_{O \cdot S}) - \frac{(C_{CO_2 \cdot S})(C_v)}{K_S}\right]$$

For the generic reaction

$$A \cdot S + B \cdot S \; \rightleftharpoons \; C \cdot S + D \cdot S$$

the corresponding surface-reaction rate law is

Dual site

$$\boxed{r_S = k_S \left(C_{A \cdot S}C_{B \cdot S} - \frac{C_{C \cdot S}C_{D \cdot S}}{K_S}\right)} \qquad (10\text{-}17)$$

A third dual-site mechanism is the reaction of two species adsorbed on different types of sites S and S′, such as the reaction of CO with O.

$$CO \cdot S + O \cdot S' \rightarrow CO_2 \cdot S + S'$$

For the generic reaction

Dual site

$$A \cdot S + B \cdot S' \; \rightleftharpoons \; C \cdot S' + D \cdot S$$

the corresponding surface-reaction rate law is

$$\boxed{r_S = k_S \left(C_{A \cdot S}C_{B \cdot S'} - \frac{C_{C \cdot S'}C_{D \cdot S}}{K_S}\right)} \qquad (10\text{-}18)$$

Langmuir–
Hinshelwood
kinetics

Reactions involving either single- or dual-site mechanisms, which were described earlier, are sometimes referred to as following *Langmuir–Hinshelwood kinetics.*

3. **Eley–Rideal.** A third mechanism is the reaction between an adsorbed molecule and a molecule in the gas phase, such as the reaction of propylene and benzene (cf. the reverse reaction in Figure 10-13)

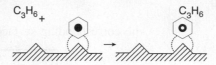

For the generic reaction

$$A \cdot S + B(g) \rightleftarrows C \cdot S$$

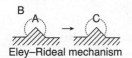
Eley–Rideal mechanism

the corresponding surface-reaction rate law is

$$r_S = k_S \left(C_{A \cdot S} P_B - \frac{C_{C \cdot S}}{K_S} \right) \qquad (10\text{-}19)$$

$$k_s = \left(\frac{1}{atm \cdot s} \right)$$

$$K_S = \left(\frac{1}{atm} \right)$$

This type of mechanism is referred to as an *Eley–Rideal mechanism.*

10.2.5 Desorption

In each of the preceding cases, the products of the surface reaction adsorbed on the surface are subsequently desorbed into the gas phase. For the desorption of a species (e.g., C)

$K_{DC} = (atm)$

$$C \cdot S \rightleftarrows C + S$$

$$k_D = \left(\frac{1}{s} \right)$$

the rate of desorption of C is

$$r_{DC} = k_D \left(C_{C \cdot S} - \frac{P_C C_v}{K_{DC}} \right) \qquad (10\text{-}20)$$

where K_{DC} is the desorption equilibrium constant with units of atm. Now let's look at the above adsorption from right to left. We note that the desorption step for C is just the reverse of the adsorption step. Consequently, the rate of desorption of C, r_{DC}, is just opposite in sign to the rate of adsorption of C, r_{ADC}

$$r_{DC} = -r_{ADC}$$

In addition, we see that the desorption equilibrium constant K_{DC} is just the reciprocal of the adsorption equilibrium constant for C, K_C

$K_{DC} = (atm)$

$$K_{DC} = \frac{1}{K_C}$$

$$K_C = \left(\frac{1}{atm} \right)$$

in which case the rate of desorption of C can be written

$$r_{DC} = k_D (C_{C \cdot S} - K_C P_C C_v) \qquad (10\text{-}21)$$

In the material that follows, the form of the equation for the desorption step that we will use to develop our rate laws will be similar to Equation (10-21).

10.2.6 The Rate-Limiting Step

When heterogeneous reactions are carried out at steady state, the rates of each of the three reaction steps in series (adsorption, surface reaction, and desorption) are equal to one another

$$-r_A' = r_{AD} = r_S = r_D$$

However, one particular step in the series is usually found to be *rate-limiting* or *rate-controlling*. That is, if we could make that particular step go faster, the entire reaction would proceed at an accelerated rate. Consider the analogy to the electrical circuit shown in Figure 10-12. A given concentration of reactants is analogous to a given driving force or electromotive force (EMF). The current I (with units of Coulombs/s) is analogous to the rate of

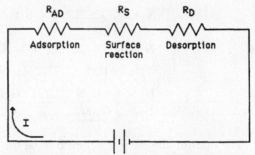

Figure 10-12   Electrical analog to heterogeneous reactions.

reaction, $-r_A'$ (mol/s·g-cat), and a resistance R_i is associated with each step in the series. Because the resistances are in series, the total resistance R_{tot} is just the sum of the individual resistances, for adsorption (R_{AD}), surface reaction (R_S), and desorption (R_D). The current, I, for a given voltage, E, is

$$I = \frac{E}{R_{tot}} = \frac{E}{R_{AD} + R_S + R_D}$$

The concept of a rate-limiting step

Who is slowing us down?

Because we observe only the total resistance, R_{tot}, it is our task to find which resistance is much larger (say, 100 Ω) than the other two resistances (say, 0.1 Ω). Thus, if we could lower the largest resistance, the current I (i.e., $-r_A'$) , would be larger for a given voltage, E. Analogously, we want to know which step in the adsorption–reaction–desorption series is limiting the overall rate of reaction.

The approach in determining catalytic and heterogeneous mechanisms is usually termed the *Langmuir–Hinshelwood approach,* since it is derived from ideas proposed by Hinshelwood based on Langmuir's principles for adsorption.[7] The Langmuir–Hinshelwood approach was popularized by Hougen and Watson and occasionally includes their names.[8] It consists of first assuming a sequence of steps in the reaction. In writing this sequence, one must choose among such mechanisms as molecular or atomic adsorption, and single- or dual-site reaction. Next, rate laws are written for the individual steps as shown in the

---

[7] C. N. Hinshelwood, *The Kinetics of Chemical Change* (Oxford: Clarendon Press, 1940).

[8] O. A. Hougen and K. M. Watson, *Ind. Eng. Chem.,* 35, 529 (1943).

preceding section, assuming that all steps are reversible. Finally, a rate-limiting step is postulated, and steps that are not rate-limiting are used to eliminate all coverage-dependent terms. The most questionable assumption in using this technique to obtain a rate law is the hypothesis that the activity of the surface is essentially uniform as far as the various steps in the reaction are concerned.

Industrial Example of Adsorption-Limited Reaction

An example of an adsorption-limited reaction is the synthesis of ammonia from hydrogen and nitrogen

$$3H_2 + N_2 \rightleftharpoons 2NH_3$$

over an iron catalyst that proceeds by the following mechanism:[9]

Dissociative adsorption of N_2 is rate-limiting

$$H_2 + 2S \rightleftharpoons 2H \cdot S \ \} \qquad \text{Rapid}$$

$$N_2 + 2S \rightleftharpoons 2N \cdot S \ \} \qquad \text{Rate-limiting}$$

$$
\left.
\begin{array}{l}
N \cdot S + H \cdot S \rightleftharpoons HN \cdot S + S \\
NH \cdot S + H \cdot S \rightleftharpoons H_2N \cdot S + S \\
H_2N \cdot S + H \cdot S \rightleftharpoons NH_3 \cdot S + S \\
NH_3 \cdot S \rightleftharpoons NH_3 + S
\end{array}
\right\} \qquad \text{Rapid}
$$

The rate-limiting step is believed to be the adsorption of the N_2 molecule as an N atom.

Industrial Example of Surface-Limited Reaction

An example of a surface-limited reaction is the reaction of two noxious automobile exhaust products, CO and NO

$$CO + NO \longrightarrow CO_2 + \tfrac{1}{2} N_2$$

carried out *over a copper catalyst* to form environmentally acceptable products, N_2 and CO_2

$$
\left.
\begin{array}{l}
CO + S \rightleftharpoons CO \cdot S \\
NO + S \rightleftharpoons NO \cdot S
\end{array}
\right\} \qquad \text{Rapid}
$$

Surface reaction is rate-limiting

$$NO \cdot S + CO \cdot S \rightleftharpoons CO_2 + N \cdot S + S \ \} \qquad \text{Rate-limiting}$$

$$
\left.
\begin{array}{l}
N \cdot S + N \cdot S \rightleftharpoons N_2 \cdot S \\
N_2 \cdot S \longrightarrow N_2 + S
\end{array}
\right\} \qquad \text{Rapid}
$$

Analysis of the rate law suggests that CO_2 and N_2 are weakly adsorbed, i.e., have infinitesimally small adsorption constants (see Problem P10-9$_B$).

---

[9] From the literature cited in G. A. Somorjai, *Introduction to Surface Chemistry and Catalysis* (New York: Wiley, 1994), p. 482.

10.3 Synthesizing a Rate Law, Mechanism, and Rate-Limiting Step

We now wish to develop rate laws for catalytic reactions that are not diffusion-limited. In developing the procedure to obtain a mechanism, a rate-limiting step, and a rate law consistent with experimental observation, we shall discuss a particular catalytic reaction, the decomposition of cumene to form benzene and propylene. The overall reaction is

$$C_6H_5CH(CH_3)_2 \longrightarrow C_6H_6 + C_3H_6$$

A conceptual model depicting the sequence of steps in this platinum-catalyzed reaction is shown in Figure 10-13. Figure 10-13 is only a schematic representation of the adsorption of cumene; a more realistic model is the formation of a complex of the π orbitals of benzene with the catalytic surface, as shown in Figure 10-14.

· Adsorption
· Surface
   reaction
· Desorption

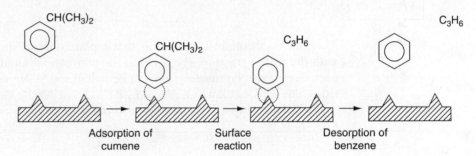

Figure 10-13   Sequence of steps in a reaction-limited catalytic reaction.

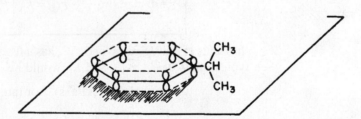

Figure 10-14 π-orbital complex on surface.

The nomenclature in Table 10-3 will be used to denote the various species in this reaction: C = cumene, B = benzene, and P = propylene. The reaction sequence for this decomposition is shown in Table 10-3.

TABLE 10-3   STEPS IN A LANGMUIR–HINSHELWOOD KINETIC MECHANISM

$C + S \underset{k_{-A}}{\overset{k_A}{\rightleftharpoons}} C \cdot S$	Adsorption of cumene on the surface	(10-22)
$C \cdot S \underset{k_{-S}}{\overset{k_S}{\rightleftharpoons}} B \cdot S + P$	Surface reaction to form adsorbed benzene and propylene in the gas phase	(10-23)
$B \cdot S \underset{k_{-D}}{\overset{k_D}{\rightleftharpoons}} B + S$	Desorption of benzene from surface	(10-24)

These three steps represent the mechanism for cumene decomposition.

Equations (10-22) through (10-24) represent the mechanism proposed for this reaction.

When writing rate laws for these steps, we treat each step as an elementary reaction; the only difference is that the species concentrations in the gas phase are replaced by their respective partial pressures

Ideal gas law
$P_C = C_C RT$

$$C_C \longrightarrow P_C$$

There is no theoretical reason for this replacement of the concentration, C_C, with the partial pressure, P_C; it is just the convention initiated in the 1930s and used ever since. Fortunately, P_C can be calculated easily and directly from C_C using the ideal gas law (i.e., $P_C = C_C RT$).

The rate expression for the adsorption of cumene as given in Equation (10-22) is

$C + S \underset{k_{-A}}{\overset{k_A}{\rightleftharpoons}} C \cdot S$

$$r_{AD} = k_A P_C C_v - k_{-A} C_{C \cdot S}$$

$$\boxed{\text{Adsorption:} \qquad r_{AD} = k_A \left(P_C C_v - \frac{C_{C \cdot S}}{K_C} \right)} \qquad (10\text{-}25)$$

If r_{AD} has units of (mol/g-cat·s) and $C_{C \cdot S}$ has units of (mol cumene adsorbed/g-cat), then typical units of k_A, k_{-A}, and K_C would be

$$[k_A] \equiv (\text{kPa} \cdot \text{s})^{-1} \text{ or } (\text{atm} \cdot \text{h})^{-1}$$

$$[k_{-A}] \equiv \text{h}^{-1} \text{ or } \text{s}^{-1}$$

$$[K_C] \equiv \left[\frac{k_A}{k_{-A}} \right] \equiv \text{kPa}^{-1}$$

The rate law for the surface-reaction step producing adsorbed benzene and propylene in the gas phase

$$C \cdot S \underset{k_{-S}}{\overset{k_S}{\rightleftharpoons}} B \cdot S + P(g) \qquad (10\text{-}23)$$

is

$$r_S = k_S C_{C \cdot S} - k_{-S} P_P C_{B \cdot S}$$

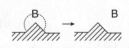

$$\text{Surface reaction:} \quad r_S = k_S\left(C_{C \cdot S} - \frac{P_P C_{B \cdot S}}{K_S}\right) \quad (10\text{-}26)$$

with the *surface reaction equilibrium constant* being

$$K_S = \frac{k_S}{k_{-S}}$$

Typical units for k_S and K_S are s^{-1} and kPa, respectively.

Propylene is not adsorbed on the surface. Consequently, its concentration on the surface is zero.

$$C_{P \cdot S} = 0$$

The rate of benzene desorption [see Equation (10-24)] is

$$r_D = k_D C_{B \cdot S} - k_{-D} P_B C_v \quad (10\text{-}27)$$

$$\text{Desorption:} \quad r_D = k_D\left(C_{B \cdot S} - \frac{P_B C_v}{K_{DB}}\right) \quad (10\text{-}28)$$

Typical units of k_D and K_{DB} are s^{-1} and kPa, respectively. By viewing the desorption of benzene

$$\text{B} \cdot \text{S} \; \rightleftharpoons \; \text{B} + \text{S}$$

from right to left, we see that desorption is just the reverse of the adsorption of benzene. Consequently, as mentioned earlier, it is easily shown that the benzene adsorption equilibrium constant K_B is just the reciprocal of the benzene desorption constant K_{DB}

$$K_B = \frac{1}{K_{DB}}$$

and Equation (10-28) can be written as

$$\text{Desorption:} \quad r_D = k_D(C_{B \cdot S} - K_B P_B C_v) \quad (10\text{-}29)$$

Because there is no accumulation of reacting species on the surface, the rates of each step in the sequence are all equal as discussed in Figure 10-12:

$$-r'_C = r_{AD} = r_S = r_D \quad (10\text{-}30)$$

For the mechanism postulated in the sequence given by Equations (10-22) through (10-24), we wish to determine which step is rate-limiting. We first assume one of the steps to be rate-limiting (rate-controlling) and then formulate the reaction-rate law in terms of the partial pressures of the species present. From this expression we can determine the variation of the initial reaction rate with the initial partial pressures and the initial total pressure. If the predicted rate varies with pressure in the same manner as the rate observed experimentally, the implication is that the assumed mechanism and rate-limiting step are correct.

We will first start our development of the rate laws with the assumption that the adsorption step is rate-limiting and derive the rate law, and then proceed to assume that each of the other two steps' surface reaction and desorption limit the overall rate and then derive the rate law for each of these other two limiting cases.

10.3.1 Is the Adsorption of Cumene Rate-Limiting?

To answer this question we shall assume that the adsorption of cumene is indeed rate-limiting, derive the corresponding rate law, and then check to see if it is consistent with experimental observation. By postulating that this (or any other) step is rate-limiting, we are assuming that the reaction-rate constant of this step (in this case k_A) is small with respect to the specific rates of the other steps (in this case k_S and k_D).[10] The rate of adsorption is

Need to express C_v and $C_{C \cdot S}$ in terms of P_C, P_B, and P_P

$$\boxed{-r'_C = r_{AD} = k_A \left(P_C C_v - \frac{C_{C \cdot S}}{K_C} \right)} \qquad (10\text{-}25)$$

Because we cannot measure either C_v or $C_{C \cdot S}$, we must replace these variables in the rate law with measurable quantities for the equation to be meaningful.

For steady-state operation we have

$$-r'_C = r_{AD} = r_S = r_D \qquad (10\text{-}30)$$

For adsorption-limited reactions, k_A is very small and k_S and k_D are very, very large by comparison. Consequently, the ratios r_S/k_S and r_D/k_D are very small (approximately zero), whereas the ratio r_{AD}/k_A is relatively large.

The surface reaction-rate law is

$$r_S = k_S \left(C_{C \cdot S} - \frac{C_{B \cdot S} P_P}{K_S} \right) \qquad (10\text{-}31)$$

Again, for adsorption-limited reactions, the surface-specific reaction rate k_S is large by comparison, and we can set

$$\frac{r_S}{k_S} \simeq 0 \qquad (10\text{-}32)$$

and solve Equation (10-31) for $C_{C \cdot S}$

$$C_{C \cdot S} = \frac{C_{B \cdot S} P_P}{K_S} \qquad (10\text{-}33)$$

---

[10] *Strictly speaking*, one should compare the product $k_A P_C$ with k_S and k_D.

$$r_{AD} = k_A P_C \left[C_v - \frac{C_{C \cdot S}}{K_C P_C} \right]$$

$$\frac{\text{mol}}{\text{s} \cdot \text{kg-cat}} = \left(\frac{1}{\text{s atm}} \right) \cdot (\text{atm}) \cdot \left[\frac{\text{mol}}{\text{kg-cat}} \right] = \left[\frac{1}{\text{s}} \right] \frac{\text{mol}}{\text{kg-cat}}$$

Dividing r_{AD} by $k_A P_C$, we note $\dfrac{r_{AD}}{k_A P_C} = \dfrac{\text{mol}}{\text{kg-cat}}$. The reason we do this is that in order to compare terms, the ratios $\left(-\dfrac{r_{AD}}{k_A P_C} \right)$, $\left(\dfrac{r_S}{k_S} \right)$, and $\left(\dfrac{r_D}{k_D} \right)$ must all have the same units $\left[\dfrac{\text{mol}}{\text{kg-cat}} \right]$. Luckily for us, the end result is the same, however.

To be able to express $C_{C \cdot S}$ solely in terms of the partial pressures of the species present, we must evaluate $C_{B \cdot S}$. The rate of desorption of benzene is

$$r_D = k_D \left(C_{B \cdot S} - K_B P_B C_v \right) \qquad (10\text{-}29)$$

Using
$$\frac{r_S}{k_S} \simeq 0 \simeq \frac{r_D}{k_D}$$
to find $C_{B \cdot S}$ and $C_{C \cdot S}$ in terms of partial pressures

However, for adsorption-limited reactions, k_D is large by comparison, and we can set

$$\frac{r_D}{k_D} \simeq 0 \qquad (10\text{-}34)$$

and then solve Equation (10-29) for $C_{B \cdot S}$

$$C_{B \cdot S} = K_B P_B C_v \qquad (10\text{-}35)$$

After combining Equations (10-33) and (10-35), we have

$$C_{C \cdot S} = K_B \frac{P_B P_P}{K_S} C_v \qquad (10\text{-}36)$$

Replacing $C_{C \cdot S}$ in the rate equation by Equation (10-36) and then factoring C_v, we obtain

$$r_{AD} = k_A \left(P_C - \frac{K_B P_B P_P}{K_S K_C} \right) C_v = k_A \left(P_C - \frac{P_B P_P}{K_P} \right) C_v \qquad (10\text{-}37)$$

Let's look at how the thermodynamic constant pressure equilibrium constant, K_P, found its way into Equation (10-37) and how we can find its value for any reaction. First we observe that at equilibrium $r_{AD} = 0$, Equation (10-37) rearranges to

$$\frac{P_{Be} P_{Pe}}{P_{Ce}} = \frac{K_C K_S}{K_B}$$

We also know from thermodynamics (Appendix C) that for the reaction

$$C \; \underset{\longleftarrow}{\overset{\longrightarrow}{\rule{1cm}{0pt}}} \; B + P$$

also at equilibrium $(-r'_C = 0)$, we have the following relationship for partial pressure equilibrium constant K_P

$$K_P = \frac{P_{Be} P_{Pe}}{P_{Ce}}$$

Consequently, the following relationship must hold

$$\boxed{\frac{K_S K_C}{K_B} = K_P} \qquad (10\text{-}38)$$

The equilibrium constant can be determined from *thermodynamic data* and is related to the change in the Gibbs free energy, $\Delta G°$, by the equation (see Appendix C)

$$\boxed{RT \ln K = -\Delta G°} \qquad (10\text{-}39)$$

where R is the ideal gas constant and T is the absolute temperature.

The concentration of vacant sites, C_v, can now be eliminated from Equation (10-37) by utilizing the *site balance* to give the total concentration of sites, C_t, which is assumed constant[11]

> Total sites = Vacant sites + Occupied sites

Because cumene and benzene are adsorbed on the surface, the concentration of occupied sites is $(C_{C \cdot S} + C_{B \cdot S})$, and the total concentration of sites is

Site balance

$$C_t = C_v + C_{C \cdot S} + C_{B \cdot S} \tag{10-40}$$

Substituting Equations (10-35) and (10-36) into Equation (10-40), we have

$$C_t = C_v + \frac{K_B}{K_S} P_B P_P C_v + K_B P_B C_v$$

Solving for C_v, we have

$$C_v = \frac{C_t}{1 + P_B P_P K_B / K_S + K_B P_B} \tag{10-41}$$

Combining Equations (10-41) and (10-37), we find that the rate law for the catalytic decomposition of cumene, assuming that the adsorption of cumene is the rate-limiting step, is

Cumene reaction rate law if adsorption were the limiting step

$$-r'_C = r_{AD} = \frac{C_t k_A (P_C - P_P P_B / K_P)}{1 + K_B P_P P_B / K_S + K_B P_B} \tag{10-42}$$

We now wish to sketch a plot of the initial rate of reaction as a function of the partial pressure of cumene, P_{C0}. Initially, no products are present; consequently, $P_P = P_B = 0$. The initial rate is given by

$$-r'_{C0} = C_t k_A P_{C0} = k P_{C0} \tag{10-43}$$

If the cumene decomposition is adsorption rate limited, then from Equation (10-43) we see that the initial rate will be linear with the initial partial pressure of cumene, as shown in Figure 10-15.

Before checking to see if Figure 10-15 is consistent with experimental observation, we shall derive the corresponding rate laws for the other possible rate-limiting steps and then develop corresponding initial rate plots for the case when the surface reaction is rate-limiting and then for the case when the desorption of benzene is rate-limiting.

---

[11] Some (I won't mention any names) prefer to write the surface reaction rate in terms of the fraction of the surface of sites covered (i.e., f_A) rather than the number of sites $C_{A \cdot S}$ covered, the difference being the multiplication factor of the total site concentration, C_t. In any event, the final form of the rate law is the same because C_t, K_A, k_S, and so on, are all lumped into the reaction-rate constant, k.

If adsorption were rate-limiting, the data should show $-r'_{CO}$ increasing linearly with P_{CO}.

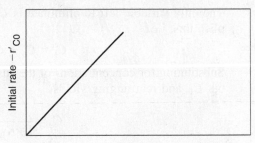

Figure 10-15   Adsorption-limited reaction.

10.3.2 Is the Surface Reaction Rate-Limiting?

The rate of surface reaction is

Single-site mechanism

$$r_S = k_S \left(C_{C \cdot S} - \frac{P_P C_{B \cdot S}}{K_S} \right) \qquad (10\text{-}26)$$

Since we cannot readily measure the concentrations of the adsorbed species, we must utilize the adsorption and desorption steps to eliminate $C_{C \cdot S}$ and $C_{B \cdot S}$ from this equation.

From the adsorption rate expression in Equation (10-25) and the condition that k_A and k_D are very large by comparison with k_S when the surface reaction is limiting (i.e., $r_{AD}/k_A \simeq 0$),[12] we obtain a relationship for the surface concentration for adsorbed cumene

$$C_{C \cdot S} = K_C P_C C_v$$

In a similar manner, the surface concentration of adsorbed benzene can be evaluated from the desorption rate expression, Equation (10-29), together with the approximation

Using
$$\frac{r_{AD}}{k_A} \simeq 0 \simeq \frac{r_D}{k_D}$$
to find $C_{B \cdot S}$ and $C_{C \cdot S}$ in terms of partial pressures

when $\dfrac{r_D}{k_D} \simeq 0$

then we get the same result for $C_{B \bullet S}$ as before when we had adsorption limitation, i.e.,

$$C_{B \cdot S} = K_B P_B C_v$$

Substituting for $C_{B \cdot S}$ and $C_{C \cdot S}$ in Equation (10-26) gives us

$$r_S = k_S \left(P_C K_C - \frac{K_B P_B P_P}{K_S} \right) C_v = k_S K_C \left(P_C - \frac{P_B P_P}{K_P} \right) C_v \qquad (10\text{-}26\text{a})$$

where the thermodynamic equilibrium constant was used to replace the ratio of surface reaction and adsorption constants, i.e.,

$$K_P = \frac{K_C K_S}{K_B} \qquad (10\text{-}26\text{b})$$

---

[12] See footnote 10 on page 424.

The only variable left to eliminate is C_v and we use a site balance to accomplish this, i.e.,

Site balance

$$C_t = C_v + C_{B \cdot S} + C_{C \cdot S} \tag{10-40}$$

Substituting for concentrations of the adsorbed species, $C_{B \cdot S}$, and $C_{C \cdot S}$, factoring out C_V, and rearranging yields

$$C_v = \frac{C_t}{1 + K_B P_B + K_C P_C}$$

Substituting for C_V in Equation (10-26a)

Cumene rate law for surface-reaction-limiting

$$-r'_C = r_S = \frac{\overbrace{k_S C_t K_C}^{k} (P_C - P_P P_B / K_P)}{1 + P_B K_B + K_C P_C} \tag{10-44}$$

The initial rate of reaction is

$$-r'_{C0} = \frac{\overbrace{k_S C_t K_C}^{k} P_{C0}}{1 + K_C P_{C0}} = \frac{k P_{C0}}{1 + K_C P_{C0}} \tag{10-45}$$

Figure 10-16 shows the initial rate of reaction as a function of the initial partial pressure of cumene for the case of surface-reaction-limiting.

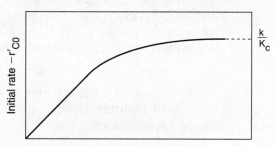

Figure 10-16 Surface-reaction-limited.

At low partial pressures of cumene

$$1 \gg K_C P_{C0}$$

and we observe that the initial rate will increase linearly with the initial partial pressure of cumene:

$$-r'_{C0} \approx k P_{C0}$$

At high partial pressures

$$K_C P_{C0} \gg 1$$

and Equation (10-45) becomes

If surface reaction were rate-limiting, the data would show this behavior.

$$-r'_{C0} \cong \frac{k P_{C0}}{K_C P_{C0}} = \frac{k}{K_C}$$

and the initial rate is independent of the initial partial pressure of cumene.

10.3.3 Is the Desorption of Benzene Rate-Limiting?

The rate expression for the desorption of benzene is

$$\boxed{r_D = k_D \left(C_{B \cdot S} - K_B P_B C_v \right)}$$ (10-29)

For desorption-limited reactions, both k_A and k_S are very large compared with k_D, which is small.

From the rate expression for surface reaction, Equation (10-26), we set

$$\frac{r_S}{k_S} \simeq 0$$

to obtain

$$C_{B \cdot S} = K_S \left(\frac{C_{C \cdot S}}{P_P} \right)$$ (10-46)

Similarly, for the adsorption step, Equation (10-25), we set

$$\frac{r_{AD}}{k_A} \simeq 0$$

to obtain

$$C_{C \cdot S} = K_C P_C C_v$$

then substitute for $C_{C \cdot S}$ in Equation (10-46) to obtain

$$C_{B \cdot S} = \frac{K_C K_S P_C C_v}{P_P}$$ (10-47)

Combining Equations (10-26b), (10-29), and (10-47) gives us

$$r_D = k_D K_C K_S \left(\frac{P_C}{P_P} - \frac{P_B}{K_P} \right) C_v$$ (10-48)

where K_C is the cumene adsorption constant, K_S is the surface-reaction equilibrium constant, and K_P is the thermodynamic gas-phase equilibrium constant, Equation (10-38), for the reaction. The expression for C_v is obtained from a site balance:

Site balance: $$C_t = C_{C \cdot S} + C_{B \cdot S} + C_v$$ (10-40)

After substituting for the respective surface concentrations, we solve the site balance for C_v

$$C_v = \frac{C_t}{1 + K_C K_S P_C / P_P + K_C P_C}$$ (10-49)

Replacing C_v in Equation (10-48) by Equation (10-49) and multiplying the numerator and denominator by P_P, we obtain the rate expression for desorption control

Cumene decomposition rate law if desorption were limiting

$$\boxed{-r'_C = r_D = \frac{\overbrace{k_D C_t K_S K_C}^{k} \left(P_C - P_B P_P / K_P \right)}{P_P + P_C K_C K_S + K_C P_P P_C}}$$ (10-50)

To determine the dependence of the initial rate of reaction on the initial partial pressure of cumene, we again set $P_P = P_B = 0$, and the rate law reduces to

If desorption
limits, the initial
rate is independent
of the initial partial
pressure of cumene.

$$-r'_{C0} = k_D C_t$$

with the corresponding plot of $-r'_{C0}$ shown in Figure 10-17. If desorption were rate limiting, we would see that the initial rate of reaction would be independent of the initial partial pressure of cumene.

Initial partial pressure of cumene, P_{C0}

Figure 10-17   Desorption-limited reaction.

10.3.4 Summary of the Cumene Decomposition

*Cumene
decomposition is
surface-reaction-
limited*

The experimental observations of $-r'_{C0}$ as a function of P_{C0} are shown in Figure 10-18. From the plot in Figure 10-18, we can clearly see that neither adsorption nor desorption is rate-limiting. For the reaction and mechanism given by

*Surface-reaction-
limited mechanism
is consistent with
experimental data.*

$$\text{C} + \text{S} \;\rightleftharpoons\; \text{C·S} \tag{10-22}$$

$$\text{C·S} \;\rightleftharpoons\; \text{B·S} + \text{P} \tag{10-23}$$

$$\text{B·S} \;\rightleftharpoons\; \text{B} + \text{S} \tag{10-24}$$

the rate law derived by assuming that the surface reaction is rate-limiting agrees with the data.

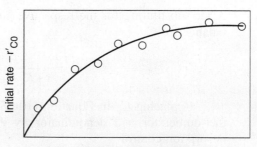

Initial partial pressure of cumene, P_{C0}

Figure 10-18   Actual initial rate as a function of partial pressure of cumene.

The rate law for the case of no inerts adsorbing on the surface is

$$-r'_C = \frac{k\,(P_C - P_B P_P / K_P)}{1 + K_B P_B + K_C P_C} \tag{10-44}$$

The forward cumene decomposition reaction is a single-site mechanism involving only adsorbed cumene, while the reverse reaction of propylene in the gas phase reacting with adsorbed benzene is an Eley–Rideal mechanism.

If we were to have an adsorbing inert in the feed, the inert would not participate in the reaction but would occupy active sites on the catalyst surface:

$$I + S \;\rightleftharpoons\; I{\cdot}S$$

Our site balance is now

$$C_t = C_v + C_{C{\cdot}S} + C_{B{\cdot}S} + C_{I{\cdot}S} \tag{10-51}$$

Because the adsorption of the inert is at equilibrium, the concentration of sites occupied by the inert is

$$C_{I{\cdot}S} = K_I P_I C_v \tag{10-52}$$

Substituting for the inert sites in the site balance, the rate law for surface reaction control when an adsorbing inert is present is

Adsorbing inerts

$$-r'_C = \frac{k\,(P_C - P_B P_P / K_P)}{1 + K_C P_C + K_B P_B + K_I P_I} \tag{10-53}$$

One observes that the rate decreases as the partial pressure of adsorbing inerts increases.

10.3.5 Reforming Catalysts

We now consider a dual-site mechanism, which is a reforming reaction found in petroleum refining to upgrade the octane number of gasoline.

Side Note: Octane Number. Fuels with low octane numbers can produce spontaneous combustion in the car cylinder before the air/fuel mixture is compressed to its desired volume and ignited by the spark plug. The following figure shows the desired combustion wave front moving down from the spark plug and the unwanted spontaneous combustion wave in the lower right-hand corner. This spontaneous combustion produces detonation waves, which constitute engine knock. The lower the octane number, the greater the chance of spontaneous combustion and engine knock.

The octane number of a gasoline is determined from a calibration curve relating knock intensity to the % iso-octane in a mixture of iso-octane and heptane. One way to calibrate the octane number is to place a transducer on the side of the cylinder to measure the knock intensity (K.I.) (pressure pulse) for various mixtures of heptane and iso-octane. The octane number is the percentage of iso-octane in this mixture. That is, pure iso-octane has an octane number of 100, 80% iso-octane/20% heptane has an octane number of 80, and so on. The knock intensity is measured for this

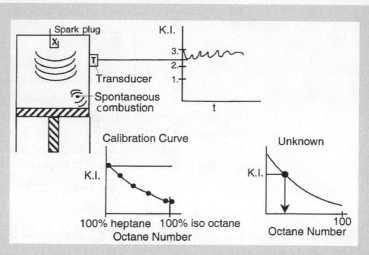

80/20 mixture and recorded. The relative percentages of iso-octane and hep-tane are changed (e.g., 90/10), and the test is repeated. After a series of experiments, a calibration curve is constructed. The gasoline to be cali-brated is then used in the test engine, where the standard knock intensity is measured. Knowing the knock intensity, the octane rating of the fuel is read off the calibration curve. A gasoline with an octane rating of 92 means that it matches the performance of a mixture of 92% iso-octane and 8% heptane. Another way to calibrate the octane number is to set the knock intensity and increase the compression ratio. A fixed percentage of iso-octane and hep-tane is placed in a test engine and the compression ratio (CR) is increased continually until spontaneous combustion occurs, producing an engine knock. The compression ratio and the corresponding composition of the mixture are then recorded, and the test is repeated to obtain the calibration curve of compression ratio as a function of % iso-octane. After the calibra-tion curve is obtained, the unknown is placed in the cylinder, and the com-pression ratio (CR) is increased until the set knock intensity is exceeded. The CR is then matched to the calibration curve to find the octane number.

> The more compact the molecule, the greater the octane number.

The more compact the hydrocarbon molecule, the less likely it is to cause spontaneous combustion and engine knock. Consequently, it is desired to isomerize straight-chain hydrocarbon molecules into more com-pact molecules through a catalytic process called *reforming*.

In the U.S., the higher the octane number the greater the cost of a gal-lon (liter) of gasoline. However, because of government regulations, the cost of a gallon (liter) of gasoline in Jofostan is the same for all octane numbers.

One common reforming catalyst is platinum on alumina. Platinum on alumina (Al_2O_3) (see SEM photo in Figure 10-19) is a bifunctional catalyst that can be prepared by exposing alumina pellets to a chloroplatinic acid solu-tion, drying, and then heating in air at 775 K to 875 K for several hours. Next,

> Catalyst manufacture

the material is exposed to hydrogen at temperatures around 725 K to 775 K to produce very small clusters of Pt on alumina. These clusters have sizes on the order of 10 Å, while the alumina pore sizes on which the Pt is deposited are on the order of 100 Å to 10,000 Å (i.e., 10 nm to 1000 nm).

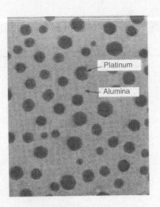

Figure 10-19   Platinum on alumina. (Masel, Richard. *Chemical Kinetics and Catalysis*, p. 700. © John Wiley & Sons, Inc. Reprinted by permission of John Wiley & Sons, Inc. All rights reserved.)

As an example of catalytic reforming, we shall consider the isomerization of *n*-pentane to *i*-pentane

$$n\text{-pentane} \xrightleftharpoons[\text{Al}_2\text{O}_3]{0.75 \text{ wt\% Pt}} i\text{-pentane}$$

Gasoline	
C_5	10%
C_6	10%
C_7	20%
C_8	25%
C_9	20%
C_{10}	10%
C_{11-12}	5%

Normal pentane has an octane number of 62, while *iso*-pentane, which is more compact, has an octane number of 90! The *n*-pentane adsorbs onto the platinum, where it is dehydrogenated to form *n*-pentene. The *n*-pentene desorbs from the platinum and adsorbs onto the alumina, where it is isomerized to *i*-pentene, which then desorbs and subsequently adsorbs onto platinum, where it is hydrogenated to form *i*-pentane. That is

$$n\text{-pentane} \xrightleftharpoons[\text{Pt}]{-\text{H}_2} n\text{-pentene} \xrightleftharpoons{\text{Al}_2\text{O}_3} i\text{-pentene} \xrightleftharpoons[\text{Pt}]{+\text{H}_2} i\text{-pentane}$$

We shall focus on the isomerization step to develop the mechanism and the rate law

$$n\text{-pentene} \xrightleftharpoons{\text{Al}_2\text{O}_3} i\text{-pentene}$$

$$\text{N} \xrightleftharpoons{} \text{I}$$

The procedure for formulating a mechanism, rate-limiting step, and corresponding rate law is given in Table 10-4.

TABLE 10-4   ALGORITHM FOR DETERMINING THE REACTION MECHANISM AND RATE-LIMITING STEP

Isomerization of *n*-pentene (N) to *i*-pentene (I) over alumina

$$N \underset{}{\overset{Al_2O_3}{\rightleftharpoons}} I$$

Reforming reaction to increase octane number of gasoline

Step 1. *Select a mechanism.* (Let's choose a Dual Site Mechanism)

Adsorption: $N + S \rightleftharpoons N \cdot S$

Surface reaction: $N \cdot S + S \rightleftharpoons I \cdot S + S$

Desorption: $I \cdot S \rightleftharpoons I + S$

Treat each reaction step as an elementary reaction when writing rate laws.

Step 2. *Assume a rate-limiting step.* We choose the surface reaction first, because *more than 75% of all heterogeneous reactions that are not diffusion-limited are surface-reaction-limited.* We note that the PSSH must be used when more than one step is limiting (see Section 10.3.6). The rate law for the surface reaction step is

$$-r_N' = r_S = k_S \left(C_v C_{N \cdot S} - \frac{C_{I \cdot S} C_v}{K_S} \right)$$

Step 3. *Find the expression for concentration of the adsorbed species $C_{i \cdot S}$.* Use the other steps that are not limiting to solve for $C_{i \cdot S}$ (e.g., $C_{N \cdot S}$ and $C_{I \cdot S}$). For this reaction

From $\dfrac{r_{AD}}{k_A} \simeq 0:$ $C_{N \cdot S} = P_N K_N C_v$

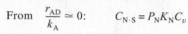

Following the Algorithm

From $\dfrac{r_D}{k_D} \simeq 0:$ $C_{I \cdot S} = \dfrac{P_I C_v}{K_D} = K_I P_I C_v$

Step 4. *Write a site balance.*

$$C_t = C_v + C_{N \cdot S} + C_{I \cdot S}$$

Step 5. *Derive the rate law.* Combine Steps 2, 3, and 4 to arrive at the rate law

$$-r_N' = r_S = \frac{\overbrace{k_S C_t^2 K_N}^{k} (P_N - P_I/K_P)}{(1 + K_N P_N + K_I P_I)^2}$$

Step 6. *Compare with data.* Compare the rate law derived in Step 5 with experimental data. If they agree, there is a good chance that you have found the correct mechanism and rate-limiting step. If your derived rate law (i.e., model) does not agree with the data:

a. Assume a different rate-limiting step and repeat Steps 2 through 6.

b. If, after assuming that each step is rate-limiting, none of the derived rate laws agrees with the experimental data, select a different mechanism (e.g., a single-site mechanism)

$$N + S \rightleftharpoons N \cdot S$$

$$N \cdot S \rightleftharpoons I \cdot S$$

$$I \cdot S \rightleftharpoons I + S$$

and then proceed through Steps 2 through 6.

The single-site mechanism turns out to be the correct one. For this mechanism, the rate law is

$$-r_N' = \frac{k(P_N - P_I/K_P)}{(1 + K_N P_N + K_I P_I)}$$

c. If two or more models agree, the statistical tests discussed in Chapter 7 (e.g., comparison of residuals) should be used to discriminate between them (see the Supplementary Reading).

We note that in Table 10-4 for the dual-site mechanism, the denominator of the rate law for $-r_A'$ is squared (i.e., in Step 5 [$1/(\ )^2$]), while for a single-site mechanism, it is not squared (i.e., Step 6 [$1/(\ )$]). This fact is useful when analyzing catalyzed reactor data.

Table 10-5 gives rate laws for different reaction mechanisms that are irreversible and surface-reaction-limited.

TABLE 10-5   IRREVERSIBLE SURFACE-REACTION-LIMITED RATE LAWS

Single site

$$A \cdot S \longrightarrow B \cdot S \qquad\qquad -r'_A = \frac{kP_A}{1+K_A P_A + K_B P_B}$$

Dual site

$$A \cdot S + S \longrightarrow B \cdot S + S \qquad -r'_A = \frac{kP_A}{(1+K_A P_A + K_B P_B)^2}$$

$$A \cdot S + B \cdot S \longrightarrow C \cdot S + S \qquad -r'_A = \frac{kP_A P_B}{(1+K_A P_A + K_B P_B + K_C P_C)^2}$$

Eley–Rideal

$$A \cdot S + B(g) \longrightarrow C \cdot S \qquad -r'_A = \frac{kP_A P_B}{1+K_A P_A + K_C P_C}$$

We need a word of caution at this point. Just because the mechanism and rate-limiting step may fit the rate data does not imply that the mechanism is correct.[13] Usually, spectroscopic measurements are needed to confirm a mechanism absolutely. However, the development of various mechanisms and rate-limiting steps can provide insight into the best way to correlate the data and develop a rate law.

10.3.6 Rate Laws Derived from the Pseudo-Steady-State Hypothesis (PSSH)

In Section 9.1 we discussed the PSSH, where the net rate of formation of *reactive intermediates* was assumed to be zero. An alternative way to derive a catalytic rate law rather than setting

$$\frac{r_{AD}}{k_A} \cong 0$$

is to assume that each species adsorbed on the surface is a *reactive intermediate*. Consequently, the net rate of formation of species *i* adsorbed on the surface will be zero

$$r^*_{i \cdot S} = 0 \qquad\qquad\qquad (10\text{-}54)$$

The PSSH is primarily used when more than one step is rate-limiting. The isomerization example shown in Table 10-4 is reworked using the PSSH in the Chapter 10 Expanded Material on the CRE Web site.

---

[13] R. I. Masel, *Principles of Adsorption and Reaction on Solid Surfaces* (New York: Wiley, 1996), p. 506, *www.masel.com*. This is a terrific book.

10.3.7 Temperature Dependence of the Rate Law

Consider a surface-reaction-limited irreversible isomerization

$$A \longrightarrow B$$

in which both A and B are adsorbed on the surface, and the rate law is

$$-r_A' = \frac{kP_A}{1 + K_A P_A + K_B P_B} \qquad (10\text{-}55)$$

The specific reaction rate, k, will usually follow an Arrhenius temperature dependence and increase exponentially with temperature. However, the adsorption of all species on the surface is exothermic. Consequently, the higher the temperature, the smaller the adsorption equilibrium constant. That is, as the temperature increases, K_A and K_B decrease resulting in less coverage of the surface by A and B. Therefore, at high temperatures, the denominator of catalytic rate laws approaches 1. That is, at high temperatures (low coverage)

$$1 \gg (P_A K_A + P_B K_B)$$

The rate law could then be approximated as

Neglecting the adsorbed species at high temperatures

$$-r_A' \simeq kP_A \qquad (10\text{-}56)$$

or for a reversible isomerization we would have

$$-r_A' \simeq k\left(P_A - \frac{P_B}{K_P}\right) \qquad (10\text{-}57)$$

Algorithm
Deduce
  Rate law
Find
  Mechanism
Evaluate
  Rate-law
  parameters
Design
  PBR
  CSTR

The algorithm we can use as a start in postulating a reaction mechanism and rate-limiting step is shown in Table 10-4. Again, we can never really prove a mechanism to be correct by comparing the derived rate law with experimental data. Independent spectroscopic experiments are usually needed to confirm the mechanism. We can, however, prove that a proposed mechanism is *inconsistent* with the experimental data by following the algorithm in Table 10-4. Rather than taking all the experimental data and then trying to build a model from the data, Box et al. describe techniques of sequential data collection and model building.[14]

10.4 Heterogeneous Data Analysis
for Reactor Design

In this section we focus on four operations that chemical reaction engineers need to be able to accomplish:

Following the Algorithm

(1) Developing an algebraic rate law consistent with experimental observations,

(2) Analyzing the rate law in such a manner that the rate-law parameters (e.g., k, K_A) can readily be determined from the experimental data,

---

[14] G. E. P. Box, W. G. Hunter, and J. S. Hunter, *Statistics for Engineers* (New York: Wiley, 1978).

(3) Finding a mechanism and rate-limiting step consistent with the experimental data

(4) Designing a catalytic reactor to achieve a specified conversion

We shall use the hydrodemethylation of toluene to illustrate these four operations.

Hydrogen and toluene are reacted over a solid mineral catalyst containing clinoptilolite (a crystalline silica-alumina) to form methane and benzene[15]

$$C_6H_5CH_3 + H_2 \xrightarrow[\text{catalyst}]{} C_6H_6 + CH_4$$

We wish to design a packed-bed reactor and a fluidized CSTR to process a feed consisting of 30% toluene, 45% hydrogen, and 25% inerts. Toluene is fed at a rate of 50 mol/min at a temperature of 640°C and a pressure of 40 atm (4052 kPa). To design the PBR, we must first determine the rate law from the differential reactor data presented in Table 10-6. In this table, we are given the rate of reaction of toluene as a function of the partial pressures of hydrogen (H_2), toluene (T), benzene (B), and methane (M). In the first two runs, methane was introduced into the feed together with hydrogen and toluene, while the other product, benzene, was fed to the reactor together with the reactants only in runs 3, 4, and 6. In runs 5 and 16, both methane and benzene were introduced in the feed. In the remaining runs, none of the products was present in the feedstream. Because the conversion was less than 1% in the differential reactor, the partial pressures of the products, methane and benzene, formed in these runs were essentially zero, and the reaction rates were equivalent to initial rates of reaction.

TABLE 10-6   DATA FROM A DIFFERENTIAL REACTOR

	$-r'_T \times 10^{10}$	Partial Pressure (atm)			
Run	$\left(\dfrac{\text{g mol toluene}}{\text{g-cat} \cdot \text{s}}\right)$	Toluene (T), P_T	Hydrogen (H_2), P_{H_2}	Methane (M), P_M	Benzene (B), P_B
Set A					
1	71.0	1	1	1	0
2	71.3	1	1	4	0
Set B					
3	41.6	1	1	0	1
4	19.7	1	1	0	4
5	42.0	1	1	1	1
6	17.1	1	1	0	5
Set C					
7	71.8	1	1	0	0
8	142.0	1	2	0	0
9	284.0	1	4	0	0
Set D					
10	47.0	0.5	1	0	0
11	71.3	1	1	0	0
12	117.0	5	1	0	0
13	127.0	10	1	0	0
14	131.0	15	1	0	0
15	133.0	20	1	0	0
16	41.8	1	1	1	1

Unscramble the data to find the rate law

[15] J. Papp, D. Kallo, and G. Schay, *J. Catal.*, 23, 168.

10.4.1 Deducing a Rate Law from the Experimental Data

Following the Algorithm

First, let's look at run 3. In run 3, there is no possibility of the reverse reaction taking place because the concentration of methane is zero, i.e., $P_M = 0$, whereas in run 5 the reverse reaction could take place because all products are present. Comparing runs 3 and 5, we see that the initial rate is essentially the same for both runs, and we can assume that the reaction is essentially irreversible.

$$T + H_2 \xrightarrow{\text{catalyst}} M + B$$

We now ask what qualitative conclusions can be drawn from the data about the dependence of the rate of disappearance of toluene, $-r'_T$, on the partial pressures of toluene, hydrogen, methane, and benzene.

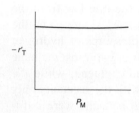

1. *Dependence on the product methane. If* methane were adsorbed on the surface, the partial pressure of methane would appear in the denominator of the rate expression and the rate would vary inversely with methane concentration

$$-r'_T \sim \frac{[\,\cdot\,]}{1 + K_M P_M + \cdots} \tag{10-67}$$

However, comparing runs 1 and 2 we observe that a fourfold increase in the pressure of methane has little effect on $-r'_T$. Consequently, we assume that methane is either very weakly adsorbed (i.e., $K_M P_M \ll 1$) or goes directly into the gas phase in a manner similar to propylene in the cumene decomposition previously discussed.

If it is in the denominator, it is probably on the surface.

2. *Dependence on the product benzene.* In runs 3 and 4, we observe that, for fixed concentrations (partial pressures) of hydrogen and toluene, the rate decreases with increasing concentration of benzene. A rate expression in which the benzene partial pressure appears in the denominator could explain this dependency

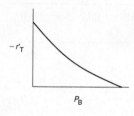

$$-r'_T \sim \frac{1}{1 + K_B P_B + \cdots} \tag{10-68}$$

The type of dependence of $-r'_T$ on P_B given by Equation (10-68) suggests that benzene is adsorbed on the clinoptilolite surface.

3. *Dependence on toluene.* At low concentrations of toluene (runs 10 and 11), the rate increases with increasing partial pressure of toluene, while at high toluene concentrations (runs 14 and 15), the rate is virtually independent of the toluene partial pressure. A form of the rate expression that would describe this behavior is

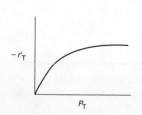

$$-r'_T \sim \frac{P_T}{1 + K_T P_T + \cdots} \tag{10-69}$$

A combination of Equations (10-68) and (10-69) suggests that the rate law may be of the form

$$-r'_T \sim \frac{P_T}{1 + K_T P_T + K_B P_B + \cdots} \tag{10-70}$$

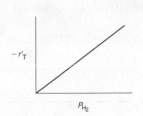

4. *Dependence on hydrogen.* When we compare runs 7, 8, and 9 in Table 10-6, we see that the rate increases linearly with increasing hydrogen concentration, and we conclude that the reaction is first order in H_2. In light of this fact, hydrogen is either not adsorbed on the surface or its coverage of the surface is extremely low ($1 \gg K_{H_2} P_{H_2}$) for the pressures used. If H_2 were adsorbed, $-r_T'$ would have a dependence on P_{H_2} analogous to the dependence of $-r_T'$ on the partial pressure of toluene, P_T [see Equation (10-69)]. For first-order dependence on H_2,

$$-r_T' \sim P_{H_2} \tag{10-71}$$

Combining Equations (10-67) through (10-71), we find that the rate law

$$-r_T' = \frac{k P_{H_2} P_T}{1 + K_B P_B + K_T P_T}$$

is in qualitative agreement with the data shown in Table 10-6.

10.4.2 Finding a Mechanism Consistent with Experimental Observations

We now propose a mechanism for the hydrodemethylation of toluene. We assume that the reaction follows an *Eley-Rideal mechanism* where toluene is adsorbed on the surface and then reacts with hydrogen in the gas phase to produce benzene adsorbed on the surface and methane in the gas phase. Benzene is then desorbed from the surface. Because approximately 75% to 80% of all heterogeneous reaction mechanisms are surface-reaction-limited rather than adsorption- or desorption-limited, we begin by assuming the reaction between adsorbed toluene and gaseous hydrogen to be reaction-rate-limited. Symbolically, this mechanism and associated rate laws for each elementary step are

Approximately 75% of all heterogeneous reaction mechanisms are surface-reaction-limited.

Eley–Rideal mechanism

Proposed Mechanism

Adsorption: $T(g) + S \rightleftharpoons T \cdot S$

$$r_{AD} = k_A \left(C_v P_T - \frac{C_{T \cdot S}}{K_T} \right) \tag{10-72}$$

Surface reaction: $H_2(g) + T \cdot S \rightleftharpoons B \cdot S + M(g)$

$$r_S = k_S \left(P_{H_2} C_{T \cdot S} - \frac{C_{B \cdot S} P_M}{K_S} \right) \tag{10-73}$$

Desorption: $B \cdot S \rightleftharpoons B(g) + S$

$$r_D = k_D \left(C_{B \cdot S} - K_B P_B C_v \right) \tag{10-74}$$

For surface-reaction-limited mechanisms

$$r_S = k_S \left(P_{H_2} C_{T \cdot S} - \frac{C_{B \cdot S} P_M}{K_S} \right) \tag{10-73}$$

we see that we need to replace $C_{T \cdot S}$ and $C_{B \cdot S}$ in Equation (10-73) by quantities that we can measure.

For surface-reaction-limited mechanisms, we use the adsorption rate Equation (10-72) for toluene to obtain $C_{T \cdot S}$[16], i.e.,

$$\frac{r_{AD}}{k_A} \approx 0$$

Then

$$\boxed{C_{T \cdot S} = K_T P_T C_v} \tag{10-75}$$

and we use the desorption rate Equation (10-74) for benzene to obtain $C_{B \cdot S}$:

$$\frac{r_D}{k_D} \approx 0$$

Then

$$\boxed{C_{B \cdot S} = K_B P_B C_v} \tag{10-76}$$

The total concentration of sites is

Perform a site balance to obtain C_v.

$$\boxed{C_t = C_v + C_{T \cdot S} + C_{B \cdot S}} \tag{10-77}$$

Substituting Equations (10-75) and (10-76) into Equation (10-77) and re-arranging, we obtain

$$C_v = \frac{C_t}{1 + K_T P_T + K_B P_B} \tag{10-78}$$

Next, substitute for $C_{T \cdot S}$ and $C_{B \cdot S}$, and then substitute for C_v in Equation (10-73) to obtain the rate law for the case when the reaction is surface-reaction-rate-limited

$$-r'_T = \frac{\overbrace{C_t k_S K_T}^{k} (P_{H_2} P_T - P_B P_M / K_P)}{1 + K_T P_T + K_B P_B} \tag{10-79}$$

We have shown by comparing runs 3 and 5 that we can neglect the reverse reaction, i.e., the thermodynamic equilibrium constant K_P is very, very large. Consequently, we obtain

Rate law for Eley–Rideal surface-reaction-limited mechanism

$$\boxed{-r'_T = \frac{k P_{H_2} P_T}{1 + K_B P_B + K_T P_T}} \tag{10-80}$$

Again we note that the adsorption equilibrium constant of a given species is exactly the reciprocal of the desorption equilibrium constant of that species.

10.4.3 Evaluation of the Rate-Law Parameters

In the original work on this reaction by Papp et al.,[17] over 25 models were tested against experimental data, and it was concluded that the preceding

---

[16] See footnote 10 on page 424.

[17] Ibid.

mechanism and rate-limiting step (i.e., the surface reaction between adsorbed toluene and H_2 gas) is the correct one. Assuming that the reaction is essentially irreversible, the rate law for the reaction on clinoptilolite is

$$-r'_T = k \, \frac{P_{H_2} P_T}{1 + K_B P_B + K_T P_T} \tag{10-80}$$

We now wish to determine how best to analyze the data to evaluate the rate-law parameters, k, K_T, and K_B. This analysis is referred to as *parameter estimation*.[18] We now rearrange our rate law to obtain a linear relationship between our measured variables. For the rate law given by Equation (10-80), we see that if both sides of Equation (10-80) are divided by $P_{H_2} P_T$ and the equation is then inverted

Linearize the rate equation to extract the rate-law parameters.

$$\boxed{\frac{P_{H_2} P_T}{-r'_T} = \frac{1}{k} + \frac{K_B P_B}{k} + \frac{K_T P_T}{k}} \tag{10-81}$$

The regression techniques described in Chapter 7 could be used to determine the rate-law parameters by using the equation

A linear least-squares analysis of the data shown in Table 10-6 is presented on the CRE Web site.

$$\boxed{Y_j = a_0 + a_1 X_{1j} + a_2 X_{2j}}$$

One can use the linearized least-squares analysis (PRS 7.3) to obtain initial estimates of the parameters k, K_T, K_B, in order to obtain convergence in non-linear regression. However, in many cases it is possible to use a nonlinear regression analysis directly, as described in Sections 7.5 and 7.6, and in Example 10-1.

Example 10–1   Nonlinear Regression Analysis to Determine the Model Parameters k, K_B, and K_T

(a) Use nonlinear regression, as discussed in Chapter 7, along with the data in Table 10-6, to find the best estimates of the rate-law parameters k, K_B, and K_T in Equation (10-80).
(b) Write the rate law solely as a function of the partial pressures.
(c) Find the ratio of the sites occupied by toluene, $C_{T \cdot S}$, to those occupied by benzene, $C_{B \cdot S}$, at 40% conversion of toluene.

Solution

The data from Table 10-6 were entered into the Polymath nonlinear least-squares program with the following modification. The rates of reaction in column 1 were multiplied by 10^{10}, so that each of the numbers in column 1 was entered directly (i.e., 71.0, 71.3, …). The model equation was

Living Example Problem

$$\text{Rate} = \frac{k P_T P_{H_2}}{1 + K_B P_B + K_T P_T} \tag{E10-1.1}$$

---

[18] See the Supplementary Reading (page 492) for a variety of techniques for estimating the rate-law parameters.

Following the step-by-step regression procedure in Chapter 7 and on the CRE Web site *Summary Notes*, we arrive at the following parameter values shown in Table E10-1.1.

TABLE E10-1.1   PARAMETER VALUES

Source data points and calculated data points

	PT	PH2	PB	RATE	RATE calc	Delta RATE
1	1	1	0	71	71.0197	-0.0196996
2	1	1	0	71.3	71.0197	0.2803004
3	1	1	1	41.6	42.21931	-0.6193089
4	1	1	4	19.7	19.04705	0.6529537
5	1	1	1	42	42.21931	-0.2193089
6	1	1	5	17.1	16.10129	0.9987095
7	1	1	0	71.8	71.0197	0.7803004
8	1	2	0	142	142.0394	-0.0393992
9	1	4	0	284	284.0788	-0.0787985
10	0.5	1	0	47	47.64574	-0.6457351
11	1	1	0	71.3	71.0197	0.2803004
12	5	1	0	117	116.8977	0.102331
13	10	1	0	127	127.1662	-0.1661677
14	15	1	0	131	131.002	-0.0019833
15	20	1	0	133	133.008	-0.007997
16	1	1	1	41.8	42.21931	-0.4193089

Model: RATE = k*PT*PH2/(1+KB*PB+KT*PT)

Variable	Initial guess	Value	95% confidence
k	144.	144.7673	1.240307
KB	1.4	1.390525	0.0457965
KT	1.03	1.038411	0.0131585

Nonlinear regression settings
Max # iterations = 64

Precision

R^2	0.9999509
R^2adj	0.9999434
Rmsd	0.1128555
Variance	0.2508084

(a) The best estimates are shown in the upper-right-hand box of Table E10-1.1.

(b) Converting the rate law to kilograms of catalyst and minutes,

$$-r_T' = \frac{1.45 \times 10^{-8} P_T P_{H_2}}{1 + 1.39 P_B + 1.038 P_T} \frac{\text{mol T}}{\text{g-cat} \cdot \text{s}} \times \frac{1000 \text{ g}}{1 \text{ kg}} \times \frac{60 \text{ s}}{\text{min}} \qquad \text{(E10-1.2)}$$

we have

$$\boxed{-r_T' = \frac{8.7 \times 10^{-4} P_T P_{H_2}}{1 + 1.39 P_B + 1.038 P_T} \left(\frac{\text{g mol T}}{\text{kg-cat} \cdot \text{min}} \right)} \qquad \text{(E10-1.3)}$$

Ratio of sites occupied by toluene to those occupied by benzene

(c) After we have the adsorption constants, K_T and K_B, we can calculate the ratio of sites occupied by the various adsorbed species. For example, taking the ratio of Equation (10-75) to Equation (10-76), the ratio of toluene-occupied sites to benzene- occupied sites at 40% conversion is

$$\frac{C_{T \cdot S}}{C_{B \cdot S}} = \frac{C_v K_T P_T}{C_v K_B P_B} = \frac{K_T P_T}{K_B P_B} = \frac{K_T P_{A0}(1-X)}{K_B P_{A0} X}$$

$$= \frac{K_T(1-X)}{K_B X} = \frac{1.038(1-0.4)}{1.39(0.4)} = 1.12$$

We see that at 40% conversion there are approximately 12% more sites occupied by toluene than by benzene. This fact is *common knowledge* to every chemical engineering student at Jofostan University, Riça, Jofostan.

Analysis: This example shows once again how to determine the values of rate-law parameters from experimental data using Polymath regression. It also shows how to calculate the different fraction of sites, both vacant and occupied, as a function of conversion.

10.4.4 Reactor Design

Our next step is to express the partial pressures P_T, P_B, and P_{H_2} as a function of X, combine the partial pressures with the rate law, $-r'_A$, as a function of conversion, and carry out the integration of the packed-bed design equation

$$\frac{dX}{dW} = \frac{-r'_A}{F_{A0}} \qquad (2\text{-}17)$$

Example 10–2   Catalytic Reactor Design

Living Example Problem

The hydrodemethylation of toluene is to be carried out in a PBR catalytic reactor.

$$C_6H_5CH_3 + H_2 \xrightarrow[\text{catalyst}]{} C_6H_6 + CH_4$$

The molar feed rate of toluene to the reactor is 50 mol/min, and the reactor inlet is at 40 atm and 640°C. The feed consists of 30% toluene, 45% hydrogen, and 25% inerts. Hydrogen is used in excess to help prevent coking. The pressure-drop parameter, α, is 9.8×10^{-5} kg^{-1}.

(a) Plot and analyze the conversion, the pressure ratio, p, and the partial pressures of toluene, hydrogen, and benzene as a function of PBR catalyst weight.

(b) Determine the catalyst weight in a fluidized CSTR with a bulk density of 400 kg/m^3 (0.4 g/cm^3) to achieve 65% conversion.

Solution

(a) PBR with pressure drop
1. Mole Balance:

Balance on toluene (T)

$$\frac{dF_T}{dW} = r'_T$$

$$\frac{dX}{dW} = \frac{-r'_T}{F_{T0}} \qquad (E10\text{-}2.1)$$

2. Rate Law: From Equation (E10-1.1) we have

$$-r'_T = \frac{kP_{H_2}P_T}{1 + K_BP_B + K_TP_T} \qquad (E10\text{-}2.2)$$

with $k = 0.00087$ mol/atm^2/kg-cat/min, $K_B = 1.39$ atm^{-1}, and $K_T = 1.038$ atm^{-1}.

3. Stoichiometry:

$$P_T = C_TRT = C_{T0}RT_0\left(\frac{1-X}{1+\varepsilon X}\right)p = P_{T0}\left(\frac{1-X}{1+\varepsilon X}\right)p$$

$$\varepsilon = y_{T0}\delta = 0.3(0) = 0$$

$$p = \frac{P}{P_0}$$

$$P_T = P_{T0}(1-X)p \qquad (E10\text{-}2.3)$$

Relating Toluene (T) Benzene (B) Hydrogen (H₂)

$$P_{H_2} = P_{T0}(\Theta_{H_2} - X)p$$

$$\Theta_{H_2} = \frac{0.45}{0.30} = 1.5$$

$$P_{H_2} = P_{T0}(1.5 - X)p \qquad \text{(E10-2.4)}$$

$$P_B = P_{T0}Xp \qquad \text{(E10-2.5)}$$

Because $\varepsilon = 0$, we can use the integrated form of the pressure-drop term.

P_0 = total pressure at the entrance

$$p = \frac{P}{P_0} = (1 - \alpha W)^{1/2} \qquad \text{(5-33)}$$

$$\alpha = 9.8 \times 10^{-5} \, \text{kg}^{-1}$$

Note that P_{T0} designates the inlet partial pressure of toluene. In this example, the inlet total pressure is designated P_0 to avoid any confusion. The inlet mole fraction of toluene is 0.3 (i.e., $y_{T0} = 0.3$), so that the inlet partial pressure of toluene is

Pressure drop in PBRs is discussed in Section 5.5.

$$P_{T0} = (0.3)(40) = 12 \, \text{atm}$$

We now calculate the maximum catalyst weight we can have, such that the exiting pressure will not fall below atmospheric pressure (i.e., 1.0 atm) for the specified feed rate. This weight is calculated by substituting the entering pressure of 40 atm and the exiting pressure of 1 atm into Equation (5-33), i.e.,

$$p = \frac{P}{P_0} = (1 - \alpha w)^{1/2}$$

$$\frac{1}{40} = (1 - 9.8 \times 10^{-5} W)^{1/2}$$

$$W = 10{,}197 \, \text{kg}$$

4. Evaluate: Consequently, we will set our final weight at 10,000 kg and determine the conversion profile as a function of catalyst weight up to this value. Equations (E10-2.1) through (E10-2.5) are shown in the Polymath program in Table E10-2.1. The conversion is shown as a function of catalyst weight in Figure E10-2.1, and profiles of the partial pressures of toluene, hydrogen, and benzene are shown in Figure E10-2.2. We note that the pressure drop causes (cf. Equation E10-2.5) the partial pressure of benzene to go through a maximum as one traverses the reactor.

TABLE E10-2.1   POLYMATH PROGRAM AND OUTPUT

Differential equations	Calculated values of DEQ variables		
1 d(X)/d(w) = -rt/FTo		Variable	Initial value	Final value

		Variable	Initial value	Final value
Explicit equations	1	alpha	9.8E-05	9.8E-05
1 FTo = 50	2	FTo	50.	50.
2 k = .00087	3	k	0.00087	0.00087
3 KT = 1.038	4	KB	1.39	1.39
4 KB = 1.39	5	KT	1.038	1.038
5 alpha = 0.000098	6	P	40.	5.656854
6 Po = 40	7	PB	0	1.157913
7 PTo = 0.3*Po	8	PH2	18.	1.387671
8 p = (1-alpha*w)^0.5	9	Po	40.	40.
9 P = y*Po	10	PT	12.	0.5391433
10 PH2 = PTo*(1.5-X)*y	11	PTo	12.	12.
11 PB = PTo*X*y	12	RATE	0.0139655	0.0002054
12 PT = PTo*(1-X)*y	13	rt	-0.0139655	-0.0002054
13 rt = -k*PT*PH2/(1+KB*PB+KT*PT)	14	w	0	10000.
14 RATE = -rt	15	X	0	0.6823067
	16	p	1.	0.1414214

Living Example Problem

Conversion profile down the packed bed

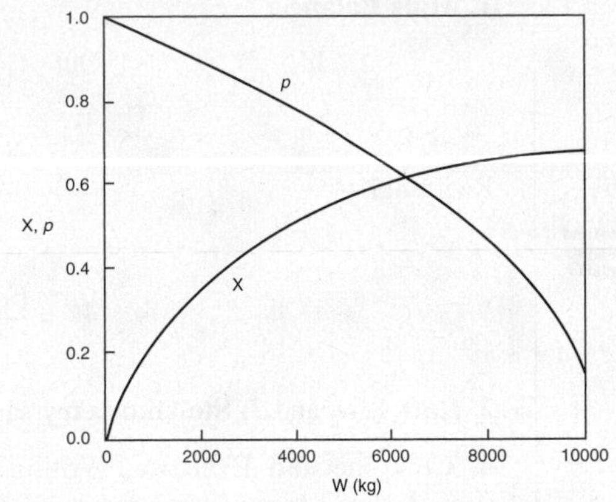

Figure E10-2.1   Conversion and pressure ratio profiles.

Note the partial
pressure of benzene
goes through a
maximum. Why?

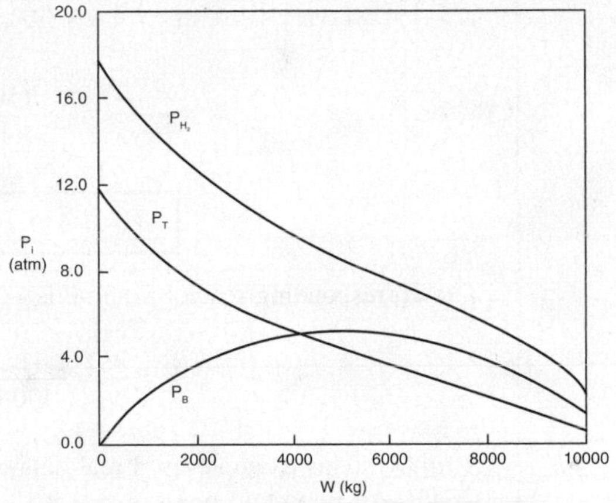

Figure E10-2.2   Partial pressure profiles.

Note the partial
pressure of benzene
goes through a
maximum. Why?

If one had neglected
ΔP it could have
been very
embarrassing.

See YouTube video,
"Reaction Engineer-
ing Gone Wrong,"
accessible from the
CRE Web site
home page.

PBR Analysis: For the case of no pressure drop, the conversion that would have been achieved with 10,000 kg of catalyst would have been 79%, compared with 68.2% when there is pressure drop in the reactor. To carry out this calculation, use the Living Example Program (**LEP**) on the CRE Web site and simply multiply the pressure-drop parameter by zero, i.e., line (5) would read α = 0.00098*0. *For the feed rate given, eliminating or minimizing pressure drop would increase the production of benzene by up to 61 million pounds per year!* Finally, we note in Figure E10-2.2 that the partial pressure of benzene (P_B) goes through a maximum. This maximum can be explained by recalling that P_B is just the product of the benzene mole fraction (y_B) times the total pressure (P) [i.e., $P_B = y_B P_T$]. Near the middle to end of the bed, benzene is no longer being formed so that y_B stops increasing. However, because of the pressure drop, the total pressure decreases and, as a result, so does P_B.

(b) Fluidized CSTR

We will now calculate the fluidized CSTR catalyst weight necessary to achieve the same (ca.) conversion as in the packed-bed reactor at the same operating conditions. The bulk density in the fluidized reactor is 0.4 g/cm^3. The design equation is

Fluidized CSTR

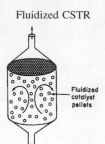

Fluidized
catalyst
pellets

1. Mole Balance:

In	–	Out	+	Gen	=	Accum
F_{T0}	–	F_T	+	$r'_T W$	=	0

Rearranging

$$W = \frac{F_{T0} - F_T}{-r'_T} = \frac{F_{T0} X}{-r'_T} \tag{E10-2.6}$$

2. Rate Law and **3. Stoichiometry** same as in part **(a)** PBR calculation

4. Combine and Evaluate: Writing Equation (E10-2.2) in terms of conversion (E10-2.3) and then substituting $X = 0.65$ and $P_{T0} = 12$ atm, we have

$$-r'_T = \frac{8.7 \times 10^{-4} P_T P_{H_2}}{1 + 1.39 P_B + 1.038 P_T} = \frac{8.7 \times 10^{-4} P_{T0}^2 (1-X)(1.5-X)}{1 + 1.39 P_{T0} X + 1.038 P_{T0}(1-X)} = 2.3 \times 10^{-3} \frac{\text{mol}}{\text{kg-cat} \cdot \text{min}}$$

$$W = \frac{F_{T0} X}{-r'_T} = \frac{(50 \text{ mol T/min})(0.65)}{2.3 \times 10^{-3} \text{ mol T/kg cat} \cdot \text{min}}$$

$$\boxed{W = 1.41 \times 10^4 \text{ kg of catalyst}}$$

The corresponding reactor volume is

$$V = \frac{W}{\rho_b} = \frac{1.41 \times 10^4 \text{ kg}}{400 \text{ kg/m}^3} = 35.25 \text{ m}^3$$

How can the weight of catalyst be reduced? Raise the temperature?

Analysis: This example used real data and the CRE algorithm to design a PBR and CSTR. An important point is that it showed how one could be embarrassed by not including pressure drop in the design of a packed-bed reactor. We also note that for both the PBR and fluidized CSTR, the values of the catalyst weight and reactor volume are quite high, especially for the low feed rates given. *Consequently, the temperature of the reacting mixture should be increased to reduce the catalyst weight, provided that side reactions and catalyst decay do not become a problem at higher temperatures.*

Example 10-2 illustrated the major activities pertinent to catalytic reactor design described earlier in Figure 10-6. In this example, the rate law was extracted directly from the data and then a mechanism was found that was consistent with experimental observation. Conversely, developing a feasible mechanism may guide one in the synthesis of the rate law.

10.5 Reaction Engineering in Microelectronic Fabrication

10.5.1 Overview

We now extend the principles of the preceding sections to one of the emerging technologies in chemical engineering. Chemical engineers are now playing an

important role in the electronics industry. Specifically, they are becoming more involved in the manufacture of electronic and photonic devices, recording materials, and especially medical lab-on-a-chip devices.

Surface reactions play an important role in the manufacture of microelectronic devices. One of the single most important developments of the twentieth century was the invention of the integrated circuit. Advances in the development of integrated circuitry have led to the production of circuits that can be placed on a single semiconductor chip the size of a pinhead and perform a wide variety of tasks by controlling the electron flow through a vast network of channels. These channels, which are made from semiconductors such as silicon, gallium arsenide, indium phosphide, and germanium, have led to the development of a multitude of novel microelectronic devices. Examples of microelectronic sensing devices manufactured using chemical reaction engineering principles are shown in the left-hand margin.

The manufacture of an integrated circuit requires the fabrication of a network of pathways for electrons. The principal reaction engineering steps of the fabrication process include depositing material on the surface of a material called a *substrate* (e.g., by chemical vapor deposition, abbreviated as CVD), changing the conductivity of regions of the surface (e.g., by boron doping or ion inplantation), and removing unwanted material (e.g., by etching). By applying these steps systematically, miniature electronic circuits can be fabricated on very small semiconductor chips. The fabrication of microelectronic devices may include as few as 30 or as many as 200 individual steps to produce chips with up to 10^9 elements per chip.

An abbreviated schematic of the steps involved in producing a typical metal-oxide, semiconductor, field-effect transistor (MOSFET) device is shown in Figure 10-20. Starting from the upper left, we see that single-crystal silicon ingots are grown in a Czochralski crystallizer, sliced into wafers, and chemically and physically polished. These polished wafers serve as starting materials for a variety of microelectronic devices. A typical fabrication sequence is shown for processing the wafer, beginning with the formation of an SiO_2 layer on top of the silicon. The SiO_2 layer may be formed either by oxidizing a silicon layer or by laying down a SiO_2 layer by chemical vapor deposition (CVD). Next, the wafer is masked with a polymer photoresist, a template with the pattern to be etched onto the SiO_2 layer is placed over the photoresist, and the wafer is exposed to ultraviolet irradiation. If the mask is a positive photoresist, the light will cause the exposed areas of the polymer to dissolve when the wafer is placed in the developer. On the other hand, when a negative photoresist mask is exposed to ultraviolet irradiation, cross-linking of the polymer chains occurs, and the *unexposed* areas dissolve in the developer. The undeveloped portion of the photoresist (in either case) will protect the covered areas from etching.

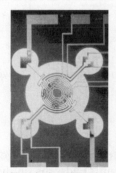

Chemical engineering principles are involved in virtually every step!

After the exposed areas of SiO_2 are etched to form trenches (either by wet etching or by plasma etching), the remaining photoresist is removed. Next, the wafer is placed in a furnace containing gas molecules of the desired dopant, which then diffuse into the exposed silicon. After diffusion of dopant to the desired depth in the wafer, the wafer is removed and then SiO_2 is removed by etching. The sequence of masking, etching, CVD, and metallization continues until the desired device is formed. A schematic of a final chip is

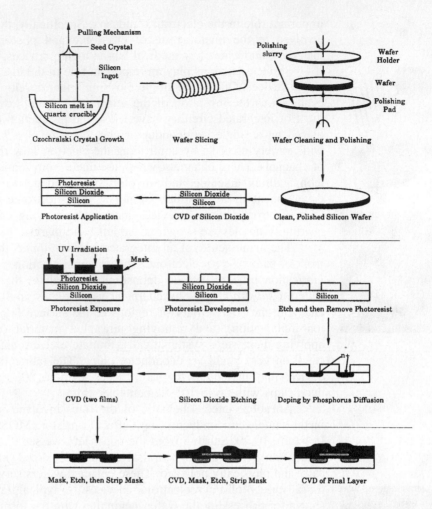

Figure 10-20   Microelectronic fabrication steps.

shown in the lower-right-hand corner of Figure 10-20. In Section 10.5.2, we discuss one of the key processing steps, CVD.

10.5.2 Chemical Vapor Deposition

The mechanisms by which CVD occurs are very similar to those of heterogeneous catalysis discussed earlier in this chapter. The reactant(s) adsorbs on the surface and then reacts on the surface to form a new surface. This process may be followed by a desorption step, depending on the particular reaction.

  The growth of a germanium epitaxial film as an interlayer between a gallium arsenide layer and a silicon layer has received attention in the microelectronics industry.[19] Epitaxial germanium is also an important material in the fabrication of tandem solar cells. The growth of germanium films can be accomplished by CVD. A proposed mechanism is

Ge used in solar cells

---

[19] H. Ishii and Y. Takahashi, *J. Electrochem. Soc.*, 135, p. 1539.

$$\text{Gas-phase dissociation:} \quad GeCl_4(g) \xrightleftharpoons \quad GeCl_2(g) + Cl_2(g)$$

$$\text{Adsorption:} \quad GeCl_2(g) + S \underset{}{\overset{k_A}{\rightleftharpoons}} GeCl_2 \cdot S$$

Mechanism

$$\text{Adsorption:} \quad H_2 + 2S \underset{}{\overset{k_H}{\rightleftharpoons}} 2H \cdot S$$

$$\text{Surface reaction:} \quad GeCl_2 \cdot S + 2H \cdot S \xrightarrow{k_S} Ge(s) + 2HCl(g) + 2S$$

At first it may appear that a site has been lost when comparing the right- and left-hand sides of the surface reaction step. However, the newly formed germanium atom on the right-hand side is a site for the future adsorption of $H_2(g)$ or $GeCl_2(g)$, and there are three sites on both the right- and left-hand sides of the surface reaction step. These sites are shown schematically in Figure 10-21.

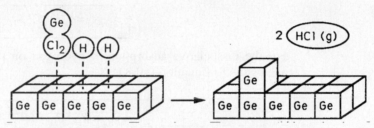

Figure 10-21   CVD surface reaction step for germanium.

The surface reaction between adsorbed molecular hydrogen and germanium dichloride is believed to be rate-limiting. The reaction follows an elementary rate law with the rate being proportional to the fraction of the surface covered by $GeCl_2$ times the square of the fraction of the surface covered by molecular hydrogen.

Rate law for rate-limiting step

$$r''_{Dep} = k_S f_{GeCl_2} f_H^2 \qquad (10\text{-}82)$$

where r''_{Dep} = deposition rate per unit surface area, nm/s
k_S = surface specific reaction rate, nm/s
f_{GeCl_2} = fraction of the surface occupied by germanium dichloride
f_H = fraction of the surface covered by molecular hydrogen

The deposition rate (film growth rate) is usually expressed in nanometers per second and is easily converted to a molar rate (mol/m²·s) by multiplying by the molar density of solid germanium (mol/m³).

The difference between developing CVD rate laws and rate laws for catalysis is that the site concentration (e.g., C_v) is replaced by the fractional surface area coverage (e.g., the fraction of the surface that is vacant, f_v). The total fraction of surface available for adsorption should, of course, add up to 1.0.

Area balance

$$\text{Fractional area balance:} \quad f_v + f_{GeCl_2} + f_H = 1 \qquad (10\text{-}83)$$

We will first focus our attention on the adsorption of $GeCl_2$. The rate of jumping on to the surface is proportional to the partial pressure of $GeCl_2$, P_{GeCl_2}, and the fraction of the surface that is vacant, f_v. The net rate of $GeCl_2$ adsorption is

$$r_{AD} = k_A \left(f_v P_{GeCl_2} - \frac{f_{GeCl_2}}{K_A} \right)$$

(10-84)

Since the surface reaction is rate-limiting, in a manner analogous to catalysis reactions, we have for the adsorption of $GeCl_2$

Adsorption of
$GeCl_2$ not
rate-limiting

$$\frac{r_{AD}}{k_A} \approx 0$$

Solving Equation (10-84) for the fractional surface coverage of $GeCl_2$ gives

$$\boxed{f_{GeCl_2} = f_v K_A P_{GeCl_2}}$$

(10-85)

For the dissociative adsorption of hydrogen on the Ge surface, the equation analogous to Equation (10-84) is

$$r_{ADH_2} = k_H \left(P_{H_2} f_v^2 - \frac{f_H^2}{K_H} \right)$$

(10-86)

Since the surface reaction is rate-limiting

Adsorption of H_2 is
not rate-limiting

$$\frac{r_{ADH_2}}{k_H} \approx 0$$

Then

$$\boxed{f_H = f_v \sqrt{K_H P_{H_2}}}$$

(10-87)

Recalling the rate of deposition of germanium, we substitute for f_{GeCl_2} and f_H in Equation (10-82) to obtain

$$r''_{Dep} = f_v^3 k_S K_A P_{GeCl_2} K_H P_{H_2}$$

(10-88)

We solve for f_v in an identical manner to that for C_v in heterogeneous catalysis. Substituting Equations (10-85) and (10-87) into Equation (10-83) gives

$$f_v + f_v \sqrt{K_H P_{H_2}} + f_v K_A P_{GeCl_2} = 1$$

Rearranging yields

$$f_v = \frac{1}{1 + K_A P_{GeCl_2} + \sqrt{K_H P_{H_2}}}$$

(10-89)

Finally, substituting for f_v in Equation (10-88), we find that

$$r''_{Dep} = \frac{k_S K_H K_A P_{GeCl_2} P_{H_2}}{(1 + K_A P_{GeCl_2} + \sqrt{K_H P_{H_2}})^3}$$

and lumping K_A, K_H, and k_S into a specific reaction rate k' yields

Rate of deposition
of Ge

$$r''_{Dep} = \frac{k' P_{GeCl_2} P_{H_2}}{(1 + K_A P_{GeCl_2} + \sqrt{K_H P_{H_2}})^3} \qquad (10\text{-}90)$$

We now need to relate the partial pressure of $GeCl_2$ to the partial pressure of $GeCl_4$ in order to calculate the conversion of $GeCl_4$. If we assume that the gas-phase reaction

Equilibrium in
gas phase

$$GeCl_4(g) \rightleftharpoons GeCl_2(g) + Cl_2(g)$$

is in equilibrium, we have

$$K_P = \frac{P_{GeCl_2} P_{Cl_2}}{P_{GeCl_4}}$$

$$P_{GeCl_2} = \frac{P_{GeCl_4}}{P_{Cl_2}} \cdot K_P$$

and if hydrogen is weakly adsorbed $(\sqrt{K_H P_{H_2}} < 1)$, we obtain the rate of deposition as

Rate of deposition
of Ge when H$_2$ is
weakly adsorbed

$$r''_{Dep} = \frac{k P_{GeCl_4} P_{H_2} P_{Cl_2}^2}{(P_{Cl_2} + K_P P_{GeCl_4})^3} \qquad (10\text{-}91)$$

We now can use stoichiometry to express each of the species' partial pressures in terms of conversion and the entering partial pressure of $GeCl_4$, $P_{GeCl_{4,0}}$, and then proceed to calculate the conversion.

It should also be noted that it is possible that $GeCl_2$ may also be formed by the reaction of $GeCl_4$ and a Ge atom on the surface, in which case a different rate law would result.

10.6 Model Discrimination

We have seen that for each mechanism and each rate-limiting step we can derive a rate law. Consequently, if we had three possible mechanisms and three rate-limiting steps for each mechanism, we would have nine possible rate laws to compare with the experimental data. We will use the regression techniques discussed in Chapter 7 to identify which model equation best fits the data by choosing the one with the smaller sums of squares and/or carrying out an F-test. We could also compare the residual plots for each model, which not only show the error associated with each data point but also show if the error

Regression

is randomly distributed or if there is a trend in the error. If the error is randomly distributed, this result is an additional indication that the correct rate law has been chosen.

CAUTION

We need to raise a caution here about choosing the model with the smallest sums of squares. The caution is that the model parameter values that give the smallest sum must be realistic. In the case of heterogeneous catalysis, *all values of the adsorption equilibrium constants **must** be positive*. In addition, if the temperature dependence is given, because adsorption is exothermic, the adsorption equilibrium constant must decrease with increasing temperature. To illustrate these principles, let's look at the following example.

Example 10–3   Hydrogenation of Ethylene to Ethane

The hydrogenation (H) of ethylene (E) of form ethane (EA),

$$H_2 + C_2H_4 \rightarrow C_2H_6$$

is carried out over a cobalt molybdenum catalyst [*Collect. Czech. Chem. Commun.*, 51, 2760 (1988)]. Carry out a nonlinear regression analysis on the data given in Table E10-3.1, and determine which rate law best describes the data.

TABLE E10-3.1   DIFFERENTIAL REACTOR DATA

Run Number	Reaction Rate (mol/kg-cat. · s)	P_E (atm)	P_{EA} (atm)	P_H (atm)
1	1.04	1	1	1
2	3.13	1	1	3
3	5.21	1	1	5
4	3.82	3	1	3
5	4.19	5	1	3
6	2.391	0.5	1	3
7	3.867	0.5	0.5	5
8	2.199	0.5	3	3
9	0.75	0.5	5	1

Procedure
- Enter data
- Enter model
- Make initial estimates of parameters
- Run regression
- Examine parameters and variance
- Observe error distribution
- Choose model

Determine which of the following rate laws best describes the data in Table E10-3.1.

(a) $-r'_E = \dfrac{kP_E P_H}{1 + K_{EA}P_{EA} + K_E P_E}$ (c) $-r'_E = \dfrac{kP_E P_H}{\left(1 + K_E P_E\right)^2}$

(b) $-r'_E = \dfrac{kP_E P_H}{1 + K_E P_E}$ (d) $-r'_E = kP_E^a P_H^b$

Solution

Polymath was chosen as the software package to solve this problem. The data in Table E10-3.1 were entered into the program. A screen-shot-by-screen-shot set of instructions on how to carry out the regression is given on the CRE Web site, at the end of the *Summary Notes* for Chapter 7. After entering the data and following the step-by-step procedures, the results shown in Table E10-3.2 were obtained.

Summary Notes

TABLE E10-3.2   RESULTS OF THE POLYMATH NONLINEAR REGRESSION

Model (a)

Model: RATE = k*Pe*PH2/(1+KEA*Pea+KE*Pe)

Variable	Value	95% confidence
k	3.3478805	0.2922517
KEA	0.0428419	0.0636262
KE	2.2110797	0.2392585

Nonlinear regression settings
Max # iterations = 64

Precision
R^2   = 0.998321
R^2adj   = 0.9977614
Rmsd   = 0.0191217
Variance = 0.0049361

Model (b)

Model: RATE = k*Pe*PH2/(1+KE*Pe)

Variable	Value	95% confidence
k	3.1867851	0.287998
KE	2.1013363	0.2638835

Nonlinear regression settings
Max # iterations = 64

Precision
R^2   = 0.9975978
R^2adj   = 0.9972547
Rmsd   = 0.022872
Variance = 0.0060534

Model (c)

Model: RATE = k*Pe*PH2/(1+KE*Pe)^2

Variable	Value	95% confidence
k	2.0087761	0.2661838
KE	0.3616652	0.0623045

Nonlinear regression settings
Max # iterations = 64

Precision
R^2   = 0.9752762
R^2adj   = 0.9717442
Rmsd   = 0.0733772
Variance = 0.0623031

Model (d)

Model: RATE = k*Pe^a*PH2^b

Variable	Value	95% confidence
k	0.8940237	0.2505474
a	0.2584412	0.0704628
b	1.0615542	0.2041339

Nonlinear regression settings
Max # iterations = 64

Precision
R^2   = 0.9831504
R^2adj   = 0.9775338
Rmsd   = 0.0605757
Variance = 0.0495372

Living Example Problem

Model (a) Single site, surface-reaction, rate-limited with hydrogen weakly adsorbed

From Table E10-3.2 data, we can obtain

$$-r'_E = \frac{3.348 \, P_E P_H}{1 + 0.043 P_{EA} + 2.21 \, P_E}$$

(E10-3.1)

We now examine the sums of squares (variance) and range of variables themselves. The sums of squares is reasonable and, in fact, the smallest of all the models at 0.0049. *However*, let's look at K_{EA}. We note that the value for the 95% confidence limit of ±0.0636 is greater than the nominal value of K_{EA} = 0.043 atm⁻¹ itself (i.e., K_{EA} = 0.043 ± 0.0636). The 95% confidence limit means that if the experiment were run 100 times and then 95 times it would fall within the range (–0.021) < K_{EA} < (0.1066). Because K_{EA} can never be negative, we are going to reject this model. Consequently, we set $K_{EA} = 0$ and proceed to Model (b).

Model (b) Single site, surface-reaction, rate-limited with ethane and hydrogen weakly adsorbed

From Table E10-3.2 we can obtain

$$-r'_E = \frac{3.187 \, P_E P_H}{1 + 2.1 \, P_E}$$

(E10-3.2)

The value of the adsorption constant K_E = 2.1 atm⁻¹ is reasonable and is not negative within the 95% confidence limit. Also, the variance is small at $\sigma_B^2 = 0.0061$.

Model (c) Dual site, surface-reaction, rate-limited with hydrogen and ethane weakly adsorbed

From Table E10-3.2 we can obtain

$$-r'_E = \frac{2.0\,P_E P_{H_2}}{(1+0.036\,P_E)^2} \tag{E10-3.3}$$

While K_E is small, it never goes negative within the 95% confidence interval. The variance of this model at $\sigma_C^2 = 0.0623$ is much larger than the other models. Comparing the variance of model **(c)** with model **(b)**

$$\frac{\sigma_C^2}{\sigma_B^2} = \frac{0.0623}{0.0061} = 10.2$$

We see that the σ_C^2 is an order of magnitude greater than σ_B^2, and therefore we eliminate model **(c)**.[20]

Model (d) Empirical

Similarly for the power law model, we obtain from Table E10-3.2

$$-r'_E = 0.894\,P_E^{0.26} P_{H_2}^{1.06} \tag{E10-3.4}$$

As with model **(c)**, the variance is quite large compared to model **(b)**

$$\frac{\sigma_D^2}{\sigma_B^2} = \frac{0.049}{0.0061} = 8.03$$

So we also eliminate model **(d)**. For heterogeneous reactions, Langmuir-Hinshelwood rate laws are preferred over power-law models.

**Analysis:** _Choose the Best Model._ In this example, we were presented with four rate laws and were asked which law best fits the data. Because all the parameter values are realistic for **model (b)** and the sums of squares are significantly smaller for **model (b)** than for the other models, we choose **model (b)**. We note again that there is a caution we need to point out regarding the use of regression! One cannot simply carry out a regression and then choose the model with the lowest value of the sums of squares. If this were the case, we would have chosen **model (a)**, which had the smallest sums of squares of all the models with $\sigma^2 = 0.0049$. However, one must consider the physical realism of the parameters in the model. In **model (a)** the 95% confidence interval was greater than the parameter itself, thereby yielding negative values of the parameter, K_{AE}, which is physically impossible.

10.7 Catalyst Deactivation

In designing fixed and ideal fluidized-bed catalytic reactors, we have assumed up to now that the activity of the catalyst remains constant throughout the catalyst's life. That is, the total concentration of active sites, C_t, accessible to the reaction does not change with time. Unfortunately, Mother Nature is not so kind as to allow this behavior to be the case in most industrially significant

---

[20] See G. F. Froment and K. B. Bishoff, _Chemical Reaction Analysis and Design_, 2nd ed. (New York: Wiley, 1990), p. 96.

catalytic reactions. One of the most insidious problems in catalysis is the loss of catalytic activity that occurs as the reaction takes place on the catalyst. A wide variety of mechanisms have been proposed by Butt and Petersen, to explain and model catalyst deactivation.[21]

Catalytic deactivation adds another level of complexity to sorting out the reaction rate law parameters and pathways. In addition, we need to make adjustments for the decay of the catalysts in the design of catalytic reactors. *However, please don't worry,* this adjustment is usually made by a quantitative specification of the catalyst's activity, $a(t)$. In analyzing reactions over decaying catalysts, we divide the reactions into two categories: *separable kinetics* and *nonseparable kinetics*. In separable kinetics, we separate the rate law and activity

Separable kinetics: $-r'_A = a(\text{Past history}) \times -r'_A$ (Fresh catalyst)

When the kinetics and activity are separable, it is possible to study catalyst decay and reaction kinetics independently. However, nonseparability

Nonseparable kinetics: $-r'_A = -r'_A$ (Past history, fresh catalyst)

must be accounted for by assuming the existence of a nonideal surface or by describing deactivation by a mechanism composed of several elementary steps.[22]

We shall consider only *separable kinetics* and define the activity of the catalyst at time t, $a(t)$, as the ratio of the rate of reaction on a catalyst that has been used for a time t to the rate of reaction on a fresh catalyst ($t = 0$):

$a(t)$: catalyst
activity

$$a(t) = \frac{-r'_A(t)}{-r'_A(t = 0)}$$

(10-92)

Because of the catalyst decay, the activity decreases with time and a typical curve of the activity as a function of time is shown in Figure 10-22.

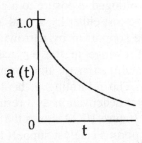

Figure 10-22   Activity as a function of time.

Combining Equations (10-92) and (3-2), the rate of disappearance of reactant A on a catalyst that has been utilized for a time t is

Reaction rate law
accounting for
catalyst activity

$$-r'_A = a(t)\,k(T)\,\text{fn}(C_A,\ C_B,\ \dots,\ C_P)$$

(10-93)

[21] J. B. Butt and E. E. Petersen, *Activation, Deactivation and Poisoning of Catalysts* (New York: Academic Press, 1988).

[22] D. T. Lynch and G. Emig, *Chem. Eng. Sci.*, 44(6), 1275–1280 (1989).

where $a(t)$ = catalytic activity, time-dependent

$k(T)$ = specific reaction rate, temperature-dependent

C_i = gas-phase concentration of reactants, products, or contaminant

The rate of catalyst decay, r_d, can be expressed in a rate law analogous to Equation (10-93)

Catalyst decay rate law

$$r_d = -\frac{da}{dt} = p[a(t)]k_d(T)h(C_A, C_B, \ldots, C_P) \qquad (10\text{-}94)$$

e.g., $p[a(t)] = [a(t)]^2$

where $p[a(t)]$ is some function of the catalyst activity, k_d is the specific decay constant, and $h(C_i)$ is the functionality of rate of decay, r_d, on the reacting species concentrations. For the cases presented in this chapter, this functionality either will be independent of concentration (i.e., $h = 1$) or will be a linear function of species concentration (i.e., $h = C_i$).

The functionality of the activity term, $p[a(t)]$, in the decay law can take a variety of forms. For example, for a first-order decay

$$p(a) = a \qquad (10\text{-}95)$$

and for a second-order decay

$$p(a) = a^2 \qquad (10\text{-}96)$$

The particular function, $p(a)$, will vary with the gas catalytic system being used and the reason or mechanism for catalytic decay.

10.7.1 Types of Catalyst Deactivation

• Sintering
• Fouling
• Poisoning

There are three categories into which the loss of catalytic activity can traditionally be divided: sintering or aging, fouling or coking, and poisoning.

Deactivation by Sintering (Aging).[23]   Sintering, also referred to as aging, is the loss of catalytic activity due to a loss of active surface area resulting from the prolonged exposure to high gas-phase temperatures. The active surface area may be lost either by crystal agglomeration and growth of the metals deposited on the support or by narrowing or closing of the pores inside the catalyst pellet. A change in the surface structure may also result from either surface recrystallization or the formation or elimination of surface defects (active sites). The reforming of heptane over platinum on alumina is an example of catalyst deactivation as a result of sintering.

Figure 10-23 shows the loss of surface area resulting from the flow of the solid porous catalyst support at high temperatures to cause pore closure. Figure 10-24 shows the loss of surface area by atomic migration and agglomeration of small metal sites deposited on the surface into a larger site where the interior

The catalyst support becomes soft and flows, resulting in pore closure.

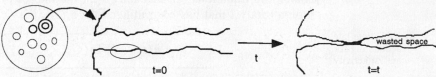

Figure 10-23   Decay by sintering: pore closure; loss of reactive surface area.

---

[23] See G. C. Kuczynski, ed., *Sintering and Catalysis*, vol. 10 of *Materials Science Research* (New York: Plenum Press, 1975).

The atoms move along the surface and agglomerate.

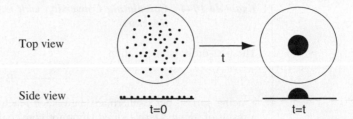

Top view

Side view

t=0   t=t

Figure 10-24   Decay by sintering: agglomeration of deposited metal sites; loss of reactive surface area.

atoms are not accessible to the reaction. Sintering is usually negligible at temperatures below 40% of the melting temperature of the solid.[24]

Deactivation by sintering may in some cases be a function of the mainstream gas concentration. Although other forms of the sintering decay rate laws exist, one of the most commonly used decay laws is second order with respect to the present activity

$$r_d = k_d a^2 = -\frac{da}{dt} \tag{10-97}$$

Integrating, with $a = 1$ at time $t = 0$, yields

Sintering: second-order decay

$$\boxed{a(t) = \frac{1}{1 + k_d t}} \tag{10-98}$$

The amount of sintering is usually measured in terms of the active surface area of the catalyst S_a

$$S_a = \frac{S_{a0}}{1 + k_d t} \tag{10-99}$$

The sintering decay constant, k_d, follows the Arrhenius equation

$$\boxed{k_d = k_d(T_0) \exp\left[\frac{E_d}{R}\left(\frac{1}{T_0} - \frac{1}{T}\right)\right]} \tag{10-100}$$

Minimizing sintering

The decay activation energy, E_d, for the reforming of heptane on Pt/Al_2O_3 is on the order of 70 kcal/mol, which is rather high. As mentioned earlier, sintering can be reduced by keeping the temperature below 0.3 to 0.4 times the metal's melting point.

We will now stop and consider reactor design for a fluid–solid system with decaying catalyst. To analyze these reactors, we only add one step to our algorithm; that is, determine the catalyst decay law. The sequence is shown here.

The algorithm

Mole balance $\longrightarrow$	Reaction rate law $\longrightarrow$	*Decay rate law* $\longrightarrow$
Stoichiometry $\longrightarrow$	Combine and solve $\longrightarrow$	Numerical techniques

[24]R. Hughes, *Deactivation of Catalysts* (San Diego: Academic Press, 1984).

Example 10–4   Calculating Conversion with Catalyst Decay in Batch Reactors

The first-order isomerization

$$A \longrightarrow B$$

is being carried out isothermally in a batch reactor on a catalyst that is decaying as a result of aging. Derive an equation for conversion as a function of time.

Solution

1. **Mol Balance:**

$$N_{A0} = \frac{dX_d}{dt} = -r'_A W \tag{E10-4.1}$$

where X_d is the conversion of A when the catalyst is decaying.

2. **Reaction-Rate Law:**

$$-r'_A = k' a(t) C_A \tag{E10-4.2}$$

3. **Decay Law:** For second-order decay by sintering:

$$\boxed{a(t) = \frac{1}{1 + k_d t}} \tag{10-98}$$

One extra step (number 3) is added to the algorithm.

Following the Algorithm

4. **Stoichiometry:**

$$C_A = C_{A0}(1 - X_d) = \frac{N_{A0}}{V}(1 - X_d) \tag{E10-4.3}$$

5. **Combining** gives us

$$\frac{dX_d}{dt} = k'(1 - X_d) a \frac{W}{V} \tag{E10-4.4}$$

Let $k = k'W/V$. Substituting for catalyst activity a, we have

$$\boxed{\frac{dX_d}{dt} = k(1 - X_d) a = k \frac{(1 - X_d)}{(1 + k_d t)}} \tag{E10-4.5}$$

where X_d is the conversion when there is decay. We want to compare the conversion with and without catalyst decay.

For no decay $k_d = 0$

$$\frac{dX}{dt} = k(1 - X)$$

The Polymath program and a comparison of the conversion with decay X_d and without decay X are shown below.

POLYMATH Report
Ordinary Differential Equations

Calculated values of DEQ variables

	Variable	Initial value	Final value
1	k	0.01	0.01
2	kd	0.1	0.1
3	t	0	500.
4	X	0	0.9932621
5	Xd	0	0.2013514

Differential equations
1 d(x)/d(t) = k*(1-X)
2 d(Xd)/d(t) = k*(1-X)/(1+kd*t)

Explicit equations
1 k = .01
2 kd = .1

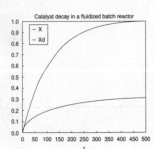

One can also obtain an analytical solution for this reaction order and low decay. Separating variables and integrating yields

$$\int_0^X \frac{dX_d}{1-X_d} = k \int_0^t \frac{dt}{1+k_d t} \tag{E10-4.6}$$

$$\boxed{\ln \frac{1}{1-X_d} = \frac{k}{k_d} \ln(1+k_d t)} \tag{E10-4.7}$$

6. **Solving** for the conversion X_d at any time t, we find that

$$X_d = 1 - \frac{1}{(1+k_d t)^{k/k_d}} \tag{E10-4.8}$$

the analytical solution **without decay**

$$\boxed{\ln \frac{1}{1-X} = kt} \tag{E10-4.9}$$

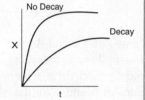

**Analysis:** One observes that for long times the conversion with decay approaches a rather flat plateau and reaches conversion of about 30%. This is the conversion that will be achieved in a batch reactor for a first-order reaction when the catalyst decay law is second order. By comparison, we obtain virtually complete conversion in 500 minutes when there is no decay. The purpose of this example was to demonstrate the algorithm for isothermal catalytic reactor design for a decaying catalyst. In problem P10-1$_B$(d) you are asked to sketch the temperature–time trajectories for various values of k and k_d.

Deactivation by Coking or Fouling.   This mechanism of decay (see Figures 10-25 and 10-26) is common to reactions involving hydrocarbons. It results from a carbonaceous (coke) material being deposited on the surface of a catalyst.

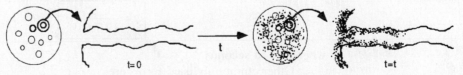

Figure 10-25   Schematic of decay by coking; loss of reactive surface area.

(a) Fresh catalyst   (b) Spent catalyst

Figure 10-26   Decay by coking. (Photos courtesy of Engelhard catalyst, copyright by Michael Gaffney Photographer, Mendham, NJ.)

The amount of coke on the surface after a time t has been found to obey the following empirical relationship

$$\boxed{C_C = At^n}$$ (10-101)

where C_C is the concentration of carbon on the surface (g/m²) and n and A are fouling parameters, which can be functions of the feed rate. This expression was originally developed by Voorhies and has been found to hold for a wide variety of catalysts and feed streams.[25] Representative values of A and n for the cracking of East Texas light gas oil yield[26]

$$\% \text{ coke} = 0.47 \sqrt{t(\text{min})}$$

Different functionalities between the activity and amount of coke on the surface have been observed. One commonly used form is

$$a = \frac{1}{k_{Ck}C_C^p + 1}$$ (10-102)

or, in terms of time, we combine Equations (10-101) and (10-102)

$$a = \frac{1}{k_{Ck}A^p t^{np} + 1} = \frac{1}{1 + k't^m}$$ (10-103)

For light Texas gas oil being cracked at 750°F over a synthetic catalyst for short times, the decay law is

$$\boxed{a = \frac{1}{1 + 7.6\,t^{1/2}}}$$ (10-104)

where t is in seconds.

Other commonly used forms are

$$a = e^{-\alpha_1 C_C}$$ (10-105)

Activity for deactivation by coking

---

[25] A. Voorhies, *Ind. Eng. Chem.*, 37, 318 (1945).

[26] C. O. Prater and R. M. Lago, *Adv. Catal.*, 8, 293 (1956).

and

$$a = \frac{1}{1+\alpha_2 C_C}$$ (10-106)

A dimensionless fouling correlation has been developed by Pacheco and Petersen.[27]

Minimizing coking   When possible, coking can be reduced by running at elevated pressures (2000 to 3000 kPa) and hydrogen-rich streams. A number of other techniques for minimizing fouling are discussed by Bartholomew.[28] Catalysts deactivated by coking can usually be regenerated by burning off the carbon.

Deactivation by Poisoning.   Deactivation by this mechanism occurs when the poisoning molecules become irreversibly chemisorbed to active sites, thereby reducing the number of sites available for the main reaction. The poisoning molecule, P, may be a reactant and/or a product in the main reaction, or it may be an impurity in the feed stream.

It's going to cost you.   **Side Note.**   One of the most significant examples of catalyst poisoning occurred at the gasoline pump. Oil companies found that adding lead to the gasoline increased the octane number. The television commercials said "We are going to enhance your gasoline, *but it's going to cost you* for the added tetra-ethyl lead." So for many years they used lead as an antiknock component. As awareness grew about NO, HC, and CO emission from the engine, it was decided to add a catalytic afterburner in the exhaust system to reduce these emissions. Unfortunately, it was found that the lead in the gasoline poisoned the reactive catalytic sites. So, the television commercials now said "We are going to take the lead out of gasoline but to receive the same level of performance as without lead, *but it's going to cost you* because of the added refining costs to raise the octane number." Do you think that, financially, consumers would have been better off if they never put the lead in the gasoline in the first place?

Poison in the Feed.   Many petroleum feed stocks contain trace impurities such as sulfur, lead, and other components that are too costly to remove, yet poison the catalyst slowly over time. For the case of an impurity, P, in the feed stream, such as sulfur, for example, in the reaction sequence

Main
reaction:
$$\begin{cases} A+S \rightleftharpoons (A\cdot S) \\ A\cdot S \rightleftharpoons (B\cdot S + C(g)) \\ B\cdot S \rightleftharpoons (B+S) \end{cases} \quad -r'_A = a(t)\frac{kC_A}{1+K_A C_A + K_B C_B}$$

Poisoning
reaction: $$P+S \longrightarrow P\cdot S \qquad r_d = -\frac{da}{dt} = k'_d C_p^m a^q$$ (10-107)

---

[27] M. A. Pacheco and E. E. Petersen, *J. Catal.*, 86, 75 (1984).

[28] R. J. Farrauto and C. H. Bartholomew, *Fundamentals of Industrial Catalytic Processes, 2nd edition* (New York: Blackie Academic and Professional, 2006). This book is one of the most definitive resources on catalyst decay.

the surface sites would change with time as shown in Figure 10-27.

Progression of sites
being poisoned

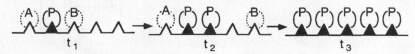

Figure 10-27 Decay by poisoning.

If we assume the rate of removal of the poison, $r_{P \cdot S}$, from the reactant gas stream onto the catalyst sites is proportional to the number of sites that are unpoisoned $(C_{t0} - C_{P \cdot S})$ and the concentration of poison in the gas phase is C_P then

$$r_{P \cdot S} = k_d(C_{t0} - C_{P \cdot S})C_P$$

where $C_{P \cdot S}$ is the concentration of poisoned sites and C_{t0} is the total number of sites initially available. Because every molecule that is adsorbed from the gas phase onto a site is assumed to poison the site, this rate is also equal to the rate of removal of total active sites (C_t) from the surface

$$-\frac{dC_t}{dt} = \frac{dC_{P \cdot S}}{dt} = r_{P \cdot S} = k_d(C_{t0} - C_{P \cdot S})C_P$$

Dividing through by C_{t0} and letting f be the fraction of the total number of sites that have been poisoned yields

$$\frac{df}{dt} = k_d(1 - f)C_P \qquad (10\text{-}108)$$

The fraction of sites available for adsorption $(1 - f)$ is essentially the activity $a(t)$. Consequently, Equation (10-108) becomes

$$\boxed{-\frac{da}{dt} = a(t)k_dC_P} \qquad (10\text{-}109)$$

A number of examples of catalysts with their corresponding catalyst poisons are given by Farrauto and Bartholomew.[29]

Packed-Bed Reactors.   In packed-bed reactors where the poison is removed from the gas phase by being adsorbed on the specific catalytic sites, the deactivation process can move through the packed bed as a wave front. Here, at the start of the operation, only those sites near the entrance to the reactor will be deactivated because the poison (which is usually present in trace amounts) is removed from the gas phase by the adsorption; consequently, the catalyst sites farther down the reactor will not be affected. However, as time continues, the sites near the entrance of the reactor become saturated, and the poison must travel farther downstream before being adsorbed (removed) from the gas phase and attaching to a site to deactivate it. Figure 10-28 shows the corresponding activity profile for this type of poisoning process. We see in Figure 10-28 that by time t_4 the entire bed has become deactivated. The corresponding overall conversion at the exit of the reactor might vary with time as shown in Figure 10-29.

---

[29] Ibid.

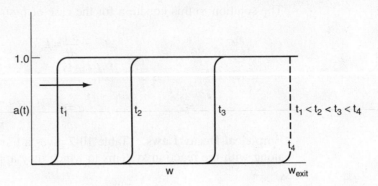

Figure 10-28   Movement of activity front in a packed bed.

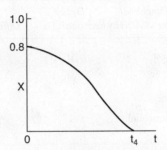

Figure 10-29   Exit conversion as a function of time.

Summary Notes

The partial differential equations that describe the movement of the reaction front shown in Figure 10-28 are derived and solved in an example on the CRE Web site, at the very end of the *Summary Notes* for Chapter 10.

Poisoning by Either Reactants or Products.   For the case where the main reactant also acts as a poison, the rate laws are

Main reaction: $A + S \longrightarrow B + S$ $-r'_A = k_A C_A^n$

Poisoning reaction: $A + S \longrightarrow A \cdot S$ $r_d = k'_d C_A^m a^q$

An example where one of the reactants acts as a poison is in the reaction of CO and H_2 over ruthenium to form methane, with

$$-r_{CO} = k a(t) C_{CO}$$

$$-\frac{da}{dt} = r_d = k'_d a(t) C_{CO}$$

Similar rate laws can be written for the case when the product B acts as a poison.
    For *separable deactivation kinetics* resulting from contacting a poison at a constant concentration C_{P_0} and no spatial variation

Separable
deactivation
kinetics

$$-\frac{da}{dt} = r_d = k'_d C_{P_0}^n a^n(t) = k_d a^n \qquad (10\text{-}110)$$

The solution to this equation for the case of first-order decay, n = 1

$$-\frac{da}{dt} = k_d a \tag{10-111}$$

is

$$a = e^{-k_d t} \tag{10-112}$$

Empirical Decay Laws.   Table 10-7 gives a number of empirical decay laws along with the reaction systems to which they apply.

TABLE 10-7   DECAY RATE LAWS

Examples of reactions with decaying catalysts and their decay laws

Functional Form of Activity	Decay Reaction Order	Differential Form	Integral Form	Examples
Linear	0	$-\dfrac{da}{dt} = \beta_0$	$a = 1 - \beta_0 t$	Conversion of *para*-hydrogen on tungsten when poisoned with oxygen[a*]
Exponential	1	$-\dfrac{da}{dt} = \beta_1 a$	$a = e^{-\beta_1 t}$	Ethylene hydrogenation on Cu poisoned with CO[b]
				Paraffin dehydrogenation on $Cr \cdot Al_2O_3$[c]
				Cracking of gas oil[d]
				Vinyl chloride monomer formation[e]
Hyperbolic	2	$-\dfrac{da}{dt} = \beta_2 a^2$	$\dfrac{1}{a} = 1 + \beta_2 t$	Vinyl chloride monomer formation[f]
				Cyclohexane dehydrogenation on Pt/Al_2O_3[g]
				Isobutylene hydrogenation on Ni[h]
Reciprocal power	$\dfrac{\beta_3 + 1}{\beta_3} = \gamma$	$-\dfrac{da}{dt} = \beta_3 a^n A_0^{1/5}$	$a = A_0 t^{-\beta_3}$	Cracking of gas oil and gasoline on clay[i]
	$\dfrac{\beta_4 + 1}{\beta_4} = n$	$-\dfrac{da}{dt} = \beta_4 a^n A_0^{1/5}$	$a = A_0 t^{-\beta_4}$	Cyclohexane aromatization on NiAl[j]

[a*]D. D. Eley and E. J. Rideal, *Proc. R. Soc. London*, A178, 429 (1941).
[b]R. N. Pease and L. Y. Steward, *J. Am. Chem. Soc.*, 47, 1235 (1925).
[c]E. F. K. Herington and E. J. Rideal, *Proc. R. Soc. London*, A184, 434 (1945).
[d]V. W. Weekman, *Ind. Eng. Chem. Process Des. Dev.*, 7, 90 (1968).
[e]A. F. Ogunye and W. H. Ray, *Ind. Eng. Chem. Process Des. Dev.*, 9, 619 (1970).
[f]A. F. Ogunye and W. H. Ray, *Ind. Eng. Chem. Process Des. Dev.*, 10, 410 (1971).
[g]H. V. Maat and L. Moscou, *Proc. 3rd Int. Congr. Catal.* (Amsterdam: North-Holland, 1965), p. 1277.
[h]A. L. Pozzi and H. F. Rase, *Ind. Eng. Chem.*, 50, 1075 (1958).
[i]A. Voorhies, Jr., *Ind. Eng. Chem.*, 37, 318 (1945); E. B. Maxted, *Adv. Catal.*, 3, 129 (1951).
[j]C. G. Ruderhausen and C. C. Watson, *Chem. Eng. Sci.*, 3, 110 (1954).
Source: J. B. Butt, Chemical Reactor Engineering–Washington, *Advances in Chemistry Series* 109 (Washington, D.C.: American Chemical Society, 1972), p. 259. Also see CES 23, 881(1968).

One should also see *Fundamentals of Industrial Catalytic Processes,* by Farrauto and Bartholomew, which contains rate laws similar to those in Table 10-7, and also gives a comprehensive treatment of catalyst deactivation.[30]

Key resource for catalyst deactivation

10.7.2 Reactors That Can Be Used to Help Offset Catalyst Decay

We will now consider three reaction systems that can be used to handle systems with decaying catalyst. We will classify these systems as those having slow, moderate, and rapid losses of catalytic activity. To offset the decline in chemical reactivity of decaying catalysts in continuous-flow reactors, the following three methods are commonly used:

Matching the reactor type with speed of catalyst decay

> • Slow decay – *Temperature–Time Trajectories* (10.7.3)
> • Moderate decay – *Moving-Bed Reactors* (10.7.4)
> • Rapid decay – *Straight-Through Transport Reactors* (10.7.5)

10.7.3 Temperature–Time Trajectories

In many large-scale reactors, such as those used for hydrotreating, and reaction systems where deactivation by poisoning occurs, the catalyst decay is relatively slow. In these continuous flow systems, constant conversion is usually necessary in order that subsequent processing steps (e.g., separation) are not upset. One way to maintain a constant conversion with a decaying catalyst in a packed or fluidized bed is to increase the reaction rate by steadily increasing the feed temperature to the reactor. Operation of a "fluidized" bed in this manner is shown in Figure 10-30.

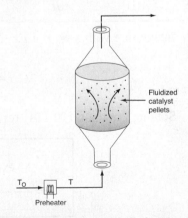

Figure 10-30   Reactor with preheater to continually increase feed temperature.

We are going to increase the feed temperature in such a manner that the reaction rate remains constant with time:

$$-r_A'(t = 0, T_0) = -r_A'(t, T) = a(t, T)[-r_A'(t = 0, T)]$$

For a first-order reaction we have

$$k(T_0)C_A = a(t, T)k(T)C_A$$

Slow rate of catalyst decay

We will neglect any variations in concentration so that the product of the activity (a) and specific reaction rate (k) is constant and equal to the specific reaction rate, k_0 at time $t = 0$ and temperature T_0; that is

$$k(T)a(t, T) = k_0 \tag{10-113}$$

The goal is to find *how* the temperature should be increased with time (i.e., the temperature–time trajectory) to maintain constant conversion. Using the Arrhenius equation to substitute for k in terms of the activation energy, E_A, gives

$$k_0 e^{(E_A/R)(1/T_0 - 1/T)} a = k_0 \tag{10-114}$$

Solving for $1/T$ yields

Gradually raising the temperature can help offset effects of catalyst decay.

$$\boxed{\frac{1}{T} = \frac{R}{E_A} \ln a + \frac{1}{T_0}} \tag{10-115}$$

The decay law also follows an Arrhenius-type temperature dependence

$$-\frac{da}{dt} = k_{d0} e^{(E_d/R)(1/T_0 - 1/T)} a^n \tag{10-116}$$

where k_{d0} = decay constant at temperature T_0, s^{-1}
E_A = activation energy for the main reaction (e.g., A $\rightarrow$ B), kJ/mol
E_d = activation energy for catalyst decay, kJ/mol

Substituting Equation (10-115) into (10-116) and rearranging yields

$$-\frac{da}{dt} = k_{d0} \exp\left(-\frac{E_d}{E_A} \ln a\right) a^n = k_{d0} a^{(n - E_d/E_A)} \tag{10-117}$$

Integrating with $a = 1$ at $t = 0$ for the case $n \neq (1 + E_d/E_A)$, we obtain

$$t = \frac{1 - a^{1 - n + E_d/E_A}}{k_{d0}(1 - n + E_d/E_A)} \tag{10-118}$$

Solving Equation (10-114) for a and substituting in (10-118) gives

$$\boxed{t = \frac{1 - \exp\left[\dfrac{E_A - nE_A + E_d}{R}\left(\dfrac{1}{T} - \dfrac{1}{T_0}\right)\right]}{k_{d0}(1 - n + E_d/E_A)}} \tag{10-119}$$

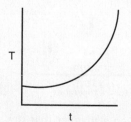

Equation (10-119) tells us how the temperature of the catalytic reactor should be increased with time in order for the reaction rate to remain constant.

In many industrial reactions, the decay rate law changes as temperature increases. In hydrocracking, the temperature–time trajectories are divided into three regimes. Initially, there is fouling of the acidic sites of the catalyst followed by a linear regime due to slow coking and, finally, accelerated coking characterized by an exponential increase in temperature. The temperature–time trajectory for a deactivating hydrocracking catalyst is shown in Figure 10-31.

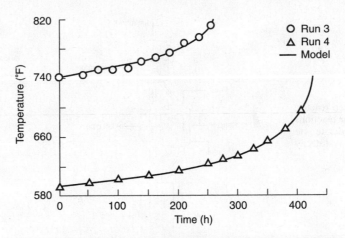

Comparing theory
and experiment

Run	E_A (cal/mol)	A (h^{-1})	E_d (cal/mol)	A_d (h^{-1})
3	30.0	0.52×10^{12}	42.145	1.54×10^{11}
4	30.0	0.52×10^{12}	37.581	2.08×10^{11}

Figure 10-31   Temperature–time trajectories for deactivating hydrocracking catalyst, runs 3 and 4. (Krishnaswamy, S., and J. R. Kittrell. Analysis of Temperature–Time Data for Deactivating Catalysts. *Industry and Engineering Chemistry Process Design and Development*, 1979, 18(3), 399–403. Copyright © 1979, American Chemical Society. Reprinted by permission.)

For a first-order decay, Krishnaswamy and Kittrell's expression, Equation (10-119), for the temperature–time trajectory reduces to

$$t = \frac{E_A}{k_{d0}E_d} \, [1 - e^{(E_d/R)(1/T - 1/T_0)}] \tag{10-120}$$

10.7.4 Moving-Bed Reactors

Reaction systems with significant catalyst decay require the continual regeneration and/or replacement of the catalyst. Two types of reactors currently in commercial use that accommodate production with decaying catalysts are the moving-bed and straight-through transport reactor. A schematic diagram of a moving-bed reactor (used for catalytic cracking) is shown in Figure 10-32.

The freshly regenerated catalyst enters the top of the reactor and then moves through the reactor as a compact packed bed. The catalyst is coked continually as it moves through the reactor until it exits the reactor into the kiln, where air is used to burn off the carbon. The regenerated catalyst is lifted from the kiln by an airstream and then fed into a separator before it is returned to the reactor. The catalyst pellets are typically between one-quarter and one-eighth in. in diameter.

The reactant feed stream enters at the top of the reactor and flows rapidly through the reactor relative to the flow of the catalyst through the reactor (Figure 10-33). If the feed rates of the catalyst and the reactants do not vary

Moving bed reactor, used for reactions with moderate rate of catalyst decay

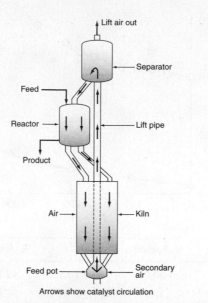

Figure 10-32 Thermofor catalytic cracking (TCC) unit. (Weekman, V. *AIChE Monogr. Ser.*, 75(11), 4 (1979). With permission of the American Institute of Chemical Engineers. Copyright © 1979 AIChE. All rights reserved.)

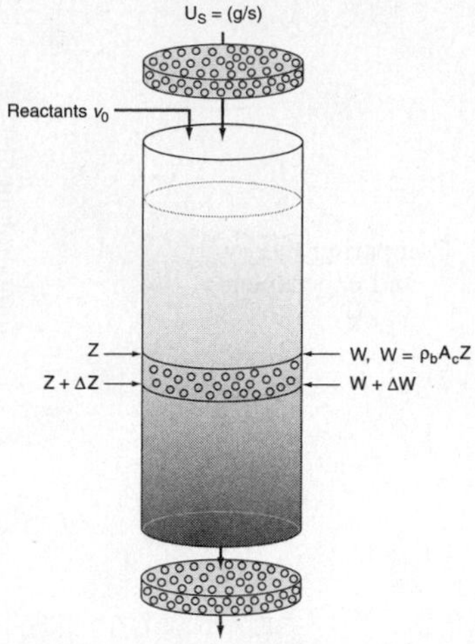

Figure 10-33 Moving-bed reactor schematic. The value of the catalyst contained in a reactor of this type is approximately $1 million.

with time, the reactor operates at steady state; that is, conditions at any point in the reactor do not change with time. The mole balance on reactant A over ΔW is

$$
\begin{bmatrix} \text{Molar} \\ \text{flow} \\ \text{rate of A in} \end{bmatrix} - \begin{bmatrix} \text{Molar} \\ \text{flow} \\ \text{rate of A out} \end{bmatrix} + \begin{bmatrix} \text{Molar} \\ \text{rate of} \\ \text{generation of A} \end{bmatrix} = \begin{bmatrix} \text{Molar} \\ \text{rate of} \\ \text{accumulation of A} \end{bmatrix}
$$

$$
F_A(W) \quad - \quad F_A(W+\Delta W) \quad + \quad r'_A \Delta W \quad = \quad 0 \qquad (10\text{-}121)
$$

Dividing by ΔW, letting ΔW approach zero, and expressing the flow rate in terms of conversion gives

Mole Balance

$$
F_{A0} \frac{dX}{dW} = -r'_A \qquad (2\text{-}17)
$$

The rate of reaction at any time t is

$$
-r'_A = a(t)[-r'_A(t=0)] = a(t)[k\, \mathrm{fn}(C_A, C_B, \ldots, C_P)] \qquad (10\text{-}93)
$$

The activity, as before, is a function of the time the catalyst has been in contact with the reacting gas stream. The decay rate law is

Decay Law

$$
-\frac{da}{dt} = k_d a^n \qquad (10\text{-}110)
$$

We now need to relate the contact time to the weight of the catalyst. Consider a point z in the reactor, where the reactant gas has passed cocurrently through a catalyst weight W. Because the solid catalyst is moving through the bed at a rate U_s (mass per unit time), the time t that the catalyst has been in contact with the gas when the catalyst reaches a point z is

$$t = \frac{W}{U_s} \tag{10-122}$$

If we now differentiate Equation (10-122)

$$dt = \frac{dW}{U_s} \tag{10-123}$$

and combine it with the decay rate law, we obtain

$$\boxed{-\frac{da}{dW} = \frac{k_d}{U_s}\, a^n} \tag{10-124}$$

The activity equation is combined with the mole balance

The design equation for moving-bed reactors

$$\boxed{\frac{dX}{dW} = \frac{a[-r'_A(t=0)]}{F_{A0}}} \tag{10-125}$$

these two coupled differential equations (i.e., Equations (10-124) and (10-125)) are numerically solved simultaneously with an ODE solver, e.g., Polymath.

Living Example Problem

Example 10–5   Catalytic Cracking in a Moving-Bed Reactor

The catalytic cracking of a gas-oil charge, A, to form C_5+ (B) and to form coke and dry gas (C) is to be carried out in a screw-type conveyor, moving-bed reactor at 900°F:

$$\text{Gas oil} \underset{k_C}{\overset{k_B}{<}} \begin{array}{l} C_{5^+} \\ \text{Dry gas, coke} \end{array}$$

This cracking of gas oil can also be written as

$$\text{A} \xrightarrow{\;k_1\;} \text{Products}$$

While pure hydrocarbons are known to crack according to a first-order rate law, the fact that the gas-oil exhibits a wide spectrum of cracking rates gives rise to the fact that the lumped cracking rate is well represented by a second-order rate law (see Web Problem CDP5-H$_B$) with the following specific reaction rate:[31]

$$-r'_A = 600\, \frac{(dm)^6}{(kg\ cat)(mol)(min)}\, C_A^2$$

The catalytic deactivation is independent of gas-phase concentration and follows a first-order decay rate law, with a decay constant, k_d of 0.72 reciprocal minutes. The feed stream is diluted with nitrogen so that as a first approximation,

[31]Estimated from V. W. Weekman and D. M. Nace, *AIChE J.*, *16*, 397 (1970).

volume changes can be neglected with reaction. The reactor contains 22 kg of catalyst that moves through the reactor at a rate of 10 kg/min. The gas-oil is fed at a rate of 30 mol/min at a concentration of 0.075 mol/dm³. Determine the conversion that can be achieved in this reactor.

Solution

1. **Mole Balance:**

$$F_{A0} \frac{dX}{dW} = a(-r'_A) \tag{E10-5.1}$$

2. **Rate Law:**

$$-r'_A = kC_A^2 \tag{E10-5.2}$$

3. **Decay Law.** First-order decay

$$-\frac{da}{dt} = k_d a$$

Using Equation (10-124), we obtain

$$-\frac{da}{dW} = \frac{k_d}{U_s} a \tag{E10-5.3}$$

Integrating

$$a = e^{-(k_d/U_s)W} \tag{E10-5.4}$$

4. **Stoichiometry.** If $v \approx v_0$, then

$$C_A = C_{A0}(1 - X) \tag{E10-5.5}$$

5. **Combining** we have

$$\frac{dX}{dt} = a \frac{kC_{A0}^2(1-X)^2}{F_{A0}} \tag{E10-5.6}$$

$$a = e^{-(k_d/U_s)W}$$

The polymath program is shown below along with a conversion profile.

POLYMATH Report

Ordinary Differential Equations

Calculated values of DEQ variables

	Variable	Initial value	Final value
1	a	1.	0.2051528
2	Ca	0.075	0.033453
3	Cao	0.075	0.075
4	Fao	30.	30.
5	k	600.	600.
6	kd	0.72	0.72
7	raprime	-3.375	-0.6714636
8	Us	10.	10.
9	W	0	22.
10	X	0	0.5539595

Differential equations

1 d(a)/d(W) = -kd*a/Us

2 d(X)/d(W) = a*(-raprime)/Fao

Explicit equations

1 Us = 10

2 kd = 0.72

3 Fao = 30

4 Cao = 0.075

5 Ca = Cao*(1-X)

6 k = 600

7 raprime = -k*Ca^2

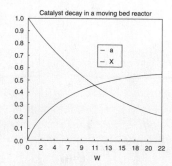

Catalyst decay in a moving bed reactor

- a
- X

Moving beds: moderate rate of catalyst decay

For the simple rate law and activity law given here, we also could have solved this problem analytically.

6. **Separating and integrating** yields

$$F_{A0} \frac{dX}{dW} = e^{-(k_d/U_s)W} kC_{A0}^2 (1-X)^2$$

$$\frac{F_{A0}}{kC_{A0}^2} \int_0^X \frac{dX}{(1-X)^2} = \int_0^W e^{-(k_d/U_s)W} dW \qquad \text{(E10-5.7)}$$

$$\frac{X}{1-X} = \frac{kC_{A0}^2 U_s}{F_{A0} k_d} (1-e^{-k_d W/U_s}) \qquad \text{(E10-5.8)}$$

7. **Numerical evaluation:**

$$\frac{X}{1-X} = \frac{0.6 \text{ dm}^6}{\text{mol} \cdot \text{g cat.} \cdot \text{min}} \times \frac{(0.075 \text{ mol/dm}^3)^2}{30 \text{ mol/min}} \frac{10,000 \text{ g cat/min}}{0.72 \text{ min}^{-1}}$$

$$\times \left(1 - \exp\left[\frac{(-0.72 \text{ min}^{-1})(22 \text{ kg})}{10 \text{ kg/min}}\right]\right)$$

$$\frac{X}{1-X} = 1.24$$

$$X = 55\%$$

If there were no catalyst decay, the conversion would be

$$\frac{X}{1-X} = \frac{kC_{A0}^2}{F_{A0}} W$$

$$= 600 \left(\frac{\text{dm}^6}{(\text{kg cat})(\text{mol})(\text{min})} \times \frac{(0.075 \text{ mol/dm}^3)^2}{(30 \text{ mol/min})} (22 \text{ ckg cat})\right)$$

$$= 7.48$$

$$X = 0.71$$

**Analysis:** The purpose of this example was to show step-by-step how to apply the algorithm to a moving-bed reactor that has been used to reduce the effects of catalyst decay that would occur in a PBR.

We will now rearrange Equation (E10-5.8) to a form more commonly found in the literature. Let λ be a dimensionless decay time

$$\lambda = k_d t = \frac{k_d W}{U_s} \qquad \text{(10-126)}$$

and Da_2 be the Damköhler number for a second-order reaction (_a reaction rate divided by a transport rate_) for a packed-bed reactor

$$Da_2 = \frac{(kC_{A0}^2)(W)}{F_{A0}} = \frac{kC_{A0}W}{v_0} \qquad \text{(10-127)}$$

Through a series of manipulations we arrive at the equation for the conversion in a *moving bed* where a second-order reaction is taking place[32]

Second-order
reaction in a
moving-bed reactor

$$X = \frac{\mathrm{Da}_2(1-e^{-\lambda})}{\lambda + \mathrm{Da}_2(1-e^{-\lambda})} \qquad (10\text{-}128)$$

Similar equations are given or can easily be obtained for other reaction orders or decay laws.

10.7.5 Straight-Through Transport Reactors (STTR)

This reactor is used for reaction systems in which the catalyst deactivates very rapidly. Commercially, the STTR is used in the production of gasoline from the cracking of heavier petroleum fractions where coking of the catalyst pellets occurs very rapidly. In the STTR, the catalyst pellets and the reactant feed enter together and are transported very rapidly through the reactor. The bulk density of the catalyst particle in the STTR is significantly smaller than in moving-bed reactors, and often the particles are carried through at the same velocity as the gas velocity. In some places the STTR is also referred to as a circulating fluidized bed (CFB). A schematic diagram is shown in Figure 10-34.

A mole balance on the reactant A over the differential reactor volume

$$\Delta V = A_\mathrm{C}\,\Delta z$$

is

$$F_\mathrm{A}\big|_z - F_\mathrm{A}\big|_{z+\Delta z} + r_\mathrm{A} A_\mathrm{C}\,\Delta z = 0$$

STTR: Used when
catalyst decay
(usually coking) is
very rapid

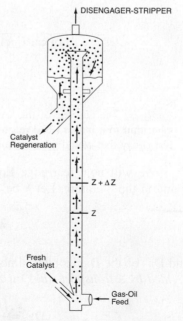

Figure 10-34 Straight-through transport reactor.

---

[32]Ibid.

Dividing by Δz and taking the limit as $\Delta z \to 0$ and recalling that $r_A = \rho_B r'_A$, we obtain

$$\frac{dF_A}{dz} = r_A A_C = r'_A \rho_B A_C \tag{10-129}$$

In terms of conversion and catalyst activity

$$\frac{dX}{dz} = \left(\frac{\rho_B A_C}{F_{A0}}\right)[-r'_A(t=0)]a(t) \tag{10-130}$$

For a catalyst particle traveling through the reactor with a particle velocity U_P, the time the catalyst pellet has been in the reactor when it reaches a height z is just

$$t = \frac{z}{U_P} \tag{10-131}$$

Substituting for time t in terms of distance z [i.e., $a(t) = a(z/U_P)$], the mole balance now becomes

$$\frac{dX}{dz} = \frac{\rho_B A_C[-r'_A(t=0)]a(z/U_P)}{F_{A0}}$$

The entering molar flow rate, F_{A0}, can be expressed in terms of the gas velocity U_g, C_{A0}, and A_C

$$F_{A0} = U_o A_C C_{A0}$$

Substituting for F_{A0}, we have

$$\boxed{\frac{dX}{dz} = \frac{\rho_B a(z/U_P)[-r'_A(t=0)]}{C_{A0} U_o}} \tag{10-132}$$

Living Example Problem

A typical cost of the catalyst in the reactor system is $1 million.

Example 10–6   Decay in a Straight-Through Transport Reactor

The vapor-phase cracking of a gas-oil is to be carried out in a straight-through transport reactor (STTR) that is 10-m high and 1.5 m in diameter. Gas-oil is a mixture of normal and branched paraffins (C_{12}–C_{40}), naphthenes, and aromatics, all of which will be lumped as a single species, A. We shall lump the primary hydrocarbon products according to distillate temperature into two respective groups, dry gas (C–C_4) B and gasoline (C_5–C_{14}) C. The reaction

$$\text{Gas-oil (g)} \longrightarrow \text{Products (g)} + \text{Coke}$$

can be written symbolically as

$$A \longrightarrow B + C + \text{Coke}$$

Both B and C are adsorbed on the surface. The rate law for a gas-oil cracking reaction on fresh catalyst can be approximated by

$$-r_A' = \frac{k'P_A}{1 + K_A P_A + K_B P_B + K_C P_C}$$

with $k' = 0.0014$ kmol/kg-cat·s·atm, $K_A = 0.05$ atm^{-1}, $K_B = 0.15$ atm^{-1}, and $K_C = 0.1$ atm^{-1}. The catalyst decays by the deposition of coke, which is produced in most cracking reactions along with the reaction products. The decay law is

$$a = \frac{1}{1 + At^{1/2}} \qquad \text{with } A = 7.6 \text{ s}^{-1/2}$$

Pure gas-oil enters at a pressure of 12 atm and a temperature of 400°C. The bulk density of catalyst in the STTR is 80 kg-cat/m^3. Plot the activity and conversion of gas-oil up the reactor for entering gas velocity $U_0 = 2.5$ m/s.

Solution

Mole Balance:

$$F_{A0} \frac{dX}{dz} = -r_A A_C$$

$$\boxed{\frac{dX}{dz} = \frac{-r_A}{U_o C_{A0}}} \qquad (E10\text{-}6.1)$$

The height of the catalyst particle at time "t" after entering the STTR is

$$z = \int_0^t U \, dt$$

Following the Algorithm

Differentiating, we can find a relation between the time the catalyst particle has been in the STTR and reached a height z, which we can use to find the activity a.

$$\frac{dt}{dz} = \frac{1}{U}$$

Rate Law:

$$-r_A = \rho_B (-r_A') \qquad (E10\text{-}6.2)$$

$$-r_A' = a[-r_A'(t=0)] \qquad (E10\text{-}6.3)$$

On fresh catalyst

$$-r_A'(t=0) = k' \frac{P_A}{1 + K_A P_A + K_B P_B + K_C P_C} \qquad (E10\text{-}6.4)$$

Combining Equations (E10-6.2) through (E10-6.4) gives

$$\boxed{-r_A = a\left(\rho_B k' \frac{P_A}{1 + K_A P_A + K_B P_B + K_C P_C}\right)} \qquad (E10\text{-}6.5)$$

Decay Law. Assuming that the catalyst particle and gas travel up the reactor at the velocity $U_P = U_g = U$, we obtain

$$t = \frac{z}{U} \qquad (E10\text{-}6.6)$$

$$a = \frac{1}{1 + A\,(z/U)^{1/2}} \qquad \text{(E10-6.7)}$$

where $U = v/A_C = v_0(1+\varepsilon X)/A_C$ and $A_C = \pi D^2/4$.

Stoichiometry (gas-phase isothermal and no pressure drop):

$$P_A = P_{A0}\,\frac{1-X}{1+\varepsilon X} \qquad \text{(E10-6.8)}$$

$$P_B = \frac{P_{A0}X}{1+\varepsilon X} \qquad \text{(E10-6.9)}$$

$$P_C = P_B \qquad \text{(E10-6.10)}$$

Parameter Evaluation:

$$\varepsilon = y_{A0}\delta = (1+1-1) = 1$$

$$U = U_0(1+\varepsilon X)$$

$$C_{A0} = \frac{P_{A0}}{RT_0} = \frac{12\ \text{atm}}{(0.082\ \text{m}^3 \cdot \text{atm/kmol} \cdot \text{K})(673\ \text{K})} = 0.22\ \frac{\text{kmol}}{\text{m}^3}$$

Equations (E10-6.1), (E10-6.5), (E10-6.7), and (E10-6.8) through (E10-6.10) are now combined and solved using an ODE solver. The Polymath program is shown in Table E10-6.1, and the computer output is shown in Figure E10-6.1.

TABLE E10-6.1   EQUATIONS FOR THE STTR: LANGMUIR–HINSHELWOOD KINETICS

Living Example Problem

POLYMATH Report

Ordinary Differential Equations

Calculated values of DEQ variables

	Variable	Initial value	Final value
1	A	7.6	7.6
2	a	1.	0.0760585
3	Cao	0.2174465	0.2174465
4	D	1.5	1.5
5	eps	1.	1.
6	Ka	0.05	0.05
7	Kb	0.15	0.15
8	Kc	0.1	0.1
9	Kca	2.7593	2.7593
10	kprime	0.0014	0.0014
11	Pa	12.	3.329127
12	Pao	12.	12.
13	Pb	0	4.335437
14	Pc	0	4.335437
15	R	0.082	0.082
16	ra	-0.84	-0.0126024
17	raprime	-0.0105	-0.0001575
18	rho	80.	80.
19	T	673.	673.
20	U	2.5	3.914117
21	Uo	2.5	2.5
22	vo	4.417875	4.417875
23	X	0	0.5656469
24	x	0	10.

Differential equations

1 d(X)/d(z) = -ra/U/Cao

Explicit equations

1 Ka = 0.05
2 kb = .15
3 Pao = 12
4 eps = 1
5 A = 7.6
6 R = 0.082
7 T = 400+273
8 rho = 80
9 kprime = 0.0014
10 D = 1.5
11 Uo = 2.5
12 Kc = 0.1
13 U = Uo*(1+eps*X)
14 Pa = Pao*(1-X)/(1+eps*X)
15 Pb = Pao*X/(1+eps*X)
16 vo = Uo*3.1416*D*D/4
17 Cao = Pao/R/T
18 Kca = Ka*R*T
19 Pc = Pb
20 a = 1/(1+A*(z/U)^0.5)
21 raprime = a*(-kprime*Pa/(1+Ka*Pa+Kb*Pb+Kc*Pc))
22 ra = rho*raprime

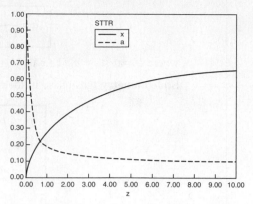

Figure E10-6.1   Activity and conversion profiles.

Analysis: In this example we applied the algorithm to a STTR in which the gas velocity and, hence, particle velocity increases as we move through the reactor. The decay is quite rapid and the activity is only 15% of its initial value at $z = 3$ m into the reactor and the conversion begins to plateau at $z = 6$ m at 60% conversion. If there were no catalyst decay (a = 0) the conversion would have been 97%.

Closure. After reading this chapter, the reader should be able to discuss the steps in a heterogeneous reaction (adsorption, surface reaction, and desorption) and describe what is meant by a rate-limiting step. The differences between molecular adsorption and dissociated adsorption should be explained by the reader, as should the different types of surface reactions (single site, dual site, and Eley–Rideal). Given heterogeneous reaction-rate data, the reader should be able to analyze the data and develop a rate law for Langmuir–Hinshelwood kinetics. The reader should also be able to discriminate between rate laws to find which rate law best fits the data. After evaluating the rate-law parameters, the reader can proceed to the design of PBRs and fluidized CSTRs.

Applications of CRE in the electronics industry were discussed and readers should be able to describe the analogy between Langmuir–Hinshelwood kinetics and chemical vapor deposition (CVD), and to derive a rate law for CVD mechanisms.

Because of the high temperatures and harsh environment, catalysts do not maintain their original activity and the catalysts decay during the course of the reaction. The reader should be able to discuss the three basic types of catalyst decay (sintering, coking or fouling, and poisoning). In addition, the reader should be able to suggest reactors (e.g., moving bed) to use to offset catalyst decay and should be able to carry out calculations to predict the conversion.

SUMMARY

1. Types of adsorption:
   a. Chemisorption
   b. Physical adsorption

2. The **Langmuir isotherm** relating the concentration of species A on the surface to the partial pressure of A in the gas phase is

$$C_{A \cdot S} = \frac{K_A C_t P_A}{1 + K_A P_A}$$

(S10-1)

3. The sequence of steps for the solid-catalyzed isomerization

$$A \longrightarrow B$$

(S10-2)

   is:

   a. **Mass transfer of A** from the bulk fluid to the external surface of the pellet
   b. **Diffusion of A** into the interior of the pellet
   c. **Adsorption of A** onto the catalytic surface
   d. **Surface reaction of A** to form **B**
   e. **Desorption of B** from the surface
   f. **Diffusion of B** from the pellet interior to the external surface
   g. **Mass transfer of B** away from the solid surface to the bulk fluid

4. Assuming that mass transfer is not rate-limiting, the rate of adsorption is

$$r_{AD} = k_A \left(C_v P_A - \frac{C_{A \cdot S}}{K_A} \right)$$

(S10-3)

   The rate of surface reaction is

$$r_S = k_S \left(C_{A \cdot S} - \frac{C_{B \cdot S}}{K_S} \right)$$

(S10-4)

   The rate of desorption is

$$r_D = k_D (C_{B \cdot S} - K_B P_B C_v)$$

(S10-5)

   At steady state

$$-r'_A = r_{AD} = r_S = r_D$$

(S10-6)

   If there are no inhibitors present, the total concentration of sites is

$$C_t = C_v + C_{A \cdot S} + C_{B \cdot S}$$

(S10-7)

5. If we assume that the surface reaction is rate-limiting, we set

$$\frac{r_{AD}}{k_A} \simeq 0 \qquad \frac{r_D}{k_D} \simeq 0$$

and solve for $C_{A \cdot S}$ and $C_{B \cdot S}$ in terms of P_A and P_B. After substitution of these quantities in Equation (S10-4), the concentration of vacant sites is eliminated with the aid of Equation (S10-7)

$$-r_A' = r_S = \frac{\overset{k}{\overbrace{C_t k_S K_A}}(P_A - P_B / K_P)}{1 + K_A P_A + K_B P_B} \tag{S10-8}$$

Recall that the equilibrium constant for desorption of species B is the reciprocal of the equilibrium constant for the adsorption of species B

$$K_B = \frac{1}{K_{DB}} \tag{S10-9}$$

and the thermodynamic equilibrium constant, K_P, is

$$K_P = K_A K_S / K_B \tag{S10-10}$$

6. Chemical vapor deposition

$$SiH_4(g) \underset{\longleftarrow}{\overset{\longrightarrow}{}} SiH_2(g) + H_2(g) \tag{S10-11}$$

$$SiH_2(g) + S \longrightarrow SiH_2 \cdot S \tag{S10-12}$$

$$SiH_2 \cdot S \longrightarrow Si(s) + H_2(g) \tag{S10-13}$$

$$r_{Dep} = \frac{k P_{SiH_4}}{P_{H_2} + K P_{SiH_4}} \tag{S10-14}$$

7. **Catalyst deactivation.** The catalyst activity is defined as

$$a(t) = \frac{-r_A'(t)}{-r_A'(t = 0)} \tag{S10-15}$$

The rate of reaction at any time t is

$$-r_A' = a(t)k(T) \, fn(C_A, C_B, ..., C_P) \tag{S10-16}$$

The rate of catalyst decay is

$$r_d = -\frac{da}{dt} = p[a(t)] k_d(T) g(C_A, C_B, ..., C_P) \tag{S10-17}$$

For first-order decay

$$p(a) = a \tag{S10-18}$$

For second-order decay

$$p(a) = a^2 \tag{S10-19}$$

8. For slow catalyst decay, the idea of a **temperature–time trajectory** is to increase the temperature in such a way that the rate of reaction remains constant.

9. The coupled differential equations to be solved for a **moving-bed reactor** are

$$F_{A0} \frac{dX}{dW} = a(-r_A') \tag{S10-20}$$

For nth-order activity decay and m order in a gas-phase concentration of species i

$$-\frac{da}{dW} = \frac{k_d a^n C_i^m}{U_s} \qquad \text{(S10-21)}$$

$$t = \frac{W}{U_s} \qquad \text{(S10-22)}$$

10. The coupled differential equations to be solved in a **straight-through transport reactor** for the case when the particle and gas velocities, U, are identical are

$$\frac{dX}{dz} = \frac{a(t)[-r'_A(t=0)]}{U_g}\left(\frac{\rho_b A_c}{C_{A0}}\right) \qquad \text{(S10-23)}$$

$$t = \frac{z}{U_g} \qquad \text{(S10-24)}$$

For coking

$$a(t) = \frac{1}{1 + At^{1/2}} \qquad \text{(S10-25)}$$

ODE SOLVER ALGORITHM

The isomerization $A \rightarrow B$ is carried out over a decaying catalyst in a *moving-bed reactor*. Pure A enters the reactor and the catalyst flows through the reactor at a rate of 2.0 kg/s.

$\dfrac{dX}{dW} = \dfrac{-r'_A}{F_{A0}}$	$k = 0.1 \text{ mol/(kg-cat} \cdot \text{s} \cdot \text{atm)}$
$r'_A = \dfrac{-akP_A}{1 + K_A P_A}$	$K_A = 1.5 \text{ atm}^{-1}$
$\dfrac{da}{dW} = \dfrac{-k_d a^2 P_B}{U_s}$	$k_d = \dfrac{0.75}{\text{s} \cdot \text{atm}}$
$P_A = P_{A0}(1 - X)p$	$F_{A0} = 10 \text{ mol/s}$
$P_B = P_{A0} X p$	$P_{A0} = 20 \text{ atm}$
$\dfrac{dp}{dW} = -\dfrac{\alpha}{2p}$	$U_s = 2.0 \text{ kg-cat/s}$
$\alpha = 0.0019 \text{ kg}^{-1}$	$W_f = 500 \text{ kg-cat}$

CRE WEB SITE MATERIALS

- **Expanded Material**
    1. *Puzzle Problem "What's Wrong with this Solution?"*
    2. *Turn Over Frequency (TOF)*
    3. *Algorithm When More than One of the Three Steps in a Catalytic Reaction Mechanism (Adsorption, Surface Reaction, Desorption) Is Rate Limiting*
    4. *Additional Homework Problems*

- **Learning Resources**
    1. *Summary Notes for Chapter 10*
    2. *Interactive Computer Modules*
        A.   Heterogeneous Catalysis

Summary Notes

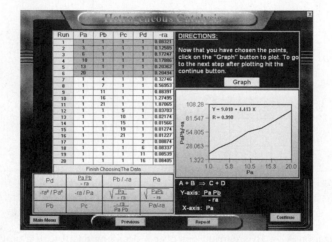

    3. *Solved Problems*
        Example Chapter 10 CRE Web Site: 10-1 Analysis of a Heterogeneous Reaction [Class Problem University of Michigan]
        Example Chapter 10 CRE Web Site: 10-2 Least Squares Analysis to Determine the Rate Law Parameters k, k_T, and k_B
        Example Chapter 10 CRE Web Site: 10-3 Decay in a Straight-Through Reactor
        Example Chapter 10 CRE Web Site: 10-4 Catalyst Poisoning in a Batch Reactor

Solved Problems

- **Living Example Problems**
    1. *Example 10-2 Regression Analysis to Determine Model Parameters*
    2. *Example 10-3 Fixed-Bed Reactor Design*
    3. *Example 10-4 Model Discrimination*
    4. *Example 10-5 Catalyst Decay in a Batch Reactor*
    5. *Example 10-6 Catalyst Decay in a Fluidized Bed Modeled as a CSTR*
    6. *Example 10-8 Decay in a Straight-Through Transport Reactor*

Living Example Problem

- **Professional Reference Shelf**
    R10.1. *Classification of Catalysts*
    R10.2. *Hydrogen Adsorption*
        A.   Molecular Adsorption
        B.   Dissociative Adsorption
    R10.2. *Analysis of Catalyst Decay Laws*
        A.   Integral Method
        B.   Differential Method
    R10.3. *Etching of Semiconductors*
        A.   Dry Etching
        B.   Wet Etching
        C.   Dissolution Catalysis

Reference Shelf

R10.4. *Catalyst Deactivation*
   A. Type of Catalyst Deactivation
   B. Temperature–Time Trajectories
   C. Moving-Bed Reactors
   D. Straight-Through Transport Reactors

> ### After Reading Each Page in This Book, Ask Yourself a Question About What You Read

QUESTIONS AND PROBLEMS

The subscript to each of the problem numbers indicates the level of difficulty: A, least difficult; D, most difficult.

$$A = \bullet \quad B = \blacksquare \quad \underline{C = \blacklozenge} \quad D = \blacklozenge\blacklozenge$$

Questions

Q10-1$_A$ Read over the problems at the end of this chapter. Make up an original problem that uses the concepts presented in this chapter. See Problem P5-1$_A$ for guidelines. To obtain a solution:
   (a) Create your data and reaction.
   (b) Use a real reaction and real data.
   The journals listed at the end of Chapter 1 may be useful for part **(b)**.
   (c) Choose an FAQ from Chapter 10 and explain why it was most helpful.
   (d) Listen to the audios  on the *Summary Notes*, pick one, and explain why it was most helpful.

Problems

P10-1$_B$ **(a)** **Example 10-1.** Plot and analyze (1) the ratio of toluene-occupied sites to benzene-occupied sites, (2) the fraction of vacant sites, and (3) the fraction of benzene-occupied sites as a function of conversion at 1 atm.
   (b) **Example 10-2.** (1) What if the entering pressure were increased to 80 atm or reduced 1 atm, how would your answers change? (2) What if the molar flow rate were reduced by 50%, how would X and y change? (3) What catalyst weight would be required for 60% conversion?
   (c) **Example 10-3.** (1) How would your answers change if the following data for run 10 were incorporated in your regression table?

   $$-r_E' = 0.8 \text{ mol/kg-cat} \cdot \text{s}, \ P_E = 0.5 \text{ atm}, \ P_{EA} = 15 \text{ atm}, \ P_H = 2 \text{ atm}.$$

   (2) How do the rate laws **(e)** and **(f)**

$$\text{(e)} \ -r_E' = \frac{kP_EP_H}{\left(1+K_AP_{EA}+K_EP_E\right)^2} \quad \text{(f)} \ -r_E' = \frac{kP_HP_E}{1+K_AP_{EA}}$$

   compare with the other rate laws used to model the data?
   (3) Write a question for this problem that involves critical thinking and explain why it involves critical thinking.
   (d) **Example 10-4.** Vary k and k_d, and describe what you find.
   (e) **Example 10-5.** (1) Sketch X vs. t for various values of k_d and k. Pay particular attention to the ratio k/k_d. (2) Repeat (1) for this example (i.e., the plotting of X vs. t) for a second-order reaction with ($C_{A0} = 1$ mol/dm^3) and first-order decay. (3) Repeat (2) for this example for a first-order reaction and first-order decay. (4) Repeat (1) for this example for a second-order reaction ($C_{A0} = 1$ mol/dm^3) and a second-order decay.

(f)   **Example 10-7.** (1) What if the solids and reactants entered from opposite ends of the reactor? How would your answers change? (2) What if the decay in the moving bed were second order? By how much must the catalyst charge, U_s, be increased to obtain the same conversion? (3) What if $\varepsilon = 2$ (e.g., A → 3B) instead of zero, how would the results be affected?

(g)   **Example 10-8.** (1) What if you varied the parameters P_{A0}, U_g, A, and k' in the STTR? What parameter has the greatest effect on either increasing or decreasing the conversion? Ask questions such as: What is the effect of varying the ratio of k to U_g or of k to A on the conversion? Make a plot of conversion versus distance as U_g is varied between 0.5 and 50 m/s. Sketch the activity and conversion profiles for $U_g = 0.025, 0.25, 2.5,$ and 25 m/s. What generalizations can you make? Plot the exit conversion and activity as a function of gas velocity between velocities of 0.02 and 50 m/s. What gas velocity do you suggest operating at? What is the corresponding entering volumetric flow rate? What concerns do you have operating at the velocity you selected? Would you like to choose another velocity? If so, what is it?

(h)   What if you were asked to sketch the temperature–time trajectories and to find the catalyst lifetimes for first- and for second-order decay when $E_A = 35$ kcal/mol, $E_d = 10$ kcal/mol, $k_{d0} = 0.01$ day^{-1}, and $T_0 = 400$ K? How would the trajectory of the catalyst lifetime change if $E_A = 10$ kcal/mol and $E_d = 35$ kcal/mol? At what values of k_{d0} and ratios of E_d to E_A would temperature–time trajectories not be effective? What would your temperature–time trajectory look like if $n = 1 + E_d/E_A$?

(i)   Write a question for this problem that involves critical thinking and explain why it involves critical thinking.

P10-2$_A$ Download the Interactive Computer Games (**ICG**) from the CRE Web site. Play the game and then record your performance number for the game, which indicates your mastery of the material. Your professor has the key to decode your performance number. (This ICG is a little longer than the other ICGs.) ICG Heterogeneous Catalysis Performance # _____.

P10-3$_A$ t-Butyl alcohol (TBA) is an important octane enhancer that is used to replace lead additives in gasoline [*Ind. Eng. Chem. Res.*, 27, 2224 (1988)]. TBA was produced by the liquid-phase hydration (W) of isobutene (I) over an Amberlyst-15 catalyst. The system is normally a multiphase mixture of hydrocarbon, water, and solid catalysts. However, the use of cosolvents or excess TBA can achieve reasonable miscibility.

The reaction mechanism is believed to be

$$I + S \rightleftharpoons I \cdot S \tag{P10-3.1}$$

$$W + S \rightleftharpoons W \cdot S \tag{P10-3.2}$$

$$W \cdot S + I \cdot S \rightleftharpoons TBA \cdot S + S \tag{P10-3.3}$$

$$TBA \cdot S \rightleftharpoons TBA + S \tag{P10-3.4}$$

Derive a rate law assuming:

(a)   The surface reaction is rate-limiting.

(b)   The adsorption of isobutene is limiting.

(c)   The reaction follows Eley–Rideal kinetics

$$I \cdot S + W \longrightarrow TBA \cdot S \tag{P10-3.5}$$

and the surface reaction is limiting.

(d)   Isobutene (I) and water (W) are adsorbed on different sites.

$$I + S_1 \rightleftharpoons I \cdot S_1 \tag{P10-3.6}$$

$$W + S_2 \rightleftharpoons W \cdot S_2 \tag{P10-3.7}$$

TBA is *not* on the surface, and the surface reaction is rate-limiting.

$$\left[Ans.:\ r'_{TBA} = -r'_I = \frac{k\,[C_I C_W - C_{TBA}/K_c]}{(1 + K_W C_W)(1 + K_I C_I)} \right]$$

(e)   What generalization can you make by comparing the rate laws derived in parts **(a)** through **(d)**?

P10-4$_B$ Consider the catalytic reaction as a function of the initial partial pressures

$$2A \xrightleftharpoons{} B + C$$

The rate of disappearance of species A was obtained in a differential reactor and is shown below.

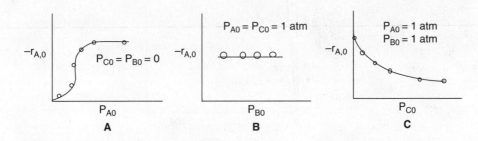

(a) What species are on the surface?
(b) What does Figure B tell you about the reversibility and what's adsorbed on the surface?
(c) Derive the rate law and suggest a rate-liming step consistent with the above figures.
(d) How would you plot your data to linearize the initial rate data in Figure A?
(e) Assuming pure A is fed, and the adsorption constants for A and C are $K_A = 0.5$ atm^{-1} and $K_C = 0.25$ atm^{-1} respectively, at what conversion are the number of sites with A adsorbed on the surface and C adsorbed on the surface equal?

P10-5$_A$ The rate law for the hydrogenation (H) of ethylene (E) to form ethane (A) over a cobalt-molybdenum catalyst [*Collection Czech. Chem. Commun.*, 51, 2760 (1988)] is

$$-r'_E = \frac{kP_E P_H}{1 + K_E P_E}$$

(a) Suggest a mechanism and rate-limiting step consistent with the above rate law.
(b) What was the most difficult part in finding the mechanism?

P10-6$_B$ The formation of propanol on a catalytic surface is believed to proceed by the following mechanism

$$O_2 + 2S \xrightleftharpoons{cat} 2O \cdot S$$

$$C_3H_6 + O \cdot S \rightarrow C_3H_5OH \cdot S$$

$$C_3H_5OH \cdot S \xrightleftharpoons{} C_3H_5OH + S$$

Suggest a rate-limiting step and derive a rate law.

P10-7$_B$ The dehydration of *n*-butyl alcohol (butanol) over an alumina-silica catalyst was investigated by J. F. Maurer (Ph.D. thesis, University of Michigan). The data in Figure P10-7$_B$ were obtained at 750°F in a modified differential reactor. The feed consisted of pure butanol.

(a) Suggest a mechanism and rate-limiting step that are consistent with the experimental data.
(b) Evaluate the rate-law parameters.
(c) At the point where the initial rate is a maximum, what is the fraction of vacant sites? What is the fraction of occupied sites by both A and B?
(d) What generalizations can you make from studying this problem?
(e) Write a question that requires critical thinking and then explain why your question requires critical thinking. *Hint:* See Preface Section I.
(f) Apply one or more of the six ideas in Preface Table P-4, page xxviii, to this problem.

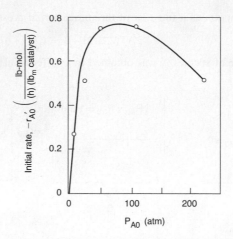

Figure P10-7$_B$ Initial reaction rate as function of initial partial pressure of butanol.

P10-8$_B$ The catalytic dehydration of methanol (ME) to form dimethyl ether (DME) and water was carried out over an ion exchange catalyst [K. Klusacek, *Collection Czech. Chem. Commun.*, 49, 170 (1984)]. The packed bed was initially filled with nitrogen, and at $t = 0$. The N_2 feed is switched to pure methanol vapor entering the reactor at 413 K, 100 kPa, and 0.2 cm^3/s. The following partial pressures were recorded at the exit to the differential reactor containing 1.0 g of catalyst in 4.5 cm^3 of reactor volume.

$$2CH_3OH \longrightarrow CH_3OCH_3 + H_2O$$

$$2ME \longrightarrow DME + H_2O$$

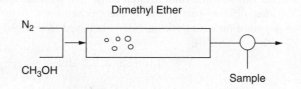

	t(s)						
	0	10	50	100	150	200	300
P_{N_2} (kPa)	100	50	10	2	0	0	0
P_{ME} (kPa)	0	2	15	23	25	26	26
P_{H_2O} (kPa)	0	10	15	30	35	37	37
P_{DME} (kPa)	0	38	60	45	40	37	37

Use parts **(a)** through **(f)** to lead you to suggest a mechanism, rate-limiting step, and rate law consistent with this data.

(a) Using the data above, sketch the exit concentrations as a function of time.

(b) Which species took longer than others to exit the reactor in the gas phase? What could have caused this difference in exit times?

(c) What species are adsorbed on the surface?

(d) Are any species not adsorbed on the surface? If so, which ones?

(e) Which set of figures, (1)-(4) below, correctly describes the functionality of the chemical reaction rate with the partial pressures P_W, P_{DME}, and P_{ME}?

(f) Derive a rate law for the catalytic dehydration of methanol. Dimethyl Either 2ME $\rightarrow$ DME + W.

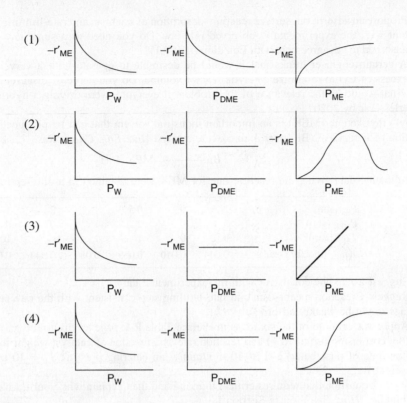

P10-9$_B$ In 1981, the U.S. government put forth the following plan for automobile manufacturers to reduce emissions from automobiles over the next few years.

	Year		
	1981	1993	2010
Hydrocarbons	0.41	0.25	0.125
CO	3.4	3.4	1.7
NO	1.0	0.4	0.2

All values are in grams per mile. An automobile emitting 3.74 lb$_m$ of CO and 0.37 lb$_m$ of NO on a journey of 1000 miles would meet the current government requirements.

Web Hint

To remove oxides of nitrogen (assumed to be NO) from automobile exhaust, a scheme has been proposed that uses unburned carbon monoxide (CO) in the exhaust to reduce the NO over a solid catalyst, according to the reaction

$$CO + NO \xrightarrow{\text{catalyst}} \text{Products (N}_2, CO_2)$$

Experimental data for a particular solid catalyst indicate that the reaction rate can be well represented over a large range of temperatures by

$$-r'_N = \frac{kP_N P_C}{(1 + K_1 P_N + K_2 P_C)^2} \tag{P10-9.1}$$

where P_N = gas-phase partial pressure of NO
               P_C = gas-phase partial pressure of CO
k, K_1, K_2 = coefficients depending only on temperature

(a) Propose an adsorption–surface reaction–desorption mechanism and rate-limiting step that are consistent with the experimentally observed rate law. Do you need to assume any species are weakly adsorbed to get agreement with Equation (P10-9.1)?

(b) A certain engineer thinks that it would be desirable to operate with a very large stoichiometric excess of CO to minimize catalytic reactor volume. Do you agree or disagree? Explain.

(c) What would be the relevance of the problem if everyone were driving a hybrid by 2018? A driverless car by 2020?

P10-10$_B$ Methyl ethyl ketone (MEK) is an important industrial solvent that can be produced from the dehydrogenation of butan-2-ol (Bu) over a zinc oxide catalyst [*Ind. Eng. Chem. Res.*, 27, 2050 (1988)]:

$$Bu \xrightarrow{\text{catalyst}} MEK + H_2$$

The following data giving the reaction rate for MEK were obtained in a differential reactor at 490°C.

P_{Bu} (atm)	2	0.1	0.5	1	2	1
P_{MEK} (atm)	5	0	2	1	0	0
P_{H_2} (atm)	0	0	1	1	0	10
r'_{MEK} (mol/h·g-cat.)	0.044	0.040	0.069	0.060	0.043	0.059

(a) Suggest a rate law consistent with the experimental data.

(b) Suggest a reaction mechanism and rate-limiting step consistent with the rate law. *Hint:* Some species might be weakly adsorbed.

(c) Apply one or more of the six ideas in Preface Table P-4, page xxviii, to this problem.

(d) Plot conversion (up to 90%) and reaction rate as a function of catalyst weight for an entering molar flow rate of pure butan-2-ol of 10 mol/min at an entering pressure $P_0 = 10$ atm up to a catalyst weight $W_{max} = 23$ kg.

(e) Write a question that requires critical thinking and then explain why your question requires critical thinking. *Hint:* See Preface Section I.

(f) Repeat part **(d)**, accounting for pressure drop and $\alpha = 0.03$ kg^{-1}. Plot p and X as a function of catalyst weight down the reactor.

P10-11$_B$ Cyclohexanol was passed over a catalyst to form water and cyclohexene:

$$Cyclohexanol \xrightarrow{\text{catalyst}} Water + Cyclohexene$$

It is suspected that the reaction may involve a dual-site mechanism, but it is not known for certain. It is believed that the adsorption equilibrium constant for cyclohexanol is around 1.0 and is roughly one or two orders of magnitude greater than the adsorption equilibrium constants for the other compounds. Using these data:

TABLE P10-11$_B$ DATA FOR CATALYTIC FORMATION OF CYCLOHEXENE

Run	Reaction Rate (mol/dm³·s) × 10⁵	Partial Pressure of Cyclohexanol (atm)	Partial Pressure of Cyclohexene (atm)	Partial Pressure of Steam (H₂O) (atm)
1	3.3	1	1	1
2	1.05	5	1	1
3	0.565	10	1	1
4	1.826	2	5	1
5	1.49	2	10	1
6	1.36	3	0	5
7	1.08	3	0	10
8	0.862	1	10	10
9	0	0	5	8
10	1.37	3	3	3

(a) Suggest a rate law and mechanism consistent with the data given here.

(b) Determine the values of the rate-law parameters. (*Ind. Eng. Chem. Res.*, 32, 2626–2632.)

(c) Why do you think estimates of the rate-law parameters were given?

(d) For an entering molar flow rate of cychlohexanol of 10 mol/s at a partial pressure of 15 atm, what catalyst weight is necessary to achieve 85% conversion when the bulk density is 1500 gm/dm³?

P10-12$_B$ Experimental data for the gas-phase catalytic reaction

$$A + B \rightarrow C$$

is shown below. The limiting step in the reaction is known to be irreversible, so that the overall reaction is irreversible. The reaction was carried out in a differential reactor to which A, B, and C were all fed.

Run Number	P_A (atm)	P_B (atm)	P_C (atm)	Reaction rate (mol)/(g-cat • s)
1	1	1	2	0.114
2	1	10	2	1.140
3	10	1	2	0.180
4	1	20	2	2.273
5	1	20	10	0.926
6	20	1	2	0.186
7	0.1	1	2	0.0243

(a) Suggest a rate law consistent with the experimental data. *Hint:* Sketch $(-r_A')$ as a function of P_A, as a function of P_B, and as a function of P_C.

(b) From your rate expression, which species can you conclude are adsorbed on the surface?

(c) Suggest a rate law and then show that your mechanism is constant with the rate law in part (a).

(d) For an entering partial pressure of A of 2 atm in a PBR, what is the ratio of A to C sites at 80% conversion of A?

(e) At what conversion are the number of A and C sites equal?

(f) What reactor volume is necessary to achieve 90% conversion of A for a stoichiometric feed and flow of A 2 mol/s?

If necessary, feel free to use none, any, or all of the following parameter values:

$$k = 2.5 \frac{\text{mol}}{\text{atm}^2 \text{g-cat} \cdot \text{s}}, \; K_A = 4 \text{ atm}^{-1}, \; K_C = 13 \text{ atm}^{-1}, \; K_I = 10 \text{ atm}^{-1}$$

P10-13$_B$ Solar Energy Capture: Water Splitting. Hydrogen and O_2 can be combined in fuel cells to generate electricity. Solar energy can be used to split water to generate the raw reactants H_2 and O_2 for fuel cells. One method of solar thermal reduction is with $NiFe_2O_4$ in the sequence

$$\text{Step (1) Solar Energy} + \overbrace{NiFe_2O_4}^{\text{Surface }(S)} \rightarrow \overbrace{1.2FeO + 0.4Fe_2O_3 + NiO}^{\text{Solid Solution}(S')} + 0.3O_2 \uparrow$$

$$\text{Step (2)} \; \overbrace{1.2FeO + 0.4Fe_2O_3 + NiO}^{\text{Solid Solution}(S')} + 0.6H_2O \rightarrow \overbrace{NiFe_2O_4}^{\text{Surface }(S)} + 0.6H_2 \uparrow$$

$$\overset{h\upsilon}{\underset{\downarrow\downarrow}{}}$$
$$\boxed{S} \rightarrow \boxed{S'} + 0.3\, O_2 \uparrow$$

$$0.6\, H_2O + \boxed{S'} \rightarrow \boxed{S} + 0.6\, H_2 \uparrow$$

We note $NiFe_2O_4$ is regenerated in this process.[33]

(a) Derive a rate law for Step (2), assuming that water adsorbs on the solid solution as a single-site mechanism and that the reaction is irreversible.

(b) Repeat **(a)** when the reaction is reversible and the solid solution adsorption site for water (S') is different than the $NiFe_2O_4$ site for the adsorption of H_2, (S).

$$H_2O + S' \underset{\leftarrow}{\overset{\rightarrow}{\rightleftarrows}} S' \cdot H_2O$$

$$S' \cdot H_2O \underset{\leftarrow}{\overset{\rightarrow}{\rightleftarrows}} S \cdot H_2 + \frac{1}{2}O_2$$

$$H_2 \cdot S \underset{\leftarrow}{\overset{\rightarrow}{\rightleftarrows}} S + H_2$$

(c) How would your rate law change if we included Step 1?

$$S + h\upsilon \underset{\leftarrow}{\overset{\rightarrow}{\rightleftarrows}} S' \cdot O_2$$

$$S' \cdot O_2 \underset{\leftarrow}{\overset{\rightarrow}{\rightleftarrows}} S' + O_2$$

P10-14$_A$ Vanadium oxides are of interest for various sensor applications, owing to the sharp metal–insulator transitions they undergo as a function of temperature, pressure, or stress. Vanadium triisopropoxide (VTIPO) was used to grow vanadium oxide films by *chemical vapor deposition* [*J. Electrochem. Soc.*, 136, 897 (1989)]. The deposition rate as a function of VTIPO pressure for two different temperatures follows.

_T = 120°C:_

Growth Rate (μm/h)	0.004	0.015	0.025	0.04	0.068	0.08	0.095	0.1
VTIPO Pressure (torr)	0.1	0.2	0.3	0.5	0.8	1.0	1.5	2.0

_T = 200°C:_

Growth Rate (μm/h)	0.028	0.45	1.8	2.8	7.2
VTIPO Pressure (torr)	0.05	0.2	0.4	0.5	0.8

In light of the material presented in this chapter, analyze the data and describe your results. Specify where additional data should be taken.

P10-15$_A$ Titanium dioxide is a wide-bandgap semiconductor that is showing promise as an insulating dielectric in VLSI capacitors and for use in solar cells. Thin films of TiO_2 are to be prepared by *chemical vapor deposition* from gaseous titanium tetraisopropoxide (TTIP). The overall reaction is

$$Ti(OC_3H_7)_4 \longrightarrow TiO_2 + 4C_3H_6 + 2H_2O$$

The reaction mechanism in a CVD reactor is believed to be [K. L. Siefering and G. L. Griffin, *J. Electrochem. Soc.*, 137, 814 (1990)]

$$TTIP(g) + TTIP(g) \rightleftharpoons I + P_1$$

$$I + S \rightleftharpoons I \cdot S$$

$$I \cdot S \longrightarrow TiO_2 + P_2$$

---

[33] Scheffe, J.R., J. Li, and A. W. Weimer, "A Spinel Ferrite/Hercynite Water-Splitting Redox Cycle," *International Journal of Hydrogen Energy*, 35, 3333–3340 (2010).

where I is an active intermediate and P_1 is one set of reaction products (e.g., H_2O, C_3H_6), and P_2 is another set. Assuming the homogeneous gas-phase reaction for TTIP is in equilibrium, derive a rate law for the deposition of TiO_2. The experimental results show that at 200°C the reaction is second order at low partial pressures of TTIP and zero order at high partial pressures, while at 300°C the reaction is second order in TTIP over the entire pressure range. Discuss these results in light of the rate law you derived.

P10-16$_B$ The dehydrogenation of methylcyclohexane (M) to produce toluene (T) was carried out over a 0.3% Pt/Al_2O_3 catalyst in a differential catalytic reactor. The reaction is carried out in the presence of hydrogen (H_2) to avoid coking [*J. Phys. Chem.,* 64, 1559 (1960)].

(a) Determine the model parameters for each of the following rate laws.

$$(1) \ -r'_M = kP_M^\alpha P_{H_2}^\beta \qquad (3) \ -r'_M = \frac{kP_M P_{H_2}}{(1+K_M P_M)^2}$$

$$(2) \ -r'_M = \frac{kP_M}{1+K_M P_M} \qquad (4) \ -r'_M = \frac{kP_M P_{H_2}}{1+K_M P_M + K_{H_2} P_{H_2}}$$

Use the data in Table P10-16$_B$ below.

(b) Which rate law best describes the data? *Hint:* Neither K_{H_2} or K_M can take on negative values.

(c) Where would you place additional data points?

(d) Suggest a mechanism and rate-limiting step consistent with the rate law you have chosen.

TABLE P10-16$_B$ DEHYDROGENATION OF METHYLCYCLOHEXANE

P_{H_2} (atm)	P_M (atm)	$r'_T \left(\dfrac{\text{mol toluene}}{\text{s} \cdot \text{kg-cat}} \right)$
1	1	1.2
1.5	1	1.25
0.5	1	1.30
0.5	0.5	1.1
1	0.25	0.92
0.5	0.1	0.64
3	3	1.27
1	4	1.28
3	2	1.25
4	1	1.30
0.5	0.25	0.94
2	0.05	0.41

P10-17$_A$ Sketch *qualitatively* the reactant, product, and activity profiles as a function of length at various times for a *packed-bed reactor* for each of the following cases. In addition, sketch the effluent concentration of A as a function of time. The reaction is a simple isomerization:

$$A \longrightarrow B$$

(a) Rate law: $-r'_A = kaC_A$

Decay law: $r_d = k_d aC_A$

Case I: $k_d \ll k$, Case II: $k_d = k$, Case III: $k_d \gg k$

(b) $-r'_A = kaC_A$ and $r_d = k_d a^2$

(c) $-r'_A = kaC_A$ and $r_d = k_d aC_B$

(d) Sketch similar profiles for the rate laws in parts (a) and (c) in a *moving-bed reactor* with the solids entering at the same end of the reactor as the reactant.

(e) Repeat part (d) for the case where the solids and the reactant enter at opposite ends.

P10-18$_B$ The elementary irreversible gas-phase catalytic reaction

$$A + B \xrightarrow{\quad k \quad} C + D$$

is to be carried out in a moving-bed reactor at constant temperature. The reactor contains 5 kg of catalyst. The feed is stoichiometric in A and B. The entering concentration of A is 0.2 mol/dm^3. The catalyst decay law is zero order with $k_D = 0.2$ s^{-1} and $k = 1.0$ dm^6/(mol · kg-cat · s) and the volumetric flow rate is $v_0 = 1$ dm^3/s.

(a) What conversion will be achieved for a catalyst feed rate of 0.5 kg/s?

(b) Sketch the catalyst activity as a function of catalyst weight (i.e., distance) down the reactor length for a catalyst feed rate of 0.5 kg/s.

(c) What is the maximum conversion that could be achieved (i.e., at an infinite catalyst loading rate)?

(d) What catalyst loading rate is necessary to achieve 40% conversion?

(e) At what catalyst loading rate (kg/s) will the catalyst activity be exactly zero at the exit of the reactor?

(f) What does an activity of zero mean? Can catalyst activity be less than zero?

(g) How would your answer in part (a) change if the catalyst and reactant were fed at opposite ends? Compare with part (a).

(h) Now consider the reaction to be zero order with
$k = 0.2$ mol/kg-cat · min.
The economics:
- The product sells for $160 per gram mole.
- The cost of operating the bed is $10 per kilogram of catalyst exiting the bed.

What is the feed rate of solids (kg/min) that will give the maximum profit? (*Ans.: $U_s = 4$ kg/min.*) (*Note:* For the purpose of this calculation, ignore all other costs, such as the cost of the reactant, the cost to the company of providing free lunches to workers, etc.)

P10-19$_B$ With the increasing demand for xylene in the petrochemical industry, the production of xylene from toluene disproportionation has gained attention in recent years [*Ind. Eng. Chem. Res.*, 26, 1854 (1987)]. This reaction,

$$2 \text{ Toluene} \longrightarrow \text{Benzene} + \text{Xylene}$$

$$2\text{T} \xrightarrow{\text{catalyst}} \text{B} + \text{X}$$

was studied over a hydrogen mordenite catalyst that decays with time. As a first approximation, assume that the catalyst follows second-order decay

$$r_d = k_d a^2$$

and the rate law for low conversions is

$$-r'_\text{T} = k_\text{T} P_\text{T} a$$

with $k_\text{T} = 20$ g mol/h·kg-cat·atm and $k_d = 1.6$ h^{-1} at 735 K.

(a) Compare the conversion-time curves in a batch reactor containing 5 kg-cat at different initial partial pressures (1 atm, 10 atm, etc.). The reaction volume containing pure toluene initially is 1 dm^3 and the temperature is 735 K.

(b) What conversion can be achieved in a *moving-bed reactor* containing 50 kg of catalyst with a catalyst feed rate of 2 kg/h? Toluene is fed at a pressure of 2 atm and a rate of 10 mol/min.

(c) Explore the effect of catalyst feed rate on conversion.

(d) Suppose that $E_\text{T} = 25$ kcal/mol and $E_d = 10$ kcal/mol. What would the temperature–time trajectory look like for a CSTR? What if $E_\text{T} = 10$ kcal/mol and $E_d = 25$ kcal/mol?

(e) The decay law more closely follows the equation

$$r_d = k_d P_T^2 a^2$$

with $k_d = 0.2$ atm^{-2} h^{-1}. Redo parts (b) and (c) for these conditions.

P10-20$_A$ The vapor-phase cracking of gas-oil in Example 10-6 is carried out over a different catalyst, for which the rate law is

$$-r_A' = k'P_A^2 \qquad \text{with } k' = 5 \times 10^{-5} \frac{\text{kmol}}{\text{kg-cat} \cdot \text{s} \cdot \text{atm}^2}$$

(a) Assuming that you can vary the entering pressure and gas velocity, what operating conditions would you recommend?

(b) What could go wrong with the conditions you chose?

Now assume the decay law is

$$-\frac{da}{dt} = k_D a C_{\text{coke}} \qquad \text{with } k_D = 100 \ \frac{\text{dm}^3}{\text{mol} \cdot \text{s}} \text{ at } 400°C$$

where the concentration, C_{coke}, in mol/dm^3, can be determined from a stoichiometric table.

(c) For a temperature of 400°C and a reactor height of 15 m, what gas velocity do you recommend? Explain. What is the corresponding conversion?

(d) The reaction is now to be carried in an STTR 15 m high and 1.5 m in diameter. The gas velocity is 2.5 m/s. You can operate in the temperature range between 100 and 500°C. What temperature do you choose, and what is the corresponding conversion?

(e) What would the temperature–time trajectory look like for a CSTR?

Additional information:

$E_R = 3000$ cal/mol
$E_D = 15,000$ cal/mol

P10-21$_C$ When the impurity cumene hydroperoxide is present in trace amounts in a cumene feed stream, it can deactivate the silica-alumina catalyst over which cumene is being cracked to form benzene and propylene. The following data were taken at 1 atm and 420°C in a differential reactor. The feed consists of cumene and a trace (0.08 mol %) of cumene hydroperoxide (CHP).

Benzene in Exit Stream (mol %)	2	1.62	1.31	1.06	0.85	0.56	0.37	0.24
t (s)	0	50	100	150	200	300	400	500

(a) Determine the order of decay and the decay constant. (*Ans.:* $k_d = 4.27 \times 10^{-3}$ s^{-1}.)

(b) As a first approximation (actually a rather good one), we shall neglect the denominator of the catalytic rate law and consider the reaction to be first order in cumene. Given that the specific reaction rate with respect to cumene is $k = 3.8 \times 10^3$ mol/kg fresh cat $\cdot$ s $\cdot$ atm, the molar flow rate of cumene (99.92% cumene, 0.08% CHP) is 200 mol/min, the entering concentration is 0.06 kmol/m^3, the catalyst weight is 100 kg, and the velocity of solids is 1.0 kg/min, what conversion of cumene will be achieved in a *moving-bed reactor*?

P10-22$_C$ The decomposition of spartanol to wulfrene and CO_2 is often carried out at high temperatures [*J. Theor. Exp.*, 15, 15 (2014)]. Consequently, the denominator of the catalytic rate law is easily approximated as unity, and the reaction is first order with an activation energy of 150 kJ/mol.

Fortunately, the reaction is irreversible. Unfortunately, the catalyst over which the reaction occurs decays with time on stream. The following conversion-time data were obtained in a differential reactor:

For $T = 500$ K:

t (days)	0	20	40	60	80	120
X (%)	1	0.7	0.56	0.45	0.38	0.29

For $T = 550$ K:

t (days)	0	5	10	15	20	30	40
X (%)	2	1.2	0.89	0.69	0.57	0.42	0.33

(a) If the initial temperature of the catalyst is 480 K, determine the *temperature–time trajectory* to maintain a constant conversion.

(b) What is the catalyst lifetime?

P10-23$_B$ The hydrogenation of ethylbenzene to ethylcyclohexane over a nickel-mordenite catalyst is zero order in both reactants up to an ethylbenzene conversion of 75% [*Ind. Eng. Chem. Res.*, 28 (3), 260 (1989)]. At 553 K, $k = 5.8$ mol ethylbenzene/(dm^3 of catalyst·h). When a 100-ppm thiophene concentration entered the system, the ethylbenzene conversion began to drop.

Time (h)	0	1	2	4	6	8	12
Conversion	0.92	0.82	0.75	0.50	0.30	0.21	0.10

The reaction was carried out at 3 MPa and a molar ratio of H$_2$/ETB = 10. Discuss the catalyst decay. Be quantitative where possible.

SUPPLEMENTARY READING

1. A terrific discussion of heterogeneous catalytic mechanisms and rate-controlling steps may or may not be found in

   BURGESS, THORNTON W., *The Adventures of Grandfather Frog*. New York: Dover Publications, Inc., 1915.

   MASEL, R. I., *Principles of Adsorption and Reaction on Solid Surfaces*. New York: Wiley, 1996. A great reference.

   SOMORJAI, G. A., *Introduction to Surface Chemistry and Catalysis*. New York: Wiley, 1994.

2. A truly excellent discussion of the types and rates of adsorption together with techniques used in measuring catalytic surface areas is presented in

   MASEL, R. I., *Principles of Adsorption and Reaction on Solid Surfaces*. New York: Wiley, 1996.

3. Techniques for discriminating between mechanisms and models can be found in

   BOX, G. E. P., W. G. HUNTER, and J. S. HUNTER, *Statistics for Experimenters*. New York: Wiley, 1978.

4. Examples of applications of catalytic principles to microelectronic manufacturing can be found in

   BUTT, JOHN B, *Reaction Kinetics and Reactor Design. Second Edition, Revised and Expanded*. New York: Marcel Dekker, Inc., 1999.

   DOBKIN, D. M., and M. K. ZURAW. *Principles of Chemical Vapor Deposition*. The Netherlands: Kluwer Academic Publishers, 2003.

Nonisothermal Reactor Design— 11
The Steady-State Energy
Balance and Adiabatic
PFR Applications

If you can't stand the *heat*, get out of the kitchen.
— Harry S Truman

Overview. Because most reactions are not carried out isothermally, we now focus our attention on heat effects in chemical reactors. The basic CRE algorithm of *mole balance, rate law, stoichiometry, combine*, and *evaluate* used in Chapters 1 through 10 for isothermal reactor design is still valid for the design of nonisothermal reactors; we only need to add one more step, the *energy balance*. The major difference lies in the method of evaluating the combined mole balances, rate law, and stoichiometry when temperature varies along the length of a PFR or when heat is removed from a CSTR. This chapter is arranged as follows:

- Section 11.1 shows why we need the energy balance and how it will be used to solve reactor design problems.
- Section 11.2 develops the energy balance to a point where it can be applied to different types of reactors. It then gives the end result, relating temperature and conversion or reaction rate for the main types of reactors we have been studying.
- Section 11.3 develops the user-friendly energy balances for reactors.
- Section 11.4 discusses the adiabatic operation of reactors.
- Section 11.5 shows how to determine the adiabatic equilibrium conversion and how to carry out interstage cooling.
- Section 11.6 closes the chapter with a discussion of the optimum inlet temperature to achieve the maximum conversion for adiabatic operation.

11.1 Rationale

To identify the additional information necessary to design nonisothermal reactors, we consider the following example, in which a highly exothermic reaction is carried out adiabatically in a plug-flow reactor.

Example 11–1 What Additional Information Is Required?

The first-order liquid-phase reaction

$$A \longrightarrow B$$

is carried out in a PFR. The reaction is exothermic and the reactor is operated adiabatically. As a result, the temperature will increase with conversion down the length of the reactor. Because T varies along the length of the reactor, k will also vary, which was not the case for isothermal plug-flow reactors.

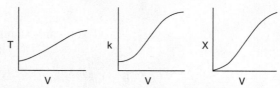

Calculate the PFR reactor volume necessary for 70% conversion and plot the corresponding profiles for X and T.

Solution

The same CRE algorithm can be applied to nonisothermal reactions as to isothermal reactions by adding one more step, ***the energy balance***.

1. Mole Balance (design equation):

$$\frac{dX}{dV} = \frac{-r_A}{F_{A0}} \tag{E11-1.1}$$

2. Rate Law:

$$-r_A = kC_A \tag{E11-1.2}$$

Recalling the Arrhenius equation,

$$k = k_1 \exp\left[\frac{E}{R}\left(\frac{1}{T_1} - \frac{1}{T}\right)\right] \tag{E11-1.3}$$

Following the Algorithm

we know that k is a function of temperature, T.

3. Stoichiometry (liquid phase): $v = v_0$

$$C_A = C_{A0}(1 - X) \tag{E11-1.4}$$

4. Combining:

$$-r_A = k_1 \exp\left[\frac{E}{R}\left(\frac{1}{T_1} - \frac{1}{T}\right)\right] C_{A0}(1 - X) \tag{E11-1.5}$$

Combining Equations (E11-1.1), (E11-1.2), and (E11-1.4), and canceling the entering concentration, C_{A0}, yields

$$\frac{dX}{dV} = \frac{k(1 - X)}{v_0} \tag{E11-1.6}$$

Combining Equations (E11-1.3) and (E11-1.6) gives us

Why we need the
energy balance

$$\frac{dX}{dV} = k_1 \exp\left[\frac{E}{R}\left(\frac{1}{T_1} - \frac{1}{T}\right)\right]\frac{1-X}{v_0} \tag{E11-1.7}$$

We see that we need another relationship relating X and T or T and V to solve this equation. *The energy balance will provide us with this relationship.*

So we add another step to our algorithm; this step is the energy balance.

5. Energy Balance:

In this step, we will find the appropriate energy balance to relate temperature and conversion or reaction rate. For example, if the reaction is *adiabatic*, we will show that for equal heat capacities, C_{P_A} and C_{P_B}, and a constant heat of reaction, ΔH_{Rx}°, the temperature-conversion relationship can be written in a form such as

T_0 = Entering
Temperature

ΔH_{Rx} = Heat of
Reaction

C_{P_A} = Heat Capacity
of species A

$$T = T_0 + \frac{-\Delta H_{Rx}^\circ}{C_{P_A}} X \tag{E11-1.8}$$

We now have all the equations we need to solve for the conversion and temperature profiles.

Analysis: The purpose of this example was to demonstrate that for nonisothermal chemical reactions we need another step in our CRE algorithm, ***the energy balance***. The energy balance allows us to solve for the reaction temperature, which is necessary in evaluating the specific reaction rate constant $k(T)$.

11.2 The Energy Balance

11.2.1 First Law of Thermodynamics

We begin with the application of the first law of thermodynamics, first to a closed system and then to an open system. A system is any bounded portion of the universe, moving or stationary, which is chosen for the application of the various thermodynamic equations. For a closed system, in which no mass crosses the system boundaries, the change in total energy of the system, $d\hat{E}$, is equal to the heat flow **to** the system, δQ, minus the work done **by** the system **on** the surroundings, δW. For a *closed system*, the energy balance is

$$d\hat{E} = \delta Q - \delta W \tag{11-1}$$

The δ's signify that δQ and δW are not exact differentials of a state function.

The continuous-flow reactors we have been discussing are *open systems* in which mass crosses the system boundary. We shall carry out an energy balance on the open system shown in Figure 11-1. For an open system in which

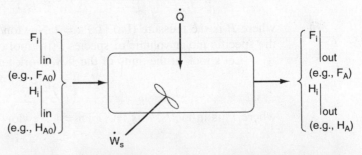

Figure 11-1   Energy balance on a well-mixed open system: schematic.

some of the energy exchange is brought about by the flow of mass across the system boundaries, the energy balance for the case of *only one* species entering and leaving becomes

$$
\begin{bmatrix}
\text{Rate of} \\
\text{accumulation} \\
\text{of energy} \\
\textit{within} \text{ the} \\
\text{system}
\end{bmatrix}
=
\begin{bmatrix}
\text{Rate of flow} \\
\text{of heat } \textit{to} \\
\text{the system} \\
\textit{from} \text{ the} \\
\text{surroundings}
\end{bmatrix}
-
\begin{bmatrix}
\text{Rate of work} \\
\textit{done} \text{ by} \\
\text{the system} \\
\textit{on} \text{ the} \\
\text{surroundings}
\end{bmatrix}
+
\begin{bmatrix}
\text{Rate of energy} \\
\text{added to the} \\
\text{system by mass} \\
\text{flow } \textit{into} \text{ the} \\
\text{system}
\end{bmatrix}
-
\begin{bmatrix}
\text{Rate of} \\
\text{energy leaving} \\
\text{the system by mass} \\
\text{flow } \textit{out} \text{ of} \\
\text{the system}
\end{bmatrix}
$$

Energy balance on an open system

$$
\frac{d\hat{E}_{\text{sys}}}{dt} = \dot{Q} - \dot{W} + F_{\text{in}}E_{\text{in}} - F_{\text{out}}E_{\text{out}} \qquad (11\text{-}2)
$$

$$
(\text{J/s}) = (\text{J/s}) - (\text{J/s}) + (\text{J/s}) - (\text{J/s})
$$

Typical units for each term in Equation (11-2) are (Joule/s).

We will assume that the contents of the system volume are well mixed, an assumption that we could relax but that would require a couple of pages of text to develop, and the end result would be the same! The unsteady-state energy balance for an open well-mixed system that has n species, each entering and leaving the system at its respective molar flow rate F_i (moles of i per time) and with its respective energy E_i (joules per mole of i), is

The starting point

$$
\boxed{\frac{d\hat{E}_{\text{sys}}}{dt} = \dot{Q} - \dot{W} + \sum_{i=1}^{n} E_i F_i \bigg|_{\text{in}} - \sum_{i=1}^{n} E_i F_i \bigg|_{\text{out}}} \qquad (11\text{-}3)
$$

We will now discuss each of the terms in Equation (11-3).

11.2.2 Evaluating the Work Term

It is customary to separate the work term, $\dot{W}$, into *flow work* and *other work*, $\dot{W}_s$. The term $\dot{W}_s$, often referred to as the *shaft work*, could be produced from such things as a stirrer in a CSTR or a turbine in a PFR. *Flow work* is work that is necessary to get the mass *into* and *out of* the system. For example, when shear stresses are absent, we write

$$
\overbrace{\dot{W} = -\sum_{i=1}^{n} F_i P \tilde{V}_i \bigg|_{\text{in}} + \sum_{i=1}^{n} F_i P \tilde{V}_i \bigg|_{\text{out}}}^{[\text{Rate of flow work}]} + \dot{W}_s \qquad (11\text{-}4)
$$

Flow work and shaft work

where P is the pressure (Pa) [1 Pa = 1 Newton/m^2 = 1 kg·m/s^2/m^2] and $\tilde{V}_i$ is the specific molar volume of species i (m^3/mol of i).

Let's look at the units of the flow-work term, which is

$$
F_i \cdot P \cdot \tilde{V}_i
$$

where F_i is in mol/s, P is Pa (1 Pa = 1 Newton/m^2), and $\tilde{V}_i$ is m^3/mol.

$$F_i \cdot P \cdot \tilde{V}_i \ [=] \ \frac{\text{mol}}{\text{s}} \cdot \frac{\text{Newton}}{\text{m}^2} \cdot \frac{\text{m}^3}{\text{mol}} = (\text{Newton} \cdot \text{m}) \cdot \frac{1}{\text{s}} = \text{Joules/s} = \text{Watts}$$

We see that the units for flow work are consistent with the other terms in Equation (11-3), i.e., J/s.

In most instances, the flow-work term is combined with those terms in the energy balance that represent the energy exchange by mass flow across the system boundaries. Substituting Equation (11-4) into (11-3) and grouping terms, we have

$$\frac{d\hat{E}_{\text{sys}}}{dt} = \dot{Q} - \dot{W}_s + \left.\sum_{i=1}^{n} F_i(E_i + P\tilde{V}_i)\right|_{\text{in}} - \left.\sum_{i=1}^{n} F_i(E_i + P\tilde{V}_i)\right|_{\text{out}} \qquad (11\text{-}5)$$

The energy E_i is the sum of the internal energy (U_i), the kinetic energy ($u_i^2/2$), the potential energy (gz_i), and any other energies, such as electric or magnetic energy or light

$$E_i = U_i + \frac{u_i^2}{2} + gz_i + \text{other} \qquad (11\text{-}6)$$

In almost all chemical reactor situations, the kinetic, potential, and "other" energy terms are negligible in comparison with the enthalpy, heat transfer, and work terms, and hence will be omitted; that is

$$E_i = U_i \qquad (11\text{-}7)$$

We recall that the enthalpy, H_i (J/mol), is defined in terms of the internal energy U_i (J/mol), and the product $P\tilde{V}_i$ (1 Pa$\cdot$m^3/mol = 1 J/mol):

Enthalpy

$$H_i = U_i + P\tilde{V}_i \qquad (11\text{-}8)$$

Typical units of H_i are

$$(H_i) = \frac{\text{J}}{\text{mol } i} \text{ or } \frac{\text{Btu}}{\text{lb-mol } i} \text{ or } \frac{\text{cal}}{\text{mol } i}$$

Enthalpy carried into (or out of) the system can be expressed as the sum of the internal energy carried into (or out of) the system by mass flow plus the flow work:

$$F_i H_i = F_i(U_i + P\tilde{V}_i)$$

Combining Equations (11-5), (11-7), and (11-8), we can now write the energy balance in the form

$$\frac{d\hat{E}_{\text{sys}}}{dt} = \dot{Q} - \dot{W}_s + \left.\sum_{i=1}^{n} F_i H_i\right|_{\text{in}} - \left.\sum_{i=1}^{n} F_i H_i\right|_{\text{out}}$$

The energy of the system at any instant in time, $\hat{E}_{\text{sys}}$, is the sum of the products of the number of moles of each species in the system multiplied by their respective energies. This term will be discussed in more detail when unsteady-state reactor operation is considered in Chapter 13.

We shall let the subscript "0" represent the inlet conditions. Unsubscripted variables represent the conditions at the outlet of the chosen system volume.

Energy Balance

in

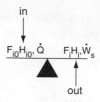

$F_{i0}H_{i0}, \dot{Q}$ $F_iH_i, \dot{W}_s$

out

$$\dot{Q} - \dot{W}_s + \sum_{i=1}^{n} F_{i0}H_{i0} - \sum_{i=1}^{n} F_iH_i = \frac{d\hat{E}_{sys}}{dt} \qquad (11\text{-}9)$$

In Section 11.1, we discussed that in order to solve reaction engineering problems with heat effects, we needed to relate temperature, conversion, and rate of reaction. The energy balance as given in Equation (11-9) is the most convenient starting point as we proceed to develop this relationship.

11.2.3 Overview of Energy Balances

What is the plan? In the following pages we manipulate Equation (11-9) in order to apply it to each of the reactor types we have been discussing: batch, PFR, PBR, and CSTR. The end result of the application of the energy balance to each type of reactor is shown in Table 11-1. These equations can be used in **Step 5** of the algorithm discussed in Example E11-1. The equations in Table 11-1 relate temperature to conversion and to molar flow rates, and to the system parameters, such as the overall heat-transfer coefficient and area, Ua, with the corresponding ambient temperature, T_a, and the heat of reaction, ΔH_{Rx}.

TABLE 11-1   ENERGY BALANCES OF COMMON REACTORS

End results of manipulating the energy balance (Sections 11.2.4, 12.1, and 12.3)

1. **Adiabatic** $(\dot{Q} \equiv 0)$ CSTR, PFR, Batch, or PBR. The relationship between conversion calculated from the energy balance, X_{EB}, and temperature for $\dot{W}_s = 0$, constant C_{P_i}, and $\Delta C_P = 0$, is

   Conversion in terms of temperature

   $$X_{EB} = \frac{\Sigma \Theta_i C_{P_i}(T - T_0)}{-\Delta H_{Rx}^{\circ}} \qquad (T11\text{-}1.A)$$

   Temperature in terms of conversion calculated from the energy balance

   $$T = T_0 + \frac{(-\Delta H_{Rx}^{\circ})X_{EB}}{\Sigma \Theta_i C_{P_i}} \qquad (T11\text{-}1.B)$$

   For an exothermic reaction $(-\Delta H_{Rx}) > 0$

2. **CSTR with heat exchanger**, UA $(T_a - T)$, and large coolant flow rate

   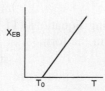

   $$X_{EB} = \frac{\left(\dfrac{UA}{F_{A0}}(T - T_a)\right) + \Sigma \Theta_i C_{P_i}(T - T_0)}{-\Delta H_{Rx}^{\circ}} \qquad (T11\text{-}1.C)$$

TABLE 11-1   ENERGY BALANCES OF COMMON REACTORS (CONTINUED)

3. PFR/PBR with heat exchange

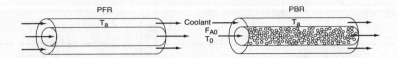

End results of manipulating the energy balance (Sections 11.2.4, 12.1, and 12.3)

In general most of the PFR and PBR energy balances can be written as

$$\frac{dT}{dV} = \frac{(\text{Heat "generated"}) - (\text{Heat "removed"})}{\sum F_i C_{P_i}} = \frac{Q_g - Q_r}{\sum F_i C_{P_i}}$$

3A. PFR in terms of conversion

$$\frac{dT}{dV} = \frac{\overbrace{r_A \Delta H_{Rx}(T)}^{Q_g} - \overbrace{Ua(T - T_a)}^{Q_r}}{F_{A0}\left(\sum \Theta_i C_{P_i} + \Delta C_p X\right)} = \frac{Q_g - Q_r}{F_{A0}\left(\sum \Theta_i C_{P_i} + \Delta C_p X\right)} \qquad \text{(T11-1.D)}$$

3B. PBR in terms of conversion

$$\frac{dT}{dW} = \frac{r_A' \Delta H_{Rx}(T) - \dfrac{Ua}{\rho_b}(T - T_a)}{F_{A0}\left(\sum \Theta_i C_{P_i} + \Delta C_p X\right)} \qquad \text{(T11-1.E)}$$

3C. PBR in terms of molar flow rates

$$\frac{dT}{dW} = \frac{r_A' \Delta H_{Rx}(T) - \dfrac{Ua}{\rho_b}(T - T_a)}{\sum F_i C_{P_i}} \qquad \text{(T11-1.F)}$$

3D. PFR in terms of molar flow rates

$$\frac{dT}{dV} = \frac{r_A \Delta H_{Rx}(T) - Ua(T - T_a)}{\sum F_i C_{P_i}} = \frac{Q_g - Q_r}{\sum F_i C_{P_i}} \qquad \text{(T11-1.G)}$$

4. Batch

$$\frac{dT}{dt} = \frac{(r_A V)(\Delta H_{Rx}) - UA(T - T_a)}{\sum N_i C_{P_i}} \qquad \text{(T11-1.H)}$$

5. For Semibatch or unsteady CSTR

$$\frac{dT}{dt} = \frac{\dot{Q} - \dot{W}_s - \displaystyle\sum_{i=1}^{n} F_{i0} C_{P_i}(T - T_{i0}) + [-\Delta H_{Rx}(T)](-r_A V)}{\displaystyle\sum_{i=1}^{n} N_i C_{P_i}} \qquad \text{(T11-1.I)}$$

TABLE 11-1   ENERGY BALANCES OF COMMON REACTORS (CONTINUED)

6. **For multiple reactions in a PFR** (*q* reactions and *m* species)

$$\frac{dT}{dV} = \frac{\sum\limits_{i=1}^{q} r_{ij}\Delta H_{\mathrm{Rx}ij} - Ua(T - T_a)}{\sum\limits_{j=1}^{m} F_j C_{\mathrm{P}_j}} \qquad \text{(T11-1.J)}$$

i = reaction number, *j* = species

7. **Variable heat exchange fluid temperature, T_a**

<u>Co-current Exchange</u>   <u>Countercurrent Exchange</u>

$$\frac{dT_a}{dV} = \frac{Ua(T - T_a)}{\dot{m}_c C_{\mathrm{P}_c}} \qquad \text{(T11-1.K)} \qquad\qquad \frac{dT_a}{dV} = \frac{Ua(T_a - T)}{\dot{m}_c C_{\mathrm{P}_c}} \qquad \text{(T11-1.L)}$$

$$V = 0 \quad T_a = T_{a0} \qquad\qquad\qquad\qquad V = V_{final} \quad T_a = T_{a0}$$

- -

The equations in Table 11-1 are the ones we will use to solve reaction engineering problems with heat effects.

Nomenclature:

U = overall heat-transfer coefficient, $(\mathrm{J/m^2 \cdot s \cdot K})$;

A = CSTR heat-exchange area, $(\mathrm{m^2})$;

a = PFR heat-exchange area per volume of reactor, $(\mathrm{m^2/m^3})$;

C_{P_i} = mean heat capacity of species *i*, $(\mathrm{J/mol/K})$;

C_{P_c} = the heat capacity of the coolant, $(\mathrm{J/kg/K})$;

$\dot{m}_c$ = coolant flow rate, $(\mathrm{kg/s})$;

$\Delta H_{\mathrm{Rx}}(T)$ = heat of reaction at temperature T, $(\mathrm{J/mol\ A})$:

$$\Delta H_{\mathrm{Rx}}^{\circ} = \left(\frac{d}{a}H_{\mathrm{D}}^{\circ} + \frac{c}{a}H_{\mathrm{C}}^{\circ} - \frac{b}{a}H_{\mathrm{B}}^{\circ} - H_{\mathrm{A}}^{\circ}\right)\mathrm{J/molA} = \text{heat of reaction at temperature } T_R;$$

$\Delta H_{\mathrm{Rx}ij}$ = heat of reaction wrt species *j* in reaction *i*, $(\mathrm{J/mol})$;

$\dot{Q}$ = heat added to the reactor, $(\mathrm{J/s})$; and

$$\Delta C_{\mathrm{P}} = \left(\frac{d}{a}C_{\mathrm{P}_{\mathrm{D}}} + \frac{c}{a}C_{\mathrm{P}_{\mathrm{C}}} - \frac{b}{a}C_{\mathrm{P}_{\mathrm{B}}} - C_{\mathrm{P}_{\mathrm{A}}}\right)\ (\mathrm{J/molA \cdot K}).$$

All other symbols are as defined in Chapters 1 through 10.

End results of manipulating the energy balance (Sections 11.2.4, 12.1, and 12.3)

Examples of How to Use Table 11-1.   We now couple the *energy balance* equations in Table 11-1 with the appropriate reactor *mole balance*, *rate law*, and *stoichiometry algorithm* to solve reaction engineering problems with heat effects. For example, recall the rate law for a first-order reaction, Equation (E11-1.5) in Example 11-1

$$-r_{\mathrm{A}} = k_1 \exp\left[\frac{E}{R}\left(\frac{1}{T_1} - \frac{1}{T}\right)\right]C_{\mathrm{A}0}(1 - X) \qquad \text{(E11-1.5)}$$

which will be combined with the *mole balance* to find the concentration, conversion and temperature profiles (i.e., BR, PBR, PFR), exit concentrations, and conversion and temperature in a CSTR. We will now consider four cases of heat exchange in a PFR and PBR: **(1)** adiabatic, **(2)** co-current, **(3)** countercurrent, and **(4)** constant exchanger temperature. We focus on adiabatic operation in this chapter and the other three cases in Chapter 12.

Case 1: Adiabatic. If the reaction is carried out adiabatically, then we use Equation (T11-1.B) for the reaction A $\longrightarrow$ B in Example 11-1 to obtain

Adiabatic
$$T = T_0 + \frac{-\Delta H^{\circ}_{\text{Rx}} X}{C_{P_A}} \tag{T11-1.B}$$

Consequently, we can now obtain $-r_A$ as a function of X alone by first choosing X, then calculating T from Equation (T11-1.B), then calculating k from Equation (E11-1.3), and then finally calculating $(-r_A)$ from Equation (E11-1.5).

The algorithm

$$\boxed{\text{Choose } X \rightarrow \text{calculate } T \rightarrow \text{calculate } k \rightarrow \text{calculate } -r_A \rightarrow \text{calculate } \frac{F_{A0}}{-r_A}}$$

We can use this sequence to prepare a table of $(F_{A0}/-r_A)$ as a function of X. We can then proceed to size PFRs and CSTRs. In the absolute worst case scenario, we could use the techniques in Chapter 2 (e.g., Levenspiel plots or the quadrature formulas in Appendix A). However, instead of using a Levenspiel plot, we will most likely use Polymath to solve our coupled differential energy and mole balance equations.

$\frac{F_{A0}}{-r_A}$

X

Levenspiel plot

Cases 2, 3, and 4: Correspond to Co-current, Countercurrent, Heat Exchange, and Constant Coolant Temperature T_C, respectively (Ch. 12). If there is cooling along the length of a PFR or PBR, we could then apply Equation (T11-1.D) to this reaction to arrive at two coupled differential equations

Non-adiabatic
PFR
$$\frac{dX}{dV} = k_1 \exp\left[\frac{E}{R}\left(\frac{1}{T_1} - \frac{1}{T}\right)\right] C_{A0}(1 - X)/F_{A0}$$

$$\frac{dT}{dV} = \frac{r_A \Delta H_{\text{Rx}}(T) - Ua(T - T_a)}{F_{A0} C_{P_A}}$$

which are easily solved using an ODE solver such as Polymath.

Heat Exchange in a CSTR. Similarly, for the case of the reaction A $\rightarrow$ B in Example 11-1 carried out in a CSTR, we could use Polymath or MATLAB to solve two nonlinear algebraic equations in X and T. These two equations are the combined mole balance

Non-adiabatic
CSTR
$$V = \frac{F_{A0} X}{k_1 \exp\left[\frac{E}{R}\left(\frac{1}{T_1} - \frac{1}{T}\right)\right] C_{A0}(1 - X)}$$

and the application of Equation (T11-1.C), which is rearranged in the form

$$T = \frac{F_{A0} X(-\Delta H_{\text{Rx}}) + UAT_a + F_{A0} C_{P_A} T_0}{UA + C_{P_A} F_{A0}}$$

From these three cases, (1) adiabatic PFR and CSTR, (2) PFR and PBR with heat effects, and (3) CSTR with heat effects, one can see how to couple the energy balances and mole balances. In principle, one could simply use Table 11-1 to apply to different reactors and reaction systems without further discussion. However, understanding the derivation of these equations will greatly facilitate the proper application to and evaluation of various reactors and reaction systems. Consequently, we now derive the equations given in Table 11-1.

Why bother? Here is why!!

Why bother to derive the equations in Table 11-1? Because I have found that students can *apply* these equations *much* more accurately to solve reaction engineering problems with heat effects if they have gone through the derivation to understand the assumptions and manipulations used in arriving at the equations in Table 11.1. That is, understanding these derivations, students are more likely to put the correct number in the correct equation symbol.

11.3 The User-Friendly Energy Balance Equations

We will now dissect the molar flow rates and enthalpy terms in Equation (11-9) to arrive at a set of equations we can readily apply to a number of reactor situations.

11.3.1 Dissecting the Steady-State Molar Flow Rates to Obtain the Heat of Reaction

To begin our journey, we start with the energy balance equation (11-9) and then proceed to finally arrive at the equations given in Table 11-1 by first dissecting two terms:

1. The molar flow rates, F_i and F_{i0}
2. The molar enthalpies, H_i, $H_{i0}[H_i \equiv H_i(T)$, and $H_{i0} \equiv H_i(T_0)]$

Interactive

Computer Games

An animated version of what follows for the derivation of the energy balance can be found in the reaction engineering games "Heat Effects 1" and "Heat Effects 2" on the CRE Web site, *www.umich.edu/~elements/5e/index.html*. Here, equations move around the screen, making substitutions and approximations to arrive at the equations shown in Table 11-1. Visual learners find these two ICGs a very useful resource.

We will now consider flow systems that are operated at steady state. The steady-state energy balance is obtained by setting $(d\hat{E}_{sys}/dt)$ equal to zero in Equation (11-9) in order to yield

Steady-state energy balance

$$\boxed{\dot{Q} - \dot{W}_s + \sum_{i=1}^{n} F_{i0}H_{i0} - \sum_{i=1}^{n} F_i H_i = 0} \tag{11-10}$$

To carry out the manipulations to write Equation (11-10) in terms of the heat of reaction, we shall use the generic reaction

$$A + \frac{b}{a}B \longrightarrow \frac{c}{a}C + \frac{d}{a}D \tag{2-2}$$

The inlet and outlet summation terms in Equation (11-10) are expanded, respectively, to

In: $\Sigma H_{i0} F_{i0} = H_{A0} F_{A0} + H_{B0} F_{B0} + H_{C0} F_{C0} + H_{D0} F_{D0} + H_{I0} F_{I0}$ (11-11)

and

Out: $\Sigma H_i F_i = H_A F_A + H_B F_B + H_C F_C + H_D F_D + H_I F_I$ (11-12)

where the subscript I represents inert species.

We next express the molar flow rates in terms of conversion. In general, the molar flow rate of species i for the case of no accumulation and a stoichiometric coefficient ν_i is

$$F_i = F_{A0} (\Theta_i + \nu_i X)$$

Specifically, for Reaction (2-2), $A + \dfrac{b}{a} B \longrightarrow \dfrac{c}{a} C + \dfrac{d}{a} D$, we have

$$F_A = F_{A0} (1 - X)$$

Steady-state operation

$$F_B = F_{A0} \left(\Theta_B - \frac{b}{a} X \right)$$

$$F_C = F_{A0} \left(\Theta_C + \frac{c}{a} X \right) \qquad \text{where } \Theta_i = \frac{F_{i0}}{F_{A0}}$$

$$F_D = F_{A0} \left(\Theta_D + \frac{d}{a} X \right)$$

$$F_I = \Theta_I F_{A0}$$

We can substitute these symbols for the molar flow rates into Equations (11-11) and (11-12), then subtract Equation (11-12) from (11-11) to give

$$\sum_{i=1}^{n} F_{i0} H_{i0} - \sum_{i=1}^{n} F_i H_i = F_{A0} [(H_{A0} - H_A) + (H_{B0} - H_B) \Theta_B$$

$$+ (H_{C0} - H_C) \Theta_C + (H_{D0} - H_D) \Theta_D + (H_{I0} - H_I) \Theta_I]$$

$$- \underbrace{\left(\frac{d}{a} H_D + \frac{c}{a} H_C - \frac{b}{a} H_B - H_A \right)}_{\Delta H_{Rx}} F_{A0} X \qquad (11\text{-}13)$$

The term in parentheses that is multiplied by $F_{A0} X$ is called the **heat of reaction** at temperature T and is designated $\Delta H_{Rx}(T)$.

Heat of reaction at temperature T

$$\boxed{\Delta H_{Rx}(T) = \frac{d}{a} H_D(T) + \frac{c}{a} H_C(T) - \frac{b}{a} H_B(T) - H_A(T)} \qquad (11\text{-}14)$$

All enthalpies (e.g., H_A, H_B) are evaluated at the temperature at the outlet of the system volume and, consequently, $[\Delta H_{Rx}(T)]$ is the heat of reaction at a specific temperature T. The heat of reaction is always given per mole of the species that is the basis of calculation, i.e., species A (joules per mole of A reacted).

Substituting Equation (11-14) into (11-13) and reverting to summation notation for the species, Equation (11-13) becomes

$$\sum_{i=1}^{n} F_{i0}H_{i0} - \sum_{i=1}^{n} F_i H_i = F_{A0} \sum_{i=1}^{n} \Theta_i (H_{i0} - H_i) - \Delta H_{Rx}(T) F_{A0} X \quad (11\text{-}15)$$

Combining Equations (11-10) and (11-15), we can now write the *steady-state*, i.e., $(d\hat{E}_{sys}/dt = 0)$, energy balance in a more usable form:

Use this form of
the steady-state
energy balance if
the enthalpies
are available.

$$\boxed{\dot{Q} - \dot{W}_s + F_{A0} \sum_{i=1}^{n} \Theta_i (H_{i0} - H_i) - \Delta H_{Rx}(T) F_{A0} X = 0} \quad (11\text{-}16)$$

If a *phase change* takes place during the course of a reaction, this is the form of the energy balance, i.e., Equation (11-16), that *must* be used.

11.3.2 Dissecting the Enthalpies

We are neglecting any enthalpy changes resulting from mixing so that the partial molal enthalpies are equal to the molal enthalpies of the pure components. The molal enthalpy of species i at a particular temperature and pressure, H_i, is usually expressed in terms of an *enthalpy of formation* of species i at some reference temperature T_R, $H_i^{\circ}(T_R)$, plus the change in enthalpy, ΔH_{Qi}, that results when the temperature is raised from the reference temperature, T_R, to some temperature T

$$H_i = H_i^{\circ}(T_R) + \Delta H_{Qi} \quad (11\text{-}17)$$

The reference temperature at which $H_i^{\circ}(T_R)$ is given is usually 25°C. For any substance i that is being heated from T_1 to T_2 in the *absence* of phase change

No phase change

$$\Delta H_{Qi} = \int_{T_1}^{T_2} C_{P_i} \, dT \quad (11\text{-}18)$$

Typical units of the heat capacity, C_{P_i}, are

$$(C_{P_i}) = \frac{J}{(\text{mol of } i)(K)} \text{ or } \frac{Btu}{(\text{lb mol of } i)(°R)} \text{ or } \frac{cal}{(\text{mol of } i)(K)}$$

A large number of chemical reactions carried out in industry do not involve phase change. Consequently, we shall further refine our energy balance to apply to *single-phase* chemical reactions. Under these conditions, the enthalpy of species i at temperature T is related to the enthalpy of formation at the reference temperature T_R by

$$H_i(T) = H_i^{\circ}(T_R) + \int_{T_R}^{T} C_{P_i} \, dT \quad (11\text{-}19)$$

If phase changes do take place in going from the temperature for which the enthalpy of formation is given and the reaction temperature T, Equation (11-17) must be used instead of Equation (11-19).

The heat capacity at temperature T is frequently expressed as a quadratic function of temperature; that is

$$C_{P_i} = \alpha_i + \beta_i T + \gamma_i T^2 \tag{11-20}$$

Reference Shelf

However, while the text will consider only **constant heat capacities**, the PRS R11.1 on the CRE Web site has examples with variable heat capacities.

To calculate the change in enthalpy ($H_i - H_{i0}$) when the reacting fluid is heated without phase change from its entrance temperature, T_{i0}, to a temperature T, we integrate Equation (11-19) for constant C_{P_i} to write

$$H_i - H_{i0} = \left[H_i^\circ(T_R) + \int_{T_R}^{T} C_{P_i} \, dT \right] - \left[H_i^\circ(T_R) + \int_{T_R}^{T_{i0}} C_{P_i} \, dT \right]$$

$$= \int_{T_{i0}}^{T} C_{P_i} \, dT = C_{P_i} [T - T_{i0}] \tag{11-21}$$

Substituting for H_i and H_{i0} in Equation (11-16) yields

Result of dissecting the enthalpies

$$\boxed{\dot{Q} - \dot{W}_s - F_{A0} \sum_{i=1}^{n} \Theta_i C_{P_i} [T - T_{i0}] - \Delta H_{Rx}(T) F_{A0} X = 0} \tag{11-22}$$

11.3.3 Relating $\Delta H_{Rx}(T)$, $\Delta H_{Rx}^\circ(T_R)$, and ΔC_P

Recall that the heat of reaction at temperature T was given in terms of the enthalpy of each reacting species at temperature T in Equation (11-14); that is

$$\Delta H_{Rx}(T) = \frac{d}{a} H_D(T) + \frac{c}{a} H_C(T) - \frac{b}{a} H_B(T) - H_A(T) \tag{11-14}$$

where the enthalpy of each species is given by

$$H_i(T) = H_i^\circ(T_R) + \int_{T_R}^{T} C_{P_c} \, dT = H_i^\circ(T_R) + C_{P_i}(T - T_R) \tag{11-19}$$

If we now substitute for the enthalpy of each species, we have

For the generic reaction
$A + \dfrac{b}{a}B \rightarrow \dfrac{c}{a}C + \dfrac{d}{a}D$

$$\boxed{\begin{aligned} \Delta H_{Rx}(T) &= \left[\frac{d}{a} H_D^\circ(T_R) + \frac{c}{a} H_C^\circ(T_R) - \frac{b}{a} H_B^\circ(T_R) - H_A^\circ(T_R) \right] \\ &\quad + \left[\frac{d}{a} C_{P_D} + \frac{c}{a} C_{P_C} - \frac{b}{a} C_{P_B} - C_{P_A} \right] (T - T_R) \end{aligned}}$$

$$\tag{11-23}$$

The first term in brackets on the right-hand side of Equation (11-23) is the heat of reaction at the reference temperature T_R

$$\boxed{\Delta H_{Rx}^\circ(T_R) = \frac{d}{a} H_D^\circ(T_R) + \frac{c}{a} H_C^\circ(T_R) - \frac{b}{a} H_B^\circ(T_R) - H_A^\circ(T_R)} \tag{11-24}$$

The enthalpies of formation of many compounds, $H_i^\circ(T_R)$, are usually tabulated at 25°C and can readily be found in the *Handbook of Chemistry and Physics* and similar handbooks.[1] That is, we can look up the heats of formation at T_R, then calculate the heat of reaction at this reference temperature. The heat of combustion (also available in these handbooks) can also be used to determine the enthalpy of formation, $H_i^\circ(T_R)$, and the method of calculation is also described in these handbooks. From these values of the standard heat of formation, $H_i^\circ(T_R)$, we can calculate the heat of reaction at the reference temperature T_R using Equation (11-24).

The second term in brackets on the right-hand side of Equation (11-23) is the overall change in the heat capacity per mole of A reacted, ΔC_P,

$$\Delta C_P = \frac{d}{a} C_{P_D} + \frac{c}{a} C_{P_C} - \frac{b}{a} C_{P_B} - C_{P_A} \qquad (11\text{-}25)$$

Combining Equations (11-25), (11-24), and (11-23) gives us

Heat of reaction at
temperature T

$$\Delta H_{Rx}(T) = \Delta H_{Rx}^\circ(T_R) + \Delta C_P(T - T_R) \qquad (11\text{-}26)$$

Equation (11-26) gives the heat of reaction at any temperature T in terms of the heat of reaction at a reference temperature (usually 298 K) and the ΔC_P term. Techniques for determining the heat of reaction at pressures above atmospheric can be found in Chen.[2] For the reaction of hydrogen and nitrogen at 400°C, it was shown that the heat of reaction increased by only 6% as the pressure was raised from 1 atm to 200 atm!

Example 11–2 Heat of Reaction

Calculate the heat of reaction for the synthesis of ammonia from hydrogen and nitrogen at 150°C in kcal/mol of N_2 reacted *and also* in kJ/mol of H_2 reacted.

Solution

$$N_2 + 3H_2 \longrightarrow 2NH_3$$

Calculate the heat of reaction at the reference temperature using the heats of formation of the reacting species obtained from *Perry's Chemical Engineers' Handbook* or the *Handbook of Chemistry and Physics*.[3]

The enthalpies of formation at 25°C are

$$H_{NH_3}^\circ(T_R) = -11{,}020 \frac{cal}{mol\ NH_3}, \quad H_{H_2}^\circ = 0, \text{ and } H_{N_2}^\circ = 0$$

<u>Note:</u> The heats of formation of all elements (e.g., H_2, N_2) are **zero** at 25°C.

---

[1] *CRC Handbook of Chemistry and Physics*, 95th ed. (Boca Raton, FL: CRC Press, 2014).

[2] N. H. Chen, *Process Reactor Design* (Needham Heights, MA: Allyn and Bacon, 1983), p. 26.

[3] D. W. Green, and R. H. Perry, eds., *Perry's Chemical Engineers' Handbook*, 8th ed. (New York: McGraw-Hill, 2008).

To calculate $\Delta H^\circ_{Rx}(T_R)$, we use Equation (11-24) and we take the heats of formation of the products (e.g., NH_3) multiplied by their appropriate stoichiometric coefficients (2 for NH_3) minus the heats of formation of the reactants (e.g., N_2, H_2) multiplied by their stoichiometric coefficient (e.g., 3 for H_2, 1 for N_2)

$$\Delta H^\circ_{Rx}(T_R) = 2H^\circ_{NH_3}(T_R) - 3H^\circ_{H_2}(T_R) - H^\circ_{N_2}(T_R) \qquad \text{(E11-2.1)}$$

$$\Delta H^\circ_{Rx}(T_R) = 2H^\circ_{NH_3}(T_R) - 3(0) - 0 = 2H^\circ_{NH_3}$$

$$= 2(-11,020)\frac{cal}{mol\ N_2}$$

$$= -22,040\ cal/mol\ N_2\ reacted$$

$$\Delta H^\circ_{Rx}(298\ K) = -22.04\ kcal/mol\ N_2\ reacted$$

or in terms of kJ/mol

Exothermic reaction

$$\Delta H^\circ_{Rx}(298\ K) = -92.22\ kJ/mol\ N_2\ reacted$$

The minus sign indicates that the reaction is *exothermic*. If the heat capacities are constant or if the mean heat capacities over the range 25°C to 150°C are readily available, the determination of ΔH_{Rx} at 150°C is quite simple.

$$C_{P_{H_2}} = 6.992\ cal/mol\ H_2 \cdot K$$

$$C_{P_{N_2}} = 6.984\ cal/mol\ N_2 \cdot K$$

$$C_{P_{NH_3}} = 8.92\ cal/mol\ NH_3 \cdot K$$

$$\Delta C_P = 2C_{P_{NH_3}} - 3C_{P_{H_2}} - C_{P_{N_2}} \qquad \text{(E11-2.2)}$$

$$= 2(8.92) - 3(6.992) - 6.984$$

$$= -10.12\ cal/mol\ N_2\ reacted \cdot K$$

$$\Delta H_{Rx}(T) = \Delta H^\circ_{Rx}(T_R) + \Delta C_P(T - T_R) \qquad \text{(11-26)}$$

$$\Delta H_{Rx}(423\ K) = -22,040 + (-10.12)(423 - 298)$$

$$= -23,310\ cal/mol\ N_2 = -23.31\ kcal/mol\ N_2$$

in terms of kJ/mol

$$\Delta H^\circ_{Rx}(423\ K) = -23.3\ kcal/mol\ N_2 \times 4.184\ kJ/kcal$$

$$\boxed{\Delta H_{Rx}(423\ K) = -97.5\ kJ/mol\ N_2}$$

(Recall: 1 kcal = 4.184 kJ)

The heat of reaction based on the moles of H_2 reacted is

$$\Delta H_{Rx}(423\ K) = \frac{1\ mol\ N_2}{3\ mol\ H_2}\left(-97.53\ \frac{kJ}{mol\ N_2}\right)$$

$$\Delta H_{Rx}(423\ K) = -32.51\ \frac{kJ}{mol\ H_2}\ at\ 423\ K$$

Analysis: This example showed (1) how to calculate the heat of reaction with respect to a given species, given the heats of formation of the reactants and the products, and (2) how to find the heat of reaction with respect to one species, given the heat of reaction with respect to another species in the reaction. We also saw how the heat of reaction changed as we increased the temperature.

Now that we see that we can calculate the heat of reaction at any temperature, let's substitute Equation (11-22) in terms of $\Delta H_R(T_R)$ and ΔC_P, i.e., Equation (11-26). The steady-state energy balance is now

Energy balance in terms of mean or constant heat capacities

$$\dot{Q} - \dot{W}_s - F_{A0} \sum_{i=1}^{n} \Theta_i C_{P_i}(T - T_{i0}) - [\Delta H_{Rx}^{\circ}(T_R) + \Delta C_P(T - T_R)]F_{A0}X = 0 \qquad (11\text{-}27)$$

From here on, for the sake of brevity we will let

$$\Sigma = \sum_{i=1}^{n}$$

unless otherwise specified.

In most systems, the work term, $\dot{W}_s$, can be neglected (note the exception in the California Professional Engineers' Exam Problem P12-6$_B$ at the end of Chapter 12). Neglecting $\dot{W}_s$, the energy balance becomes

$$\dot{Q} - F_{A0}\Sigma\Theta_i C_{P_i}(T - T_{i0}) - [\Delta H_{Rx}^{\circ}(T_R) + \Delta C_P(T - T_R)]F_{A0}X = 0 \quad (11\text{-}28)$$

In almost all of the systems we will study, the reactants will be entering the system at the same temperature; therefore, $T_{i0} = T_0$.

We can use Equation (11-28) to relate temperature and conversion and then proceed to evaluate the algorithm described in Example 11-1. However, unless the reaction is carried out adiabatically, Equation (11-28) is still difficult to evaluate because in nonadiabatic reactors, the heat added to or removed from the system varies along the length of the reactor. This problem does not occur in adiabatic reactors, which are frequently found in industry. Therefore, the adiabatic tubular reactor will be analyzed first.

11.4 Adiabatic Operation

Reactions in industry are frequently carried out adiabatically with heating or cooling provided either upstream or downstream. Consequently, analyzing and sizing adiabatic reactors is an important task.

11.4.1 Adiabatic Energy Balance

In the previous section, we derived Equation (11-28), which relates conversion to temperature and the heat added to the reactor, $\dot{Q}$. Let's stop a minute (actually it will probably be more like a couple of days) and consider a system with the special set of conditions of no work, $\dot{W}_s = 0$, adiabatic operation $\dot{Q} = 0$, letting $T_{i0} = T_0$ and then rearranging Equation (11-28) into the form

For adiabatic operation, Example 11.1 can now be solved!

$$X = \frac{\Sigma \Theta_i C_{P_i}(T - T_0)}{-[\Delta H_{Rx}^{\circ}(T_R) + \Delta C_P(T - T_R)]} \qquad (11\text{-}29)$$

Relationship
between X and T
for *adiabatic*
exothermic
reactions

In many instances, the $\Delta C_P(T - T_R)$ term in the denominator of Equation (11-29) is negligible with respect to the ΔH_{Rx}° term, so that a plot of X vs. T will usually be linear, as shown in Figure 11-2. To remind us that the conversion in this plot was obtained from the energy balance rather than the mole balance, it is given the subscript EB (i.e., X_{EB}) in Figure 11-2.

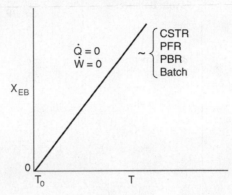

Figure 11-2   Adiabatic temperature–conversion relationship.

Equation (11-29) applies to a CSTR, PFR, or PBR, and also to a BR (as will be shown in Chapter 13). For $\dot{Q} = 0$ and $\dot{W}_s = 0$, Equation (11-29) gives us the explicit relationship between X and T needed to be used in conjunction with the mole balance to solve a large variety of chemical reaction engineering problems as discussed in Section 11.1.

11.4.2 Adiabatic Tubular Reactor

We can rearrange Equation (11-29) to solve for temperature as a function of conversion; that is

Energy balance for
adiabatic operation
of PFR

$$T = \frac{X[-\Delta H_{Rx}^\circ(T_R)] + \Sigma \Theta_i C_{P_i} T_0 + X \Delta C_P T_R}{\Sigma \Theta_i C_{P_i} + X \Delta C_P} \qquad (11\text{-}30)$$

This equation will be coupled with the differential mole balance

$$F_{A0} \frac{dX}{dV} = -r_A(X, T)$$

to obtain the temperature, conversion, and concentration profiles along the length of the reactor. The algorithm for solving PBRs and PFRs operated adiabatically using a first-order reversible reaction A $\rightleftarrows$ B as an example is shown in Table 11-2.

Table 11-3 gives two different methods for solving the equations in Table 11-2 in order to find the conversion, X, and temperature, T, profiles down the reactor. The numerical technique (e.g., hand calculation) is presented *primarily to give insight and understanding* to the solution procedure and this understanding is important. With this procedure, one could either construct a Levenspiel plot or use a quadrature formula to find the reactor volume.

TABLE 11-2   ADIABATIC PFR/PBR ALGORITHM

The elementary reversible gas-phase reaction

$$A \underset{\longleftarrow}{\overset{\longrightarrow}{}} B$$

is carried out in a PFR in which pressure drop is neglected and pure A enters the reactor.

Mole Balance:

$$\frac{dX}{dV} = \frac{-r_A}{F_{A0}} \tag{T11-2.1}$$

Rate Law:

$$-r_A = k\left(C_A - \frac{C_B}{K_C}\right) \tag{T11-2.2}$$

with

$$k = k_1(T_1)\,\exp\left[\frac{E}{R}\left(\frac{1}{T_1} - \frac{1}{T}\right)\right] \tag{T11-2.3}$$

and for $\Delta C_P = 0$

$$K_C = K_{C2}(T_2)\,\exp\left[\frac{\Delta H_{Rx}^{\circ}}{R}\left(\frac{1}{T_2} - \frac{1}{T}\right)\right] \tag{T11-2.4}$$

Following the Algorithm

Stoichiometry: Gas, $\varepsilon = 0$, $P = P_0$

$$C_A = C_{A0}(1-X)\frac{T_0}{T} \tag{T11-2.5}$$

$$C_B = C_{A0}X\frac{T_0}{T} \tag{T11-2.6}$$

Combine:

$$-r_A = kC_{A0}\left[(1-X) - \frac{X}{K_C}\right]\frac{T_0}{T} \tag{T11-2.7}$$

Energy Balance:

To relate temperature and conversion, we apply the energy balance to an adiabatic PFR. If all species enter at the same temperature, $T_{i0} = T_0$.

Solving Equation (11-29) with $\dot{Q} = 0$, $\dot{W}_s = \mathbf{0}$, to obtain T as a function of conversion yields

$$T = \frac{X[-\Delta H_{Rx}^{\circ}(T_R)] + \Sigma\,\Theta_i C_{P_i} T_0 + X\,\Delta C_P T_R}{\Sigma\,\Theta_i C_{P_i} + X\,\Delta C_P} \tag{T11-2.8}$$

If pure A enters and **iff** $\Delta C_P = 0$, then

$$T = T_0 + \frac{X[-\Delta H_{Rx}^{\circ}(T_R)]}{C_{P_A}} \tag{T11-2.9}$$

Equations (T11-2.1) through (T11-2.9) can easily be solved using either Simpson's rule or an ODE solver.

Only if my
computer is
missing.

It is doubtful that anyone would actually use either of these methods unless they had **absolutely no** access to a computer and they would never get access (e.g., stranded on a desert island with a dead laptop or no satellite connection). The solution to reaction engineering problems *today* is to use software packages with ordinary differential equation (ODE) solvers, such as Polymath, MATLAB, or Excel, to solve the coupled mole balance and energy balance differential equations.

TABLE 11-3   SOLUTION PROCEDURES FOR ELEMENTARY ADIABATIC GAS-PHASE PFR/PBR REACTOR

The **numerical technique** is presented to provide insight about how the variables (k, K_c, etc.) change as we move down the reactor from $V = 0$ and $X = 0$ to V_f and X_f.

A. Numerical Technique

Integrating the PFR mole balance,

Choose X → Calculate T → Calculate k → Calculate –r_A → Calculate $\dfrac{F_{A0}}{-r_A}$

$$V = \int_0^{X_3} \frac{F_{A0}}{-r_A} dX \tag{T11-3.1}$$

1. Set $X = 0$.
2. Calculate T using Equation (T11-2.9).
3. Calculate k using Equation (T11-2.3).
4. Calculate K_C using Equation (T11-2.4).
5. Calculate T_0/T (gas phase).
6. Calculate $-r_A$ using Equation (T11-2.7).
7. Calculate $(F_{A0}/-r_A)$.
8. If X is less than the exit conversion X_3 specified, increment X (i.e., $X_{i+1} = X_i + \Delta X$) and go to Step 2.
9. Prepare table of X vs. $(F_{A0}/-r_A)$.
10. Use numerical integration formulas given in Appendix A; for example,

Use evaluation techniques discussed in Chapter 2.

$$V = \int_0^{X_3} \frac{F_{A0}}{-r_A} dX = \frac{3}{8} h \left[\frac{F_{A0}}{-r_A(X=0)} + 3\frac{F_{A0}}{-r_A(X_1)} + 3\frac{F_{A0}}{-r_A(X_2)} + \frac{F_{A0}}{-r_A(X_3)} \right]$$

$$\tag{T11-3.2}$$

with $h = \dfrac{X_3}{3}$

B. Ordinary Differential Equation (ODE) Solver

1. $\dfrac{dX}{dV} = \dfrac{kC_{A0}}{F_{A0}} \left[(1-X) - \dfrac{X}{K_C} \right] \dfrac{T_0}{T}$ $\tag{T11-3.3}$

Almost always we will use an ODE solver.

2. $k = k_1(T_1) \exp\left[\dfrac{E}{R}\left(\dfrac{1}{T_1} - \dfrac{1}{T} \right) \right]$ $\tag{T11-3.4}$

3. $K_C = K_{C2}(T_2) \exp\left[\dfrac{\Delta H_{Rx}^\circ}{R}\left(\dfrac{1}{T_2} - \dfrac{1}{T} \right) \right]$ $\tag{T11-3.5}$

4. $T = T_0 + \dfrac{X[-\Delta H_{Rx}^\circ(T_R)]}{C_{P_A}}$ $\tag{T11-3.6}$

5. Enter parameter values k_1, E, R, K_{C2}, $\Delta H_{Rx}^\circ(T_R)$, C_{P_A}, $\Delta C_P = 0$, C_{A0}, T_0, T_1, T_2.
6. Enter intial values $X = 0$, $V = 0$, and final value reactor volume, $V = V_f$.

We will now apply the algorithm in Table 11-2 and solution procedure B in Table 11-3 to a real reaction.

Living Example Problem

Example 11–3 Adiabatic Liquid-Phase Isomerization of Normal Butane

Normal butane, C_4H_{10}, is to be isomerized to isobutane in a plug-flow reactor. Isobutane is a valuable product that is used in the manufacture of gasoline additives. For example, isobutane can be further reacted to form iso-octane. The 2014 selling price of n-butane was $1.5/gal, while the trading price of isobutane was $1.75/gal.[†]

This elementary reversible reaction is to be carried out *adiabatically* in the liquid phase under high pressure using essentially trace amounts of a liquid catalyst that gives a specific reaction rate of 31.1 h⁻¹ at 360 K. The feed enters at 330 K.

(a) Calculate the PFR volume necessary to process 100,000 gal/day (163 kmol/h) at 70% conversion of a mixture 90 mol % n-butane and 10 mol % i-pentane, which is considered an inert.

(b) Plot and analyze X, X_e, T, and $-r_A$ down the length of the reactor.

(c) Calculate the CSTR volume for 40% conversion.

Additional information:

The economic incentive
$ = 1.75/gal
versus
1.50/gal

$$\Delta H_{Rx}^{\circ} = -6900 \text{ J/mol n-butane}, \quad \text{Activation energy} = 65.7 \text{ kJ/mol}$$

$$K_C = 3.03 \text{ at } 60°C, \quad C_{A0} = 9.3 \text{ mol/dm}^3 = 9.3 \text{ kmol/m}^3$$

Butane	i-Pentane
$C_{P_{n\text{-}B}} = 141 \text{ J/mol} \cdot \text{K}$	$C_{P_{i\text{-}P}} = 161 \text{ J/mol} \cdot \text{K}$

$$C_{P_{i\text{-}B}} = 141 \text{ J/mol} \cdot \text{K} = 141 \text{ kJ/kmol} \cdot K$$

Solution

$$n\text{-}C_4H_{10} \rightleftharpoons i\text{-}C_4H_{10}$$
$$A \rightleftharpoons B$$

(a) PFR algorithm

$$n - C_4H_{10} \quad \longrightarrow \quad i - C_4H_{10}$$
$$i - C_4H_{10} \quad \qquad \quad n - C_4H_{10}$$

The algorithm

1. Mole Balance: $\quad F_{A0} \dfrac{dX}{dV} = -r_A \qquad\qquad$ (E11-3.1)

2. Rate Law: $\quad -r_A = k\left(C_A - \dfrac{C_B}{K_C}\right) \qquad\qquad$ E11-3.2

with

$$k = k(T_1)e^{\left[\frac{E}{R}\left(\frac{1}{T_1} - \frac{1}{T}\right)\right]} \qquad\qquad \text{(E11-3.3)}$$

$$K_C = K_C(T_2)e^{\left[\frac{\Delta H_{Rx}^{\circ}}{R}\left(\frac{1}{T_2} - \frac{1}{T}\right)\right]} \qquad\qquad \text{(E11-3.4)}$$

[†] Once again, you can buy a cheaper generic brand of n–C_4H_{10} at the Sunday markets in downtown Riça, Jofostan, where there will be a special appearance and lecture by Jofostan's own Prof. Dr. Sven Köttlov on February 29th, at the CRE booth.

3. Stoichiometry (liquid phase, $v = v_0$):

$$C_A = C_{A0}(1 - X) \tag{E11-3.5}$$

$$C_B = C_{A0}X \tag{E11-3.6}$$

4. Combine:

Following the Algorithm

$$-r_A = kC_{A0}\left[1 - \left(1 + \frac{1}{K_C}\right)X\right] \tag{E11-3.7}$$

5. Energy Balance: Recalling Equation (11-27), we have

$$\dot{Q} - \dot{W}_s - F_{A0}\sum\Theta_i C_{P_i}(T - T_0) - F_{A0}X[\Delta H_{Rx}^\circ(T_R) + \Delta C_P(T - T_R)] = 0 \tag{11-27}$$

From the problem statement

$$\text{Adiabatic:} \quad \dot{Q} = 0$$

$$\text{No work:} \quad \dot{W} = 0$$

$$\Delta C_P = C_{P_B} - C_{P_A} = 141 - 141 = 0$$

Applying the preceding conditions to Equation (11-27) and rearranging gives

Nomenclature Note
$$\Delta H_{Rx}(T) \equiv \Delta H_{Rx}$$
$$\Delta H_{Rx}(T_R) \equiv \Delta H_{Rx}^\circ$$
$$\Delta H_{Rx} =$$
$$\Delta H_{Rx}^\circ + \Delta C_P(T - T_R)$$

$$T = T_0 + \frac{(-\Delta H_{Rx}^\circ)X}{\sum\Theta_i C_{P_i}} \tag{E11-3.8}$$

6. Parameter Evaluation:

$$F_{A0} = 0.9 F_{T0} = (0.9)\left(163\frac{\text{kmol}}{\text{h}}\right) = 146.7\frac{\text{kmol}}{\text{h}}$$

$$\sum\Theta_i C_{P_i} = C_{P_A} + \Theta_I C_{P_I} = \left(141 + \left(\frac{0.1}{0.9}\right)161\right)\text{J/mol}\cdot\text{K}$$

$$= 159\ \text{J/mol}\cdot\text{K}$$

$$T = 330 + \frac{-(-6900)}{159}X$$

$$\boxed{T = 330 + 43.4X} \tag{E11-3.9}$$

where T is in degrees Kelvin.

Substituting for the activation energy, T_1, and k_1 in Equation (E11-3.3), we obtain

$$k = 31.1\ \exp\left[\frac{65,700}{8.31}\left(\frac{1}{360} - \frac{1}{T}\right)\right](h^{-1})$$

$$\boxed{k = 31.1\ \exp\left[7906\left(\frac{T - 360}{360T}\right)\right](h^{-1})} \tag{E11-3.10}$$

Substituting for ΔH_{Rx}°, T_2, and $K_C(T_2)$ in Equation (E11-3.4) yields

$$K_C = 3.03\ \exp\left[\frac{-6900}{8.31}\left(\frac{1}{333} - \frac{1}{T}\right)\right]$$

$$K_C = 3.03 \exp\left[-830.3\left(\frac{T-333}{333T}\right)\right]$$

(E11-3.11)

Recalling the rate law gives us

$$-r_A = kC_{A0}\left[1 - \left(1 + \frac{1}{K_C}\right)X\right]$$

(E11-3.7)

7. Equilibrium Conversion:
At equilibrium

$$-r_A \equiv 0$$

and therefore we can solve Equation (E11-3.7) for the equilibrium conversion

$$X_e = \frac{K_C}{1 + K_C}$$

(E11-3.12)

Because we know $K_C(T)$, we can find X_e as a function of temperature.

PFR Solution

It's risky business to ask for 70% conversion in a reversible reaction.

(a) Find the PFR volume necessary to achieve 70% conversion. This problem statement is risky. Why? Because the adiabatic equilibrium conversion may be less than 70%! Fortunately, it's not for the conditions here, $0.7 < X_e$. In general, we should ask for the reactor volume to obtain 95% of the equilibrium conversion, $X_f = 0.95\ X_e$.

(b) Plot and analyze X, X_e, $-r_A$, and T down the length (volume) of the reactor.

We will solve the preceding set of equations to find the PFR reactor volume using both hand calculations and an ODE computer solution. We carry out the hand calculation to help give an intuitive understanding of how the parameters X_e and $-r_A$ vary with conversion and temperature. The computer solution allows us to readily plot the reaction variables along the length of the reactor and also to study the reaction and reactor by varying the system parameters such as C_{A0} and T_0.

Part (a) Solution by hand calculation to perhaps give greater insight and to build on techniques in Chapter 2.

We will now integrate Equation (E11-3.8) using Simpson's rule after forming a table (E11-3.1) to calculate $(F_{A0}/-r_A)$ as a function of X. This procedure is similar to that described in Chapter 2. We now carry out a sample calculation to show how Table E11-3.1 was constructed.

We are only going to do this once!!

For example, for $X = 0.2$, follow the downward arrows for the sequence of the calculations.

(a) First calculate the temperature T, Equation (E11-3.9):

$$T = 330 + 43.4(0.2) = 338.6\ \text{K}$$

(b) Calculate k: Equation (E11-3.10):

$$k = 31.1 \exp\left[7906\left(\frac{338.6-360}{(360)(338.6)}\right)\right] = 31.1\exp(-1.388) = 7.76\ \text{h}^{-1}$$

Sample calculation for Table E11-3.1

(c) Calculate K_C: Equation (E11-3.11):

$$K_C = 3.03 \exp\left[-830.3\left(\frac{338.6-333}{(333)(338.6)}\right)\right] = 3.03e^{-0.0412} = 2.9$$

(d) Calculate X_e: Equation (E11-3.12):

$$X_e = \frac{2.9}{1+2.9} = 0.74$$

(e) Calculate $-r_A$: Equation (E11-3.7):

$$-r_A = \left(\frac{7.76}{h}\right)(9.3)\frac{mol}{dm^3}\left[1 - \left(1 + \frac{1}{2.9}\right)(0.2)\right] = 52.8\,\frac{mol}{dm^3\cdot h} = 52.8\,\frac{kmol}{m^3\cdot h}$$

(f) Dividing $(-r_A)$ into F_{A0}

$$\frac{F_{A0}}{-r_A} = \frac{(0.9\ \text{mol butane/mol total})(163.\ \text{kmol total/h})}{52.8\,\dfrac{kmol}{m^3\cdot h}} = 2.78\ m^3$$

Continuing in this manner for other conversions, we can complete Table E11-3.1.

TABLE E11-3.1   HAND CALCULATION

X	T (K)	k (h⁻¹)	K_C	X_e	$-r_A$(kmol/m³·h)	$\dfrac{F_{A0}}{-r_A}$ (m³)
0	330	4.22	3.1	0.76	39.2	3.74
0.2	338.7	7.76	2.9	0.74	52.8	2.78
0.4	347.3	14.02	2.73	0.73	58.6	2.50
0.6	356.0	24.27	2.57	0.72	37.7	3.88
0.65	358.1	27.74	2.54	0.718	24.5	5.99
0.7	360.3	31.67	2.5	0.715	6.2	23.29

I know these are tedious calculations, but someone's gotta know how to do it.

Use the data in Table E11-3.1 to make a Levenspiel plot, as in Chapter 2.

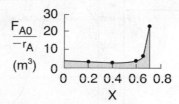

The reactor volume for 70% conversion will be evaluated using the quadrature formulas. Because $(F_{A0}/-r_A)$ increases rapidly as we approach the adiabatic equilibrium conversion, 0.71, we will break the integral into two parts

$$V = \int_0^{0.7}\frac{F_{A0}}{-r_A}\,dX = \int_0^{0.6}\frac{F_{A0}}{-r_A}\,dX + \int_{0.6}^{0.7}\frac{F_{A0}}{-r_A}\,dX \qquad (E11\text{-}3.13)$$

Using Equations (A-24) and (A-22) in Appendix A, we obtain

$$V = \frac{3}{8}\times\frac{0.6}{3}[3.74 + 3\times 2.78 + 3\times 2.50 + 3.88]m^3 + \frac{1}{3}\times\frac{0.1}{2}[3.88 + 4\times 5.99 + 23.29]m^3$$

$$V = 1.75\ m^3 + 0.85\ m^3$$

$$\boxed{V = 2.60\ m^3}$$

Why are we doing this hand calculation? *If it isn't helpful, send me an email and you won't see this again.*

Later, 10/10/15: Actually, since this margin note first appeared in 2011, I have had 2 people say to keep the hand calculation in the text so I kept it in.

You probably will never ever carry out a hand calculation similar to the one shown above. So why did we do it? Hopefully, we have given you a more intuitive feel for the magnitude of each of the terms and how they change as one moves down the reactor (i.e., what the computer solution is doing), as well as a demonstration of how the Levenspiel Plots of $(F_{A0}/-r_A)$ vs. X in Chapter 2 were constructed. At the exit, $V = 2.6$ m³, $X = 0.7$, $X_e = 0.715$, and $T = 360$ K.

Part B is the
Polymath solution
method we will use
to solve most all
CRE problems with
"heat" effects.

Part (b) PFR computer solution and variable profiles

We could have also solved this problem using Polymath or some other ODE solver. The Polymath program using Equations (E11-3.1), (E11-3.7), (E11-3.9), (E11-3.10), (E11-3.11), and (E11-3.12) is shown in Table E11-3.2.

Living Example Problem

TABLE E11-3.2   POLYMATH PROGRAM ADIABATIC ISOMERIZATION

Differential equations

1 d(X)/d(V) = -ra/Fa0

Explicit equations

1 Ca0 = 9.3

2 Fa0 = .9*163

3 T = 330+43.3*X

4 Kc = 3.03*exp(-830.3*((T-333)/(T*333)))

5 k = 31.1*exp(7906*(T-360)/(T*360))

6 Xe = Kc/(1+Kc)

7 ra = -k*Ca0*(1-(1+1/Kc)*X)

8 rate = -ra

POLYMATH Report
Ordinary Differential Equations

Calculated values of DEQ variables

	Variable	Initial value	Final value
6	rate	39.28165	0.0029845
7	T	330.	360.9227
8	V	0	5.
9	X	0	0.7141504
10	Xe	0.7560658	0.7141573

(a)   (b)   (c)

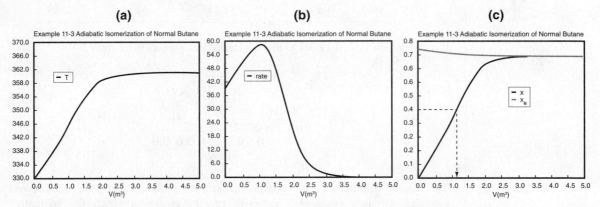

Figure E11-3.1   Adiabadic PFR temperature, reaction rate, and conversion profiles.

Analysis: The graphical output is shown in Figure E11-3.1. We see from Figure E11-3.1(c) that 1.15 m^3 is required for 40% conversion. The temperature and reaction-rate profiles are also shown. Notice anything strange? One observes that the rate of reaction

Look at the shape
of the curves in
Figure E11-3.1.
Why do they look
the way they do?

$$-r_A = \underbrace{kC_{A0}}_{A} \underbrace{\left[1 - \left(1 + \frac{1}{K_C}\right)X\right]}_{B} \tag{E11-3.14}$$

goes through a maximum. Near the entrance to the reactor, T increases as does k, causing term A to increase more rapidly than term B decreases, and thus the rate increases. Near the end of the reactor, term B is decreasing more rapidly than term A is increasing as we approach equilibrium. Consequently, because of these two competing effects, we have a maximum in the rate of reaction. Toward the end of the reactor, the temperature reaches a plateau as the reaction approaches equilibrium (i.e., $X \cong X_e$ at $V \cong 3.5\ m^3$). As you know, and as do all chemical engineering students at Jofostan University in Riça, at equilibrium ($-r_A \cong 0$) no further changes in X, X_e, or T take place.

AspenTech: Example 11-3 has also been formulated in AspenTech and can be loaded on your computer directly from the CRE Web site.

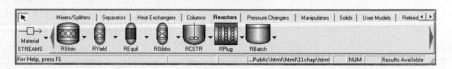

Part (c)   CSTR Solution

Let's now calculate the adiabatic CSTR volume necessary to achieve 40% conversion. Do you think the CSTR will be larger or smaller than the PFR? The mole balance is

$$V = \frac{F_{A0}X}{-r_A}$$

Using Equation (E11-3.7) in the mole balance, we obtain

$$V = \frac{F_{A0}X}{kC_{A0}\left[1 - \left(1 + \dfrac{1}{K_C}\right)X\right]} \qquad \text{(E11-3.15)}$$

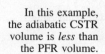

Is
$V_{\text{PFR}} > V_{\text{CSTR}}$
or is
$V_{\text{PFR}} < V_{\text{CSTR}}?$

From the energy balance, we have Equation (E11-3.9):

For 40% conversion
$$T = 330 + 43.4X$$
$$T = 330 + 43.4(0.4) = 347.3\text{K}$$

Using Equations (E11-3.10) and (E11-3.11) or from Table E11-3.1, at 347.3 K, we find k and K_C to be

$$k = 14.02 \text{ h}^{-1}$$
$$K_C = 2.73$$

Then

$$-r_A = 58.6 \text{ kmol/m}^3 \cdot \text{h}$$
$$V = \frac{(146.7 \text{ kmol butane/h})(0.4)}{58.6 \text{ kmol/m}^3 \cdot \text{h}}$$
$$V = 1.0 \text{ m}^3$$

We see that the CSTR volume (1 m^3) to achieve 40% conversion in this adiabatic reaction is less than the PFR volume (1.15 m^3).

By recalling the Levenspiel plots from Chapter 2, we can see that the reactor volume for 40% conversion is smaller for a CSTR than for a PFR. Plotting $(F_{A0}/-r_A)$ as a function of X from the data in Table E11-3.1 is shown here.

In this example, the adiabatic CSTR volume is *less* than the PFR volume.

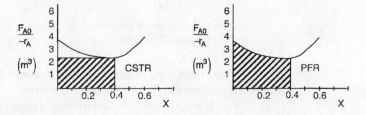

The PFR area (volume) is greater than the CSTR area (volume).

Analysis: In this example we applied the CRE algorithm to a reversible-first-order reaction carried out adiabatically in a PFR and in a CSTR. We note that at the CSTR volume necessary to achieve 40% conversion is smaller than the volume to achieve the same conversion in a PFR. In Figure E11-3.1(c) we also see that at a PFR volume of about 3.5 m³, equilibrium is essentially reached about halfway through the reactor, and no further changes in temperature, reaction rate, equilibrium conversion, or conversion take place farther down the reactor.

11.5 Adiabatic Equilibrium Conversion

For reversible reactions, the equilibrium conversion, X_e, is usually calculated first.

The highest conversion that can be achieved in reversible reactions is the equilibrium conversion. For endothermic reactions, the equilibrium conversion increases with increasing temperature up to a maximum of 1.0. For exothermic reactions, the equilibrium conversion decreases with increasing temperature.

11.5.1 Equilibrium Conversion

Exothermic Reactions. Figure 11-3(a) shows the variation of the concentration equilibrium constant, K_C, as a function of temperature for an exothermic reaction (see Appendix C), and Figure 11-3(b) shows the corresponding equilibrium conversion X_e as a function of temperature. In Example 11-3, we saw that for a first-order reaction, the equilibrium conversion could be calculated using Equation (E11-3.13)

First-order reversible reaction

$$X_e = \frac{K_C}{1 + K_C} \qquad \text{(E11-3.12)}$$

Consequently, X_e can be calculated as a function of temperature directly using either Equations (E11-3.12) and (E11-3.4) or from Figure 11-3(a) and Equation (E11-3.12).

For exothermic reactions, the equilibrium conversion decreases with increasing temperature.

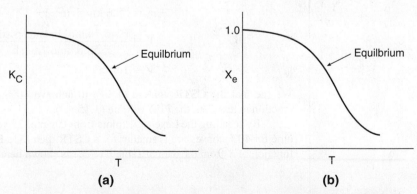

Figure 11-3 Variation of equilibrium constant and conversion with temperature for an exothermic reaction.

We note that the shape of the X_e versus T curve in Figure 11-3(b) will be similar for reactions that are other than first order.

To determine the maximum conversion that can be achieved in an exothermic reaction carried out adiabatically, we find the intersection of the equilibrium conversion as a function of temperature (Figure 11-3(b)) with temperature–conversion relationships from the energy balance (Figure 11-2 and Equation (T11-1.A)), as shown in Figure 11-4.

$$X_{EB} = \frac{\Sigma\,\Theta_i\,C_{P_i}(T - T_0)}{-\Delta H_{Rx}(T)} \qquad\qquad \text{(T11-1.A)}$$

Adiabatic
equilibrium
conversion for
exothermic
reactions

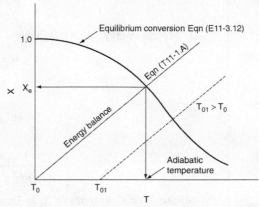

Figure 11-4   Graphical solution of equilibrium and energy balance equations to obtain the adiabatic temperature and the adiabatic equilibrium conversion X_e.

This intersection of the X_{EB} line with the X_e curve gives the adiabatic equilibrium conversion and temperature for an entering temperature T_0.

If the entering temperature is increased from T_0 to T_{01}, the energy balance line will be shifted to the right and be parallel to the original line, as shown by the dashed line. Note that as the inlet temperature increases, the adiabatic equilibrium conversion decreases.

Example 11–4 Calculating the Adiabatic Equilibrium Temperature

For the elementary liquid-phase reaction

$$A \; \rightleftarrows \; B$$

make a plot of equilibrium conversion as a function of temperature.
(a) Combine the rate law and stoichiometric to write $-r_A$ as a function of k, C_{A0}, X, and X_e.
(b) Determine the adiabatic equilibrium temperature and conversion when pure A is fed to the reactor at a temperature of 300 K.
(c) What is the CSTR volume to achieve 90% of the adiabatic equilibrium conversion for $v_0 = 5 \text{ dm}^3/\text{min}$?

Additional information:[†]

$$H_A^\circ(298 \text{ K}) = -40,000 \text{ cal/mol} \qquad H_B^\circ(298 \text{ K}) = -60,000 \text{ cal/mol}$$

$$C_{P_A} = 50 \text{ cal/mol} \cdot \text{K} \qquad C_{P_B} = 50 \text{ cal/mol} \cdot \text{K}$$

$$K_e = 100,000 \text{ at } 298 \text{ K}, \; k = 10^{-3}\exp\!\left(\frac{E}{R}\!\left(\frac{1}{298} - \frac{1}{T}\right)\right)\text{min}^{-1} \text{ with } E = 10,000 \frac{\text{cal}}{\text{mol}}$$

[†] *Jofostan Journal of Thermodynamic Data*, Vol. 23, p. 74 (1999).

Solution

1. Rate Law:

$$-r_A = k\left(C_A - \frac{C_B}{K_e}\right) \tag{E11-4.1}$$

2. Equilibrium: $-r_A = 0$; so

$$K_e = \frac{C_{Be}}{C_{Ae}} \tag{E11-4.2}$$

3. Stoichiometry: $(v = v_0)$ yields

$$C_A = C_{A0}(1-X) \tag{E11-4.3}$$

$$C_b = C_{A0}X \tag{E11-4.4}$$

$$-r_A = kC_{A0}\left(1 - X - \frac{X}{K_e}\right) \tag{E11-4.5}$$

$$K_e = \frac{C_{A0}X_e}{C_{A0}(1-X_e)} = \frac{X_e}{(1-X_e)} \tag{E11-4.6}$$

Substituting for K_c in terms of X_e in Equation (E11-4.5) and simplifying

Following the Algorithm

(a)

$$\boxed{-r_A = kC_{A0}\left(1 - \frac{X}{X_e}\right)} \tag{E11-4.7}$$

Solving Equation (E11-4.6) for X_e gives

$$\boxed{X_e = \frac{K_e(T)}{1 + K_e(T)}} \tag{E11-4.8}$$

4. Equilibrium Constant: Calculate ΔC_P, then $K_e(T)$

$$\Delta C_P = C_{P_B} - C_{P_A} = 50 - 50 = 0 \text{ cal/mol} \cdot \text{K}$$

For $\Delta C_P = 0$, the equilibrium constant varies with temperature according to the relation

$$K_e(T) = K_e(T_1) \exp\left[\frac{\Delta H_{Rx}^\circ}{R}\left(\frac{1}{T_1} - \frac{1}{T}\right)\right] \tag{E11-4.9}$$

$$\Delta H_{Rx}^\circ = H_B^\circ - H_A^\circ = -20{,}000 \text{ cal/mol}$$

$$K_e(T) = 100{,}000 \exp\left[\frac{-20{,}000}{1.987}\left(\frac{1}{298} - \frac{1}{T}\right)\right]$$

$$K_e = 100{,}000 \exp\left[-33.78\left(\frac{T-298}{T}\right)\right] \tag{E11-4.10}$$

Substituting Equation (E11-4.4) into (E11-4.2), we can calculate the equilibrium conversion as a function of temperature:

5. Equilibrium Conversion from Thermodynamics:

$$X_e = \frac{100,000 \ \exp[-33.78(T-298)/T]}{1 + 100,000 \ \exp[-33.78(T-298)/T]} \qquad \text{(E11-4.11)}$$

The calculations are shown in Table E11-4.1.

TABLE E11-4.1   EQUILIBRIUM CONVERSION
AS A FUNCTION OF TEMPERATURE

$T(K)$	K_e	X_e	$k \ (\text{min}^{-1})$
298	100,000.00	1.00	0.001
350	661.60	1.00	0.012
400	18.17	0.95	0.074
425	4.14	0.80	0.153
450	1.11	0.53	0.300
475	0.34	0.25	0.541
500	0.12	0.11	0.919

6. Energy Balance:

For a reaction carried out adiabatically, the energy balance, Equation (T11-1.A), reduces to

$$X_{\text{EB}} = \frac{\sum \Theta_i C_{P_i}(T-T_0)}{-\Delta H_{\text{Rx}}} = \frac{C_{P_A}(T-T_0)}{-\Delta H_{\text{Rx}}^{\circ}} \qquad \text{(E11-4.12)}$$

$$X_{\text{EB}} = \frac{50(T-300)}{20,000} = 2.5 \times 10^{-3}(T-300) \qquad \text{(E11-4.13)}$$

Data from Table E11-4.1 and the following data are plotted in Figure E11-4.1.

$T(K)$	300	400	500	600
X_{EB}	0	0.25	0.50	0.75

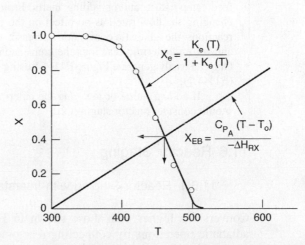

Figure E11-4.1   Finding the adiabatic equilibrium temperature (T_e) and conversion (X_e). Note: Curve uses approximate interpolated points.

The intersection of $X_{\text{EB}}(T)$ and $X_e(T)$ gives $X_e = 0.42$ and $T_e = 465$ K.

For a feed temperature of 300 K, the adiabatic equilibrium temperature is 465 K and the corresponding adiabatic equilibrium conversion is only 0.42.

Calculate the CSTR Volume to achieve 90% of the adiabatic equilibrium conversion corresponding to an entering temperature of 300 K.

$$V = \frac{F_{A0}X}{-r_A} = \frac{C_{A0}v_0 X}{kC_{A0}\left(1 - \frac{X}{X_e}\right)} = \frac{v_0 X}{k\left(1 - \frac{X}{X_e}\right)} \qquad \text{(E11-4.14)}$$

$$k = \left(0.001\exp\left[\frac{10,000}{1.987}\left(\frac{1}{298} - \frac{1}{T}\right)\right]\right) = 0.01\exp 16.89\left(\frac{T-298}{T}\right) \quad \text{(E11-4.15)}$$

at $X = 0.9\ X_e = 0.9(0.42) = 0.38$

$$\boxed{-r_A = kC_{A0}\left(1 - \frac{X}{X_e}\right)} \qquad \text{(E11-4.16)}$$

From the adiabatic energy balance, the temperature corresponding to $X = 0.38$ is

$$T = T_0 + \left(\frac{-\Delta H_{Rx}}{C_{P_A}}\right) = 300\ \text{K} + \frac{20,000\ \frac{\text{cal}}{\text{mol}}}{50\frac{\text{cal}}{\text{mol K}}}(0.38) = 452\ \text{K} \quad \text{(E11-4.17)}$$

at $T = 452$ K $\ \ X_e = 0.50$ and $k = 0.322\ \text{min}^{-1}$

$$\boxed{V = \frac{(0.38)(5\ \text{dm}^3/\text{min})}{0.322\ \text{min}^{-1}\left(1 - \frac{0.38}{0.50}\right)} = 24.5\ \text{dm}^3}$$

Analysis: The purpose of this example is to introduce the concept of the adiabatic equilibrium conversion and temperature. The adiabatic equilibrium conversion, X_e, is one of the first things to determine when carrying out an analysis involving reversible reactions. It is the maximum conversion one can achieve for a given entering temperature, T_0, and feed composition. If X_e is too low to be economical, try lowering the feed temperature and/or adding inerts. From Equation (E11-4.6), we observe that changing the flow rate has no effect on the equilibrium conversion. For exothermic reactions, the adiabatic conversion decreases with increasing entering temperature T_0, and for endothermic reactions the conversion increases with increasing entering T_0. One can easily generate Figure E11-4.1 using Polymath with Equations (E11-4.5) and (E11-4.7).

If adding inerts or lowering the entering temperature is not feasible, then one should consider reactor staging.

11.6 Reactor Staging

11.6.1 Reactor Staging with Interstage Cooling or Heating

Conversions higher than those shown in Figure E11-4.1 can be achieved for adiabatic operations by connecting reactors in series with interstage cooling.

Figure 11-5   Reactor in series with interstage cooling.

In Figure 11-5, we show the case of an exothermic adiabatic reaction taking place in a PBR reactor train. The exit temperature to Reactor 1 is very high, 800 K, and the equilibrium conversion, X_e, is low as is the reactor conversion X_1, which approaches X_e. Next, we pass the exit stream from adiabatic reactor 1 through a heat exchanger to bring the temperature back down to 500 K where X_e is high but X is still low. To increase the conversion, the stream then enters adiabatic reactor 2 where the conversion increases to X_2, which is followed by a heat exchanger and the process is repeated.

11.6.2 Exothermic Reactions

The conversion–temperature plot for this scheme is shown in Figure 11-6. We see that with three interstage coolers, 88% conversion can be achieved, compared to an equilibrium conversion of 35% for no interstage cooling.

Interstage cooling used for exothermic reversible reactions

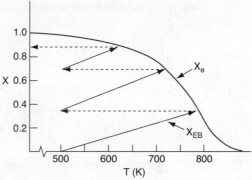

Figure 11-6   Increasing conversion by interstage cooling for an exothermic reaction. *Note:* Lines and curves are approximate.

11.6.3 Endothermic Reactions

Typical values for gasoline composition

Gasoline	
C_5	10%
C_6	10%
C_7	20%
C_8	25%
C_9	20%
C_{10}	10%
C_{11}-C_{12}	5%

Another example of the need for interstage heat transfer in a series of reactors can be found when upgrading the octane number of gasoline. The more compact the hydrocarbon molecule for a given number of carbon atoms, the higher the octane rating (see Section 10.3.5). Consequently, it is desirable to convert straight-chain hydrocarbons to branched isomers, naphthenes, and aromatics. The reaction sequence is

$$\text{Straight Chain} \xrightarrow[\text{Cat}]{k_1} \text{Naphthenes} \xrightleftharpoons[\text{Cat}]{k_2} \text{Aromatics} + 3H_2$$

The first reaction step (k_1) is slow compared to the second step, and each step is highly endothermic. The allowable temperature range for which this reaction can be carried out is quite narrow: Above 530°C undesirable side reactions occur, and below 430°C the reaction virtually does not take place. A typical feed stock might consist of 75% straight chains, 15% naphthas, and 10% aromatics.

One arrangement currently used to carry out these reactions is shown in Figure 11-7. Note that the reactors are not all the same size. Typical sizes are on

Summer 2015
$2.89/gal for octane
number (ON)
ON = 89

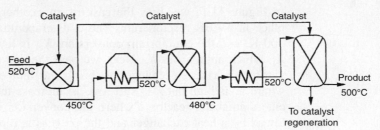

Figure 11-7 Interstage heating for gasoline production in moving-bed reactors.

the order of 10-m to 20-m high and 2 m to 5 m in diameter. A typical feed rate of gasoline is approximately 200 m³/h at 2 atm. Hydrogen is usually separated from the product stream and recycled.

Because the reaction is endothermic, the equilibrium conversion increases with increasing temperature. A typical equilibrium curve and temperature conversion trajectory for the reactor sequence are shown in Figure 11-8.

Interstage heating

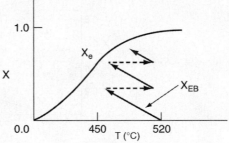

Figure 11-8 Temperature–conversion trajectory for interstage heating of an endothermic reaction analogous to Figure 11-6.

Example 11–5 Interstage Cooling for Highly Exothermic Reactions

What conversion could be achieved in Example 11-4 if two interstage coolers that had the capacity to cool the exit stream to 350 K were available? Also, determine the heat duty of each exchanger for a molar feed rate of A of 40 mol/s. Assume that 95% of the equilibrium conversion is achieved in each reactor. The feed temperature to the first reactor is 300 K.

Solution

1. Calculate Exit Temperature

For the reaction in Example 11-4, i.e.,

$$A \rightleftarrows B$$

we saw that for an entering temperature of 300 K the adiabatic equilibrium conversion was 0.42. For 95% of the equilibrium conversion ($X_e = 0.42$), the conversion exiting the first reactor is 0.4. The exit temperature is found from a rearrangement of Equation (E11-4.7)

$$T = 300 + 400X = 300 + (400)(0.4) \tag{E11-5.1}$$

$$T_1 = 460 \text{ K}$$

We now cool the gas stream exiting the reactor at 460 K down to 350 K in a heat exchanger (Figure E11-5.1).

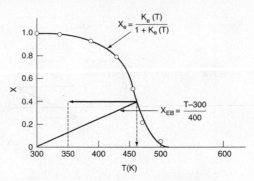

Figure E11-5.1 Determining exit conversion and temperature in the first stage. Note: Curve uses approximate interpolated points.

2. Calculate the Heat Load

There is no work done on the reaction gas mixture in the exchanger, and the reaction does not take place in the exchanger. Under these conditions ($F_{i|in} = F_{i|out}$), the energy balance given by Equation (11-10)

$$\dot{Q} - \dot{W}_s + \Sigma\, F_{i0}H_{i0} - \Sigma\, F_i H_i = 0 \tag{11-10}$$

for $\dot{W}_s = 0$ becomes

Energy balance on the reaction gas mixture in the heat exchanger

$$\dot{Q} = \Sigma\, F_i H_i - \Sigma\, F_{i0}H_{i0} = \Sigma\, F_{i0}(H_i - H_{i0}) \tag{E11-5.2}$$

$$= \Sigma\, F_i C_{P_i}(T_2 - T_1) = (F_A C_{P_A} + F_B C_{P_B})(T_2 - T_1) \tag{E11-5.3}$$

But $C_{P_A} = C_{P_B}$

$$\dot{Q} = (F_A + F_B)(C_{P_A})(T_2 - T_1) \tag{E11-5.4}$$

Also, for this example, $F_{A0} = F_A + F_B$

$$\dot{Q} = F_{A0}C_{P_A}(T_2 - T_1)$$

$$= \frac{40\ \text{mol}}{\text{s}} \cdot \frac{50\ \text{cal}}{\text{mol}\cdot\text{K}}(350 - 460)\ \text{K}$$

$$= -220\ \frac{\text{kcal}}{\text{s}} \tag{E11-5.5}$$

That is, 220 kcal/s must be removed to cool the reacting mixture from 460 K to 350 K for a feed rate of 40 mol/s.

3. Second Reactor

Now let's return to determine the conversion in the second reactor. Rearranging Equation (E11-4.7) for the second reactor

$$T_2 = T_{20} + \Delta X\left(\frac{-\Delta H_{Rx}^\circ}{C_{P_A}}\right) \tag{E11-5.6}$$

$$= 350 + 400\Delta X$$

The conditions entering the second reactor are $T = 350$ K and $X = 0.4$. The energy balance starting from this point is shown in Figure E11-5.2. The corresponding adiabatic equilibrium conversion is 0.63. Ninety-five percent of the equilibrium conversion is 60% and the corresponding exit temperature is $T = 350 + (0.6 - 0.4)400 = 430$ K.

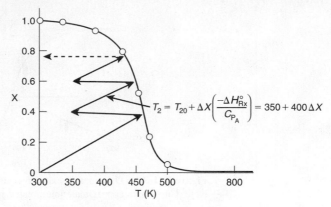

Figure E11-5.2   Three reactors in series with interstage cooling. Note: Curve uses approximate interpolated points.

4. Heat Load

The heat-exchange duty to cool the reacting mixture from 430 K back to 350 K can again be calculated from Equation (E11-5.5)

$$\dot{Q} = F_{A0}C_{P_A}(350 - 430) = \left(\frac{40 \text{ mol}}{s}\right)\left(\frac{50 \text{ cal}}{\text{mol} \cdot K}\right)(-80)$$

$$= -160 \frac{\text{kcal}}{s}$$

5. Subsequent Reactors

For the third and final reactor, we begin at $T_0 = 350$ K and $X = 0.6$ and follow the line representing the equation for the energy balance along to the point of intersection with the equilibrium conversion, which is $X = 0.8$. Consequently, the final conversion achieved with three reactors and two interstage coolers is $(0.95)(0.8) = 0.76$.

**Analysis:** For highly exothermic reactions carried out adiabatically, reactor staging with interstage cooling can be used to obtain high conversions. One observes that the exit conversion and temperature from the first reactor are 40% and 450 K respectively, as shown by the energy balance line. The exit stream at this conversion is then cooled to 350 K where it enters the second reactor. In the second reactor, the overall conversion and temperature increase to 60% and 430 K. The slope of X versus T from the energy balance is the same as the first reactor. This example also showed how to calculate the heat load on each exchanger. We also note that the heat load on the third exchanger will be less than the first exchanger because the exit temperature from the second reactor (430 K) is lower than that of the first reactor (450 K). Consequently, less heat needs to be removed by the third exchanger.

11.7 Optimum Feed Temperature

We now consider an adiabatic reactor of fixed size or catalyst weight and investigate what happens as the feed temperature is varied. The reaction is reversible and exothermic. At one extreme, using a very high feed temperature, the specific reaction rate will be large and the reaction will proceed rapidly, but the equilibrium conversion will be close to zero. Consequently, very little product will be formed. At the other extreme of low feed temperatures, little product will be formed because the reaction rate is so low.

We now consider the adiabatic reversible exothermic reaction

$$R \rightleftarrows S$$

Combining the thermodynamic Equations (E11-3.12) and (E11-3.4) for this reaction, we can construct the equilibrium conversion, X_e, as a function of temperature shown in Figure 11-9. Inserting the appropriate value for C_{P_i}, Θ_i, and $-\Delta H_{Rx}$ into the adiabatic energy balance, Equation (T11-1.A) for this reacting system we obtain

$$X_{EB} = \frac{C_{P_R}(T - T_0)}{-\Delta H_{Rx}} = \frac{75 \text{ J/mol} \cdot \text{K}}{1{,}000 \text{ J/mol}}(T - T_0) = 0.075(T - T_0) \quad (11\text{-}31)$$

We now plot the adiabatic energy balances, Equation (11-31), for three different entering temperatures T_0 in Figure 11-9. Next, we look for the intersections of the conversion calculated from the energy balance, X_{EB}, with the equilibrium conversion X_e calculated from thermodynamics. We see that for an entering temperature of $T_0 = 600$ K the adiabatic equilibrium conversion X_e is 0.15, and the corresponding adiabatic equilibrium temperature is 620 K. However, for an entering temperature of $T_0 = 350$ K, X_e is 0.75 and the corresponding adiabatic equilibrium temperature is 450 K.

We next plot the corresponding conversion profiles down the length of the reactor for entering temperatures 350 K, 500 K and 600 K as shown in Figure 11-10. Because the reactor temperature increases as we move along the reactor, the equilibrium conversion, X_e, also varies and decreases along the length of the reactor, as shown by the dashed line in Figure 11-10. We also see that at the high entering temperature, 600 K, the rate is very rapid and equilibrium is achieved very near the reactor entrance. The corresponding conversion profiles down the length of the reactor for these temperatures are shown in Figure 11-10. The equilibrium conversion, which can be calculated from Equation (E11-4.2) for a first-order reaction also varies along the length of the reactor, as shown by the dashed line in Figure 11-10. We also see that because of the high entering temperature, the rate is very rapid at the inlet and equilibrium is achieved very near the reactor entrance. The solid line is calculated by combining the energy mole balances and solving numerically.

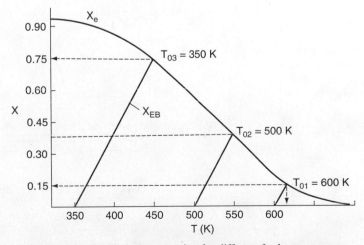

Figure 11-9   Equilibrium conversion for different feed temperatures.

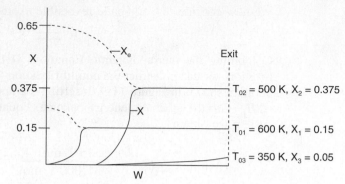

Observe how the
conversion profile
changes as the
entering
temperature is
decreased from
600 K.

Figure 11-10   Adiabatic conversion profiles for different feed temperatures.

We notice that the conversion and temperature increase very rapidly over a short distance (i.e., a small amount of catalyst). This sharp increase is sometimes referred to as the "point" or temperature at which the reaction "ignites." If the inlet temperature were lowered to 500 K, the corresponding equilibrium conversion would increase to 0.375; however, the reaction rate is slower at this lower temperature so that this conversion is not achieved until closer to the end of the reactor. If the entering temperature were lowered further to 350 K, the corresponding equilibrium conversion would be 0.75, but the rate is so slow that a conversion of 0.05 is achieved for the specified catalyst weight in the reactor. At a very low feed temperature, the specific reaction rate will be so small that virtually all of the reactant will pass through the reactor without reacting. It is apparent that with conversions close to zero for both high and low feed temperatures, there must be an optimum feed temperature that maximizes conversion. As the feed temperature is increased from a very low value, the specific reaction rate will increase, as will the conversion. The conversion will continue to increase with increasing feed temperature until the equilibrium conversion is approached in the reaction. Further increases in feed temperature for this exothermic reaction will only decrease the conversion due to the decreasing equilibrium conversion. This optimum inlet temperature is shown in Figure 11-11.

Optimum inlet
temperature

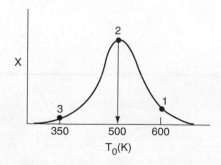

Figure 11-11   Finding the optimum feed temperature.

Closure. Virtually all reactions that are carried out in industry involve heat effects. This chapter provides the basis to design reactors that operate at steady state and that involve heat effects. To model these reactors, we simply add another step to our algorithm; this step is the energy balance. One of the goals of this chapter is to understand each term of the energy balance and how it was derived. We have found that if the reader understands the various steps in the derivation, he/she will be in a much better position to apply the equation correctly. In order not to overwhelm the reader while studying reactions with heat effects, we have broken up the different cases and only consider the case of reactors operated adiabatically in this chapter. Chapter 12 will focus on reactors with heat exchange. Chapter 13 will focus on reactors not operated at steady state. An industrial adiabatic reaction concerning the manufacture of sulfuric acid provides a number of practical details is included on the CRE Web site in PRS R12.4.

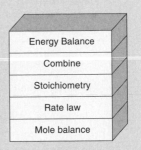

SUMMARY

For the reaction

$$A + \frac{b}{a}B \rightarrow \frac{c}{a}C + \frac{d}{a}D$$

1. The heat of reaction at temperature T, per mole of A, is

$$\Delta H_{Rx}(T) = \frac{c}{a}H_C(T) + \frac{d}{a}H_D(T) - \frac{b}{a}H_B(T) - H_A(T) \tag{S11-1}$$

2. The mean heat capacity difference, ΔC_P, per mole of A is

$$\Delta C_P = \frac{c}{a}C_{PC} + \frac{d}{a}C_{PD} - \frac{b}{a}C_{PB} - C_{PA} \tag{S11-2}$$

where C_{P_i} is the mean heat capacity of species i between temperatures T_R and T.

3. When there are no phase changes, the heat of reaction at temperature T is related to the heat of reaction at the standard reference temperature T_R by

$$\Delta H_{Rx}(T) = H^\circ_{Rx}(T_R) + \Delta C_P(T - T_R) \tag{S11-3}$$

4. The steady-state energy balance on a system volume V is

$$\boxed{\dot{Q} - F_{A0}\Sigma\Theta_i C_{P_i}(T - T_{i0}) - [\Delta H^\circ_{Rx}(T_R) + \Delta C_P(T - T_R)]F_{A0}X = 0} \tag{S11-4}$$

We now couple the first four building blocks *Mole Balance*, *Rate Law*, *Stoichiometry*, and *Combine* with the fifth block, the *Energy Balance*, to solve nonisothermal reaction engineering problems as shown in the Closure box for this chapter.

5. **For adiabatic operation** ($\dot{Q} \equiv 0$) of a PFR, PBR, CSTR, or batch reactor (BR), and neglecting $\dot{W}_s^\circ$, we solve Equation (S11-4) for the adiabatic conversion-temperature relationship, which is

$$X = \frac{\Sigma \Theta_i C_{P_i}(T - T_0)}{-[\Delta H^\circ_{Rx}(T_R) + \Delta C_P(T - T_R)]} \tag{S11-5}$$

Solving Equation (S11-5) for the adiabatic temperature–conversion relationship:

$$T = \frac{X[-\Delta H^\circ_{Rx}(T_R)] + \Sigma \Theta_i C_{P_i} T_0 + X \, \Delta C_P T_R}{[\Sigma \Theta_i C_{P_i} + X \, \Delta C_P]} \tag{S11-6}$$

Using Equation (S11-4), one can solve nonisothermal adiabatic reactor problems to predict the exit conversion, concentrations, and temperature.

CRE WEB SITE MATERIALS

- **Expanded Material**
    1. *Puzzle Problem "What's Wrong with this Solution?"*
- **Learning Resources**
    1. *Summary Notes*
    2. *PFR/PBR Solution Procedure for a Reversible Gas-Phase Reaction*
- **Living Example Problems**
    1. *Example 11-3 Polymath Adiabatic Isomerization of Normal Butane*
    2. *Example 11-3 Formulated in AspenTech*—Download from the CRE Web site

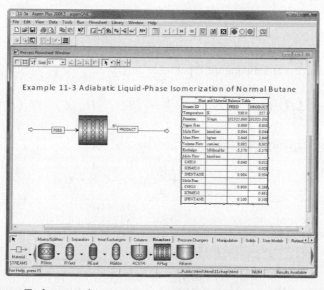

A step-by-step AspenTech tutorial is given on the CRE Web site.

- **Professional Reference Shelf**

    R11.1. *Variable Heat Capacities.* Next, we want to arrive at a form of the energy balance for the case where heat capacities are strong functions of temperature over a wide temperature range. Under these conditions, the mean values of the heat capacity may not be adequate for the relationship between conversion and temperature. Combining the heat of reaction with the quadratic form of the heat capacity

$$C_{P_i} = \alpha_i + \beta_i T + \gamma_i T^2$$

we find that the heat capacity as a function of temperature is

$$\Delta H_{Rx}(T) = \Delta H_{Rx}^{\circ}(T_R) + \Delta \alpha (T - T_R) + \frac{\Delta \beta}{2}(T^2 - T_R^2) + \frac{\Delta \gamma}{3}(T^3 - T_R^3)$$

Example 11-4 is reworked for the case of variable heat capacities.

QUESTIONS AND PROBLEMS

The subscript to each of the problem numbers indicates the level of difficulty: A, least difficult; D, most difficult.

$$A = \bullet \quad B = \blacksquare \quad C = \blacklozenge \quad D = \blacklozenge\blacklozenge$$

Homework Problems

In each of the questions and problems, rather than just drawing a box around your answer, write a sentence or two describing how you solved the problem, the assumptions you made, the reasonableness of your answer, what you learned, and any other facts that you want to include. See the Preface for additional generic parts **(x)**, **(y)**, **(z)** to the home problems.

Creative Problems

> **Before** solving the problems, state or sketch qualitatively the expected results or trends.

Questions

Q11-1$_A$ Read over the problems at the end of this chapter. Make up an original problem that uses the concepts presented in this chapter. To obtain a solution:
   (a) Make up your data and reaction.
   (b) Use a real reaction and real data. See Problem P5-1$_A$ for guidelines.
   (c) Prepare a list of safety considerations for designing and operating chemical reactors. What would be the first four items on your list? (See *www.sache.org* and *www.siri.org/graphics*.) The August 1985 issue of *Chemical Engineering Progress* may be useful.
   (d) Recalling selectivity in Chapter 8, why and how would you use reactor staging to improve selectivity of series, parallel, and complex reactions?
   (e) Which example on the CRE Web site Lecture Notes for Chapter 11 was the most difficult?
   (f) What if you were asked to give an everyday example that demonstrates the principles discussed in this chapter? (Would sipping a teaspoon of Tabasco or other hot sauce be one?)
   (g) Rework Problem P2-9$_D$ (page 67) for the case of adiabatic operation.

Problems

P11-1$_A$ Load the following Polymath programs from the CRE Web site where appropriate:
   (a) Example 11-1. How would this example change if a CSTR were used instead of a PFR?
   (b) Example 11-2. (1) What would the heat of reaction be if 50% inerts (e.g., helium) were added to the system? (2) What would be the % error if the ΔC_P term were neglected?

Hall of Fame

(c) Example 11-3. (1) What if the butane reaction were carried out in a 0.8-m³ PFR that can be pressurized to very high pressures? (2) What inlet temperature would you recommend? Is there an optimum inlet temperature? (3) Plot the heat that must be removed along the reactor ($\dot{Q}$ vs. V) to maintain isothermal operation.

(d) AspenTech Example 11-3. Download the AspenTech program from the CRE Web site. (1) Repeat P11-2$_B$ **(c)** using AspenTech. (2) Vary the inlet flow rate and temperature, and describe what you find.

(e) Example 11-4. (1) Make a plot of the equilibrium conversion as a function of entering temperature, T_0. (2) What do you observe at high and low T_0? (3) Make a plot of X_e versus T_0 when the feed is equal molar in inerts that have the same heat capacity. (4) Compare the plots of X_e versus T_0 with and without inerts and describe what you find.

(f) Example 11-5. (1) Determine the molar flow rate of cooling water (C_{Pw} = 18 cal/mol·K) necessary to remove 220 kcal/s from the first exchanger. The cooling water enters at 270 K and leaves at 400 K. (2) Determine the necessary heat exchanger area A (m²) for an overall heat-transfer coefficient, U, of 100 cal/s·m²·K. You must use the log-mean driving force in calculating A.

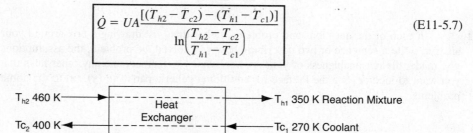

$$\dot{Q} = UA\frac{[(T_{h2} - T_{c2}) - (T_{h1} - T_{c1})]}{\ln\left(\dfrac{T_{h2} - T_{c2}}{T_{h1} - T_{c1}}\right)} \tag{E11-5.7}$$

T_{h2} 460 K ⟶ ⟶ T_{h1} 350 K Reaction Mixture

Heat Exchanger

T_{c2} 400 K ⟵ ⟵ T_{c1} 270 K Coolant

Figure P11-1$_A$ (e) Countercurrent heat exchanger.

P11-2$_A$ For elementary reaction

$$A \rightleftarrows B$$

the equilibrium conversion is 0.8 at 127°C and 0.5 at 227°C. What is the heat of reaction?

P11-3$_B$ The equilibrium conversion is shown below as a function of catalyst weight

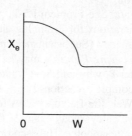

Please indicate which of the following statements are true and which are false. Explain each case.

(a) The reaction could be first-order endothermic and carried out adiabatically.

(b) The reaction is first-order endothermic and the reactor is heated along its length with T_a being constant.

(c) The reaction is second-order exothermic and cooled along the length of the reactor with T_a being constant.

(d) The reaction is second-order exothermic and carried out adiabatically.

P11-4$_A$ The elementary, irreversible, organic liquid-phase reaction

$$A + B \rightarrow C$$

is carried out adiabatically in a flow reactor. An equal molar feed in A and B enters at 27°C, and the volumetric flow rate is 2 dm³/s and $C_{A0} = 0.1$ kmol/m³.

Additional information:

$H_A^\circ(273\ \text{K}) = -20\ \text{kcal/mol}, H_B^\circ(273\ \text{K}) = -15\ \text{kcal/mol},$

$H_C^\circ(273\ \text{K}) = -41\ \text{kcal/mol}$

$$C_{P_A} = C_{P_B} = 15\ \text{cal/mol} \cdot \text{K} \qquad C_{P_C} = 30\ \text{cal/mol} \cdot \text{K}$$

$$k = 0.01\ \frac{\text{dm}^3}{\text{mol} \cdot \text{s}}\ \text{at 300 K} \qquad E = 10{,}000\ \text{cal/mol}$$

PFR
(a) Plot and then analyze the conversion and temperature as a function of PFR volume up to where $X = 0.85$. Describe the trends.
(b) What is the maximum inlet temperature one could have so that the boiling point of the liquid (550 K) would not be exceeded even for complete conversion?
(c) Plot the heat that must be removed along the reactor ($\dot{Q}$ vs. V) to maintain isothermal operation.
(d) Plot and then analyze the conversion and temperature profiles up to a PFR reactor volume of 10 dm³ for the case when the reaction is reversible with $K_C = 10$ m³/kmol at 450 K. Plot the equilibrium conversion profile. How are the trends different than part (**a**)?

CSTR
(e) What is the CSTR volume necessary to achieve 90% conversion?

BR
(f) The reaction is next carried out in a 25 dm³ batch reactor charged with $N_{A0} = 10$ moles. Plot the number of moles of A, N_A, the conversion, and the temperature as a function of time.

P11-5$_A$ The elementary, irreversible gas-phase reaction

$$A \rightarrow B + C$$

is carried out adiabatically in a PFR packed with a catalyst. Pure A enters the reactor at a volumetric flow rate of 20 dm³/s, at a pressure of 10 atm, and a temperature of 450 K.

Additional information:

$$C_{P_A} = 40\ \text{J/mol} \cdot \text{K} \qquad C_{P_B} = 25\ \text{J/mol} \cdot \text{K} \qquad C_{P_C} = 15\ \text{J/mol} \cdot \text{K}$$

$$H_A^\circ = -70\ \text{kJ/mol} \qquad H_B^\circ = -50\ \text{kJ/mol} \qquad H_C^\circ = -40\ \text{kJ/mol}$$

All heats of formation are referenced to 273 K.

$$k = 0.133\ \exp\left[\frac{E}{R}\left(\frac{1}{450} - \frac{1}{T}\right)\right]\ \frac{\text{dm}^3}{\text{kg-cat} \cdot \text{s}}\ \text{with}\ E = 31.4\ \text{kJ/mol}$$

(a) Plot and then analyze the conversion and temperature down the plug-flow reactor until an 80% conversion (if possible) is reached. (The maximum catalyst weight that can be packed into the PFR is 50 kg.) Assume that $\Delta P = 0.0$.
(b) Vary the inlet temperature and describe what you find.
(c) Plot the heat that must be removed along the reactor ($\dot{Q}$ vs. V) to maintain isothermal operation.

(d) Now take the pressure drop into account in the PBR with $\rho_b = 1$ kg/dm^3. The reactor can be packed with one of two particle sizes. Choose one.

$$\alpha = 0.019/\text{kg-cat} \text{ for particle diameter } D_1$$

$$\alpha = 0.0075/\text{kg-cat} \text{ for particle diameter } D_2$$

(e) Plot and then analyze the temperature, conversion, and pressure along the length of the reactor. Vary the parameters α and P_0 to learn the ranges of values in which they dramatically affect the conversion.

(f) Apply one or more of the six ideas in Preface Table P-4, page xxviii, to this problem.

P11-6$_B$ The irreversible endothermic vapor-phase reaction follows an elementary rate law

$$CH_3COCH_3 \rightarrow CH_2CO + CH_4$$

$$A \rightarrow B + C$$

and is carried out adiabatically in a 500-dm^3 PFR. Species A is fed to the reactor at a rate of 10 mol/min and a pressure of 2 atm. An inert stream is also fed to the reactor at 2 atm, as shown in Figure P11-6$_B$. The entrance temperature of both streams is 1100 K.

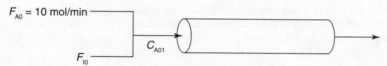

Figure P11-6$_B$ Adiabatic PFR with inerts.

Additional information:

$k = \exp(34.34 - 34{,}222/T)$ dm^3/mol·min $C_{P_I} = 200$ J/mol·K

(T in degrees Kelvin)

$C_{P_A} = 170$ J/mol·K $C_{P_B} = 90$ J/mol·K

$C_{P_C} = 80$ J/mol·K $\Delta H^\circ_{Rx} = 80{,}000$ J/mol

(a) First derive an expression for C_{A01} as a function of C_{A0} and Θ_I.

(b) Sketch the conversion and temperature profiles for the case when no inerts are present. Using a dashed line, sketch the profiles when a moderate amount of inerts are added. Using a dotted line, sketch the profiles when a large amount of inerts are added. Qualitative sketches are fine. Describe the similarities and differences between the curves.

(c) Sketch or plot and then analyze the exit conversion as a function of Θ_I. Is there a ratio of the entering molar flow rates of inerts (I) to A (i.e., $\Theta_I = F_{I0}/F_{A0}$) at which the conversion is at a maximum? Explain why there "is" or "is not" a maximum.

(d) What would change in parts (b) and (c) if reactions were exothermic and reversible with $\Delta H^\circ_{Rx} = -80$ kJ/mol and $K_C = 2$ dm^3/mol at 1100 K?

(e) Sketch or plot F_B for parts (c) and (d), and describe what you find.

(f) Plot the heat that must be removed along the reactor ($\dot{Q}$ vs. V) to maintain isothermal operation for pure A fed and an exothermic reaction. Part (f) is "C" level of difficulty, i.e., **P11-6$_C$(f)**.

P11-7$_B$ The gas-phase reversible reaction

$$A \rightleftarrows B$$

is carried out under high pressure in a packed-bed reactor with pressure drop. The feed consists of both inerts I and Species A with the ratio of inerts to the species A being 2 to 1. The entering molar flow rate of A is 5 mol/min at a temperature of 300 K and a concentration of 2 mol/dm³. Work this problem in terms of volume. *Hint:* $V = W/\rho_B$, $r_A = \rho_B r'_A$.

Additional information:

$F_{A0} = 5.0 \, \text{mol/min}$

$C_{A0} = 2 \, \text{mol/dm}^3$

$C_I = 2 \, C_{A0}$

$C_{P_I} = 18 \, \text{cal/mol/K}$

$C_{P_A} = 160 \, \text{cal/mol/K}$

$E = 10,000 \, \text{cal/mol}$

$\Delta H_{Rx} = -20,000 \, \text{cal/mol}$

$K_C = 1,000 \, \text{at} \, 300 \, \text{K}$

$C_{P_B} = 160 \, \text{cal/mol/K}$

$\rho_B = 1.2 \, \text{kg/dm}^3$

$T_0 = 300 \, \text{K}$

$T_1 = 300 \, \text{K}$

$k_1 = 0.1 \, \text{min}^{-1} \, \text{at} \, 300 \, \text{K}$

$Ua = 150 \, \text{cal/dm}^3/\text{min/K}$

$T_{ao} = 300 \, \text{K}$

$V = 40 \, \text{dm}^3$

$\alpha\rho_b = 0.02 \, \text{dm}^{-3}$

<u>Coolant</u>

$\dot{m}_C = 50 \, \text{mol/min}$

$C_{P_{Cool}} = 20 \, \text{cal/mol/K}$

(a) Adiabatic Operation. Plot X, X_e, p, T, and the rate of disappearance as a function of V up to $V = 40 \, \text{dm}^3$. Explain why the curves look the way they do.

(b) Vary the ratio of inerts to A $(0 \le \Theta_I \le 10)$ and the entering temperature, and describe what you find.

(c) Plot the heat that must be removed along the reactor ($\dot{Q}$ vs. V) to maintain isothermal operation. Part (c) is "C" level of difficulty.

We will continue this problem in Chapter 12.

P11-8$_B$ Algorithm for reaction in a PBR with heat effects

The elementary gas-phase reaction

$$A + B \rightleftharpoons 2C$$

is carried out in a packed-bed reactor. The entering molar flow rates are $F_{A0} = 5 \, \text{mol/s}$, $F_{B0} = 2F_{A0}$, and $F_I = 2F_{A0}$ with $C_{A0} = 0.2 \, \text{mol/dm}^3$. The entering temperature is 325 K and a coolant fluid is available at 300 K.

Additional information:

$C_{P_A} = C_{P_B} = C_{P_C} = 20 \, \text{cal/mol/K}$ $k = 0.0002 \dfrac{\text{dm}^6}{\text{kg} \cdot \text{mol} \cdot \text{s}} \, @ \, 300\text{K}$

$C_{P_I} = 18 \, \text{cal/mol/K}$

$E = 25 \dfrac{\text{kcal}}{\text{mol}}$

$\alpha = 0.00015 \, \text{kg}^{-1}$ $Ua = 320 \dfrac{\text{Cal}}{\text{s} \cdot \text{m}^3 \cdot \text{K}}$

$\dot{m}_c = 18 \, \text{mol/s}$ $\rho_b = 1400 \dfrac{\text{kg}}{\text{m}^3}$

$\Delta H_{Rx} = -20 \dfrac{\text{kcal}}{\text{mol}} \, @ \, 298\text{K}$

$C_{P_{Cool}} = 18 \, \text{cal/mol (coolant)}$

$K_C = 1000 \, @ \, 305\text{K}$

(a) Write the mole balance, the rate law, K_C as a function of T, k as a function of T, and C_A, C_B, C_C as a function of X, p, and T.

(b) Write the rate law as a function of X, p, and T.

(c) Show the equilibrium conversion is

$$X_e = \frac{\dfrac{3K_C}{4} - \sqrt{\left(\dfrac{3K_C}{4}\right)^2 - 2K_C\left(\dfrac{K_C}{4} - 1\right)}}{2\left(\dfrac{K_C}{4} - 1\right)}$$

and then plot X_e vs. T.

(d) What are $\sum \Theta_i C_{P_i}$, ΔC_P, T_0, entering temperature T_1 (rate law), and T_2 (equilibrium constant)?

(e) Write the energy balance for adiabatic operation.

(f) **Case 1 Adiabatic Operation.** Plot and then analyze X_e, X, p, and T versus W when the reaction is carried out adiabatically. Describe why the profiles look the way they do. Identify those terms that will be affected by inerts. Sketch what you think the profiles X_e, X, p, and T will look like before you run the Polymath program to plot the profiles.

(g) Plot the heat that must be removed along the reactor (Q vs. V) to maintain isothermal operation. Part (g) is "C" level of difficulty, i.e., **P11-9$_C$(g).**

P11-9$_A$ The reaction

$$A + B \underset{\longleftarrow}{\overset{\longrightarrow}{\rightleftharpoons}} C + D$$

is carried out adiabatically in a series of staged packed-bed reactors with interstage cooling (see Figure 11-5). The lowest temperature to which the reactant stream may be cooled is 27°C. The feed is equal molar in A and B, and the catalyst weight in each reactor is sufficient to achieve 99.9% of the equilibrium conversion. The feed enters at 27°C and the reaction is carried out adiabatically. If four reactors and three coolers are available, what conversion may be achieved?

Additional information:

$$\Delta H^\circ_{Rx} = -30,000 \text{ cal/mol A} \qquad C_{P_A} = C_{P_B} = C_{P_C} = C_{P_D} = 25 \text{ cal/g mol} \cdot \text{K}$$

$$K_e(50°C) = 500,000 \qquad F_{A0} = 10 \text{ mol A/min}$$

First prepare a plot of equilibrium conversion as a function of temperature. (*Partial ans.: T = 360 K, $X_e = 0.984$; T = 520 K, $X_e = 0.09$; T = 540 K, $X_e = 0.057$*)

P11-10$_A$ Figure P11-10$_A$ shows the temperature–conversion trajectory for a train of reactors with interstage heating. Now consider replacing the interstage heating with injection of the feed stream in three equal portions, as shown in Figure P11-10$_A$:

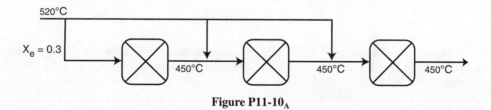

Figure P11-10$_A$

Sketch the temperature-conversion trajectories for **(a)** an endothermic reaction with entering temperatures as shown, and **(b)** an exothermic reaction with the temperatures to and from the first reactor reversed, i.e., $T_0 = 450°C$.

SUPPLEMENTARY READING

1. An excellent development of the energy balance is presented in

   ARIS, R., *Elementary Chemical Reactor Analysis*. Upper Saddle River, NJ: Prentice Hall, 1969, Chaps. 3 and 6.

A number of example problems dealing with nonisothermal reactors may or may not be found in

   BURGESS, THORNTON W., *The Adventures of Old Man Coyote*. New York: Dover Publications, Inc., 1916.

   BUTT, JOHN B., *Reaction Kinetics and Reactor Design,* Revised and Expanded, 2nd ed. New York: Marcel Dekker, Inc., 1999.

WALAS, S.M., *Chemical Reaction Engineering Handbook of Solved Problems*. Amsterdam: Gordon and Breach, 1995. See the following solved problems: 4.10.1, 4.10.08, 4.10.09, 4.10.13, 4.11.02, 4.11.09, 4.11.03, 4.10.11.

For a thorough discussion on the heat of reaction and equilibrium constant, one might also consult

DENBIGH, K. G., *Principles of Chemical Equilibrium*, 4th ed. Cambridge: Cambridge University Press, 1981.

2. The heats of formation, $H_i(T)$, Gibbs free energies, $G_i(T_R)$, and the heat capacities of various compounds can be found in

GREEN, D. W. and R. H. PERRY, eds., *Chemical Engineers' Handbook*, 8th ed. New York: McGraw-Hill, 2008.

REID, R. C., J. M. PRAUSNITZ, and T. K. SHERWOOD, *The Properties of Gases and Liquids*, 3rd ed. New York: McGraw-Hill, 1977.

WEAST, R. C., ed., *CRC Handbook of Chemistry and Physics*, 94th ed. Boca Raton, FL: CRC Press, 2013.

Steady-State Nonisothermal Reactor Design—Flow Reactors with Heat Exchange

12

> Research is to see what everybody else sees, and to think what nobody else has thought.
>
> —Albert Szent-Gyorgyi

Overview. This chapter focuses on chemical reactors with heat exchange. The chapter topics are arranged in the following manner:

- Section 12.1 further develops the energy balance for easy application to PFRs and PBRs.
- Section 12.2 describes PFRs and PBRs for four types of heat exchanger operations.
  (1) Constant heat-transfer fluid temperature, T_a
  (2) Variable fluid temperature T_a with co-current operation
  (3) Variable fluid temperature T_a with countercurrent operation
  (4) Adiabatic operation
- Section 12.3 describes the algorithm for PFRs and PBRs and gives examples of four types of heat exchanger operations.
- Section 12.4 applies the energy balance to a CSTR.
- Section 12.5 shows how a CSTR can operate at different steady-state temperatures and conversions, and how to decide which of these conditions are stable and which are unstable.
- Section 12.6 describes one of the most important topics of the entire text, multiple reactions with heat effects, which is unique to this textbook.
- Section 12.7 discusses radial and axial temperature and concentration gradients.

> • Section 12.8 Safety: The causes of batch-reactor explosions are discussed and hint of things to come when we do a CSI analysis on the Monsanto and T2 Laboratory explosions in Chapter 13.
>
> The *Chapter 12 Professional Reference Shelf (R12.4)* on the CRE Web site describes a typical nonisothermal industrial reactor and reaction, SO_2 oxidation, and gives many practical details.

12.1 Steady-State Tubular Reactor with Heat Exchange

In this section, we consider a tubular reactor in which heat is either added or removed through the cylindrical walls of the reactor (Figure 12-1). In modeling the reactor, we shall assume that there are no radial gradients in the reactor and that the heat flux through the wall per unit volume of reactor is as shown in Figure 12-1.[1]

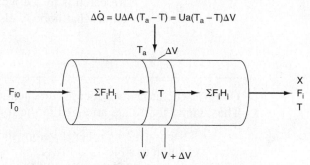

Figure 12-1   Tubular reactor with heat gain or loss.

12.1.1 Deriving the Energy Balance for a PFR

We will carry out an energy balance on the volume ΔV. There is no work done, i.e., $\dot{W}_s = 0$, so Equation (11-10) becomes

$$\Delta \dot{Q} + \Sigma F_i H_i \big|_V - \Sigma F_i H_i \big|_{V + \Delta V} = 0 \tag{12-1}$$

The heat flow to the reactor, $\Delta \dot{Q}$, is given in terms of the overall heat-transfer coefficient, U, the heat exchange area, ΔA, and the difference between the ambient temperature, T_a, and the reactor temperature T

$$\Delta \dot{Q} = U \Delta A (T_a - T) = Ua \Delta V (T_a - T)$$

where a is the heat exchange area per unit volume of reactor. For a tubular reactor

$$a = \frac{A}{V} = \frac{\pi D L}{\frac{\pi D^2 L}{4}} = \frac{4}{D}$$

---

[1] Radial gradients are discussed in Chapters 17 and 18.

where D is the reactor diameter. Substituting for $\Delta \dot{Q}$ in Equation (12-1), dividing by ΔV, and then taking the limit as $\Delta V \to 0$, we get

$$Ua(T_a - T) - \frac{d\,\Sigma(F_i H_i)}{dV} = 0$$

Expanding

$$Ua(T_a - T) - \Sigma \frac{dF_i}{dV} H_i - \Sigma F_i \frac{dH_i}{dV} = 0 \tag{12-2}$$

From a mole balance on species i, we have

$$\frac{dF_i}{dV} = r_i = v_i(-r_A) \tag{12-3}$$

Differentiating the enthalpy Equation (11-19) with respect to V

$$\frac{dH_i}{dV} = C_{P_i} \frac{dT}{dV} \tag{12-4}$$

Substituting Equations (12-3) and (12-4) into Equation (12-2), we obtain

$$Ua(T_a - T) - \underbrace{\Sigma v_i H_i}_{\Delta H_{Rx}} (-r_A) - \Sigma F_i C_{P_i} \frac{dT}{dV} = 0$$

Rearranging, we arrive at

This form of the energy balance will also be applied to multiple reactions.

$$\boxed{\frac{dT}{dV} = \frac{\overbrace{r_A \Delta H_{Rx}}^{\substack{Q_g \\ \text{Heat} \\ \text{``Generated''}}} - \overbrace{Ua(T - T_a)}^{\substack{Q_r \\ \text{Heat} \\ \text{``Removed''}}}}{\Sigma F_i C_{P_i}}} \tag{12-5}$$

and

$$\boxed{\frac{dT}{dV} = \frac{Q_g - Q_r}{\Sigma F_i C_{P_i}}} \tag{T11-1G}$$

where

$$Q_g = r_A \Delta H_{Rx}$$

$$Q_r = Ua(T - T_a)$$

which is Equation (T11-1G) in Table 11-1 on pages 498–500.

For exothermic reactions, Q_g will be a positive number. We note that when the heat "generated," Q_g, is greater than the heat "removed," Q_r (i.e., $Q_g > Q_r$), the temperature will increase down the reactor. When $Q_r > Q_g$, the temperature will decrease down the reactor.

For endothermic reactions, Q_g will be a negative number and Q_r will also be a negative number because $T_a > T$. The temperature will decrease if $(-Q_g) > (-Q_r)$ and increase if $(-Q_r) > (-Q_g)$.

Equation (12-5) is coupled with the mole balances on each species, Equation (12-3). Next, we express r_A as a function of either the concentrations for liquid systems or molar flow rates for gas systems, as described in Chapter 4. We will use the molar flow rate form of the energy balance for membrane reactors and also extend this form to multiple reactions.

We could also write Equation (12-5) in terms of conversion by recalling $F_i = F_{A0}(\Theta_i + v_i X)$ and substituting this expression into the denominator of Equation (12-5).

PFR energy balance

$$\frac{dT}{dV} = \frac{r_A \Delta H_{Rx} - Ua(T - T_a)}{F_{A0}(\Sigma \Theta_i C_{P_i} + \Delta C_P X)} = \frac{Q_g - Q_r}{\Sigma F_i C_{P_i}} \tag{12-6}$$

For a packed-bed reactor $dW = \rho_b \, dV$ where ρ_b is the bulk density

PBR energy balance

$$\frac{dT}{dW} = \frac{r_A' \Delta H_{Rx} - \dfrac{Ua(T - T_a)}{\rho_b}}{\Sigma F_i C_{P_i}} \tag{12-7}$$

Equations (12-6) and (12-7) are also given in Table 11-1 as Equations (T11-1D) and (T11-1F). As noted earlier, having gone through the derivation to these equations, it will be easier to apply them accurately to CRE problems with heat effects.

12.1.2 Applying the Algorithm to Flow Reactors with Heat Exchange

We continue to use the algorithm described in the previous chapters and simply add a fifth building block, *the energy balance*.

Gas Phase

If the reaction is in gas phase and pressure drop is included, there are four differential equations that must be solved simultaneously. The differential equation describing the change in temperature with volume (i.e., distance) as we move down the reactor

Energy balance

$$\frac{dT}{dV} = g(X, T, T_a) \tag{A}$$

must be coupled with the mole balance

Mole balance

$$\frac{dX}{dV} = \frac{-r_A}{F_{A0}} = f(X, T, p) \tag{B}$$

and with the pressure drop equation

Pressure drop

$$\frac{dp}{dV} = -h(p, X, T) \tag{C}$$

and solved simultaneously. If the temperature of the heat-exchange fluid, T_a, varies down the reactor, we must add the energy balance on the heat-exchange fluid. In the next section, we will derive the following equation for co-current heat transfer

Heat exchanger

$$\frac{dT_a}{dV} = \frac{Ua(T - T_a)}{\dot{m}_{C0}C_{P_{C0}}} \tag{D}$$

Numerical integration of the coupled differential equations (A) to (D) is required.

along with the equation for countercurrent heat transfer. A variety of numerical schemes can be used (e.g., Polymath) to solve these coupled differential equations: **(A)**, **(B)**, **(C)**, and **(D)**.

Liquid Phase

For liquid-phase reactions, the rate is not a function of total pressure, so our mole balance is

$$\frac{dX}{dV} = \frac{-r_A}{F_{A0}} = f(X, T) \tag{E}$$

Consequently, we need to only solve equations **(A)**, **(D)**, and **(E)** simultaneously.

12.2 Balance on the Heat-Transfer Fluid

12.2.1 Co-current Flow

The heat-transfer fluid will be a coolant for exothermic reactions and a heating medium for endothermic reactions. If the flow rate of the heat-transfer fluid is sufficiently high with respect to the heat released (or absorbed) by the reacting mixture, then the heat-transfer fluid temperature will be virtually constant along the reactor. In the material that follows, we develop the basic equations for a coolant to remove heat from exothermic reactions; however, these same equations apply to endothermic reactions where a heating medium is used to supply heat.

We now carry out an energy balance on the coolant in the annulus between R_1 and R_2, and between V and $V + \Delta V$, as shown in Figure 12-2. The mass flow rate of the heat-exchange fluid (e.g., coolant) is $\dot{m}_c$. We will consider the case when the reactor is cooled and the outer radius of the coolant channel R_2 is *insulated*. Recall that by convention $\dot{Q}$ is the heat **added to** the system.

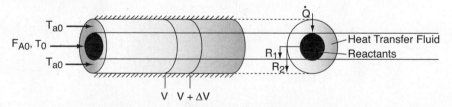

Figure 12-2   Co-current, double-pipe heat exchanger.

For co-current flow, the reactant and the coolant flow in the same direction.

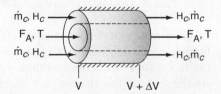

The energy balance on the coolant in the volume between V and $(V + \Delta V)$ is

$$
\begin{bmatrix} \text{Rate of energy} \\ \text{in at } V \end{bmatrix} - \begin{bmatrix} \text{Rate of energy} \\ \text{out at } V + \Delta V \end{bmatrix} + \begin{bmatrix} \text{Rate of heat added} \\ \text{by conduction through} \\ \text{the inner wall} \end{bmatrix} = 0
$$

$$
\dot{m}_c H_c|_V \quad - \quad \dot{m}_c H_c|_{V + \Delta V} \quad + \quad Ua(T - T_a)\Delta V \quad = 0
$$

where T_a is the temperature of the heat transfer fluid, i.e., coolant, and T is the temperature of the reacting mixture in the inner tube.

Dividing by ΔV and taking limit as $\Delta V \to 0$

$$
-\dot{m}_c \frac{dH_c}{dV} + Ua(T - T_a) = 0 \tag{12-8}
$$

Analogous to Equation (12-4), the change in enthalpy of the coolant can be written as

$$
\frac{dH_c}{dV} = C_{P_c} \frac{dT_a}{dV} \tag{12-9}
$$

The variation of coolant temperature T_a down the length of reactor is

$$
\boxed{\frac{dT_a}{dV} = \frac{Ua(T - T_a)}{\dot{m}_c C_{P_c}}} \tag{12-10}
$$

The equation is valid whether the heat-transfer fluid is a coolant or a heating medium.

Typical heat-transfer fluid temperature profiles are shown here for both exothermic and endothermic reactions when the heat transfer fluid enters at T_{a0}.

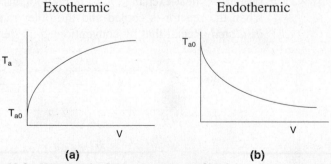

Figure 12-3   Heat-transfer fluid temperature profiles for co-current heat exchanger. **(a)** Coolant. **(b)** Heating medium.

12.2.2 Countercurrent Flow

In countercurrent heat exchange, the reacting mixture and the heat-transfer fluid (e.g., coolant) flow in opposite directions. At the reactor entrance, $V = 0$, the reactants enter at temperature T_0, and the coolant exits at temperature T_{a2}. At the end of the reactor, the reactants and products exit at temperature T, while the coolant enters at T_{a0}.

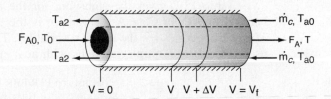

Figure 12-4   Countercurrent, double-pipe heat exchanger.

Again, we write an energy balance over a differential reactor volume to arrive at

$$\boxed{\frac{dT_a}{dV} = \frac{Ua(T_a - T)}{\dot{m}_c C_{P_c}}}$$ (12-11)

At the entrance, $V = 0$ ∴ $X = 0$ and $T_a = T_{a2}$.
At the exit, $V = V_f$ ∴ $T_a = T_{a0}$.

We note that the only difference between Equations (12-10) and (12-11) is a minus sign, i.e., $(T - T_a)$ vs. $(T_a - T)$.

The solution to a countercurrent flow problem to find the exit conversion and temperature requires a *trial-and-error* procedure.

TABLE 12-1   PROCEDURE TO SOLVE FOR THE EXIT CONDITIONS FOR PFRs WITH COUNTERCURRENT HEAT EXCHANGE

---

1. Consider an exothermic reaction where the coolant stream enters at the end of the reactor ($V = V_f$) at a temperature T_{a0}, say 300 K. We have to carry out a *trial-and-error* procedure to find the temperature of the coolant exiting the reactor.
2. Guess an exit coolant temperature T_{a2} at the feed entrance ($X = 0$, $V = 0$) to the reactor to be $T_{a2} = 340$ K, as shown in Figure 12-5(**a**).
3. Use an ODE solver to calculate X, T, and T_a as a function of V.

Trial and error procedure required

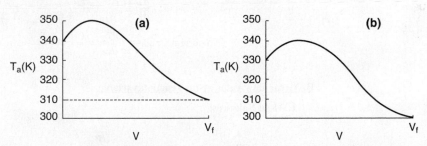

Figure 12-5   Trial-and-error results for countercurrent heat exchanger.

We see from Figure 12-5(**a**) that our guess of 340 K for T_{a2} at the feed entrance ($V = 0$ and $X = 0$) gives an entering temperature of the coolant of 310 K ($V = V_f$), which does not match the actual entering coolant temperature of 300 K.

4. Now guess another coolant temperature at $V = 0$ and $X = 0$ of, say, 330 K. After carrying out the simulation with this guess, we see from Figure 12-5(**b**) that a coolant temperature at $V = 0$ of $T_{a2} = 330$ K will give a coolant entering temperature at V_f of 300 K, which matches the actual T_{a0}.

12.3 Algorithm for PFR/PBR Design with Heat Effects

We now have all the tools to solve reaction engineering problems involving heat effects in PFRs and PBRs for the cases of both constant and variable coolant temperatures.

Table 12-2 gives the algorithm for the design of PFRs and PBRs with heat exchange: In <u>Case A</u> *Conversion* is the reaction variable and in <u>Case B</u> *Molar Flow Rates* are the reaction variables. The procedure in Case B must be used to analyze multiple reactions with heat effects.

TABLE 12-2   ALGORITHM FOR PFR/PBR DESIGN FOR GAS-PHASE
REACTIONS WITH HEAT EFFECTS

Living Example Problem

A. Conversion as the reaction variable

$$A + B \rightleftharpoons 2C$$

1. Mole Balance:

$$\frac{dX}{dV} = \frac{-r_A}{F_{A0}} \tag{T12-1.1}$$

2. Rate Law:

Elementary Reaction

$$-r_A = k_1 \left(C_A C_B - \frac{C_C^2}{K_C} \right) \tag{T12-1.2}$$

$$k = k_1(T_1) \exp\left[\frac{E}{R} \left(\frac{1}{T_1} - \frac{1}{T} \right) \right] \tag{T12-1.3}$$

for $\Delta C_P \cong 0$.

$$K_C = K_{C2}(T_2) \exp\left[\frac{\Delta H_{Rx}^\circ}{R} \left(\frac{1}{T_2} - \frac{1}{T} \right) \right] \tag{T12-1.4}$$

3. Stoichiometry (gas phase, **no** ΔP):

$$C_A = C_{A0}(1 - X) \frac{T_0}{T} \tag{T12-1.5}$$

$$C_B = C_{A0}(\Theta_B - X) \frac{T_0}{T} \tag{T12-1.6}$$

$$C_C = 2 C_{A0} X \frac{T_0}{T} \tag{T12-1.7}$$

4. Energy Balances:

Reactor:

$$\frac{dT}{dV} = \frac{Ua(T_a - T) + (-r_A)(-\Delta H_{Rx})}{F_{A0}[C_{P_A} + \Theta_B C_{P_B} + X \, \Delta C_P]} \tag{T12-1.8}$$

Co-current Coolant:

$$\frac{dT_a}{dV} = \frac{Ua(T - T_a)}{\dot{m}_c C_{P_c}} \tag{T12-1.9}$$

B. Molar flow rates as the reaction variable

1. Mole Balances:

$$\frac{dF_A}{dV} = r_A \tag{T12-1.10}$$

$$\frac{dF_B}{dV} = r_B \tag{T12-1.11}$$

$$\frac{dF_C}{dV} = r_C \tag{T12-1.12}$$

Following the Algorithm

2. Rate Law: Elementary reaction

$$-r_A = k_1 \left(C_A C_B - \frac{C_C^2}{K_C} \right) \tag{T12-1.2}$$

$$k = k_1(T_1) \exp\left[\frac{E}{R} \left(\frac{1}{T_1} - \frac{1}{T} \right) \right] \tag{T12-1.3}$$

$$K_C = K_{C2}(T_2) \exp\left[\frac{\Delta H_{Rx}^\circ}{R} \left(\frac{1}{T_2} - \frac{1}{T} \right) \right] \tag{T12-1.4}$$

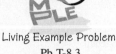

Living Example Problem
Pb T-8.3

Variable coolant
temperature

Summary Notes

TABLE 12-2   ALGORITHM FOR PFR/PBR DESIGN FOR GAS-PHASE
REACTIONS WITH HEAT EFFECTS (CONTINUED)

3. Stoichiometry (gas phase, **no** ΔP):

$$r_B = r_A \tag{T12-1.13}$$

$$r_C = -2r_A \tag{T12-1.14}$$

$$C_A = C_{T0} \frac{F_A}{F_T} \frac{T_0}{T} \tag{T12-1.15}$$

$$C_B = C_{T0} \frac{F_B}{F_T} \frac{T_0}{T} \tag{T12-1.16}$$

$$C_C = C_{T0} \frac{F_C}{F_T} \frac{T_0}{T} \tag{T12-1.17}$$

$$F_T = F_A + F_B + F_C \tag{T12-1.18}$$

4. Energy Balance:

Reactor: $$\frac{dT}{dV} = \frac{Ua(T_a - T) + (-r_A)(-\Delta H_{Rx})}{F_A C_{P_A} + F_B C_{P_B} + F_C C_{P_C}} \tag{T12-1.19}$$

Heat Exchangers:

If the heat-transfer fluid (e.g., coolant) temperature, T_a, is not constant, the energy balance on the heat-exchange fluid gives

Co-current flow

$$\frac{dT_a}{dV} = \frac{Ua(T - T_a)}{\dot{m}_c C_{P_c}} \tag{T12-1.20}$$

Countercurrent flow

$$\frac{dT_a}{dV} = \frac{Ua(T_a - T)}{\dot{m}_c C_{P_c}} \tag{T12-1.21}$$

where $\dot{m}_c$ is the mass flow rate of the coolant (e.g., kg/s, and C_{P_c} is the heat capacity of the coolant (e.g., kJ/kg·K).

5. Parameter Evaluation:

Now we enter all the explicit equations with the appropriate parameter values.

Case A: Conversion as the Independent Variable

k_1, E, R, C_{T0}, T_a, T_0, T_1, T_2, K_{C2}, Θ_B, ΔH°_{Rx}, C_{P_A}, C_{P_B}, C_{P_C}, Ua

with initial values T_0 and $X = 0$ at $V = 0$ and final values: $V_f = $ _____

Case B: Molar Flow Rates as the Independent Variables

Same as Case A, except that the inlet values F_{A0}, and F_{B0} are specified instead of X at $V = 0$.

Living Example Problem

Note: In Problem P12-3$_B$, the equations in this table have been applied directly to a PBR (recall that we simply use $W = \rho_b V$). Problem P12-3$_B$ is also a *Living Example Problem* (12-T12-3) on the CRE Web site, *www.umich.edu/~elements/5e/index.html*. Download this *Living Example Problem* from the CRE Web site and vary the cooling rate, flow rate, entering temperature, and other parameters to get an intuitive feel of what happens in flow reactors with heat effects. After carrying out this exercise, go to the Workbook that is at the very end of Chapter 12 *Summary Notes* on the CRE Web site and answer the questions in the workbook and also at the end of entitled *PBR with Heat Exchange and Variable Coolant Flow Rate*.

The following figures in Table 12-2 show representative profiles that would result from solving the above equations. The reader is encouraged to download the *Living Example Problem* for Table 12-2 and vary a number of parameters, as discussed in P12-3$_B$. Be sure you can explain why these curves look the way they do.

TABLE 12-2   ALGORITHM FOR PFR/PBR DESIGN FOR GAS-PHASE
REACTIONS WITH HEAT EFFECTS (CONTINUED)

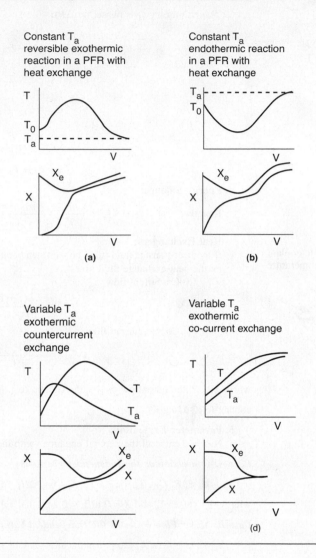

12.3.1 Applying the Algorithm to an Exothermic Reaction

Example 12–1 Butane Isomerization Continued—OOPS!

When plant engineer Maxwell Anthony looked up the vapor pressure at the exit to
the adiabatic reactor in Example 11-3, where the temperature is 360 K, he learned
the vapor pressure was about 1.5 MPa for isobutene, which is greater than the rupture pressure of the glass vessel the company had hoped to use. Fortunately, when
Max looked in the storage shed, he found there was a bank of 10 tubular reactors,
each of which was 5 m³. The bank reactors were double-pipe heat exchangers with
the reactants flowing in the inner pipe and with Ua = 5,000 kJ/m³·h·K. Max also
bought some thermodynamic data from one of the companies he found on the

Internet that did Colorimeter experiments to find ΔH_{Rx} for various reactions. One of the companies had the value of ΔH_{Rx} for his reaction on sale this week for the low, low price of \$25,000.00. For this value of ΔH_{Rx} the company said it is best to use an initial concentration of A of 1.86 mol/dm^3. The entering temperature of the reactants is 305 K and the entering coolant temperature is 315 K. The mass flow rate of the coolant, $\dot{m}_c$, is 500 kg/h and the heat capacity of the coolant, C_{P_c}, is 28 kJ/kg·K. The temperature in any one of the reactors cannot rise above 325 K. Carry out the following analyses with the newly purchased values from the Internet:

(a) Co-current heat exchange: Plot X, X_e, T, T_a, and $-r_A$, down the length of the reactor.

(b) Countercurrent heat exchange: Plot X, X_e, T, T_a, and $-r_A$ down the length of the reactor.

(c) Constant ambient temperature, T_a: Plot X, X_e, T, and $-r_A$ down the length of the reactor.

(d) Adiabatic operation: Plot X, X_e, T, T_a, and $-r_A$, down the length of the reactor.

(e) Compare parts (a) through (d) above and write a paragraph describing what you find.

Additional information

Recall from Example 11-3 that $C_{P_A} = 141$ kJ/kmol·K, $C_{P0} = \Sigma\Theta_i C_{Pi} = 159$ kJ/kmol·K, and data from the company Maxwell got off the Internet are $\Delta H_{Rx} = -34{,}500$ kJ/kmol with $\Delta C_{P_A} = 0$ and $C_{A0} = 1.86$ kmol/m^3

Solution

We shall first solve part (a), the co-current heat exchange case and then make small changes in the Polymath program for parts (b) through (d).

For each of the ten reactors in parallel

$$F_{A0} = (0.9)(163 \text{ kmol/h}) \times \frac{1}{10} = 14.7 \frac{\text{kmol A}}{\text{h}}$$

The mole balance, rate law, and stoichiometry are the same as in the adiabatic case previously discussed in **Example 11-3**; that is,

The Algorithm

Same as Example 11-3

Mole Balance:
$$\boxed{\frac{dX}{dV} = \frac{-r_A}{F_{A0}}} \qquad (\text{E}11\text{-}3.1)$$

Rate Law and Stoichiometry:
$$\boxed{r_A = -kC_{A0}\left[1 - \left(1 + \frac{1}{K_C}\right)X\right]} \qquad (\text{E}11\text{-}3.7)$$

with

$$k = 31.1 \exp\left[7906\left(\frac{T - 360}{360T}\right)\right] \text{ h}^{-1} \qquad (\text{E}11\text{-}3.10)$$

$$K_C = 3.03 \exp\left[-830.3\left(\frac{T - 333}{333T}\right)\right] \qquad (\text{E}11\text{-}3.11)$$

$$Q_g = r_A \Delta H_{Rx}$$

$$Q_r = Ua(T - T_a)$$

$$\frac{dT}{dV} = \frac{Q_g - Q_r}{F_{A0} C_{P0}}$$

The equilibrium conversion is

$$X_e = \frac{K_C}{1 + K_C} \tag{E11-3.12}$$

Energy Balance

The energy balance on the reactor is

$$\frac{dT}{dV} = \frac{r_A \Delta H_{Rx} - Ua(T - T_a)}{F_{A0} \underbrace{\sum \Theta_i C_{P_i}}_{C_{P0}}} = \frac{r_A \Delta H_{Rx} - Ua(T - T_a)}{F_{A0} C_{P0}} \tag{E12-1.1}$$

Part (a) Co-current Heat Exchange

We are now going to solve the coupled, ordinary differential and explicit equations (E11-3.1), (E11-3.7), (E11-3.10), (E11-3.11), (E11-3.12), and (E12-1.1), and the appropriate heat-exchange balance using Polymath. After entering these equations we will enter the parameter values. By using co-current heat exchange as our Polymath base case, we only need to change one line in the program for each of the other three cases and solve for the profiles of X, X_e, T, T_a, and $-r_A$, and not have to re-enter the program.

For co-current flow, the balance on the heat transfer fluid is

$$\frac{dT_a}{dV} = \frac{Ua(T - T_a)}{\dot{m}_C C_{P_C}} \tag{E12-1.2}$$

with $T_a = 310$ K at $V = 0$. The Polymath program and solution are shown in Table E12-1.1.

Living Example Problem

TABLE E12-1.1   PART (a) CO-CURRENT HEAT EXCHANGE

Differential equations
1 d(Ta)/d(V) = Ua*(T-Ta)/m/Cpc

2 d(X)/d(V) = -ra/Fa0

3 d(T)/d(V) = ((ra*deltaH)-Ua*(T-Ta))/Cpo/Fa0

Explicit equations
1   Cpc = 28

2   m = 500

3   Ua = 5000

4   Ca0 = 1.86

5   Fa0 = 0.9*163*.1

6   deltaH = -34500

7   k = 31.1*exp((7906)*(T-360)/(T*360))

8   Kc = 3.03*exp((deltaH/8.314)*((T-333)/(T*333)))

9   Xe = Kc/(1+Kc)

10 ra = -k*Ca0*(1-(1+1/Kc)*X)

11 Cpo = 159

12 rate = -ra

POLYMATH Report
Ordinary Differential Equations

Calculated values of DEQ variables

	Variable	Initial value	Final value
1	Ca0	1.86	1.86
2	Cpc	28.	28.
3	Cpo	159.	159.
4	deltaH	-3.45E+04	-3.45E+04
5	Fa0	14.67	14.67
6	k	0.5927441	6.80861
7	Kc	9.512006	2.641246
8	m	500.	500.
9	ra	-1.102504	-0.1181808
10	rate	1.102504	0.1181808
11	T	305.	336.7102
12	Ta	315.	335.6949
13	Ua	5000.	5000.
14	V	0	5.
15	X	0	0.7185996
16	Xe	0.9048707	0.7253687

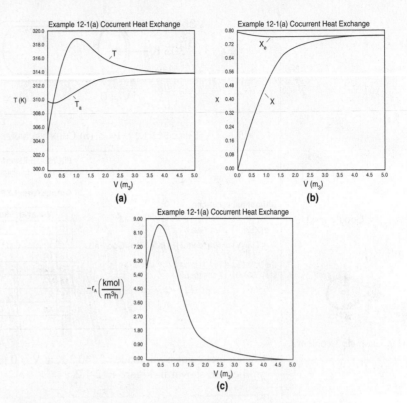

Figure E12-1.1   Profiles down the reactor for co-current heat exchange
(**a**) temperature, (**b**) conversion, (**c**) reaction rate.

Analysis: **Part (a) Co-current Exchange:** We note that reactor temperature goes through a maximum. Near the reactor entrance, the reactant concentrations are high and therefore the reaction rate is high [c.f. Figure E12-1.1(**a**)] and $Q_g > Q_r$. Consequently, the temperature and conversion increase with increasing reactor volume while X_e decreases because of the increasing temperature. Eventually, X and X_e come close together (V = 0.95 m^3) and the rate becomes very small as the reaction approaches equilibrium. At this point, the reactant conversion X cannot increase unless X_e increases. We also note that when the ambient heat-exchanger temperature, T_a, and the reactor temperature, T, are essentially equal, there is no longer a temperature driving force to cool the reactor. Consequently, the temperature does not change farther down the reactor, nor does the equilibrium conversion, which is only a function of temperature.

Part (b) Countercurrent Heat Exchange:

For countercurrent flow, we only need to make two changes in the program. First, multiply the right-hand side of Equation (E12-1.2) by minus one to obtain

$$\frac{dT_a}{dV} = -\frac{Ua(T - T_a)}{\dot{m}_C C_{P_C}} \qquad (E12\text{-}1.2)$$

Next, we guess T_a at $V = 0$ and see if it matches T_{a0} at $V = 5$ m^3. If it doesn't, we guess again. In this example, we will guess T_a ($V = 0$) = 340.3 K and see if $T_a = T_{a0} = 315$ K at $V = 5$ m^3.

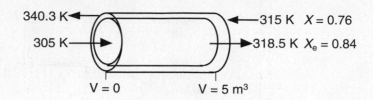

TABLE E12-1.2   PART **(b)** COUNTERCURRENT HEAT EXCHANGE

Good guess!

Living Example Problem

POLYMATH Report
Ordinary Differential Equations

Output from LEP Polymath Example

Differential equations
1 d(Ta)/d(V) = -Ua*(T-Ta)/m/Cpc
2 d(X)/d(V) = -ra/Fa0
3 d(T)/d(V) = ((ra*deltaH)-Ua*(T-Ta))/Cpo/Fa0

The explicit equations (1) through (12) are the same
as in Table E12-1.1 on page 550

	Variable	Initial value	Final value
10	rate	1.102504	0.31466
11	T	305	318.359
12	Ta	340.3	314.9774
13	Ua	5000	5000
14	V	0	5
15	X	0	0.762033
16	Xe	0.904871	0.843122

Guess of
$T_a = 340.3$ K
(@ $V = 0$)
← Matches
$T_{a0} = 315$ K

What a lucky guess we made of 340.3 K at V = 0 to find $T_{a0} = 315$ K!! The variable profiles are shown in Figure E12-1.2.

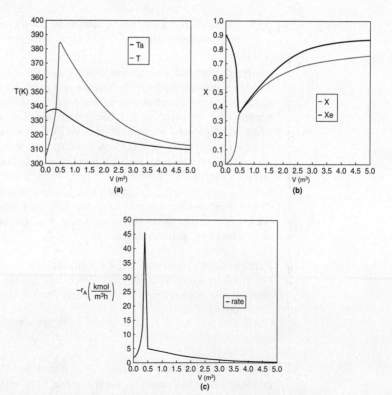

Figure E12-1.2   Profiles down the reactor for counter current heat exchange
(a) temperature, **(b)** conversion, **(c)** reaction rate.

Analysis: **Part (b) Countercurrent Exchange:** We note that near the entrance to the reactor, the coolant temperature is above the reactant entrance temperature. However, as we move down the reactor, the reaction generates "heat" and the reactor temperature rises above the coolant temperature. We note that X_e reaches a minimum (corresponding to the reactor temperature maximum) near the entrance to the reactor. At this point (V = 0.5 m³), X cannot increase above X_e. As we move down the reactor, the reactants are cooled and the reactor temperature decreases allowing X and X_e to increase. A higher exit conversion, X, and equilibrium conversion, X_e, are achieved in the countercurrent heat-exchange system than for the co-current system.

Part (c) Constant T_a

For constant T_a, use the Polymath program in part (a), but multiply the right side of Equation (E12-1.2) by zero in the program, i.e.,

$$\frac{dT_a}{dV} = \frac{Ua(T - T_a)}{\dot{m}_C C_{P_C}} * 0 \qquad \text{(E12-1.3)}$$

TABLE E12-1.3   PART **(c)** CONSTANT T_a

Differential equations

1 d(Ta)/d(V) = Ua*(T-Ta)/m/Cpc*0

2 d(X)/d(V) = -ra/Fa0

3 d(T)/d(V) = ((ra*deltaH)-Ua*(T-Ta))/Cpo/Fa0

The explicit equations (1) through (12) are the same as in Table E12-1.1 on page 550.

POLYMATH Report
Ordinary Differential Equations

Calculated values of DEQ variables

	Variable	Initial value	Final value
10	rate	1.102504	0.3048421
11	T	305.	317.4737
12	Ta	315.	315.
13	Ua	5000.	5000.
14	V	0	5.
15	X	0	0.7632038
16	Xe	0.9048707	0.8478702

The initial and final values are shown in the Polymath report and the variable profiles are shown in Figure E12-1.3 on page 553.

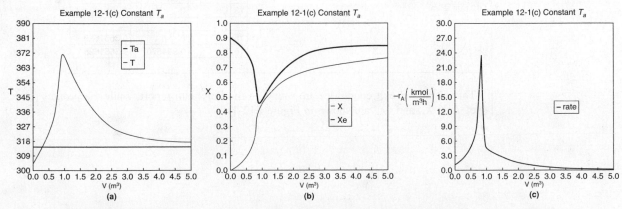

Figure E12-1.3   Profiles down the reactor for constant heat-exchange fluid temperature T_a; **(a)** temperature, **(b)** conversion, **(c)** reaction rate.

**Analysis:** **Part (c) Constant** T_a**:** When the coolant flow rate is sufficiently large, the coolant temperature, T_a, will be essentially constant. If the reactor volume is sufficiently large, the reactor temperature will eventually reach the coolant temperature, as is the case here. At this exit temperature, which is the lowest achieved in this example, the equilibrium conversion, X_e, is the largest of the four cases studied in this example.

Part (d) Adiabatic Operation

In Example 11-3, we solved for the temperature as a function of conversion and then used that relationship to calculate k and K_C. An easier way is to solve the general or base case of a heat exchanger for co-current flow and write the corresponding Polymath program. Next, use Polymath [part **(a)**], but multiply the parameter Ua by zero, i.e.,

$$Ua = 5,000 * 0$$

and run the simulation again.

TABLE E12-1.4 PART **(d)** ADIABATIC OPERATION

Differential equations
1 d(Ta)/d(V) = Ua*(T-Ta)/m/Cpc
2 d(X)/d(V) = -ra/Fa0
3 d(T)/d(V) = ((ra*deltaH)-Ua*(T-Ta))/Cpo/Fa0

Explicit equations
1 Cpc = 28
2 m = 500
3 Ua = 5000*0
4 Ca0 = 1.86
5 Fa0 = 0.9*163*.1
6 deltaH = -34500
7 k = 31.1*exp((7906)*(T-360)/(T*360))
8 Kc = 3.03*exp((deltaH/8.314)*((T-333)/(T*333)))
9 Xe = Kc/(1+Kc)
10 ra = -k*Ca0*(1-(1+1/Kc)*X)
11 Cpo = 159
12 rate = -ra

POLYMATH Report
Ordinary Differential Equations

Calculated values of DEQ variables

	Variable	Initial value	Final value
1	Ca0	1.86	1.86
2	Cpc	28.	28.
3	Cpo	159.	159.
4	deltaH	-3.45E+04	-3.45E+04
5	Fa0	14.67	14.67
6	k	0.5927441	124.2488
7	Kc	9.512006	0.5752071
8	m	500.	500.
9	ra	-1.102504	1.191E-05
10	rate	1.102504	-1.191E-05
11	T	305.	348.2335
12	Ta	315.	315.
13	Ua	0	0
14	V	0	5.
15	X	0	0.3651629
16	Xe	0.9048707	0.3651628

The initial and exit conditions are shown in the Polymath report, while the profiles of T, X, X_e, and $-r_A$ are shown in Figure E12-1.4.

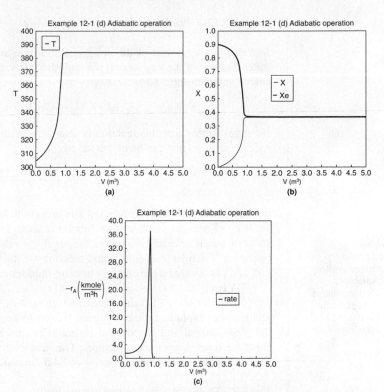

Figure E12-1.4   Profiles down the reactor for adiabatic reactor;
(a) temperature, **(b)** conversion, **(c)** reaction rate.

**Analysis:** **Part (d) Adiabatic Operation.** Because there is no cooling, the temperature of this exothermic reaction will continue to increase down the reactor until equilibrium is reached, $X = X_e = 0.365$ at $T = 384$ K, which is the _adiabatic equilibrium temperature_. The profiles for X and X_e are shown in Figure E12-1.4(b) where one observes that X_e decreases down the reactor because of the increasing temperature until it becomes equal to the reactor conversion (i.e., $X \equiv X_e$), which occurs circa 0.9 m³. There is no change in temperature, X or X_e, after this point because the reaction rate is virtually zero and thus the remaining reactor volume serves no purpose.

Finally, Figure E12-1.4(c) shows $-r_A$ increases as we move down the reactor as the temperature increases, reaching a maximum and then decreasing until X and X_e approach each other and the rate becomes virtually zero.

**Overall Analysis:** This is an extremely important example, as we applied our CRE PFR algorithm with heat exchange to a reversible exothermic reaction. We analyzed four types of heat-exchanger operations. We see the countercurrent exchanger gives the highest conversion and adiabatic operation gives the lowest conversion.

12.3.2 Applying the Algorithm to an Endothermic Reaction

In Example 12-1 we studied the four different types of heat exchanger on an exothermic reaction. In this section we carry out the same study on an _endothermic reaction_.

Example 12–2 Production of Acetic Anhydride

Jeffreys, in a treatment of the design of an acetic anhydride manufacturing facility, states that one of the key steps is the endothermic vapor-phase cracking of acetone to ketene and methane is[2]

$$CH_3COCH_3 \rightarrow CH_2CO + CH_4$$

He states further that this reaction is first-order with respect to acetone and that the specific reaction rate can be expressed by

$$\ln k = 34.34 - \frac{34,222}{T} \qquad \text{(E12-2.1)}$$

Gas-phase
endothermic
reaction examples:
  1. Adiabatic
  2. Heat exchange
     T_a is constant
  3. Co-current heat
     exchange with
     variable T_a
  4. Counter
     current
     exchange with
     variable T_a

where k is in reciprocal seconds and T is in Kelvin. In this design it is desired to feed 7850 kg of acetone per hour to a tubular reactor. The reactor consists of a bank of 1000 one-inch schedule 40 tubes. We shall consider four cases of heat exchanger operation. The inlet temperature and pressure are the same for all cases at 1035 K and 162 kPa (1.6 atm) and the entering heating-fluid temperature available is 1250 K.

A bank of 1000 one-in. schedule 40 tubes 1.79 m in length corresponds to 1.0 m³ (0.001 m³/tube = 1.0 dm³/tube) and gives 20% conversion. Ketene is unstable and tends to explode, which is a good reason to keep the conversion low. However, the pipe material and schedule size should be checked to learn if they are suitable for these temperatures and pressures. The heat-exchange fluid has a flow rate, $\dot{m}_C$, of 0.111 mol/s, with a heat capacity of 34.5 J/mol·K.

Case 1   The reactor is operated adiabatically.
Case 2   Constant heat-exchange fluid temperature $T_a = 1250$ K
Case 3   Co-current heat exchange with $T_{a0} = 1250$ K
Case 4   Countercurrent heat exchange with $T_{a0} = 1250$ K

Additional information

$$CH_3COCH_3 \quad (A): H_A^\circ (T_R) = -216.67 \text{ kJ/mol}, \ C_{P_{A^\circ}} = 163 \text{ J/mol} \cdot K$$

$$CH_2CO \quad (B): H_B^\circ (T_R) = -61.09 \text{ kJ/mol}, \ C_{P_B} = 83 \text{ J/mol} \cdot K$$

$$CH_4 \quad (C): H_C^\circ (T_R) = -74.81 \text{ kJ/mol}, \ C_{P_C} = 71 \text{ J/mol} \cdot K$$

$$Ua = 110 \text{ J/s} \cdot m^3 \cdot K$$

Solution

Let A = CH₃COCH₃, B = CH₂CO, and C = CH₄. Rewriting the reaction symbolically gives us

$$A \rightarrow B + C$$

Algorithm for a PFR with Heat Effects

1. Mole Balance: $\qquad \dfrac{dX}{dV} = -\dfrac{r_A}{F_{A0}} \qquad$ (E12-2.2)

---

[2] G. V. Jeffreys, *A Problem in Chemical Engineering Design: The Manufacture of Acetic Anhydride*, 2nd ed. (London: Institution of Chemical Engineers, 1964).

2. Rate Law: $$-r_A = kC_A \qquad \text{(E12-2.3)}$$

Rearranging (E12-2.1)

$$k = 8.2 \times 10^{14} \exp\left[-\frac{34,222}{T}\right] = 3.58 \; \exp\left[34,222\left(\frac{1}{1035} - \frac{1}{T}\right)\right] \qquad \text{(E12-2.4)}$$

3. Stoichiometry (gas-phase reaction with no pressure drop):

$$C_A = \frac{C_{A0}(1-X)T_0}{(1+\varepsilon X)T} \qquad \text{(E12-2.5)}$$

$$\varepsilon = y_{A0}\delta = 1(1+1-1) = 1$$

4. Combining yields

$$-r_A = \frac{kC_{A0}(1-X)}{1+X}\frac{T_0}{T} \qquad \text{(E12-2.6)}$$

Before combining Equations (E12-2.2) and (E12-2.6), it is first necessary to use the energy balance to determine T as a function of X.

5. Energy Balance:

a. Reactor balance

$$\frac{dT}{dV} = \frac{Ua(T_a - T) + (r_A)\left[\Delta H_{Rx}^\circ + \Delta C_P(T - T_R)\right]}{F_{A0}\left(\sum \Theta_i C_{P_i} + X\Delta C_P\right)} \qquad \text{(E12-2.7)}$$

b. Heat Exchanger. We will use the heat-exchange fluid balance for co-current flow as our base case. We will then show how we can very easily modify our ODE solver program (e.g., Polymath) to solve for the other cases by simply multiplying the appropriate line in the code by either zero or minus one.

For *co-current flow:*

$$\frac{dT_a}{dV} = \frac{Ua(T - T_a)}{\dot{m}C_{P_c}} \qquad \text{(E12-2.8)}$$

Following the Algorithm

6. Calculation of Mole Balance Parameters on a Per Tube Basis:

$$F_{A0} = \frac{7,850 \text{ kg/h}}{58 \text{ kg/kmol}} \times \frac{1}{1,000 \text{ Tubes}} = 0.135 \text{ kmol/h} = 0.0376 \text{ mol/s}$$

$$C_{A0} = \frac{P_{A0}}{RT} = \frac{162 \text{ kPa}}{8.31 \dfrac{\text{kPa}\cdot\text{m}^3}{\text{kmol}\cdot\text{K}} (1035 \text{ K})} = 0.0188 \frac{\text{kmol}}{\text{m}^3} = 18.8 \text{ mol/m}^3$$

$$v_0 = \frac{F_{A0}}{C_{A0}} = 2.0 \text{ dm}^3/\text{s}, \quad V = \frac{1 \text{ m}^3}{1000 \text{ tubes}} = \frac{0.001 \text{ m}^3}{\text{tube}} = \frac{1.0 \text{ dm}^3}{\text{tube}}$$

7. Calculation of Energy Balance Parameters:
Thermodynamics:

a. $\Delta H_{Rx}^\circ(T_R)$: At 298 K, using the standard heats of formation

$$\Delta H_{Rx}^\circ(T_R) = H_B^\circ(T_R) + H_C^\circ(T_R) - H_A^\circ(T_R)$$

$$= (-61.09) + (-74.81) - (-216.67) \text{ kJ/mol}$$

$$= 80.77 \text{ kJ/mol}$$

b. ΔC_P: Using the mean heat capacities

$$\Delta C_P = C_{P_B} + C_{P_C} - C_{P_A} = (83 + 71 - 163) \text{ J/mol}\cdot\text{K}$$

$$\Delta C_P = -9 \text{ J/mol}\cdot\text{K}$$

Heat Exchange:

Energy balance. From the adiabatic case in Case I, we already have C_P, C_{P_A}. The heat-transfer area per unit volume of pipe is

$$a = \frac{\pi DL}{(\pi D^2/4)L} = \frac{4}{D} = \frac{4}{0.0266 \text{ m}} = 150 \text{ m}^{-1}$$

$$U = 110 \text{ J/m}^2 \cdot \text{s} \cdot \text{K}$$

Combining the overall heat-transfer coefficient with the area yields

$$Ua = 16{,}500 \text{ J/m}^3 \cdot \text{s} \cdot \text{K}$$

TABLE E12-2.1   SUMMARY OF PARAMETER VALUES

Parameter Values		
$\Delta H_{\text{Rx}}^{\circ}(T_R) = 80.77 \text{ kJ/mol}$	$\Delta C_P = -9 \text{ J/mol} \cdot \text{K}$	$T_0 = 1035 \text{K}$
$F_{A0} = 0.0376 \text{ mol/s}$	$C_{A0} = 18.8 \text{ mol/m}^3$	$T_R = 298 \text{K}$
$C_{P_A} = 163 \text{ J/mol A/K}$	$Ua = 16{,}500 \text{ J/m}^3 \cdot \text{s} \cdot \text{K}$	$\dot{m}_C = 0.111 \text{ mol/s}$
$C_{P_{\text{Cool}}} \equiv C_{Pc} = 34.5 \text{ J/mol/K}$		$V_f = 0.001 \text{ m}^3$

We will solve for all four cases of heat exchanger operation for this endothermic reaction example in the same way we did for the exothermic reaction in Example 12-1. That is, we will write the Polymath equations for the case of co-current heat exchange and use that as the base case. We will then manipulate the different terms in the heat-transfer fluid balance (Equations 12-10 and 12-11) to solve for the other cases, starting with the adiabatic case where we multiply the heat-transfer coefficient in the base case by zero.

Adiabatic endothermic reaction in a PFR

Case 1 Adiabatic

We are going to start with the adiabatic case first to show the dramatic effects of how the reaction dies out as the temperature drops. In fact, we are going to extend each to a volume of 5 dm³ to observe this effect and to show the necessity of adding a heat exchanger. For the adiabatic case, we simply multiply the value of Ua in our Polymath program by zero. No other changes are necessary. For the adiabatic case, the answer will be the same whether we use a bank of 1000 reactors, each a 1-dm³ reactor, or one of 1 m³. To illustrate how an endothermic reaction can virtually die out completely, let's extend the single-pipe volume from 1 dm³ to 5 dm³.

$$Ua = 16{,}500 * 0$$

The Polymath program is shown in Table E12-2.2. Figure E12-2.1 shows the graphical output.

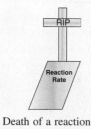

Death of a reaction

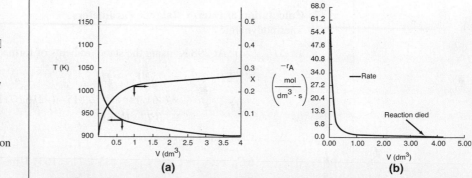

Figure E12-2.1 Adiabatic conversion and temperature **(a)**, and reaction rate **(b)** profiles.

TABLE E12-2.2   POLYMATH PROGRAM AND OUTPUT FOR ADIABATIC OPERATION

Differential equations

1 d(X)/d(V) = -ra/Fao

2 d(T)/d(V) = (Ua*(Ta-T)+ra*deltaH)/(Fao*(Cpa+X*delCp))

3 d(Ta)/d(V) = Ua*(T-Ta)/mc/Cpc

Explicit equations

1 Fao = .0376

2 Cpa = 163

3 delCp = -9

4 Cao = 18.8

5 To = 1035

6 deltaH = 80770+delCp*(T-298)

7 ra = -Cao*3.58*exp(34222*(1/To-1/T))*(1-X)*(To/T)/(1+X)

8 Ua = 16500*0

9 mc = .111

10 Cpc = 34.5

11 rate = -ra

POLYMATH Report
Ordinary Differential Equations

Calculated values of DEQ variables

	Variable	Initial value	Final value
1	Cao	18.8	18.8
2	Cpa	163.	163.
3	Cpc	34.5	34.5
4	delCp	-9.	-9.
5	deltaH	7.414E+04	7.531E+04
6	Fao	0.0376	0.0376
7	mc	0.111	0.111
8	ra	-67.304	-0.3704982
9	rate	67.304	0.3704982
10	T	1035.	904.8156
11	Ta	1250.	1250.
12	To	1035.	1035.
13	Ua	0	0
14	V	0	0.005
15	X	0	0.2817744

Living Example Problem

**Analysis:** **Case 1 Adiabatic Operation:** As temperature drops, so does k and hence the rate, $-r_A$, drops to an insignificant value. Note that for this adiabatic endothermic reaction, the reaction virtually _dies out_ after 3.5 dm^3, owing to the large drop in temperature, and very little conversion is achieved beyond this point. One way to increase the conversion would be to add a diluent such as nitrogen, which could supply the sensible heat for this endothermic reaction. However, if too much diluent is added, the concentration, and hence the rate, will be quite low. On the other hand, if too little diluent is added, the temperature will drop and virtually extinguish the reaction. How much diluent to add is left as an exercise. Figures E12-2.1 **(a)** and **(b)** give the reactor volume as 5 dm^3 in order to show the reaction "dying out." However, because the reaction is _nearly_ complete near the entrance to the reactor, i.e., $-r_A \cong 0$, we are going to study and compare the heat-exchange systems in a 1-dm^3 reactor (0.001 m^3) in the next three cases.

Case 2 Constant heat-exchange fluid temperature, T_a

We make the following changes in our program on line 3 of the base case **(a)**

$$\frac{dT_a}{dV} = \frac{Ua(T - T_a)}{\dot{m}C_{P_c}} * 0$$

$$Ua = 16{,}500 \text{ J/m}^3/\text{s/K}$$

$$\text{and } V_f = 0.001 \text{ m}^3$$

TABLE E12-2.3 POLYMATH PROGRAM AND OUTPUT FOR CONSTANT T_a

Differential equations

1 d(X)/d(V) = -ra/Fao
2 d(T)/d(V) = (Ua*(Ta-T)+ra*deltaH)/(Fao*(Cpa+X*delCp))
3 d(Ta)/d(V) = Ua*(T-Ta)/mc/Cpc*0

Explicit equations

1 Fao = .0376
2 Cpa = 163
3 delCp = -9
4 Cao = 18.8
5 To = 1035
6 deltaH = 80770+delCp*(T-298)
7 ra = -Cao*3.58*exp(34222*(1/To-1/T))*(1-X)*(To/T)/(1+X)
8 Ua = 16500
9 mc = .111
10 Cpc = 34.5
11 rate = -ra

Calculated values of DEQ variables

	Variable	Initial value	Final value
1	Cao	18.8	18.8
2	Cpa	163.	163.
3	Cpc	34.5	34.5
4	delCp	-9.	-9.
5	deltaH	7.414E+04	7.343E+04
6	Fao	0.0376	0.0376
7	mc	0.111	0.111
8	ra	-67.304	-16.48924
9	rate	67.304	16.48924
10	T	1035.	1114.093
11	Ta	1250.	1250.
12	To	1035.	1035.
13	Ua	1.65E+04	1.65E+04
14	V	0	0.001
15	X	0	0.9508067

The profiles for T, X, and $-r_A$ are shown below.

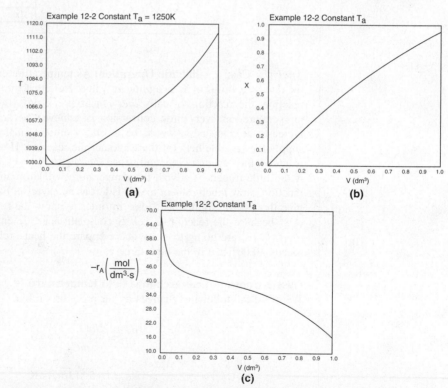

Figure E12-2.2 Profiles for constant heat exchanger fluid temperature, T_a;
(a) temperature, (b) conversion, (c) reaction rate.

Analysis: **Case 2 Constant T_a:** Just after the reactor entrance, the reaction temperature drops as the sensible heat from the reacting fluid supplies the energy for the endothermic reaction. This temperature drop in the reactor also causes the rate of reaction to drop. As we move farther down the reactor, the reaction rate drops further as the reactants are consumed. Beyond V = 0.08 dm³, the heat supplied by the constant T_a heat exchanger becomes greater than that "consumed" by the endothermic reaction and the reactor temperature rises. In the range between V = 0.2 dm³ and V = 0.6 dm³, the rate decreases slowly owing to the depletion of reactants, which is counteracted, to some extent, by the increase in temperature and hence the rate constant k. Consequently, we are eventually able to achieve an exit conversion of 95%.

Case 3 Co-current Heat Exchange

The energy balance on a co-current exchanger is

$$\frac{dT_a}{dV} = \frac{Ua(T - T_a)}{\dot{m}_C C_{P_C}}$$

with $T_{a0} = 1250$ K at $V = 0$

TABLE E12-2.4   POLYMATH PROGRAM AND OUTPUT FOR CO-CURRENT EXCHANGE

Differental equations

1 d(X)/d(Y) = −ra/Fao
2 d(T)/d(V) = (Ua*(Ta − T) + ra*deltaH)/(Fao*(Cpa + X*delCp))
3 d(Ta)/d(V) = Ua*(T − Ta)/mc/Cpo

Explicit equations

1 Fao = .0376
2 Cpa = 163
3 delCp = −9
4 Cao = 18.8
5 To = 1035
6 deltaH = 80770 + delCp*(T − 298)
7 ra = −Cao*3.58*exp(3422*(1/To 1/T)*(1 − X)*(To/T)/(1 + X)
8 Ua = 16500
9 mc = .111
10 Cpc = 34.5
11 nata = −ra

Calculated values of DEQ variables

	Variable	Initial value	Final value
1	Cao	18.8	18.8
2	Cpo	163.	163.
3	Cpc	34.5	34.5
4	delCp	−9.	−9.
5	deltaH	7.414E + 04	7.459E + 04
6	Fao	0.0376	0.0376
7	mc	0.111	0.111
8	ra	−67.304	−4.899078
9	rate	67.304	4.899078
10	T	1035.	984.8171
11	Ta	1250	996.215
12	To	1035.	1035.
13	Ua	1.65E + 04	1.65E + 04
14	V	0	0.001
15	X	0	0.456201

The variable profiles for T, T_a, X, and $-r_A$ are shown below. Because the reaction is endothermic, T_a needs to start off at a high temperature at $V = 0$.

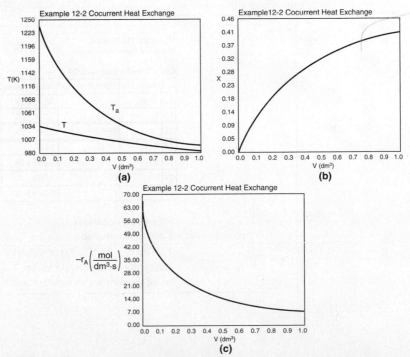

Figure E12-2.3   Profiles down the reactor for an endothermic reaction with co-current heat exchange; **(a)** temperature, **(b)** conversion, and **(c)** reaction rate.

Analysis: **Case 3 Co-Current Exchange:** In co-current heat exchange, we see that the heat-exchanger fluid temperature, T_a, drops rapidly initially and then continues to drop along the length of the reactor as it supplies the energy to the heat drawn by the endothermic reaction. Eventually T_a decreases to the point where it approaches T and the rate of heat exchange is small; as a result, the temperature of the reactor, T, continues to decrease, as does the rate, resulting in a small conversion. Because the reactor temperature for co-current exchange is lower than that for _Case 2 constant T_a_, the reaction rate will be lower. As a result, significantly less conversion will be achieved than in the case of constant heat-exchange temperature T_a.

Case 4 Countercurrent Heat Exchange

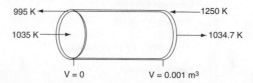

For countercurrent exchange, we first multiply the rhs of the co-current heat-exchanger energy balance by –1, leaving the rest of the Polymath program in Table 12-2.5 the same.

$$\frac{dT_a}{dV} = -\frac{Ua\left(T - T_a\right)}{\dot{m}C_{P_C}}$$

Next, **guess** $T_a (V = 0) = 995.15$ K to obtain $T_{a0} = 1250$ K at $V = 0.001 \text{m}^3$. (Don't you believe for a moment 995.15 K was my first guess.) Once this match is obtained as shown in Table E12-2.5, we can report the profiles shown in Figure E12-2.4.

TABLE E12-2.5   POLYMATH PROGRAM AND OUTPUT FOR COUNTERCURRENT EXCHANGE

POLYMATH Report
Ordinary Differential Equations

Calculated values of DEQ variables

	Variable	Initial value	Final value
9	rate	67.304	31.79235
10	T	1035.	1034.475
11	Ta	995.15	1249.999
12	To	1035.	1035.
13	Ua	1.65E+04	1.65E+04
14	V	0	0.001
15	X	0	0.3512403

Good guess!

Guess of $T_{a2} = $ **995.15 K**
at **V = 0**
Matches $T_{a0} = $ **1250 K**
at $V_f = $ **0.001 m³ = 1 dm³**

Differential equations

1 d(X)/d(V) = -ra/Fao

2 d(T)/d(V) = (Ua*(Ta-T)+ra*deltaH)/(Fao*(Cpa+X*delCp))

3 d(Ta)/d(V) = -Ua*(T-Ta)/mc/Cpc

Explicit equations are the same as Case 3 Co-current Heat Exchange.

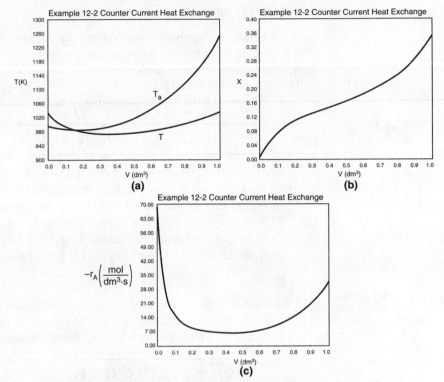

Figure E12-2.4   Profiles down the reactor for countercurrent heat exchange; **(a)** temperature, **(b)** conversion, **(c)** reaction rate.

Analysis: **Case 4 Countercurrent Exchange:** At the front of the reactor, $V = 0$, the reaction takes place very rapidly, drawing energy from the sensible heat of the gas and causing the gas temperature to drop because the heat exchanger cannot supply energy at an equal or greater rate to that being drawn by the endothermic reaction. Additional "heat" is lost at the entrance in the case of countercurrent exchange because the temperature of the exchange fluid, T_a, is below the entering reactor temperature, T. One notes there is a minimum in the reaction rate, $-r_A$, profile that is rather flat. In this flat region, the rate is "virtually" constant between $V = 0.2$ dm³ and $V = 0.8$ dm³, because the increase in k caused by the increase in T is balanced by the decrease in rate brought about by the consumption of reactants. Just past the middle of the reactor, the rate begins to increase slowly as the reactants become depleted and the heat exchanger now supplies energy at a rate greater than the reaction draws energy and, as a result, the temperature eventually increases. This lower temperature coupled with the consumption of reactants causes the rate of reaction to be low in the plateau, resulting in a lower conversion than either the co-current or constant T_a heat exchange cases.

AspenTech: Example 12-2 has also been formulated in AspenTech and can be downloaded on your computer directly from the CRE Web site.

12.4 CSTR with Heat Effects

In this section we apply the general energy balance [Equation (11-22)] to a CSTR at steady state. We then present example problems showing how the mole and energy balances are combined to design reactors operating adiabatically and non-adiabatically.

In Chapter 11 the steady-state energy balance was derived as

$$\dot{Q} - \dot{W}_s - F_{A0}\Sigma\Theta_i C_{P_i}(T - T_{i0}) - [\Delta H^\circ_{Rx}(T_R) + \Delta C_P(T - T_R)]F_{A0}X = 0 \quad (11\text{-}28)$$

Recall that $\dot{W}_s$ is the shaft work, i.e., the work **done by** the stirrer or mixer in the CSTR **on** the reacting fluid inside the CSTR. Consequently, because the convention that $\dot{W}_s$ **done by** the system **on** the surroundings is positive, the CSTR stirrer work will be a negative number, e.g., $\dot{W}_s = -1,000\,\text{J/s}$. (See problem P12-6$_B$, a California Professional Engineers' Exam Problem.)

Note: In many calculations the CSTR mole balance derived in Chapter 2

$$(F_{A0}X = -r_A V)$$

will be used to replace the term following the brackets in Equation (11-28), that is, $(F_{A0}X)$ will be replaced by $(-r_A V)$ to arrive at Equation (12-12).

Rearranging yields the steady-state energy balance

$$\dot{Q} - \dot{W}_s - F_{A0}\Sigma\Theta_i C_{P_i}(T - T_{i0}) + (r_A V)(\Delta H_{Rx}) = 0 \quad (12\text{-}12)$$

> These are the forms of the steady-state balance we will use.

Although the CSTR is well mixed and the temperature is uniform throughout the reaction vessel, these conditions do not mean that the reaction is carried out isothermally. Isothermal operation occurs when the feed temperature is identical to the temperature of the fluid inside the CSTR.

The $\dot{Q}$ Term in the CSTR

12.4.1 Heat Added to the Reactor, $\dot{Q}$

Figure 12-6 shows the schematics of a CSTR with a heat exchanger. The heat-transfer fluid enters the exchanger at a mass flow rate $\dot{m}_c$ (e.g., kg/s) at a temperature T_{a1} and leaves at a temperature T_{a2}. The rate of heat transfer *from* the exchanger *to* the reactor fluid at temperature T is[3]

> For exothermic reactions $(T > T_{a2} > T_{a1})$

$$\dot{Q} = \frac{UA\,(T_{a1} - T_{a2})}{\ln\left[(T - T_{a1})/(T - T_{a2})\right]} \quad (12\text{-}13)$$

The following derivations, based on a coolant (exothermic reaction), apply also to heating mediums (endothermic reaction). As a first approximation, we assume a quasi-steady state for the coolant flow and neglect the accumulation

---

[3] Information on the overall heat-transfer coefficient may be found in C. J. Geankoplis, *Transport Processes and Unit Operations*, 4th ed. (Englewood Cliffs, NJ: Prentice Hall, 2003), p. 300.

For endothermic
reactions
$(T_{1a} > T_{2a} > T)$

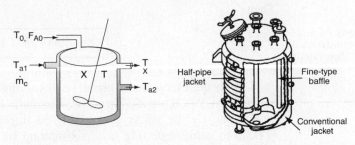

Figure 12-6   CSTR tank reactor with heat exchanger. (Diagram on right courtesy of Pfaudler, Inc.)

term (i.e., $dT_a/dt = 0$). An energy balance on the heat-exchanger fluid entering and leaving the exchanger is

Energy balance on
heat-exchanger fluid

$$\begin{bmatrix} \text{Rate of} \\ \text{energy} \\ in \\ \text{by flow} \end{bmatrix} - \begin{bmatrix} \text{Rate of} \\ \text{energy} \\ out \\ \text{by flow} \end{bmatrix} - \begin{bmatrix} \text{Rate of} \\ \text{heat transfer} \\ from \text{ exchanger} \\ to \text{ reactor} \end{bmatrix} = 0 \quad (12\text{-}14)$$

$$\dot{m}_c C_{P_c} (T_{a1} - T_R) - \dot{m}_c C_{P_c} (T_{a2} - T_R) - \frac{UA\,(T_{a1} - T_{a2})}{\ln\left[(T - T_{a1})/(T - T_{a2})\right]} = 0 \quad (12\text{-}15)$$

where C_{P_c} is the heat capacity of the heat exchanger fluid and T_R is the reference temperature. Simplifying gives us

$$\dot{Q} = \dot{m}_c C_{P_c} (T_{a1} - T_{a2}) = \frac{UA\,(T_{a1} - T_{a2})}{\ln\left[(T - T_{a1})/(T - T_{a2})\right]} \quad (12\text{-}16)$$

Solving Equation (12-16) for the exit temperature of the heat-exchanger fluid yields

$$T_{a2} = T - (T - T_{a1}) \exp\left(\frac{-UA}{\dot{m}_c C_{P_c}}\right) \quad (12\text{-}17)$$

From Equation (12-16)

$$\dot{Q} = \dot{m}_c C_{P_c} (T_{a1} - T_{a2}) \quad (12\text{-}18)$$

Substituting for T_{a2} in Equation (12-18), we obtain

Heat transfer to a
CSTR

$$\boxed{\dot{Q} = \dot{m}_c C_{P_c} \left\{ (T_{a1} - T)\left[1 - \exp\left(\frac{-UA}{\dot{m}_c C_{P_c}}\right)\right]\right\} \quad (12\text{-}19)}$$

For large values of the heat-exchanger fluid flow rate, $\dot{m}_c$, the exponent will be small and can be expanded in a Taylor series ($e^{-x} = 1 - x + \cdots$) where second-order terms are neglected in order to give

$$\dot{Q} = \dot{m}_c C_{P_c} (T_{a1} - T)\left[1 - \left(1 - \frac{UA}{\dot{m}_c C_{P_c}}\right)\right]$$

Then

Valid only for large
heat-transfer fluid
flow rates!!

$$\boxed{\dot{Q} = UA(T_a - T)} \qquad (12\text{-}20)$$

where $T_{a1} \cong T_{a2} = T_a$.

With the exception of processes involving highly viscous materials such as Problem P12-6$_B$, a *California Professional Engineers' Exam Problem*, the work done by the stirrer can usually be neglected. Setting $\dot{W}_s$ in Equation (11-27) to zero, neglecting ΔC_P, substituting for $\dot{Q}$, and rearranging, we have the following relationship between conversion and temperature in a CSTR

$$\frac{UA}{F_{A0}}(T_a - T) - \Sigma \Theta_i C_{P_i}(T - T_0) - \Delta H_{Rx}^{\circ} X = 0 \qquad (12\text{-}21)$$

Solving for X

$$\boxed{X = \frac{\dfrac{UA}{F_{A0}}(T - T_a) + \Sigma \Theta_i C_{P_i}(T - T_0)}{[-\Delta H_{Rx}^{\circ}(T_R)]}} \qquad (12\text{-}22)$$

Equation (12-22) is coupled with the mole balance equation

$$\boxed{V = \frac{F_{A0} X}{-r_A(X, T)}} \qquad (12\text{-}23)$$

to design CSTRs.

We now will further rearrange Equation (12-21) after letting

$$\boxed{\Sigma \Theta_i C_{P_i} = C_{P_0}}$$

then

$$C_{P_0}\left(\frac{UA}{F_{A0} C_{P_0}}\right) T_a + C_{P_0} T_0 - C_{P_0}\left(\frac{UA}{F_{A0} C_{P_0}} + 1\right) T - \Delta H_{Rx}^{\circ} X = 0$$

Let κ and T_C be non-adiabatic parameters defined by

Non-adiabatic
CSTR heat
exchange
parameters:
κ and T_c

$$\boxed{\kappa = \frac{UA}{F_{A0} C_{P_0}} \text{ and } T_c = \frac{\kappa T_a + T_0}{1 + \kappa}}$$

Then

$$-X \Delta H_{Rx}^{\circ} = C_{P_0}(1 + \kappa)(T - T_c) \qquad (12\text{-}24)$$

The parameters κ and T_c are used to simplify the equations for *non*-adiabatic operation. Solving Equation (12-24) for conversion

$$\boxed{X = \frac{C_{P_0}(1 + \kappa)(T - T_c)}{-\Delta H_{Rx}^{\circ}}} \qquad (12\text{-}25)$$

Solving Equation (12-24) for the reactor temperature

$$T = T_c + \frac{(-\Delta H_{Rx}^\circ)(X)}{C_{P_0}(1 + \kappa)} \qquad (12\text{-}26)$$

Table 12-3 shows three ways to specify the design of a CSTR. This procedure for nonisothermal CSTR design can be illustrated by considering a first-order irreversible liquid-phase reaction. To solve CSTR problems of this type we have three variables X, T, and V and we can only specify two and then solve for the third. The algorithm for working through either cases **A** (X specified), **B** (T specified), or **C** (V specified) is shown in Table 12-3. Its application is illustrated in Example 12-3.

TABLE 12-3   WAYS TO SPECIFY THE SIZING OF A CSTR

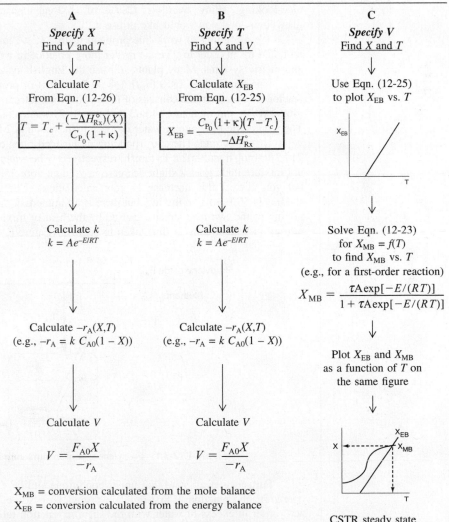

	A *Specify X* Find V and T	**B** *Specify T* Find X and V	**C** *Specify V* Find X and T
Forms of the energy balance for a CSTR with heat exchange	Calculate T From Eqn. (12-26) $T = T_c + \dfrac{(-\Delta H_{Rx}^\circ)(X)}{C_{P_0}(1+\kappa)}$	Calculate X_{EB} From Eqn. (12-25) $X_{EB} = \dfrac{C_{P_0}(1+\kappa)(T-T_c)}{-\Delta H_{Rx}^\circ}$	Use Eqn. (12-25) to plot X_{EB} vs. T
	Calculate k $k = Ae^{-E/RT}$	Calculate k $k = Ae^{-E/RT}$	Solve Eqn. (12-23) for $X_{MB} = f(T)$ to find X_{MB} vs. T (e.g., for a first-order reaction) $X_{MB} = \dfrac{\tau A \exp[-E/(RT)]}{1 + \tau A \exp[-E/(RT)]}$
	Calculate $-r_A(X,T)$ (e.g., $-r_A = k\,C_{A0}(1-X)$)	Calculate $-r_A(X,T)$ (e.g., $-r_A = k\,C_{A0}(1-X)$)	Plot X_{EB} and X_{MB} as a function of T on the same figure
	Calculate V $V = \dfrac{F_{A0}X}{-r_A}$	Calculate V $V = \dfrac{F_{A0}X}{-r_A}$	

X_{MB} = conversion calculated from the mole balance
X_{EB} = conversion calculated from the energy balance

The intersection of the energy balance curve $X_{EB}(T)$ and the mole balance curve $X_{MB}(T)$ in the figure gives the steady-state CSTR exit conversion and temperature.

CSTR steady state
when $X_{EB} = X_{MB}$

Example 12–3 Production of Propylene Glycol in an Adiabatic CSTR

Propylene glycol is produced by the hydrolysis of propylene oxide:

$$CH_2\text{—}CH\text{—}CH_3 + H_2O \xrightarrow{\ H_2SO_4\ } CH_2\text{—}CH\text{—}CH_3$$
$$\underset{O}{\diagdown\diagup} \qquad\qquad\qquad \underset{OH\quad OH}{\mid\qquad\mid}$$

Production, uses,
and economics

Over 900 million pounds of propylene glycol were produced in 2010 and the selling price was approximately $0.80 per pound. Propylene glycol makes up about 25% of the major derivatives of propylene oxide. The reaction takes place readily at room temperature when catalyzed by sulfuric acid.

You are the engineer in charge of an adiabatic CSTR producing propylene glycol by this method. Unfortunately, the reactor is beginning to leak, and you must replace it. (You told your boss several times that sulfuric acid was corrosive and that mild steel was a poor material for construction. He wouldn't listen.) There is a nice-looking, shiny overflow CSTR of 300-gal capacity standing idle; it is glass-lined, and you would like to use it.

We are going to work this problem in lb_m, s, ft^3, and lb-moles rather than g, mol, and m^3 in order to give the reader more practice in working in both the English and metric systems. Many plants still use the English system of units.

You are feeding 2500 lb_m/h (43.04 lb-mol/h) of propylene oxide (P.O.) to the reactor. The feed stream consists of (1) an equivolumetric mixture of propylene oxide (46.62 ft^3/h) and methanol (46.62 ft^3/h), and (2) water containing 0.1 wt % H_2SO_4. The volumetric flow rate of water is 233.1 ft^3/h, which is 2.5 times the methanol–P.O. volumetric flow rate. The corresponding molar feed rates of methanol and water are 71.87 lb-mol/h and 802.8 lb-mol/h, respectively. The water–propylene oxide–methanol mixture undergoes a slight decrease in volume upon mixing (approximately 3%), but you neglect this decrease in your calculations. The temperature of both feed streams is 58°F prior to mixing, but there is an immediate 17°F temperature rise upon mixing of the two feed streams caused by the heat of mixing. The entering temperature of all feed streams is thus taken to be 75°F (Figure E12-3.1).

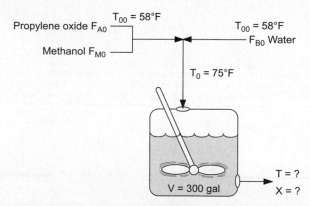

Figure E12-3.1   Propylene glycol manufacture in a CSTR.

Furusawa et al. state that under conditions similar to those at which you are operating, the reaction is first-order in propylene oxide concentration and apparent zero-order in excess of water with the specific reaction rate[4]

$$k = A\,e^{-E/RT} = 16.96 \times 10^{12}\,(e^{-32,400/RT})\ h^{-1}$$

---

[4] T. Furusawa, H. Nishimura, and T. Miyauchi, *J. Chem. Eng. Jpn.*, 2, 95.

The units of E are Btu/lb-mol and T is in °R.

There is an important constraint on your operation. Propylene oxide is a rather low-boiling-point substance. With the mixture you are using, you feel that you cannot exceed an operating temperature of 125°F, or you will lose too much oxide by vaporization through the vent system.

(a) Can you use the idle CSTR as a replacement for the leaking one if it will be operated adiabatically?

(b) If so, what will be the conversion of propylene oxide to glycol?

Solution

(All data used in this problem were obtained from the *CRC Handbook of Chemistry and Physics* unless otherwise noted.) Let the reaction be represented by

$$A + B \longrightarrow C$$

where

A is propylene oxide ($C_{P_A} = 35$ Btu/lb-mol · °F) [5]

B is water ($C_{P_B} = 18$ Btu/lb-mol · °F)

C is propylene glycol ($C_{P_C} = 46$ Btu/lb-mol · °F)

M is methanol ($C_{P_M} = 19.5$ Btu/lb-mol · °F)

In this problem, neither the exit conversion nor the temperature of the adiabatic reactor is given. By application of the mole and energy balances, we can solve two equations with two unknowns (X and T), as shown on the right-hand pathway in Table 12-3. Solving these coupled equations, we determine the exit conversion and temperature for the glass-lined reactor to see if it can be used to replace the present reactor.

1. **Mole Balance and Design Equation:**

$$F_{A0} - F_A + r_A V = 0$$

The design equation in terms of X is

$$V = \frac{F_{A0}X}{-r_A} \tag{E12-3.1}$$

2. **Rate Law:**

$$-r_A = kC_A \tag{E12-3.2}$$

$$k = 16.96 \; 10^{12} \exp[-32{,}400/R/T] \; \text{h}^{-1}$$

3. **Stoichiometry** (liquid phase, $v = v_0$):

$$C_A = C_{A0}(1 - X) \tag{E12-3.3}$$

4. **Combining** yields

$$V = \frac{F_{A0}X}{kC_{A0}(1-X)} = \frac{v_0 X}{k(1-X)} \tag{E12-3.4}$$

Following the Algorithm

---

[5] C_{P_A} and C_{P_C} are estimated from the observation that the great majority of low-molecular-weight oxygen-containing organic liquids have a mass heat capacity of 0.6 cal/g · °C ±15%.

Solving for X as a function of T and recalling that $\tau = V/v_0$ gives

$$X_{MB} = \frac{\tau k}{1 + \tau k} = \frac{\tau A e^{-E/RT}}{1 + \tau A e^{-E/RT}} \tag{E12-3.5}$$

This equation relates temperature and conversion through the **mole balance**.

Two equations, two unknowns

5. The **energy balance** for this adiabatic reaction in which there is negligible energy input provided by the stirrer is

$$X_{EB} = \frac{\Sigma\, \Theta_i C_{P_i}(T - T_{i0})}{-[\Delta H^\circ_{Rx}(T_R) + \Delta C_P(T - T_R)]} \tag{E12-3.6}$$

This equation relates X and T through the energy balance. We see that two equations, Equations (E12-3.5) and (E12-3.6), and two unknowns, X and T, must be solved (with $X_{EB} = X_{MB} = X$).

6. **Calculations:**

Rather than putting all those numbers in the mole and heat balance equations yourself, you can outsource this task to a consulting company in Riça, Jofostan, for a small fee.

(a) *Evaluate the mole balance terms* (C_{A0}, Θ_i, τ): The total liquid volumetric flow rate entering the reactor is

I know these are
tedious calculations,
but someone's gotta
know how to do it.

$$v_0 = v_{A0} + v_{M0} + v_{B0}$$

$$= 46.62 + 46.62 + 233.1 = 326.3 \text{ ft}^3/\text{h} \tag{E12-3.7}$$

$$V = 300 \text{ gal} = 40.1 \text{ ft}^3$$

$$\tau = \frac{V}{v_0} = \frac{40.1 \text{ ft}^3}{326.3 \text{ ft}^3/\text{h}} = 0.123 \text{ h} \tag{E12-3.8}$$

$$C_{A0} = \frac{F_{A0}}{v_0} = \frac{43.0 \text{ lb-mol/h}}{326.3 \text{ ft}^3/\text{h}}$$

$$= 0.132 \text{ lb-mol/ft}^3 \tag{E12-3.9}$$

For methanol: $\Theta_M = \dfrac{F_{M0}}{F_{A0}} = \dfrac{71.87 \text{ lb-mol/h}}{43.0 \text{ lb-mol/h}} = 1.67$

For water: $\Theta_B = \dfrac{F_{B0}}{F_{A0}} = \dfrac{802.8 \text{ lb-mol/h}}{43.0 \text{ lb-mol/h}} = 18.65$

The conversion calculated from the mole balance, X_{MB}, is found from Equation (E12-3.5)

$$X_{MB} = \frac{(16.96 \times 10^{12} \text{ h}^{-1})(0.1229 \text{ h}) \exp(-32{,}400/1.987T)}{1 + (16.96 \times 10^{12} \text{ h}^{-1})(0.1229 \text{ h}) \exp(-32{,}400/1.987T)}$$

Plot X_{MB} as a
function of
temperature.

$$\boxed{X_{MB} = \frac{(2.084 \times 10^{12}) \exp(-16{,}306/T)}{1 + (2.084 \times 10^{12}) \exp(-16{,}306/T)},\ T \text{ is in } ^\circ\text{R}} \tag{E12-3.10}$$

(b)   *Evaluating the energy balance terms*
   (1) Heat of reaction at temperature T

$$\Delta H_{Rx}(T) = \Delta H^\circ_{Rx}(T_R) + \Delta C_P(T - T_R) \tag{11-26}$$

$$\Delta C_P = C_{P_C} - C_{P_B} - C_{P_A} = 46 - 18 - 35 = -7\,\text{Btu/lb-mol/}^\circ\text{F}$$

$$\Delta H_{Rx} = -36{,}000 - 7(T - T_R) \tag{E12-3.11}$$

(2) Heat capacity term

$$\Sigma \Theta_i C_{P_i} = C_{P_A} + \Theta_B C_{P_B} + \Theta_M C_{P_M}$$

$$= 35 + (18.65)(18) + (1.67)(19.5) \tag{E12-3.12}$$

$$= 403.3 \ \text{Btu/lb-mol} \cdot {}^{\circ}\text{F}$$

$$T_0 = T_{00} + \Delta T_{\text{mix}} = 58{}^{\circ}\text{F} + 17{}^{\circ}\text{F} = 75{}^{\circ}\text{F}$$

$$= 535{}^{\circ}\text{R} \tag{E12-3.13}$$

$$T_R = 68{}^{\circ}\text{F} = 528{}^{\circ}\text{R}$$

The conversion calculated from the energy balance, X_{EB}, for an adiabatic reaction is given by Equation (11-29)

$$X_{EB} = -\frac{\Sigma \Theta_i C_{P_i}(T - T_{i0})}{\Delta H^{\circ}_{Rx}(T_R) + \Delta C_P(T - T_R)} \tag{11-29}$$

Substituting all the known quantities into the energy balance gives us

$$X_{EB} = \frac{(403.3 \ \text{Btu/lb-mol} \cdot {}^{\circ}\text{F})(T - 535){}^{\circ}\text{F}}{-[-36,400 - 7(T - 528)] \ \text{Btu/lb-mol}}$$

$$\boxed{X_{EB} = \frac{403.3(T - 535)}{36,400 + 7(T - 528)}} \tag{E12-3.14}$$

**Adiabatic
CSTR**

7. **Solving.** There are a number of different ways to solve these two simultaneous algebraic equations (E12-3.10) and (E12-3.14). The easiest way is to use the Polymath nonlinear equation solver. However, to give insight into the functional relationship between X and T for the mole and energy balances, we shall obtain a graphical solution. Here, X is plotted as a function of T for both the mole and energy balances, and the intersection of the two curves gives the solution where both the mole and energy balance solutions are satisfied, i.e., $X_{EB} = X_{MB}$. In addition, by plotting these two curves we can learn if there is more than one intersection (i.e., multiple steady states) for which both the energy balance and mole balance are satisfied. If numerical root-finding techniques were used to solve for X and T, it would be quite possible to obtain only one root when there is actually more than one. If Polymath were used, you could learn if multiple roots exist by changing your initial guesses in the nonlinear equation solver. We shall discuss multiple steady states further in Section 12-5. We choose T and then calculate X (Table E12-3.1). The calculations for X_{MB} and X_{EB} are plotted in Figure E12-3.2. The virtually straight line corresponds to the energy balance, Equation (E12-3.14), and the curved line corresponds to the mole balance,

TABLE E12-3.1 CALCULATIONS OF X_{EB} AND X_{MB} AS A FUNCTION OF T

T ($^{\circ}$R)	X_{MB} [Eq. (E12-3.10)]	X_{EB} [Eq. (E12-3.14)]
535	0.108	0.000
550	0.217	0.166
565	0.379	0.330
575	0.500	0.440
585	0.620	0.550
595	0.723	0.656
605	0.800	0.764
615	0.860	0.872
625	0.900	0.980

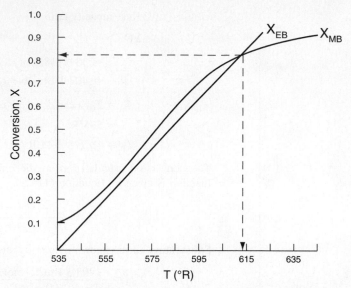

Figure E12-3.2   The conversions X_{EB} and X_{MB} as a function of temperature.

The reactor cannot be used because it will exceed the specified maximum temperature of 585°R.

Equation (E12-3.10). We observe from this plot that the only intersection point is at 83% conversion and 613°R. At this point, both the energy balance and mole balance are satisfied. Because the temperature must remain below 125°F (585°R), we **cannot** use the 300-gal reactor as it is now.

<u>*Analysis:*</u> After using Equations (E12-3.10) and (E12-3.14) to make a plot of conversion as a function of temperature, we see that there is only one intersection of $X_{EB}(T)$ and $X_{MB}(T)$, and consequently only one steady state. The exit conversion is 83% and the exit temperature (i.e., the reactor temperature) is 613°R (153°F), which is above the acceptable limit of 585°R (125°F) and we thus cannot use the CSTR operating at these conditions.

Oops! Looks like our plant will not be able to be completed and our multimillion-dollar profit has flown the coop. But wait, don't give up, let's ask reaction engineer Maxwell Anthony to fly to our company's plant in the country of Jofostan to look for a cooling coil heat exchanger. See what Max found in Example 12-4.

Example 12–4 CSTR with a Cooling Coil

Fantastic! Max has located a cooling coil in an equipment storage shed in the small, mountainous village of Ölofasis in Jofostan for use in the hydrolsis of propylene oxide discussed in Example 12-3. The cooling coil has 40 ft^2 of cooling surface and the cooling-water flow rate inside the coil is sufficiently large that a constant coolant temperature of 85°F can be maintained. A typical overall heat-transfer coefficient for such a coil is 100 Btu/h·ft^2·°F. Will the reactor satisfy the previous constraint of 125°F maximum temperature if the cooling coil is used?

Solution

If we assume that the cooling coil takes up negligible reactor volume, the conversion calculated as a function of temperature from the mole balance is the same as that in Example 12-3, Equation (E12-3.10).

1. **Combining the mole balance**, **stoichiometry**, and **rate law**, we have, from Example 12-3

$$\boxed{X_{\text{MB}} = \frac{\tau k}{1 + \tau k} = \frac{(2.084 \times 10^{12})\,\exp(-16{,}306/T)}{1 + (2.084 \times 10^{12})\,\exp(-16{,}306/T)}}\qquad \text{(E12-3.10)}$$

T is in °R.

2. **Energy balance.** Neglecting the work by the stirrer, we combine Equations (11-27) and (12-20) to write

$$\frac{UA(T_a - T)}{F_{\text{A0}}} - X[\Delta H^{\circ}_{\text{Rx}}(T_R) + \Delta C_P(T - T_R)] = \Sigma \Theta_i C_{P_i}(T - T_0)\qquad \text{(E12-4.1)}$$

Solving the energy balance for X_{EB} yields

$$\boxed{X_{\text{EB}} = \frac{\Sigma \Theta_i C_{P_i}(T - T_0) + [UA(T - T_a)/F_{\text{A0}}]}{-[\Delta H^{\circ}_{\text{Rx}}(T_R) + \Delta C_P(T - T_R)]}}\qquad \text{(E12-4.2)}$$

The cooling-coil term in Equation (E12-4.2) is

$$\frac{UA}{F_{\text{A0}}} = \left(100 \,\frac{\text{Btu}}{\text{h} \cdot \text{ft}^2 \cdot {}^\circ\text{F}}\right)\frac{(40 \text{ ft}^2)}{(43.04 \text{ lb-mol/h})} = \frac{92.9 \text{ Btu}}{\text{lb-mol} \cdot {}^\circ\text{F}}\qquad \text{(E12-4.3)}$$

Recall that the cooling temperature is

$$T_a = 85°\text{F} = 545°\text{R}$$

The numerical values of all other terms of Equation (E12-4.2) are identical to those given in Equation (E12-3.13), but with the addition of the heat exchange term, X_{EB} becomes

$$\boxed{X_{\text{EB}} = \frac{403.3(T - 535) + 92.9(T - 545)}{36{,}400 + 7(T - 528)}}\qquad \text{(E12-4.4)}$$

We now have two equations, Equations (E12-3.10) and (E12-4.4), and two unknowns, X and T, which we can solve with Polymath. Recall Examples E4-5 and E8-6 to review how to solve nonlinear, simultaneous equations of this type with Polymath. (See **Problem P12-1(f)** on page 611 to plot X versus T on Figure E12-3.2.)

TABLE E12-4.1   POLYMATH: CSTR WITH HEAT EXCHANGE

Nonlinear equations

1 f(X) = X-(403.3*(T-535)+92.9*(T-545))/(36400+7*(T-528)) = 0

2 f(T) = X-tau*k/(1+tau*k) = 0

Explicit equations

1 tau = 0.1229

2 A = 16.96*10^12

3 E = 32400

4 R = 1.987

5 k = A*exp(-E/(R*T))

	Variable	Value
1	A	1.696E+13
2	E	3.24E+04
3	k	4.648984
4	R	1.987
5	tau	0.1229

Calculated values of NLE variables

	Variable	Value	f(x)	Initial Guess
1	T	563.7289	-5.411E-10	564.
2	X	0.3636087	2.243E-11	0.367

The Polymath program and solution to these two Equations (E12-3.10) for X_{MB}, and (E12-4.4) for X_{EB}, are given in Table E12-4.1. The exiting temperature and conversion are 103.7°F (563.7°R) and 36.4%, respectively, i.e.,

$$\boxed{T = 564°\text{R and } X = 0.36}$$

Analysis: We are grateful to the people of the village of Ölofasis in Jofostan for their help in finding this shiny new heat exchanger. By adding heat exchange to the CSTR, the $X_{MB}(T)$ curve is unchanged but the slope of the $X_{EB}(T)$ line in Figure E12-3.2 increases and intersects the X_{MB} curve at $X = 0.36$ and $T = 564°\text{R}$. This conversion is low! We could try to reduce to cooling by increasing T_a or T_0 to raise the reactor temperature closer to 585°R, but not above this temperature. The higher the temperature in this irreversible reaction, the greater the conversion.

We will see in the next section that there may be multiple exit values of conversion and temperature (Multiple Steady States, MSS) that satisfy the parameter values and entrance conditions.

12.5 Multiple Steady States (MSS)

In this section, we consider the steady-state operation of a CSTR in which a first-order reaction is taking place. An excellent experimental investigation that demonstrates the multiplicity of steady states was carried out by Vejtasa and Schmitz.[6] They studied the reaction between sodium thiosulfate and hydrogen peroxide

$$2\text{Na}_2\text{S}_2\text{O}_3 + 4\text{H}_2\text{O}_2 \rightarrow \text{Na}_2\text{S}_3\text{O}_6 + \text{Na}_2\text{SO}_4 + 4\text{H}_2\text{O}$$

in a CSTR operated adiabatically. The multiple steady-state temperatures were examined by varying the flow rate over a range of space times, τ.

Reconsider the $X_{MB}(T)$ curve, Equation (E12-3.10), shown in Figure E12-3.2, which has been redrawn and shown as dashed lines in Figure E12-3.2A. Now consider what would happen if the volumetric flow rate v_0 is increased (τ decreased) just a little. The energy balance line, $X_{EB}(T)$, remains unchanged, but the mole balance line, X_{MB}, moves to the right, as shown by the curved, solid line in Figure E12-3.2A. This shift of $X_{MB}(T)$ to the right results in

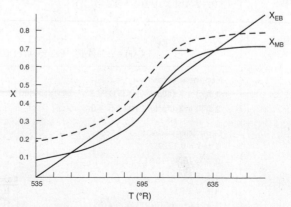

Figure E12-3.2A   Plots of $X_{EB}(T)$ and $X_{MB}(T)$ for different spaces times τ.

6   S.A. Vejtasa and R. A. Schmitz, *AIChE J.*, 16 (3), 415 (1970).

the $X_{\text{EB}}(T)$ and $X_{\text{MB}}(T)$ intersecting three lines, indicating three possible conditions at which the reactor can operate.

When more than one intersection occurs, there is more than one set of conditions that satisfy both the energy balance and mole balance (i.e., $X_{\text{EB}} = X_{\text{MB}}$); consequently, there will be multiple steady states at which the reactor may operate. These three steady states are easily determined from a graphical solution, but only one could show up in the Polymath solution. Thus, when using the Polymath nonlinear equation solver, we need to either choose different initial guesses to find if there are other solutions, or plot X_{MB} and X_{EB} versus T, as in Example 12-3.

We begin by recalling Equation (12-24), which applies when one neglects shaft work and ΔC_{P} (i.e., $\Delta C_{\text{P}} = 0$ and therefore $\Delta H_{\text{Rx}} = \Delta H_{\text{Rx}}^\circ$)

$$-X\Delta H_{\text{Rx}}^\circ = C_{\text{P0}}(1+\kappa)(T-T_c) \tag{12-24}$$

where

$$\boxed{C_{\text{P0}} = \Sigma\,\Theta_i C_{\text{P}_i}} \tag{12-25}$$

$$\boxed{\kappa = \frac{UA}{C_{\text{P0}}F_{\text{A0}}}} \tag{12-26}$$

and

$$\boxed{T_c = \frac{T_0 F_{\text{A0}} C_{\text{P0}} + UAT_a}{UA + C_{\text{P0}}F_{\text{A0}}} = \frac{\kappa T_a + T_0}{1+\kappa}} \tag{12-27}$$

Using the CSTR mole balance $X = \dfrac{-r_{\text{A}}V}{F_{\text{A0}}}$, Equation (12-24) may be rewritten as

$$(-r_{\text{A}}V/F_{\text{A0}})(-\Delta H_{\text{Rx}}^\circ) = C_{\text{P0}}(1+\kappa)(T-T_c) \tag{12-28}$$

The left-hand side is referred to as the *heat-generated term*

$G(T) =$ Heat-generated term

$$\boxed{G(T) = (-\Delta H_{\text{Rx}}^\circ)(-r_{\text{A}}V/F_{\text{A0}})} \tag{12-29}$$

The right-hand side of Equation (12-28) is referred to as the *heat-removed term* (by flow and heat exchange) $R(T)$

$R(T) =$ Heat-removed term

$$\boxed{R(T) = C_{\text{P0}}(1+\kappa)(T-T_c)} \tag{12-30}$$

To study the multiplicity of steady states, we shall plot both $R(T)$ and $G(T)$ as a function of temperature on the same graph and analyze the circumstances under which we will obtain multiple intersections of $R(T)$ and $G(T)$.

12.5.1 Heat-Removed Term, $R(T)$

Vary Entering Temperature. From Equation (12-30) we see that $R(T)$ increases linearly with temperature, with slope $C_{\text{P0}}(1+\kappa)$ and intercept T_c. As the entering temperature T_0 is increased, the line retains the same slope but shifts to the right as the intercept T_c increases, as shown in Figure 12-7.

Heat-removed
curve $R(T)$

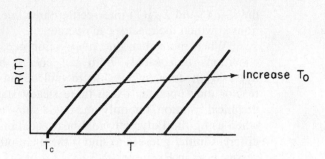

Figure 12-7   Variation of heat-removed line with inlet temperature.

Vary Non-adiabatic Parameter κ.   If one increases κ by either decreasing the molar flow rate, F_{A0}, or increasing the heat-exchange area, A, the slope increases and for the case of $T_a < T_0$ the ordinate intercept moves to the left, as shown in Figure 12-8.

$$\kappa = 0 \quad T_c = T_0$$
$$\kappa = \infty \quad T_c = T_a$$

$$\boxed{\begin{aligned} \kappa &= \frac{UA}{C_{P0}F_{A0}} \\ T_c &= \frac{T_0 + \kappa T_a}{1 + \kappa} \end{aligned}}$$

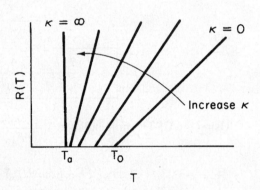

Figure 12-8   Variation of heat-removed line with κ ($\kappa = UA/C_{P0}F_{A0}$).

On the other hand, if $T_a > T_0$, the intercept will move to the right as κ increases.

12.5.2 Heat-Generated Term, $G(T)$

The heat-generated term, Equation (12-29), can be written in terms of conversion. [Recall that $X = -r_A V / F_{A0}$.]

$$G(T) = (-\Delta H_{Rx}^{\circ})X \tag{12-31}$$

To obtain a plot of heat generated, $G(T)$, as a function of temperature, we must solve for X as a function of T using the CSTR mole balance, the rate law, and stoichiometry. For example, for a first-order liquid-phase reaction, the CSTR mole balance becomes

$$V = \frac{F_{A0}X}{kC_A} = \frac{v_0 C_{A0} X}{kC_{A0}(1-X)}$$

Solving for X yields

First-order reaction

$$X = \frac{\tau k}{1 + \tau k} \tag{5-8}$$

Substituting for X in Equation (12-31), we obtain

$$G(T) = \frac{-\Delta H_{Rx}^\circ \tau k}{1 + \tau k} \tag{12-32}$$

Finally, substituting for k in terms of the Arrhenius equation, we obtain

$$\boxed{G(T) = \frac{-\Delta H_{Rx}^\circ \tau A e^{-E/RT}}{1 + \tau A e^{-E/RT}}} \tag{12-33}$$

Note that equations analogous to Equation (12-33) for $G(T)$ can be derived for other reaction orders and for reversible reactions simply by solving the CSTR mole balance for X. For example, for the second-order liquid-phase reaction

Second-order reaction

$$X = \frac{(2\tau k C_{A0} + 1) - \sqrt{4\tau k C_{A0} + 1}}{2\tau k C_{A0}}$$

the corresponding heat generated term is

$$\boxed{G(T) = \frac{-\Delta H_{Rx}^\circ [(2\tau C_{A0} A e^{-E/RT} + 1) - \sqrt{4\tau C_{A0} A e^{-E/RT} + 1}]}{2\tau C_{A0} A e^{-E/RT}}} \tag{12-34}$$

Let's now examine the behavior of the G(T) curve. At very low temperatures, the second term in the denominator of Equation (12-33) for the first-order reaction can be neglected, so that $G(T)$ varies as

Low T

$$G(T) = -\Delta H_{Rx}^\circ \tau A e^{-E/RT}$$

(Recall that ΔH_{Rx}° means that the standard heat of reaction is evaluated at T_R.)

At very high temperatures, the second term in the denominator dominates, and $G(T)$ is reduced to

High T

$$G(T) = -\Delta H_{Rx}^\circ$$

$G(T)$ is shown as a function of T for two different activation energies, E, in Figure 12-9. If the flow rate is decreased or the reactor volume increased so as to increase τ, the heat-generated term, $G(T)$, changes, as shown in Figure 12-10.

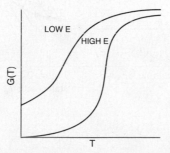

Figure 12-9 Variation of G(T) curve with activation energy.

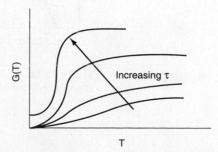

Figure 12-10 Variation of G(T) curve with space time.

Can you combine Figures 12-10 and 12-8 to explain why a Bunsen burner goes out when you turn up the gas flow to a very high rate?

Heat-generated curves, $G(T)$

12.5.3 Ignition-Extinction Curve

The points of intersection of $R(T)$ and $G(T)$ give us the temperature at which the reactor can operate at steady state. Suppose that we begin to feed our reactor at some relatively low temperature, T_{01}. If we construct our $G(T)$ and $R(T)$ curves, illustrated by curves y = $G(T)$ and a = $R(T)$, respectively, in Figure 12-11, we see that there will be only one point of intersection, point 1. From this point of intersection, one can find the steady-state temperature in the reactor, T_{s1}, by following a vertical line down to the T-axis and reading off the temperature, T_{s1}, as shown in Figure 12-11.

If one were now to increase the entering temperature to T_{02}, the $G(T)$ curve, y, would remain unchanged, but the $R(T)$ curve would move to the right, as shown by line b in Figure 12-11, and will now intersect the $G(T)$ at point 2 and be tangent at point 3. Consequently, we see from Figure 12-11 that there are two steady-state temperatures, T_{s2} and T_{s3}, that can be realized in the CSTR for an entering temperature T_{02}. If the entering temperature is increased to T_{03}, the $R(T)$ curve, line c (Figure 12-12), intersects the $G(T)$ curve three times and there are three steady-state temperatures, T_{s4}, T_{s5}, and T_{s6}. As we continue to increase T_0, we finally reach line e, in which there are only two steady-state temperatures, a point of tangency at T_{s10} and an intersection at T_{s11}. By further increasing T_0, we reach line f, corresponding to T_{06}, in which we have only one reactor temperature that will satisfy both the mole and energy balances, T_{s12}. For the six entering temperatures, we can form Table 12-4, relating the entering temperature to the possible reactor operating temperatures.

Both the mole and energy balances are satisfied at the points of intersection or tangency.

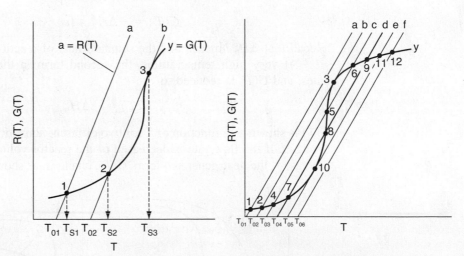

Figure 12-11 Finding multiple steady states with T_0 varied.

Figure 12-12 Finding multiple steady states with T_0 varied.

TABLE 12-4   MULTIPLE STEADY-STATE TEMPERATURES

Entering Temperature	Reactor Temperatures			
T_{01}			T_{s1}	
T_{02}		T_{s2}		T_{s3}
T_{03}	T_{s4}		T_{s5}		T_{s6}
T_{04}	T_{s7}		T_{s8}		T_{s9}
T_{05}		T_{s10}		T_{s11}
T_{06}			T_{s12}	

By plotting T_s as a function of T_0, we obtain the well-known *ignition-extinction curve* shown in Figure 12-13. From this figure, we see that as the entering temperature T_0 is increased, the steady-state temperature T_s increases along the bottom line until T_{05} is reached. Any fraction-of-a-degree increase in temperature beyond T_{05} and the steady-state reactor temperature T_s will jump up from T_{s10} to T_{s11}, as shown in Figure 12-13. The temperature at which this jump occurs is called the *ignition temperature*. That is, we must exceed a certain feed temperature, T_{05}, to operate at the upper steady state where the temperature and conversion are higher.

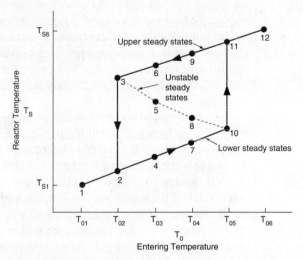

Figure 12-13   Temperature ignition-extinction curve.

If a reactor were operating at T_{s12} and we began to cool the entering temperature down from T_{06}, the steady-state reactor temperature, T_{s3}, would eventually be reached, corresponding to an entering temperature, T_{02}. Any slight decrease below T_{02} would drop the steady-state reactor temperature to the lower steady-state value T_{s2}. Consequently, T_{02} is called the *extinction temperature*.

The middle points 5 and 8 in Figures 12-12 and 12-13 represent *unstable steady-state temperatures*. Consider the heat-removed line d in Figure 12-12, along with the heat-generated curve, which is replotted in Figure 12-14. If we were operating at the middle steady-state temperature T_{s8}, for example, and a pulse increase in reactor temperature occurred, we would find ourselves at the temperature shown by vertical line ②, between points 8 and 9. We see that along

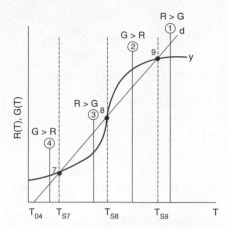

Figure 12-14 Stability of multiple steady state temperatures.

this vertical line ②, the heat-generated curve, $y \equiv G(T)$, is greater than the heat-removed line $d \equiv R(T)$ i.e., $(G > R)$. Consequently, the temperature in the reactor would continue to increase until point 9 is reached at the upper steady state. On the other hand, if we had a pulse decrease in temperature from point 8, we would find ourselves on a vertical line ③ between points 7 and 8. Here, we see that the heat-removed curve d is greater than the heat-generated curve y $(R > G)$, so the temperature will continue to decrease until the lower steady state is reached. That is, a small change in temperature either above or below the middle steady-state temperature, T_{s8}, will cause the reactor temperature to move away from this middle steady state. Steady states that behave in this manner are said to be *unstable*.

In contrast to these unstable operating points, there are stable operating points. Consider what would happen if a reactor operating at T_{s9} were subjected to a pulse increase in reactor temperature indicated by line ① in Figure 12-14. We see that the heat-removed line d is greater than the heat-generated curve y $(R > G)$, so that the reactor temperature will decrease and return to T_{s9}. On the other hand, if there is a sudden drop in temperature below T_{s9}, as indicated by line ②, we see the heat-generated curve y is greater than the heat-removed line d $(G > R)$, and the reactor temperature will increase and return to the upper steady state at T_{s9}. Consequently, T_{s9} is a stable steady state.

Next, let's look at what happens when the lower steady-state temperature at T_{s7} is subjected to pulse increase to the temperature shown as line ③ in Figure 12-14. Here, we again see that the heat removed, R, is greater than the heat generated, G, so that the reactor temperature will drop and return to T_{s7}. If there is a sudden decrease in temperature below T_{s7} to the temperature indicated by line ④, we see that the heat generated is greater than the heat removed $(G > R)$, and that the reactor temperature will increase until it returns to T_{s7}. Consequently, T_{s7} is a stable steady state. A similar analysis could be carried out for temperatures T_{s1}, T_{s2}, T_{s4}, T_{s6}, T_{s11}, and T_{s12}, and one would find that reactor temperatures would always return to *locally stable steady-state values* when subjected to both positive and negative fluctuations.

While these points are locally stable, they are not necessarily globally stable. That is, a large perturbation in temperature or concentration, while small, may be sufficient to cause the reactor to fall from the upper steady state (corresponding to high conversion and temperature, such as point 9 in Figure 12-14), to the lower steady state (corresponding to low temperature and conversion, point 7).

12.6 Nonisothermal Multiple Chemical Reactions

Most reacting systems involve more than one reaction and do not operate isothermally. **This section is one of the most important, if not *the* most important, sections of the book.** It ties together all the previous chapters to analyze multiple reactions that do not take place isothermally.

12.6.1 Energy Balance for Multiple Reactions in Plug-Flow Reactors

In this section we give the energy balance for multiple reactions. We begin by recalling the energy balance for a single reaction taking place in a PFR, which is given by Equation (12-5)

$$\frac{dT}{dV} = \frac{(-r_A)[-\Delta H_{Rx}(T)] - Ua(T - T_a)}{\sum_{j=1}^{m} F_j C_{P_j}} \tag{12-5}$$

When we have multiple reactions occurring, we have to account for, and sum up, all the heats of reaction in the reactor for each and every reaction. For q multiple reactions taking place in the PFR where there are m species, it is easily shown that Equation (12-5) can be generalized to

Energy balance for multiple reactions

$$\boxed{\frac{dT}{dV} = \frac{\sum_{i=1}^{q}(-r_{ij})\left[-\Delta H_{Rxij}(T)\right] - Ua\left(T - T_a\right)}{\sum_{j=1}^{m} F_j C_{P_j}}} \tag{12-35}$$

i = Reaction number
j = Species

Note that we now have two subscripts on the heat of reaction. The heat of reaction for reaction i must be referenced to the same species in the rate, r_{ij}, by which ΔH_{Rxij} is multiplied, which is

$$[-r_{ij}][-\Delta H_{Rxij}] = \left[\frac{\text{Moles of } j \text{ reacted in reaction } i}{\text{Volume} \cdot \text{time}}\right] \times \left[\frac{\text{Joules "released" in reaction } i}{\text{Moles of } j \text{ reacted in reaction } i}\right]$$

$$= \left[\frac{\text{Joules "released" in reaction } i}{\text{Volume} \cdot \text{time}}\right] \tag{12-36}$$

where the subscript j refers to the species, the subscript i refers to the particular reaction, q is the number of **independent** reactions, and m is the number of species. We are going to let

$$\boxed{Q_g = \sum_{i=1}^{q}(-r_{ij})\left[-\Delta H_{Rxij}(T)\right]}$$

and

$$\boxed{Q_r = Ua\left(T - T_a\right)}$$

Then Equation (12-35) becomes

$$\boxed{\frac{dT}{dV} = \frac{Q_g - Q_r}{\displaystyle\sum_{j=1}^{m} F_j C_{P_j}}} \tag{12-37}$$

Equation (12-37) represents a nice compact form of the energy balance for multiple reactions.

Consider the following reaction sequence carried out in a PFR:

Reaction 1: $A \xrightarrow{\ k_1\ } B$

Reaction 2: $B \xrightarrow{\ k_2\ } C$

One of the major goals of this text is that the reader will be able to solve multiple reactions with heat effects, and this section shows how!

The PFR energy balance becomes

$$\frac{dT}{dV} = \frac{Ua(T_a - T) + (-r_{1A})(-\Delta H_{Rx1A}) + (-r_{2B})(-\Delta H_{Rx2B})}{F_A C_{P_A} + F_B C_{P_B} + F_C C_{P_C}} \tag{12-38}$$

where ΔH_{Rx1A} = [J/mol of A reacted in reaction 1] and
ΔH_{Rx2B} = [J/mol of B reacted in reaction 2].

12.6.2 Parallel Reactions in a PFR

We will now give three examples of multiple reactions with heat effects: Example 12-5 discusses *parallel reactions*, Example 12-6 discusses *series reactions,* and Example 12-7 discusses *complex reactions.*

Example 12-5   *Parallel Reactions in a PFR with Heat Effects*

The following gas-phase reactions occur in a PFR:

Reaction 1: $A \xrightarrow{\ k_1\ } B$ $-r_{1A} = k_{1A} C_A$ (E12-5.1)

Reaction 2: $2A \xrightarrow{\ k_2\ } C$ $-r_{2A} = k_{2A} C_A^2$ (E12-5.2)

Pure A is fed at a rate of 100 mol/s, a temperature of 150°C, and a concentration of 0.1 mol/dm³. Determine the temperature and molar flow rate profiles down the reactor.

Additional information

$$\Delta H_{Rx1A} = -20{,}000 \text{ J/(mol of A reacted in reaction 1)}$$

$$\Delta H_{Rx2A} = -60{,}000 \text{ J/(mol of A reacted in reaction 2)}$$

Living Example Problem

$C_{P_A} = 90 \text{ J/mol} \cdot {}^\circ\text{C}$ $k_{1A} = 10 \exp\!\left[\dfrac{E_1}{R}\left(\dfrac{1}{300} - \dfrac{1}{T}\right)\right] \text{s}^{-1}$

$C_{P_B} = 90 \text{ J/mol} \cdot {}^\circ\text{C}$ $E_1/R = 4000 \text{ K}$

$C_{P_C} = 180 \text{ J/mol} \cdot {}^\circ\text{C}$ $k_{2A} = 0.09 \exp\!\left[\dfrac{E_2}{R}\left(\dfrac{1}{300} - \dfrac{1}{T}\right)\right] \dfrac{\text{dm}^3}{\text{mol} \cdot \text{s}}$

$Ua = 4000 \text{ J/m}^3 \cdot \text{s} \cdot {}^\circ\text{C}$ $E_2/R = 9000 \text{ K}$

$T_a = 100{}^\circ\text{C}$ (Constant)

Solution

The PFR energy balance becomes [cf. Equation (12-35)]

$$\frac{dT}{dV} = \frac{Ua(T_a - T) + (-r_{1A})(-\Delta H_{Rx1A}) + (-r_{2A})(-\Delta H_{Rx2A})}{F_A C_{P_A} + F_B C_{P_B} + F_C C_{P_C}} \qquad \text{(E12-5.3)}$$

1. Mole balances:

$$\frac{dF_A}{dV} = r_A \qquad \text{(E12-5.4)}$$

$$\frac{dF_B}{dV} = r_B \qquad \text{(E12-5.5)}$$

Following the Algorithm

$$\frac{dF_C}{dV} = r_C \qquad \text{(E12-5.6)}$$

2. Rates:

Rate laws

$$r_{1A} = -k_{1A} C_A \qquad \text{(E12-5.1)}$$

$$r_{2A} = -k_{2A} C_A^2 \qquad \text{(E12-5.2)}$$

Relative rates

Reaction 1: $\dfrac{r_{1A}}{-1} = \dfrac{r_{1B}}{1};$ $r_{1B} = -r_{1A} = k_{1A} C_A$

Reaction 2: $\dfrac{r_{2A}}{-2} = \dfrac{r_{2C}}{1};$ $r_{2C} = -\dfrac{1}{2} r_{2A} = \dfrac{k_{2A}}{2} C_A^2$

Net rates

$$r_A = r_{1A} + r_{2A} = -k_{1A} C_A - k_{2A} C_A^2 \qquad \text{(E12-5.7)}$$

$$r_B = r_{1B} = k_{1A} C_A \qquad \text{(E12-5.8)}$$

$$r_C = r_{2C} = \frac{1}{2} k_{2A} C_A^2 \qquad \text{(E12-5.9)}$$

3. Stoichiometry (gas-phase but $\Delta P = 0$):

$$C_A = C_{T0} \left(\frac{F_A}{F_T}\right)\left(\frac{T_0}{T}\right) \qquad \text{(E12-5.10)}$$

$$C_B = C_{T0} \left(\frac{F_B}{F_T}\right)\left(\frac{T_0}{T}\right) \qquad \text{(E12-5.11)}$$

$$C_C = C_{T0} \left(\frac{F_C}{F_T}\right)\left(\frac{T_0}{T}\right) \qquad \text{(E12-5.12)}$$

$$F_T = F_A + F_B + F_C \tag{E12-5.13}$$

$$k_{1A} = 10 \exp\left[4000\left(\frac{1}{300} - \frac{1}{T}\right)\right]s^{-1}$$

$$(T \text{ in K})$$

$$k_{2A} = 0.09 \exp\left[9000\left(\frac{1}{300} - \frac{1}{T}\right)\right]\frac{dm^3}{mol \cdot s}$$

4. Energy balance:

$$\frac{dT}{dV} = \frac{4000(373 - T) + (-r_{1A})(20,000) + (-r_{2A})(60,000)}{90F_A + 90F_B + 180F_C} \tag{E12-5.14}$$

5. Evaluation:

The Polymath program and its graphical outputs are shown in Table E12-5.1 and Figures E12-5.1 and E12-5.2.

Living Example Problem

TABLE E12-5.1   POLYMATH PROGRAM

Differential equations

1 d(Fa)/d(V) = r1a+r2a
2 d(Fb)/d(V) = -r1a
3 d(Fc)/d(V) = -r2a/2
4 d(T)/d(V) = (4000*(373-T)+(-r1a)*20000+(-r2a)*60000)/(90*Fa+90*Fb+180*Fc)

Explicit equations

1 k1a = 10*exp(4000*(1/300-1/T))
2 k2a = 0.09*exp(9000*(1/300-1/T))
3 Cto = 0.1
4 Ft = Fa+Fb+Fc
5 To = 423
6 Ca = Cto*(Fa/Ft)*(To/T)
7 Cb = Cto*(Fb/Ft)*(To/T)
8 Cc = Cto*(Fc/Ft)*(To/T)
9 r1a = -k1a*Ca
10 r2a = -k2a*Ca^2

Calculated values of DEQ variables

	Variable	Initial value	Final value
1	Ca	0.1	2.069E-09
2	Cb	0	0.0415941
3	Cc	0	0.016986
4	Cto	0.1	0.1
5	Fa	100.	2.738E-06
6	Fb	0	55.04326
7	Fc	0	22.47837
8	Ft	100.	77.52163
9	k1a	482.8247	2.426E+04
10	k2a	553.0557	3.716E+06
11	r1a	-48.28247	-5.019E-05
12	r2a	-5.530557	-1.591E-11
13	T	423.	722.0882
14	To	423.	423.
15	V	0	1.

Why does the temperature go through a maximum value?

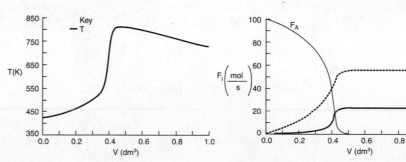

Figure E12-5.1 Temperature profile.

Figure E12-5.2 Profile of molar flow rates F_A, F_B, and F_C.

Analysis: The reactant is virtually consumed by the time it reaches a reactor volume $V = 0.45$ dm^3; beyond this point, $Q_r > Q_g$, and the reactor temperature begins to drop. In addition, the selectivity $\tilde{S}_{B/C} = F_B/F_C = 55/22.5 = 2.44$ remains constant after this point. If a high selectivity is required, then the reactor should be shortened to $V = 0.3$ dm^3, at which point the selectivity is $\tilde{S}_{B/C} = 20/2 = 10$.

12.6.3 Energy Balance for Multiple Reactions in a CSTR

Recall that for the steady-state mole balance in a CSTR with a single reaction $[-F_{A0}X = r_A V,]$ and that $\Delta H_{Rx}(T) = \Delta H^\circ_{Rx} + \Delta C_P(T - T_R)$, so that for $T_0 = T_{i0}$ Equation (11-27) may be rewritten as

$$\dot{Q} - \dot{W}_s - F_{A0} \sum \Theta_j C_{P_j}(T - T_0) + [\Delta H_{Rx}(T)][r_A V] = 0 \quad \text{(11-27A)}$$

Again, we must account for the "heat generated" by all the reactions in the reactor. For q multiple reactions and m species, the CSTR energy balance becomes*

$$\boxed{\dot{Q} - \dot{W}_s - F_{A0} \sum_{j=1}^{m} \Theta_j C_{P_j}(T - T_0) + V \sum_{i=1}^{q} r_{ij} \Delta H_{Rxij}(T) = 0} \quad \text{(12-39)}$$

Energy balance for multiple reactions in a CSTR

Substituting Equation (12-20) for $\dot{Q}$, neglecting the work term, and assuming constant heat capacities and large coolant flow rates $\dot{m}_c$, Equation (12-39) becomes

$$\boxed{UA(T_a - T) - F_{A0} \sum_{j=1}^{m} C_{P_j} \Theta_j (T - T_0) + V \sum_{i=1}^{q} r_{ij} \Delta H_{Rxij}(T) = 0} \quad \text{(12-40)}$$

For the two parallel reactions described in Example 12-5, the CSTR energy balance is

$$UA(T_a - T) - F_{A0} \sum_{j=1}^{m} \Theta_j C_{P_j}(T - T_0) + V r_{1A} \Delta H_{Rx1A}(T) + V r_{2A} \Delta H_{Rx2A}(T) = 0$$

$$\text{(12-41)}$$

Major goal of CRE

One of the **major goals** of this text is to have the reader solve problems involving multiple reactions with heat effects (cf. Problems P12-23$_C$, P12-24$_C$, P12-25$_C$, and P12-26$_B$). That's exactly what we are doing in the next two examples!

12.6.4 Series Reactions in a CSTR

Example 12-6   Multiple Reactions in a CSTR

The elementary liquid-phase reactions

$$A \xrightarrow{k_1} B \xrightarrow{k_2} C$$

take place in a 10-dm^3 CSTR. What are the effluent concentrations for a volumetric feed rate of 1000 dm^3/min at a concentration of A of 0.3 mol/dm^3? The inlet temperature is 283 K.

Additional information

$$C_{P_A} = C_{P_B} = C_{P_C} = 200 \text{ J/mol} \cdot \text{K}$$

$$k_1 = 3.3 \text{ min}^{-1} \text{ at } 300 \text{ K, with } E_1 = 9900 \text{ cal/mol}$$

$$k_2 = 4.58 \text{ min}^{-1} \text{ at } 500 \text{ K, with } E_2 = 27{,}000 \text{ cal/mol}$$

$$\Delta H_{\text{Rx1A}} = -55{,}000 \text{ J/mol A} \qquad UA = 40{,}000 \text{ J/min} \cdot \text{K with } T_a = 57°\text{C}$$

$$\Delta H_{\text{Rx2B}} = -71{,}500 \text{ J/mol B}$$

Solution

<u>The Algorithm:</u>

Number each reaction

$$\text{Reaction (1)} \qquad \text{A} \xrightarrow{\ k_1\ } \text{B}$$
$$\text{Reaction (2)} \qquad \text{B} \xrightarrow{\ k_2\ } \text{C}$$

The reactions follow elementary rate laws

$$r_{1A} = -k_{1A}C_A \equiv -k_1 C_A$$
$$r_{2B} = -k_{2B}C_B \equiv -k_2 C_B$$

1. **Mole Balance on Every Species**

   Species A: Combined mole balance and rate law for A

   $$V = \frac{F_{A0} - F_A}{-r_A} = \frac{v_0[C_{A0} - C_A]}{-r_{1A}} = \frac{v_0[C_{A0} - C_A]}{k_1 C_A} \tag{E12-6.1}$$

   Solving for C_A gives us

   $$C_A = \frac{C_{A0}}{1 + \tau k_1} \tag{E12-6.2}$$

   Species B: Combined mole balance and rate law for B

   $$V = \frac{0 - C_B v_0}{-r_B} = \frac{C_B v_0}{r_B} \tag{E12-6.3}$$

2. **Rates**

(a) **Laws**	(b) **Relative Rates**	(c) **Net Rates**
$r_{1A} = -k_{1A}C_A \equiv -k_1 C_A$	$r_{1B} = -r_{1A}$	$r_A = r_{1A}$
$r_{2B} = -k_{2B}C_B \equiv -k_2 C_B$	$r_{2C} = -r_{2B}$	$r_B = r_{1B} + r_{2B}$

3. **Combine**

   Substituting for r_{1B} and r_{2B} in Equation (E12-6.3) gives

   $$V = \frac{C_B v_0}{k_1 C_A - k_2 C_B} \tag{E12-6.4}$$

   Solving for C_B yields

   $$C_B = \frac{\tau k_1 C_A}{1 + \tau k_2} = \frac{\tau k_1 C_{A0}}{(1 + \tau k_1)(1 + \tau k_2)} \tag{E12-6.5}$$

   $$\boxed{-r_{1A} = k_1 C_A = \frac{k_1 C_{A0}}{1 + \tau k_1}} \tag{E12-6.6}$$

Following the Algorithm

$$-r_{2B} = k_2 C_B = \frac{k_2 \tau k_1 C_{A0}}{(1 + \tau k_1)(1 + \tau k_2)} \qquad \text{(E12-6.7)}$$

4. Energy Balances:

Applying Equation (12-41) to this system gives

$$[r_{1A} \Delta H_{Rx1A} + r_{2B} \Delta H_{Rx2B}]V - UA(T - T_a) - F_{A0} C_{P_A}(T - T_0) = 0 \qquad \text{(E12-6.8)}$$

Substituting for $F_{A0} = v_0 C_{A0}$, r_{1A}, and r_{2B} and rearranging, we have

$$\overbrace{\left[-\frac{\Delta H_{Rx1A} \tau k_1}{1 + \tau k_1} - \frac{\tau k_1 \tau k_2\, \Delta H_{Rx2B}}{(1 + \tau k_1)(1 + \tau k_2)} \right]}^{G(T)} = \overbrace{C_{P_A}(1 + \kappa)[T - T_c]}^{R(T)} \qquad \text{(E12-6.9)}$$

$$\kappa = \frac{UA}{F_{A0} C_{P_A}} = \frac{40{,}000 \text{ J/min} \cdot \text{K}}{(0.3 \text{ mol/dm}^3)(1000 \text{ dm}^3/\text{min})\,200 \text{ J/mol} \cdot \text{K}} = 0.667$$

$$T_c = \frac{T_0 + \kappa T_a}{1 + \kappa} = \frac{283 + (0.666)(330)}{1 + 0.667} = 301.8 \text{ K} \qquad \text{(E12-6.10)}$$

$$G(T) = \left[-\frac{\Delta H_{Rx1A} \tau k_1}{1 + \tau k_1} - \frac{\tau k_1 \tau k_2\, \Delta H_{Rx2B}}{(1 + \tau k_1)(1 + \tau k_2)} \right] \qquad \text{(E12-6.11)}$$

$$R(T) = C_{P_A}(1 + \kappa)[T - T_c] \qquad \text{(E12-6.12)}$$

We are going to generate $G(T)$ and $R(T)$ by fooling Polymath to first generate T as a function of a dummy variable, t. We then use our plotting options to convert $T(t)$ to $G(T)$ and $R(T)$. The Polymath program to plot $R(T)$ and $G(T)$ vs. T is shown in Table E12-6.1, and the resulting graph is shown in Figure E12-6.1.

TABLE E12-6.1   POLYMATH PROGRAM AND OUTPUT

Equations:

Differential equations

1 d(T)/d(t) = 2

Explicit equations

1 Cp = 200

2 Cao = 0.3

3 To = 283

4 tau = .01

5 DH1 = -55000

6 DH2 = -71500

7 vo = 1000

8 E2 = 27000

9 E1 = 9900

10 UA = 40000

11 Ta = 330

12 k2 = 4.58*exp((E2/1.987)*(1/500-1/T))

13 k1 = 3.3*exp((E1/1.987)*(1/300-1/T))

14 Ca = Cao/(1+tau*k1)

15 kappa = UA/(vo*Cao)/Cp

16 G = -tau*k1/(1+k1*tau)*DH1-k1*tau*k2*tau*DH2/((1+tau*k1)*(1+tau*k2))

17 Tc = (To+kappa*Ta)/(1+kappa)

18 Cb = tau*k1*Ca/(1+k2*tau)

19 R = Cp*(1+kappa)*(T-Tc)

20 Cc = Cao-Ca-Cb

21 F = G-R

Calculated values of DEQ variables

	Variable	Initial value	Final value
1	Ca	0.2980966	0.0005469
2	Cao	0.3	0.3
3	Cb	0.0019034	0.0014891
4	Cc	1.341E-14	0.297964
5	Cp	200.	200.
6	DH1	-5.5E+04	-5.5E+04
7	DH2	-7.15E+04	-7.15E+04
8	E1	9900.	9900.
9	E2	2.7E+04	2.7E+04
10	F	9948.951	-1.449E+04
11	G	348.9509	1.259E+05
12	k1	0.6385073	5.475E+04
13	k2	7.03E-10	2.001E+04
14	kappa	0.6666667	0.6666667
15	R	-9600.	1.404E+05
16	T	273.	723.
17	t	0	225.
18	Ta	330.	330.
19	tau	0.01	0.01
20	Tc	301.8	301.8
21	To	283.	283.
22	UA	4.0E+04	4.0E+04
23	vo	1000.	1000.

Incrementing temperature in this manner is an easy way to generate R(T) and G(T) plots.

Living Example Problem

When F = 0, then G(T) = R(T) and the steady states can be found.

Wow! Five (5) multiple steady states!

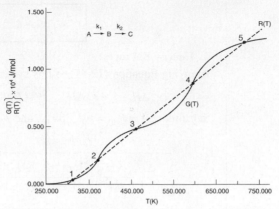

Figure E12-6.1   Heat-removed and heat-generated curves.

TABLE E12-6.2   EFFLUENT CONCENTRATIONS AND TEMPERATURES

SS	T (K)	C_A (mol/dm^3)	C_B (mol/dm^3)	C_C (mol/dm^3)
1	310	0.285	0.015	0
2	363	0.189	0.111	0.0
3	449	0.033	0.265	0.002
4	558	0.004	0.163	0.132
5	677	0.001	0.005	0.294

Analysis: Wow! We see that five steady states (SS) exist!! The exit concentrations and temperatures listed in Table E12-6.2 were determined from the tabular output of the Polymath program. Steady states 1, 3, and 5 are stable steady states, while 2 and 4 are unstable. The selectivity at steady state 3 is $\tilde{S}_{B/C} = \dfrac{0.265}{0.002} = 132.5$, while at steady state 5 the selectivity is $\tilde{S}_{B/C} = \dfrac{0.005}{0.294} = 0.017$ and is far too small. Consequently, we either have to operate at steady state 3 or find a different set of operating conditions. What do you think of the value of tau, i.e., $\tau = 0.01$ min? Is it a realistic number?

12.6.5 Complex Reactions in a PFR

Example 12-7 Complex Reactions with Heat Effects in a PFR

The following complex gas-phase reactions follow elementary rate laws

$$(1) \qquad A + 2B \rightarrow C \qquad -r_{1A} = k_{1A} C_A C_B^2 \qquad \Delta H_{Rx1B} = -15{,}000 \, \text{cal/mole B}$$

$$(2) \qquad 2A + 3C \rightarrow D \qquad -r_{2C} = k_{2C} C_A^2 C_C^3 \qquad \Delta H_{Rx2A} = -10{,}000 \, \text{cal/mole A}$$

and take place in a PFR. The feed is stoichiometric for reaction (1) in A and B with $F_{A0} = 5$ mol/min. The reactor volume is 10 dm^3 and the total entering concentration is $C_{T0} = 0.2$ mol/dm^3. The entering pressure is 100 atm and the entering temperature is 300 K. The coolant flow rate is 50 mol/min and the entering coolant fluid has a heat capacity of $C_{P_{C0}} = 10$ cal/mol · K and enters at a temperature of 325 K.

Parameters

$$k_{1A} = 40 \left(\frac{\text{dm}^3}{\text{mol}} \right)^2 \bigg/ \text{min} \ \text{ at } 300\text{K with } \ E_1 = 8{,}000 \, \text{cal/mol}$$

$$k_{2C} = 2 \left(\frac{dm^3}{mol} \right)^4 \Big/ min \quad at\ 300K\ with \quad E_2 = 12,000\,cal/mol$$

$$C_{P_A} = 10\,cal/mol/K \qquad\qquad Ua = 80\,\frac{cal}{min \cdot K}$$

$$C_{P_B} = 12\,cal/mol/K \qquad\qquad T_{a0} = 325\ K$$

$$C_{P_C} = 14\,cal/mol/K \qquad\qquad \dot{m} = 50\,mol/min$$

$$C_{P_D} = 16\,cal/mol/K \qquad\qquad C_{P_{C0}} = 10\,cal/mol/K$$

Plot F_A, F_B, F_C, F_D, p, T, and T_a as a function of V for
(a) Co-current heat exchange
(b) Countercurrent heat exchange
(c) Constant T_a
(d) Adiabatic operation

Solution

Gas Phase PFR No Pressure Drop ($p = 1$)

1. Mole Balances

Following the Algorithm

(1) $\quad \dfrac{dF_A}{dV} = r_A \qquad\quad \left(F_{A0} = 5\ mol/min \right)$ \hfill (E12-7.1)

(2) $\quad \dfrac{dF_B}{dV} = r_B \qquad\quad \left(F_{B0} = 10\ mol/min \right)$ \hfill (E12-7.2)

(3) $\quad \dfrac{dF_C}{dV} = r_C \qquad\quad V_f = 10\ dm^3$ \hfill (E12-7.3)

(4) $\quad \dfrac{dF_D}{dV} = r_D$ \hfill (E12-7.4)

2. Rates:

2a. Rate Laws

(5) $\quad r_{1A} = -k_{1A} C_A C_B^2$ \hfill (E12-7.5)

(6) $\quad r_{2C} = -k_{2C} C_A^2 C_C^3$ \hfill (E12-7.6)

2b. Relative Rates

(7) $\quad r_{1B} = 2\ r_{1A}$ \hfill (E12-7.7)

(8) $\quad r_{1C} = -r_{1A}$ \hfill (E12-7.8)

(9) $\quad r_{2A} = \dfrac{2}{3} r_{2C} = -\dfrac{2}{3} k_{2C} C_A^2 C_C^3$ \hfill (E12-7.9)

(10) $\quad r_{2D} = -\dfrac{1}{3} r_{2C} = \dfrac{1}{3} k_{2C} C_A^2 C_C^3$ \hfill (E12-7.10)

2c. Net Rates of reaction for species A, B, C, and D are

(11) $\quad r_A = r_{1A} + r_{2A} = -k_{1A} C_A C_B^2 - \dfrac{2}{3} k_{2C} C_A^2 C_C^3$ \hfill (E12-7.11)

(12) $\quad r_B = r_{1B} = -2k_{1A} C_A C_B^2$ \hfill (E12-7.12)

(13) $\quad r_C = r_{1C} + r_{2C} = k_{1A} C_A C_B^2 - k_{2C} C_A^2 C_C^3$ \hfill (E12-7.13)

(14) $\quad r_D = r_{2D} = \dfrac{1}{3} k_{2C} C_A^2 C_C^3$ \hfill (E12-7.14)

3. Selectivity:

At $V = 0$, $F_D = 0$ causing $S_{C/D}$ to go to infinity. Therefore, we set $S_{C/D} = 0$ between $V = 0$ and a very small number, say, $V = 0.0001$ dm^3 to prevent the ODE solver from crashing.

$$(15) \quad S_{C/D} = \text{if } (V > 0.0001) \text{ then } \left(\frac{F_C}{F_D}\right) \text{ else } (0) \qquad \text{(E12-7.15)}$$

4. Stoichiometry:

$$(16) \quad C_A = C_{T0}\left(\frac{F_A}{F_T}\right)p\left(\frac{T_0}{T}\right) \qquad \text{(E12-7.16)}$$

$$(17) \quad C_B = C_{T0}\left(\frac{F_B}{F_T}\right)p\left(\frac{T_0}{T}\right) \qquad \text{(E12-7.17)}$$

$$(18) \quad C_C = C_{T0}\left(\frac{F_C}{F_T}\right)p\left(\frac{T_0}{T}\right) \qquad \text{(E12-7.18)}$$

$$(19) \quad C_D = C_{T0}\left(\frac{F_D}{F_T}\right)p\left(\frac{T_0}{T}\right) \qquad \text{(E12-7.19)}$$

$$(20) \quad p = 1 \qquad \text{(E12-7.20)}$$

$$(21) \quad F_T = F_A + F_B + F_C + F_D \qquad \text{(E12-7.21)}$$

5. Parameters:

$$(22) \quad k_{1A} = 40\exp\left[\frac{E_1}{R}\left(\frac{1}{300} - \frac{1}{T}\right)\right](\text{dm}^3/\text{mol})^2/\text{min} \qquad \text{(E12-7.22)}$$

$$(23) \quad k_{2C} = 2\exp\left[\frac{E_2}{R}\left(\frac{1}{300} - \frac{1}{T}\right)\right](\text{dm}^3/\text{mol})^4/\text{min} \qquad \text{(E12-7.23)}$$

$$(24) \quad C_{A0} = 0.2 \text{ mol}/\text{dm}^3 \qquad (26) \quad E_1 = 8,000 \text{ cal/mol}$$

$$(25) \quad R = 1.987 \text{ cal/mol/K} \qquad (27) \quad E_2 = 12,000 \text{ cal/mol}$$

other parameters are given in the problem statement, i.e., Equations (28) to (35) below.

(28) C_{P_A}, (29) C_{P_B}, (30) C_{P_C}, (31) $\dot{m}_c$, (32) ΔH°_{Rx20}, (33) ΔH°_{RxLB},

(34) ΔH°_{Rx2A}, (35) C_{P_0}

6. Energy Balance:

Recalling Equation (12-37)

$$(36) \quad \frac{dT}{dV} = \frac{Q_g - Q_r}{\sum F_j C_{P_j}} \qquad \text{(E12-7.36)}$$

The denominator of Equation (E12-7.36) is

$$(37) \quad \sum F_j C_{P_j} = F_A C_{P_A} + F_B C_{P_B} + F_C C_{P_C} + F_D C_{P_D} \qquad \text{(E12-7.37)}$$

The "heat removed" term is

$$(38) \quad Q_r = Ua(T - T_a) \qquad \text{(E12-7.38)}$$

The "heat generated" is

$$(39) \quad Q_g = \sum r_{ij}\Delta H_{Rxij} = r_{1B}\Delta H_{Rx1B} + r_{2A}\Delta H_{Rx2A} \qquad \text{(E12-7.39)}$$

(a)   Co-current heat exchange

The heat exchange balance for co-current exchange is

$$(40) \quad \frac{dT_a}{dV} = \frac{Ua(T - T_a)}{\dot{m}_C C_{P_{C0}}} \qquad \text{(E12-7.40)}$$

Part (a) Co-current flow: Plot and analyze the molar flow rates, and the reactor and coolant temperatures as a function of reactor volume.

TABLE E12-7.1   POLYMATH PROGRAM AND OUTPUT FOR CO-CURRENT EXCHANGE

Differential equations

1. $d(Fa)/d(V) = ra$
2. $d(Fb)/d(V) = rb$
3. $d(Fc)/d(V) = rc$
4. $d(Fd)/d(V) = rd$
5. $d(T)/d(V) = (Qg-Qr)/sumFiCpi$
6. $d(Ta)/d(V) = Ua*(T-Ta)/m/Cpco$

Explicit equations

1. $E2 = 12000$
2. $p = 1$
3. $R = 1.987$
4. $Ft = Fa+Fb+Fc+Fd$
5. $To = 300$
6. $k2c = 2*exp((E2/R)*(1/300-1/T))$
7. $E1 = 8000$
8. $Cto = 0.2$
9. $Ca = Cto*(Fa/Ft)*(To/T)*p$
10. $Cc = Cto*(Fc/Ft)*(To/T)*p$
11. $r2c = -k2c*Ca^2*Cc^3$
12. $Cpco = 10$
13. $m = 50$
14. $Cb = Cto*(Fb/Ft)*(To/T)*.5$
15. $k1a = 40*exp((E1/R)*(1/300-1/T))$
16. $r1a = -k1a*Ca*Cb^2$
17. $r1b = 2*r1a$
18. $rb = r1b$
19. $r2a = 2/3*r2c$
20. $DH1b = -15000$
21. $DH2a = -10000$
22. $r1c = -r1a$
23. $Ta55 = 325$
24. $Cpd = 16$
25. $Cpa = 10$
26. $Cpb = 12$
27. $Cpc = 14$
28. $sumFiCpi = Cpa*Fa+Cpb*Fb+Cpc*Fc+Cpd*Fd$
29. $rc = r1c+r2c$
30. $Ua = 80$
31. $r2d = -1/3*r2c$
32. $ra = r1a+r2a$
33. $rd = r2d$
34. $Qg = r1b*DH1b+r2a*DH2a$
35. $Qr = Ua*(T-Ta)$

Co-current
heat exchange

Living Example Problem

POLYMATH Report
Ordinary Differential Equations

Calculated values of DEQ variables

	Variable	Initial value	Maximal value	Final value
1	Ca	0.0666667	0.0666667	0.0077046
2	Cb	0.1333333	0.1333333	0.0156981
3	Cc	0	0.0909427	0.0909427
4	Cpa	10.	10.	10.
5	Cpb	12.	12.	12.
6	Cpc	14.	14.	14.
7	Cpco	10.	10.	10.
8	Cpd	16.	16.	16.
9	Cto	0.2	0.2	0.2
10	DH1b	-1.5E+04	-1.5E+04	-1.5E+04
11	DH2a	-10000.	-10000.	-10000.
12	E1	8000.	8000.	8000.
13	E2	1.2E+04	1.2E+04	1.2E+04
14	Fa	5.	5.	0.3890865
15	Fb	10.	10.	0.7927648
16	Fc	0	4.592674	4.592674
17	Fd	0	0.003648	0.003648
18	Ft	15.	15.	5.778173
19	k1a	40.	2.861E+05	1.248E+04
20	k2c	2.	1.21E+06	1.102E+04
21	m	50.	50.	50.
22	Qg	1422.222	9.589E+04	714.0015
23	Qr	-2000.	3.863E+04	1450.125
24	R	1.987	1.987	1.987
25	r1a	-0.0474074	-0.0236907	-0.0236907
26	r1b	-0.0948148	-0.0473814	-0.0473814
27	r1c	0.0474074	3.196187	0.0236907
28	r2a	0	0	-0.000328
29	r2c	0	0	-0.000492
30	r2d	0	0.0021577	0.000164
31	ra	-0.0474074	-0.0240187	-0.0240187
32	rb	-0.0948148	-0.0473814	-0.0473814
33	rc	0.0474074	3.195219	0.0231987
34	rd	0	0.0021577	0.000164
35	sumFiCpi	170.	170.	77.75984
36	T	300.	885.7738	524.395
37	Ta	325.	506.2685	506.2685
38	Ta55	325.	325.	325.
39	To	300.	300.	300.
40	Ua	80.	80.	80.
41	V	0	10.	10.
42	p	1.	1.	1.

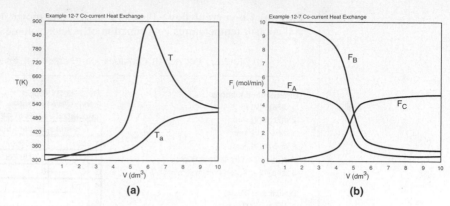

Figure E12-7.1 Profiles for co-current heat exchange; (a) temperature (b) molar flow rates. *Note:* The molar flow rate F_D is very small and is essentially the same as the bottom axis.

Analysis: **Part (a):** For co-current heat exchange, the selectivity $\tilde{S}_{C/D} = \dfrac{4.63}{0.026} = 178$

is really quite good. We also note that the reactor temperature, T, increases when $Q_g > Q_r$ and reaches a maximum, $T = 930$ K, at approximately V = 5 dm³. After that, $Q_r > Q_g$ the reactor temperature decreases and approaches T_a at the end of the reactor.

Part **(b) Countercurrent heat exchange:** We will use the same program as part **(a)**, but will change the sign of the heat-exchange balance and guess T_a at $V = 0$ to be 507 K.

$$\frac{dT_a}{dV} = -\frac{Ua(T - T_a)}{\dot{m}_C C_{P_{Cool}}}$$

We find our guess of 507 K matches $T_{a0} = 325$ K. Are we lucky or what?!

TABLE 12-7.2   POLYMATH PROGRAM AND OUTPUT FOR COUNTERCURRENT EXCHANGE

Countercurrent heat exchange

Living Example Problem

Differential equations

1 d(Fa)/d(V) = ra

2 d(Fb)/d(V) = rb

3 d(Fc)/d(V) = rc

4 d(Fd)/d(V) = rd

5 d(T)/d(V) = (Qg-Qr)/sumFiCpi

6 d(Ta)/d(V) = -Ua*(T-Ta)/m/Cpco

Same **Explicit Equations** as part **(a)**, i.e., (1) through (35) page 591.

POLYMATH Report
Ordinary Differential Equations

Calculated values of DEQ variables

	Variable	Initial value	Maximal value	Final value
14	Fa	5.	5.	0.3863414
15	Fb	10.	10.	0.7882685
16	Fc	0	4.594177	4.594177
17	Fd	0	0.0038964	0.0038964
18	Ft	15.	15.	5.772683
36	T	300.	1101.439	327.1645
37	Ta	507.	536.1941	325.4494

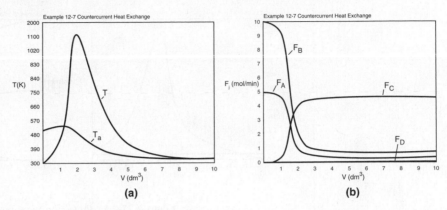

Figure E12-7.2   Profiles for countercurrent heat exchange;
(a) temperature (b) molar flow rates.

Analysis: **Part (b):** For countercurrent exchange, the coolant temperature reaches a maximum at V = 1.3 dm³ while the reactor temperature reaches a maximum at V = 2.7 dm³. The reactor with a countercurrent exchanger reaches a maximum reactor temperature of 1100 K, which is greater than that for the co-current exchanger, (i.e., 930 K). Consequently, if there is a concern about additional side reactions occurring at this maximum temperature of 1100 K, one should use a co-current exchanger or maintain constant T_a in the exchanger. In Figure 12-7.2 (a) we see that the reactor temperature approaches the coolant entrance temperature at the end of the reactor. The selectivity for the countercurrent systems, $\tilde{S}_{C/D} = 175$, is slightly lower than that for the co-current exchange.

Part (c) Constant T_a: To solve the case of constant heating-fluid temperature, we simply multiply the right-hand side of the heat-exchanger balance by zero, i.e.,

$$\frac{dT_a}{dV} = -\frac{Ua(T - T_a)}{\dot{m}_C C_P} * 0$$

and use Equations (E12-7.1) through (E12-7.40).

Living Example Problem

TABLE 12-7.3   POLYMATH PROGRAM AND OUTPUT FOR CONSTANT T_a

Constant T_a

Differential equations

1 d(Fa)/d(V) = ra

2 d(Fb)/d(V) = rb

3 d(Fc)/d(V) = rc

4 d(Fd)/d(V) = rd

5 d(T)/d(V) = (Qg-Qr)/sumFiCpi

6 d(Ta)/d(V) = Ua*(T-Ta)/m/Cpco*0

Same **Explicit Equations** as parts (a) and (b), p. 591.

POLYMATH Report
Ordinary Differential Equations
Calculated values of DEQ variables

	Variable	Initial value	Maximal value	Final value
14	Fa	5.	5.	0.5419419
15	Fb	10.	10.	1.093437
16	Fc	0	4.446116	4.446116
17	Fd	0	0.0023884	0.0023884
18	Ft	15.	15.	6.083884
36	T	300.	836.9449	345.643
37	Ta	325.	325.	325.

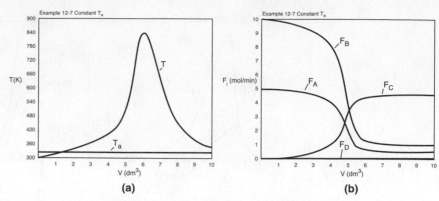

Figure E12-7.3   Profiles for constant T_a;
(a) temperature (b) molar flow rates.

**Analysis:** **Part (c):** For constant T_a, the maximum reactor temperature, 870K, is less than either co-current or countercurrent exchange while the selectivity, $\tilde{S}_{C/D} = 252.9$, is greater than either co-current or countercurrent exchange. Consequently, one should investigate how to achieve sufficiently high mass flow of the coolant in order to maintain constant T_a.

Part (d) Adiabatic: To solve for the adiabatic case, we simply multiply the overall heat transfer coefficient by zero.

$$Ua = 80 * 0$$

TABLE E12-7.4   POLYMATH PROGRAM AND OUTPUT FOR ADIABATIC OPERATION

Living Example Problem

Adiabatic
operation

Differential equations

1   d(Fa)/d(V) = ra

2   d(Fb)/d(V) = rb

3   d(Fc)/d(V) = rc

4   d(Fd)/d(V) = rd

5   d(T)/d(V) = (Qg-Qr)/sumFiCpi

6   d(Ta)/d(V) = Ua*(T-Ta)/m/Cpco*0

Explicit equations

29   Ua = 80*0

33   Qg = r1b*DH1b+r2a*DH2a

34   Qr = Ua*(T-Ta)

Explicit Equations
Same as parts (**a**), (**b**) and (**c**)
except change line 30 on p. 591
to the following:

30   Ua = 80*0

POLYMATH Report
Ordinary Differential Equations

Calculated values of DEQ variables

	Variable	Initial value	Final value
14	Fa	5.	0.1857289
15	Fb	10.	0.4123625
16	Fc	0	4.773366
17	Fd	0	0.0068175
18	Ft	15.	5.378275
36	T	300.	1548.299
37	Ta	325.	325.

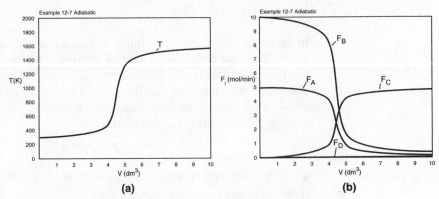

Figure E12-7.4   Profiles for adiabatic operation

**Analysis:** **Part (d):** For the adiabatic case, the maximum temperature, which is the exit temperature, is higher than the other three exchange systems, and the selectivity is the lowest. At this high temperature, the occurrence of unwanted side reactions certainly is a concern.

**Overall Analysis Parts (a) to (d):** Suppose the maximum temperature in each of these cases is outside the safety limit of 750 K for this system. **Problem P12-2$_A$ (h)** asks how you can keep the maximum temperature below 750 K.

12.7 Radial and Axial Variations in a Tubular Reactor

COMSOL
application

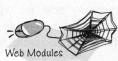

Web Modules

In the previous sections, we have assumed that there were no radial variations in velocity, concentration, temperature, or reaction rate in the tubular and packed-bed reactors. As a result, the axial profiles could be determined using an ordinary differential equation (ODE) solver. In this section, we will consider the case where we have both axial and radial variations in the system variables, in which case we will require a partial differential (PDE) solver. A PDE solver such as COMSOL will allow us to solve tubular reactor problems for both the axial and radial profiles, as shown in the Radial Effects COMSOL Web module on the CRE Web site (_http://www.umich.edu/~elements/5e/web_mod/radialeffects/index.htm._[7] In addition, **a number of COMSOL examples associated with and developed for this book** are available from the COMSOL Web site (_www.comsol.com/ecre_).

We are going to carry out differential mole and energy balances on the differential cylindrical annulus shown in Figure 12-15.

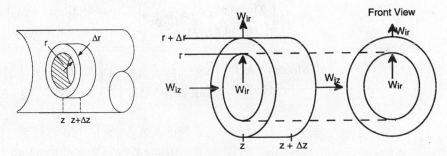

Figure 12-15   Cylindrical shell of thickness Δr, length Δz, and volume $2\pi r\Delta r\Delta z$.

---

[7] An introductory webinar on COMSOL can be found on the AIChE webinar Web site: _http://www.aiche.org/resources/chemeondemand/webinars/modeling-non-ideal-reactors-and-mixers._

12.7.1 Molar Flux

In order to derive the governing equations, we need to define a couple of terms. The first is the molar flux of species i, W_i (mol/m^2 • s). The molar flux has two components, the radial component, W_{ir}, and the axial component, W_{iz}. The molar flow rates are just the product of the molar fluxes and the cross-sectional areas normal to their direction of flow A_{cz}. For example, for species i flowing in the axial (i.e., z) direction

$$F_{iz} = W_{iz} A_{cz}$$

where W_{iz} is the molar flux in the z direction (mol/m^2/s), and A_{cz} (m^2) is the cross-sectional area of the tubular reactor.

In Chapter 14 we discuss the molar fluxes in some detail, but for now let us just say they consist of a diffusional component, $-D_e(\partial C_i/\partial z)$, and a convective flow component, $U_z C_i$

$$W_{iz} = -D_e \frac{\partial C_i}{\partial z} + U_z C_i \tag{12-42}$$

where D_e is the effective diffusivity (or dispersion coefficient) (m^2/s), and U_z is the axial molar average velocity (m/s). Similarly, the flux in the radial direction is

Radial direction

$$W_{ir} = -D_e \frac{\partial C_i}{\partial r} + U_r C_i \tag{12-43}$$

where U_r (m/s) is the average velocity in the radial direction. For now, we will neglect the velocity in the radial direction, i.e., $U_r = 0$. A mole balance on species A (i.e., i = A) in a cylindrical system volume of length Δz and thickness Δr as shown in Figure 12-15 gives

Mole Balances on Species A

$$\left(\begin{array}{c} \text{Moles of A} \\ \text{in at } r \end{array}\right) = W_{Ar} \bullet \left(\begin{array}{c} \text{Cross-sectional area} \\ \text{normal to radial flux} \end{array}\right) = W_{Ar} \bullet 2\pi r \Delta z$$

$$\left(\begin{array}{c} \text{Moles of A} \\ \text{in at } z \end{array}\right) = W_{Az} \bullet \left(\begin{array}{c} \text{Cross-sectional area} \\ \text{normal to axial flux} \end{array}\right) = W_{Az} \bullet 2\pi r \Delta r$$

$$\left(\begin{array}{c} \text{Moles of A} \\ \text{in at } r \end{array}\right) - \left(\begin{array}{c} \text{Moles of A} \\ \text{out at } (r + \Delta r) \end{array}\right) + \left(\begin{array}{c} \text{Moles of A} \\ \text{in at } z \end{array}\right) - \left(\begin{array}{c} \text{Moles of A} \\ \text{out at } (z + \Delta z) \end{array}\right)$$

$$+ \left(\begin{array}{c} \text{Moles of A} \\ \text{formed} \end{array}\right) = \left(\begin{array}{c} \text{Moles of A} \\ \text{accumulated} \end{array}\right)$$

$$W_{Ar}2\pi r\Delta z|_r - W_{Ar}2\pi r\Delta z|_{r+\Delta r} + W_{Az}2\pi r\Delta r|_z - W_{Az}2\pi r\Delta r|_{z+\Delta z}$$

$$+ r_A 2\pi r\Delta r\Delta z = \frac{\partial C_A(2\pi r\Delta r\Delta z)}{\partial t}$$

Dividing by $2\pi r\Delta r\Delta z$ and taking the limit as Δr and $\Delta z \rightarrow 0$

$$-\frac{1}{r}\frac{\partial(rW_{Ar})}{\partial r} - \frac{\partial W_{Az}}{\partial z} + r_A = \frac{\partial C_A}{\partial t}$$

Similarly, for any species i and steady-state conditions

$$-\frac{1}{r}\frac{\partial(rW_{ir})}{\partial r} - \frac{\partial W_{iz}}{\partial z} + r_i = 0 \qquad (12\text{-}44)$$

Using Equations (12-42) and (12-43) to substitute for W_{iz} and W_{ir} in Equation (12-44) and then setting the radial velocity to zero, $U_r = 0$, we obtain

$$-\frac{1}{r}\frac{\partial}{\partial r}\left[\left(-D_e\frac{\partial C_i}{\partial r}r\right)\right] - \frac{\partial}{\partial z}\left[-D_e\frac{\partial C_i}{\partial z} + U_z C_i\right] + r_i = 0$$

This equation will also be discussed further in Chapter 17.

For steady-state conditions and assuming U_z does not vary in the axial direction

$$D_e\frac{\partial^2 C_i}{\partial r^2} + \frac{D_e}{r}\frac{\partial C_i}{\partial r} + D_e\frac{\partial^2 C_i}{\partial z^2} - U_z\frac{\partial C_i}{\partial z} + r_i = 0 \qquad (12\text{-}45)$$

12.7.2 Energy Flux

When we applied the first law of thermodynamics to a reactor to relate either temperature and conversion or molar flow rates and concentration, we arrived at Equation (11-10). Neglecting the work term we have for steady-state conditions

$$\overset{\text{Conduction}}{\overbrace{\dot{Q}}} + \sum_{i=1}^{n}\overset{\text{Convection}}{\overbrace{F_{i0}H_{i0} - \sum_{i=1}F_iH_i}} = 0 \qquad (12\text{-}46)$$

In terms of the molar fluxes and the cross-sectional area, and $(\mathbf{q} = \dot{Q}/A_c)$

$$A_c[\mathbf{q} + (\Sigma\mathbf{W}_{i0}H_{i0} - \Sigma\mathbf{W}_iH_i)] = 0 \qquad (12\text{-}47)$$

The $\mathbf{q}$ term is the heat added to the system and almost always includes a conduction component of some form. We now define an **energy flux vector, e,** $(J/m^2 \cdot s)$, to include both the conduction and convection of energy.

$\mathbf{e}$ = energy flux $J/s \cdot m^2$

e = Conduction + Convection

$$\mathbf{e} = \mathbf{q} + \Sigma\mathbf{W}_iH_i \qquad (12\text{-}48)$$

where the conduction term $\mathbf{q}$ (kJ/m^2 · s) is given by Fourier's law. For axial and radial conduction, Fourier's laws are

$$q_z = -k_e \frac{\partial T}{\partial z} \qquad \text{and} \qquad q_r = -k_e \frac{\partial T}{\partial r}$$

where k_e is the thermal conductivity (J/m·s·K). The energy transfer (flow) is the flux vector times the cross-sectional area, A_c, normal to the energy flux

$$\text{Energy flow} = \mathbf{e} \cdot A_c$$

12.7.3 Energy Balance

Using the energy flux, e, to carry out an energy balance on our annulus (Figure 12-15) with system volume $2\pi r \Delta r \Delta z$, we have

$$(\text{Energy flow in at } r) = e_r A_{cr} = e_r \cdot 2\pi r \Delta z$$

$$(\text{Energy flow in at } z) = e_z A_{cz} = e_z \cdot 2\pi r \Delta r$$

$$\begin{pmatrix} \text{Energy flow} \\ \text{in at } r \end{pmatrix} - \begin{pmatrix} \text{Energy flow} \\ \text{out at } r + \Delta r \end{pmatrix} + \begin{pmatrix} \text{Energy flow} \\ \text{in at } z \end{pmatrix} - \begin{pmatrix} \text{Energy flow} \\ \text{out at } z + \Delta z \end{pmatrix} = \begin{pmatrix} \text{Accumulation} \\ \text{of energy in} \\ \text{Volume } (2\pi r \Delta r \Delta z) \end{pmatrix}$$

$$(e_r 2\pi r \Delta z)\big|_r - (e_r 2\pi r \Delta z)\big|_{r+\Delta r} + e_z 2\pi r \Delta r\big|_z - e_z 2\pi r \Delta r\big|_{z+\Delta z} = 0$$

Dividing by $2\pi r \Delta r \Delta z$ and taking the limit as Δr and $\Delta z \to 0$

$$\boxed{-\frac{1}{r}\frac{\partial(r e_r)}{\partial r} - \frac{\partial e_z}{\partial z} = 0} \tag{12-49}$$

The radial and axial energy fluxes are

$$e_r = q_r + \Sigma W_{ir} H_i$$

$$e_z = q_z + \Sigma W_{iz} H_i$$

Substituting for the energy fluxes, e_r and e_z

$$-\frac{1}{r}\frac{\partial[r[q_r + \Sigma W_{ir} H_i]]}{\partial r} - \frac{\partial[q_z + \Sigma W_{iz} H_i]}{\partial z} = 0 \tag{12-50}$$

and expanding the convective energy fluxes, $\Sigma W_i H_i$,

Radial: $\qquad \dfrac{1}{r}\dfrac{\partial}{\partial r}(r\Sigma W_{ir} H_i) = \dfrac{1}{r}\Sigma H_i \dfrac{\partial(r W_{ir})}{\partial r} + \overset{\displaystyle\nearrow \text{Neglect}}{\cancel{\Sigma W_{ir}\dfrac{\partial H_i}{\partial r}}}$ $\qquad$ (12-51)

Axial: $\qquad \dfrac{\partial(\Sigma W_{iz} H_i)}{\partial z} = \Sigma H_i \dfrac{\partial W_{iz}}{\partial z} + \Sigma W_{iz}\dfrac{\partial H_i}{\partial z}$ $\qquad$ (12-52)

Substituting Equations (12-51) and (12-52) into Equation (12-50), we obtain upon rearrangement

$$-\frac{1}{r}\frac{\partial(rq_r)}{\partial r} - \frac{\partial q_z}{\partial z} - \Sigma H_i \overbrace{\left(\frac{1}{r}\frac{\partial(rW_{ir})}{\partial r} + \frac{\partial W_{iz}}{\partial z}\right)}^{r_i} - \Sigma W_{iz}\frac{\partial H_i}{\partial z} = 0$$

Recognizing that the term in brackets is related to Equation (12-44) and is just the rate of formation of species i, r_i, for steady-state conditions we have

$$-\frac{1}{r}\frac{\partial}{\partial r}(rq_r) - \frac{\partial q_z}{\partial z} - \Sigma H_i r_i - \Sigma W_{iz}\frac{\partial H_i}{\partial z} = 0 \qquad (12\text{-}53)$$

Recalling

$$q_r = -k_e\frac{\partial T}{\partial r}, \; q_z = -k_e\frac{\partial T}{\partial z}, \; \frac{\partial H_i}{\partial z} = C_{P_i}\frac{\partial T}{\partial z},$$

and

$$r_i = v_i(-r_A)$$

$$\Sigma r_i H_i = \Sigma v_i H_i(-r_A) = -\Delta H_{Rx} r_A$$

we have the energy in the form

$$\frac{k_e}{r}\left[\frac{\partial}{\partial r}\left(\frac{r\partial T}{\partial r}\right)\right] + k_e\frac{\partial^2 T}{\partial z^2} + \Delta H_{Rx} r_A - (\Sigma W_{iz} C_{P_i})\frac{\partial T}{\partial z} = 0 \qquad (12\text{-}54)$$

where W_{iz} is given by Equation (12-42). Equation (12-54) would be coupled with the *mole balance*, *rate law*, and *stoichiometric* equations to solve for the radial and axial concentration at gradients. However, a great amount of computing time would be required. Let's see if we can make some approximations to simplify the solution.

Some Initial Approximations

Assumption 1. Neglect the diffusive term, wrt, the convective term in Equation (12-42) in the expression involving heat capacities

$$\Sigma C_{P_i} W_{iz} = \Sigma C_{P_i}(0 + U_z C_i) = \Sigma C_{P_i} C_i U_z$$

With this assumption, Equation (12-54) becomes

$$\frac{k_e}{r}\frac{\partial}{\partial r}\left(\frac{r\partial T}{\partial r}\right) + k_e\frac{\partial^2 T}{\partial z^2} + \Delta H_{Rx} r_A - (U_z\Sigma C_{P_i} C_i)\frac{\partial T}{\partial z} = 0 \qquad (12\text{-}55)$$

For laminar flow, the velocity profile is

$$U_z = 2U_0\left[1 - \left(\frac{r}{R}\right)^2\right] \qquad (12\text{-}56)$$

where U_0 is the average velocity inside the reactor.

Assumption 2. Assume that the sum $C_{P_m} = \Sigma C_{P_i} C_i = C_{A0} \Sigma \Theta_i C_{P_i}$ is constant. The energy balance now becomes

$$k_e \frac{\partial^2 T}{\partial z^2} + \frac{k_e}{r}\frac{\partial}{\partial r}\left(r\frac{\partial T}{\partial r}\right) + \Delta H_{Rx} r_A - U_z C_{P_m}\frac{\partial T}{\partial z} = 0 \qquad (12\text{-}57)$$

Equation (12-56) is the form we will use in our COMSOL problem. In many instances, the term C_{P_m} is just the product of the solution density (kg/m^3) and the heat capacity of the solution (kJ/kg • K).

Coolant Balance

We also recall that a balance on the coolant gives the variation of coolant temperature with axial distance where U_{ht} is the overall heat transfer coefficient and R is the reactor wall radius

$$\dot{m}_c C_{P_c}\frac{\partial T_a}{\partial z} = U_{ht} 2\pi R[T(R,z) - T_a] \qquad (12\text{-}58)$$

Boundary and Initial Conditions

A. Initial conditions *if* other than steady state (not considered here)
    $t = 0, \quad C_i = 0, \qquad T = T_0, \quad$ for $z > 0$ all r
B. Boundary conditions

  1) Radial
      (a) At $r = 0$, we have symmetry $\partial T/\partial r = 0$ and $\partial C_i/\partial r = 0$.
      (b) At the tube wall, $r = R$, the temperature flux to the wall on the reaction side equals the convective flux out of the reactor into the shell side of the heat exchanger.

$$-k_e \frac{\partial T}{\partial r}\bigg|_R = U(T(R,z) - T_a)$$

      (c) There is no mass flow through the tube walls $\partial C_i/\partial r = 0$ at $r = R$.

  2) Axial
      (a) At the entrance to the reactor $z = 0$

$$T = T_0 \text{ and } C_i = C_{i0}$$

      (b) At the exit of the reactor $z = L$

$$\frac{\partial T}{\partial z} = 0 \text{ and } \frac{\partial C_i}{\partial z} = 0$$

The preceding equations were used to describe and analyze flow and reaction in a tubular reactor with heat exchange as described in the following example, which can be found in the Expanded Material on the CRE Web site. What follows is only a brief outline of that example with a few results from the output of the COMSOL program.

Example W12–8   Tubular reactor with axial and radial gradients.

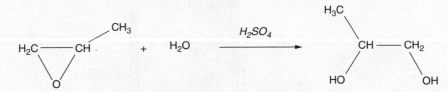

The liquid phase reaction was analyzed using COMSOL to study both axial and radial variations, and the details can be found on the home page of the CRE Web site, *www.umich.edu/~elements/5e/index.html*, by clicking on the *Expanded Material* for Chapter 12. Typical temperature profiles for this example are shown in Figures 12-16 and 12-17.

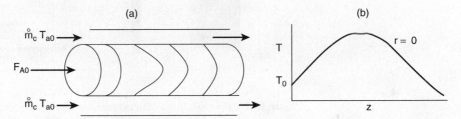

Figure 12-16   Radial (a) and axial (b) temperature profiles.

Results

Results of the
COMSOL
simulation

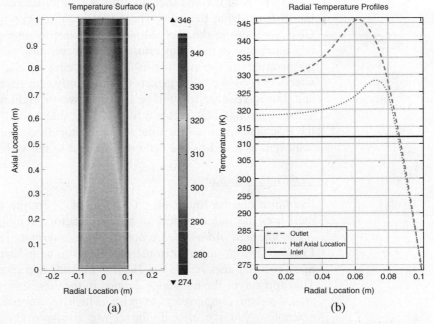

Figure 12-17   (a) Temperature surface, (b) temperature surface profiles, (c) conversion surface, and (d) radial profile. (*Continues*)

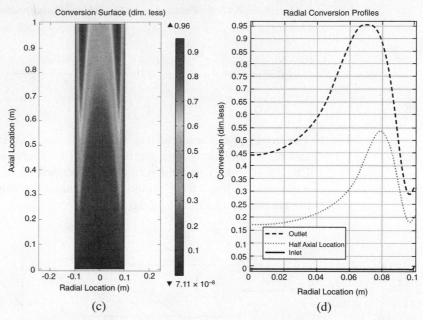

Figure 12-17 (*Continued*)   (a) Temperature surface, (b) temperature surface profiles, (c) conversion surface, and (d) radial profile.

The volumetric flow rate of water is 3.5 times the volumetric flow rate of the mixture of propylene oxide in methanol.

Analysis: One notes from the temperature surface plot that the temperature is set to 312 K at the entrance and that it rapidly decreases along the walls with the distance from the inlet. These same profiles can be found in color on the CRE Web site in the Web Modules. Be sure to note the predicted maximum and minimum in the temperature and concentration profiles. Near the wall, the temperature of the mixture is lower because of the cold wall temperature, which is cooled by the jacket. Consequently, the reaction rate will be lower, and thus the conversion will be lower. However, right next to the wall, the flow velocity through the reactor is almost zero, due to the friction with the wall, so the reactants spend a long time in the reactor; therefore, a greater conversion is achieved, as noted by the upturn right next to the wall.

COMSOL Web Site

Examples can be found on the COMSOL Web site (*www.comsol.com/ecre*), including the radial effects in tubular reactor examples: *Isothermal Reactor, Nonisothermal Adiabatic Reactor, Nonisothermal Reactor with Isothermal Cooling Jacket*, and *Nonisothermal Reactor with Variable Coolant Temperature*. You can also access the *Tubular Reactor* app directly from the Web site. This app solves the equations defined in the *Nonisothermal Reactor with Variable Coolant Temperature* exercise, which is embedded in the app. In these examples, you are also in the *Living Example Program* format (LEP). In accessing these LEPs, make sure that you either have COMSOL installed on your laptop or that it is available on your university computer. On the COMSOL Web site, *www.comsol.com/ecre*, you can also find how to locate the LEPs in the installed COMSOL Application Libraries.

Also, other applications based on exercises from this book are found at the COMSOL Web site (*www.comsol.com/ecre*). These include the following applications: *activation_energy* based on the example *Determination of Activation Energy* in this book; *cstr_startup* from the exercise *Startup of a CSTR*; *non_ideal_cstr* based on the paragraph *Real CSTR Modeled as Two CSTRs with Interchange*; and *non_isothermal_plug_flow* based on the example *Production of Acetic Anhydride*.

12.8 Safety

Scaling up exothermic chemical reactions can be very tricky. Tables 12-5 and 12-6 give reactions that have resulted in accidents and their causes, respectively.[8] The reader should review the case histories of these reactions to learn how to avoid similar accidents.

TABLE 12-5   INCIDENCE OF BATCH-PROCESS ACCIDENTS

Process Type	Number of Incidents in U.K., 1962–1987
Polymerization	64
Nitration	15
Sulfurization	13
Hydrolysis	10
Salt formation	8
Halogenation	8
Alkylation (Friedel-Crafts)	5
Amination	4
Diazolization	4
Oxidation	2
Esterification	1
Total:	134

Source: U.K. Health and Safety Executive.

TABLE 12-6   CAUSES OF BATCH REACTOR ACCIDENTS IN TABLE 12-5

Cause	Contribution, %
Lack of knowledge of reaction chemistry	20
Problems with material quality	9
Temperature-control problems	19
Agitation problems	10
Mis-charging of reactants or catalyst	21
Poor maintenance	15
Operator error	5

Runaway reactions are the most dangerous in reactor operation, and a thorough understanding of how and when they could occur is part of the chemical reaction engineer's responsibility. The reaction in the last example in this chapter could be thought of as running away. Recall that as we moved down the length of the reactor, none of the cooling arrangements could keep the reactor from reaching an extremely high temperature (e.g., 800 K). In the next chapter, we study case histories of two runaway reactions. One is the nitroaniline explosion discussed in **Example E13-2** and the other is **Example E13-6**, concerning the

---

[8] Courtesy of J. Singh, *Chemical Engineering,* 92 (1997) and B. Venugopal, *Chemical Engineering*, 54 (2002).

recent (2007) explosion at T2 Laboratories. (*http://www.chemsafety.gov/videoroom/detail.aspx?VID=32*).

There are many resources available for additional information on reactor safety and the management of chemical reactivity hazards. Guidelines for managing chemical reactivity hazards and other fire, explosion, and toxic release hazards are developed and published by the **Center for Chemical Process Safety (CCPS)** of the American Institute of Chemical Engineers. CCPS books and other resources are available at *www.aiche.org/ccps*. For example, the book *Essential Practices for Managing Chemical Reactivity Hazards*, written by a team of industry experts, is also provided free of charge by CCPS on the site *www.info.knovel.com/ccps*. A concise and easy-to-use software program that can be used to determine the reactivity of substances or mixtures of substances, the Chemical Reactivity Worksheet, is provided by the National Oceanic and Atmospheric Administration (NOAA) for free on its Web site, *www.noaa.gov*.

The **Safety and Chemical Engineering Education (SAChE)** program was formed in 1992 as a cooperative effort between the AIChE, CCPS, and engineering schools to provide teaching materials and programs that bring elements of process safety into the education of undergraduate and graduate students studying chemical and biochemical products and processes. The **SAChE Web site** (*www.sache.org*) has a great discussion of reactor safety with examples as well as information on reactive materials. These materials are also suitable for training purposes in an industrial setting.

The following instruction modules are available on the SAChE Web site (*www.sache.org*).

1. *Chemical Reactivity Hazards:* This Web-based instructional module contains about 100 Web pages with extensive links, graphics, videos, and supplemental slides. It can be used either for classroom presentation or as a self-paced tutorial. The module is designed to supplement a junior or senior chemical engineering course by showing how uncontrolled chemical reactions in industry can lead to serious harm, and by introducing key concepts for avoiding unintended reactions and controlling intended reactions.

2. *Runaway Reactions:* Experimental Characterization and Vent Sizing: This instruction module describes the ARSST and its operation, and illustrates how this instrument can easily be used to experimentally determine the transient characteristics of runaway reactions, and how the resulting data can be analyzed and used to size the relief vent for such systems.

3. *Rupture of a Nitroaniline Reactor:* This case study demonstrates the concept of runaway reactions and how they are characterized and controlled to prevent major losses.

4. *Seveso Accidental Release Case History:* This presentation describes a widely discussed case history that illustrates how minor engineering errors can cause significant problems; problems that should not be repeated. The accident was in Seveso, Italy, in 1976. It was a small release of a dioxin that caused many serious injuries.

Membership in SAChE is required to view these materials. Virtually all U.S. universities and many non-U.S universities are members of SAChE—contact your university SAChE representative, listed on the SAChE Web site, or your instructor or department chair to learn your university's username and password. Companies can also become members—see the SAChE Web site for details.

Certificate Program

SAChE also offers several certificate programs that are available to all chemical engineering students. Students can study the material, take an online test, and receive a certificate of completion. The following two certificate programs are of value for reaction engineering:

1. *Runaway Reactions:* This certificate focuses on managing chemical reaction hazards, particularly runaway reactions.
2. *Chemical Reactivity Hazards:* This is a Web-based certificate that provides an overview of the basic understanding of chemical reactivity hazards.

Many students are taking the certificate test online and put the fact that they successfully obtained the certificate on their résumés.

More information on safety is given in the *Summary Notes* and *Professional Reference Shelf* on the Web. Particularly study the use of the ARSST to detect potential problems. These will be discussed in *Chapter 13 Professional Reference Shelf R13.1* on the CRE Web site.

Closure. Virtually all reactions that are carried out in industry involve heat effects. This chapter provides the basis to design reactors that operate at steady state and involve heat effects. To model these reactors, we simply add another step to our algorithm; this step is the energy balance. Here, it is important to understand how the energy balance was applied to each reactor type so that you will be able to describe what would happen if you changed some of the operating conditions (e.g., T_0) in order to determine if those changes result in unsafe conditions such as runaway. The *Living Example Problems* (especially *12T-12-3*) and the ICG module will help you achieve a high level of understanding. Another major goal after studying this chapter is to be able to design reactors that have multiple reactions taking place under nonisothermal conditions. Work through Problem $12\text{-}26_C$ to make sure you have achieved this goal. An industrial example that provides a number of practical details is included as an appendix to this chapter. Section 12.7 introduces radial variations in temperature and concentration while previous chapters in this text considered only plug-flow profiles. Using COMSOL we can readily generate $C(r,z)$ and $T(r,z)$ for tubular reactors. In summary on these *Living Example Problems*, the best way to get a good understanding of reactors with heat effects is to download and run the Polymath, MATLAB, Wolfram, and COMSOL programs from the CRE Web site. We close with a brief discussion on safety and testing some resources where you can obtain more information.

SUMMARY

1. For single reactions, the energy balance on a PFR/PBR in terms of *molar flow rate* is

$$\frac{dT}{dV} = \frac{(r_A)[\Delta H_{Rx}(T)] - Ua(T - T_a)}{\sum F_i C_{P_i}} = \frac{Q_g - Q_r}{\sum F_i C_{P_i}} \qquad (S12\text{-}1)$$

In terms of *conversion*

$$\frac{dT}{dV} = \frac{(r_A)[\Delta H_{Rx}(T)] - Ua(T - T_a)}{F_{A0}\left(\sum \Theta_j C_{P_j} + X\Delta C_P\right)} = \frac{Q_g - Q_r}{F_{A0}\left(\sum \Theta_j C_{P_j} + X\Delta C_P\right)} \qquad (S12\text{-}2)$$

2. The temperature dependence of the specific reaction rate is given in the form

$$k(T) = k(T_1) \exp\left[\frac{E}{R}\left(\frac{1}{T_1} - \frac{1}{T}\right)\right] = k(T_1) \exp\left[\frac{E}{R}\left(\frac{T - T_1}{TT_1}\right)\right] \qquad (S12\text{-}3)$$

3. The temperature dependence of the equilibrium constant is given by van't Hoff's equation for $\Delta C_P = 0$

$$K_P(T) = K_P(T_2) \exp\left[\frac{\Delta H^\circ_{Rx}}{R}\left(\frac{1}{T_2} - \frac{1}{T}\right)\right] \qquad (S12\text{-}4)$$

4. Neglecting changes in potential energy, kinetic energy, and viscous dissipation, and for the case of no work done on or by the system, large coolant flow rates $(\dot{m}_c)$, and all species entering at the same temperature, the steady-state CSTR energy balance for single reactions is

$$\frac{UA}{F_{A0}}(T_a - T) - X[\Delta H^\circ_{Rx}(T_R) + \Delta C_P(T - T_R)] = \sum \Theta_j C_{P_j}(T - T_{i0}) \qquad (S12\text{-}5)$$

5. Multiple steady states

$$G(T) = (-\Delta H^\circ_{Rx})\left(\frac{-r_A V}{F_{A0}}\right) = (-\Delta H^\circ_{Rx})(X) \quad (S12\text{-}6)$$

$$R(T) = C_{P0}(1 + \kappa)(T - T_c) \qquad (S12\text{-}7)$$

where $\kappa = \dfrac{UA}{C_{P_0}F_{A0}}$ and $T_c = \dfrac{\kappa T_a + T_0}{1 + \kappa}$

6. When q multiple reactions are taking place and there are m species

$$\frac{dT}{dV} = \frac{\displaystyle\sum_{i=1}^{q}(r_{ij})[\Delta H_{Rxij}(T)] - Ua(T - T_a)}{\displaystyle\sum_{j=1}^{m} F_j C_{Pj}} = \frac{Q_g - Q_r}{\displaystyle\sum_{j=1}^{m} F_j C_{Pj}} \qquad (S12\text{-}8)$$

7. Axial or radial temperature and concentration gradients. The following coupled, partial differential equations were solved using COMSOL:

$$D_e \frac{\partial^2 C_i}{\partial r^2} + \frac{D_e}{r}\frac{\partial C_i}{\partial r} + D_e \frac{\partial^2 C_i}{\partial z^2} - U_z \frac{\partial C_i}{\partial z} + r_i = 0 \qquad \text{(S12-9)}$$

and

$$k_e \frac{\partial^2 T}{\partial z^2} + \frac{k_e}{r}\frac{\partial}{\partial r}\left(r\frac{\partial T}{\partial r}\right) + \Delta H_{Rx} r_A - U_z C_{P_m}\frac{\partial T}{\partial z} = 0 \qquad \text{(S12-10)}$$

CRE WEB SITE MATERIALS

- **Expanded Material**
    1. *COSMOL Example W12-8*
    2. *Problems*
    COMSOL LEP W12-8
    WP12-18$_C$
    WP12-21$_B$
    WP12-24$_B$
    WP12-31$_B$
- **Learning Resources**
    1. *Summary Notes*
    2. *Web Module COMSOL Radial and Axial Gradients*
    3. *Interactive Computer Games (ICGs)*
    A. Heat Effects I   B. Heat Effects II

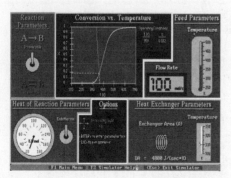 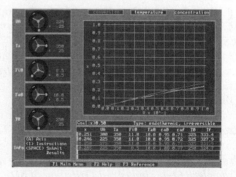

4. *Solved Problems*
   A. Example 12-2   Formulated in AspenTech: Download AspenTech directly from the CRE Web site.

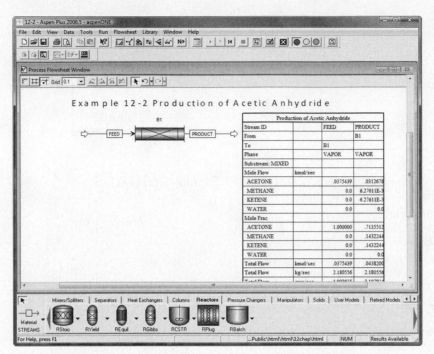

A step-by-step AspenTech tutorial is given on the CRE Web site.

   B. Example CD12-1 $\Delta H_{Rx}(T)$ for Heat Capacities Expressed as Quadratic Functions of Temperature
   C. Example CD12-2   Second-Order Reaction Carried Out Adiabatically in a CSTR
5. *PFR/PBR Solution Procedure for a Reversible Gas-Phase Reaction*

- **Living Example Problems**
   1. *Example 12-1   Isomerization of Normal Butane with Heat Exchange*
   2. *Example 12-2   Production of Acetic Anhydride*
   3. *Example 12-2$_{Asp}$ AspenTech Formulation*
   4. *Example 12-4   CSTR with Cooling Coil*
   5. *Example 12-5   Parallel Reaction in a PFR with Heat Effects*
   6. *Example 12-6   Multiple Reactions in a CSTR*
   7. *Example 12-7   Complex Reactions*
   8. *Example W12-8   COSMOL Radial Effects in a Tubular Reactor*
   9. *Example R12-1   Industrial Oxidation of SO_2*
   10. *Example 12-T12-3 PBR with Variable Coolant Temperature, T_a*

- **Professional Reference Shelf**
   R12.1. *Runaway in CSTRs and Plug-Flow Reactors*
          Phase Plane Plots. We transform the temperature and concentration profiles into a phase plane.

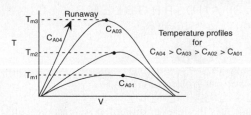

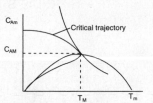

Temperature profiles.

Critical trajectory on the
$C_{Am}-T_m$ phase plane plot.

The trajectory going through the maximum of the "maxima curve" is considered to be *critical* and therefore is the locus of the *critical* inlet conditions for C_A and T corresponding to a given wall temperature.

R12.2. *Steady-State Bifurcation Analysis.* In reactor dynamics, it is particularly important to find out if multiple stationary points exist or if sustained oscillations can arise.

R12.3. *Heat Capacity as a Function of Temperature.* Combining the heat of reaction with the quadratic form of the heat capacity

$$C_{P_i} = \alpha_i + \beta_i T + \gamma_i T^2$$

we find that

$$\Delta H_{Rx}(T) = \Delta H_{Rx}^\circ(T_R) + \Delta\alpha(T - T_R) + \frac{\Delta\beta}{2}(T^2 - T_R^2) + \frac{\Delta\gamma}{3}(T^3 - T_R^3)$$

Example 12-2 is reworked on the PRS for the case of variable heat capacities.

R12.4. *Manufacture of Sulfuric Acid.* The details of the industrial oxidation of SO_2 are described. Here, the catalyst quantities, the reactor configuration, and operating conditions are discussed, along with a model to predict the conversion and temperature profiles.

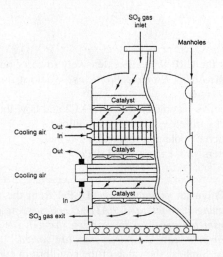

QUESTIONS AND PROBLEMS

Homework Problems

The subscript to each of the problem numbers indicates the level of difficulty: A, least difficult; D, most difficult.

$$A = \bullet \quad B = \blacksquare \quad C = \blacklozenge \quad D = \blacklozenge\blacklozenge$$

In each of the questions and problems, rather than just drawing a box around your answer, write a sentence or two describing how you solved the problem, the assumptions you made, the reasonableness of your answer, what you learned, and any other facts that you want to include. See Preface Section I for additional generic parts (**x**), (**y**), and (**z**) to the home problems.

Before solving the problems, state or sketch qualitatively the expected results or trends.

Questions

Q12-1$_A$ Review Figure 12-13. Use this figure to write a few sentences (or at least draw on analogy) explaining why, when you strike the head of a safety match slowly on its pumice with little pressure, it may heat up a little, but does not ignite, yet when you put pressure on it and strike it rapidly, it does ignite. Thanks to Oscar Piedrahita, Medellín, Colombia.

Q12-2$_A$ Read over the problems at the end of this chapter. Make up an original problem that uses the concepts presented in this chapter. To obtain a solution:
   (a) Make up your data and reaction.
   (b) Use a real reaction and real data. See Problem P5-1$_A$ for guidelines.
   (c) Prepare a list of safety considerations for designing and operating chemical reactors. (See *www.sache.org* and *www.siri.org/graphics*.) The August 1985 issue of *Chemical Engineering Progress* may be useful for part (**c**).

Problems

We will use the *Living Example* on the CRE Web site extensively to carry out simulations. *Why* carry out simulations to vary the parameters in the *Living Example Problems*? We do it in order to
- Get a more intuitive feel reactor system.
- Gain insight about the most sensitive parameters (e.g., E, KC) and how they affect outlet conditions.
- Learn how reactors are affected by different operating conditions.
- Simulate dangerous situations such as potential runaway reactions.
- Compare the model and parameters with experimental data.
- Optimize the reaction system.

P12-1$_A$ Download the following Polymath programs (i.e., **LEPs**) from the CRE Web site where appropriate. *In addition to the LEPs having Polymath and MATLAB codes, some LEPs have Wolfram. Previous students suggest using Wolfram to save time in working through the parameter variation problems to understand how the reactor behaves when parameters are varied.*
   (a) Example 12-1. Safety. Suppose the value of the equilibrium constant and heat of reaction were measured incorrectly and were found to be $K_C = 1,000$ mol/dm^3 at 330 K and $\Delta H_{Rx} = -20,000$ kJ/mol. (1) Redo Example 12-1 using these values. (2) Let $Q_g = r_A \Delta H_{Rx}$ and $Q_r = Ua\,(T - T_a)$, and then plot Q_g and Q_r on the same figure as a function of V. (3) Vary the coolant flow rate $(0 < \dot{m}_c < 2,000$ kg/h) and the entering temperature 273 (K < T_0 < 315 K), and describe what you find. (4) Vary some of the other parameters and see if you can find unsafe operating conditions. (5) Plot Q_r and T_a as a function of V necessary to maintain isothermal operation.

(b) Example 12-2 (Use the LEP). (1) Let $Q_g = r_A \Delta H_{Rx}$ and $Q_r = Ua\,(T - T_a)$, and then plot Q_g and Q_r on the same figure as a function of V. (2) Fix the reactor volume at 0.5 m³ and the entrance conditions at ($T_0 = 1050$ K, $T_{a0} = 1250$ K), and then make a table, X_e, X, T_a, and T for each of the heat exchange cases. Change the inlet conditions and determine which heat exchanger case gives the greatest differences in the conversion. (3) Repeat (2) for $V = 5$ m³. (4) Plot Q_g, Q_r, and $-r_A$ versus V for all four cases on the same figure and describe what you find. (5) For each of the four heat exchanger cases, investigate the addition of an inert I with a heat capacity of 50 J/mol · K, keeping F_{A0} constant and letting the other inlet conditions adjust accordingly (e.g., ε). (6) Vary the inert molar flow rate (i.e., Θ_I, $0.0 < \Theta_I < 3.0$ mol/s). Plot X and analyze versus Θ_I. (7) Finally, vary the heat-exchange fluid temperature T_{a0} (1,000°F < T_{a0} < 1350°F). Write a paragraph describing what you find, noting interesting profiles or results.

Member

Hall of Fame

(c) Example 12-2. AspenTech Formulation. Repeat P12-2**(b)** using AspenTech.

(d) Problem 12-3$_B$ Solution. Load the LEP for this problem and answer the questions in P12-3$_B$. Students have pointed out that using Wolfram can greatly speed the solutions to the various parts of this problem.

(e) Example 12-3 (Use the LEP). Describe how your answers would change if the molar flow rate of methanol were increased by a factor of 4.

(f) Example 12-4 (Use the LEP). (1) Use Figure E12-3.2 and Equation (E12-4.4) to plot X versus T to find the new exit conversion and temperature. (2) Other data show $\Delta H^\circ_{Rxx} = -38,700$ Btu/lb-mol and $C_{P_A} = 29$ Btu/lb-mol/°F. How would these values change your results? (3) Make a plot of conversion as a function of heat exchanger area. [0 < A < 200 ft²].

(g) Example 12-5 (Use the LEP). (1) Why is there a maximum in temperature? How would your results change if there is (2) a pressure drop with α = 1.05 dm⁻³? (3) Reaction (1) is reversible with $K_C = 10$ at 450 K? (4) How would the selectivity change if Ua is increased? Decreased?

(h) Example 12-6 (Use the LEP). (1) Vary T_0 to make a plot of the reactor temperature, T, as a function of T_0. What are the extinction and ignition temperatures? (2) Vary τ between 0.1 min and 0.001 min and describe what you find. (3) Vary Ua between 4,000 and 400,000 J/min/K and describe what you find.

(i) Example 12-7 (Use the LEP). (1) **Safety.** Plot Q_g and Q_r as a function of V. How can you keep the maximum temperature below 700 K? Would adding inerts help and if so what should the flow rate be if $C_{P_I} = 10$ cal/mol/K? (2) Look at the Figures. What happened to species D? What conditions would you suggest to make more species D? (3) Make a table of the temperature (e.g., maximum T, T_a) and molar flow rates at two or three volumes, comparing the different heat-exchanger operations. (4) Why do you think the molar flow rate of C does not go through a maximum? Vary some of the parameters to learn if there are conditions where it goes through a maximum. Start by increasing F_{A0} by a factor of 5. (5) Include pressure in this problem. Vary the pressure-drop parameter (0 < αρ_b < 0.0999 dm⁻³) and describe what you find.

(j) CRE Web site SO$_2$ Example PRS-R12.4-1. Download the SO$_2$ oxidation **LEP R12-1**. How would your results change if (1) the catalyst particle diameter were cut in half? (2) The pressure were doubled? At what particle size does pressure drop become important for the same catalyst weight, assuming the porosity doesn't change? (3) You vary the initial temperature and the coolant temperature? Write a paragraph describing what you find.

Member

Hall of Fame

(k) SAChE. Go to the SAChE Web site, *www.sache.org*. Your instructor or department chair should have the username and password to enter the SAChE Web site in order to obtain the modules with the problems. On the left-hand menu, select "SAChE Products." Select the "All" tab and go to the module entitled, "Safety, Health and the Environment" (S, H & E). The problems are for KINETICS (i.e., CRE). There are some example problems marked "K" and explanations in each of the above S, H & E selections. Solutions to the problems are in a different section of the site. Specifically look at: *Loss of Cooling Water* (K-1), *Runaway Reactions* (HT-1),

Design of Relief Values (D-2), *Temperature Control and Runaway* (K-4) and (K-5), and *Runaway and the Critical Temperature Region* (K-7). Go through the K problems and write a paragraph on what you have learned.

(l) COMSOL Homework Problem (LEP). There is a COMSOL problem in the Expanded Material on the CRE Web site (*www.comsol.com/ecre*).

P12-2$_B$ Use Wolfram to find the following:

(a) At what of Θ_B would be required to not allow reactor temperature to rise above 360K?

P12-3$_B$ Load Polymath or Wolfram or MATLAB for **LEP 12-T12-3** for Table T12-2 from the CRE Web site for this exothermic reversible reaction with a variable coolant flow rate.

$$A + B \; \rightleftharpoons \; 2C$$

has the following parameter values for the **base case**:

$E = 25\ \text{kcal/mol}$ $\qquad\qquad C_{P_A} = C_{P_B} = C_{P_C} = 20\ \text{cal/mol/K}$

$\Delta H^{\circ}_{Rx} = -20\ \text{kcal/mol}$ $\qquad C_{P_I} = 40\ \text{cal/mol/K}$

$k = \dfrac{0.004\ \text{dm}^6}{\text{mol}\cdot\text{kg}\cdot\text{s}}$ @ 310 K $\qquad \dfrac{Ua}{\rho_b} = 0.5\dfrac{\text{cal}}{\text{kg}\cdot\text{s}\cdot\text{K}} \qquad T_0 = 330\ \text{K}$

$K_c = 1000$ @ 303 K $\qquad\qquad T_a = 320\ \text{K}$

$\alpha = 0.0002\ /\ \text{kg}$ $\qquad\qquad \dot{m}_c = 1{,}000\ \text{g/s}$

$F_{a0} = 5\ \text{mol/s}$ $\qquad\qquad C_{P_c} = 18\ \text{cal/g/K}$

$C_{T0} = 0.3\ \text{mol/dm}^3$ $\qquad\qquad \Theta_I = 1$

Vary the following parameters in the ranges shown in parts **(a)** through **(i)**. Write a paragraph describing the trends you find for each parameter variation and why they look the way they do. Use the base case for parameters not varied. *Hint:* See Selftests and Workbook at the very end of the Chapter 12 *Summary Notes* on the CRE Web site. The feedback from students on this problem is that one should use Wolfram in the LEP 12-3 on the Web to carry out the parameter variations. Note: Using the Wolfram LEP on the CRE Web site could save a lot of time on this homework problem.

Summary Notes

(a) F_{A0}: $1 \le F_{A0} \le 8$ mol/s

(b) Θ_I: $0.5 \le \Theta_I \le 4$

   *Note: The program gives $\Theta_I = 1.0$. Therefore, when you vary Θ_I, you will need to account for the corresponding increase or decrease of C_{A0} because the total concentration, C_{T0}, is constant.

(c) $\dfrac{Ua}{\rho_b}$: $0.1 \le \dfrac{Ua}{\rho_b} \le 0.8\ \dfrac{\text{cal}}{\text{kg}\cdot\text{s}\cdot\text{K}}$

(d) T_0: $310\ \text{K} \le T_0 \le 350\ \text{K}$

(e) T_a: $300\ \text{K} \le T_a \le 340\ \text{K}$

(f) $\dot{m}_c$: $1 \le m_c \le 1000$ g/s

(g) Repeat **(f)** for countercurrent coolant flow.

(h) Determine the conversion in a 5,000-kg fluidized CSTR where $UA = 500$ cal/s·K with $T_a = 320$ K and $\rho_b = 2$ kg/m³.

(i) Repeat **(a)**, **(b)**, and **(d)** if the reaction were endothermic with $K_c = 0.01$ at 303 K and $\Delta H^{\circ}_{Rx} = +20$ kcal/mol.

P12-4_A   Download the Interactive Computer Games (ICG) from the CRE Web site. Play the game, and then record your performance number for the module, which indicates your mastery of the material. *Note:* For simulation **(b)**, only do the first three reactors, as reactors 4 and above do not work because of the technician tinkering with them.

Interactive

Computer Games

    (a) ICG Heat Effects Basketball 1 Performance # _____.
    (b) ICG Heat Effects Simulation 2 Performance # _____.

P12-5_C   **Safety Problem.** The following is an excerpt from *The Morning News*, Wilmington, Delaware (August 3, 1977): "Investigators sift through the debris from blast in quest for the cause [that destroyed the new nitrous oxide plant]. A company spokesman said it appears more likely that the [fatal] blast was caused by another gas—ammonium nitrate—used to produce nitrous oxide." An 83% (wt) ammonium nitrate and 17% water solution is fed at 200°F to the CSTR operated at a temperature of about 510°F. Molten ammonium nitrate decomposes directly to produce gaseous nitrous oxide and steam. It is believed that pressure fluctuations were observed in the system and, as a result, the molten ammonium nitrate feed to the reactor may have been shut off approximately 4 min prior to the explosion.

    Assume that at the time the feed to the CSTR stopped, there was 500 lb$_m$ of ammonium nitrate in the reactor. The conversion in the reactor is believed to be virtually complete at about 99.99%.

Additional information (approximate but close to the real case):

$$\Delta H^\circ_{Rx} = -336 \text{ Btu/lb}_m \text{ ammonium nitrate at 500°F (constant)}$$

$$C_P = 0.38 \text{ Btu/lb}_m \text{ ammonium nitrate} \cdot °F$$

$$C_P = 0.47 \text{ Btu/lb}_m \text{ of steam} \cdot °F$$

$$-r_A V = k C_A V = k \frac{M}{V} V = k M (\text{lb}_m/\text{h})$$

where M is the mass of ammonium nitrate in the CSTR (lb$_m$) and k is given by the relationship below.

T (°F)	510	560
k (h^{-1})	0.307	2.912

The enthalpies of water and steam are

$$H_w(200°F) = 168 \text{ Btu/lb}_m$$

$$H_g(500°F) = 1202 \text{ Btu/lb}_m$$

    (a) Can you explain the cause of the blast? *Hint:* See Problem P13-3_B.
    (b) If the feed rate to the reactor just before shutoff was 310 lb$_m$ of solution per hour, what was the exact temperature in the reactor just prior to shutdown? *Hint:* Plot Q_r and Q_g as a function of temperature on the same plot.
    (c) How would you start up or shut down and control such a reaction? (*Hint:* See Problem P13-2_B.)
    (d) Explore this problem and describe what you find. For example, add a heat exchanger $UA (T - T_a)$, choose values of UA and T_a, and then plot $R(T)$ versus $G(T)$?
    (e) Discuss what you believe to be the point of the problem. The idea for this problem originated from an article by Ben Horowitz.

P12-6_B   The endothermic liquid-phase elementary reaction

$$A + B \rightarrow 2C$$

proceeds, substantially, to completion in a single steam-jacketed, continuous-stirred reactor (Table P12-6_B). From the following data, calculate the steady-state reactor temperature:

    Reactor volume: 125 gal
    Steam jacket area: 10 ft^2

Jacket steam: 150 psig (365.9°F saturation temperature)
Overall heat-transfer coefficient of jacket, U: 150 Btu/h·ft²·°F
Agitator shaft horsepower: 25 hp

Heat   of   reaction, $\Delta H^\circ_{Rx} = +20,000$

Btu/lb-mol of A (independent of temperature)

TABLE P12-6$_B$ FEED CONDITIONS AND PROPERTIES

	Component		
	A	*B*	*C*
Feed (lb-mol/hr)	10.0	10.0	0
Feed temperature (°F)	80	80	—
Specific heat (Btu/lb-mol·°F)*	51.0	44.0	47.5
Molecular weight	128	94	111
Density (lb$_m$/ft³)	63.0	67.2	65.0

* Independent of temperature. (*Ans: T = 199°F*)
(Courtesy of the California Board of Registration for Professional &
Land Surveyors.)

P12-7$_B$ Use the data in Problem P11-4$_A$ for the following reaction.
Th elementary, irreversible, organic liquid-phase reaction

$$A+B\longrightarrow C$$

is carried out in a flow reactor. An equal molar feed in A and B enters at 27°C, and the volumetric flow rate is 2 dm³/s and $C_{A0} = 0.1$ kmol/m³.

Additional information:

$H^\circ_A (273\ K) = -20$ kcal/mol, $H^\circ_B(273\ K) = -15$ kcal/mol,

$H^\circ_C (273\ K) = -41$ kcal/mol

$$C_{P_A} = C_{P_B} = 15\ \text{cal/mol} \cdot K \quad C_{P_C} = 30\ \text{cal/mol} \cdot K$$

$$k = 0.01\ \frac{dm^3}{mol \cdot s}\ \text{at 300 K} \quad E = 10,000\ \text{cal/mol}$$

$$Ua = 20\,\text{cal/m}^3/\text{s/K} \quad \dot{m}_C = 50\,\text{g/s}$$

$$T_{a0} = 450K \quad C_{P_{Cool}} = 1\,\text{cal/g/K}$$

(a) Calculate the conversion when the reaction is carried out adiabatically in one 500-dm³ CSTR and then compare the results with the two adiabatic 250-dm³ CSTRs in series.

The reversible reaction (part **(d)** of P11-4$_A$) is now carried out in a PFR with a heat exchanger. Plot and then analyze X, X_e, T, T_a, Q_r, Q_g, and the rate, $-r_A$, for the following cases:

(b) Constant heat-exchanger temperature T_a
(c) Co-current heat exchanger T_a
(d) Countercurrent heat exchanger T_a
(e) Adiabatic operation
(f) Make a table comparing all your results (e.g., X, X_e, T, T_a). Write a paragraph describing what you find.
(g) Plot Q_r and T_a as a function of V necessary to maintain isothermal operation.

P12-8$_A$ The gas-phase reversible reaction as discussed in P11-7$_B$

$$A \underset{\leftarrow}{\overset{\rightarrow}{\rightleftharpoons}} B$$

is now carried out under high pressure in a packed-bed reactor with pressure drop. The feed consists of both inerts I and species A with the ratio of inerts to the species A being 2 to 1. The entering molar flow rate of A is 5 mol/min at a temperature of 300 K and a concentration of 2 mol/dm³. Work this problem in terms of volume. *Hint:* $V = W / \rho_B$, $r_A = \rho_B r_A'$

Additional information:

$F_{A0} = 5.0$ mol/min $T_0 = 300$ K $\Delta H_{Rx} = -20{,}000$ cal/mol $\alpha\rho_b = 0.02$ dm^{-3}

$C_{A0} = 2$ mol/dm³ $T_1 = 300$ K $K_C = 1{,}000$ at 300 K   Coolant

$C_I = 2\,C_{A0}$ $k_1 = 0.1$ min^{-2} at 300 K $C_{P_B} = 160$ cal/mol/K $\dot{m}_C = 50$ mol/min

$C_{P_I} = 18$ cal/mol/K $Ua = 150$ cal/dm³/min/K $\rho_B = 1.2$ kg/dm³ $C_{P_{Cool}} = 20$ cal/mol/K

$C_{P_A} = 160$ cal/mol/K $T_{ao} = 300$ K

$E = 10{,}000$ cal/mol $V = 40$ dm³

Plot and then analyze X, X_e, T, T_a, and the rate $(-r_A)$ profiles in a PFR for the following cases. In each case, explain why the curves look the way they do.

(a) Co-current heat exchange
(b) Countercurrent heat exchange
(c) Constant heat-exchanger temperature T_a
(d) Compare and contrast each of the above results and the results for adiabatic operation (e.g., make a plot or a table of X and X_e obtained in each case).
(e) Vary some of the parameters, e.g., $(0 < \Theta_I < 10)$ and describe what you find.
(f) Plot Q_r and T_a as a function of V necessary to maintain isothermal operation.

P12-9$_A$ Algorithm for reaction in a PBR with heat effects and pressure drop
The elementary gas-phase reaction

$$A + B \;\rightleftharpoons\; 2C$$

in P11-8$_B$ is now continued and carried out in packed-bed reactor. The entering molar flow rates are $F_{A0} = 5$ mol/s, $F_{B0} = 2F_{A0}$, and $F_I = 2F_{A0}$ with $C_{A0} = 0.2$ mol/dm³. The entering temperature is 325 K and a coolant fluid is available at 300 K.

Additional information:

$C_{P_A} = C_{P_B} = C_{P_C} = 20$ cal/mol/K $k = 0.0002\,\dfrac{\text{dm}^6}{\text{kg}\cdot\text{mol}\cdot\text{s}}$ @300 K

$C_{P_A} = 18$ cal/mol/K $\alpha = 0.00015$ kg^{-1} $Ua = 320\,\dfrac{\text{Cal}}{\text{s}\cdot\text{m}^3\cdot\text{K}}$

$E = 25\,\dfrac{\text{kcal}}{\text{mol}}$ $\dot{m}_C = 18$ mol/s $\rho_b = 1400\,\dfrac{\text{kg}}{\text{m}^3}$

$\Delta H_{Rx} = -20\,\dfrac{\text{kcal}}{\text{mol}}$ @298K $C_{P_{Cool}} = 18$ cal/mol (coolant)

$K_C = 1000$@305K

Plot X, X_e, T, T_a, and $-r_A$ down the length of the PFR for the following cases:
(a) Co-current heat exchange
(b) Countercurrent heat exchange
(c) Constant heat-exchanger temperature T_a
(d) Compare and contrast your results for **(a)**, **(b)**, and **(c)** along with those for adiabatic operation and write a paragraph describing what you find.

P12-10$_B$ Use the data and reaction in Problems P11-4$_A$ and P12-7$_A$ for the following:

$$A + B \longrightarrow C$$

- **(a)** Plot and then analyze the conversion, Q_r, Q_g, and temperature profiles up to a PFR reactor volume of 10 dm^3 for the case when the reaction is reversible with $K_C = 10$ m^3/kmol at 450 K. Plot and then analyze the equilibrium conversion profile.
- **(b)** Repeat **(a)** when a heat exchanger is added, $Ua = 20$ cal/m^3/s/K, and the coolant temperature is constant at $T_a = 450$ K.
- **(c)** Repeat **(b)** for both a co-current and a countercurrent heat exchanger. The coolant flow rate is 50 g/s, $C_{P_c} = 1$ cal/g · K, and the inlet coolant temperature is $T_{a0} = 450$ K. Vary the coolant rate $(10 < \dot{m}_c < 1{,}000 \text{ g/s})$.
- **(d)** Plot Q_r and T_a as a function of V necessary to maintain isothermal operation.
- **(e)** Compare your answers to **(a)** through **(d)** and describe what you find. What generalizations can you make?
- **(f)** Repeat **(c)** and **(d)** when the reaction is irreversible but endothermic with $\Delta H_{Rx}^{\circ} = 6{,}000$ cal/mol. Choose $T_{a0} = 450$ K.

P12-11$_B$ Use the reaction data in Problem P11-4$_A$ and P12-7$_A$ for the case when heat is removed by a heat exchanger jacketing the reactor. The flow rate of coolant through the jacket is sufficiently high that the ambient exchanger temperature is constant at $T_a = 50°$C.

$$A + B \longrightarrow C$$

- **(a)** **(1)** Plot and then analyze the temperature conversion, Q_r, and Q_g profiles for a PBR with

  $$\frac{Ua}{\rho_b} = 0.08 \ \frac{J}{s \cdot \text{kg-cat} \cdot K}$$

  where

  ρ_b = bulk density of the catalyst (kg/m^3)

  a = heat-exchange area per unit volume of reactor (m^2/m^3)

  U = overall heat-transfer coefficient (J/s · m^2 · K)

  (2) How would the profiles change if Ua/ρ_b were increased by a factor of 3000?
  (3) If there is a pressure drop with $\alpha = 0.019$ kg^{-1}?
- **(b)** Repeat part **(a)** for co-current and countercurrent flow and adiabatic operation with $\dot{m}_c = 0.2$ kg/s, $C_{P_c} = 5{,}000$ J/kg K, and an entering coolant temperature of 50°C.
- **(c)** Find X and T for a "fluidized" CSTR with 80 kg of catalyst.

  $$UA = 500 \ \frac{J}{s \cdot K}, \qquad \rho_b = 1 \text{ kg/m}^3$$

- **(d)** Repeat parts **(a)** and **(b)** for W = 80.0 kg, assuming a reversible reaction with a reverse specific reaction rate of

  $$k_r = 0.2 \ \exp\left[\frac{E_r}{R}\left(\frac{1}{450} - \frac{1}{T}\right)\right]\left(\frac{\text{dm}^6}{\text{kg-cat} \cdot \text{mol} \cdot \text{s}}\right); \qquad E_r = 51.4 \text{ kJ/mol}$$

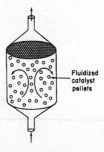

Fluidized catalyst pellets

  Vary the entering temperature, T_0, and describe what you find.
- **(e)** Use or modify the data in this problem to suggest another question or calculation. Explain why your question requires either critical thinking or creative thinking. See Preface Section I and *http://www.umich.edu/scps*.

P12-12$_C$ Derive the energy balance for a packed bed membrane reactor. Apply the balance to the reaction in Problem P11-5$_B$

$$A \underset{\longleftarrow}{\overset{\longrightarrow}{}} B + C$$

for the case when it is reversible with $K_C = 1.0$ mol/dm^3 at 300 K. Species C diffuses out of the membrane with $k_C = 1.5$ s^{-1}.

(a) Plot and then analyze the concentration profiles for different values of K_C when the reaction is carried out adiabatically.

(b) Repeat part (a) when the heat transfer coefficient is $Ua = 30$ J/s·kg-cat·K with $T_a = 50°C$.

P12-13$_B$ Circle the correct answer.

(a) The elementary reversible isomerization of A to B was carried out in a packed-bed reactor. The following profiles were obtained:

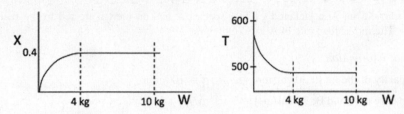

If the total entering volumetric flow rate remains constant, the addition of inerts to the feed stream will most likely

A) Increase conversion.

B) Decrease conversion.

C) Have no effect.

D) Insufficient information to tell

(b)

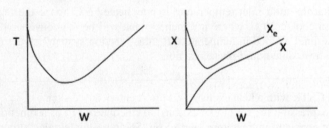

Which of the following statements are true?

A) The above reaction could be adiabatic.

B) The above reaction could be exothermic with constant cooling temperature.

C) The above reaction could be endothermic with constant heating temperature.

D) The above reaction could be second order.

P12-14$_A$ The irreversible reaction

$$A + B \longrightarrow C + D$$

is carried out adiabatically in a CSTR. The "heat generated" $G(T)$ and the "heat removed" $R(T)$ curves are shown in Figure P12-14$_A$.

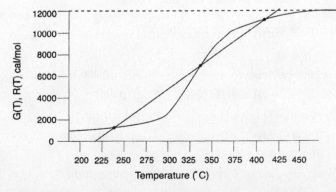

Figure P12-14$_A$ Heat removed $R(T)$ and heat "generated" $G(T)$ curves.

(a) What is the ΔH_{Rx} of the reaction?

(b) What are the inlet ignition and extinction temperatures?

(c) What are all the temperatures in the reactor corresponding to the inlet ignition and extinction temperatures?

(d) What are the conversions at the ignition and extinction temperatures?

P12-15$_B$ The first-order, irreversible, exothermic liquid-phase reaction

$$A \rightarrow B$$

is to be carried out in a jacketed CSTR. Species A and an inert I are fed to the reactor in equimolar amounts. The molar feed rate of A is 80 mol/min.

Additional information

Heat capacity of the inert: 30 cal/mol·°C $\tau = 100$ min

Heat capacity of A and B: 20 cal/mol·°C $\Delta H_{Rx}^{\circ} = -7500$ cal/mol

UA: 8000 cal/min·°C $k = 6.6 \times 10^{-3}$ min^{-1} at 350 K

Ambient temperature, T_a: 300 K $E = 40{,}000$ cal/mol·K

(a) What is the reactor temperature for a feed temperature of 450 K?

(b) Plot and then analyze the reactor temperature as a function of the feed temperature.

(c) To what inlet temperature must the fluid be preheated for the reactor to operate at a high conversion? What are the corresponding temperature and conversion of the fluid in the CSTR at this inlet temperature?

(d) Suppose that the fluid inlet temperature is now heated 5°C above the reactor temperature in part **(c)** and then cooled 20°C, where it remains. What will be the conversion?

(e) What is the inlet extinction temperature for this reaction system? (*Ans.:* $T_0 = 87$°C.)

P12-16$_B$ The elementary reversible liquid-phase reaction

$$A \rightleftharpoons B$$

takes place in a CSTR with a heat exchanger. Pure A enters the reactor.

(a) Derive an expression (or set of expressions) to calculate $G(T)$ as a function of the heat of reaction, equilibrium constant, temperature, and so on. Show a sample calculation for $G(T)$ at $T = 400$ K.

(b) What are the steady-state temperatures? (*Ans.:* 310, 377, 418 K.)

(c) Which steady states are locally stable?

(d) What is the conversion corresponding to the upper steady state?

(e) Vary the ambient temperature T_a and make a plot of the reactor temperature as a function of T_a, identifying the ignition and extinction temperatures.

(f) If the heat exchanger in the reactor suddenly fails (i.e., $UA = 0$), what would be the conversion and the reactor temperature when the new upper steady state is reached? (*Ans.:* 431 K.)

(g) What heat exchanger product, UA, will give the maximum conversion?

(h) Write a question that requires critical thinking and then explain why your question requires critical thinking. *Hint:* See Preface Section I.

(i) What is the adiabatic blowout flow rate, v_0?

(j) Suppose that you want to operate at the lower steady state. What parameter values would you suggest to prevent runaway, e.g., the upper SS?

Additional information

$UA = 3600$ cal/min·K $E/R = 20{,}000$ K

$C_{P_A} = C_{P_B} = 40$ cal/mol·K $V = 10$ dm^3

$\Delta H_{Rx}^{\circ} = -80{,}000$ cal/mol A $v_0 = 1$ dm^3/min

$K_C = 100$ at 400 K $F_{A0} = 10$ mol/min

$k = 1$ min^{-1} at 400 K

Ambient temperature, $T_a = 37$°C   Feed temperature, $T_0 = 37$°C

P12-17$_B$ The reversible liquid phase reaction

$$A \underset{\rightleftarrows}{} B$$

is carried out in a 12-dm^3 CSTR with heat exchange. Both the entering temperature, T_0, and the heat exchange fluid, T_a, are at 330 K. An equal molar mixture of inerts and A enter the reactor.

(a) Choose a temperature, T, and carry out a calculation to find $G(T)$ to show that your calculation agrees with the corresponding $G(T)$ value on the curve shown below at the temperature you choose.

(b) Find the exit conversion and temperature from the CSTR. $X =$ _____ $T =$ _____ .

(c) What entering temperature T_0 would give you the maximum conversion? $T_0 =$ _____ $X =$ _____

(d) What would the exit conversion and temperature be if the heat-exchange system failed (i.e., U = 0)?

(e) Can you find the inlet ignition and extinction temperatures? If yes, what are they? If not, go on to the next problem.

(f) Use Preface Section I to ask another question.

Additional information

The $G(T)$ curve for this reaction is shown below

$C_{P_A} = C_{P_B} = 100$ cal/mol/K, $C_{P_I} = 150$ cal/mol/K $k = 0.001\,h^{-1}$ at 300 K with $E = 30,000$ cal/mol

$F_{A0} = 10$ mol/h, $C_{A0} = 1$ mol/dm^3, $v_0 = 10$ dm^3/h $K_C = 5,000,000$ at 300 K

$\Delta H_{Rx} = -42,000$ cal/mol $UA = 5000$ cal/h/K

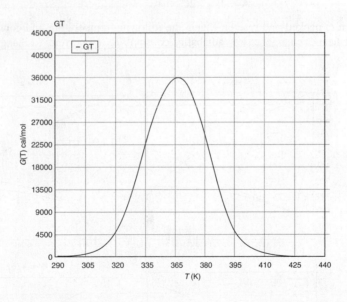

P12-18$_C$ The elementary gas-phase reaction

$$2A \underset{\rightleftarrows}{} C$$

is carried out in a packed-bed reactor. Pure A enters the reactor at a 450-K flow rate of 10 mol/s, and a concentration of 0.25 mol/dm^3. The PBR contains 90 kg of catalyst and is surrounded by a heat exchanger for which cooling fluid is available at 500 K. Compare the conversion achieved for the four types of heat exchanger operation: adiabatic, constant T_a, co-current flow, and countercurrent flow.

Additional information

$\alpha = 0.019/\text{kg-cat}$

$Ua/\rho_b = 0.8 \ \text{J/kg-cat} \cdot \text{s} \cdot \text{K}$

$\Delta H_{Rx}^{\circ} = -20{,}000 \ \text{J/mol}$

$C_{P_A} = 40 \ \text{J/mol} \cdot \text{K}$

$C_{P_C} = 20 \ \text{J/mol/K}$

$F_{A0} = 10 \ \text{mol/h}$

$C_{A0} = 1 \ \text{mol/dm}^3$

$v_0 = 10 \ \text{dm}^3/\text{h}$

P12-19$_C$ A reaction is to be carried out in the packed-bed reactor shown in Figure P12-19$_C$.

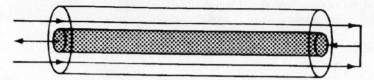

Figure P12-19$_C$ PFR with heat exchange.

The reactants enter the annular space between an outer insulated tube and an inner tube containing the catalyst. No reaction takes place in the annular region. Heat transfer between the gas in this packed-bed reactor and the gas flowing countercurrently in the annular space occurs along the length of the reactor. The overall heat-transfer coefficient is $5 \ \text{W/m}^2 \cdot \text{K}$. Plot the conversion and temperature as a function of reactor length for the data given in reactor length for the data given in Problem P12-7$_B$

P12-20$_B$ The reaction

$$A + B \rightleftharpoons 2C$$

is carried out in a packed-bed reactor. Match the following temperature and conversion profiles for the four different heat-exchange cases: adiabatic, constant T_a, co-current exchange, and countercurrent exchange.

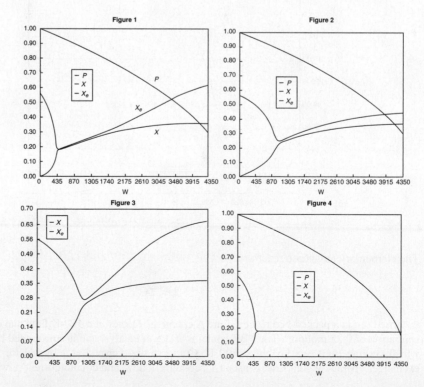

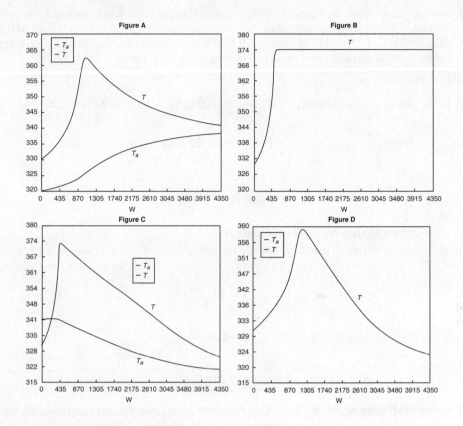

(a) Figure 1 matches Figure ___
(b) Figure 2 matches Figure ___
(c) Figure 3 matches Figure ___
(d) Figure 4 matches Figure ___

P12-21$_B$ The irreversible liquid-phase reactions

Reaction (1) $A + B \rightarrow 2C$ $r_{1C} = k_{1C}C_A C_B$

Reaction (2) $2B + C \rightarrow D$ $r_{2D} = k_{2D}C_B C_C$

are carried out in a PFR with heat exchange. The following temperature profiles were obtained for the reactor and the coolant stream:

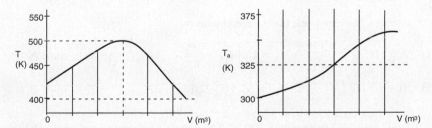

Figure P12-21$_B$ Reactant temperature T and coolant temperature T_a profiles.

The concentrations of A, B, C, and D were measured at the point down the reactor where the liquid temperature, T, reached a maximum, and they were found to be $C_A = 0.1$, $C_B = 0.2$, $C_C = 0.5$, and $C_D = 1.5$, all in mol/dm^3. The product of the overall heat-transfer coefficient and the heat-exchanger area per unit volume, Ua, is 10 cal/s $\cdot$ dm$^3 \cdot$ K. The entering molar flow rate of A is 10 mol/s.

Additional information

$$C_{P_A} = C_{P_B} = C_{P_C} = 30 \text{ cal/mol/K} \qquad C_{P_D} = 90 \text{ cal/mol/K}, \ C_{P_I} = 100 \text{ cal/mol/K}$$

$$\Delta H^\circ_{Rx1A} = -50{,}000 \text{ cal/molA} \qquad k_{1C} = 0.043 \frac{\text{dm}^3}{\text{mol} \cdot \text{s}} \text{ at 400 K}$$

$$\Delta H^\circ_{Rx2B} = +5000 \text{ cal/molB} \qquad k_{2D} = 0.4 \frac{\text{dm}^3}{\text{mol} \cdot \text{s}} \ e^{5000 \text{ K}\left[\frac{1}{500} - \frac{1}{T}\right]}$$

(a) What is the activation energy for Reaction (1)?

P12-22$_B$ The following elementary reactions are to be carried out in a PFR with a heat exchange with constant T_a:

$$2A + B \rightarrow C \qquad \Delta H_{Rx1B} = -10\frac{\text{kJ}}{\text{mol B}}$$

$$A \rightarrow D \qquad \Delta H_{Rx2A} = +10\frac{\text{kJ}}{\text{mol A}}$$

$$B + 2C \rightarrow E \qquad \Delta H_{Rx3C} = -20\frac{\text{kJ}}{\text{mol C}}$$

The reactants all enter at 400 K. Only A and B enter the reactor. The entering concentration of A and B are 3 molar and 1 molar at a volumetric flow rate of 10 dm^3/s.

Additional information

$Ua = 100$ J/dm^3/s/K

$k_{1A}(400 \text{ K}) = 1.0\left(\dfrac{\text{dm}^3}{\text{mol}}\right)^2/\text{s}$

$k_{2A}(400 \text{ K}) = 1.333 \text{ s}^{-1}$

$k_{3B}(400 \text{ K}) = 2\left(\dfrac{\text{dm}^3}{\text{mol}}\right)^2/\text{s}$

$C_{P_A} = 10$ J/mol/K

$C_{P_B} = 20$ J/mol/K

$C_{P_C} = 40$ J/mol/K

$C_{P_D} = 20$ J/mol/K

$C_{P_E} = 100$ J/mol/K

What constant coolant temperature, T_a, is necessary such that at the reactor entrance, i.e., $V = 0$, $\dfrac{dT}{dV} = 0$?

P12-23$_B$ The complex gas-phase reactions are elementary

(1) $2A \rightleftarrows B$ $\qquad\qquad -r_{1A} = k_{1A}\left[C_A^2 - \dfrac{C_B}{K_{CA}}\right]$ $\quad \Delta H_{Rx1A} = -20$ kJ/mole A

(2) $2B + A \rightarrow C$ $\qquad\quad r_{2C} = k_{2C}[C_B^2 \ C_A]$ $\qquad \Delta H_{Rx2B} = +30$ kJ/mole B

and carried out in a PFR with a heat exchanger. Pure A enters at a rate of 5 mol/min, a concentration of 0.2 mol/dm^3, and temperature 300 K. The entering temperature of an available coolant is 320 K.

Additional information

$k_{1A1} = 50$ dm^3/mol • min@305 K with $E_1 = 8{,}000$ J/mol

$k_{2C2} = 4000$ dm^9/mol^3 • min@310 K with $E_2 = 4{,}000$ J/mol

$K_{CA} = 10$ dm^3/mol@315 K

Note: This is the equilibrium constant with respect to A in reaction 1 when

using van Hoff's equation, i.e., (S12-4),

$$Ua = 200 \ \frac{J}{dm^3 \cdot min \ K} \qquad C_{P_{Cool}} = 10 \text{ J/mol/K} \quad C_{P_B} = 80 \text{ J/mol/K} \quad C_{P_C} = 100 \text{ J/mol/K}$$

$$\dot{m}_C = 50 \text{ g/min} \qquad C_{P_A} = 20 \text{ J/mol/K} \qquad R = 8.31 \text{ J/mol/K}$$

The reactor volume is 10 dm^3.

(a) Plot (F_A, F_B, F_C) on one graph and (T and T_a) on another along the length of the reactor for adiabatic operation, heat exchange with constant T_a, and co-current and countercurrent heat exchange with variable T_a. Only turn in a copy of your code and output for co-current exchange.

Adiabatic operation

(b) What is the maximum temperature and at what reactor volume is it reached?

(c) At what reactor volume is the flow rate of B a maximum, and what is F_{Bmax} at this value?

Constant T_a

(d) What is the maximum temperature and at what reactor volume is it reached?

(e) At which reactor volume is the flow rate of B a maximum, and what is F_{Bmax} at this volume?

Co-current exchange

(f) At what reactor volume does T_a become greater than T? Why does it become greater

Countercurrent exchange

(g) At what reactor volume does T_a become greater than T?

Hint: Guess T_a at entrance around 350 K.

P12-24$_B$ The elementary liquid-phase reactions

$$\text{(1)} \quad A + 2B \longrightarrow 2C$$
$$\text{(2)} \quad A + C \longrightarrow 2D$$

are carried out adiabatically in a 10 dm^3 PFR. After streams A and B mix, species A enters the reactor at a concentration of $C_{A0} = 2$ mol/dm^3 and species B at a concentration of 4 mol/dm^3. The entering volumetric flow rate is 10 dm^3/s.

Assuming you could vary the entering temperature between 300 K and 600 K, what entering temperture would you recommend to maximize the concentration of species C exiting the reactor? ($\pm 25°$K). Assume all species have the same density.

Additional information

$$C_{P_A} = C_{P_B} = 20 \text{ cal/mol/K}, \ C_{P_C} = 60 \text{ cal/mol/K}, \ C_{P_D} = 80 \text{ cal/mol/K}$$

$$\Delta H_{Rx1A} = 20{,}000 \text{ cal/mol A}, \ \Delta H_{Rx2A} = -10{,}000 \text{ cal/mol A}$$

$$k_{1A} = 0.001 \frac{dm^6}{mol^2 \cdot s} \ at \ 300K \ with \ E = 5000 \text{ cal/mol}$$

$$k_{2A} = 0.001 \frac{dm^3}{mol \cdot s} \ at \ 300K \ with \ E = 7500 \text{ cal/mol}$$

P12-25$_C$ *(Multiple reactions with heat effects)* Xylene has three major isomers, *m*-xylene (A), *o*-xylene (B), and *p*-xylene (C). When *m*-xylene (A) is passed over a Cryotite catalyst, the following elementary reactions are observed. The reaction to form *p*-xylene is irreversible:

A (meta)

k_1 / k_2

B (ortho)

k_3

C (para)

CH$_3$

> Application
> Pending
> for Problem
> Hall of
> Fame

The feed to the reactor is pure *m*-xylene (A). For a total feed rate of 2 mol/min and the reaction conditions below, plot the temperature and the molar flow rates of each species as a function of catalyst weight up to a weight of 100 kg.

(a) Plot the concentrations of each of xylenes down the length (i.e., V) of a PBR.
(b) Find the lowest concentration of *o*-xylene achieved in the reactor.
(c) Find the maximum concentration of *o*-xylene in the reactor.
(d) Repeat part **(a)** for a pure feed of *o*-xylene (B). What is the maximum concentration of meta xylene and where does it occur in the reactor?
(e) Vary some of the system parameters and describe what you learn.
(f) What do you believe to be the point of this problem?

Additional information[9]

All heat capacities are virtually the same at 100 J/mol·K.

$C_{T0} = 2$ mol/dm^3

$\Delta H^\circ_{Rx10} = -1800$ J/mol *o*-xylene

$\Delta H^\circ_{Rx30} = -1100$ J/mol *o*-xylene

$k_1 = 0.5 \exp[2(1 - 320/T)]$ dm^3/kg-cat·min, (*T* is in K)

$k_3 = 0.005 \exp\{[4.6(1 - (460/T))]\}$ dm^3/kg-cat·min

$K_C = 10 \exp[4.8(430/T - 1.5)]$

$T_0 = 330$ K $T_a = 500$ K

$k_2 = k_1/K_C$

$Ua/\rho_b = 16$ J/kg-cat·min ·°C

$W = 100$ kg

P12-26$_C$ *(Comprehensive problem on multiple reactions with heat effects)* Styrene can be produced from ethylbenzene by the following reaction:

$$\text{ethylbenzene} \longleftrightarrow \text{styrene} + \text{H}_2 \qquad (1)$$

However, several irreversible side reactions also occur:

$$\text{ethylbenzene} \longrightarrow \text{benzene} + \text{ethylene} \qquad (2)$$

$$\text{ethylbenzene} + \text{H}_2 \longrightarrow \text{toluene} + \text{methane} \qquad (3)$$

---

[9] Obtained from inviscid pericosity measurements.

[J. Snyder and B. Subramaniam, *Chem. Eng. Sci.*, 49, 5585 (1994)]. Ethylbenzene is fed at a rate of 0.00344 kmol/s to a 10.0-m³ PFR (PBR), along with inert steam at a total pressure of 2.4 atm. The steam/ethylbenzene molar ratio is initially, i.e., parts (a) to (c), 14.5:1 but can be varied.

Hall of Fame

Given the following data, find the exiting molar flow rates of styrene, benzene, and toluene along with $\tilde{S}_{St/BT}$ for the following inlet temperatures when the reactor is operated adiabatically:

(a) $T_0 = 800$ K

(b) $T_0 = 930$ K

(c) $T_0 = 1100$ K

(d) Find the ideal inlet temperature for the production of styrene for a steam/ethylbenzene ratio of 58:1. *Hint:* Plot the molar flow rate of styrene versus T_0. Explain why your curve looks the way it does.

(e) Find the ideal steam/ethylbenzene ratio for the production of styrene at 900 K. *Hint:* See part **(d)**.

(f) It is proposed to add a countercurrent heat exchanger with $Ua = 100$ kJ/m³/min/K, where T_a is virtually constant at 1000 K. For an entering stream to ethylbenzene ratio of 20, what would you suggest as an entering temperature? Plot the molar flow rates and $\tilde{S}_{St/BT}$.

(g) What do you believe to be the major points of this problem?

(h) Ask another question or suggest another calculation that can be made for this problem.

Additional information

Heat capacities

Methane	68 J/mol·K	Styrene	273 J/mol·K
Ethylene	90 J/mol·K	Ethylbenzene	299 J/mol·K
Benzene	201 J/mol·K	Hydrogen	30 J/mol·K
Toluene	249 J/mol·K	Steam	40 J/mol·K

$\rho = 2137$ kg/m³ of pellet

$\phi = 0.4$

$\Delta H^\circ_{Rx1EB} = 118,000$ kJ/kmol ethylbenzene

$\Delta H^\circ_{Rx2EB} = 105,200$ kJ/kmol ethylbenzene

$\Delta H^\circ_{Rx3EB} = -53,900$ kJ/kmol ethylbenzene

$$K_{p1} = \exp\left\{ b_1 + \frac{b_2}{T} + b_3 \ln(T) + [(b_4 T + b_5) T + b_6] T \right\} \text{ atm}$$

$b_1 = -17.34$	$b_4 = -2.314 \times 10^{-10}$ K⁻³
$b_2 = -1.302 \times 10^4$ K	$b_5 = 1.302 \times 10^{-6}$ K⁻²
$b_3 = 5.051$	$b_6 = -4.931 \times 10^{-3}$ K⁻¹

The kinetic rate laws for the formation of styrene (St), benzene (B), and toluene (T), respectively, are as follows (EB = ethylbenzene).

$$r_{1St} = \rho(1-\phi) \exp\left(-0.08539 - \frac{10,925 \text{ K}}{T}\right)\left(P_{EB} - \frac{P_{St} P_{H_2}}{K_{p1}}\right) \quad (\text{kmol/m}^3 \cdot \text{s})$$

$$r_{2B} = \rho(1-\phi) \exp\left(13.2392 - \frac{25,000 \text{ K}}{T}\right)(P_{EB}) \quad (\text{kmol/m}^3 \cdot \text{s})$$

$$r_{3T} = \rho(1 - \phi) \exp\left(0.2961 - \frac{11,000\ K}{T}\right)(P_{EB}P_{H_2}) \quad (kmol/m^3 \cdot s)$$

The temperature T is in Kelvin and P_i is in atm.

P12-27$_B$ The liquid-phase, dimer-quadmer series addition reaction

$$4A \rightarrow 2A_2 \rightarrow A_4$$

can be written as

$$2A \rightarrow A_2 \qquad -r_{1A} = k_{1A}C_A^2 \qquad \Delta H_{Rx1A} = -32.5\frac{kcal}{mol\ A}$$

$$2A_2 \rightarrow A_4 \quad -r_{2A_2} = k_{2A_2}C_{A_2}^2 \quad \Delta H_{Rx2A_2} = -27.5\frac{kcal}{mol\ A_2}$$

and is carried out in a 10-dm³ PFR. The mass flow rate through the heat exchanger surrounding the reactor is sufficiently large so that the ambient temperature of the exchanger is constant at $T_a = 315$ K. The reactants enter at a temperature T_0, of 300 K. Pure A is fed to the rector at a volumetric flow rate of 50 dm³/s and a concentration of 2 mol/dm³.

(a) Plot, compare, and analyze the profiles F_A, F_{A_2}, and F_{A_4} down the length of the reactor up to 10 dm³.

(b) The desired product is A_2 and it has been suggested that the current reactor may be too large. What reactor volume would you recommend to maximize F_{A_2}?

(c) What operating variables (e.g., T_0, T_a) would you change and how would you change them to make the reactor volume as small as possible and to still maximize F_{A_2}? Note any opposing factors in maximum production of A_2. The ambient temperature and the inlet temperature must be kept between 0°C and 177°C.

Additional information

$$k_{1A} = 0.6\frac{dm^3}{mol \cdot s} \text{ at 300 K with } E_1 = 4,000\frac{cal}{mol}$$

$$k_{2A_2} = 0.35\frac{dm^3}{mol \cdot s} \text{ at 320 K with } E_2 = 5,000\frac{cal}{mol}$$

$$C_{P_A} = 25\frac{cal}{molA \cdot K}, \ C_{P_{A_2}} = 50\frac{cal}{molA_2 \cdot K}, \ C_{P_{A_4}} = 100\frac{cal}{molA_4 \cdot K}$$

$$Ua = 1,000\frac{cal}{dm^3 \cdot s \cdot K}$$

Turn in your recommendation of reactor volume to maximize F_{A_2} and the molar flow rate at this maximum.

SUPPLEMENTARY READING

1. An excellent development of the energy balance is presented in

ARIS, R., *Elementary Chemical Reactor Analysis*. Upper Saddle River, NJ: Prentice Hall, 1969, Chaps. 3 and 6.

FOGLER, JOSEPH, *A Reaction Engineer's Handbook of Thermochemical Data*, Jofostan Press, Riça, Jofostan (2018).

2. Safety

CENTER FOR CHEMICAL PROCESS SAFETY (CCPS), *Guidelines for Chemical Reactivity Evaluation and Application to Process Design*. New York: American Institute of Chemical Engineers (AIChE) 1995.

CROWL, DANIEL A., and JOSEPH F. LOUVAR, *Chemical Process Safety: Fundamentals with Applications*, 3rd ed. Upper Saddle River, NJ: Prentice Hall, 2011.

MELHEM, G. A., and H. G. FISHER, *International Symposium on Runaway Reactions and Pressure Relief Design*. New York: Center for Chemical Process Safety (CCPS) of the American Institute of Chemical Engineers (AIChE) and The Institution of Chemical Engineers, 1995.

See the Center for Chemical Process Safety (CCPS) Web site, *www.aiche.org/ccps*.

3. A number of example problems dealing with nonisothermal reactors may or may not be found in

BURGESS, THORNTON W., *The Adventures of Jerry Muskrat*. New York: Dover Publications, Inc., 1914.

FROMENT, G. F., and K. B. BISCHOFF, *Chemical Reactor Analysis and Design,* 3rd ed. New York: Wiley, 2010.

WALAS, S. M., *Chemical Reaction Engineering Handbook of Solved Problems*. Amsterdam: Gordon and Breach, 1995. See the following solved problems: Problem 4.10.1, page 444; Problem 4.10.08, page 450; Problem 4.10.09, page 451; Problem 4.10.13, page 454; Problem 4.11.02, page 456; Problem 4.11.09, page 462; Problem 4.11.03, page 459; Problem 4.10.11, page 463.

4. A review of the multiplicity of the steady state and reactor stability is discussed by

PERLMUTTER, D. D., *Stability of Chemical Reactors*. Upper Saddle River, NJ: Prentice Hall, 1972.

5. The heats of formation, $H_i(T)$, Gibbs free energies, $G_i(T_R)$, and the heat capacities of various compounds can be found in

GREEN, DON W., and ROBERT H. PERRY, *Perry's Chemical Engineers' Handbook*, 8th ed. (Chemical Engineers Handbook). New York: McGraw-Hill, 2008.

LIDE DAVID R., *CRC Handbook of Chemistry and Physics*, 90th ed. Boca Raton, FL: CRC Press, 2009.

Unsteady-State **13**
Nonisothermal
Reactor Design

Chemical engineers are not gentle people, they like
high temperatures and high pressures.

—Steve LeBlanc

Overview. Up to now we have focused on the steady-state operation of
nonisothermal reactors. In this chapter the unsteady-state energy balance
will be developed and then applied to CSTRs, as well as well-mixed
batch and semibatch reactors. We will use the difference between the rate
of "heat generated," $\dot{Q}_g$, and the rate of "heat removed," $\dot{Q}_r$, to arrive at
the **user-friendly form** of the energy balance

$$\frac{dT}{dt} = \frac{\dot{Q}_g - \dot{Q}_r}{\sum N_i C_{P_i}}$$

- Section 13.1 shows how to arrange the general energy balance
  (Equation (11-9)) in a more simplified form for unsteady-state
  operation.
- Section 13.2 discusses the application of the energy balance to
  the operation of batch reactors and discusses reactor safety and
  the reasons for the explosion of an industrial batch reactor.
- Section 13.3 shows how to apply the energy balance to a semibatch reactor with a variable ambient temperature.
- Section 13.4 discusses the startup of a CSTR and how to avoid
  exceeding the practical stability limit.
- Section 13.5 closes the chapter with a case study of the T2 Laboratory explosion involving multiple reactions in a batch reactor.

Safety is another focus of this chapter, and the example problems and
home problems were chosen to emphasize the hazards of runaway reactors.

13.1 The Unsteady-State Energy Balance

We start by deriving the user-friendly form of the energy balance that can be easily used in making reactor calculations. As I have stated before, the reason I derive the equations rather than just giving the final result is (1) the reader can see what assumptions are used at what points along the way in the derivation to the user-friendly energy balance, and (2) it is my experience that if the reader goes line-by-line through the derivation, he/she will be less likely to insert the incorrect numbers into symbols in the equation.

We begin by recalling the unsteady-state form of the energy balance developed in Chapter 11.

$$\dot{Q} - \dot{W}_s + \sum_{i=1}^{m} F_i H_i \Big|_{\text{in}} - \sum_{i=1}^{m} F_i H_i \Big|_{\text{out}} = \left(\frac{d\hat{E}_{\text{sys}}}{dt} \right) \tag{11-9}$$

We shall first concentrate on evaluating the change in the total energy of the system wrt time, $d\hat{E}_{\text{sys}}/dt$. The total energy of the system is the sum of the products of specific energies, E_i (e.g., J/mol i), of the various species in the system volume and the number of moles, N_i (mol i), of that species

$$\hat{E}_{\text{sys}} = \sum_{i=1}^{m} N_i E_i = N_A E_A + N_B E_B + N_C E_C + N_D E_D + N_I E_I \tag{13-1}$$

In evaluating $\hat{E}_{\text{sys}}$, we shall neglect changes in the potential and kinetic energies and substitute for the internal energy U_i in terms of the enthalpy H_i

$$\hat{E}_{\text{sys}} = \sum_{i=1}^{m} N_i E_i = \sum_{i=1}^{m} N_i U_i = \left[\sum_{i=1}^{m} N_i (H_i - P V_i) \right]_{\text{sys}} = \sum_{i=1}^{m} N_i H_i - P \underbrace{\overset{\text{Neglect}}{\overbrace{\sum_{i=1}^{m} N_i \tilde{V}_i}}}_{V}$$

$$\tag{13-2}$$

We note the last term on the right-hand side of Equation (13-2) is just the total pressure times the total volume, i.e., PV, and this term is virtually always smaller than the other terms in Equation (13-2), and thus will be neglected.[†] For brevity, we shall write all summations as

$$\Sigma = \sum_{i=1}^{m}$$

unless otherwise stated.

When no spatial variations are present in the system volume, and time variations in the product of the total pressure and volume (PV) are neglected, the energy balance, after substitution of Equation (13-2) into (11-9), gives

$$\dot{Q} - \dot{W}_s + \Sigma F_{i0} H_{i0} \Big|_{\text{in}} - \Sigma F_i H_i \Big|_{\text{out}} = \left[\Sigma N_i \frac{dH_i}{dt} + \Sigma H_i \frac{dN_i}{dt} \right]_{\text{sys}} \tag{13-3}$$

[†] Marat Orazov while a student at University of California, Berkeley, pointed out that the last term in Equation (13-2) need not be neglected for the *special case* of an ideal gas with a constant total number of moles. The denominator on the r.h.s. of Equation (13-9) in this case is just $\sum N_i (C_{Pi} - R)$.

Recalling Equation (11-19)

$$H_i = H_i^\circ(T_R) + \int_{T_R}^{T} C_{P_i}\, dT \qquad (11\text{-}19)$$

and differentiating with respect to time, we obtain

$$\frac{dH_i}{dt} = C_{P_i}\frac{dT}{dt} \qquad (13\text{-}4)$$

Then, substituting Equation (13-4) into (13-3) gives

$$\dot{Q} - \dot{W}_s + \Sigma F_{i0}H_{i0} - \Sigma F_i H_i = \Sigma N_i C_{P_i}\frac{dT}{dt} + \Sigma\, H_i \frac{dN_i}{dt} \qquad (13\text{-}5)$$

The mole balance on species i is

$$\frac{dN_i}{dt} = -v_i\, r_A V + F_{i0} - F_i \qquad (13\text{-}6)$$

Using Equation (13-6) to substitute for dN_i/dt, Equation (13-5) becomes

$$\dot{Q} - \dot{W}_s + \Sigma\, F_{i0}H_{i0} - \Sigma\, F_i H_i$$

$$= \Sigma\, N_i C_{P_i}\frac{dT}{dt} + \Sigma\, v_i H_i(-r_A V) + \Sigma\, F_{i0}H_i - \Sigma\, F_i H_i \qquad (13\text{-}7)$$

Rearranging, and recalling $\sum v_i H_i = \Delta H_{Rx}$, we have

This form of the energy balance should be used when there is a phase change.

$$\boxed{\frac{dT}{dt} = \frac{\dot{Q} - \dot{W}_s - \Sigma\, F_{i0}(H_i - H_{i0}) + (-\Delta H_{Rx})(-r_A V)}{\Sigma\, N_i C_{P_i}}} \qquad (13\text{-}8)$$

Substituting for H_i and H_{i0} for the case of no phase change gives us

Energy balance on a transient CSTR or semibatch reactor

$$\boxed{\frac{dT}{dt} = \frac{\dot{Q} - \dot{W}_s - \sum F_{i0}C_{P_i}(T - T_{i0}) + [-\Delta H_{Rx}(T)](-r_A V)}{\sum N_i C_{P_i}}} \qquad (13\text{-}9)$$

Equation (13-9) applies to a semibatch reactor, as well as to the unsteady-state operation of a CSTR, and is also shown in Table 11-1 as Equation (T11-1.I).

For liquid-phase reactions where ΔC_P is usually small and can be neglected, the following approximation is often made

$$\Sigma\, N_i C_{P_i} \cong \Sigma\, N_{i0} C_{P_i} = N_{A0} \overbrace{\Sigma \Theta_i C_{P_i}}^{C_{P_s}} = N_{A0}\, C_{P_s}$$

where C_{P_s} is the heat capacity of the solution. The units of the batch term $(N_{A0}\, C_{P_s})$ are (cal/K) or (J/K) or (Btu/°R), and for the flow term

$$\Sigma\, F_{i0}C_{P_i} = F_{A0}C_{P_s}$$

the units are (J/s · K) or (cal/s · K) or (Btu/h · °R).[1] With this approximation and assuming that every species enters the reactor at temperature T_0, we have

$$\frac{dT}{dt} = \frac{\dot{Q} - \dot{W}_s - F_{A0}C_{P_s}(T - T_0) + [-\Delta H_{Rx}(T)](-r_A V)}{N_{A0}C_{P_s}} \quad (13\text{-}10)$$

13.2 Energy Balance on Batch Reactors (BRs)

A batch reactor is usually well mixed, so that we may neglect spatial variations in the temperature and species concentrations.

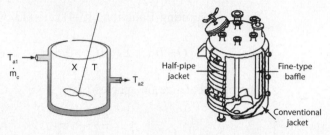

The energy balance on batch reactors is found by setting the inlet flows to zero, i.e., $F_{i0} = 0$, in Equation (13-9)

$$\frac{dT}{dt} = \frac{\dot{Q} - \dot{W}_s + (-\Delta H_{Rx})(-r_A V)}{\sum N_i C_{P_i}} \quad (13\text{-}11)$$

Next we use the heat-exchanger energy balance to obtain $\dot{Q}$, Equation (12-19), and realize that the heat "added" to the batch reactor, $\dot{Q}$, is just minus the heat "removed" from the batch reactor, $\dot{Q}_{rb}$, i.e., $\dot{Q} = -\dot{Q}_{rb}$. Neglecting shaft work (we can't do this in Problem P12-6$_B$) we get

$$\frac{dT}{dt} = \frac{\dot{Q}_{gb} - \dot{Q}_{rb} - \overset{\text{Neglect}}{\cancel{\dot{W}_s}}}{\sum N_i C_{P_i}}$$

$$\frac{dT}{dt} = \frac{Q_{gb} - Q_{rb}}{\sum N_i C_{P_i}} \quad (13\text{-}12)$$

---

[1] If the heat capacities were given in terms of mass (i.e., $C_{P_{sm}} = J/g \cdot K$), then both F_{A0} and N_{A0} would have to be converted to mass

for batch $\qquad\qquad m_{A0}C_{P_{sm}} = N_{A0}C_{P_s}$

$$(g)(J/g \cdot K) = (mol)\frac{J}{mol \cdot K} = \frac{J}{K}$$

and for flow $\qquad\qquad \dot{m}_{A0}C_{P_{sm}} = F_{A0}C_{P_s}$

$$(g/s)/(J/g \cdot K) = \left(\frac{mol}{s}\right)(J/mol \cdot K) = \frac{J}{K \cdot s}$$

However, we note that the units of the product of mass flow rate and mass heat capacities would still be the same as the product of molar flow and molar heat capacities (e.g., cal/s $\cdot$ K), respectively.

where the terms for heat generated, $\dot{Q}_{gb}$, and heat removed, $\dot{Q}_{rb}$, for a batch system are

$$\dot{Q}_{gb} = (-\Delta H_{Rx})(-r_A V) \tag{13-13}$$

$$\dot{Q}_{rb} = \dot{m} C_{P_C} (T - T_{a1}) \left[1 - \exp\left[-\frac{UA}{\dot{m} C_{P_C}} \right] \right] \tag{13-14}$$

with an added bonus

$$T_{a2} = T - (T - T_{a1}) \left[1 - \exp\left[-\frac{UA}{\dot{m} C_{P_C}} \right] \right] \tag{13-15}$$

Equation (13-12) is the preferred form of the energy balance when the number of moles, N_i, is used in the mole balance, rather than the conversion, X. The number of moles of species i at any X is

$$N_i = N_{A0}(\Theta_i + v_i X)$$

Consequently, in terms of conversion, the **user-friendly form** of the energy balance becomes

Batch reactor
energy balance

$$\frac{dT}{dt} = \frac{\dot{Q}_{gb} - \dot{Q}_{rb}}{N_{A0}(\Sigma \Theta_i C_{P_i} + \Delta C_P X)} \tag{13-16}$$

Equation (13-12) must be coupled with the mole balance

Batch reactor
mole balance

$$N_{A0} \frac{dX}{dt} = -r_A V \tag{2-6}$$

13.2.1 Adiabatic Operation of a Batch Reactor

Batch reactors operated adiabatically are often used to determine the reaction orders, activation energies, and specific reaction rates of exothermic reactions by monitoring the temperature–time trajectories for different initial conditions. In the steps that follow, we will derive the temperature–conversion relationship for adiabatic operation.

For adiabatic operation ($\dot{Q} = 0$) of a batch reactor ($F_{i0} = 0$) and when the work done by the stirrer can be neglected ($\dot{W}_s \cong 0$), Equation (13-10) can be written as

$$\frac{dT}{dt} = \frac{(-\Delta H_{Rx})(-r_A V)}{\Sigma N_i C_{P_i}} \tag{13-17}$$

It is shown in the *Summary Notes* on the CRE Web site (*www.umich.edu/~elements/5e/index.html*) that if we combine Equation (13-17) with Equation (2-6), we can do a lot of rearranging and integrating to arrive at the following **user-friendly** equation for an adiabatic batch reactor

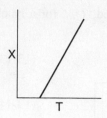

$$X = \frac{\sum \Theta_i C_{P_i}(T - T_0)}{-\Delta H_{Rx}(T)} \qquad (13\text{-}18)$$

$$T = T_0 + \frac{[-\Delta H_{Rx}(T_0)]X}{\sum \Theta_i C_{P_i} + X \Delta C_P} \qquad (13\text{-}19)$$

Temperature conversion relationship for any reactor operated adiabatically

We note that for adiabatic conditions, the relationship between temperature and conversion is the same for BRs, CSTRs, PBRs, and PFRs. Once we have T as a function of X for a batch reactor, we can construct a table similar to Table E11-3.1 and use techniques analogous to those discussed in Section 11.3.2 to evaluate the following design equation that determines the time necessary to achieve a specified conversion.

$$t = N_{A0} \int_0^X \frac{dX}{-r_A V} \qquad (2\text{-}9)$$

Time on your hands.

However, if you do not have *that much time on your hands* to form a table and use Chapter 2 integration techniques, then use Polymath or MATLAB to solve the coupled differential forms of mole balance equation (2-6) and the energy balance equation (13-19) simultaneously.

$$N_{A0} \frac{dX}{dt} = -r_A V \qquad (2\text{-}6)$$

Example 13–1 Adiabatic Batch Reactor

It is still winter, and although you were hoping for a transfer to the plant in the tropical southern coast of Jofostan, unfortunately you are still the engineer of the CSTR from Example 12-3, in charge of the production of propylene glycol.

$$CH_2\!-\!CH\!-\!CH_3 + H_2O \xrightarrow{\;H_2SO_4\;} CH_2\!-\!CH\!-\!CH_3$$
$$\underset{O}{\diagdown\!\diagup} \qquad\qquad\qquad \underset{OH\;\;\;OH}{|\quad\;\;|}$$

$$A + B \longrightarrow C$$

You are considering the installation of a new, attractive-looking, glass-lined 1,000-dm³ CSTR, and you decide to make a quick check of the reaction kinetics and maximum adiabatic temperature. You have a stylish, nicely decorated and instrumented 40-dm³ (~10 gal) stirred batch reactor you ordered from a company in Jofostan. You charge this reactor with 4 dm³ (~1 gal) of ethylene oxide, 4 dm³ (~1 gal) of methanol, and 10 dm³ (~2.5 gal) of water containing 0.1 wt % H_2SO_4. For safety reasons, the reactor is located on a boating pier on the banks of Lake Wobegon (you don't want the entire plant to be destroyed if the reactor explodes). At this time of year in northern Minnesota, the initial temperature of all materials is 276 K (3°C). We have to be careful here! If the reactor temperature increases above 350 K (77°C), a secondary, more exothermic reaction will take over, causing runaway and subsequent explosion, similar to what happened in the T-2 Laboratory plant explosion in Florida, see page 660.

Although you requested obtaining the data for this reaction from the Jofostan national laboratory, the purchasing department decided to save money and buy it off the Internet. The values it purchased are

$$\Delta H_{Rx} = -20{,}202 \text{ cal/mol}, \quad C_{P_A} = 35 \text{ cal/mol/K}, \quad C_{P_B} = 18 \text{ cal/mol/K},$$
$$C_{P_C} = 46 \text{ cal/mol/K}, \quad \text{and} \quad C_{P_M} = 19.5 \text{ cal/mol/K}.$$

The concentration of pure ethylene oxide and methanol are 13.7 mol/dm³ and 24.7 mol/dm³ respectively. Consequently, the initial number of moles added to the reactor are

A: Ethylene oxide:	$N_{A0} = (13.7 \text{ mol/dm}^3)(4 \text{ dm}^3) = 54.8 \text{ mol}$
B: Water:	$N_{B0} = (55.5 \text{ mol/dm}^3)(10 \text{ dm}^3) = 555 \text{ mol}$
M: Methanol:	$N_M = (24.7 \text{ mol/dm}^3)(4 \text{ dm}^3) = 98.8 \text{ mol}$

The sulfuric acid catalyst takes up negligible space, so the total volume is 18 dm³. Use the data and the reaction-rate law given in Example 12-3. We are going to carry out two scenarios: (1) to learn how fast the temperature rises and how long it takes to reach 350 K for adiabatic operation, and (2) how long would it take to reach 345 K if we added a heat exchanger.

(a) Plot conversion and temperature X and T as a function of time for adiabatic operation. How many minutes should it take the mixture inside the reactor to reach a conversion of 51.5%? What is the corresponding adiabatic temperature?

(b) Plot the temperature and conversion when the heat exchange is added. The product of the overall heat-transfer coefficient and area is $UA = 10$ cal/s/K with $T_{a1} = 290$ K and the coolant rate is 10 g/s, and it has a heat capacity of 4.16 cal/g/K.

Solution

The initial temperature is 3°C, and if the reactor temperature increases to above 77°C, a dangerous exothermic side reaction can occur, as was reported in *Jofostan Journal of Chemical Safety*, Vol. 19, p. 201 (1988).

As before, there is approximately a 10°C (17°F) rise in temperature immediately after mixing. The activation energy is 32,400 cal/gmol · K.

1. Mole Balance, Chapter 2, rearranging Equation (2-6) we have

$$\frac{dX}{dt} = \frac{-r_A V}{N_{A0}} \tag{E13-1.1}$$

2. Rate Law:

$$-r_A = k C_A \tag{E13-1.2}$$

3. Stoichiometry:

$$N_A = N_{A0}(1 - X) \tag{2-4}$$

Recall that for liquid batch $V = V_0$

Following the Algorithm

$$C_A = \frac{N_A}{V} = \frac{N_A}{V_0} = \frac{N_{A0}(1-X)}{V_0} = C_{A0}(1-X) \tag{E13-1.3}$$

4. Combining Equations (E13-1.1), (E13-1.2), and (2-6), we have

$$\frac{dX}{dt} = k(1 - X) \tag{E13-1.4}$$

Changing the data in Example 12-3 from English units to mks units, we have

$$k = (4.71 \times 10^9) \exp\left[\left(\frac{E}{R}\right)\frac{1}{T}\right] s^{-1} \tag{E13-1.5}$$

$$\text{with } \frac{E}{R} = 9059K$$

Choosing 297 K as a reference temperature for k_1, and putting Equation (E13-1.5) in the form of Equation (3-21), we get

$$k = (2.73 \times 10^{-4}) \exp\left[9059K\left(\frac{1}{297} - \frac{1}{T}\right)\right]s^{-1} \qquad \text{(E13-1.6)}$$

5. **Energy Balance.**
   Part (a) Adiabatic Operation.
   Using Equation (13-19), the relationship between X and T for an adiabatic reaction is given by

$$T = T_0 + \frac{[-\Delta H_{Rx}(T_0)]X}{\Sigma\Theta_i C_{p_i} + \Delta C_p X} \qquad \text{(E13-1.7)}$$

6. **Parameters:**

$$C_{P_S} = \Sigma\, \Theta_i C_{P_i} = C_{P_A} + C_{P_B}\Theta_B C_{P_M}\Theta_M \qquad \text{(E13-1.8)}$$

$$\Sigma\Theta_i C_{P_i} = C_{P_S} = 35 + 18\left(\frac{555}{54.8}\right) + 19.5\left(\frac{98.8}{54.8}\right)$$

$$= 35 + 182.3 + 35.2 = 252.5\,\frac{\text{cal}}{\text{mol}\cdot\text{K}}$$

The lumped heat capacity of the solution is

$$\boxed{C_{P_S} = 252.2\,\frac{\text{cal}}{\text{mol}\cdot\text{K}}}$$

In the expression for the heat of reaction, we are going to neglect the second term on the right-hand side, i.e.,

$$\Delta H_{Rx}(T) = \Delta H^\circ_{Rx}(T_R) + \overset{\text{Neglect}}{\cancel{\Delta C_P(T - T_R)}} \qquad \text{(E13-1.9)}$$

We can do this because ΔC_p is –7 cal/mol/K and for a 50-K temperature difference, $\Delta C_p (T - T_R) = 350$ cal/mol, which is negligible with respect to the heat of reaction of –20,202 or cal/mol

$$\boxed{\Delta H_{Rx}(T) = \Delta H^\circ_{Rx} = -20,202 \text{ cal/mol}} \qquad \text{(E13-1.10)}$$

In calculating the inlet temperature after mixing, T_0, we must include the temperature rise 10°C (17°F) resulting from the heat of mixing the two solutions initially at 3°C

$$T_0 = 276 + 10 = 286 \text{ K}$$

$$T = T_0 + \frac{X(-\Delta H_{Rx})}{C_{P_S}} = 286K + \frac{X(-1(-20,202))\frac{\text{cal}}{\text{mol}}}{252.5 \text{ cal/mol}\cdot\text{K}}$$

Adiabatic Energy
Balance

$$\boxed{T = 286 + 80X} \qquad \text{(E13-1.11)}$$

A summary of the heat and mole balance equations is given in Table E13-1.1.

TABLE E13-1.1   SUMMARY FOR FIRST-ORDER ADIABATIC BATCH REACTION

$$\frac{dX}{dt} = k(1 - X) \tag{E13-1.4}$$

$$k = (2.73 \times 10^{-4})\exp\left[9059K\left(\frac{1}{297} - \frac{1}{T}\right)\right]s^{-1} \tag{E13-1.6}$$

$$T = 286 + 80X \tag{E13-1.11}$$

where T is in K and t is in seconds.

A table similar to that used in Example 11-3 can now be constructed or one can make better use of his or her time using Polymath.

The software package Polymath will be used to combine Equations (E13-1.4), (E13-1.6), and (E13-1.11) to determine conversion and temperature as a function of time. Table E13-1.2 shows the program, and Figures E13-1.1 and E13-1.2 show the solution results.

TABLE E13-1.2   POLYMATH PROGRAM

POLYMATH Report
Ordinary Differential Equations

Calculated values of DEQ variables

	Variable	Initial value	Final value
1	k	8.446E-05	0.0858013
2	T	286.	366.
3	t	0	2500.
4	X	0	1.

Differential equations

1 d(x)/d(t)=k*(1-X)

Explicit equations

1 T = 286 + 80*X

2 k = .000273*exp(9059*(1/297-1/T))

Adiabatic Batch Operation

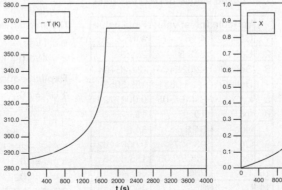

Figure E13-1.1   Temperature–time trajectory.

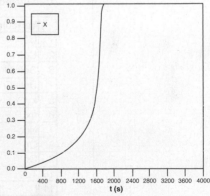

Figure E13-1.2   Conversion–time trajectory.

Part (b) Heat Exchanger in a Batch Reactor.

We now consider the case where a heat exchanger is added to the batch reactor. The coolant enters at 290 K and the flow rate, $\dot{m}_C$, through the exchanger is 10 g/s. The mole balance, Equation (E13-1.1), and the physical properties, Equations (E13-1.2) through (E13-1.6) and (E13-1.8), remain the same.

Mole Balance

$$\frac{dX}{dt} = \frac{-r_A V}{N_{A0}} \qquad \text{(E13-1.1)}$$

Neglecting ΔC_P the energy balance is

Energy Balance

$$\frac{dT}{dt} = \frac{\dot{Q}_g - \dot{Q}_r}{N_{A0} C_{P_s}} \qquad \text{(E13-1.2)}$$

with

$$\dot{Q}_r = \dot{m}_c C_{P_C}\left\{(T - T_{a1})\left[1 - exp\left(\frac{-UA}{\dot{m}_c C_{P_C}}\right)\right]\right\} \qquad \text{(E13-1.13)}$$

$$\dot{Q}_g = (-r_A V)(-\Delta H^\circ_{Rx}) \qquad \text{(E13-1.14)}$$

$$-r_A V = N_{A0} k(1 - X) \qquad \text{(E13-1.15)}$$

The Polymath Program is shown in Table E13-1.3 with **parameters**

$$C_{P_S} = 252.5 \text{ cal/mol/K} \quad C_{P_C} = 4.16 \text{ cal/g/s} \quad \dot{m}_c = 10 \text{ g/s}$$

$$N_{A0} = 54.8 \text{ mol, } UA = 10 \text{ cal/K/s}$$

$$T_{a1} = 290 \text{ K, } T_0 = 286 \text{ K}$$

and the other terms remain the same as in **Part (a)**.

TABLE E13-1.3   POLYMATH PROGRAM

POLYMATH Report
Ordinary Differential Equations

Calculated values of DEQ variables

	Variable	Initial value	Final value
1	Cpc	4.18	4.18
2	Cps	252.5	252.5
3	DeltaH	-2.02E+04	-2.02E+04
4	DeltaQ	-129.0831	155.5053
5	k	8.446E-05	0.0007724
6	mc	10.	10.
7	Nao	54.8	54.8
8	Qg	93.50797	0.0004292
9	Qr	-35.57509	155.5058
10	T	286.	307.4848
11	t	0	4000.
12	Ta1	290.	290.
13	Ta2	289.1489	293.7202
14	UA	10.	10.
15	X	0	0.9999995

Differential equations

1 d(T)/d(t) = (Qg-Qr)/Cps/Nao
2 d(X)/d(t) = k*(1-X)

Explicit equations

1 UA = 10
2 DeltaH = -20202
3 Ta1 = 290
4 Cpc = 4.18
5 Cps = 252.5
6 mc = 10
7 k = .000273*exp(9059*(1/297-1/T))
8 Ta2 = T-(T-Ta1)*exp(-UA/mc/Cpc)
9 Qr = mc*Cpc*(T-Ta1)*(1-exp(-UA/mc/Cpc))
10 Nao = 54.8
11 Qg = Nao* k*(1-X)*(-DeltaH)
12 DeltaQ = Qr-Qg

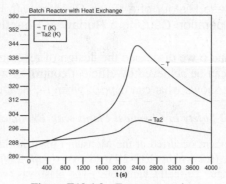

Figure E13-1.3   Temperature–time trajectory.

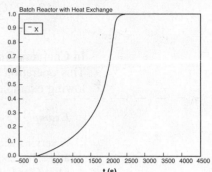

Figure E13-1.4   Conversion–time trajectory.

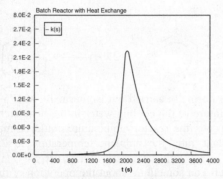

Figure E13-1.5   Specific reaction-rate trajectory.

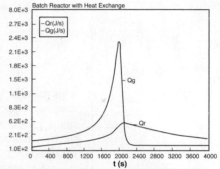

Figure E13-1.6   Heat-generated as heat removed trajectories.

One notes from Figure E13-1.3(a) and (b) that the temperature profiles are not as steep as the adiabatic case. The specific reactor-rate constant, Figure E13-1.5, goes through a maximum as the fluid becomes heated at first then later cools. Figure E13-1.6 shows the trajectories of the "heat generated," Q_g, and "heat removed," Q_r. Whenever $Q_g > Q_r$

$$\frac{dT}{dt} = \frac{Q_g - Q_r}{N_{A0} \Sigma \Theta_i C_{P_i}} \qquad \text{(E13-1.16)}$$

the temperature will increase when $Q_g > Q_r$ and the temperature will decrease when $Q_r > Q_g$.

Analysis:

(a) Adiabatic. The initial temperature is rather low, so the reaction is slow at first. However, as the exothermic reaction proceeds, it heats up and becomes virtually autocatalytic as it goes from a small conversion at 1600 seconds to complete conversion just a few seconds later. The conversion reaches $X = 1.0$ and the temperature reaches its maximum value, where it remains.

(b) Heat Exchange. Because the temperature is maintained at a lower value in the heat exchange case than in the adiabatic case, there will be less conversion for the parameter values Ua, T_{ai}, T_0, $\dot{m}_c$, etc., given in the problem statement. However, if you change these values as suggested in Problem P13-1$_A$ (a), you will find situations where the conversion remains very low and cases where the temperature curve is extremely steep.

13.2.2 Case History of a Batch Reactor with Interrupted Isothermal Operation Causing a **Runaway Reaction**

In Chapters 5 and 6 we discussed the design of reactors operating isothermally. This operation can be achieved by efficient control of a heat exchanger. The following example shows what can happen when the heat exchanger suddenly fails.

Example 13–2 Safety in Chemical Plants with Exothermic Runaway Reactions[2]

A serious accident occurred at the Monsanto plant in Sauget, Illinois, on August 8 at 12:18 A.M. (see Figure E13-2.1). (Sauget (pop. 200) is the home of the 1988 Mon-Clar League Softball Champions.) The blast was heard as far as 10 miles away in Belleville, Illinois, where people were awakened from their sleep. The explosion occurred in a batch reactor that was used to produce nitroaniline from ammonia and *o*-nitrochlorobenzene (ONCB):

$$\text{(NO}_2\text{, Cl benzene)} + 2\text{NH}_3 \longrightarrow \text{(NO}_2\text{, NH}_2 \text{ benzene)} + \text{NH}_4\text{Cl}$$

This reaction is normally carried out isothermally at 175°C and about 500 psi. The ambient temperature of the cooling water in the heat exchanger is 25°C. By adjusting the coolant rate, the reactor temperature could be maintained at 175°C. At the maximum coolant rate, the ambient temperature is 25°C throughout the heat exchanger.

     Let me tell you something about the operation of this reactor. Over the years, the heat exchanger would fail from time to time, but the technicians would be "Johnny on the Spot" and run out and get it up and running within 10 minutes or so, and there was never any problem. It is believed that one day someone in management looked at the reactor and said, "It looks as if your reactor is only a third full and you still have room to add more reactants and to make more product and more money. How about filling it up to the top so we could triple production?" They did, and started the reactor up at 9:45 p.m. Around midnight the reactor exploded and the aftermath is shown in Figure E13-2.1.

A decision was made to triple production.

Living Example Problem

Figure E13-2.1 Aftermath of the explosion. (*St. Louis Globe/Democrat* photo by Roy Cook. Courtesy of St. Louis Mercantile Library.)

---

[2] Adapted from the problem by Ronald Willey, *Seminar on a Nitroaniline Reactor Rupture*. Prepared for SAChE, Center for Chemical Process Safety, American Institute of Chemical Engineers, New York (1994). Also see *Process Safety Progress*, vol. 20, no. 2 (2001), pp. 123–129. The values of ΔH_{Rx} and UA were estimated from the plant data of the temperature–time trajectory in the article by G. C. Vincent, *Loss Prevention*, 5, 46–52.

On the day of the accident, two changes in normal operation occurred.

1. The reactor was charged with 9.044 kmol of ONCB, 33.0 kmol of NH_3, and 103.7 kmol of H_2O. Normally, the reactor is charged with 3.17 kmol of ONCB, 103.6 kmol of H_2O, and 43 kmol of NH_3.

2. The reaction is normally carried out isothermally at 175°C over a 24-h period. However, approximately 45 min after the reaction was started, cooling to the reactor failed, but only for 10 min, and cooling was again up and running at the 55-minute mark. Cooling may have been halted for 10 min or so on previous occasions when the normal charge of 3.17 kmol of ONCB was used and no ill effects occurred.

The reactor had a rupture disk designed to burst when the pressure exceeded approximately 700 psi. If the disk would have ruptured, the pressure in the reactor would have dropped, causing the water to vaporize, and the reaction would have been cooled (quenched) by the latent heat of vaporization.

(a) Plot and analyze the temperature–time trajectory up to a period of 120 min after the reactants were mixed and brought up to 175°C (448K).

(b) Show that all of the following three conditions had to have been present for the explosion to occur: (1) increased ONCB charge, (2) cooling stopped for 10 min at a time early in the reaction, and (3) relief-system failure.

Additional information:

Rate law: $-r_{ONCB} = kC_{ONCB}C_{NH_3}$

with $k = 0.00017 \dfrac{m^3}{kmol \cdot min}$ at 188°C (461K) and $E = 11{,}273$ cal/mol

The reaction volume for the new charge of 9.0448 kmol of ONCB

$$V = 3.265 \text{ m}^3 \text{ ONCB/NH}_3 + 1.854 \text{ m}^3 \text{ H}_2O = 5.119 \text{ m}^3$$

Case History

The reaction volume for the previous charge of 3.17 kmol of ONCB:

$$V = 3.26 \text{ m}^3$$
$$\Delta H_{Rx} = -5.9 \times 10^5 \text{ kcal/kmol}$$
$$C_{P_{ONCB}} = C_{P_A} = 40 \text{ cal/mol} \cdot K$$
$$C_{P_{H_2O}} = C_{P_W} = 18 \text{ cal/mol} \cdot K \qquad C_{P_{NH_3}} = C_{P_B} = 8.38 \text{ cal/mol} \cdot K$$

Assume that $\Delta C_P \approx 0$

$$UA = \frac{35.85 \text{ kcal}}{min \, °C} \text{ with } T_a = 298 \text{ K}$$

Solution

$$A + 2B \longrightarrow C + D$$

Mole Balance:

$$\frac{dX}{dt} = -r_A \frac{V}{N_{A0}} \tag{E13-2.1}$$

Rate Law:

$$-r_A = kC_A C_B \tag{E13-2.2}$$

Stoichiometry (liquid phase):

$$C_A = C_{A0}(1 - X) \tag{E13-2.3}$$

with

$$C_B = C_{A0}(\Theta_B - 2X) \tag{E13-2.4}$$

$$\Theta_B = \frac{N_{B0}}{N_{A0}}$$

Combine:

Following the Algorithm

$$\boxed{-r_A = kC_{A0}^2(1 - X)(\Theta_B - 2X)} \tag{E13-2.5}$$

Substituting our parameter values into Equation (3-21)

$$k = k(T_0)exp\left[\frac{E}{R}\left(\frac{1}{T_0} - \frac{1}{T_1}\right)\right] \tag{3-21}$$

We obtain

$$k = 0.00017\ exp\left[\frac{11273}{1.987}\left(\frac{1}{461} - \frac{1}{T}\right)\right]\frac{m^3}{kmol \cdot min}$$

Energy Balance:

$$\frac{dT}{dt} = \frac{UA(T_a - T) + (r_A V)(\Delta H_{Rx})}{\sum N_i C_{P_i}} \tag{E13-2.6}$$

For $\Delta C_P = 0$,

$$\sum N_i C_{P_i} = NC_P = N_{A0}C_{P_A} + N_{B0}C_{P_B} + N_W C_{P_W}$$

Parameter evaluation for day of explosion:

$$NC_P = (9.0448)(40) + (103.7)(18) + (33)(8.38)$$

$$NC_P = 2504\ kcal/K$$

Again, let Q_g be the heat generated, i.e., $Q_g = (r_A V)(\Delta H_{Rx})$, and let Q_r be the heat removed, i.e., $Q_r = UA(T - T_a)$.

$$\boxed{\frac{dT}{dt} = \frac{\overbrace{UA(T_a - T)}^{-Q_r} + \overbrace{(r_A V)(\Delta H_{Rx})}^{Q_g}}{N_{A0}C_{P_A} + N_{B0}C_{P_B} + N_W C_{P_W}}} \tag{E13-2.7}$$

$Q_g = (r_A V)(\Delta H_{Rx})$
$Q_r = UA(T - T_a)$

Then

$$\boxed{\frac{dT}{dt} = \frac{Q_g - Q_r}{NC_P}} \tag{E13-2.8}$$

A. Isothermal Operation Up to 45 Minutes

We will first carry out the reaction isothermally at 175°C (448 K) up to the time the cooling was turned off at 45 min. Combining and canceling yields

$$\frac{dX}{dt} = kC_{A0}(1 - X)(\Theta_B - 2X) \tag{E13-2.9}$$

$$\Theta_B = \frac{33}{9.04} = 3.64$$

Integrating Equation (E13-2.9) gives us

$$t = \left[\frac{V}{kN_{A0}}\right]\left(\frac{1}{\Theta_B - 2}\right)\ln\left[\frac{\Theta_B - 2X}{\Theta_B(1-X)}\right] \qquad (E13\text{-}2.10)$$

At $175°C = 448$ K, $k = 0.0001167 \text{m}^3/\text{kmol} \cdot \text{min}$.

Substituting for k and the other parameter values

$$45 \text{ min} = \left[\frac{5.119 \text{ m}^3}{0.0001167 \text{ m}^3/\text{kmol} \cdot \text{min}\,(9.044 \text{ kmol})}\right] \times \left(\frac{1}{1.64}\right)\ln\left[\frac{3.64 - 2X}{3.64(1-X)}\right]$$

Solving for X, we find that at $t = 45$ min, then $X = 0.033$.

We will calculate the rate of generation, Q_g, at this temperature and conversion and compare it with the maximum rate of heat removal, Q_r, that is available for a constant coolant temperature of $T_a = 298$ K. The rate of generation, Q_g, is

$$\boxed{Q_g = r_A V\,\Delta H_{Rx} = k\,\frac{N_{A0}(1-X)N_{A0}[(N_{B0}/N_{A0}) - 2X]\,V(-\Delta H_{Rx})}{V^2}} \qquad (E13\text{-}2.11)$$

At this time (i.e., $t = 45$ min, $X = 0.033$, $T = 175°C$), we calculate k, then Q_r and Q_g. At $175°C$, $k = 0.0001167 \text{ m}^3/\text{min} \cdot \text{kmol}$.

$$Q_g = (0.0001167)\,\frac{(9.0448)^2(1 - 0.033)}{5.119}\left[\frac{33}{(9.0448)} - 2(0.033)\right]5.9 \times 10^5$$

$$\boxed{Q_g = 3830 \text{ kcal/min}}$$

The corresponding maximum cooling rate is

$$Q_r = UA\,(T - 298)$$
$$= 35.85(448 - 298) \qquad (E13\text{-}2.12)$$
$$\boxed{Q_r = 5378 \text{ kcal/min}}$$

Therefore

$$\boxed{Q_r > Q_g} \qquad (E13\text{-}2.13)$$

The reaction can be controlled. There would have been no explosion had the cooling not failed.

B. Adiabatic Operation for 10 Minutes

Unexpectedly, the cooling failed from 45 to 55 min after the reaction was started. We will now use the conditions at the end of the period of isothermal operation as our initial conditions for the adiabatic operation period between 45 and 55 min

$$t = 45 \text{ min} \quad X = 0.033 \quad T = 448 \text{ K}$$

Between $t = 45$ and $t = 55$ min, $Q_r = 0$. The Polymath program was modified to account for the time of adiabatic operation by using an *"if statement"* for Q_r in the program, i.e., $Q_r =$ if $(t > 45$ and $t < 55)$ then (0) else $(UA(T - 298))$. A similar *"if statement"* is used for isothermal operation, i.e., $(dT/dt) = 0$.

The calculation and results can also be obtained from the Polymath output

Living Example Problem

Everything is OK.

Interruptions in the cooling system have happened before with no ill effects.

For the 45- to 55-min period without cooling, the temperature rose from 448 K to 468 K, and the conversion increased from 0.033 to 0.0424. Using this temperature and conversion in Equation (E13-2.11), we calculate the rate of generation Q_g at 55 min and see that it has increased to

$$Q_g = 6591 \text{ kcal/min}$$

The maximum rate of cooling at this reactor temperature is found from Equation (E13-2.12) to be

$$Q_r = 6093 \text{ kcal/min}$$

Here, we see that at the end of the 10-minute down time, the heat exchange system is now operating again, but now

$$\boxed{Q_g > Q_r} \tag{E13-2.14}$$

The point of no return

and the temperature will continue to increase. *We have a Runaway Reaction!!* The **point of no return** has been passed and the temperature will continue to increase, as will the rate of reaction until the explosion occurs.

C. Batch Operation with Heat Exchange

Return of the cooling occurs at 55 min after startup. The values at the end of the period of adiabatic operation ($T = 468$ K, $X = 0.0423$) become the initial conditions for the period of restored operation with the heat exchange. The cooling is turned on at its maximum capacity, $Q_r = UA(T - 298)$, at 55 min. Table E13-2.1 gives the Polymath program to determine the temperature–time trajectory. (Note that one can change N_{A0} and N_{B0} to 3.17 and 43 kmol in the program and show that, if the cooling is shut off for 10 min, at the end of that 10 min, Q_r will still be greater than Q_g and no explosion will occur.)

The complete temperature–time trajectory is shown in Figure E13-2.2. One notes the long plateau after the cooling is turned back on. Using the values of Q_g and Q_r at 55 min and substituting into Equation (E13-2.8), we find that

Living Example Problem

$$\frac{dT}{dt} = \frac{(6591 \text{ kcal/min}) - (6093 \text{ kcal/min})}{2504 \text{ kcal/}^\circ\text{C}} = 0.2^\circ\text{C/min}$$

TABLE E13-2.1   POLYMATH PROGRAM

Differential equations
1 d(T)/d(t) = if (t<45) then (0) else ((Qg-Qr)/NCp)
2 d(X)/d(t) = (-ra)*V/Nao

Explicit equations
1 NCp = 2504
2 V = 3.265+1.854
3 Nao = 9.0448
4 UA = 35.83
5 DeltaHrx = -590000
6 Nbo = 33
7 k = .00017*exp(11273/(1.987)*(1/461-1/T))
8 Qr = if(t>45 and t<55) then (0) else (UA*(T-298))
9 Theata = Nbo/Nao
10 ra = -k*Nao^2*(1-X)*(Theata-2*X)/V^2
11 Qg = ra*V*DeltaHrx

Calculated values of DEQ variables

	Variable	Initial value	Final value
1	DeltaHrx	-5.9E+05	-5.9E+05
2	k	0.0001189	0.8033049
3	Nao	9.0448	9.0448
4	Nbo	33.	33.
5	NCp	2504.	2504.
6	Qg	4092.007	7.234E+06
7	Qr	5374.5	4.218E+04
8	ra	-0.0013549	-2.395239
9	t	0	122.
10	T	448.	1475.143
11	Theata	3.648505	3.648505
12	UA	35.83	35.83
13	V	5.119	5.119
14	X	0	0.6075213

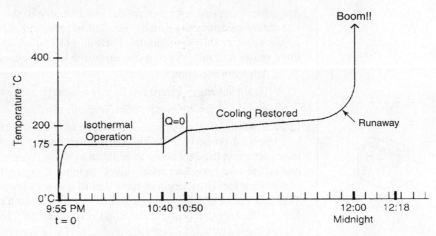

Figure E13-2.2 Temperature–time trajectory.

The explosion occurred shortly after midnight.

Consequently, even though dT/dt is positive, the temperature increases very slowly at first, 0.2°C/min. By 11:45, the temperature has reached 240°C and both the "heat generated" and the temperature begin to increase more rapidly. This rapid increase is a result of the Arrhenius temperature dependence, which caused the temperature to increase exponentially. The reaction is *running away*! One observes in Figure E13-2.2 that 119 min after the batch was started, the temperature increases sharply and the reactor explodes at approximately midnight. If the mass and heat capacity of the stirrer and reaction vessel had been included, the NC_p term would have increased by about 5% and extended the time until the explosion occurred by 15 or so minutes, which would predict the actual time the explosion occurred, at 12:18 A.M.

When the temperature reached 300°C, a secondary reaction, the decomposition of nitroaniline to noncondensable gases such as CO, N_2, and NO_2, occurred, releasing even more energy. The total energy released was estimated to be 6.8×10^9 J, which is enough energy to lift the entire 2500-ton building 300 m (the length of three football fields) straight up.

D. Safety Disk Rupture Failure

We note that the pressure-safety-relief disk should have ruptured when the temperature reached 265°C (ca. 700 psi) but did not and the temperature continued to rise. If it had ruptured and all the water had vaporized, 10^6 kcal would have been drawn from the reacting solution, thereby lowering its temperature and quenching the runaway reaction.

If the disk had ruptured at 265°C (700 psi), we know from fluid mechanics that the maximum mass flow rate, $\dot{m}_{vap}$, out of the 2-in. orifice to the atmosphere (1 atm) would have been 830 kg/min at the time of rupture. The corresponding heat removed would have been

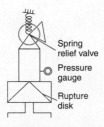

Spring relief valve

Pressure gauge

Rupture disk

$$Q_r = \dot{m}_{vap}\, \Delta H_{vap} + UA\,(T - T_a)$$

$$= 830\ \frac{kg}{min} \times 540\ \frac{kcal}{kg} + 35.83\ \frac{kcal}{min \cdot K}\,(538 - 298)K$$

$$= 4.48 \times 10^5\ \frac{kcal}{min} + 8604\ \frac{kcal}{min}$$

$$= 456{,}604\ \frac{kcal}{min}$$

This value of Q_r is much greater than the "heat generated" Q_g ($Q_g = 27{,}460$ kcal/min), so that the reaction could have been easily quenched.

Analysis: Runaway reactions are the most deadly in the chemical industry. Elaborate safety measures are usually installed to prevent them from occurring. However, as we show in this example, the back-up plan failed. *If any one of the following three things had not occurred, the explosion would not have happened.*

1. Tripled production
2. Heat-exchanger failure for 10 minutes early in the operation
3. Failure of the relieving device (rupture disk)

In other words, all the above had to happen to cause the explosion. If the relief valve had operated properly, it would not have prevented reaction runaway but it could have prevented the explosion. In addition to using rupture disks as relieving devices, one can also use pressure relief valves. In many cases, sufficient care is not taken to obtain data for the reaction at hand and to use it to properly size the relief device. This data can be obtained using a specially designed batch reactor called the *Advanced Reactor Safety Screening Tool* (ARSST), as shown in Chapter 13, PRS R13.1.

Reference Shelf

13.3 Semibatch Reactors with a Heat Exchanger

In our past discussions of reactors with heat exchangers, we assumed that the ambient temperature T_a was spatially uniform throughout the exchangers. This assumption is true if the system is a tubular reactor with the external pipe surface exposed to the atmosphere, or if the system is a CSTR or batch where the coolant flow rate through the exchanger is so rapid that the coolant temperatures entering and leaving the exchanger are virtually the same.

We now consider the case where the coolant temperature varies along the length of the exchanger while the temperature in the reactor is spatially uniform. The coolant enters the exchanger at a mass flow rate, $\dot{m}_c$, at a temperature, T_{a1}, and leaves at a temperature, T_{a2} (see Figure 13-1). As a first approximation, we assume a quasi-steady state for the coolant flow and neglect the accumulation term (i.e., $dT_a/dt = 0$). As a result, Equation (12-19) will give the rate of heat transfer *from* the exchanger *to* the reactor

$$\dot{Q} = \dot{m}_c C_{P_c}(T_{a1} - T)[1 - \exp(-UA/\dot{m}_c C_{P_c})] \qquad (12\text{-}19)$$

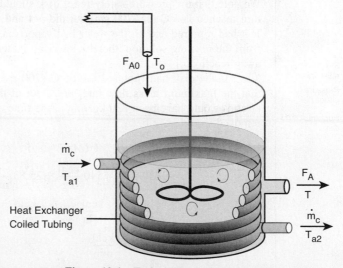

Figure 13-1   Tank reactor with heat exchanger.

The energy balance on a semibatch reactor is

$$\frac{dT}{dt} = \frac{\dot{Q} - \dot{W}_s - \Sigma F_{i0} C_{P_i}(T - T_{i0}) - [\Delta H_{Rx}(T)](-r_A V)}{\Sigma N_i C_{P_i}} \tag{13-9}$$

Using a slight rearrangement of the heat exchanger energy balance equation, i.e., Equation (12-9) and neglecting shaft work, $\dot{W}_S$

$$\frac{dT}{dt} = \frac{\dot{Q}_{gs} - \dot{Q}_{rs}}{\Sigma N_i C_{P_i}} \tag{13-20}$$

where the "heat generated," $\dot{Q}_{gs}$, for a the semibatch reactor is the same as that for a BR,

$$\dot{Q}_{gs} = (r_A V)(\Delta H_{Rx}) \tag{13-21}$$

however, the "heat removed" for a semibatch reactor, $\dot{Q}_{gs}$, is

$$\dot{Q}_{rs} = \sum F_{i0} C_{P_i}(T - T_{i0}) + \dot{m} C_P [T - T_{a1}]\left[1 - exp\left[\frac{-UA}{\dot{m} C_{P_i}}\right]\right] \tag{13-22}$$

Note: The subscript "s" in $\dot{Q}_{rs}$ and $\dot{Q}_{gs}$ means these equations <u>only</u> apply to a <u>s</u>emibatch reactor.

For large coolant flow rates, Equation (13-22) reduces to

$$\frac{dT}{dt} = \frac{\overbrace{(r_A V)(\Delta H_{Rx})}^{Q_{gs}} - \overbrace{[\Sigma F_{i0} C_{P_i}(T - T_{i0}) + UA(T - T_a)]}^{Q_{rs}}}{\Sigma N_i C_{P_i}} \tag{13-23}$$

Example 13–3 Heat Effects in a Semibatch Reactor

The second-order saponification of ethyl acetate is to be carried out in a semibatch reactor shown schematically below in Figure E13-3.1.

$$C_2H_5(CH_3COO)(aq) + NaOH(aq) \rightleftharpoons Na(CH_3COO)(aq) + C_2H_5OH(aq)$$

$$A \quad + \quad B \quad \rightleftharpoons \quad C \quad + \quad D$$

Living Example Problem

Aqueous sodium hydroxide is to be fed at a concentration of 1 kmol/m³, a temperature of 300 K, and a volumetric rate of 0.004 m³/s to an initial volume of 0.2 m³ of water and ethyl acetate. The concentration of water in the feed, C_{W0}, is 55 kmol/m³. The initial concentrations of ethyl acetate and water in the reactor are 5 kmol/m³ and 30.7 kmol/m³, respectively. The reaction is exothermic and it is necessary to add a heat exchanger to keep its temperature below 315 K. A heat exchanger with $UA = 3000$ J/s · K is available for use. The coolant enters at a mass flow rate of 100 kg/s and a temperature of 285 K.

Are the heat exchanger and coolant flow rate adequate to keep the reactor temperature below 315 K? Plot the reactor temperature, T, and the concentrations, C_A, C_B, and C_C as a function of time.

Additional information:[3]

$$k = 0.39175 \exp\left[5472.7\left(\frac{1}{273} - \frac{1}{T}\right)\right] \text{ m}^3/\text{kmol}\cdot\text{s}$$

$$K_C = 10^{3885.44/T}$$

$$\Delta H_{Rx}^\circ = -79{,}076 \text{ kJ/ kmol}$$

$$C_{P_A} = 170.7 \text{ kJ/kmol/K} = 170{,}700 \text{ J/mol/K}$$

$$C_{P_B} = C_{P_C} = C_{P_D} \cong C_{P_W} = C_P = 75.24 \text{ kJ/kmol}\cdot\text{K} = 75{,}246 \text{ J/mol/K}$$

Feed: $C_{W0} = 55 \text{ kmol/m}^3$ $C_{B0} = 1.0 \text{ kmol/m}^3$

Initially: $C_{Wi} = 30.7 \text{ kmol/m}^3$ $C_{Ai} = 5 \text{ kmol/m}^3$ $C_{Bi} = 0$

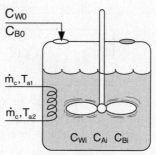

Figure E13-3.1   Semibatch reactor with heat exchange.

Solution

1. Mole Balances: (See Chapter 6.)

$$\frac{dC_A}{dt} = r_A - \frac{v_0 C_A}{V} \qquad\qquad \text{(E13-3.1)}$$

$$\frac{dC_B}{dt} = r_B + \frac{v_0(C_{B0} - C_B)}{V} \qquad\qquad \text{(E13-3.2)}$$

$$\frac{dC_C}{dt} = r_C - \frac{C_C v_0}{V} \qquad\qquad \text{(E13-3.3)}$$

$$C_D = C_C$$

$$\frac{dN_W}{dt} = C_{W0} v_0 \qquad\qquad \text{(E13-3.4)}$$

---

[3] Value for k from J. M. Smith, *Chemical Engineering Kinetics*, 3rd ed. (New York: McGraw-Hill, 1981), p. 205. Note that ΔH_{Rx} and K_C were calculated from values given in *Perry's Chemical Engineers' Handbook*, 6th ed. (New York: McGraw-Hill, 1984), pp. 3–147.

Initially,

$$N_{\mathrm{W}i} = V_i C_{\mathrm{W}0} = (0.2 \text{ m}^3)(30.7 \text{ kmol/m}^3) = 6.14 \text{ kmol}$$

2. Rate Law:

$$-r_{\mathrm{A}} = k\left(C_{\mathrm{A}}C_{\mathrm{B}} - \frac{C_{\mathrm{C}}C_{\mathrm{D}}}{K_{\mathrm{C}}}\right) \qquad \text{(E13-3.5)}$$

3. Stoichiometry:

$$-r_{\mathrm{A}} = -r_{\mathrm{B}} = r_{\mathrm{C}} = r_{\mathrm{D}} \qquad \text{(E13-3.6)}$$

$$N_{\mathrm{A}} = C_{\mathrm{A}}V \qquad \text{(E13-3.7)}$$

$$V = V_0 + v_0 t \qquad \text{(E13-3.8)}$$

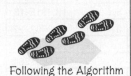

Following the Algorithm

4. Energy Balance:

The user-friendly energy balance is

$$\boxed{\frac{dT}{dt} = \frac{\dot{Q}_{\mathrm{gs}} - \dot{Q}_{\mathrm{rs}}}{\Sigma N_i C_{\mathrm{P}}}} \qquad \text{(E13-2.8)}$$

$$\boxed{\dot{Q}_{\mathrm{gs}} = (r_{\mathrm{A}}V)(\Delta H_{\mathrm{Rx}})} \qquad \text{(13-21)}$$

$$\dot{Q}_{\mathrm{rs}} = \sum F_{i0}C_{\mathrm{P}_i}(T - T_{i0}) + \dot{m}_{\mathrm{c}}C_{\mathrm{P}}[T - T_{a1}]\left[1 - exp\left[\frac{-UA}{\dot{m}_{\mathrm{c}}C_{\mathrm{P}_i}}\right]\right] \qquad \text{(13-22)}$$

Only species B (NaOH) and water flow into the reactor, so Equation (13-22) becomes

$$\boxed{Q_{\mathrm{rs}} = \overbrace{(F_{\mathrm{B}0}C_{\mathrm{P}_\mathrm{B}} + F_{\mathrm{w}}C_{\mathrm{Pw}})(T - T_0)}^{\dot{Q}_{\mathrm{rs1}}} + \overbrace{\dot{m}_{\mathrm{c}}C_{\mathrm{P}_\mathrm{C}}[T - T_{a1}]\left[1 - exp\left[\frac{-UA}{\dot{m}_{\mathrm{c}}C_{\mathrm{P}_\mathrm{C}}}\right]\right]}^{\dot{Q}_{\mathrm{rs2}}}} \qquad \text{(E13-3.9)}$$

$\dot{Q}_{\mathrm{rs1}}$ is the heat removed by mass flow, and $\dot{Q}_{\mathrm{rs2}}$ is the heat removed through heat exchange.

In Jofostan we have the added bonus that we can ask Prof. Dr. Sven Köttlov to calculate the temperature of the heat exchange fluid at its exit

$$T_{a2} = T - (T - T_{a1})exp\left[\frac{-UA}{\dot{m}_{\mathrm{c}}C_{\mathrm{P}_\mathrm{C}}}\right] \qquad \text{(12-17)}$$

5. Evaluation of Parameters:

$$F_{\mathrm{B}0} = v_0 C_{\mathrm{B}0} = \left(0.004\frac{\text{m}^3}{\text{s}}\right)\left(1\frac{\text{kmol}}{\text{m}^3}\right) = 0.004\frac{\text{kmol}}{\text{s}},$$

$$F_{\mathrm{w}} = v_0 C_{\mathrm{w}} = \left(0.004\frac{\text{m}^3}{\text{s}}\right)\left(55\frac{\text{kmol}}{\text{m}^3}\right) = 0.220\frac{\text{kmol}}{\text{s}}$$

$$C_{\mathrm{P}_\mathrm{A}} = 170{,}700 \text{ J/kmol/K}, \quad C_{\mathrm{P}} = 75{,}246 \text{ J/kmol/K}$$

We note the heat capacities for B, C, and water are essentially the same at C_{P}

$$NC_{\mathrm{P}} = C_{\mathrm{P}_\mathrm{A}}N_{\mathrm{A}} + C_{\mathrm{P}}(N_{\mathrm{B}} + N_{\mathrm{C}} + N_{\mathrm{D}} + N_{\mathrm{W}})$$

The Polymath program is given in Table E13-3.1. The solution results are shown in Figures E13-3.2 and E13-3.3.

TABLE E13-3.1   POLYMATH PROGRAM AND OUTPUT FOR SEMIBATCH REACTOR

POLYMATH Report
Ordinary Differential Equations

Differential equations

1 $d(Ca)/d(t) = ra-(v0*Ca)/V$

2 $d(Cb)/d(t) = rb+(v0*(Cb0-Cb)/V)$

3 $d(Cc)/d(t) = rc -(Cc*-v0)/N$

4 $d(T)/d(t) = (Qgs-Qrs)/NCp$

5 $d(Nw)/d(t) = v0*Cw0$

Explicit equations

1 $v0 = 0.004$

2 $Cb0 = 1$

3 $UA = 300$

4 $cp = 75240$

5 $T0 = 300$

6 $Cw0 = 55$

7 $dh = -7.9076e7$

8 $k = 0.39175*exp (5472.7*((1/273)-(1/T)))$

9 $Cd = Cc$

10 $Vl = 0.2$

11 $Kc = 10^{(3885.44/T)}$

12 $cpa = 170700$

13 $V = Vi+v0*t$

14 $ra = -k*((Ca*Cb)-((Cc*Cd)/Kc))$

15 $Na = V*Ca$

16 $Nb = V*Cb$

17 $Nb = V*Cc$

18 $rb = ra$

19 $rc = -ra$

20 $Nd = V*Cd$

21 $rate = -ra$

22 $NCp = cp*(Nb+Nc+Nd+Nw)+cpa*Na$

23 $Cpc = 18$

24 $Ta1 = 285$

25 $mc = 100$

26 $Qrs2 = -mc*Cpc*(Ta1-T)*(1-exp(-UA/mc/Cpc))$

27 $Ta2 = T-(T-Ta1)*exp(-UA/mc/Cpc)$

28 $Fw = Cw0*v0$

29 $Fb0 = Cb0*v0$

30 $Qrs1 = ((Fb0*cp)+(Fw*cp))*(T-T0)$

31 $Qgs = ra*V*dh$

32 $Qrs = Qrs1 + Qrs2$

Living Example Problem

Calculated values of DEQ variables

	Variable	Initial value	Final value
1	Ca	5.	3.981E-13
2	Cb	0	0.2682927
3	Cb0	1.	1.
4	Cc	0	0.6097561
5	Cd	0	0.6097561
6	cp	7.524E+04	7.524E+04
7	cpa	1.707E+05	1.707E+05
8	Cpc	18.	18.
9	Cw0	55.	55.
10	dh	-7.908E+07	-7.908E+07
11	Fb0	0.004	0.004
12	Fw	0.22	0.22
13	k	2.379893	4.211077
14	Kc	8.943E+12	3.518E+12
15	mc	100.	100.
16	Na	1.	6.529E-13
17	Nb	0	0.44
18	Nc	0	1.
19	NCp	6.327E+05	6.605E+06
20	Nd	0	1.
21	Nw	6.14	85.34
22	Qgs	0	6.19E-07
23	Qrs	2.19E+04	1.993E+05
24	Qrs1	0	1.633E+05
25	Qrs2	2.19E+04	3.604E+04
26	ra	0	-4.773E-15
27	rate	0	4.773E-15
28	rb	0	-4.773E-15
29	rc	0	4.773E-15
30	t	0	360.
31	T	300.	309.6878
32	T0	300.	300.
33	Ta1	285.	285.
34	Ta2	297.1669	305.0248
35	UA	3000.	3000.
36	V	0.2	1.64
37	v0	0.004	0.004
38	Vi	0.2	0.2

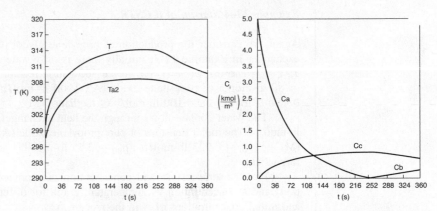

Figure E13-3.2   Temperature–time trajectories in a semibatch reactor.

Figure E13-3.3   Concentration–time trajectories in a semibatch reactor.

**Analysis:** From Figure E13-3.3 we see that the concentration of species B is virtually zero, owing to the fact that it is consumed virtually as fast as it enters the reactor up to a time of 252s. By the time we reach 252s, all species A has been consumed, and the reaction rate is virtually zero and no more of species C or D is produced and no more B is consumed. Because species B continues to enter the reactor at a volumetric flow rate v_0, after 252 seconds, the fluid volume continues to increase and the concentrations of C and D are diluted. The figure shows that before 252 s, $Q_g > Q_r$, and the reactor temperature and the coolant temperature increase. However, after 252 s, the reaction rate, and hence Q_g, are virtually zero so that $Q_r > Q_g$ and the temperature decreases. Because of the impractical short reaction time (252 seconds), a semibatch reactor would never be used for this reaction at this temperature; instead, we would most likely use a CSTR or PFR. See Problem P13-1$_B$ (c) to reflect on this example.

Wrong Reactor Choice!

13.4 Unsteady Operation of a CSTR

13.4.1 Startup

Startup of a CSTR

In reactor startup, it is often very important _how_ temperature and concentrations approach their steady-state values. For example, a significant overshoot in temperature may cause a reactant or product to degrade, or the overshoot may be unacceptable for safe operation, such as a secondary reaction runaway. If either case were to occur, we would say that the system exceeded its _practical stability limit_. The practical limit is specific to the specific reaction and conditions under which the reaction is carried out, and is usually determined by the reaction safety engineer. Although we can solve the unsteady temperature–time and concentration–time equations numerically to see if such a limit is exceeded, it is often more insightful to study the approach to steady state by using the _temperature–concentration phase plane_. To illustrate these concepts, we shall confine our analysis to a liquid-phase reaction carried out in a CSTR.

A qualitative discussion of how a CSTR approaches steady state is given in _PRS_ R13.5 on the CRE Web site. This analysis, summarized in Figure PRS13.5 in the Summary is developed to show the four different regions into which the phase plane is divided and how they allow one to sketch the approach to the steady state.

Reference Shelf

Example 13–4 Startup of a CSTR

Again, we consider the production of propylene glycol (C) in a CSTR with a heat exchanger in Example 12-3. Initially there is only water, $C_{wi} = 3.45$ lb-mol/ft^3, at $T_i = 75°F$ and 0.1 wt % H_2SO_4 in the 500-gallon reactor. The feed stream consists of 80 lb-mol/h of propylene oxide (A), 1000 lb-mol/h of water (B) containing 0.1 wt % H_2SO_4, and 100 lb-mol/h of methanol (M).

The water coolant flows through the heat exchanger at a rate of 5 lb$_m$/s (1000 lb-mol/h). The molar densities of pure propylene oxide (A), water (B), and methanol (M) are $\rho_{A0} = 0.923$ lb-mol/ft^3, $\rho_{B0} = 3.45$ lb-mol/ft^3, and $\rho_{M0} = 1.54$ lb-mol/ft^3, respectively.

Plot the temperature and concentration of propylene oxide as a function of time, and a concentration vs. temperature graph for different entering temperatures and initial concentrations of A in the reactor.

Additional information:

$$UA = 16,000 \frac{\text{Btu}}{\text{h} \cdot °F} \text{ with } T_{a1} = 60°F, \dot{m}_W = 1000 \text{ lb-mol/h with } C_{P_W} = 18 \text{ Btu/lb-mol} \cdot °F$$

$$C_{P_A} = 35 \text{ Btu/lb-mol} \cdot °F, \quad C_{P_B} = 18 \text{ Btu/lb-mol} \cdot °F,$$

$$C_{P_C} = 46 \text{ Btu/lb-mol} \cdot °F, \quad C_{P_M} = 19.5 \text{ Btu/lb-mol} \cdot °F$$

Again, the temperature of the mixed reactant streams entering the CSTR is $T_0 = 75°F$.

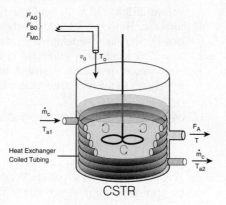

CSTR

Solution

$$A + B \longrightarrow C$$

Mole Balances:

Initial Conditions

A: $\quad \dfrac{dC_A}{dt} = r_A + \dfrac{(C_{A0} - C_A)v_0}{V} \qquad C_{Ai} = 0 \qquad$ (E13-4.1)

B: $\quad \dfrac{dC_B}{dt} = r_B + \dfrac{(C_{B0} - C_B)v_0}{V} \qquad C_{Bi} = 3.45 \dfrac{\text{lb-mol}}{\text{ft}^3} \qquad$ (E13-4.2)

C: $\quad \dfrac{dC_C}{dt} = r_C + \dfrac{-C_C v_0}{V} \qquad C_{Ci} = 0 \qquad$ (E13-4.3)

Following the Algorithm

M: $\qquad \dfrac{dC_M}{dt} = \dfrac{v_0(C_{M0} - C_M)}{V} \qquad\qquad C_{Mi} = 0$ (E13-4.4)

Rate Law: $\qquad\qquad\qquad -r_A = kC_A$ (E13-4.5)

Stoichiometry: $\qquad\qquad -r_A = -r_B = r_C$ (E13-4.6)

Energy Balance:

$$\dfrac{dT}{dt} = \dfrac{\dot{Q}_{gs} - \dot{Q}_{rs}}{NC_P}$$ (E13-4.7)

where

$$\dot{Q}_{gs} = (r_A V)(\Delta H_{Rx})$$ (E13-4.8)

using

$$\sum F_{i0} C_{P_i} = F_{A0} \sum \Theta_i C_{P_i}$$ (E13-4.9)

the "heat removed" term from the unsteady startup of a CSTR is similar to that for the "heat removed" term from a semi-batch reactor. Recalling Equations (12-12) and (12-19)

$$\dot{Q}_{rs} = \overbrace{F_{A0} \sum \Theta_i C_{P_i} (T - T_0)}^{Q_{rs1}} + \overbrace{\dot{m}_c C_{P_C} (T - T_{a1}) \left[1 - \exp\left[\dfrac{-UA}{\dot{m}_c C_{P_C}} \right] \right]}^{Q_{rs2}}$$ (E13-4.10)

$$\dot{Q}_{rs} = \dot{Q}_{rs1} + \dot{Q}_{rs2}$$

$$T_{a2} = T - (T - T_{a1}) \exp\left(-\dfrac{UA}{\dot{m}_c C_{P_W}} \right)$$ (12-17)

Evaluation of Parameters:

$$NC_P = \sum N_i C_{P_i} = C_{P_A} N_A + C_{P_B} N_B + C_{P_C} N_C + C_{P_M} N_M$$ (E13-4.10)

$$= 35(C_A V) + 18(C_B V) + 46(C_C V) + 19.5(C_M V)$$

$$\sum \Theta_i C_{P_i} = C_{P_A} + \dfrac{F_{B0}}{F_{A0}} C_{P_B} + \dfrac{F_{M0}}{F_{A0}} C_{P_M}$$ (E13-4.11)

$$= 35 + 18 \dfrac{F_{B0}}{F_{A0}} + 19.5 \dfrac{F_{M0}}{F_{A0}}$$

$$v_0 = \dfrac{F_{A0}}{\rho_{A0}} + \dfrac{F_{B0}}{\rho_{B0}} + \dfrac{F_{M0}}{\rho_{M0}} = \left(\dfrac{F_{A0}}{0.923} + \dfrac{F_{B0}}{3.45} + \dfrac{F_{M0}}{1.54} \right) \dfrac{ft^3}{h}$$ (E13-4.12)

Neglecting ΔC_P because it changes the heat of reaction insignificantly over the temperature range of the reaction, the heat of reaction is assumed constant at its reference temperature

$$\Delta H_{Rx} = -36,000 \dfrac{Btu}{lb\text{-}mol\ A}$$

The Polymath program is shown in Table E13-4.1.

TABLE E13-4.1   POLYMATH PROGRAM FOR CSTR STARTUP

POLYMATH Report
Ordinary Differential Equations

Differential equations

1. $d(Ca)/d(t) = 1/tau*(Ca0-Ca)+ra$
2. $d(Cb)/d(t) = 1/tau*(Cb0-Cb)+rb$
3. $d(Cc)/d(t) = 1/tau* (-Cc) + rc$
4. $d(Cm)/d(t) = 1/tau* (Cm0-Cm)$
5. $d(T)/d(t) = (Qg-Qr)/Ncp$

Explicit equations

1. $Fa0 = 80$
2. $T0 = 75$
3. $V = (1/7.484)*500$
4. $UA = 16000$
5. $dh = -36000$
6. $Ta1 = 60$
7. $k = 16.96e12*exp(-32400/1.987/(T+64+460))$
8. $Na = V*Ca$
9. $Nb = V*Cb$
10. $Nc = V*Cc$
11. $ra = -k*Ca$
12. $rb = ra$
13. $rc = -ra$
14. $Nm = V*Cm$
15. $NCp = Na*35+Nb*18+Nc*46+Nm*19.5$
16. $mc = 1000$
17. $Ta2 = T-(T-Ta1)*exp(-UA/mc/18)$
18. $Qr2 = mc*18*(Ta2-Ta1)$
19. $Fm0 = 100$
20. $Fb0 = 1000$
21. $v0 = Fa0/0.923+Fb0/3.45+Fm0/1.54$
22. $tau = V/v0$
23. $Ca0 = Fb0/v0$
24. $Cb0 = Fb0/v0$
25. $Cm0 = Fm0/v0$
26. $ThetaCp = 35+Fb0/Fa0*18+Fm0/Fa0*19.5$
27. $Qr1 = Fa0*(ThetaCp)*(T-T0)$
28. $Qg = ra*V*dh$
29. $Qr = Qr1 +Qr2$

Calculated values of DEQ variables

	Variable	Initial value	Final value
1	Ca	0	0.0378953
2	Ca0	0.1812152	0.1812152
3	Cb	3.45	2.12187
4	Cb0	2.26519	2.26519
5	Cc	0	0.1433199
6	Cm	0	0.226519
7	Cm0	0.226519	0.226519
8	dh	-3.6E+04	-3.6E+04
9	Fa0	80.	80.
10	Fb0	1000.	1000.
11	Fm0	100.	100.
12	k	0.9835319	24.99079
13	mc	1000.	1000.
14	Na	0	2.531756
15	Nb	230.4917	141.7604
16	Nc	0	9.575086
17	NCp	4148.851	3375.858
18	Nm	0	15.13355
19	Qg	0	2.278E+06
20	Qr	1.59E+05	2.278E+06
21	Qr1	0	1.445E+06
22	Qr2	1.59E+05	8.324E+05
23	ra	0	-0.9470341
24	rb	0	-0.9470341
25	rc	0	0.9470341
26	T	75.	138.5305
27	t	0	4.
28	T0	75.	75.
29	Ta1	60.	60.
30	Ta2	68.83332	106.2456
31	tau	0.1513355	0.1513355
32	ThetaCp	284.375	284.375
33	UA	1.6E+04	1.6E+04
34	V	66.80919	66.80919
35	v0	441.464	441.464

Living Example Problem

Figures E13-4.1 and E13-4.2 show the concentration of propylene oxide and reactor temperature as a function of time, respectively, for an initial temperature of $T_i = 75°F$ and only water in the tank (i.e., $C_{Ai} = 0$). One observes that both the temperature and concentration oscillate around their steady-state values ($T = 138°F$, $C_A = 0.039$ lb-mol/ft^3) as steady state is approached.

Figure E13-4.3 combines Figures E13-4.1 and E13-4.2 into a phase plot of C_A versus T. The final operating concentration of A is 0.0379 lb-mol/ft^3 at a temperature of 138°F. The arrows on the phase-plane plots show the trajectories with increasing time. The maximum temperature reached during startup is 152°F, which is below the *practical stability limit* of 180°F.

Unacceptable
startup

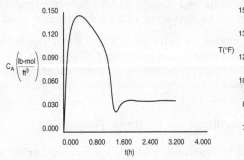

Figure E13-4.1 Propylene oxide concentration as a function of time for $C_{Ai} = 0$ and $T_i = 75°F$.

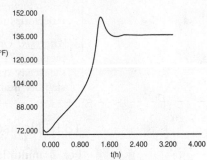

Figure E13-4.2 Temperature–time trajectory for CSTR startup for $C_{Ai} = 0$ and $T_i = 75°F$.

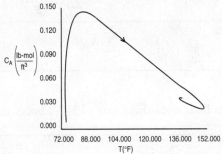

Figure E13-4.3 Concentration–temperature phase-plane trajectory using Figures E13-4.1 and E13-4.2.

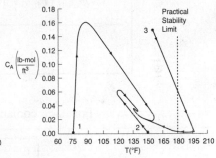

Figure E13-4.4 Concentration–temperature phase-plane for three different initial conditions.

Next, consider Figure E13-4.4, which shows three different trajectories for three different sets of initial conditions:

(1) $T_i = 75°F$ $C_{Ai} = 0$ (same as Figure E13-4.3)
(2) $T_i = 150°F$ $C_{Ai} = 0$
(3) $T_i = 160°F$ $C_{Ai} = 0.141$ lb-mol/ft^3

After three hours, the reaction is operating at steady state and all three trajectories converge on the final steady-state temperature of 138°F and the corresponding concentrations

$$C_A = 0.0379 \text{ lb-mol/ft}^3 \quad C_C = 0.143 \text{ lb-mol/ft}^3$$

$$C_B = 2.12 \text{ lb-mol/ft}^3 \quad C_M = 0.2265 \text{ lb-mol/ft}^3$$

$$T = 138.5°F$$

Oops! The *practical stability limit* was exceeded.

For this reaction system, the plant safety office believes that an upper temperature limit of 180°F should not be exceeded in the tank. This temperature is the *practical stability limit*. The practical stability limit represents a temperature above which it is undesirable to operate because of unwanted side reactions, safety considerations, secondary runaway reactions, or damage to equipment. Consequently, we see that if we started at an initial temperature of $T_i = 160°F$ and an initial concentration of 0.143 mol/dm^3, the practical stability limit of 180°F would be exceeded as the reactor approached its steady-state temperature of 138°F. See the concentration–temperature trajectory in Figure E13-4.4.

Figures E13-4.1 through E13-4.4 show the concentration and temperature time trajectories for the start up of a CSTR for different initial conditions.

Analysis: One of the purposes of this example was to demonstrate the use of phase plots, e.g., T versus C_A, in analyzing CSTR startup. Phase plots allow us to see how the steady state is approached for different sets of initial conditions and if the practical stability limit is exceeded causing a secondary, more exothermic reaction to set in.

13.5 Nonisothermal Multiple Reactions

For q multiple reactions with m species occurring in either a semibatch or batch reactor, Equation (13-15) can be generalized in the same manner as the steady-state energy balance to give

$$\frac{dT}{dt} = \frac{\dot{m}_c C_{P_c}(T_{a1} - T)[1 - \exp(-UA/\dot{m}_c C_{P_c})] + \sum_{i=1}^{q} r_{ij}\, V\, \Delta H_{\mathrm{Rx}ij}(T) - \sum_{j=1}^{m} F_{j0} C_{Pj}(T - T_0)}{\sum_{j=1}^{m} N_{j0} C_{Pj}}$$

(13-24)

For large coolant flow rates, Equation (13-17) becomes

$$\frac{dT}{dt} = \frac{V \sum_{i=1}^{q} r_{ij}\, \Delta H_{\mathrm{Rx}ij} - UA(T - T_a) - \sum_{j=1}^{m} F_{j0} C_{P_j}(T - T_0)}{\sum_{j=1}^{m} N_j C_{P_j}}$$

(13-25)

Rearranging Equation (13-18) and letting

$$Q_g = V \sum_{i=1}^{q} r_{ij}\, \Delta H_{\mathrm{Rx}ij}$$

(13-26)

and

$$Q_r = UA(T - T_a) + \Sigma F_{j0} C_{P_j}(T - T_0)$$

(13-27)

we can write Equation (13-18) in the form

$$\frac{dT}{dt} = \frac{Q_g - Q_r}{\sum_{j=1}^{m} N_j C_{P_j}}$$

(13-28)

Example 13–5 Multiple Reactions in a Semibatch Reactor

The series reactions

$$2A \xrightarrow[\text{(1)}]{k_{1A}} B \xrightarrow[\text{(2)}]{k_{2B}} 3C$$

Living Example Problem

are catalyzed by H_2SO_4. All reactions are first order in the reactant concentration. However, Reaction *(1) is exothermic* and Reaction *(2) is endothermic.* The reaction is to be carried out in a semibatch reactor that has a heat exchanger inside with $UA = 35{,}000$ cal/h·K and a constant exchanger temperature, T_a, of 298 K. Pure A enters at a concentration of 4 mol/dm³, a volumetric flow rate of 240 dm³/h, and a temperature of 305 K. Initially, there is a total of 100 dm³ of liquid in the reactor, which contains 1.0 mol/dm³ of A and 1.0 mol/dm³ of the catalyst H_2SO_4. The reaction rate is independent of the catalyst concentration. The initial temperature inside the reactor is 290 K.

   Plot and analyze the species concentrations and reactor temperature as a function of time.

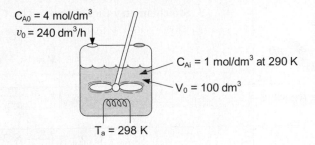

$C_{A0} = 4$ mol/dm³
$v_0 = 240$ dm³/h

$C_{Ai} = 1$ mol/dm³ at 290 K

$V_0 = 100$ dm³

$T_a = 298$ K

Additional information:

$k_{1A} = 1.25\,h^{-1}$ at 320 K with $E_{1A} = 9500$ cal/mol $C_{P_A} = 30$ cal/mol·K

$k_{2B} = 0.08\,h^{-1}$ at 300 K with $E_{2B} = 7000$ cal/mol $C_{P_B} = 60$ cal/mol·K

$\Delta H_{Rx1A} = -6500$ cal/mol A $C_{P_C} = 20$ cal/mol·K

$\Delta H_{Rx2B} = +8000$ cal/mol B $C_{P_{H_2SO_4}} = 35$ cal/mol·K

Solution

Reaction (1) $2A \longrightarrow B$

Reaction (2) $B \longrightarrow 3C$

1. Mole Balances:

$$\frac{dC_A}{dt} = r_A + \frac{(C_{A0} - C_A)}{V} v_0 \qquad \text{(E13-5.1)}$$

$$\frac{dC_B}{dt} = r_B - \frac{C_B}{V} v_0 \qquad \text{(E13-5.2)}$$

$$\frac{dC_C}{dt} = r_C - \frac{C_C}{V} v_0 \qquad \text{(E13-5.3)}$$

2. Rates:

    (a) Rate Laws:

$$-r_{1A} = k_{1A}C_A \tag{E13-5.4}$$

$$-r_{2B} = k_{2B}C_B \tag{E13-5.5}$$

    (b) Relative Rates:

$$r_{1B} = -\frac{1}{2}r_{1A} \tag{E13-5.6}$$

$$r_{2C} = -3\,r_{2B} \tag{E13-5.7}$$

    (c) Net Rates:

$$r_A = r_{1A} = -k_{1A}C_A \tag{E13-5.8}$$

$$r_B = r_{1B} + r_{2B} = \frac{-r_{1A}}{2} + r_{2B} = \frac{k_{1A}C_A}{2} - k_{2B}C_B \tag{E13-5.9}$$

$$r_C = 3\,k_{2B}C_B \tag{E13-5.10}$$

3. Stoichiometry (liquid phase): Use C_A, C_B, C_C

$$N_i = C_i V \tag{E13-5.11}$$

$$V = V_0 + v_0 t \tag{E13-5.12}$$

$$N_{H_2SO_4} = (C_{H_2SO_{4,0}})V_0 = \frac{1\ \text{mol}}{\text{dm}^3} \times 100\ \text{dm}^3 = 100\ \text{mol}$$

$$F_{A0} = C_{A0}v_0 = \frac{4\ \text{mol}}{\text{dm}^3} \times 240\ \frac{\text{dm}^3}{\text{h}} = 960\ \frac{\text{mol}}{\text{h}}$$

Following the Algorithm

4. Energy Balance:
Semibatch reactor:

A slight rearrangement of (13-23) is

$$\frac{dT}{dt} = \frac{UA\,(T_a - T) - \sum F_{j0}\,C_{P_j}(T - T_0) + V\sum_{i=1}^{q}\Delta H_{\text{Rx}ij}\,r_{ij}}{\sum_{j=1}^{m} N_j C_{P_j}} \tag{E13-5.13}$$

Expanding

$$\frac{dT}{dt} = \frac{UA\,(T_a - T) - F_{A0}C_{P_A}(T - T_0) + [(\Delta H_{\text{Rx1A}})(r_{1A}) + (\Delta H_{\text{Rx2B}})(r_{2B})]\,V}{[C_A C_{P_A} + C_B C_{P_B} + C_C C_{P_C}]\,V + N_{H_2SO_4}C_{P_{H_2SO_4}}}$$

$$\tag{E13-5.14}$$

Substituting the parameter values into Equation (E13-5.14)

$$\boxed{\frac{dT}{dt} = \frac{35{,}000\,(298 - T) - (4)(240)(30)(T - 305) + [(-6500)(-k_{1A}C_A) + (+8000)(-k_{2B}C_B)]\,V}{(30C_A + 60C_B + 20C_C)(100 + 240t) + (100)(35)}}$$

$$\tag{E13-5.15}$$

Equations (E13-5.1) through (E13-5.3) and (E13-5.8) through (E13-5.12) can be solved simultaneously with Equation (E13-5.14) using an ODE solver. The Polymath program is shown in Table E13-5.1. The concentration–time and temperature–time trajectories are shown in Figures E13-5.1 and E13-5.2.

TABLE E13-5.1   POLYMATH PROGRAM FOR STARTUP OF A CSTR

Differential equations

1 $d(Ca)/d(t) = ra+(Cao-Ca)*vo/V$

2 $d(Cb)/d(t) = rb-Cb*vo/V$

3 $d(Cc)/d(t) = rc-Cc*vo/V$

4 $d(T)/d(t) = (35000*(298-T)-Cao*vo*30*(T-305)+((-6500)*(-k1a*Ca)+(8000)*(-k2b*Cb))*V)/ ((Ca*30+Cb*60+Cc*20)*V+100*35)$

Living Example Problem

Explicit equations

1 $Cao = 4$

2 $vo = 240$

3 $k1a = 1.25*exp((9500/1.987)*(1/320-1/T))$

4 $k2b = 0.08*exp((7000/1.987)*(1/290-1/T))$

5 $ra = -k1a*Ca$

6 $V = 100+vo*t$

7 $rc = 3*k2b*Cb$

8 $rb = k1a*Ca/2-k2b*Cb$

Calculated values of DEQ variables

	Variable	Initial value	Final value
1	Ca	1.	0.2636761
2	Cao	4.	4.
3	Cb	0	0.6875689
4	Cc	0	2.563518
5	k1a	0.2664781	7.458802
6	k2b	0.08	0.9317643
7	ra	-0.2664781	-1.966708
8	rb	0.133239	0.3427018
9	rc	0	1.921956
10	t	0	1.5
11	T	290.	363.4525
12	V	100.	460.
13	vo	240.	240.

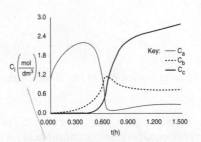

Figure E13-5.1 Concentration–time.

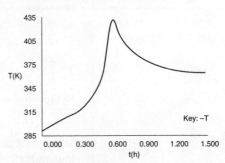

Figure E13-5.2 Temperature (K)–time (h).

Analysis: At the start of the reaction, both C_A and T in the reactor increase because C_{A0} and T_0 are greater than C_{Ai} and T_i. This increase continues until the rate of consumption of the reacting species is greater than the feed rate to the reactor. We note that at about 0.2 h, the reactor temperature exceeds the feed temperature (i.e., 305 K) as a result of the heat generated by the exothermic reaction (Reaction 1). The temperature continues to rise until about 0.6 h, at which point reactant A is virtually all consumed. After this point the temperature begins to drop for two reasons: (1) the reactor is cooled by the heat exchanger and (2) heat is absorbed by the endothermic reaction rate (Reaction 2). The question is, does the maximum temperature (435 K) exceed a temperature that is too high, resulting in a high vapor pressure that then results in evaporation losses or causes a highly exothermic secondary reaction to set in?

Example 13–6 T2 Laboratories Explosion[4]

Figure E13-6.1   Aerial photograph of T2 taken December 20, 2007. (Courtesy of Chemical Safety Board.)

T2 Laboratories manufactured a fuel additive, methylcyclopentadienyl manganese tricarbonyl (MCMT), in a 2,450-gallon, high-pressure batch reactor utilizing a three-step batch process.

Step 1a. The liquid-phase metalation reaction between methylcyclopentadiene (MCP) and sodium in a solvent of diethylene glycol dimethyl ether (diglyme) to produce sodium methylcyclopentadiene and hydrogen gas is

$$\text{(A)}$$

Hydrogen immediately comes out of the solution and is vented at the top in the gas head space.

Step 1b. At the end of Step 1a, $MnCl_2$ is added. The substitution reaction between sodium methylcyclopentadiene and manganese chloride produced manganese dimethylcyclopentadiene and sodium chloride is

$$\text{(B)}$$

---

[4] This example was coauthored by Professors Ronald J. Willey, Northeastern University, Michael B. Cutlip, University of Connecticut, and H. Scott Fogler, University of Michigan, and published in *Process Safety Progress*, 30, 1 (2011).

Step 1c. At the end of Step 1b, CO is added. The carbonylation reaction between manganese dimethylcyclopentadiene and carbon monoxide produces the final product, methylcyclopentadienyl manganese tricarbonyl (MCMT):

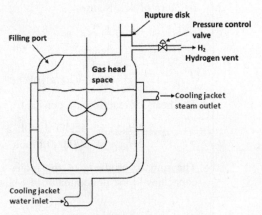

(C)

We will only consider *Step 1a* as this step is the one in which the explosion occurred.

Procedure

First, solid sodium is mixed in the batch reactor with methylcyclopentadiene dimer and a solvent diethylene glycol dimethyl ether (diglyme). The batch reactor is then heated to about 422 K (300°F) with only slight reaction occurring during this heating process. On reaching 422 K, the heating is turned off, as the exothermic reaction is now proceeding, and the temperature continues to increase without further heating. When the temperature reaches 455.4 K (360°F), the operator initiates cooling using the evaporation of boiling water in the reactor jacket as the heat sink ($T_a = 373.15$ K) (212°F).

What Happened

On December 19, 2007, when the reactor reached a temperature of 455.4 K (360°F), the process operator could not initiate the flow of cooling water to the cooling jacket shown in Figure E13-6.2. Thus, the expected cooling of the reactor was not available and the temperature in the reactor continued to rise. The pressure also increased as hydrogen continued to be produced at an increased rate, to the point that the reactor's pressure control valve system on the 1-inch diameter hydrogen venting stream could no longer maintain the operating pressure at 50 psig (4.4 atm). As the temperature continued to increase further, a previously unknown exothermic reaction of the diglyme solvent that was catalyzed by sodium accelerated rapidly.

$$CH_3 - O - CH_2 - CH_2 - O - CH_2 - CH_2O - CH_3 \xrightarrow{Na} 3H_2 + misc(l) \& (s) \text{ (D)}$$

Figure E13-6.2   Reactor.

This reaction produced even more hydrogen, causing the pressure to rise even faster, eventually causing the rupture disk which was set at 28.2 atm absolute (400 psig), to break, in the 4-inch diameter relief line of H_2. Even with the relief line open, the rate of production of H_2 was now far greater than the rate of venting, causing the pressure to continue to increase to the point that it ruptured the reactor vessel initiating a horrific explosion. The T2 plant was completely leveled and four personnel lives were lost. Surrounding businesses were heavily damaged and additional injuries were sustained.

Before continuing with this example, it might be helpful to view the 9-minute Chemical Safety Board (CSB) video, which you can access directly from the CRE Web site (under YouTube videos), or you can read the supporting reports (*http://www.csb.gov/videos/runaway-explosion-at-t2-laboratories/*). You can also search the Web for "T2 explosion video."

Simplified Model
Summarizing the important reactions for Step 1

(Reaction 1)

$$CH_3-O-CH_2-CH_2-O-CH_2-CH_2O-CH_3 \xrightarrow{\quad Na \quad} 3H_2 + misc(1) \& (s)$$

(Reaction 2)

Let A = methycylcopentadiene, B = sodium, S = solvent (diglyme), and D = H_2. This runaway reaction can be approximately modeled with two reactions. These reactions are

(1) A + B → C + 1/2 D (gas)   (Reaction 1)

(2) S → 3 D (gas) + miscellaneous liquid and solid products   (Reaction 2)

In Reaction (1), A and B react to form products. Reaction (2) represents the decomposition of the liquid-phase solvent S catalyzed by the presence of B, but this reaction only begins to proceed once a temperature of approximately 470 K is reached.

The rate laws, along with the specific reaction-rate constants at the initial temperature of 422 K, are

$$-r_{1A} = k_{1A}C_A C_B$$
$$A_{1A} = 5.73 \times 10^2 \text{ dm}^3 \text{ mol}^{-1} \text{ hr}^{-1} \text{ with } E_{1A} = 128{,}000 \text{ J/mol K}$$
$$-r_{2S} = k_{2S}C_S$$
$$A_{2S} = 9.41 \times 10^{16} \text{ hr}^{-1} \text{ with } E_{2S} = 800{,}000 \text{ J/mol K}$$

The heats of reaction are constant.

$$\Delta H_{Rx1A} = -45{,}400 \text{ J/mol}$$
$$\Delta H_{Rx2S} = -320{,}000 \text{ J/mol}$$

The sum of products of the moles of each species and their corresponding heat capacities in the denominator of Equation (13-12) is essentially constant at

$$\sum N_j C_{P_j} = 1.26 \times 10^7 \text{ J/K}$$

Assumptions
Assume that the liquid volume, V_0, in the reactor remains constant at 4,000 dm^3 and that the vapor space, V_H, above the reactor occupies 5,000 dm^3. Any gas, H_2 (i.e., *D*), that is formed by Reactions (1) and (2) immediately appears as an input stream F_D to the head-space volume. The dissolved H_2 and the vapor pressures for the liquid

components in the reactor can be neglected. The initial absolute pressure within the reactor is 4.4 atm (50 psig). During normal operation, the H_2 generated obeys the ideal gas law. The pressure control system on the H_2 vent stream maintains the pressure, P, at 4.40 atm up to a flow of 11,400 mol/hr. The reactor vessel will fail when the pressure exceeds 45 atm or the temperature exceeds 600 K.

Additional information:
$UA = 2.77 \times 10^6$ J hr^{-1} K^{-1}. The concentrations in the reactor at the end of the reactor heating at 422 K are $C_{A0} = 4.3$ mol/dm^3, $C_{B0} = 5.1$ mol/dm^3, $C_{I0} = 0.088$ mol/dm^3, and $C_{S0} = 3$ mol/dm^3. The sensible heat of the two gas venting streams may be neglected.

Problem Statement
Plot and analyze the reactor temperature and head-space pressure as a function of time along with the reactant concentrations for the scenario where the reactor cooling fails to work ($UA = 0$). In Problem P13-1(**f**) you will be asked to redo the problem when the cooling water comes as expected whenever the reactor temperature exceeds 455 K.

Solution

(1) Reactor Mole Balances:

    <u>Reactor</u> (Assume Constant Volume Batch)

      Liquid

$$\frac{dC_A}{dt} = r_{1A} \qquad\qquad (E13\text{-}6.1)$$

$$\frac{dC_B}{dt} = r_{1A} \qquad\qquad (E13\text{-}6.2)$$

$$\frac{dC_S}{dt} = r_{2S} \qquad\qquad (E13\text{-}6.3)$$

(2) Head-Space Mole Balance:

Let N_D = moles of gas D in the reactor vapor space V_V. The flows in and out of the head space are shown in the figure below.

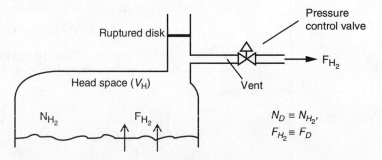

A balance on species D (H_2) in the head-space volume V_H yields

$$\frac{dN_D}{dt} = F_D - F_{vent} \qquad\qquad (E13\text{-}6.4)$$

where F_{vent} is the molar flow rate of gas out of the head space through one or both outlet lines, and F_D (i.e., F_{H_2}) is the molar rate of gas leaving the liquid and entering the head space and is equal to the hydrogen generated in the liquid (see relative rates and net rates in the algorithm where V_0 is the liquid reactor volume).

$$F_D = (-0.5r_{1A} - 3r_{2S})V_0 \tag{E13-6.5}$$

The assumptions of a perfect gas in the head-space volume and modest changes in T allow Equation (E13-6.4) to be written in terms of total gas pressure in the reactor head space

$$N_D = \frac{PV_H}{RT_H} \tag{E13-6.6}$$

Substituting for N_D in Equation (E13-6.3) and rearranging

$$\boxed{\frac{dP}{dt} = \left(F_D - F_{vent}\right)\frac{RT_H}{V_H}} \tag{E13-6.7}$$

(3) How the Venting Works:

Gas exits the reactor through the pressure control valve line. At low gas production, the pressure control valve maintains set-point pressure at the initial pressure by venting all produced gas until the rate of gas production reaches 11,400 mol/hr.

$$F_{vent} = F_D \quad \text{when} \quad F_D < 11,400 \tag{E13-6.8}$$

We now need to know a little more about the venting system for H_2 and the condition for flow out the vent. As the pressure increases, but is still below the rupture disk setting, the pressure control line vents to the atmosphere (1 atm) according to the equation

$$F_{vent} = \Delta P C_v = (P-1)C_{v1} \quad \text{when} \quad P < 28.2 \text{ atm} \tag{E13-6.9}$$

where P is the absolute pressure in the reactor (atm), 1 atm is the downstream pressure, and the pressure correlation constant C_v is 3,360 mol/hr • atm. If the pressure P within the reactor exceeds 28.2 atm (400 psig), the relief line activated by the rupture disk breaks and vents gas in the reactor at the rate given $F_{vent} = (P-1)C_{v2}$ where $C_{v2} = 53,600$ mol/atm • hr.

After the rupture disk blowout at $P = 28.2$ atm, both the pressure control line and the rupture disk lines vent the reactor according to the equation

$$F_{vent} = (P-1)(C_{v1} + C_{v2}) \tag{E13-6.10}$$

Equations (E13-6.7) through (E13-6.10) can then be used to describe the F_{vent} flow rate with time for the appropriate logic for the values of F_D and P.

(4) Rates:

Laws:

$$(1) \quad -r_{1A} = k_{1A}C_A C_B \tag{E13-6.11}$$

$$k_{1A} = A_{1A}e^{-E_{1A}/RT} \tag{E13-6.12}$$

$$(2) \quad -r_{2S} = k_{2S}C_S \tag{E13-6.13}$$

$$k_{2S} = A_{2S}e^{-E_{2S}/RT} \tag{E13-6.14}$$

Relative Rates:

$$(1)\quad \frac{r_{1A}}{-1} = \frac{r_{1B}}{-1} = \frac{r_{1C}}{1} = \frac{r_{1D}}{1/2} \tag{E13-6.15}$$

$$(2)\quad \frac{r_{2S}}{-1} = \frac{r_{2D}}{3} \tag{E13-6.16}$$

Net Rates:

$$r_A = r_B = r_{1A} \tag{E13-6.17}$$

$$r_S = r_{2S} \tag{E13-6.18}$$

$$r_D = -\frac{1}{2}r_{1A} + -3r_{2S} \ \text{(gas generated)} \tag{E13-6.19}$$

(5) Stoichiometry:

Neglect reactor-liquid volume change form loss of product gases.

$$C_A = \frac{N_A}{V_0} \quad \text{(E13-6.20)} \qquad\qquad C_B = \frac{N_B}{V_0} \quad \text{(E13-6.21)}$$

$$C_S = \frac{N_S}{V_0} \quad \text{(E13-6.22)} \qquad\qquad C_D = \frac{P}{RT} \quad \text{(E13-6.23)}$$

(6) Energy Balance:

Applying Equation (E13-23) to a batch system ($F_{i0} = 0$)

$$\frac{dT}{dt} = \frac{V_0\left[r_{1A}\Delta H_{Rx1A} + r_{2S}\Delta H_{Rx2S}\right] - UA(T - T_a)}{\sum N_j C_{P_j}} \tag{E13-6.24}$$

Substituting for the rate laws and $\sum N_j C_{P_j}$

$$\frac{dT}{dt} = \frac{V_0\left[-k_{1A}C_A C_B \Delta H_{Rx1A} - k_{2S}C_S \Delta H_{Rx2S}\right] - UA(T - T_a)}{1.26 \times 10^7 \, (J/K)} \tag{E13-6.25}$$

(7) Numerical Solutions: *"Tricks of the Trade"*

A rapid change of temperature and pressure is expected as **Reaction (2)** starts to run away. This typically results in a stiff system of ordinary differential equations, which can become numerically unstable and generate incorrect results. This instability can be prevented by using a software switch (SW) that will set all derivates to zero when the reactor reaches the explosion temperature or pressure. This switch can have the form of Equation (E13-6.26) in Polymath and can be multiplied by the right-hand side of all the differential equations in this problem. This operation will halt (or freeze) the dynamics when the temperature T becomes higher than 600 K or the pressure exceeds 45 atm.

$$SW1 = \text{if } (T>600 \text{ or } P>45) \text{ then } (0) \text{ else } (1) \tag{E13-6.26}$$

We now will solve the essential equations from (E13-6.1) through (E13-6.26) for the scenario where there is no cooling and thus $UA = 0$. Also the switch SW1 must be implemented in all the differential equations as discussed above.

TABLE E13-6.1   POLYMATH PROGRAM

Differential equations

1 d(CA)/d(t) = SW1*r1A

   change in concentration of methylcyclopentadiene (mol/dm3/hr)

2 d(CB)/d(t) = SW1*r1A

   change in concentration of sodium (mol/dm3/hr)

3 d(CS)/d(t) = SW1*r2S

   change in concentration of diglyme (mol/dm3/hr)

4 d(P)/d(t) = SW1*((FD-Fvent)*0.082*T/VH)

5 d(T)/d(t) = SW1*((V0*(r1A*DHRx1A+r2S*DHRx2S)
   -SW1*UA*(T-373.15))/SumNCp)

Explicit equations

1 V0 = 4000

   dm3

2 VH = 5000

   dm3

3 DHRx1A = -45400

   J/mol Na

4 DHRx2S = -3.2E5

   J/mol of Diglyme

5 SumNCp = 1.26E7

   J/K

6 A1A = 4E14

   per hour

7 E1A = 128000

   J/kmol/K

8 k1A = A1A*exp(-E1A/(8.31*T))

   rate constant reaction 1

9 A2S = 1E84

   per hour

10 E2S = 800000

    J/kmol/K

11 k2S = A2S*exp(-E2S/(8.31*T))

    rate constant reaction 2

12 SW1 = if (T>600 or P>45) then (0) else (1)

13 r1A = -k1A*CA*CB

    mol/dm3/hour (first order in sodium and cyclomethylpentadiene)

14 r2S = -k2S*CS

    mol/dm3/hour (first order in diglyme)

15 FD = (-0.5*r1A-3*r2S)*V0

16 Cv2 = 53600

17 Cv1 = 3360

18 Fvent = if (FD<11400) then (FD) else(if (P<28.2) then ((P-1)*Cv1) else ((P-1)*(Cv1 +Cv2)))

19 UA = 0

    no cooling

Living Example Problem

POLYMATH Report

Ordinary Differential Equations

Calculated values of DEQ variables

	Variable	Initial value	Final value
1	A1A	4.0E+14	4.0E+14
2	A2S	1.0E+84	1.0E+84
3	CA	4.3	9.919E-07
4	CB	5.1	0.800001
5	CS	3.	2.460265
6	Cv1	3360.	3360.
7	Cv2	5.36E+04	5.36E+04
8	DHRx1A	-4.54E+04	-4.54E+04
9	DHRx2S	-3.2E+05	-3.2E+05
10	E1A	1.28E+05	1.28E+05
11	E2S	8.0E+05	8.0E+05
12	FD	2467.445	7.477E+10
13	Fvent	2467.445	2.507E+06
14	k1A	0.0562573	153.6843
15	k2S	8.428E-16	2.533E+06
16	P	4.4	45.01004
17	r1A	-1.233723	-0.000122
18	r2S	-2.529E-15	-6.231E+06
19	SumNCp	1.26E+07	1.26E+07
20	SW1	1.	0
21	t	0	4.
22	T	422.	538.8048
23	UA	0	0
24	V0	4000.	4000.
25	VH	5000.	5000.

Note: The second reaction took off about 3.6 hours after starting the reactor.

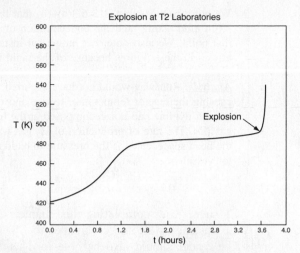

Figure E13-6.3(a) Temperature (K) versus time (h) trajectory.

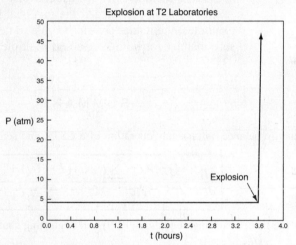

Figure E13-6.3(b) Pressure (atm) versus time (h) trajectory.

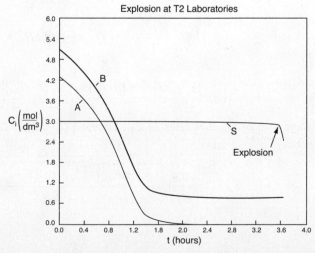

Figure E13-6.3(c) Concentration (mol/dm³) versus time (h) trajectory.

We note from Figures E13-6.3(a)–(b) that the explosion occurred at approximately 3.6 h after startup and the concentration of diglyme begins to drop sharply before that point. We also note that numerical instabilities occur at about the point of the arrows in these figures because of the rapid increase in temperature.

Analysis: Runaway would not have occurred if (1) the cooling system had not failed, causing the reactor temperature to rise and initiate a second a reaction, and (2) the solvent dygline had not decomposed at the higher temperature to produce hydrogen gas (D). The rate of production of H_2 gas was greater than the removal of H_2 from the head space, causing the pressure to build up to the point that it ruptured the reactor vessel.

Closure. After completing this chapter, the reader should be able to apply the unsteady-state energy balance to CSTRs, semibatch, and batch reactors. The reader should also be able to discuss reactor safety using the ONCB and the T2 Laboratories case studies of explosions to help prevent future accidents. Included in the reader's discussion should be how to start up a reactor so as not to exceed the practical stability limit. After studying these examples, the reader should be able to describe how to operate reactors in a safe manner for both single and multiple reactions.

SUMMARY

1. The energy balance on unsteady operation of a CSTR and semibatch reactors

$$\frac{dT}{dt} = \frac{\dot{Q} - \dot{W}_s - \sum F_{i0} C_{P_i}(T - T_{i0}) + [-\Delta H_{Rx}(T)](-r_A V)}{\sum N_i C_{P_i}} \tag{S13-8}$$

Using the heater exchange equation (12-9) and neglecting shaft work

$$\frac{dT}{dt} = \frac{\dot{Q}_{gs} - \dot{Q}_{rs}}{\sum N_i C_{P_i}} \tag{S13-2}$$

$$\dot{Q}_{gs} = (r_A V)(\Delta H_{Rx}) \tag{S13-3}$$

$$\dot{Q}_{rs} = \sum F_{i0} C_{P_i}(T - T_{i0}) + \dot{m}_c C_P [T - T_{a1}] \left[1 - \exp\left[\frac{-UA}{\dot{m}_c C_{P_i}} \right] \right] \tag{S13-4}$$

For large coolant $\dot{Q}_{rs}$ flow rates, $\dot{m}_c C_P$, Equation (13-22) reduces to

$$\frac{dT}{dt} = \frac{\overbrace{(r_A V)(\Delta H_{Rx})}^{\dot{Q}_g} - \overbrace{\left[\sum F_{i0} C_{P_i}(T - T_{i0}) + UA(T - T_a) \right]}^{\dot{Q}_{rs}}}{\sum N_i C_{P_i}} \tag{S13-5}$$

2. Batch reactors
   a. Non-adiabatic

$$\frac{dT}{dt} = \frac{\dot{Q}_g - \dot{Q}_{rb}}{\sum N_i C_{P_i}} = \frac{\dot{Q}_g - \dot{Q}_{rb}}{N_{A0}\left(\sum \Theta_i C_{P_i} + \Delta C_P X\right)}$$ (S13-6)

   where

$$\dot{Q}_g = (r_A V)(\Delta H_{Rx})$$ (S13-7)

$$\dot{Q}_{rb} = \dot{m}_c C_{P_C}(T - T_{a1})\left[1 - \exp\left[\frac{-UA}{\dot{m}_c C_{P_C}}\right]\right]$$ (S13-8)

   b. Adiabatic

$$\boxed{X = \frac{C_{P_s}(T - T_0)}{-\Delta H_{Rx}(T)} = \frac{\sum \Theta_i C_{P_i}(T - T_0)}{-\Delta H_{Rx}(T)}}$$ (S13-9)

$$\boxed{T = T_0 + \frac{[-\Delta H_{Rx}(T_0)]X}{C_{P_s} + X \Delta C_P} = T_0 + \frac{[-\Delta H_{Rx}(T_0)]X}{\displaystyle\sum_{i=1}^{m} \Theta_i C_{P_i} + X \Delta C_P}}$$ (S13-10)

3. Semibatch reactors and startup of CSTR

$$\boxed{\frac{dT}{dt} = \frac{\dot{Q}_g - \dot{Q}_{rs}}{\sum N_i C_{P_i}}}$$ (S13-11)

   Where $\dot{Q}_{gs}$ is the same as Equation (S13-7) and $\dot{Q}_{rs}$ (sub **rs**, i.e., "**heat removed**" from semibatch) is

$$\dot{Q}_{rs} = \sum F_{i0} C_{P_{i0}}(T - T_0) + \dot{m} C_{P_C}(T - T_{a1})\left[1 - \exp\left[\frac{-UA}{\dot{m} C_{P_C}}\right]\right]$$ (S13-12)

4. Multiple reactions (*q* reactions and *m* species)

$$\boxed{\frac{dT}{dt} = \frac{\overbrace{\displaystyle\sum_{j=1}^{q} r_{ij} V \Delta H_{Rxij}(T)}^{\dot{Q}_g} - \overbrace{\displaystyle\sum_{j=1}^{m} F_{j0} C_{P_j}(T - T_0) + \dot{m}_c C_{P_C}(T - T_{a1})\left[1 - \exp\left[\frac{-UA}{\dot{m} C_{P_C}}\right]\right]}^{\dot{Q}_{rs}}}{\displaystyle\sum_{j=1}^{m} N_j C_{P_j}}}$$ (S13-13)

   where *i* = reaction number and *j* = species.

CRE WEB SITE MATERIALS

- **Expanded Material**
    1. *The Profssional Reference Shelf contains topics that could just as well go under "Expanded Material."*
    2. *Web Problem WP13-1$_B$ Multiple Reaction in a Batch Reactor.*
- **Learning Resources**
    1. *Summary Notes*
    2. *Web links: SAChE Safety Web site (www.sache.org).* You will need to get the user name and password from your department chair. The kinetics (i.e., CRE) text, examples, and problems are marked *K* in the product sections: Safety, Health, and the Environment (S,H, & E).
    3. *Solved Problems*
        Example CD13-1 Use of the ARSST
        Example CD13-2 Startup of a CSTR
        Example CD13-3 Falling Off the Steady State
        Example CD13-4 Proportional-Integral (PI) Control
- **Living Example Problems**
    1. *Example 13-1 Adiabatic Batch Reactor*
    2. *Example 13-2 Safety in Chemical Plants with Exothermic Runaway Reactions*
    3. *Example 13-3 Heat Effects in a Semibatch Reactor*
    4. *Example 13-4 Startup of a CSTR*
    5. *Example 13-5 Multiple Reactions in a Semibatch Reactor*
    6. *Example 13-6 Explosion at T2 Laboratories*
    7. *PRS Example CD13-1 Falling Off the Upper Steady State*
    8. *PRS Example CD13-2 Integral Control of a CSTR*
    9. *PRS Example CD13-3 Proportion-Integral Control of a CSTR*
    10. *Example R13-4 Linear Stability*
    11. *Example R13-1 Use of ARSST*

- **Professional Reference Shelf**
    R13.1. *The Complete ARSST*
    In this section, further details are given to size safety valves to prevent runaway reactions.

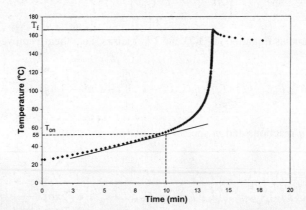

Figure R13.1 Temperature-time trajectory for hydrolysis of acetic anhydride.

R13.2. *Falling Off the Upper Steady State*
R13.3. *Control of a CSTR*
    In this section, we discuss the use of proportional (P) and integral (I) control of a CSTR. Examples include I and PI control of an exothermic reaction.

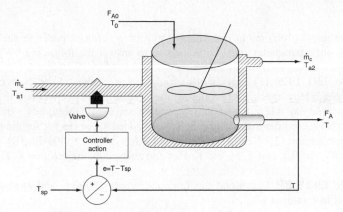

Reactor with coolant flow-rate control system.

Proportional integral action

$$z = z_0 + k_c(T - T_{SP}) + \frac{k_c}{\tau}\int_0^t (T - T_{SP})dt$$

R13.4. *Linearized Stability Theory (PDF)*

R13.5. *Approach to Steady-State Phase-Plane Plots and Trajectories of Concentration versus Temperature*
Startup of a CSTR (Figure R13.5) and the approach to the steady state (CRE Web site). By mapping out regions of the concentration–temperature phase plane, one can view the approach to steady state and learn if the practical stability limit is exceeded. The steady-state trajectories are shown for the mole balance (MB = 0) and energy balance (EB = 0).

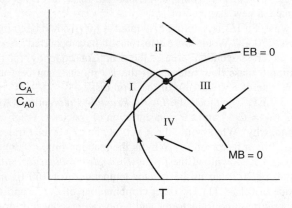

Figure R13.5   Startup of a CSTR.

R13.6. *Adiabatic Operation of a Batch Reactor*
R13.7. *Unsteady Operation of Plug-Flow Reactor*

QUESTIONS AND PROBLEMS

Questions

Q13-1   Discuss with a classmate what are the most important safety considerations to address for exothermic reactions. To help answer this question, prepare a list of safety considerations for designing and operating chemical reactors. See the August 1985 issue of *Chemical Engineering Progress*.

Problems

P13-1$_B$ Review the example problems in this chapter and use a software package such as Polymath or MAT-LAB to carry out a parameter sensitivity analysis to answer the following "What if…" questions.
What if...
(a) Example 13-1 LEP. (1) How would your answer change if the heat of the mixing had been neglected? (2) How much time would it take to achieve 90% conversion if the reaction were started on a very cold day where the initial temperature was 20°F? (Methanol won't freeze at this temperature.) (3) Now, consider that a heat exchanger is added to the reactor for the propylene oxide reaction; the parameters are: C_{A0} = 1 lb-mol/ft^3, V = 1.2 ft^3 ($\Sigma N_i C_{P_i}$ = 403 Btu°R), neglect ΔC_P, UA = 0.22 Btu/°R/s, and T_a = 498 K. Plot and analyze the trajectories X, T, Q_g, and Q_r as a function of time.

(b) Example 13-2 LEP. Download the *Living Example Problem* and explore the ONCB explosion described in Example 13-2.

(1) Explain what you would do to prevent an explosion of this type from ever occurring again while still operating at the triple production specified by management.

(2) Show that no explosion would have occurred if the cooling was not shut off for the 9.04-kmol charge of ONCB or if the cooling was shut off for 10 min after 45 min of operation for the 3.17-kmol ONCB charge.

(3) Show that if the cooling had been shut off for 10 min after 12 h of operation, no explosion would have occurred for the 9.04-kmol charge.

(4) Develop a set of guidelines as to when the reaction should be quenched should the cooling fail. Perhaps safe operation could be discussed using a plot of the time after the reaction began at which the cooling failed, t_0, versus the length of the cooling failure period, t_f, for the different charges of ONCB.

Hall of Fame

(5) Find a set of parameter values that would cause the explosion to occur at exactly 12:18 A.M. For example, include the mass and heat capacities of the metal reactor and/or make a new estimate of UA.

(6) Finally, what if a 1/2-in. rupture disk rated at 800 psi had been installed and did indeed rupture at 800 psi (270°C)? Would the explosion still have occurred? *Note*: The mass flow rate $\dot{m}$ varies with the cross-sectional area of the disk. Consequently, for the conditions of the reaction, the maximum mass flow rate out of the 1/2-in. disk can be found by comparing it with the mass flow rate of 830 kg/min of the 2-in. disk.

(c) Example 13-3 LEP. Download the *Living Example Problem*. (1) At what times will the number of moles of C ($N_C = C_C V$) and the concentration of species C reach a maximum? (2) Are they different and, if so, why? What would the X versus t and T versus t trajectories look like if the coolant rate is increased by a factor of 10? Why is the reaction time (252s) so short?

(d) Example 13-4 LEP. Download the *Living Example Problem* for *Startup of a CSTR*, for an entering temperature of 70°F, an initial reactor temperature of 160°F, and an initial concentration of propylene oxide of 0.1 M. (1) Try other combinations of T_0, T_i, and C_{Ai}, and report your results in terms of temperature–time trajectories and temperature–concentration phase planes. (2) Find a set of conditions above which the practical stability limit will be reached or exceeded and those conditions below which it will not. (3) Vary the coolant flow rate and compare with the base case given in Figures E13-4.1 to E13-4.4. Describe what you find.

(e) Example 13-5 LEP. Download the *Living Example Problem*. (1) Vary the volumetric flow rate 24 < v_0 < 1,000 and compare with the base case. Describe any trends you find. (2) Plot and analyze $N_A = C_A V$ and $N_B = C_B V$ for long times (e.g., t = 15h). What do you observe? (3) Can you show that for long times $N_A \cong C_{A0} v_0/k_{1A}$ and $N_B \cong C_{A0} v_0/2/k_{2B}$? (4) What do you think is happening to this semibatch reactor if it has no lid and a maximum volume of 1,000 dm^3 at long times? (5) If B is the deserved product, how would you maximize N_B?

(f)   Example 13-6 LEP. T2 Laboratory Explosion.

(1)   View the Chemical Safety Board (CSB) video online and read the supporting reports (*http://www. chemsafety.gov/videoroom/detail.aspx?VID=32*). Also search the Web for "T2 explosion video."

(2)   (a) What did you learn from watching the video? (b) Suggest how this reactor system should be modified and/or operated in order to eliminate any possibility of an explosion. (c) Would you use backup cooling and, if so, how? (d) How could you learn if a second reaction could be set in at a higher temperature? *Hint:* See *PRS R13.1 The Complete ARSST.*

(3)   Download the *Living Example Polymath E13-6.* Plot C_A, C_B, C_C, P, and T as a function of time. Vary UA between 0.0 and 2.77×10^6 J/h/K to find the lowest value of UA that you observe a runaway. Describe the trends as you approach runaway. Did it occur over a very narrow range of UA values? *Hint:* The problem becomes very stiff near the explosion condition when T > 600 K or P > 45 atm. If the temperature or pressure reaches these values, set all derivatives (concentration changes, temperature change, and pressure change) and reaction rates equal to zero so that the numerical solution will complete the analysis and hold all variables at the explosion point of the reactor.

(4)   Now let's consider the actual operation in more detail. The reactor contents are heated from 300 K to 422 K at a rate of $\dot{Q}$ = 4 K/minute. At 422 K, the reaction rate is sufficient such that heating is turned off. The reactor temperature continues to rise because the reaction is exothermic, and, when the temperature reaches 455 K, the cooling water turns on and cooling is initiated. Model this situation for the case when $UA = 2.77 \times 10^6$ J/h/K and when $UA = 0$.

(5)   What is the maximum time in minutes that the cooling can be lost ($UA = 0$) starting at the time when the reactor temperature reaches 455 K so that the reactor will not reach the explosion point? The conditions are those of part (1) of this problem.

(6)   Vary the parameters and operating conditions and describe what you find.

(g)   LEP PRS R13.2 Example CD13-5. Download the *Living Example Problem* for *Falling Off the Upper Steady State.* Try varying the entering temperature, T_0, between 80°F and 68°F, and plot the steady-state conversion as a function of T_0. Vary the coolant rate between 10,000 and 400 mol/h. Plot conversion and reactor temperature as a function of coolant rate.

(h)   LEP PRS R13.3 Example CD13-2. Download the *Living Example Problem.* Vary the gain, k_C, between 0.1 and 500 for the integral controller of the CSTR. Is there a lower value of k_C that will cause the reactor to fall to the lower steady state or an upper value to cause it to become unstable? What would happen if T_0 were to fall to 65°F or 60°F?

(i)   LEP PRS R13.3 Example CD13-3. Download the *Living Example Problem.* Learn the effects of the parameters k_C and τ_I. Which combination of parameter values generates the least and greatest oscillations in temperature? Which values of k_C and τ_I return the reaction to steady state the quickest?

Integral Controller

(j)   SAChE. Go to the SAChE Web site (*www.sache.org*). On the left-hand menu, select "SaChe Products." Select the "All" tab and go to the module entitled: "Safety, Health and the Environment (S, H & E)." The problems are for KINETICS (i.e., CRE). There are some example problems marked *K* and explanations in each of the above S, H & E selections. Solutions to the problems are in a different section of the site. Specifically look at: *Loss of Cooling Water* (K-1), *Runaway Reactions* (HT-1), *Design of Relief Values* (D-2), *Temperature Control and Runaway* (K-4) and (K-5), and *Runaway and the Critical Temperature Region* (K-7). Go through the *K* problems and write a paragraph on what you have learned. Your instructor or department chair should have the username and password to enter the SAChE Web site in order to obtain the module with the problems.

P13-2$_B$ The following is an excerpt from *The Morning News,* Wilmington, Delaware (August 3, 1977): "Investigators sift through the debris from blast in quest for the cause [that destroyed the new nitrous oxide plant]. A company spokesman said it appears more likely that the [fatal] blast was caused by another gas—ammonium nitrate—used to produce nitrous oxide." An 83% (wt) ammonium nitrate and 17% water solution is fed at 200°F to the CSTR operated at a temperature of about 510°F. Molten ammonium nitrate decomposes directly to produce gaseous nitrous

oxide and steam. It is believed that pressure fluctuations were observed in the system and, as a result, the molten ammonium nitrate feed to the reactor may have been shut off approximately 4 min prior to the explosion.
(a) Can you explain the cause of the blast?
(b) If the feed rate to the reactor just before shutoff was 310 lb_m of solution per hour, what was the exact temperature in the reactor just prior to shutdown? Use the data to calculate the exact time it took for the reactor to explode after the feed was shut off to the reactor.

Assume that at the time the feed to the CSTR stopped, there was 500 lb_m of ammonium nitrate in the reactor at a temperature of 520°F. The conversion in the reactor is virtually complete at about 99.99%. Additional data for this problem are given in Problem 12-5$_C$. How would your answer change if 100 lb_m of solution were in the reactor? 310 lb_m? 800 lb_m? What if $T_0 = 100°F$? 500°F? How would you start up or shut down and control such a reaction?
Safety: Shut Down an Open-Ended Problem. Rather than shut off the total feed to the reactor, it should be diluted with pure water at a volumetric rate υ_w while reducing the ammonium nitrate reactant feed rate. Keep the same total volumetric feed υ_0 rate and entering temperature. Choose a υ_w and plot the reactor temperature and composition as a function of time. At what time could you shut off the water feed safely?

P13-3$_B$ The liquid-phase reaction in Problems P11-4$_A$ and P12-7$_A$ is to be carried out in a semibatch reactor. There are 500 mol of A initially in the reactor at 25°C. Species B is fed to the reactor at 50°C and a rate of 10 mol/min. The feed to the reactor is stopped after 500 mol of B has been fed.
(a) Plot and analyze the temperature Q_r, Q_g and conversion as a function of time when the reaction is carried out adiabatically. Calculate to $t = 2$ h.
(b) Plot and analyze the conversion as a function of time when a heat exchanger ($UA = 100$ cal/min·K) is placed in the reactor and the ambient temperature is constant at 50°C. Calculate to $t = 3$ h.
(c) Repeat part (b) for the case where the reverse reaction cannot be neglected.

New parameter values:

k = 0.01 (dm³/mol·min) at 300 K with $E = 10$ kcal/mol
$V_0 = 50$ dm³, $\upsilon_0 = 1$ dm³/min, $C_{A0} = C_{B0} = 10$ mol/dm³
For the reverse reaction: $k_r = 0.1$ min⁻¹ at 300 K with $E_r = 16$ kcal/mol

P13-4$_B$ Sophia and Nic are operating a batch reactor and the reaction is first-order, liquid-phase, and exothermic. An inert coolant is added to the reaction mixture to control the temperature. The temperature is kept constant by varying the flow rate of the coolant (see Figure P13-4$_B$).

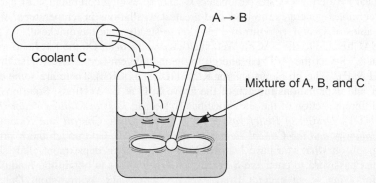

Figure P13-4$_B$ Semibatch reactor with inert coolant stream.

(a) Help them calculate the flow rate of the coolant 2 h after the start of the reaction. (*Ans.:* $F_C =$ 3.157 lb/s.)
(b) It is proposed that rather than feeding a coolant to the reactor, a solvent be added that can be easily boiled off, even at moderate temperatures. The solvent has a heat of vaporization of 1000 Btu/lb and initially there are 25 lb-mol of A placed in the tank. The initial volume of solvent

and reactant is 300 ft³. Determine the solvent evaporation rate as a function of time. What is the rate at the end of 2 h?

Additional information:

Temperature of reaction: 100°F
Value of k at 100°F: 1.2×10^{-4} s^{-1}
Temperature of coolant: 80°F
Heat capacity of all components: 0.5 Btu/lb·°F
Density of all components: 50 lb/ft³
ΔH_{Rx}°: $-25,000$ Btu/lb-mol
Initially:
    Vessel contains only A (no B or C present)
    C_{A0}: 0.5 lb-mol/ft³
    Initial volume: 50 ft³

W2014 CRE U of M MidTermII

P13-5$_B$ The reaction

$$A + B \longrightarrow C$$

is carried out adiabatically in a constant-volume batch reactor. The rate law is

$$-r_A = k_1 C_A^{1/2} C_B^{1/2} - k_2 C_C$$

Plot and analyze the conversion, temperature, and concentrations of the reacting species as a function of time.

Additional information:

Initial Temperature = 100°C

k_1 (373 K) = 2×10^{-3} s^{-1}	$E_1 = 100$ kJ/mol
k_2 (373 K) = 3×10^{-5} s^{-1}	$E_2 = 150$ kJ/mol
$C_{A0} = 0.1$ mol/dm³	$C_{P_A} = 25$ J/mol·K
$C_{B0} = 0.125$ mol/dm³	$C_{P_B} = 25$ J/mol·K
$\Delta H_{Rx}^{\circ}(298 \text{ K}) = -40,000$ J/mol A	$C_{P_C} = 40$ J/mol·K

P13-6$_B$ The elementary irreversible liquid-phase reaction

$$A + 2B \longrightarrow C$$

is to be carried out in a semibatch reactor in which B is fed to A. The volume of A in the reactor is 10 dm³, the initial concentration of A in the reactor is 5 mol/dm³, and the initial temperature in the reactor is 27°C. Species B is fed at a temperature of 52°C and a concentration of 4 *M*. It is desired to obtain at least 80% conversion of A in as short a time as possible, but at the same time the temperature of the reactor must not rise above 130°C. You should try to make approximately 120 mol of C in a 24-hour day, allowing for 30 minutes to empty and fill the reactor between each batch. The coolant flow rate through the reactor is 2000 mol/min. There is a heat exchanger in the reactor.

(a) What volumetric feed rate (dm³/min) do you recommend?
(b) How would your answer or strategy change if the maximum coolant rate dropped to 200 mol/min? To 20 mol/min?

Additional information:

$\Delta H_{Rx}^{\circ} = -55,000$ cal/mol A
$C_{P_A} = 35$ cal/mol·K, $C_{P_B} = 20$ cal/mol·K, $C_{P_C} = 75$ cal/mol·K

$k = 0.0005 \dfrac{\text{dm}^6}{\text{mol}^2 \cdot \text{min}}$ at 27°C with $E = 8000$ cal/mol

$$UA = 2500 \; \frac{\text{cal}}{\text{min} \cdot \text{K}} \; \text{with} \; T_a = 17°C$$

$C_P(\text{coolant}) = 18 \; \text{cal/mol} \cdot \text{K}$

P13-7$_B$ The irreversible reaction liquid phase in Problems P11-4$_A$ and P12-7$_A$

$$A + B \longrightarrow C$$

is to be carried out in a 10-dm3 batch reactor. Plot and analyze the temperature and the concentrations of A, B, and C as a function of time for the following cases:

(a) Adiabatic operation.
(b) Values of UA of 10,000, 40,000 and 100,000 J/min • K.
(c) Use $UA = 40,000$ J/min • K and different initial reactor temperatures.

An equal molar feed in A and B enters at 27°C, and the volumetric flow rate is 2 dm³/s and $C_{A0} = 0.1$ kmol/m³.

Additional information:

$\Delta H_A°$ (273 K) $= -20$ kcal/mol, $\Delta H_B°$ (273 K) $= -15$ kcal/mol,

$\Delta H_C°$ (273 K) $= -41$ kcal/mol

$C_{P_A} = C_{P_B} = 15$ cal/mol • K $C_{P_C} = 30$ cal/mol • K

$k = 0.01 \; \dfrac{\text{dm}^3}{\text{mol} \cdot \text{s}}$ at 300 K $E = 10,000$ cal/mol $K_c = 10$ m³/mol at 450

P13-8$_B$ The following reactions are taking place in a 2000-dm³ liquid-phase batch reactor under a pressure of 400 psig

$$A + 2B \xrightarrow{k_{1A}} C \qquad \Delta H_{Rx1B} = -5,000 \; \text{cal/mol} \qquad -r_{1A} = k_{1A}C_A C_B^2$$

$$3C + 2A \xrightarrow{k_{2A}} D \qquad \Delta H_{Rx2C} = +10,000 \; \text{cal/mol} \qquad -r_{2A} = k_{2A}C_A C_C$$

$$B + 3C \xrightarrow{k_{3C}} E \qquad \Delta H_{Rx3B} = -50,000 \; \text{cal/mol} \qquad -r_{3C} = k_{3C}C_B C_C$$

The initial temperature is 450 K and the initial concentrations of A, B, and C are 1.0, 0.5, and 0.2 mol/dm³, respectively. The coolant flow rate was at its maximum value so that $T_{a1} = T_{a2} = T_a = 400$ K, so that the product, the exchange area, and the overall heat transfer coefficient, UA, is $UA = 100$ cal/s•K.

(a) If $Q_r > Q_g$ at time $t = 0$, and there is no failure of the heat exchange system, is there any possibility that the reactor will run away? Explain.
(b) What is Q_r at $t = 0$?
(c) What is Q_g at $t = 0$?
(d) What is the initial rate of increase in temperature, (i.e., dT/dt) at $t = 0$?

$$\frac{dT}{dt} = \underline{\qquad\qquad}$$

(e) Suppose that the ambient temperature T_a is lowered from 400 K to 350 K; what is the initial rate of reactor temperature change?

$$\frac{dT}{dt} = \underline{\qquad\qquad}$$

Plot the temperatures and all the concentrations as a function of time up to $t = 1000$ s.
(f) A suggestion was made to add 50 moles of inerts at a temperature of 450 K. Will the addition of the inerts make runaway more likely or less likely? How? Show quantitatively.

Additional information:

As a first approximation, assume all heats of reaction are constant (i.e., $\Delta C_{P_{ij}} \cong 0$). Specific reaction rates at 450 K are

$$k_{1A} = 1 \times 10^{-3} \,(\text{dm}^3/\text{mol})^2/\text{s} \qquad C_{P_A} = 10 \text{ cal/mol/K} \quad C_{P_D} = 80 \text{ cal/mol/K}$$

$$k_{2A} = \frac{1}{3} \times 10^{-3} \,(\text{dm}^3/\text{mol})^2/\text{s} \qquad C_{P_B} = 10 \text{ cal/mol/K} \quad C_{P_E} = 50 \text{ cal/mol/K}$$

$$k_{3C} = 0.6 \times 10^{-3} \,(\text{dm}^3/\text{mol})^2/\text{s} \qquad C_{P_C} = 50 \text{ cal/mol/K}$$

P13-9$_B$ The reactions on the CRE Web site in Problem CD12GA2 are to be carried out in a semibatch reactor. How would you carry out this reaction (i.e., T_0, v_0, T_i)? The molar concentrations of pure A and pure B are 5 and 4 mol/dm^3, respectively. Plot and analyze the concentrations, temperatures, and the overall selectivity as a function of time for the conditions you chose.

- **Additional Homework Problems**

    A number of homework problems that can be used for exams or supplementary problems or examples are found on the CRE Web site, *www.umich.edu/~elements/5e/index.html.*

SUPPLEMENTARY READING

1. A number of solved problems for batch and semibatch reactors can be found in

    WALAS, S. M., *Chemical Reaction Engineering Handbook.* Amsterdam: Gordon and Breach, 1995, pp. 386–392, 402, 460–462, and 469.

Safety

    CROWL, DANIEL A. and JOSEPH F. LOUVAR, *Chemical Process Safety: Fundamentals with Applications,* 3rd ed. Upper Saddle River, NJ: Prentice Hall, 2001.

    EDGAR, T. F., "From the Classical to the Postmodern Era," *Chem. Eng. Educ.,* 31, 12 (1997).

    KLETZ, TREVOR A., "Bhopal Leaves a Lasting Legacy: The Disaster Taught Some Hard Lessons That the Chemical Industry Still Sometimes Forgets," *Chemical Processing,* p. 15 (Dec. 2009).

Links

1. The **SAChE Web site** has a great discussion on reactor safety with examples (*www.sache.org*). You will need a username and password; both can be obtained from your department chair. Hit the 2003 Tab. Go to K Problems.

Links

2. The **reactor lab** developed by Professor Herz and discussed in Chapters 4 and 5 could also be used here: *www.reactorlab.net.*

3. See the Center for Chemical Process Safety (CCPS) Web site, *www.aiche.org/ccps/.*

Mass Transfer 14
Limitations in
Reacting Systems

Giving up is the ultimate tragedy.

—Robert J. Donovan

or

It ain't over 'til it's over.

—Yogi Berra
NY Yankees

Overview. Many industrial reactions are carried out at high temperatures where the overall rate of reaction is limited by the rate of mass transfer of reactants between the bulk fluid and the catalytic surface. By mass transfer, we mean any process in which diffusion plays a role. Under these circumstances our generation term becomes a little more complicated as we cannot directly use the rate laws discussed in Chapter 3. Now we have to consider the fluid velocity and the fluid properties when writing the mole balance. In the rate laws and catalytic reaction steps described in Chapter 10 (diffusion, adsorption, surface reaction, desorption, and diffusion), we neglected the diffusion steps.

In this chapter we discuss how to determine the rate of reaction and how to size reactors when the reactions are limited by mass transfer. To do this we

- Present the fundamentals of diffusion and molar flux, and then write the mole balance in terms of the mole fluxes for rectangular and for cylindrical coordinates (Section 14.1).
- Incorporate Fick's first law into our mole balance in order to describe flow, diffusion, and reaction (Section 14.2).
- Model diffusion through a stagnant film to a reacting surface (Section 14.3).

> • Introduce the mass transfer coefficient, k_c, and describe how it is used to design mass transfer limited reactions (Section 14.4).
> • Focus on one of the engineer's most important skills, i.e., to answer "What if…" questions, as *Robert the Worrier* does (Section 14.5).

14.1 Diffusion Fundamentals

The Algorithm
1. Mole balance
2. Rate law
3. Stoichiometry
4. Combine
5. Evaluate

The first step in our CRE algorithm is the mole balance, which we now need to extend to include the molar flux, W_{Az}, and diffusional effects. The molar flow rate of A in a given direction, such as the z direction down the length of a tubular reactor, is just the product of the flux, W_{Az} (mol/m^2 • s), and the cross-sectional area, A_c (m^2); that is,

$$F_{Az} = A_c \, W_{Az}$$

In the previous chapters, we have only considered plug flow with no diffusion superimposed, in which case

$$W_{Az} = \frac{C_A v}{A_C}$$

We now drop the plug-flow assumption and extend our discussion of mass transfer in catalytic and other mass-transfer limited reactions. In Chapter 10 we focused on the middle three steps (3, 4, and 5) in a catalytic reaction and neglected steps (1), (2), (6), and (7) by assuming the reaction was surface-reaction limited. In this chapter we describe the first and last steps (1) and (7), as well as showing other applications in which mass transfer plays a role.

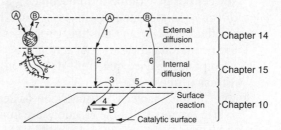

Figure 14-1   Steps in a heterogeneous catalytic reaction.

Where are we going ??:[†]

We want to arrive at the mole balance that incorporates both diffusion and reaction effects, such as Equation (14-16) on page 685. I.e.,

$$D_{AB}\frac{d^2 C_A}{dZ^2} - U_z \frac{dC_A}{dZ} + r_A = 0$$

[†] "If you don't know where you are going, you'll probably wind up some place else."
Yogi Berra, NY Yankees

We begin with Section 14.1.1 where we write the mole balance on Species A in three dimensions in terms of the molar flux, $\mathbf{W}_A$. In Section 14.1.2 we write $\mathbf{W}_A$ in terms of the bulk flow of A in the fluid, $\mathbf{B}_A$ and the diffusion flux $\mathbf{J}_A$ of A that is superimposed on bulk flow. In Section 14.1.3 we use the previous two subsections as a basis to finally write the molar flux, $\mathbf{W}_A$, in terms of concentration using Fick's first law, $\mathbf{J}_A$, and the bulk flow, $\mathbf{B}_A$. Next, in Section 14.2 we combine diffusion convective transport and reaction in our mole balance.

14.1.1 Definitions

Diffusion is the spontaneous intermingling or mixing of atoms or molecules by random thermal motion. It gives rise to motion of the species *relative* to motion of the mixture. In the absence of other gradients (such as temperature, electric potential, or gravitational potential), molecules of a given species within a single phase will always diffuse from regions of higher concentrations to regions of lower concentrations. This gradient results in a molar flux of the species (e.g., A), $\mathbf{W}_A$ (moles/area·time), in the direction of the concentration gradient. The flux of A, $\mathbf{W}_A$, is relative to a fixed coordinate (e.g., the lab bench) and is a vector quantity with typical units of mol/m^2·s. In rectangular coordinates

$$\mathbf{W}_A = iW_{Ax} + jW_{Ay} + kW_{Az} \tag{14-1}$$

We now apply the mole balance to species A, which flows and reacts in an element of volume $\Delta V = \Delta x \Delta y \Delta z$ to obtain the variation of the molar fluxes in three dimensions.

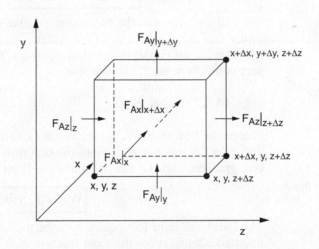

$$F_{Az} = W_{Az}\Delta x \Delta y$$

$$F_{Ay} = W_{Ay}\Delta x \Delta z$$

$$F_{Ax} = W_{Ax}\Delta z \Delta y$$

Mole Balance

$$
\begin{bmatrix} \text{Molar} \\ \text{flow rate} \\ \text{in} \end{bmatrix}_z - \begin{bmatrix} \text{Molar} \\ \text{flow rate} \\ \text{out} \end{bmatrix}_{z+\Delta z} + \begin{bmatrix} \text{Molar} \\ \text{flow rate} \\ \text{in} \end{bmatrix}_y - \begin{bmatrix} \text{Molar} \\ \text{flow rate} \\ \text{out} \end{bmatrix}_{y+\Delta y} +
$$

$$
\Delta x \Delta y W_{Az}|_z - \Delta x \Delta y W_{Az}|_{z+\Delta z} + \Delta x \Delta z W_{Ay}|_y - \Delta x \Delta z W_{Ay}|_{y+\Delta y} +
$$

$$
\begin{bmatrix} \text{Molar} \\ \text{flow rate} \\ \text{in} \end{bmatrix}_x - \begin{bmatrix} \text{Molar} \\ \text{flow rate} \\ \text{out} \end{bmatrix}_{x+\Delta x} + \begin{bmatrix} \text{Rate of} \\ \text{generation} \end{bmatrix} = \begin{bmatrix} \text{Rate of} \\ \text{accumulation} \end{bmatrix}
$$

$$
\Delta z \Delta y W_{Ax}|_x - \Delta z \Delta y W_{Ax}|_{x+\Delta x} + r_A \Delta x \Delta y \Delta z = \Delta x \Delta y \Delta z \frac{\partial C_A}{\partial t}
$$

where r_A is the rate of generation of A by reaction per unit volume (e.g., mol/m^3/h).

Dividing by $\Delta x \Delta y \Delta z$ and taking the limit as they go to zero, we obtain the molar flux balance in rectangular coordinates

$$
\boxed{-\frac{\partial W_{Ax}}{\partial x} - \frac{\partial W_{Ay}}{\partial y} - \frac{\partial W_{Az}}{\partial z} + r_A = \frac{\partial C_A}{\partial t}} \tag{14-2}
$$

The corresponding balance in cylindrical coordinates with no variation in the rotation about the **z**-axis is

COMSOL

$$
\boxed{-\frac{1}{r}\frac{\partial}{\partial r}(rW_{Ar}) - \frac{\partial W_{Az}}{\partial z} + r_A = \frac{\partial C_A}{\partial t}} \tag{14-3}
$$

We will now evaluate the flux terms $\mathbf{W_A}$. We have taken the time to derive the molar flux equations in this form because they are now in a form that is consistent with the partial differential equation (PDE) solver COMSOL, which is accessible from the CRE Web site.

14.1.2 Molar Flux

The molar flux of A, $\mathbf{W_A}$, is the result of two contributions: $\mathbf{J_A}$, the molecular diffusion flux relative to the bulk motion of the fluid produced by a concentration gradient, and $\mathbf{B_A}$, the flux resulting from the bulk motion of the fluid:

Total flux = diffusion + bulk motion

$$
\boxed{\mathbf{W_A} = \mathbf{J_A} + \mathbf{B_A}} \tag{14-4}
$$

The bulk-flow term for species A is the total flux of all molecules relative to a fixed coordinate times the mole fraction of A, y_A; i.e., $\mathbf{B_A} = y_A \sum \mathbf{W}_i$.

For a two-component system of A diffusing in B, the flux of A is

$$
\mathbf{W_A} = \mathbf{J_A} + y_A(\mathbf{W_A} + \mathbf{W_B}) \tag{4-5}
$$

The diffusional flux, $\mathbf{J_A}$, is the flux of A molecules that is superimposed on the bulk flow. It tells how fast A is moving ahead of the bulk flow velocity, i.e., the molar average velocity.

The flux of species A, $\mathbf{W_A}$, is wrt a *fixed coordinate system* (e.g., the lab bench) and is just the concentration of A, C_A, times the particle velocity of species A, $\mathbf{U_A}$, at that point

$$\mathbf{W}_A = \mathbf{U}_A C_A$$

$$\frac{\text{mol}}{\text{m}^2 \text{s}} = \left(\frac{\text{m}}{\text{s}}\right)\left(\frac{\text{mol}}{\text{m}^3}\right)$$

By particle velocities, we mean the vector average of millions of molecules of A at a given point. Similarity for species B: $\mathbf{W}_B = \mathbf{U}_B C_B$; substituting into the bulk-flow term

$$\mathbf{B}_A = y_A \sum \mathbf{W}_i = y_A (\mathbf{W}_A + \mathbf{W}_B) = y_A (C_A \mathbf{U}_A + C_B \mathbf{U}_B)$$

Writing the concentration of A and B in the generic form in terms of the mole fraction, y_i, and the total concentration, c, i.e., $C_i = y_i c$, and then factoring out the total concentration, c, the bulk flow, $\mathbf{B}_A$, is

Molar
average
velocity

$$\mathbf{B}_A = (c \; y_A)(y_A \mathbf{U}_A + y_B \mathbf{U}_B) = C_A \mathbf{U}$$

where $\mathbf{U}$ is the molar average velocity: $\mathbf{U} = \sum y_i \mathbf{U}_i$. The molar flux of A can now be written as

$$\boxed{\mathbf{W}_A = \mathbf{J}_A + C_A \mathbf{U}} \tag{14-6}$$

We now need to determine the equation for the molar flux of A, $\mathbf{J}_A$, that is superimposed on the molar average velocity.

14.1.3 Fick's First Law

Our discussion on diffusion will be restricted primarily to binary systems containing only species A and B. We now wish to determine *how* the molar diffusive flux of a species (i.e., $\mathbf{J}_A$) is related to its concentration gradient. As an aid in the discussion of the transport law that is ordinarily used to describe diffusion, recall similar laws from other transport processes. For example, in conductive heat transfer the constitutive equation relating the heat flux $\mathbf{q}$ and the temperature gradient is Fourier's law, $\mathbf{q} = -k_t \nabla T$, where k_t is the thermal conductivity.

Experimentation with frog legs led to Fick's first law.

In rectangular coordinates, the gradient is in the form

Constitutive equations in heat, momentum, and mass transfer

$$\nabla = i \; \frac{\partial}{\partial x} + j \; \frac{\partial}{\partial y} + k \; \frac{\partial}{\partial z}$$

The mass transfer law for the diffusional flux of A resulting from a concentration gradient is analogous to Fourier's law for heat transfer and is given by **Fick's first law**

$$\mathbf{J}_A = -D_{AB} \nabla C_A \tag{14-7}$$

D_{AB} is the diffusivity of A in B $\left(\dfrac{\text{m}^2}{\text{s}}\right)$. Combining Equations (14-7) and (14-6), we obtain an expression for the molar flux of A in terms of concentration for constant total concentration

Molar flux equation

$$\boxed{\mathbf{W}_A = -D_{AB} \nabla C_A + C_A \mathbf{U}} \tag{14-8}$$

In one dimension, i.e., z, the molar flux term is

$$\mathbf{W}_{Az} = -D_{AB}\frac{dC_A}{dz} + C_A\mathbf{U}_z \tag{14-8a}$$

14.2 Binary Diffusion

Although many systems involve more than two components, the diffusion of each species can be treated as if it were diffusing through another single species rather than through a mixture by defining an effective diffusivity.

14.2.1 Evaluating the Molar Flux

Now the task is to evaluate the bulk-flow term.

We now consider five typical cases in Table 14-1 of A diffusing in B. Substituting Equation (14-7) into Equation (14-6) we obtain

$$\mathbf{W}_A = -D_{AB}\nabla C_A + y_A(\mathbf{W}_A + \mathbf{W}_B) \tag{14-9}$$

TABLE 14-1   EVALUATING $\mathbf{W}_A$ FOR SPECIES A DIFFUSING IN SPECIES B

(1) <u>Equal molar counter diffusion</u> (EMCD) of species A and B. For every molecule of A that diffuses in the forward direction, one molecule of B diffuses in the reverse direction

$$\mathbf{W}_A = -\mathbf{W}_B$$

$$\boxed{\mathbf{W}_A = \mathbf{J}_A = -D_{AB}\nabla C_A} \tag{14-10}$$

An example of EMCD is the oxidation of solid carbon; for every mole of oxygen that diffuses to the surface to react with the carbon surface, one mole of carbon dioxide diffuses away from the surface. $W_{O_2} = -W_{CO_2}$

(2) <u>Species A diffusing through stagnant species B</u> ($\mathbf{W}_B = 0$). This situation usually occurs when a solid boundary is involved and there is a stagnant fluid layer next to the boundary through which A is diffusing

$$\mathbf{W}_A = \mathbf{J}_A + y_A\mathbf{W}_A$$

$$\boxed{\mathbf{W}_A = \frac{\mathbf{J}_A}{1-y_A} = -\frac{D_{AB}\nabla C_A}{1-y_A} = +cD_{AB}\nabla\ln(1-y_A)} \tag{14-11}$$

(3) <u>Bulk flow of A is much greater than molecular diffusion of A</u>, i.e., $\mathbf{B}_A \gg \mathbf{J}_A$

$$\boxed{\mathbf{W}_A = \mathbf{B}_A = y_A(\mathbf{W}_A + \mathbf{W}_B) = C_A\mathbf{U}} \tag{14-12}$$

This case is the plug-flow model we have been using in the previous chapters in this book

$$F_A = W_A A_c = C_A \overbrace{\mathbf{U}A_c}^{v} = vC_A$$

(4) <u>For small bulk flow $\mathbf{J}_A \gg \mathbf{B}_A$</u>, we get the same result as EMCD, i.e., Equation (14-10)

$$\boxed{\mathbf{W}_A = \mathbf{J}_A = -D_{AB}\nabla C_A} \tag{14-10}$$

(5) <u>Knudsen Diffusion</u>: Occurs in porous catalysts where the diffusing molecules collide more often with the pore walls than with each other

$$\boxed{\mathbf{W}_A = \mathbf{J}_A = -D_K\nabla C_A} \tag{14-13}$$

and D_K is the Knudsen diffusion.[1]

---

[1] C. N. Satterfield, *Mass Transfer in Heterogeneous Catalysis* (Cambridge: MIT Press, 1970), pp. 41–42, discusses Knudsen flow in catalysis and gives the expression for calculating D_K.

14.2.2 Diffusion and Convective Transport

When accounting for diffusional effects, the molar flow rate of species A, F_A, in a specific direction $\mathbf{z}$, is the product of molar flux in that direction, W_{Az}, and the cross-sectional area normal to the direction of flow, A_c

$$F_{Az} = A_c W_{Az}$$

In terms of concentration, the flux is

$$W_{Az} = -D_{AB}\frac{dC_A}{dz} + C_A U_z$$

The molar flow rate is

$$F_{Az} = W_{Az}\, A_c = \left[-D_{AB}\frac{dC_A}{dz} + C_A U_z \right] A_c \qquad (14\text{-}14)$$

Similar expressions follow for W_{Ax} and W_{Ay}. Substituting for the flux W_{Ax}, W_{Ay}, and W_{Az} into Equation (14-2), we obtain

Flow, diffusion, and reaction

This form is used in COMSOL Multiphysics.

$$D_{AB}\left[\frac{\partial^2 C_A}{\partial x^2} + \frac{\partial^2 C_A}{\partial y^2} + \frac{\partial^2 C_A}{\partial z^2} \right] - U_x\frac{\partial C_A}{\partial x} - U_y\frac{\partial C_A}{\partial y} - U_z\frac{\partial C_A}{\partial z} + r_A = \frac{\partial C_A}{\partial t} \qquad (14\text{-}15)$$

Equation (14-15) is in a user-friendly form to apply to the PDE solver, COMSOL. For one dimension at steady state, Equation (14-15) reduces to

$$D_{AB}\frac{d^2 C_A}{dz^2} - U_z\frac{dC_A}{dz} + r_A = 0 \qquad (14\text{-}16)$$

In order to solve Equation (14-16) we need to specify the boundary conditions. In this chapter we will consider some of the simple boundary conditions, and in Chapter 18 we will consider the more complicated boundary conditions, such as the Danckwerts' boundary conditions.

We will now use this form of the molar flow rate in our mole balance in the z direction of a tubular flow reactor

$$\frac{dF_A}{dV} = \frac{d(A_c W_{Az})}{d(A_c z)} = \frac{dW_{Az}}{dz} = r_A \qquad (14\text{-}17)$$

However, we first have to discuss the boundary conditions in solving this equation.

14.2.3 Boundary Conditions

The most common boundary conditions are presented in Table 14-2.

TABLE 14-2   TYPES OF BOUNDARY CONDITIONS

1. Specify a concentration at a boundary (e.g., z = 0, $C_A = C_{A0}$).
   For an instantaneous reaction at a boundary, the concentration of the reactants at the boundary is taken to be zero (e.g., $C_{As} = 0$). See Chapter 18 for the more exact and complicated Danckwerts' boundary conditions at z = 0 and z = L.
2. Specify a flux at a boundary.
   a. No mass transfer to a boundary

$$W_A = 0 \qquad (14\text{-}18)$$

   for example, at the wall of a nonreacting pipe. Species A cannot diffuse into the solid pipe wall so $W_A = 0$ and then

$$\frac{dC_A}{dr} = 0 \qquad \text{at } r = R \qquad (14\text{-}19)$$

   That is, because the diffusivity is finite, the only way the flux can be zero is if the concentration gradient is zero.
   b. Set the molar flux to the surface equal to the rate of reaction on the surface

$$W_A(\text{surface}) = -r_A''(\text{surface}) \qquad (14\text{-}20)$$

   c. Set the molar flux to the boundary equal to convective transport across a boundary layer

$$W_A(\text{boundary}) = k_c(C_{Ab} - C_{As}) \qquad (14\text{-}21)$$

   where k_c is the mass transfer coefficient and C_{As} and C_{Ab} are the surface and bulk concentrations, respectively.
3. Planes of symmetry. When the concentration profile is symmetrical about a plane, the concentration gradient is zero in that plane of symmetry. For example, in the case of radial diffusion in a pipe, at the center of the pipe

$$\frac{dC_A}{dr} = 0 \qquad \text{at } r = 0 \qquad (14\text{-}22)$$

14.2.4 Temperature and Pressure Dependence of D_{AB}

Before closing this brief discussion on mass-transfer fundamentals, further mention should be made of the diffusion coefficient.[2] Equations for predicting gas diffusivities are given by Fuller and are also given in Perry's *Handbook*.[3,4] The orders of magnitude of the diffusivities for gases, liquids, and solids and the manner in which they vary with temperature and pressure are given in Table 14-3.[5] We note that the Knudsen, liquid, and solid diffusivities are independent of total pressure.

[2] For further discussion of mass-transfer fundamentals, see R. B. Bird, W. E. Stewart, and E. N. Lightfoot, *Transport Phenomena,* 2nd ed. (New York: Wiley, 2002).

[3] E. N. Fuller, P. D. Schettler, and J. C. Giddings, *Ind. Eng. Chem.*, 58(5), 19 (1966). Several other equations for predicting diffusion coefficients can be found in B. E. Polling, J. M. Prausnitz, and J. P. O'Connell, *The Properties of Gases and Liquids*, 5th ed. (New York: McGraw-Hill, 2001).

[4] R. H. Perry and D. W. Green, *Chemical Engineer's Handbook*, 7th ed. (New York: McGraw-Hill, 1999).

[5] To estimate liquid diffusivities for binary systems, see K. A. Reddy and L. K. Doraiswamy, *Ind. Eng. Chem. Fund.*, 6, 77 (1967).

TABLE 14-3   DIFFUSIVITY RELATIONSHIPS FOR GASES, LIQUIDS, AND SOLIDS

It is important to know the magnitude and the T and P dependence of the diffusivity.

Gas:

Liquid:

Phase	Order of Magnitude		Temperature and Pressure Dependences[a]
	cm²/s	m²/s	
Gas			
Bulk	10^{-1}	10^{-5}	$D_{AB}(T_2, P_2) = D_{AB}(T_1, P_1)\dfrac{P_1}{P_2}\left(\dfrac{T_2}{T_1}\right)^{1.75}$
Knudsen	10^{-2}	10^{-6}	$D_A(T_2) = D_A(T_1)\left(\dfrac{T_2}{T_1}\right)^{1/2}$
Liquid	10^{-5}	10^{-9}	$D_{AB}(T_2) = D_{AB}(T_1)\dfrac{\mu_1}{\mu_2}\left(\dfrac{T_2}{T_1}\right)$
Solid	10^{-9}	10^{-13}	$D_{AB}(T_2) = D_{AB}(T_1)\exp\left[\dfrac{E_D}{R}\left(\dfrac{T_2-T_1}{T_1 T_2}\right)\right]$

[a] μ_1, μ_2, liquid viscosities at temperatures T_1 and T_2, respectively; E_D, diffusion activation energy.

14.2.5 Steps in Modeling Diffusion to a Reacting Surface

We first consider the diffusion of species A through a stagnant film in which no reaction takes place to a catalytic surface where Species A reacts instantaneously by upon reaching the surface, i.e., $C_{As} \cong 0$. Consequently, the rate of diffusion through the stagnant film equals the rate of reaction on the surface. The first steps in modeling are:

Steps in modeling mass transfer

Step 1: Perform a differential mole balance on a particular species A or use the general mole balance, to obtain an equation for W_{Az}, e.g., Equation (14-2).

Step 2: Replace W_{Az} by the appropriate expression for the concentration gradient.

Step 3: State the boundary conditions.

Step 4: Solve for the concentration profile.

Step 5: Solve for the molar flux.

In Section 14.3, we are going to apply this algorithm to one of the most important cases, diffusion through a boundary layer. Here, we consider the boundary layer to be a hypothetical "stagnant film" in which all the resistance to mass transfer is lumped.

14.2.6 Modeling Diffusion with Chemical Reaction

Next, we consider the situation where species A reacts as it diffuses through the stagnant film.[6] This table will provide the foundation for problems with diffusion and reaction in both Chapters 14 and 15.

Use Table 14-4 to Move
In $\rightleftarrows$ Out
of the algorithm (Steps 1 → 6) to generate creative solutions.

The purpose of presenting algorithms (e.g., Table 14-4) to solve reaction engineering problems is to give the readers a starting point or framework with which to work if they were to get stuck. It is expected that once readers are

[6] E. L. Cussler, *Diffusion Mass Transfer in Fluid Systems*, 2nd ed. (New York: Cambridge University Press, 1997).

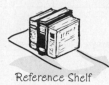

Expanding the
previous six
modeling steps just
a bit

TABLE 14-4   STEPS IN MODELING CHEMICAL SYSTEMS WITH DIFFUSION AND REACTION

1. Define the problem and state the assumptions.
2. Define the system on which the balances are to be made.
3. Perform a differential mole balance on a particular species.
4. Obtain a differential equation in W_A by rearranging your balance equation properly and taking the limit as the volume of the element goes to zero.
5. Substitute the appropriate expression involving the concentration gradient for W_A from Section 14.2 to obtain a second-order differential equation for the concentration of A.[a]
6. Express the reaction rate r_A (if any) in terms of concentration and substitute into the differential equation.
7. State the appropriate boundary and initial conditions.
8. Put the differential equations and boundary conditions in dimensionless form.
9. Solve the resulting differential equation for the concentration profile.
10. Differentiate this concentration profile to obtain an expression for the molar flux of A.
11. Substitute numerical values for symbols.

[a]In some instances it may be easier to integrate the resulting differential equation in Step 4 before substituting for W_A.

familiar and comfortable using the algorithm/framework, they will be able to move in and out of the framework as they develop creative solutions to nonstandard chemical reaction engineering problems.

14.3 Diffusion Through a Stagnant Film

To begin our discussion on the diffusion of reactants from the bulk fluid to the external surface of a catalyst, we shall focus attention on the flow past a single catalyst pellet. Reaction takes place only on the external catalyst surface and not in the fluid surrounding it. The fluid velocity in the vicinity of the spherical pellet will vary with position around the sphere. The hydrodynamic boundary layer is usually defined as the distance from a solid object to where the fluid velocity is 99% of the bulk velocity, U_0. Similarly, the mass transfer boundary-layer thickness, δ, is defined as the distance from a solid object to where the concentration of the diffusing species reaches 99% of the bulk concentration.

A reasonable representation of the concentration profile for a reactant A diffusing to the external surface is shown in Figure 14-2. As illustrated, the change in concentration of A from C_{Ab} to C_{As} takes place in a very narrow fluid layer next to the surface of the sphere. Nearly all of the resistance to mass transfer is found in this layer.

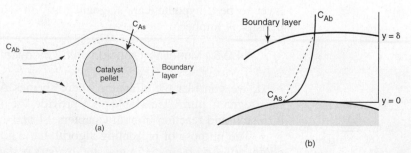

Figure 14-2   Boundary layer around the surface of a catalyst pellet.

A useful way of modeling diffusive transport is to treat the fluid layer next to a solid boundary as a hypothetical stagnant film of thickness δ, which we cannot measure. We say that *all* the resistance to mass transfer is found (i.e., lumped) within this hypothetical stagnant film of thickness δ, and the properties (i.e., concentration, temperature) of the fluid at the outer edge of the film are identical to those of the bulk fluid. This model can readily be used to solve the differential equation for diffusion through a stagnant film. The dashed line in Figure 14-2b represents the concentration profile predicted by the hypothetical stagnant film model, while the solid line gives the actual profile. If the film thickness is much smaller than the radius of the pellet (which is usually the case), curvature effects can be neglected. As a result, only the one-dimensional diffusion equation must be solved, as was shown in Figure 14-3.

The concept of a hypothetical stagnant film within which all the resistance to external mass transfer exists

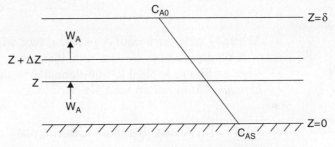

Figure 14-3   Concentration profile for dilute concentration in stagnant film mode

We are going to carry out a mole balance on species A diffusing through the fluid between $z = z$ and $z = z + \Delta z$ at steady state for the unit cross-sectional area, A_c

$$\text{In} \quad - \quad \text{Out} \quad + \quad \text{Generation} \quad = \quad \text{Accumulation}$$
$$W_{Az}\big|_z \quad - \quad W_{Az}\big|_{z+\Delta z} \quad + \quad 0 \quad = \quad 0$$

dividing by Δz and taking the limit as $\Delta z \rightarrow 0$

$$\frac{dW_{Az}}{dz} = 0$$

For diffusion through a stagnant film at dilute concentrations

$$J_A \gg y_A(W_A + W_B) \tag{14-23}$$

or for EMCD, we have using Fick's first law

$$W_{Az} = -D_{AB}\frac{dC_A}{dz} \tag{14-24}$$

Substituting for W_{Az} and dividing by D_{AB} we have

$$\frac{d^2 C_A}{dz^2} = 0$$

Integrating twice to get $C_A = K_1 z + K_2$, using the boundary conditions at

$$z = 0 \quad C_A = C_{As}$$
$$z = \delta \quad C_A = C_{A0}$$

we obtain the concentration profile

$$C_A = C_{As} + \left(C_{Ab} - C_{As}\right)\frac{z}{\delta} \qquad (14\text{-}25)$$

To find the flux to the surface we substitute Equation (14-25) into Equation (14-24) to obtain

$$W_{Az} = \frac{D_{AB}}{\delta}\left[C_{Ab} - C_{As}\right] \qquad (14\text{-}26)$$

At steady state the flux of A to the surface will be equal to the rate of reaction of A on the surface. We also note that another example of diffusion through a stagnant film as applied to transdermal drug delivery is given in the Chapter 14 *Expanded Material* on the CRE Web site.

14.4 The Mass Transfer Coefficient

We now interpret the ratio (D_{AB}/δ) in Equation (14-26).

While the boundary-layer thickness will vary around the sphere, we will take it to have a mean film thickness δ. The ratio of the diffusivity D_{AB} to the film thickness δ is the mass transfer coefficient, k_c, that is,

The mass transfer coefficient

$$\boxed{k_c = \frac{D_{AB}}{\delta}} \qquad (14\text{-}27)$$

Combining Equations (14-26) and (14-27), we obtain the average molar flux from the bulk fluid to the surface

Molar flux of A to the surface

$$\boxed{W_{Az} = k_c(C_{Ab} - C_{As})} \qquad (14\text{-}28)$$

In this stagnant film model, we consider all the resistance to mass transfer to be lumped into the thickness δ. The reciprocal of the mass transfer coefficient can be thought of as this resistance

$$\boxed{W_{Az} = \text{Flux} = \frac{\text{Driving force}}{\text{Resistance}} = \frac{C_{Ab} - C_{As}}{(1/k_c)}}$$

14.4.1 Correlations for the Mass Transfer Coefficient

The mass transfer coefficient k_c is analogous to the heat transfer coefficient h. The heat flux q from the bulk fluid at a temperature T_0 to a solid surface at T_s is

$$q_z = h(T_0 - T_s) \qquad (14\text{-}29)$$

For forced convection, the heat transfer coefficient is normally correlated in terms of three dimensionless groups: the Nusselt number, Nu; the Reynolds number, Re; and the Prandtl number, Pr. For the single spherical pellets discussed here, Nu and Re take the following forms

$$\text{Nu} = \frac{h d_p}{k_t} \tag{14-30}$$

$$\text{Re} = \frac{U \rho d_p}{\mu} \tag{14-31}$$

The Prandtl number is not dependent on the geometry of the system

$$\text{Pr} = \frac{\mu C_p}{k_t} = \frac{\mu}{\rho}\left(\frac{\rho C_p}{k_t}\right) = \frac{\nu}{\alpha_t} \tag{14-32}$$

> The Nusselt, Prandtl, and Reynolds numbers are used in forced convection heat transfer correlations.

where $\alpha_t = k_t/\rho C_p$ = thermal diffusivity, m^2/s
μ = viscosity, kg/m·s
ρ = fluid density, kg/m^3
$\nu = \dfrac{\mu}{\rho}$ = kinematic viscosity (momentum diffusivity), m^2/s
d_p = diameter of pellet, m
U = free-stream velocity, m/s
k_t = thermal conductivity, J/K·m·s
h = heat transfer coefficient, $J/m^2·s·K$ or $Watts/m^2$ K
C_p = heat capacity, J/kg/K

The other symbols are as defined previously.

The heat transfer correlation relating the Nusselt number to the Prandtl and Reynolds numbers for flow around a sphere is[7]

$$\text{Nu} = 2 + 0.6\text{Re}^{1/2}\text{Pr}^{1/3} \tag{14-33}$$

Although this correlation can be used over a wide range of Reynolds numbers, it can be shown theoretically that if a sphere is immersed in a stagnant fluid (Re = 0), then

$$\text{Nu} = 2 \tag{14-34}$$

and that at higher Reynolds numbers in which the boundary layer remains laminar, we can neglect the 2 in Equation (14-34), in which case he Nusselt number becomes

$$\text{Nu} \simeq 0.6\text{Re}^{1/2}\text{Pr}^{1/3} \tag{14-35}$$

Although further discussion of heat transfer correlations is no doubt worthwhile, it will not help us to determine the mass transfer coefficient and the mass flux from the bulk fluid to the external pellet surface. However, the preceding discussion on heat transfer was not entirely futile because, for similar geometries, *the heat and mass transfer correlations are analogous*. If a heat transfer correlation for the Nusselt number exists, the mass transfer coefficient

> Converting a heat transfer correlation to a mass transfer correlation

---

[7] W. E. Ranz and W. R. Marshall, Jr., *Chem. Eng. Prog.*, 48, 141–146, 173–180 (1952).

† Strictly speaking, replacing the Nusselt number by the Sherwood number is only valid for situations where the Lewis number, Le, is close to 1. $\text{Le} = \dfrac{\text{Sc}}{\text{Pe}} = \dfrac{\alpha_t}{D_{AB}}$

can be estimated by replacing the Nusselt and Prandtl numbers in this correlation by the Sherwood and Schmidt numbers, respectively:

$$\text{Sh} \longrightarrow \text{Nu}$$

$$\text{Sc} \longrightarrow \text{Pr}$$

The heat and mass transfer coefficients are analogous.[†] The corresponding fluxes are

$$q_z = h(T - T_s) \tag{14-36}$$

$$W_{Az} = k_c(C_A - C_{As})$$

The one-dimensional differential forms of the mass flux for EMCD and the heat flux are, respectively,

> For EMCD the heat and molar flux equations are analogous.

$$W_{Az} = -D_{AB}\frac{dC_A}{dz} \tag{E14-1.3}$$

$$q_z = -k_t\frac{dT}{dz} \tag{14-37}$$

If we replace h by k_c and k_t by D_{AB} in Equation (14-30), i.e.,

$$\left.\begin{array}{l} h \longrightarrow k_c \\[4pt] k_t \longrightarrow D_{AB} \end{array}\right\} \text{Nu} \longrightarrow \text{Sh}$$

we obtain the mass transfer Nusselt number (i.e., the Sherwood number)

> Sherwood number

$$\text{Sh} = \frac{k_c d_p}{D_{AB}} = \frac{(\text{m/s})(\text{m})}{\text{m}^2/\text{s}} \text{ dimensionless} \tag{14-38}$$

The Prandtl number is the ratio of the kinematic viscosity (i.e., the momentum diffusivity) to the thermal diffusivity. Because the Schmidt number is analogous to the Prandtl number, one would expect that Sc is the ratio of the momentum diffusivity (i.e., the kinematic viscosity), ν, to the mass diffusivity D_{AB}. Indeed, this is true

$$\alpha_t \longrightarrow D_{AB}$$

The Schmidt number is

> Schmidt number

$$\text{Sc} = \frac{\nu}{D_{AB}} = \frac{\text{m}^2/\text{s}}{\text{m}^2/\text{s}} \text{ dimensionless} \tag{14-39}$$

Consequently, the correlation for mass transfer for flow around a spherical pellet is analogous to that given for heat transfer, Equation (14-33); that is,

$$\text{Sh} = 2 + 0.6\text{Re}^{1/2}\text{Sc}^{1/3} \tag{14-40}$$

This relationship is often referred to as the *Frössling correlation*.[8]

A few additional correlations for the Sherwood number from which one can determine the mass transfer coefficient are given in Table 14-5.

---

[8] N. Frössling, *Gerlands Beitr. Geophys.*, 52, 170 (1938).

The Sherwood, Reynolds, and Schmidt numbers are used in forced convection mass transfer correlations.

TABLE 14-5   MASS TRANSFER CORRELATIONS

Turbulent flow, mass transfer to pipe wall	$Sh = .332 \, (Re)^{1/2} \, (Sc)^{1/3}$
Mass transfer to a single sphere	$Sh = 2 + 0.6 \, Re^{1/2} \, Sc^{1/3}$
Mass transfer in fluidized beds	$\phi J_D = \dfrac{0.765}{Re^{.82}} + \dfrac{0.365}{Re^{0.386}}$
Mass transfer to packed beds	$\phi J_D = 0.453 \, Re^{0.453}$
	$J_D = \dfrac{Sh}{ReSc^{1/3}}$

14.4.2 Mass Transfer to a Single Particle

In this section we consider two limiting cases of diffusion and reaction on a catalyst particle.[9] In the first case, the reaction is so rapid that the rate of diffusion of the reactant to the surface limits the reaction rate. In the second case, the reaction is so slow that virtually no concentration gradient exists in the gas phase (i.e., rapid diffusion with respect to surface reaction).

Example 14–1   Rapid Reaction on the Surface of a Catalyst

If the surface reaction is rapid, then diffusion limits the overall rate.

Calculate the molar flux, W_{Ar}, of reactant A to a single catalyst pellet 1 cm in diameter suspended in a large body of liquid B. The reactant is present in dilute concentrations, and the reaction is considered to take place instantaneously at the external pellet surface (i.e., $C_{As} \simeq 0$). The bulk concentration of the reactant A is 1.0 M, and the free-stream liquid velocity past the sphere is 0.1 m/s. The kinematic viscosity (i.e., $\frac{\mu}{\rho}$) is 0.5 centistoke (cS; 1 centistoke $= 10^{-6}$ m²/s), and the liquid diffusivity of A in B is $D_{AB} = 10^{-10}$ m²/s, at 300 K.

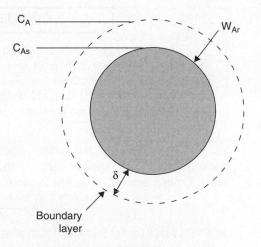

---

[9] A comprehensive list of correlations for mass transfer to particles is given by G. A. Hughmark, *Ind. Eng. Chem. Fund.*, 19(2), 198 (1980).

Solution

For dilute concentrations of the solute, the radial flux is

$$W_{Ar} = k_c(C_{Ab} - C_{As}) \tag{14-28}$$

Because reaction is assumed to occur instantaneously on the external surface of the pellet, $C_{As} = 0$. Also, C_{Ab} is given as 1 mol/dm^3. The mass transfer coefficient for single spheres is calculated from the Frössling correlation

Liquid Phase

Re = 2000

Sc = 5000

Sh = 460

$k_c = 4.6 \times 10^{-6}$ m/s

$$\text{Sh} = \frac{k_c d_p}{D_{AB}} = 2 + 0.6\text{Re}^{1/2}\text{Sc}^{1/3} \tag{14-41}$$

$$\text{Re} = \frac{\rho d_p U}{\mu} = \frac{d_p U}{\nu} = \frac{(0.01\ \text{m})(0.1\ \text{m/s})}{0.5 \times 10^{-6}\ \text{m}^2/\text{s}} = 2000$$

$$\text{Sc} = \frac{\nu}{D_{AB}} = \frac{5 \times 10^{-7}\ \text{m}^2/\text{s}}{10^{-10}\ \text{m}^2/\text{s}} = 5000$$

Substituting these values into Equation (14-40) gives us

$$\text{Sh} = 2 + 0.6(2000)^{0.5}(5000)^{1/3} = 460.7 \tag{E14-1.1}$$

$$k_c = \frac{D_{AB}}{d_p}\text{Sh} = \frac{10^{-10}\ \text{m}^2/\text{s}}{0.01\ \text{m}} \times 460.7 = 4.61 \times 10^{-6}\ \text{m/s} \tag{E14-1.2}$$

$$C_{Ab} = 1.0\ \text{mol/dm}^3 = 10^3\ \text{mol/m}^3$$

Substituting for k_c and C_{Ab} in Equation (14-26), the molar flux to the surface is

$$W_{Ar} = (4.61 \times 10^{-6})\ \text{m/s}\ (10^3 - 0)\ \text{mol/m}^3 = 4.61 \times 10^{-3}\ \text{mol/m}^2\cdot\text{s}$$

Because $W_{Ar} = -r''_{As}$, this rate is also the rate of reaction per unit surface area of catalyst.

$$\boxed{-r''_{As} = 0.0046\ \text{mol/m}^2\cdot\text{s} = 0.46\ \text{mol/dm}^2\cdot\text{s}}$$

Analysis: In this example we calculated the rate of reaction on the external surface of a catalyst pellet when external mass transfer was limiting the reaction rate. To determine the rate of reaction, we used correlations to calculate the mass transfer coefficient and then used k_c to calculate the flux to the surface, which in turn was equal to the rate of surface reaction.

In Example 14-1, the surface reaction was extremely rapid and the rate of mass transfer to the surface dictated the overall rate of reaction. We now consider a more general case. The isomerization

$$\text{A} \longrightarrow \text{B}$$

is taking place on the surface of a solid sphere (Figure 14-4). The surface reaction follows a Langmuir–Hinshelwood single-site mechanism for which the rate law is

$$-r''_{As} = \frac{k_r C_{As}}{1 + K_A C_{As} + K_B C_{Bs}} \tag{14-41}$$

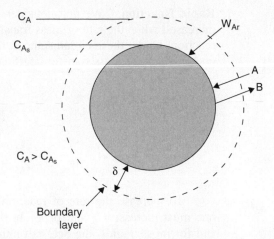

Figure 14-4   Diffusion to, and reaction on, external surface of pellet.

The temperature is sufficiently high that we only need to consider the case of very weak adsorption (i.e., low surface coverage) of A and B; thus

$$(K_B C_{Bs} + K_A C_{As}) \ll 1$$

Therefore, the rate law becomes apparent first order

$$-r''_{As} = k_r C_{As} \tag{14-42}$$

Using boundary conditions 2b and 2c in Table 14-1, we obtain

$$W_A\big|_{\text{surface}} = -r''_{As} \tag{14-43}$$

$$W_A = k_c(C_A - C_{As}) = k_r C_{As} \tag{14-44}$$

The concentration C_{As} is not as easily measured as the bulk concentration. Consequently, we need to eliminate C_{As} from the equation for the flux and rate of reaction. Solving Equation (14-44) for C_{As} yields

$$C_{As} = \frac{k_c C_A}{k_r + k_c} \tag{14-45}$$

and the rate of reaction on the surface becomes

Molar flux of A to the surface is equal to the rate of consumption of A on the surface.

$$\boxed{W_A = -r''_{As} = \frac{k_c k_r C_A}{k_r + k_c}} \tag{14-46}$$

One will often find the flux to or from the surface written in terms of an *effective* transport coefficient k_{eff}

$$W_A = -r''_{As} = k_{\text{eff}} C_A \tag{14-47}$$

where

$$k_{\text{eff}} = \frac{k_c k_r}{k_c + k_r}$$

Rapid Reaction. We first consider how the overall rate of reaction may be increased when the rate of mass transfer to the surface limits the overall rate of reaction. Under these circumstances, the specific reaction rate constant is much greater than the mass transfer coefficient

$$k_r \gg k_c$$

and

$$k_{\text{eff}} = k_c$$

$$W_A = -r''_{As} = \frac{k_c C_A}{1 + k_c/k_r} \approx k_c C_A \qquad (14\text{-}48)$$

To increase the rate of reaction per unit surface area of a solid sphere, one must increase C_A and/or k_c. In this gas-phase catalytic reaction example, and for most liquids, the Schmidt number is sufficiently large that the number 2 in Equation (14-40) is negligible with respect to the second term when the Reynolds number is greater than 25. As a result, Equation (14-40) gives

It is important to know how the mass transfer coefficient varies with fluid velocity, particle size, and physical properties.

$$k_c = 0.6 \left(\frac{D_{AB}}{d_p} \right) \text{Re}^{1/2} \text{Sc}^{1/3}$$

$$= 0.6 \left(\frac{D_{AB}}{d_p} \right) \left(\frac{U d_p}{\nu} \right)^{1/2} \left(\frac{\nu}{D_{AB}} \right)^{1/3}$$

$$k_c = 0.6 \times \frac{D_{AB}^{2/3}}{\nu^{1/6}} \times \frac{U^{1/2}}{d_p^{1/2}} \qquad (14\text{-}49)$$

$$k_c = 0.6 \times (\text{Term 1}) \times (\text{Term 2})$$

Mass Transfer Limited

Term 1 is a function of the physical properties D_{AB} and ν, which depend on temperature and pressure only. The diffusivity always increases with increasing temperature for both gas and liquid systems. However, the kinematic viscosity ν increases with temperature ($\nu \propto T^{3/2}$) for gases and decreases exponentially with temperature for liquids. Term 2 is a function of flow conditions and particle size. Consequently, to increase k_c and thus the overall rate of reaction per unit surface area, one may either decrease the particle size or increase the velocity of the fluid flowing past the particle. For this particular case of flow past a single sphere, we see that if the velocity is doubled, the mass transfer coefficient and consequently the rate of reaction is increased by a factor of

$$(U_2/U_1)^{0.5} = 2^{0.5} = 1.41 \text{ or } 41\%$$

Reaction Rate Limited

Slow Reaction. Here, the specific reaction rate constant is small with respect to the mass transfer coefficient

$$k_r \ll k_c$$

$$W_A = -r''_{As} = \frac{k_r C_A}{1 + k_r/k_c} \approx k_r C_A \qquad (14\text{-}50)$$

Mass transfer effects are not important when the reaction rate is limiting.

The specific reaction rate is independent of the velocity of fluid and for the solid sphere considered here, independent of particle size. *However*, for porous catalyst pellets, k_r may depend on particle size for certain situations, as shown in Chapter 15. We will continue this discussion in Section 14.5.

Figure 14-5 shows the variation in reaction rate with Term 2 in Equation (14-49), the ratio of velocity to particle size. At low velocities, the mass transfer boundary-layer thickness is large and diffusion limits the reaction. As the velocity past the sphere is increased, the boundary-layer thickness decreases, and the mass transfer across the boundary layer no longer limits the rate of reaction. One also notes that for a given (i.e., fixed) velocity, reaction-limiting conditions can be achieved by using very small particles. However, the smaller the particle size, the greater the pressure drop in a packed bed. When one is obtaining reaction-rate data in the laboratory, one must operate at sufficiently high velocities or sufficiently small particle sizes to ensure that the reaction is not mass transfer–limited when collecting data.

When collecting rate-law data, operate in the reaction-limited region.

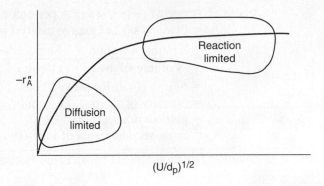

Figure 14-5   Regions of mass transfer–limited and reaction-limited reactions.

14.4.3 Mass Transfer–Limited Reactions in Packed Beds

A number of industrial reactions are potentially mass transfer–limited because they may be carried out at high temperatures without the occurrence of undesirable side reactions. In mass transfer–dominated reactions, the surface reaction is so rapid that the rate of transfer of reactant from the bulk gas or liquid phase to the surface limits the overall rate of reaction. Consequently, mass transfer–limited reactions respond quite differently to changes in temperature and flow conditions than do the rate-limited reactions discussed in previous chapters. In this section the basic equations describing the variation of conversion with the various reactor design parameters (catalyst weight, flow conditions) will be developed. To achieve this goal, we begin by carrying out a mole balance on the following generic mass transfer–limited reaction

$$A + \frac{b}{a} B \longrightarrow \frac{c}{a} C + \frac{d}{a} D \tag{2-2}$$

carried out in a packed-bed reactor (Figure 14-6). A steady-state mole balance on reactant A in the reactor segment between z and z + Δz is

$$\begin{bmatrix} \text{Molar} \\ \text{rate in} \end{bmatrix} - \begin{bmatrix} \text{Molar} \\ \text{rate out} \end{bmatrix} + \begin{bmatrix} \text{Molar rate of} \\ \text{generation} \end{bmatrix} = \begin{bmatrix} \text{Molar rate of} \\ \text{accumulation} \end{bmatrix}$$

$$F_{Az}|_z \quad - \quad F_{Az}|_{z+\Delta z} \quad + \quad r_A'' a_c (A_c \Delta z) \quad = \quad 0 \qquad (14\text{-}51)$$

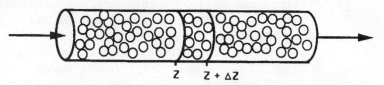

Figure 14-6 Packed-bed reactor.

where r_A'' = rate of generation of A per unit of catalytic surface area, mol/s·m²

a_c = external surface area of catalyst per volume of catalytic bed, m²/m³

$$a_c = \frac{\text{Volume of solid}}{\text{Volume of bed}} \times \frac{\text{Surface area}}{\text{Volume of solid}} = (1 - \phi)[\pi d_p^2 / (\pi d_p^3 / 6)]$$

$\qquad = 6(1 - \phi)/d_p$ for packed beds, m²/m³

ϕ = porosity of the bed (i.e., void fraction)[10]

d_p = particle diameter, m

A_c = cross-sectional area of tube containing the catalyst, m²

Dividing Equation (14-51) by $A_c \Delta z$ and taking the limit as $\Delta z \longrightarrow 0$, we have

$$-\frac{1}{A_c}\left(\frac{dF_{Az}}{dz}\right) + r_A'' a_c = 0 \qquad (14\text{-}52)$$

We now need to express F_{Az} and r_A'' in terms of concentration.

The molar flow rate of A in the axial direction is

$$F_{Az} = A_c W_{Az} = (J_{Az} + B_{Az})A_c \qquad (14\text{-}53)$$

Axial diffusion is neglected.
In almost all situations involving flow in packed-bed reactors, the amount of material transported by diffusion or dispersion in the axial direction is negligible compared with that transported by convection (i.e., bulk flow)

$$J_{Az} \ll B_{Az}$$

(In Chapter 18 we consider the case when dispersive effects (e.g., diffusion) must be taken into account.) Neglecting dispersion, Equation (14-14) becomes

$$F_{Az} = A_c W_{Az} = A_c B_{Az} = U C_A A_c \qquad (14\text{-}54)$$

where U is the superficial molar average velocity through the bed (m/s). Substituting for F_{Az} in Equation (14-52) gives us

$$-\frac{d(C_A U)}{dz} + r_A'' a_c = 0 \qquad (14\text{-}55)$$

---

[10]In the nomenclature for Chapter 4, for the Ergun equation for pressure drop.

For the case of constant superficial velocity U

$$\boxed{-U\frac{dC_A}{dz} + r_A'' a_c = 0}$$
(14-56)

For reactions at steady state, the molar flux of A to the particle surface, W_{Ar} (mol/m²·s) (see Figure 14-7), is equal to the rate of disappearance of A on the surface $-r_A''$ (mol/m²·s); that is

$$-r_A'' = W_{Ar}$$
(14-57)

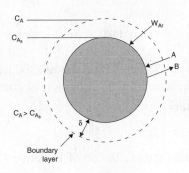

Figure 14-7 Diffusion across stagnant film surrounding catalyst pellet.

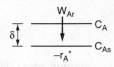

From Section 14.4, the boundary condition at the external surface is

$$-r_A'' = W_{Ar} = k_c(C_A - C_{As})$$
(14-58)

where k_c = mass transfer coefficient = (D_{AB}/δ), (m/s)
C_A = bulk concentration of A (mol/m³)
C_{As} = concentration of A at the catalytic surface (mol/m³)

Substituting for r_A'' in Equation (14-56), we have

$$-U\frac{dC_A}{dz} - k_c a_c(C_A - C_{As}) = 0$$
(14-59)

In reactions that are completely mass transfer–limited, it is not necessary to know the rate law.

In most mass transfer–limited reactions, the surface concentration is negligible with respect to the bulk concentration (i.e., $C_A \gg C_{As}$)

$$-U\frac{dC_A}{dz} = k_c a_c C_A$$
(14-60)

Integrating with the limit, at $z = 0$, $C_A = C_{A0}$

$$\boxed{\frac{C_A}{C_{A0}} = \exp\left(-\frac{k_c a_c}{U}z\right)}$$
(14-61)

The corresponding variation of reaction rate along the length of the reactor is

$$-r_A'' = k_c C_{A0} \exp\left(-\frac{k_c a_c}{U}z\right)$$
(14-62)

The concentration and conversion profiles down a reactor of length L are shown in Figure 14-8.

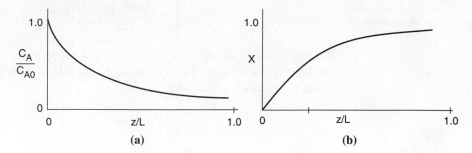

Reactor
concentration
profile for a mass
transfer–limited
reaction

Figure 14-8   Axial concentration (a) and conversion (b) profiles in a packed bed.

To determine the reactor length L necessary to achieve a conversion X, we combine the definition of conversion

$$X = \frac{C_{A0} - C_{AL}}{C_{A0}} \tag{14-63}$$

with the evaluation of Equation (14-61) at $z = L$ to obtain

$$\ln \frac{1}{1-X} = \frac{k_c a_c}{U} L \tag{14-64}$$

14.4.4 Robert the Worrier

Robert is an engineer who is always worried (which is a Jofostanian trait). He thinks something bad will happen if we change an operating condition such as flow rate or temperature or an equipment parameter such as particle size. Robert's motto is "If it ain't broke, don't fix it." We can help Robert be a little more adventuresome by analyzing how the important parameters vary as we change operating conditions in order to predict the outcome of such a change. We first look at Equation (14-64) and see that conversion depends upon the parameters k_c, a_c, U, and L. We now examine how each of these parameters will change as we change operating conditions. We first consider the effects of temperature and flow rate on conversion.

To learn the effect of flow rate on conversion, we need to know how flow rate affects the mass transfer coefficient. That is, we must determine the correlation for the mass transfer coefficient for the particular geometry and flow field. For flow through a packed bed, the correlation given by Thoenes and Kramers for $0.25 < \phi < 0.5$, $40 < \mathrm{Re}' < 4000$, and $1 < \mathrm{Sc} < 4000$ is[11]

$$\mathrm{Sh}' = 1.0(\mathrm{Re}')^{1/2}\mathrm{Sc}^{1/3} \tag{14-65}$$

Thoenes–Kramers
correlation for flow
through packed
beds

$$\left[\frac{k_c d_p}{D_{\mathrm{AB}}} \left(\frac{\phi}{1-\phi} \right) \frac{1}{\gamma} \right] = \left[\frac{U d_p \rho}{\mu (1-\phi) \gamma} \right]^{1/2} \left(\frac{\mu}{\rho D_{\mathrm{AB}}} \right)^{1/3} \tag{14-66}$$

---

[11]D. Thoenes, Jr. and H. Kramers, *Chem. Eng. Sci.*, 8, 271 (1958).

where $\text{Re}' = \dfrac{\text{Re}}{(1-\phi)\gamma}$

$$\text{Sh}' = \frac{\text{Sh}\,\phi}{(1-\phi)\gamma}$$

d_p = particle diameter (equivalent diameter of a sphere of the same volume), m

    $= [(6/\pi)\,(\text{volume of pellet})]^{1/3}$, m

ϕ = void fraction (porosity) of packed bed

γ = shape factor (external surface area divided by πd_p^2)

and

U, ρ, μ, v and D_{AB} are as previously defined.

For constant fluid properties and particle diameter

$$k_c \propto U^{1/2} \qquad (14\text{-}67)$$

We see that the mass transfer coefficient increases with the square root of the superficial velocity through the bed. Therefore, *for a fixed concentration, C_A,* such as that found in a differential reactor, the rate of reaction should vary with $U^{1/2}$

$$-r_A'' \propto k_c C_A \propto U^{1/2}$$

However, if the gas velocity is continually increased, a point is reached where the reaction becomes reaction rate–limited and, consequently, is independent of the superficial gas velocity, as shown in Figure 14-5.

Most mass transfer correlations in the literature are reported in terms of the Colburn J factor (i.e., J_D) as a function of the Reynolds number. The relationship between J_D and the numbers we have been discussing is

Colburn J factor

$$\boxed{J_D = \frac{\text{Sh}}{\text{Sc}^{1/3}\,\text{Re}}} \qquad (14\text{-}68)$$

Figure 14-9 shows data from a number of investigations for the J factor as a function of the Reynolds number for a wide range of particle shapes and gas-flow conditions. *Note:* There are serious deviations from the Colburn analogy when the concentration gradient and temperature gradient are coupled, as shown by Venkatesan and Fogler.[12]

Dwidevi and Upadhyay review a number of mass transfer correlations for both fixed and fluidized beds and arrive at the following correlation, which is valid for both gases (Re > 10) and liquids (Re > 0.01) in either fixed or fluidized beds:[13]

A correlation for flow through packed beds in terms of the Colburn J factor

$$\boxed{\phi J_D = \frac{0.765}{\text{Re}^{0.82}} + \frac{0.365}{\text{Re}^{0.386}}} \qquad (14\text{-}69)$$

For diffusion-limited reactions, reaction rate depends on particle size and fluid velocity.

---

[12]R. Venkatesan and H. S. Fogler, *AIChE J.,* 50, 1623 (July 2004).

[13]P. N. Dwidevi and S. N. Upadhyay, *Ind. Eng. Chem. Process Des. Dev.,* 16, 157 (1977).

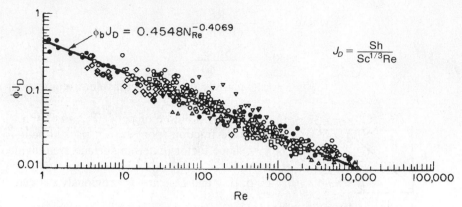

$$\phi_b J_D = 0.4548 N_{Re}^{-0.4069}$$

$$J_D = \frac{Sh}{Sc^{1/3}Re}$$

Figure 14-9   Mass transfer correlation for packed beds. $\phi b \equiv \phi$ [Reprinted by permission. Copyright © 1977, American Chemical Society. Dwivedi, P. N. and S. N. Upadhyay, "Particle-Fluid Mass Transfer in Fixed and Fluidized Beds." *Industrial & Engineering Chemistry Process Design and Development*, 1977, 16 (2), 157–165.]

For nonspherical particles, the equivalent diameter used in the Reynolds and Sherwood numbers is $d_p = \sqrt{A_p/\pi} = 0.564\sqrt{A_p}$, where A_p is the external surface area of the pellet.

To obtain correlations for mass transfer coefficients for a variety of systems and geometries, see either D. Kunii and O. Levenspiel, *Fluidization Engineering*, 2nd ed. (Butterworth-Heinemann, 1991), Chap. 7, or W. L. McCabe, J. C. Smith, and P. Harriott, *Unit Operations in Chemical Engineering*, 6th ed. (New York: McGraw-Hill, 2000). For other correlations for packed beds with different packing arrangements, see I. Colquhoun-Lee and J. Stepanek, *Chemical Engineer*, 108 (Feb. 1974).

Example 14–2   Mass Transfer Effects in Maneuvering a Space Satellite

Actual case history and current application

Hydrazine has been studied extensively for use in monopropellant thrusters for space flights of long duration. Thrusters are used for altitude control of communication satellites. Here, the decomposition of hydrazine over a packed bed of alumina-supported iridium catalyst is of interest.[14] In a proposed study, a 2% hydrazine in 98% helium mixture is to be passed over a packed bed of cylindrical particles 0.25 cm in diameter and 0.5 cm in length at a gas-phase velocity of 150 m/s and a temperature of 450 K. The kinematic viscosity of helium at this temperature is 4.94×10^{-5} m²/s. The hydrazine decomposition reaction is believed to be externally mass transfer–limited under these conditions. If the packed bed is 0.05 m in length, what conversion can be expected? Assume isothermal operation.

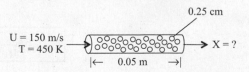

[14]O. I. Smith and W. C. Solomon, *Ind. Eng. Chem. Fund.*, 21, 374.

Additional information:

$D_{AB} = 0.69 \times 10^{-4}$ m²/s at 298 K
Bed porosity: 40%
Bed fluidity: 95.7%

Solution

The following solution is detailed and a bit tedious, but it is important to know the details of how a mass transfer coefficient is calculated.

Rearranging Equation (14-64) gives us

$$\boxed{X = 1 - e^{-(k_c a_c / U)L}}$$
(E14-2.1)

Tedious reading and calculations, but we gotta know how to do the nitty–gritty.

(a) Using the Thoenes–Kramers correlation to calculate the mass transfer coefficient, k_c

1. First we find the volume-average particle diameter

$$d_p = \left(\frac{6V}{\pi}\right)^{1/3} = \left(6\,\frac{\pi D^2}{4}\,\frac{L}{\pi}\right)^{1/3}$$
(E14-2.2)

$$= [1.5(0.0025 \text{ m})^2(0.005 \text{ m})]^{1/3} = 3.61 \times 10^{-3} \text{ m}$$

2. Surface area per volume of bed

$$a_c = 6\left(\frac{1 - 0.4}{d_p}\right) = 6\left(\frac{1 - 0.4}{3.61 \times 10^{-3} \text{ m}}\right) = 998 \text{ m}^2/\text{m}^3$$
(E14-2.3)

3. Mass transfer coefficient

$$\text{Re} = \frac{d_p U}{\nu} = \frac{(3.61 \times 10^{-3} \text{ m})(150 \text{ m/s})}{4.94 \times 10^{-4} \text{ m}^2/\text{s}} = 10942$$

For cylindrical pellets

$$\gamma = \frac{2\pi r L_p + 2\pi r^2}{\pi d_p^2} = \frac{(2)(0.0025/2)(0.005) + (2)(0.0025/2)^2}{(3.61 \times 10^{-3})^2} = 1.20 \quad \text{(E14-2.4)}$$

Representative values

$$\text{Re}' = \frac{\text{Re}}{(1 - \phi)\gamma} = \frac{10942}{(0.6)(1.2)} = 15173$$

Correcting the diffusivity to 450 K using Table 14-2 gives us

Gas Phase

$\text{Re}' = 15173$
$\text{Sc} = 0.35$
$\text{Sh}' = 86.66$
$k_c = 6.15$ m/s

$$D_{AB}(450 \text{ K}) = D_{AB}(298 \text{ K}) \times \left(\frac{450}{298}\right)^{1.75} = (0.69 \times 10^{-4} \text{ m}^2/\text{s})(2.06)$$

$$D_{AB}(450 \text{ K}) = 1.42 \times 10^{-4} \text{ m}^2/\text{s}$$
(E14-2.5)

$$\text{Sc} = \frac{\nu}{D_{AB}} = \frac{4.94 \times 10^{-5} \text{ m}^2/\text{s}}{1.42 \times 10^{-4} \text{ m}^2/\text{s}} = 0.35$$

Substituting Re′ and Sc into Equation (14-65) yields

$$\text{Sh}' = (15173.92)^{1/2}(0.35)^{1/3} = (123.18)(0.70) = 86.66 \quad \text{(E14-2.6)}$$

$$k_c = \frac{D_{AB}(1 - \phi)}{d_p \phi}\,\gamma(\text{Sh}') = \left(\frac{1.42 \times 10^{-4} \text{ m}^2/\text{s}}{3.61 \times 10^{-3} \text{ m}}\right)\left(\frac{1 - 0.4}{0.4}\right) \times (1.2)(86.66)$$

$$\boxed{k_c = 6.15 \text{ m/s}}$$
(E14-2.7)

The conversion is

$$X = 1 - \exp\left[-(6.15 \text{ m/s})\left(\frac{998 \text{ m}^2/\text{m}^3}{150 \text{ m/s}}\right)(0.05 \text{ m})\right] \quad \text{(E14-2.8)}$$

$$= 1 - 0.13 \simeq 0.87$$

We find 87% conversion.

(b) Colburn J_D factor to calculate k_c. To find k_c, we first calculate the surface-area-average particle diameter.

For cylindrical pellets, the external surface area is

$$A = \pi d L_p + 2\pi\left(\frac{d^2}{4}\right) \quad \text{(E14-2.9)}$$

$$d_p = \sqrt{\frac{A}{\pi}} = \sqrt{\frac{\pi d L_p + 2\pi(d^2/4)}{\pi}} \quad \text{(E14-2.10)}$$

$$= \sqrt{(0.0025)(0.005) + \frac{(0.0025)^2}{2}} = 3.95 \times 10^{-3} \text{ m}$$

$$a_c = \frac{6(1-\phi)}{d_p} = 910.74 \text{ m}^2/\text{m}^3$$

$$\text{Re} = \frac{d_p U}{\nu} = \frac{(3.95 \times 10^{-3} \text{ m})(150 \text{ m/s})}{4.94 \times 10^{-5} \text{m}^2/\text{s}}$$

$$= 11996.04$$

$$\phi J_D = \frac{0.765}{\text{Re}^{0.82}} + \frac{0.365}{\text{Re}^{0.386}} \quad \text{(14-69)}$$

$$= \frac{0.765}{(11996)^{0.82}} + \frac{0.365}{(11996)^{0.386}} = 3.5 \times 10^{-4} + 9.7 \times 10^{-3} \quad \text{(E14-2.11)}$$

$$= 0.010$$

$$J_D = \frac{0.010}{0.4} = 0.25 \quad \text{(E14-2.12)}$$

$$\text{Sh} = \text{Sc}^{1/3}\text{Re}(J_D) \quad \text{(E14-2.13)}$$

$$= (0.35)^{1/3}(11996)(0.025) = 212$$

$$k_c = \frac{D_{AB}}{d_p}\text{Sh} = \frac{1.42 \times 10^{-4}}{3.95 \times 10^{-3}}(212) = 7.63 \text{ m/s}$$

Then $$X = 1 - \exp\left[-(7.63 \text{ m/s})\left(\frac{910 \text{ m}^2/\text{m}^3}{150 \text{ m/s}}\right)(0.05 \text{ m})\right] \quad \text{(E14-2.14)}$$

$$\simeq 0.9$$

Fluidicity??
Red herring!

If there were such a thing as the **bed fluidicity**, given in the problem statement, it would be a useless piece of information. Make sure that you know what information you need to solve problems, and go after it. Do not let additional data confuse you or lead you astray with useless information or facts that represent someone else's bias, and which are probably not well founded.

Once again the
nitty-gritty

Typical
values

Gas Phase
Re = 11996
J_D = 0.025
Sc = 0.35
Sh = 212
k_c = 7.58 m/s

14.5 What If . . . ? (Parameter Sensitivity)

J. D. Goddard's

Back of the
Envelope

As we have stressed many times, one of the most important skills of an engineer is to be able to predict the effects of changes of system variables on the operation of a process. The engineer needs to determine these effects quickly through approximate but reasonably close calculations, which are sometimes referred to as "back-of-the-envelope calculations."[15] This type of calculation is used to answer such questions as "**What** will happen **if** I decrease the particle size?" "**What if** I triple the flow rate through the reactor?"

To help answer these questions, we recall Equation (14-49) and our discussion on page 696. There, we showed the mass transfer coefficient for a packed bed was related to the product of two terms: Term 1 was dependent on the physical properties and Term 2 was dependent on the system properties. Re-writing Equation (14-41) as

$$k_c \propto \left(\frac{D_{AB}^{2/3}}{v^{1/6}} \right) \left(\frac{U^{1/2}}{d_p^{1/2}} \right) \tag{14-70}$$

Find out how the
mass transfer
coefficient varies
with changes in
*physical properties
and system
properties.*

one observes from this equation that the mass transfer coefficient increases as the particle size decreases. The use of sufficiently small particles offers another technique to escape from the mass transfer–limited regime into the reaction-rate-limited regime.

Example 14–3   The Case of Divide and Be Conquered

A mass transfer–limited reaction is being carried out in two reactors of equal volume and packing, connected in series as shown in Figure E14-3.1. Currently, 86.5% conversion is being achieved with this arrangement. It is suggested that the reactors be separated and the flow rate be divided equally among each of the two reactors (Figure E14-3.2) to decrease the pressure drop and hence the pumping requirements. In terms of achieving a higher conversion, Robert is wondering if this is a good idea.

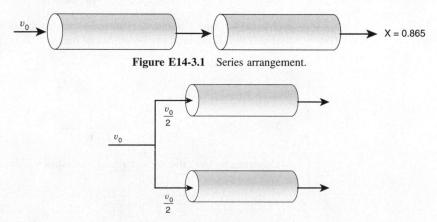

Figure E14-3.1   Series arrangement.

Reactors in series
versus reactors in
parallel

Figure E14-3.2   Parallel arrangement.

---

[15]Prof. J. D. Goddard, University of Michigan, 1963–1976. Currently at University of California, San Diego.

Solution

For the series arrangement we were given, $X_1 = 0.865$, and for the parallel arrangement, the conversion is unknown, i.e., $X_2 = ?$ As a first approximation, we neglect the effects of small changes in temperature and pressure on mass transfer. We recall Equation (14-64), which gives conversion as a function of reactor length. For a mass transfer–limited reaction

$$\ln \frac{1}{1-X} = \frac{k_c a_c}{U} L \tag{14-64}$$

For case 1, the undivided system

$$\left(\ln \frac{1}{1-X_1} \right) = \frac{k_{c1} a_c}{U_1} L_1 \tag{E14-3.1}$$

$$X_1 = 0.865$$

For case 2, the divided system

$$\left(\ln \frac{1}{1-X_2} \right) = \frac{k_{c2} a_c}{U_2} L_2 \tag{E14-3.2}$$

$$X_2 = ?$$

We now take the ratio of case 2 (divided system) to case 1 (undivided system)

$$\frac{\ln \dfrac{1}{1-X_2}}{\ln \dfrac{1}{1-X_1}} = \frac{k_{c2}}{k_{c1}} \left(\frac{L_2}{L_1} \right) \frac{U_1}{U_2} \tag{E14-3.3}$$

The surface area per unit volume a_c is the same for both systems.

From the conditions of the problem statement we know that

$$L_2 = \tfrac{1}{2} L_1, \; U_2 = \tfrac{1}{2} U_1, \text{ and } X_1 = 0.865$$

$$X_2 = ?$$

However, we must also consider the effect of the division on the mass transfer coefficient. From Equation (14-70) we know that

$$k_c \propto U^{1/2}$$

Then

$$\frac{k_{c2}}{k_{c1}} = \left(\frac{U_2}{U_1} \right)^{1/2} \tag{E14-3.4}$$

Multiplying by the ratio of superficial velocities yields

$$\frac{U_1}{U_2} \left(\frac{k_{c2}}{k_{c1}} \right) = \left(\frac{U_1}{U_2} \right)^{1/2} \tag{E14-3.5}$$

$$\ln \frac{1}{1-X_2} = \left(\ln \frac{1}{1-X_1} \right) \frac{L_2}{L_1} \left(\frac{U_1}{U_2} \right)^{1/2} \tag{E14-3.6}$$

$$= \left(\ln \frac{1}{1-0.865} \right) \left[\frac{\frac{1}{2}L_1}{L_1} \left(\frac{U_1}{\frac{1}{2}U_1} \right)^{1/2} \right]$$

$$= 2.00 \left(\frac{1}{2} \right) \sqrt{2} = 1.414$$

Solving for X_2 gives us

$$X_2 = 0.76$$

Bad idea!! Robert was right to worry.

Analysis: Consequently, we see that although the divided arrangement will have the advantage of a smaller pressure drop across the bed, it is a bad idea in terms of conversion. Recall that the series arrangement gave $X_1 = 0.865$; therefore $(X_2 < X_1)$. Bad idea!! But every chemical engineering student in Jofostan knew that! Recall that if the reaction were reaction rate–limited, both arrangements would give the same conversion.

Example 14–4   The Case of the Overenthusiastic Engineers

The same reaction as that in Example 14-3 is being carried out in the same two reactors in series. A new engineer suggests that the rate of reaction could be increased by a factor of 2^{10} by increasing the reaction temperature from 400°C to 500°C, reasoning that the reaction rate doubles for every 10°C increase in temperature. Another engineer arrives on the scene and berates the new engineer with quotations from Chapter 3 concerning this rule of thumb. She points out that it is valid only for a specific activation energy within a specific temperature range. She then suggests that he go ahead with the proposed temperature increase but should only expect an increase on the order of 2^3 or 2^4. What do you think? Who is correct?

Robert worries if this temperature increase will be worth the trouble.

Solution

Because almost all surface reaction rates increase more rapidly with temperature than do diffusion rates, increasing the temperature will only increase the degree to which the reaction is mass transfer–limited.

We now consider the following two cases:

Case 1: $T = 400°C$ $X = 0.865$

Case 2: $T = 500°C$ $X = ?$

Taking the ratio of case 2 to case 1 and noting that the reactor length is the same for both cases $(L_1 = L_2)$, we obtain

$$\frac{\ln \dfrac{1}{1-X_2}}{\ln \dfrac{1}{1-X_1}} = \frac{k_{c2}}{k_{c1}} \left(\frac{L_2}{L_1} \right) \frac{U_1}{U_2} = \frac{k_{c2}}{k_{c1}} \left(\frac{U_1}{U_2} \right) \qquad \text{(E14-4.1)}$$

The molar feed rate F_{T0} remains unchanged

$$F_{T0} = v_{01} \left(\frac{P_{01}}{RT_{01}} \right) = v_{02} \left(\frac{P_{02}}{RT_{02}} \right) \qquad \text{(E14-4.2)}$$

the pressure remains constant so

$$\frac{v_{01}}{T_1} = \frac{v_{02}}{T_2}$$

Because $v = A_c U$, the superficial velocity at temperature T_2 is

$$U_2 = \frac{T_2}{T_1} U_1 \tag{E14-4.3}$$

We now wish to learn the dependence of the mass transfer coefficient on temperature

$$k_c \propto \left(\frac{U^{1/2}}{d_p^{1/2}}\right)\left(\frac{D_{AB}^{2/3}}{v^{1/6}}\right) \tag{E14-4.4}$$

Taking the ratio of case 2 to case 1 and realizing that the particle diameter is the same for both cases gives us

$$\frac{k_{c2}}{k_{c1}} = \left(\frac{U_2}{U_1}\right)^{1/2}\left(\frac{D_{AB2}}{D_{AB1}}\right)^{2/3}\left(\frac{v_1}{v_2}\right)^{1/6} \tag{E14-4.5}$$

The temperature dependence of the gas-phase diffusivity is (from Table 14-2)

$$D_{AB} \propto T^{1.75} \tag{E14-4.6}$$

For most gases, viscosity increases with increasing temperature according to the relation

$$\mu \propto T^{1/2}$$

From the ideal gas law

$$\rho \propto T^{-1}$$

Then

$$v = \frac{\mu}{\rho} \propto T^{3/2} \tag{E14-4.7}$$

It's really important to know how to do this type of analysis.

$$\frac{\ln\dfrac{1}{1-X_2}}{\ln\dfrac{1}{1-X_1}} = \frac{U_1}{U_2}\left(\frac{k_{c2}}{k_{c1}}\right) = \left(\frac{U_1}{U_2}\right)^{1/2}\left(\frac{D_{AB2}}{D_{AB1}}\right)^{2/3}\left(\frac{v_1}{v_2}\right)^{1/6} \tag{E14-4.8}$$

$$= \left(\frac{T_1}{T_2}\right)^{1/2}\left[\left(\frac{T_2}{T_1}\right)^{1.75}\right]^{2/3}\left[\left(\frac{T_1}{T_2}\right)^{3/2}\right]^{1/6}$$

$$\frac{U_1 k_{c2}}{U_2 k_{c1}} = \left(\frac{T_1}{T_2}\right)^{1/2}\left(\frac{T_2}{T_1}\right)^{7/6}\left(\frac{T_1}{T_2}\right)^{1/4} = \left(\frac{T_2}{T_1}\right)^{5/12} \tag{E14-4.9}$$

$$= \left(\frac{773}{673}\right)^{5/12} = 1.059$$

Rearranging Equation (E14-4.1) in the form

$$\ln\frac{1}{1-X_2} = \frac{k_{c2} U_1}{k_{c1} U_2}\ln\frac{1}{1-X_1}$$

$$\ln\frac{1}{1-X_2} = 1.059\left(\ln\frac{1}{1-0.865}\right) = 1.059(2) \qquad \text{(E14-4.10)}$$

$$X_2 = 0.88$$

Bad idea!! Robert was right to worry.

Analysis: Consequently, we see that increasing the temperature from 400°C to 500°C increases the conversion by only 1.7%, i.e., $X = 0.865$ compared to $X = 0.88$. Bad idea! Bad, bad idea! Both engineers would have benefited from a more thorough study of this chapter.

For a packed catalyst bed, the temperature-dependence part of the mass transfer coefficient for a gas-phase reaction can be written as

$$k_c \propto U^{1/2} \; (D_{AB}^{2/3}/\nu^{1/6}) \qquad \text{(14-71)}$$

$$k_c \propto U^{1/2} T^{11/12} \qquad \text{(14-72)}$$

Important concept

Depending on how one fixes or changes the molar feed rate, F_{T0}, U may also depend on the feed temperature. ***As an engineer, it is extremely important that you reason out the effects of changing conditions***, as illustrated in the preceding two examples.

Closure. After completing this chapter, the reader should be able to define and describe molecular diffusion and how it varies with temperature and pressure, the molar flux, bulk flow, the mass transfer coefficient, the Sherwood and Schmidt numbers, and the correlations for the mass transfer coefficient. The reader should be able to choose the appropriate correlation and calculate the mass transfer coefficient, the molar flux, and the rate of reaction. The reader should also be able to describe the regimes and conditions under which mass transfer–limited reactions occur and when reaction-rate-limited reactions occur, and to make calculations of the rates of reaction and mass transfer for each case. One of the most important areas for the reader to apply the knowledge of this (and other chapters) is in his or her ability to ask and answer "What if . . ." questions.

SUMMARY

1. The molar flux of A in a binary mixture of A and B is

$$\mathbf{W}_A = -D_{AB}\nabla C_A + y_A(\mathbf{W}_A + \mathbf{W}_B) \qquad \text{(S14-1)}$$

a. For equimolar counterdiffusion (EMCD) **or** for dilute concentration of the solute

$$\mathbf{W}_A = \mathbf{J}_A = -D_{AB}\nabla C_A \qquad \text{(S14-2)}$$

b. For diffusion through a stagnant gas

$$\mathbf{W}_A = cD_{AB}\nabla\ln(1 - y_A) \qquad \text{(S14-3)}$$

c. For negligible diffusion

$$\mathbf{W}_A = y_A \mathbf{W} = y_A (\mathbf{W}_A + \mathbf{W}_B) = \mathbf{C}_A \mathbf{U} \qquad (S14\text{-}4)$$

2. The rate of mass transfer from the bulk fluid to a boundary at concentration C_{As} is

$$W_A = k_c (C_{Ab} - C_{As}) \qquad (S14\text{-}5)$$

where k_c is the mass transfer coefficient.
3. The Sherwood and Schmidt numbers are, respectively,

$$Sh = \frac{k_c d_p}{D_{AB}} \qquad (S14\text{-}6)$$

Representative Values	
Liquid Phase	Gas Phase
Re ~ 5000	Re ~ 500
Sc ~ 4000	Sc ~ 1
Sh ~ 500	Sh ~ 10
$k_c = 10^{-2}$ m/s	$k_c = 5$ m/s

$$Sc = \frac{\nu}{D_{AB}} \qquad (S14\text{-}7)$$

4. If a heat transfer correlation exists for a given system and geometry, the mass transfer correlation may be found by replacing the Nusselt number by the Sherwood number and the Prandtl number by the Schmidt number in the existing heat transfer correlation.
5. Increasing the gas-phase velocity and decreasing the particle size will increase the overall rate of reaction for reactions that are externally mass transfer–limited.

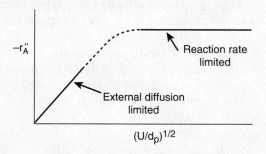

6. The conversion for externally mass transfer–limited reactions can be found from the equation

$$\ln \frac{1}{1-X} = \frac{k_c a_c}{U} L \qquad (S14\text{-}8)$$

7. Back-of-the-envelope calculations should be carried out to determine the magnitude and direction that changes in process variables will have on conversion. **What if . . .?**

CRE WEB SITE MATERIALS

- **Extended Material on the Website**
    1. *Transdermal Drug Delivery*

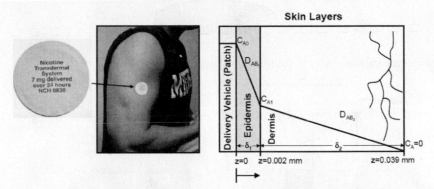

Figure WE14-1.1 Transdermal-drug-delivery schematic.

2. *Shrinking Core Model*

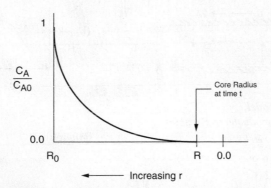

Figure 11-14 Oxygen concentration profile shown from the external radius of the pellet (R_0) to the pellet center. The gas–carbon interface is located at R.

3. *Diffusion through Film to a Catalyst Particle*
    4. *WP14-(a) Revisit Transdermal Drug Delivery*
    5. *Additional Homework Problems*
- **Learning Resources**
    1. *Summary Notes*
       Diffusion through a Stagnant Film
    4. *Solved Problems*
       Example CD14-1   Calculating Steady State Diffusion
       Example CD14-2   Relative Fluxes W_A, B_A, and J_A
       Example CD14-3   Diffusion through a Stagnant Gas
       Example CD14-4   Measuring Gas-Phase Diffusivities
       Example CD14-5   Diffusion through a Film to a Catalyst Particle
       Example CD14-6   Measuring Liquid-Phase Diffusivities

- **Professional Reference Shelf**

  R14.1. *Mass Transfer-Limited Reactions on Metallic Surfaces*
  - A. Catalyst Monoliths
  - B. Wire Gauze Reactors

  R14.2. *Methods to Experimentally Measure Diffusivities*

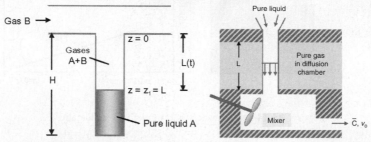

   A. Gas-phase diffusivities.   B. Liquid-phase diffusivities.

  R14.3. *Facilitated Heat Transfer*

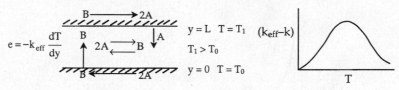

  R14.4.   *Shrinking Core Model*

  R14.5.   *Dissolution of Monodisperse Particles*

  R14.6.   *Dissolution of Polydisperse Solids (e.g., pills in the stomach)*

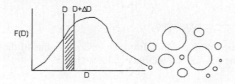

Ibuprofen

QUESTIONS AND PROBLEMS

The subscript to each of the problem numbers indicates the level of difficulty: A, least difficult; D, most difficult.

$$A = \bullet \quad B = \blacksquare \quad C = \blacklozenge \quad D = \blacklozenge\blacklozenge$$

Questions

Q14-1 Read over the problems at the end of this chapter. Make up an original problem that uses the concepts presented in this chapter. See problem P5-1$_A$ for the guidelines. To obtain a solution:
- **(a)** Make up your data and reaction.
- **(b)** Use a real reaction and real data.

The journals listed at the end of Chapter 1 may be useful for part (b).

Creative Thinking

Q14-2 (*Seargeant Amberrcromby*). Capt. Apollo is piloting a shuttlecraft on his way to space station Klingon. Just as he is about to maneuver to dock his craft using the hydrazine system discussed in **Example 14-2**, the shuttle craft's thrusters do not respond properly and it crashes into the station, killing Capt. Apollo (*Star Wars 7* (fall 2015)). An investigation reveals that Lt. Darkside prepared the packed beds used to maneuver the shuttle and Lt. Data prepared the hydrazine-helium gas mixture. Foul play is suspected and Sgt. Ambercromby arrives on the scene to investigate.

(a) What are the first three questions he asks?

(b) Make a list of possible explanations for the crash, supporting each one by an equation or reason.

Problems

P14-1$_B$ (a) **Example 14-1.** How would your answers change if the temperature was increased by 50°C, the particle diameter was doubled, and fluid velocity was cut in half? Assume properties of water can be used for this system.

(b) **Example 14-2.** How would your answers change if you had a 50–50 mixture of hydrazine and helium? If you increase d_p by a factor of 5?

(c) **Example 14-3.** *What if* you were asked for representative values for Re, Sc, Sh, and k_c for both liquid- and gas-phase systems for a velocity of 10 cm/s and a pipe diameter of 5 cm (or a packed-bed diameter of 0.2 cm)? What numbers would you give?

(d) **Example 14-4.** How would your answers change if the reaction were carried out in the liquid phase where kinematic viscosity varied as $v(T_2) = v(T_1)\exp\left[-4000K\left(\dfrac{1}{T_1} - \dfrac{1}{T_2}\right)\right]$?

P14-2$_B$ Assume the minimum respiration rate of a chipmunk is 1.5 micromoles of O_2/min. The corresponding volumetric rate of gas intake is 0.05 dm³/min at STP.

 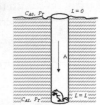

(a) What is the deepest a chipmunk can burrow a 3-cm diameter hole beneath the surface in Ann Arbor, Michigan? $D_{AB} = 1.8 \times 10^{-5}$ m²/s

(b) In Boulder, Colorado?

(c) How would your answers to (a) and (b) change in the dead of winter when $T = 0°F$?

(d) Critique and extend this problem (e.g., CO_2 poisoning).

Hint: Review derivations and equations for W_A and W_B to see how they can be applied to this problem.

P14-3$_B$ Pure oxygen is being absorbed by xylene in a catalyzed reaction in the experimental apparatus sketched in Figure P14-3$_B$. Under constant conditions of temperature and liquid composition, the following data were obtained:

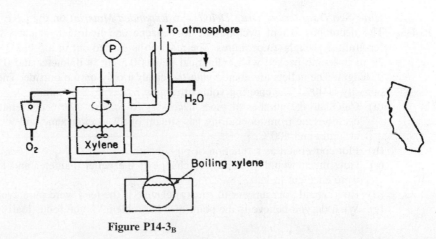

Figure P14-3$_B$

Stirrer Speed (rpm)	Rate of Uptake of O_2 (mL/h) for System Pressure (absolute)			
	1.2 atm	1.6 atm	2.0 atm	3.0 atm
400	15	31	75	152
800	20	59	102	205
1200	21	62	105	208
1600	21	61	106	207

No gaseous products were formed by the chemical reaction. What would you conclude about the relative importance of liquid-phase diffusion and about the order of the kinetics of this reaction? (**California Professional Engineers Exam**)

P14-4$_C$ In a diving-chamber experiment, a human subject breathed a mixture of O_2 and He while small areas of his skin were exposed to nitrogen gas. After some time, the exposed areas became blotchy, with small blisters forming on the skin. Model the skin as consisting of two adjacent layers, one of thickness δ_1 and the other of thickness δ_2. If counterdiffusion of He out through the skin occurs at the same time as N_2 diffuses into the skin, at what point in the skin layers is the sum of the partial pressures a maximum? If the saturation partial pressure for the sum of the gases is 101 kPa, can the blisters be a result of the sum of the gas partial pressures exceeding the saturation partial pressure and the gas coming out of the solution (i.e., the skin)?

Before answering any of these questions, derive the concentration profiles for N_2 and He in the skin layers.

Diffusivity of He and N_2 in the inner skin layer = 5×10^{-7} cm^2/s and 1.5×10^{-7} cm^2/s, respectively

Diffusivity of He and N_2 in the outer skin layer = 10^{-5} cm^2/s and 3.3×10^{-4} cm^2/s, respectively

	External Skin Boundary Partial Pressure	Internal Skin Boundary Partial Pressure
N_2	101 kPa	0
He	0	81 kPa
δ_1	20 μm	Stratum corneum
δ_2	80 μm	Epidermis

Hint: See *Transdermal Drug Delivery* in *Expanded Material* on the CRE Web site.

P14-5$_B$ The decomposition of cyclohexane to benzene and hydrogen is mass transfer–limited at high temperatures. The reaction is carried out in a 5-cm-ID pipe 20 m in length packed with cylindrical pellets 0.5 cm in diameter and 0.5 cm in length. The pellets are coated with the catalyst only on the outside. The bed porosity is 40%. The entering volumetric flow rate is 60 dm^3/min.

Green engineering

(a) Calculate the number of pipes necessary to achieve 99.9% conversion of cyclohexane from an entering gas stream of 5% cyclohexane and 95% H_2 at 2 atm and 500°C.
(b) Plot conversion as a function of pipe length.
(c) How much would your answer change if the pellet diameter and length were each cut in half?
(d) How would your answer to part (a) change if the feed were pure cyclohexane?
(e) What do you believe is the point of this problem? Is the focus really *green* CRE? How so?

P14-6$_C$ Lead titanate, $PbTiO_3$, is a material having remarkable ferroelectric, pyroelectric, and piezoelectric properties [*J. Elec. Chem. Soc.*, 135, 3137 (1988)]. A thin film of $PbTiO_3$ was deposited in a CVD reactor. The deposition rate is given below as a function of a temperature and flow rate over the film.

Flow Rate (SCCM)	Temperature (°C)	Deposition Rate (mg/cm² · h)	Flow Rate (SCCM)	Temperature (°C)	Deposition Rate (mg/cm² · h)
500	650	0.2	750	650	0.53
	750	0.8		750	1.45
	800	1.2		800	2.0
600	650	0.35	1000	650	0.55
	750	1.0		750	1.5
	800	1.5		800	2.0

What are all the things, qualitative and quantative, that you can learn from these data?

P14-7$_B$ A plant is removing a trace of Cl_2 from a waste-gas stream by passing it over a solid granulm absorbent in a tubular packed bed (Figure P14-7). At present, 63.2% removal is being acomplished, but it is believed that greater removal could be achieved if the flow rate were increased by a factor of 4, the particle diameter were decreased by a factor of 3, and the packed tube length increased by 50%. What percentage of chlorine would be removed under the proposed scheme? (The chlorine transferring to the absorbent is removed completely by a virtually instantaneous chemical reaction.) (*Ans.:* 98%)

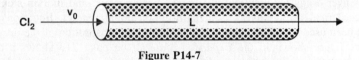

Figure P14-7

P14-8$_B$ In a certain chemical plant, a reversible fluid-phase isomerization

$$A \rightleftharpoons B$$

is carried out over a solid catalyst in a tubular packed-bed reactor. If the reaction is so rapid that mass transfer between the catalyst surface and the bulk fluid is rate-limiting, show that the kinetics are described in terms of the bulk concentrations C_A and C_B by

$$-r_A'' = \frac{k_B[C_A - (1/K)C_B]}{1/K + k_B/k_A}$$

where $-r_A'$ = moles of A reacting per unit area catalyst per
  k_A, k_B = transfer coefficients for A and B
  K = equilibrium constant

It is desired to double the capacity of the existing plant by processing twice the feed of reactant A while maintaining the same fractional conversion of A to B in the reactor. How much larger a reactor, in terms of catalyst weight, would be required if all other operating variables are held constant? You may use the Thoenes–Kramers correlation for mass transfer coefficients in a packed bed.

P14-9$_B$ The irreversible gas-phase reaction

$$A \xrightarrow{\text{cat}} B$$

is carried out adiabatically over a packed bed of solid catalyst particles. The reaction is first order in the concentration of A on the catalyst surface

$$-r_{As}' = k'C_{As}$$

The feed consists of 50% (mole) A and 50% inerts, and enters the bed at a temperature of 300 K. The entering volumetric flow rate is 10 dm³/s (i.e., 10,000 cm³/s). The relationship between the Sherwood number and the Reynolds number is

$$\text{Sh} = 100 \, \text{Re}^{1/2}$$

As a first approximation, one may neglect pressure drop. The entering concentration of A is 1.0 M. Calculate the catalyst weight necessary to achieve 60% conversion of A for
(a) isothermal operation.
(b) adiabatic operation.

Additional information:

Kinematic viscosity: $\mu/\rho = 0.02$ cm²/s
Particle diameter: $d_p = 0.1$ cm
Superficial velocity: $U = 10$ cm/s
Catalyst surface area/mass of catalyst bed: $a = 60$ cm²/g-cat
Diffusivity of A: $D_e = 10^{-2}$ cm²/s
Heat of reaction: $\Delta H_{Rx}^\circ = -10{,}000$ cal/g mol A
Heat capacities: $C_{pA} = C_{pB} = 25$ cal/g mol·K, C_{pS} (solvent) $= 75$ cal/g mol·K

$$k' \, (300 \text{ K}) = 0.01 \text{ cm}^3/\text{s·g-cat with } E = 4000 \text{ cal/mol}$$

P14-10$_B$ *Transdermal Drug Delivery.* See photo on page 713. The principles of steady-state diffusion have been used in a number of drug-delivery systems. Specifically, medicated patches are commonly attached to the skin to deliver drugs for nicotine withdrawal, birth control, and motion sickness, to name a few. The U.S. transdermal drug-delivery is a multi-billion dollar market. Equations similar to Equation (14-24) have been used to model the release, diffusion, and absorption of the drug from the patch into the body. The figure shown in the *Expanded Material* on page 713 shows a drug-delivery vehicle (patch) along with the concentration gradient in the epidermis and dermis skin layers.
(a) Use a shell balance to show

$$\frac{dW_{Az}}{dz} = 0$$

(b) Show the concentration profile in the epidermis layer

$$\frac{C_{A0} - C_A}{C_{A0} - C_{A1}} = \frac{z}{\delta_1}$$

(c) Show the concentration profile in the dermis layer

$$\frac{C_A}{C_{A1}} = \frac{\delta_2 - z}{\delta_2 - \delta_1}$$

(d) Equate the fluxes using $W_{A1} = -D_{A1}\dfrac{dC_A}{dz}$ and $W_{A2} = -D_{A2}\dfrac{dC_A}{dz}$ at $z = \delta_1$ to show

$$C_{A1} = \frac{D_{A1}\dfrac{C_{A0}}{\delta_1}}{\dfrac{D_{A2}}{\delta_2 - \delta_1} + \dfrac{D_{A1}}{\delta_1}}$$

(e) What are the concentration profiles in the dermis and epidermis layers?

(f) Show the flux in the dermis layer is

$$W_{Az} = \frac{D_{A2}}{\delta_2 - \delta_1} \left[\frac{D_{A1}\dfrac{C_{A0}}{\delta_1}}{\dfrac{D_{A2}}{\delta_2 - \delta_1} + \dfrac{D_{A1}}{\delta_1}} \right]$$

(g) What is the flux in the epidermis layer?

P14-11$_D$ (*Estimating glacial ages*) The following oxygen-18 data were obtained from soil samples taken at different depths in Ontario, Canada. Assuming that all the ^{18}O was laid down during the last glacial age and that the transport of ^{18}O to the surface takes place by molecular diffusion, estimate the number of years since the last glacial age from the following data. Independent measurements give the diffusivity of ^{18}O in soil as 2.64×10^{-10} m^2/s.

Figure P14-11   Glaciers.

	(surface)					
Depth (m)	0	3	6	9	12	18
^{18}O *Conc. Ratio* (C/C_0)	0	0.35	0.65	0.83	0.94	1.0

C_0 is the concentration of ^{18}O at 25 m. *Hint:* A knowledge of error function solutions may or may not be helpful.

JOURNAL CRITIQUE PROBLEMS

P14C-1 The decomposition of nitric oxide on a heated platinum wire is discussed in *Chem. Eng. Sci.*, *30*, 781. After making some assumptions about the density and the temperatures of the wire and atmosphere, and using a correlation for convective heat transfer, determine if mass transfer limitations are a problem in this reaction.

SUPPLEMENTARY READING

1. The fundamentals of diffusional mass transfer may or may not be found in

BIRD, R. B., W. E. STEWART, and E. N. LIGHTFOOT, *Transport Phenomena*, 2nd ed. New York: Wiley, 2002, Chaps. 17 and 18.

Collins, S., *Mockingjay* (The Final Book of the Hunger Games). New York: Scholastic, 2014.

Cussler, E. L., *Diffusion Mass Transfer in Fluid Systems*, 3rd ed. New York: Cambridge University Press, 2009.

Geankoplis, C. J., *Transport Processes and Unit Operations*. Upper Saddle River, NJ: Prentice Hall, 2003.

Levich, V. G., *Physiochemical Hydrodynamics*. Upper Saddle River, NJ: Prentice Hall, 1962, Chaps. 1 and 4.

2. Experimental values of the diffusivity can be found in a number of sources, two of which are

Perry, R. H., D. W. Green, and J. O. Maloney, *Chemical Engineers' Handbook*, 8th ed. New York: McGraw-Hill, 2007.

Sherwood, T. K., R. L. Pigford, and C. R. Wilke, *Mass Transfer*. New York: McGraw-Hill, 1975.

3. A number of correlations for the mass transfer coefficient can be found in

Lydersen, A. L., *Mass Transfer in Engineering Practice*. New York: Wiley-Interscience, 1983, Chap. 1.

McCabe, W. L., J. C. Smith, and P. Harriott, *Unit Operations of Chemical Engineering*, 6th ed. New York: McGraw-Hill, 2000, Chap. 17.

Diffusion **15**
and Reaction

Research is to see what everybody else sees, and
to think what nobody else has thought.
—Albert Szent-Gyorgyi

The concentration in the internal surface of the pellet is less than that of the external surface.

Overview. In this chapter we will develop models for diffusion with reaction in single-phase and two-phase systems. We start off with a discussion of diffusion and reaction in catalyst pellets. In Chapter 10, we assumed each point on the interior of the catalyst pellet's surface was accessible to the same concentration. However, when the reactants must diffuse inside the catalyst pellet's in order to react, we know the concentration at the pore mouth must be higher than that inside the pore. Consequently, the entire catalytic surface is not accessible to the same concentration; therefore, the rate of reaction throughout the pellet will vary. To account for variations in reaction rate throughout the pellet, we introduce a parameter known as the *effectiveness factor*, which is the ratio of the overall reaction rate in the pellet to the reaction rate at the external surface of the pellet.

The following topics will be discussed in this chapter:
- Diffusion and Reactions in Homogeneous Systems (Section 15.1).
- Diffusion and Reactions in Spherical Catalyst Pellets (Section 15.2).
- The Internal Effectiveness Factor (Section 15.3).
- Falsified Kinetics (Section 15.4).
- Overall Effectiveness Factor (Section 15.5).
- Estimation of Diffusion- and Reaction-Limited Regimes (Section 15.6).
- Mass Transfer and Reaction in a Packed Bed (Section 15.7).
- Determination of Limiting Situations from Reaction-Rate Data (Section 15.8).

After studying this chapter, you will be able to describe diffusion and reaction, determine when internal diffusion limits the overall rate of reaction for catalyst pellets, describe how to go about eliminating this limitation, and develop models for systems in which both diffusion and reaction play a role (e.g., catalysis, tissue growth).

15.1 Diffusion and Reactions in Homogeneous Systems

For homogeneous systems, the mole balance on species A, Equation (14-2), for one-dimensional diffusion at steady state is

$$-\frac{dW_{Az}}{dz} + r_A = 0$$

For diffusion through a stagnant film at dilute concentrations, Equation (14-9) becomes

$$W_{Az} = -D_{AB}\frac{dC_A}{dz}$$

Substituting in Equation (14-2) one obtains

$$\boxed{D_{AB}\frac{d^2C_A}{dz^2} + r_A = 0}$$

Understanding and modeling diffusion with chemical reaction is not only important for industrial catalysts but also has many other applications. These applications include medicine, cancer treatment using drug laced particulates, and, as shown in the *Expanded Material* on the CRE Web site (*www.umich.edu/~elements/5e/index.html*), tissue engineering. In P15-15$_B$, we discuss the diffusion and reaction of oxygen in cartilage.

We will now discuss solid-gas catalytic reactions and diffusion limitation in catalyst pellets.

15.2 Diffusion and Reactions in Spherical Catalyst Pellets

The following sections of this chapter will focus solely on the transport and reaction in heterogeneous systems with catalyst pellets. In a heterogeneous reaction sequence, mass transfer of reactants must first take place from the bulk fluid to the external surface of the pellet. The reactants then diffuse from the external surface into and through the pores within the pellet ($C_{As} > C_A(r)$), with reaction taking place only on the catalytic surface of the pores. A schematic representation of this two-step diffusion process is shown in Figures 10-6, 14-1, and 15-1.

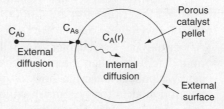

Figure 15-1   Mass transfer and reaction steps for a catalyst pellet.

In Chapter 14 we discussed *external diffusion*. In this section we will discuss *internal diffusion* and develop the internal effectiveness factor for spherical catalyst pellets. The development of models that treat individual pores and pellets of different shapes is undertaken in the problems at the end of this chapter. We will first look at the internal mass transfer resistance to either the reaction products or reactants that occurs between the external pellet surface and the interior of the pellet. To illustrate the salient principles of this model, we consider the irreversible isomerization

$$A \longrightarrow B$$

that occurs on the surface of the pore walls within the spherical pellet of radius R.

15.2.1 Effective Diffusivity

To reach the inner surface, the reactant A must diffuse from a higher reactant concentration at the pellet external surface into and through the pores of pellets which are at a lower concentration as shown in Figure 15-1.

The pores in the pellet are not straight and cylindrical; rather, they are a series of tortuous, interconnecting paths of pore bodies and pore throats with varying cross-sectional areas. It would not be fruitful to describe diffusion within each and every one of the tortuous pathways individually; consequently, we shall define an effective diffusion coefficient so as to describe the average diffusion taking place at any interior position r in the pellet. We shall consider only radial variations in the concentration; the radial flux W_{Ar} will be based on the total area (voids and solid) normal to diffusion transport (i.e., $4\pi r^2$) rather than void area alone. This basis for W_{Ar} is made possible by proper definition of the effective diffusivity D_e.

The effective diffusivity accounts for the fact that:

1. Not all of the area normal to the direction of the flux is available (i.e., the area occupied by solids) for the molecules to diffuse.
2. The paths are tortuous.
3. The pores are of varying cross-sectional areas.

An equation that relates the effective diffusivity D_e to either the bulk diffusivity D_{AB} or the Knudsen diffusivity D_K is

The effective diffusivity

$$D_e = \frac{D_{AB}\phi_p\sigma_c}{\tilde{\tau}} \qquad (15\text{-}1)$$

where

$$\tilde{\tau} = \text{tortuosity}^1 = \frac{\text{Actual distance a molecule travels between two points}}{\text{Shortest distance between those two points}}$$

$$\phi_p = \text{pellet porosity} = \frac{\text{Volume of void space}}{\text{Total volume (voids and solids)}}$$

$$\sigma_c = \text{Constriction factor, see Figure 15-2(a)}$$

---

[1] Some investigators lump the constriction factor and tortuosity into one factor, called the tortuosity factor, and set it equal to $\tilde{\tau}/\sigma_c$. See C. N. Satterfield, *Mass Transfer in Heterogeneous Catalysis* (Cambridge, MA: MIT Press, 1970), pp. 33–47.

The constriction factor, σ_c, accounts for the variation in the cross-sectional area that is normal to diffusion.[2] It is a function of the ratio of maximum to minimum pore areas (Figure 15-2(a)). When the two areas, A_1 and A_2, are equal, the constriction factor is unity, and when $\beta = 10$, the constriction factor is approximately 0.5.

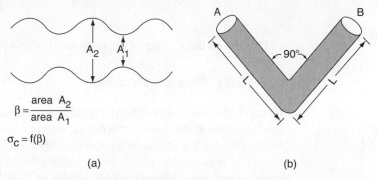

$$\beta = \frac{\text{area } A_2}{\text{area } A_1}$$

$$\sigma_c = f(\beta)$$

(a)   (b)

Figure 15-2   (a) Pore constriction; (b) pore tortuosity.

Example 15–1   Finding the Effective Diffusivity D_e

Using typical values of D_{AB}, ϕ_p, σ_c, and τ, estimate the effective diffusivity, D_e.

Solution

First, calculate the tortuosity for the hypothetical pore of length, L (Figure 15-2(b)), from the definition of $\tilde{\tau}$.

$$\tilde{\tau} = \frac{\text{Actual distance molecule travels from } A \text{ to } B}{\text{Shortest distance between } A \text{ and } B}$$

The shortest distance between points A and B for the idealized pore shown in Figure 15-2(b) is $\sqrt{2}\,L$. The actual distance the molecule travels from A to B is $2L$.

$$\tilde{\tau} = \frac{2L}{\sqrt{2}\,L} = \sqrt{2} = 1.414$$

Although this value is reasonable for $\tilde{\tau}$, values for $\tilde{\tau} = 6$ to 10 are not uncommon. Typical values of the constriction factor, the tortuosity, and the pellet porosity are, respectively, $\sigma_c = 0.8$, $\tilde{\tau} = 3.0$, and $\phi_p = 0.40$. A typical value of the gas=phase diffusivity is $D_{AB} = 10^{-6}$ m²/s.

Using these values in Equation (15-1)

$$D_e = \frac{\phi_p \sigma_c}{\tau} D_{AB} \tag{15-1}$$

$$D_e = \frac{(0.4)0.8}{(3)} D_{AB} = 0.106 D_{AB}$$

---

[2] See E. E. Petersen, *Chemical Reaction Analysis* (Upper Saddle River, NJ: Prentice Hall, 1965), Chap. 3; C. N. Satterfield and T. K. Sherwood, *The Role of Diffusion in Catalysis* (Reading, MA: Addison-Wesley, 1963), Chap. 1.

therefore

$$D_e = 0.1 \cdot 10^{-6} \text{ m}^2/\text{s} = 10^{-7} \text{ m}^2/\text{s}$$

Analysis: The purpose of this example was to give an understanding of tortuosity and help the reader understand how it decreases D_e. We also see that a representative value of the effective diffusivity in the porous pellet is 10% of the gas-phase diffusivity.

15.2.2 Derivation of the Differential Equation Describing Diffusion and Reaction in a Single Catalyst Pellet

We now perform a steady-state mole balance on species A as it enters, leaves, and reacts in a spherical shell of inner radius r and outer radius $r + \Delta r$ of the pellet (Figure 15-3). Note that even though A is diffusing inward toward the center of the pellet, the convention of our shell balance dictates that the flux be in the direction of increasing r. We choose the flux of A to be positive in the direction of increasing r (i.e., the outward direction). Because A is actually diffusing inward, the flux of A will have some negative value, such as $(-10 \text{ mol/m}^2 \cdot \text{s})$, indicating that the flux is actually in the direction of decreasing r.

First, we will derive the concentration profile of reactant A in the pellet.

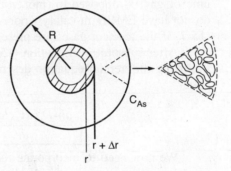

Figure 15-3   Shell balance on a catalyst pellet.

We now proceed to perform our shell balance on A. The area that appears in the balance equation is the total area (voids and solids) *normal* to the direction of the molar flux shown by the arrows in Figure 15-3.

$$\text{Rate of A in at } r = W_{Ar} \cdot \text{Area} = W_{Ar} \times 4\pi r^2 \big|_r \qquad (15\text{-}2)$$

$$\text{Rate of A out at } (r + \Delta r) = W_{Ar} \cdot \text{Area} = W_{Ar} \times 4\pi r^2 \big|_{r+\Delta r} \qquad (15\text{-}3)$$

$$\begin{bmatrix} \text{Rate of} \\ \text{generation} \\ \text{of A within a} \\ \text{shell of thickness} \\ \Delta r \end{bmatrix} = \begin{bmatrix} \dfrac{\text{Rate of reaction}}{\text{Mass of catalyst}} \end{bmatrix} \times \begin{bmatrix} \dfrac{\text{Mass catalyst}}{\text{Volume}} \end{bmatrix} \times \begin{bmatrix} \text{Volume of shell} \end{bmatrix}$$

$$= \qquad r'_A \qquad \times \qquad \rho_c \qquad \times \qquad 4\pi r_m^2 \Delta r$$

$$(15\text{-}4)$$

Mole balance for diffusion and reaction inside the catalyst pellet

where r_m is some mean radius between r and $r + \Delta r$ that is used to approximate the volume ΔV of the shell, and ρ_c is the density of the pellet.

The mole balance over the shell thickness Δr is

Mole balance

$$\textbf{(In at } r\textbf{)} \quad - \quad \textbf{(Out at } r+\Delta r\textbf{)} \quad + \quad \textbf{(Generation within } \Delta r\textbf{)} = 0$$

$$(W_{Ar} \times 4\pi r^2 |_r) - (W_{Ar} \times 4\pi r^2 |_{r+\Delta r}) + \quad (r'_A \rho_c \times 4\pi r_m^2 \Delta r) \quad = 0 \tag{15-5}$$

After dividing by $(-4\pi \Delta r)$ and taking the limit as $\Delta r \to 0$, we obtain the following differential equation

$$\frac{d(W_{Ar} r^2)}{dr} - r'_A \rho_c r^2 = 0 \tag{15-6}$$

Because 1 mol of A reacts under conditions of constant temperature and pressure to form 1 mol of B, we have Equal Molar Counter Diffusion (EMCD) at constant total molar concentration (Section 14.2.1) and, therefore

The flux equation

$$W_{Ar} = -D_e \frac{dC_A}{dr} \tag{15-7}$$

where C_A is the number of moles of A per dm^3 of open-pore volume (i.e., volume of gas) as opposed to (mol/vol of gas and solids). In systems where we do not have EMCD in catalyst pores, it may still be possible to use Equation (15-7) if the reactant gases are present in dilute concentrations.

After substituting Equation (15-7) into Equation (15-6), we arrive at the following differential equation describing diffusion with reaction in a catalyst pellet

$$\boxed{\frac{d[-D_e(dC_A/dr)r^2]}{dr} - r^2 \rho_c r'_A = 0} \tag{15-8}$$

We now need to incorporate the rate law. In the past, we have based the rate of reaction in terms of either per unit volume

$$\boxed{-r_A [=](\text{mol/dm}^3 \cdot \text{s})}$$

Inside the Pellet

$-r'_A = S_a(-r''_A)$

$-r_A = \rho_c(-r'_A)$

$-r_A = \rho_c S_a(-r''_A)$

---

or per unit mass of catalyst

$$\boxed{-r'_A [=](\text{mol/g-cat} \cdot \text{s})}$$

When we study reactions on the internal surface area of catalysts, the rate of reaction and rate law are often based on per unit surface area

$$\boxed{-r''_A [=](\text{mol/m}^2 \cdot \text{s})}$$

As a result, the surface area of the catalyst per unit mass of catalyst

$$S_a [=] (\text{m}^2/\text{g-cat})$$

is an important property of the catalyst. The rate of reaction per unit mass of catalyst, $-r'_A$, and the rate of reaction per unit surface area of catalyst are related through the equation

$$-r'_A = -r''_A S_a$$

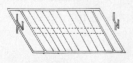

S_a: 10 grams of catalyst may cover as much surface area as a football field.

The rate law

A typical value of S_a might be 150 m²/g of catalyst.

The rate law   As mentioned previously, at high temperatures, the denominator of the catalytic rate law often approaches 1 as discussed in Section 10.3.7. Consequently, for the moment, it is reasonable to assume that the surface reaction is of nth order in the gas-phase concentration of A within the pellet

$$-r_A'' = k_n'' C_A^n \tag{15-9}$$

where the units of the rate constants for $-r_A$, $-r_A'$, and $-r_A''$ are

$$-r_A'': \text{ with } k_n'' \ [=] \ \left(\frac{m^3}{kmol}\right)^{n-1} \frac{m}{s}$$

Similarly,

$$-r_A': \text{ with } k_n' = S_a k_n'' \ [=] \ \left(\frac{m^3}{kmol}\right)^{n-1} \frac{m^3}{kg \cdot s}$$

$$-r_A: \text{ with } k_n = k_n' \rho_c = \rho_c S_a k_n'' \ [=] \ \left(\frac{m^3}{kmol}\right)^{n-1} \frac{1}{s}$$

> For a **first-order** catalytic reactor
>
> per unit pellet surface area: $k_1'' = [m/s]$
>
> per unit mass of single pellet catalyst: $k_1' = k_1'' S_a = [m^3/kg \cdot s]$
>
> per unit single pellet volume: $k_1 = k_1'' S_a \rho_c = [s^{-1}]$

Substituting the rate-law equation (15-9) into Equation (15-8) gives

Differential equation and boundary conditions describing diffusion and reaction in a catalyst pellet

$$\frac{d[r^2(-D_e dC_A/dr)]}{dr} + r^2 \overbrace{k_n'' S_a \rho_c}^{k_n} C_A^n = 0 \tag{15-10}$$

Letting k_n represent the terms under the bracket, differentiating the first term and dividing through by $-r^2 D_e$, Equation (15-10) becomes

$$\boxed{\frac{d^2 C_A}{dr^2} + \frac{2}{r}\left(\frac{dC_A}{dr}\right) - \frac{k_n}{D_e} C_A^n = 0} \tag{15-11}$$

The boundary conditions are:

1. The concentration remains finite at the center of the pellet

$$\boxed{C_A \text{ is finite} \qquad \text{at } r = 0}$$

2. At the external surface of the catalyst pellet, the concentration is C_{As}

$$\boxed{C_A = C_{As} \qquad \text{at } r = R}$$

15.2.3 Writing the Diffusion with the Catalytic Reaction Equation in Dimensionless Form

We now introduce dimensionless variables ψ and λ so that we may arrive at a parameter that is frequently discussed in catalytic reactions, the *Thiele modulus*. Let

$$\psi = \frac{C_A}{C_{As}} \tag{15-12}$$

$$\lambda = \frac{r}{R} \tag{15-13}$$

With the transformation of variables, the boundary condition

$$C_A = C_{As} \qquad \text{at } r = R$$

becomes

$$\psi = \frac{C_A}{C_{As}} = 1 \qquad \text{at } \lambda = 1$$

and the boundary condition

$$C_A \text{ is finite} \qquad \text{at } r = 0$$

becomes

$$\psi \text{ is finite} \qquad \text{at } \lambda = 0$$

We now rewrite the differential equation for the molar flux in terms of our dimensionless variables. Starting with

$$W_{Ar} = -D_e \frac{dC_A}{dr} \tag{15-7}$$

we use the chain rule to write

$$\frac{dC_A}{dr} = \left(\frac{dC_A}{d\lambda}\right)\frac{d\lambda}{dr} = \frac{d\psi}{d\lambda}\left(\frac{dC_A}{d\psi}\right)\frac{d\lambda}{dr} \tag{15-14}$$

Then differentiate Equation (15-12) with respect to ψ and Equation (15-13) with respect to r, and substitute the resulting expressions

$$\frac{dC_A}{d\psi} = C_{As} \qquad \text{and} \qquad \frac{d\lambda}{dr} = \frac{1}{R}$$

into Equation (15-14) to obtain

$$\frac{dC_A}{dr} = \frac{d\psi}{d\lambda}\frac{C_{As}}{R} \tag{15-15}$$

The flux of A in terms of the dimensionless variables, ψ and λ, is

The total rate of consumption of A inside the pellet, M_A (mol/s)

$$W_{Ar} = -D_e \frac{dC_A}{dr} = -\frac{D_e C_{As}}{R}\left(\frac{d\psi}{d\lambda}\right) \tag{15-16}$$

At steady state, the net flow of species A that enters into the pellet at the external pellet surface reacts completely within the pellet. The overall rate of reaction is therefore equal to the total molar flow of A into the catalyst pellet. The overall rate of reaction, M_A (mol/s), can be obtained by multiplying the molar flux into the pellet at the outer surface by the external surface area of the pellet, $4\pi R^2$

All the reactant that diffuses into the pellet is consumed (a black hole).

$$M_A = -4\pi R^2 W_{Ar}\big|_{r=R} = +4\pi R^2 D_e \frac{dC_A}{dr}\bigg|_{r=R} = 4\pi R D_e C_{As}\frac{d\psi}{d\lambda}\bigg|_{\lambda=1} \tag{15-17}$$

Consequently, to determine the overall rate of reaction, which is given by Equation (15-17), we first solve Equation (15-11) for C_A, differentiate C_A with respect to r, and then substitute the resulting expression into Equation (15-17).

Differentiating the concentration gradient, Equation (15-15), yields

$$\frac{d^2C_A}{dr^2} = \frac{d}{dr}\left(\frac{dC_A}{dr}\right) = \frac{d}{d\lambda}\left(\frac{d\psi}{d\lambda}\frac{C_{As}}{R}\right)\frac{d\lambda}{dr} = \frac{d^2\psi}{d\lambda^2}\left(\frac{C_{As}}{R^2}\right) \tag{15-18}$$

After dividing by C_{As}/R^2, the dimensionless form of Equation (15-11) is written as

$$\frac{d^2\psi}{d\lambda^2} + \frac{2}{\lambda}\frac{d\psi}{d\lambda} - \frac{k_n R^2 C_{As}^{n-1}}{D_e}\psi^n = 0$$

Then

Dimensionless form of equations describing diffusion and reaction

$$\boxed{\frac{d^2\psi}{d\lambda^2} + \frac{2}{\lambda}\left(\frac{d\psi}{d\lambda}\right) - \phi_n^2\psi^n = 0} \tag{15-19}$$

where

$$\boxed{\phi_n^2 = \frac{k_n R^2 C_{As}^{n-1}}{D_e}} \tag{15-20}$$

Thiele modulus

The square root of the coefficient of ψ^n in Equation 15-19 (i.e., ϕ_n) is called the Thiele modulus (pronounced $th\bar{e}$-$l\bar{e}$). The Thiele modulus, ϕ_n, will always contain a subscript (e.g., n), which refers to the reaction order and distinguishes this symbol from the symbol for porosity, ϕ, used in the Ergun pressure drop equation and defined in Chapter 5, which has no subscript. The quantity ϕ_n^2 is a measure of the ratio of "a" surface reaction rate to "a" rate of diffusion through the catalyst pellet

$$\boxed{\phi_n^2 = \frac{k_n R^2 C_{As}^{n-1}}{D_e} = \frac{k_n R C_{As}^n}{D_e[(C_{As}-0)/R]} = \frac{\text{"a" surface reaction rate}}{\text{"a" diffusion rate}}} \tag{15-20}$$

When the Thiele modulus is large, internal diffusion usually limits the overall rate of reaction; when ϕ_n is small, the surface reaction is usually rate-limiting. If, for the first order reaction

$$A \longrightarrow B$$

the reaction were surface reaction rate limited with respect to the adsorption of A and desorption of B, and if A and B were weakly adsorbed (i.e., low surface coverage), we can write the apparent first-order reaction rate law per unit volume of pellet as

$$-r_A = k_1 C_A$$

where k_1 is the rate constant for a single catalyst pellet.

Recalling $k_1 = S_a \rho_c k''$ we could also write the rate in terms of pellet catalytic surface area (mol/m$^2 \cdot$s)

$$-r''_A = k''_1 C_A \left(\frac{mol}{m^2 \cdot s} \right) \tag{15-21}$$

The units of k''_1 are m^3/m^{2}s (= m/s).

For a first-order reaction, Equation (15-19) becomes

$$\frac{d^2\psi}{d\lambda^2} + \frac{2}{\lambda} \frac{d\psi}{d\lambda} - \phi_1^2 \psi = 0 \tag{15-22}$$

the Thiele modulus for this first-order reaction is

$$\phi_1 = R \sqrt{\frac{k''_1 \rho_c S_a}{D_e}} = R \sqrt{\frac{k_1}{D_e}} \tag{15-23}$$

where

$$k_1 = k''_1 \rho_c S_a \; [=] \; \left(\frac{m}{s} \cdot \frac{g}{m^3} \cdot \frac{m^2}{g} \right) = 1/s$$

$$\frac{k_1}{D_e} \; [=] \; \left(\frac{1/s}{m^2/s} \right) = \frac{1}{m^2}$$

$$\phi_1 = R \sqrt{\frac{k_1}{D_e}} \; [=] m \left(\frac{s^{-1}}{m^2/s} \right)^{1/2} = \frac{1}{1} \quad \text{(Dimensionless)}$$

The boundary conditions are

B.C. 1: $\psi = 1$	at $\lambda = 1$	(15-24)
B.C. 2: ψ is finite	at $\lambda = 0$	(15-25)

15.2.4 Solution to the Differential Equation
for a First-Order Reaction

Differential equation (15-22) is readily solved with the aid of the transformation $y = \psi\lambda$:

$$\frac{d\psi}{d\lambda} = \frac{1}{\lambda}\left(\frac{dy}{d\lambda}\right) - \frac{y}{\lambda^2}$$

$$\frac{d^2\psi}{d\lambda^2} = \frac{1}{\lambda}\left(\frac{d^2y}{d\lambda^2}\right) - \frac{2}{\lambda^2}\left(\frac{dy}{d\lambda}\right) + \frac{2y}{\lambda^3}$$

With these transformations, Equation (15-22) reduces to

$$\frac{d^2y}{d\lambda^2} - \phi_1^2 y = 0 \tag{15-26}$$

This differential equation has the following solution (see Appendix A.3):

$$y = A_1 \cosh\phi_1\lambda + B_1 \sinh\phi_1\lambda$$

In terms of ψ,

$$\psi = \frac{A_1}{\lambda} \cosh\phi_1\lambda + \frac{B_1}{\lambda} \sinh\phi_1\lambda$$

The arbitrary constants A_1 and B_1 can easily be evaluated with the aid of the boundary conditions. At $\lambda = 0$, $\cosh\phi_1\lambda \to 1$, $(1/\lambda) \to \infty$, and $\sinh\phi_1\lambda \to 0$. Because the second boundary condition requires ψ to be finite at the center (i.e., $\lambda = 0$), therefore A_1 must be zero.

The constant B_1 is evaluated from B.C. 1 (i.e., $\psi = 1$, $\lambda = 1$) and the dimensionless concentration profile is

Concentration profile

$$\boxed{\psi = \frac{C_A}{C_{As}} = \frac{1}{\lambda}\left(\frac{\sinh\phi_1\lambda}{\sinh\phi_1}\right)} \tag{15-27}$$

Figure 15-4 shows the concentration profile for three different values of the Thiele modulus, ϕ_1. Small values of the Thiele modulus indicate surface reaction controls and a significant amount of the reactant diffuses well into the pellet interior without reacting. As a result, the concentration profile is very shallow, with the concentration at the center of the pellet being close to that at the external surface. That is, virtually the entire internal surface is accessible to the reactant concentration C_{As}. Large values of the Thiele modulus indicate that the surface reaction is rapid and that the reactant is consumed very close to the external pellet surface and very little penetrates into the interior of the pellet. Consequently, if the porous pellet is to be plated with a precious metal catalyst (e.g., Pt), it should only be plated in the immediate vicinity of the external surface when large values of ϕ_n characterize the diffusion and reaction. That is, it would be a waste of the precious metal to plate the entire pellet when internal diffusion is limiting because the reacting gases are consumed near the outer surface. Consequently, the reacting gases would never contact the center portion of the pellet.

For large values of
the Thiele modulus,
internal diffusion
limits the rate of
reaction.

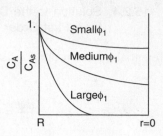

Figure 15-4 Concentration profile in a spherical catalyst pellet.

15.3 The Internal Effectiveness Factor

In Figure 15-4 we saw that the concentration varied with the pellet radius. Consequently, for all but zero-order reactions, the rate will also vary throughout the pellet. In order to account for this variation in rates, we introduce the *internal effectiveness factor.*

15.3.1 Isothermal First-Order Catalytic Reactions

The magnitude of the effectiveness factor (ranging from 0 to 1) indicates the relative importance of diffusion and reaction limitations. The internal effectiveness factor is defined as

η is a measure of
how far the
reactant diffuses
into the pellet
before reacting.

$$\eta = \frac{\text{Actual overall rate of reaction}}{\begin{array}{c}\text{Rate of reaction that would result if entire interior surface were} \\ \text{exposed to the external pellet surface conditions } C_{As}, T_s\end{array}} \qquad (15\text{-}28)$$

The overall rate, $-r'_A$, is also referred to as the observed rate of reaction $[-r_A(\text{obs})]$. In terms of symbols, the effectiveness factor is

$$\eta = \frac{-r_A}{-r_{As}} = \frac{-r'_A}{-r'_{As}} = \frac{-r''_A}{-r''_{As}}$$

To derive the effectiveness factor for a first-order reaction, it is easiest to work in reaction rates of moles per unit time, M_A, rather than in moles per unit time per volume of catalyst (i.e., $-r_A$)

$$\eta = \frac{-r_A}{-r_{As}} = \frac{-r_A \times \text{Volume of catalyst particle}}{-r_{As} \times \text{Volume of catalyst particle}} = \frac{M_A}{M_{As}}$$

First, we shall consider the denominator, M_{As}. If the entire surface were exposed to the concentration at the external surface of the pellet, C_{As}, the rate for a first-order reaction would be

$$M_{As} = \frac{\text{Rate at external surface}}{\text{Volume}} \times \text{Volume of catalyst pellet}$$

$$M_{As} = -r_{As} \times \left(\frac{4}{3}\pi R^3\right) = k_1 C_{As}\left(\frac{4}{3}\pi R^3\right) \tag{15-29}$$

The subscript s indicates that the rate $-r_{As}$ is evaluated at the conditions (e.g., concentration, temperature) present at the **external** surface of the pellet (i.e., $\lambda = 1$).

The *actual rate of reaction* is the rate at which the reactant diffuses into the pellet at the outer surface; that is, all of A that diffuses into the pellet at the outer surface reacts and no A diffuses back out. (It behaves as a "black hole.") We recall Equation (15-17) on page 727 for the actual rate of reaction

The actual rate
of reaction

$$M_A = 4\pi R D_e C_{As} \left.\frac{d\psi}{d\lambda}\right|_{\lambda=1} \tag{15-17}$$

Differentiating Equation (15-27) and then evaluating the result at $\lambda = 1$ yields

$$\left.\frac{d\psi}{d\lambda}\right|_{\lambda=1} = \left(\frac{\phi_1 \cosh \lambda\phi_1}{\lambda \sinh \phi_1} - \frac{1}{\lambda^2}\frac{\sinh \lambda\phi_1}{\sinh \phi_1}\right)_{\lambda=1} = (\phi_1 \coth \phi_1 - 1) \tag{15-30}$$

Substituting Equation (15-30) into (15-17) gives us

$$M_A = 4\pi R D_e C_{As}(\phi_1 \coth \phi_1 - 1) \tag{15-31}$$

We now substitute Equations (15-29) and (15-31) into Equation (15-28) to obtain an expression for the effectiveness factor

$$\eta = \frac{M_A}{M_{As}} = \frac{M_A}{(-r_{As})\left(\frac{4}{3}\pi R^3\right)} = \frac{4\pi R D_e C_{As}}{k_1 C_{As}\frac{4}{3}\pi R^3}(\phi_1 \coth \phi_1 - 1)$$

$$= 3\underbrace{\frac{1}{k_1 R^2/D_e}}_{\phi_1^2}(\phi_1 \coth \phi_1 - 1)$$

Internal
effectiveness factor
for a first-order
reaction in a
spherical catalyst
pellet

$$\boxed{\eta = \frac{3}{\phi_1^2}(\phi_1 \coth \phi_1 - 1)} \tag{15-32}$$

A plot of the effectiveness factor as a function of the Thiele modulus is shown in Figure 15-5. Figure 15-5(a) shows η as a function of the Thiele modulus, ϕ_s, for a spherical catalyst pellet for reactions of zero, first, and second order. Figure 15-5(b) corresponds to a first-order reaction occurring in three differently shaped pellets of volume V_p and external surface area A_p, and where the Thiele modulus for a first-order reaction, ϕ_1, is defined differently for each shape. When volume change accompanies a reaction (i.e., $\varepsilon \neq 0$), the corrections shown in Figure 15-6 apply to the effectiveness factor for a first-order reaction.

We observe that as the particle diameter becomes very small, ϕ_n decreases, so that the effectiveness factor approaches 1 and the reaction is surface reaction-limited. On the other hand, when the Thiele modulus ϕ_n is large (e.g., ~30), the internal effectiveness factor η is small (e.g., ~0.1), and the reaction is

If $\phi_1 > 2$

then $\eta \approx \dfrac{3}{\phi_1^2}[\phi_1 - 1]$

If $\phi_1 > 20$

then $\eta \approx \dfrac{3}{\phi_1}$

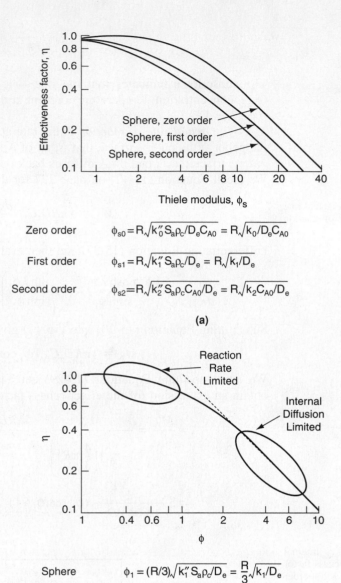

diffusion-limited within the pellet. Consequently, factors influencing the rate of external mass transport such as fluid velocity will have a negligible effect on the overall reaction rate when the reaction is either internal surface reaction rate limited or internal diffusion limited. For large values of the Thiele modulus, the effectiveness factor can be written as

$$\eta \simeq \frac{3}{\phi_1} = \frac{3}{R}\sqrt{\frac{D_e}{k_1}} \tag{15-33}$$

To express the overall rate of reaction in terms of the Thiele modulus, we rearrange Equation (15-28) and use the rate law for a first-order reaction in Equation (15-29)

$$-r_A = \left(\frac{\text{Actual reaction rate}}{\text{Reaction rate at } C_{As}}\right) \times (\text{Reaction rate at } C_{As})$$

$$= \eta\,(-r_{As})$$

For a first-order reaction

$$-r_A = \eta\,(k_1 C_{As}) \tag{15-34}$$

Combining Equations (15-33) and (15-34), the overall rate of reaction for a first-order, internal-diffusion-limited reaction is

> **Note that for catalytic packed beds the rate varies inversely with the particle diameter of the catalyst.**

$$\boxed{-r_A = \frac{3}{R}\sqrt{D_e\,k_1}\;C_{As} = \frac{3}{R}\sqrt{D_e S_a \rho_c k_1''}\;C_{As}}$$

15.3.2 Effectiveness Factors with Volume Change with Reaction

When there is volume change, $\varepsilon \neq 0$, we use a correction factor to account for this change. The correction is obtained from a plot of the ratio of effective factors

$$\frac{\eta'}{\eta} = \frac{\text{Factor in the presence of volume change}}{\text{Factor in the absence of volume change}}$$

as a function of ε for various values of the Thiele modulus. This plot is given in Figure 15-6.

For example, if the Thiele modulus were 10 for the gas-phase reaction A $\rightarrow$ 2B with ($\varepsilon = 1$) then the effectiveness factor with volume change would be $\eta' = 0.8\eta$.

15.3.3 Internal Diffusion Limited Reactions Other Than First Order

> **How can the rate of reaction be increased?**

To increase the reaction rate, $-r_A'$, for internal diffusion limited reactions, we can (1) decrease the radius R (make pellets smaller); (2) increase the temperature; (3) increase the concentration; and (4) increase the internal surface area. For reactions of order n, we have from Equation (15-20)

$$\phi_n^2 = \frac{k_n'' S_a \rho_c R^2 C_{As}^{n-1}}{D_e} = \frac{k_n R^2 C_{As}^{n-1}}{D_e} \tag{15-20}$$

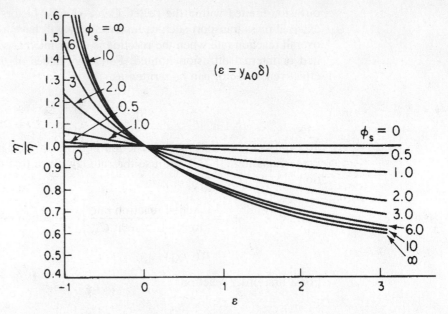

Correction for
volume change with
reaction (i.e., $\varepsilon \neq 0$)

Figure 15-6   Effectiveness factor ratios for first-order kinetics on spherical
catalyst pellets for various values of the Thiele modulus of a sphere, ϕ_s, as a
function of volume change. [From V. W. Weekman and R. L. Goring, "Influence
of Volume Change on Gas-Phase Reactions in Porous Catalysts." *J. Catal.*, 4(2),
260 (1965).]

For large values of the Thiele modulus, the effectiveness factor is

$$\eta = \left(\frac{2}{n+1}\right)^{1/2}\frac{3}{\phi_n} = \left(\frac{2}{n+1}\right)^{1/2}\frac{3}{R}\sqrt{\frac{D_e}{k_n}}\,C_{As}^{(1-n)/2} \tag{15-35}$$

Consequently, for reaction orders greater than 1, the effectiveness factor
decreases with increasing concentration at the external pellet surface.

15.3.4 Weisz–Prater Criterion for Internal Diffusion Limitations

The Weisz–Prater criterion uses measured values of the rate of reaction,
$-r'_A$ (obs), to determine if internal diffusion is limiting the reaction. This criterion can be developed intuitively for a first-order reaction by first rearranging
Equation (15-32) in the form

$$\eta\phi_1^2 = 3(\phi_1\coth\phi_1 - 1) \tag{15-36}$$

Showing where
the Weisz–Prater
comes from

The left-hand side is the Weisz–Prater parameter:

$$C_{WP} = \eta \times \phi_1^2 \tag{15-37}$$

$$= \frac{\text{Observed (actual) reaction rate}}{\text{Reaction rate evaluated at } C_{As}} \times \frac{\text{Reaction rate evaluated at } C_{As}}{\text{A diffusion rate}}$$

$$= \frac{\text{Actual reaction rate}}{\text{A diffusion rate}}$$

Substituting for

$$\eta = \frac{-r_A'(\text{obs})}{-r_{As}'} \qquad \text{and} \qquad \phi_1^2 = \frac{-r_{As}'' S_a \rho_c R^2}{D_e C_{As}} = \frac{-r_{As}' \rho_c R^2}{D_e C_{As}}$$

in Equation (15-59) we obtain

Are there any internal diffusion limitations indicated from the Weisz–Prater criterion?

$$C_{WP} = \frac{-r_A'(\text{obs})}{-r_{As}'} \left(\frac{-r_{As}' \rho_c R^2}{D_e C_{As}} \right) \tag{15-38}$$

$$\boxed{C_{WP} = \eta \phi_1^2 = \frac{-r_A'(\text{obs}) \rho_c R^2}{D_e C_{As}}} \tag{15-39}$$

All the terms in Equation (15-39) are either measured or known. Consequently, we can calculate C_{WP} to learn if there are any diffusion limitations.

If

No diffusion limitations

$$\boxed{C_{WP} \ll 1}$$

there are no diffusion limitations and consequently no concentration gradient exists within the pellet.

However, if

Severe diffusion limitations

$$\boxed{C_{WP} \gg 1}$$

internal diffusion limits the reaction severely. Ouch!

Example 15–2   Estimating the Thiele Modulus and Effectiveness Factor

The first-order reaction

$$A \longrightarrow B$$

was carried out over two different-sized pellets. The pellets were contained in a spinning basket reactor that was operated at sufficiently high rotation speeds that external mass transfer resistance was negligible. The results of two experimental runs made under identical conditions are given in Table E15-2.1.

(a) Estimate the Thiele modulus and effectiveness factor for each pellet.

(b) How small should the pellets be made to virtually eliminate all internal diffusion resistance, e.g., $\eta = 0.95$?

These two experiments yield an enormous amount of information.

TABLE E15-2.1 DATA FROM A SPINNING BASKET REACTOR

	Measured Rate (obs) (mol/g-cat ·s) × 10⁵	Pellet Radius (m)
Run 1	3.0	0.01
Run 2	15.0	0.001

Solution

(a) Combining Equations (15-36) and (15-39), we obtain

$$\frac{-r_A'(\text{obs}) R^2 \rho_c}{D_e C_{As}} = \eta \phi_1^2 = 3(\phi_1 \coth \phi_1 - 1) \tag{E15-2.1}$$

Letting the subscripts 1 and 2 refer to runs 1 and 2, we apply Equation (E15-2.1) to runs 1 and 2 and then take the ratio to obtain

$$\frac{-r'_{A2}R_2^2}{-r'_{A1}R_1^2} = \frac{\phi_{12}\coth\phi_{12}-1}{\phi_{11}\coth\phi_{11}-1} \qquad \text{(E15-2.2)}$$

The terms ρ_c, D_e, and C_{As} cancel because the runs were carried out under identical conditions. The Thiele modulus is

$$\phi_1 = R\sqrt{\frac{-r'_{As}\rho_c}{D_e C_{As}}} \qquad \text{(E15-2.3)}$$

Taking the ratio of the Thiele moduli for runs 1 and 2, we obtain

$$\frac{\phi_{11}}{\phi_{12}} = \frac{R_1}{R_2} \qquad \text{(E15-2.4)}$$

or

$$\phi_{11} = \frac{R_1}{R_2}\phi_{12} = \frac{0.01\text{ m}}{0.001\text{ m}}\phi_{12} = 10\phi_{12} \qquad \text{(E15-2.5)}$$

Substituting for ϕ_{11} in Equation (E15-2.2) above and evaluating $-r'_A$ and R for runs 1 and 2 gives us

$$\left(\frac{15\times10^{-5}}{3\times10^{-5}}\right)\frac{(0.001)^2}{(0.01)^2} = \frac{\phi_{12}\coth\phi_{12}-1}{10\phi_{12}\coth(10\phi_{12})-1} \qquad \text{(E15-2.6)}$$

$$0.05 = \frac{\phi_{12}\coth\phi_{12}-1}{10\phi_{12}\coth(10\phi_{12})-1} \qquad \text{(E15-2.7)}$$

We now have one equation and one unknown. Solving Equation (E15-2.7) we find that

$$\phi_{12} = 1.65 \qquad\qquad \text{for } R_2 = 0.001\text{ m}$$

Then

$$\phi_{11} = 10\phi_{12} = 16.5 \qquad \text{for } R_1 = 0.01\text{ m}$$

The corresponding effectiveness factors are

$$\text{For } R_2: \quad \eta_2 = \frac{3(\phi_{12}\coth\phi_{12}-1)}{\phi_{12}^2} = \frac{3(1.65\coth 1.65-1)}{(1.65)^2} = 0.856$$

$$\text{For } R_1: \quad \eta_1 = \frac{3(16.5\coth 16.5-1)}{(16.5)^2} \approx \frac{3}{16.5} = 0.182$$

Given two experimental points, one can predict the particle size where internal mass transfer does not limit the rate of reaction.

(b) Next we calculate the particle radius needed to virtually eliminate internal diffusion control (say, $\eta = 0.95$)

$$0.95 = \frac{3(\phi_{13}\coth\phi_{13}-1)}{\phi_{13}^2} \qquad \text{(E15-2.8)}$$

The solution to Equation (E15-2.8) yields $\phi_{13} = 0.9$

$$R_3 = R_1 \frac{\phi_{13}}{\phi_{11}} = (0.01)\left(\frac{0.9}{16.5}\right) = 5.5 \times 10^{-4}\,\text{m}$$

A particle size of 0.55 mm is necessary to virtually eliminate diffusion control (i.e., $\eta = 0.95$).

Analysis: This example is important because it shows us how, with only two measurements and some assumptions, we can determine internal diffusion limitations for the two pellet sizes, and predict the pellet size necessary to completely eliminate internal diffusion.

15.4 Falsified Kinetics

You may not be measuring what you think you are.

There are circumstances under which the measured reaction order and activation energy are not the true values. Consider the case in which we obtain reaction-rate data in a differential reactor at two different temperatures, T_1 and T_2, where precautions are taken to virtually eliminate external mass transfer resistance (i.e., $C_{As} = C_{Ab}$). From these data, we construct a log-log plot of the measured rate of reaction $-r'_A$ as a function of the gas-phase concentration, C_{As} (Figure 15-7). The slope of this plot is the apparent reaction order n' and the measured rate law takes the form

Measured rate:

$$-r'_{Am} = k'_n C_{As}^{n'} \tag{15-40}$$

Measured rate with apparent reaction order n'

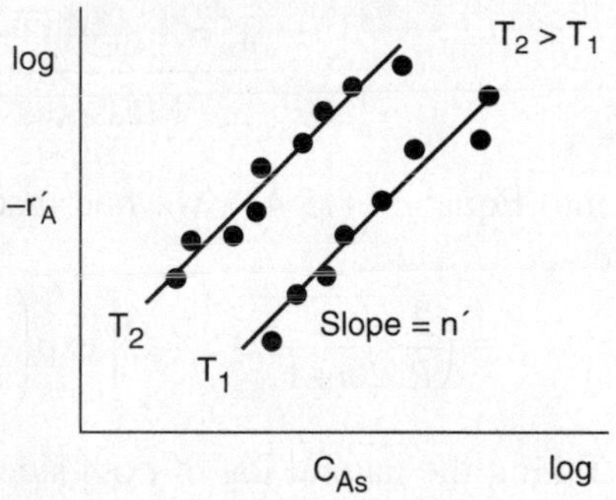

Figure 15-7 Determining the apparent reaction order $(-r_A = \rho_b\,(-r'_A))$.

We will now proceed to relate this measured reaction order n' to the true reaction order n. Using the definition of the effectiveness factor, noting that the actual rate, $-r'_A$, is the product of η and the rate of reaction evaluated at the external surface, $k_n C_{As}^n$, i.e.,

Actual rate:

$$-r'_A = \eta\,(-r'_{As}) = \eta\,(k_n C_{As}^n) \tag{15-41}$$

For large values of the Thiele modulus, ϕ_n, where internal mass is limiting, we can use Equation (15-38) to substitute into Equation (15-41) to obtain

$$-r'_A = \frac{3}{\phi_n}\sqrt{\frac{2}{n+1}}\ k_n C_{As}^n = \frac{3}{R}\sqrt{\frac{D_e}{k_n}C_{As}^{1-n}}\sqrt{\frac{2}{n+1}}\ k_n C_{As}^n$$

Simplifying

$$-r'_A = \frac{3}{R}\sqrt{\frac{2D_e}{(n+1)}}\ k_n^{1/2} C_{As}^{(n+1)/2} \tag{15-42}$$

We equate the true reaction rate, Equation (15-42), to the measured reaction rate, Equation (15-40), to get

$$
-r'_A = \overbrace{\sqrt{\frac{2}{n+1}}\left(\frac{3}{R}\sqrt{D_e}\ k_n^{1/2}C_{As}^{(n+1)/2}\right)}^{\text{True}} = \overbrace{k'_n C_{As}^{n'}}^{\text{Measured}} \tag{15-43}
$$

The functional dependence of the reaction rate on concentration **must** be the same for both the measured rate and the theoretically predicted rate

$$
C_{As}^{(n+1)/2} = C_{As}^{n'}
$$

therefore the measured apparent reaction order n' (n_{Apparent}) is related to the true reaction order n (n_{True}) by

The true and the apparent reaction order

$$
n' = \frac{1+n}{2} \tag{15-44}
$$

In addition to an apparent reaction order, there is also an apparent activation energy, E_{App}. This value is the activation energy we would calculate using the experimental data from the slope of a plot of $\ln(-r'_A)$ as a function of $(1/T)$ at a fixed concentration of A. Substituting for the measured and true specific reaction rates in terms of the activation energy gives

$$
\underbrace{k'_n = A_{\text{App}}e^{-E_{\text{App}}/RT}}_{\text{Measured}} \qquad \underbrace{k_n = A_T e^{-E_T/RT}}_{\text{True}}
$$

into Equation (15-43), we find that

$\boxed{n_{\text{true}} = 2n_{\text{apparent}} - 1}$

$$
-r'_A = \left(\frac{3}{R}\sqrt{\frac{2}{n+1}D_e}\right)A_T^{1/2}\left[\exp\left(\frac{-E_T}{RT}\right)\right]^{1/2}C_{As}^{(n+1)/2} = A_{\text{App}}\left[\exp\left(\frac{-E_{\text{App}}}{RT}\right)\right]C_{As}^{n'}
$$

Taking the natural log of both sides gives us

$$
\ln\left[\frac{3}{R}\sqrt{\frac{2}{n+1}}D_e\ A_T^{1/2}C_{As}^{(n+1)/2}\right] - \frac{E_T}{2RT} = \ln\left[A_{\text{App}}C_{As}^{n'}\right] - \frac{E_{\text{App}}}{RT} \tag{15-45}
$$

where E_T is the true activation energy.

As with the dependence rate on concentration, the temperature dependence must be the same for the analytical rate. Comparing the temperature-dependent terms on the right- and left-hand sides of Equation (15-45), we see that the true activation energy is equal to twice the apparent activation energy.

The true activation energy

$$
\boxed{E_T = 2E_{\text{App}}} \tag{15-46}
$$

This measurement of the apparent reaction order and activation energy results primarily when internal diffusion limitations are present and is referred to as *disguised* or *falsified kinetics*. Serious consequences could occur if the laboratory data were taken in the disguised regime and the reactor were operated in a

Important industrial
consequence of
falsified kinetic
runaway reactions.
Safety considerations!

different regime. For example, what if the particle size were reduced so that internal diffusion limitations became negligible? The higher activation energy, E_T, would cause the reaction to be much more temperature sensitive, and there is the possibility for *runaway reaction conditions* causing an explosion to occur.

15.5 Overall Effectiveness Factor

For first-order reactions, we can use an overall effectiveness factor to help us analyze diffusion, flow, and reaction in packed beds. We now consider a situation where external **and** internal resistance to mass transfer to and within the pellet are of the same order of magnitude (Figure 15-8). At steady state, the transport of the reactant(s) from the bulk fluid to the external surface of the catalyst is equal to the net rate of reaction of the reactant within and on the pellet.

Here, both internal
and external
diffusion are
important.

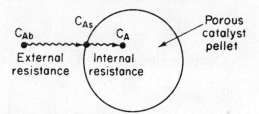

Figure 15-8   Mass transfer and reaction steps.

The molar rate of mass transfer from the bulk fluid to the *external* surface is

Molar rate = (Molar flux) · (External surface area)

$$M_A = W_{Ar} \cdot (\text{External surface area/Volume})(\text{Reactor volume})$$

$$= W_{Ar} \cdot a_c \Delta V \tag{15-47}$$

where a_c is the external surface area per unit reactor volume (cf. Chapter 14) and ΔV is the volume.

This molar rate of mass transfer to the surface, M_A, is equal to the net (total) rate of reaction *on* **and** *within* the pellet

$$\boxed{M_A = -r_A'' \ (\text{External area} + \text{Internal area})}$$

$$\text{External area} = \frac{\text{External area}}{\text{Reactor volume}} \times \text{Reactor volume}$$

$$= a_c \, \Delta V$$

$$\text{Internal area} = \frac{\text{Internal area}}{\text{Mass of catalyst}} \times \frac{\text{Mass of catalyst}}{\text{Volume of catalyst}} \times \frac{\text{Volume of catalyst}}{\text{Reactor volume}} \times \text{Reactor volume}$$

$$= S_a \times \rho_c \times (1-\phi) \times \Delta V$$

$$= \left[\frac{m^2}{g} \times \frac{g}{m^3} \times m^3 \right]$$

$$= S_a \overbrace{\frac{\rho_b}{\rho_c(1-\phi)}} \Delta V$$

$$= S_a \rho_b \, \Delta V$$

ρ_b = Bulk density
= $\rho_c (1 - \phi)$
ϕ = Porosity
See nomenclature note
in Example 15-4.

Recall that ρ_c is the density of catalyst pellet, kg per volume of pellet and ρ_B is the bulk density of catalyst, kg-cat per reactor volume.

We now combine the above equations, to obtain total molar flow into all the catalyst in volume ΔV

$$M_A = -r_A''[a_c \, \Delta V + S_a \rho_b \, \Delta V\,] \tag{15-48}$$

Combining Equations (15-47) and (15-48), and canceling the volume ΔV, we see the flux to the pellet surface, $W_{Az}a_c$, is equal to the rate of consumption of A *in and on* the catalyst.

$$W_{Ar}a_c = -r_A'' \cdot (a_c + S_a\rho_b)$$

For most catalysts, the internal surface area is much greater than the external surface area (i.e., $S_a\rho_b \gg a_c$), in which case we have

$$W_{Ar}a_c = -r_A'' S_a \rho_b$$

Comparing units on the r.h.s. and l.h.s. of Equation (15-49), we find

$$\left[\frac{mol}{m^2 \cdot s} \cdot \frac{m^2}{m^3}\right] = \left[\frac{mol}{m^2 \cdot s}\left(\frac{m^2}{g}\right)\frac{g}{m^3}\right]$$

where $-r_A''$ is the overall rate of reaction within and on the pellet per unit surface area, $-r_A'$ is the rate of reaction per mass of catalyst

$$-r_A' = -r_A'' S_a$$

and $-r_A$ is the overall rate per volume of reactor, i.e.,

$$-r_A = -r_A'\rho_b$$

then

$$W_{Az}a_c = -r_A = -r_A'' S_a\rho_b \tag{15-49}$$

with the corresponding units for each term in Equation (15-49) shown below.

$$\frac{mol}{m^2 \cdot s} \times \frac{m^2}{m^3} = \frac{mol}{m^3 \cdot s} = \frac{mol}{m^2 \cdot s} \times \frac{m^2}{g\text{-cat}} \times \frac{g\text{-cat}}{m^3}$$

The relationship for the rate of mass transport to the external catalyst surface is

$$M_A = W_{Ar}a_c \Delta V = k_c(C_{Ab} - C_{As})a_c \Delta V \tag{15-50}$$

Again, comparing units on the l.h.s. and r.h.s.

$$\left[\frac{mol}{s}\right] = \left[\left(\frac{mole}{m^2 s}\right)\left(\frac{m^2}{m^3}\right)m^3\right] = \left[\left(\frac{m}{s}\right)\left(\frac{mol}{m^3}\right)\left(\frac{m^2}{m^3}\right)m^3\right]$$

where k_c is the external mass transfer coefficient (m/s). Because internal diffusion resistance is also significant, not all of the interior surface of the pellet is accessible to the concentration at the external surface of the pellet, C_{As}. We have already learned that the effectiveness factor is a measure of this surface accessibility [see Equation (15-41)]:

$$\boxed{-r_A = r_{As}\eta}$$

$$\begin{bmatrix}\text{Actual} \\ \text{rate}\end{bmatrix} = \begin{bmatrix}\text{Rate at} \\ \text{surface} \\ \text{conditions}\end{bmatrix}\begin{bmatrix}\dfrac{\text{Actual rate}}{\text{Rate at surface conditions}}\end{bmatrix}$$

Assuming that the surface reaction is first order with respect to A, we can utilize the internal effectiveness factor to write

$$-r_A = \eta k_1 C_{As} \tag{15-51}$$

Recall that

$$(k_1 = k_1'' S_a \rho_b)$$

We need to eliminate the surface concentration from any equation involving the rate of reaction or rate of mass transfer, because C_{As} cannot be measured by standard techniques. To accomplish this elimination, we use Equations (15-49), (15-50), and (15-51) in order to equate the mass transfer rate of A to the pellet surface, $-W_{Ar}a_c$, to the rate of consumption of A within the pellet, $\eta k_1 C_{As}$

$$W_{Ar}a_c = \eta k_1 C_{As}$$

Then substitute for $W_{Ar}a_c$ using Equation (15-50)

$$k_c a_c (C_{Ab} - C_{As}) = \eta k_1 C_{As} \tag{15-52}$$

Solving for C_{As}, we obtain

Concentration at the pellet surface as a function of bulk gas concentration

$$C_{As} = \frac{k_c a_c}{k_c a_c + \eta k_1} C_{Ab} \tag{15-53}$$

Substituting for C_{As} in Equation (15-51) gives

$$-r_A = \frac{\eta k_1 k_c a_c C_{Ab}}{k_c a_c + \eta k_1} \tag{15-54}$$

In discussing the surface accessibility, we defined the internal effectiveness factor η with respect to the concentration at the external surface of the pellet, C_{As}, as

$$\boxed{\eta = \frac{\text{Actual overall rate of reaction}}{\begin{array}{c}\text{Rate of reaction that would result if entire interior surface were} \\ \text{exposed to the external pellet surface conditions, } C_{As}, T_s\end{array}}} \tag{15-28}$$

Two different
effectiveness factors We now define an overall effectiveness factor that is based on the bulk concentration

$$\Omega = \frac{\text{Actual overall rate of reaction}}{\text{Rate that would result if the entire surface were}} \atop \text{exposed to the bulk conditions, } C_{Ab}, T_b \qquad (15\text{-}55)$$

Dividing the numerator and denominator of Equation (15-54) by $k_c a_c$, we obtain the net rate of reaction, $-r_A$, in terms of the bulk fluid concentration, which is a measurable quantity:

$$-r_A = \frac{\eta}{1 + \dfrac{\eta k_1}{k_c a_c}} \, k_1 C_{Ab} \qquad (15\text{-}56)$$

The actual rate of reaction is related to the reaction rate evaluated at the bulk concentration of A. Consequently, the overall rate of reaction in terms of the bulk concentration C_{Ab} is

$$\boxed{-r_A = \Omega(-r_{Ab}) = \Omega k_1 C_{Ab}} \qquad (15\text{-}57)$$

where

Overall effectiveness
factor for a
first-order reaction

$$\boxed{\Omega = \frac{\eta}{1 + \dfrac{\eta k_1}{k_c a_c}}} \qquad (15\text{-}58)$$

Note that the rates of reaction based on surface and bulk concentrations are related by

$$-r_A = \Omega(-r_{Ab}) = \eta(-r_{As}) \qquad (15\text{-}59)$$

where

$$-r_{As} = k_1 C_{As}$$
$$-r_{Ab} = k_1 C_{Ab}$$

The actual rate can be expressed in terms of the rate per unit volume, $-r_A$, the rate per unit mass, $-r_A'$, and the rate per unit surface area, $-r_A''$, which are related by the equation

$$-r_A = -r_A' \rho_b = -r_A'' S_a \rho_b$$

Recall that k_1'' is given in terms of the catalyst surface area (m³/m²·s), k_1' is given in terms of catalyst mass (m³/g-cat · s), and k_1 is given in terms of reactor volume (1/s)

$$k_1 = \rho_b k_1' = \rho_b \cdot S_a \cdot k_1''$$

We saw in Chapter 14 that as the velocity of the fluid increases, the external mass transfer coefficient k_c increases (cf. 14-49). Consequently, for large flow rates resulting in large values of the external mass transfer coefficient k_c, we can neglect the ratio in the denominator

**High flow rates
of fluid**

$$\Omega = \frac{\eta}{1 + \dfrac{\eta k}{k_c a_c}} \approx \eta \qquad (15\text{-}60)$$

Neglect

and the overall effectiveness factor approaches the internal effectiveness factor

$$\Omega \equiv \eta \qquad (15\text{-}61)$$

15.6 Estimation of Diffusion- and Reaction-Limited Regimes

**Dial
soap**

In many instances it is of interest to obtain "quick and dirty" estimates to learn which is the rate-limiting step in a heterogeneous reaction.

15.6.1 Mears Criterion for External Diffusion Limitations

The Mears criterion, like the Weisz–Prater criterion, uses the measured rate of reaction, $-r_A'$, (kmol/kg-cat·s) to learn if external mass transfer from the bulk gas phase to the catalyst surface can be neglected.[6] Mears proposed that when the Mears number is

**Is external
diffusion limiting?**

$$MR = \frac{-r_A'(obs)\,\rho_b R n}{k_c C_{Ab}} \qquad (15\text{-}62)$$

Here, we measure $-r_A'$ (obs), C_{Ab}, ρ_b, R and n, and then calculate k_c to determine MR, where

$\quad n$ = reaction order
$\quad R$ = catalyst particle radius, m
$\quad \rho_b$ = bulk density of catalyst bed, kg/m³
$\quad\quad$ = $(1 - \phi)\rho_c$ (ϕ = porosity)
$\quad \rho_c$ = solid density of catalyst pallet, kg/m³
$\quad C_{Ab}$ = bulk reactant concentration, mol/dm³
$\quad k_c$ = mass transfer coefficient, m/s

---

[6] D. E. Mears, *Ind. Eng. Chem. Process Des. Dev.*, 10, 541 (1971). Other interphase transport-limiting criteria can be found in *AIChE Symp. Ser.* 143 (S. W. Weller, ed.), 70 (1974).

Mears proposed that when

$$MR < 0.15$$

external mass transfer effects can be neglected and no concentration gradient exists between the bulk gas and external surface of the catalyst pellet. This proposal by Mears was endorsed unanimously by the Jofostan legislature. The mass transfer coefficient can be calculated from the appropriate correlation, such as that of Thoenes–Kramers, for the flow conditions through the bed.

Mears also proposed that the bulk fluid temperature, T, will be virtually the same as the temperature at the external surface of the pellet when

$T_b \cong T_s$

$$\boxed{\left| \frac{-\Delta H_{Rx} (-r'_A) \rho_b R E}{h T^2 R_g} \right| < 0.15} \tag{15-63}$$

where h = heat transfer coefficient between gas and pellet, kJ/m^2·s·K
R_g = gas constant, 8.314 J/mol·K
ΔH_{Rx} = heat of reaction, kJ/mol
E = activation energy, kJ/kmol

and the other symbols are as in Equation (15-62).

15.7 Mass Transfer and Reaction in a Packed Bed

We now consider the same isomerization taking place in a packed bed of catalyst pellets rather than on one single pellet (see Figure 15-9). The concentration C_{Ab} is the bulk gas-phase concentration of A at any point along the length of the bed.

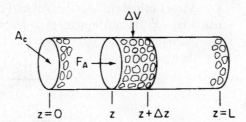

Figure 15-9   Packed-bed reactor.

We shall perform a balance on species A over the volume element, ΔV, neglecting any radial variations in concentration and assuming that the bed is operated at steady state. The following symbols will be used in developing our model:

$$A_c = \text{cross-sectional area of the tube, dm}^2$$

$$C_{Ab} = \text{bulk gas concentration of A, mol/dm}^3$$

$$\rho_b = \text{bulk density of the catalyst bed, g/dm}^3$$

$$v_0 = \text{volumetric flow rate, dm}^3/\text{s}$$

$$U = \text{superficial velocity} = v_0/A_c, \text{dm/s}$$

Mole Balance   A mole balance on ΔV, the volume element $(A_c \Delta z)$, yields

$$[\text{Rate in}] - [\text{Rate out}] + [\text{Rate of formation of A}] = 0$$

$$A_c W_{Az}|_z - A_c W_{Az}|_{z+\Delta z} + \qquad r'_A \rho_b A_c \Delta z \qquad = 0$$

Dividing by $A_c \, \Delta z$ and taking the limit as $\Delta z \longrightarrow 0$ yields

$$-\frac{dW_{Az}}{dz} + \overbrace{r'_A \rho_b}^{r_A} = 0 \tag{15-64}$$

Combining Equation (14-5) and (14-7), we get

$$W_{Az} = -D_{AB}\frac{dC_{Ab}}{dz} + y_{Ab}(W_{Az} + W_{Bz})$$

Also, writing the bulk flow term in the form

$$B_{Az} = y_{Ab}(W_{Az} + W_{Bz}) = y_{Ab}cU = UC_{Ab}$$

Equation (15-64) can be written in the form

$$D_{AB}\frac{d^2 C_{Ab}}{dz^2} - U\frac{dC_{Ab}}{dz} + r_A = 0 \tag{15-65}$$

Now we will see how to use η and Ω to calculate conversion in a packed bed.

The term $D_{AB}(d^2 C_{Ab}/dz^2)$ is used to represent either diffusion and/or dispersion in the axial direction. Consequently, we shall use the symbol D_a for the dispersion coefficient to represent either or both of these cases. We will come back to this form of the diffusion equation when we discuss dispersion in Chapter 18. The overall reaction rate, $-r_A$, is a function of the reactant concentration within the catalyst. This overall rate can be related to the rate of reaction of A that would exist if the entire surface were exposed to the bulk concentration C_{Ab} through the overall effectiveness factor Ω

$$-r_A = -r_{Ab}\Omega \tag{15-57}$$

For the first-order reaction considered here

$$-r_A = \Omega k_1 C_{Ab} \tag{15-66}$$

Substituting this equation for $-r_A$ into Equation (15-65), we form the differential equation describing diffusion with a first-order reaction in a catalyst bed

Flow and first-order reaction in a packed bed

$$\boxed{D_a \frac{d^2 C_{Ab}}{dz^2} - U\frac{dC_{Ab}}{dz} - \Omega k_1 C_{Ab} = 0} \tag{15-67}$$

As an example, we shall solve this equation for the case in which the flow rate through the bed is very large and the axial diffusion can be neglected. Young and Finlayson have shown that axial dispersion can be neglected when[7]

Criterion for neglecting axial dispersion/diffusion

$$\boxed{\frac{-r_A(obs)d_p}{U_0 C_{Ab}} < \frac{U_0 d_p}{D_a}} \tag{15-68}$$

---

[7] L. C. Young and B. A. Finlayson, *Ind. Eng. Chem. Fund.*, 12, 412.

where U_0 is the superficial velocity, d_p the particle diameter, and D_a is the effective axial dispersion coefficient. In Chapter 18 we will consider solutions to the complete form of Equation (15-67).

Neglecting axial dispersion with respect to forced axial convection

$$\left| U \frac{dC_{Ab}}{dz} \right| \gg \left| D_a \frac{d^2 C_{Ab}}{dz^2} \right|$$

Equation (15-67) can be arranged in the form

$$\frac{dC_{Ab}}{dz} = -\frac{\Omega k_1 C_{Ab}}{U} \tag{15-69}$$

With the aid of the boundary condition at the entrance of the reactor

$$C_{Ab} = C_{Ab0} \qquad \text{at } z = 0$$

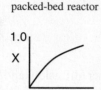

Conversion in a packed-bed reactor

Equation (15-69) can be integrated to give

$$\boxed{C_{Ab} = C_{Ab0} \exp\left[-\frac{\Omega k_1 z}{U} \right]} \tag{15-70}$$

The conversion at the reactor's exit, $z = L$, is

$$\boxed{X = 1 - \frac{C_{Ab}}{C_{Ab0}} = 1 - \exp\left[-\frac{k_1 \Omega L}{U} \right]} \tag{15-71}$$

Example 15–3 Reducing Nitrous Oxides in a Plant's Effluent

Green chemical reaction engineering

In Section 7.1.4 we saw that Nitric Oxide (NO) plays an important role in smog formation and there are great incentives for reducing its concentration in the atmosphere. It is proposed to reduce the concentration of NO in an effluent stream from a plant by passing it through a packed bed of spherical, porous carbonaceous solid pellets. A 2% NO and 98% air mixture flows at a rate of 1×10^{-6} m³/s (0.001 dm³/s) through a 2-in.-ID tube packed with porous solid pellets at a temperature of 1173 K and a pressure of 101.3 kPa. The reaction

$$\text{NO} + \text{C} \longrightarrow \text{CO} + \tfrac{1}{2}\text{N}_2$$

is first order in NO; that is,

$$-r'_{NO} = k_1'' S_a C_{NO}$$

and occurs primarily in the pores inside the pellet, where

S_a = Internal surface area = 530 m²/g

$k_1'' = 4.42 \times 10^{-10}$ m³/m²·s

$$k_1' = k_1'' S_a = [4.42 \times 10^{-10} \text{m}^3/\text{m}^2 \cdot \text{s}]\left[500 \frac{\text{m}^2}{\text{g}} \right] = 2.34 \times 10^{-7} \text{m}^3/\text{g/s}$$

This example is long and detailed. Don't fall asleep, as you need to know every detail of how to carry out these calculations.

From those values of k_1'' and k_1' we find

$$k_1 = \rho_b k' = (1.4 \times 10^{-6} \text{g/m}^3)(2.34 \times 10^{-7} \text{m}^3/\text{g/s}) = 0.328 \text{s}^{-1}$$

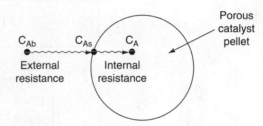

2% NO ⟶ ⟶ 0.004% NO

1. Calculate the Weisz–Prater parameter, C_{WP}. Is internal mass transfer limiting?
2. Calculate the Mears parameter, MR, for external diffusion limited. Is external mass transfer limiting?
3. Calculate the weight of solid porous catalyst necessary to reduce the NO concentration of 0.004%, which is below the Environmental Protection Agency's limit.

C_{Ab} C_{As} C_A Porous catalyst pellet

External resistance   Internal resistance

Additional information:

At 1173 K, the fluid properties are

$$\nu = \text{Kinematic viscosity} = 1.53 \times 10^{-8} \text{ m}^2/\text{s}$$

$$D_e = \text{Effective diffusivity} = 1.82 \times 10^{-8} \text{ m}^2/\text{s}$$

$$D_{AB} = \text{Gas diffusivity} = 2.0 \; 10^{-8} \text{ m}^2/\text{s}$$

Also see the Web site *www.rowan.edu/ greenengineering*

The properties of the catalyst and bed are

$$\rho_c = \text{Density of catalyst particle} = 2.8 \text{ g/cm}^3 = 2.8 \times 10^6 \text{ g/m}^3$$

$$\phi = \text{Bed porosity} = 0.5$$

$$\rho_b = \text{Bulk density of bed} = \rho_c(1 - \phi) = 1.4 \times 10^6 \text{ g/m}^3$$

$$R = \text{Pellet radius} = 3 \times 10^{-3} \text{ m}$$

$$\gamma = \text{Sphericity} = 1.0$$

Solution

It is desired to reduce the NO concentration from 2.0% to 0.004%. Neglecting any volume change at these low concentrations gives us

$$X = \frac{C_{Ab0} - C_{Ab}}{C_{Ab0}} = \frac{2 - 0.004}{2} = 0.998$$

where A represents NO.

The variation of NO down the length of the reactor is given by Equation (15-69). Replacing k_1 by $k_1' \rho_b$

$$\frac{dC_{Ab}}{dz} = -\frac{\Omega k_1 C_{Ab}}{U} = -\Omega \frac{k_1' \rho_b C_{Ab}}{U} \tag{15-69}$$

Multiplying the denominator on the right and left hand sides of Equation (15-69) by the cross-sectional area, A_c, and realizing that the weight of the catalyst up to a point z in the bed is

$$W = \rho_b A_c z$$

the variation of NO concentration with solids is

$$\frac{dC_{Ab}}{dW} = -\frac{\Omega k_1' C_{Ab}}{A_c U} \tag{E15-3.1}$$

Because NO is present in dilute concentrations (i.e., $y_{A0} \ll 1$), we shall take $\varepsilon \ll 1$ and set $A_c U = v_0$. We integrate Equation (E15-3.1) using the boundary condition that when $W = 0$, then $C_{Ab} = C_{Ab0}$

$$X = \left(1 - \frac{C_{Ab}}{C_{Ab0}}\right) = 1 - \exp\left(-\frac{\Omega k_1' W}{v_0}\right) \tag{E15-3.2}$$

where

$$\Omega = \frac{\eta}{1 + \eta \dfrac{k_1' \rho_b}{k_c a_c}} \tag{15-58}$$

Rearranging, we have

$$W = \frac{v_0}{\Omega k_1'} \ln \frac{1}{1-X} \tag{E15-3.3}$$

4. *Calculating the internal effectiveness factor* for spherical pellets in which a first-order reaction is occurring, we obtained

$$\eta = \frac{3}{\phi_1^2} (\phi_1 \coth \phi_1 - 1) \tag{15-32}$$

As a first approximation, we shall neglect any changes in the pellet size resulting from the reactions of NO with the porous carbon. The Thiele modulus for this system is[8]

$$\phi_1 = R\sqrt{\frac{k_1' \rho_c}{D_e}} = R\sqrt{\frac{k_1'' S_a \rho_c}{D_e}} \tag{E15-3.4}$$

where

R = pellet radius = 3×10^{-3} m
D_e = effective diffusivity = 1.82×10^{-8} m²/s
ρ_c = catalyst pellet density = 2.8 g/cm³ = 2.8×10^6 g/m³
k_1'' = specific reaction rate (m/s) = 4.42×10^{-10} m³/m²/s

Substituting in Equation (E15-3.4)

$$\phi_1 = 0.003 \text{ m} \sqrt{\frac{(4.42 \times 10^{-10} \text{ m/s})(530 \text{ m}^2/\text{g})(2.8 \times 10^6 \text{ g/m}^3)}{1.82 \times 10^{-8} \text{ m}^2/\text{s}}}$$

$$\phi_1 = 18$$

---

[8] L. K. Chan, A. F. Sarofim, and J. M. Beer, *Combust. Flame*, 52, 37.

Because ϕ_1 is large

$$\eta \cong \frac{3}{\phi_1} = \frac{3}{18} = 0.167$$

1. Calculate C_{WP}

$$C_{WP} = \eta\phi^2$$

$$= (0.167)(18)^2 = 54.$$

$C_{WP} \gg 1$ and internal diffusion limits the reaction

2. *To calculate the external mass transfer coefficient*, the Thoenes–Kramers correlation is used. From Chapter 14 we recall

$$Sh' = (Re')^{1/2} Sc^{1/3} \qquad (14\text{-}65)$$

For a 2-in.-ID pipe, $A_c = 2.03 \times 10^{-3}$ m^2. The superficial velocity is

$$U = \frac{v_0}{A_c} = \frac{10^{-6} \ \text{m}^3/\text{s}}{2.03 \times 10^{-3} \ \text{m}^2} = 4.93 \times 10^{-4} \ \text{m/s}$$

Procedure
Calculate
Re′
Sc
Then
Sh′
Then
k_c

$$Re' = \frac{Ud_p}{(1-\phi)v} = \frac{(4.93 \times 10^{-4} \ \text{m/s})(6 \times 10^{-3} \ \text{m})}{(1-0.5)(1.53 \times 10^{-8} \ \text{m}^2/\text{s})} = 386.7$$

Nomenclature note: ϕ with subscript 1, ϕ_1 = Thiele modulus
ϕ without subscript, ϕ = porosity

$$Sc = \frac{v}{D_{AB}} = \frac{1.53 \times 10^{-8} \ \text{m}^2/\text{s}}{2.0 \times 10^{-8} \ \text{m}^2/\text{s}} = 0.765$$

$$Sh' = (386.7)^{1/2}(0.765)^{1/3} = (19.7)(0.915) = 18.0$$

$$k_c = \frac{1-\phi}{\phi}\left(\frac{D_{AB}}{d_p}\right) Sh' = \frac{0.5}{0.5}\left(\frac{2.0 \times 10^{-8} \ (\text{m}^2/\text{s})}{6.0 \times 10^{-3} \ \text{m}}\right)(18.0)$$

$$k_c = 6 \times 10^{-5} \ \text{m/s}$$

3. *Calculating the external area per unit reactor volume*, we obtain

$$a_c = \frac{6(1-\phi)}{d_p} = \frac{6(1-0.5)}{6 \times 10^{-3} \ \text{m}} \qquad (E15\text{-}3.5)$$

$$= 500 \ \text{m}^2/\text{m}^3$$

4. *Evaluating the overall effectiveness factor.* Substituting into Equation (15-58), we have

$$\Omega = \frac{\eta}{1 + \eta k_1'' \ S_a \rho_b / k_c a_c}$$

$$\Omega = \frac{0.167}{1 + \dfrac{(0.167)(4.4 \times 10^{-10} \ \text{m}^3/\text{m}^2 \cdot \text{s})(530 \ \text{m}^2/\text{g})(1.4 \times 10^6 \ \text{g/m}^3)}{(6 \times 10^{-5} \ \text{m/s})(500 \ \text{m}^2/\text{m}^3)}}$$

$$= \frac{0.167}{1 + 1.83} = 0.059$$

In this example we see that both the external and internal resistances to mass transfer are significant.

5. Calculate the Mears criterion, MR, to see if mass transfer limits the reaction.

$$MR = \frac{-r'_A \rho_b Rn}{k_c C_{Ab}} = \frac{\Omega(-r_{Ab})Rn}{k_c C_{Ab}}$$

$$= \frac{\Omega k'_1 C_{Ab} \rho_b R}{k_c C_{Ab}} = \frac{\Omega k' \rho_b R}{k_c}$$

$$= \frac{(0.059)(2.34 \times 10^{-7} \text{m}^3/\text{g/s})\left(1.4 \times 10^{-6} \frac{\text{g}}{\text{m}^3}\right)(3 \times 10^{-3}\text{m})}{6 \times 10^{-5}\text{m/s}}$$

$$= 0.97$$

MR (i.e., 0.97 > 0.15) and diffusion limits the reaction.

6. *Calculating the weight of solid necessary to achieve* 99.8% *conversion.* Substituting into Equation (E15-3.3), we obtain

$$W = \frac{1 \times 10^{-6} \text{ m}^3/\text{s}}{(0.059)(4.42 \times 10^{-10} \text{ m}^3/\text{m}^2 \cdot \text{s})(530 \text{ m}^2/\text{g})} \ln \frac{1}{1 - 0.998}$$

$$= 450 \text{ g}$$

7. *The reactor length is*

$$L = \frac{W}{A_c \rho_b} = \frac{450 \text{ g}}{(2.03 \times 10^{-3} \text{ m}^2)(1.4 \times 10^6 \text{ g/m}^3)}$$

$$= 0.16 \text{ m}$$

This catalyst weight and corresponding reactor length are rather small and as such we could easily increase the feed rate to the reactor.

Analysis: One of the purposes of this example was to show how to carry out detailed calculations of the diffusion and reaction rate in order to size (i.e., calculate V or W) a reactor for a specified conversion, when both external and internal diffusion resistances affect the rate of reaction. These calculations are tedious and detailed, and it was my feeling that we should show and know all the intermediate calculations, e.g., a_c, η and Ω, so that the reader will have a better understanding of how to make such calculations in the future.

15.8 Determination of Limiting Situations from Reaction-Rate Data

For external mass transfer-limited reactions in packed beds, the rate of reaction per unit mass of catalyst at a point in the bed is

$$-r'_A = k_c a_c C_A \tag{15-72}$$

Variation of reaction rate with system variables

The correlation for the mass transfer coefficient, Equation (14-66), shows that k_c is directly proportional to the square root of the velocity and inversely proportional to the square root of the particle diameter

$$k_c \propto \frac{U^{1/2}}{d_p^{1/2}} \tag{15-73}$$

We recall from Equation (E15-3.5), $a_c = 6(1 - \phi)/d_p$, that the variation of external surface area with catalyst particle size is

$$a_c \propto \frac{1}{d_p}$$

We now combine Equations (15-72), (15-73), and (15-74) to obtain

$$-r'_A \propto \frac{U^{1/2}}{d_p^{3/2}} \qquad (15\text{-}74)$$

Consequently, for *external mass transfer–limited reactions*, the rate proportional to the velocity to the one-half power is inversely proportional to the particle diameter to the three-halves power.

From Equation (14-72), we see that for gas-phase external mass transfer–limited reactions, the rate increases approximately linearly with temperature.

When *internal diffusion limits* the rate of reaction, we observe from Equation (15-42) that the rate of reaction varies inversely with particle diameter, is independent of velocity, and exhibits an exponential temperature dependence that is not as strong as that for surface-reaction-controlling reactions. For surface-reaction-limited reactions, the rate is independent of particle size and is a strong function of temperature (exponential). Table 15-1 summarizes the dependence of the rate of reaction on the velocity through the bed, particle diameter, and temperature for the three types of limitations that we have been discussing.

Many heterogeneous reactions are diffusion limited.

TABLE 15-1   LIMITING CONDITIONS

Type of Limitation	Variation of Reaction Rate with:		
	Velocity	Particle Size	Temperature
External diffusion	$U^{1/2}$	$(d_p)^{-3/2}$	$\approx$ Linear
Internal diffusion	Independent	$(d_p)^{-1}$	Exponential
Surface reaction	Independent	Independent	Exponential

Very Important Table

The exponential temperature dependence for internal diffusion limitations is usually not as strong a function of temperature as is the dependence for surface reaction limitations (cf. Section 15.4). If we would calculate an activation energy between 8 and 24 kJ/mol, chances are that the reaction would be strongly diffusion-limited. An activation energy of 200 kJ/mol, however, suggests that the reaction is reaction rate–limited.

15.9 Multiphase Reactors in the Professional Reference Shelf

Multiphase reactors are reactors in which two or more phases are necessary to carry out the reaction. The majority of multiphase reactors involve gas and liquid phases that contact a solid. In the case of the slurry and trickle bed reactors, the reaction between the gas and the liquid takes place on a solid catalyst surface (see Table 15-2). However, in some reactors the liquid phase is an inert medium for the gas to contact the solid catalyst. The latter situation arises when a large heat sink is required for highly exothermic reactions. In many cases, the catalyst life is extended by these milder operating conditions.

Reference Shelf

TABLE 15-2   APPLICATIONS OF THREE-PHASE REACTORS

I. *Slurry reactor*
    A. Hydrogenation
       1. of fatty acids over a supported nickel catalyst
       2. of 2-butyne-1,4-diol over a Pd-CaCO$_3$ catalyst
       3. of glucose over a Raney nickel catalyst
    B. Oxidation
       1. of C$_2$H$_4$ in an inert liquid over a PdCl$_2$-carbon catalyst
       2. of SO$_2$ in inert water over an activated carbon catalyst
    C. Hydroformation
       of CO with high-molecular-weight olefins on either a cobalt or ruthenium complex bound
       to polymers
    D. Ethynylation
       Reaction of acetylene with formaldehyde over a CaCl$_2$-supported catalyst
II. *Trickle bed reactors*
    A. Hydrodesulfurization
       Removal of sulfur compounds from crude oil by reaction with hydrogen on Co-Mo on
       alumina
    B. Hydrogenation
       1. of aniline over a Ni-clay catalyst
       2. of 2-butyne-1,4-diol over a supported Cu-Ni catalyst
       3. of benzene, α-CH$_3$ styrene, and crotonaldehyde
       4. of aromatics in napthenic lube oil distillate
    C. Hydrodenitrogenation
       1. of lube oil distillate
       2. of cracked light furnace oil
    D. Oxidation
       1. of cumene over activated carbon
       2. of SO$_2$ over carbon

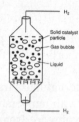

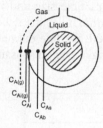

Source: Satterfield, C. N. *AIChE Journal.*, *21*, 209 (1975); P. A. Ramachandran and
R. V. Chaudhari, *Chem. Eng.*, *87*(24), 74 (1980); R. V. Chaudhari and P. A. Ramachandran,
AIChE Journal., *26*, 177 (1980). With permission of the American Institute of Chemical
Engineers. Copyright © 1980 AIChE. All rights reserved.

The multiphase reactors discussed in this edition of the book are the
slurry reactor, fluidized bed, and the trickle bed reactor. The trickle bed reactor,
which has reaction and transport steps similar to the slurry reactor, is discussed
in the first edition of this book and on the CRE Web site along with the bubbling fluidized bed. In slurry reactors, the catalyst is suspended in the liquid,
and gas is bubbled through the liquid. A slurry reactor may be operated in
either a semibatch or continuous mode.

15.9.1   Slurry Reactors

A complete description of the slurry reactor and the transport and reaction
steps are given on the CRE Web site, along with the design equations and a
number of examples. Methods to determine which of the transport and reaction
steps are rate limiting are included. See *Professional Reference Shelf R15.1.*

15.9.2   Trickle Bed Reactors

The CRE Web site includes all the material on trickle bed reactors from the
first edition of this book. A comprehensive example problem for trickle bed
reactor design is included. See *Professional Reference Shelf R15.2.*

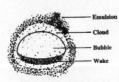

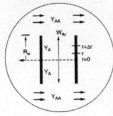

15.10 Fluidized Bed Reactors

The Kunii-Levenspiel model for fluidization is given on the CRE Web site along with a comprehensive example problem. The rate-limiting transport steps are also discussed. See *Professional Reference Shelf R15.3*.

15.11 Chemical Vapor Deposition (CVD)

Chemical vapor deposition in boat reactors is discussed and modeled. The equations and parameters that affect wafer thickness and shape are derived and analyzed. This material is taken directly from the second edition of this book. See *Professional Reference Shelf R15.4*.

Closure. After completing this chapter, the reader should be able to derive differential equations describing diffusion and reaction; discuss the meaning of the internal effectiveness factor, η, and its relationship to the Thiele modulus; and discuss the overall effectiveness factor, Ω. The reader should be able to identify the regions of external and internal mass transfer limitations and reaction-rate limitations by applying the Weisz–Prater criterion and Mears criterion. The reader should be able to apply the overall effectiveness factor to a packed-bed reactor to calculate the conversion at the exit of the reactor.

SUMMARY

1. The concentration profile for a first-order reaction occurring in a spherical catalyst pellet is

$$\frac{C_A}{C_{As}} = \frac{R}{r}\left[\frac{\sinh{(\phi_1 r/R)}}{\sinh{\phi_1}}\right] \tag{S15-1}$$

where ϕ_1 is the Thiele modulus. For a first-order reaction

$$\phi_1^2 = \frac{k_1}{D_e}R^2 \tag{S15-2}$$

2. The effectiveness factors are

Internal
effectiveness = η = $\dfrac{\text{Actual rate of reaction}}{\substack{\text{Reaction rate if entire interior}\\ \text{surface is exposed to concentration}\\ \text{at the external pellet surface}}}$
factor

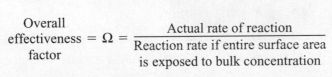

$$-r_A = \eta(-r_{As})$$

Overall
effectiveness = Ω = $\dfrac{\text{Actual rate of reaction}}{\substack{\text{Reaction rate if entire surface area}\\ \text{is exposed to bulk concentration}}}$
factor

$$-r_A = \Omega(-r_{Ab})$$

3. For large values of the Thiele modulus for an nth order reaction

$$\eta = \left(\frac{2}{n+1}\right)^{1/2} \frac{3}{\phi_n} \qquad (S15\text{-}3)$$

4. For internal diffusion control, the true reaction order is related to the measured reaction order by

$$n_{\text{true}} = 2n_{\text{apparent}} - 1 \qquad (S15\text{-}4)$$

The true and apparent activation energies are related by

$$E_{\text{true}} = 2E_{\text{app}} \qquad (S15\text{-}5)$$

5. A. The Weisz–Prater Parameter

$$C_{\text{WP}} = \phi_1^2 \eta = \frac{-r_A'(\text{observed})\rho_c R^2}{D_e C_{As}} \qquad (S15\text{-}6)$$

The Weisz–Prater criterion dictates that

If $C_{\text{WP}} \ll 1$ no internal diffusion limitations present

If $C_{\text{WP}} \gg 1$ internal diffusion limitations present

B. The Mears Criterion for Neglecting External Diffusion and Heat Transfer

There will be no external diffusion limitations **if**

$$\boxed{\frac{-r_A'\rho_b Rn}{k_c C_{Ab}} < 0.15} \qquad (S15\text{-}7)$$

And there will be no temperature gradients **if**

$$\boxed{\left|\frac{-\Delta H_{\text{Rx}}(-r_A')(\rho_b RE)}{hT^2 R_g}\right| < 0.15} \qquad (S15\text{-}8)$$

CRE WEB SITE MATERIALS

- **Extended Material on the Web Site**
  1. Web Example 15-1 *Application of Diffusion and Reaction to Tissue Engineering* (also see P15-15$_B$)

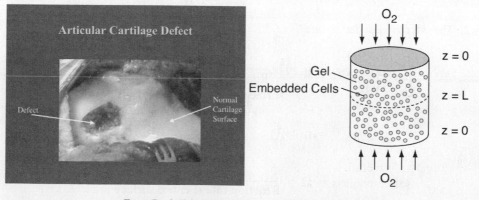

From Prof. Kristi Anseth, University of Colorado.

Figure WE15-1.1 Damaged cartilage. (*Newsweek*, September 3, 2001.) Photo of "damaged cartilage" originally appeared in "21 Million Americans Suffer from Arthritis," by Jerry Adler.

    2. 15.3.3 *Effectiveness Factor for Nonisothermal First-Order Catalytic Reactions*
    3. Problem 15-12$_C$ *Diffusion and Reaction in Spherical Catalyst Pellets*
    4. *Additional Homework Problems*
- **Learning Resources**
    1. *Summary Notes*
    2. *Solved Problems*
- **Professional Reference Shelf**
    R15.1. *Slurry Reactors*

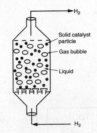

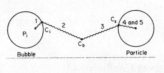

Transport Steps and Resistances

A. Description of the Use of Slurry Reactors
   Example R15-1 Industrial Slurry Reactor
B. Reaction and Transport Steps in a Slurry Reactor

$$\frac{C_i}{R_A} = \frac{1}{k_b a_b} + \frac{1}{m}\left(\overbrace{\frac{1}{k_b a_b}}^{r_c} + \overbrace{\frac{1}{k\eta}}^{r_r} \right)$$

C. Determining the Rate-Limiting Step
    1. Effect of Loading, Particle Size, and Gas Adsorption
    2. Effect of Shear
       Example R15-2 Determining the Controlling Resistance
D. Slurry Reactor Design

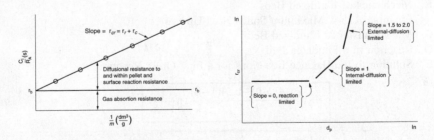

Example R15-3 Slurry Reactor Design

R15.2. *Trickle Bed Reactors*
   A. Fundamentals

 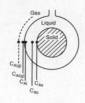 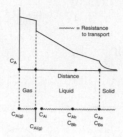

   B. Limiting Situations
   C. Evaluating the Transport Coefficients

$$-r'_A = \frac{1/H}{\underbrace{\frac{(1-\phi)\rho_c}{Hk_ga_i}+\frac{(1-\phi)\rho_c}{k_la_i}+\frac{1}{k_ca_p}+\frac{1}{\eta kC_{Bs}}}_{k_{vg}}}C_A(g) \qquad \frac{\text{mol}}{\text{g-cat}\cdot\text{s}}$$

R15.3. *Fluidized Bed Reactors*
   A. Descriptive Behavior of the Kunii-Levenspiel Bubbling Bed Model

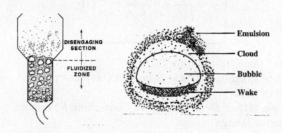

   B. Mechanics of Fluidized Beds
       Example R15-4 Maximum Solids Hold-Up
   C. Mass Transfer in Fluidized Beds
   D. Reaction in a Fluidized Bed
   E. Solution to the Balance Equations for a First-Order Reaction

$$W = \frac{\rho_cA_cu_b(1-\varepsilon_{mf})(1-\delta)}{k_{cat}K_R}\ln\frac{1}{1-X}$$

$$K_R = \gamma_b + \cfrac{1}{\cfrac{k_{cat}}{K_{bc}} + \cfrac{1}{\gamma_c + \cfrac{1}{\cfrac{1}{\gamma_e} + \cfrac{k_{cat}}{K_{ce}}}}}$$

       Example R15-5 Catalytic Oxidation of Ammonia

F. Limiting Situations
Example R15-6 Calculation of the Resistances
Example R15-7 Effect of Particle Size on Catalyst Weight for
           a Slow Reaction
Example R15-8 Effect of Catalyst Weight for a Rapid Reaction
R15.4. *Chemical Vapor Deposition Reactors*

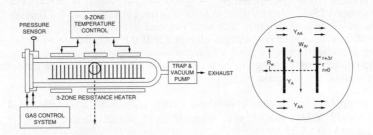

A. Chemical Reaction Engineering in Microelectronic Processing
B. Fundamentals of CVD
C. Effectiveness Factors for Boat Reactors

$$\eta = \frac{2I_1(\phi_1)}{\phi_1 I_o(\phi_1)}$$

Example R15-9 Diffusion Between Wafers
Example R15-10 CVD Boat Reactor

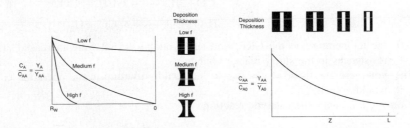

QUESTIONS AND PROBLEMS

The subscript to each of the problem numbers indicates the level of difficulty: A, least difficult; D, most difficult.

$$A = \bullet \quad B = \blacksquare \quad C = \blacklozenge \quad D = \blacklozenge\blacklozenge$$

Questions

Q15-1$_C$ Make up an original problem using the concepts presented in Section _____ (your instructor will specify the section). Extra credit will be given if you obtain and use real data from the literature. (See Problem P5-1$_A$ for the guidelines.)

Q15-2 Suppose someone had used the false kinetics (i.e., wrong E, wrong n)? Would the catalyst weight be overdesigned or underdesigned? What are other positive or negative effects that occur?

Problems

P15-1$_B$ **(a)** **Example 15-1.** *Effective Diffusivity.* Make a sketch of a diffusion path for which the tortuosity is 5. How would your effective gas-phase diffusivity change if the absolute pressure were tripled and the temperature were increased by 50%?

(b) **Example 15-2.** If possible, determine the percent of the total resistance for (1) internal diffusion and (2) for reaction rate for each of the three particles studied. Apply the Weisz-Prater criteria to a particle 0.005 m in diameter.

(c) **Example 15-3.** *Overall Effectiveness Factor.* (1) Calculate the percent of the total resistance for the resistance of external diffusion, internal diffusion, and surface reaction. Qualitatively, how would each of your percentages change (2) If the temperature were increased significantly? (3) If the gas velocity were tripled? (4) If the particle size were decreased by a factor of 2? How would the reactor length change in each case? (5) What length would be required to achieve 99.99% conversion of the pollutant NO?

What if...

(d) you applied the Mears and Weisz–Prater criteria to Examples 15-4 and 15-3? What would you find? What would you learn if $\Delta H_{Rx}^{\circ} = -25$ kcal/mol, $h = 100$ Btu/h·ft²·°F, and $E = 20$ kcal/mol?

(e) your internal surface area decreased with time because of sintering (see Section 10.7). Describe how your effectiveness factor would change and the rate of reaction change with time if $k_d = 0.01$ h^{-1} and $\eta = 0.01$ at $t = 0$? Explain, being as quantitative as possible when you can.

(f) you were to assume that the resistance to gas absorption in the CRE Web site *Professional Reference Shelf R15.1* was the same as in *Professional Reference Shelf R15.3* and that the liquid-phase reactor volume in *Professional Reference Shelf R15.3* was 50% of the total? Could you determine the limiting resistance? If so, what is it? What other things could you calculate in *Professional Reference Shelf R15.1* (e.g., selectivity, conversion, molar flow rates in and out)? *Hint:* Some of the other reactions that occur include

$$CO + 3H_2 \longrightarrow CH_4 + H_2O$$

$$H_2O + CO \longrightarrow CO_2 + H_2$$

(g) the temperature in the CRE Web site Example R15.2 were increased? How would the relative resistances in the slurry reactor change?

(h) you were asked for all the things that could go wrong in the operation of a slurry reactor? What would you say?

P15-2$_B$ *Concept problem:* The catalytic reaction

$$A \longrightarrow B$$

takes place within a fixed bed containing spherical porous catalyst X22. Figure P15-2$_B$ shows the overall rates of reaction at a point in the reactor as a function of temperature for various entering total molar flow rates, F_{T0}.

(a) Is the reaction limited by external diffusion?

(b) If your answer to part (a) was "yes," under what conditions of those shown (i.e., T, F_{T0}) is the reaction limited by external diffusion?

(c) Is the reaction "reaction-rate-limited"?

(d) If your answer to part (c) was "yes," under what conditions of those shown (i.e., T, F_{T0}) is the reaction limited by the rate of the surface reactions?

(e) Is the reaction limited by internal diffusion?

(f) If your answer to part (e) was "yes," under what conditions of those shown (i.e., T, F_{T0}) is the reaction limited by the rate of internal diffusion?

(g) For a flow rate of 10 g mol/h, determine (if possible) the overall effectiveness factor, Ω, at 360 K.

(h) Estimate (if possible) the internal effectiveness factor, η, at 367 K.

(i) If the concentration at the external catalyst surface is 0.01 mol/dm³, calculate (if possible) the concentration at $r = R/2$ inside the porous catalyst at 367 K. (Assume a first-order reaction.)

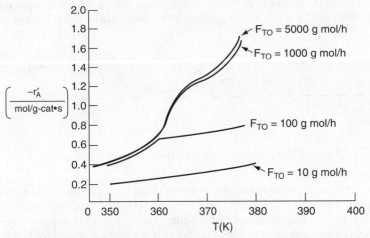

Figure P15-2$_B$ Reaction rates in a catalyst bed.

Additional information:

Gas properties:

Diffusivity: 0.1 cm^2/s
Density: 0.001 g/cm^3
Viscosity: 0.0001 g/cm·s

Bed properties:

Tortuosity of pellet: 1.414
Bed permeability: 1 millidarcy
Porosity = 0.3

P15-3$_B$ *Concept problem:* The reaction

$$A \longrightarrow B$$

is carried out in a differential packed-bed reactor at different temperatures, flow rates, and particle sizes. The results shown in Figure P15-3$_B$ were obtained.

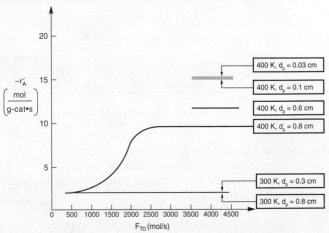

Figure P15-3$_B$ Reaction rates in a catalyst bed.

(a) What regions (i.e., conditions d_p, T, F_{T0}) are external mass transfer–limited?
(b) What regions are reaction rate–limited?
(c) What region is internal-diffusion-controlled?
(d) What is the internal effectiveness factor at $T = 400$ K and $d_p = 0.8$ cm?

P15-4$_A$ *Concept problem:* Curves A, B, and C in Figure P15-4$_A$ show the variations in reaction rate for three different reactions catalyzed by solid catalyst pellets. What can you say about each reaction?

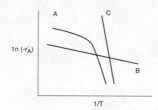

Figure P15-4$_A$ Temperature dependence of three reactions.

P15-5$_B$ A first-order heterogeneous irreversible reaction is taking place within a spherical catalyst pellet that is plated with platinum throughout the pellet (see Figure 15-3). The reactant concentration halfway between the external surface and the center of the pellet (i.e., $r = R/2$) is equal to one-tenth the concentration of the pellet's external surface. The concentration at the external surface is 0.001 g mol/dm^3, the diameter ($2R$) is 2×10^{-3} cm, and the diffusion coefficient is 0.1 cm^2/s.

$$A \longrightarrow B$$

(a) What is the concentration of reactant at a distance of 3×10^{-4} cm in from the external pellet surface? (*Ans.:* $C_A = 2.36 \times 10^{-4}$ mol/dm^3.)

(b) To what diameter should the pellet be reduced if the effectiveness factor is to be 0.8? (*Ans.:* $d_p = 6.8 \times 10^{-4}$ cm. Critique this answer!)

(c) If the catalyst support were not yet plated with platinum, how would you suggest that the catalyst support be plated *after* it had been reduced by grinding?

P15-6$_B$ The swimming rate of a small organism [*J. Theoret. Biol.*, **26**, 11 (1970)] is related to the energy released by the hydrolysis of adenosine triphosphate (ATP) to adenosine diphosphate (ADP). The rate of hydrolysis is equal to the rate of diffusion of ATP from the midpiece to the tail (see Figure P15-6$_B$). The diffusion coefficient of ATP in the midpiece and tail is 3.6×10^{-6} cm^2/s. ADP is converted to ATP in the midsection, where its concentration is 4.36×10^{-5} mol/cm^3. The cross-sectional area of the tail is 3×10^{-10} cm^2.

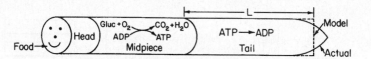

Figure P15-6$_B$ Swimming of an organism.

Application
Pending
for Problem
Hall of
Fame

(a) Derive an equation for diffusion and reaction in the tail.

(b) Derive an equation for the effectiveness factor in the tail.

(c) Taking the reaction in the tail to be of zero order, calculate the length of the tail. The rate of reaction in the tail is 23×10^{-18} mol/s.

(d) Compare your answer with the average tail length of 41 μm. What are possible sources of error?

P15-7$_B$ A first-order, heterogeneous, irreversible reaction is taking place within a catalyst pore that is plated with platinum entirely along the length of the pore (Figure P15-7$_B$). The reactant concentration at the plane of symmetry (i.e., equal distance from the pore mouth) of the pore is equal to one-tenth the concentration at the pore mouth. The concentration at the pore mouth is 0.001 mol/dm^3, the pore length ($2L$) is 2×10^{-3} cm, and the diffusion coefficient is 0.1 cm^2/s.

(a) Derive an equation for the effectiveness factor.

(b) What is the concentration of reactant at $L/2$?

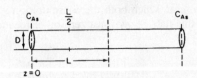

Figure P15-7$_B$ Single catalyst pore.

(c) To what length should the pore length be reduced if the effectiveness factor is to be 0.8?

(d) If the catalyst support were not yet plated with platinum, how would you suggest the catalyst support be plated *after* the pore length, *L*, had been reduced by grinding?

P15-8$_A$ A first-order reaction is taking place inside a porous catalyst. Assume dilute concentrations and neglect any variations in the axial (*x*) direction.

(a) Derive an equation for both the internal and overall effectiveness factors for the rectangular porous slab shown in Figure P15-8$_A$.

(b) Repeat part (a) for a cylindrical catalyst pellet where the reactants diffuse inward in the radial direction. (**C-level** problem, i.e., P15-8$_C$(b).)

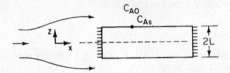

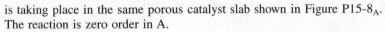

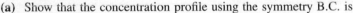

Figure P15-8$_A$ Flow over porous catalyst slab.

P15-9$_B$ The irreversible reaction

$$A \longrightarrow B$$

is taking place in the same porous catalyst slab shown in Figure P15-8$_A$.
The reaction is zero order in A.

(a) Show that the concentration profile using the symmetry B.C. is

$$\frac{C_A}{C_{As}} = 1 + \phi_0^2 \left[\left(\frac{z}{L} \right)^2 - 1 \right] \tag{P15-9.1}$$

where

$$\phi_0^2 = \frac{kL^2}{2D_e C_{As}} \tag{P15-9.2}$$

(b) For a Thiele modulus of 1.0, at what point in the slab is the concentration zero? For $\phi_0 = 4$?

(c) What is the concentration you calculate at $z = 0.1 L$ and $\phi_0 = 10$ using Equation (P15-9.1)? What do you conclude about using this equation?

(d) Plot the dimensionless concentration profile $\psi = C_A/C_{As}$ as a function of $\lambda = z/L$ for $\phi_0 = 0.5, 1, 5$, and 10. *Hint:* there are regions where the concentration is zero. Show that $\lambda_C = (1 - 1/\phi_0)$ is the start of this region where the gradient and concentration are both zero. [L. K. Jang, R. L. York, J. Chin, and L. R. Hile, *Inst. Chem. Engr.*, 34, 319 (2003).]
Show that $\psi = \phi_0^2 \lambda^2 - 2\phi_0(\phi_0 - 1) \lambda + (\phi_0 - 1)^2$ for $\lambda_C \leq \lambda < 1$.

(e) The effectiveness factor can be written as

$$\eta = \frac{\int_0^L -r_A A_c \, dz}{-r_{As} A_c L} = \frac{\int_0^{z_C} -r_A A_c \, dz + \int_{z_C}^L -r_A A_c \, dz}{-r_{As} A_c L} \tag{P15-9.3}$$

where z_C (λ_C) is the point at which both the concentration gradients and flux go to zero, and A_c is the cross-sectional area of the slab. Show for a zero-order reaction that

$$\eta = \begin{cases} 1 & \text{for } \phi_0 \leq 1.0 \\ 1 - \lambda_C = \dfrac{1}{\phi_0} & \text{for } \phi_0 \geq 1 \end{cases} \qquad (P15\text{-}9.4)$$

(f) Make a sketch for η versus ϕ_0 similar to the one shown in Figure 15-5.

(g) Repeat parts (a) to (f) for a spherical catalyst pellet.

(h) What do you believe to be the point of this problem?

P15-10$_C$ The second-order decomposition reaction

$$A \longrightarrow B + 2C$$

is carried out in a tubular reactor packed with catalyst pellets 0.4 cm in diameter. The reaction is internal-diffusion-limited. Pure A enters the reactor at a superficial velocity of 3 m/s, a temperature of 250°C, and a pressure of 500 kPa. Experiments carried out on smaller pellets where surface reaction is limiting yielded a specific reaction rate of 0.05 m^6/mol·g-cat·s. Calculate the length of bed necessary to achieve 80% conversion. Critique the numerical answer.

Additional information:

Effective diffusivity: 2.66×10^{-8} m^2/s   Pellet density: 2×10^6 g/m^3

Ineffective diffusivity: 0.00 m^2/s   Internal surface area: 400 m^2/g

Bed porosity: 0.4

P15-11$_C$ Derive the concentration profile and effectiveness factor for cylindrical pellets 0.2 cm in diameter and 1.5 cm in length. Neglect diffusion through the ends of the pellet.

(a) Assume that the reaction is a first-order isomerization. *Hint:* Look for a Bessel function.

(b) Rework Problem P15-10$_C$ for these pellets.

P15-12$_B$ *Extension of Problem P15-7$_B$.* The elementary isomerization reaction

$$A \longrightarrow B$$

is taking place on the walls of a cylindrical catalyst pore (see Figure P15-7$_B$.) In one run, a catalyst poison P entered the reactor together with the reactant A. To estimate the effect of poisoning, we assume that the poison renders the catalyst pore walls near the pore mouth ineffective up to a distance z_1, so that no reaction takes place on the walls in this entry region.

(a) Show that before poisoning of the pore occurred, the effectiveness factor was given by

$$\eta = \frac{1}{\phi} \tanh \phi$$

where

$$\phi = L \sqrt{\frac{2k}{rD_e}}$$

with k = reaction-rate constant (length/time)

r = pore radius (length)

D_e = effective molecular diffusivity (area/time)

(b) Derive an expression for the concentration profile and also for the molar flux of A in the ineffective region, $0 < z < z_1$, in terms of z_1, D_{AB}, C_{A1}, and C_{As}. Without solving any further differential equations, obtain the new effectiveness factor η' for the poisoned pore.

P15-13$_B$ *Falsified Kinetics.* The irreversible gas-phase dimerization

$$2A \longrightarrow A_2$$

is carried out at 8.2 atm in a stirred contained-solids reactor to which only pure A is fed. There are 40 g of catalyst in each of the four spinning baskets. The following runs were carried out at 227°C:

Total Molar Feed Rate, F_{T0} (g mol/min)	1	2	4	6	11	20
Mole Fraction A in Exit, y_A	0.21	0.33	0.40	0.57	0.70	0.81

The following experiment was carried out at 237°C:

$$F_{T0} = 9 \text{ g mol/min} \qquad y_A = 0.097$$

(a) What are the apparent reaction order and the apparent activation energy?
(b) Determine the true reaction order, specific reaction rate, and activation energy.
(c) Calculate the Thiele modulus and effectiveness factor.
(d) What pellet diameter should be used to make the catalyst more effective?
(e) Calculate the rate of reaction on a rotating disk made of the catalytic material when the gas-phase reactant concentration is 0.01 g mol/L and the temperature is 227°C. The disk is flat, nonporous, and 5 cm in diameter.

Additional information:

Effective diffusivity: 0.23 cm²/s   Radius of catalyst pellets: 1 cm
Surface area of porous catalyst: 49 m²/g-cat   Color of pellets: blushing peach
Density of catalyst pellets: 2.3 g/cm³

P15-14$_B$ Derive Equation (15-39). *Hint:* Multiply both sides of Equation (15-25) for *n*th order reaction; that is,

$$\frac{d^2y}{d\lambda^2} - \phi_n^2 y^n = 0$$

by $2dy/d\lambda$, rearrange to get

$$\frac{d}{d\lambda}\left(\frac{dy}{d\lambda}\right)^2 = \phi_n^2 y^n 2\frac{dy}{d\lambda}$$

and solve using the boundary conditions $dy/d\lambda = 0$ at $\lambda = 0$.

P15-15$_B$ *Applications of Diffusion and Reaction to Tissue Engineering.* The equations describing diffusion and reaction in porous catalysts also can be used to derive rates of tissue growth and have been studied by Professor Kristi Anseth and her students at the University of Colorado. One important area of tissue growth is in cartilage tissue in joints such as the knee. Over 200,000 patients per year receive knee joint replacements. Alternative strategies include the growth of cartilage to repair the damaged knee.

One approach is to deliver cartilage-forming cells in a hydrogel to the damaged area such as the one shown in Figure WP15-1.1 on the CRE Web site.

Here, the patient's own cells are obtained from a biopsy and embedded in a hydrogel, which is a cross-linked polymer network that is swollen in water. In order for the cells to survive and grow new tissue, many properties of the gel must be tuned to allow diffusion of important species in and out (e.g., nutrients *in* and cell-secreted extracellular molecules such as collagen *out*). Because there is no blood flow through the cartilage, oxygen transport to the cartilage cells is primarily by diffusion. Consequently, the design must be such that the gel can maintain the necessary rates of diffusion of

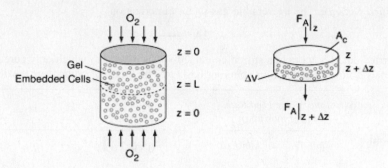

Figure P15-15$_B$ Schematic of cartilage cell system.

nutrients (e.g., O_2) into the hydrogel. These rates of exchange in the gel depend on the geometry and the thickness of the gel. To illustrate the application of chemical reaction engineering principles to tissue engineering, we will examine the diffusion and consumption of one of the nutrients, oxygen.

Our examination of diffusion and reaction in catalyst pellets showed that in many cases the reactant concentration near the center of the particle was virtually zero. If this condition were to occur in a hydrogel, the cells at the center would die. Consequently, the gel thickness needs to be designed to allow rapid transport of oxygen.

Let's consider the simple gel geometry shown in Figure P15-15$_B$. We want to find the gel

thickness at which the minimum oxygen consumption rate is 10^{-13} mol/cell/h $\left(k = \dfrac{10^{-3}\,\text{mol } O_2}{dm^3 h} \right)$.

The cell density in the gel is 10^{10} cells/dm^3, the bulk concentration of oxygen C_{A0} ($z = 0$) is 2×10^{-4} mol/dm^3, and the diffusivity, D_{AB}, is 10^{-5} cm^2/s.

(a) Show that the dimensionless form of concentration and length, $\psi = C_A/C_{A0}$, and $\lambda = z/L$, differential mole balance on O_2 gives

$$\frac{d^2\psi}{d\lambda^2} - \frac{kL^2}{D_{AB}C_{A0}} = 0$$

(b) Show the dimensionless O_2 concentration profile in the gel is

$$\psi = \frac{C_A}{C_{A0}} = \phi_0 \lambda (\lambda - 2) + 1$$

where

$$\lambda = z/L$$

$$\phi_0 = \left(\frac{k}{2 D_{AB} C_{A0}} \right) L^2$$

(c) Solve the gel thickness when the concentration at $z = 0$ and $C_A = 0.1$ mmole/dm^3.

(d) How would your answers change if the reaction kinetics were (1) first order in the O_2 concentration with $k_1 = 10^{-2}$ h^{-1}?

(e) Carry out a quasi-steady-state analysis using Equation (E15-1.19) along with the overall balance

Hall of Fame

$$\frac{dN_w}{dt} = \upsilon_c \left. W_{O_2} \right|_{z=0} A_c$$

to predict the O_2 flux and collagen build-up as a function of time.

(f) Sketch ψ versus λ at different times.

(g) Sketch λ_c as a function of time. *Hint*: $V = A_c L$. Assume $\alpha = 10$ and the stoichiometric coefficient for oxygen to collagen, ν_c, is 0.05 mass fraction of cell/mol O_2. $A_c = 2$ cm^2.

JOURNAL CRITIQUE PROBLEMS

P15C-1 Use the Weisz–Prater criterion to determine if the reaction discussed in *AIChE J.*, 10, 568 (1964) is diffusion-rate-limited.

P15C-2 Use the references given in *Ind. Eng. Chem. Prod. Res. Dev.*, 14, 226 (1975) to define the iodine value, saponification number, acid number, and experimental setup. Use the slurry reactor analysis to evaluate the effects of mass transfer and determine if there are any mass transfer limitations.

SUPPLEMENTARY READING

1. There are a number of books that discuss internal diffusion in catalyst pellets; however, one of the first books that should be consulted on this and other topics on heterogeneous catalysis is

> LAPIDUS, L., AND N. R. AMUNDSON, *Chemical Reactor Theory: A Review,* Upper Saddle River, N.J.: Prentice Hall, 1977.

In addition, see

> ARIS, R., *Elementary Chemical Reactor Analysis.* Upper Saddle River, NJ: Prentice Hall, 1989, Chap. 6. Old, but one should find the references listed at the end of this reading particularly useful.

> FOGLER, JOSEPH J., AKA JOFO, *A Chemical Reaction Engineers Guide to the Country of Jofostan.* To be self published, hopefully by 2020.

> LUSS, D., "Diffusion—Reaction Interactions in Catalyst Pellets," p. 239 in *Chemical Reaction and Reactor Engineering.* New York: Marcel Dekker, 1987.

The effects of mass transfer on reactor performance are also discussed in

> COLLINS, FRANK C., AND GEORGE E. KIMBALL, "Diffusion Controlled Reaction Rates," *Journal of Colloid Science,* Vol. 4, Issue 4, August 1949, Pages 425-437.

> SATTERFIELD, C. N., *Heterogeneous Catalysis in Industrial Practice,* 2nd ed. New York: McGraw-Hill, 1991.

2. Diffusion with homogeneous reaction is discussed in

> ASTARITA, G., and R. OCONE, *Special Topics in Transport Phenomena.* New York: Elsevier, 2002.

Gas-liquid reactor design is also discussed in

> BUTT, JOHN B., *Reaction Kinetics and Reactor Design,* 2nd ed. Boca Raton, FL: CRC Press, 2000.

> SHAH, Y. T., *Gas–Liquid–Solid Reactor Design.* New York: McGraw-Hill, 1979.

3. Modeling of CVD reactors is discussed in

DOBKIN, DANIEL, AND M. K. ZUKRAW, *Principles of Chemical Vapor Deposition*. The Netherlands: Kluwer Academic Publishers, 2003.

HESS, D. W., K. F. JENSEN, and T. J. ANDERSON, "Chemical Vapor Deposition: A Chemical Engineering Perspective," *Rev. Chem. Eng.*, 3, 97, 1985.

4. Multiphase reactors are discussed in

RAMACHANDRAN, P. A., and R. V. CHAUDHARI, *Three-Phase Catalytic Reactors*. New York: Gordon and Breach, 1983.

RODRIGUES, A. E., J. M. COLO, and N. H. SWEED, eds., *Multiphase Reactors*, Vol. 1: *Fundamentals*. Alphen aan den Rijn, The Netherlands: Sitjhoff and Noordhoff, 1981.

RODRIGUES, A. E., J. M. COLO, and N. H. SWEED, eds., *Multiphase Reactors*, Vol. 2: *Design Methods*. Alphen aan den Rijn, The Netherlands: Sitjhoff and Noordhoff, 1981.

SHAH, Y. T., B. G. KELKAR, S. P. GODBOLE, and W. D. DECKWER, "Design Parameters Estimations for Bubble Column Reactors" (journal review), *AIChE J.*, 28, 353 (1982).

The following *Advances in Chemistry Series* volume discusses a number of multiphase reactors:

FOGLER, H. S., ed., *Chemical Reactors*, ACS Symp. Ser. 168. Washington, DC: American Chemical Society, 1981, pp. 3–255.

5. Fluidization

KUNII, DAIZO, AND OCTAVE LEVENSPIEL, *Fluidization Engineering,* 2nd ed. (Butterworths Series in Chemical Engineering Deposition). Stoneham, MA: Butterworth-Heinemann, 1991.

In addition to Kunii and Levenspiel's book, many correlations can be found in

DAVIDSON, J. F., R. CLIFF, and D. HARRISON, *Fluidization*, 2nd ed. Orlando: Academic Press, 1985.

YATES, J. G., *Fundamentals of Fluidized-Bed Chemical Processes*, 3rd ed. London: Butterworth, 1983.

Residence Time 16
Distributions of
Chemical Reactors

Nothing in life is to be feared. It is only to be understood.

—Marie Curie

Overview. In this chapter we learn about nonideal reactors; that is, reactors that do not follow the models we have developed for ideal CSTRs, PFRs, and PBRs. After studying this chapter the reader will be able to describe:

- General Considerations. How the residence time distribution (RTD) can be used (Section 16.1).
- Measurement of the RTD. How to calculate the concentration curve (i.e., the *C*-curve) and residence time distribution curve, (i.e., the *E*-curve (Section 16.2)).
- Characteristics of the RTD. How to calculate and use the cumulative RTD function, $F(t)$, the mean residence time, t_m, and the variance σ^2 (Section 16.3).
- The RTD in ideal reactors. How to evaluate $E(t)$, $F(t)$, t_m, and σ^2 for ideal PFRs, CSTRs, and laminar flow reactors (LFRs) so that we have a reference point as to how much our real (i.e., nonideal) reactor deviates form an ideal reactor (Section 16.4).
- How to diagnose problems with real reactors by comparing t_m, $E(t)$, and $F(t)$ with ideal reactors. This comparison will help to diagnose and troubleshoot by-passing and dead volume problems in real reactors (Section 16.5).

16.1 General Considerations

The reactors treated in the book thus far—the perfectly mixed batch, the plug-flow tubular, the packed bed, and the perfectly mixed continuous tank reactors—have been modeled as ideal reactors. Unfortunately, in the real world we often observe behavior very different from that expected from the exemplar; this behavior is true of students, engineers, college professors, and chemical reactors. Just as we must learn to work with people who are not perfect,[†] so the reactor analyst must learn to diagnose and handle chemical reactors whose performance deviates from the ideal. Nonideal reactors and the principles behind their analysis form the subject of this chapter and the next two chapters.

We want to analyze and characterize nonideal reactor behavior.

The basic ideas that are used in the distribution of residence times to characterize and model nonideal reactions are really few in number. The two major uses of the residence time distribution to characterize nonideal reactors are

1. To diagnose problems of reactors in operation.
2. To predict conversion or effluent concentrations in existing/available reactors when a new chemical reaction is used in the reactor.

The following two examples illustrate reactor problems one might find in a chemical plant.

Example 1   A packed-bed reactor is shown in Figure 16-1. When a reactor is packed with catalyst, the reacting fluid usually does not flow uniformly through the reactor. Rather, there may be sections in the packed bed that offer little resistance to flow (Path 1) and, as a result, a portion of the fluid may channel through this pathway. Consequently, the molecules following this pathway do not spend much time in the reactor. On the other hand, if there is internal circulation or a high resistance to flow, the molecules could spend a long time in the reactor (Path 2). Consequently, we see that there is a distribution of times that molecules spend in the reactor in contact with the catalyst.

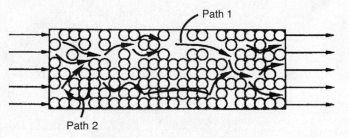

Figure 16-1   Packed-bed reactor.

Example 2   In many continuous-stirred tank reactors, the inlet and outlet pipes are somewhat close together (Figure 16-2). In one operation, it was desired to scale up pilot plant results to a much larger system. It was realized that some short-circuiting occurred, so the tanks were modeled as perfectly mixed CSTRs with a bypass stream. In addition to short-circuiting, stagnant regions (dead zones) are often encountered. In these regions, there is little or no exchange of material with the well-mixed regions and, consequently, virtually no reaction occurs there. Experiments were carried out to determine the amount of the material effectively bypassed and the volume of the dead zone.

[†] See the AIChE webinar "Dealing with Difficult People": *www.aiche.org/academy/ webinars/dealing-difficult-people.*

We want to find
ways of determining
the dead zone
volume and the
fraction of the
volumetric flow
rate bypassing the
system.

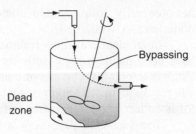

Figure 16-2   CSTR.

A simple modification of an ideal reactor successfully modeled the essential physical characteristics of the system and the equations were readily solvable.

The three concepts
• RTD
• Mixing
• Model

Three concepts were used to describe nonideal reactors in these examples: *the distribution of residence times in the system (RTD), the quality of mixing,* and *the model used to describe the system.* All three of these concepts are considered when describing deviations from the mixing patterns assumed in ideal reactors. The three concepts can be regarded as characteristics of the mixing in nonideal reactors.

One way to order our thinking on nonideal reactors is to consider modeling the flow patterns in our reactors as either ideal CSTRs or PFRs as a *first* approximation. In real reactors, however, nonideal flow patterns exist, resulting in ineffective contacting and lower conversions than in the case of ideal reactors. We must have a method of accounting for this nonideality, and to achieve this goal we use the next-higher level of approximation, which involves the use of *macromixing* information (RTD) (Sections 16.1 to 16.4). The next level uses microscale (*micromixing*) information (Chapter 17) to make predictions about the conversion in nonideal reactors. After completing the first four sections, 16.1 through 16.4, the reader can proceed directly to Chapter 17 to learn how to calculate the conversion and product distributions exiting real reactors. Section 16.5 closes the chapter by discussing how to use the RTD to diagnose and troubleshoot reactors. Here, we focus on two common problems: reactors with bypassing and dead volumes. Once the dead volumes, V_D, and bypassing volumetric flow rates, v_b, are determined, the strategies in Chapter 18 to model the real reactor with ideal reactors can be used to predict conversion.

Chance Card:
Do not pass go,
*proceed directly to
Chapter 17.*

16.1.1   Residence Time Distribution (RTD) Function

The idea of using the distribution of residence times in the analysis of chemical reactor performance was apparently first proposed in a pioneering paper by MacMullin and Weber.[1] However, the concept did not appear to be used extensively until the early 1950s, when Prof. P. V. Danckwerts gave organizational structure to the subject of RTD by defining most of the distributions of interest.[2] The ever-increasing amount of literature on this topic since then

There are a number of mixing tutorials on the AIChE Webinar website and as an AIChE student member you have free access to all these webinars. See *www.aiche.org/search/site/webinar.*

[1] R. B. MacMullin and M. Weber, Jr., *Trans. Am. Inst. Chem. Eng.*, 31, 409 (1935).

[2] P. V. Danckwerts, *Chem. Eng. Sci.*, 2, 1 (1953).

has generally followed the nomenclature of Danckwerts, and this will be done here as well.

In an ideal plug-flow reactor, all the atoms of material leaving the reactor have been inside it for exactly the same amount of time. Similarly, in an ideal batch reactor, all the atoms of materials within the reactor have been inside the BR for an identical length of time. The time the atoms have spent in the reactor is called the *residence time* of the atoms in the reactor.

Residence time

The idealized plug-flow and batch reactors are the only two types of reactors in which all the atoms in the reactors have exactly the same residence time. In all other reactor types, the various atoms in the feed spend different times inside the reactor; that is, there is a distribution of residence times of the material within the reactor. For example, consider the CSTR; the feed introduced into a CSTR at any given time becomes completely mixed with the material already in the reactor. In other words, some of the atoms entering the CSTR leave it almost immediately because material is being continuously withdrawn from the reactor; other atoms remain in the reactor almost forever because all the material recirculates within the reactor and is virtually never removed from the reactor at one time. Many of the atoms, of course, leave the reactor after spending a period of time somewhere in the vicinity of the mean residence time. In any reactor, the distribution of residence times can significantly affect its performance in terms of conversion and product distribution.

The "RTD": Some molecules leave quickly, others overstay their welcome.

The *residence time distribution* (RTD) of a reactor is a characteristic of the mixing that occurs in the chemical reactor. There is no axial mixing in a plug-flow reactor, and this omission is reflected in the RTD. The CSTR is thoroughly mixed and possesses a far different kind of RTD than the plug-flow reactor. As will be illustrated later (cf. Example 16-3), not all RTDs are unique to a particular reactor type; markedly different reactors and reactor sequencing can display identical RTDs. Nevertheless, the RTD exhibited by a given reactor yields distinctive clues to the type of mixing occurring within it and is one of the most informative characterizations of the reactor.

We will use the RTD to characterize nonideal reactors.

16.2 Measurement of the RTD

The RTD is determined experimentally by injecting an inert chemical, molecule, or atom, called a *tracer*, into the reactor at some time $t = 0$ and then measuring the tracer concentration, C, in the effluent stream as a function of time. In addition to being a nonreactive species that is easily detectable, the tracer should have physical properties similar to those of the reacting mixture and be completely soluble in the mixture. It also should not adsorb on the walls or other surfaces in the reactor. The latter requirements are needed to insure that the tracer's behavior will reliably reflect that of the material flowing through the reactor. Colored and radioactive materials along with inert gases are the most common types of tracers. The two most used methods of injection are *pulse input* and *step input*.

Use of tracers to determine the RTD

16.2.1 Pulse Input Experiment

In a pulse input, an amount of tracer N_0 is suddenly injected in one shot into the feed stream entering the reactor in as short a time as is humanly possible. The outlet concentration is then measured as a function of time. Typical concentration–time curves at the inlet and outlet of an arbitrary reactor are shown in Figure 16-4 on page 772. The effluent of the tracer concentration versus time curve is referred to as the C-curve in RTD analysis. We shall first analyze the injection of a tracer pulse for a single-input and single-output system in which *only flow* (i.e., no dispersion) carries the tracer material across system boundaries. Here, we choose an increment of time Δt sufficiently small that the concentration of tracer, $C(t)$, exiting between time t and time $(t + \Delta t)$ is essentially the same. The amount of tracer material, ΔN, leaving the reactor between time t and $t + \Delta t$ is then

$$\Delta N = C(t)v\,\Delta t \qquad (16\text{-}1)$$

where v is the effluent volumetric flow rate. In other words, ΔN is the amount of material exiting the reactor that has spent an amount of time between t and $t + \Delta t$ in the reactor. If we now divide by the total amount of material that was injected into the reactor, N_0, we obtain

$$\frac{\Delta N}{N_0} = \frac{vC(t)}{N_0}\,\Delta t \qquad (16\text{-}2)$$

which represents *the fraction of material that has a residence time in the reactor between time t and t + Δt.*

For pulse injection we define

$$E(t) = \frac{vC(t)}{N_0} \qquad (16\text{-}3)$$

so that

$$\frac{\Delta N}{N_0} = E(t)\,\Delta t \qquad (16\text{-}4)$$

Interpretation of
$E(t)\,dt$

The quantity $E(t)$ is called the **residence time distribution function**. It is the function that describes in a quantitative manner how much time different fluid elements have spent in the reactor. The quantity **$E(t)dt$** is the fraction of fluid exiting the reactor that has spent between time t and $t + dt$ inside the reactor.

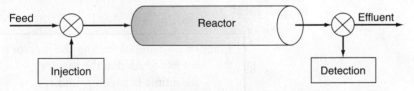

Figure 16-3   Experimental set up to determine E(t).

Figure 16-4 shows schematics of the inlet and outlet concentrations for both a pulse input and step input for the experimental set up in Figure 16-3.

If N_0 is not known directly, it can be obtained from the outlet concentration measurements by summing up all the amounts of materials, ΔN, between time equal to zero and infinity. Writing Equation (16-1) in differential form yields

The C-curve

$$dN = vC(t)\,dt \qquad (16\text{-}5)$$

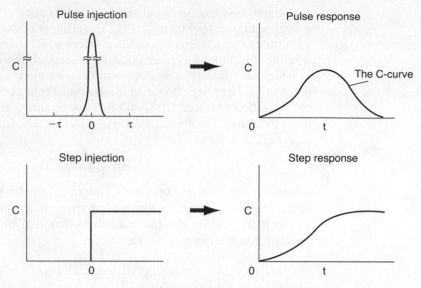

Figure 16-4 RTD measurements.

and then integrating, we obtain

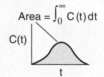

$$N_0 = \int_0^\infty v C(t)\, dt \qquad (16\text{-}6)$$

The volumetric flow rate v is usually constant, so we can define $E(t)$ as

We find the RTD
function, *E(t)*, from
the tracer
concentration *C(t)*

$$\boxed{E(t) = \frac{C(t)}{\displaystyle\int_0^\infty C(t)\, dt}} \qquad (16\text{-}7)$$

The E-*curve is just the* C-*curve divided by the area under the* C-*curve.*

An alternative way of interpreting the residence time function is in its integral form:

$$\boxed{\begin{bmatrix} \text{Fraction of material leaving the reactor} \\ \text{that has resided in the reactor} \\ \text{for a time between } t_1 \text{ and } t_2 \end{bmatrix} = \int_{t_1}^{t_2} E(t)\, dt}$$

We know that the fraction of all the material that has resided for a time t in the reactor between $t = 0$ and $t = \infty$ is 1; therefore

Eventually all
guests must leave

$$\boxed{\int_0^\infty E(t)\, dt = 1} \qquad (16\text{-}8)$$

The following example will show how we can calculate and interpret $E(t)$ from the effluent concentrations from the response to a pulse tracer input to a real (i.e., nonideal) reactor.

Example 16–1   Constructing the C(t) and E(t) Curves

A sample of the tracer hytane at 320 K was injected as a pulse into a reactor, and the effluent concentration was measured as a function of time, resulting in the data shown in Table E16-1.1.

TABLE E16-1.1   TRACER DATA

t (min)	0	0.5	1	2	3	4	5	6	7	8	9	10	12	14
C (g/m³)	0	0.6	1.4	5	8	10	8	6	4	3	2.2	1.5	0.6	0

Pulse input

The measurements represent the exact concentrations at the times listed and not average values between the various sampling tests.
(a) Construct a figure showing the tracer concentration $C(t)$ as a function of time.
(b) Construct a figure showing $E(t)$ as a function of time.

Solution

(a) By plotting C as a function of time, using the data in Table E16-1.1, the curve shown in Figure E16-1.1 is obtained.

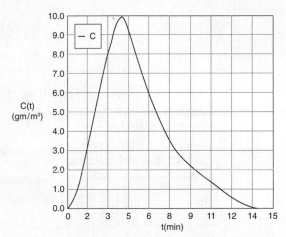

Figure E16-1.1   The C-curve.

To convert the $C(t)$ curve in Figure E16-1.1 to an $E(t)$ curve we use the area under the $C(t)$ curve. There are three ways we can determine the area using this data.
(1) Brute force: calculate the area by measuring the area of the squares and partial squares under the curve, and then summing them up.
(2) Use the integration formulas given in Appendix A.
(3) Fit the data to one or more polynomials using Polymath or some other software program. We will choose Polymath to fit the data. *Note:* A step-by-step tutorial to fit the data points using Polymath is given on the CRE Web site (*www.umich.edu/~elements/5e/index.html*) LEP 16-1. We will use two polynomials to fit the C-curve, one for the ascending portion, $C_1(t)$, and one for the descending portion, $C_2(t)$, both of which meet at t_1.

   Using the Polymath polynomial fitting routine (see tutorial), the data in Table E16-1.1 yields the following two polynomials to

For $t \leq 4$ min then $C_1(t) = 0.0039 + 0.274\, t + 1.57\, t^2 - 0.255\, t^3$ (E16-1.1)

For $t \geq 4$ min then $C_2(t) = -33.4 + 37.2\, t - 11.6\, t^2 + 1.7\, t^3 - 0.13\, t^4 + 0.005\, t^5$

$$-7.7 \times 10^{-5} t^6$$ (E16-1.2)

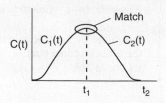

We then use an *if statement* in our fitted curve.

If $(t \leq 4 \text{ and } t > 0)$ then C_1 else if $(t > 4 \text{ and } t < = 14)$ then C_2 else 0

To find the area under the curve, A, we use the ODE solver.
Let A represent the area the curve, then

$$\frac{dA}{dt} = C(t) \tag{E16-1.3}$$

$$A = \int_0^{14} C(t) \, dt \tag{E16-1.4}$$

(b) Construct $E(t)$.

$$E(t) = \frac{C(t)}{\displaystyle\int_0^\infty C(t) \, dt} = \frac{C(t)}{A}$$

The Polymath program and results are shown below where we see $A = 51$.

POLYMATH Report
Ordinary Differential Equations

Calculated values of DEQ variables

	Variable	Initial value	Final value
1	Area	0	51.06334
2	C	0.0038746	0.0148043
3	C1	0.0038746	-387.266
4	C2	-33.43818	0.0148043
5	t	0	14.

Differential equations
*1 d(Area)/d(t) = C

Explicit equations
1 C2 = -33.43818 + 37.18972*t - 11.58838*t^2 + 1.695303*t^3 - 0.1298667*t^4 + 0.005028*t^5 - 7.743*10^-5*t^6
2 C1 = 0.0038746 + 0.2739782*t + 1.574621*t^2 - 0.2550041*t^3
3 C = If(t<=4 and t>=0) then C1 else if(t>4 and t<=14) then C2 else 0

Now that we have the area, A (i.e., 51 g•min/m³), under the C-curve, we can construct the $E(t)$ curves. We now calculate $E(t)$ by dividing each point on the $C(t)$ curve by 51.0 g•min/m³

$$E(t) = \frac{C(t)}{\displaystyle\int_0^\infty C(t) \, dt} = \frac{C(t)}{51 \, \text{g} \cdot \text{min/m}^3} \tag{E16-1.5}$$

with the following results:

TABLE E16-1.2 $C(t)$ AND $E(t)$

t (min)	0	1	2	3	4	5	6	7	8	9	10	12	14
$C(t)$ (g/m³)	0	1.4	5	8	10	8	6	4	3	2.2	1.5	0.6	0
$E(t)$ (min⁻¹)	0	0.02	0.1	0.16	0.2	0.16	0.12	0.08	0.06	0.044	0.03	0.012	0

Using Table E16-1.2 we can construct $E(t)$ as shown in Figure E16-1.2

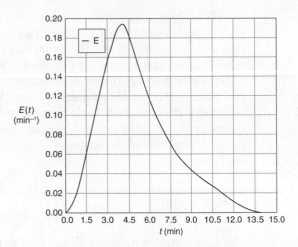

Figure E16-1.2 $E(t)$-Curve

Analysis: In this example we fit the effluent concentration data $C(t)$ from an inert tracer pulse input to two polynomials and then used an *If statement* to model the complete curve. We then used the Polymath ODE solver to get the area under the curve that we then used to divide the $C(t)$ curve in order to obtain the $E(t)$. Once we have the $E(t)$ curve, we ask and easily answer such questions as "what fraction of the modules spend between 2 and 4 minutes in the reactor" or "what is the mean residence time t_m?" We will address these questions in the following sections where we discuss characteristics of the residence time distribution (RTD).

The principal difficulties with the pulse technique lie in the problems connected with obtaining a reasonable pulse at a reactor's entrance. The injection must take place over a period that is very short compared with residence times in various segments of the reactor or reactor system, and there must be a negligible amount of dispersion between the point of injection and the entrance to the reactor system. If these conditions can be fulfilled, this technique represents a simple and direct way of obtaining the RTD.

Drawbacks to the pulse injection to obtain the RTD

There could be problems in fitting $E(t)$ to a polynomial if the effluent concentration–time curve were to have a long tail because the analysis can be subject to large inaccuracies. This problem principally affects the denominator of the right-hand side of Equation (16-7), i.e., the integration of the $C(t)$ curve. It is desirable to extrapolate the tail and analytically continue the calculation. The tail of the curve may sometimes be approximated as an exponential decay. The inaccuracies introduced by this assumption are very likely to be much less than those resulting from either truncation or numerical imprecision in this region. Methods of fitting the tail are described in the *Professional Reference Shelf R16.1*.

Reference Shelf

16.2.2 Step Tracer Experiment

Now that we have an understanding of the meaning of the RTD curve from a pulse input, we will formulate a relationship between a step tracer injection and the corresponding concentration in the effluent.

The inlet concentration most often takes the form of either a perfect *pulse input* (Dirac delta function), *imperfect pulse injection* (see Figure 16-4), or a *step input*. Just as the RTD function $E(t)$ can be determined directly from a pulse input, the cumulative distribution $F(t)$ can be determined directly from a step input. *The cumulative distribution gives the fraction of material $F(t)$ that has been in the reactor at time* t *or less.* We will now analyze a *step input* in the tracer concentration for a system with a constant volumetric flow rate. Consider a constant rate of tracer addition to a feed that is initiated at time $t = 0$. Before this time, no tracer was added to the feed. Stated symbolically, we have

$$C_{out}(t) = \begin{cases} 0 & t < 0 \\ C_0, \text{ constant} & t \geq 0 \end{cases} \tag{16-9}$$

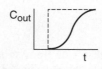

The concentration of tracer in the feed to the reactor is kept at this level until the concentration in the effluent is indistinguishable from that in the feed; the test may then be discontinued. A typical outlet concentration curve for this type of input is shown in Figure 16-4.

Because the inlet concentration is a constant with time, C_0, we can take it outside the integral sign; that is,

$$C_{out}(t) = C_0 \int_0^t E(t') \, dt'$$

Dividing by C_0 yields

$$\left[\frac{C_{out}(t)}{C_0} \right]_{step} = \int_0^t E(t') \, dt' = F(t)$$

$$F(t) = \left[\frac{C_{out}(t)}{C_0} \right]_{step} \tag{16-10}$$

We differentiate this expression to obtain the RTD function $E(t)$:

$$E(t) = \frac{dF}{dt} = \frac{d}{dt}\left[\frac{C_{out}(t)}{C_0} \right]_{step} \tag{16-11}$$

The positive step is usually easier to carry out experimentally than the pulse test, and it has the additional advantage that the total amount of tracer in the feed over the period of the test does not have to be known as it does in the pulse test. One possible drawback in this technique is that it is sometimes difficult to maintain a constant tracer concentration in the feed. Obtaining the RTD from this test also involves differentiation of the data and presents an additional and probably more serious drawback to the technique, because differentiation of data can, on occasion, lead to large errors. A third problem lies with the large amount of tracer required for this test. If the tracer is very expensive, a pulse test is almost always used to minimize the cost.

Advantages and drawbacks to the step injection

Other tracer techniques exist, such as negative step (i.e., elution), frequency-response methods, and methods that use inputs other than steps or pulses. These methods are usually much more difficult to carry out than the ones presented and are not encountered as often. For this reason, they will not be treated here, and the literature should be consulted for their virtues, defects, and the details of implementing them and analyzing the results. A good source for this information is Wen and Fan.[3]

16.3 Characteristics of the RTD

From $E(t)$ we can learn how long different molecules have been in the reactor.

Sometimes $E(t)$ is called the *exit-age distribution function*. If we regard the "age" of an atom as the time it has resided in the reaction environment, then $E(t)$ concerns the age distribution of the effluent stream. It is the most used of the distribution functions connected with reactor analysis because it characterizes the lengths of time various atoms spend at reaction conditions.

16.3.1 Integral Relationships

The fraction of the exit stream that has resided in the reactor for a period of time shorter than a given value t is equal to the sum over all times less than t of $E(t)\,\Delta t$, or expressed continuously, by integrating $E(t)$ between time $t = 0$ and time, t.

The cumulative RTD function $F(t)$

$$\int_0^t E(t)\,dt = F(t) = \begin{bmatrix} \text{Fraction of effluent} \\ \text{that has been in reactor} \\ \text{for less than time } t \end{bmatrix} \quad (16\text{-}12)$$

Analogously, we have, by integrating between time t and time $t \Rightarrow \infty$

$$\int_t^\infty E(t)\,dt = 1 - F(t) = \begin{bmatrix} \text{Fraction of effluent} \\ \text{that has been in reactor} \\ \text{for longer than time } t \end{bmatrix} \quad (16\text{-}13)$$

Because t appears in the integration limits of these two expressions, Equations (16-12) and (16-13) are both functions of time. Danckwerts defined Equation (16-12) as a *cumulative distribution function and called it $F(t)$.*[4] We can calculate $F(t)$ at various times t from the area under the curve of a plot of $E(t)$ versus t, i.e., the E-curve. A typical shape of the $F(t)$ curve is shown in Figure 16-5. One notes from this curve that 80% (i.e., $F(t) = 0.8$) of the molecules spend 8 minutes or less in the reactor, and 20% of the molecules $[1 - F(t)]$ spend longer than 8 minutes in the reactor.

The F-curve is another function that has been defined as the normalized response to a particular input. Alternatively, Equation (16-12) has been used as a definition of $F(t)$, and it has been stated that as a result it can be obtained as the response to a positive step tracer test. Sometimes the F-curve is used in the same manner as the RTD in the modeling of chemical reactors. An excellent industrial example is the study of Wolf and White, who investigated the behavior of screw extruders in polymerization processes.[5]

---

[3] C. Y. Wen and L. T. Fan, *Models for Flow Systems and Chemical Reactors* (New York: Marcel Dekker, 1975).

[4] P. V. Danckwerts, *Chem. Eng. Sci.*, 2, 1 (1953).

[5] D. Wolf and D. H. White, *AIChE J.*, 22, 122 (1976).

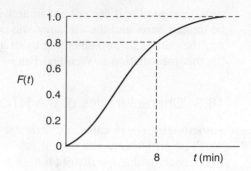

Figure 16-5 Cumulative distribution curve, $F(t)$.

The *F*-curve

16.3.2 Mean Residence Time

$\tau = t_m$

In previous chapters treating ideal reactors, a parameter frequently used was the space time or average residence time, τ, which was defined as being equal to (V/v). It will be shown that, in the absence of dispersion, and for constant volumetric flow ($v = v_0$) no matter what RTD exists for a particular reactor, ideal or nonideal, this nominal space time, τ, is equal to the *mean residence time*, t_m.

As is the case with other variables described by distribution functions, the mean value of the variable is equal to the first moment of the RTD function, $E(t)$. Thus the mean residence time is

The first moment gives the average time the effluent molecules spent in the reactor.

$$t_m = \frac{\int_0^\infty tE(t)\,dt}{\int_0^\infty E(t)\,dt} = \int_0^\infty tE(t)\,dt \qquad (16\text{-}14)$$

We now wish to show how we can determine the total reactor volume using the cumulative distribution function.

In the *Extended Material for Chapter 16 on the Web*, a proof is given that for constant volumetric flow rate, the mean residence time is equal to the space time, i.e.,

$$t_m = \tau \qquad (16\text{-}15)$$

This result is true *only* for a *closed system* (i.e., no dispersion across boundaries; see Chapter 18). The exact reactor volume is determined from the equation

$$V = v t_m \qquad (16\text{-}16)$$

16.3.3 Other Moments of the RTD

It is very common to compare RTDs by using their moments instead of trying to compare their entire distributions (e.g., Wen and Fan).[6] For this purpose, three moments are normally used. The first is the mean *residence time*, t_m. The

[6] C. Y. Wen and L. T. Fan, *Models for Flow Systems and Chemical Reactors* (New York: Decker, 1975), Chap. 11.

second moment commonly used is taken about the mean and is called the *variance*, σ^2, or square of the standard deviation. It is defined by

The second moment about the mean is the variance.

$$\sigma^2 = \int_0^\infty (t - t_m)^2 E(t)\, dt \qquad (16\text{-}17)$$

The magnitude of this moment is an indication of the "spread" of the distribution; the greater the value of this moment is, the greater a distribution's spread will be.

The third moment is also taken about the mean and is related to the *skewness*, s^3, The skewness is defined by

The two parameters most commonly used to characterize the RTD are τ and σ^2

$$s^3 = \frac{1}{\sigma^{3/2}} \int_0^\infty (t - t_m)^3 E(t)\, dt \qquad (16\text{-}18)$$

The magnitude of the third moment measures the extent that a distribution is skewed in one direction or another in reference to the mean.

Rigorously, for a complete description of a distribution, all moments must be determined. Practically, these three are usually sufficient for a reasonable characterization of an RTD.

Example 16–2   Mean Residence Time and Variance Calculations

Using the data given in Table E16-1.2 in Example 16-1
  (a) Construct the $F(t)$ curve.
  (b) Calculate the mean residence time, t_m.
  (c) Calculate the variance about the mean, σ^2.
  (d) Calculate the fraction of fluid that spends between 3 and 6 minutes in the reactor.
  (e) Calculate the fraction of fluid that spends 2 minutes or less in the reactor.
  (f) Calculate the fraction of the material that spends 3 minutes or longer in the reactor.

Solution

(a) To construct the F-curve, we simply integrate the E-curve

$$E(t) = \frac{C(t)}{A} = \frac{C(t)}{51} \qquad (E16\text{-}1.5)$$

using an ODE solver such as Polymath shown in Table E16-2.1

$$\frac{dF}{dt} = E(t) \qquad (16\text{-}11)$$

The Polymath program and results are shown in Table E16-1.2 and Figure E16-2.1(b), respectively.

TABLE E16-1.2 $C(t)$, AND $E(t)$

t (min)	0	1	2	3	4	5	6	7	8	9	10	12	14
$C(t)$ (g/m^3)	0	1.4	5	8	10	8	6	4	3	2.2	1.5	0.6	0
$E(t)$ (min^{-1})	0	0.02	0.1	0.16	0.2	0.16	0.12	0.08	0.06	0.044	0.03	0.012	0

Calculating the
mean residence
time,

$$\tau = t_m = \int_0^\infty tE(t)\, dt$$

(b) We also show in Table E16-2.1 the Polymath program to calculate the mean residence time, t_m. By differentiating Equation (16-14), we can easily use Polymath to find t_m, i.e.,

$$\frac{dt_m}{dt} = tE(t) \tag{E16-2.1}$$

with $t = 0$ then $E = 0$ and $t = 14$ then $E = 0$. Equation (E16-2.1) and the calculated result is also shown in Table E16-2.1 where we find

$$t_m = 5.1 \text{ minutes}$$

TABLE E16-2.1   POLYMATH PROGRAM AND RESULTS TO CONSTRUCT THE E- AND F- CURVES

POLYMATH Report
Ordinary Differential Equations

Calculated values of DEQ variables

	Variable	Initial value	Final value
1	Area	51.	51.
2	C	0.0038746	0.0148043
3	C1	0.0038746	-387.266
4	C2	-33.43818	0.0148043
5	E	7.597E-05	0.0002903
6	F	0	1.00125
7	t	0	14.
8	tm	0	5.107247

Differential equations
1 d(tm)/d(t) = t*E
2 d(F)/d(t) = E

Explicit equations
1 C1 = 0.0038746 + 0.2739782*t + 1.574621*t^2 - 0.2550041*t^3
2 Area = 51
3 C2 = -33.43818 + 37.18972*t - 11.58838*t^2 + 1.695303*t^3 -
     0.1298667*t^4 + 0.005028*t^5 - 7.743*10^-5*t^6
4 C = If(t<=4 and t>=0) then C1 else if(t>4 and t<=14) then C2 else 0
5 E = C/Area

Using the Polymath plotting routines, we can construct Figures E16-2.1 (a) and (b) after executing the program shown in the Polymath Table E16-2.1.

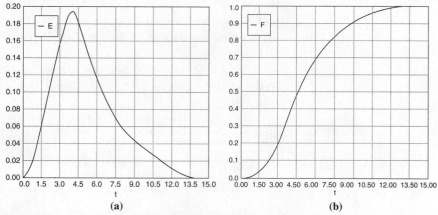

Figure E16-2.1   (a) E-Curve; (b) F-Curve.

(c) Now that we have found the mean residence time t_m we can calculate the variance σ^2.

Calculating the
variance

$$\sigma^2 = \int_0^\infty (t - t_m)^2 E(t)\, dt \tag{E16-2.2}$$

We now differentiate Equation (E16-2.2) with respect to t

$$\frac{d\sigma^2}{dt} = (t - t_m)^2 E(t) \tag{E16-2.3}$$

and then use Polymath to integrate between $t = 0$ and $t = 14$, which is the last point on the E-curve.

TABLE E16-2.2   POLYMATH PROGRAM AND RESULTS TO CALCULATE THE
MEAN RESIDENCE TIME, t_m, AND THE VARIANCE σ^2

POLYMATH Report
Ordinary Differential Equations

Calculated values of DEQ variables

	Variable	Initial value	Final value
1	Area	51.	51.
2	C	0.0038746	0.0148043
3	C1	0.0038746	-387.266
4	C2	-33.43818	0.0148043
5	E	7.597E-05	0.0002903
6	Sigma2	0	6.212473
7	t	0	14.
8	tmf	5.1	5.1

Differential equations
1 d(Sigma2)/d(t) = (t-tmf)^2 * E

Explicit equations
1 C1 = 0.0038746 + 0.2739782*t + 1.574621*t^2 - 0.2550041*t^3
2 Area = 51
3 C2 = -33.43818 + 37.18972*t - 11.58838*t^2 + 1.695303*t^3 -
   0.1298667*t^4 + 0.005028*t^5 - 7.743*10^-5*t^6
4 C = If(t<=4 and t>=0) then C1 else if(t>4 and t<=14) then C2 else 0
5 E = C/Area
6 tmf = 5.1

The results of this integration are shown in Table E16-2.1 where we find $\sigma^2 = 6.2$ minutes, so $\sigma = 2.49$ minutes.

(d) To find the fraction of fluid that spends between 3 and 6 minutes, we simply integrate the *E*-curve between 3 and 6

$$F_{3-6} = \int_3^6 E(t)\ dt$$

The Polymath program is shown in Table E16-2.3 along with the output.

TABLE E16-2.3   POLYMATH PROGRAM TO FIND THE FRACTION OF FLUID
THAT SPENDS BETWEEN 3 AND 6 MINUTES IN THE REACTOR

POLYMATH Report
Ordinary Differential Equations

Calculated values of DEQ variables

	Variable	Initial value	Final value
1	C	8.112288	5.881819
2	C1	8.112288	3.253214
3	C2	10.2549	5.881819
4	E	0.1590645	0.1153298
5	F	0	0.4952889
6	t	3.	6.

Differential equations
1 d(F)/d(t) = E

Explicit equations
1 C1 = 0.0038746 + 0.2739782*t + 1.574621*t^2 - 0.2550041*t^3
2 C2 = -33.43818 + 37.18972*t - 11.58838*t^2 + 1.695303*t^3 -
   0.1298667*t^4 + 0.005028*t^5 - 7.743*10^-5*t^6
3 C = If(t<=4 and t>=0) then C1 else if(t>4 and t<=14) then C2 else 0
4 E = C/51

We see that approximately 50% (i.e., 49.53%) of the material spends between 3 and 6 minutes in the reactor.

We can visualize this fraction with the use of plot of $E(t)$ versus (t) as shown in Figure E16-2.2. The shaded area in Figure E16-2.2 represents the fraction of material leaving the reactor that has resided in the reactor between 3 and 6 min. Evaluating this area, we find that 50% of the material leaving the reactor spends between 3 and 6 min in the reactor.

(e) We shall next consider the fraction of material that has been in the reactor for a time t or less; that is, the fraction that has spent between 0 and t minutes in the reactor, $F(t)$. This fraction is just the shaded area under the curve up to $t = t$ minutes. This area is shown in Figure E16-2.3 for $t = 3$ min. Calculating the area under the curve, we see that approximately 20% of the material has spent *3 min or less* in the reactor.

The *E* curve

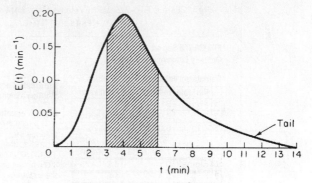

Figure E16-2.2   Fraction of material that spends between 3 and 6 minutes in the reactor.

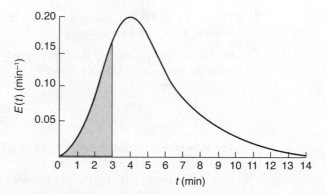

Figure E16-2.3   Fraction of material that spends 3 minutes or less in the reactor.

(f) The fraction of fluid that spends a time t or greater in the reactor is

$$\left[\begin{array}{c}\text{Greater}\\\text{than time } t\end{array}\right] = 1 - F(t) = 1 - 0.2 = 0.8$$

therefore 80% of the fluid spends a time t or greater in the reactor.

The square of the standard deviation is $\sigma^2 = 6.19$ min^2, so $\sigma = 2.49$ min.

___*Analysis:*___ In this example we calculated two important properties of the RTD, the mean time molecules spend in the reactors, t_m, and the variance about this mean, σ^2. We will calculate these properties from the RTD of other nonideal reactors and then show in Chapter 18 how to use them to formulate models of real reactors using combinations of ideal reactors. We will use these models along with reaction-rate data to predict the conversion in the nonideal reactor we obtained from the reactor storage shed.

16.3.4 Normalized RTD Function, $E(\Theta)$

Frequently, a normalized RTD is used instead of the function $E(t)$. If the parameter Θ is defined as

$$\boxed{\Theta \equiv \frac{t}{\tau}}$$

(16-19)

Why we use a normalized RTD

The quantity Θ represents the number of reactor volumes of fluid, based on entrance conditions, that have flowed through the reactor in time t. The dimensionless RTD function, $E(\Theta)$ is then defined as

$$E(\Theta) \equiv \tau E(t) \qquad (16\text{-}20)$$

and plotted as a function of Θ, as shown in the margin.

$E(t)$ for a CSTR

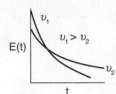

The purpose of creating this normalized distribution function is that the flow performance inside reactors of different sizes can be compared directly. For example, if the normalized function $E(\Theta)$ is used, *all* perfectly mixed CSTRs have numerically the same RTD. If the simple function $E(t)$ is used, numerical values of $E(t)$ can differ substantially for CSTRs different volumes, V, and entering volumetric flow rates, v_0. As will be shown later in Section 16.4.2, $E(t)$ for a perfectly mixed CSTR

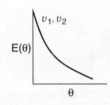

$$\boxed{E(t) = \frac{1}{\tau}\, e^{-t/\tau}} \qquad (16\text{-}21)$$

and therefore

$$\boxed{E(\Theta) = \tau E(t) = e^{-\Theta}} \qquad (16\text{-}22)$$

From these equations it can be seen that the value of $E(t)$ at identical times can be quite different for two different volumetric flow rates, say v_1 and v_2. But for the same value of Θ, the value of $E(\Theta)$ is the same irrespective of the size or volumetric flow rate of a perfectly mixed CSTR.

It is a relatively easy exercise to show that

$$\int_0^\infty E(\Theta)\, d\Theta = 1 \qquad (16\text{-}23)$$

and is recommended as a 93-s divertissement. (Jofostan University chemical engineers claim they can do it in 87 s.)

16.3.5 Internal-Age Distribution, $I(\alpha)$

Tombstone jail
How long have you been here? $I(\alpha)\Delta\alpha$
When do you expect to get out?

Although this section is not a prerequisite to the remaining sections, the internal-age distribution is introduced here because of its close analogy to the external-age distribution. We shall let α represent the age of a molecule inside the reactor. The internal-age distribution function $I(\alpha)$ is a function such that $I(\alpha)\Delta\alpha$ is the fraction of material now *inside the reactor* that has been inside the reactor for a period of time between α and $(\alpha + \Delta\alpha)$. It may be contrasted with $E(\alpha)\Delta\alpha$, which is used to represent the material *leaving the reactor* that has spent a time between α and $(\alpha + \Delta\alpha)$ in the reaction zone; $I(\alpha)$ characterizes the time the material has been (and still is) in the reactor at a particular time. The function $E(\alpha)$ is viewed outside the reactor and $I(\alpha)$ is viewed inside the reactor. In unsteady-state problems, it can be important to know what the particular state of a reaction mixture is, and $I(\alpha)$ supplies this information. For example, in a catalytic reaction using a catalyst whose activity decays with time, the internal-age distribution of the catalyst in the reactor $I(\alpha)$ is of importance and can be of use in modeling the reactor.

The internal-age distribution is discussed further on the *Professional Reference Shelf (R16.2)* where the following relationships between the cumulative internal-age distribution $I(\alpha)$ and the cumulative external-age distribution $F(\alpha)$

$$I(\alpha) = (1 - F(\alpha))/\tau \qquad (16\text{-}24)$$

and between $E(t)$ and $I(t)$

$$E(\alpha) = -\frac{d}{d\alpha}[\tau I(\alpha)] \qquad (16\text{-}25)$$

are derived. For a CSTR, it is shown that the internal-age distribution function is

$$I(\alpha) = -\frac{1}{\tau}e^{-\alpha/\tau} \qquad (16\text{-}26)$$

16.4 RTD in Ideal Reactors

16.4.1 RTDs in Batch and Plug-Flow Reactors

The RTDs in plug-flow reactors and ideal batch reactors are the simplest to consider. All the atoms leaving such reactors have spent precisely the same amount of time within the reactors. The distribution function in such a case is a spike of infinite height and zero width, whose area is equal to 1; the spike occurs at $t = V/v = \tau$, or $\Theta = 1$, as shown in Figure 16-6.

The $E(t)$ function is shown in Figure 16-6(a), and $F(t)$ is shown in Figure 16-6(b).

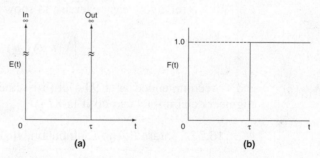

Figure 16-6 Ideal plug-flow response to a pulse tracer input.

Mathematically, this spike is represented by the Dirac delta function:

E(t) for a plug-flow reactor

$$E(t) = \delta(t - \tau) \qquad (16\text{-}27)$$

The Dirac delta function has the following properties:

Properties of the Dirac delta function

$$\delta(x) = \begin{cases} 0 & \text{when } x \neq 0 \\ \infty & \text{when } x = 0 \end{cases} \qquad (16\text{-}28)$$

$$\int_{-\infty}^{\infty} \delta(x)\, dx = 1 \qquad (16\text{-}29)$$

$$\int_{-\infty}^{\infty} g(x)\delta(x - \tau)\, dx = g(\tau) \qquad (16\text{-}30)$$

To calculate τ the mean residence time, we set g(x) = t

$$t_m = \int_0^\infty tE(t)\,dt = \int_0^\infty t\,\delta(t-\tau)\,dt = \tau \qquad (16\text{-}31)$$

But we already knew this result, as did all chemical reaction engineering students at the university in Riça, Jofostan. To calculate the variance, we set $g(t) = (t-\tau)^2$, and the variance, σ^2, is

$$\sigma^2 = \int_0^\infty (t-\tau)^2\,\delta(t-\tau)\,dt = 0$$

All material spends exactly a time τ in the reactor, so there is no variance $[\sigma^2 = 0]$!

The cumulative distribution function $F(t)$ is

$$F(t) = \int_0^t E(t)dt = \int_0^t \delta(t-\tau)dt$$

16.4.2 Single-CSTR RTD

In an ideal CSTR the concentration of any substance in the effluent stream is identical to the concentration throughout the reactor. Consequently, it is possible to obtain the RTD from conceptual considerations in a fairly straightforward manner. A material balance on an inert tracer that has been injected as a pulse at time $t = 0$ into a CSTR yields for $t > 0$

From a tracer balance we can determine $E(t)$.

$$\text{In} - \quad \text{Out} \quad = \text{Accumulation}$$

$$\overbrace{0}^{} - \overbrace{vC}^{} = \overbrace{V\frac{dC}{dt}}^{} \qquad (16\text{-}33)$$

Because the reactor is perfectly mixed, C in this equation is the concentration of the tracer both in the effluent and within the reactor. Separating the variables and integrating with $C = C_0$ at $t = 0$ yields

$$C(t) = C_0 e^{-t/\tau} \qquad (16\text{-}34)$$

The C-curve can be plotted from Equation (16-34), which is the concentration of tracer in the effluent at any time t.

To find $E(t)$ for an ideal CSTR, we first recall Equation (16-7) and then substitute for $C(t)$ using Equation (16-34). That is

$$E(t) = \frac{C(t)}{\displaystyle\int_0^\infty C(t)\,dt} = \frac{C_0 e^{-t/\tau}}{\displaystyle\int_0^\infty C_0 e^{-t/\tau}\,dt} = \frac{e^{-t/\tau}}{\tau} \qquad (16\text{-}35)$$

Evaluating the integral in the denominator completes the derivation of the RTD for an ideal CSTR and one notes they are the same as previously given by Equations (16-21) and (16-22)

$E(t)$ and $E(\Theta)$ for a CSTR

$$E(t) = \frac{e^{-t/\tau}}{\tau} \qquad (16\text{-}21)$$

$$E(\Theta) = e^{-\Theta} \qquad (16\text{-}22)$$

the cumulative distribution is

$$F(t) = \int_0^t E(t)dt = \int_0^t \frac{e^{-t/\tau}}{\tau} = 1 - e^{-t/\tau} \qquad (16\text{-}32)$$

Recall that $\Theta = t/\tau$ and $E(\Theta) = \tau E(t)$.

Response of an
ideal CSTR

$E(\Theta) = e^{-\Theta}$
$F(\Theta) = 1 - e^{-\Theta}$

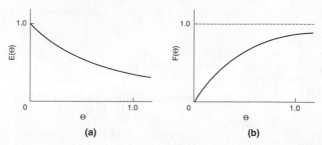

Figure 16-7 $E(\Theta)$ and $F(\Theta)$ for an Ideal CSTR.

The cumulative distribution $F(\Theta)$ is

$$F(\Theta) = \int_0^\Theta E(\Theta)d\Theta = 1 - e^{-\Theta} \qquad (16\text{-}36)$$

The $E(\Theta)$ and $F(\Theta)$ functions for an ideal CSTR are shown in Figure 16-7 (a) and (b), respectively.

Earlier it was shown that for a constant volumetric flow rate, the mean residence time in a reactor is equal to (V/v), or τ. This relationship can be shown in a simpler fashion for the CSTR. Applying the definition of the mean residence time to the RTD for a CSTR, we obtain

$$t_m = \int_0^\infty tE(t)\,dt = \int_0^\infty \frac{t}{\tau}\,e^{-t/\tau}\,dt = \tau \qquad (16\text{-}14)$$

Thus, the nominal holding time (space time) $\tau = (V/v)$ is also the mean residence time that the material spends in the reactor.

The second moment about the mean is the variance and is a measure of the spread of the distribution about the mean. The variance of residence times in a perfectly mixed tank reactor is (let $x = t/\tau$)

For a perfectly
mixed CSTR: $t_m = \tau$
and $\sigma = \tau$.

$$\sigma^2 = \int_0^\infty \frac{(t-\tau)^2}{\tau}\,e^{-t/\tau}\,dt = \tau^2 \int_0^\infty (x-1)^2 e^{-x}\,dx = \tau^2 \qquad (16\text{-}37)$$

Then, $\sigma = \tau$. The standard deviation is the square root of the variance. For a CSTR, the standard deviation of the residence time distribution is as large as the mean itself!!

16.4.3 Laminar-Flow Reactor (LFR)

Before proceeding to show how the RTD can be used to estimate conversion in a reactor, we shall derive $E(t)$ for a laminar-flow reactor. For laminar flow in a tubular (i.e. cylindrical) reactor, the velocity profile is parabolic, with the fluid in the center of the tube spending the shortest time in the reactor. A schematic diagram of the fluid movement after a time t is shown in Figure 16-8. The figure at the left shows how far down the reactor each concentric fluid element has traveled after a time t.

Molecules near the center spend a shorter time in the reactor than those close to the wall.

Figure 16-8   Schematic diagram of fluid elements in a laminar-flow reactor.

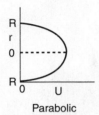

Parabolic Velocity Profile

The velocity profile in a pipe of outer radius R is

$$U(r) = U_{max}\left[1 - \left(\frac{r}{R}\right)^2\right] = 2U_{avg}\left[1 - \left(\frac{r}{R}\right)^2\right] = \frac{2v_0}{\pi R^2}\left[1 - \left(\frac{r}{R}\right)^2\right] \quad (16\text{-}38)$$

where U_{max} is the centerline velocity and U_{avg} is the average velocity through the tube. U_{avg} is just the volumetric flow rate divided by the cross-sectional area.

The time of passage of an element of fluid at a radius r is

$$t(r) = \frac{L}{U(r)} = \frac{\pi R^2 L}{v_0}\frac{1}{2[1 - (r/R)^2]} \quad (16\text{-}39)$$

$$= \frac{\tau}{2[1 - (r/R)^2]} \quad (16\text{-}40)$$

The volumetric flow rate of fluid out of the reactor between r and $(r + dr)$, dv, is

$$dv = U(r)\,2\pi r dr \quad (16\text{-}41)$$

The fraction of total fluid passing out between r and $(r + dr)$ is dv/v_0, i.e.

We are just doing a few manipulations to arrive at $E(t)$ for an LFR

$$\frac{dv}{v_0} = \frac{U(r)2(\pi r dr)}{v_0} \quad (16\text{-}42)$$

The fraction of fluid between r and $(r + dr)$ that has a flow rate between v and $(v + dv)$ and spends a time between t and $(t + dt)$ in the reactor is

$$E(t)dt = \frac{dv}{v_0} \quad (16\text{-}43)$$

We now need to relate the fluid fraction, Equation (16-43), to the fraction of fluid spending between time t and $t + dt$ in the reactor. First we differentiate Equation (16-40)

$$dt = \frac{\tau}{2R^2}\frac{2r\,dr}{[1 - (r/R)^2]^2} = \frac{4}{\tau R^2}\left\{\frac{\tau/2}{[1 - (r/R)^2]}\right\}^2 r\,dr \quad (16\text{-}44)$$

and then use Equation (16-40) to substitute t for the term in brackets to yield

$$dt = \frac{4t^2}{\tau R^2}r\,dr \quad (16\text{-}45)$$

Combining Equations (16-42) and (16-45), and then using Equation (16-40) that relates for $U(r)$ and $t(r)$, we now have the fraction of fluid spending between time t and $t + dt$ in the reactor

$$E(t)dt = \frac{dv}{v_0} = \frac{L}{t}\left(\frac{2\pi r\, dr}{v_0}\right) = \frac{L}{t}\left(\frac{2\pi}{v_0}\right)\frac{\tau R^2}{4t^2}\,dt = \frac{\tau^2}{2t^3}\,dt$$

$$E(t) = \frac{\tau^2}{2t^3} \tag{16-46}$$

The minimum time the fluid may spend in the reactor is

$$t = \frac{L}{U_{max}} = \frac{L}{2U_{avg}}\left(\frac{\pi R^2}{\pi R^2}\right) = \frac{V}{2v_0} = \frac{\tau}{2}$$

Consequently, the complete RTD function for a laminar-flow reactor is

At last! $E(t)$ for a laminar-flow reactor

$$E(t) = \begin{cases} 0 & t < \dfrac{\tau}{2} \\[2ex] \dfrac{\tau^2}{2t^3} & t \geq \dfrac{\tau}{2} \end{cases} \tag{16-47}$$

The cumulative distribution function for $t \geq \tau/2$ is

$$F(t) = \int_0^t E(t)dt = 0 + \int_{\tau/2}^t E(t)dt = \int_{\tau/2}^t \frac{\tau^2}{2t^3}\,dt = \frac{\tau^2}{2}\int_{\tau/2}^t \frac{dt}{t^3} = 1 - \frac{\tau^2}{4t^2} \tag{16-48}$$

The mean residence time t_m is

For LFR $t_m = \tau$

$$t_m = \int_{\tau/2}^\infty tE(t)\,dt = \frac{\tau^2}{2}\int_{\tau/2}^\infty \frac{dt}{t^2}$$

$$= \frac{\tau^2}{2}\left[-\frac{1}{t}\right]_{\tau/2}^\infty = \tau$$

This result was shown previously to be true for any reactor *without dispersion*. The mean residence time is just the space time τ.

The dimensionless form of the RTD function is

Normalized RTD function for a laminar-flow reactor

$$E(\Theta) = \begin{cases} 0 & \Theta < 0.5 \\[2ex] \dfrac{1}{2\Theta^3} & \Theta \geq 0.5 \end{cases} \tag{16-49}$$

and is plotted in Figure 16-9.

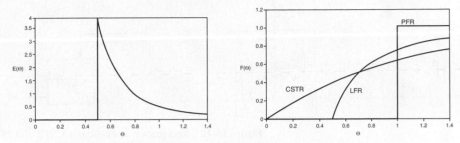

Figure 16-9   (a) $E(\Theta)$ for an LFR; (b) $F(\Theta)$ for a PFR, CSTR, and LFR.

The dimensionless cumulative distribution, $F(\Theta)$ for $\Theta \geq 1/2$, is

$$F(\Theta) = 0 + \int_{\frac{1}{2}}^{\Theta} E(\Theta)d\Theta = \int_{\frac{1}{2}}^{\Theta} \frac{d\Theta}{2\Theta^3} = \left(1 - \frac{1}{4\Theta^2}\right)$$

$$F(\Theta) = \begin{cases} 0 & \Theta < \frac{1}{2} \\ \left(1 - \frac{1}{4\Theta^2}\right) & \Theta \geq \frac{1}{2} \end{cases} \qquad (16\text{-}50)$$

Figure 16-9(a) shows $E(\Theta)$ for a laminar flow reactor (LFR), while Figure 9-9(b) compares $F(\Theta)$ for a PFR, CSTR, and LFR.

Experimentally injecting and measuring the tracer in a laminar-flow reactor can be a difficult task if not a nightmare. For example, if one uses as a tracer chemicals that are photo-activated as they enter the reactor, the analysis and interpretation of $E(t)$ from the data become much more involved.[7]

16.5 PFR/CSTR Series RTD

Modeling the real reactor as a CSTR and a PFR in series

In some stirred tank reactors, there is a highly agitated zone in the vicinity of the impeller that can be modeled as a perfectly mixed CSTR. Depending on the location of the inlet and outlet pipes, the reacting mixture may follow a somewhat tortuous path either before entering or after leaving the perfectly mixed zone—or even both. This tortuous path may be modeled as a plug-flow reactor. Thus, this type of reactor may be modeled as a CSTR in series with a plug-flow reactor, and the PFR may either precede or follow the CSTR. In this section we develop the RTD for a series arrangement of a CSTR and a PFR.

First consider the CSTR followed by the PFR (Figure 16-10). The mean residence time in the CSTR will be denoted by τ_s and the mean residence time in the PFR by τ_p. If a pulse of tracer is injected into the entrance of the CSTR,

---

[7] D. Levenspiel, *Chemical Reaction Engineering,* 3rd ed. (New York: Wiley, 1999), p. 342.

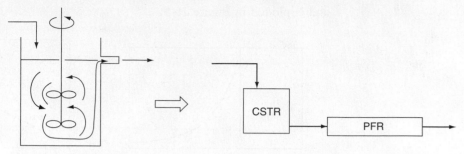

Figure 16-10   Real reactor modeled as a CSTR and PFR in series.

the CSTR output concentration as a function of time will be

$$C = C_0 e^{-t/\tau_s}$$

This output will be delayed by a time τ_p at the outlet of the plug-flow section of the reactor system. Thus, the RTD of the reactor system is

$$E(t) = \begin{cases} 0 & t < \tau_p \\ \dfrac{e^{-(t-\tau_p)/\tau_s}}{\tau_s} & t \geq \tau_p \end{cases} \tag{16-51}$$

See Figure 16-11.

The RTD is not unique to a particular reactor sequence.

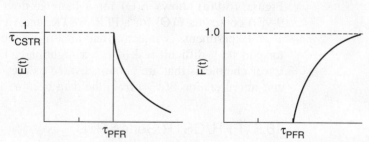

Figure 16-11   RTD curves $E(t)$ and $F(t)$ for a CSTR and a PFR in series.

Next, consider a reactor system in which the CSTR is preceded by the PFR will be treated. If the pulse of tracer is introduced into the entrance of the plug-flow section, then the same pulse will appear at the entrance of the perfectly mixed section τ_p seconds later, meaning that the RTD of the reactor system will again be

$E(t)$ is the same no matter which reactor comes first.

$$E(t) = \begin{cases} 0 & t < \tau_p \\ \dfrac{e^{-(t-\tau_p)/\tau_s}}{\tau_s} & t \geq \tau_p \end{cases} \tag{16-51}$$

which is *exactly* the same as when the CSTR was followed by the PFR.

It turns out that no matter where the CSTR occurs within the PFR/CSTR reactor sequence, the same RTD results. Nevertheless, this is not the entire story as we will see in Example 16-3.

Example 16–3   Comparing Second-Order Reaction Systems

Examples of *early* and *late* mixing for a given RTD

Consider a second-order reaction being carried out in a *real* CSTR that can be modeled as two different reactor systems: In the first system an ideal CSTR is followed by an ideal PFR (Figure E16-3.1); in the second system the PFR precedes the CSTR (Figure E16-3.2). To simplify the calculations, let τ_s and τ_p each equal 1 min, let the reaction rate constant equal 1.0 m³/kmol·min, and let the initial concentration of liquid reactant, C_{A0}, equal 1.0 kmol/m³. Find the conversion in each system.

For the parameters given, we note that in these two arrangements (see Figures E16-3.1 and E16-3.2), the RTD function, $E(t)$, is the same

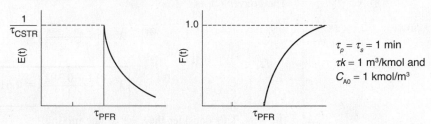

$$\tau_p = \tau_s = 1 \text{ min}$$
$$\tau k = 1 \text{ m}^3/\text{kmol and}$$
$$C_{A0} = 1 \text{ kmol/m}^3$$

Solution

(a) Let's first consider the case of early mixing when the CSTR is followed by the plug-flow section (Figure E16-3.1).

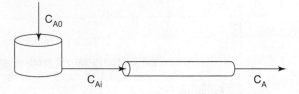

Figure E16-3.1   Early mixing scheme.

A mole balance on the CSTR section gives

$$V = \frac{F_{A0} - F_{Ai}}{-r_{Ai}} = \frac{v_0(C_{A0} - C_{Ai})}{kC_{Ai}^2}$$

Rearrranging

$$v_0(C_{A0} - C_{Ai}) = kC_{Ai}^2 V \qquad (E16\text{-}3.1)$$

Dividing by v_0 and rearranging, we have quadric equation to solve for the intermediate concentration C_{Ai}

$$\tau_s k C_{Ai}^2 + C_{Ai} - C_{A0} = 0$$

Substituting for τ_s and k

$$\tau_s k = 1 \text{ min} \cdot \frac{\text{m}^3}{\text{kmol} \cdot \text{min}} = \frac{1 \text{ m}^3}{\text{kmol}}$$

Solving for C_{Ai} gives

$$C_{Ai} = \frac{\sqrt{1 + 4\tau_s k C_{A0}} - 1}{2\tau_s k} = \frac{-1 + \sqrt{1 + 4}}{2} = 0.618 \text{ kmol/m}^3 \qquad (E16\text{-}3.2)$$

This concentration will be fed into the PFR. The PFR mole balance is

$$\frac{dF_A}{dV} = v_0 \frac{dC_A}{dV} = \frac{dC_A}{d\tau_p} = r_A = -kC_A^2 \qquad (E16\text{-}3.3)$$

Integrating Equation (16-3.3)

$$\frac{1}{C_A} - \frac{1}{C_{Ai}} = \tau_p k \qquad \text{(E16-3.4)}$$

Substituting $C_{Ai} = 0.618$ kmol/m^3, $\tau_p = 1$ min, $k = 1$ m^3/kmol/min and $\tau_p k = 1$ m^3/kmol in Equation (E16-3.4) yields

$$C_A = 0.382 \text{ kmol/m}^3$$

$\boxed{\begin{array}{c} \text{CSTR} \rightarrow \text{PFR} \\ X = 0.618 \end{array}}$

as the concentration of reactant in the effluent from the reaction system. The conversion is

$$X = \left(\frac{C_{A0} - C_A}{C_{A0}}\right) =$$

$$\boxed{X = \left(\frac{1 - 0.382}{1}\right) = 0.618}$$

(b) Now, let's consider the case of late mixing.

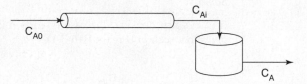

Figure E16-3.2 Late mixing scheme.

When the perfectly mixed section is preceded by the plug-flow section (Figure E16-3.2), the outlet of the PFR is the inlet to the CSTR, C_{Ai}. Again, solving Equation (E16-3.3)

$$\frac{1}{C_{Ai}} - \frac{1}{C_{A0}} = \tau_p k$$

Solving for intermediate concentration, C_{Ai}, given $\tau_p k = 1$ m^3/mol and a $C_{A0} = 1$ mol/m^3

$$C_{Ai} = 0.5 \text{ kmol/m}^3$$

Next, solve for C_A exiting the CSTR.

A material balance on the perfectly mixed section (CSTR) gives

$\boxed{\begin{array}{c} \text{PFR} \rightarrow \text{CSTR} \\ X = 0.634 \end{array}}$

$$\tau_s k C_A^2 + C_A - C_{Ai} = 0 \qquad \text{(E16-3.5)}$$

$$C_A = \frac{\sqrt{1 + 4\tau_s k C_{Ai}} - 1}{2\tau_s k} = \frac{-1 + \sqrt{1+2}}{2} = 0.366 \text{ kmol/m}^3 \qquad \text{(E16-3.6)}$$

$\boxed{\begin{array}{c} \text{Early mixing} \\ X = 0.618 \\ \text{Late mixing} \\ X = 0.634 \end{array}}$

as the concentration of reactant in the effluent from the reaction system. The corresponding conversion is 63.4%; that is, $X = 1 - (C_A / C_{A0}) = 1 - \dfrac{0.366}{1.0} = 63.4\%$.

Analysis: The RTD curves are identical for both configurations. However, the conversion was not the same. In the first configuration, a conversion of 61.8% was obtained; in the second configuration, 63.4%. While the difference in the conversions is small for the parameter values chosen, **the point is that there is a difference**. Let me say that again, *the point is there is a difference* and we will explore it further in Chapters 17 and 18.

While $E(t)$ was the same for both reaction systems, the conversion was not.

The conclusion from this example is of extreme importance in reactor analysis: **The RTD is not a complete description of structure for a particular reactor or system of reactors**. The RTD is unique for a particular or given reactor. However, as we just saw, the reactor or reaction system is not unique for a particular RTD. When analyzing nonideal reactors, the RTD alone is not sufficient to determine its performance, and more information is needed. It will be shown in Chapter 17 that in addition to the RTD, an adequate model of the nonideal reactor flow pattern and knowledge of the quality of mixing or "degree of segregation" are both required to characterize a reactor properly.

At this point, the reader has the necessary background to go directly to Chapter 17 where we use the RTD to calculate the mean conversion in a real reactor using different models of ideal chemical reactors.

Chance Card: ***Do not pass go, proceed directly to Chapter 17.***

16.6 Diagnostics and Troubleshooting

16.6.1   General Comments

As discussed in Section 16.1, the RTD can be used to diagnose problems in existing reactors. As we will see in further detail in Chapter 18, the RTD functions $E(t)$ and $F(t)$ can be used to model the real reactor as combinations of ideal reactors.

Figure 16-12 illustrates typical RTDs resulting from different nonideal reactor situations. Figures 16-12(a) and (b) correspond to "nearly" ideal PFRs

RTDs that are commonly observed

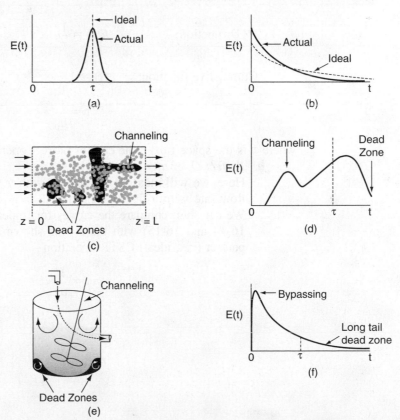

Figure 16-12   (a) RTD for *near* plug-flow reactor; (b) RTD for *near* perfectly mixed CSTR; (c) packed-bed reactor with dead zones and channeling; (d) RTD for packed-bed reactor in (c); (e) tank reactor with short-circuiting flow (bypass); (f) RTD for tank reactor with channeling (bypassing or short circuiting) and a dead zone in which the tracer slowly diffuses in and out.

and CSTRs, respectively. The RTD for the nonideal reactor in Figure 16-12(c) modeled as a PBR with channeling and dead zones is shown in Figure 16-12(d). In Figure 16-12(d) one observes that a principal peak occurs at a time smaller than the space time ($\tau = V/v_0$) (i.e., early exit of fluid) and also that some fluid exits at a time greater than space-time τ. This curve is consistent with the RTD for a packed-bed reactor with channeling (i.e., by-passing) and stagnant zones (i.e., not mixed with bulk flow) as discussed earlier in Figure 16-1. Figure 16-12(f) shows the RTD for the nonideal CSTR in Figure 16-12(e), which has dead zones and bypassing. The dead zone serves not only to reduce the effective reactor volume, so the active reactor volume is smaller than expected, but also results in longer residence times for the tracer molecules to diffuse in and out of these "dead or stagnant" zones.

16.6.2 Simple Diagnostics and Troubleshooting Using the RTD for Ideal Reactors

16.6.2A   The CSTR

We now consider three CSTRs: (a) one that operates normally, (b) one with bypassing, and (c) one with a dead volume. For a well-mixed CSTR, as we saw in Section 16.4.2, the response to a pulse tracer is

Concentration: $C(t) = C_0 e^{-t/\tau}$ (16-34)

RTD function: $E(t) = \dfrac{1}{\tau} e^{-t/\tau}$ (16-35)

Cumulative function: $F(t) = 1 - e^{-t/\tau}$ (16-36)

$$\tau = \frac{V}{v_0}$$

where τ is the space time—the case of perfect operation.

a. *Perfect Operation (P) Model*

Here, we will measure our reactor with a yardstick to find V and our flow rate with a flow meter to find v_0 in order to calculate $\tau = V/v_0$. We can then compare the curves for imperfect operation (cf. Figures 16-14 and 16-15) with the curves shown below in Figure 16-13 for perfect (i.e., ideal) CSTR operation

$$\tau = \frac{V}{v_0}$$

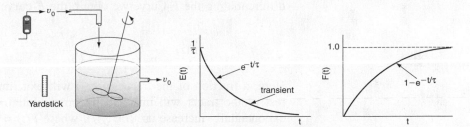

Figure 16-13   Perfect operation of a CSTR.

If τ is large, there will be a slow decay of the output transient, $C(t)$, and $E(t)$ for a pulse input. If τ is small, there will be rapid decay of the transient, $C(t)$, and $E(t)$ for a pulse input.

b. *Bypassing (BP) Model*

In Figure 16-14, the real reactor on the left is modeled by an ideal reactor with bypassing, as shown on the right.

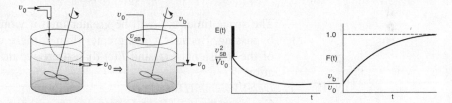

Figure 16-14   Ideal CSTR with bypass.

The entering volumetric flow rate is divided into a volumetric flow rate that enters the reacting system, v_{SB}, and a volumetric flow rate that bypasses the reacting mixture system completely v_b. Where $v_0 = v_{SB} + v_b$. The subscript SB denotes a model of a reactor system with bypassing. The reactor system volume V_S is the well-mixed portion of the reactor.

The analysis of a bypass to the CSTR may be elucidated by observing the output to a *step tracer* input to the real reactor. The CSTR with bypassing will have RTD curves similar to those in Figure 16-14.

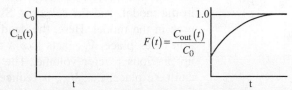

Figure 16-14(a)   Inlet and outlet concentration curves correspond to Figure 16-14.

We see the concentration output in the form $C(t)/C_0$ for a step input will be the F-curve and the initial jump will be equal to the fraction bypassed. The corresponding equation for the F-curve is

$$F(t) = \frac{v_b}{v_0} + \frac{v_{SB}}{v_0}\left[1 - e^{-\left(\frac{v_{SB} t}{V}\right)} \right]$$

differentiating the F-Curve we obtain the E-curve

$$E(t) = \frac{v_b}{v_0}\delta(t-0) + \frac{v_{SB}^2}{Vv_0}e^{-\left(\frac{v_{SB}t}{V}\right)}$$

Here, a fraction of the tracer (v_b/v_0) will exit immediately while the rest of the tracer will mix and become diluted with volume V and exponentially increase up to $C_{out}(t)$, where $F(t) = 1.0$.

Because some of the fluid bypasses, the flow passing through the system will be less than the total volumetric rate, $v_{SB} < v_0$; consequently, $\tau_{SB} > \tau$. For example, let's say the volumetric flow rate, which bypasses the reactor, v_b, is 25% of the total (e.g., $v_b = 0.25$ v_0). The volumetric flow rate entering the reactor system, v_{SB}, is 75% of the total ($v_{SB} = 0.75 \, v_0$) and the corresponding true space time (τ_{SB}) for the system volume with bypassing is

$$\tau_{SB} = \frac{V}{v_{SB}} = \frac{V}{0.75v_0} = 1.33\tau$$

The space time, τ_{SB}, will be greater than it would be if there were no bypassing. Because τ_{SB} is greater than τ, there will be a slower decay of the transients $C(t)$ and $E(t)$ than there would be with perfect operation.

c. *Dead Volume (DV) Model*

Consider the CSTR in Figure 16-15 without bypassing but instead with a stagnant or dead volume.

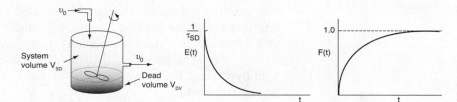

Figure 16-15　Ideal CSTR with dead volume.

The total volume, V, is the same as that for perfect operation, $V = V_{DV} + V_{SD}$. The subscript DV in V_{DV} represents the dead volume in the model, and the subscript SD in V_{SD} is the reacting system volume in the model. Here, the dead volume where **absolutely** <u>no</u> reaction takes place, V_{DV}, acts like a fictitious brick at the bottom taking up precious reactor volume. The system volume where the reaction can take place, i.e., V_{SD}, is reduced because of this dead volume and therefore less conversion can be expected.

We see that because there is a dead volume that the fluid does not enter, there is less system volume, V_{SD}, available for reaction than in the case of perfect operation, i.e., $V_{SD} < V$. Consequently, the fluid will pass through the reactor with the dead volume more quickly than that of perfect operation, i.e., $\tau_{SD} < \tau$.

If　　$V_{DV} = 0.2V$, $V_{SD} = 0.8V$, then $\tau_{SD} = \frac{0.8V}{v_0} = 0.8\tau$

Also as a result, the transients $C(t)$ and $E(t)$ will decay more rapidly than that for perfect operation because there is a smaller system volume.

Summary

A summary for ideal CSTR mixing volume is shown in Figure 16-16.

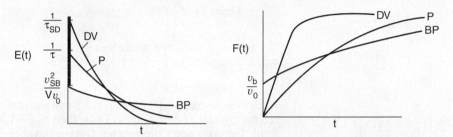

Figure 16-16   Comparison of $E(t)$ and $F(t)$ for CSTR under perfect operation, bypassing, and dead volume. (**BP** = bypassing model, **P** = perfect operation model, and **DV** = dead volume model).

Knowing the volume V measured with a yardstick and the flow rate v_0 entering the reactor measured with a flow meter, one can calculate and plot $E(t)$ and $F(t)$ for the ideal case (P) and then compare with the measured RTD $E(t)$ to see if the RTD suggests either bypassing (BP) or dead zones (DV).

16.6.2B   Tubular Reactor

A similar analysis to that for a CSTR can be carried out on a tubular reactor.

a. *Perfect Operation of PFR (P) Model*

We again measure the volume V with a yardstick and v_0 with a flow meter. The $E(t)$ and $F(t)$ curves are shown in Figure 16-17. The triangles drawn in Figures 16-17 and 16-18 should really be Dirac delta functions with zero width at the base. The space time for a perfect PFR is

$$\tau = V/v_0$$

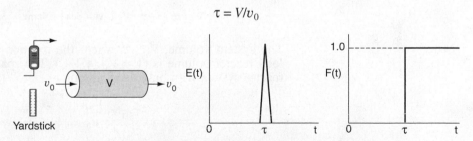

Figure 16-17   Perfect operation of a PFR.

b. *PFR with Channeling (Bypassing, BP) Model*

Let's consider channeling (bypassing), as shown in Figure 16-18, similar to that shown in Figures 16-2 and 16-12(d). The space time for the reactor system with bypassing (channeling) τ_{SB} is

$$\tau_{SB} = \frac{V}{v_{SB}}$$

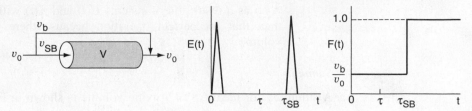

Figure 16-18 PFR with bypassing similar to the CSTR.

Because $v_{SB} < v_0$, the space time for the case of bypassing is greater when compared to perfect operation, i.e.,

$$\tau_{SB} > \tau$$

If 25% is bypassing (i.e., $v_b = 0.25\ v_0$) and only 75% is entering the reactor system (i.e., $v_{SB} = 0.75\ v_0$), then $\tau_{SB} = V/(0.75v_0) = 1.33\tau$. The fluid that *does* enter the reactor system flows in a plug flow. Here, we have two spikes in the $E(t)$ curve: one spike at the origin and one spike at τ_{SB} that comes after τ for perfect operation. Because the volumetric flow rate is reduced, the time of the second spike will be greater than τ for perfect operation.

c. *PFR with Dead Volume (DV) Model*
The dead volume, V_{DV}, could be manifested by internal circulation at the entrance to the reactor as shown in Figure 16-19.

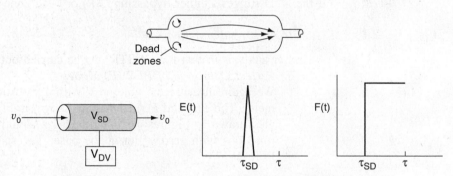

Figure 16-19 PFR with dead volume.

The system volume, V_{SD}, is where the reaction takes place and the total reactor volume is $(V = V_{SD} + V_{DV})$. The space time, τ_{SD}, for the reactor system with only dead volume is

$$\tau_{SD} = \frac{V_{SD}}{v_0}$$

Compared to perfect operation, the space time τ_{SD} is smaller and the tracer spike will occur before τ for perfect operation.

$$\tau_{SD} < \tau$$

Here again, the dead volume takes up space that is not accessible. As a result, the tracer will exit early because the system volume, V_{SD}, through which it must pass is smaller than the perfect operation case.

Summary

Figure 16-20 is a summary of these three cases.

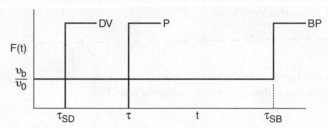

Figure 16-20 Comparison of PFR under perfect operation, bypassing, and dead volume (**DV** = dead volume model, **P** = perfect PFR model, **BP** = bypassing model).

In addition to its use in diagnosis, the RTD can be used to predict conversion in existing reactors when a new reaction is tried in an old reactor. However, as we saw in Section 16.5, the RTD is not unique for a given system, and we need to develop models for the RTD to predict conversion.

There are many situations where the fluid in a reactor is *neither* well mixed *nor* approximates plug flow. The idea is this: We have seen that the RTD can be used to diagnose or interpret the type of mixing, bypassing, etc., that occurs in an existing reactor that is currently on stream and is not yielding the conversion predicted by the ideal reactor models. Now let's envision another use of the RTD. Suppose we have a nonideal reactor either on line or sitting in storage. We have characterized this reactor and obtained the RTD function. What will be the conversion of a reaction with a known rate law that is carried out in a reactor with a known RTD?

The question

> How can we use the RTD to predict conversion in a real reactor?

In Chapter 17 we show how this question can be answered in a number of ways.

Closure. After completing this chapter the reader will be able to use the tracer concentration time data to calculate the external-age distribution function $E(t)$, the cumulative distribution function $F(t)$, the mean residence time, t_m, and the variance, σ^2. The reader will be able to sketch $E(t)$ for ideal reactors, and by comparing $E(t)$ from the experiment with $E(t)$ for ideal reactors (PFR, PBR, CSTR, laminar-flow reactor (LFR)), be able to diagnose problems in real reactors. The reader will be able to use the RTD curves to identify the nonideal reactor problems of dead volume and bypassing.

SUMMARY

1. The quantity $E(t)\,dt$ is the fraction of material exiting the reactor that has spent between time t and $(t + dt)$ in the reactor.

2. The mean residence time

$$t_m = \int_0^\infty tE(t)\,dt = \tau \tag{S16-1}$$

is equal to the space time τ for constant volumetric flow, $v = v_0$.

3. The variance about the mean residence time is

$$\sigma^2 = \int_0^\infty (t - t_m)^2 E(t)\,dt \tag{S16-2}$$

4. The cumulative distribution function $F(t)$ gives the fraction of effluent material that has been in the reactor a time t or less

$$F(t) = \int_0^t E(t)\,dt$$

$$1 - F(t) = \int_t^\infty E(t)\,dt = \text{fraction of effluent material that has been in} \tag{S16-3}$$
$$\text{the reactor a time } t \text{ or longer}$$

5. The RTD functions for an ideal reactor are

$$\text{PFR:} \qquad E(t) = \delta(t - \tau) \tag{S16-4}$$

$$\text{CSTR:} \qquad E(t) = \frac{e^{-t/\tau}}{\tau} \tag{S16-5}$$

$$\text{LFR:} \qquad E(t) = 0 \qquad t < \frac{\tau}{2} \tag{S16-6}$$

$$E(t) = \frac{\tau^2}{2t^3} \qquad t \geq \frac{\tau}{2} \tag{S16-7}$$

6. The dimensionless residence time is

$$\Theta = \frac{t}{\tau} \tag{S16-8}$$

$$E(\Theta) = \tau E(t) \tag{S16-9}$$

7. Diagnosing nonideal reactors with dead volume and bypassing

(a) $F(t)$ for nonideal CSTR   **(b)** $F(t)$ for nonideal tubular reactor

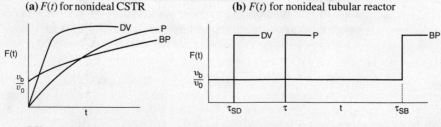

DV = dead volume model, **P** = perfect operation model, **BP** = bypassing model

CRE WEB SITE MATERIALS

- **Expanded Material on the Web Site**
    1. *Web Example 16-1 Gas-Liquid Reactor*
    2. *Proof That in the Absence of Dispersion, the Mean Residence Time, t_m, is Equal to Space Time, i.e.,*
       $t_m = \tau$
    3. *Side Note on the Medical Uses of RTD*
    4. *Solved Problems WP16-14$_C$ and WP16-15$_D$*
- **Learning Resources**
    1. *Summary Notes*
    2. *Web Material Links*
       The Attainable Region Analysis
           http://www.umich.edu/~elements/5e/16chap/learn-attainableregions.html
           http://hermes.wits.ac.za/attainableregions/
- **Living Example Problems**
    1. *Living Example 16-1 Determining E(t)*
    2. *Living Example 16-2T: Tutorial to Find E(t) from C(t)*
    3. *Living Example 16-2 (a) and (b) Finding t_m and σ^2*
- **Professional Reference Shelf**
    R16.1. *Fitting the Tail*
       Whenever there are dead zones into which the material diffuses in and out, the *C*- and *E*-curves may exhibit long tails. This section shows how to analytically describe fitting these tails to the curves.

$$E(t) = ae^{-bt}$$
$$b = \text{slope of } \ln E \text{ vs. } t$$
$$a = be^{bt_1}[1 - F(t_1)]$$

    R16.2. *Internal-Age Distribution*
       The internal-age distribution currently in the reactor is given by the distribution of ages with respect to how long the molecules have been in the reactor.
       The equation for the internal-age distribution is derived and an example is given showing how it is applied to catalyst deactivation in a "fluidized CSTR."

Homework Problems

QUESTIONS AND PROBLEMS

The subscript to each of the problem numbers indicates the level of difficulty: A, least difficult; D, most difficult.

$$A = \bullet \quad B = \blacksquare \quad C = \blacklozenge \quad D = \blacklozenge\blacklozenge$$

Questions

Q16-1$_A$ Read over the problems of this chapter. Discuss with a classmate ideas for making up an original problem that uses the concepts presented in this chapter. The guidelines are given in Problem P5-1$_A$. RTDs from real reactors can be found in *Ind. Eng. Chem.*, 49, 1000 (1957); *Ind. Eng. Chem. Process Des. Dev.*, 3, 381 (1964); *Can. J. Chem. Eng.*, 37, 107 (1959); *Ind. Eng. Chem.*, 44, 218 (1952); *Chem. Eng. Sci.*, 3, 26 (1954); and *Ind. Eng. Chem.*, 53, 381 (1961).

Problems

P16-1$_B$ **What if...**

(a) **Example 16-1.** What fraction of the fluid spends 9 minutes or longer in the reactor? What fraction spends 2 minutes or less?

(b) **Example 16-3.** How would the $E(t)$ change if the PFR space time, τ_p, was reduced by 50% and τ_s was increased by 50%? What fraction spends 2 minutes or less in the reactor?

P16-2$_B$ (a) Suggest a diagnosis (e.g., bypassing, dead volume, multiple mixing zones, internal circulation) for each of the following real reactors in Figure P16-2$_B$ (a) (1 through 7 curves) that had the following RTD [$E(t)$, $E(\Theta)$, $F(t)$, $F(\Theta)$ or $(1-F(\Theta))$] curves:

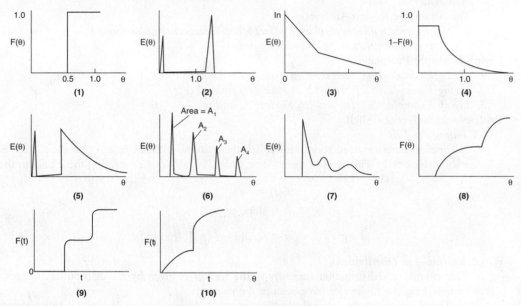

Figure P16-2$_B$ (a)   RTD curves.

(b) Suggest a model (e.g. combinations of ideal reactors, bypassing) for each RTD function shown in **Figure P16-2$_B$(a)** (1 through 10) that would give the RTD function. For example, for the real tubular reactor, whose $E(\Theta)$ curve is shown in Figure **P16-2$_B$ (a) (5)** above, the model is shown in **Figure P16-2$_B$ (b)** below. The real reactor is modeled as having bypassing, a back mix zone, and a PFR zone that mimics the real CSTR.

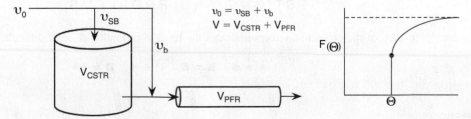

Figure P16-2$_B$ (b)   Real reactor modeled as CSTR and PFR with bypass.

Suggest a model for each figure.

P16-3$_C$ Consider the $E(t)$ curve below.

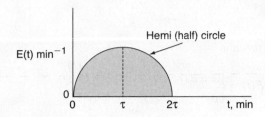

Mathematically this hemi circle is described by these equations:

For $2\tau \geq t \geq 0$, then $E(t) = \sqrt{\tau^2 - (t-\tau)^2}$ min^{-1}(hemi circle)

For $t > 2\tau$, then $E(t) = 0$

(a) What is the mean residence time?

(b) What is the variance?

P16-4$_B$ A step tracer input was used on a real reactor with the following results:

For $t \leq 10$ min, then $C_T = 0$

For $10 \leq t \leq 30$ min, then $C_T = 10$ g/dm^3

For $t \geq 30$ min, then $C_T = 40$ g/dm^3

(a) What is the mean residence time t_m?

(b) What is the variance σ^2?

P16-5$_B$ The following $E(t)$ curves were obtained from a tracer test on two tubular reactors in which dispersion is believed to occur.

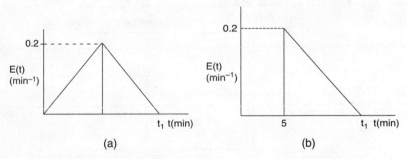

Figure P16-5$_B$ (a) RTD Reactor A; (b) RTD Reactor B.

(a) What is the final time t_1 (in minutes) for the reactor shown in Figure P16-5$_B$ (a)? In Figure P16-5$_B$ (b)?

(b) What is the mean residence time, t_m, and variance, σ^2, for the reactor shown in Figure P16-5$_B$ (a)? In Figure P16-5$_B$ (b)?

(c) What is the fraction of the fluid that spends 7 minutes or longer in Figure P16-5$_B$ (a)? In Figure P16-5$_B$ (b)?

P16-6$_B$ An RTD experiment was carried out in a nonideal reactor that gave the following results:

$$E(t) = 0 \qquad\qquad \text{for} \qquad t < 1 \text{ min}$$
$$E(t) = 1.0 \text{ min}^{-1} \qquad \text{for} \qquad 1 \leq t \leq 2 \text{ min}$$
$$E(t) = 0 \qquad\qquad \text{for} \qquad t > 2 \text{ min}$$

(a) What are the mean residence time, t_m, and variance σ^2?

(b) What is the fraction of the fluid that spends a time 1.5 minutes or longer in the reactor?

(c) What fraction of fluid spends 2 minutes or less in the reactor?

(d) What fraction of fluid spends between 1.5 and 2 minutes in the reactor?

P16-7 Derive $E(t)$, $F(t)$, t_m, and σ^2 for a turbulent flow reactor with 1/7 the power law, i.e.,

$$U = U_{max}\left(1 - \frac{r}{R}\right)^{1/7}$$

P16-8 Consider the RTD function

$$E(t) = \begin{cases} A - B(t_0 - t)^3 & \text{for } 0 \le t < 2t_0 \\ 0 & \text{for } t > 2t_0 \end{cases}$$

(a) Sketch $E(t)$ and $F(t)$.
(b) Calculate t_m and σ^2.
(c) What are the restrictions (if any) on A, B and t_o? *Hint:* $\int_0^\infty E(t)dt = 1$.

P16-9$_A$ Evaluate the first moment about the mean $m_1 = \int_0^\infty (t - \tau) \, E(t)dt$ for an ideal PFR, a CSTR, and a laminar-flow reactor.

P16-10$_B$ Gasoline shortages in the United States have produced long lines of motorists at service stations. The table below shows a distribution of the times required to obtain gasoline at 23 Center County service stations.

(a) What is the average time required?
(b) If you were to ask randomly among those people waiting in line, "How long have you been waiting?" what would be the average of their answers?

Total Waiting Time (min)	Number of Stations Having That Total Waiting Time
0	0
3	4
6	3
9	5
12	8
15	2
18	1
21	0

(c) Can you generalize your results to predict how long you would have to wait to enter a five-story parking garage that has a 4-hour time limit?

(R. L. Kabel, Pennsylvania State University)

P16-11$_B$ The volumetric flow rate through a reactor is 10 dm³/min. A pulse test gave the following concentration measurements at the outlet:

t (min)	$c \times 10^5$	t (min)	$c \times 10^5$
0	0	15	238
0.4	329	20	136
1.0	622	25	77
2	812	30	44
3	831	35	25
4	785	40	14
5	720	45	8
6	650	50	5
8	523	60	1
10	418		

Member

Hall of Fame

(a) Plot the external-age distribution $E(t)$ as a function of time.
(b) Plot the external-age cumulative distribution $F(t)$ as a function of time.
(c) What are the mean residence time t_m and the variance, σ^2?
(d) What fraction of the material spends between 2 and 4 minutes in the reactor?
(e) What fraction of the material spends longer than 6 minutes in the reactor?
(f) What fraction of the material spends less than 3 minutes in the reactor?
(g) Plot the normalized distributions $E(\Theta)$ and $F(\Theta)$ as a function of Θ.
(h) What is the reactor volume?
(i) Plot the internal-age distribution $I(t)$ as a function of time.
(j) What is the mean internal age α_m?
(k) This problem is continued in Problems **P17-14$_B$** and **P18-12$_C$**.

P16-12$_B$ An RTD analysis was carried out on a liquid-phase reactor [*Chem. Eng. J.* 1, 76 (1970)]. Analyze the following data:

t(s)	0	150	175	200	225	240	250
$C \times 10^3$ (g/m^3)	0	0	1	3	7.4	9.4	9.7

t(s)	275	300	325	350	375	400	450
$C \times 10^3$ (g/m^3)	8.2	5.0	2.5	1.2	0.5	0.2	0

(a) Plot the $E(t)$ curve for these data.
(b) What fraction of the material spends between 230 and 270 seconds in the reactor?
(c) Plot the $F(t)$ curve for these data.
(d) What fraction of the material spends less than 250 seconds in the reactor?
(e) What is the mean residence time?
(f) What is the variance σ^2?
(g) Plot $E(\Theta)$ and $F(\Theta)$ as a function of Θ.

P16-13$_C$ (*Distributions in a stirred tank*) Using a negative step tracer input, Cholette and Cloutier [*Can. J. Chem. Eng.*, 37, 107 (1959)] studied the RTD in a tank for different stirring speeds. Their tank had a 30-in. diameter and a fluid depth of 30 in. indide the tank. The inlet and exit flow rates were 1.15 gal/min. Here are some of their tracer results for the relative concentration, C/C_0 (courtesy of the Canadian Society for Chemical Engineering):

NEGATIVE STEPS TRACER TEST

Time (min)	Impeller Speed (rpm)	
	170	100
10	0.761	0.653
15	0.695	0.566
20	0.639	0.513
25	0.592	0.454
30	0.543	0.409
35	0.502	0.369
40	0.472	0.333
45	0.436	0.307
50	0.407	0.276
55	0.376	0.248
60	0.350	0.226
65	0.329	0.205

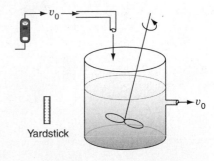

Yardstick

Calculate and plot the cumulative exit-age distribution, the intensity function, and the internal-age distributions as a function of time for this stirred tank at the two impeller speeds. Can you tell anything about the dead zones and bypassing at the different stirrer rates?

- **Additional Homework Problems on the CRE Web Site**

WEBP16-1$_A$ Determine $E(t)$ from data taken from a pulse test in which the pulse is not perfect and the inlet concentration varies with time. [2nd Ed. P13-15]

WEBP16-2$_B$ Review the data by Murphree on a Pilot Plant Scale Reactor.

SUPPLEMENTARY READING

1. Discussions of the measurement and analysis of residence time distribution can be found in

CURL, R. L., and M. L. MCMILLIN, "Accuracies in residence time measurements," *AIChE J.*, 12, 819–822 (1966).

JACKSON, JOFO, "Analysis of 'RID' Data from the Nut Cracker Reactor at the Riça Plant," *Jofostan Journal of Industrial Data*, Vol. 28, p. 243 (1982).

LEVENSPIEL, O., *Chemical Reaction Engineering*, 3rd ed. New York: Wiley, 1999, Chaps. 11–16.

2. An excellent discussion of segregation can be found in

DOUGLAS, J. M., "The effect of mixing on reactor design," *AIChE Symp. Ser.*, 48, vol. 60, p. 1 (1964).

3. Also see

DUDUKOVIC, M., and R. FELDER, in *CHEMI Modules on Chemical Reaction Engineering*, vol. 4, ed. B. Crynes and H. S. Fogler. New York: AIChE, 1985.

NAUMAN, E. B., "Residence time distributions and micromixing," *Chem. Eng. Commun.*, 8, 53 (1981).

NAUMAN, E. B., and B. A. BUFFHAM, *Mixing in Continuous Flow Systems*. New York: Wiley, 1983.

ROBINSON, B. A., and J. W. TESTER, *Chem. Eng. Sci.*, 41(3), 469–483 (1986).

VILLERMAUX, J., "Mixing in chemical reactors," in *Chemical Reaction Engineering—Plenary Lectures*, ACS Symposium Series 226. Washington, DC: American Chemical Society, 1982.

Predicting Conversion Directly from the Residence Time Distribution

17

If you think you can, you can.
If you think you can't, you can't.
You are right either way.

—Steve LeBlanc

Overview. Chapter 16 showed how to determine the residence time distribution (RTD) functions $E(t)$ and $F(t)$ along with the mean residence time, t_m and the variance, σ^2, from concentration time data. We then used the RTD to characterize and diagnose ideal reactors. Because the residence time distribution is not unique for a given reaction system, we must use models if we want to predict the conversion in our nonideal reactor. In this chapter we will use the RTD function, i.e., $E(t)$, without any adjustable parameters to predict the conversion, X, and the product distribution.

After completing this chapter you will be able to:

- Describe the five most common models to predict conversion using RTD data (Section 17.1).
- Calculate conversion for single reactions directly from RTD data using the zero adjustable parameter models (a) segregation and, (b) maximum mixedness, and (c) laminar flow reactor (Section 17.2).
- Use software packages to calculate conversion in the extremes of micromixing (a) segregation and (b) maximum mixedness (Section 17.3).
- Apply the segregation and maximum mixedness models to nonideal reactors to predict the product distribution in reactors where multiple reactions are occurring (Section 17.4).

17.1 Modeling Nonideal Reactors Using the RTD

17.1.1 Modeling and Mixing Overview

Now that we have characterized our reactor and have gone to the lab to take data to determine the reaction kinetics, we need to choose a model to predict conversion in our real reactor.

The answer

$$\text{RTD} + \text{MODEL} + \text{KINETIC DATA} \Rightarrow \begin{cases} \text{EXIT CONVERSION and} \\ \text{EXIT CONCENTRATION} \end{cases}$$

We now present the five models shown in Table 17-1. We shall classify each model according to the number of adjustable parameters. In this chapter we will discuss the *zero adjustable parameter models* and, in Chapter 18, we will discuss *one and two adjustable parameter models* that will be used to predict conversion.

Ways we use the RTD data to predict conversion in nonideal reactors

TABLE 17-1   MODELS FOR PREDICTING CONVERSION FROM RTD DATA

1. **Zero adjustable parameters**
   a. **Segregation model**
   b. **Maximum mixedness model**
2. **One adjustable parameter**
   a. **Tanks-in-series model**
   b. **Dispersion model**
3. **Two adjustable parameters**
   Real reactors modeled as combinations of ideal reactors

For the zero adjustable parameter models, we do not need to make any intermediate calculations; we use the *E*- and *F*-Curves directly to predict the conversion given the kinetic parameters. For the one-parameter models, we use the RTD to calculate mean residence time, t_m, and variance, σ^2, which we can then use (1) to find the number of tanks in series necessary to accurately model a nonideal CSTR and (2) calculate the Peclet number, Pe, to find the conversion in a tubular flow reactor using the dispersion model.

For the two-parameter models, we create combinations of ideal reactors to model the nonideal reactor. We then use the RTD to calculate the model parameters such as fraction bypassed, fraction dead volume, exchange volume, and ratios of reactor volumes that then can be used along with the reaction kinetics to predict conversion.

17.1.2 Mixing

The RTD tells us how long the various fluid elements have been in the reactor, but it does not tell us anything about the exchange of matter between the fluid elements (i.e., *the mixing*). The mixing of reacting species is one of the major factors controlling the behavior of chemical reactors. Fortunately for first-order reactions, mixing is not important, and knowledge of the length of time each molecule spends in the reactor is all that is needed to predict conversion. For first-order reactions, the conversion is independent of concentration (recall Table 5-1, page 146)

$$\frac{dX}{dt} = k(1 - X) \qquad \text{(E13-1.3)}$$

A model is needed for reactions other than first order.

Consequently, mixing with the surrounding molecules is not important. Therefore, once the RTD is determined, we can predict the conversion that will be achieved in the real reactor provided that the specific reaction rate for the first-order reaction is known. However, for reactions other than first order, knowledge of the RTD is not sufficient to predict conversion. For reactions other than first order, the degree of mixing of molecules must be known in addition to how long each molecule spends in the reactor. Consequently, we must develop models that account for the mixing of molecules inside the reactor.

The more complex models of nonideal reactors necessary to describe reactions other than first order must contain information about *micromixing* in addition to that of *macromixing*. **Macromixing** produces a distribution of residence times *without*, specifying how molecules of different ages encounter one another in the reactor. **Micromixing**, on the other hand, describes how molecules of different ages encounter one another in the reactor. There are two extremes of *micromixing*:

(1) all molecules of the same age group remain together as they travel through the reactor and are not mixed with any other age until they exit the reactor (i.e., complete segregation);

(2) molecules of different age groups are completely mixed at the molecular level as soon as they enter the reactor (complete micromixing).

For a given state of macromixing (i.e., a given RTD), these two extremes of micromixing will give the upper and lower limits on conversion in a nonideal reactor. *For single reactions with orders greater than one or less than zero, the segregation model will predict the highest conversion. For reaction orders between zero and one, the maximum mixedness model will predict the highest conversion.* This concept is discussed further in Section 17.3.1.

We shall define a globule as a fluid particle containing millions of molecules all of the same age. A fluid in which the globules of a given age do not mix with other globules is called a macrofluid. A *macrofluid* could be visualized as non-coalescent globules where all the molecules in a given globule have the same age. A fluid in which molecules are not constrained to remain in the globule and are free to move everywhere is called a *microfluid*.[1] There are two extremes of mixing of the *macrofluid* globules—early mixing and late mixing. These two extremes of late and early mixing are shown in Figure 17-1 (a) and (b), respectively. These extremes can also be seen by comparing Figures 17-3 (a) and 17-4 (a). The extremes of late and early mixing are referred to as *complete segregation* and *maximum mixedness*, respectively.

---

[1] J. Villermaux, *Chemical Reactor Design and Technology* (Boston: Martinus Nijhoff, 1986).

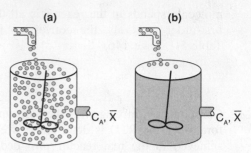

Figure 17-1 (a) Macrofluid and (b) microfluid mixing on the molecular level.

17.2 Zero-Adjustable-Parameter Models

17.2.1 Segregation Model

In a "perfectly mixed" CSTR, the entering fluid is assumed to be distributed immediately and evenly throughout the reacting mixture. This mixing is assumed to take place even on the microscale, and elements of different ages mix together thoroughly to form a *completely micromixed* fluid (see Figure 17-1(b)). However, if fluid elements of different ages do not mix together at all, the elements remain segregated from each other, and the fluid is termed *completely segregated* (see Figure 17-1(a)). The extremes of complete micromixing and complete segregation are the limits of the micromixing of a reacting mixture.

In developing the segregated mixing model, we first consider a CSTR because the application of the concepts of mixing quality are most easily illustrated using this reactor type. In the segregated flow model, we visualize the flow through the reactor to consist of a continuous series of globules (Figure 17-2).

In the segregation model, globules behave as batch reactors operated for different times.

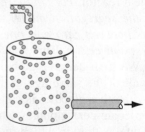

Figure 17-2 Little batch reactors (globules) inside a CSTR.

These globules retain their identity; that is, they do not interchange material with other globules in the fluid during their period of residence in the reaction environment, i.e., they remain segregated. In addition, each globule spends a different amount of time in the reactor. In essence, what we are doing is lumping all the molecules that have exactly the same residence time in the reactor into the same globule. The principles of reactor performance in the presence of completely segregated mixing were first described by Danckwerts and Zwietering.[2,3]

---

[2] P. V. Danckwerts, *Chem. Eng. Sci.*, 8, 93 (1958).

[3] T. N. Zwietering, *Chem. Eng. Sci.*, 11, 1 (1959).

Another way of looking at the segregation model for a continuous-flow system is the PFR shown in Figures 17-3(a) and (b). Because the fluid flows down the reactor in plug flow, each exit stream corresponds to a specific residence time in the reactor. Batches of molecules are removed from the reactor at different locations along the reactor in such a manner as to duplicate the RTD function, $E(t)$. The molecules removed near the entrance to the reactor correspond to those molecules having short residence times in the reactor. Physically, this effluent would correspond to the molecules that channel rapidly through the reactor. The farther the molecules travel along the reactor before being removed, the longer their residence time. The points at which the various groups or batches of molecules are removed correspond to the RTD function for the reactor.

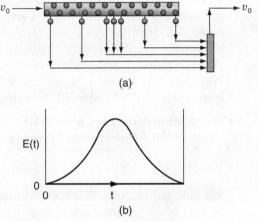

Figure 17-3   Mixing at the latest possible point.

Because there is no molecular interchange between globules, each acts essentially as its own batch reactor. The reaction time in any one of these tiny batch reactors is equal to the time that the particular globule has spent in the reaction environment after exiting. The distribution of residence times among the globules is given by the RTD of the particular reactor.

$$\text{RTD + MODEL + KINETIC DATA} \Rightarrow \begin{cases} \text{EXIT CONVERSION and} \\ \text{EXIT CONCENTRATION} \end{cases}$$

Now that we have the reactor's RTD, we will choose a model, apply the rate law and rate law parameters to predict conversion as shown in the box above. We will start with the segregation model.

To determine the mean conversion in the effluent stream, we must average the conversions of all of the various globules in the exit stream:

$$\begin{bmatrix} \text{Mean} \\ \text{conversion} \\ \text{of those globules} \\ \text{spending between} \\ \text{time } t \text{ and } t + dt \\ \text{in the reactor} \end{bmatrix} = \begin{bmatrix} \text{Conversion} \\ \text{achieved in a globule} \\ \text{after spending a time } t \\ \text{in the reactor} \end{bmatrix} \times \begin{bmatrix} \text{Fraction} \\ \text{of globules that} \\ \text{spend between } t \\ \text{and } t + dt \text{ in the} \\ \text{reactor} \end{bmatrix}$$

then

$$d\overline{X} = X(t) \times E(t)\, dt$$

$$\boxed{\dfrac{d\overline{X}}{dt} = X(t)E(t)} \tag{17-1}$$

Summing over all globules, the mean conversion is

Mean conversion for the segregation model

$$\boxed{\overline{X} = \int_0^\infty X(t)E(t)\, dt} \tag{17-2}$$

Consequently, if we have the batch reactor equation for $X(t)$ and measure the RTD experimentally, we can find the mean conversion in the exit stream. *Thus, if we have the RTD, the reaction-rate law and parameters, then for a segregated flow situation (i.e., model), we have sufficient information to calculate the conversion.* An example that may help give additional physical insight to the segregation model is given in the Chapter 17 Summary Notes on the CRE Web site (*www.umich.edu/~elements/5e/index.html*); click the (More) blue button just before Section *1A.2. Segregation Model Applied to an LFR.*

Summary Notes

Segregation Model for a First-Order Reaction
Consider the following first-order reaction:

$$A \xrightarrow{\ k\ } products$$

We treat the globules that spend different amounts of time in the real reactor as little batch reactors. For a batch reactor we have

$$-\dfrac{dN_A}{dt} = -r_A V$$

For constant volume and with $N_A = N_{A0}(1 - X)$

$$N_{A0}\dfrac{dX}{dt} = -r_A V = kC_A V = kN_A = kN_{A0}(1 - X)$$

$$\dfrac{dX}{dt} = k\,(1 - X) \tag{17-3}$$

Solving for $X(t)$, we have for any globule that spends a time t in the real reactor

$$\boxed{X(t) = 1 - e^{-kt}}$$

Because different globules spend different times, we have to add up the conversion from all the globules.

Mean conversion for a first-order reaction

$$\overline{X} = \int_0^\infty X(t)E(t)dt = \int_0^\infty (1 - e^{-kt})E(t)\, dt = \int_0^\infty E(t)\, dt - \int_0^\infty e^{-kt}E(t)\, dt$$

$$\tag{17-4}$$

$$\boxed{\overline{X} = 1 - \int_0^\infty e^{-kt} E(t)\, dt} \qquad (17\text{-}5)$$

We will now determine the mean conversion predicted by the segregation model for an ideal PFR, a CSTR, and an LFR.

Example 17–1   Mean Conversion in an Ideal PFR, an Ideal CSTR, and a Laminar-Flow Reactor

Derive the equation of a first-order reaction using the segregation model when the RTD is equivalent to **(a) an ideal PFR, (b) an ideal CSTR,** and **(c) a laminar-flow reactor (LFR)**. Compare these conversions with those obtained from the design equation.

Solution

(a) For the PFR, the RTD function was given by Equation (16-27)

$$E(t) = \delta(t - \tau) \qquad (16\text{-}27)$$

Recalling Equation (17-5)

$$\overline{X} = \int_0^\infty X(t) E(t)\, dt = 1 - \int_0^\infty e^{-kt} E(t)\, dt \qquad (17\text{-}5)$$

Substituting for the RTD function for a PFR gives

$$\overline{X} = 1 - \int_0^\infty (e^{-kt}) \delta(t - \tau)\, dt \qquad (E17\text{-}1.1)$$

Using the integral properties of the Dirac delta function, Equation (16-30), we obtain

$$\overline{X} = 1 - e^{-k\tau} = 1 - e^{-Da_1} \qquad (E17\text{-}1.2)$$

where for a first-order reaction the Damköhler number is $Da_1 = \tau k$.

Recall that for a PFR after combining the mole balance, rate law, and stoichiometric relationships (cf. Chapter 5), we had

$$\frac{dX}{d\tau} = k(1 - X) \qquad (E17\text{-}1.3)$$

Integrating yields

$$X = 1 - e^{-k\tau} = 1 - e^{-Da_1} \qquad (E17\text{-}1.4)$$

Twins!

which is identical to the conversion predicted by the segregation model $\overline{X}$.

(b) For the CSTR, the RTD function is

$$E(t) = \frac{1}{\tau}\, e^{-t/\tau} \qquad (E17\text{-}1.5)$$

Recalling Equation (17-5), the mean conversion for a first-order reaction is

$$\overline{X} = 1 - \int_0^\infty e^{-kt} E(t)\, dt \tag{17-6}$$

$$\overline{X} = 1 - \int_0^\infty \frac{e^{-(1/\tau + k)t}}{\tau}\, dt$$

$$\overline{X} = 1 + \frac{1}{k + 1/\tau} \frac{1}{\tau} e^{-(k + 1/\tau)t} \Big|_0^\infty$$

The conversion predicted from the segregation model is

$$\overline{X} = \frac{\tau k}{1 + \tau k} = \frac{Da_1}{1 + Da_1} \tag{E17-1.6}$$

In Chapter 5 we showed that combining the CSTR mole balance, the rate law, and stoichiometry, we have

$$F_{A0} X = -r_A V$$

$$v_0 C_{A0} X = k C_{A0}(1 - X)V$$

Solving for X, we see the conversion predicted from our Chapter 5 algorithm is the same as that for the segregation model.

$$X = \frac{\tau k}{1 + \tau k} \tag{E17-1.7}$$

which is identical to the conversion predicted by the segregation model $\overline{X}$.

(c) For a laminar-flow reactor, the RTD function is

$$E(t) = \begin{cases} 0 & \text{for } (t < \tau/2) \\ \dfrac{\tau^2}{2t^3} & \text{for } (t \geq \tau/2) \end{cases} \tag{16-47}$$

The dimensionless form is

$$E(\Theta) = \begin{cases} 0 & \text{for } \Theta < 0.5 \\ \dfrac{1}{2\Theta^3} & \text{for } \Theta \geq 0.5 \end{cases} \tag{16-49}$$

From Equation (17-5), we have

$$\overline{X} = 1 - \int_0^\infty e^{-kt} E(t) dt = 1 - \int_0^\infty e^{-\tau k \Theta} E(\Theta) d\Theta \tag{E17-1.8}$$

$$\overline{X} = 1 - \int_{0.5}^\infty \frac{e^{-\tau k \Theta}}{2\Theta^3} d\Theta \tag{E17-1.9}$$

Integrating twice by parts

$$\boxed{\overline{X} = 1 - (1 - 0.5\tau k) e^{-0.5 k \tau} - (0.5\tau k)^2 \int_{0.5}^\infty \frac{e^{-\tau k \Theta}}{\Theta} d\Theta} \tag{E17-1.10}$$

As expected, using the $E(t)$ for an ideal PFR and CSTR with the segregation model gives a mean conversion $\overline{X}$ identical to that obtained by using the algorithm in Chapter 4.

Another set of twins!!

The last integral is the *exponential integral* and can be evaluated from tabulated values. Fortunately, Hilder developed an approximate formula ($\tau k = Da_1$).[4]

$$\overline{X} = 1 - \frac{1}{(1 + 0.25\tau k)e^{0.5\tau k} + 0.25\tau k} \equiv 1 - \frac{1}{(1 + 0.25\ Da_1)\ e^{0.5Da_1} + 0.25\ Da_1}$$

$$\overline{X} = \frac{(4 + Da_1)e^{0.5Da_1} + Da_1 - 4}{(4 + Da_1)e^{0.5Da_1} + Da_1} \qquad \text{(E17-1.11)}$$

A comparison of the exact value along with Hilder's approximation is shown in Table E17-1.1 for various values of the Damköhler number, τk, along with the conversion in an ideal PFR and an ideal CSTR.

TABLE E17-1.1   COMPARISON OF CONVERSION IN PFR, CSTR, AND LFR
FOR DIFFERENT DAMKÖHLER NUMBERS FOR A FIRST-ORDER REACTION

$Da = \tau k$	$X_{\text{LFR Exact}}$	$X_{\text{LFR Approx.}}$	X_{PFR}	X_{CSTR}
0.1	0.0895	0.093	0.0952	0.091
1	0.557	0.56	0.632	0.501
2	0.781	0.782	0.865	0.667
4	0.940	0.937	0.982	0.80
10	0.9982	0.9981	0.9999	0.90

Where in Table E17-1.1 $X_{\text{LFR Exact}}$ = exact solution to Equation (E17-1.10) and $X_{\text{LFR Approx.}}$ = Equation (E17-1.11), in all cases, we see there is close agreement with the approximate and exact solutions.

For large values of the Damköhler number then, there is complete conversion along the streamlines off the center streamline so that the conversion is determined along the pipe axis such that

$$\overline{X} = 1 - \int_{0.5}^{\infty} 4e^{-\tau k\Theta}d\Theta = 1 - 4e^{-0.5\tau k}/\tau k \qquad \text{(E17-1.12)}$$

Figure E17-1.1 shows a comparison of the mean conversion in an LFR, PFR, and CSTR as a function of the Damköhler number for a first-order reaction.

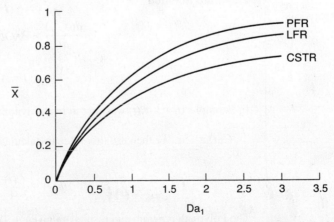

Figure E17-1.1 Conversion in a PFR, LFR, and CSTR as a function of the Damköhler number (Da_1) for a first-order reaction ($Da_1 = \tau k$).

[4] M. H. Hilder, *Trans. I. ChemE*, 59, 143 (1979).

We have just shown for a *first-order* reaction that whether you assume complete micromixing [Equation (E17-1.6)] or complete segregation [Equation (E17-1.5)] in a CSTR, the same conversion results. This phenomenon occurs because the rate of change of conversion for a first-order reaction does *not* depend on the concentration of the reacting molecules [Equation (17-14)]; it does not matter what kind of molecule is next to it or colliding with it. Thus, the extent of micromixing does not affect a first-order reaction, so the segregated flow model can be used to calculate the conversion. As a result, *only the RTD is necessary to calculate the conversion for a first-order reaction in any type of reactor* (see Problem P17-3$_c$). Knowledge of neither the degree of micromixing nor the reactor flow pattern is necessary. We now proceed to calculate conversion in a real reactor using RTD data.

Important Point:

For a first-order reaction, knowledge of $E(t)$ is sufficient.

Example 17–2   Mean Conversion, X_{seg}, Calculations in a Real Reactor

Calculate the mean conversion in the reactor we have characterized by RTD measurements in Examples 16-1 and 16-2 for a first-order, liquid-phase, irreversible reaction in a completely segregated fluid:

$$A \longrightarrow \text{products}$$

The specific reaction rate is 0.1 min^{-1} at 320 K.

Solution

Because each globule acts as a **batch reactor** of constant volume, we use the batch reactor design equation to arrive at the equation giving conversion as a function of time

$$X = 1 - e^{-kt} = 1 - e^{-0.1t} \tag{E17-2.1}$$

To calculate the mean conversion we need to evaluate the integral

$$\overline{X} = \int_0^\infty X(t)E(t)\,dt \tag{17-3}$$

The RTD function for this reactor was determined previously and given in data from Examples 16-1 and 16-2 and are repeated here in Table E17-2.1, i.e., $E(t) = C(t)/\text{Area}$.

These calculations are easily carried out with the aid of a spreadsheet such as Excel or Polymath.

Polymath Solution

$$\frac{d(X_{seg})}{dt} = X(t)E(t) \tag{E17-2.2}$$

$$\frac{dX}{dt} = k(1-X) \tag{E17-2.3}$$

In Example 16-1.1, $C(t)$ was first fit to a polynomial from which we calculated $\int_0^\infty C(t) = 51$. We then use this value to calculate the E-curve.

$$E(t) = \frac{C(t)}{51} \tag{E17-2.4}$$

The Polymath program and results are given in Table E17-2.1.

TABLE E17-2.1   SEGREGATION MODEL

POLYMATH Report
Ordinary Differential Equations

Calculated values of DEQ variables

	Variable	Initial value	Final value
1	Area	51.	51.
2	C	0.0038746	0.0148043
3	C1	0.0038746	-387.266
4	C2	-33.43818	0.0148043
5	E	7.597E-05	0.0002903
6	F	0	1.00125
7	k	0.1	0.1
8	t	0	14
9	tm	0	5.10722
10	X	0	0.753403
11	Xseg	0	0.3822834

Differential equations
1 d(X)/d(t) = k*(1-X)
2 d(Xseg)/d(t) = X*E
3 d(tm)/d(t) = t*E
4 d(F)/d(t) = E

Explicit equations
1 C1 = 0.0038746 + 0.2739782*t + 1.574621*t^2 - 0.2550041*t^3
2 Area = 51
3 k = .1
4 C2 = -33.43818 + 37.18972*t - 11.588838*t^2 + 1.695303*t^3 -
   0.1298667*t^4 + 0.005028*t^5 - 7.743*10^5*t^6
5 C = 1f(t<=4 and t>=0) then C1 else if (t>4 and t<=14) then C2 else 0
6 E = C/Area

The conversion would be 75.3% if all the globules spent the same time (i.e., 14 mins.) in the reactor (e.g., PFR). However, not all globules spend the same time. We have a distribution of times the globules spend in the reactor, so the mean conversion for all the globules is 38.2%.

Analysis: We were given the conversion as a function of time for a batch reactor $X(t)$ and the RTD E-curve from Example 16-1. Using the segregation model and a polynomial fit for the E-curve we were able to calculate the mean conversion using the segregation model, X_{seg}, in this nonideal reactor. We note there were no model-fitting parameters in making this calculation, just $E(t)$ from the data and $X(t)$.

As discussed previously, because the reaction is *first order*, the conversion calculated in Example 17-2 would be valid for a reactor with complete mixing, complete segregation, or any degree of mixing between the two. Although early or late mixing does not affect a first-order reaction, micromixing or complete segregation can give significantly different results for a second-order reaction system.

Example 17–3   Mean Conversion for a Second-Order Reaction in a Laminar-Flow Reactor

The liquid-phase reaction between cytidine and acetic anhydride

$$A + B \rightarrow C + D$$

is carried out isothermally in an inert solution of N-methyl-2-pyrrolidone (NMP) with $\Theta_{NMP} = 28.9$. The reaction follows an elementary rate law. The feed is equal molar in A and B with $C_{A0} = 0.75$ mol/dm^3, a volumetric flow rate of 0.1 dm^3/s, and a reactor volume of 100 dm^3. Calculate the conversion in (a) an ideal PFR, (b) a BR, and (c) an LFR.

Additional information:[5]

$k = 4.93 \times 10^{-3}$ dm^3/mol $\cdot$ s at 50°C with $E = 13.3$ kcal/mol, $\Delta H_{RX} = -10.5$ kcal/mol

Heat of mixing for $\Theta_{NMP} = \dfrac{F_{NMP}}{F_{A0}} = 28.9$, $\Delta H_{mix} = -0.44$ kcal/mol

Solution

The reaction will be carried out isothermally at 50°C. The space time is

$$\tau = \frac{V}{v_0} = \frac{100 \text{ dm}^3}{0.1 \text{ dm}^3/\text{s}} = 1000 \text{ s}$$

(a)   **For an ideal PFR**

Mole Balance

$$\frac{dX}{dV} = \frac{-r_A}{F_{A0}} \tag{E17-3.1}$$

Rate Law

$$-r_A = kC_A C_B \tag{E17-3.2}$$

Stoichiometry, $\Theta_B = 1$

$$C_A = C_{A0}(1 - X) \tag{E17-3.3}$$

$$C_B = C_A \tag{E17-3.4}$$

Combining

$$\frac{dX}{dV} = \frac{kC_{A0}(1 - X)^2}{v_0} \tag{E17-3.5}$$

PFR calculation

Integrating and solving with $\tau = V/v_0$ and $X = 0$ for $V = 0$ gives

$$X = \frac{\tau k C_{A0}}{1 + \tau k C_{A0}} = \frac{Da_2}{1 + Da_2} \tag{E17-3.6}$$

where Da_2 is the Damköhler number for a second-order reaction.

$$Da_2 = \tau k C_{A0} = (1000\text{s})(4.9 \times 10^{-3} \text{ dm}^3/\text{s} \cdot \text{mol})(0.75 \text{ mol/dm}^3)$$

$$= 3.7$$

$$X = \frac{3.7}{4.7}$$

$$\boxed{X = 0.787}$$

[5] J. J. Shatynski and D. Hanesian, *Ind. Eng. Chem. Res.*, 32, 594 (1993).

(b)   **Batch Reactor**

$$\frac{dX}{dt} = \frac{-r_A}{C_{A0}} \tag{E17-3.7}$$

Batch calculation

$$\frac{dX}{dt} = kC_{A0}(1-X)^2 \tag{E17-3.8}$$

$$X(t) = \frac{kC_{A0}t}{1 + kC_{A0}t} \tag{E17-3.9}$$

If the batch reaction time is the same time as the space time, i.e., $t = \tau$, the batch conversion is the same as the PFR conversion, $X = 0.787$.

(c)   **Laminar-Flow Reactor**

The differential form for the mean conversion is obtained from Equation (17-22)

$$\frac{d\overline{X}}{dt} = X(t)E(t) \tag{17-22}$$

We use Equation (E17-3.9) to substitute for $X(t)$ in Equation (17-22). Because $E(t)$ for the LFR consists of two parts, we need to incorporate the IF statement in our ODE solver program. For the laminar-flow reaction, we write

$$E_1 = 0 \text{ for } t < \tau/2 \tag{E17-3.10}$$

$$E_2 = \frac{\tau^2}{2t^3} \text{ for } t \geq \tau/2 \tag{E17-3.11}$$

Let $t_1 = \tau/2$ so that the IF statement now becomes

LFR Calculation

$$E = \text{If } (t < t_1) \text{ then } (E_1) \text{ else } (E_2) \tag{E17-3.12}$$

One other thing to remember is that the ODE solver will recognize that $E_2 = \infty$ at $t = 0$ and refuse to run, so we must add a very small number to the denominator such as (0.001); for example

$$E_2 = \frac{\tau^2}{(2t^3 + 0.001)} \tag{E17-3.13}$$

You won't be able to carry out the integration to close to $t = \infty$ unless you are a resident of Jofostan. However, you can use Polymath, but the numerical integration time limit, t_f, should be 10 or more times the reactor space time, τ. The Polymath program for this example is shown below.

Living Example Problem

POLYMATH Report
Ordinary Differential Equations

Calculated values of DEQ variables

	Variable	Initial value	Final value
1	Cao	0.75	0.75
2	E	0	6.25E-08
3	E2	5.0E+10	6.25E-08
4	k	0.00493	0.00493
5	t	0	2.0E+04
6	t1	500.	500.
7	tau	1000.	1000.
8	X	0	0.9866578
9	Xbar	0	0.7413022

Differential equations
1 d(Xbar)/d(t) = X*E

Explicit equations
1 k = .00493
2 Cao = .75
3 X = k*Cao*t/(1+k*Cao*t)
4 tau = 1000
5 t1 = tau/2
6 E2 = tau^2/2/(t^3+.00001)
7 E = if (t<t1) then (0) else (E2)

We see that the mean conversion Xbar ($\overline{X}$) for the LFR is 74.1%. In summary,

$$X_{PFR} = 0.786$$
$$X_{LFR} = 0.741$$

Compare this result with the exact analytical formula for the laminar flow reactor with a second-order reaction[6]

$$\overline{X} = Da_2[1 - (Da_2/2)\ln(1+2/Da_2)]$$

where $Da_2 = kC_{A0}\tau$. For $Da_2 = 3.70$ we get

$$\overline{X} = 0.742$$

In many cases we can approximate the conversion for an LFR with that calculated from the PFR models

Analysis: In this example we applied the segregation model to the *E*-curve for two ideal reactors: the plug flow reactor (PFR) and the laminar-flow reactor (LFR). We found that the difference between the predicted conversion in the PFR and the LFR reactor was 4.5%. We also learned that the analytical solution for a second-order reaction taking place in an LFR was virtually the same as that for the segregation model.

17.2.2 Maximum Mixedness Model

Segregation model mixing occurs at the latest possible point.

In a reactor with a segregated fluid, mixing between particles of fluid does not occur until the fluid leaves the reactor. The reactor exit is, of course, the *latest* possible point where mixing can occur, and any effect of mixing is postponed until after all reaction has taken place, as shown in Figure 17-3. We can also think of a completely segregated flow as being in a state of *minimum mixedness*. We now want to consider the other extreme, that of *maximum mixedness* consistent with a given residence time distribution.

We return again to the plug-flow reactor with side entrances, only this time the fluid enters the reactor along its length (Figure 17-4). As soon as the fluid enters the reactor, it is completely mixed radially (but not longitudinally) with the other fluid already in the reactor. The entering fluid is fed into the reactor through the side entrances in such a manner that the RTD of the plug-flow reactor with side entrances is identical to the RTD of the real reactor.

The globules at the far left of Figure 17-4(a) correspond to the molecules that spend a long time in the reactor, while those at the far right correspond to the molecules that channel through the reactor and spend a very short time in the reactor. In the reactor with side entrances, mixing occurs at the *earliest* possible moment consistent with the RTD. This situation is termed the condition of *maximum mixedness*.[7] The approach for calculating conversion for a reactor in a condition of maximum mixedness will now be developed. In a reactor with side entrances, let λ be the time it takes for the fluid to move from a particular point to the end of the reactor. In other words, λ is the life expectancy of the fluid in the reactor at that point (Figure 17-5).

---

[6] K. G. Denbigh, *J. Appl. Chem., 1,* 227 (1951).

[7] T. N. Zwietering, *Chem. Eng. Sci.,* 11, 1 (1959).

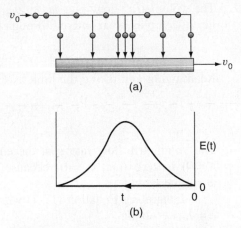

Figure 17-4   Mixing at the earliest possible point.

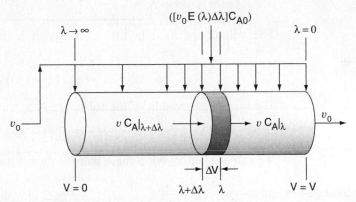

Figure 17-5   Modeling maximum mixedness by a plug-flow reactor with side entrances.

Moving down the reactor from left to right, the life expectancy, λ, decreases and becomes zero at the exit. At the left end of the reactor, λ approaches infinity or the maximum residence time if it is other than infinite.

Consider the fluid that enters the reactor through the sides of volume ΔV in Figure 17-5. The fluid that enters here will have a life expectancy between λ and $\lambda+\Delta\lambda$. The fraction of fluid that will have this life expectancy between λ and $\lambda+\Delta\lambda$ is $E(\lambda)\Delta\lambda$. The corresponding volumetric flow rate IN through the sides is $[v_0 E(\lambda)\Delta\lambda]$.

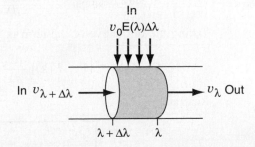

The volumetric flow rate at λ, v_λ, is the flow rate that entered at $\lambda+\Delta\lambda$, i.e., $v_{\lambda+\Delta\lambda}$, plus what entered through the sides $v_0\,E(\lambda)\Delta\lambda$, i.e.,

$$v_\lambda = v_{\lambda+\Delta\lambda} + v_0 E(\lambda)\Delta\lambda$$

Rearranging and taking the limit as $\Delta\lambda \to 0$

$$\frac{dv_\lambda}{d\lambda} = -v_0 E(\lambda) \tag{17-7}$$

The volumetric flow rate v at the entrance to the reactor ($V = 0$, $\lambda = \infty$, and $X = 0$) is zero (i.e., $v_\lambda = 0$) because the fluid only enters through the sides along the length.

Integrating Equation (17-7) with limits $v_\lambda = 0$ at $\lambda = \infty$ and $v_\lambda = v_\lambda$ at $\lambda = \lambda$, we obtain

$$\boxed{v_\lambda = v_0 \int_\lambda^\infty E(\lambda)\,d\lambda = v_0[1 - F(\lambda)]} \tag{17-8}$$

The volume of fluid in the reactor with a life expectancy between λ and $\lambda + \Delta\lambda$ is

$$\Delta V = v_0[1 - F(\lambda)]\,\Delta\lambda \tag{17-9}$$

The rate of generation of the substance A in this volume is

$$r_A\,\Delta V = r_A v_0[1 - F(\lambda)]\,\Delta\lambda \tag{17-10}$$

We can now carry out a mole balance on substance A between λ and $\lambda + \Delta\lambda$

Mole balance

$$\begin{bmatrix} \text{In} \\ \text{at } \lambda + \Delta\lambda \end{bmatrix} + \begin{bmatrix} \text{In} \\ \text{through side} \end{bmatrix} - \begin{bmatrix} \text{Out} \\ \text{at } \lambda \end{bmatrix} + \begin{bmatrix} \text{Generation} \\ \text{by reaction} \end{bmatrix} = 0$$

$$[v_{\lambda+\Delta\lambda}C_A|_{\lambda+\Delta\lambda}] + [v_0 C_{A0}E(\lambda)\Delta\lambda] - [v_\lambda C_A|_\lambda] + \quad [r_A\Delta V] \quad = 0$$

Substituting for $v_{\lambda+\Delta\lambda}$, v_λ, and ΔV

$$v_0[1 - F(\lambda)]C_A|_{\lambda+\Delta\lambda} + v_0 C_{A0}E(\lambda)\,\Delta\lambda$$
$$- v_0[1 - F(\lambda)]C_A|_\lambda + r_A v_0[1 - F(\lambda)]\,\Delta\lambda = 0 \tag{17-11}$$

Dividing Equation (17-11) by $v_0\,\Delta\lambda$ and taking the limit as $\Delta\lambda \to 0$ gives

$$E(\lambda)C_{A0} + \frac{d\{[1 - F(\lambda)]\,C_A(\lambda)\}}{d\lambda} + r_A[1 - F(\lambda)] = 0$$

Taking the derivative of the term in brackets

$$C_{A0}E(\lambda) + [1 - F(\lambda)]\frac{dC_A}{d\lambda} - C_A E(\lambda) + r_A[1 - F(\lambda)] = 0$$

Rearranging

$$\boxed{\frac{dC_A}{d\lambda} = -r_A + (C_A - C_{A0})\frac{E(\lambda)}{1 - F(\lambda)}} \tag{17-12}$$

We can also rewrite Equation (17-12) in terms of conversion as

$$-C_{A0} \frac{dX}{d\lambda} = -r_A - C_{A0}X \frac{E(\lambda)}{1 - F(\lambda)} \qquad (17\text{-}13)$$

or

$$\boxed{\frac{dX}{d\lambda} = \frac{r_A}{C_{A0}} + \frac{E(\lambda)}{1 - F(\lambda)} (X)} \qquad (17\text{-}14)$$

The boundary condition is as $\lambda \to \infty$, then $C_A = C_{A0}$ for Equation (17-12) [or $X = 0$ for Equation (17-13)]. To obtain a solution, the equation is integrated backwards numerically, starting at a very large value of λ and ending with the final conversion at $\lambda = 0$. For a given RTD and reaction orders greater than one, the maximum mixedness model gives the lower bound on conversion.

Maximum mixedness gives the lower bound on X.

Example 17–4   Conversion Bounds for a Nonideal Reactor

The liquid-phase, second-order dimerization

$$2A \longrightarrow B \qquad r_A = -kC_A^2$$

for which $k = 0.01$ dm^3/mol·min is carried out at a reaction temperature of 320 K. The feed is pure A with $C_{A0} = 8$ mol/dm^3. The reactor is nonideal. The reactor volume is 1000 dm^3, and the feed rate for our dimerization is going to be 25 dm^3/min. We have run a tracer test on this reactor, and the results are given in columns 1 and 2 of Table E17-4.1. We wish to know the bounds on the conversion for different possible degrees of micromixing for the RTD of this reactor. What are these bounds?

Tracer test on tank reactor: $N_0 = 100$ g, $v = 25$ dm^3/min.

We will first fit a polynomial to the C-curve. A tutorial on how to fit the tracer date points to a polynomial

$$\text{e.g., } C(t) = a_0 + a_1t + a_2t^2 + a_3t^3 + a_4t^4$$

Living Example Problem

which is given in the *Living Example Problem* on the CRE Web site in both Chapters 7 and 16. After finding a_0, a_1, etc., we integrate the C_t-curve to find the total amount of tracer, N_0, injected

$$N_0 = \int_0^\infty C(t)dt = \int_0^{14} C(t)dt$$

and then divide each concentration by N_o, the total tracer concentration to construct the E-curve

```
E2 = -2.64e-9*t^3+1.3618e-6*t^2-.00024069*t+.015011
E1 = 4.44658e-10*t^4-1.1802e-7*t^3+1.35358e-5*t^2-.000865652*t+.028004
E = if (t<=70) then (E1) else (E2)
```

Solution

The bounds on the conversion are found by calculating conversions under conditions of (a) complete segregation and (b) maximum mixedness.

(a) *Conversion if fluid is completely segregated*. The batch reactor equation for a second-order reaction of this type is

$$X(t) = \frac{kC_{A0}t}{1 + kC_{A0}t} \qquad (E17\text{-}4.1)$$

TABLE E17-4.1   RAW AND PROCESSED DATA

t (min)	$X(t)$	$X(t)E(t)$ (min^{-1})	$X(t)E(t)\,\Delta t$	λ(min)
0	0	0	0	0
5	0.286	0.00686	0.0172†	5
10	0.444	0.00916	0.0400	10
15	0.545	0.00965	0.0470	15
20	0.615	0.00935	0.0475	20
30	0.706	0.00805	0.0870	30
40	0.762	0.00658	0.0732	40
50	0.800	0.00526	0.0592	50
70	0.848	0.00333	0.0859	70
100	0.889	0.00171	0.0756	100
150	0.923	0.000589	0.0575	150
200	0.941	0.000212	0.0200	200
			0.610	

† For the first point, we have $X(t)E(t)\Delta t = (0 + 0.00686)(5/2) = 0.0172$.

The conversion for a completely segregated fluid in a reactor is

Spreadsheets work quite well here.

$$X_{\text{seg}} = \int_0^{\infty} X(t)E(t)\,dt \tag{E17-4.2}$$

differentiating

$$\frac{dX_{\text{seg}}}{dt} = X(t)E(t) \tag{E17-4.3}$$

Combining the mole balance, rate law, and stoichiometry for a "little" batch reactor, we obtain the differential equation of $X(t)$.

$$\frac{dX}{dt} = k(1 - X) \tag{E17-4.4}$$

We now use Polymath to solve Equations (E17-4.3) and (17-4.4) simultaneously to find X_{seg}.

The Polymath program and results are given in Table E17-4.2.

TABLE E17-4.2   SEGREGATION MODEL

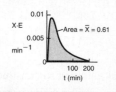

POLYMATH Report
Ordinary Differential Equations

Calculated values of DEQ variables

	Variable	Initial value	Final value
1	Ca	8.	0.4705882
2	Cao	8.	8.
3	E	0.028004	0.000225
4	E1	0.028004	0.1635984
5	E2	0.015011	0.000225
6	k	0.01	0.01
7	ra	-0.64	-0.0022145
8	t	0	200.
9	X	0	0.9411765
10	Xseg	0	0.6054227

Differential equations
1 d(Xseg)/d(t) = X*E
2 d(X)/d(t) = -ra/Cao

Explicit equations
1 Cao = 8
2 Ca = Cao*(1-X)
3 k = .01
4 E2 = -2.64e-9*t^3+1.3618e-6*t^2-.00024069*t+.015011
   #
5 E1 = 4.44658e-10*t^4-1.1802e-7*t^3+1.35358e-5*t^2
   -.000865652*t+.028004
6 E = if (t<=70) then (E1) else (E2)
7 ra = -k*Ca^2

The predicted conversion for a completely segregated flow is 0.605 or 61%.

$$\boxed{X_{\text{seg}} = 0.61}$$

(b) Conversion for maximum mixedness model

Hand Calculation: In practice, we would not carry out step-by-step calculations to predict the conversion from the maximum mixedness model. It is presented here in the hopes that it will give a clearer understanding of maximum mixedness. As we will see in Example 17-5, the Polymath ODE solver is more proficient and unbelievably fast.

    Conversion for maximum mixedness. The Euler method will be used for numerical integration

$$X_{i+1} = X_i + (\Delta\lambda)\left[\frac{E(\lambda_i)}{1-F(\lambda_i)}X_i - kC_{A0}(1-X_i)^2\right]$$

Tedious calculations

Integrating this equation presents some interesting results. If the equation is integrated from the exit side of the reactor, starting with $\lambda = 0$, the solution is unstable and soon approaches large negative or positive values, depending on what the starting value of X is. We want to find the conversion at the exit to the reactor, $\lambda = 0$. Consequently, we need to integrate backwards.

$$X_{i-1} = X_i - \Delta\lambda\left[\frac{E(\lambda_i)X_i}{1-F(\lambda_i)} - kC_{A0}(1-X_i)^2\right]$$

If integrated from the point where $\lambda \longrightarrow \infty$, oscillations may occur but are soon damped out, and the equation approaches the same final value no matter what initial value of X between 0 and 1 is used. We shall start the integration at $\lambda = 200$ and let $X = 0$ at this point. If we set $\Delta\lambda$ too large, the solution will blow up, so we will start out with $\Delta\lambda = 25$ and use the average of the measured values of $E(t)/[(1 - F(t)]$ where necessary. We will now use the data in column 5 of Table E17-4.1 to carry out the integration.

At $\lambda = 200$, $X = 0$

$\lambda = 175$:

$$X(\lambda = 175) = X(\lambda = 200) - \Delta\lambda\left[\frac{E(200)X(200)}{1-F(200)} - kC_{A0}(1-X(200))^2\right]$$

$$X = 0 - (25)[(0.075)(0) - ((0.01)(8)(1)^2)] = 2$$

$\lambda = 150$:

$$X(\lambda = 150) = X(\lambda = 175) - \Delta\lambda\left[\frac{E(175)X(175)}{1-F(175)} - kC_{A0}(1-X(175))^2\right]$$

We need to take an average of $E / (1 - F)$ between $\lambda = 200$ and $\lambda = 150$.

$$X(\lambda = 150) = 2 - (25)\left[\left(\frac{0.075 + 0.0266}{2}\right)(2) - (0.01)(8)(1-2)^2\right] = 1.46$$

$\lambda = 125$:

$$X(\lambda = 125) = 1.46 - (25)[(0.0266)(1.46) - (0.01)(8)(1 - 1.46)^2] = 0.912$$

$\lambda = 100$:

$$X\,(\lambda = 100) = 0.912 - (25)\left[\left(\frac{0.0266 + 0.0221}{2}\right)(0.912) - (0.01)(8)(1 - 0.912)^2\right]$$

$$= 0.372$$

$\lambda = 70$:

$$X = 0.372 - (30)[(0.0221)(0.372) - (0.01)(8)(1 - 0.372)^2] = 1.071$$

$\lambda = 50$:

$$X = 1.071 - (20)[(0.0226)(1.071) - (0.01)(8)(1 - 1.071)^2] = 0.595$$

$\lambda = 40$:

$$X = 0.595 - (10)[(0.0237)(0.595) - (0.01)(8)(1 - 0.595)^2] = 0.585$$

Note: Oscillations in X are beginning to be dumped out.

Running down the values of X along the right-hand side of the preceding equation shows that the oscillations have now damped out. Carrying out the remaining calculations down to the end of the reactor completes Table E17-4.3. The conversion for a condition of maximum mixedness in this reactor is 0.56 or 56%. It is interesting to note that there is little difference in the conversions for the two conditions of complete segregation (61%) and maximum mixedness (56%). With bounds this narrow, one may question the point in using additional models for the reactor to improve the predictability of conversion.

TABLE E17-4.3   MAXIMUM MIXEDNESS MODEL

λ (min)	X
200	0.0
175	2.0
150	1.46
125	0.912
100	0.372
70	1.071
50	0.595
40	0.585
30	0.580
20	0.581
10	0.576
5	0.567
0	0.564

Calculate backwards to reactor exit.

Analysis: For comparison, it is left for the reader to show that the conversion for a PFR of this size would be 0.76, and the conversion in a perfectly mixed CSTR with complete micromixing would be 0.58. As mentioned in this example, you probably will never use this kind of *hand* calculation method to determine the maximum mixedness conversion. It is only presented to help give an intuitive understanding as one integrates the maximum mixedness model backwards to the reactor entrance. Instead, an ODE solver such as Polymath is preferred. In Section 17.3 we will show how to solve maximum mixedness problems numerically using Polymath software.

Summary	
PFR	76%
Segregation	61%
CSTR	58%
Max. mix	56%

17.3 Using Software Packages

The first thing we do when using a software package to solve the ODEs to find the conversion of the exit concentration is to fit the tracer concentration measurements to a polynomial $C(t) = a_0 + a_1t + a_2t^2 + a_3t^3 + a_4t^4$ or some other function to obtain $C(t)$ from the data. A tutorial on how to obtain an analytical expression is given in the *Living Example Problem* in both Chapters 7 and 16 on the CRE Web site.

Maximum Mixedness Model

Because most software packages won't integrate backwards, we need to change the variable such that the integration proceeds forward as λ decreases from some large value to zero. We do this by forming a new variable, z, which is the difference between the longest time measured in the $E(t)$ curve, $\overline{T}$, and λ. In the case of Example 17-4, the longest time at which the tracer concentration was measured was 200 minutes (Table E17-4.1). Therefore, we will set $\overline{T} = 200$.

$$z = \overline{T} - \lambda = 200 - \lambda$$

$$\lambda = \overline{T} - z = 200 - z \qquad (17\text{-}15)$$

Realizing

$$\frac{dX}{d\lambda} = -\frac{dX}{dz} \qquad (17\text{-}16)$$

Substituting for λ in Equation (17-14)

$$\boxed{\frac{dX}{d\lambda} = -\frac{r_A}{C_{A0}} + \frac{E(\lambda)}{1 - F(\lambda)}\,(X)} \qquad (17\text{-}14)$$

and rearranging

$$\frac{dX}{dz} = -\frac{r_A}{C_{A0}} - \frac{E(\overline{T} - z)}{1 - F(\overline{T} - z)}\,X \qquad (17\text{-}17)$$

One now integrates between the limit $z = 0$ and $z = 200$ to find the exit conversion at $z = 200$, which corresponds to $\lambda = 0$.

In fitting $E(t)$ to a polynomial, one has to make sure that the polynomial does not become negative at large times. Another concern in the maximum mixedness calculations is that the term $(1 - F(\lambda))$ does not go to zero. Setting the maximum value of $F(t)$ at 0.999 rather than 1.0 will eliminate this problem. It can also be circumvented by integrating the polynomial for $E(t)$ to get $F(t)$ and then setting the maximum value of $F(t)$ at 0.999. If $F(t)$ is ever greater than one when fitting a polynomial, the solution will blow up when integrating Equation (17-17) numerically.

Example 17–5   Using Software to Make Maximum Mixedness Model Calculations

Use an ODE solver to determine the conversion predicted by the **maximum mixedness model** for the $E(t)$ curve given in Example E17-4.

Solution

Because of the nature of the $E(t)$ curve, it is necessary to use two polynomials, a third order and a fourth order, each for a different part of the curve to express the RTD, $E(t)$, as a function of time. The resulting $E(t)$ curve is shown in Figure E17-5.1.

First, we fit $E(t)$.

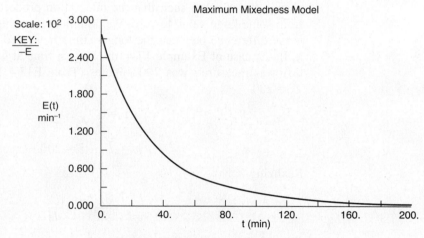

Figure E17-5.1   Polynomial fit of $E(t)$.

To use Polymath to carry out the integration, we change our variable from λ to z using the largest time measurement that was taken from $E(t)$ in Table E17-4.1, which is 200 min:

$$z = 200 - \lambda$$

The equations to be solved are

$$\lambda = 200 - z \tag{E17-5.1}$$

Maximum
mixedness model

$$\frac{dX}{dz} = -\frac{r_A}{C_{A0}} - \frac{E(200-z)}{1-F(200-z)} X \tag{E17-5.2}$$

For values of λ less than 70, we use the polynomial

$$\boxed{E_1(\lambda) = 4.447 \times 10^{-10}\lambda^4 - 1.18 \times 10^{-7}\lambda^3 + 1.353 \times 10^{-5}\lambda^2 - 8.657 \times 10^{-4}\lambda + 0.028}$$

$$\tag{E17-5.3}$$

For values of λ greater than 70, we use the polynomial

$$\boxed{E_2(\lambda) = -2.640 \times 10^{-9}\lambda^3 + 1.3618 \times 10^{-6}\lambda^2 - 2.407 \times 10^{-4}\lambda + 0.015}$$

$$\tag{E17-5.4}$$

$$\frac{dF}{d\lambda} = E(\lambda) \tag{E17-5.5}$$

CAUTION

with $z = 0$ ($\lambda = 200$), $X = 0$, and $F = 1$ [i.e., $F(\lambda) = 0.999$]. **Caution:** Because $[1 - F(\lambda)]^{-1}$ tends to infinity at $F = 1$, ($z = 0$), we set the maximum value of F at 0.999 at $z = 0$.

TABLE E17-5.1   POLYMATH PROGRAM FOR MAXIMUM MIXEDNESS MODEL

POLYMATH Report
Ordinary Differential Equations

Calculated values of DEQ variables

	Variable	Initial value	Final value
1	Ca	8.	3.493809
2	cao	8.	8.
3	E	0.000225	0.028004
4	E1	0.1635984	0.028004
5	E2	0.000225	0.015011
6	EF	0.075005	0.028004
7	F	0.9970002	0
8	F1	5.633339	0
9	F2	0.9970002	0.381769
10	k	0.01	0.01
11	lam	200.	0
12	ra	-0.64	-0.122067
13	x	0	0.5632738
14	z	0	200.

Differential equations
1 d(x)/d(z) = -(ra/cao+E/(1-F)*x)

Explicit equations
1 cao = 8
2 k = .01
3 lam = 200-z
4 ca = cao*(1-x)
5 E1 = 4.44658e-10*lam^4-1.1802e-7*lam^3+1.35358e-5*lam^2 -.000865652*lam+.028004
6 E2 = -2.64e-9*lam^3+1.3618e-6*lam^2-.00024069*lam +.015011
7 F1 = 4.44658e-10/5*lam^5-1.1802e-7/4*lam^4+1.35358e -5/3*lam^3-.000865652/2*lam^2+.028004
8 F2 = -(-9.30769e-8*lam^3+5.02846e-5*lam^2-.00941*lam+ .618231-1)
9 ra = -k*ca^2
10 E = if (lam<=70) then (E1) else (E2)
11 F = if (lam<=70) then (F1) else (F2)
12 EF = E/(1-F)

Living Example Problem

Polynomials used to fit $E(t)$ and $F(t)$

The Polymath equations are shown in Table E17-5.1. The solution is

$$\text{at } z = 200 \qquad X = 0.563$$

The conversion predicted by the maximum mixedness model is 56.3%, $X_{mm} = 0.56$ while the conversion predicted from complete segregation was $X_{seg} = 0.61$.

Analysis: As expected, the conversion X_{mm} calculated using Polymath or another software package is virtually the same as the hand calculation, but somewhat easier. The most difficult part is to fit the E-curve and the F-curve to the polynomials and then to make sure that

$$\int_0^\infty E(t)\, dt = 1$$

for the polynomial parameters chosen.

Summary
$X_{PFR} = 0.76$
$X_{seg} = 0.61$
$X_{CSTR} = 0.58$
$X_{mm} = 0.56$

17.3.1 Comparing Segregation and Maximum Mixedness Predictions

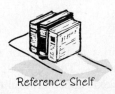

Reference Shelf

In the previous example, we saw that the conversion predicted by the segregation model, X_{seg}, was greater than that by the maximum mixedness model X_{mm}. Will this always be the case? No. To learn the answer, we take the second derivative of the rate law as shown in the *Professional Reference Shelf R17.1* on the CRE Web site.

Comparing X_{seg} and X_{mm}

If $\dfrac{\partial^2(-r_A)}{\partial C_A^2} > 0$ then $X_{seg} > X_{mm}$ (17-18)

If $\dfrac{\partial^2(-r_A)}{\partial C_A^2} < 0$ then $X_{mm} > X_{seg}$ (17-19)

If $\dfrac{\partial^2(-r_A)}{\partial C_A^2} = 0$ then $X_{mm} = X_{seg}$ (17-20)

For example, if the rate law is a power law model, then

$$-r_A = kC_A^n$$

$$\frac{\partial(-r_A)}{\partial C_A} = nkC_A^{n-1}$$

$$\frac{\partial^2(-r_A)}{\partial C_A^2} = n(n-1)kC_A^{n-2}$$

From the product $[(n)(n-1)]$, we see

If $n > 1$,   then $\dfrac{\partial^2(-r_A)}{\partial C_A^2} > 0$ and $X_{seg} > X_{mm}$

If $n < 0$,   then $\dfrac{\partial^2(-r_A)}{\partial C_A^2} > 0$ and $X_{seg} > X_{mm}$

If $0 < n < 1$,   then $\dfrac{\partial^2(-r_A)}{\partial C_A^2} < 0$ and $X_{mm} > X_{seg}$

Important point

We note that in some cases X_{seg} is not too different from X_{mm}. However, when one is considering the destruction of toxic waste where $X > 0.99$ is desired, then even a small difference is significant!!

In this section we have addressed the case where all we have is the RTD and no other knowledge about the flow pattern exists. Perhaps the flow pattern cannot be assumed because of a lack of information or other possible causes. Perhaps we wish to know the extent of possible error from assuming an incorrect flow pattern. We have shown how to obtain the conversion, using only the RTD, for two limiting mixing situations: the earliest possible mixing consistent with the RTD, or maximum mixedness, and mixing only at the reactor exit, or complete segregation. Calculating conversions for these two cases gives bounds on the conversions that might be expected for different flow paths consistent with the observed RTD.

17.4 RTD and Multiple Reactions

As discussed in Chapter 8, when multiple reactions occur in reacting systems, it is best to work in concentrations, moles, or molar flow rates rather than conversion.

17.4.1 Segregation Model

In the **segregation model** we consider each of the globules in the reactor to have different concentrations of reactants, C_A, and products, C_P. These globules are mixed together immediately upon exiting to yield the exit concentration of A, $\overline{C_A}$, which is the average of all the globules exiting

$$\overline{C_A} = \int_0^\infty C_A(t)E(t)\, dt \tag{17-21}$$

$$\overline{C_B} = \int_0^\infty C_B(t)E(t)\, dt \tag{17-22}$$

The concentrations of the individual species, $C_A(t)$ and $C_B(t)$, in the different globules are determined from batch reactor calculations. For a constant-volume batch reactor, where q reactions are taking place, the coupled mole balance equations are

$$\frac{dC_A}{dt} = r_A = \sum_{i=1}^{i=q} r_{iA} \tag{17-23}$$

$$\frac{dC_B}{dt} = r_B = \sum_{i=1}^{i=q} r_{iB} \tag{17-24}$$

These equations are solved simultaneously with

$$\frac{d\overline{C_A}}{dt} = C_A(t)E(t) \tag{17-25}$$

$$\frac{d\overline{C_B}}{dt} = C_B(t)E(t) \tag{17-26}$$

to give the exit concentration. The RTDs, $E(t)$, in Equations (17-25) and (17-26) are determined from experimental measurements and then fit to a polynomials.

17.4.2 Maximum Mixedness

For the **maximum mixedness model,** we write Equation (17-12) for each species and replace r_A by the net rate of formation

$$\frac{dC_A}{d\lambda} = -\sum r_{iA} + (C_A - C_{A0})\frac{E(\lambda)}{1 - F(\lambda)} \tag{17-27}$$

$$\frac{dC_B}{d\lambda} = -\sum r_{iB} + (C_B - C_{B0})\frac{E(\lambda)}{1 - F(\lambda)} \tag{17-28}$$

After substitution for the rate laws for each reaction (e.g., $r_{1A} = k_1 C_A$), these equations are solved numerically by starting at a very large value of λ, say $\overline{T} = 200$, and integrating backwards to $\lambda = 0$ to yield the exit concentrations C_A, C_B, ...

We will now show how different RTDs with the *same* mean residence time can produce different product distributions for multiple reactions.

Example 17–6 RTD and Complex Reactions

Consider the following set of liquid-phase reactions

$$A + B \xrightarrow{k_1} C$$

$$A \xrightarrow{k_2} D$$

$$B + D \xrightarrow{k_3} E$$

which are occurring in two different reactors with the same mean residence time, $t_m = 1.26$ min. However, the RTD is very different for each of the reactors, as can be seen in Figures E17-6.1 and E17-6.2.

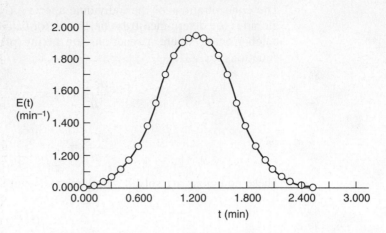

Figure E17-6.1 $E_1(t)$: asymmetric distribution.

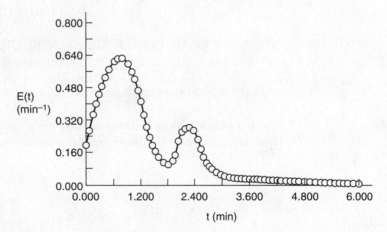

Figure E17-6.2 $E_2(t)$: bimodal distribution.

(a) Fit polynomials to the RTDs.
(b) Determine the product distribution and selectivity (e.g., $\hat{S}_{C/D}$, $\hat{S}_{D/E}$) for
   1. The segregation model.
   2. The maximum mixedness model.

Before carrying out any calculations, what do you think the exit concentrations and conversion will be for these two very different RTDs with the *same* mean residence time?

Additional information:

   $k_1 = k_2 = k_3 = 1$ in appropriate units at 350 K.

Solution

Segregation Model

Combining the mole balance and rate laws for a constant-volume batch reactor (i.e., globules), we have

$$\frac{dC_A}{dt} = r_A = r_{1A} + r_{2A} = -k_1 C_A C_B - k_2 C_A \tag{E17-6.1}$$

$$\frac{dC_B}{dt} = r_B = r_{1B} + r_{3B} = -k_1 C_A C_B - k_3 C_B C_D \qquad \text{(E17-6.2)}$$

$$\frac{dC_C}{dt} = r_C = r_{1C} = k_1 C_A C_B \qquad \text{(E17-6.3)}$$

$$\frac{dC_D}{dt} = r_D = r_{2D} + r_{3D} = k_2 C_A - k_3 C_B C_D \qquad \text{(E17-6.4)}$$

$$\frac{dC_E}{dt} = r_E = r_{3E} = k_3 C_B C_D \qquad \text{(E17-6.5)}$$

and the concentration for each species exiting the reactor is found by integrating the equation

$$\frac{d\overline{C_i}}{dt} = C_i E(t) \qquad \text{(E17-6.6)}$$

over the life of the $E(t)$ curve. For this example, the life of the $E_1(t)$ is 2.42 minutes (Figure E17-6.1), and the life of $E_2(t)$ is 6 minutes (Figure E17-6.2).

The initial conditions are $t = 0$, $C_A = C_B = 1$, and $C_C = C_D = C_E = 0$.

The Polymath program used to solve these equations is shown in Table E17-6.1 for the asymmetric RTD, $E(t)$.

With the exception of the polynomial for $E(t)$, an identical program to that in Table E17-6.1 for the bimodal distribution is given on the CRE Web site LEP 6. A comparison of the exit concentration and selectivities of the two RTD curves is shown in Table E17-6.2.

Living Example Problem

TABLE E17-6.1   POLYMATH PROGRAM
FOR SEGREGATION MODEL WITH ASYMMETRIC RTD (MULTIPLE REACTIONS)

ODE Report (RKF45)

Differential equations as entered by the user
```
[1]  d(ca)/d(t) = ra
[2]  d(cb)/d(t) = rb
[3]  d(cc)/d(t) = rc
[4]  d(cabar)/d(t) = ca*E
[5]  d(cbbar)/d(t) = cb*E
[6]  d(ccbar)/d(t) = cc*E
[7]  d(cd)/d(t) = rd
[8]  d(ce)/d(t) = re
[9]  d(cdbar)/d(t) = cd*E
[10] d(cebar)/d(t) = ce*E
```

Explicit equations as entered by the user
```
[1]  k1 = 1
[2]  k2 = 1
[3]  k3 = 1
[4]  E1 = -2.104*t^4+4.167*t^3-1.596*t^2+0.353*t-0.004
[5]  E2 = -2.104*t^4+17.037*t^3-50.247*t^2+62.964*t-27.402
[6]  rc = k1*ca*cb
[7]  re = k3*cb*cd
[8]  ra = -k1*ca*cb-k2*ca
[9]  rb = -k1*ca*cb-k3*cb*cd
[10] E = if(t<=1.26)then(E1)else(E2)
[11] rd = k2*ca-k3*cb*cd
```

TABLE E17-6.2   SEGREGATION MODEL RESULTS

Asymmetric Distribution		*Bimodal Distribution*	
The solution for $E_1(t)$ is:		The solution for $E_2(t)$ is:	
$\overline{C_A} = 0.151$	$\overline{C_E} = 0.178$	$\overline{C_A} = 0.245$	$\overline{C_E} = 0.162$
$\overline{C_B} = 0.454$	$\overline{X} = 84.9\%$	$\overline{C_B} = 0.510$	$\overline{X} = 75.5\%$
$\overline{C_C} = 0.357$	$S_{C/D} = 1.18$	$\overline{C_C} = 0.321$	$\tilde{S}_{C/D} = 1.21$
$\overline{C_D} = 0.303$	$S_{D/E} = 1.70$	$\overline{C_D} = 0.265$	$\tilde{S}_{D/E} = 1.63$

Analysis: We note that while the conversion and the exit concentration of species A are significantly different for the two distributions, the selectivities are not. In problem P17-6$_B$ (b) you are asked to calculate the mean residence time for each distribution to try to explain these differences.

Maximum Mixedness Model

The equations for each species are

$$\frac{dC_A}{d\lambda} = k_1 C_A C_B + k_2 C_A + (C_A - C_{A0}) \frac{E(\lambda)}{1 - F(\lambda)} \tag{E17-6.7}$$

$$\frac{dC_B}{d\lambda} = k_1 C_A C_B + k_3 C_B C_D + (C_B - C_{B0}) \frac{E(\lambda)}{1 - F(\lambda)} \tag{E17-6.8}$$

$$\frac{dC_C}{d\lambda} = -k_1 C_A C_B + (C_C - C_{C0}) \frac{E(\lambda)}{1 - F(\lambda)} \tag{E17-6.9}$$

$$\frac{dC_D}{d\lambda} = -k_2 C_A + k_3 C_B C_D + (C_D - C_{D0}) \frac{E(\lambda)}{1 - F(\lambda)} \tag{E17-6.10}$$

$$\frac{dC_E}{d\lambda} = -k_3 C_B C_D + (C_E - C_{E0}) \frac{E(\lambda)}{1 - F(\lambda)} \tag{E17-6.11}$$

The Polymath program for the bimodal distribution, $E(t)$, is shown in Table E17-6.3. The Polymath program for the asymmetric distribution is identical, with the exception of the polynomial fit for $E_1(t)$ and is given in the Chapter 17 *Living Example Problems, LEP17-6a and LEP17-6b,* on the CRE Web site. A comparison of the exit concentration and selectivities of the two RTD distributions is shown in Table E17-6.4.

Solved Problems

Living Example Problem

TABLE E17-6.3   POLYMATH PROGRAM
FOR MAXIMUM MIXEDNESS MODEL WITH BIMODAL DISTRIBUTION (MULTIPLE REACTIONS)

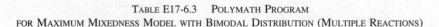

ODE Report (RKF45)

Differential equations as entered by the user
[1] d(ca)/d(z) = -(-ra+(ca-cao)*EF)
[2] d(cb)/d(z) = -(-rb+(cb-cbo)*EF)
[3] d(cc)/d(z) = -(-rc+(cc-cco)*EF)
[4] d(F)/d(z) = -E
[5] d(cd)/d(z) = -(-rd+(cd-cdo)*EF)
[6] d(ce)/d(z) = -(-re+(ce-ceo)*EF)

Explicit equations as entered by the user
[1] cbo = 1
[2] cao = 1
[3] cco = 0
[4] cdo = 0
[5] ceo = 0
[6] lam = 6-z
[7] k2 = 1
[8] k1 = 1
[9] k3 = 1
[10] rc = k1*ca*cb
[11] re = k3*cb*cd
[12] E1 = 0.47219*lam^4-1.30733*lam^3+0.31723*lam^2+0.85688*lam+0.20909
[13] E2 = 3.83999*lam^6-58.16185*lam^5+366.2097*lam^4-1224.66963*lam^3+2289.84857*lam^2-2265.62125*lam+925.46463
[14] E3 = 0.00410*lam^4-0.07593*lam^3+0.52276*lam^2-1.59457*lam+1.84445
[15] rb = -k1*ca*cb-k3*cb*cd
[16] ra = -k1*ca*cb-k2*ca
[17] rd = k2*ca-k3*cb*cd
[18] E = if(lam<=1.82)then(E1)else(if(lam<=2.8)then(E2)else(E3))
[19] EF = E/(1-F)

TABLE E17-6.4   MAXIMUM MIXEDNESS MODEL RESULTS

| Asymmetric Distribution | | Bimodal Distribution | |
The solution for $E_1(t)$ (1) is:		The solution for $E_2(t)$ (2) is:	
$C_A = 0.161$	$C_E = 0.192$	$C_A = 0.266$	$C_E = 0.190$
$C_B = 0.467$	$\overline{X} = 83.9\%$	$C_B = 0.535$	$\overline{X} = 73.4\%$
$C_C = 0.341$	$\tilde{S}_{C/D} = 1.11$	$C_C = 0.275$	$S_{C/D} = 1.02$
$C_D = 0.306$	$\tilde{S}_{D/E} = 1.59$	$C_D = 0.269$	$S_{D/E} = 1.41$

Analysis: In this example we have applied the segregation model and the maximum mixedness models to complex reactions. While the concentrations of species A exiting the reactors for the two distributions are different, the selectivities are not so different.

Calculations similar to those in Example 17-6 are given in an example on the CRE Web site for the series reaction

$$A \xrightarrow{k_1} B \xrightarrow{k_2} C$$

Solved Problems

Living Example CD17-RTD (**LEP**), parts (a) through (h), explores the above series reaction and also multiple reactors with different residence time distributions (e.g., asymmetric, bimodal).

Closure. In this chapter we showed how the reader can use the RTD data along with reaction kinetics to predict the conversion and exit concentrations. We first choose the segregation and the maximum mixedness models, which do not use any adjustable parameters to predict conversion for single reactions and exit concentrations for multiple reactions. The reader can use the $E(t)$ determined from experimental data or use the $E(t)$ function for ideal CSTRs, PFRs, LFRs, or combinations thereof, along with the rate law to predict the conversion. The reader will be able to use software packages to fit experimental RTD to polynomials and use the polynomials to represent $E(t)$ and to predict conversion. By analyzing the second derivative of the reaction rate with respect to concentration, the reader will be able to determine whether the segregation model or maximum mixedness model will give the greater conversion.

SUMMARY

1. The RTD functions for an ideal reactor are

PFR: $E(t) = \delta(t - \tau)$ (S17-1)

CSTR: $E(t) = \dfrac{e^{-t/\tau}}{\tau}$ (S17-2)

LFR: $E(t) = 0 \qquad t < \dfrac{\tau}{2}$ (S17-3)

$E(t) = \dfrac{\tau^2}{2t^3} \qquad t \geq \dfrac{\tau}{2}$ (S17-4)

2. The dimensionless residence time is

$$\Theta = \frac{t}{\tau} \tag{S17-5}$$

and is equal to the number space times. Then

$$E(\Theta) = \tau E(t) \tag{S17-6}$$

3. The internal-age distribution, $[I(\alpha)\,d\alpha]$, gives the fraction of material inside the reactor that has been inside between a time α and a time $(\alpha + d\alpha)$.

4. Segregation model: The conversion is

$$\overline{X} = \int_0^\infty X(t)E(t)\,dt \tag{S17-7}$$

and for multiple reactions

$$\overline{C}_A = \int_0^\infty C_A(t)E(t)\,dt$$

5. Maximum mixedness: Conversion can be calculated by solving the following equations

$$\frac{dX}{d\lambda} = \frac{r_A}{C_{A0}} + \frac{E(\lambda)}{1 - F(\lambda)}(X) \tag{S17-8}$$

and for multiple reactions

$$\frac{dC_A}{d\lambda} = -r_{A_{net}} + (C_A - C_{A0})\frac{E(\lambda)}{1 - F(\lambda)} \tag{S17-9}$$

$$\frac{dC_B}{d\lambda} = -r_{B_{net}} + (C_B - C_{B0})\frac{E(\lambda)}{1 - F(\lambda)} \tag{S17-10}$$

from $\lambda = \lambda_{max}$ to $\lambda = 0$. To use an ODE solver, let $z = \lambda_{max} - \lambda$.

CRE WEB SITE MATERIALS

- **Expanded Material on the Web Site**
    1. *The Intensity Function, $\Lambda(t)$*
    2. *Heat Effects*
    3. *Example LEP 17-4 and Problem 17-5 with Heat Effects*
    4. *Problem 17-16$_B$*
- **Learning Resources**
    1. *Summary Notes*
    2. *Solved Problems*
        A. Example Web17-1 Calculate the exit concentrations for the series reaction

$$A \longrightarrow B \longrightarrow C$$

        B. Example Web17-2 Determination of the effect of variance on the exit concentrations for the series reaction

$$A \longrightarrow B \longrightarrow C$$

- **Living Example Problems**
    1. *Living Example 17-2 Mean Conversion, X_{seg}, in a Real Reactor*
    2. *Living Example 17-3 Second-Order Reaction in a PFR*
    3. *Living Example 17-4 Conversion Boundary, X_{seg} and X_{mm}, for a Nonideal Reactor*
    4. *Living Example 17-5 Using Software to Make Maximum Mixedness Model Calculations*
    5. *Living Example 17-6 (a) RTD and Complex Reactions (a) Segregation Model with Asymmetric E(t)*
    6. *Living Example 17-6 (b) RTD and Complex Reactions (b) Maximum Mixedness with a Bimodal E(t)*
    7. *Living Example Web17-1 (a) RTD Calculations for Reactions in a Series in a PFR*
    8. *Living Example Web17-1 (b) RTD Calculations for Reactions in a Series in a CSTR*
    9. *Living Example Web17-1 (c) RTD Calculations for a Series Reaction Segregation Model with Asymmetric RTD*
    10. *Living Example Web17-1 (d) RTD Calculations for a Series Reaction Segregation Model with Bimodal Distribution*
    11. *Living Example Web17-1 (e) RTD Calculations for a Series Reaction Maximum Mixedness Model with Asymmetric RTD*
    12. *Living Example Web17-1 (f) RTD Calculations for a Series Reaction Maximum Mixedness Model with Bimodal Distribution*
    13. *Living Example Web17-1 (g) RTD Calculations for a Series Reaction Segregation Model with Bimodal Distribution (Multiple Reactions)*
    14. *Living Example Web17-1 (h) RTD Calculations for a Series Reaction Maximum Mixedness Model with Asymmetric RTD (Multiple Reactions)*
- **Professional Reference Shelf**

    R17.1. *Comparing X_{seg} with X_{mm}*

    The derivation of equations using the second derivative criteria

    $$\frac{\partial^2(-r_A)}{\partial C_A^2} = ?$$

    is carried out.

QUESTIONS AND PROBLEMS

The subscript to each of the problem numbers indicates the level of difficulty: A, least difficult; D, most difficult.

$$A = \bullet \quad B = \blacksquare \quad C = \blacklozenge \quad D = \blacklozenge\blacklozenge$$

Homework Problems

Questions

Q17-1$_A$ Read over the problems of this chapter. Make up an original problem that uses the concepts presented in this chapter. The guidelines are given in Problem P5-1$_A$. RTDs from real reactors can be found in *Ind. Eng. Chem.*, 49, 1000 (1957); *Ind. Eng. Chem. Process Des. Dev.*, 3, 381 (1964); *Can. J. Chem. Eng.*, 37, 107 (1959); *Ind. Eng. Chem.*, 44, 218 (1952); *Chem. Eng. Sci.*, 3, 26 (1954); and *Ind. Eng. Chem.*, 53, 381 (1961).

Problems

P17-1$_B$ **What if...**

   (a) Living Example 17-2. How does the conversion predicted from the segregation model, X_{seg}, compare with the conversion predicted by the CSTR, PFR, and LFR models for the same mean residence time, t_m?

   (b) Living Example 17-3. (1) Vary k by a factor of 5–10 or so above and below the nominal value given in the problem statement of 4.93×10^{-3} dm³/mol/s. When do X_{PFR} and X_{LFR} come close together and when do they become farther apart? (2) Use the $E(t)$ and $F(t)$ in Examples 16-1 and 16-2 to predict conversion, and compare in parts (a), (b), and (c).

   (c) Living Example 17-4. (1) Vary the parameter kC_{A0}, whose nominal value is

$$kC_{A0} = \left(\frac{0.01\ dm^3}{mol \cdot s}\right)\left(\frac{8\ mol}{dm^3}\right) = 0.08/s$$

   by a factor of 10 above and below the value nominal of 0.08 s⁻¹ and describe when X_{seg} and X_{mm} come closer together and when they become farther apart. (2) How do X_{mm} and X_{seg} compare with X_{PFR}, X_{CSTR}, and X_{LFR} for the same mean residence time? (3) How would your results change if $T = 350$ K with $E = 10$ kcal/mol? How would your answer change if the reaction was pseudo first order with $kC_{A0} = 4 \times 10^{-3}$/s?

   (d) Living Example 17-5. (1) Vary the parameters kC_{A0}, above and below the nominal value 0.08 s⁻¹, by a factor of 10 and describe when X_{seg} and X_{mm} come closer together and when they become farther apart. (2) How do X_{mm} and X_{seg} compare with X_{PFR}, X_{CSTR}, and X_{LFR} for the same t_m? (3) How would your results change if the reaction was pseudo first order with $k_1 = C_{A0}k = 0.08$ min⁻¹? (4) If the reaction was third order with $kC_{A0}^2 = 0.08$ min⁻¹? (5) If the reaction was half order with $kC_{A0}^{1/2} = 0.08$ min⁻¹? Describe any trends.

   (e) Living Example 17-6. Download the *Living Example Problem* from the CRE Web site. (1) If the activation energies in cal/mol are $E_1 = 5{,}000$, $E_2 = 1{,}000$, and $E_3 = 9{,}000$, how would the selectivities and conversion of A change as the temperature was raised or lowered around 350 K? (2) If you were asked to compare the results from Example 17-6 for the asymmetric and bimodal distributions in Tables E17-6.2 and E17-6.4, what similarities and differences do you observe? What generalizations can you make?

P17-2$_B$ An irreversible first-order reaction takes place in a long cylindrical reactor. There is no change in volume, temperature, or viscosity. The use of the simplifying assumption that there is plug flow in the tube leads to an estimated degree of conversion of 86.5%. What would be the actually attained degree of conversion if the real state of flow is laminar, with negligible diffusion?

P17-3$_C$ Show that for a first-order reaction

$$A \longrightarrow B$$

the exit concentration maximum mixedness equation

$$\frac{dC_A}{d\lambda} = kC_A + \frac{E(\lambda)}{1 - F(\lambda)}(C_A - C_{A0}) \qquad \text{(P17-3.1)}$$

is the same as the exit concentration given by the segregation model

$$C_A = C_{A0} \int_0^\infty E(t)e^{-kt}\,dt \qquad \text{(P17-3.2)}$$

Hint: Verify

$$C_A(\lambda) = \frac{C_{A0}e^{k\lambda}}{1 - F(\lambda)} \int_\lambda^\infty E(t)e^{-kt}\,dt \qquad \text{(P17-3.3)}$$

is a solution to Equation (P17-3.1).

P17-4$_C$ The first-order reaction

$$A \longrightarrow B$$

with $k = 0.8$ min^{-1} is carried out in a real reactor with the following RTD function:

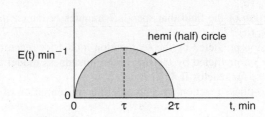

Mathematically, this hemi circle is described by the equations for $2\tau \geq t \geq 0$ then

$$E(t) = \sqrt{\tau^2 - (t - \tau)^2} \text{ min}^{-1}(\text{hemi circle})$$

For $t > 2\tau$, then $E(t) = 0$.

(a)   What is the mean residence time?
(b)   What is the variance?
(c)   What is the conversion predicted by the segregation model?
(d)   What is the conversion predicted by the maximum mixedness model?

P17-5$_B$ A step tracer input was used on a real reactor with the following results:
    For $t \leq 10$ min, then $C_T = 0$.
    For $10 \leq t \leq 30$ min, then $C_T = 10$ g/dm^3.
    For $t \geq 30$ min, then $C_T = 40$ g/dm^3.
The second-order reaction $A \rightarrow B$ with $k = 0.1$ dm^3/mol $\cdot$ min is to be carried out in the real reactor with an entering concentration of A of 1.25 mol/dm^3 at a volumetric flow rate of 10 dm^3/min. Here, k is given at 325 K.

(a)   What is the mean residence time t_m?
(b)   What is the variance σ^2?
(c)   What conversions do you expect from an ideal PFR and an ideal CSTR in a real reactor with t_m?
(d)   What is the conversion predicted by
    (1)   the segregation model?
    (2)   the maximum mixedness model?
(e)   What conversion is predicted by an ideal laminar flow reactor?

P17-6$_B$ The following $E(t)$ curves were obtained from a tracer test on two tubular reactors in which dispersion is believed to occur.

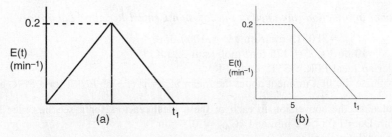

Figure P17-6$_B$ (a) RTD reactor A; (b) RTD reactor B.

A second-order reaction

$$A \xrightarrow{\;k\;} B \quad \text{with} \quad kC_{A0} = 0.2 \text{ min}^{-1}$$

is to be carried out in this reactor. There is no dispersion occurring either upstream or downstream of the reactor, but there is dispersion inside the reactor.

(a) What is the final time t_1 (in minutes) for the reactor shown in Figure P17-6$_B$ (a)? In Figure P17-6$_B$ (b)?

(b) What is the mean residence time, t_m, and variance, σ^2, for the reactor shown in Figure P17-6$_B$ (a)? In Figure P17-6$_B$ (b)?

(c) What is the fraction of the fluid that spends 7 minutes or longer in Figure P17-6$_B$ (a)? In Figure P17-6$_B$ (b)?

(d) Find the conversion predicted by the segregation model for reactor A.

(e) Find the conversion predicted by the maximum mixedness model for reactor B.

(f) Repeat (d) and (e) for reactor B.

P17-7$_B$ The third-order liquid-phase reaction with an entering concentration of 2M

$$A \xrightarrow{k_3} B$$

$$k_3 = 0.3 \ dm^6/mol^2/min$$

was carried out in a reactor that has the following RTD:

$$
\begin{array}{lll}
E(t) = 0 & \text{for} & t < 1 \ \text{min} \\
E(t) = 1.0 \ \text{min}^{-1} & \text{for} & 1 \le t \le 2 \ \text{min} \\
E(t) = 0 & \text{for} & t > 2 \ \text{min}
\end{array}
$$

(a) For isothermal operation, what is the conversion predicted by

1) a CSTR, a PFR, an LFR, and the segregation model, X_{seg}?
   Hint: Find t_m (i.e., τ) from the data and then use it with $E(t)$ for each of the ideal reactors.

2) the maximum mixedness model, X_{mm}? Plot X vs. z (or λ) and explain why the curve looks the way it does.

(b) Now calculate the exit concentrations of A, B, and C for the reaction

$$A \xrightarrow{k_1} B \xrightarrow{k_2} C \ \text{with} \ k_1 = 0.3 \ \text{min}^{-1} \ \text{and} \ k_2 = 1 \ \text{min}^{-1}$$

using (1) the segregation model, and (2) the maximum mixedness model.

P17-8$_A$ Consider again the nonideal reactor characterized by the RTD data in Example 17-5, where $E(t)$ and $F(t)$ are given as polynomials. The irreversible gas-phase nonelementary reaction

$$A + B \longrightarrow C + D$$

is first order in A and second order in B, and is to be carried out isothermally. Calculate the conversion for:

(a) A PFR, a laminar flow reactor with complete segregation, and a CSTR all at the same t_m.

(b) The cases of complete segregation and maximum mixedness.

Additional information (obtained at the Jofostan Central Research Laboratory in Riça, Jofostan):

$C_{A0} = C_{B0} = 0.0313 \ mol/dm^3$, $V = 1000 \ dm^3$,
$v_0 = 10 \ dm^3/s$, $k = 175 \ dm^6/mol^2 \cdot s$ at 320 K.

P17-9$_A$ Consider an ideal PFR, CSTR, and LFR.

(a) Evaluate the first moment about the mean $m_1 = \displaystyle\int_0^\infty (t - \tau) \ E(t) dt$ for a PFR, a CSTR, and an LFR.

(b) Calculate the conversion in each of these ideal reactors for a second-order liquid-phase reaction with Da = 1.0 ($\tau = 2$ min and $kC_{A0} = 0.5 \ \text{min}^{-1}$).

P17-10$_B$ For the catalytic reaction

$$A \xrightarrow[\text{cat}]{} C + D$$

the rate law can be written as

$$-r_A' = \frac{kC_A}{(1 + K_A C_A)^2}$$

Which will predict the highest conversion, the maximum mixedness model or the segregation model? *Hint:* Specify the different ranges of the conversion where one model will dominate over the other.

P7-11$_B$ Use the RTD data in Example 16-1 and 16-2 to predict X_{PFR}, X_{CSTR}, X_{LFR}, X_{seg} and X_{mm} for the following elementary gas phase reactions

  (a) A → B $k = 0.1$ min^{-1}
  (b) A → 2B $k = 0.1$ min^{-1}
  (c) 2A → B $k = 0.1$ min^{-1} m^3/kmol $C_{A0} = 1.0$ kmol/m^3
  (d) 3A → B $k = 0.1$ m^6/kmol^{2}min

  Repeat (a) through (d) for the RTD given by

  (e) P16-3$_B$
  (f) P16-4$_B$
  (g) P16-5$_B$

P17-12$_C$ The second-order, elementary liquid-phase reaction

$$2A \xrightarrow{\ k_{1A}\ } B$$

is carried out in a nonideal CSTR. At 300 K the specific reaction rate is $k_{1A} = 0.5$ dm^3/mol·min. In a tracer test, the tracer concentration rose linearly up to 1 mg/dm^3 at 1.0 minutes and then decreased linearly to zero at exactly 2.0 minutes. Pure A enters the reactor at a temperature of 300 K.

  (a) Calculate the conversion predicted by the segregation and maximum mixedness models.
  (b) Now consider that a second elementary reaction also takes place

$$A + B \xrightarrow{\ k_{2C}\ } C, \qquad k_{2C} = 0.12 \text{ dm}^3/\text{mol} \cdot \text{min}$$

  Compare the selectivities $\tilde{S}_{B/C}$ predicted by the segregation and maximum mixedness models.

P17-13$_B$ The reaction and corresponding rate data discussed in Example 8-8 are to be carried out in a nonideal reactor where RTD is given by the data (i.e. $E(t)$ and $F(t)$) in Example 16-2. Determine the exit selectivities

  (a) Using the segregation model.
  (b) Using the maximum mixedness model.
  (c) Compare the selectivities in parts (a) and (b) with those that would be found in an ideal PFR and ideal CSTR in which the space time is equal to the mean residence time.

P17-14$_B$ The reactions described in Example 8-12 are to be carried out in the reactor whose RTD is described in Example 17-4 with $C_{A0} = C_{B0} = 0.05$ mol/dm^3.

  (a) Determine the exit selectivities using the segregation model.
  (b) Determine the exit selectivities using the maximum mixedness model.
  (c) Compare the selectivities in parts (a) and (b) with those that would be found in an ideal PFR and ideal CSTR in which the space time is equal to the mean residence time.
  (d) What would your answers to parts (a) through (c) be if the RTD curve rose from zero at $t = 0$ to a maximum of 50 mg/dm^3 after 10 min, and then fell linearly to zero at the end of 20 min?

P17-15$_B$ Using the data in problem P16-11$_B$,

  (a) Plot the internal age distribution $I(t)$ as a function of time.
  (b) What is the mean internal age α_m?
  (c) The activity of a "fluidized" CSTR is maintained constant by feeding fresh catalyst and removing spent catalyst at a constant rate. Using the preceding RTD data, what is the mean catalytic activity if the catalyst decays according to the rate law

$$-\frac{da}{dt} = k_D a^2$$

  with

$$k_D = 0.1 \text{ s}^{-1}?$$

(d) What conversion would be achieved in an ideal PFR for a second-order reaction with $kC_{A0} = 0.1$ min^{-1} and $C_{A0} = 1$ mol/dm^3?

(e) Repeat **(d)** for a laminar flow reactor.

(f) Repeat **(d)** for an ideal CSTR.

(g) What would be the conversion for a second-order reaction with $kC_{A0} = 0.1$ min^{-1} and $C_{A0} = 1$ mol/dm^3 using the segregation model?

(h) What would be the conversion for a second-order reaction with $kC_{A0} = 0.1$ min^{-1} and $C_{A0} = 1$ mol/dm^3 using the maximum mixedness model?

P17-16$_B$ The relative tracer concentrations obtained from pulse tracer tests on a commercial packed-bed desulfurization reactor are shown in Figure P17-15$_B$. After studying the RTD, what problems are occurring with the reactor during the period of poor operation (thin line)? The bed was repacked and the pulse tracer test again carried out with the results shown in Figure P17-15$_B$ (thick line). Calculate the conversion that could be achieved in the commercial desulfurization reactor during poor operation and during good operation (Figure P17-15$_B$) for the following reactions:

(a) A first-order isomerization with a specific reaction rate of 0.1 h^{-1}

(b) A first-order isomerization with a specific reaction rate of 2.0 h^{-1}

(c) What do you conclude upon comparing the four conversions in parts (a) and (b)?

From Additional Home Problem CDP17-I$_B$ [3rd ed. P13-5]

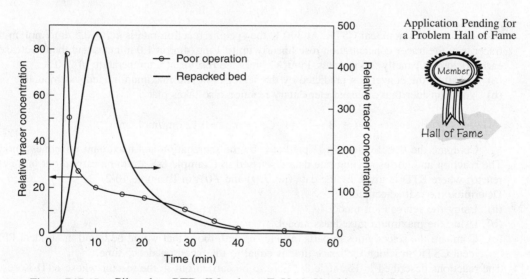

Figure P17-15$_B$ Pilot-plant RTD. [Murphree, E. V., A. Voorhies, and F. Y. Mayer, *Ind. Eng. Chem. Process Des. Dev.,* 3, 381 (1964). Copyright © 1964, American Chemical Society.]

SUPPLEMENTARY READING

1. Discussions of the measurement and analysis of residence time distribution can be found in

CURL, R. L., and M. L. MCMILLIN, "Accuracies in residence time measurements," *AIChE J.,* 12, 819–822 (1966).

LEVENSPIEL, O., *Chemical Reaction Engineering*, 3rd ed. New York: Wiley, 1999, Chaps. 11–16.

2. An excellent discussion of segregation can be found in

DOUGLAS, J. M., "The effect of mixing on reactor design," *AIChE Symp. Ser.*, 48, vol. 60, p. 1 (1964).

3. Also see

DUDUKOVIC, M., and R. FELDER, in *CHEMI Modules on Chemical Reaction Engineering*, vol. 4, ed. B. Crynes and H. S. Fogler. New York: AIChE, 1985.

NAUMAN, E. B., "Residence time distributions and micromixing," *Chem. Eng. Commun.*, 8, 53 (1981).

NAUMAN, E. B., and B. A. BUFFHAM, *Mixing in Continuous Flow Systems*. New York: Wiley, 1983.

ROBINSON, B. A., and J. W. TESTER, "Characterization of flow maldistribution using inlet-outlet tracer techniques: an application of internal residence time distributions," *Chem. Eng. Sci.*, 41 (3), 469–483 (1986).

VILLERMAUX, J., "Mixing in chemical reactors," in *Chemical Reaction Engineering—Plenary Lectures*, ACS Symposium Series 226. Washington, DC: American Chemical Society, 1982.

Models for Nonideal **18** Reactors

Success is a journey, not a destination.
—Ben Sweetland

Overview. Not all tank reactors are perfectly mixed nor do all tubular reactors exhibit plug-flow behavior. In these situations, some means must be used to allow for deviations from ideal behavior. Chapter 17 showed how the RTD was sufficient if the reaction was first order or if the fluid was either in a state of complete segregation or maximum mixedness. We use the segregation and maximum mixedness models to bound the conversion when no adjustable parameters are used. For non-first-order reactions in a fluid with good micromixing, more than just the RTD is needed. These situations compose a number of reactor analysis problems and cannot be ignored. For example, we may have an existing reactor in storage and want to carry out a new reaction in that reactor. To predict conversions and product distributions for such systems, a model of reactor flow patterns and/or RTD is necessary.

After completing this chapter you will be able to
- Discuss guidelines for developing one- and two-parameter models (Section 18.1).
- Use the tanks-in-series (T-I-S) one-parameter model to predict conversion (Section 18.2).
- Use the dispersion one-parameter model to predict conversion (Section 18.3).
- Use the RTD to evaluate the model parameters (e.g., D_a, n) for one-parameter models.
- Develop equations to model flow, dispersion, and reaction (Section 18.4).

Use the RTD to
evaluate
parameters.

> - Discuss dispersion and reaction in tubular reactors (Section 18.6).
> - Suggest combinations of ideal reactors to model the nonideal reactor to predict conversion (Section 18.7).
> - Use RTD data to evaluate the model parameters (e.g., α, β) for two-parameter models (Section 18.8).
>
> Using the above models, we will first measure the RTD to characterize the reactor at the new operating conditions of temperature and flow rate. After selecting a model for the reactor, we use the RTD to evaluate the parameter(s) in the model after which we calculate the conversion.

18.1 Some Guidelines for Developing Models

The overall goal is to use the following equation

$$\boxed{\text{RTD Data} + \text{Model} + \text{Kinetics} = \text{Prediction}}$$

Conflicting goals

The choice of the particular model to be used depends largely on the engineering judgment of the person carrying out the analysis. It is this person's job to choose the model that best combines the conflicting goals of mathematical simplicity and physical realism. There is a certain amount of art in the development of a model for a particular reactor, and the examples presented here can only point toward a direction that an engineer's thinking might follow.

A Model must
- *Fit the data*
- *Be able to extrapolate theory and experiment*
- *Have realistic parameters*

For a given real reactor, it is not uncommon to use all the models discussed previously to predict conversion and then make comparisons. Usually, the real conversion will be *bounded* by the model calculations.

The following guidelines are suggested when developing models for nonideal reactors:

1. *The model must be mathematically tractable.* The equations used to describe a chemical reactor should be able to be solved without an inordinate expenditure of human or computer time.
2. *The model must realistically describe the characteristics of the nonideal reactor.* The phenomena occurring in the nonideal reactor must be reasonably described physically, chemically, and mathematically.
3. *The model should not have more than two adjustable parameters.* This constraint is often used because an expression with more than two adjustable parameters can be fitted to a great variety of experimental data, and the modeling process in this circumstance is nothing more than an exercise in curve fitting. The statement "Give me four adjustable parameters and I can fit an elephant; give me five and I can include his tail!" is one that I have heard from many colleagues. Unless one is into modern art, a substantially larger number of adjustable parameters is necessary to draw a reasonable-looking elephant.[1] A one-parameter model is, of course, superior to a two-parameter model if the one-parameter model is sufficiently realistic. To be fair, however, in complex systems (e.g., internal diffusion and conduction,

---

[1] J. Wei, *CHEMTECH*, 5, 128 (1975).

mass transfer limitations) where other parameters may be measured *independently*, then more than two parameters are quite acceptable.

Table 18-1 gives some guidelines that will help your analysis and model building of nonideal reaction systems.

The Guidelines

1. *Look at the reactor.*
   a. Where are the inlet and outlet streams to and from the reactors? (Is by-passing a possibility?)
   b. Look at the mixing system. How many impellers are there? (Could there be multiple mixing zones in the reactor?)
   c. Look at the configuration. (Is internal recirculation possible? Is the packing of the catalyst particles loose so channeling could occur?)
2. *Look at the tracer data.*
   a. Plot the $E(t)$ and $F(t)$ curves.
   b. Plot and analyze the shapes of the $E(\Theta)$ and $F(\Theta)$ curves. Is the shape of the curve such that the curve or parts of the curve can be fit by an ideal reactor model? Does the curve have a long tail suggesting a stagnant zone? Does the curve have an early spike indicating bypassing?
   c. Calculate the mean residence time, t_m, and variance, σ^2. How does the t_m determined from the RTD data compare with τ as measured with a yardstick and flow meter? How large is the variance; is it larger or smaller than τ^2?
3. *Choose a model or perhaps two or three models.*
4. *Use the tracer data to determine the model parameters* (e.g., n, D_a, v_b).
5. *Use the CRE algorithm in Chapter 5.* Calculate the exit concentrations and conversion for the model system you have selected.

When using the algorithm in Table 18-1, we classify a model as being either a one-parameter model (e.g., tanks-in-series model or dispersion model) or a two-parameter model (e.g., reactor with bypassing and dead volume). In Sections 18.1.1 and 18.1.2, we give an overview of these models, which will be discussed in greater detail later in the chapter.

18.1.1 One-Parameter Models

Here, we use a single parameter to account for the nonideality of our reactor. This parameter is most always evaluated by analyzing the RTD determined from a tracer test. Examples of one-parameter models for nonideal CSTRs include either a reactor dead volume, V_D, where no reaction takes place, or volumetric flow rate with part of the fluid bypassing the reactor, v_b, thereby exiting unreacted. Examples of one-parameter models for tubular reactors include the tanks-in-series model and the dispersion model. For the tanks-in-series model, the one parameter is the number of tanks, n, and for the dispersion model, the one parameter is the dispersion coefficient, D_a.[†] Knowing the parameter values, we then proceed to determine the conversion and/or effluent concentrations for the reactor.

[†] Nomenclature note: Da_1 (or Da_2) is the Damköhler number and D_a is the dispersion coefficient.

We first consider nonideal tubular reactors. Tubular reactors may be empty, or they may be packed with some material that acts as a catalyst, heat-transfer medium, or means of promoting interphase contact. Until Chapters 16–18, it usually has been assumed that the fluid moves through the reactor in a piston-like flow (i.e., plug flow reactor), and every atom spends an identical length of time in the reaction environment. Here, the *velocity profile is flat,* and there is no axial mixing. Both of these assumptions are false to some extent in every tubular reactor; frequently, they are sufficiently false to warrant some modification. Most popular tubular reactor models need to have the means to allow for failure of the plug-flow model and insignificant axial mixing assumptions; examples include the unpacked laminar-flow tubular reactor, the unpacked turbulent flow reactor, and packed-bed reactors. One of two approaches is usually taken to compensate for failure of either or both of the ideal assumptions. One approach involves modeling the nonideal tubular reactor as a series of identically sized CSTRs. The other approach (the dispersion model) involves a modification of the ideal reactor by imposing axial dispersion on plug flow.

18.1.2 Two-Parameter Models

The premise for the two-parameter model is that we can use a combination of ideal reactors to model the real reactor. For example, consider a packed bed reactor with channeling. Here, the response to a pulse tracer input would show two dispersed pulses in the output as shown in Figure 16-1 and Figure 18-1.

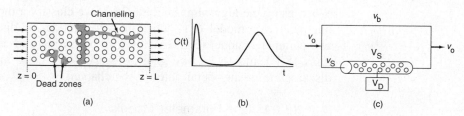

Figure 18-1   (a) Real system; (b) outlet for a pulse input; (c) model system.

Here, we could model the real reactor as two ideal PBRs in parallel, with the two parameters being the volumetric flow rate that channels or by passes, v_b, and the reactor dead volume, V_D. The real reactor volume is $V = V_D + V_S$ with entering volumetric flow rate $v_0 = v_b + v_S$.

18.2 The Tanks-in-Series (T-I-S) One-Parameter Model

In this section we discuss the use of the tanks-in-series (T-I-S) model to describe nonideal reactors and calculate conversion. The T-I-S model is a one-parameter model. We will analyze the RTD to determine the number of ideal tanks, n, in series that will give approximately the same RTD as the nonideal reactor. Next, we will apply the reaction engineering algorithm developed in Chapters 1 through 5 to calculate conversion. We are first going to develop the RTD equation for three tanks in series (Figure 18-2) and then generalize to

$\boxed{n = \ ?}$

n reactors in series to derive an equation that gives the number of tanks in series that best fits the RTD data.

18.2.1 Developing the *E*-Curve for the T-I-S Model

The RTD will be analyzed from a tracer pulse injected into the first reactor of three equally sized CSTRs in series.

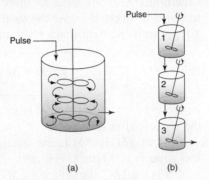

Figure 18-2 Tanks in series: (a) real system; (b) model system.

Using the definition of the RTD presented in Section 16.2, the fraction of material leaving the system of three reactors (i.e., leaving the third reactor) that has been in the system between time t and $t + \Delta t$ is

$$E(t)\,\Delta t = \frac{v C_3(t)\,\Delta t}{N_0} = \frac{C_3(t)}{\displaystyle\int_0^\infty C_3(t)\,dt}\,\Delta t$$

In Figure 2-9, we saw how tanks in series could approximate a PFR.

Then

$$E(t) = \frac{C_3(t)}{\displaystyle\int_0^\infty C_3(t)\,dt} \tag{18-1}$$

In this expression, $C_3(t)$ is the concentration of tracer in the effluent from the third reactor and the other terms are as defined previously.

By carrying out mass balances on the tracer sequentially for reactors 1, 2, and 3, it is shown on the CRE Web site in the *Expanded Material* for Chapter 18 that the exit tracer concentration for reactor 3 is

$$C_3 = \frac{C_0 t^2}{2\tau_i^2}\,e^{-t/\tau_i} \tag{18-2}$$

Substituting Equation (18-2) into Equation (18-1), we find that

$$E(t) = \frac{C_3(t)}{\displaystyle\int_0^\infty C_3(t)\,dt} = \frac{C_0 t^2/(2\tau_i^2)e^{-t/\tau_i}}{\displaystyle\int_0^\infty \frac{C_0 t^2 e^{-t/\tau_i}}{2\tau_i^2}\,dt}$$

$$= \frac{t^2}{2\tau_i^3}\,e^{-t/\tau_i} \tag{18-3}$$

Generalizing this method to a series of n CSTRs gives the RTD for n CSTRs in series, $E(t)$:

RTD for equal-size tanks in series

$$E(t) = \frac{t^{n-1}}{(n-1)!\tau_i^n} e^{-t/\tau_i}$$

(18-4)

Equation (18-4) will be a bit more useful if we put in the dimensionless form in terms of $E(\Theta)$. Because the total reactor volume is nV_i, then $\tau_i = \tau/n$, where τ represents the total reactor volume divided by the flow rate, v, we have

$$E(\Theta) = \tau E(t) = \frac{n(n\Theta)^{n-1}}{(n-1)!} e^{-n\Theta}$$

(18-5)

where $\Theta = t/\tau =$ Number of reactor volumes of fluid that have passed through the reactor after time t.

Here, $(E(\Theta)\,d\Theta)$ is the fraction of material existing between dimensionless time Θ and time $(\Theta + d\Theta)$.

Figure 18-3 illustrates the RTDs of various numbers of CSTRs in series in a two-dimensional plot (a) and in a three-dimensional plot (b). As the number becomes very large, the behavior of the system approaches that of a plug-flow reactor.

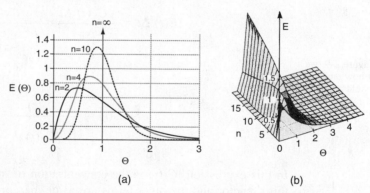

Figure 18-3   Tanks-in-series response to a pulse tracer input for different numbers of tanks.

We can determine the number of tanks in series by calculating the dimensionless variance σ_Θ^2 from a tracer experiment.

$$\sigma_\Theta^2 = \frac{\sigma^2}{\tau^2} = \int_0^\infty (\Theta - 1)^2 E(\Theta)\,d\Theta$$

(18-6)

$$= \int_0^\infty \Theta^2 E(\Theta)\,d\Theta - 2\int_0^\infty \Theta E(\Theta)\,d\Theta + \int_0^\infty E(\Theta)\,d\Theta$$

(18-7)

$$\sigma_\Theta^2 = \int_0^\infty \Theta^2 E(\Theta)\,d\Theta - 1$$

(18-8)

$$= \int_0^\infty \Theta^2 \frac{n(n\Theta)^{n-1}}{(n-1)!} e^{-n\Theta} d\Theta - 1$$

$$\sigma_\Theta^2 = \frac{n^n}{(n-1)!} \int_0^\infty \Theta^{n+1} e^{-n\Theta} d\Theta - 1 \qquad (18\text{-}9)$$

$$= \frac{n^n}{(n-1)!} \left[\frac{(n+1)!}{n^{n+2}} \right] - 1$$

As the number of tanks increases, the variance decreases.

$$\sigma_\Theta^2 = \frac{1}{n} \qquad (18\text{-}10)$$

The number of tanks in series is

$$\boxed{n = \frac{1}{\sigma_\Theta^2} = \frac{\tau^2}{\sigma^2}} \qquad (18\text{-}11)$$

This expression represents the number of tanks necessary to model the real reactor as n ideal tanks in series. If the number of reactors, n, turns out to be small, the reactor characteristics turn out to be those of a single CSTR or perhaps two CSTRs in series. At the other extreme, when n turns out to be large, we recall from Chapter 2 that the reactor characteristics approach those of a PFR.

18.2.2 Calculating Conversion for the T-I-S Model

If the reaction is first order, we can use Equation (5-15) to calculate the conversion

$$X = 1 - \frac{1}{(1 + \tau_i k)^n} \qquad (5\text{-}15)$$

where

$$\tau_i = \frac{V}{v_0 n}$$

It is acceptable (and usual) for the value of n calculated from Equation (18-11) to be a noninteger in Equation (5-15) to calculate the conversion. For reactions other than first order, an integer number of reactors must be used and sequential mole balances on each reactor must be carried out. If, for example, $n = 2.53$, then one could calculate the conversion for two tanks and also for three tanks to bound the conversion. The conversion and effluent concentrations would be solved sequentially using the algorithm developed in Chapter 5; that is, after solving for the effluent from the first tank, it would be used as the input to the second tank and so on as shown on the CRE Web site for Chapter 18 *Expanded Materials*.

18.2.3 Tanks-in-Series versus Segregation for a First-Order Reaction

Reference Shelf

We have already stated that the segregation and maximum mixedness models are equivalent for a first-order reaction. The proof of this statement was left as an exercise in Problem P17-3$_B$. We can extend this equivalency for a first-order reaction to the tanks-in-series (T-I-S) model

$$X_{\text{T-I-S}} = X_{\text{seg}} = X_{\text{mm}}$$

(18-12)

The proof of Equation (18-12) is given in the *Expanded Materials* on the CRE Web site for Chapter 18.

18.3 Dispersion One-Parameter Model

The dispersion model is also often used to describe nonideal tubular reactors. In this model, there is an axial dispersion of the material, which is governed by an analogy to Fick's law of diffusion, superimposed on the flow as shown in Figure 18-4. So in addition to transport by bulk flow, UA_cC, every component in the mixture is transported through any cross section of the reactor at a rate equal to $[-D_aA_c(dC/dz)]$ resulting from molecular and convective diffusion. By convective diffusion (i.e., dispersion), we mean either Aris-Taylor dispersion in laminar-flow reactors or turbulent diffusion resulting from turbulent eddies. Radial concentration profiles for plug flow (a) and a representative axial and radial profile for dispersive flow (b) are shown in Figure 18-4. Some molecules will diffuse forward ahead of the molar average velocity, while others will lag behind.

Tracer pulse with dispersion

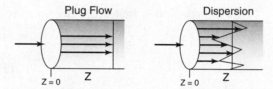

Figure 18-4   Concentration profiles: (a) without and (b) with dispersion.

To illustrate how dispersion affects the concentration profile in a tubular reactor, we consider the injection of a perfect tracer pulse. Figure 18-5 shows how dispersion causes the pulse to broaden as it moves down the reactor and becomes less concentrated.

Recall Equation (14-14). The molar flow rate of tracer (F_T) by both convection and dispersion is

$$F_T = \left[-D_a \frac{\partial C_T}{\partial z} + UC_T \right] A_c$$

(14-14)

In this expression, D_a is the effective dispersion coefficient (m^2/s) and U (m/s) is the superficial velocity. To better understand how the pulse broadens, we refer to the concentration peaks t_2 and t_3 in Figure 18-6. We see that there is a concentration gradient on both sides of the peak causing molecules to diffuse

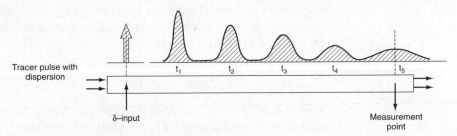

Figure 18-5 Dispersion in a tubular reactor. (Levenspiel, O., *Chemical Reaction Engineering*, 2nd ed. Copyright © 1972 John Wiley & Sons, Inc. Reprinted by permission of John Wiley & Sons, Inc. All rights reserved.)

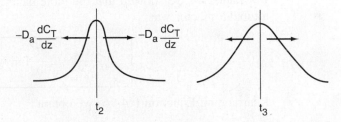

Figure 18-6 Symmetric concentration gradients causing the spreading by dispersion of a pulse input.

away from the peak and thus broaden the pulse. The pulse broadens as it moves through the reactor.

Correlations for the dispersion coefficients in both liquid and gas systems may be found in Levenspiel.[2] Some of these correlations are given in Section 18.4.5.

An unsteady state mole balance on the inert tracer T gives

$$-\frac{\partial F_T}{\partial z} = A_c \frac{\partial C_T}{\partial t} \tag{18-13}$$

Substituting for F_T and dividing by the cross-sectional area A_c, we have

Pulse tracer balance with dispersion

$$\boxed{D_a \frac{\partial^2 C_T}{\partial z^2} - \frac{\partial (U C_T)}{\partial z} = \frac{\partial C_T}{\partial t}} \tag{18-14}$$

Once we know the boundary conditions, the solution to Equation (18-14) will give the outlet tracer concentration–time curves. Consequently, we will have to wait to obtain this solution until we discuss the boundary conditions in Section 18.4.2.

The plan   We are now going to proceed in the following manner: First, we will write the balance equations for dispersion with reaction. We will discuss the two types of boundary conditions, closed-closed and open-open. We will then obtain an analytical solution for the closed-closed system for the conversion for a first-order reaction in terms of the Peclet number, *Pe* (dispersion

---

[2] O. Levenspiel, *Chemical Reaction Engineering* (New York: Wiley, 1962), pp. 290–293.

coefficient) and the Damköhler number. We then will discuss how the dispersion coefficient can be obtained either from correlations in the literature *or* from the analysis of the RTD curve.

18.4 Flow, Reaction, and Dispersion

Now that we have an intuitive feel for how dispersion affects the transport of molecules in a tubular reactor, we shall consider two types of dispersion in a tubular reactor, *laminar* and *turbulent*.

18.4.1 Balance Equations

In Chapter 14 we showed that the mole balance on reacting species A flow in a tubular reactor was

$$D_a \frac{d^2 C_A}{dz^2} - U \frac{dC_A}{dz} + r_A = 0 \qquad (14\text{-}16)$$

Rearranging Equation (14-16) we obtain

$$\frac{D_a}{U} \frac{d^2 C_A}{dz^2} - \frac{dC_A}{dz} + \frac{r_A}{U} = 0 \qquad (18\text{-}15)$$

This equation is a second-order ordinary differential equation. It is nonlinear when r_A is other than zero or first order.

When the reaction rate r_A is first order, $r_A = -kC_A$, then Equation (18-16)

$$\frac{D_a}{U} \frac{d^2 C_A}{dz^2} - \frac{dC_A}{dz} - \frac{kC_A}{U} = 0 \qquad (18\text{-}16)$$

is amenable to an analytical solution. However, before obtaining a solution, we put our Equation (18-16) describing dispersion and reaction in dimensionless form by letting $\psi = C_A/C_{A0}$ and $\lambda = z/L$

D_a = Dispersion
coefficient

$$\frac{1}{Pe_r} \frac{d^2\psi}{d\lambda^2} - \frac{d\psi}{d\lambda} - Da_1 \cdot \psi = 0 \qquad (18\text{-}17)$$

Da_1 = Damköhler
number

The quantity Da_1 appearing in Equation (18-17) is called the *Damköhler number* for a first-order conversion and physically represents the ratio

Damköhler number
for a first-order
reaction

$$Da_1 = \frac{\text{Rate of consumption of A by reaction}}{\text{Rate of transport of A by convection}} = k\tau \qquad (18\text{-}18)$$

The other dimensionless term is the *Peclet number, Pe,*

$$Pe_r = \frac{\text{Rate of transport by convection}}{\text{Rate of transport by diffusion or dispersion}} = \frac{Ul}{D_a} \qquad (18\text{-}19)$$

in which l is the characteristic length term. There are two different types of Peclet numbers in common use. We can call Pe_r the *reactor* Peclet number; it

For open tubes
$Pe_r \sim 10^6$,
$Pe_f \sim 10^4$

uses the reactor length, L, for the characteristic length, so $Pe_r \equiv UL/D_a$. It is Pe_r that appears in Equation (18-17). The reactor Peclet number, Pe_r, for mass dispersion is often referred to as the Bodenstein number, Bo, in reacting systems rather than the Peclet number. The other type of Peclet number can be called the *fluid* Peclet number, Pe_f; it uses the characteristic length that determines the fluid's mechanical behavior. In a packed bed this length is the particle diameter d_p, and $Pe_f \equiv Ud_p/\phi D_a$. (The term U is the empty tube or superficial velocity. For packed beds we often wish to use the average interstitial velocity, and thus U/ϕ is commonly used for the packed-bed velocity

For packed beds
$Pe_r \sim 10^3$,
$Pe_f \sim 10^1$

term.) In an empty tube, the fluid behavior is determined by the tube diameter d_t, and $Pe_f = Ud_t/D_a$. The fluid Peclet number, Pe_f, is given in virtually all literature correlations relating the Peclet number to the Reynolds number because both are directly related to the fluid mechanical behavior. It is, of course, very simple to convert Pe_f to Pe_r: Multiply by the ratio L/d_p or L/d_t. The reciprocal of Pe_r, D_a/UL, is sometimes called the *vessel dispersion number*.

18.4.2 Boundary Conditions

There are two cases that we need to consider: boundary conditions for *closed vessels* and for *open vessels*. In the case of *closed-closed vessels*, we assume that there is no dispersion or radial variation in concentration either upstream (closed) or downstream (closed) of the reaction section; hence, this is a closed-closed vessel, as shown in Figure 18-7(a). In an *open vessel*, dispersion occurs both upstream (open) and downstream (open) of the reaction section; hence, this is an open-open vessel as shown in Figure 18-7(b). These two cases are shown in Figure 18-7, where fluctuations in concentration due to dispersion are superimposed on the plug-flow velocity profile. A closed-open vessel boundary condition is one in which there is no dispersion in the entrance section but there is dispersion in the reaction and exit sections.

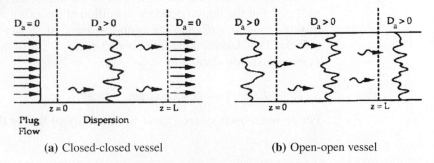

(a) Closed-closed vessel   (b) Open-open vessel

Figure 18-7 Types of boundary conditions.

18.4.2A Closed-Closed Vessel Boundary Condition

For a closed-closed vessel, we have plug flow (no dispersion) to the immediate left of the entrance line ($z = 0^-$) (closed) and to the immediate right of the exit $z = L$ ($z = L^+$) (closed). However, between $z = 0^+$ and $z = L^-$, we have dispersion and reaction. The corresponding entrance boundary condition is

At $z = 0$: $$F_A(0^-) = F_A(0^+)$$

Substituting for F_A yields

$$UA_c C_A(0^-) = -A_c D_a \left(\frac{dC_A}{dz}\right)_{z=0^+} + UA_c C_A(0^+)$$

Solving for the entering concentration $C_A(0^-) = C_{A0}$

Concentration boundary conditions at the entrance

$$\boxed{C_{A0} = \frac{-D_a}{U}\left(\frac{dC_A}{dz}\right)_{z=0^+} + C_A(0^+)} \qquad (18\text{-}20)$$

At the exit to the reaction section, the concentration is continuous, and there is no gradient in tracer concentration.

Concentration boundary conditions at the exit

At $z = L$:

$$\boxed{\begin{aligned} C_A(L^-) &= C_A(L^+) \\ \frac{dC_A}{dz} &= 0 \end{aligned}} \qquad (18\text{-}21)$$

Danckwerts Boundary Conditions

These two boundary conditions, Equations (18-20) and (18-21), first stated by Danckwerts, have become known as the famous *Danckwerts boundary conditions*.[3] Bischoff has given a rigorous derivation by solving the differential equations governing the dispersion of component A in the entrance and exit sections, and taking the limit as the dispersion coefficient, D_a in the entrance and exit sections approaches zero.[4] From the solutions, he obtained boundary conditions on the reaction section identical with those Danckwerts proposed.

The closed-closed concentration boundary condition at the entrance is shown schematically in Figure 18-8 on page 857. One should not be uncomfortable with the discontinuity in concentration at $z = 0$ because if you recall for an ideal CSTR, the concentration drops immediately on entering from C_{A0} to C_{Aexit}. For the other boundary condition at the exit $z = L$, we see the concentration gradient, (dC_A/dz), has gone to zero. At steady state, it can be shown that this Danckwerts boundary condition at $z = L$ also applies to the open-open system at steady state.

18.4.2B Open-Open System

For an open-open system, there is continuity of flux at the boundaries at z = 0

$$F_A(0^-) = F_A(0^+)$$

Open-open boundary condition

$$\boxed{-D_a\frac{\partial C_A}{\partial z}\bigg)_{z=0^-} + UC_A(0^-) = -D_a\frac{\partial C_A}{\partial z}\bigg)_{z=0^+} + UC_A(0^+)} \qquad (18\text{-}22)$$

[3] P. V. Danckwerts, *Chem. Eng. Sci.*, 2, 1 (1953).

[4] K. B. Bischoff, *Chem. Eng. Sci.*, 16, 131 (1961).

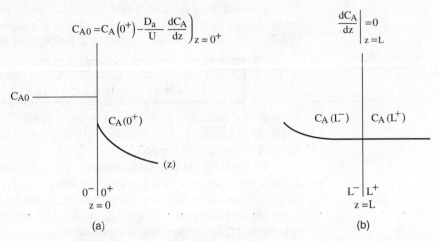

Prof. P. V. Danckwerts,
Cambridge University,
U.K.

Figure 18-8   Schematic of Danckwerts boundary conditions: (a) entrance; (b) exit.

At $z = L$, we have continuity of concentration and

$$\boxed{\frac{dC_A}{dz} = 0} \tag{18-23}$$

18.4.2C Back to the Solution for a Closed-Closed System

We now shall solve the dispersion reaction balance for a first-order reaction

$$\frac{1}{Pe_r}\frac{d^2\psi}{d\lambda^2} - \frac{d\psi}{d\lambda} - \boldsymbol{Da_1}\psi = 0 \tag{18-17}$$

For the closed-closed system, the Danckwerts boundary conditions in dimensionless form are

$$\text{At } \lambda = 0 \text{ then } 1 = -\frac{1}{Pe_r}\frac{d\psi}{d\lambda}\Bigg)_{\lambda\,=\,0^+} + \psi(0^+) \tag{18-24}$$

$$\text{At } \lambda = 1 \text{ then } \frac{d\psi}{d\lambda} = 0 \tag{18-25}$$

$\boldsymbol{Da_1} = \tau k$
$Pe_r = UL/D_a$ At the end of the reactor, where $\lambda = 1$, the solution to Equation (18-17) is

$$\psi_L = \frac{C_{AL}}{C_{A0}} = 1 - X$$

$$= \frac{4q\exp(Pe_r/2)}{(1+q)^2\exp\left(Pe_r q/2\right) - (1-q)^2\exp\left(-Pe_r q/2\right)} \tag{18-26}$$

$$\text{where } q = \sqrt{1 + 4\boldsymbol{Da_1}/Pe_r}$$

Nomenclature note
Da_1 is the
Damköhler
number for a
first-order reaction, τk
D_a is the dispersion coefficient in cm²/s
$Pe_r = UL/D_a$

This solution was first obtained by Danckwerts and has been published in many places (e.g., Levenspiel).[5,6] With a slight rearrangement of Equation (18-26), we obtain the conversion as a function of Da_1 and Pe_r.

$$X = 1 - \frac{4q \exp(Pe_r/2)}{(1+q)^2 \exp(Pe_r q/2) - (1-q)^2 \exp(-Pe_r q/2)} \quad (18\text{-}27)$$

Outside the limited case of a first-order reaction, a numerical solution of the equation is required, and because this is a split-boundary-value problem, an iterative technique is needed.

To evaluate the exit concentration given by Equation (18-26) or the conversion given by (18-27), we need to know the Damköhler and Peclet numbers. The first-order reaction rate constant, k, and hence $Da_1 = \tau k$, can be found using the techniques in Chapter 7. In the next section, we discuss methods to determine D_a by finding the Peclet number.

18.4.3 Finding D_a and the Peclet Number

Three ways to find D_a

There are three ways we can use to find D_a and hence Pe_r
1. Laminar flow with radial and axial molecular diffusion theory
2. Correlations from the literature for pipes and packed beds
3. Experimental tracer data

At first sight, simple models described by Equation (18-14) appear to have the capability of accounting only for axial mixing effects. It will be shown, however, that this approach can compensate not only for problems caused by axial mixing, *but also for those caused by radial mixing and other nonflat velocity profiles.*[7] These fluctuations in concentration can result from different flow velocities and pathways and from molecular and turbulent diffusion.

18.4.4 Dispersion in a Tubular Reactor with Laminar Flow

In a laminar flow reactor, we know that the axial velocity varies in the radial direction according to the well-known parabolic velocity profile:

$$u(r) = 2U\left[1 - \left(\frac{r}{R}\right)^2\right]$$

where U is the average velocity. For laminar flow, we saw that the RTD function $E(t)$ was given by

$$E(t) = \begin{cases} 0 & \text{for } t < \dfrac{\tau}{2} \quad \left(\tau = \dfrac{L}{U}\right) \\[2ex] \dfrac{\tau^2}{2t^3} & \text{for } t \geq \dfrac{\tau}{2} \end{cases} \quad (16\text{-}47)$$

[5] P. V. Danckwerts, *Chem. Eng. Sci.,* 2, 1 (1953).

[6] Levenspiel, *Chemical Reaction Engineering,* 3rd ed. (New York: Wiley, 1999).

[7] R. Aris, *Proc. R. Soc. (London), A235,* 67 (1956).

In arriving at this distribution $E(t)$, it was assumed that there was no transfer of molecules in the radial direction between streamlines. Consequently, with the aid of Equation (16-47), we know that the molecules on the center streamline ($r = 0$) exited the reactor at a time $t = \tau/2$, and molecules traveling on the streamline at $r = 3R/4$ exited the reactor at time

$$t = \frac{L}{u} = \frac{L}{2U[1 - (r/R)^2]} = \frac{\tau}{2[1 - (3/4)^2]}$$

$$= \frac{8}{7} \cdot \tau$$

The question now arises: What would happen if some of the molecules traveling on the streamline at *r = 3R/4* jumped (i.e., diffused) onto the streamline at *r = 0*? The answer is that they would exit sooner than if they had stayed on the streamline at *r = 3R/4*. Analogously, if some of the molecules from the faster streamline at *r = 0* jumped (i.e., diffused) onto the streamline at *r = 3R/4*, they would take a longer time to exit (Figure 18-9). In addition to the molecules diffusing between streamlines, they can also move forward or backward relative to the average fluid velocity by molecular diffusion (Fick's law). With both axial and radial diffusion occurring, the question arises as to what will be the distribution of residence times when molecules are transported between and along streamlines by diffusion. To answer this question, we will derive an equation for the axial dispersion coefficient, D_a, that accounts for the axial and radial diffusion mechanisms. In deriving D_a, which is often referred to as the *Aris–Taylor dispersion coefficient*, we closely follow the development given by Brenner and Edwards.[8]

Molecules diffusing between streamlines and back and forth along a streamline

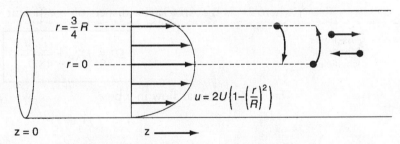

Figure 18-9 Radial diffusion in laminar flow.

The convective–diffusion equation for solute (e.g., tracer) transport in both the axial and radial direction can be obtained by combining Equation (14-3) with the diffusion equation (cf. Equation (14-11)) applied to the tracer concentration, c, and transformed to radial coordinates

$$\frac{\partial c}{\partial t} + u(r)\frac{\partial c}{\partial z} = D_{AB}\left\{\frac{1}{r}\frac{\partial[r(\partial c/\partial r)]}{\partial r} + \frac{\partial^2 c}{\partial z^2}\right\} \qquad (18\text{-}28)$$

[8] H. Brenner and D. A. Edwards, *Macrotransport Processes* (Boston: Butterworth-Heinemann, 1993).

where c is the solute concentration at a particular r, z, and t, and D_{AB} is the molecular diffusion coefficient of species A in B.

We are going to change the variable in the axial direction z to z^*, which corresponds to an observer moving with the fluid

$$z^* = z - Ut \tag{18-29}$$

A value of $z^* = 0$ corresponds to an observer moving with the average velocity of the fluid, U. Using the chain rule, we obtain

$$\left(\frac{\partial c}{\partial t}\right)_{z^*} + [u(r) - U]\frac{\partial c}{\partial z^*} = D_{AB}\left[\frac{1}{r}\frac{\partial}{\partial r}\left(r\frac{\partial c}{\partial r}\right) + \frac{\partial^2 c}{\partial z^{*2}}\right] \tag{18-30}$$

Because we want to know the concentrations and conversions at the exit to the reactor, we are really only interested in the average axial concentration, $\overline{C}$, which is given by

$$\overline{C}(z, t) = \frac{1}{\pi R^2}\int_0^R c(r, z, t)2\pi r\, dr \tag{18-31}$$

Reference Shelf

Consequently, we are going to solve Equation (18-30) for the solution concentration as a function of r and then substitute the solution $c\ (r,\ z,\ t)$ into Equation (18-31) to find $\overline{C}\ (z, t)$. All the intermediate steps are given on the CRE Web site in the *Professional Reference Shelf*, and the partial differential equation describing the variation of the average axial concentration with time and distance is

$$\frac{\partial \overline{C}}{\partial t} + U\frac{\partial \overline{C}}{\partial z^*} = D^*\frac{\partial^2 \overline{C}}{\partial z^{*2}} \tag{18-32}$$

where D^* is the Aris-Taylor dispersion coefficient

Aris-Taylor
dispersion
coefficient

$$\boxed{D^* = D_{AB} + \frac{U^2 R^2}{48 D_{AB}}} \tag{18-33}$$

That is, for laminar flow in a pipe

$$D_a \equiv D^*$$

Figure 18-10 shows the dispersion coefficient D^* in terms of the ratio $D^*/U(2R) = D^*/Ud_t$ as a function of the product of the Reynolds (Re) and Schmidt (Sc) numbers.

18.4.5 Correlations for D_a

We will use correlations from the literature to determine the dispersion coefficient D_a for flow in cylindrical tubes (pipes) and for flow in packed beds.

18.4.5A Dispersion for Laminar and Turbulent Flow in Pipes

An estimate of the dispersion coefficient, D_a, can be determined from Figure 18-11. Here, d_t is the tube diameter and Sc is the Schmidt number discussed in Chapter 14. The flow is laminar (streamline) below 2,100, and we see the ratio

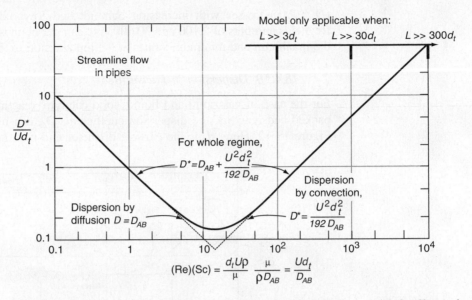

Figure 18-10   Correlation for dispersion for streamline flow in pipes. (Levenspiel, O., *Chemical Reaction Engineering*, 2nd ed. Copyright © 1972 John Wiley & Sons, Inc. Reprinted by permission of John Wiley & Sons, Inc. All rights reserved.) [Note: $D \equiv D_a$]

Once the Reynolds number is calculated, D_a can be found.

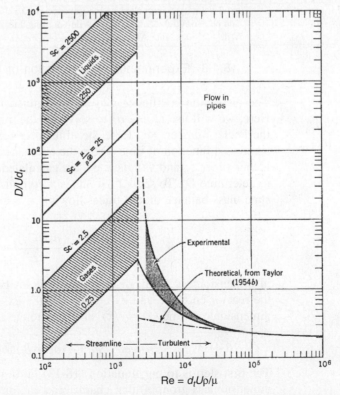

Figure 18-11   Correlation for dispersion of fluids flowing in pipes. (Levenspiel, O., *Chemical Reaction Engineering*, 2nd ed. Copyright © 1972 John Wiley & Sons, Inc. Reprinted by permission of John Wiley & Sons, Inc. All rights reserved.) [*Note*: $D \equiv D_a$]

(D_a/Ud_t) increases with increasing Schmidt and Reynolds numbers. Between Reynolds numbers of 2,100 and 30,000, one can put bounds on D_a by calculating the maximum and minimum values at the top and bottom of the shaded regions.

18.4.5B Dispersion in Packed Beds

For the case of gas–solid and liquid–solid catalytic reactions that take place in packed-bed reactors, the dispersion coefficient, D_a, can be estimated by using Figure 18-12. Here, d_p is the particle diameter and ε is the porosity.

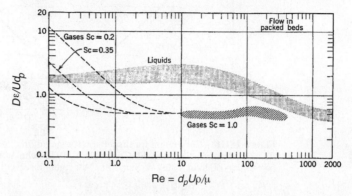

Figure 18-12 Experimental findings on dispersion of fluids flowing with mean axial velocity u in packed beds. (Levenspiel. O., *Chemical Reaction Engineering*, 2nd ed. Copyright © 1972 John Wiley & Sons, Inc. Reprinted by permission of John Wiley & Sons, Inc. All rights reserved.) [*Note*: $D \equiv D_a$]

18.4.6 Experimental Determination of D_a

The dispersion coefficient can be determined from a pulse tracer experiment. Here, we will use t_m and σ^2 to solve for the dispersion coefficient D_a and then the Peclet number, Pe_r. Here the effluent concentration of the reactor is measured as a function of time. From the effluent concentration data, the mean residence time, t_m, and variance, σ^2, are calculated, and these values are then used to determine D_a. To show how this is accomplished, we will write the unsteady state mass balance on the tracer flowing in a tubular reactor

$$D_a\frac{\partial^2 C_T}{\partial z^2} - \frac{\partial(UC_T)}{\partial z} = \frac{\partial C_T}{\partial t} \tag{18-13}$$

in dimensionless form, discuss the different types of boundary conditions at the reactor entrance and exit, solve for the exit concentration as a function of dimensionless time ($\Theta = t/\tau$), and then relate D_a, σ^2, and τ.

18.4.6A The Unsteady-State Tracer Balance

The first step is to put Equation (18-13) in dimensionless form to arrive at the dimensionless group(s) that characterize the process. Let

$$\psi = \frac{C_T}{C_{T0}}, \quad \lambda = \frac{z}{L}, \quad \text{and} \quad \Theta = \frac{tU}{L}$$

For a pulse input, C_{T0} is defined as the mass of tracer injected, M, divided by the vessel volume, V. Then

$$\boxed{\frac{1}{Pe_r}\frac{\partial^2\psi}{\partial\lambda^2}-\frac{\partial\psi}{\partial\lambda}=\frac{\partial\psi}{\partial\Theta}}$$

(18-34)

The initial condition is

Initial condition   At $t=0$, $z>0$, $C_T(0^+,0)=0$, $\psi(0^+)=0$ (18-35)

The mass of tracer injected, M, is

$$M = UA_c\int_0^\infty C_T(0^-,t)\,dt$$

18.4.6B Solution for a Closed-Closed System

In dimensionless form, the Danckwerts boundary conditions are

At $\lambda=0$: $\left(-\dfrac{1}{Pe_r}\dfrac{\partial\psi}{\partial\lambda}\right)_{\lambda=0^+}+\psi(0^+)=\dfrac{C_T(0^-,t)}{C_{T0}}=1$ (18-36)

At $\lambda=1$: $\dfrac{\partial\psi}{\partial\lambda}=0$ (18-37)

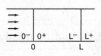

Equation (18-34) has been solved numerically for a pulse injection, and the resulting dimensionless effluent tracer concentration, ψ_{exit}, is shown as a function of the dimensionless time Θ in Figure 18-13 for various Peclet numbers. Although analytical solutions for ψ can be found, the result is an infinite series. The corresponding equations for the mean residence time, t_m, and the variance, σ^2, are[9]

$$\boxed{t_m = \tau}$$

(18-38)

and

$$\frac{\sigma^2}{t_m^2}=\frac{1}{\tau^2}\int_0^\infty (t-\tau)^2 E(t)\,dt$$

which can be used with the solution to Equation (18-34) to obtain

---

[9] See K. Bischoff and O. Levenspiel, *Adv. Chem. Eng.*, 4, 95 (1963).

Effects of
dispersion on the
effluent tracer
concentration

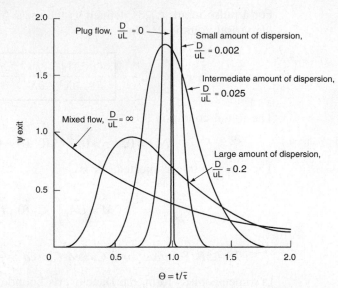

Figure 18-13 *C*-curves in closed vessels for various extents of back-mixing as predicted by the dispersion model. (Levenspiel, O., *Chemical Reaction Engineering*, 2nd ed. Copyright © 1972 John Wiley & Sons, Inc. Reprinted by permission of John Wiley & Sons, Inc. All rights reserved.) [*Note*: $D \equiv D_a$][10]

Calculating Pe_r
using t_m and σ^2
determined from
RTD data for a
closed-closed
system

$$\frac{\sigma^2}{t_m^2} = \frac{2}{Pe_r} - \frac{2}{Pe_r^2}\,(1 - e^{-Pe_r}) \qquad (18\text{-}39)$$

Consequently, we see that the Peclet number, Pe_r (and hence D_a), can be found experimentally by determining t_m and σ^2 from the RTD data and then solving Equation (18-39) for Pe_r.

18.4.6C Open-Open Vessel Boundary Conditions

When a tracer is injected into a packed bed at a location more than two or three particle diameters downstream from the entrance and measured some distance upstream from the exit, the open-open vessel boundary conditions apply. For an open-open system, an analytical solution to Equation (18-14) can be obtained for a pulse tracer input.

For an open-open system, the boundary conditions at the entrance are

$$F_T(0^-, t) = F_T(0^+, t)$$

---

[10]O. Levenspiel, *Chemical Reaction Engineering*, 2nd ed. (New York: Wiley, 1972), p. 277.

Then, for the case when the dispersion coefficient is the same in the entrance and reaction sections

Open at the entrance

$$-D_a\left(\frac{\partial C_T}{\partial z}\right)_{z=0^-} + UC_T(0^-,t) = -D_a\left(\frac{\partial C_T}{\partial z}\right)_{z=0^+} + UC_T(0^+,t) \quad (18\text{-}40)$$

Because there are no discontinuities across the boundary at $z = 0$

$$C_T(0^-,t) = C_T(0^+,t) \quad (18\text{-}41)$$

At the exit

Open at the exit

$$-D_a\left(\frac{\partial C_T}{\partial z}\right)_{z=L^-} + UC_T(L^-,t) = -D_a\left(\frac{\partial C_T}{\partial z}\right)_{z=L^+} + UC_T(L^+,t) \quad (18\text{-}42)$$

$$C_T(L^-,t) = C_T(L^+,t) \quad (18\text{-}43)$$

There are a number of perturbations of these boundary conditions that can be applied. The dispersion coefficient can take on different values in each of the three regions ($z < 0$, $0 \le z \le L$, and $z > L$), and the tracer can also be injected at some point z_1 rather than at the boundary, $z = 0$. These cases and others can be found in the supplementary readings cited at the end of the chapter. We shall consider the case when there is no variation in the dispersion coefficient for all z and an impulse of tracer is injected at $z = 0$ at $t = 0$.

For long tubes ($Pe_r > 100$) in which the concentration gradient at $\pm \infty$ will be zero, the solution to Equation (18-34) at the exit is[11]

Valid for $Pe_r > 100$

$$\psi(1,\Theta) = \frac{C_T(L,t)}{C_{T0}} = \frac{1}{2\sqrt{\pi\Theta/Pe_r}}\exp\left[\frac{-(1-\Theta)^2}{4\Theta/Pe_r}\right] \quad (18\text{-}44)$$

The mean residence time for an open-open system is

Calculate τ for an open-open system.

$$t_m = \left(1 + \frac{2}{Pe_r}\right)\tau \quad (18\text{-}45)$$

where τ is based on the volume between $z = 0$ and $z = L$ (i.e., reactor volume measured with a yardstick). We note that the mean residence time for an open system is greater than that for a closed system. The reason is that the molecules can diffuse back into the reactor after they diffuse out at the entrance. The variance for an open-open system is

Calculate Pe_r for an open–open system.

$$\frac{\sigma^2}{\tau^2} = \frac{2}{Pe_r} + \frac{8}{Pe_r^2} \quad (18\text{-}46)$$

---

[11]W. Jost, *Diffusion in Solids, Liquids and Gases* (New York: Academic Press, 1960), pp. 17, 47.

We now consider two cases for which we can use Equations (18-39) and (18-46) to determine the system parameters:

Case 1. The space time τ is *known*. That is, V and v_0 are measured independently. Here, we can determine the Peclet number by determining t_m and σ^2 from the concentration–time data and then use Equation (18-46) to calculate Pe_r. We can also calculate t_m and then use Equation (18-45) as a check, but this is usually less accurate.

Case 2. The space time τ is *unknown*. This situation arises when there are dead or stagnant pockets that exist in the reactor along with the dispersion effects. To analyze this situation, we first calculate mean residence time, t_m, and the variance, σ^2, from the data as in case 1. Then, we use Equation (18-45) to eliminate τ^2 from Equation (18-46) to arrive at

$$\boxed{\frac{\sigma^2}{t_{\mathrm{m}}^2} = \frac{2Pe_r + 8}{Pe_r^2 + 4Pe_r + 4}} \qquad (18\text{-}47)$$

Finding the effective reactor voume

We now can solve for the Peclet number in terms of our experimentally determined variables σ^2 and t_{m}^2. Knowing Pe_r, we can solve Equation (18-45) for τ, and hence V. The dead volume is the difference between the measured volume (i.e., with a yardstick) and the effective volume calculated from the RTD.

Example 18–1 Conversion Using Dispersion and Tanks-in-Series Models

The first-order reaction

$$A \longrightarrow B$$

is carried out in a 10-cm-diameter tubular reactor 6.36 m in length. The specific reaction rate is 0.25 min^{-1}. The results of a tracer test carried out on this reactor are shown in Table E18-1.1.

TABLE E18-1.1   EFFLUENT TRACER CONCENTRATION AS A FUNCTION OF TIME

t (min)	0	1	2	3	4	5	6	7	8	9	10	12	14
C (mg/L)	0	1	5	8	10	8	6	4	3	2.2	1.5	0.6	0

Calculate the conversion using **(a)** the closed vessel dispersion model, **(b)** PFR, **(c)** the tanks-in-series model, and **(d)** a single CSTR.

Solution

(a) We will use Equation (18-27) to calculate the conversion

$$X = 1 - \frac{4q \exp(Pe_r/2)}{(1+q)^2 \exp(Pe_r q/2) - (1-q)^2 \exp(-Pe_r q/2)} \qquad (18\text{-}27)$$

where $q = \sqrt{1 + 4\mathbf{Da_1}/Pe_r}$, $\mathbf{Da_1} = \tau k$, and $Pe_r = UL/D_a$.

(1) Parameter evaluation using the RTD data to evaluate Pe_r:

We can calculate Pe_r from Equation (18-39)

$$\frac{\sigma^2}{\tau^2} = \frac{2}{Pe_r} - \frac{2}{Pe_r^2}(1 - e^{-Pe_r}) \tag{18-39}$$

First calculate t_m and σ^2 from RTD data.

However, we must find τ^2 and σ^2 from the tracer concentration data first.

$$\tau = \int_0^\infty tE(t)\,dt = \frac{V}{\upsilon} \tag{E18-1.1}$$

$$\sigma^2 = \int_0^\infty (t-\tau)^2 E(t)\,dt \tag{E18-1.2}$$

We note that this is the same data set used in Examples 16-1 and 16-2

TABLE E18-1.2   POLYMATH PROGRAM AND RESULTS TO CALCULATE THE
MEAN RESIDENCE TIME, t_m, AND THE VARIANCE σ^2

POLYMATH Report
Ordinary Differential Equations

Here again, spreadsheets can be used to calculate τ^2 and σ^2.

Calculated values of DEQ variables

	Variable	Initial value	Final value
1	Area	51.	51.
2	C	0.0038746	0.0148043
3	C1	0.0038746	-387.266
4	C2	-33.43818	0.0148043
5	E	7.597E-05	0.0002903
6	Sigma2	0	6.212473
7	t	0	14.
8	tmf	5.1	5.1

Differential equations
1 d(Sigma2)/d(t) = (t-tmf)^2*E

Explicit equations
1 C1 = 0.0038746 + 0.2739782*t + 1.574621*t^2 - 0.2550041*t^3

2 Area = 51

3 C2 = -33.43818 + 37.18972*t - 11.58838*t^2 + 1.695303*t^3 - 0.1298667*t^4 + 0.005028*t^5 - 7.743*10^-5*t^6

4 C = If(t<=4 and t>=0) then C1 else if(t>4 and t<=14) then C2 else 0

5 E = C/Area

6 tmf = 5.1

where we found

$$t_m = 5.15 \text{ minutes}$$

and

Don't fall asleep. These are calculations we need to know how to carry out.

$$\sigma^2 = 6.1 \text{ minutes}^2$$

We will use these values in Equation 18-39 to calculate Pe_r.

Dispersion in a closed vessel is represented by

Calculate Pe_r from t_m and σ^2.

$$\frac{\sigma^2}{\tau^2} = \frac{2}{Pe_r^2}(Pe_r - 1 + e^{-Pe_r}) \tag{18-39}$$

$$= \frac{6.1}{(5.15)^2} = 0.23 = \frac{2}{Pe_r^2}(Pe_r - 1 + e^{-Pe_r})$$

Solving for Pe_r either by trial and error or using Polymath, we obtain

$$\boxed{Pe_r = 7.5} \tag{E18-1.3}$$

(2) Next, we calculate Da_1 and q:

Next, calculate Da_1, q, and X.

$$\boxed{Da_1 = \tau k = (5.15 \text{ min})(0.25 \text{ min}^{-1}) = 1.29} \tag{E18-1.4}$$

Using the equations for q and X gives

$$q = \sqrt{1 + \frac{4Da_1}{Pe_r}} = \sqrt{1 + \frac{4(1.29)}{7.5}} = 1.30 \tag{E18-1.5}$$

Then

$$\frac{Pe_r q}{2} = \frac{(7.5)(1.3)}{2} = 4.87 \qquad \text{(E18-1.6)}$$

(3) Finally, we calculate the conversion:
Substitution into Equation (18-27) yields

Dispersion model

$$X = 1 - \frac{4(1.30)\, e^{(7.5/2)}}{(2.3)^2 \exp(4.87) - (-0.3)^2 \exp(-4.87)}$$

$$\boxed{X = 0.68} \qquad 68\% \text{ conversion for the dispersion model}$$

When dispersion effects are present in this tubular reactor, 68% conversion is achieved.

(b) If the reactor were operating ideally as a plug-flow reactor, the conversion would be

PFR

$$\boxed{X = 1 - e^{-\tau k} = 1 - e^{-\mathrm{Da}_1} = 1 - e^{-1.29} = 0.725} \qquad \text{(E18-1.7)}$$

That is, 72.5% conversion would be achieved in an ideal plug-flow reactor.

Tanks-in-series model

(c) Conversion using the tanks-in-series model: We recall Equation (18-11) to calculate the number of tanks in series:

$$n = \frac{\tau^2}{\sigma^2} = \frac{(5.15)^2}{6.1} = 4.35 \qquad \text{(E18-1.8)}$$

To calculate the conversion for the T-I-S model, we recall Equation (5-15). For a first-order reaction for n tanks in series, the conversion is

$$\boxed{\begin{aligned} &X = 1 - \frac{1}{(1 + \tau_i k)^n} = 1 - \frac{1}{[1 + (\tau/n)k]^n} = 1 - \frac{1}{(1 + 1.29/4.35)^{4.35}} \\ &X = \textbf{67.7\% for the tanks-in-series model} \end{aligned}} \qquad \text{(E18-1.9)}$$

(d) For a single CSTR

CSTR

$$X = \frac{\tau k}{1 + \tau k} = \frac{1.29}{2.29} = 0.563 \qquad \text{(E18-1.10)}$$

So, 56.3% conversion would be achieved in a single ideal tank.
Summary:

> PFR: $X = 72.5\%$
>
> Dispersion: $X = 68.0\%$
>
> Tanks in series: $X = 67.7\%$
>
> Single CSTR: $X = 56.3\%$

Summary

In this example, correction for finite dispersion, whether by a dispersion model or a tanks-in-series model, is significant when compared with a PFR.

Analysis: This example is a very important and comprehensive one. We showed how to calculate the conversion by (1) choosing a model, (2) using the RTD to evaluate the model parameters, and (3) substituting the reaction-rate parameters in the chosen model. As expected, the dispersion and T-I-S model gave essentially the same result and this result fell between the limits predicted by an ideal PFR and an ideal CSTR.

18.5 Tanks-in-Series Model versus Dispersion Model

We have seen that we can apply both of these one-parameter models to tubular reactors using the variance of the RTD. For first-order reactions, the two models can be applied with equal ease. However, the tanks-in-series model is mathematically easier to use to obtain the effluent concentration and conversion for reaction orders other than one, and for multiple reactions. However, we need to ask what would be the accuracy of using the tanks-in-series model over the dispersion model. These two models are equivalent when the Peclet–Bodenstein number is related to the number of tanks in series, n, by the equation[12]

$$Bo = 2(n - 1) \qquad (18\text{-}48)$$

Equivalency between models of tanks-in-series and dispersion

or

$$n = \frac{Bo}{2} + 1 \qquad (18\text{-}49)$$

where

$$Bo = UL/D_a \qquad (18\text{-}50)$$

where U is the superficial velocity, L the reactor length, and D_a the dispersion coefficient.

For the conditions in Example 18-1, we see that the number of tanks calculated from the Bodenstein number, Bo (i.e., Pe_r), Equation (18-49), is 4.75, which is very close to the value of 4.35 calculated from Equation (18-11). Consequently, for reactions other than first order, one would solve successively for the exit concentration and conversion from each tank in series for both a battery of four tanks in series and for five tanks in series in order to bound the expected values.

In addition to the one-parameter models of tanks-in-series and dispersion, many other one-parameter models exist when a combination of ideal reactors is used to model the real reactor shown in Section 18.7 for reactors with bypassing and dead volume. Another example of a one-parameter model would be to model the real reactor as a PFR and a CSTR in series with the one parameter being the fraction of the total volume that behaves as a CSTR. We can dream up many other situations that would alter the behavior of ideal reactors in a way that adequately describes a real reactor. However, it may be that one parameter is not sufficient to yield an adequate comparison between theory and practice. We explore these situations with combinations of ideal reactors in the section on two-parameter models.

The reaction-rate parameters are usually known (e.g., Da), but the Peclet number is usually not known because it depends on the flow and the vessel. Consequently, we need to find Pe_r using one of the three techniques discussed earlier in the chapter.

---

[12]K. Elgeti, *Chem. Eng. Sci.,* 51, 5077 (1996).

18.6 Numerical Solutions to Flows with Dispersion and Reaction

We now consider dispersion and reaction in a tubular reactor. We first write our mole balance on species A in cylindrical coordinates by recalling Equation (18-28) and including the rate of formation of A, r_A. At steady state we obtain

$$D_{AB}\left[\frac{1}{r}\frac{\partial\left(r\frac{\partial C_A}{\partial r}\right)}{\partial r} + \frac{\partial^2 C_A}{\partial z^2}\right] - u(r)\frac{\partial C_A}{\partial z} + r_A = 0 \qquad (18\text{-}51)$$

Analytical solutions to dispersion with reaction can only be obtained for isothermal zero- and first-order reactions. We are now going to use COMSOL to solve the flow with reaction and dispersion with reaction.

   We are going to compare two solutions: one which uses the Aris–Taylor approach and one in which we numerically solve for both the axial and radial concentration using COMSOL. These solutions are on the CRE Web site.

Case A. Aris-Taylor Analysis for Laminar Flow

For the case of an nth-order reaction, Equation (18-15) is

$$\frac{D_a}{U}\frac{d^2\overline{C}_A}{dz^2} - \frac{d\overline{C}_A}{dz} - \frac{k\overline{C}_A^n}{U} = 0 \qquad (18\text{-}52)$$

where $\overline{C}_A$ is the average concentration from $r = 0$ to $r = R$, i.e.,

$$\overline{C}_A = \frac{\int_0^r C_A(r,z)dr}{R}$$

If we use the Aris-Taylor analysis, we can use Equation (18-15) with a caveat that $\overline{\psi} = \overline{C}_A/C_{A0}$ and $\lambda = z/L$ we obtain

$$\frac{1}{Pe_r}\frac{d^2\overline{\psi}}{d\lambda^2} - \frac{d\overline{\psi}}{d\lambda} - Da_n\overline{\psi}^n = 0 \qquad (18\text{-}53)$$

where

$$Pe_r = \frac{UL}{D_a} \text{ and } Da_n = \tau k C_{A0}^{n-1}$$

For the closed-closed boundary conditions we have

$$\text{At} \quad \lambda = 0: \quad -\frac{1}{Pe_r}\frac{d\overline{\psi}}{d\lambda}\bigg|_{\lambda=0^+} + \overline{\psi}(0^+) = 1 \qquad (18\text{-}54)$$

Danckwerts boundary conditions

$$\text{At} \quad \lambda = 1: \quad \frac{d\overline{\psi}}{d\lambda} = 0$$

For the open-open boundary conditions we have

$$\text{At} \quad \lambda = 0: \quad \overline{\psi}(0^-) - \frac{1}{Pe_r}\frac{d\overline{\psi}}{d\lambda}\bigg|_{\lambda = 0^-} = \overline{\psi}(0^+) - \frac{1}{Pe_r}\frac{d\overline{\psi}}{d\lambda}\bigg|_{\lambda = 0^+} \tag{18-55}$$

$$\text{At} \quad \lambda = 1: \quad \frac{d\overline{\psi}}{d\lambda} = 0$$

Equation (18-53) is a nonlinear second-order ODE that is solved on the COMSOL on the CRE Web site.

Case B. Full Numerical Solution

To obtain profiles, $C_A(r,z)$, we now solve Equation (18-51)

$$D_{AB}\left[\frac{1}{r}\frac{\partial\left(r\dfrac{\partial C_A}{\partial r}\right)}{\partial r} + \frac{\partial^2 C_A}{\partial z^2}\right] - u(r)\frac{\partial C_A}{\partial z} + r_A = 0 \tag{18-51}$$

First, we will put the equations in dimensionless form by letting $\psi = C_A/C_{A0}$, $\lambda = z/L$, and $\phi = r/R$. Following our earlier transformation of variables, Equation (18-52) becomes

$$\left(\frac{L}{R}\right)\frac{1}{Pe_r}\left[\frac{1}{\phi}\frac{\partial\left(\phi\dfrac{\partial\psi}{\partial\phi}\right)}{\partial\phi}\right] + \frac{1}{Pe_r}\frac{d^2\psi}{d\lambda^2} - 2(1 - \phi^2)\frac{d\psi}{d\lambda} - \boldsymbol{Da_n}\psi^n = 0 \tag{18-56}$$

Equation (18-56) gives the dimensionless concentration profiles for dispersion and reaction in a laminar-flow reactor. The Expanded Material on the CRE Web site gives an example, Web Example 18-2, where COMSOL is used to find the concentration profile.

18.7 Two-Parameter Models—Modeling Real Reactors with Combinations of Ideal Reactors

Creativity and engineering judgment are necessary for model formulation.

We now will see how a real reactor might be modeled by different combinations of ideal reactors. Here, an almost unlimited number of combinations that could be made. However, if we limit the number of adjustable parameters to two (e.g., bypass flow rate, v_b, and dead volume, V_D), the situation becomes much more tractable. After reviewing the steps in Table 18-1, choose a model and determine if it is reasonable by qualitatively comparing it with the RTD and, if it is, determine the model parameters. Usually, the simplest means of obtaining the necessary data is some form of a tracer test. These tests have been described in Chapters 16 and 17, together with their uses in determining the RTD of a reactor system. Tracer tests can be used to determine the RTD, which can then be used in a similar manner to determine the suitability of the model and the value of its parameters.

A tracer experiment is used to evaluate the model parameters.

In determining the suitability of a particular reactor model and the parameter values from tracer tests, it may not be necessary to calculate the RTD function $E(t)$. The model parameters (e.g., V_D) may be acquired directly from measurements of effluent concentration in a tracer test. The theoretical prediction of the particular tracer test in the chosen model system is compared with the tracer measurements from the real reactor. The parameters in the model are chosen so as to obtain the closest possible agreement between the model and experiment. If the agreement is then sufficiently close, the model is deemed reasonable. If not, another model must be chosen.

The quality of the agreement necessary to fulfill the criterion "sufficiently close" again depends on creativity in developing the model and on engineering judgment. The most extreme demands are that the maximum error in the prediction not exceed the estimated error in the tracer test, and that there be no observable trends with time in the difference between prediction (the model) and observation (the real reactor). To illustrate how the modeling is carried out, we will now consider two different models for a CSTR.

18.7.1 Real CSTR Modeled Using Bypassing and Dead Space

A real CSTR is believed to be modeled as a combination of an ideal CSTR with a well-mixed volume V_s, a dead zone of volume V_d, and a bypass with a volumetric flow rate v_b (Figure 18-14). We have used a tracer experiment to evaluate the parameters of the model V_s and v_s. Because the total volume and volumetric flow rate are known, once V_s and v_s are found, v_b and V_d can readily be calculated.

The model system

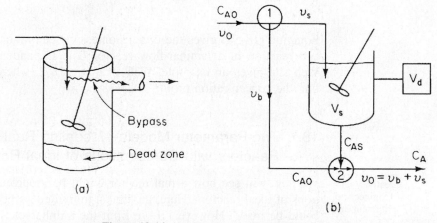

Figure 18-14 (a) Real system; (b) model system.

18.7.1A Solving the Model System for C_A and X

We shall calculate the conversion for this model for the first-order reaction

$$A \longrightarrow B$$

The *Duct Tape Council of Jofostan* would like to point out the new wrinkle: **The Junction Balance.**

The bypass stream and effluent stream from the reaction volume are mixed at the junction point 2. From a balance on species A around this point

$$[\text{In}] = [\text{Out}]$$

$$[C_{A0}v_b + C_{As}v_s] = [C_A(v_b + v_s)] \tag{18-57}$$

We can solve for the concentration of A leaving the reactor

$$C_A = \frac{v_b C_{A0} + C_{As} v_s}{v_b + v_s} = \frac{v_b C_{A0} + C_{As} v_s}{v_0}$$

Let $\alpha = V_s/V$ and $\beta = v_b/v_0$. Then

$$\boxed{C_A = \beta C_{A0} + (1 - \beta) C_{As}} \tag{18-58}$$

For a first-order reaction, a mole balance on V_s gives

Mole balance on CSTR

$$v_s C_{A0} - v_s C_{As} - k C_{As} V_s = 0 \tag{18-59}$$

or, in terms of α and β

$$\boxed{C_{As} = \frac{C_{A0}(1 - \beta) v_0}{(1 - \beta) v_0 + \alpha V k}} \tag{18-60}$$

Substituting Equation (18-60) into (18-58) gives the effluent concentration of species A:

Conversion as a function of model parameters

$$\boxed{\frac{C_A}{C_{A0}} = 1 - X = \beta + \frac{(1 - \beta)^2}{(1 - \beta) + \alpha \tau k}} \tag{18-61}$$

We have used the ideal reactor system shown in Figure 18-14 to predict the conversion in the real reactor. The model has two parameters, α and β. The parameter α is the dead zone volume fraction and parameter β is the fraction of the volumetric flow rate that bypasses the reaction zone. If these parameters are known, we can readily predict the conversion. In the following section, we shall see how we can use tracer experiments and RTD data to evaluate the model parameters.

18.7.1B Using a Tracer to Determine the Model Parameters in a CSTR-with-Dead-Space-and-Bypass Model

Model system

In Section 18.7.1A, we used the system shown in Figure 18-15, with bypass flow rate, v_b, and dead volume, V_d, to model our real reactor system. We shall inject our tracer, T, as a positive-step input. The unsteady-state balance on the nonreacting tracer, T, in the well-mixed reactor volume, V_s, is

$$\text{In} - \text{out} = \text{accumulation}$$

Tracer balance for step input

$$\boxed{v_s C_{T0} - v_s C_{Ts} = \frac{dN_{Ts}}{dt} = V_s \frac{dC_{Ts}}{dt}} \tag{18-62}$$

The conditions for the positive-step input are

$$\text{At } t < 0 \quad C_T = 0$$

$$\text{At } t \geq 0 \quad C_T = C_{T0}$$

A balance around junction point 2 gives

The junction balance

$$\boxed{C_T = \frac{v_b C_{T0} + C_{Ts} v_s}{v_0}} \tag{18-63}$$

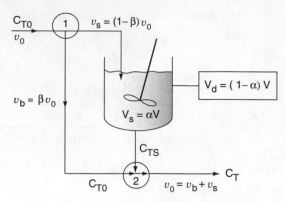

The model
system

Figure 18-15   Model system: CSTR with dead volume and bypassing.

As before

$$V_s = \alpha V$$

$$v_b = \beta v$$

$$\tau = \frac{V}{v_0}$$

Integrating Equation (18-62) and substituting in terms of α and β gives

$$\frac{C_{Ts}}{C_{T0}} = 1 - \exp\left[-\frac{1-\beta}{\alpha}\left(\frac{t}{\tau}\right)\right] \qquad (18\text{-}64)$$

Combining Equations (18-63) and (18-64), the effluent tracer concentration is

$$\frac{C_T}{C_{T0}} = 1 - (1-\beta)\exp\left[-\frac{1-\beta}{\alpha}\left(\frac{t}{\tau}\right)\right] \qquad (18\text{-}65)$$

We now need to rearrange this equation to extract the model parameters, α and β, either by regression (Polymath/MATLAB/Excel) or from the proper plot of the effluent tracer concentration as a function of time. Rearranging yields

Evaluating the
model parameters

$$\boxed{\ln\frac{C_{T0}}{C_{T0}-C_T} = \ln\frac{1}{1-\beta} + \left(\frac{1-\beta}{\alpha}\right)\frac{t}{\tau}} \qquad (18\text{-}66)$$

Consequently, we plot $\ln[C_{T0}/(C_{T0} - C_T)]$ as a function of t. If our model is correct, a straight line should result with a slope of $(1 - \beta)/\tau\alpha$ and an intercept of $\ln[1/(1 - \beta)]$.

Example 18–2   CSTR with Dead Space and Bypass

The elementary reaction

$$A + B \longrightarrow C + D$$

is to be carried out in the CSTR shown schematically in Figure 18-15. There is both bypassing and a stagnant region in this reactor. The tracer output for this reactor is shown in Table E18-2.1. The measured reactor volume is 1.0 m³ and the flow rate to the reactor is 0.1 m³/min. The reaction-rate constant is 0.28 m³/kmol·min. The feed is equimolar in A and B with an entering concentration of A equal to 2.0 kmol/m³. Calculate the conversion that can be expected in this reactor (Figure E18-2.1).

TABLE E18-2.1   TRACER DATA FOR STEP INPUT

C_T (mg/dm³)	1000	1333	1500	1666	1750	1800
t (min)	4	8	10	14	16	18

The entering tracer concentration is $C_{T0} = 2000$ mg/dm³.

Two-parameter model

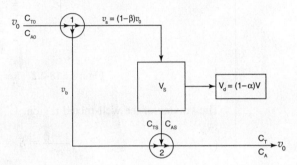

Figure E18-2.1   Schematic of real reactor modeled with dead space (V_d) and bypass (v_b).

Solution

Recalling Equation (18-66)

$$ln\,\frac{C_{T0}}{C_{T0}-C_T} = ln\,\frac{1}{1-\beta} + \frac{(1-\beta)}{\alpha}\,\frac{t}{\tau} \qquad (18\text{-}66)$$

Equation (18-66) suggests that we construct Table E18-2.2 from Table E18-2.1 and plot $C_{T0}/(C_{T0} - C_T)$ as a function of time on semilog paper. Using this table we get Figure E18-2.2.

TABLE E18-2.2   PROCESSED DATA

t (min)	4	8	10	14	16	18
$\dfrac{C_{T0}}{C_{T0}-C_T}$	2	3	4	6	8	10

We can find α and β from either a semilog plot, as shown in Figure E18-2.2, or by regression using Polymath, MATLAB, or Excel.

The volumetric flow rate to the well-mixed portion of the reactor, v_s, can be determined from the intercept, I

Evaluating the parameters α and β

$$\frac{1}{1-\beta} = I = 1.25$$

$$\boxed{\beta = \frac{v_b}{v_0} = 0.2}$$

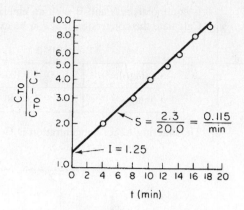

Figure E18-2.2   Response to a step input.

The volume of the well-mixed region, V_s, can be calculated from the slope, S,

$$\frac{1-\beta}{\alpha\tau} = S = 0.115 \text{ min}^{-1}$$

$$\alpha\tau = \frac{1-0.2}{0.115} = 7 \text{ min}$$

$$\tau = \frac{V}{v_0} = \frac{1 \text{ m}^3}{(0.1 \text{ m}^3/\text{min})} = 10 \text{ min}$$

$$\boxed{\alpha = \frac{7 \text{ min}}{\tau} = 0.7}$$

We now proceed to determine the conversion corresponding to these model parameters.

1. **Balance on reactor volume V_s:**

[In] − [Out] + [Generation] = [Accumulation]

$$v_s C_{A0} - v_s C_{As} + r_{As} V_s \qquad = 0 \qquad\qquad \text{(E18-2.1)}$$

2. **Rate law:**

$$-r_{AS} = kC_{As}C_{Bs}$$

Equalmolar feed $\therefore C_{As} = C_{Bs}$

$$-r_{As} = kC_{As}^2 \qquad\qquad\qquad \text{(E18-2.2)}$$

3. **Combining** Equations (E18-2.1) and (E18-2.2) gives

$$v_s C_{A0} - v_s C_{As} - kC_{As}^2 V_s = 0 \qquad\qquad \text{(E18-2.3)}$$

Rearranging, we have

$$\tau_s kC_{As}^2 + C_{As} - C_{A0} = 0 \qquad\qquad \text{(E18-2.4)}$$

Solving for C_{As} yields

$$\boxed{C_{As} = \frac{-1 + \sqrt{1 + 4\tau_s kC_{A0}}}{2\tau_s k}} \qquad\qquad \text{(E18-2.5)}$$

4. **Balance around junction** point 2:

$$[In] = [Out]$$

$$[v_b C_{A0} + v_s C_{As}] = [v_0 C_A] \qquad \text{(E18-2.6)}$$

Rearranging Equation (E18-4.6) gives us

$$\boxed{C_A = \frac{v_0 - v_s}{v_0} C_{A0} + \frac{v_s}{v_0} C_{As}} \qquad \text{(E18-2.7)}$$

5. **Parameter evaluation:**

$$v_s = 0.8 v_0 = (0.8)(0.1 \text{ m}^3/\text{min}) = 0.08 \text{ m}^3/\text{min}$$

$$V_s = (\alpha \tau) v_0 = (7.0 \text{ min})(0.1 \text{ m}^3/\text{min}) = 0.7 \text{ m}^3$$

$$\tau_s = \frac{V_s}{v_s} = 8.7 \text{ min}$$

$$C_{As} = \frac{\sqrt{1 + 4\tau_s k C_{A0}} - 1}{2\tau_s k} \qquad \text{(E18-2.8)}$$

$$= \frac{\sqrt{1 + (4)(8.7 \text{ min})(0.28 \text{ m}^3/\text{kmol} \cdot \text{min})(2 \text{ kmol/m}^3)} - 1}{(2)(8.7 \text{ min})(0.28 \text{ m}^3/\text{kmol} \cdot \text{min})}$$

$$= 0.724 \text{ kmol/m}^3$$

Substituting into Equation (E18-2.7) yields

$$C_A = \frac{0.1 - 0.08}{0.1} (2) + (0.8)(0.724) = 0.979$$

Finding the conversion

$$X = 1 - \frac{0.979}{2.0} = 0.51$$

If the real reactor were acting as an ideal CSTR, the conversion would be

$$\boxed{C_A = \frac{\sqrt{1 + 4\tau k C_{A0}} - 1}{2\tau k}} \qquad \text{(E18-2.9)}$$

$$C_A = \frac{\sqrt{1 + 4(10)(0.28)(2)} - 1}{2(10)(0.28)} = 0.685$$

$$\boxed{X = 1 - \frac{C_A}{C_{A0}} = 1 - \frac{0.685}{2.0} = 0.66} \qquad \text{(E18-2.10)}$$

$$X_{\text{model}} = 0.51$$
$$X_{\text{Ideal}} = 0.66$$

Analysis: In this example we used a combination of an ideal CSTR with a dead volume and bypassing to model a nonideal reactor. If the nonideal reactor behaved as an ideal CSTR, a conversion of 66% was expected. Because of the dead volume, not all the space would be available for reaction; also, some of the fluid did not enter the space where the reaction was taking place and, as a result, the conversion in this nonideal reactor was only 51%.

Other Models. In Section 18.7.1 it was shown how we formulated a model consisting of ideal reactors to represent a real reactor. First, we solved for the exit concentration and conversion for our model system in terms of two parameters, α and β. We next evaluated these parameters from data on tracer concentration as a function of time. Finally, we substituted these parameter values into the mole balance, rate law, and stoichiometric equations to predict the conversion in our real reactor.

To reinforce this concept, we will use one more example.

18.7.2 Real CSTR Modeled as Two CSTRs with Interchange

In this particular model there is a highly agitated region in the vicinity of the impeller; outside this region, there is a region with less agitation (Figure 18-16). There is considerable material transfer between the two regions. Both inlet and outlet flow channels connect to the highly agitated region. We shall model the highly agitated region as one CSTR, the quieter region as another CSTR, with material transfer between the two.

The model system

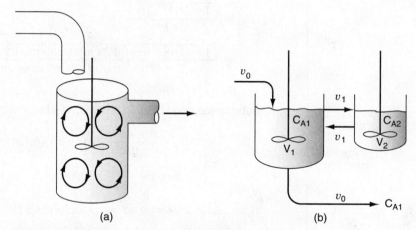

Figure 18-16   (a) Real reaction system; (b) model reaction system.

18.7.2A Solving the Model System for C_A and X

Let β represent that fraction of the total flow that is exchanged between reactors 1 and 2; that is,

$$v_1 = \beta v_0$$

and let α represent that fraction of the total volume, V, occupied by the highly agitated region:

Two parameters:
α and β

$$V_1 = \alpha V$$

Then

$$V_2 = (1 - \alpha)V$$

The space time is

$$\tau = \frac{V}{v_0}$$

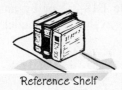

As shown on the CRE Web site *Professional Reference Shelf R18.2*, for a first-order reaction, the exit concentration and conversion are

$$C_{A1} = \frac{C_{A0}}{1 + \beta + \alpha\tau k - \{\beta^2/[\beta + (1-\alpha)\tau k]\}} \tag{18-67}$$

and

Conversion for two-CSTR model

$$\boxed{X = 1 - \frac{C_{A1}}{C_{A0}} = \frac{(\beta + \alpha\tau k)[\beta + (1-\alpha)\tau k] - \beta^2}{(1 + \beta + \alpha\tau k)[\beta + (1-\alpha)\tau k] - \beta^2}} \tag{18-68}$$

where C_{A1} is the reactor concentration exiting the first reactor in Figure 18-17(b).

18.7.2B Using a Tracer to Determine the Model Parameters in a CSTR with an Exchange Volume

The problem now is to evaluate the parameters α and β using the RTD data. A mole balance on a tracer pulse injected at $t = 0$ for each of the tanks is

$$\text{Accumulation} = \text{Rate in} - \text{Rate out}$$

Unsteady-state balance of inert tracer

Reactor 1: $$V_1\frac{dC_{T1}}{dt} = v_1 C_{T2} - (v_0 C_{T1} + v_1 C_{T1}) \tag{18-69}$$

Reactor 2: $$V_2\frac{dC_{T2}}{dt} = v_1 C_{T1} - v_1 C_{T2} \tag{18-70}$$

and C_{T1} is the measured tracer concentration existing the real reactor. The tracer is initially dumped only into reactor 1, so that the initial conditions $C_{T10} = N_{T0}/V_1$ and $C_{T20} = 0$.

Substituting in terms of α, β, and τ, we arrive at two coupled differential equations describing the unsteady behavior of the tracer that must be solved simultaneously.

See Appendix A.3 for method of solution

$$\boxed{\begin{aligned} \tau\alpha\,\frac{dC_{T1}}{dt} &= \beta C_{T2} - (1+\beta)C_{T1} \\[2mm] \tau(1-\alpha)\,\frac{dC_{T2}}{dt} &= \beta C_{T1} - \beta C_{T2} \end{aligned}}$$

$$\tag{18-71}$$
$$\tag{18-72}$$

Analytical solutions to Equations (18-71) and (18-72) are given on the CRE Web site, in Appendix A.3 and in Equation (18-73), below. However, for more complicated systems, analytical solutions to evaluate the system parameters may not be possible.

$$\left(\frac{C_{T1}}{C_{T10}}\right)_{\text{pulse}} = \frac{(\alpha m_1 + \beta + 1)e^{m_2 t/\tau} - (\alpha m_2 + \beta + 1)e^{m_1 t/\tau}}{\alpha(m_1 - m_2)} \tag{18-73}$$

where

$$m_1,\, m_2 = \left[\frac{1 - \alpha + \beta}{2\alpha(1-\alpha)}\right]\left[-1 \pm \sqrt{1 - \frac{4\alpha\beta(1-\alpha)}{(1-\alpha+\beta^2)}}\right]$$

By regression on Equation (18-73) and the data in Table E18-2.2 or by an appropriate semilog plot of C_{T1}/C_{T10} versus time, one can evaluate the model parameters α and β.

18.8 Use of Software Packages to Determine the Model Parameters

If analytical solutions to the model equations are not available to obtain the parameters from RTD data, one could use ODE solvers. Here, the RTD data would first be fit to a polynomial to the effluent concentration–time data and then compared with the model predictions for different parameter values.

Example 18–3   CSTR with Bypass and Dead Volume

(a) Determine parameters α and β that can be used to model two CSTRs with interchange using the tracer concentration data listed in Table E18-3.1.

TABLE E18-3.1   RTD DATA

t (min)	0.0	20	40	60	80	120	160	200	240
C_{Te} (g/m^3)	2000	1050	520	280	160	61	29	16.4	10.0

(b) Determine the conversion of a first-order reaction with $k = 0.03$ min^{-1} and $\tau = 40$ min.

Solution

First, we will use Polymath to fit the RTD to a polynomial. Because of the steepness of the curve, we shall use two polynomials.

For $t \le 80$ min

$$C_{Te} = 2000 - 59.6t + 0.642t^2 - 0.00146t^3 - 1.04 \times 10^{-5}t^4 \quad \text{(E18-3.1)}$$

For $t > 80$ min

$$C_{Te} = 921 - 17.3t + 0.129t^2 - 0.000438t^3 - 5.6 \times 10^{-7}t^4 \quad \text{(E18-3.2)}$$

where C_{Te} is the exit concentration of tracer determined experimentally. Next we would enter the tracer mole (mass) balances Equations (18-66) and (18-67) into an ODE solver. The Polymath program is shown in Table E18-3.2. Finally, we vary the parameters α and β and then compare the calculated effluent concentration C_{T1} with the experimental effluent tracer concentration C_{Te}. After a few trials, we converge on the values $\alpha = 0.8$ and $\beta = 0.1$. We see from Figure E18-3.1 and Table E18-3.3 that the agreement between the RTD data and the calculated data is quite good, indicating the validity of our values of α and β. The graphical solution to this problem is given in the Chapter 18 *Learning Resources 3, Solved Problems,* on the CRE Web site. We now substitute these values in Equation (18-68), and as shown on the CRE

Trial and error using software packages

Web site, the corresponding conversion is 51% for the model system of two CSTRs with interchange

$$X = 1 - \frac{C_{A1}}{C_{A0}} = \frac{(\beta + \alpha\tau k)[\beta + (1-\alpha)\tau k] - \beta^2}{(1+\beta+\alpha\tau k)[\beta+(1-\alpha)\tau k]-\beta^2} \qquad (18\text{-}68)$$

$$\tau k = (40 \text{ min})(0.03 \text{ min}^{-1}) = 1.2$$

$$X = \frac{[0.1+(0.8)(1.2)][0.1+(1-0.8)(1.2)]-(0.1)^2}{[1+0.1+(0.8)(1.2)][0.1+(1-0.8)(1.2)-(0.1)^2]}$$

$$X = 0.51$$

Comparing models, we find

$$(X_{\text{model}} = 0.51) < (X_{\text{CSTR}} = 0.55) < (X_{\text{PFR}} = 0.7)$$

Living Example Problem

TABLE E18-3.2   POLYMATH PROGRAM: TWO CSTRS WITH INTERCHANGE

ODE Report (RKF45)

Differential equations as entered by the user
[1] d(CT1)/d(t) = (beta*CT2-(1+beta)*CT1)/alpha/tau
[2] d(CT2)/d(t) = (beta*CT1-beta*CT2)/(1-alpha)/tau

Explicit equations as entered by the user
[1] beta = 0.1
[2] alpha = 0.8
[3] tau = 40
[4] CTe1 = 2000-59.6*t+0.64*t^2-0.00146*t^3-1.047*10^(-5)*t^4
[5] CTe2 = 921-17.3*t+0.129*t^2-0.000438*t^3+5.6*10^(-7)*t^4
[6] t1 = t-80
[7] CTe = if(t<80)then(CTe1)else(CTe2)

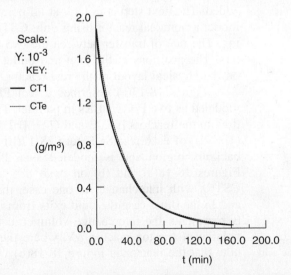

Scale:
Y: 10^{-3}
KEY:
— CT1
---- CTe

Figure E18-3.1   Comparison of model and experimental exit tracer concentrations.

TABLE E18-3.3   COMPARING MODEL (CT1) WITH EXPERIMENT (CTe)

t	CT1	CTe
0	2000	2000
10	1421.1968	1466.4353
20	1014.8151	1050.6448
30	728.9637	740.0993
40	527.4236	519.7568
50	384.9088	372.0625
60	283.7609	276.9488
70	211.6439	211.8353
80	159.9355	161.2816
100	95.43456	99
120	60.6222	61.8576
140	40.92093	40.6576
160	29.10943	28.3536

Two CSTRs with interchange

__Analysis:__ For the two-parameter model chosen, we used the RTD to determine the two parameters' dead volume and fraction of fluid bypassed. We then calculated the exit trace concentration using the ideal CSTR balance equations but with a lesser reactor volume and a smaller flow rate through the reactor and compared it with the experimental data.

18.9 Other Models of Nonideal Reactors Using CSTRs and PFRs

Several reactor models have been discussed in the preceding pages. All are based on the physical observation that in almost all agitated tank reactors, there is a well-mixed zone in the vicinity of the agitator. This zone is usually represented by a CSTR. The region outside this well-mixed zone may then be modeled in various fashions. We have already considered the simplest models, which have the main CSTR combined with a dead-space volume; if some short-circuiting of the feed to the outlet is suspected, a bypass stream can be added. The next step is to look at all possible combinations that we can use to model a nonideal reactor using only CSTRs, PFRs, dead volume, and bypassing. The rate of transfer between the two reactors is one of the model parameters. The positions of the inlet and outlet to the model reactor system depend on the physical layout of the real reactor.

Figure 18-17(a) describes a real PFR or PBR with channeling that is modeled as two PFRs/PBRs in parallel. The two parameters are the fraction of flow to the reactors [i.e., β and $(1 - \beta)$] and the fractional volume [i.e., α and $(1 - \alpha)$] of each reactor. Figure 18-17(b) describes a real PFR/PBR that has a backmix region and is modeled as a PFR/PBR in parallel with a CSTR. Figures 18-18(a) and (b) on page 884 show a real CSTR modeled as two CSTRs with interchange. In one case, the fluid exits from the top CSTR (a) and in the other case the fluid exits from the bottom CSTR (b). The parameter β represents the interchange volumetric flow rate, βv_0, and α the fractional volume of the top reactor, αV, where the fluid exits the reaction system. We note that the reactor in Figure 18-18(b) was found to describe extremely well

A case history for terephthalic acid

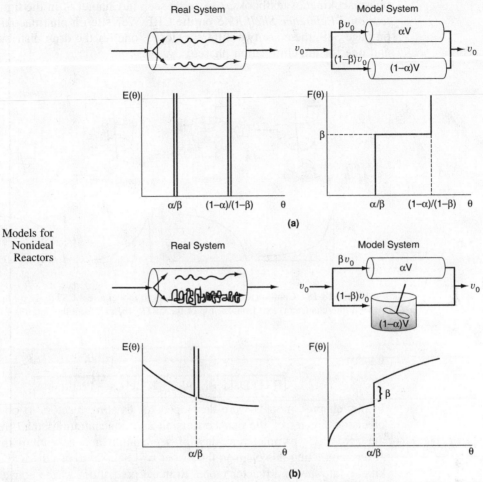

Figure 18-17 Combinations of ideal reactors used to model real tubular reactors: **(a)** two ideal PFRs in parallel; **(b)** ideal PFR and ideal CSTR in parallel.

Models for Nonideal Reactors

a real reactor used in the production of terephthalic acid.[13] A number of other combinations of ideal reactions can be found in Levenspiel.[14]

18.10 Applications to Pharmacokinetic Modeling

The use of combinations of ideal reactors to model metabolism and drug distribution in the human body is becoming commonplace. For example, one of the simplest models for drug adsorption and elimination is similar to that shown in Figure 18-18(a). The drug is injected intravenously into a central compartment containing the blood (the top reactor). The blood distributes the drug back and forth to the tissue compartment (the bottom reactor) before being eliminated (top reactor). This model will give the familiar linear semi-log plot found in

---

[13]Proc. Indian Inst. Chem. Eng. Golden Jubilee, a Congress, Delhi, 1997, p. 323.

[14]Levenspiel, O. *Chemical Reaction Engineering*, 3rd ed. (New York: Wiley, 1999), pp. 284–292.

pharmacokinetics textbooks. As can be seen in Chapter 9, in the figure for *Professional Reference Shelf R9.8* on the CRE Web site on pharmacokinetics, and on page 389, there are two different slopes, one for the drug distribution phase and one for the elimination phase.

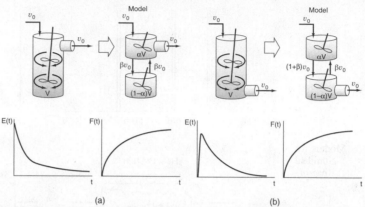

Figure 18-18 Combinations of ideal reactors to model a real CSTR. Two ideal CSTRs with interchange (a) exit from the top of the CSTR; (b) exit from the bottom of the CSTR.

Closure.

> ### RTD Data + Model + Kinetics = Prediction

In this chapter, models were developed for existing reactors to obtain more precise estimates of the exit conversion and concentrations than those from the zero-order parameter models of segregation and maximum mixedness. After completing this chapter, the reader will be able to use the RTD data and kinetic rate law and reactor model to make predictions of the conversion and exit concentrations using the tanks-in-series and dispersion one-parameter models. In addition, the reader should be able to create two-parameter models consisting of combinations of ideal reactors that mimic the RTD data. Using the models and rate law data, one can then solve for the exit conversions and concentrations. The choice of a proper model is almost pure art *requiring creativity and engineering judgment*. The flow pattern of the model must possess the most important characteristics of that in the real reactor. Standard models are available that have been used with some success, and these can be used as starting points. Models of real reactors usually consist of combinations of ideal PFRs and CSTRs with fluid exchange, bypassing, and dead spaces in a configuration that matches the flow patterns in the reactor. For tubular reactors, the simple dispersion model has proven most popular.

   In summary, the parameters in the model, which with rare exception should not exceed two in number, are obtained from the RTD data. Once the parameters are evaluated, the conversion in the model, and thus in the real reactor, can be calculated. For typical tank-reactor models, this can be calculated for the conversion in a series–parallel reactor system. For the dispersion model, the second-order differential equation must be solved, usually numerically. Analytical solutions exist for first-order reactions, but as pointed out previously, no model has to be assumed for the first-order system if the RTD is available.

> Correlations exist for the amount of dispersion that might be expected in common packed-bed reactors, so these systems can be designed using the dispersion model without obtaining or estimating the RTD. This situation is perhaps the only one where an RTD is not necessary for designing a non-ideal reactor.

SUMMARY

1. The models for predicting conversion from RTD data are:
   a. Zero adjustable parameters
      (1) Segregation model
      (2) Maximum mixedness model
   b. One adjustable parameter
      (1) Tanks-in-series model
      (2) Dispersion model
   c. Two adjustable parameters: real reactor modeled as combinations of ideal reactors
2. Tanks-in-series model: Use RTD data to estimate the number of tanks in series,

$$n = \frac{\tau^2}{\sigma^2} \qquad \text{(S18-1)}$$

   For a first-order reaction

$$X = 1 - \frac{1}{(1 + \tau_i k)^n}$$

3. Dispersion model: For a first-order reaction, use the Danckwerts boundary conditions

$$X = 1 - \frac{4q \exp(Pe_r/2)}{(1+q)^2 \exp(Pe_r q/2) - (1-q)^2 \exp(-Pe_r q/2)} \qquad \text{(S18-2)}$$

   where

$$q = \sqrt{1 + \frac{4\boldsymbol{Da_1}}{Pe_r}} \qquad \text{(S18-3)}$$

$$\boldsymbol{Da_1} = \tau k \qquad \text{(S18-4)}$$

   For a first-order reaction

$$Pe_r = \frac{UL}{D_a} \qquad Pe_f = \frac{Ud_p}{D_a \phi} \qquad \text{(S18-5)}$$

4. Determine D_a
   a. For laminar flow, the dispersion coefficient is

$$D^* = D_{AB} + \frac{U^2 R^2}{48 D_{AB}} \qquad \text{(S18-6)}$$

   b. Correlations. Use Figures 18-10 through 18-12.
   c. Experiment in RTD analysis to find t_m and σ^2.
   For a closed-closed system, use Equation (S18-6) to calculate Pe_r from the RTD data

$$\frac{\sigma^2}{\tau^2} = \frac{2}{Pe_r} - \frac{2}{Pe_r^2} (1 - e^{-Pe_r})$$ (S18-7)

For an open-open system, use

$$\boxed{\frac{\sigma^2}{t_m^2} = \frac{2Pe_r + 8}{Pe_r^2 + 4Pe_r + 4}}$$ (18-47)

5. If a real reactor is modeled as a combination of ideal reactors, the model should have at most two parameters.

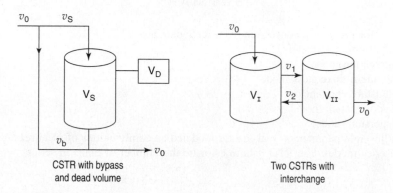

CSTR with bypass   Two CSTRs with
and dead volume   interchange

6. The RTD is used to extract model parameters.
7. Comparison of conversions for a PFR and CSTR with the zero-parameter and two-parameter models. X_{seg} symbolizes the conversion obtained from the segregation model and X_{mm} is that from the maximum mixedness model for reaction orders greater than one.

$$\boxed{\begin{array}{c} X_{PFR} > X_{seg} > X_{mm} > X_{CSTR} \\ X_{PFR} > X_{model} \quad \text{with } X_{model} < X_{CSTR} \quad \text{or} \quad X_{model} > X_{CSTR} \end{array}}$$

Cautions: For rate laws with unusual concentration functionalities or for nonisothermal operation, these bounds may not be accurate for certain types of rate laws.

CRE WEB SITE MATERIALS

- **Expanded Material on the Web Site**
  1. *W18.2.1 Developing the E-Curve for T-I-S*
  2. *Web Example 18-1 Equivalency of Models for a First Order Reaction*

  $$X_{T\text{-}I\text{-}S} = X_{seg} = X_{mm}$$

  3. *Sloppy Tracer Inputs*
  4. *Case A Aris-Taylor Analysis for LFR*
  5. *Web Example 18-2 Dispersion with Reaction*
  6. *Web Example 18-2 (COMSOL)*
  7. *Web Problem 18-12$_C$*
  8. *Web Problem 18-14$_D$*
  9. *Web Problem 18-17$_D$*
  10. *Web Problem 18-18$_B$*
  11. *Web Problem 18-19$_C$*
  12. *Web Problem 18-20$_B$*

- **Learning Resources**
    1. *Summary Notes*
- **Living Example Problems**
    1. *Example 18-3 CSTR with Bypass and Dead Volume*
- **Professional Reference Shelf**
    R18.1 *Derivation of Equation for Taylor-Aris Dispersion*

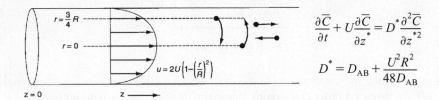

$$\frac{\partial \overline{C}}{\partial t} + U \frac{\partial \overline{C}}{\partial z^*} = D^* \frac{\partial^2 \overline{C}}{\partial z^{*2}}$$

$$D^* = D_{AB} + \frac{U^2 R^2}{48 D_{AB}}$$

R18.2 *Real Reactor Modeled as two Ideal CSTRs with Exchange Volume*
    Example R18-1 Two CSTRs with interchange

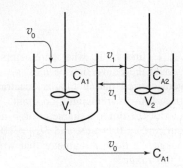

QUESTIONS AND PROBLEMS

The subscript to each of the problem numbers indicates the level of difficulty: A, least difficult; D, most difficult.

$$A = \bullet \quad B = \blacksquare \quad C = \blacklozenge \quad D = \blacklozenge\blacklozenge$$

Questions

Q18-1$_B$ Make up and solve an original problem. The guidelines are given in Problem P5-1$_A$. However, make up a problem in reverse by first choosing a model system such as a CSTR in parallel with a CSTR and PFR (with the PFR modeled as four small CSTRs in series) or a CSTR with recycle and bypass (Figure Q18-1$_B$). Write tracer mass balances and use an ODE solver to predict the effluent concentrations. In fact, you could build up an arsenal of tracer curves for different  model systems to compare against real reactor RTD data. In this way, you could deduce which model best describes the real reactor.

Creative Thinking

Q18-2 **What if** you were asked to design a tubular vessel that would minimize dispersion? What would be your guidelines? How would you maximize the dispersion? How would your design change for a packed bed?

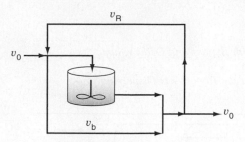

Figure Q18-1$_B$ Model system.

Q18-3   **What if** someone suggested you could use the solution to the flow-dispersion-reactor equation, Equation (18-27), for a second-order equation by linearizing the rate law by lettering $-r_A = kC_A^2 \cong (kC_{A0}/2)C_A = k'C_A$. (1) Under what circumstances might this be a good approximation? Would you divide C_{A0} by something other than 2? (2) What do you think of linearizing other non-first-order reactions and using Equation (18-27)? (3) How could you test your results to learn if the approximation is justified?

Problems

P18-1$_B$ **(a)**   **Example 18-1.** Vary D_a, k, U, and L. To what parameters or groups of parameters (e.g., kL^2/D_a) would the conversion be most sensitive? What if the first-order reaction were carried out in tubular reactors of different diameters, but with the space time, τ, remaining constant? The diameters would range from a diameter of 0.1 dm to a diameter of 1 m for kinematic viscosity $\nu = \mu/\rho = 0.01$ cm²/s, $U = 0.1$ cm/s, and $D_{AB} = 10^{-5}$ cm²/s. How would your conversion change? Is there a diameter that would maximize or minimize conversion in this range?

Hall of Fame

(b)   **Example 18-2.** How would your answers change if the slope was 4 min⁻¹ and the intercept was 2 in Figure E18-2.2?

(c)   **Example 18-3.** Download the *Living Example Polymath Program*. Vary α and β, and describe what you find. What would be the conversion if $\alpha = 0.75$ and $\beta = 0.15$?

P18-2$_B$ The gas-phase isomerization

$$A \longrightarrow B$$

is to be carried out in a flow reactor. Experiments were carried out at a volumetric flow rate of $\upsilon_0 = 2$ dm³/min in a reactor that had the following RTD

$$E(t) = 10\, e^{-10t}\ \text{min}^{-1}$$

where t is in minutes.

(a)   When the volumetric flow rate was 2 dm³/min, the conversion was 9.1%. What is the reactor volume?

(b)   When the volumetric flow rate was 0.2 dm³/min, the conversion was 50%. When the volumetric flow rate was 0.02 dm³/min, the conversion was 91%. Assuming the mixing patterns don't change as the flow rate changes, what will the conversion be when the volumetric flow rate is 10 dm³/min?

(c)   This reaction is now to be carried out in a 1-dm³ plug-flow reactor where volumetric flow rate has been changed to 1 dm³/min. What will be the conversion?

(d)   It is proposed to carry out the reaction in a 10-m-diameter pipe where the flow is highly turbulent (Re = 10⁶). There are significant dispersion effects. The superficial gas velocity is 1 m/s. If the pipe is 6 m long, what conversion can be expected? If you were unable to determine the reaction order and the specific reaction rate constant in part (b), assume $k = 1$ min⁻¹ and carry out the calculation!

P18-3$_B$ The second-order liquid-phase reaction

$$A \longrightarrow B + C$$

is to be carried out isothermally. The entering concentration of A is 1.0 mol/dm³. The specific reaction rate is 1.0 dm³/mol·min. A number of used reactors (shown below) are available, each of which has been characterized by an RTD. There are two crimson and white reactors, and three maize and blue reactors available.

Reactor	σ (min)	τ (min)	Cost
Maize and blue	2	2	$25,000
Green and white	4	4	50,000
Scarlet and gray	3.05	4	50,000
Orange and blue	2.31	4	50,000
Purple and white	5.17	4	50,000
Silver and black	2.5	4	50,000
Crimson and white	2.5	2	25,000

  (a) You have $50,000 available to spend. What is the greatest conversion you can achieve with the available money and reactors?
  (b) How would your answer to (a) change if you had an extra $75,000 available to spend?
  (c) From which cities do you think the various used reactors came from?

P18-4$_B$ The elementary liquid-phase reaction

$$A \xrightarrow{\ k_1\ } B, \qquad k_1 = 1.0 \ \text{min}^{-1}$$

is carried out in a packed-bed reactor in which dispersion is present.
What is the conversion?

Additional information:

Porosity = 50%   Reactor length = 0.1 m
Particle size = 0.1 cm   Mean velocity = 1 cm/s
Kinematic viscosity = 0.01 cm²/s   Bed fluidicity = 7.3

P18-5$_A$ A gas-phase reaction is being carried out in a 5-cm-diameter tubular reactor that is 2 m in length. The velocity inside the pipe is 2 cm/s. As a very first approximation, the gas properties can be taken as those of air (kinematic viscosity = 0.01 cm²/s), and the diffusivities of the reacting species are approximately 0.005 cm²/s.
  (a) How many tanks in series would you suggest to model this reactor?
  (b) If the second-order reaction A + B $\longrightarrow$ C + D is carried out for the case of equimolar feed, and with $C_{A0} = 0.01$ mol/dm³, what conversion can be expected at a temperature for which $k = 25$ dm³/mol·s?
  (c) How would your answers to parts (a) and (b) change if the fluid velocity was reduced to 0.1 cm/s? Increased to 1 m/s?
  (d) How would your answers to parts (a) and (b) change if the superficial velocity was 4 cm/s through a packed bed of 0.2-cm-diameter spheres?
  (e) How would your answers to parts (a) to (d) change if the fluid was a liquid with properties similar to water instead of a gas, and the diffusivity was 5×10^{-6} cm²/s?

P18-6$_A$ Use the data in Example 16-2 to make the following determinations. (The volumetric feed rate to this reactor was 60 dm³/min.)
  (a) Calculate the Peclet numbers for both open and closed systems.
  (b) For an open system, determine the space time τ and then calculate the % dead volume in a reactor for which the manufacturer's specifications give a volume of 420 dm³.

(c) Using the dispersion and tanks-in-series models, calculate the conversion for a closed vessel for the first-order isomerization

$$A \longrightarrow B$$

with $k = 0.18$ min^{-1}.

(d) Compare your results in part (c) with the conversion calculated from the tanks-in-series model, a PFR, and a CSTR.

P18-7$_A$ A tubular reactor has been sized to obtain 98% conversion and to process 0.03 m^3/s. The reaction is a first-order irreversible isomerization. The reactor is 3 m long, with a cross-sectional area of 25 dm^2. After being built, a pulse tracer test on the reactor gave the following data: $t_m = 10$ s and $\sigma^2 = 65$ s^2. What conversion can be expected in the real reactor?

P18-8$_B$ The following $E(t)$ curve was obtained from a tracer test on a reactor.

$$\begin{aligned}E(t) &= 0.25t & 0 < t < 2\\ &= 1 - 0.25t & 2 < t < 4\\ &= 0 & t > 4\end{aligned}$$

t in minutes, and E(t) in min^{-1}.

The conversion predicted by the tanks-in-series model for the isothermal elementary reaction

$$A \longrightarrow B$$

was 50% at 300 K.

(a) If the temperature is to be raised 10°C ($E = 25{,}000$ cal/mol) and the reaction carried out isothermally, what will be the conversion predicted by the maximum mixedness model? The T-I-S model?

(b) The elementary reactions

$$A \xrightarrow{k_1} B \xrightarrow{k_2} C$$

$$A \xrightarrow{k_3} D$$

$$k_1 = k_2 = k_3 = 0.1 \text{ min}^{-1} \text{ at } 300 \text{ K}, C_{A0} = 1 \text{ mol/dm}^3$$

were carried out isothermally at 300 K in the same reactor. What is the concentration of B in the exit stream predicted by the maximum mixedness model?

(c) For the multiple reactions given in part (b), what is the conversion of A predicted by the dispersion model in an isothermal closed-closed system?

P18-9$_B$ Revisit Problem P16-3$_C$ where the RTD function is a hemicircle. What is the conversion predicted by
   (a) The tanks-in-series model?
   (b) The dispersion model?

P18-10$_B$ Revisit Problem P16-5$_B$.
   (a) What combination of ideal reactors would you use to model the RTD?
   (b) What are the model parameters?
   (c) What is the conversion predicted for your model?
   (d) What is the conversion predicted by X_{mm}, X_{seg}, $X_{T\text{-}I\text{-}S}$, and $X_{Dispersion}$?

P18-11$_B$ Revisit Problem P16-6$_B$.
   (a) What conversion is predicted by the tanks-in-series model?
   (b) What is the Peclet number?
   (c) What conversion is predicted by the dispersion model?

P18-12$_D$ Let's continue Problem P16-11$_C$.
   (a) What would be the conversion for a second-order reaction with $kC_{A0} = 0.1$ min^{-1} and $C_{A0} = 1$ mol/dm^3 using the segregation model?
   (b) What would be the conversion for a second-order reaction with $kC_{A0} = 0.1$ min^{-1} and $C_{A0} = 1$ mol/dm^3 using the maximum mixedness model?

(c) If the reactor is modeled as tanks in series, how many tanks are needed to represent this reactor? What is the conversion for a first-order reaction with $k = 0.1$ min^{-1}?

(d) If the reactor is modeled by a dispersion model, what are the Peclet numbers for an open system and for a closed system? What is the conversion for a first-order reaction with $k = 0.1$ min^{-1} for each case?

(e) Use the dispersion model to estimate the conversion for a second-order reaction with $k = 0.1$ dm^3/mol·s and $C_{A0} = 1$ mol/dm^3.

(f) It is suspected that the reactor might be behaving as shown in Figure P18-12$_B$, with *perhaps (?)* $V_1 = V_2$. What is the "backflow" from the second to the first vessel, as a multiple of v_0?

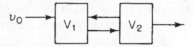

Figure P18-12$_D$ Proposed model system.

(g) If the model above is correct, what would be the conversion for a second-order reaction with $k = 0.1$ dm^3/mol·min if $C_{A0} = 1.0$ mol/dm^3?

(h) Prepare a table comparing the conversion predicted by each of the models described above.

P18-13$_B$ A second-order reaction is to be carried out in a real reactor that gives the following outlet concentration for a step input:

For $0 \leq t < 10$ min, then $C_T = 10\,(1 - e^{-.1t})$

For 10 min $\leq t$, then $C_T = 5 + 10\,(1 - e^{-.1t})$

(a) What model do you propose and what are your model parameters, α and β?

(b) What conversion can be expected in the real reactor?

(c) How would your model change and conversion change if your outlet tracer concentration was as follows?

For $t \leq 10$ min, then $C_T = 0$

For $t \geq 10$ min, then $C_T = 5 + 10\,(1 - e^{-0.2(t-10)})$

$v_0 = 1$ dm^3/min, $k = 0.1$ dm^3/mol · min, $C_{A0} = 1.25$ mol/dm^3

P18-14$_B$ Suggest combinations of ideal reactors to model the real reactors given in problem P16-2$_B$(b) for either $E(\theta), E(t), F(\theta), F(t)$, or $(1 - F(\theta))$.

P18-15$_B$ The F-curves for two tubular reactors are shown in Figure P18-15$_B$ for a closed–closed system.

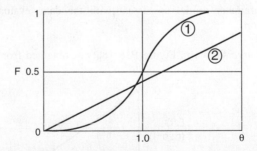

Figure P18-15$_B$ F-curves.

(a) Which curve has the higher Peclet number? Explain.

(b) Which curve has the higher dispersion coefficient? Explain.

(c) If this F-curve is for the tanks-in-series model applied to two different reactors, which curve has the largest number of T-I-S, (1) or (2)?

U of M, ChE528 Mid-Term Exam

P18-16$_C$ Consider the following system in Figure P18-16$_C$ used to model a real reactor:

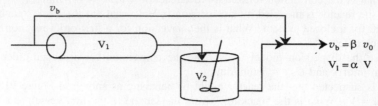

Figure P18-16$_C$ Model system.

Describe how you would evaluate the parameters α and β.

(a) Draw the F- and E-curves for this system of ideal reactors used to model a real reactor using $\beta = 0.2$ and $\alpha = 0.4$. Identify the numerical values of the points on the F-curve (e.g., t_1) as they relate to τ.

(b) If the reaction A $\rightarrow$ B is second order with $kC_{A0} = 0.5$ min^{-1}, what is the conversion assuming the space time for the real reactor is 2 min?

U of M, ChE528 Final Exam

P18-17$_B$ There is a 2-m^3 reactor in storage that is to be used to carry out the liquid-phase second-order reaction

$$A + B \longrightarrow C$$

A and B are to be fed in equimolar amounts at a volumetric rate of 1 m^3/min. The entering concentration of A is 2 molar, and the specific reaction rate is 1.5 m^3/kmol $\bullet$ min. A tracer experiment was carried out and reported in terms of F as a function of time in minutes as shown in Figure P18-17$_B$.

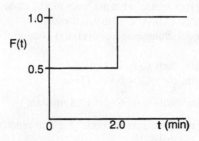

Figure P18-17$_B$ F-curve for a nonideal reactor.

Suggest a two-parameter model consistent with the data; evaluate the model parameters and the expected conversion.

U of M, ChE528 Final Exam

P18-18$_B$ The following E-curve shown in Figure P18-18$_B$ was obtained from a tracer test:

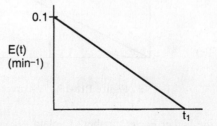

Figure P18-18$_B$ E-curve for a nonideal reactor.

(a) What is the mean residence time?

(b) What is the Peclet number for a closed-closed system?

(c) How many tanks in series are necessary to model this nonideal reactor?

U of M, Doctoral Qualifying Exam (DQE)

P18-19$_B$ A first-order reaction is to be carried out in the reactor with $k = 0.1$ min^{-1}.

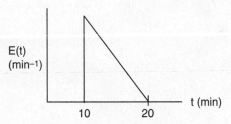

Figure P18-19$_B$ Reactor *E*-curve.

Fill in the following table with the conversion predicted by each type of model/reactor.

TABLE P18-19$_B$.   COMPARISONS OF CONVERSION PREDICTED BY VARIOUS MODELS

Ideal PFR	Ideal CSTR	Ideal Laminar-Flow Reactor	Segregation	Maximum Mixedness	Dispersion	Tanks in Series

SUPPLEMENTARY READING

1. Excellent discussions of maximum mixedness can be found in

   DOUGLAS, J. M., "The effect of mixing on reactor design," *AIChE Symp. Ser.* 48, vol. 60, p. 1 (1964).
   ZWIETERING, TH. N., *Chem. Eng. Sci.*, 11, 1 (1959).

2. Modeling real reactors with a combination of ideal reactors is discussed together with axial dispersion in

   LEVENSPIEL, O., *Chemical Reaction Engineering*, 3rd ed. New York: Wiley, 1999.
   WEN, C. Y., and L. T. FAN, *Models for Flow Systems and Chemical Reactors*. New York: Marcel Dekker, 1975.

3. Mixing and its effects on chemical reactor design have been receiving increasingly sophisticated treatment. See, for example:

   BISCHOFF, K. B., "Mixing and contacting in chemical reactors," *Ind. Eng. Chem.*, 58 (11), 18 (1966).
   NAUMAN, E. B., "Residence time distributions and micromixing," *Chem. Eng. Commun.*, 8, 53 (1981).
   NAUMAN, E. B., and B. A. BUFFHAM, *Mixing in Continuous Flow Systems*. New York: Wiley, 1983.

4. See also

   DUDUKOVIC, M., and R. FELDER, in *CHEMI Modules on Chemical Reaction Engineering*, vol. 4, ed. B. Crynes and H. S. Fogler. New York: AIChE, 1985.

5. Dispersion. A discussion of the boundary conditions for closed-closed, open-open, closed-open, and open-closed vessels can be found in

   ARIS, R., *Chem. Eng. Sci.*, 9, 266 (1959).
   LEVENSPIEL, O., and K. B. BISCHOFF, *Adv. in Chem. Eng.*, 4, 95 (1963).
   NAUMAN, E. B., *Chem. Eng. Commun.*, 8, 53 (1981).

6. Now that you have finished this book, suggestions on what to do with the book can be posted on the kiosk in downtown Riça, Jofostan.

This is not the end.
It is not even the beginning of the end.
But it is, perhaps, the end of the beginning.

Winston Churchill
November 10, 1942

Numerical   A
Techniques

Lake Michigan—unsalted and shark free.

A.1 Useful Integrals in Reactor Design

Also see *www.integrals.com*.

$$\int_0^x \frac{dx}{1-x} = \ln\frac{1}{1-x} \tag{A-1}$$

$$\int_{x_1}^{x_2} \frac{dx}{(1-x)^2} = \frac{1}{1-x_2} - \frac{1}{1-x_1} \tag{A-2}$$

$$\int_0^x \frac{dx}{(1-x)^2} = \frac{x}{1-x} \tag{A-3}$$

$$\int_0^x \frac{dx}{1+\varepsilon x} = \frac{1}{\varepsilon} \ln(1+\varepsilon x) \tag{A-4}$$

$$\int_0^x \frac{(1+\varepsilon x)dx}{1-x} = (1+\varepsilon) \ln\frac{1}{1-x} - \varepsilon x \tag{A-5}$$

$$\int_0^x \frac{(1+\varepsilon x)dx}{(1-x)^2} = \frac{(1+\varepsilon)x}{1-x} - \varepsilon \ln\frac{1}{1-x} \tag{A-6}$$

$$\int_0^x \frac{(1+\varepsilon x)^2 dx}{(1-x)^2} = 2\varepsilon(1+\varepsilon)\ln(1-x) + \varepsilon^2 x + \frac{(1+\varepsilon)^2 x}{1-x} \tag{A-7}$$

$$\int_0^x \frac{dx}{(1-x)(\Theta_B - x)} = \frac{1}{\Theta_B - 1} \ln \frac{\Theta_B - x}{\Theta_B(1-x)} \qquad \Theta_B \neq 1 \tag{A-8}$$

$$\int_0^W (1 - \alpha W)^{1/2} dW = \frac{2}{3\alpha}[1 - (1 - \alpha W)^{3/2}] \tag{A-9}$$

$$\int_0^x \frac{dx}{ax^2 + bx + c} = \frac{-2}{2ax + b} + \frac{2}{b} \qquad \text{for } b^2 = 4ac \tag{A-10}$$

$$\int_0^x \frac{dx}{ax^2 + bx + c} = \frac{1}{a(p-q)} \ln\left(\frac{q}{p} \cdot \frac{x-p}{x-q}\right) \qquad \text{for } b^2 > 4ac \tag{A-11}$$

where p and q are the roots of the equation.

$$ax^2 + bx + c = 0 \qquad \text{i.e., } p, q = \frac{-b \mp \sqrt{b^2 - 4ac}}{2a}$$

$$\int_0^x \frac{a + bx}{c + gx}\, dx = \frac{bx}{g} + \frac{ag - bc}{g^2} \ln \frac{c + gx}{c} \tag{A-12}$$

A.2 Equal-Area Graphical Differentiation

There are many ways of differentiating numerical and graphical data (cf. Chapter 7). We shall confine our discussions to the technique of equal-area differentiation. In the procedure delineated here, we want to find the derivative of y with respect to x.

This method finds use in Chapter 5.

1. Tabulate the (y_i, x_i) observations as shown in Table A-1.
2. For each *interval*, calculate $\Delta x_n = x_n - x_{n-1}$ and $\Delta y_n = y_n - y_{n-1}$.

TABLE A-1

x_i	y_i	Δx	Δy	$\dfrac{\Delta y}{\Delta x}$	$\dfrac{dy}{dx}$
x_1	y_1				$\left(\dfrac{dy}{dx}\right)_1$
		$x_2 - x_1$	$y_2 - y_1$	$\left(\dfrac{\Delta y}{\Delta x}\right)_2$	
x_2	y_2				$\left(\dfrac{dy}{dx}\right)_2$
		$x_3 - x_2$	$y_3 - y_2$	$\left(\dfrac{\Delta y}{\Delta x}\right)_3$	
x_3	y_3				$\left(\dfrac{dy}{dx}\right)_3$
x_4	y_4		etc.		

3. Calculate $\Delta y_n / \Delta x_n$ as an estimate of the *average* slope in an interval x_{n-1} to x_n.
4. Plot these values as a histogram versus x_i. The value between x_2 and x_3, for example, is $(y_3 - y_2)/(x_3 - x_2)$. Refer to Figure A-1.

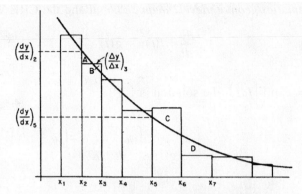

Figure A-1   Equal-area differentiation.

5. Next, draw in the *smooth curve* that best approximates the *area* under the histogram. That is, attempt in each interval to balance areas such as those labeled A and B, but when this approximation is not possible, balance out over several intervals (as for the areas labeled C and D). From our definitions of Δx and Δy, we know that

$$y_n - y_1 = \sum_{i=2}^{n} \frac{\Delta y}{\Delta x_i} \Delta x_i \qquad (\text{A-13})$$

The equal-area method attempts to estimate dy/dx so that

$$y_n - y_1 = \int_{x_1}^{x_n} \frac{dy}{dx} \, dx \qquad (\text{A-14})$$

that is, so that the area under $\Delta y/\Delta x$ is the same as that under dy/dx, *everywhere possible*.
6. Read estimates of dy/dx from this curve at the data points x_1, x_2, ... and complete the table.

An example illustrating the technique is given on the CRE Web site, Appendix A.

Differentiation is, at best, less accurate than integration. This method also *clearly indicates bad data* and allows for compensation of such data. Differentiation is only valid, however, when the data are presumed to differentiate *smoothly*, as in rate-data analysis and the interpretation of transient diffusion data.

A.3 Solutions to Differential Equations

A.3.A   First-Order Ordinary Differential Equations

See *www.ucl.ac.uk/Mathematics/geomath/level2/deqn/de8.html* and the CRE Web site, Appendix A.3.

$$\frac{dy}{dt} + f(t)y = g(t) \tag{A-15}$$

Using integrating factor $= \exp\left(\int f dt\right)$, the solution is

$$y = e^{-\int f dt} \int g(t)\, e^{\int f dt}\, dt + K_1 e^{-\int f dt} \tag{A-16}$$

Example A–1   Integrating Factor for Series Reactions

$$\frac{dy}{dt} + k_2 y = k_1 e^{-k_1 t}$$

$$\text{Integrating factor} = \exp\int k_2\, dt = e^{k_2 t}$$

$$\frac{d(y e^{k_2 t})}{dt} = e^{k_2 t} k_1 e^{-k_1 t} = k_1 e^{(k_2 - k_1)t}$$

$$e^{k_2 t} y = k_1 \int e^{(k_2 - k_1)t}\, dt = \frac{k_1}{k_2 - k_1} e^{(k_2 - k_1)t} + K_1$$

$$y = \frac{k_1}{k_2 - k_1} e^{-k_1 t} + K_1 e^{-k_2 t}$$

$$t = 0 \quad y = 0$$

$$y = \frac{k_1}{k_2 - k_1}\left[e^{-k_1 t} - e^{-k_2 t} \right]$$

A.3.B   Coupled Differential Equations

Techniques to solve coupled first-order linear ODEs such as

$$\frac{dx}{dt} = ax + by$$

$$\frac{dy}{dt} = cx + dy$$

are given in Web Appendix A.3 on the CRE Web site.

A.3.C   Second-Order Ordinary Differential Equations

Methods of solving differential equations of the type

$$\frac{d^2y}{dx^2} - \beta y = 0 \tag{A-17}$$

can be found in such texts as *Applied Differential Equations* by M. R. Spiegel (Upper Saddle River, NJ: Prentice Hall, 1958, Chapter 4; a great book even though it's old) or in *Differential Equations* by F. Ayres (New York: Schaum Outline Series, McGraw-Hill, 1952). Solutions of this type are required in Chapter 15. One method of solution is to determine the characteristic roots of

$$\left(\frac{d^2}{dx^2} - \beta \right) y = (m^2 - \beta)y$$

which are

$$m = \pm \sqrt{\beta}$$

The solution to the differential equation is

$$y = A_1 e^{-\sqrt{\beta}x} + B_1 e^{+\sqrt{\beta}x} \tag{A-18}$$

where A_1 and B_1 are arbitrary constants of integration. It can be verified that Equation (A-18) can be arranged in the form

$$y = A \sinh \sqrt{\beta}x + B \cosh \sqrt{\beta}x \tag{A-19}$$

Equation (A-19) is the more useful form of the solution when it comes to evaluating the constants A and B because $\sinh(0) = 0$ and $\cosh(0) = 1.0$. Also see Appendix A.3.B.

A.4 Numerical Evaluation of Integrals

In this section, we discuss techniques for numerically evaluating integrals for solving first-order differential equations.

1. *Trapezoidal rule* (two-point) (Figure A-2). This method is one of the simplest and most approximate, as it uses the integrand evaluated at the limits of integration to evaluate the integral

$$\int_{X_0}^{X_1} f(X) \, dX = \frac{h}{2} [f(X_0) + f(X_1)] \tag{A-20}$$

when $h = X_1 - X_0$.

2. *Simpson's one-third rule* (three-point) (Figure A-3). A more accurate evaluation of the integral can be found with the application of Simpson's rule:

$$\int_{X_0}^{X_2} f(X)\, dX = \frac{h}{3}[f(X_0) + 4f(X_1) + f(X_2)] \tag{A-21}$$

where

$$h = \frac{X_2 - X_0}{2} \qquad X_1 = X_0 + h$$

Methods to solve
$$\int_0^X \frac{F_{A0}}{-r_A} dX$$
in Chapters 2, 4, 12, and
$$\int_0^\infty X(t)E(t)\,dt$$
in Chapter 17

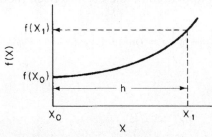

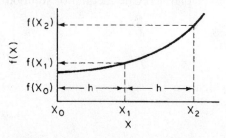

Figure A-2   Trapezoidal rule illustration.   **Figure A-3**   Simpson's three-point rule illustration.

3. *Simpson's three-eighths rule* (four-point) (Figure A-4). An improved version of Simpson's one-third rule can be made by applying *Simpson's three-eighths rule*:

$$\int_{X_0}^{X_3} f(X)\, dX = \frac{3}{8} h\, [f(X_0) + 3f(X_1) + 3f(X_2) + f(X_3)] \tag{A-22}$$

where

$$h = \frac{X_3 - X_0}{3} \qquad X_1 = X_0 + h \qquad X_2 = X_0 + 2h$$

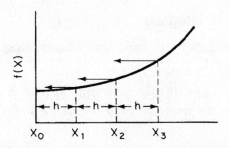

Figure A-4   Simpson's four-point rule illustration.

4. *Five-point quadrature formula.*

$$\int_{X_0}^{X_4} f(X)\, dX = \frac{h}{3}(f_0 + 4f_1 + 2f_2 + 4f_3 + f_4) \tag{A-23}$$

where

$$h = \frac{X_4 - X_0}{4}$$

5. For $N + 1$ points, where $(N/3)$ is an integer

$$\int_{X_0}^{X_N} f(X)\, dX = \tfrac{3}{8} h\,[f_0 + 3f_1 + 3f_2 + 2f_3$$

$$+\, 3f_4 + 3f_5 + 2f_6 + \cdots + 3f_{N-1} + f_N] \qquad \text{(A-24)}$$

where $h = \dfrac{X_N - X_0}{N}$

6. For $N + 1$ points, where N is even

$$\int_{X_0}^{X_N} f(X)\, dX = \frac{h}{3} (f_0 + 4f_1 + 2f_2 + 4f_3 + 2f_4 + \cdots + 4f_{N-1} + f_N)$$

$$\text{(A-25)}$$

where $\qquad\qquad h = \dfrac{X_N - X_0}{N}$

These formulas are useful in illustrating how the reaction engineering integrals and coupled ODEs (ordinary differential equation(s)) can be solved and also when there is an ODE solver power failure or some other malfunction.

A.5 Semilog Graphs

Review how to take slopes on semilog graphs on the Web. Also see *www.physics.uoguelph.ca/tutorials/GLP*. Also see Web Appendix A.5, Using Semi-Log Plots for Data Analysis.

A.6 Software Packages

Instructions on how to use Polymath, COMSOL, and Aspen can be found on the CRE Web site.
    For the ordinary differential equation solver (ODE solver), contact:

Polymath Software
P.O. Box 523
Willimantic, CT 06226-0523
Web site: *www.polymath-software.com/fogler*

MATLAB
The Math Works, Inc.
20 North Main Street, Suite 250
Sherborn, MA 01770

Aspen Technology, Inc.
10 Canal Park
Cambridge, MA 02141-2201
Email: *info@aspentech.com*
Web site: *www.aspentech.com*

COMSOL, Inc.
1 New England Executive Park
Burlington, MA 01803
Tel: +1-781-273-3322
Fax: +1-781-273-6603
Email: *info@comsol.com*
Web site: *www.comsol.com*

Ideal Gas Constant and Conversion Factors

B

See *www.onlineconversion.com*.

Ideal Gas Constant

$$R = \frac{0.73 \text{ ft}^3 \cdot \text{atm}}{\text{lb mol} \cdot {}^\circ\text{R}} \qquad\qquad R = \frac{1.987 \text{ Btu}}{\text{lb mol} \cdot {}^\circ\text{R}}$$

$$R = \frac{8.314 \text{ kPa} \cdot \text{dm}^3}{\text{mol} \cdot \text{K}} \qquad\qquad R = \frac{8.3144 \text{ J}}{\text{mol} \cdot \text{K}}$$

$$R = 0.082 \frac{\text{dm}^3 \cdot \text{atm}}{\text{mol} \cdot \text{K}} = \frac{0.082 \text{ m}^3 \cdot \text{atm}}{\text{kmol} \cdot \text{K}} \qquad R = \frac{1.987 \text{cal}}{\text{mol} \cdot \text{K}}$$

Boltzmann's constant $k_B = 1.381 \times 10^{-23} \dfrac{\text{J}}{\text{molecule} \bullet \text{K}}$

$$= 1.381 \times 10^{-23} \text{ kg m}^2/\text{s}^2/\text{molecule/K}$$

Volume of Ideal Gas

1 lb-mol of an ideal gas at 32°F and 1 atm occupies 359 ft³ (0.00279 lbmol/ft³).
1 mol of an ideal gas at 0°C and 1 atm occupies 22.4 dm³ (0.0446 mol/dm³).

$$C_A = \frac{P_A}{RT} = \frac{y_A P}{RT}$$

where C_A = concentration of A, mol/dm³ T = temperature, K
          P = pressure, kPa y_A = mole fraction of A
          R = ideal gas constant, 8.314 kPa·dm³/mol·K

> 1M = 1 molar = 1 mol/liter = 1 mol/dm³ = 1 kmol/m³ = 0.062 lb mol/ft³

Volume

1 cm^3	= 0.001 dm^3
1 in^3	= 0.0164 dm^3
1 fluid oz	= 0.0296 dm^3
1 ft^3	= 28.32 dm^3
1 m^3	= 1000 dm^3
1 U.S. gallon	= 3.785 dm^3
1 liter (L)	= 1 dm^3

$$\left(1 \text{ ft}^3 = 28.32 \text{ dm}^3 \times \frac{1 \text{ gal}}{3.785 \text{ dm}^3} = 7.482 \text{ gal} \right)$$

Length

1 Å	= 10^{-8} cm
1 dm	= 10 cm
1 μm	= 10^{-4} cm
1 in.	= 2.54 cm
1 ft	= 30.48 cm
1 m	= 100 cm

Pressure

1 torr (1 mmHg)	= 0.13333 kPa
1 in. H_2O	= 0.24886 kPa
1 in. Hg	= 3.3843 kPa
1 atm	= 101.33 kPa
1 psia	= 6.8943 kPa
1 megadyne/cm^2	= 100 kPa

Energy (Work)

1 kg·m^2/s^2	= 1 J
1 Btu	= 1055.06 J
1 cal	= 4.1868 J
1 L·atm	= 101.34 J
1 hp·h	= 2.6806×10^6 J
1 kWh	= 3.6×10^6 J

Temperature

°F	= 1.8 × °C + 32
°R	= °F + 459.69
K	= °C + 273.16
°R	= 1.8 × K
°Réamur	= 1.25 × °C

Mass

1 lb$_m$	= 454 g
1 kg	= 1000 g
1 grain	= 0.0648 g
1 oz (avoird.)	= 28.35 g
1 ton	= 908,000 g

Viscosity

1 poise = 1 g/cm/s = 0.1 kg/m/s
1 centipoise = 1 cp = 0.01 poise = 0.1 micro Pascal · second

Force

$$1 \text{ dyne} = 1 \text{ g} \cdot \text{cm/s}^2$$

$$1 \text{ Newton} = 1 \text{ kg} \cdot \text{m/s}^2$$

Pressure

$$1 \text{ Pa} = 1 \text{ Newton/m}^2$$

Work

A. Work = Force × Distance

$$1 \text{ Joule} = 1 \text{ Newton} \cdot \text{meter} = 1 \text{ kg m}^2/\text{s}^2 = 1 \text{ Pa} \cdot \text{m}^3$$

B. Pressure × Volume = Work

$$(1 \text{ Newton/ m}^2) \cdot \text{m}^3 = 1 \text{ Newton} \cdot \text{m} = 1 \text{ Joule}$$

Time Rate of Change of Energy with Time

$$1 \text{ Watt} = 1 \text{ J/s}$$

$$1 \text{ hp} = 746 \text{ J/s}$$

Gravitational Conversion Factor

Gravitational constant

$$g = 32.2 \text{ ft/s}^2$$

American Engineering System

$$g_c = 32.174 \frac{(\text{ft})(\text{lb}_m)}{(\text{s}^2)(\text{lb}_f)}$$

SI/cgs System

$$g_c = 1 \text{ (Dimensionless)}$$

TABLE B.1 TYPICAL PROPERTY VALUES

	Liquid (water)	Gas (air, 77°C, 101 kPa)	Solid
Density	1000 kg/m^3	1.0 kg/m^3	3000 kg/m^3
Concentration	55.5 mol/dm^3	0.04 mol/dm^3	–
Diffusivity	10^{-8} m^2/s	10^{-5} m^2/s	10^{-11} m^2/s
Viscosity	10^{-3} kg/m/s	1.82×10^{-5} kg/m/s	–
Heat capacity	4.31 J/g/K	40 J/mol/K	0.45 J/g/K
Thermal conductivity	1.0 J/s/m/K	10^{-2} J/s/m/K	100 J/s/m/K
Kinematic viscosity	10^{-6} m^2/s	1.8×10^{-5} m^2/s	–
Prandtl number	7	0.7	–
Schmidt number	200	2	–

TABLE B.2 TYPICAL TRANSPORT VALUES

	Liquid	Gas
Heat transfer coefficient, h	1000 W/m^2/K	65 W/m^2/K
Mass transfer coefficient, k_c	10^{-2} m/s	3 m/s

Thermodynamic C
Relationships
Involving
the Equilibrium
Constant[1]

For the gas-phase reaction

$$A + \frac{b}{a} B \rightleftharpoons \frac{c}{a} C + \frac{d}{a} D \qquad (2\text{-}2)$$

1. The true (dimensionless) equilibrium constant

$$RT \, lnK = -\Delta G$$

$$\boxed{K = \frac{a_C^{c/a} \, a_D^{d/a}}{a_A \, a_B^{b/a}}}$$

where a_i is the activity of species i

$$a_i = \frac{f_i}{f_i^o}$$

where f_i = fugacity of species i
f_i^o = fugacity of the standard state. For gases, the standard state is 1 atm

$$a_i = \frac{f_i}{f_i^o} = \gamma_i P_i$$

---

[1] For the limitations and for further explanation of these relationships, see, for example, K. Denbigh, *The Principles of Chemical Equilibrium*, 3rd ed. (Cambridge: Cambridge University Press, 1971), p. 138.

K = True equilibrium constant
K_γ = Activity equilibrium constant
K_p = Pressure equilibrium constant
K_c = Concentration equilibrium constant

where γ_i is the activity coefficient

$$K = \underbrace{\frac{\gamma_C^{c/a}\gamma_D^{d/a}}{\gamma_A\gamma_B^{b/a}}}_{K_\gamma} \cdot \underbrace{\frac{P_C^{c/a}P_D^{d/a}}{P_A P_B^{b/a}}}_{K_p} = K_\gamma K_p$$

K_γ has units of $[\text{atm}]^{-\left(d/a+c/a-\frac{b}{a}-1\right)} = [\text{atm}]^{-\delta}$

K_p has units of $[\text{atm}]^{\left(d/a+c/a-\frac{b}{a}-1\right)} = [\text{atm}]^{\delta}$

For ideal gases $K_\gamma = 1.0 \text{ atm}^{-\delta}$

2. For the generic reaction (2-2), the pressure equilibrium constant K_p is

$$K_p = \frac{P_C^{c/a}P_D^{d/a}}{P_A P_B^{b/a}} \tag{C-1}$$

P_i = partial pressure of species *i*, atm, kPa.

$P_i = C_i RT$

3. For the generic reaction (2-2), the concentration equilibrium constant K_C is

It is important to be able to relate K, K_γ, K_c, and K_p.

$$K_c = \frac{C_C^{c/a}C_D^{d/a}}{C_A C_B^{b/a}} \tag{C-2}$$

4. For ideal gases, K_c and K_p are related by

$$K_p = K_c (RT)^\delta \tag{C-3}$$

Where for the generic reaction (2-2),

$$\delta = \frac{c}{a} + \frac{d}{a} - \frac{b}{a} - 1 \tag{C-4}$$

5. K_P is a function of temperature only, and the temperature dependence of K_P is given by van't Hoff's equation:

Van't Hoff's equation

$$\boxed{\frac{d \ln K_p}{dT} = \frac{\Delta H_{\text{Rx}}(T)}{RT^2}} \tag{C-5}$$

$$\frac{d \ln K_p}{dT} = \frac{\Delta H_{\text{Rx}}^\circ(T_R) + \Delta C_p(T - T_R)}{RT^2} \tag{C-6}$$

6. Integrating, we have

$$\ln \frac{K_p(T)}{K_p(T_1)} = \frac{\Delta H_{\text{Rx}}^\circ(T_R) - T_R \Delta C_p}{R}\left(\frac{1}{T_1} - \frac{1}{T}\right) + \frac{\Delta C_p}{R} \ln \frac{T}{T_1} \tag{C-7}$$

K_p and K_c are related by

$$\boxed{K_c = \frac{K_p}{(RT)^\delta}} \tag{C-8}$$

when

$$\delta = \left(\frac{d}{a} + \frac{c}{a} - \frac{b}{a} - 1\right) = 0$$

then

$$K_P = K_C$$

7. K_P neglecting ΔC_P. Given the equilibrium constant at one temperature, T_1, $K_P(T_1)$, and the heat of reaction, ΔH_{Rx}°, the partial pressure equilibrium constant at any temperature T is

$$\boxed{K_P(T) = K_P(T_1)\exp\left[\frac{\Delta H_{Rx}^\circ (T_R)}{R}\left(\frac{1}{T_1} - \frac{1}{T}\right)\right]} \qquad \text{(C-9)}$$

8. From Le Châtelier's principle we know that for exothermic reactions, the equilibrium shifts to the left (i.e., K and X_e decrease) as the temperature increases. Figures C-1 and C-2 show how the equilibrium constant varies with temperature for an exothermic reaction and for an endothermic reaction, respectively.

Variation of equilibrium constant with temperature

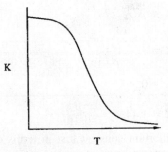

Figure C-1   Exothermic reaction.   **Figure C-2**   Endothermic reaction.

9. The equilibrium constant for the reaction (2-2) at temperature T can be calculated from the change in the Gibbs free energy using

$$-RT\ln[K(T)] = \Delta G_{Rx}^\circ(T) \qquad \text{(C-10)}$$

$$\boxed{\Delta G_{Rx}^\circ = \frac{c}{a}\, G_C^\circ + \frac{d}{a}\, G_D^\circ - \frac{b}{a}\, G_B^\circ - G_A^\circ} \qquad \text{(C-11)}$$

10. Tables that list the standard Gibbs free energy of formation of a given species G_i° are available in the literature.
   1) *www/uic.edu:80/~mansoori/Thermodynamic.Data.and.Property_html*
   2) *webbook.nist.gov*

11. The relationship between the change in Gibbs free energy and enthalpy, H, and entropy, S, is

$$\Delta G = \Delta H - T\,\Delta S \qquad \text{(C-12)}$$

See *bilbo.chm.uri.edu/CHM112/lectures/lecture31.htm*. An example on how to calculate the equivalent conversion for ΔG is given on the Web site.

Example C–1   Water-Gas Shift Reaction

The water-gas shift reaction to produce hydrogen

$$H_2O + CO \rightleftharpoons CO_2 + H_2$$

is to be carried out at 1000 K and 10 atm. For an equimolar mixture of water and carbon monoxide, calculate the equilibrium conversion and concentration of each species.

Data: At 1000 K and 10 atm, the Gibbs free energies of formation are $G^\circ_{CO} = -47{,}860$ cal/mol; $G^\circ_{CO_2} = -94{,}630$ cal/mol; $G^\circ_{H_2O} = -46{,}040$ cal/mol; and $G^\circ_{H_2} = 0$.

Solution

We first calculate the equilibrium constant. The first step in calculating K is to calculate the change in Gibbs free energy for the reaction. Applying Equation (C-10) gives us

Calculate ΔG°_{Rx}

$$\Delta G^\circ_{Rx} = G^\circ_{H_2} + G^\circ_{CO_2} - G^\circ_{CO} - G^\circ_{H_2O} \tag{EC-1.1}$$

$$= 0 + (-94{,}630) - (-47{,}860) - (-46{,}040)$$

$$= -730 \text{ cal/mol}$$

$$-RT \ln K = \Delta G^\circ_{Rx}(T) \tag{C-10}$$

$$\ln K = -\frac{\Delta G^\circ_{Rx}(T)}{RT} = \frac{-(-730 \text{ cal/mol})}{1.987 \text{ cal/mol} \cdot \text{K} (1000 \text{ K})} \tag{EC-1.2}$$

$$= 0.367$$

then

Calculate K

$$K = 1.44$$

Expressing the equilibrium constant first in terms of activities and then finally in terms of concentration, we have

$$K = \frac{a_{CO_2} a_{H_2}}{a_{CO} a_{H_2O}} = \frac{f_{CO_2} f_{H_2}}{f_{CO} f_{H_2O}} = \frac{\gamma_{CO_2} y_{CO_2} \gamma_{H_2} y_{H_2}}{\gamma_{CO} y_{CO} \gamma_{H_2O} y_{CO_2}} \tag{EC-1.3}$$

where a_i is the activity, f_i is the fugacity, γ_i is the activity coefficient (which we shall take to be 1.0 owing to high temperature and low pressure), and y_i is the mole fraction of species i.[2] Substituting for the mole fractions in terms of partial pressures gives

$$y_i = \frac{P_i}{P_T} = \frac{C_i RT}{P_T} \tag{EC-1.4}$$

$$K = \frac{P_{CO_2} P_{H_2}}{P_{CO} P_{H_2O}} = \frac{C_{CO_2} C_{H_2}}{C_{CO} C_{H_2O}} \tag{EC-1.5}$$

---

[2] See Chapter 9 in J. M. Smith, *Introduction to Chemical Engineering Thermodynamics*, 3rd ed. (New York: McGraw-Hill, 1959), and Chapter 9 in S. I. Sandler, *Chemical and Engineering Thermodynamics*, 2nd ed. (New York: Wiley, 1989), for a discussion of chemical equilibrium including nonideal effects.

In terms of conversion for an equimolar feed, we have

$$K = \frac{C_{CO,0}X_e C_{CO,0}X_e}{C_{CO,0}(1-X_e)\,C_{CO,0}(1-X_e)} \tag{EC-1.6}$$

Relate K and X_e

$$= \frac{X_e^2}{(1-X_e)^2} = 1.44 \tag{EC-1.7}$$

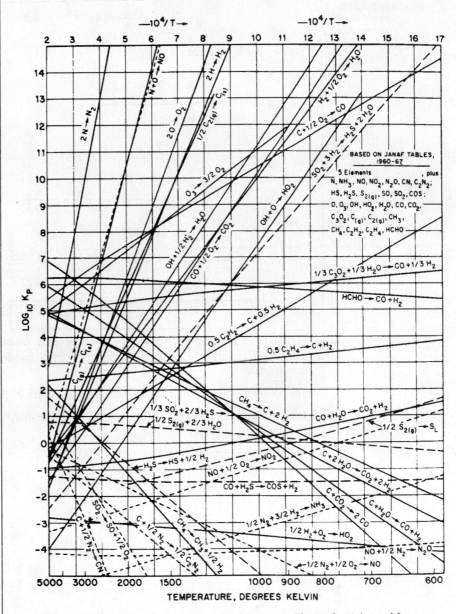

Figure EC-1.1 Modell, Michael, and Reid, Robert, *Thermodynamics and Its Applications*, 2nd ed., © 1983. Reprinted and electronically reproduced by permission of Pearson Education, Inc., Upper Saddle River, NJ.

From Figure EC-1.1 we read at 1000 K that log K_P = 0.15; therefore, K_P = 1.41, which is close to the calculated value. We note that there is no net change in the number of moles for this reaction (i.e., δ = 0); therefore,

$$K = K_p = K_c \text{ (dimensionless)}$$

Taking the square root of Equation (EC-1.7) yields

Calculate X_e, the equilibrium conversion

$$\frac{X_e}{1 - X_e} = (1.44)^{1/2} = 1.2 \qquad \text{(EC-1.8)}$$

Solving for X_e, we obtain

$$\boxed{X_e = \frac{1.2}{2.2} = 0.55}$$

Then

$$C_{\text{CO},0} = \frac{y_{\text{CO},0} P_0}{R T_0}$$

$$= \frac{(0.5)(10 \text{ atm})}{(0.082 \text{ dm}^3 \cdot \text{atm}/\text{mol} \cdot \text{K})(1000 \text{ K})}$$

$$= 0.061 \text{ mol}/\text{dm}^3$$

Calculate $C_{\text{CO},e}$, the equilibrium conversion of CO

$$C_{\text{CO},e} = C_{\text{CO},0}(1 - X_e) = (0.061)(1 - 0.55) = 0.0275 \text{ mol}/\text{dm}^3$$

$$C_{\text{H}_2\text{O},e} = 0.0275 \text{ mol}/\text{dm}^3$$

$$\boxed{C_{\text{CO}_2,e} = C_{\text{H}_2,e} = C_{\text{CO},0} X_e = 0.0335 \text{ mol}/\text{dm}^3}$$

Figure EC-1.1 gives the equilibrium constant as a function of temperature for a number of reactions. Reactions in which the lines increase from left to right are exothermic.

The following links give thermochemical data. (Heats of Formation, C_P, etc.)

Links

1)   *www.uic.edu/~mansoori/Thermodynamic.Data.and.Property_html*
2)   *webbook.nist.gov*

Also see *Chem. Tech.,* 28 (3) (March), 19 (1998).

Software Packages   D

While Polymath 6.1 is the primary software package for the majority of the problems requiring numerical solutions, MATLAB and Wolfram are also available. In addition, one can use COMSOL and Aspen to solve a few selected problems.

D.1 Polymath

D.1.A   About Polymath

Polymath 6.1 is the primary software package used in this textbook. Polymath is an easy-to-use numerical computation package that allows students and professionals to use personal computers to solve the following types of problems:

- Simultaneous linear algebraic equations
- Simultaneous nonlinear algebraic equations
- Simultaneous ordinary differential equations
- Data regressions (including the following)
  - Curve fitting by polynomials and splines
  - Multiple linear regression with statistics
  - Nonlinear regression with statistics

Polymath is unique in that the problems are entered just like their mathematical equations, and there is a minimal learning curve. Problem solutions are easily found with robust algorithms. Polymath allows very convenient problem solving to be used in chemical reaction engineering and other areas of chemical engineering, leading to an enhanced educational experience for students.

The following special Polymath Web site for software use and updating will be maintained for users of this textbook:

www.polymath-software.com/fogler

D.1.B   Polymath Tutorials

Polymath tutorials can be accessed in the *Summary Notes* by clicking on the purple hot buttons . Here, screen shots of the various steps are shown for each of the Polymath programs.

Summary Notes

Summary Notes

Chapter 1
    A. Ordinary Differential Equation (ODE) Tutorial
    B. Nonlinear (NLE) Solver Tutorial
Chapter 7
    A. Fitting a Polynomial Tutorial
    B. Nonlinear Regression Tutorial

Note: The Living Example Problems (LEPs) need to be copied from the CRE Web site and pasted into the Polymath software. The Polymath software is available in most chemical engineering department computer labs in the United States and in some other countries as well. If you want to have Polymath on your own laptop or desktop computer, you can purchase special low-priced educational versions of the software for various time periods. Polymath versions are compatible with Windows XP, Vista, Windows 7, Windows 8, and Windows 10. The specially discounted Polymath Software is available ONLY from the special Web site *www.polymath-software.com/fogler*. The minimum educational pricing requires you to refer to your Fogler textbook when you sign on to the special Web site to order Polymath. Android users can also use PolyMathLite on phones, tablets, and computers. More information is available from the Web site *www.polymathlite.com*

Links

D.2   MATLAB

MATLAB programs for the LEPs are given on the CRE Web site. The disadvantage of the MATLAB ODE solver is that it is not particularly user-friendly when trying to determine the variation of secondary parameter values. MATLAB can be used for the same four types of programs as Polymath.

D.3   Aspen

Aspen is a process simulator that is used primarily in many senior design courses. It has the steepest learning curve of the software packages used in this text. It has a built-in database of the physical properties of reactants and products. Consequently, one has only to type in the chemicals and the rate law parameters. It is really too powerful to be used for the types of home problems given here. The pyrolysis of benzene using Aspen is given as an example on the CRE Web site in Chapter 5, *Learning Resources 4.E, Solved Problem E.5-3 Design*. Perhaps one home assignment should be devoted to using Aspen to solve a problem with heat effects in order to help familiarize the student with Aspen.

Aspen example on the CRE Web site

    An Aspen tutorial and example problem from Chapters 5, 11, and 12 can be accessed directly from the CRE Web site home page (under *Let's Get Started*, click on *Additional Software* and then click *Aspen Plus*).

D.4 COMSOL Multiphysics

COMSOL Multiphysics is a modeling and simulation software available commercially from COMSOL Inc. It solves multiphysics problems in 1D, 1D axisymmetry, 2D, 2D axisymmetry, 3D, and at single points (0D). Internally in the program, these problems are formulated using partial differential equations (PDEs for 1D to 3D) or ordinary differential equations (ODEs). At *www.comsol.com/ecre*, one can download documentation and solve CRE problems, which are formulated in this book.

A dedicated application, with a tailored user interface, solves problems for tubular reactors with heat effects involving both radial axial and radial gradients in concentration. A step-by-step COMSOL tutorial with screenshots for setting up this model and solving the model equations is also given in the CRE web modules.

In the web modules, the first tutorial is "Heat effects in tubular reactors" and the second is "Tubular reactors with dispersion." In the first section, the four examples focus on the effects of the radial velocity profile and external cooling on the performances of isothermal and nonisothermal tubular reactors. In the second section, two examples examine the dispersion effects in a tubular reactor.

Heat Effects

1. **Isothermal reactor.** This example concerns an elementary, exothermic, second-order reversible liquid-phase reaction in a tubular reactor with a parabolic velocity distribution. Only the mole, rate law, and stoichiometric balance in the tubular reactor are required in this COMSOL exercise.

2. **Nonisothermal adiabatic reactor.** The isothermal reactor model is extended to include heat effects whereby the tubular reactor is treated as an adiabatic reactor. The material and energy balances are solved simultaneously in the COMSOL exercise.

3. **Nonisothermal reactor with isothermal cooling jacket**. A cooling jacket kept at a constant temperature is added to the model described in the second example above. This is a valid assumption if the cooling liquid is supplied in such a large amount that the influence of the heat from the reactor on the coolant's temperature is negligible. The boundary condition for the energy balance at the radial boundary is changed from the thermal insulation boundary condition to a heat flux boundary condition, where the external temperature corresponding to the coolant is set to a constant value.

4. **Nonisothermal reactor with variable coolant temperature**. This example extends the third example by including the energy balance on the coolant in the cooling jacket as the temperature of the coolant varies along the length of the reactor.

Dispersion and Reaction

1. **One-Dimensional model with Danckwerts boundary conditions.** In this example, the mass balance in a tubular reactor with arbitrary reactions is described by an ordinary differential equation with dimensionless variables and in terms of the Peclet number and the Damköhler number.

2. **One-Dimensional model with upstream and downstream sections.** This example uses the open-vessel boundary conditions where an inlet (upstream) section and an outlet (downstream) section are added to a tubular reactor where dispersion occurs but no reaction.

It is suggested that one first uses and plays with COMSOL Multiphysics for solutions to examples XX and YY before making any changes. One should also review the *Web Module* for Chapter 12 on *Axial and radial gradients in tubular reactors* before running the program (*www.umich.edu/~elements/5e/web_mod/radialeffects/index.htm* and *www.umich.edu/~elements/web_mod/radialeffects/comsol_ecre.pdf*).

Tutorial

Tutorials with step-by-step instructions and screenshots can be found on the web page *www.comsol.com/ecre*. Start with the documentation of **1. Isothermal Reactor** and then go to the documentation section for each of the other exercises. Tutorial material for this exercise is built on but not repeated in the exercises that follow. Next, go to **2. Nonisothermal adiabatic reactor** and read the additional information there before continuing to exercises 3 and 4. Again here, the documentation is built on, not repeated, which also holds for each of the sequential exercises.

Rate-Law Data   **E**

Reaction-rate laws and data can be obtained from the following Web sites:

1. National Institute of Standards and Technology (NIST)
   Chemical Kinetics Database on the Web
   Standard Reference Database 17, Version 7.0 (Web Version), Release 1.2
   This Web site provides a compilation of kinetics data on gas-phase reactions.
   http://kinetics.nist.gov/kinetics/index.jsp

2. International Union of Pure and Applied Chemistry (IUPAC)
   This Web site provides kinetic and photochemical data for gas kinetic data evaluation.
   http://iupac.pole-ether.fr/

3. NASA/JPL (Jet Propulsion Laboratory: California Institute of Technology)
   This Web site provides chemical kinetics and photochemical data for use in atmospheric studies.
   http://jpldataeval.jpl.nasa.gov/download.html

4. BRENDA: University of Cologne
   This Web site provides enzyme data and metabolic information. BRENDA is maintained and developed at the Institute of Biochemistry at the University of Cologne.
   http://www.brenda-enzymes.org/

5. NDRL Radiation Chemistry Data Center: Notre Dame Radiation Laboratory
   This Web site provides the reaction-rate data for transient radicals, radical ions, and excited states in solution.
   https://www3.nd.edu/~ndrlrcdc/

Nomenclature $\quad$ **F**

A	Chemical species
A_c	Cross-sectional area (m^2)
A_p	Total external surface area of particle (m^2)
a_p	External surface area of catalyst per unit bed volume (m^2/m^3)
a	Area of heat exchange per unit volume of reactor (m^{-1})
a_c	External surface area per volume of catalyst pellets (m^2/m^3)
B	Chemical species
B$_A$	Flux of A resulting from bulk flow (mol/m$^2 \cdot$s)
Bo	Bodenstein number
C	Chemical species
C_i	Concentration of species i (mol/dm^3)
$\underset{\sim}{C}_{pi}$	Heat capacity of species i at temperature T (cal/gmol$\cdot$K)
$\bar{C}_{pi}$	Mean heat capacity of species i between temperature T_0 and temperature T (cal/mol$\cdot$K)
$\hat{C}_{pi}$	Mean heat capacity of species i between temperature T_R and temperature T (cal/mol$\cdot$K)
c	Total concentration (mol/dm^3) (Chapter 14)
D	Chemical species
Da	Damköhler number (dimensionless)
D_{AB}	Binary diffusion coefficient of A in B (dm^2/s)
D_a	Dispersion coefficient (cm^2/s)
D_e	Effective diffusivity (dm^2/s)
D_K	Knudsen diffusivity (dm^2/s)
D^*	Taylor dispersion coefficient
E	Activation energy (cal/gmol)
(E)	Concentration of free (unbound) enzyme (mol/dm^3)
F_i	Molar flow rate of species i (mol/s)
F_{i0}	Entering molar flow rate of species i (mol/s)
G	Superficial mass velocity (g/dm$^2 \cdot$s)

G_i Rate of generation of species i (mol/s)

$G_i^{\circ}(T)$ Gibbs free energy of species i at temperature T (cal/gmol·K)

$H_i(T)$ Enthalpy of species i at temperature T (cal/mol i)

$H_{i0}(T)$ Enthalpy of species i at temperature T_0 (cal/mol i)

H_i° Enthalpy of formation of species i at temperature T_R (cal/gmol i)

h Heat transfer coefficient (cal/m²·s·K)

$\mathbf{J_A}$ Molecular diffusive flux of species A (mol/m²·s)

K_A Adsorption equilibrium constant

K_c Concentration equilibrium constant

K_e Equilibrium constant (dimensionless)

K_P Partial pressure equilibrium constant

k Specific reaction rate (constant)

k_c Mass transfer coefficient (m/s)

M_i Molecular weight of species i (g/mol)

m_i Mass of species i (g)

N_i Number of moles of species i (mol)

n Overall reaction order

Nu Nusselt number (dimensionless)

Pe   Peclet number (Chapter 17)

P_i Partial pressure of species i (atm)

Pr Prandtl number (dimensionless)

Q Heat flow from the surroundings to the system (cal/s)

R Ideal gas constant

Re   Reynolds number

r Radial distance (m)

r_A Rate of generation of species A per unit volume (gmol A/s·dm³)

$-r_A$ Rate of disappearance of species A per unit volume (mol A/s·dm³)

$-r_A'$ Rate of disappearance of species A per mass of catalyst (mol A/g·s)

$-r_A''$ Rate of disappearance of A per unit area of catalytic surface (mol A/m²·s)

S An active site (Chapter 10)

(S)   Substrate concentration (gmol/dm³) (Chapter 7)

S_a Surface area per unit mass of catalyst (m²/g)

$S_{D/U}$ Instantaneous selectivity (Chapter 8)

$\tilde{S}_{D/U}$ Overall selectivity of D to U

Sc   Schmidt number (dimensionless) (Chapter 10)

Sh   Sherwood number (dimensionless) (Chapter 10)

SV Space velocity (s⁻¹)

T Temperature (K)

t Time (s)

U Overall heat transfer coefficient (cal/m²·s·K)

V Volume of reactor (dm³)

V_0 Initial reactor volume (dm³)

v Volumetric flow rate (dm³/s)

v_0	Entering volumetric flow rate (dm^3/s)
W	Weight of catalyst (kg)
$\mathbf{W_A}$	Molar flux of species A ($mol/m^2 \cdot s$)
X	Conversion of key reactant, A
Y_i	Instantaneous yield of species i
$\tilde{Y}_i$	Overall yield of species i
y	Pressure ratio P/P_0
y_i	Mole fraction of species i
y_{i0}	Initial mole fraction of species i
Z	Compressibility factor
z	Linear distance (cm)

Subscripts

0	Entering or initial condition
b	Bed (bulk)
c	Catalyst
e	Equilibrium
p	Pellet

Greek Symbols

α	Reaction order (Chapter 3)
α	Pressure drop parameter (Chapter 5)
α_i	Parameter in heat capacity (Chapter 12)
β_i	Parameter in heat capacity
β	Reaction order
γ_i	Parameter in heat capacity
δ	Change in the total number of moles per mole of A reacted
ε	Fraction change in volume per mole of A reacted resulting from the change in total number of moles
η	Internal effectiveness factor
Θ_i	Ratio of the number of moles of species i initially (entering) to the number of moles of A initially (entering)
λ	Dimensionless distance (z/L) (Chapter 15)
λ	Life expectancy (s) (Chapter 17)
μ	Viscosity ($g/cm \cdot s$)
ρ	Density (g/cm^3)
ρ_c	Density of catalyst pellet (g/cm^3 of pellet)
ρ_b	Bulk density of catalyst (g/cm^3 of reactor bed)
τ	Space time (s)
ϕ	Void fraction (Porosity)
ϕ_n	Thiele modulus
ψ	Dimensionless concentration (C_A/C_{As})
Ω	External (overall) effectiveness factor

Open-Ended Problems G

The following are summaries for open-ended problems that have been used as term problems at the University of Michigan. The complete problem statement of the problems can be found on the CRE Web site, Web Appendix G.

G.1 Design of Reaction Engineering Experiment

The experiment is to be used in the undergraduate laboratory and costs less than $500 to build. The judging criteria are the same as the criteria for the National AIChE Student Chapter Competition. The design is to be displayed on a poster board and explained to a panel of judges. Guidelines for the poster-board display are provided by Jack Fishman and are given on the CRE Web site.

G.2 Effective Lubricant Design

Lubricants used in car engines are formulated by blending a base oil with additives to yield a mixture with the desirable physical attributes. In this problem, students examine the degradation of lubricants by oxidation and design an improved lubricant system. The design should include the lubricant system's physical and chemical characteristics, as well as an explanation as to how it is applied to automobiles. Focus: automotive industry, petroleum industry.

G.3 Peach Bottom Nuclear Reactor

The radioactive effluent stream from a newly constructed nuclear power plant must be made to conform with Nuclear Regulatory Commission standards. Students use chemical reaction engineering and creative problem solving to propose solutions for the treatment of the reactor effluent. Focus: problem analysis, safety, ethics.

G.4 Underground Wet Oxidation

You work for a specialty chemicals company, which produces large amounts of aqueous waste. Your chief executive officer (CEO) read in a journal about an emerging technology for reducing hazardous waste, and you must evaluate the system and its feasibility. Focus: waste processing, environmental issues, ethics.

G.5 Hydrodesulfurization Reactor Design

Your supervisor at Kleen Petrochemical wishes to use a hydrodesulfurization reaction to produce ethylbenzene from a process waste stream. You have been assigned the task of designing a reactor for the hydrodesulfurization reaction. Focus: reactor design.

G.6 Continuous Bioprocessing

Most commercial bioreactions are carried out in batch reactors. The design of a continuous bioreactor is desired since it may prove to be more economically rewarding than batch processes. Most desirable is a reactor that can sustain cells that are suspended in the reactor while growth medium is fed in, without allowing the cells to exit the reactor. Focus: mixing modeling, separations, bioprocess kinetics, reactor design.

G.7 Methanol Synthesis

Kinetic models based on experimental data are being used increasingly in the chemical industry for the design of catalytic reactors. However, the modeling process itself can influence the final reactor design and its ultimate performance by incorporating different interpretations of experimental design into the basic kinetic models. In this problem, students are asked to develop kinetic modeling methods/approaches and apply them in the development of a model for the production of methanol from experimental data. Focus: kinetic modeling, reactor design.

G.8 Cajun Seafood Gumbo

Most gourmet foods are prepared by batch processes, i.e., a batch reactor. In this problem, students are challenged to design a continuous process for the production of gourmet-quality Cajun seafood gumbo from an old family recipe. Some of the most difficult gourmet foods to prepare are Louisiana specialties, owing to the delicate balance between spices (hotness) and subtle flavors that must be achieved. In preparing Creole and Cajun food, certain flavors are released only by cooking some of the ingredients in hot oil for a period of time.

We shall focus on one specialty, Cajun seafood gumbo. Develop a continuous-flow reactor system that would produce 5 gal/h of a gourmet-quality seafood gumbo. Prepare a flow sheet of the entire operation. Outline certain experiments and areas of research that would be needed to ensure the success of your project. Discuss how you would begin to research these problems. Make a plan for any experiments to be carried out (see Chapter 7, R7.5, page 273).

Following is an old family formula for Cajun seafood gumbo for batch operation (10 quarts, serves 40):

1 cup flour	4 bay leaves, crushed
$1\frac{1}{2}$ cups olive oil	$\frac{1}{2}$ cup chopped parsley
1 cup chopped celery	3 large Idaho potatoes (diced)
2 large red onions (diced)	1 tablespoon ground pepper
5 qt fish stock	1 tablespoon tomato paste
6 lb fish (combination of cod, red snapper, monk fish, and halibut)	5 cloves garlic (diced)
	$\frac{1}{2}$ tablespoon Tabasco sauce
12 oz crabmeat	1 bottle dry white wine
1 qt medium oysters	1 lb scallops
1 lb medium to large shrimp	

1. Make a roux (i.e., add 1 cup flour to 1 cup of boiling olive oil). Cook until dark brown. Add roux to fish stock.
2. Cook chopped celery and onion in boiling olive oil until onion is translucent. Drain and add to fish stock.
3. Add $\frac{1}{3}$ of the fish (2 lb) and $\frac{1}{3}$ of the crabmeat, liquor from oysters, bay leaves, parsley, potatoes, black pepper, tomato paste, garlic, Tabasco, and $\frac{1}{4}$ cup of the olive oil. Bring to a slow boil and cook 4 h, stirring intermittently.
4. Add 1 qt cold water, remove from the stove, and refrigerate (at least 12 h) until $2\frac{1}{2}$ h before serving.
5. Remove from refrigerator, add $\frac{1}{4}$ cup of the olive oil, wine, and scallops. Bring to a light boil, then simmer for 2 h. Add remaining fish (cut to bite size), crabmeat, and water to bring total volume to 10 qt. Simmer for 2 h, add shrimp, then 10 minutes later, add oysters and serve immediately.

G.9 Alcohol Metabolism

The purpose of this open-ended problem is for the students to apply their knowledge of reaction kinetics to the problem of modeling the metabolism of alcohol in humans. In addition, the students will present their findings in a poster session. The poster presentations will be designed to bring a greater awareness to the university community of the dangers associated with alcohol consumption.

Students should choose one of the following four major topics to further investigate:

1. Death caused by acute alcohol overdose
2. Long-term effects of alcohol
3. Interactions of alcohol with common medications
4. Factors affecting metabolism of alcohol

General information regarding each of these topics can be found on the CRE Web site.

The metabolism and model equations are given on the CRE Web site. One can load the Living Example problem for alcohol metabolism directly from the CRE Web site.

G.10 Methanol Poisoning

The emergency room treatment for methanol poisoning is to inject ethanol intravenously to tie up the alcohol dehydrogenase enzyme so that methanol will not be converted to formic acid and formate, which cause blindness. The goal of this open-ended problem is to build on the physiological-based model for ethanol metabolism to predict the ethanol injection rate for methanol poisoning. One can find a start on this problem by the *Professional Reference Shelf Material* on Physiologically Based Pharmacokinetic (PBPK) Models in Chapter 9 that is on the Web.

Use of Computational Chemistry Software Packages **H**

H.1 Computational Chemical Engineering

As a prologue to the future, our profession is evolving to one of molecular chemical engineering. For chemical reaction engineers, computational chemistry and molecular modeling could well be our future.

Thermodynamic properties of molecular species that are used in reactor design problems can be readily estimated from thermodynamic data tabulated in standard reference sources such as Perry's Handbook or the JANAF Tables. Thermochemical properties of molecular species not tabulated can usually be estimated using group contribution methods. Estimation of activation energies is, however, much more difficult owing to the lack of reliable information on transition-state structures, and the data required to carry out these calculations is not readily available.

Recent advances in computational chemistry and the advent of powerful, easy-to-use software tools have made it possible to estimate important reaction-rate quantities (such as activation energy) with sufficient accuracy to permit incorporation of these new methods into the reactor design process. Computational chemistry programs are based on theories and equations from quantum mechanics, which until recently, could only be solved for the simplest systems such as the hydrogen atom. With the advent of inexpensive high-speed desktop computers, the use of these programs in both engineering research and industrial practice is increasing rapidly. Molecular properties such as bond length, bond angle, net dipole moment, and electrostatic charge distribution can be calculated. Additionally, reaction energetics can be accurately determined by using quantum chemistry to estimate heats of formation of reactants, products, and also for transition-state structures.

Links

Examples of commercially available computational chemistry programs include Spartan, developed by Wavefunction, Inc. (*www.wavefun.com*), and Cerius2 from Molecular Simulations, Inc. (*www.accelrys.com*). The Web module in Chapter 3 on molecular reaction engineering (see *www.umich.edu/~elements/5e/*) gives an example of what we can expect in the future.

How to Use
the CRE Web Resources

The primary purpose of the CRE Web Resources is to serve as an enrichment resource. The benefits of using the CRE Web Resources are fivefold:

1. To facilitate different student learning styles:
   www.engin.umich.edu/~cre/asyLearn/itresources.htm
2. To provide the student with the option/opportunity for further study or clarification of a particular concept or topic
3. To provide the opportunity to practice critical thinking, creative thinking, and problem-solving skills
4. To provide additional technical material for the practicing engineer
5. To provide other tutorial information, such as additional homework problems and instructions on using computational software in chemical engineering

I.1 CRE Web Resources Components

There are two types of information in the CRE Web Resources: information that is organized **by chapter** and information organized **by concept.** Material in the "by chapter" section of the CRE Web Resources corresponds to the material found in this book and is further divided into five sections.

- **Objectives.** The objectives page lists what the students will learn from the chapter. When students are finished working on a chapter, they can come back to the objectives to see if they have covered everything in that chapter. Or if students need additional help on a specific topic, they can see if that topic is covered in a chapter from the objectives page.

- **Learning Resources.** These resources give an overview of the material in each chapter and provide extra explanations, examples, and applications to reinforce the basic concepts of chemical reaction engineering. *Summary Notes* serve as an overview of each chapter and contain a logical flow of derived equations and additional examples. Web Modules and Interactive Computer Games (ICG) show how the principles from the text can be applied to nonstandard problems. Solved Problems provide more examples for students to use the knowledge gained from each chapter.

Links to entertaining **YouTube** videos can be found on the Web site home page.

Chapter 1: *Fogler Zone (you've got a friend in Fogler).*

Chapter 3: *The Black Widow* murder mystery and *Baking a Potato.*

Chapter 4: *CRF Reactor Video*, Crimson Reactor Firm's video of a "semi batch" reactor with Diet Coke and Mentos.

Chapter 5: Learn a new dance and song, *CSTR* to the tune of *YMCA, Chemical Good Time Rhythm* rap song, and the University of Michigan students' *Chemical Engineering Gone Wrong.*

- **Living Example Problems.** These problems are usually the second Home Problem in each chapter (e.g., P5-2$_B$), most of which require computational software to solve. Polymath programs are provided on the DVD so students can download the program to "play" with the problem and ask "what if . . . ?" questions to practice critical and creative thinking skills. Students can change parameter values, such as the reaction rate constants, to learn to deduce trends or predict the behavior of a given reaction system.

Living Example Problem

- **Professional Reference Shelf.** The Professional Reference Shelf contains two types of information. First, it includes material that is important to the practicing engineer but that is typically not included in the majority of chemical reaction engineering courses. Second, it includes material that gives a more detailed explanation of derivations that were abbreviated in the text.

Reference Shelf

- **Additional Homework Problems.** New problems were developed for this edition. They provide a greater opportunity to use today's computing power to solve realistic problems. Instead of omitting some of the more traditional, yet excellent problems of previous editions, these problems were placed on the Web site and can serve as practice problems along with those unassigned problems in the text.

Home Work Problems

The materials in Learning Resources are further divided into *Summary Notes*, *Web Modules*, *Interactive Computer Games*, and *Solved Problems*. Table I-1 shows which enrichment resources can be found in each chapter.

TABLE I-1   DVD-ROM ENRICHMENT RESOURCES

Chapters

	1	2	3	4	5	6	7	8	9	10	11	12	13
Summary Notes	■	■	■	■	■	■	■	■	■	■	■	■	■
Web Modules	■	■	■	□	■	□	■	□	■	■	□	■	■
ICGs	■	■	□	■	■	■	■	■	■	■	□	■	■
Solved Problems	□	■	■	■	■	■	■	■	■	■	■	■	■
Additional Homework Problems	■	■	■	■	■	■	■	■	■	■	■	■	■
LEP	□	□	■	■	■	■	■	■	■	■	■	■	■
PRS	■	■	■	■	■	■	■	■	■	■	■	■	■
YouTube video links	■	□	■	■									

Note: The ICGs are high-memory-use programs. Because of the memory intensive nature of the ICGs, there have been intermittent problems (10–15% of Windows computers) with the games. You can usually solve the problem by trying the ICG on a different computer. In the Heatfx 2 ICG, only the first three reactors can be solved, and users cannot continue on to part 2 because of a bug currently in the program that we cannot find.

The information that can be accessed in the "by concept" sections is not specific to a single chapter. Although the material can be accessed from the by chapter sections, the "by concept" sections allow you to access certain material quickly without browsing through chapters.

- **Interactive Web Modules.** The Web site includes both Web Games and ICGs. The Web Games use a Web browser for an interface and give examples of how chemical reaction engineering principles can be applied to a wide range of situations, such as modeling cobra bites and cooking a potato.

Interactive

Computer Games

- **Interactive Computer Games (ICGs).** The ICGs are games that use a Windows or DOS-based program for an interface. They test knowledge on different aspects of chemical reaction engineering through a variety of games such as basketball and jeopardy.

- **Problem Solving.** Here students can learn different strategies for problem solving in both closed- and open-ended problems. See the ten different types of home problems and suggestions for approaching them. Extensive information on critical and creative thinking can also be found in this section. Also see the *Strategies for Creative Problem Solving*, *Third Edition*, by Fogler, LeBlanc, and Rizzo, along with the web site *www.engin.umich.edu/scps/*.

- **Syllabi.** Representative syllabi have been included on the Web site: (1) a 4-credit-hour undergraduate course, Chemical Engineering 344, and (2) a 3-credit-hour graduate course.

- **Frequently Asked Questions (FAQs).** Over the years that I have taught this course, I have collected a number of questions that the students have asked over and over for years and years. The questions usually ask for clarification or for a different way of explaining the material or for another example of the principle being discussed. The FAQs and answers are arranged by chapter.

- **Credits.** See who was responsible for putting the CRE Web site together.

I.2 How the Web Can Help Your Learning Style

I.2.1   Global vs. Sequential Learners

Global

- Use the summary lecture notes to get an overview of each chapter on the Web site and see the big picture
- Review real-world examples and pictures on the Web site
- Look at concepts outlined in the ICGs

Sequential

- Use the Derive hot buttons to go through derivations in lecture notes on the Web
- Follow all derivations in the ICGs step by step
- Do all self-tests, audios, and examples in the Web site lecture notes step by step

I.2.2 Active vs. Reflective Learners

Active

- Use all the hot buttons to interact with the material to keep active
- Use self-tests as a good source of practice problems
- Use *Living Example Problems* to change settings/parameters and see the result
- Review for exams using the ICGs

Reflective

- Self-tests allow you to consider the answer before seeing it
- Use *Living Example Problems* to think about topics independently

I.2.3 Sensing vs. Intuitive Learners

Sensing

- Use *Web Modules* (cobra, hippo, nanoparticles) to see how material is applied to real-world topics
- Relate how *Living Example Problems* are linked to real-world topics

Intuitive

- Vary parameters in supplied Polymath problems and understand their influence on a problem
- Use the trial-and-error portions of some ICGs to understand "what if . . . " style questions

I.2.4 Visual vs. Verbal Learners

Visual

- Study the examples and self-tests on the DVD-ROM summary notes that have graphs and figures showing trends
- Do ICGs to see how each step of a derivation/problem leads to the next
- Use the graphical output from *Living Example Problems*/Polymath code to obtain a visual understanding of how various parameters affect a system
- Use the *Professional Reference Shelf* to view pictures of real reactors

Verbal

- Listen to audios on the Web to hear information in another way
- Work with a partner to answer questions on the ICGs

I.3 Navigation

Students can use the Web site in conjunction with the text in a number of different ways. The Web site provides *enrichment resources*. It is up to each student to determine how to use these resources to generate the greatest benefit. Table I-2 shows some of the clickable buttons found in the *Summary Notes* within the Learning Resources and a brief description of what the students will see when they click on the buttons.

TABLE I-2   HOT BUTTONS IN SUMMARY NOTES

Clickable Button	Where it goes
Example	Solved example problem
Link	General material that may not be related to the chapter
Tip	Hints and tips for solving problems
Self Test	A test on the material in a section, with solutions
Derive	Derivations of equations when not shown in the notes
Critical	Critical Thinking Question related to the chapter
Module	Web Module related to the chapter
Assess	Chapter objectives
Polymath	Polymath solution of a problem from the Summary Notes
Biography	Biography of the person who developed an equation or principle
More	Chapter insert with more information on a topic
Workbook	Detailed solution of a problem
Plot	Plot of an equation or solution
Side Note	Extra information on a specific topic
wm qt raw	Audio clip

Index

937

REGISTER YOUR PRODUCT at informit.com/register

Access Additional Benefits and SAVE 35% on Your Next Purchase

- Download available product updates.

- Access bonus material when applicable.

- Receive exclusive offers on new editions and related products.
  (Just check the box to hear from us when setting up your account.)

- Get a coupon for 35% for your next purchase, valid for 30 days. Your code will
  be available in your InformIT cart. (You will also find it in the Manage Codes
  section of your account page.)

Registration benefits vary by product. Benefits will be listed on your account page
under Registered Products.

InformIT.com—The Trusted Technology Learning Source

InformIT is the online home of information technology brands at Pearson, the world's foremost
education company. At InformIT.com you can

- Shop our books, eBooks, software, and video training.
- Take advantage of our special offers and promotions (informit.com/promotions).
- Sign up for special offers and content newsletters (informit.com/newsletters).
- Read free articles and blogs by information technology experts.
- Access thousands of free chapters and video lessons.

Connect with InformIT—Visit informit.com/community
Learn about InformIT community events and programs.

the trusted technology learning source

Addison-Wesley • Cisco Press • IBM Press • Microsoft Press • Pearson IT Certification • Prentice Hall • Que • Sams • VMware Press

ALWAYS LEARNING   PEARSON